The Ultimate Digital Study Tool

Self-Test

A comprehensive quizzing section lets you test your ability to identify anatomical structures in a simulated practical exam. Additional quizzes provide review of structure functions and features.

Diagnostics displayed throughout the test track your progress.

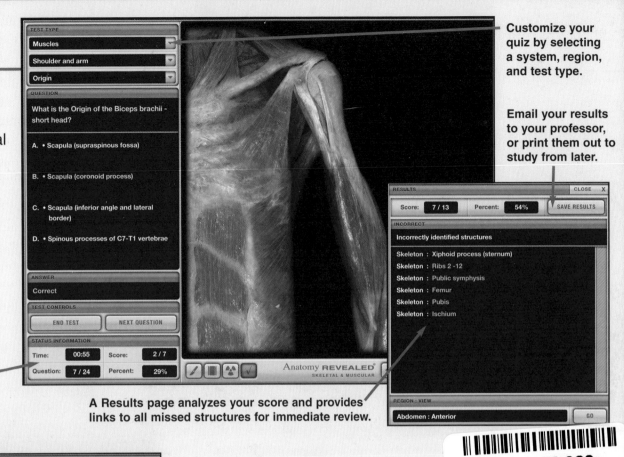

Customize your quiz by selecting a system, region, and test type.

Email your results to your professor, or print them out to study from later.

A Results page analyzes your score and provides links to all missed structures for immediate review.

Animation

Compelling animations help clarify anatomical relationships or explain difficult physiological concepts.

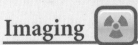

Pin and label important structures.

Imaging

Labeled X-ray, MRI, and CT images help you learn to recognize key anatomical structures as seen through various medical imaging techniques.

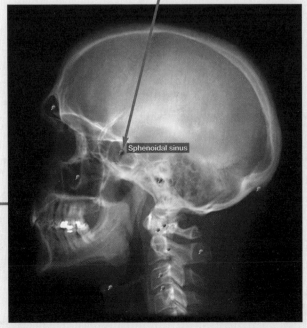

ESSENTIALS OF
Anatomy & Physiology

sixth edition

Rod R. Seeley
Idaho State University

Trent D. Stephens
Idaho State University

Philip Tate
Phoenix College

D E D I C A T I O N

This text is dedicated to the students of human anatomy and physiology. Helping students develop a working knowledge of anatomy and physiology is a satisfying challenge, and we have a great appreciation for the effort and enthusiasm of so many who want to know more. It is difficult to imagine anything more exciting, or more important, than being involved in the process of helping people learn about the subject we love.

Boston   Burr Ridge, IL   Dubuque, IA   Madison, WI   New York   San Francisco   St. Louis
Bangkok   Bogotá   Caracas   Kuala Lumpur   Lisbon   London   Madrid   Mexico City
Milan   Montreal   New Delhi   Santiago   Seoul   Singapore   Sydney   Taipei   Toronto

Higher Education

ESSENTIALS OF ANATOMY AND PHYSIOLOGY, SIXTH EDITION

Some ancillaries, including electronic and print components, may not be available to customers outside the United States.

This book is printed on acid-free paper.
Printed in China

4 5 6 7 8 9 0 CTP/CTP 0 9 8

ISBN-13   978–0–07–294369–6
ISBN-10   0–07–294369–6

Publisher: *Michelle Watnick*
Senior Developmental Editor: *Kathleen R. Loewenberg*
Marketing Manager: *Lynn M. Kalb*
Senior Project Manager: *Mary E. Powers*
Senior Production Supervisor: *Laura Fuller*
Senior Media Project Manager: *Tammy Juran*
Designer: *Rick D. Noel*
Cover/Interior Designer: *Christopher Reese*
(USE) Cover Image: © *Getty Images, Women's Beach Volleyball, Ian Waldie*
Senior Photo Research Coordinator: *John C. Leland*
Photo Research: *Jerry Marshall/pictureresearching.com*
Supplement Producer: *Tracy L. Konrardy*
Compositor: *The GTS Companies*
Typeface: *10.5/12 Minion*
Printer: *CTPS*

The credits section for this book begins on page C-1 and is considered an extension of the copyright page.

Library of Congress Cataloging-in-Publication Data

Seeley, Rod R.
   Essentials of anatomy and physiology / Rod R. Seeley, Trent D. Stephens, Philip Tate. — 6th ed.
      p.   cm.
   Includes index.
   ISBN 978–0–07–294369–6 — ISBN 0–07–294369–6 (hard copy : alk. paper)
   1. Human physiology. 2. Human anatomy. I. Stephens, Trent D. II. Tate, Philip.
   III. Title.

QP34.5.S418   2007
612—dc22

                                        2006000006
                                        CIP

www.mhhe.com

About the Authors

Rod Seeley, Trent Stephens, and Phil Tate in Driggs, Idaho, where they often retreat to collaborate on their textbooks. The Grand Tetons are pictured in the background.

Rod R. Seeley
Professor of Physiology at Idaho State University

Rod has extensive experience teaching introductory biology, anatomy and physiology, pathobiology, endocrinology, and more advanced physiology courses. He has won numerous teaching awards and is actively involved in supervision of doctoral students in biological education. With a B.S. in zoology from Idaho State University and an M.S. and Ph.D. in zoology from Utah State University, Rod has built a solid reputation as an author of journal and other professionally related articles as well as a public lecturer.

Trent D. Stephens
Professor of Anatomy and Embryology at Idaho State University

An award-winning educator and researcher, Trent Stephens teaches human anatomy, human head and neck anatomy, and human embryology. He also has many years of experience teaching neurobiology. His skill as a biological illustrator has greatly influenced the illustrations in this textbook. He has a B.S. in microbiology and a B.S. in zoology, as well as an M.S. in zoology from Brigham Young University. His Ph.D. in anatomy is from the University of Pennsylvania. Trent is actively involved in research on limb development and birth defects caused by thalidomide. He has authored numerous papers in these fields.

Philip Tate
Instructor of Anatomy and Physiology at Phoenix College

Phil Tate earned a B.S. in zoology, a B.S. in mathematics, and an M.S. in ecology at San Diego State University; and a Doctor of Arts (D.A.) in biological education from Idaho State University. He is an award-winning instructor who has taught a wide spectrum of students at the four-year and community college levels. Phil has served as the annual conference coordinator, president-elect, president, and past president of the Human Anatomy and Physiology Society (HAPS).

Brief Contents

Appendices

Contents

Preface

"In this sixth edition, we have continued the tradition of combining clear and accurate descriptions of anatomy, precise explanations of how structures function, and examples of how structures work together to maintain life."

Essentials of Anatomy & Physiology is designed to help students develop a solid, basic understanding of essential concepts in anatomy and physiology without an encyclopedic presentation of detail. It is not possible to include all that is known into a book of this type. However, it is important that a textbook provide enough information to allow students to understand basic concepts and to make reasonable predictions and analyses. We have taken great care to select critically important information and present it in a way that maximizes understanding.

We have provided explanations of how the systems respond to physical activity, disease, and aging, with a special focus on how regulatory systems control functions so that homeostasis is maintained. We have included timely and interesting examples to demonstrate the application of knowledge, often in a clinical context. For example, fundamental information is presented to allow students to understand the normal structure and function of the heart, how the heart reacts to changes in activity levels, and how the heart responds to age-related changes. It is also important for students to predict the response of the heart to blood loss or heart attack. Enough additional material is included to help students understand the effects of transfusions, treatments that reduce damage to heart muscle, and methods that help the heart continue to pump blood.

We've designed the figures to support the explanations in the text. Illustrations of structures are accurate, attractive, and presented clearly. They reflect a contemporary style and are coordinated so that colors and styles of structures in multiple figures are consistent with one another throughout the book. Many of the figures not only depict structures but also their relationships to surrounding structures. Care has been taken so that figure labels and legends are consistent with the terms used in the body of the text.

What Sets This Book Apart?

Clarity—*Just Enough Information Presented Concisely*

Essentials of Anatomy & Physiology is written in succinct, understandable language. We continue to improve this aspect of the text because we believe that basic content must be presented and explained clearly in order to support critical thinking. Therefore, whether or not critical thinking is a major emphasis in your course, this text is a valuable asset for students because of its clarity.

> "I like the way the authors explain the basic information needed to understand the complexity of the endocrine system. This is definitely one chapter that should not be oversimplified or have topics omitted. The authors have done a good job of clearly presenting complex material."
> —Michael Vitale, *Daytona Beach Community College*

Emphasis on Critical Thinking—*Building a Knowledge Base for Solving Problems*

An emphasis on critical thinking is integrated throughout this textbook. This approach can be found in questions embedded in the narrative; in clinical material that is designed to bridge concepts explained in the text with real-life applications and scenarios; in end-of-chapter questions that go beyond rote memorization; and in a visual program that presents material in logical, relevant images.

Ease of Comprehension—*Carefully Designed Pedagogy Supports Key Themes*

The pedagogical framework of *Essentials of Anatomy & Physiology* provides a multi-pronged approach to learning. This well-structured foundation is centered on clarity, ease of comprehension, and critical thinking. Knowledge- and comprehension-level questions are balanced with questions that require more complex reasoning in both the narrative of the text and in the end-of-chapter exercises. Clinical information is presented in several ways that support and restate concepts learned in the textual material.

Exceptional Art Program—*Simplicity, Consistency, and Logic Underscore Visuals*

The text's visual program is precisely planned to enhance comprehension in a number of ways: tables combined with illustrations, relevant photos side-by-side with drawings, cadaver photos where appropriate, step-by-step Process Figures, intuitive homeostasis figures that include a "Start" icon and color-coded directional arrows, and helpful flow diagrams.

> "The homeostasis figures in the Seeley *Essentials* text are very well presented. These figures are key to understanding the relationship between structure and function, and bring the anatomy and physiology in perspective for the whole organism."
> —James Foster, *Pikeville College School of Osteopathic Medicine*

Clinical Emphasis—*New "A Case in Point" Readings Add to Rich Base of Clinical Material*

When problems in structure and/or function of the human body occur, this is often the best time to comprehend how the two are related. *Essentials of Anatomy & Physiology* has always had a strong emphasis on clinical material. In this sixth edition, we strengthen that emphasis with the addition of A Case in Point readings. These brief, real-life scenarios combine with the popular brief asides, and Clinical Focus boxes to provide a thorough clinical education that fully supports the textual material.

Sixth Edition Changes— *What's New?*

The sixth edition of *Essentials of Anatomy & Physiology* is the result of extensive analysis of the text and evaluation of input from anatomy and physiology instructors who have thoroughly reviewed chapters. We are grateful to these professionals and have used their constructive comments in our continuing efforts to enhance the strengths of the book.

Improved Art—*Visualizing Relationships Between Structure and Function*

The effectiveness of the art has been improved in many ways in this edition. Over 100 figures have been changed. Some of the figures have been heavily revised to help students more easily understand concepts:

- The histology figures in chapter 4 have been changed to a table format to present both descriptions and illustrations of the major tissue types found in the body.
- Homeostasis figures have been improved by adding a "Start Here" icon and color-coded arrows to make it easier to follow the events occurring when a variable increases or decreases. These simplified flow charts succinctly map out key homeostatic events, giving students a quick and effective summary of the mechanisms described in the text.
- Figure 7.11 is just one example of a much improved illustration. The figure has been reorganized to better illustrate the relationship between ATP use in a muscle and the sources of ATP, such as creatine phosphate and aerobic metabolism. Another example is figure 12.17, which has been changed to more clearly show the events that occur during the cardiac cycle.
- Many of the Process Figures have been fine-tuned to make them clearer and easier to follow. These figures provide well-organized, self-contained visual explanations of how physiological mechanisms work. They help students learn physiological processes by combining illustrations with parallel descriptions of the major features of each process.
- Labels of orientation have been added to various illustrations so students immediately understand what perspective they are viewing (see figure 6.19 as an example).

Numerous other figures have been changed to *present more current information, achieve consistency, coordinate better with the text,* or *improve comprehension.* Just a few additional examples include:

- Figure 17.1 is a presentation of the U.S. government's new food pyramid correlating with new descriptions of macronutrient distribution ranges.
- Figure 18.15 from the fifth edition has been redesigned to form three separate, small figures in this sixth edition.
- Micrographs were added to figure 3.26 to further illustrate the phases of mitosis.
- Figures 13.14 and 13.15 have been redrawn to make them more consistent with the arm veins presented in figure 13.16.

- Figures 3.10 and 3.11 have also been reconfigured and relabeled for easier comparison.
- The anterior and posterior interventricular sulci of the heart have now been labeled in figure 12.5 because it makes describing the branches of the coronary arteries clearer.
- Explanatory labels were added to figure 9.19 to aid in understanding of the effect of sound waves on the structures of the ear.

Finally, some of the changes are more subtle, yet still important. We've adjusted leader lines, rewritten or reorganized labels, and made terminology more consistent between the text and the figures. *The end result is a beautiful, accurate, up-to-date art program that has been scrutinized and approved by our peers, and supports the hallmarks of clarity, ease of comprehension, and critical thinking that this text emphasizes.*

New! "A Case in Point" Readings—*Brief 'Snapshots' Bring Concepts to Life*

We are excited about these short essays, new to the sixth edition. Each A Case in Point reading explains how material just presented in the text can be used in understanding key anatomical and physiological concepts, especially in a clinical setting. They are designed to be interesting as well as instructional because they present information as small, news-like clips taken from real-life scenarios. Each chapter presents one or more A Case in Point essays.

The A Case in Point essays, along with clinical topics in each chapter, are designed to be consistent with the Predict and Critical Thinking Questions. All of these elements are designed to encourage students to learn to think critically. In some cases, they bring together information from more than one chapter to address sample problems and encourage more than just rote memorization.

Refined and Updated Textual Copy—*Organizing Information in a Logical Sequence*

We have kept the well-received order of topics in each chapter. Additionally, we have researched reports of new discoveries, used reviews from instructors who teach anatomy and physiology, and carefully examined the existing text to improve the book in the following ways:

- Many explanations have been made more clear.
- Terminology has been made more consistent throughout the text and the illustrations.
- New information applicable to the discipline has been researched and included where appropriate.
- Factual data has been checked and updated, or corrected, if necessary.
- Several explanations have been expanded to make the topics easier for students to understand.
- Predict Questions and Critical Thinking Questions have been updated.

See the Guided Tour starting on p. **x** for more details on each of the learning features in *Essentials of Anatomy and Physiology.*

Acknowledgments

No modern textbook is solely the work of the authors. To adequately acknowledge the support of loved ones is not possible. They have had the patience and understanding to tolerate our frustrations and absence, and they have been willing to provide assistance and undying encouragement.

Many hands besides our own have touched this text, guiding it through various stages of development and production. We wish to express our gratitude to the staff of McGraw-Hill for their help and encouragement. We sincerely appreciate Sponsoring Editor Michelle Watnick and Developmental Editor Kathy Loewenberg for their many hours of work, suggestions, and tremendous patience and encouragement. We also thank Project Manager Mary Powers, Photo Editor John Leland, Production Supervisor Laura Fuller, and Designer Rick Noel for their hours spent turning manuscript into a book; Media Producer Eric Weber, Media Project Manager Tammy Juran, and Supplement Producer Tracy Konrady for their assistance in building the various products that support our text; and Marketing Managers Jim Connely and Lynn Kalb for their enthusiasm in promoting this book. The McGraw-Hill employees with whom we have worked are excellent professionals. They have been consistently helpful and their efforts are appreciated. Their commitment to this project has clearly been more than a job to them.

Thanks are gratefully offered to Copy Editor Sue Dillon for carefully polishing our words.

We also thank the many illustrators who worked on the development and execution of the illustration program and who provided photographs and photomicrographs for the sixth edition of *Essentials of Anatomy and Physiology.* The art program for this text represents a monumental effort, and we appreciate their contribution to the overall appearance and pedagogical value of the illustrations.

Finally, we sincerely thank the reviewers and the teachers who have provided us with excellent constructive criticism. The remuneration they received represents only a token payment for their efforts. To conscientiously review a textbook requires a true commitment and dedication to excellence in teaching. Their helpful criticisms and suggestions for improvement were significant contributions that we greatly appreciate. We acknowledge them by name in the next section.

Rod Seeley
Trent Stephens
Phil Tate

Reviewers

Lena Ballard
Rock Valley College

Willodean D. S. Burton
Austin Peay State University

Bridget A. Falkenstein
Sierra College

Katherine A. Foreman
Moraine Valley Community College

James D. Foster
Western University of Health Sciences

Andrew Goliszek
N.C.A & T State University

Jennifer Hertz
Iowa Western Community College

Peggy S. M. Hill
University of Tulsa

Gary W. Hunt
Tulsa Community College

Kamal Kamal
Valencia Community College—West Campus

Charles Matsuda
Kapiolani Community College

Carla S. Murray
Carl Sandburg College

Michael Scott
Lincoln University

Elliot S. Stern
Everett Community College

Mary Elizabeth Torrano
American River College

William L. Trotter
Des Moines Area Community College

Michael A. Vitale
Daytona Beach Community College

Kennette Stiles Weatherford
Pierce College

Guided Tour

A Sound Learning System

Essentials of Anatomy and Physiology is designed to help you learn in a systematic fashion. Simple facts are the building blocks for developing explanations of more complex concepts. The text discussion is presented within a supporting framework of learning aids that help organize studying, reinforce learning, and promote problem-solving skills.

Chapter Outline and Objectives

An outline that combines chapter topics with their associated learning objectives is presented at the beginning of each chapter to introduce key concepts you will encounter in your reading.

Opening Photomicrograph

A photomicrograph depicting a structure related to the chapter content provides an interesting and relevant microscopic preview of what's to come.

Chapter Outline and Objectives

Nutrition (p. 484)
1. Define nutrition, essential nutrient, and kilocalorie.
2. For carbohydrates, lipids, and proteins, describe their dietary sources, their uses in the body, and the daily recommended amounts of each in the diet.
3. List the common vitamins and minerals and give a function for each.

Metabolism (p. 492)
4. Define metabolism, anabolism, and catabolism.
5. List three ways in which enzyme activity is controlled.
6. Describe glycolysis and name its products.
7. Describe the citric acid cycle and its products.
8. Describe the electron-transport chain and how ATP is produced in the process.
9. Explain how the breakdown of glucose yields two ATP molecules in anaerobic respiration and 38 ATP molecules in aerobic respiration.
10. Describe the basic steps involved in using lipids and amino acids as an energy source.
11. Differentiate between the absorptive and postabsorptive metabolic states.
12. Define metabolic rate.

Body Temperature Regulation (p. 502)
13. Describe heat production and regulation in the body.

CHAPTER

17

Nutrition, Metabolism, and Body Temperature Regulation

This colorized scanning electron micrograph shows a mitochondrion (pink) in the cytoplasm of an intestinal epithelial cell. The mitochondrion has an outer and inner membrane. The inner membrane has numerous folds that project into the interior of the mitochondrion. Enzymes, necessary for producing ATP, are located in these folds.

483

Predict Questions

These innovative critical thinking exercises encourage you to become an active learner as you read. Predict questions challenge you to use your understanding of new concepts to solve a problem. Answers to Predict Questions are given at the end of the text, allowing you to evaluate your response and discover the logic used to arrive at the correct answer.

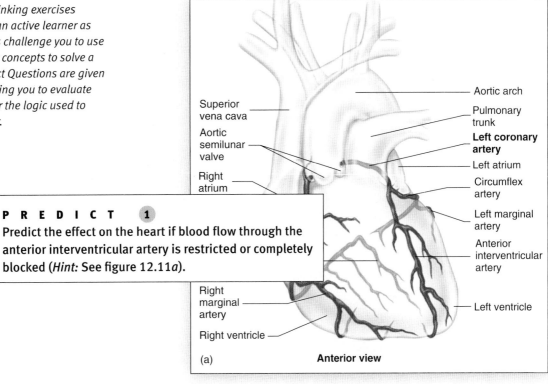

P R E D I C T   1

Predict the effect on the heart if blood flow through the anterior interventricular artery is restricted or completely blocked (*Hint:* See figure 12.11*a*).

Anterior view

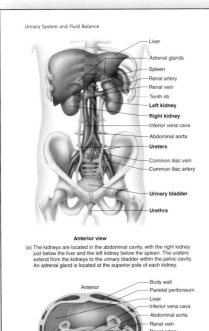

Figure 18.2   Anatomy of the Urinary System

(a) The kidneys are located in the abdominal cavity, with the right kidney just below the liver and the left kidney below the spleen. The ureters extend from the kidneys to the urinary bladder within the pelvic cavity. An adrenal gland is located at the superior pole of each kidney.

(b) The kidneys are located behind the parietal peritoneum. Fat surrounds each kidney. The renal arteries extend from the abdominal aorta to each kidney, and the renal veins extend from the kidneys to the inferior vena cava.

Urinary System and Fluid Balance   **509**

vessels, part of the system for collecting urine (see following discussion of calyces and renal pelvis), and fat.

The kidney is divided into an outer **cortex** and an inner **medulla,** which surround the renal sinus. The bases of several cone-shaped **renal pyramids** are located at the boundary between the cortex and the medulla, and the tips of the renal pyramids project toward the center of the kidney. A funnel-shaped structure called a **calyx** (kā′liks, cup of a flower) surrounds the tip of each renal pyramid. The calyces from all the renal pyramids join to form a larger funnel called the **renal pelvis.** The renal pelvis narrows to form a small tube, the **ureter** (ū-rē′ter or ū′re-ter, urinary canal), which exits the kidney and connects to the urinary bladder. Urine passes from the tips of the renal pyramids into the calyces. From the calyces, urine collects in the renal pelvis and exits the kidney through the ureter (see figure 18.3).

The functional unit of the kidney is the **nephron** (nef′ron, Greek for kidney), and there are approximately 1.3 million of them in each kidney. Each nephron consists of a **renal corpuscle,** a **proximal tubule,** a **loop of Henle,** or nephronic loop, and a **distal tubule** (figure 18.4). Fluid enters the renal corpuscle and then flows into the proximal tubule. From there, it flows into the loop of Henle. Each loop of Henle has a descending limb, which extends toward the renal sinus, and an ascending limb, which extends back toward the cortex. The fluid flows through the ascending limb of the loop of Henle to the distal tubule. Many distal tubules empty into a **collecting duct,** which carries the fluid from the cortex, through the medulla. Many collecting ducts empty into a **papillary duct,** and the papillary ducts empty their contents into a calyx.

The renal corpuscle and both convoluted tubules are in the renal cortex (see figure 18.4). The collecting duct and loop of Henle enter the medulla. Approximately 15% of the nephrons, called **juxtamedullary** (next to the medulla) **nephrons,** have loops of Henle that extend deep into the medulla of the kidney. The other nephrons (85%), called **cortical nephrons,** have loops of Henle that do not extend deep into the medulla.

The renal corpuscle of the nephron consists of Bowman's capsule and the glomerulus (see figure 18.4; figure 18.5). **Bowman's capsule** consists of the enlarged end of the nephron, which is indented to form a double-walled chamber. The indentation is occupied by a tuft of capillaries called the **glomerulus** (glō-mār′ū-lŭs, *glomus,* a ball of yarn), which resembles a ball of yarn. The cavity of Bowman's capsule opens into the proximal tubule, which carries fluid away from the capsule. The inner layer of Bowman's capsule consists of specialized cells called **podocytes** (pod′ō-sits, *pod,* foot + *kytos,* a hollow cell), which wrap around the glomerular capillaries. The outer layer of Bowman's capsule consists of simple squamous epithelial cells.

The glomerular capillaries have pores in their walls, and the podocytes have numerous cell processes with gaps between them. The endothelium of the glomerular capillaries, the podocytes, and the basement membrane between them form a **filtration membrane** (see figure 18.5*d*). In the first step of urine formation, fluid, consisting of water and solutes smaller than proteins, pass from the blood in the glomerular capillaries through the **filtration membrane** into Bowman's capsule (see figure 18.5*d*). The fluid that passes across the filtration membrane is called **filtrate.**

fat, which protects the kidney from mechanical shock. On the medial side of each kidney is the **hilum** (hī′lŭm, a small amount), where the renal artery and nerves enter and where the renal vein and ureter exit the kidney (figure 18.3). The hilum opens into a cavity called the **renal sinus,** which contains blood

Vocabulary Aids

Learning anatomy and physiology is, in many ways, like learning a new language. Mastering the terminology is key to building your knowledge base.

Key terms are set in boldface where they are defined in the chapter, and most terms are included in the glossary at the end of the book. Pronunciation guides are included for difficult words.

Because knowing the origin of a term can enhance understanding and retention, the meanings of the words or word roots that form a term are given when they are relevant. Furthermore, a handy list of prefixes, suffixes, and combining forms is printed on the inside back cover as a quick reference to help you identify commonly used word roots.

Instructive Artwork Makes the Difference

A picture is worth a thousand words—especially when you're learning anatomy and physiology. Because words alone cannot convey the nuances of anatomy or the intricacies of physiology, the sixth edition of *Essentials of Anatomy and Physiology* employs a dynamic program of full-color illustrations and photographs that support and further clarify the text explanations. Brilliantly rendered and carefully reviewed for accuracy and consistency, the precisely labeled illustrations and photos provide concrete, visual reinforcement of the topics discussed throughout the text.

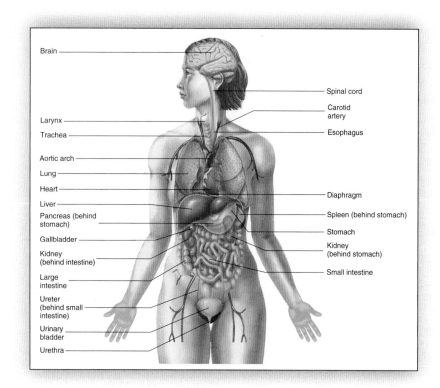

Realistic Anatomical Art

The anatomical figures in Essentials of Anatomy and Physiology *have been carefully rendered to convey realistic, three-dimensional detail. Richly textured bones and artfully shaded muscles and vessels lend a sense of realism to the figures that helps you envision the appearance of actual structures within in the body.*

The colors used to represent different anatomical structures have been applied consistently throughout the book. This reliable pattern of color consistency helps you easily identify the structures in every figure and promotes continuity between figures.

Reference diagrams orient you to the view or plane an illustration represents.

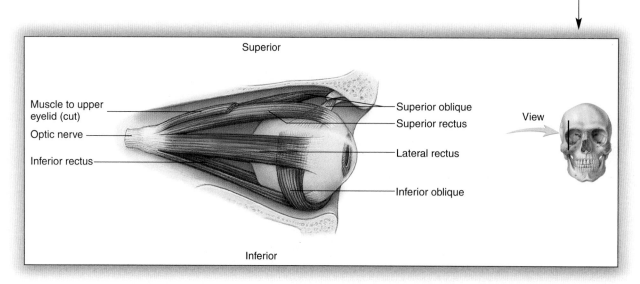

Multilevel Perspective

Illustrations depicting complex structures or processes combine macroscopic and microscopic views to help you see the relationships between increasingly detailed drawings.

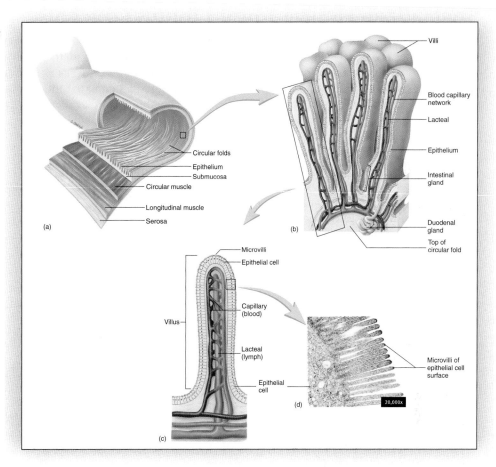

Combination Art

Drawings are often paired with photographs to enhance visualization of structures.

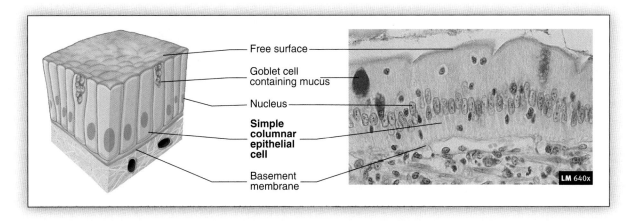

Histology Micrographs

Light micrographs, as well as scanning and transmission electron micrographs, are used in conjunction with illustrations to present a true picture of anatomy and physiology from the cellular level.

Specialized Figures Clarify Tough Concepts

Studying anatomy and physiology does not have to be an intimidating task mired in memorization. *Essentials of Anatomy and Physiology* uses two special types of illustrations to help you not only to learn the steps involved in specific processes, but also to apply this knowledge as you predict outcomes in similar situations. Process Figures organize the key occurrences of physiological processes in an easy-to-follow format. Homeostasis figures summarize the mechanisms of homeostasis by diagramming the means by which a given system regulates a parameter within a narrow range of values.

Process Figures

Process Figures break down physiological processes into a series of smaller steps, allowing you to track the key occurrences and learn them as you go.

Sequence indicators within the artwork correspond to the numbered steps along the side. These colored circles help you zero in on the site where the action described in each step takes place.

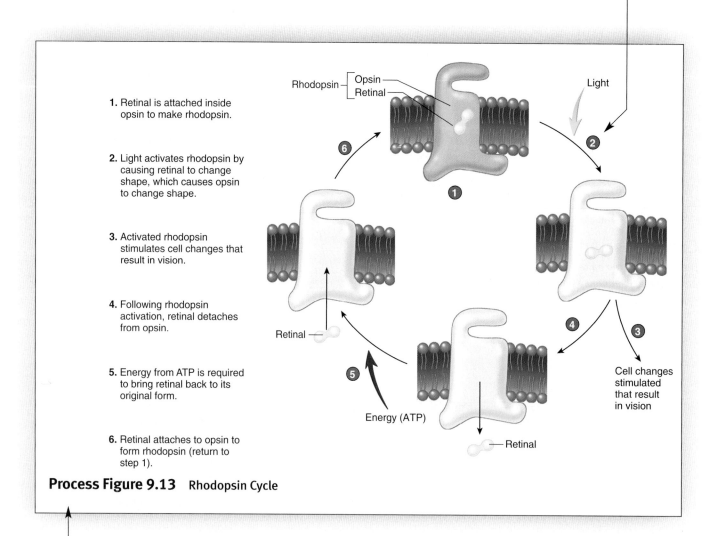

1. Retinal is attached inside opsin to make rhodopsin.

2. Light activates rhodopsin by causing retinal to change shape, which causes opsin to change shape.

3. Activated rhodopsin stimulates cell changes that result in vision.

4. Following rhodopsin activation, retinal detaches from opsin.

5. Energy from ATP is required to bring retinal back to its original form.

6. Retinal attaches to opsin to form rhodopsin (return to step 1).

Process Figure 9.13 Rhodopsin Cycle

Process Figures and Homeostasis Figures are identified next to the figure number. The accompanying caption provides additional explanation.

Homeostasis Figures

These specialized flowcharts diagram the mechanisms body systems employ to maintain homeostasis.

Changes caused by an increase of a variable toward the outside of its normal range are shown in the boxes outlined in green.

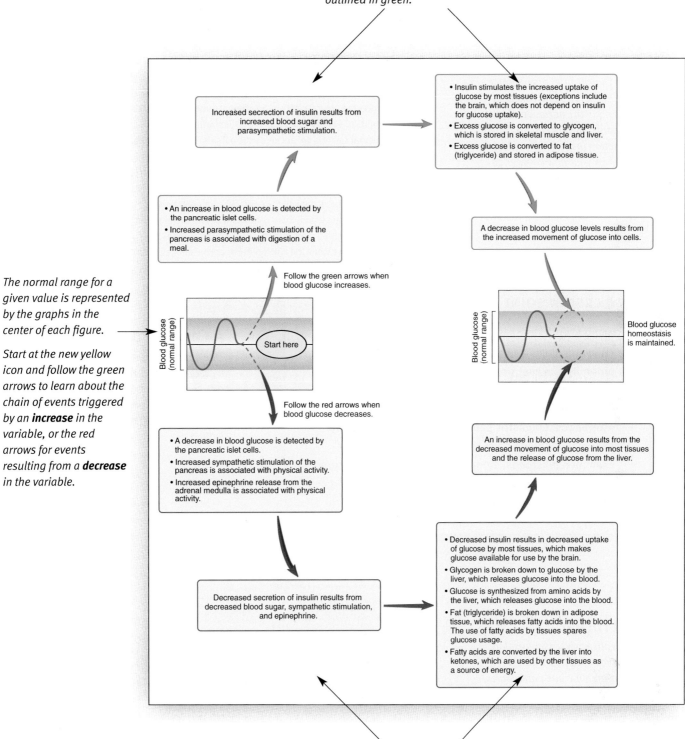

The normal range for a given value is represented by the graphs in the center of each figure.

*Start at the new yellow icon and follow the green arrows to learn about the chain of events triggered by an **increase** in the variable, or the red arrows for events resulting from a **decrease** in the variable.*

Increased secrection of insulin results from increased blood sugar and parasympathetic stimulation.

• Insulin stimulates the increased uptake of glucose by most tissues (exceptions include the brain, which does not depend on insulin for glucose uptake).
• Excess glucose is converted to glycogen, which is stored in skeletal muscle and liver.
• Excess glucose is converted to fat (triglyceride) and stored in adipose tissue.

• An increase in blood glucose is detected by the pancreatic islet cells.
• Increased parasympathetic stimulation of the pancreas is associated with digestion of a meal.

A decrease in blood glucose levels results from the increased movement of glucose into cells.

Follow the green arrows when blood glucose increases.

Blood glucose (normal range)

Start here

Blood glucose (normal range)

Blood glucose homeostasis is maintained.

Follow the red arrows when blood glucose decreases.

• A decrease in blood glucose is detected by the pancreatic islet cells.
• Increased sympathetic stimulation of the pancreas is associated with physical activity.
• Increased epinephrine release from the adrenal medulla is associated with physical activity.

An increase in blood glucose results from the decreased movement of glucose into most tissues and the release of glucose from the liver.

Decreased secretion of insulin results from decreased blood sugar, sympathetic stimulation, and epinephrine.

• Decreased insulin results in decreased uptake of glucose by most tissues, which makes glucose available for use by the brain.
• Glycogen is broken down to glucose by the liver, which releases glucose into the blood.
• Glucose is synthesized from amino acids by the liver, which releases glucose into the blood.
• Fat (triglyceride) is broken down in adipose tissue, which releases fatty acids into the blood. The use of fatty acids by tissues spares glucose usage.
• Fatty acids are converted by the liver into ketones, which are used by other tissues as a source of energy.

Changes caused by a decrease of a variable toward the outside of its normal range are shown in boxes outlined in red.

Clinical Content Puts Knowledge into Practice

Essentials of Anatomy and Physiology provides clinical examples to promote interest and demonstrate relevance, but clinical information is used primarily to illustrate the application of basic knowledge. Exposure to clinical information is especially beneficial if you are planning on using your knowledge of anatomy and physiology in a health-related career.

Clinical Topics

Interesting clinical sidebars reinforce or expand on the facts and concepts discussed within the narrative. Once you have learned a concept, applying that information in a clinical context shows you how your new knowledge can be put into practice.

Pneumothorax

A **pneumothorax** (noo-mō-thōr′aks) is the introduction of air into the pleural cavity. Air can enter by an external route when a sharp object, such as a bullet or broken rib, penetrates the thoracic wall; or air can enter the pleural cavity by an internal route if alveoli at the lung surface rupture, such as can occur in a patient with emphysema. When the pleural cavity is connected to the outside by such openings, the pressure in the pleural cavity increases and becomes equal to the air pressure outside the body. Thus, pleural pressure is also equal to alveolar pressure because pressure in the alveoli at the end of expiration is equal to air pressure outside the body. When pleural pressure and alveolar pressure are equal, there is no tendency for the alveoli to expand, lung recoil is unopposed, and the lungs collapse. A pneumothorax can occur in one lung while the lung on the opposite side remains inflated because ... separated by the mediastinum.

A CASE IN POINT | Epigastric Pain

Wilby Hurtt has epigastric pain that is most noticeable following meals and at night when he is lying in bed. He probably has gastroesophageal reflux disease (GERD), in which stomach acid improperly moves into the esophagus, damaging and irritating the lining of the esophagus. Epigastric pain, however, can have many causes and should be evaluated by a physician. For example, gallstones, stomach or small intestine ulcers, inflammation of the pancreas, and he ... cause epigastric pain.

A Case in Point

These brief essays explain how material just presented in the text can be used to understand important anatomical and physiological concepts, particularly in a clinical setting.

Clinical Focus

These in-depth boxed essays explore relevant topics of clinical interest. Subjects covered include pathologies, current research, sports medicine, exercise physiology, pharmacology, and clinical applications.

Clinical Focus   Skeletal Disorders

Growth and Developmental Disorders

Giantism (jī′an-tizm) is a condition of abnormally increased size that usually involves excessive endochondral growth at the epiphyseal plates of long bones. **Dwarfism,** the condition in which a person is abnormally small, may result from improper growth in the epiphyseal plates (figure A; see also chapter 10).

Osteogenesis imperfecta (os′tē-ō-jen′ĕ-sis im-per-fek′tă, *osteo,* bone + *genesis,* production; imperfect), a group of genetic disorders producing very brittle bones that are easily fractured, occurs because insufficient collagen or abnormal collagen is formed. Collagen normally strengthens bones and makes them flexible. In severe cases, prenatal fractures of the limbs often occur in the fetus. These fractures usually heal in poor alignment, causing the limbs to appear bent and shortened. In less severe cases, the disease first becomes apparent during childhood.

Rickets (rik′ets, *wrick,* to twist; bones become twisted in the disease) is a condition involving growth retardation resulting from nutritional deficiencies either in minerals (calcium and phosphate) necessary for normal ossification or in vitamin D, which is necessary for calcium and phosphate absorption. The condition results in bones that are soft, weak, and easily broken.

Rickets most often occurs in children who receive inadequate amounts of sunlight (necessary for vitamin D production by the body) and whose diets are deficient in vitamin D.

Bacterial Infections

Osteomyelitis (os′tē-ō-mī-e-lī′tis, *osteo,* bone + *myelos,* marrow + *itis,* inflammation) is bone inflammation that often results from bacterial infection, and it can lead to complete destruction of the bone. *Staphylococcus* (staf′i-lō-kok′ŭs, *staphyle,* a bunch of grapes + *kokkos,* a berry; these terms describe the organization and shape of the bacterium) (staph) infections, introduced into the body through wounds, are the most common cause of osteomyelitis. Tuberculosis is primarily a lung disease, but it can also affect bones. Because of milk pasteurization and other improvements in hygiene, tuberculosis became rare in the United States. Because tuberculosis can be a complication in AIDS and because a drug-resistant form of tuberculosis has emerged, tuberculosis has once again become a clinical problem in the United States.

Tumors

There are many types of bone **tumors** with a wide range of resultant bone defects. Tumors may be benign or malignant. Malignant bone tumors may metastasize (spread)

to other parts of the body or may result from metastasizing tumors elsewhere.

Decalcification

Osteomalacia (os′tē-ō-mă-lā′shē-ă, *osteo,* bone + *malakia,* softness), or the softening of bones, results from calcium depletion from bones. If the body has an unusual need for calcium (e.g., during pregnancy when fetal growth requires large amounts of calcium), it may be removed from the mother's bones, which consequently soften and weaken. Osteomalacia is sometimes called adult rickets and can result from vitamin D deficiency.

Osteoporosis (os′tē-ō-pō-rō′sis, *osteo,* bone + *poros,* pore + *osis,* condition), or porous bone, results from reduction in the overall quantity of bone tissue (see Systems Pathology: Osteoporosis on p. 152).

Figure A   Giant and Dwarf (both are adults)

Systems Pathology

These boxes explore a specific disorder or condition related to a particular body system. Presented in a modified case study format, each Systems Pathology box begins with a patient history followed by background information about the featured topic.

Systems Pathology
Burns

Mr. S is a 23-year-old man who had difficulty falling asleep at night. He often stayed up late watching TV or reading until he fell asleep. Mr. S was also a chain smoker. One night he took several sleeping pills. Unfortunately, he fell asleep before putting out his cigarette, which started a fire. As a result, Mr. S was severely burned, receiving full-thickness and partial-thickness burns (figure A*a*). He was rushed to the emergency room and was eventually transferred to a burn unit.

For the first day after his accident, his condition was critical because he went into shock. Administration of large volumes of intravenous fluid stabilized his condition. As part of his treatment, Mr. S was also given a high-protein, high-calorie diet.

A week later, dead tissue was removed from the most serious burns (figure A*b*), and skin grafts were performed. Despite the use of topical antimicrobial drugs and sterile bandages, however, some of the burns became infected. An additional complication was the development of a venous thrombosis (blood clot) in his leg.

Although the burns were painful and the treatment was prolonged, Mr. S made a full recovery. He no longer smokes.

Background Information

When large areas of skin are severely burned, systemic effects are produced that can be life-threatening. Within minutes of a major burn injury, there is increased permeability of capillaries, which are the small blood vessels in which fluid, gases, nutrients, and waste products are normally exchanged between the blood and tissues. This increased permeability occurs at the burn site and throughout the body. As a result, fluid and ions are lost from the burn wound and into tissue spaces. The loss of fluid decreases blood volume, which decreases the ability of the heart to pump blood. The resulting decrease in blood delivery to tissues can cause tissue damage, shock, and even death. Treatment consists of administering intravenous fluid at a faster rate than it leaks out of the capillaries. Although this can reverse the shock and prevent death, fluid continues to leak into tissue spaces, causing pronounced edema of the tissues.

Typically, after 24 hours, capillary permeability returns to normal, and the amount of intravenous fluid administered can be greatly decreased. How burns result in capillary permeability changes is not well understood. It is clear that following a burn, immunological and metabolic changes occur that affect not only capillaries, but the rest of the body as well. For example, mediators of inflammation (see chapter 4), which are released in response to the tissue damage, contribute to changes in capillary permeability throughout the body.

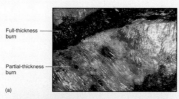

Full-thickness burn

Partial-thickness burn

(a)

(b)

Figure A Burn Victim

(*a*) Partial- and full-thickness burns. (*b*) Patient

Substances released from the bur[n] causing cells to function abnormally. Bu[r] most immediate hypermetabolic state tha[t] sure. Two other factors contributing to the[e] (1) a resetting of the temperature contro[l] higher temperature and (2) hormones rele[ase] tem (e.g., epinephrine and norepinephrin[e] increase cell metabolism). Compared wit[h] ture of approximately 37°C (98.6°F), a b[o]

A System Interactions table at the end of every box summarizes how the condition profiled affects each body system.

System Interactions	Effects of Burns on Other Systems
System	**Interactions**
Skeletal	Red bone marrow replaces red blood cells destroyed in the burnt skin.
Muscular	Loss of muscle mass resulting from the hypermetabolic state caused by the burn.
Nervous	Pain is sensed in the partial-thickness burns. The temperature-regulatory center in the brain is set to a higher temperature, contributing to increased body temperature. Abnormal K⁺ concentrations disturb normal nervous system activity: elevated levels are caused by release of K⁺ from damaged cells; low levels can be caused by rapid loss of these ions in fluid from the burn.
Endocrine	Increased secretion of epinephrine and norepinephrine from the adrenal glands in response to the injury contributes to increased body temperature by increasing metabolism.
Cardiovascular	Increased capillary permeability causes decreased blood volume, resulting in decreased blood delivery to tissues, edema, and shock. The pumping effectiveness of the heart is impaired by ion imbalance and substances released from the burn. Increased blood clotting causes venous thrombosis. Preferential delivery of blood to the injury promotes healing.
Lymphatic and immune	Increased inflammation in response to tissue damage. Later, depression of the immune system can result in infection.
Respiratory	Airway obstruction is caused by edema. Increased respiration rate results from increased metabolism and lactic acid buildup.
Digestive	Decreased blood delivery as a result of the burn causes degeneration of the intestinal lining and liver. Bacteria from the intestine can cause systemic infections. The liver releases blood-clotting factors in response to the injury. Increased nutrients necessary to support increased metabolism and for repair of the integumentary system are absorbed.
Urinary	The kidneys compensate for the increased fluid loss caused by the burn by greatly reducing or even stopping urine production. Decreased blood volume causes decreased blood flow to the kidneys, which reduces urine output, but can cause kidney tissue damage. Hemoglobin, released from red blood cells damaged in the burnt skin, can decrease urine production by blocking fluid movement from the blood into the kidney.

(101.3°F) is typical in burn patients, despite the higher loss of water by evaporation from the burn.

In severe burns, the increased metabolic rate can result in weight loss as great as 30–40% of the patient's preburn weight. To help compensate, caloric intake may double or even triple. In addition, the need for protein, which is necessary for tissue repair, is greater.

Normal skin maintains homeostasis by preventing the entry of microorganisms. Because burns damage and even completely destroy the skin, microorganisms can cause infections. For this reason, burn patients are maintained in an aseptic environment, which attempts to prevent the entry of microorganisms into the wound. They are also given antimicrobial drugs, which kill microorganisms or suppress their growth. **Debridement** (dā-brēd-mon′), the removal of dead tissue from the burn, helps to prevent infections by cleaning the wound and removing tissue in which infections could develop. Skin grafts, performed within a week of the injury, also prevent infections by closing the wound and preventing the entry of microorganisms.

Despite these efforts, however, infections are still the major cause of death of burn victims. Depression of the immune system

during the first or second week after the injury contributes to the high infection rate. The thermally altered tissue is recognized as a foreign substance that can stimulate the immune system. As a result, the immune system is overwhelmed as immune system cells become less effective and production of the chemicals that normally provide resistance to infections decreases (see chapter 14). The greater the magnitude of the burn, the greater the depression of the immune system, and the greater the risk of infection.

Venous thrombosis (throm-bō′sis), the development of a clot in a vein, is also a complication of burns. Blood normally forms a clot when exposed to damaged tissue, such as at a burn site, but clotting can also occur elsewhere, such as in veins. Clots can block blood flow, resulting in tissue destruction. The concentration of chemicals in the blood that cause clotting increases for two reasons: loss of fluid from the burn patient concentrates the chemicals and there is an increased release of clotting factors from the liver.

P R E D I C T 5
When Mr. S is first admitted to the burn unit, the nurses carefully monitor his urine output. Why does that make sense in light of his injuries?

Every Systems Pathology box includes a Predict Question specific to the case study.

Study Features Ensure Success

A carefully devised set of learning aids at the end of each chapter helps you review the chapter content, evaluate your grasp of key concepts, and utilize what you have learned. Reading the chapter summary and completing the practice questions and critical thinking exercises will greatly improve your understanding of each chapter and is also a beneficial way to study for exams.

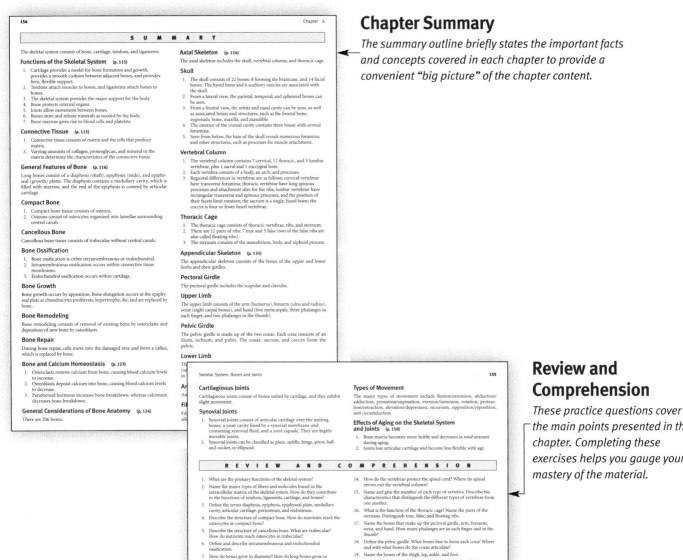

Chapter Summary

The summary outline briefly states the important facts and concepts covered in each chapter to provide a convenient "big picture" of the chapter content.

Review and Comprehension

These practice questions cover the main points presented in the chapter. Completing these exercises helps you gauge your mastery of the material.

Answers to Predict Questions

The Predict questions that appear throughout the text are answered in Appendix E, allowing you to evaluate your responses and understand the logic used to arrive at the correct answer.

Critical Thinking

These innovative exercises encourage you to apply chapter concepts to solve a problem. Answering these questions helps build your working knowledge of anatomy and physiology while developing reasoning skills. Answers are provided in Appendix D.

Teaching and Learning Supplements

McGraw-Hill offers various tools and technology products to support the sixth edition of *Essentials of Anatomy and Physiology*. Students can order supplemental study materials by contacting their campus bookstore. Instructors can obtain teaching aids by calling the McGraw-Hill Customer Service Department at 1-800-338-3987, visiting our A&P catalog at www.mhhe.com/ap, or contacting your local McGraw-Hill sales representative.

For Instructors

Digital Content Manager

This multimedia collection of visual resources allows instructors to browse, select, and export files containing artwork from the text in multiple formats to create customized classroom presentations, visually based tests and quizzes, dynamic course website content, or attractive printed support materials. The digital assets on this cross-platform CD-ROM are grouped by chapter within easy-to-use folders.

- **Art** Full-color digital files of all illustrations in the book, plus grayscale, and unlabeled versions of the same artwork, are housed in the Art folder. These files can be readily incorporated into lecture presentations, exams, or custom-made classroom materials. In addition, all files are pre-inserted into blank PowerPoint slides for ease of lecture preparation.

- **Photos** The Photos folder contains digital files of instructionally significant photographs from the text, which can be reproduced for multiple classroom uses.

- **Tables** The Tables folder includes every table that appears in the text saved in electronic form.

In addition to the content found within each chapter, the Digital Content Manager for *Essentials of Anatomy and Physiology* contains the following multimedia instructional materials.

- **Active Art** Active Art consists of art files that have been converted to a format that allows the artwork to be edited inside of PowerPoint. Each piece can be broken down to its core elements, grouped or ungrouped, and edited to create customized illustrations.

- **Animations** Numerous full-color animations illustrating physiological processes are provided. Harness the visual impact of processes in motion by importing these files into classroom presentations or online course materials.

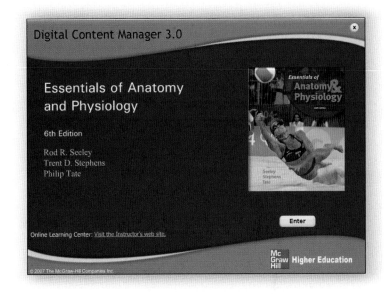

Digital Content Manager

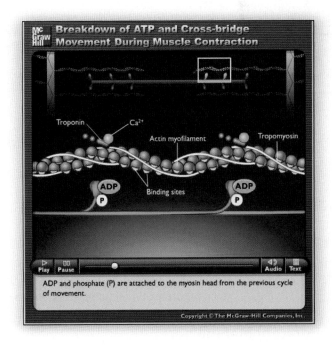

Animation

e-Instruction

The Classroom Performance System (CPS) is an interactive system that allows the instructor to administer in-class questions electronically. Students answer questions via hand-held, remote control keypads, and their individual responses are logged into a gradebook. Aggregated responses can be displayed in graphical form. Using this immediate feedback, the instructor can quickly determine if students understand the lecture topic, or if more clarification is needed. CPS promotes student participation, class productivity, and individual student accountability.

Instructor's Testing and Resource CD-ROM

This cross-platform CD-ROM provides a wealth of resources for the instructor. One of the supplements featured on this CD is EZ Test, a flexible and easy-to-use electronic testing program. This program allows instructors to create tests from book-specific items, and accommodates a wide range of question types, including the option for instructors to add their own questions. Multiple versions of the test can be created and any test can be exported for use with course management systems such as WebCT, BlackBoard, or PageOut. EZ Test Online is a new service that gives you a place to easily administer your EZ Test-created exams and quizzes online. The program is available for both Windows and Macintosh environments.

Other assets on the Instructor's Testing and Resource CD-ROM include the book's Instructor's Manual and the Instructor's Manual for the accompanying Laboratory Manual.

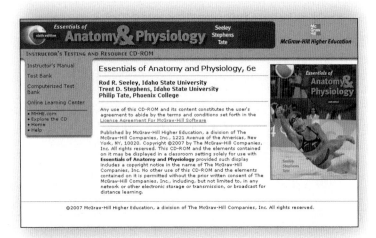

Transparencies

The set of transparency acetates that accompanies the text includes 250 full-color images identified by the authors as the most useful figures to incorporate into lecture presentations.

Laboratory Manual

The *Laboratory Manual* to accompany *Essentials of Anatomy and Physiology,* authored by Kevin Patton of St. Charles Community College and St. Louis University School of Medicine, divides the material typically covered in anatomy and physiology labs into 43 subunits. Selection of the subunits and the sequence of their use permits the design of a laboratory course that is integrated with the emphasis and sequence of the lecture material. Basic content is introduced first, and gradually more complex activities are developed. This laboratory manual also contains coloring and labeling exercises, optional computer exercises, boxed hints, safety alerts, separate lab reports, and a full-color histology minireference. Access to a comprehensive instructor's manual containing answers to the lab reports is provided to instructors.

Online Learning Center

The *Essentials of Anatomy and Physiology* Online Learning Center at **www.mhhe.com/seeleyess6** is a comprehensive website created for instructors and students using the Seeley, Stephens, and Tate textbook. The Online Learning Center allows instructors complete access to all student features, as well as exclusive access to a separate Instructor Center that houses downloadable and printable versions of traditional ancillaries, plus additional instructor content. Contact your McGraw-Hill sales representative for your instructor user name and password.

Course Delivery Systems

With help from our partners WebCT, Blackboard, Top-Class, eCollege, and other course management systems, professors can take complete control over their course content. Course cartridges containing Online Learning Center content, online testing, and powerful student tracking features are readily available for use within these platforms.

For Students
Online Learning Center

The *Essentials of Anatomy and Physiology* Online Learning Center at **www.mhhe.com/seeleyess6** offers access to a vast array of premium online content to fortify the learning experience.

Essential Study Partner   A collection of interactive study modules that contains hundreds of animations, learning activities, and quizzes designed to help students grasp complex concepts.

Online Tutoring   A 24-hour tutorial service moderated by qualified instructors. Help with difficult concepts is only an email away.

In addition to these outstanding online tools, the OLC features quizzes, interactive learning games, and study tools tailored to correlate with each chapter of the text.

Anatomy & Physiology Revealed (APR) CD-ROM

Anatomy & Physiology Revealed is a unique multimedia study aid designed to help you learn and review human anatomy using cadaver specimens. Dissections, animations, imaging, and self-tests all work together as an exceptional tool for the study of structure and function.

The complete Anatomy & Physiology Revealed series includes:

Volume 1—Skeletal and Muscular Systems
Volume 2—Nervous System
Volume 3—Cardiovascular, Respiratory, and Lymphatic Systems
Volume 4—Digestive, Urinary, Reproductive, and Endocrine Systems

Visit **www.aprevealed.com** for more information.

Physiology Interactive Lab Simulations (Ph.I.L.S) 3.0

This CD-ROM contains 26 lab simulations (**15 of which are new**) that allow students to perform experiments without using expensive lab equipment or live animals. This easy-to-use software offers students the flexibility to change the parameters of every lab experiment, with no limit to the amount of times they can repeat experiments or modify variables. This power to manipulate each experiment reinforces key physiology concepts by helping students to view outcomes, make predictions, and draw conclusions.

MediaPhys

This interactive CD-ROM offers detailed explanations, high-quality illustrations, and animations to provide a thorough introduction to the world of physiology. MediaPhys is filled with interactive activities and quizzes to help reinforce physiology concepts that are often difficult to understand.

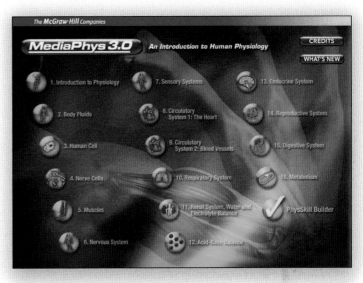

Virtual Anatomy Dissection Review CD-ROM

This multimedia program contains high-quality cat dissection photographs that are correlated to illustrations and photos of human structures. This CD makes it easy to identify and review cat anatomy, and also relate the cat specimen to corresponding human structures. Key features include:

- Vivid cat dissection photos
- Photos of human surface anatomy paired with detailed drawings
- Audio pronunciations of muscle names
- Concise listings of the actions, origins, and insertions of skeletal muscles
- Video clips demonstrating muscle actions
- Quizzing activities

Chapter Outline and Objectives

The Human Organism

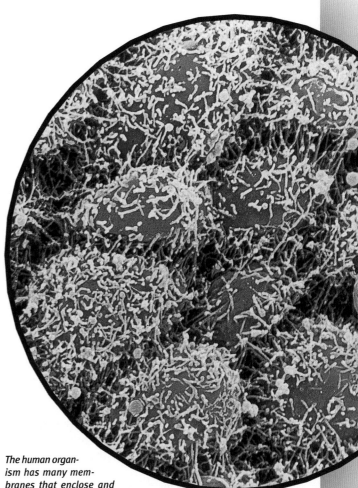

The human organism has many membranes that enclose and protect underlying structures. The colorized scanning electron micrograph shows the cells forming the peritoneum, a membrane covering abdominopelvic organs and the inside wall of the abdominopelvic cavity. These cells have many short, hairlike projections called microvilli. The microvilli increase the surface area of the cells, enabling them to secrete a slippery lubricating fluid that protects organs from friction as they rub against one another or the inside of the abdominopelvic wall.

human anatomy and physiology is the study of the structure and function of the human body. The human body has many intricate parts with coordinated functions maintained by a complex system of checks and balances. The coordinated function of all the parts of the human body allows us to detect stimuli, respond to stimuli, and perform many other functions.

Knowing the structure and function of the human body allows us to understand how the body responds to a stimulus. For example, eating a candy bar results in an increase in blood sugar (the stimulus). Knowledge of the pancreas allows us to predict that the pancreas will secrete insulin (the response). Insulin moves into blood vessels and is transported to cells, where it increases the movement of sugar from the blood into the cells, providing them with a source of energy. As glucose moves into cells, blood sugar levels decrease.

Knowing human anatomy and physiology also provides the basis for understanding disease. In one type of diabetes mellitus, for example, the pancreas does not secrete adequate amounts of insulin. Without adequate insulin, not enough sugar moves into cells, depriving them of a needed source of energy and resulting in their malfunction.

The study of human anatomy and physiology is important for those who plan a career in the health sciences because a sound knowledge of structure and function is necessary for health professionals to perform their duties. Understanding anatomy and physiology also prepares anyone to evaluate recommended treatments, critically review advertisements and reports in the popular literature, and rationally discuss the human body with health professionals and nonprofessionals.

Anatomy

Anatomy (ă-nat′ŏ-mē) is the scientific discipline that investigates the structure of the body. The word "anatomy" means to dissect, or cut apart and separate, the parts of the body for study. Anatomy covers a wide range of studies, including the structure of body parts, their microscopic organization, and the processes by which they develop. In addition, anatomy examines the relationship between the structure of a body part and its function. Just as the structure of a hammer makes it well suited for pounding nails, the structure of body parts allows them to perform specific functions effectively. For example, bones can provide strength and support because bone cells surround themselves with a hard, mineralized substance. Understanding the relationship between structure and function makes it easier to understand and appreciate anatomy.

Systemic and regional anatomy are two basic approaches to the study of anatomy. **Systemic anatomy** is the study of the body by systems and is the approach taken in this and most other introductory textbooks. Examples of systems are the circulatory, nervous, skeletal, and muscular systems. **Regional anatomy** is the study of the organization of the body by areas. Within each region, such as the head, abdomen, or arm, all systems are studied simultaneously. It is the approach taken in most medical and dental schools.

Surface anatomy and anatomical imaging are used to examine the internal structures of a living person. **Surface anatomy** is the study of external features, such as bony projections, which serve as landmarks for locating deeper structures (for examples of external landmarks, see chapters 6 and 7). **Anatomic imaging** involves the use of x-rays, ultrasound, magnetic resonance imaging (MRI), and other technologies to create pictures of internal structures. Both surface anatomy and anatomic imaging provide important information useful in diagnosing disease.

Cadavers and the Law

For much of history, public sentiment made it difficult for anatomists to obtain human bodies for dissection. In the early 1800s, the benefits of human dissection for training physicians had become very apparent, and the need for cadavers increased beyond the ability to acquire them legally. Thus arose the resurrectionists, or body snatchers. For a fee and no questions asked, they removed bodies from graves and provided them to medical schools. Because the bodies were not easy to obtain and were not always in the best condition, two enterprising men named William Burke and William Hare went one step further. Over a period of time, they murdered 17 people in Scotland and sold their bodies to a medical school. When discovered, Hare testified against Burke and went free. Burke was convicted, hanged, and publicly dissected. Discovery of Burke's activities so outraged the public that sensible laws regulating the acquisition of cadavers were soon passed, and this dark chapter in the history of anatomy was closed.

Physiology

Physiology (fiz-ē-ol′ō-jē, the study of nature) is the scientific discipline that deals with the processes or functions of living things. It is important in physiology to recognize structures as dynamic rather than static, or unchanging. The major goals of physiology are (1) to understand and predict the body's responses to stimuli, and (2) to understand how the body maintains conditions within a narrow range of values in the presence of a continually changing environment.

Physiology is divided according to (1) the organisms involved or (2) the levels of organization within a given organism. **Human physiology** is the study of a specific organism, the human, whereas **cellular** and **systemic physiology** are examples of physiology that emphasize specific organizational levels.

Structural and Functional Organization

The body can be studied at six structural levels: the chemical, cell, tissue, organ, organ system, and organism (figure 1.1).

Chemical

The structural and functional characteristics of all organisms are determined by their chemical makeup. The **chemical** level of organization involves interactions among atoms and their combinations into molecules. The function of a molecule is related intimately to its structure. For example, collagen molecules are strong, ropelike fibers that give skin structural strength and flexibility. With old age, the structure of collagen changes, and the skin becomes fragile and is torn more easily. A brief overview of chemistry is presented in chapter 2.

Cell

Cells are the basic structural and functional units of organisms, such as plants and animals. Molecules can combine to form **organelles** (or′gă-nelz, *organon*, a tool + *-elle*, small), which are the small structures that make up cells. For example, the nucleus contains the cell's hereditary information and mitochondria manufacture adenosine triphosphate (ATP), which is a molecule used by cells for a source of energy. Although cell types differ in their structure and function, they have many characteristics in common. Knowledge of these characteristics and their variations is essential to a basic understanding of anatomy and physiology. The cell is discussed in chapter 3.

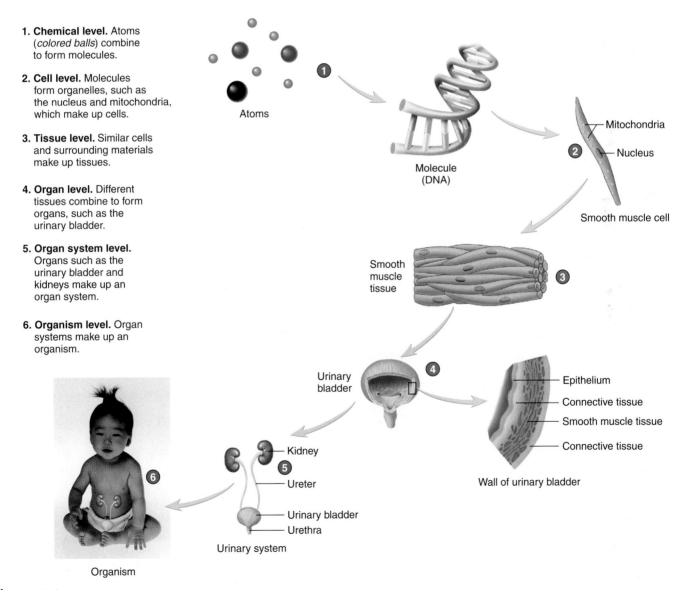

1. **Chemical level.** Atoms (*colored balls*) combine to form molecules.

2. **Cell level.** Molecules form organelles, such as the nucleus and mitochondria, which make up cells.

3. **Tissue level.** Similar cells and surrounding materials make up tissues.

4. **Organ level.** Different tissues combine to form organs, such as the urinary bladder.

5. **Organ system level.** Organs such as the urinary bladder and kidneys make up an organ system.

6. **Organism level.** Organ systems make up an organism.

Atoms

Molecule (DNA)

Mitochondria
Nucleus

Smooth muscle cell

Smooth muscle tissue

Urinary bladder

Epithelium
Connective tissue
Smooth muscle tissue
Connective tissue

Wall of urinary bladder

Kidney
Ureter
Urinary bladder
Urethra

Urinary system

Organism

Figure 1.1   Levels of Organization

Six levels of organization for the human body are the chemical, cell, tissue, organ, organ system, and organism.

Tissue

A **tissue** (tish'ū, *texo,* to weave) is a group of similar cells and the materials surrounding them. The characteristics of the cells and surrounding materials determine the functions of the tissue. The many tissues that make up the body are classified into four primary tissue types: epithelial, connective, muscle, and nervous. Tissues are discussed in chapter 4.

Organ

An **organ** (ōr′gǎn, a tool) is composed of two or more tissue types that together perform one or more common functions. The urinary bladder, skin, stomach, eye, and heart are examples of organs (figure 1.2).

Organ System

An **organ system** is a group of organs classified as a unit because of a common function or set of functions. For example, the urinary system consists of the kidneys, ureter, urinary bladder, and urethra. The kidneys produce urine, which is transported by the ureters to the urinary bladder, where it is stored until eliminated from the body by passing through the urethra. In this text the body is considered to have 11 major organ systems: integumentary, skeletal, muscular, lymphatic, respiratory, digestive, nervous, endocrine, cardiovascular, urinary, and reproductive (figure 1.3).

The coordinated activity of the organ systems is necessary for normal function. For example, the digestive system takes in and processes food, which is carried by the blood of the cardiovascular system to the cells of the other systems. These cells use the food and produce waste products that are carried by the blood to the kidneys of the urinary system, which removes waste products from the blood. Because the organ systems are so interrelated, dysfunction of one organ system can have profound effects on other systems. For example, a heart attack can result in inadequate circulation of blood. Consequently, the organs of other systems, such as the brain and kidneys, can malfunction. Throughout this text, the interactions of the organ systems are considered in the Systems Pathology essays.

Organism

An **organism** is any living thing considered as a whole, whether composed of one cell, such as a bacterium, or trillions of cells, such as a human. The human organism is a complex of organ systems that are mutually dependent on one another (see figure 1.2).

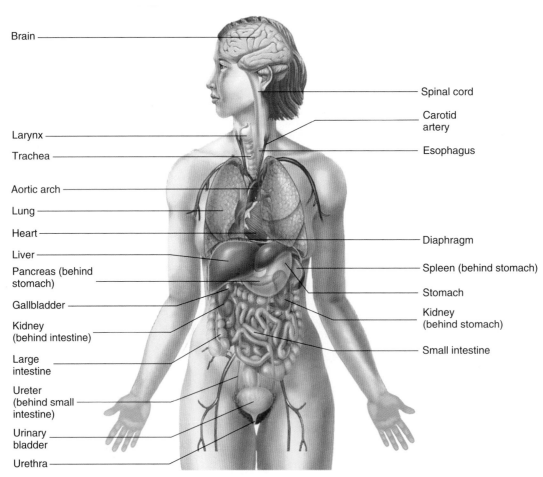

Figure 1.2   **Organs of the Body**

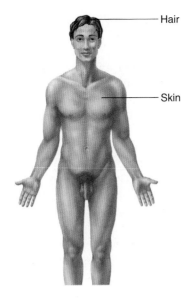

Integumentary System

Provides protection, regulates temperature, prevents water loss, and produces vitamin D precursors. Consists of skin, hair, nails, and sweat glands.

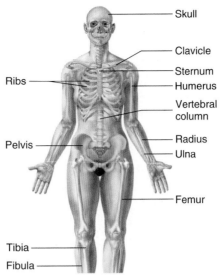

Skeletal System

Provides protection and support, allows body movements, produces blood cells, and stores minerals and fat. Consists of bones, associated cartilages, ligaments, and joints.

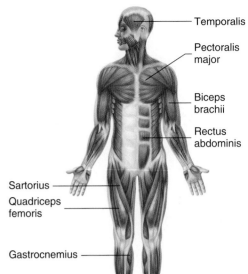

Muscular System

Produces body movements, maintains posture, and produces body heat. Consists of muscles attached to the skeleton by tendons.

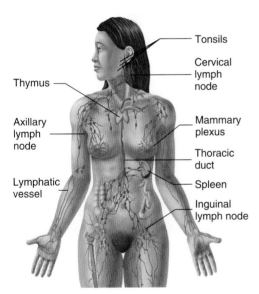

Lymphatic System

Removes foreign substances from the blood and lymph, combats disease, maintains tissue fluid balance, and absorbs fats from the digestive tract. Consists of the lymphatic vessels, lymph nodes, and other lymphatic organs.

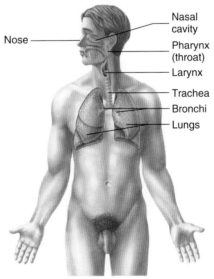

Respiratory System

Exchanges oxygen and carbon dioxide between the blood and air and regulates blood pH. Consists of the lungs and respiratory passages.

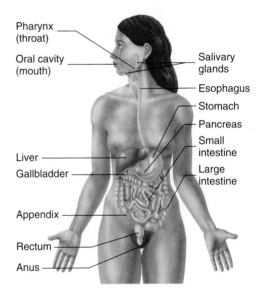

Digestive System

Performs the mechanical and chemical processes of digestion, absorption of nutrients, and elimination of wastes. Consists of the mouth, esophagus, stomach, intestines, and accessory organs.

Figure 1.3 Organ Systems of the Body

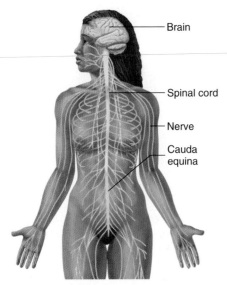

Nervous System

A major regulatory system that detects sensations and controls movements, physiologic processes, and intellectual functions. Consists of the brain, spinal cord, nerves, and sensory receptors.

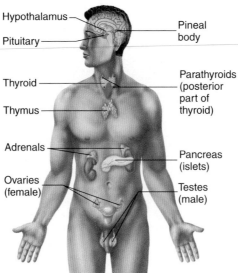

Endocrine System

A major regulatory system that influences metabolism, growth, reproduction, and many other functions. Consists of glands, such as the pituitary, that secrete hormones.

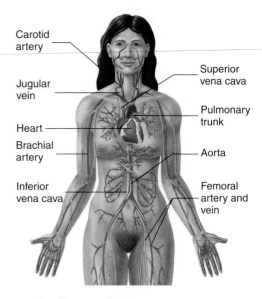

Cardiovascular System

Transports nutrients, waste products, gases, and hormones throughout the body; plays a role in the immune response and the regulation of body temperature. Consists of the heart, blood vessels, and blood.

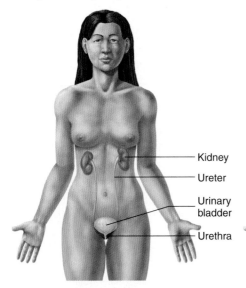

Urinary System

Removes waste products from the blood and regulates blood pH, ion balance, and water balance. Consists of the kidneys, urinary bladder, and ducts that carry urine.

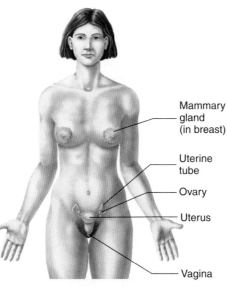

Female Reproductive System

Produces oocytes and is the site of fertilization and fetal development; produces milk for the newborn; produces hormones that influence sexual function and behaviors. Consists of the ovaries, vagina, uterus, mammary glands, and associated structures.

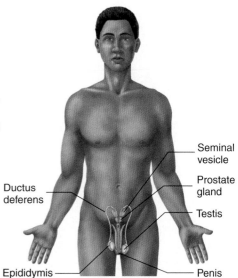

Male Reproductive System

Produces and transfers sperm cells to the female and produces hormones that influence sexual functions and behaviors. Consists of the testes, accessory structures, ducts, and penis.

Figure 1.3   Organ Systems of the Body *(continued)*

Characteristics of Life

Humans are organisms and have many characteristics in common with other organisms. The most important common feature of all organisms is life. Essential characteristics of life are organization, metabolism, responsiveness, growth, development, and reproduction.

1. **Organization** is the condition in which the parts of an organism have specific relationships to each other and the parts interact to perform specific functions. Living things are highly organized. All organisms are composed of one or more cells. Cells, in turn, are composed of highly specialized organelles, which depend on the precise functions of large molecules. Disruption of this organized state can result in loss of functions and death.

2. **Metabolism** (mĕ-tab′ō-lizm, *metabolē,* change) is the ability to use energy to perform vital functions, such as growth, movement, and reproduction. Plants can capture energy from sunlight, and humans obtain energy from food.

3. **Responsiveness** is the ability of an organism to sense changes in the environment and make the adjustments that help maintain its life. Responses include movement toward food or water and away from danger or poor environmental conditions. Organisms can also make adjustments that maintain their internal environment. For example, if body temperature increases in a hot environment, sweat glands produce sweat, which can lower body temperature back toward normal levels.

4. **Growth** results in an increase in size of all or part of the organism. It can result from an increase in cell number, cell size, or the amount of substance surrounding cells. For example, bones become larger as the number of bone cells increases and they surround themselves with bone matrix.

5. **Development** includes the changes an organism undergoes through time; it begins with fertilization and ends at death. The greatest developmental changes occur before birth, but many changes continue after birth, and some continue throughout life. Development usually involves growth, but it also involves differentiation. Differentiation is change in cell structure and function from generalized to specialized. For example, following fertilization, generalized cells specialize to become specific cell types, such as skin, bone, muscle, or nerve cells. These differentiated cells form the tissues and organs.

6. **Reproduction** is the formation of new cells or new organisms. Without reproduction of cells, growth and tissue repair are impossible. Without reproduction of the organism, the species becomes extinct.

Human Versus Animal-Based Knowledge

Humans share many characteristics with other organisms, and much of the knowledge about humans has come from studying other organisms. For example, the study of bacteria (single-celled organisms) has provided much information about human cells; and great progress in open heart surgery was made possible by perfecting techniques on other mammals before attempting them on humans. Because other organisms are also different from humans, the ultimate answers to questions about humans can be obtained only from humans.

Homeostasis

Homeostasis (hō′mē-ō-stā′sis, homeo-, the same + *stasis,* standing) is the existence and maintenance of a relatively constant environment within the body. Most cells of the body are surrounded by a small amount of fluid, and normal cell functions depend on the maintenance of the cells' fluid environment within a narrow range of conditions, including temperature, volume, and chemical content. These conditions are called **variables** because their values can change. For example, body temperature is a variable that can increase in a hot environment or decrease in a cold environment.

Homeostatic mechanisms, such as sweating or shivering, normally maintain body temperature near an ideal normal value, or **set point** (figure 1.4). Note that these mechanisms are not able to maintain body temperature precisely at the set point. Instead, body temperature increases and decreases slightly around the set point, producing a **normal range** of values. As long as body temperatures remain within this normal range, homeostasis is maintained.

The organ systems help control the internal environment so that it remains relatively constant. For example, the digestive, respiratory, circulatory, and urinary systems function together so that each cell in the body receives adequate oxygen and nutrients and so that waste products do not accumulate to a toxic level. If the fluid surrounding cells deviates from homeostasis,

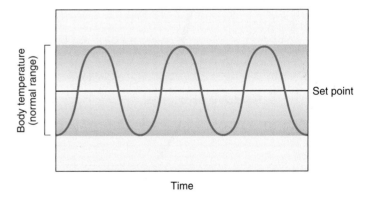

Figure 1.4 Homeostasis

Homeostasis is the maintenance of a variable, such as body temperature, around an ideal normal value, or set point. The value of the variable fluctuates around the set point to establish a normal range of values.

the cells do not function normally and may even die. Disease disrupts homeostasis and sometimes results in death.

Negative Feedback

Most systems of the body are regulated by **negative-feedback mechanisms,** which function to maintain homeostasis. *Negative* means that any deviation from the set point is made smaller or is resisted. Negative feedback does not prevent variation but maintains variation within a normal range.

The maintenance of normal blood pressure is an example of a negative-feedback mechanism. Normal blood pressure is important because it is responsible for moving blood from the heart to tissues. The blood supplies the tissues with oxygen and nutrients and removes waste products. Thus, normal blood pressure is required to ensure that tissue homeostasis is maintained.

Many negative-feedback mechanisms, such as the one maintaining normal blood pressure, have three components:

(1) a **receptor** (rē-sep′tŏr, rē-sep′tōr) monitors the value of a variable such as blood pressure; (2) a **control center,** such as part of the brain, establishes the set point around which the variable is maintained; and (3) an **effector** (ē-fek′tŏr), such as the heart, can change the value of the variable. Blood pressure depends in part on contraction (beating) of the heart: as heart rate increases, blood pressure increases, and as heart rate decreases, blood pressure decreases.

The receptors that monitor blood pressure are located within large blood vessels near the heart. If blood pressure increases slightly, the receptors detect the increased blood pressure and send that information to the control center in the brain. The control center causes heart rate to decrease, resulting in a decrease in blood pressure. If blood pressure decreases slightly, the receptors inform the control center, which increases heart rate, producing an increase in blood pressure. As a result, blood pressure is maintained within a normal range (figure 1.5).

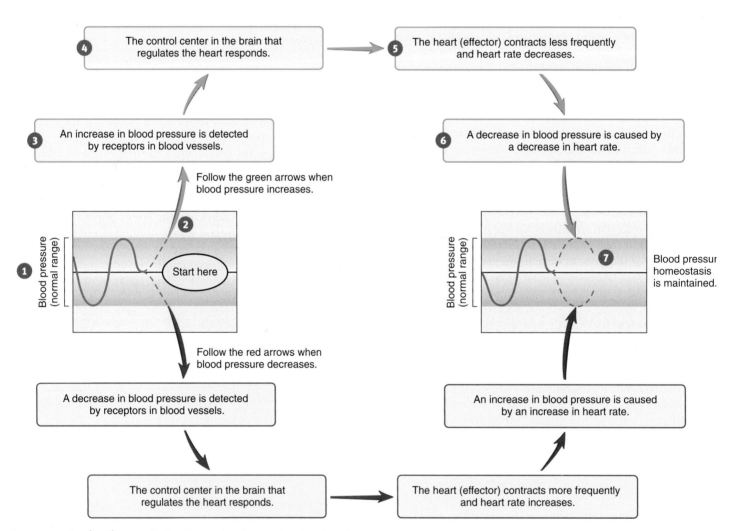

Homeostasis Figure 1.5   Example of Negative Feedback

Throughout the text, all homeostasis figures have the same format as shown here. The changes caused by an increase of a variable outside the normal range are shown in the green boxes, and the changes caused by a decrease are shown in the red boxes. To help you learn how to interpret homeostasis figures, some of the steps in this figure are numbered: (1) Blood pressure is within its normal range. (2) Blood pressure increases and is outside the normal range. (3) The increase in blood pressure is detected by receptors. (4) The blood pressure control center responds to the change in blood pressure detected by the receptors. (5) The control center causes heart rate to decrease. (6) The decrease in heart rate causes blood pressure to decrease. (7) Blood pressure returns to its normal range, and homeostasis is maintained. Follow the responses to a decrease in blood pressure outside its normal range by following the red boxes.

P R E D I C T   **1**
Donating a pint of blood reduces blood volume, which results in a decrease in blood pressure (just as air pressure in a tire decreases as air is let out of the tire). What effect does donating blood have on heart rate? What would happen if a negative-feedback mechanism did not return the value of some parameter such as blood pressure to its normal range?

Positive Feedback

Positive-feedback mechanisms are *not* homeostatic and are rare in healthy individuals. *Positive* implies that when a deviation from a normal value occurs, the response of the system is to make the deviation even greater. Positive feedback therefore usually creates a cycle leading away from homeostasis and in some cases results in death.

Inadequate delivery of blood to cardiac (heart) muscle is an example of positive feedback. Contraction of cardiac muscle generates blood pressure and moves blood through blood vessels to tissues. A system of blood vessels on the outside of the heart provides cardiac muscle with a blood supply sufficient to allow normal contractions to occur. In effect, the heart pumps blood to itself. Just as with other tissues, blood pressure must be maintained to ensure adequate delivery of blood to cardiac muscle. Following extreme blood loss, blood pressure decreases to the point at which the delivery of blood to cardiac muscle is inadequate. As a result, cardiac muscle homeostasis is disrupted, and cardiac muscle does not function normally. The heart pumps less blood, which causes the blood pressure to drop even further. The additional decrease in blood pressure causes less blood delivery to cardiac muscle, and the heart pumps even less blood, which again decreases the blood pressure (figure 1.6). The process continues until the blood pressure is too low to sustain the cardiac muscle, the heart stops beating, and death results.

Following a moderate amount of blood loss (e.g., after donating a pint of blood), negative-feedback mechanisms result in an increase in heart rate that restores blood pressure. If blood loss is severe, however, negative-feedback mechanisms may not be able to maintain homeostasis, and the positive-feedback effect of an ever-decreasing blood pressure can develop.

Circumstances in which negative-feedback mechanisms are not adequate to maintain homeostasis illustrate a basic principle. Many disease states result from failure of negative-feedback mechanisms to maintain homeostasis. The purpose of medical therapy is to overcome illness by aiding negative-feedback mechanisms. For example, a transfusion reverses a constantly decreasing blood pressure and restores homeostasis.

A few positive-feedback mechanisms do operate in the body under normal conditions, but in all cases they are eventually limited in some way. Birth is an example of a normally occurring positive-feedback mechanism. Near the end of pregnancy, the uterus is stretched by the baby's large size. This stretching, especially around the opening of the uterus, stimulates contractions of the uterine muscles. The uterine contractions push the baby against the opening of the uterus,

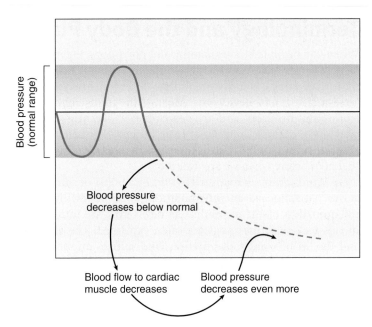

Figure 1.6   **Example of Harmful Positive Feedback**
A decrease in blood pressure below the normal range causes decreased blood flow to the heart. The heart is unable to pump enough blood to maintain blood pressure, and blood flow to the cardiac muscle decreases. Thus, the ability of the heart to pump decreases further, and blood pressure decreases even more.

stretching it further. This stimulates additional contractions that result in additional stretching. This positive-feedback sequence ends only when the baby is delivered from the uterus and the stretching stimulus is eliminated.

Humors and Homeostasis

The idea that the body maintains a balance (homeostasis) can be traced back to ancient Greece. It was believed that the body supported four juices, or humors: the red juice of blood, the yellow juice of bile, the white juice secreted from the nose and lungs, and the black juice in the pancreas. It was also thought that health resulted from a proper balance of these juices and that disease was caused by an excess of one of them. Normally the body would attempt to heal itself by expelling the excess juice. An example was thought to be the expulsion of mucus from the nose of a person with a cold. This led to the practice of bloodletting to restore the body's normal balance of juices. Tragically, in the eighteenth and nineteenth centuries, bloodletting went to extremes. When bloodletting did not result in improvement of the patient, it was taken as evidence that not enough blood had been removed to restore a healthy balance of the body's juices. The obvious solution was to let more blood, undoubtedly causing many deaths. Eventually the failure of this approach became obvious, and the practice was abandoned. Fortunately, we now understand more about how the body maintains homeostasis.

P R E D I C T   **2**
Is the sensation of thirst associated with a negative- or a positive-feedback mechanism? Explain. (*Hint:* What is being regulated when one becomes thirsty?)

Terminology and the Body Plan

When you begin to study anatomy and physiology, the number of new words may seem overwhelming. Learning is easier and more interesting if you pay attention to the origin, or **etymology** (et′ĕ-mol′o-jē), of new words. Most of the terms are derived from Latin or Greek and are descriptive in the original languages. For example, anterior in Latin means "to go before." The anterior surface of the body is therefore the surface of the body that goes before when we are walking.

Words are often modified by adding a prefix or suffix. The suffix "-itis" means an inflammation; so appendicitis is an inflammation of the appendix. As new terms are introduced in this text, their meanings are often explained. The glossary and the list of word roots, prefixes, and suffixes on the inside back cover of the textbook also provide additional information about the new terms.

Body Positions

The **anatomic position** refers to a person standing erect with the face directed forward, the upper limbs hanging to the sides, and the palms of the hands facing forward (figure 1.7). A person is **supine** when lying face upward and **prone** when lying face downward.

The position of the body can affect the description of body parts relative to each other. In the anatomic position, the elbow is above the hand, but in the supine or prone position, the elbow and hand are at the same level. To avoid confusion, relational descriptions are always based on the anatomic position, no matter the actual position of the body. Thus, the elbow is always described as being above the wrist.

Directional Terms

Directional terms describe parts of the body relative to each other (see figure 1.7 and table 1.1). It is important to become familiar with these directional terms as soon as possible because you will see them repeatedly throughout the text. Right and left are retained as directional terms in anatomic terminology. Up is replaced by **superior,** down by **inferior,** front by **anterior,** and back by **posterior.**

The word *anterior* means that which goes before, and **ventral** means belly. The anterior surface of the human body is therefore the ventral surface, or belly, because the belly "goes first" when we are walking. The word *posterior* means that which follows, and **dorsal** means back. The posterior surface of the body is the dorsal surface, or back, which follows as we are walking.

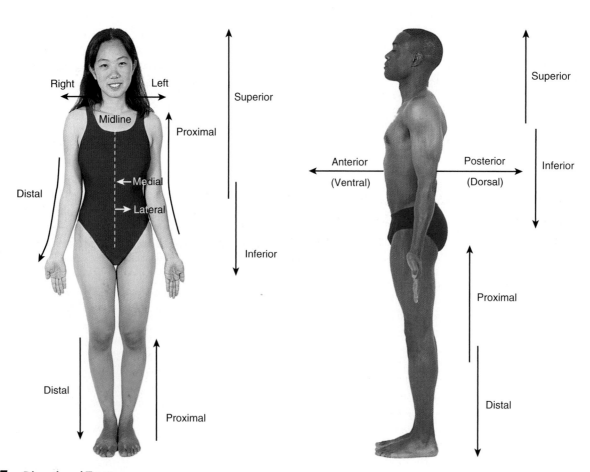

Figure 1.7   Directional Terms

All directional terms are in relation to a person in the anatomic position: a person standing erect with the face directed forward, the arms hanging to the sides, and the palms of the hands facing forward.

Proximal means nearest, whereas **distal** means distant. These terms are used to refer to linear structures, such as the limbs, in which one end is near some other structure and the other end is farther away. Each limb is attached at its proximal end to the body, and the distal end, such as the hand, is farther away.

Medial means toward the midline, and **lateral** means away from the midline. The nose is located in a medial position in the face, and the eyes are lateral to the nose. The term **superficial** refers to a structure close to the surface of the body, and **deep** is toward the interior of the body. The skin is superficial to muscle and bone.

P R E D I C T   3
Provide the correct directional term for the following statement. When a boy is standing on his head, his nose is _____ to his mouth.

Body Parts and Regions

A number of terms are used when referring to different regions or parts of the body (figure 1.8). The upper limb is divided into the arm, forearm, wrist, and hand. The **arm** extends from the shoulder to the elbow, and the **forearm** extends from the elbow to the wrist. The lower limb is divided into the thigh, leg, ankle, and foot. The **thigh** extends from the hip to the knee, and the **leg** extends from the knee to the ankle. Note that, contrary to popular usage, the terms *arm* and *leg* refer to only a part of the respective limb.

The central region of the body consists of the **head, neck,** and **trunk.** The trunk can be divided into the **thorax** (chest), **abdomen** (region between the thorax and pelvis), and **pelvis** (the inferior end of the trunk associated with the hips).

The abdomen is often subdivided superficially into four **quadrants** by two imaginary lines—one horizontal and one vertical—that intersect at the navel (figure 1.9a). The quadrants

formed are the right upper, left upper, right lower, and left lower quadrants. In addition to these quadrants, the abdomen is sometimes subdivided into nine **regions** by four imaginary lines—two horizontal and two vertical. These four lines create an imaginary tic-tac-toe figure on the abdomen, resulting in nine regions: epigastric (ep-i-gas′trik), right and left hypochondriac (hī-pō-kon′drē-ak), umbilical (ŭm-bil′i-kăl), right and left lumbar (lŭm′bar), hypogastric (hī-pō-gas′trik), and right and left iliac (il′ē-ak) (figure 1.9b). Clinicians use the quadrants or regions as reference points for locating the underlying organs. For example, the appendix is located in the right lower quadrant, and the pain of an acute appendicitis is usually felt there.

P R E D I C T   4
Using figures 1.2 and 1.9a, determine in which quadrant each of the following organs is located: spleen, gallbladder, kidneys, most of the stomach, and most of the liver.

A CASE IN POINT | Epigastric Pain

Wilby Hurtt has epigastric pain that is most noticeable following meals and at night when he is lying in bed. He probably has gastroesophageal reflux disease (GERD), in which stomach acid improperly moves into the esophagus, damaging and irritating the lining of the esophagus. Epigastric pain, however, can have many causes and should be evaluated by a physician. For example, gallstones, stomach or small intestine ulcers, inflammation of the pancreas, and heart disease can cause epigastric pain.

Table 1.1	Directional Terms for Humans		
Terms	**Etymology**	**Definition***	**Example**
Right		Toward the body's right side	The right ear
Left		Toward the body's left side	The left ear
Inferior	Lower	Below	The nose is inferior to the forehead
Superior	Higher	Above	The mouth is superior to the chin
Anterior	To go before	Toward the front of the body	The teeth are anterior to the throat
Posterior	*Posterus,* following	Toward the back of the body	The brain is posterior to the eyes
Dorsal	*Dorsum,* back	Toward the back (synonymous with posterior)	The spine is dorsal to the breastbone
Ventral	*Venter,* belly	Toward the belly (synonymous with anterior)	The navel is ventral to the spine
Proximal	*Proximus,* nearest	Closer to a point of attachment	The elbow is proximal to the wrist
Distal	*di + sto,* to be distant	Farther from a point of attachment	The knee is distal to the hip
Lateral	*Latus,* side	Away from the midline of the body	The nipple is lateral to the breastbone
Medial	*Medialis,* middle	Toward the middle or midline of the body	The bridge of the nose is medial to the eye
Superficial	*Superficialis,* surface	Toward or on the surface	The skin is superficial to muscle
Deep	*Deop,* deep	Away from the surface, internal	The lungs are deep to the ribs

* All directional terms refer to a human in the anatomic position.

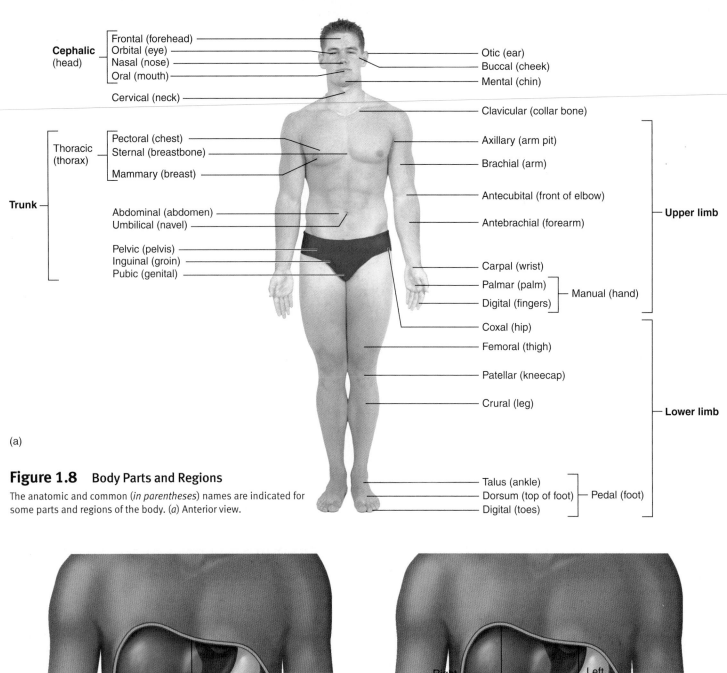

Cephalic (head)
- Frontal (forehead)
- Orbital (eye)
- Nasal (nose)
- Oral (mouth)

Cervical (neck)

Otic (ear)
Buccal (cheek)
Mental (chin)
Clavicular (collar bone)

Trunk

Thoracic (thorax)
- Pectoral (chest)
- Sternal (breastbone)
- Mammary (breast)

Abdominal (abdomen)
Umbilical (navel)

Pelvic (pelvis)
Inguinal (groin)
Pubic (genital)

Axillary (arm pit)
Brachial (arm)
Antecubital (front of elbow)
Antebrachial (forearm)

Carpal (wrist)
Palmar (palm)
Digital (fingers)
— Manual (hand)

Upper limb

Coxal (hip)
Femoral (thigh)
Patellar (kneecap)
Crural (leg)

Lower limb

Talus (ankle)
Dorsum (top of foot) — Pedal (foot)
Digital (toes)

(a)

Figure 1.8 Body Parts and Regions

The anatomic and common (*in parentheses*) names are indicated for some parts and regions of the body. (*a*) Anterior view.

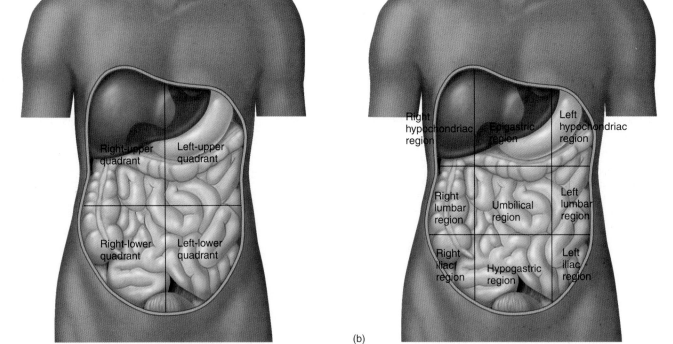

(a)
- Right-upper quadrant
- Left-upper quadrant
- Right-lower quadrant
- Left-lower quadrant

(b)
- Right hypochondriac region
- Epigastric region
- Left hypochondriac region
- Right lumbar region
- Umbilical region
- Left lumbar region
- Right iliac region
- Hypogastric region
- Left iliac region

Figure 1.9 Subdivisions of the Abdomen

Lines are superimposed over internal organs to demonstrate the relationship of the organs to the subdivisions. (*a*) Abdominal quadrants consist of four subdivisions. (*b*) Abdominal regions consist of nine subdivisions.

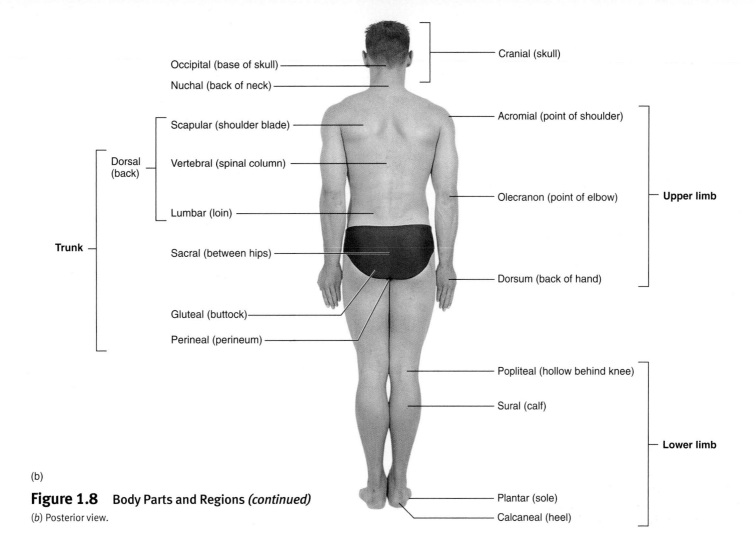

(b)

Figure 1.8 Body Parts and Regions *(continued)*

(b) Posterior view.

Planes

At times it is conceptually useful to discuss the body in reference to a series of planes (imaginary flat surfaces) passing through it (figure 1.10). Sectioning the body is a way to "look inside" and observe the body's structures. A **sagittal** (saj′i-tăl)) **plane** runs vertically through the body and separates it into right and left parts. The word *sagittal* literally means "the flight of an arrow" and refers to the way the body would be split by an arrow passing anteriorly to posteriorly. A **median plane** is a sagittal plane that passes through the midline of the body and divides it into equal right and left halves. A **transverse** (trans-vers′), or **horizontal, plane** runs parallel to the surface of the ground and divides the body into superior and inferior parts. A **frontal,** or **coronal** (kōr′ŏ-năl, kō-rō′nal), **plane** runs vertically from right to left and divides the body into anterior and posterior parts.

Organs are often sectioned to reveal their internal structure (figure 1.11). A cut through the long axis of the organ is a **longitudinal section,** and a cut at a right angle to the long axis is a **transverse,** or **cross, section.** If a cut is made across the long axis at other than a right angle, it is called an **oblique section.**

Body Cavities

The body contains many cavities. Some of these cavities such as the nasal cavity open to the outside of the body, and some do not. The trunk contains three large cavities that do not open to the outside of the body: the thoracic cavity, the abdominal cavity, and the pelvic cavity (figure 1.12). The **thoracic cavity** is surrounded by the rib cage and is separated from the abdominal cavity by the muscular diaphragm. It is divided into right and left parts by a median structure called the **mediastinum** (me′dē-as-tī′nŭm, wall). The mediastinum is a partition containing the heart, thymus, trachea, esophagus, and other structures. The two lungs are located on either side of the mediastinum.

The **abdominal cavity** is bounded primarily by the abdominal muscles and contains the stomach, intestines, liver, spleen, pancreas, and kidneys. The **pelvic** (pel′vik) **cavity** is a small space enclosed by the bones of the pelvis and contains the urinary bladder, part of the large intestine, and the internal reproductive organs. The abdominal and pelvic cavities are not physically separated and sometimes are called the **abdominopelvic** (ab-dom′i-nō-pel′vik) **cavity.**

Serous Membranes

Serous (sēr′ŭs) **membranes** line the trunk cavities and cover the organs of these cavities. To understand the relationship between serous membranes and an organ, imagine an inflated balloon into which a fist has been pushed. The inner balloon wall in contact with the fist (organ) represents the **visceral** (vis′er-ăl, organ) **serous membrane,** and the outer part of the balloon

13

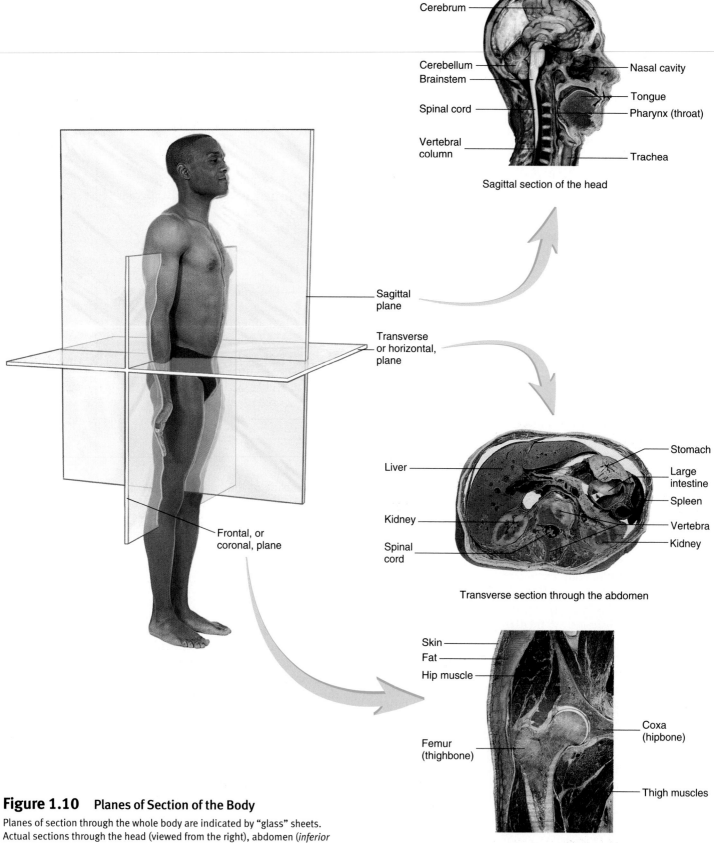

Cerebrum

Cerebellum

Brainstem

Spinal cord

Vertebral
column

Nasal cavity

Tongue

Pharynx (throat)

Trachea

Sagittal section of the head

Sagittal
plane

Transverse
or horizontal,
plane

Frontal, or
coronal, plane

Liver

Kidney

Spinal
cord

Stomach

Large
intestine

Spleen

Vertebra

Kidney

Transverse section through the abdomen

Skin

Fat

Hip muscle

Femur
(thighbone)

Coxa
(hipbone)

Thigh muscles

Frontal section through the right hip

Figure 1.10   Planes of Section of the Body

Planes of section through the whole body are indicated by "glass" sheets.
Actual sections through the head (viewed from the right), abdomen (*inferior
view*), and hip (*anterior view*) are also shown.

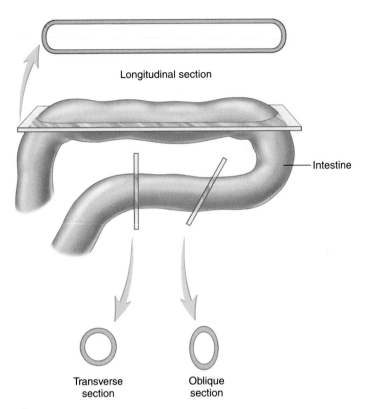

Figure 1.11 **Planes of Section Through an Organ**

Planes of section through the small intestine are indicated by "glass" sheets. The views of the small intestine after sectioning are also shown. Although the small intestine is basically a tube, the sections appear quite different in shape.

wall represents the **parietal** (pă-rī′ĕ-tăl, wall) **serous membrane** (figure 1.13). The cavity or space between the visceral and parietal serous membranes is normally filled with a thin, lubricating film of serous fluid produced by the membranes. As an organ rubs against another organ or against the body wall, the serous fluid and smooth serous membranes function to reduce friction.

The thoracic cavity contains three serous membrane-lined cavities: a pericardial cavity and two pleural cavities. The **pericardial** (per-i-kar′dē-ăl, around the heart) **cavity** surrounds the heart (figure 1.14a). The visceral pericardium covers the heart, which is contained within a connective tissue sac lined with the parietal pericardium. The pericardial cavity, which contains pericardial fluid, is located between the visceral and parietal pericardia.

A **pleural** (ploor′ăl, associated with the ribs) **cavity** surrounds each lung, which is covered by visceral pleura (figure 1.14b). Parietal pleura lines the inner surface of the thoracic wall, the lateral surfaces of the mediastinum, and the superior surface of the diaphragm. The pleural cavity is located between the visceral and parietal pleurae and contains pleural fluid.

The abdominopelvic cavity contains a serous membrane-lined cavity called the **peritoneal** (per′i-tō-nē′ăl, to stretch over) **cavity** (figure 1.14c). Visceral peritoneum covers many of the organs of the abdominopelvic cavity. Parietal peritoneum lines the wall of the abdominopelvic cavity and the inferior surface of the diaphragm. The peritoneal cavity is located between the visceral and parietal peritoneum and contains peritoneal fluid.

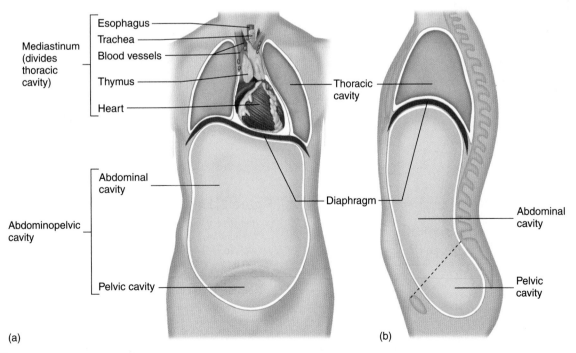

Figure 1.12 **Trunk Cavities**

(a) Anterior view showing the major trunk cavities. The diaphragm separates the thoracic cavity from the abdominal cavity. The mediastinum, which includes the heart, is a partition of organs dividing the thoracic cavity. (b) Sagittal section of the trunk cavities viewed from the left. The *dashed line* shows the division between the abdominal and pelvic cavities. The mediastinum has been removed to show the thoracic cavity.

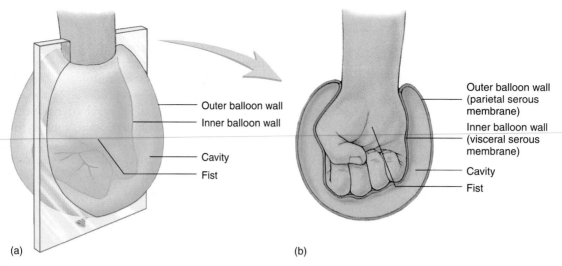

(a)

(b)

Figure 1.13   Serous Membranes

(a) Fist pushing into a balloon. A "glass" sheet indicates the location of a cross section through the balloon. (b) Interior view produced by the section in (a). The fist represents an organ, and the walls of the balloon the serous membranes. The inner wall of the balloon represents a visceral serous membrane in contact with the fist (organ). The outer wall of the balloon represents a parietal serous membrane.

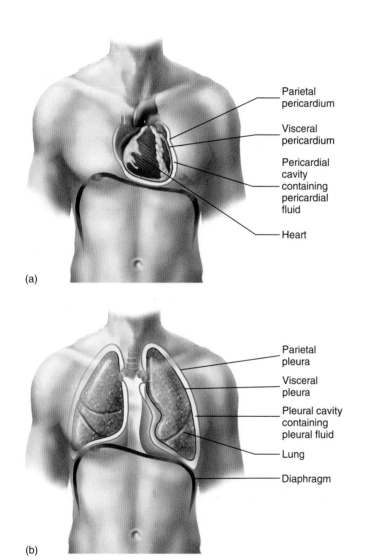

(a)

(b)

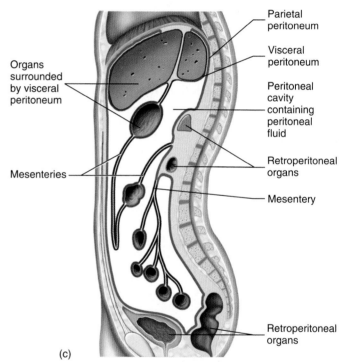

(c)

Figure 1.14   Location of Serous Membranes

(a) Frontal section showing the parietal pericardium (blue), visceral pericardium (red), and pericardial cavity. (b) Frontal section showing the parietal pleura (blue), visceral pleura (red), and pleural cavities. (c) Sagittal section through the abdominopelvic cavity showing the parietal peritoneum (blue), visceral peritoneum (red), peritoneal cavity, mesenteries (purple), and retroperitoneal organs.

Inflammation of Serous Membranes

The serous membranes can become inflamed—usually as a result of an infection. **Pericarditis** (per'i-kar-dī'tis) is inflammation of the pericardium, **pleurisy** (ploor'i-sē) is inflammation of the pleura, and **peritonitis** (per'i-tō-nī'tis) is inflammation of the peritoneum.

A CASE IN POINT | Peritonitis

May Day was rushed to the hospital emergency room. Earlier that day she experienced diffuse abdominal pain, but no fever. Then the pain became more intense and shifted to her right lower quadrant. She also developed fever. It was concluded that May Day had **appendicitis**, an inflammation of the appendix that is usually caused by an infection. The appendix is a small, wormlike sac attached to the large intestine. The outer surface of the appendix is visceral peritoneum. An infection of the appendix can rupture its wall, releasing bacteria into the peritoneal cavity, resulting in peritonitis. Appendicitis is the most common cause of emergency abdominal surgery in children and it often leads to peritonitis. May had her appendix removed, was treated with antibiotics, and made a full recovery.

Mesenteries (mes′en-ter-ē z), which consist of two layers of peritoneum fused together (see figure 1.14c), connect the visceral peritoneum of some abdominopelvic organs to the parietal peritoneum on the body wall or to the visceral peritoneum of other abdominopelvic organs. The mesenteries anchor the organs to the body wall and provide a pathway for nerves and blood vessels to reach the organs. Other abdominopelvic organs are more closely attached to the body wall and do not have mesenteries. Parietal peritoneum covers these other organs, which are said to be **retroperitoneal** (re′trō-per′i-tō-nē′ăl, *retro*, behind + peritoneum). The retroperitoneal organs include the kidneys, adrenal glands, pancreas, parts of the intestines, and the urinary bladder (see figure 1.14c)

P R E D I C T
Explain how an organ can be located within the abdominopelvic cavity but not be within the peritoneal cavity.

S U M M A R Y

A knowledge of anatomy and physiology can be used to predict the body's responses to stimuli when healthy or diseased.

Anatomy   (p. 2)

1. Anatomy is the study of the structures of the body.
2. Systemic anatomy is the study of the body by organ systems. Regional anatomy is the study of the body by areas.
3. Surface anatomy uses superficial structures to locate deeper structures, and anatomic imaging is a noninvasive method for examining deep structures.

Physiology   (p. 2)

Physiology is the study of the processes and functions of the body.

Structural and Functional Organization   (p. 3)

1. The human body can be organized into six levels: chemical (atoms and molecules), cell, tissue (groups of similar cells and the materials surrounding them), organ (two or more tissues that perform one or more common functions), organ system (groups of organs with common functions), and organism.
2. The 11 organ systems are the integumentary, skeletal, muscular, nervous, endocrine, cardiovascular, lymphatic, respiratory, digestive, urinary, and reproductive systems (see figure 1.3).

Characteristics of Life   (p. 7)

The characteristics of life include organization, metabolism, responsiveness, growth, development, and reproduction.

Homeostasis   (p. 7)

1. Homeostasis is the condition in which body functions, fluids, and other factors of the internal environment are maintained within a range of values suitable to support life.
2. Negative-feedback mechanisms operate to maintain homeostasis.
3. Positive-feedback mechanisms make deviations from normal even greater. Although a few positive-feedback mechanisms normally exist in the body, most positive-feedback mechanisms are harmful.

Terminology and the Body Plan   (p. 10)
Body Positions

1. A human standing erect with the face directed forward, the arms hanging to the sides, and the palms facing forward is in the anatomic position.
2. A person lying face upward is supine and face downward is prone.

Directional Terms

Directional terms always refer to the anatomic position, regardless of the body's actual position (see table 1.1).

Body Parts and Regions

1. The body can be divided into the limbs, head, neck, and trunk.
2. The abdomen can be divided superficially into four quadrants or nine regions that are useful for locating internal organs or describing the location of a pain.

Planes

1. A sagittal plane divides the body into left and right parts, a transverse plane divides the body into superior and inferior parts, and a frontal plane divides the body into anterior and posterior parts.
2. A longitudinal section divides an organ along its long axis, a transverse section cuts an organ at a right angle to the long axis, and an oblique section cuts across the long axis at an angle other than a right angle.

Body Cavities

1. The thoracic cavity is bounded by the ribs and the diaphragm. The mediastinum divides the thoracic cavity into two parts.
2. The abdominal cavity is bounded by the diaphragm and the abdominal muscles.
3. The pelvic cavity is surrounded by the pelvic bones.

Serous Membranes

1. The trunk cavities are lined by serous membranes. The parietal part of a serous membrane lines the wall of the cavity, and the visceral part covers the internal organs.

2. The serous membranes secrete fluid that fills the space between the parietal and visceral membranes. The serous membranes protect organs from friction.
3. The pericardial cavity surrounds the heart, the pleural cavities surround the lungs, and the peritoneal cavity surrounds certain abdominal and pelvic organs.
4. Mesenteries are parts of the peritoneum that hold the abdominal organs in place and provide a passageway for blood vessels and nerves to organs.
5. Retroperitoneal organs are found "behind" the parietal peritoneum. The kidneys, adrenal glands, pancreas, parts of the intestines, and the urinary bladder are examples of retroperitoneal organs.

R E V I E W   A N D   C O M P R E H E N S I O N

1. Define anatomy, surface anatomy, anatomic imaging, and physiology.
2. List six structural levels at which the body can be studied conceptually.
3. Define tissue. What are the four primary tissue types?
4. Define organ and organ system. What are the 11 organ systems of the body and their functions?
5. Name six characteristics of life.
6. What does the term homeostasis mean? If a deviation from homeostasis occurs, what kind of mechanism restores homeostasis?
7. Describe a negative-feedback mechanism in terms of receptor, control center, and effector. Give an example of a negative-feedback mechanism.
8. Define positive feedback. Why are positive-feedback mechanisms generally harmful? Give an example of a harmful and a beneficial positive-feedback mechanism.
9. Why is knowledge of the etymology of anatomical and physiological terms useful?
10. What is the anatomic position? Why is it important to remember the anatomic position when using directional terms?
11. Define and give an example of the following directional terms: inferior, superior, anterior, posterior, dorsal, ventral, proximal, distal, lateral, medial, superficial, and deep.
12. List the subdivisions of the upper limb, lower limb, and trunk.
13. Describe the four-quadrant and nine-region methods of subdividing the abdominal region. What is the purpose of these methods?
14. Define the sagittal, median, transverse, and frontal planes of the body.
15. Define the longitudinal, transverse, and oblique sections of an organ.
16. Define the following cavities: thoracic, abdominal, pelvic, and abdominopelvic. What is the mediastinum?
17. What is the difference between the visceral and parietal layers of a serous membrane? What function do serous membranes perform?
18. Name the serous membranes associated with the heart, lungs, and abdominopelvic organs.
19. Define mesentery. What does the term retroperitoneal mean? Give an example of a retroperitoneal organ.

C R I T I C A L   T H I N K I N G

1. A male has lost blood as a result of a gunshot wound. Even though bleeding has been stopped, his blood pressure is low and dropping and his heart rate is elevated. Following a blood transfusion, his blood pressure increases and his heart rate decreases. Which of the following statement(s) is (are) consistent with these observations?
   a. Negative-feedback mechanisms can be inadequate without medical intervention.
   b. The transfusion interrupted a positive-feedback mechanism.
   c. The increased heart rate after the gunshot wound and before the transfusion is a result of a positive-feedback mechanism.
   d. a and b
   e. a, b, and c
2. During physical exercise, respiration rate increases. Two students are discussing the mechanisms involved: Student A claims they are positive-feedback mechanisms, and student B claims they are negative-feedback mechanisms. Do you agree with student A or student B, and why?
3. Complete the following statements using the correct directional terms for a human being.
   a. The navel is _____ to the nose.
   b. The heart is _____ to the breastbone (sternum).
   c. The forearm is _____ to the arm.
   d. The ear is _____ to the brain.
4. Describe in as many directional terms as you can the relationship between your kneecap and your heel.
5. According to "Dear Abby," a wedding band should be worn closest to the heart, and an engagement ring should be worn as a "guard" on the outside. Should a woman's wedding band be worn proximal or distal to her engagement ring?
6. In which quadrants and regions are the pancreas and the urinary bladder located?
7. During pregnancy, which would increase more in size, the mother's abdominal or pelvic cavity? Explain.
8. A bullet enters the left side of a male, passes through the left lung, and lodges in the heart. Name in order the serous membranes and their cavities through which the bullet passes.
9. Can a kidney be removed without cutting through parietal peritoneum? Explain.

Answers in Appendix D

Visit this textbook's website at www.mhhe.com/seeleyess6 for practice quizzes, animations, interactive learning exercises, and other study tools.

Chapter Outline and Objectives

Basic Chemistry (p. 20)

1. Define matter, mass, and weight.
2. Define element and atom.
3. Name the subatomic particles of an atom, and describe how they are organized.
4. Describe two types of chemical bonds.
5. Define hydrogen bond, and explain its importance.
6. Distinguish between a molecule and a compound.
7. Describe the process of dissociation.

Chemical Reactions (p. 25)

8. Using symbols, explain synthesis, decomposition, and exchange reactions.
9. Explain how reversible reactions produce chemical equilibrium.
10. Distinguish between chemical reactions that release or take in energy.
11. List the factors that affect the rate of chemical reactions.

Acids and Bases (p. 29)

12. Describe the pH scale and its relationship to acidity and alkalinity.
13. Explain why buffers are important.

Inorganic Chemistry (p. 30)

14. Distinguish between inorganic and organic molecules.
15. State the importance of oxygen, carbon dioxide, and water for living organisms.

Organic Chemistry (p. 31)

16. Describe four important types of organic molecules and their functions.
17. Explain how enzymes work.

The Chemical Basis of Life

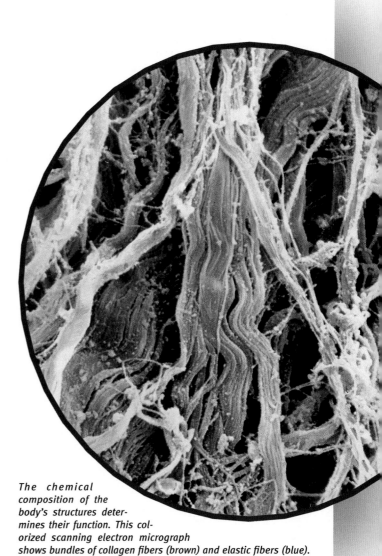

The chemical composition of the body's structures determines their function. This colorized scanning electron micrograph shows bundles of collagen fibers (brown) and elastic fibers (blue). The molecules forming collagen fibers are like tiny ropes, and collagen fibers make tissues strong yet flexible. The molecules forming elastic fibers resemble microscopic coiled springs. The elastic fibers allow tissues to be stretched and then recoil back to their original shape.

hemicals compose the structures of the body, and the interactions of chemicals with one another are responsible for the functions of the body. Nerve impulse generation, digestion, muscle contraction, and metabolism can be described in chemical terms. Many abnormal conditions and their treatments can also be explained in chemical terms, even though their symptoms appear as malfunctions in organ systems. For example, Parkinson's disease, which causes uncontrolled shaking movements, results from a shortage of a chemical called dopamine in certain nerve cells in the brain. It can be treated by giving patients another chemical that is converted to dopamine by brain cells.

A basic knowledge of chemistry is essential for understanding anatomy and physiology. **Chemistry** is the scientific discipline concerned with the atomic composition and structure of substances and the reactions they undergo. This chapter outlines some basic chemical principles and emphasizes their relationship to living organisms. Refer back to this chapter when chemical phenomena are discussed later in the text.

Basic Chemistry

Matter, Mass, and Weight

All living and nonliving things are composed of **matter,** which is anything that occupies space and has mass. **Mass** is the amount of matter in an object, and **weight** is the gravitational force acting on an object of a given mass. For example, the weight of an apple results from the force of gravity "pulling" on the apple's mass.

P R E D I C T   ①
The difference between mass and weight can be illustrated by considering an astronaut. How would an astronaut's mass and weight in outer space compare with his mass and weight on the earth's surface?

The international unit for mass is the **kilogram (kg),** which is the mass of a platinum–iridium cylinder kept at the International Bureau of Weights and Measurements in France. The mass of all other objects is compared with this cylinder. For example, a 2.2-lb lead weight or 1 liter (L) (1.06 qt) of water has a mass of approximately 1 kg. An object with 1/1000 the mass of the standard kilogram cylinder is defined to have a mass of 1 **gram (g).**

Chemists use a balance to determine the mass of objects. Although we commonly refer to weighing an object on a balance, we actually are "massing" the object because the balance compares objects of unknown mass with objects of known mass. When the unknown and known masses are exactly balanced, the gravitational pull of earth on both of them is the same. Thus, the effect of gravity on the unknown mass is counteracted by the effect of gravity on the known mass. A balance produces the same results at sea level as on a mountaintop because it does not matter that the gravitational pull at sea level is stronger than on a mountaintop. It only matters that the effect of gravity on both the unknown and known masses is the same.

Elements and Atoms

An **element** is the simplest type of matter with unique chemical properties. As of March 1996, 112 elements are known. A list of the elements commonly found in the human body is given in table 2.1. About 96% of the weight of the body results from the elements oxygen, carbon, hydrogen, and nitrogen.

An **atom** (at′ŏm, indivisible) is the smallest particle of an element that has the chemical characteristics of that element. An element is composed of atoms of only one kind. For example,

Table 2.1	Some Common Elements				
Element	Symbol	Atomic Number	Mass Number	Percent in Human Body by Weight	Percent in Human Body by Number of Atoms
Hydrogen	H	1	1	9.5	63.0
Carbon	C	6	12	18.5	9.5
Nitrogen	N	7	14	3.3	1.4
Oxygen	O	8	16	65.0	25.5
Sodium	Na	11	23	0.2	0.3
Phosphorus	P	15	31	1.0	0.22
Sulfur	S	16	32	0.3	0.05
Chlorine	Cl	17	35	0.2	0.03
Potassium	K	19	39	0.4	0.06
Calcium	Ca	20	40	1.5	0.31
Iron	Fe	26	56	Trace	Trace
Iodine	I	53	127	Trace	Trace

the element carbon is composed of only carbon atoms, and the element oxygen is composed of only oxygen atoms.

An element, or an atom of that element, often is represented by a symbol. Usually the first letter or letters of the element's name are used—for example, C for carbon, H for hydrogen, Ca for calcium, and Cl for chlorine. Occasionally the symbol is taken from the Latin, Greek, or Arabic name for the element—for example, Na from the Latin word *natrium* is the symbol for sodium.

Atomic Structure

The characteristics of matter result from the structure, organization, and behavior of atoms. Atoms are composed of subatomic particles, some of which have an electric charge. The three major types of subatomic particles are neutrons, protons, and electrons. **Neutrons** (noo′tronz) have no electrical charge, **protons** (prō′tonz) have positive charges, and **electrons** (e-lek′tronz) have negative charges. The positive charge of a proton is equal in magnitude to the negative charge of an electron. The number of protons and electrons in each atom is equal, and the individual charges cancel each other. Therefore, each atom is electrically neutral.

Protons and neutrons form the **nucleus** of the atom, and electrons are moving around the nucleus (figure 2.1). The nucleus accounts for 99.97% of an atom's mass, but only 1 ten-trillionth of its volume. Most of the volume of an atom is occupied by the electrons. Although it is impossible to know precisely where any given electron is located at any particular moment, the region where electrons are most likely to be found can be represented by an **electron cloud** (see figure 2.1). The darker the color in each small volume of the diagram, the greater the likelihood of finding an electron there at any given moment.

The **atomic number** of an element is equal to the number of protons in each atom, and because the number of electrons and protons is equal, the atomic number is also the number of electrons. Each element is uniquely defined by the number of protons in the atoms of that element. For example, only hydrogen atoms have one proton, only carbon atoms have six protons, and only oxygen atoms have eight protons (figure 2.2; see table 2.1).

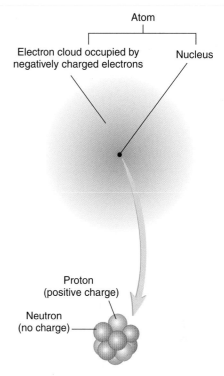

Figure 2.1   Model of an Atom

The tiny, dense nucleus consists of positively charged protons and uncharged neutrons. Most of the volume of an atom is occupied by rapidly moving, negatively charged electrons, which can be represented as an electron cloud. The probable location of an electron is indicated by the color of the electron cloud. The darker the color in each small part of the electron cloud, the more likely the electron is located there.

Making New Elements

Scientists have been able to create new elements by changing the number of protons in the nuclei of existing elements. Protons, neutrons, or electrons from one atom are accelerated to very high speeds and then smashed into the nucleus of another atom. The resulting changes in the nucleus produces a new element with a new atomic number. Twenty six elements with an atomic number greater than 92 have been synthesized in this fashion. These artificially produced elements are usually unstable and quickly convert back to more stable elements.

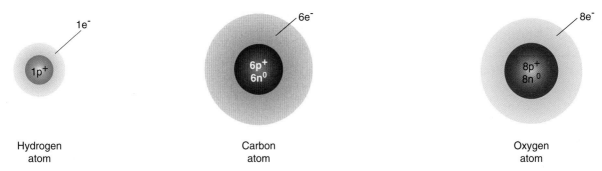

Figure 2.2   Hydrogen, Carbon, and Oxygen Atoms

Within the nucleus, the number of positively charged protons (p^+) and uncharged neutrons (n^0) is indicated. The negatively charged electrons (e^-) are around the nucleus. Atoms are electrically neutral because the number of protons and electrons within an atom is equal.

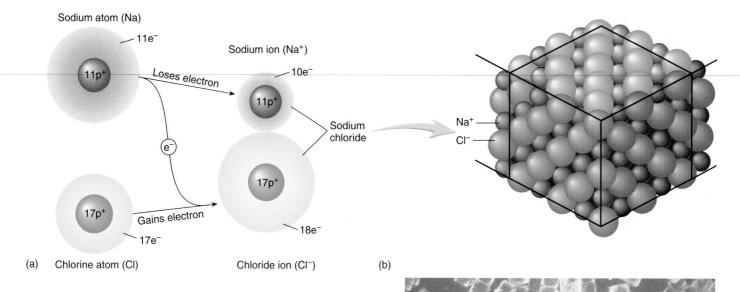

Figure 2.3   Ionic Bonding

(*a*) A sodium atom loses an electron to become a smaller-sized, positively charged ion, and a chlorine atom gains an electron to become a larger-sized, negatively charged ion. The attraction between the oppositely charged ions results in ionic bonding and the formation of sodium chloride. (*b*) The Na^+ and Cl^- are organized to form a cube-shaped array. (*c*) Microphotograph of salt crystals reflects the cubic arrangement of the ions.

Protons and neutrons have about the same mass, and they are responsible for most of the mass of atoms. Electrons, on the other hand, have very little mass. The **mass number** of an element is the number of protons plus the number of neutrons in each atom. For example, the mass number for carbon is 12 because it has six protons and six neutrons.

P R E D I C T   **2**

The atomic number of fluorine is 9, and the mass number is 19. What is the number of protons, neutrons, and electrons in an atom of fluorine?

Electrons and Chemical Bonding

The chemical behavior of an atom is determined largely by its outermost electrons. **Chemical bonding** occurs when the outermost electrons are transferred or shared between atoms. Two major types of chemical bonding are ionic and covalent bonding.

Ionic Bonding

An atom is electrically neutral because it has an equal number of protons and electrons. If an atom loses or gains electrons, the number of protons and electrons are no longer equal, and a charged particle called an **ion** (ī′on) is formed.

After an atom loses an electron, it has one more proton than it has electrons and is positively charged. For example, a sodium atom (Na) can lose an electron to become a positively charged sodium ion (Na^+) (figure 2.3*a*). After an atom gains an electron, it has one more electron than it has protons and is negatively charged. For example, a chlorine atom (Cl) can accept an electron to become a negatively charged chloride ion (Cl^-). Because oppositely charged ions are attracted to each other, positively charged ions tend to remain close to negatively charged ions, which is called **ionic** (ī-on′ik) **bonding.** Thus, Na^+ and Cl^- are held together by ionic bonding to form an array of ions called sodium chloride (NaCl), or table salt (figure 2.3*b* and *c*).

Ions are denoted by using the symbol of the atom from which the ion was formed. The charge of the ion is indicated by a superscripted plus (+) or minus (−) sign. For example, a sodium ion is Na^+, and a chloride ion is Cl^-. If more than one electron has been lost or gained, a number is used with the plus or minus sign. Thus, Ca^{2+} is a calcium ion formed by the loss of two electrons. Some ions commonly found in the body are listed in table 2.2.

P R E D I C T   **3**

If an iron (Fe) atom loses three electrons, what is the charge of the resulting ion? Write the symbol for this ion.

Table 2.2	Important Ions	
Common Ions	**Symbols**	**Significance**
Calcium	Ca^{2+}	Part of bones and teeth, blood clotting, muscle contraction, release of neurotransmitters
Sodium	Na^+	Membrane potentials, water balance
Potassium	K^+	Membrane potentials
Hydrogen	H^+	Acid–base balance
Hydroxide	OH^-	Acid–base balance
Chloride	Cl^-	Water balance
Bicarbonate	HCO_3^-	Acid–base balance
Ammonium	NH_4^+	Acid–base balance
Phosphate	PO_4^{3-}	Part of bones and teeth, energy exchange, acid–base balance
Iron	Fe^{2+}	Red blood cell formation
Magnesium	Mg^{2+}	Necessary for enzymes
Iodide	I^-	Present in thyroid hormones

The ions are part of the structures or play important roles in the processes listed.

Covalent Bonding

Covalent bonding results when atoms share one or more pairs of electrons. The resulting combination of atoms is called a molecule. An example is the covalent bond between two hydrogen atoms to form a hydrogen molecule (figure 2.4). Each hydrogen atom has one electron. As the atoms get closer together, the positively charged nucleus of each atom begins to attract the electron of the other atom. At an optimal distance, the two nuclei mutually attract the two electrons, and each electron is shared by both nuclei. The two hydrogen atoms are now held together by a covalent bond.

When an electron pair is shared between two atoms, a **single covalent bond** results. A single covalent bond can be represented by a single line between the symbols of the atoms involved (for example, H—H). A **double covalent bond** results when two atoms share two pairs of electrons. When a carbon atom combines with two oxygen atoms to form carbon dioxide, two double covalent bonds are formed. Double covalent bonds are indicated by a double line between the atoms (O=C=O).

Two hydrogen atoms can share their electrons with an oxygen atom to form a water molecule (H_2O, figure 2.5). The hydrogen atoms do not share the electrons equally with the oxygen atom, however, and the electrons tend to spend more time around the oxygen atom than around the hydrogen atoms. This unequal, asymmetrical sharing of electrons is called a **polar covalent bond** because the unequal sharing of electrons results in one end (pole) of the molecule having a small electrical charge opposite to that of the other end. Molecules with this asymmetrical electric charge are called **polar molecules,** whereas molecules with a symmetrical electric charge are called **nonpolar molecules.**

No interaction between the two hydrogen atoms because they are too far apart.

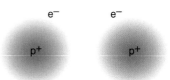

The positively charged nucleus of each hydrogen atom begins to attract the electron of the other.

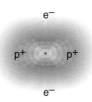

A covalent bond is formed when the electrons are shared between the nuclei because the electrons are equally attracted to each nucleus.

Figure 2.4   Covalent Bonding

Hydrogen Bonds

A polar molecule has a positive and a negative "end." The positive "end" of one polar molecule can be weakly attracted to the negative "end" of another polar molecule. Although this attraction is called a **hydrogen bond,** it is not a chemical bond because electrons are not transferred or shared between the atoms of the different polar molecules. The attraction between molecules resulting from hydrogen bonds is much weaker than ionic or covalent bonds. For example, the positively charged hydrogen of one water molecule is weakly attracted to a negatively charged oxygen of another water molecule (figure 2.6). Thus, the water molecules are held together by hydrogen bonds.

Hydrogen bonds also play an important role in determining the shape of complex molecules because the hydrogen bonds between different polar parts of a single large molecule hold the molecule in its normal three-dimensional shape (see the sections on Proteins and Nucleic Acids on pp. 34–37).

Molecules and Compounds

A **molecule** (mol′ĕ-kūl) is formed when two or more atoms chemically combine to form a structure that behaves as an independent unit. The atoms that combine to form a molecule can be of the same type, such as two hydrogen atoms combining to form a hydrogen molecule. More typically, a molecule consists of two or more different types of atoms, such as two

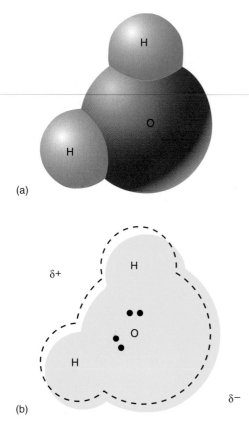

(a)

(b)

δ+

δ−

Figure 2.5   Polar Covalent Bonds

(*a*) A water molecule forms when two hydrogen atoms form covalent bonds with an oxygen atom. (*b*) Electron pairs (*indicated by the black dots*) are shared between the hydrogen atoms and oxygen. The dashed outline shows the expected location of the electron cloud if the electrons are shared equally. The electrons are shared unequally, as shown by the electron cloud (*yellow*) not coinciding with the dashed outline. Consequently, the oxygen side of the molecule has a slight negative charge (*indicated by* δ⁻), and the hydrogen side of the molecule has a slight positve charge (*indicated by* δ⁺).

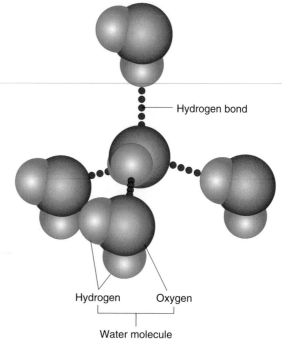

Hydrogen bond

Hydrogen   Oxygen

Water molecule

Figure 2.6   Hydrogen Bonds

The positive hydrogen part of one water molecule forms a hydrogen bond (*red dotted line*) with the negative oxygen part of another water molecule. As a result, hydrogen bonds hold the water molecules together.

hydrogen atoms and an oxygen atom forming water. Thus, a glass of water consists of a collection of individual water molecules positioned next to one another.

A **compound** (kom′pownd, to place together) is a substance composed of two or more *different* types of atoms that are chemically combined. Not all molecules are compounds. For example, a hydrogen molecule is not a compound because it does not consist of different types of atoms. Some compounds are molecules and some are not. Covalent compounds, in which different types of atoms are held together by covalent bonds, are molecules because the sharing of electrons results in the formation of distinct, independent units. Water is an example of a substance that is a compound and a molecule.

On the other hand, ionic compounds, in which ions are held together by the force of attraction between opposite charges, are not molecules because they do not consist of distinct units. A piece of sodium chloride does not consist of individual sodium chloride molecules positioned next to one another. Instead, it is an organized array of individual Na⁺ and individual Cl⁻ in which each charged ion is surrounded by several ions of the opposite charge (see figure 2.3*b*). Sodium

chloride is an example of a substance that is a compound but not a molecule.

Molecules and compounds can be represented by the symbols of the atoms forming the molecule or compound plus subscripts denoting the number of each type of atom. For example, glucose (a sugar) can be represented as $C_6H_{12}O_6$, indicating that glucose has 6 carbon, 12 hydrogen, and 6 oxygen atoms.

Dissociation

When ionic compounds dissolve in water, their ions **dissociate** (di-sō′sē-āt′, to separate), or separate, from each other because the positively charged ions are attracted to the negative ends of the water molecules, and the negatively charged ions are attracted to the positive ends of the water molecules. For example, when sodium chloride dissociates in water, the Na⁺ and Cl⁻ separate, and water molecules surround and isolate the ions, keeping them in solution (figure 2.7).

When molecules dissolve in water, the molecules usually remain intact even though they are surrounded by water molecules. Thus, in a glucose solution, glucose molecules are surrounded by water molecules.

Ions and Electricity

Ions that dissociate in water are sometimes called **electrolytes** (ē-lek′trō-lītz) because they have the capacity to conduct an electrical current, which is the flow of charged particles. An electrocardiogram (ECG) is a recording of electric currents produced by the heart. These currents can be detected by electrodes on the surface of the body because the ions in the body fluids conduct electric currents.

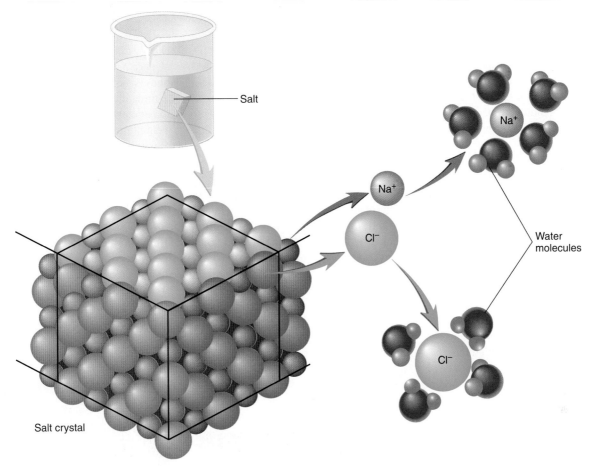

Figure 2.7 Dissociation

Sodium chloride (table salt) dissociating in water. The positively charged Na$^+$ are attracted to the negatively charged oxygen (*red*) end of the water molecule, and the negatively charged Cl$^-$ are attracted to the positively charged hydrogen (*blue*) end of the water molecule.

Chemical Reactions

In a **chemical reaction,** atoms, ions, molecules, or compounds interact either to form or to break chemical bonds. The substances that enter into a chemical reaction are called the **reactants,** and the substances that result from the chemical reaction are called the **products.**

Classification of Chemical Reactions

For our purposes, chemical reactions can be classified as synthesis, decomposition, or exchange reactions.

Synthesis Reactions

When two or more reactants combine to form a larger, more complex product, the process is called a **synthesis reaction.** This can be represented symbolically as follows:

$$A + B \rightarrow AB$$

Examples of synthesis reactions include the synthesis of the complex molecules of the human body from the basic "building blocks" obtained in food and the synthesis of **adenosine triphosphate (ATP)** (ă-den′ō-sēn trī-fos′făt) molecules. In ATP, A stands for adenosine, T stands for tri- (or

three), and P stands for a phosphate group (PO$_4$$^{3-}$). Thus, ATP consists of adenosine and three phosphate groups. ATP is synthesized when adenosine diphosphate (ADP), which has two (di-) phosphate groups, combines with a phosphate group to form the larger ATP molecule. The phosphate group that reacts with ADP is often denoted as P$_i$, where the "i" indicates that the phosphate group is associated with an inorganic substance (see the section on Inorganic Chemistry on p. 30).

$$
\begin{array}{cccc}
\text{A-P-P} & + & \text{P}_i & \rightarrow & \text{A-P-P-P} \\
\text{(ADP)} & & \text{(phosphate} & & \text{(ATP)} \\
& & \text{group)} & &
\end{array}
$$

Decomposition Reactions

In a **decomposition reaction,** reactants are broken down into smaller, less complex products. A decomposition reaction is the reverse of a synthesis reaction and can be represented in this way:

$$AB \rightarrow A + B$$

Examples of decomposition reactions include the breakdown of food molecules into basic building blocks, and the breakdown of ATP to ADP and a phosphate group.

$$
\begin{array}{cccc}
\text{A-P-P-P} & \rightarrow & \text{A-P-P} & + & \text{P}_i \\
\text{(ATP)} & & \text{(ADP)} & & \text{(phosphate group)}
\end{array}
$$

Protons, neutrons, and electrons are responsible for the chemical properties of atoms. They also have other properties that can be useful in a clinical setting. For example, some of these properties have enabled the development of methods for examining the inside of the body.

Isotopes (ī′sō-tōpz, *isos,* equal + *topos,* part) are two or more forms of the same element that have the same number of protons and electrons but a different number of neutrons. Thus, isotopes have the same atomic number (i.e., number of protons), but different mass numbers (i.e., sum of the protons and neutrons). For example, hydrogen and its isotope deuterium each have an atomic number of one because they both have one proton. However, hydrogen has no neutrons, whereas deuterium has one neutron. Therefore the mass number of hydrogen is one and deuterium is two. Water made with deuterium is called heavy water because of the weight of the "extra" neutron. Because isotopes of the same atom have the same number of electrons, they are very similar in their chemical behavior. The nuclei of some isotopes are stable and do not change. Radioactive isotopes, however, have unstable nuclei that lose neutrons or protons. Several different kinds of radiation can be produced when neutrons and protons, or the products formed by their breakdown, are released from the nucleus of the isotope.

The radiation given off by some radioactive isotopes can penetrate and destroy tissues. Rapidly dividing cells are more sensitive to radiation than are slowly dividing cells. Radiation is used to treat cancerous (malignant) tumors because cancer cells divide rapidly. If the treatment is effective, few healthy cells are destroyed, but the cancerous cells are killed.

Radioactive isotopes also are used in medical diagnosis. The radiation can be detected, and the movement of the radioactive isotopes throughout the body can be traced. For example, the thyroid gland normally takes up iodine and uses it in the formation of thyroid hormones. Radioactive iodine can be used to determine if iodine uptake is normal in the thyroid gland.

Radiation can be produced in ways other than changing the nucleus of atoms. X-rays are a type of radiation formed when electrons lose energy by moving from a higher energy state to a lower one. X-rays are used in examination of bones to determine if they are broken, and of teeth to see if they have caries (cavities). Mammograms, which are low-energy radiographs (x-ray films) of the breast, can be used to detect tumors because the tumors are slightly denser than normal tissue.

Computers can be used to analyze a series of radiographs, each made at a slightly different body location. The picture of each radiographic "slice" through the body is assembled by the computer to form a three-dimensional image. A **computed tomography** (tō-mog′ră-fē) **(CT) scan** is an example of this technique (figure A). CT scans are used to detect tumors and other abnormalities in the body.

Magnetic resonance imaging (MRI) is another method for looking into the body (figure B). The patient is placed into a very powerful magnetic field, which aligns the hydrogen nuclei. Radiowaves given off by the hydrogen nuclei are monitored, and the data are used by a computer to make an image of the body. Because MRI detects hydrogen, it is very effective for visualizing soft tissues that contain a lot of water. MRI technology is used to detect tumors and other abnormalities in the body.

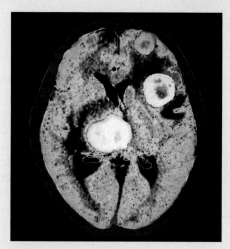

Figure A   CT Scan

CT scan of the brain with iodine injection, showing three brain tumors (ovals) that have metastasized (spread) to the brain from cancer in the large intestine.

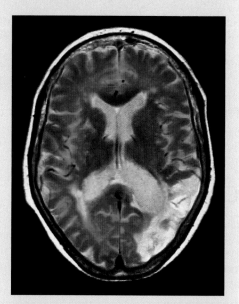

Figure B   MRI

Colorized MRI brain scan showing a stroke. The whitish area in the lower right part of the MRI is blood that has leaked into the surrounding tissue.

Exchange Reactions

An **exchange reaction** is a combination of a decomposition and a synthesis reaction. In decomposition, reactants are broken down. In synthesis, the products of the decomposition reaction are combined. The symbolic representation of an exchange reaction is:

$$AB + CD \rightarrow AC + BD$$

The reaction of hydrochloric acid (HCl) with sodium hydroxide (NaOH) to form table salt (NaCl) and water (H_2O) is an exchange reaction.

$$HCl + NaOH \rightarrow NaCl + H_2O$$

Reversible Reactions

A **reversible reaction** is a chemical reaction in which the reaction can proceed from reactants to products and from products to reactants. When the rate of product formation is equal to the rate of reactant formation, the reaction is said to be at **equilibrium.** At equilibrium the amount of the reactants relative to the amount of products remains constant.

The following analogy may help to clarify the concept of reversible reactions and equilibrium. Imagine a trough containing water. The trough is divided into two compartments by a partition, but the partition contains holes that allow water to move freely between the compartments. Because water can move in either direction, this is like a reversible reaction. Let the amount of water in the left compartment represent the amount of reactant, and the amount of water in the right compartment represent the amount of product. At equilibrium, the amount of reactant relative to the amount of product in each compartment is always the same because the partition allows water to pass between the two compartments until the level of water is the same in both compartments. If the amount of reactant is increased by adding water to the left compartment, water flows from the left compartment through the partition to the right compartment until the level of water is the same in both. Thus, the amounts of reactant and product are once again equal. Unlike this analogy, however, the amount of reactants relative to the products in most reversible reactions is not one to one. Depending on the specific reversible reaction, there can be one part reactant to two parts product, two parts reactants to one part product, or many other possibilities.

An important reversible reaction in the human body is the reaction between carbon dioxide (CO_2) and water (H_2O) to form hydrogen ions (H^+) and bicarbonate ions (HCO_3^-) (the reversibility of the reaction is indicated by two arrows pointing in opposite directions):

$$CO_2 + H_2O \rightleftarrows H^+ + HCO_3^-$$

If CO_2 is added to H_2O, the amount of CO_2 relative to the amount of H^+ increases. The reaction of CO_2 with H_2O produces more H^+, however, and the amount of CO_2 relative to the amount of H^+ returns to equilibrium. Conversely, adding H^+ results in the formation of more CO_2, and the equilibrium is restored.

Maintaining a constant level of H^+ in body fluids is necessary for the nervous system to function properly. This level can be maintained, in part, by controlling blood CO_2 levels. For example, slowing the respiration rate causes blood CO_2 levels to increase, which in turn causes an increase in H^+ concentration in the blood.

P R E D I C T

If the respiration rate increases, carbon dioxide is removed from the blood. What effect does this have on blood hydrogen ion levels?

Energy and Chemical Reactions

Energy, unlike matter, does not occupy space and it has no mass. **Energy** is defined as the capacity to do **work**—that is, to move matter. Energy can be subdivided into potential energy and kinetic energy. **Potential energy** is stored energy that could do work but is not doing so. For example, a coiled spring has potential energy. It could push against an object and move the object, but as long as the spring does not uncoil, no work is accomplished. **Kinetic** (ki-net′ik, of motion) **energy** is energy caused by the movement of an object and is the form of energy that actually does work. An uncoiling spring pushing an object causing it to move is an example. When potential energy is released, it becomes kinetic energy, thus doing work.

Potential and kinetic energy can be found in many different forms. **Mechanical energy** is energy resulting from the position or movement of objects. Many of the activities of the human body, such as moving a limb, breathing, or circulating blood, involve mechanical energy. Other forms of energy are chemical energy, heat energy, electric energy, and electromagnetic (radiant) energy.

According to the law of conservation of energy, the total energy of the universe is constant. Therefore, energy is neither created nor destroyed. One type of energy can be changed into another, however. For example, as a moving object slows down and comes to rest, its kinetic energy is converted into heat energy by friction.

The **chemical energy** of a substance is a form of stored (potential) energy that results from the relative positions and interactions among its charged subatomic particles. Consider two balls attached by a relaxed spring. In order to push the balls together and compress the spring, energy must be put into this system. As the spring is compressed, potential energy increases. When the compressed spring expands, potential energy decreases. Similarly charged particles, such as two negatively charged electrons or two positively charged nuclei, repel each other. As similarly charged particles move closer together, their potential energy increases, much like compression of a spring, and as they move farther apart, their potential energy decreases. Chemical bonding is a form of potential energy because of the charges and positions of the subatomic particles bound together.

Chemical reactions are important because of the products they form and the energy changes that result as the relative position of subatomic particles changes. If the products of a chemical reaction contain less potential energy than the reactants, energy is released. For example, food molecules contain more potential energy than waste products. The difference in potential energy between food and waste products is used by living systems for many activities, such as growth, repair, movement, and heat production, as the potential energy in the food molecules changes into other forms of energy.

An example of a reaction that releases energy is the breakdown of ATP to ADP and a phosphate group (figure 2.8a). The phosphate group is attached to the ADP molecule by a covalent bond, which has potential energy. After the breakdown of ATP, some of that energy is released as heat and some is available for use by cells for activities such as synthesizing new molecules or for muscle contraction.

$$ATP \rightarrow ADP + P_i + Heat + Energy \text{ (used by cells)}$$

P R E D I C T   5
Why does body temperature increase during exercise?

Energy must be added from another source if the products of a chemical reaction contain more energy than the reactants (figure 2.8b). The energy released during the breakdown of food molecules is the source of energy for this kind of reaction in the body. The energy from food molecules is used to synthesize molecules such as ATP, fats, and proteins.

$$ADP + P_i + Energy \text{ (from food molecules)} \rightarrow ATP$$

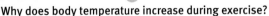

Energy Flow
The energy that makes almost all life on earth possible ultimately comes from the sun. In the process of photosynthesis, plants capture the energy in sunlight and convert it into chemical bonds in glucose. The plants and the organisms that eat plants use the energy from glucose to form ATP. The energy from the breakdown of ATP fuels the chemical reactions of life.

Rate of Chemical Reactions
The rate at which a chemical reaction proceeds is influenced by several factors, including how easily the substances react with one another, their concentrations, the temperature, and the presence of a catalyst.

Reactants
Reactants differ from one another in their ability to undergo chemical reactions. For example, iron corrodes much more rapidly than does stainless steel. For this reason, during its refurbishment the iron bars forming the skeleton of the Statue of Liberty were replaced with stainless steel bars.

Concentration
Within limits, the greater the concentration of reactants, the greater the rate at which a chemical reaction will occur because, as the concentration increases, the reacting molecules are more likely to come into contact with one another. For example, the normal concentration of oxygen inside cells enables it to come into contact with other molecules, producing the chemical reactions necessary for life. If the oxygen concentration decreases, the rate of chemical reactions decreases. A decrease in oxygen in cells can impair cell function and even result in cell death.

Temperature
The rate of chemical reactions also increases when the temperature is increased. When a person has a fever of only a few degrees, reactions occur throughout the body at a faster rate. The

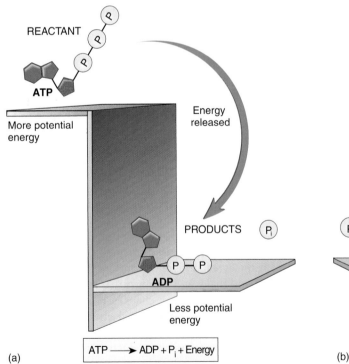

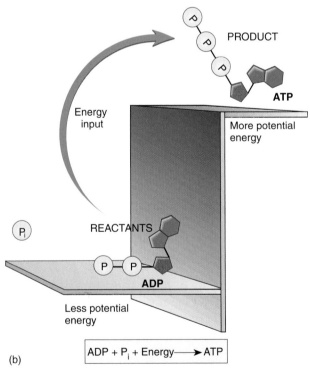

$$ATP \longrightarrow ADP + P_i + Energy$$

(a)

$$ADP + P_i + Energy \longrightarrow ATP$$

(b)

Figure 2.8   Energy and Chemical Reactions
In each figure the upper shelf represents a higher energy level, and the lower shelf represents a lower energy level. (*a*) Reaction in which energy is released as a result of the breakdown of ATP. (*b*) Reaction in which the input of energy is required for the synthesis of ATP.

result is increased activity in most organ systems, such as increased heart and respiratory rates. When body temperature drops, the rate of reactions decreases. The clumsy movement of very cold fingers results largely from the reduced rate of chemical reactions in cold muscle tissue.

Catalysts

At normal body temperatures, most chemical reactions would take place too slowly to sustain life if it were not for the body's catalysts. A **catalyst** (kat′ă-list) is a substance that increases the rate of a chemical reaction, without itself being permanently changed or depleted. An **enzyme** (en′zīm) is a protein molecule that acts as a catalyst. Many of the chemical reactions that occur in the body require enzymes. Enzymes are considered in greater detail later in the section on Enzymes (see p. 36).

Acids and Bases

An **acid** is a proton donor. Because a hydrogen atom without its electron is a proton, any substance that releases hydrogen ions in water is an acid. For example, hydrochloric acid (HCl) in the stomach forms hydrogen ions (H^+) and chloride ions (Cl^-):

$$HCl \rightarrow H^+ + Cl^-$$

A **base** is a proton acceptor. For example, sodium hydroxide (NaOH) forms sodium ions (Na^+) and hydroxide ions (OH^-). It is a base because the OH^- is a proton acceptor that binds with a H^+ to form water.

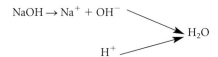

The pH Scale

The **pH scale** (figure 2.9), which ranges from 0 to 14, indicates the H^+ concentration of a solution. Pure water is defined as a neutral solution. A **neutral solution** has an equal number of H^+ and OH^- and has a pH of 7.0. An **acidic solution** has a pH less than 7.0 and has a greater concentration of H^+ than OH^-. An **alkaline** (al′kă-līn), or **basic, solution** has a pH greater than 7.0 and has fewer H^+ than OH^-.

As the pH value becomes smaller, the solution is more acidic; as the pH value becomes larger, the solution is more basic. A change of one unit on the pH scale represents a 10-fold change in the H^+ concentration. For example, a solution of pH 6.0 has 10 times more H^+ than a solution with a pH of 7.0. Thus small changes in pH represent large changes in H^+ concentration.

Acidosis and Alkalosis

The normal pH range for human blood is 7.35 to 7.45. The condition of **acidosis** (as-i-dō′sis) results if blood pH drops below 7.35. The nervous system becomes depressed, and the individual becomes disoriented and possibly comatose. **Alkalosis** (al-kă-lō′sis) results if blood pH rises above 7.45. The nervous system becomes overexcitable and the individual can be extremely nervous or have convulsions. Both acidosis and alkalosis can result in death.

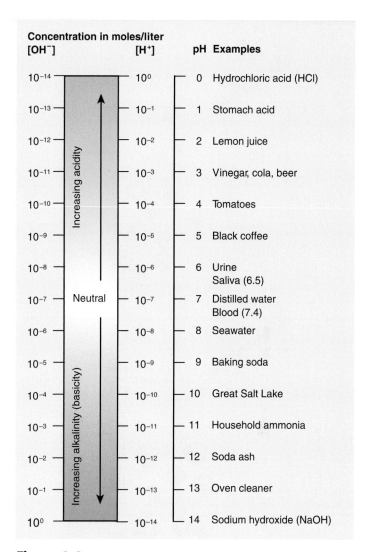

Figure 2.9   The pH Scale

A pH of 7 is neutral. Values less than 7 are acidic (the lower the number, the more acidic). Values greater than 7 are basic (the higher the number, the more basic). Representative fluids and their approximate pH values are listed.

Salts

A **salt** is a compound consisting of a positive ion other than a H^+ and a negative ion other than a OH^-. Salts are formed by the reaction of an acid and a base. For example, hydrochloric acid (HCl) combines with sodium hydroxide (NaOH) to form the salt sodium chloride (NaCl).

HCl	+	NaOH	→	NaCl	+	H_2O
(acid)		(base)		(salt)		(water)

Buffers

The chemical behavior of many molecules changes as the pH of the solution in which they are dissolved changes. The survival of an organism depends on its ability to regulate body fluid pH within a narrow range. One way normal body fluid pH is maintained is through the use of buffers. A **buffer** (bŭf′er) is a chemical that resists changes in pH when either an acid or a base is added to a solution containing the buffer. When an acid

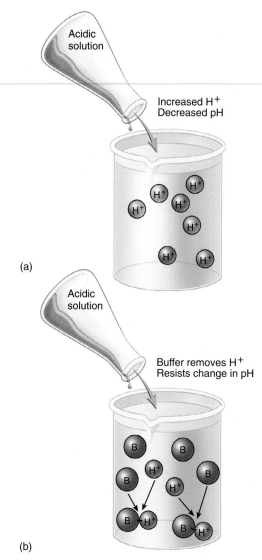

Acidic
solution

Increased H$^+$
Decreased pH

H$^+$ H$^+$
H$^+$ H$^+$
H$^+$
H$^+$ H$^+$

(a)

Acidic
solution

Buffer removes H$^+$
Resists change in pH

B B
B H$^+$ B
H$^+$
B H$^+$ B H$^+$

(b)

Figure 2.10   Buffers

(a) The addition of an acid to a nonbuffered solution results in an increase of H$^+$ and a decrease in pH. (b) The addition of an acid to a buffered solution results in a much smaller change in pH. The added H$^+$ bind to the buffer (symbolized by the letter "B").

is added to a buffered solution, the buffer binds to the H$^+$, preventing them from causing a decrease in the pH of the solution (figure 2.10).

> **PREDICT   6**
> If a base is added to a solution, would the pH of the solution increase or decrease? If the solution is buffered, what response from the buffer prevents the change in pH?

Inorganic Chemistry

Originally it was believed that inorganic substances were those that came from nonliving sources and organic substances were those extracted from living organisms. As the science of chemistry developed, however, it became apparent that organic substances could be manufactured in the laboratory. As defined currently, **inorganic chemistry** deals with those substances that do not contain carbon, whereas **organic chemistry** is the study of carbon-containing substances. These definitions have a few exceptions. For example, carbon dioxide and carbon monoxide are classified as inorganic molecules, even though they contain carbon.

Oxygen and Carbon Dioxide

Oxygen (O$_2$) is an inorganic molecule consisting of two oxygen atoms bound together by a double covalent bond. About 21% of the gas in the atmosphere is oxygen, and it is essential for most living organisms. Oxygen is required by humans in the final step of a series of chemical reactions in which energy is extracted from food molecules (see chapters 3 and 17).

   Carbon dioxide (CO$_2$) consists of one carbon atom bound to two oxygen atoms. Each oxygen atom is bound to the carbon atom by a double covalent bond. Carbon dioxide is produced when food molecules such as glucose are metabolized within the cells of the body (see chapters 3 and 17). Once carbon dioxide is produced, it is eliminated from the cell as a metabolic by-product, transferred to the lungs by the blood, and exhaled during respiration. If carbon dioxide is allowed to accumulate within cells, it becomes toxic.

Water

Water (H$_2$O) is an inorganic molecule that consists of one atom of oxygen joined by polar covalent bonds to two atoms of hydrogen. It has many important properties for living organisms.

1. *Stabilizing body temperature.* Water can absorb large amounts of heat and remain at a stable temperature. Blood, which is mostly water, can transfer heat effectively from deep within the body to the body's surface. Blood is warmed deep in the body and then flows to the surface, where the heat is released. In addition, water evaporation in the form of sweat results in significant heat loss from the body.
2. *Protection.* Water is an effective lubricant. For example, tears protect the surface of the eye from the rubbing of the eyelids. Water also forms a fluid cushion around organs that helps to protect them from damage. The cerebrospinal fluid that surrounds the brain is an example.
3. *Chemical reactions.* Most of the chemical reactions necessary for life do not take place unless the reacting molecules are dissolved in water. For example, sodium chloride must dissociate in water into Na$^+$ and Cl$^-$ before they can react with other ions. Water also directly participates in many chemical reactions. For example, during the digestion of food, large molecules and water react to form smaller molecules.
4. *Transport.* Many substances dissolve in water and can be moved from place to place as the water moves. For example, blood transports nutrients, gases, and waste products within the body.

Organic Chemistry

The ability of carbon to form covalent bonds with other atoms makes possible the formation of the large, diverse, complicated molecules necessary for life. A series of carbon atoms bound together by covalent bonds constitute the "backbone" of many large molecules. Variation in the length of the carbon chains and the combination of atoms bound to the carbon backbone allow the formation of a wide variety of molecules. For example, some protein molecules have thousands of carbon atoms bound by covalent bonds to one another or to other atoms such as nitrogen, sulfur, hydrogen, and oxygen.

The four major groups of organic molecules essential to living organisms are carbohydrates, lipids, proteins, and nucleic acids. Each of these groups has specific structural and functional characteristics (table 2.3).

Carbohydrates

Carbohydrates are composed of carbon, hydrogen, and oxygen atoms. In most carbohydrates, for each carbon atom there are two hydrogen atoms and one oxygen atom. Note that this ratio of hydrogen atoms to oxygen atoms is two to one, the same as in water (H_2O). They are called carbohydrates because each carbon (carbo) is combined with the same atoms that form water (hydrated). For example, the chemical formula for glucose is $C_6H_{12}O_6$.

The smallest carbohydrates are **monosaccharides** (mon-ō-sak′ă-rīdz, one sugar) or simple sugars. Glucose (blood sugar) and fructose (fruit sugar) are important monosaccharide energy sources for many of the body's cells. Larger carbohydrates are formed by chemically binding monosaccharides together. For this reason, monosaccharides are considered the building blocks of carbohydrates. **Disaccharides** (dī-sak′ă-rīdz, two sugars) are formed when two monosaccharides join. For example, glucose and fructose combine to form the disaccharide sucrose (table sugar) (figure 2.11a). **Polysaccharides** (pol-ē-sak′ă-rīdz, many sugars) consist of many monosaccharides bound in long chains. Glycogen, or animal starch, is a polysaccharide of glucose (figure 2.11b). When cells containing glycogen need energy, the glycogen is broken down into individual glucose molecules, which can be used as energy sources. Plant starch, also a polysaccharide of glucose, can be ingested and broken down into glucose. Cellulose, another polysaccharide of glucose, is an important structural component of plant cell walls. Humans cannot digest cellulose, however, and it is eliminated in the feces, where the cellulose fibers provide bulk.

Lipids

Lipids are substances that dissolve in nonpolar solvents, such as alcohol or acetone, but not in polar solvents, such as water. Lipids are composed mainly of carbon, hydrogen, and oxygen; but other elements such as phosphorus and nitrogen are minor

Table 2.3	Important Organic Molecules and Their Functions			
Molecule	**Elements**	**Building Blocks**	**Function**	**Examples**
Carbohydrate	C, H, O	Monosaccharides	Energy	Monosaccharides can be used as energy sources. Glycogen (polysaccharide) is an energy-storage molecule.
Lipid	C, H, O (P, N in some)	Glycerol and fatty acids (for fats)	Energy	Fats can be stored and broken down later for energy; per unit of weight fats yield twice as much energy as carbohydrates.
			Structure	Phospholipids and cholesterol are important components of cell membranes.
			Regulation	Steroid hormones regulate many physiological processes (e.g., estrogen and testosterone are responsible for many of the differences between males and females).
Protein	C, H, O, N (S in most)	Amino acids	Regulation	Enzymes control the rate of chemical reactions. Hormones regulate many physiological processes (e.g., insulin affects glucose transport into cells).
			Structure	Collagen fibers form a structural framework in many parts of the body.
			Energy	Proteins can be broken down for energy; per unit of weight they yield the same energy as carbohydrates.
			Contraction	Actin and myosin in muscle are responsible for muscle contraction.
			Transport	Hemoglobin transports oxygen in the blood.
			Protection	Antibodies and complement protect against microorganisms and other foreign substances.
Nucleic acid	C, H, O, N, P	Nucleotides	Regulation	DNA directs the activities of the cell.
			Heredity	Genes are pieces of DNA that can be passed from one generation to the next generation.
			Protein synthesis	RNA is involved in protein synthesis.

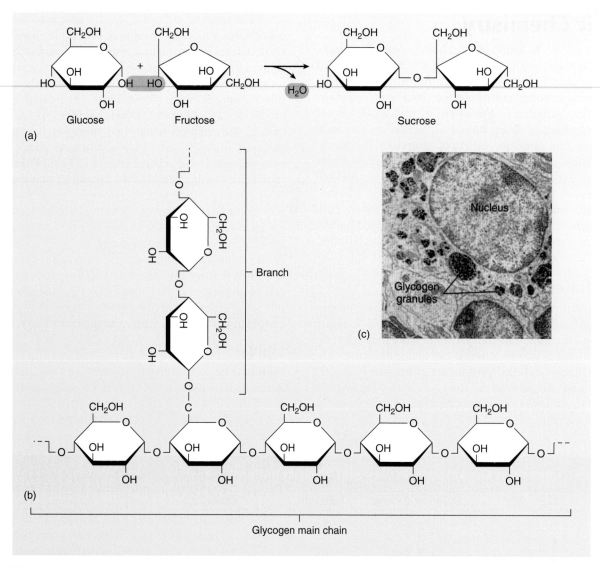

Figure 2.11   Carbohydrates

(*a*) Glucose and fructose are monosaccharides that combine to form the disaccharide sucrose. (*b*) Glycogen is a polysaccharide formed by combining many glucose molecules. (*c*) The photomicrograph shows glycogen granules in a liver cell.

components of some lipids. Lipids contain a lower proportion of oxygen to carbon than do carbohydrates.

Fats, phospholipids, and steroids are examples of lipids. **Fats** are important energy-storage molecules; they also pad and insulate the body. The building blocks of fats are **glycerol** (glis′er-ol) and **fatty acids** (figure 2.12). Glycerol is a three-carbon molecule with a **hydroxyl** (hī-drok′sil) **group** (—OH) attached to each carbon atom, and fatty acids consist of a carbon chain with a **carboxyl** (kar-bok′sil) **group** attached at one end. A carboxyl group consists of both an oxygen atom and a hydroxyl group attached to a carbon atom (—COOH).

$$\underset{\text{—C—OH}}{\overset{\displaystyle\overset{O}{\|}}{}} \qquad \text{or} \qquad \underset{\text{HO—C—}}{\overset{\displaystyle\overset{O}{\|}}{}}$$

The carboxyl group is responsible for the acidic nature of the molecule because it releases hydrogen ions into solution. **Triglycerides** (trī-glis′er-īdz), which are also called **triacylglycerols**

(trī-as′il-glis′er-olz), are the most common type of fat molecules. Triglycerides have three fatty acids bound to a glycerol molecule.

Fatty acids differ from one another according to the length and degree of saturation of their carbon chains. Most naturally occurring fatty acids contain 14–18 carbon atoms. A fatty acid is **saturated** if it contains only single covalent bonds between the carbon atoms (figure 2.13*a*). Sources of saturated fats include beef, pork, whole milk, cheese, butter, eggs, coconut oil, and palm oil. The carbon chain is **unsaturated** if it has one or more double covalent bonds (figure 2.13*b*). Because the double covalent bonds can occur anywhere along the carbon chain, many types of unsaturated fatty acids with an equal degree of unsaturation are possible. **Monounsaturated** fats, such as olive and peanut oils, have one double covalent bond between carbon atoms. **Polyunsaturated fats,** such as safflower, sunflower, corn, or fish oils, have two or more double covalent bonds between carbon atoms. Unsaturated fats are the best type of fats in

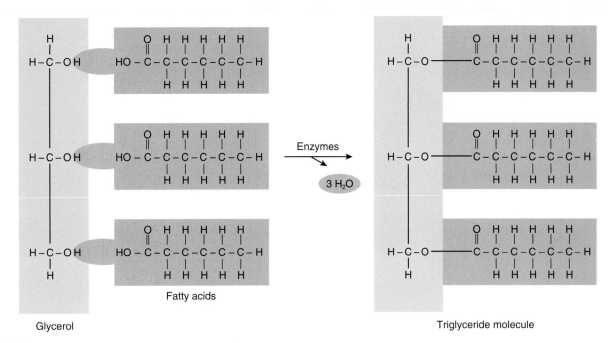

Figure 2.12 Triglyceride

Production of a triglyceride from one glycerol molecule and three fatty acids.

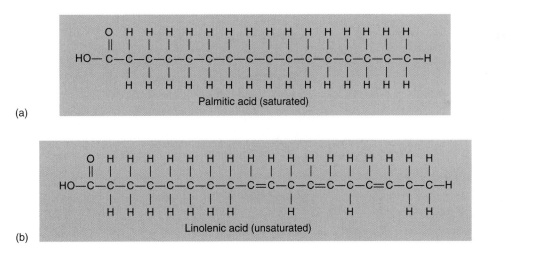

Figure 2.13 Fatty Acids

(a) Palmitic acid (saturated with no double bonds between the carbons). (b) Linolenic acid (unsaturated with three double bonds between the carbons).

the diet because, unlike saturated fats, they do not contribute to the development of cardiovascular disease.

Phospholipids are similar to triglycerides, except that one of the fatty acids bound to the glycerol is replaced by a molecule containing phosphorus (figure 2.14). They are polar at the end of the molecule to which the phosphate is bound and nonpolar at the other end. The polar end of the molecule is attracted to water and is said to be **hydrophilic** (water loving). The nonpolar end is repelled by water and is said to be **hydrophobic** (water fearing). Phospholipids are important structural components of cell membranes (see chapter 3).

The **eicosanoids** (ī′kō-să-noydz) are a group of important chemicals derived from fatty acids. They include **prostaglandins**

(pros′tă-glan′dinz), **thromboxanes** (throm′bok-zānz), and **leukotrienes** (loo-kō-trī′ēnz). Eicosanoids are made in most cells and are important regulatory molecules. Among their numerous effects is their role in the response of tissues to injuries. Prostaglandins have been implicated in regulating the secretion of some hormones, blood clotting, some reproductive functions, and many other processes. Many of the therapeutic effects of aspirin and other anti-inflammatory drugs result from their ability to inhibit prostaglandin synthesis.

Steroids are composed of carbon atoms bound together into four ringlike structures. Important steroid molecules include cholesterol, bile salts, estrogen, progesterone, and testosterone (figure 2.15). Cholesterol is an important steroid because

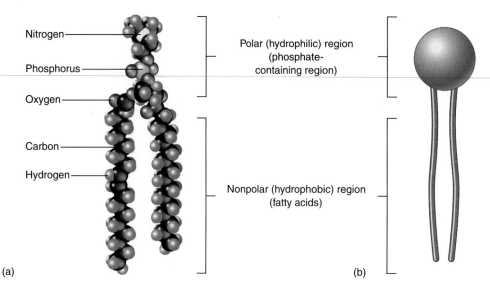

Figure 2.14   **Phospholipids**

(*a*) Molecular model of a phospholipid. (*b*) Simplified way in which phospholipids are often depicted.

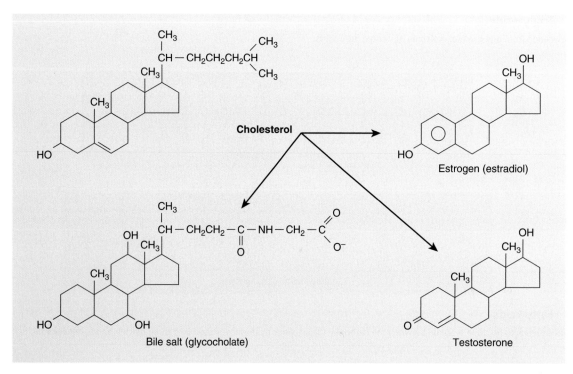

Figure 2.15   **Steroids**

Steroids are four-ringed molecules that differ from one another according to the groups attached to the rings. Cholesterol, the most common steroid, can be modified to produce other steroids.

other molecules are synthesized from it. For example, bile salts, which increase fat absorption in the intestines, are derived from cholesterol, as are the reproductive hormones estrogen, progesterone, and testosterone. In addition, cholesterol is an important component of cell membranes. Although high levels of cholesterol in the blood increase the risk of cardiovascular disease, a certain amount of cholesterol is vital for normal function.

Proteins

All **proteins** contain carbon, hydrogen, oxygen, and nitrogen, and most have some sulfur. The building blocks of proteins are **amino** (ă-mē′nō) **acids,** which are organic acids containing an **amine** (ă-mēn′) **group** ($-NH_2$) and a carboxyl group (figure 2.16). There are 20 basic types of amino acids. Humans can synthesize 12 of these from simple organic molecules, but the remaining 8 "essential amino acids" must be obtained in the diet.

(a) Two examples of amino acids. Each amino acid has an amine group (—NH₂) and a carboxyl group (—COOH).

Amino acid alanine

Amino acid glycine

(b) The individual amino acids are joined.

H_2O

(c) A protein consists of a chain of different amino acids (represented by different colored spheres).

(d) A three-dimensional representation of the amino acid chain showing hydrogen bonds (*dotted red lines*) between different amino acids. The hydrogen bonds cause the amino acid chain to become folded or coiled.

Folded sheet

Coiled

(e) An entire protein showing its complex three-dimensional shape.

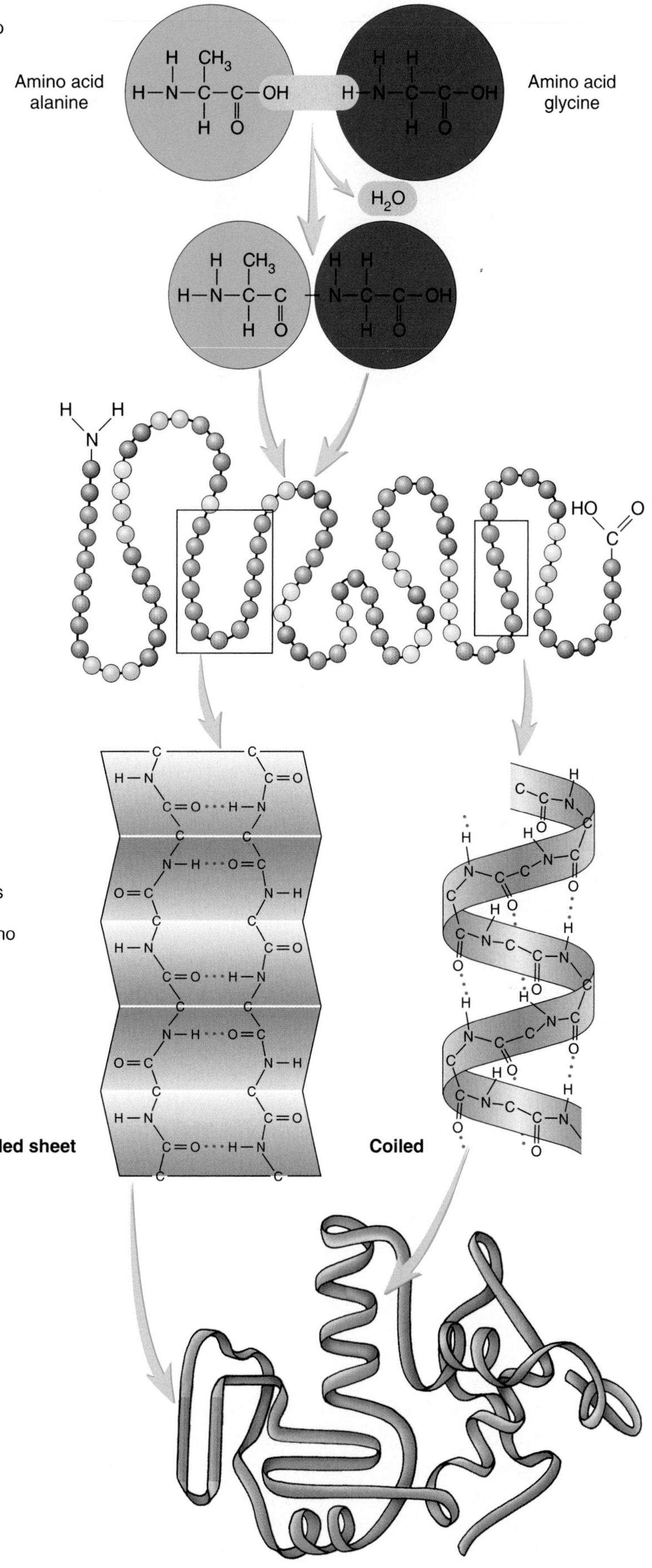

Figure 2.16 Proteins

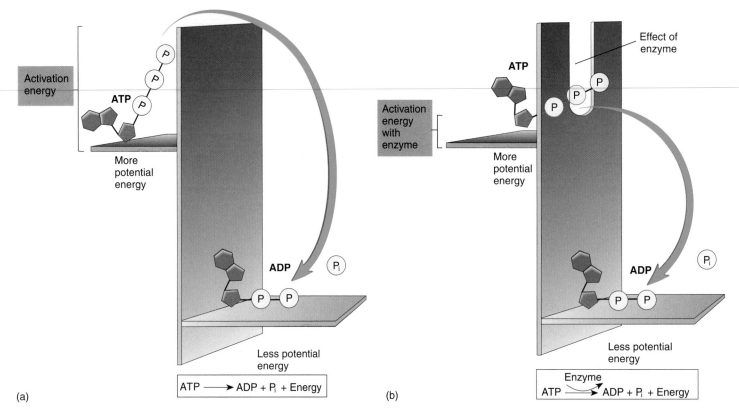

Figure 2.17 **Activation Energy and Enzymes**

(*a*) Activation energy is needed to change ATP to ADP. The upper shelf represents a higher energy level, and the lower shelf represents a lower energy level. The "wall" extending above the upper shelf represents the activation energy. Even though energy is given up moving from the upper to the lower shelf, the activation energy "wall" must be overcome before the reaction can proceed. (*b*) The enzyme lowers the activation energy, making it easier for the reaction to proceed.

A protein consists of many amino acids joined together to form a chain of amino acids (see figure 2.16). Although there are only 20 amino acids, they can combine to form numerous types of proteins with unique structures and functions. Different proteins have different kinds and numbers of amino acids. Hydrogen bonds between amino acids in the chain cause it to fold or coil to assume a specific three-dimensional shape (see figure 2.16). The ability of proteins to perform their functions depends on their shape. If the hydrogen bonds that maintain the shape of the protein are broken, the protein becomes nonfunctional. This change in shape is called **denaturation,** and it can be caused by abnormally high temperatures or changes in pH.

Proteins perform many important functions. For example, enzymes are proteins that regulate the rate of chemical reactions, structural proteins provide the framework for many of the body's tissues, and muscles contain proteins that are responsible for muscle contraction.

Enzymes

An **enzyme** (en′zīm) is a protein catalyst that increases the rate at which a chemical reaction proceeds without the enzyme being permanently changed. Enzymes increase the rate of chemical reactions by lowering the **activation energy,** which is the energy necessary to start a chemical reaction. For example, heat in the form of a spark is required to start the reaction between oxygen and gasoline. Most of the chemical reactions that

occur in the body have high activation energies, which are decreased by enzymes (figure 2.17). The lowered activation energies enable reactions to proceed at rates that sustain life.

Consider this analogy, in which paper clips represent amino acids and your hands represent enzymes. Paper clips in a box only occasionally join together. Using your hands, however, a chain of paper clips can be rapidly formed. In a similar fashion, enzymes can quickly join amino acids into a chain, forming a protein. With an enzyme, the rate of a chemical reaction can take place more than a million times faster than without the enzyme.

The three-dimensional shape of enzymes is critical for their normal function. According to the **"lock-and-key" model** of enzyme action, the shape of an enzyme and that of the reactants allow the enzyme to bind easily to the reactants. Bringing the reactants very close to one another reduces the activation energy for the reaction. Because the enzyme and the reactants must fit together, enzymes are very specific for the reactions they control, and each enzyme controls only one type of chemical reaction. After the reaction takes place, the enzyme is released and can be used again (figure 2.18).

The chemical events of the body are regulated primarily by mechanisms that control either the concentration or activity of enzymes. The rate at which enzymes are produced in cells or whether the enzymes are in an active or inactive form determines the rate of each chemical reaction.

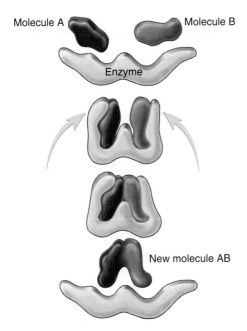

Figure 2.18 Enzyme Action

The enzyme brings two reacting molecules together. This is possible because the reacting molecules "fit" the shape of the enzyme (lock-and-key model). After the reaction, the unaltered enzyme can be used again.

Nucleic Acids: DNA and RNA

Deoxyribonucleic (dē-oks′ē-rī′bō-noo-klē′ik) **acid (DNA)** is the genetic material of cells, and copies of DNA are transferred from one generation of cells to the next. DNA contains the information that determines the structure of proteins. **Ribonucleic** (rī′bō-noo-klē′ik) **acid (RNA)** is structurally related to DNA, and three types of RNA also play important roles in protein synthesis. In chapter 3, the means by which DNA and RNA direct the functions of the cell are described.

The **nucleic** (noo-klē′ik, noo-klā′ik) **acids** are large molecules composed of carbon, hydrogen, oxygen, nitrogen, and phosphorus. Both DNA and RNA consist of basic building blocks called **nucleotides** (noo′klē-ō-tīdz). Each nucleotide is composed of a sugar (monosaccharide) to which a nitrogenous organic base and a phosphate group are attached (figure 2.19). The sugar is deoxyribose for DNA, and ribose for RNA. The organic bases are thymine (thī′mēn, thī′min), cytosine (sī′tō-sēn), and uracil (ūr′ă-sil), which are single-ringed molecules; and adenine (ad′ĕ-nēn) and guanine (gwahn′ēn), which are double-ringed molecules.

DNA has two strands of nucleotides joined together to form a twisted, ladderlike structure called a double helix. The uprights of the ladder are formed by covalent bonds between the sugar molecules and phosphate groups of adjacent nucleotides. The rungs of the ladder are formed by the bases of the nucleotides of one upright connected to the bases of the other upright by hydrogen bonds. Each nucleotide of DNA contains one of the organic bases: adenine, thymine, cytosine, or guanine. Adenine binds only to thymine because the structure of these organic bases allows two hydrogen bonds to form between them. Cytosine binds only to guanine because the structure of these organic bases allows three hydrogen bonds to form between them.

The sequence of organic bases in DNA molecules stores genetic information. Each DNA molecule consists of millions of organic bases, and their sequence ultimately determines the type and sequence of amino acids found in protein molecules. Because enzymes are proteins, DNA structure determines the rate and type of chemical reactions that occur in cells by controlling enzyme structure. The information contained in DNA, therefore, ultimately defines all cellular activities. Other proteins, such as collagen, that are coded by DNA determine many of the structural features of humans.

RNA has a structure similar to a single strand of DNA. Like DNA, four different nucleotides make up the RNA molecule, and the organic bases are the same, except that thymine is replaced with uracil. Uracil can bind only to adenine.

Adenosine Triphosphate

Adenosine triphosphate (ă-den′ō-sēn trī-fos′fāt) **(ATP)** is an especially important organic molecule found in all living organisms. It consists of adenosine and three phosphate groups (figure 2.20). Adenosine is the sugar ribose with the organic base adenine. The potential energy stored in the covalent bond between the second and third phosphate groups is important to living organisms because it provides the energy used in nearly all of the chemical reactions within cells.

ATP is often called the energy currency of cells because it is capable of both storing and providing energy. The concentration of ATP is maintained within a narrow range of values, and essentially all energy-requiring chemical reactions stop when there is an inadequate quantity of ATP.

A CASE IN POINT | Cyanide Poisoning

Although Justin Hale was rescued from his burning home, he subsequently died from cyanide poisoning. Inhalation of smoke released by the burning of rubber and plastic by household fires is the most common cause of cyanide poisoning. Cyanide compounds can be lethal to humans because they interfere with the production of ATP in mitochondria (see aerobic respiration in chapter 3). Without adequate ATP, cells malfunction and can die. The heart and brain are especially susceptible to cyanide poisoning.

Cyanide poisoning by inhalation or absorption through the skin can also occur in certain manufacturing industries, and cyanide gas was used during the Holocaust to kill people. Deliberate suicide by ingesting cyanide is rare, but has been made famous by the suicide capsules in spy movies. In 1982, seven people in the Chicago area died after taking Tylenol that some unknown person laced with cyanide. Subsequent copycat tamperings occurred and led to the widespread use of tamper-proof capsules and packaging.

1. The building blocks of nucleic acids are nucleotides, which consist of a phosphate group, a sugar, and a nitrogen base.

2. The phosphate groups connect the sugars to form two strands of nucleotides (*purple columns*).

3. Hydrogen bonds (*dotted red lines*) between the nucleotides join the two nucleotide strands together. Adenine binds to thymine and cytosine binds to guanine.

4. The two nucleotide strands coil to form a double-stranded helix.

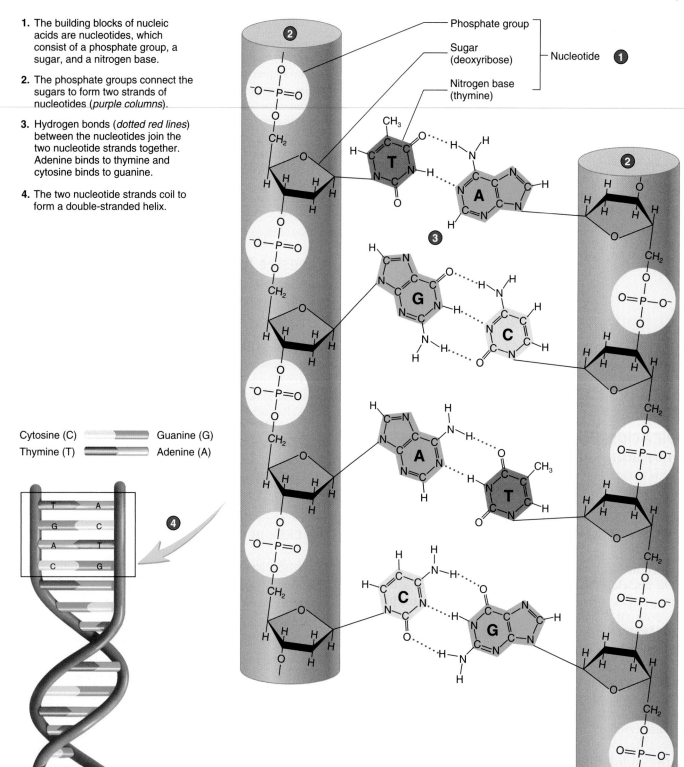

Figure 2.19   Structure of DNA

1. Adenosine is adenine, which is one of the nitrogenous bases in DNA, combined with the sugar ribose.

2. Adenosine diphosphate (ADP) is adenosine with two phosphate groups.

3. Adenosine triphosphate (ATP) is adenosine with three phosphates.

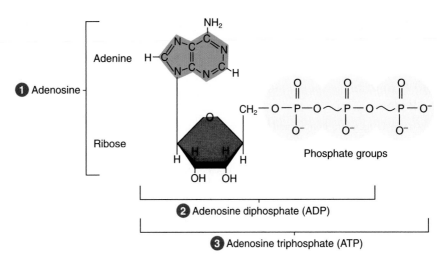

Figure 2.20 Adenosine Triphosphate (ATP) Molecule

S U M M A R Y

Chemistry is the study of the composition and structure of substances and the reactions they undergo.

Basic Chemistry (p. 20)

Matter, Mass, and Weight

1. Matter is anything that occupies space.
2. Mass is the amount of matter in an object, and weight results from the gravitational attraction between earth and matter.

Elements and Atoms

1. An element is the simplest type of matter with unique chemical and physical properties.
2. An atom is the smallest particle of an element that has the chemical characteristics of that element. An element is composed of only one kind of atom.

Atomic Structure

1. Atoms consist of neutrons, positively charged protons, and negatively charged electrons.
2. Atoms are electrically neutral because the number of protons in atoms equals the number of electrons.
3. Protons and neutrons are found in the nucleus, and electrons, which are located around the nucleus, can be represented by an electron cloud.
4. The atomic number is the unique number of protons in each atom of an element. The mass number is the number of protons and neutrons.

Electrons and Chemical Bonding

1. Ionic bonding results when an electron is transferred from one atom to another.
2. Covalent bonding results when a pair of electrons are shared between atoms. A polar covalent bond is an unequal sharing of electron pairs.

Hydrogen Bonds

A hydrogen bond is the weak attraction that occurs between the oppositely charged regions of polar molecules. Hydrogen bonds are important in determining the three-dimensional structure of large molecules.

Molecules and Compounds

1. A molecule is two or more atoms chemically combined to form a structure that behaves as an independent unit.
2. A compound is two or more different types of atoms chemically combined. A compound can be a molecule (covalent compound) or an organized array of ions (ionic compound).

Dissociation

Dissociation is the separation of ions in an ionic compound by polar water molecules.

Chemical Reactions (p. 25)

Classification of Chemical Reactions

1. A synthesis reaction is the combination of reactants to form a new, larger product.
2. A decomposition reaction is the breakdown of larger reactants into smaller products.
3. An exchange reaction is a decomposition reaction, in which reactants are broken down, and a synthesis reaction, in which the products of the decomposition reaction combine.

Reversible Reactions

1. In a reversible reaction, the reactants can form products, or the products can form reactants.
2. The amount of reactants relative to products is constant at equilibrium.

Energy and Chemical Reactions

1. Energy is the capacity to do work. Potential energy is stored energy that could do work, and kinetic energy does work by causing the movement of an object.
2. Energy can be neither created nor destroyed, but one type of energy can be changed into another.
3. Energy exists in chemical bonds as potential energy.
4. Energy is released in chemical reactions when the products contain less potential energy than the reactants. The energy can be lost as heat, be used to synthesize molecules, or can do work.
5. Energy is absorbed in reactions when the products contain more potential energy than the reactants.

Rate of Chemical Reactions

1. The rate of chemical reactions increases when the concentration of the reactants increases, temperature increases, or a catalyst is present.
2. A catalyst (enzyme) increases the rate of chemical reactions without being altered permanently.

Acids and Bases   (p. 29)

Acids are proton (hydrogen ion) donors, and bases are proton acceptors.

The pH Scale

1. A neutral solution has an equal number of H^+ and OH^- and a pH of 7.0.
2. An acidic solution has more H^+ than OH^- and a pH of less than 7.0.
3. A basic solution has fewer H^+ than OH^- and a pH greater than 7.0.

Salts

A salt is formed when an acid reacts with a base.

Buffers

Buffers are chemicals that resist changes in pH when acids or bases are added.

Inorganic Chemistry   (p. 30)

Inorganic chemistry is mostly concerned with non-carbon-containing substances, but does include such carbon-containing substances as carbon dioxide and carbon monoxide.

Oxygen and Carbon Dioxide

1. Oxygen is involved with the extraction of energy from food molecules.
2. Carbon dioxide is a by-product of the breakdown of food molecules.

Water

1. Water stabilizes body temperature.
2. Water provides protection by acting as a lubricant or cushion.
3. Water is necessary for many chemical reactions.
4. Water transports many substances.

Organic Chemistry   (p. 31)

Organic molecules contain carbon atoms bound together by covalent bonds.

Carbohydrates

1. Carbohydrates provide the body with energy.
2. Monosaccharides are the building blocks that form more complex carbohydrates, such as disaccharides and polysaccharides.

Lipids

1. Lipids are substances that dissolve in nonpolar solvents, such as alcohol or acetone, but not in polar solvents, such as water. Fats, phospholipids, and steroids are examples of lipids.
2. Lipids provide energy (fats), are structural components (phospholipids), and regulate physiological processes (steroids).
3. The building blocks of triglycerides (fats) are glycerol and fatty acids.
4. Fatty acids can be saturated (have only single covalent bonds between carbon atoms) or unsaturated (have one or more double covalent bonds between carbon atoms).

Proteins

1. Proteins regulate chemical reactions (enzymes), are structural components, and cause muscle contraction.
2. The building blocks of proteins are amino acids.
3. Denaturation of proteins disrupts hydrogen bonds, which changes the shape of proteins and makes them nonfunctional.
4. Enzymes are specific, bind to reactants according to the lock-and-key model, and function by lowering activation energy.

Nucleic Acids: DNA and RNA

1. The basic unit of nucleic acids is the nucleotide, which is a monosaccharide with an attached phosphate and organic base.
2. DNA nucleotides contain the monosaccharide deoxyribose and the organic bases adenine, thymine, guanine, or cytosine. DNA occurs as a double strand of joined nucleotides and is the genetic material of cells.
3. RNA nucleotides are composed of the monosaccharide ribose. The organic bases are the same as for DNA, except that thymine is replaced with uracil.

Adenosine Triphosphate

ATP stores energy derived from catabolism. The energy is released from ATP and is used in anabolism and other cell processes.

R E V I E W   A N D   C O M P R E H E N S I O N

1. Define chemistry. Why is an understanding of chemistry important?
2. Define matter. What is the difference between mass and weight?
3. Define element and atom. How many different kinds of atoms are found in a specific element?
4. List the components of an atom, and explain how they are organized to form an atom. Compare the charges of the subatomic particles.
5. Define the atomic number and the mass number of an element.
6. Distinguish between ionic, covalent, polar covalent, and hydrogen bonds. Define ion.
7. What is the difference between a molecule and a compound?
8. What happens to ionic and covalent compounds when they dissolve in water?
9. Define chemical reaction. Describe synthesis, decomposition, and exchange reactions, giving an example of each.
10. What is meant by the equilibrium condition in a reversible reaction?
11. Define potential and kinetic energy.
12. Give an example of a chemical reaction that releases energy and an example of a chemical reaction that requires the input of energy.
13. Name three ways that the rate of chemical reactions can be increased.
14. What is an acid and what is a base? Describe the pH scale.
15. Define salt. What is a buffer, and why are buffers important?
16. Distinguish between inorganic and organic chemistry.
17. Why is oxygen necessary for human life? Where does the carbon dioxide we breathe out come from?
18. List four functions that water performs in the human body.
19. Name the four major types of organic molecules. Give a function for each.
20. Describe the action of enzymes in terms of activation energy and the lock-and-key model.

C R I T I C A L   T H I N K I N G

1. If an atom of iodine (I) gains an electron, what is the charge of the resulting ion? Write the symbol for this ion.

2. For each of the following chemical equations, determine if a synthesis reaction, a decomposition reaction, or dissociation has taken place:
   a. $HCl \rightarrow H^+ + Cl^-$
   b. Glucose + Fructose $\rightarrow$ Sucrose (table sugar)
   c. $2 H_2O \rightarrow 2 H_2 + O_2$

3. In terms of the energy in chemical bonds, explain why eating food is necessary for increasing muscle mass.

4. Given that the hydrogen ion concentration in a solution is based on the following reversible reaction:

$$CO_2 + H_2O \rightleftarrows H^+ + HCO_3^-$$

What happens to the pH of the solution when $NaHCO_3$ (sodium bicarbonate) is added to the solution? (*Hint:* The sodium bicarbonate dissociates to form Na^+ and HCO_3^-.)

5. A mixture of chemicals is warmed slightly. As a consequence, although little heat is added, the solution becomes very hot. Explain what happens to make the solution hot.

6. Two solutions, when mixed together at room temperature, produce a chemical reaction. When the solutions are boiled, however, and allowed to cool to room temperature before mixing, no chemical reaction takes place. Explain.

Answers in Appendix D

Visit this textbook's website at www.mhhe.com/seeleyess6 for practice quizzes, animations, interactive learning exercises, and other study tools.

McGraw-Hill offers a study CD that features interactive cadaver dissection. *Anatomy & Physiology Revealed* includes cadaver photos that allow you to peel away layers of the human body to reveal structures beneath the surface. This program also includes animations, radiologic imaging, audio pronunciations, and practice quizzing.

3

Cell Structures and Their Functions

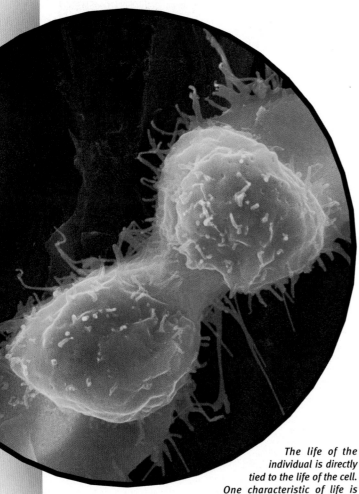

The life of the individual is directly tied to the life of the cell. One characteristic of life is growth, and growth involves cell division.
This colorized scanning electron micrograph shows a cell dividing into two new cells.

Chapter Outline and Objectives

the cell is the basic living unit of all organisms. The simplest organisms consist of single cells, whereas humans are composed of trillions of cells. An average-sized cell is one-fifth the size of the smallest dot you can make on a sheet of paper with a sharp pencil! In spite of their extremely small size, cells are complex living structures. Cells of the human body have many characteristics in common.

However, most cells are also specialized to perform specific functions. The human body is made up of many populations of these specialized cells. The coordinated functions of these populations are critical for complex organisms, such as humans, to survive.

The study of cells is an important link between the study of chemistry in chapter 2 and tissues in chapter 4. Knowledge of chem-

istry makes it possible to understand cells because cells are composed of chemicals responsible for many of the characteristics of cells. Cells, in turn, determine the form and functions of the tissues of the body. In addition, a great many diseases and other human disorders have a cellular basis. This chapter considers the structure of cells and how cells perform the activities necessary for life.

Cell Structure

Each cell is a highly organized unit. Within cells, specialized structures called **organelles** (or′gă-nelz, little organs) perform specific functions (figure 3.1 and table 3.1). The nucleus is an organelle containing the cell's genetic material. The living material surrounding the nucleus is called **cytoplasm** (sī′tō-plazm,

cyto-, cell + plasma, a thing formed), which contains many types of organelles. The cytoplasm is enclosed by the cell, or plasma, membrane.

The number and type of organelles within each cell determine the cell's specific structure and functions. For example, cells secreting large amounts of protein contain well-developed organelles that synthesize and secrete protein, whereas muscle

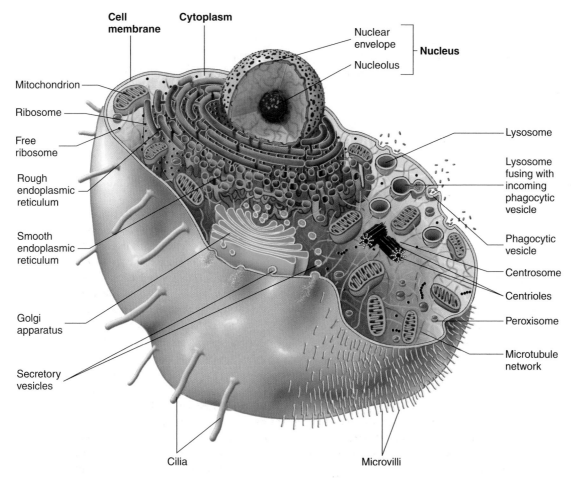

Figure 3.1   Generalized Cell Showing the Major Organelles

This generalized cell shows organelles contained in cells. No single cell, however, contains all organelle types. In addition, some kinds of the cells contain many organelles of one type, and another kind of cell contains very few.

Table 3.1   Organelles and Their Locations and Functions

Organelles	Location	Function(s)
Nucleus	Often near center of the cell	Contains genetic material of cell (DNA) and nucleoli; site of ribosome and messenger RNA synthesis
Nucleolus	In the nucleus	Site of ribosomal RNA synthesis and ribosomal subunit assembly
Rough endoplasmic reticulum (rough ER)	In cytoplasm	Many ribosomes attached to rough ER; site of protein synthesis
Smooth endoplasmic reticulum (smooth ER)	In cytoplasm	Site of lipid synthesis; detoxification
Golgi apparatus	In cytoplasm	Modifies protein structure and packages proteins in secretory vesicles
Secretory vesicle	In cytoplasm	Contains materials produced in the cell; formed by the Golgi apparatus; secreted by exocytosis
Lysosome	In cytoplasm	Contains enzymes that digest material taken into the cell
Mitochondrion	In cytoplasm	Site of aerobic respiration and the major site of ATP synthesis
Microtubule	In cytoplasm	Supports cytoplasm; assists in cell division and forms components of cilia and flagella
Cilia	On cell surface with many on each cell	Cilia move substances over surfaces of certain cells
Flagella	On sperm cell surface with one per cell	Propels sperm cells
Microvilli	Extensions of cell surface with many on each cell	Increase surface area of certain cells

cells contain proteins and organelles that enable them to contract. The following sections describe the structure and main functions of the major organelles found in cells.

Functions of the Cell

The main functions of the cell include

1. *Basic unit of life.* The cell is the smallest part of an organism that still retains all the characteristics of life.
2. *Protection and support.* Cells produce and secrete various molecules that provide protection and support of the body. For example, bone cells produce a mineralized material, making bone a hard tissue that protects vital organs and supports the weight of the body.
3. *Movement.* Movements of the body occur because of molecules located within specific cells such as muscle cells.
4. *Communication.* Cells produce and receive chemical and electrical signals that allow them to communicate with one another. For example, nerve cells communicate with one another and with muscle cells, causing muscle cells to contract.
5. *Cell metabolism and energy release.* The chemical reactions that occur within cells are referred to collectively as cell metabolism. Energy released during metabolism is used for cell activities, such as the synthesis of new molecules, muscle contraction, and heat production, which helps maintain body temperature.

6. *Inheritance.* Each cell contains a copy of the genetic information of the individual. Specialized cells, sperm cells and oocytes, transmit that genetic information to the next generation.

Cell Membrane

The **cell membrane,** or **plasma** (plaz′mă) **membrane,** is the outermost component of a cell. The cell membrane encloses the cytoplasm and forms the boundary between material inside the cell and material outside it. Substances outside the cell are called **extracellular substances,** and those inside the cell are called **intracellular substances.** The cell membrane encloses the cell, supports the cell contents, is a selective barrier that determines what moves into and out of the cell, and plays a role in communication between cells.

The major molecules that make up the cell membrane are phospholipids and proteins (figure 3.2). In addition, the membrane contains other molecules, such as cholesterol, carbohydrates, water, and ions. The phospholipids form a double layer of molecules. The polar, phosphate-containing ends of the phospholipids are hydrophilic (water loving) and therefore face the water inside and outside the cell. The nonpolar, fatty acid ends of the phospholipids are hydrophobic (water fearing) and therefore face away from the water on either side of the membrane, toward the center of the double layer of phospholipids. The double layer of phospholipids forms a lipid barrier between the inside and outside of the cell.

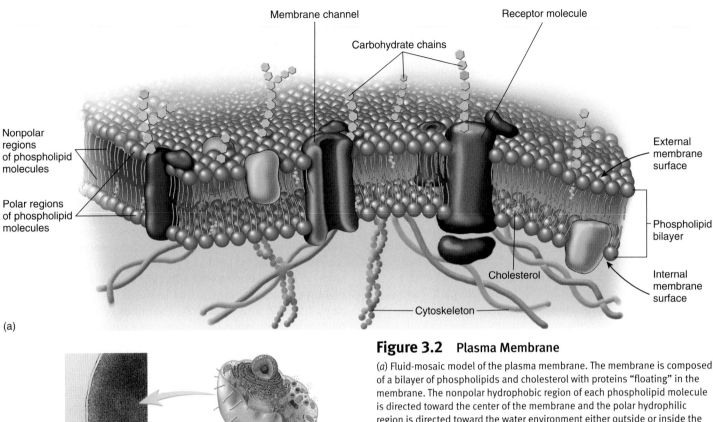

Membrane channel

Receptor molecule

Carbohydrate chains

Nonpolar
regions
of phospholipid
molecules

Polar regions
of phospholipid
molecules

External
membrane
surface

Phospholipid
bilayer

Cholesterol

Cytoskeleton

Internal
membrane
surface

(a)

(b)

TEM 100,000x

Figure 3.2   Plasma Membrane

(*a*) Fluid-mosaic model of the plasma membrane. The membrane is composed of a bilayer of phospholipids and cholesterol with proteins "floating" in the membrane. The nonpolar hydrophobic region of each phospholipid molecule is directed toward the center of the membrane and the polar hydrophilic region is directed toward the water environment either outside or inside the cell. (*b*) Transmission electron micrograph of a plasma membrane, with the membrane indicated by the *blue arrows*. Proteins at either surface of the lipid bilayer stain more readily than the lipid bilayer does and give the membrane the appearance of consisting of three parts: the two dark outer parts are proteins and the phospholipid heads, and the lighter central part is the phospholipid tails and cholesterol.

Studies of the arrangement of molecules in the cell membrane have given rise to a model of its structure called the **fluid-mosaic model.** The double layer of phospholipid molecules has a liquid quality. Cholesterol within the phospholipid membrane gives it added strength and flexibility. Protein molecules "float" among the phospholipid molecules and, in some cases, may extend from the inner to the outer surface of the cell membrane. Carbohydrates may be bound to some protein molecules, modifying their functions. The proteins function as membrane channels, carrier molecules, receptor molecules, enzymes, or structural supports in the membrane. **Membrane channels** and **carrier molecules** are involved with the movement of substances through the cell membrane. **Receptor molecules** are part of an intercellular communication system that enables cell recognition and coordination of the activities of cells. For example, a nerve cell can release a chemical messenger that moves to a muscle cell and temporarily binds to its receptor. The binding acts as a signal that triggers a response such as contraction of the muscle cell.

Movement Through the Cell Membrane

Cell membranes are **selectively permeable,** allowing some substances, but not others, to pass into or out of the cells. Intracellular material has a different composition from extracellular material, and the survival of cells depends on maintaining the difference. Substances such as enzymes, glycogen, and potassium ions are found at higher concentrations intracellularly; and Na^+, Ca^{2+}, and Cl^- are found in greater concentrations extracellularly. In addition, nutrients must enter cells continually, and waste products must exit. Because of the permeability characteristics of cell membranes and their ability to transport certain molecules, cells are able to maintain proper intracellular concentrations of molecules. Rupture of the membrane, alteration of its permeability characteristics, or inhibition of transport processes disrupts the normal intracellular concentration of molecules and can lead to cell death.

Molecules pass through cell membranes in four ways:

1. *Directly through the phospholipid membrane.* Molecules that are soluble in lipids, such as oxygen, carbon dioxide, and steroids, pass through cell membranes readily by dissolving in the phospholipid bilayer. The phospholipid bilayer acts as a barrier to most substances that are not lipid-soluble; but certain small, non-lipid-soluble molecules, such as water and urea, can diffuse between the phospholipid molecules of cell membranes.

2. *Membrane channels.* Cell membrane **channels,** consisting of large protein molecules, extend from one surface of cell membranes to the other (see figure 3.2). There are several channel types, each of which allows only certain molecules to pass through it. The size, shape, and charge of molecules determine whether they can pass through each kind of channel. For example, Na^+ pass through Na^+ channels, and K^+ and Cl^- pass through K^+ and Cl^- channels, respectively. Rapid movement of water across the cell membrane apparently occurs through membrane channels.

3. *Carrier molecules.* Large polar molecules that are not lipid-soluble, such as glucose and amino acids, cannot pass through cell membranes in significant amounts unless they are transported by special carrier molecules. Substances that are transported across cell membranes by carrier molecules are said to be transported by **carrier-mediated processes. Carrier molecules** are proteins that extend from one side of a cell membrane to the other. They bind to molecules to be transported and move them across the cell membrane. Each carrier molecule transports a specific type of molecule. For

example, carrier molecules that transport glucose across cell membranes do not transport amino acids, and carrier molecules that transport amino acids do not transport glucose.

4. *Vesicles.* Large non-lipid-soluble molecules, small pieces of matter, and even whole cells can be transported across cell membranes in **vesicles,** which are membrane-bound sacs. Because of the fluid nature of membranes, vesicles and cell membranes can fuse, allowing the contents of the vesicles to cross the cell membrane.

Diffusion

A **solution** is generally thought of as a liquid or gas and consists of one or more substances called **solutes** dissolved in the predominant liquid or gas, which is called the **solvent. Diffusion** can be viewed as the tendency for solutes, such as ions or molecules, to move from an area of higher concentration to an area of lower concentration of that solute in solution (figure 3.3, steps 1 and 2, and table 3.2). Examples of diffusion are the tendencies for smoke or perfume to distribute throughout a room in which there are no air currents, or that of a dye or salt throughout a beaker of still water.

Diffusion results from the constant random motion of all solutes in a solution. More solute particles occur in an area of higher concentration than in one of lower concentration. Because particles move randomly, the chances are greater that solute particles will move from the higher toward the lower concentration than from a lower to higher concentration. At equilibrium, the net movement of solutes stops, although the random motion continues, and the movement of solutes in any one direction is balanced by an equal movement in the opposite direction (figure 3.3, step 3).

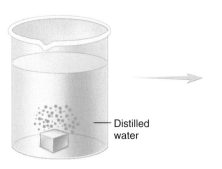 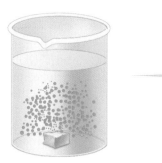 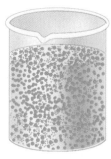

1. When a salt crystal (*green*) is placed into a beaker of water, there is a concentration gradient for salt from the salt crystal to the water that surrounds it.

2. Salt ions (*green*) move down their concentration gradient into the water.

3. Salt ions and water molecules are distributed evenly throughout the solution. Even though the salt ions and water molecules continue to move randomly, an equilibrium exists, and no net movement occurs because no concentration gradient exists.

Process Figure 3.3   Diffusion

Table 3.2 Types and Characteristics of Movement Across Membranes

Type	Transport	Requires ATP	Examples
Diffusion	With the concentration gradient through the lipid portion of the cell membrane or through membrane channels	No	Oxygen, carbon dioxide, chloride ions, and urea
Osmosis	With the concentration gradient (for water) through the lipid portion of the cell membrane or through membrane channels	No	Water
Filtration	Movement of liquid and substances by pressure through a partition containing holes	No	In the kidneys, filtration of everything in blood smaller than proteins and blood cells
Facilitated diffusion	With the concentration gradient by carrier molecules	No	Glucose in most cells
Active transport	Against the concentration gradient* by carrier molecules	Yes	Na^+, K^+, Ca^{2+}, and H^+; amino acids
Secondary active transport	Against the concentration gradient by carrier molecules; the energy for secondary active transport of one substance comes from the concentration gradient of another	Yes	Glucose, amino acids
Endocytosis	Movement into cells by vesicles	Yes	Ingestion of particles by phagocytosis or receptor-mediated endocytosis and liquids by pinocytosis
Exocytosis	Movement out of cells by vesicles	Yes	Secretion of proteins

*Active transport normally moves substances against their concentration gradient, but it can also move substances with their concentration gradient.

A **concentration gradient** is a measure of the difference in the concentration of a solute in a solvent between two points divided by the distance between the two points. The concentration gradient is said to be steeper when the concentration difference between the two points is large and/or the distance is small. Movement down, or with, a concentration gradient, describes the diffusion of solutes from a higher toward a lower concentration of solutes. Movement up, or against, a concentration gradient, describes the movement of solutes from a lower toward a higher concentration of solutes. This second type of movement does not occur by diffusion and requires energy to move solutes against their concentration gradient.

Diffusion is an important means of transporting substances through the extracellular and intracellular fluids in the body. In addition, substances that can pass either through the lipid layers of the cell membrane or through membrane channels diffuse through the cell membrane (figure 3.4). Some nutrients enter and some waste products leave the cell by diffusion. The normal intracellular concentrations of many substances depend on diffusion. For example, if the extracellular concentration of oxygen is reduced, not enough oxygen diffuses into the cell, and normal cell function cannot occur.

PREDICT 1

Urea is a toxic waste produced inside liver cells. It diffuses from those cells into the blood and is eliminated from the body by the kidneys. What would happen to the intracellular and extracellular concentration of urea if the kidneys stopped functioning?

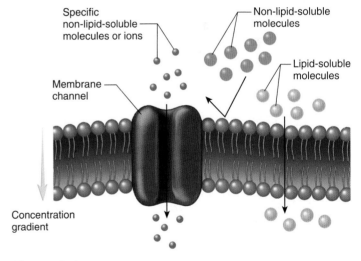

Specific non-lipid-soluble molecules or ions

Non-lipid-soluble molecules

Lipid-soluble molecules

Membrane channel

Concentration gradient

Figure 3.4 Diffusion Through the Cell Membrane

Non-lipid-soluble molecules diffuse through membrane channels. Lipid-soluble molecules diffuse directly through the cell membrane.

Osmosis

Osmosis (os-mō′sis, a thrusting) is the diffusion of water (a solvent) across a selectively permeable membrane, such as the cell membrane, from a region of higher water concentration to one of lower water concentration (figure 3.5; see table 3.2). Osmosis is important to cells because large volume changes caused by water movement can disrupt normal cell functions.

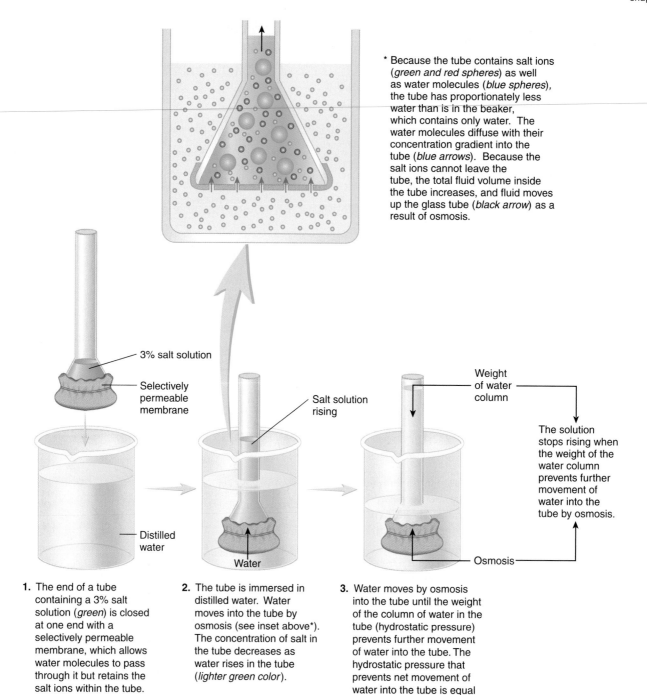

* Because the tube contains salt ions (*green and red spheres*) as well as water molecules (*blue spheres*), the tube has proportionately less water than is in the beaker, which contains only water. The water molecules diffuse with their concentration gradient into the tube (*blue arrows*). Because the salt ions cannot leave the tube, the total fluid volume inside the tube increases, and fluid moves up the glass tube (*black arrow*) as a result of osmosis.

3% salt solution

Selectively permeable membrane

Distilled water

Salt solution rising

Water

Weight of water column

The solution stops rising when the weight of the water column prevents further movement of water into the tube by osmosis.

Osmosis

1. The end of a tube containing a 3% salt solution (*green*) is closed at one end with a selectively permeable membrane, which allows water molecules to pass through it but retains the salt ions within the tube.

2. The tube is immersed in distilled water. Water moves into the tube by osmosis (see inset above*). The concentration of salt in the tube decreases as water rises in the tube (*lighter green color*).

3. Water moves by osmosis into the tube until the weight of the column of water in the tube (hydrostatic pressure) prevents further movement of water into the tube. The hydrostatic pressure that prevents net movement of water into the tube is equal to the osmotic pressure of the solution in the tube.

Process Figure 3.5   Osmosis

Osmosis occurs when the cell membrane is less permeable, selectively permeable, or not permeable to solutes *and* a concentration gradient for water exists across the cell membrane. Water diffuses from a solution with a higher concentration of water across the cell membrane into a solution with a lower water concentration. The ability to predict the direction of water movement across the cell membrane depends on knowing which solution on either side of the membrane has the highest water concentration.

The concentration of a solution, however, is not expressed in terms of water, but in terms of solute concentration. For example, if sugar solution A is more concentrated than sugar solution B, then solution A has more sugar (solute) than solution B. As the concentration of a solution increases, the amount of water (solvent) proportionately decreases. Water diffuses from the less concentrated solution, which has fewer solute molecules and more water molecules, into the more concentrated solution, which has more solute molecules and fewer water molecules.

Osmotic pressure is the force required to prevent the movement of water across a selectively permeable membrane. Thus, osmotic pressure is a measure of the tendency of water to move by osmosis across a selectively permeable membrane. It can be measured by placing a solution into a tube that is closed at one end by a selectively permeable membrane and immersing the tube in distilled water (figure 3.5, step 1). Water molecules move by osmosis through the membrane into the tube, forcing the solution to move up the tube (figure 3.5, step 2). As the solution rises, its weight produces **hydrostatic pressure** (figure 3.5, step 3), which moves water out of the tube back into the distilled water surrounding the tube. Net movement of water into the tube stops when the hydrostatic pressure in the tube causes water to move out of the tube at the same rate that it diffuses into the tube by osmosis. The osmotic pressure of the solution in the tube is equal to the hydrostatic pressure that prevents net movement of water into the tube.

The greater the concentration of a solution, the greater its osmotic pressure, and the greater the tendency for water to move into the solution. This occurs because water moves from less concentrated solutions (less solute, more water) into more concentrated solutions (more solute, less water). The greater the concentration of a solution, the greater the tendency for water to move into the solution, and the greater the osmotic pressure must be to prevent that movement.

Cells will either swell, remain unchanged, or shrink when placed into a solution. When a cell is placed into a **hypotonic** (hī′pō-ton′ik, *hypo,* under + *tonos,* tone) solution, the solution usually has a lower concentration of solutes and a higher concentration of water than the cytoplasm of the cell. Thus, the solution has less tone or osmotic pressure than the cell. Water moves by osmosis into the cell, causing it to swell. If the cell swells enough, it can rupture, a process called **lysis** (lī′sis, loosening) (figure 3.6*a*). When a cell is immersed in an **isotonic** (ī′sō-ton′ik, *iso,* equal) solution, the concentrations of various solutes and water are the same on both sides of the cell membrane. The cell therefore neither shrinks nor swells (figure 3.6*b*). When a cell is immersed in a **hypertonic** (hī′per-ton′ik, *hyper,* above) solution, the solution usually has a higher concentration of solutes and a lower concentration of water than the cytoplasm of the cell. Water moves by osmosis from the cell into the hypertonic solution, resulting in cell shrinkage, or **crenation** (krē-nā′shŭn, *crena,* a notch) (figure 3.6*c*). In general, solutions injected into the circulatory system or into tissues must be isotonic because swelling or shrinking disrupts normal cell function and can lead to cell death.

Filtration

Filtration is the movement of fluid through a partition containing small holes (see table 3.2). The fluid movement results from the pressure or weight of the fluid pushing against the partition. The fluid and substances small enough to pass through the holes move through the partition, but substances larger than the holes do not pass through it. For example, in a

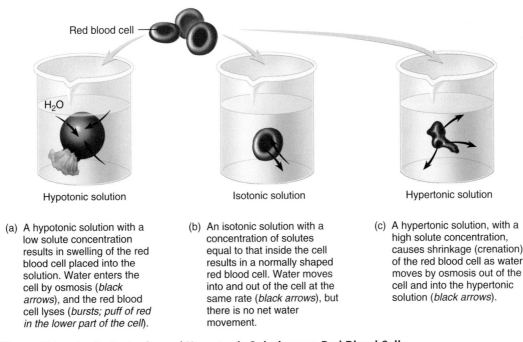

Red blood cell

H₂O

Hypotonic solution

Isotonic solution

Hypertonic solution

(a) A hypotonic solution with a low solute concentration results in swelling of the red blood cell placed into the solution. Water enters the cell by osmosis (*black arrows*), and the red blood cell lyses (*bursts; puff of red in the lower part of the cell*).

(b) An isotonic solution with a concentration of solutes equal to that inside the cell results in a normally shaped red blood cell. Water moves into and out of the cell at the same rate (*black arrows*), but there is no net water movement.

(c) A hypertonic solution, with a high solute concentration, causes shrinkage (crenation) of the red blood cell as water moves by osmosis out of the cell and into the hypertonic solution (*black arrows*).

Figure 3.6 Effects of Hypotonic, Isotonic, and Hypertonic Solutions on Red Blood Cells

car, oil but not dirt particles passes through an oil filter. In the body, filtration occurs in the kidneys as a first step in urine production. Blood pressure moves fluid from the blood through a partition, or filtration membrane. Water, ions, and small molecules pass through the filtration membrane as the first step in urine formation, whereas larger substances, such as proteins and blood cells, remain in the blood (see chapter 18).

Mediated Transport Mechanisms

Many nutrient molecules, such as amino acids and glucose, cannot enter the cell by the process of diffusion, and many substances, such as proteins, produced in cells one cannot leave the cell by diffusion. Carrier molecules within the cell membrane are involved in **carrier-mediated transport mechanisms,** which function to move large, water-soluble molecules or electrically charged ions across the cell membrane. After a molecule to be transported binds to a specific carrier molecule on one side of the membrane, the three-dimensional shape of the carrier molecule changes, and the transported molecule is moved to the opposite side of the cell membrane. The transported molecule is then released by the carrier molecule, which resumes its original shape and is available to transport another molecule. Carrier-mediated transport mechanisms exhibit **specificity;** that is, only specific molecules are transported by the carriers. There are three kinds of carrier-mediated transport: facilitated diffusion, active transport, and secondary active transport.

Facilitated Diffusion

Facilitated diffusion is a carrier-mediated transport process that moves substances into or out of cells from a higher to a lower concentration of that substance (figure 3.7; see table 3.2). Because movement is with the concentration gradient, metabolic energy in the form of ATP is not required.

> **P R E D I C T   2**
> The transport of glucose into most cells occurs by facilitated diffusion. Because diffusion occurs from a higher to a lower concentration, glucose cannot accumulate within these cells at a higher concentration than is found outside the cell. Once glucose enters a cell, it is rapidly converted to other molecules, such as glucose phosphate or glycogen. What effect does this conversion have on the ability of the cell to transport glucose?

Active Transport

Active transport is a carrier-mediated process that moves substances across the cell membrane from regions of lower concentration to those of higher concentration against a concentration gradient (see table 3.2). Consequently, active transport processes accumulate substances on one side of the cell membrane at concentrations many times greater than those on the other side. Active transport requires energy in the form of ATP, and if ATP is not available, active transport stops. Examples of active transport include the movement of various amino acids from the small intestine into the blood.

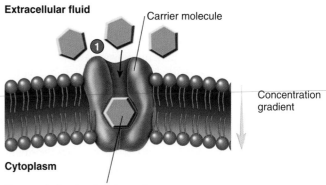

Extracellular fluid · Carrier molecule · Concentration gradient · Cytoplasm · Transported molecule (glucose)

1. The carrier molecule binds with a molecule, such as glucose, on the outside of the cell membrane.

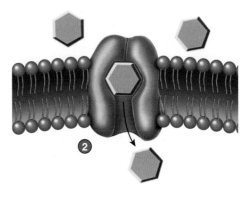

2. The carrier molecule changes shape and releases the molecule on the inside of the cell membrane.

Process Figure 3.7   Facilitated Diffusion

In some cases, the active transport mechanism can exchange one substance for another. For example, the **sodium–potassium pump** moves Na^+ out of cells and K^+ into cells (figure 3.8). The result is a higher concentration of Na^+ outside the cell and a higher concentration of K^+ inside the cell. The concentration gradients for Na^+ and K^+, established by the sodium–potassium pump, are essential in maintaining the resting membrane potential (see chapter 8).

Cystic fibrosis is a genetic disorder that affects the active transport of chlorine ions into cells. This disorder is discussed in the *Systems Pathology* essay on p. 66.

A CASE IN POINT | Addison's Disease

Lowe Blood experienced unexplained weight loss, fatigue, and low blood pressure. After running several tests, his physician concluded he had **Addison's disease.** The disease involves a decrease in aldosterone production by the adrenal cortex. When aldosterone production falls too low, certain kidney cells are unable to transport Na^+. This results in Na^+ and water loss in the urine, causing blood volume and blood pressure to drop.

1. Three sodium ions (Na⁺) and adenosine triphosphate (ATP) bind to the sodium–potassium (Na⁺–K⁺) pump .

2. The ATP breaks down to adenosine diphosphate (ADP) and a phosphate (P) and releases energy. That energy is used to power the shape change in the Na⁺–K⁺ pump. Phosphate remains bound to the Na⁺–K⁺ pump–ATP binding site.

3. The Na⁺–K⁺ pump changes shape, and the Na⁺ are transported across the membrane.

4. The Na⁺ diffuse away from the Na⁺–K⁺ pump.

5. Two potassium ions (K⁺) bind to the Na⁺–K⁺ pump.

6. The phosphate is released from the Na⁺–K⁺ pump binding site.

7. The Na⁺–K⁺ pump changes shape, transporting K⁺ across the membrane, and the K⁺ diffuse away from the pump. The Na⁺–K⁺ pump can again bind to Na⁺ and ATP.

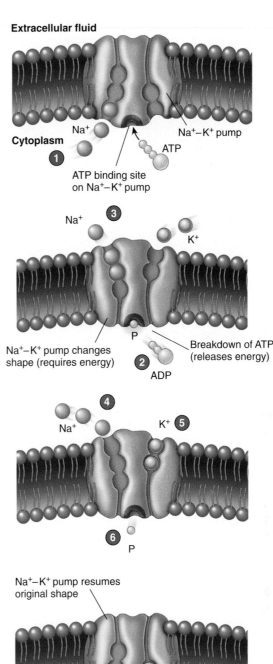

Process Figure 3.8 Active Transport: Sodium–Potassium Pump

Secondary Active Transport

Secondary active transport involves the active transport of one substance, such as Na⁺, across the cell membrane, establishing a concentration gradient. The diffusion of that transported substance down its concentration gradient provides the energy to transport a second substance, such as glucose, across the cell membrane (figure 3.9). In **cotransport,** the diffusing substance moves in the same direction as the transported substance. **Countertransport** is movement of the diffusing substance in the opposite direction as the transported substance.

Endocytosis and Exocytosis

Endocytosis (en′dō-sī-tō′sis, *endon,* within + *kytos,* cell + *-osis,* condition) is the uptake of material through the cell membrane by the formation of a membrane-bound sac called a vesicle (see table 3.2). The cell membrane invaginates to form a vesicle containing the material to be taken into the cell. The vesicle is then taken into the cell.

Endocytosis usually exhibits specificity. The cell membrane contains specific receptor molecules that bind to specific substances. When a specific substance binds to the receptor

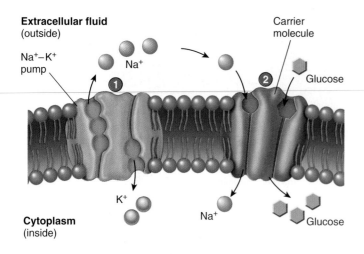

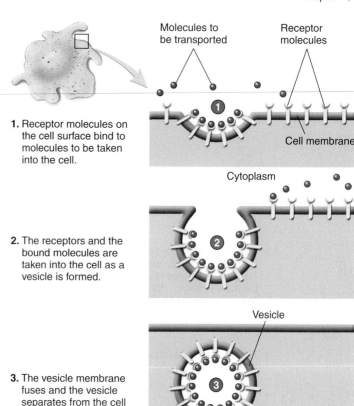

1. Receptor molecules on the cell surface bind to molecules to be taken into the cell.

2. The receptors and the bound molecules are taken into the cell as a vesicle is formed.

3. The vesicle membrane fuses and the vesicle separates from the cell membrane.

1. A sodium–potassium (Na⁺–K⁺) pump maintains a concentration of Na⁺ that is higher outside the cell than inside.

2. Na⁺ move back into the cell by a carrier molecule that also moves glucose. The concentration gradient for Na⁺ provides the energy required to move glucose against its concentration gradient.

Process Figure 3.9   Cotransport

Process Figure 3.10   Receptor-Mediated Endocytosis

molecule, endocytosis is triggered, and the substance is transported into the cell. This process is called **receptor-mediated endocytosis** (figure 3.10). Cholesterol and growth factors are examples of molecules that can be taken into a cell by receptor-mediated endocytosis. Bacterial phagocytosis is also receptor mediated.

Hypercholesterolemia

Hypercholesterolemia is a common genetic disorder affecting 1 in every 500 adults in the United States. It consists of a reduction in or absence of low-density lipoprotein (LDL) receptors on cell surfaces. This interferes with receptor-mediated endocytosis of LDL cholesterol. As a result of inadequate cholesterol uptake, cholesterol synthesis within these cells is not regulated, and too much cholesterol is produced. The excess cholesterol accumulates in blood vessels, resulting in atherosclerosis. Atherosclerosis can result in heart attacks or strokes.

   Phagocytosis (fag′ō-sī-tō′sis, "cell eating") is a term often used for endocytosis when solid particles are **ingested.** A part of the cell membrane extends around a particle and fuses so that the particle is surrounded by the membrane. That part of the membrane then "pinches off" to form a vesicle containing the particle. The vesicle is within the cytoplasm of the cell, and the cell membrane is left intact. White blood cells and some other cell types phagocytize bacteria, cell debris, and foreign particles. Phagocytosis is an important means by which white blood cells take up and destroy harmful substances that have entered the body. **Pinocytosis** (pin′ō-sī-tō′sis, "cell drinking") is distinguished from phagocytosis in that much smaller vesicles are formed that contain liquid rather than particles.

   In some cells, membrane-bound sacs called **secretory vesicles** accumulate materials for release from the cell. The secretory vesicles move to the cell membrane, where the vesicle

membrane fuses with the cell membrane, and the material in the vesicle is eliminated from the cell. This process is called **exocytosis** (ek′sō-sī-tō′sis, *exo,* outside) (figure 3.11 and see table 3.2). Secretion of digestive enzymes by the pancreas, of mucus by the salivary glands, and of milk from the mammary glands are examples of exocytosis. In many respects the process is similar to that of endocytosis, but it occurs in an opposite direction. Endocytosis results in the uptake of materials by cells, and exocytosis in the release of materials from cells. Both endocytosis and exocytosis require energy in the form of ATP to form vesicles.

Organelles
Nucleus

The **nucleus** (noo′klē-ŭs, a little nut, stone of fruit) is a large organelle usually located near the center of the cell (see figure 3.1). All cells of the body have a nucleus at some point in their life cycle, although some cells, such as red blood cells, lose their nuclei as they mature. Other cells, such as osteoclasts (a type of bone cell) and skeletal muscle cells, contain more than one nucleus.

   The nucleus is bounded by a **nuclear envelope,** which consists of outer and inner membranes with a narrow space between them (figure 3.12). At many points on the surface of the nucleus, the inner and outer membranes come together to form

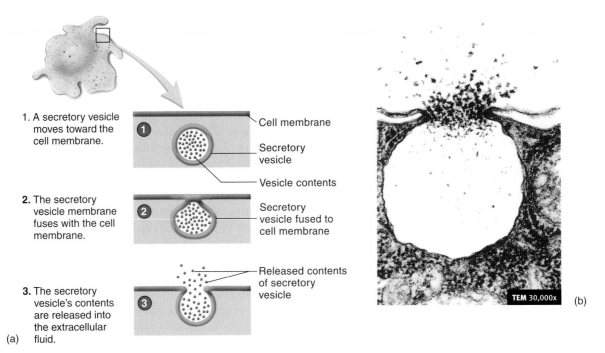

1. A secretory vesicle moves toward the cell membrane.

Cell membrane

Secretory vesicle

Vesicle contents

2. The secretory vesicle membrane fuses with the cell membrane.

Secretory vesicle fused to cell membrane

3. The secretory vesicle's contents are released into the extracellular fluid.

Released contents of secretory vesicle

(a)

TEM 30,000x (b)

Process Figure 3.11   Exocytosis

(a) Diagram of exocytosis. (b) Transmission electron micrograph of exocytosis.

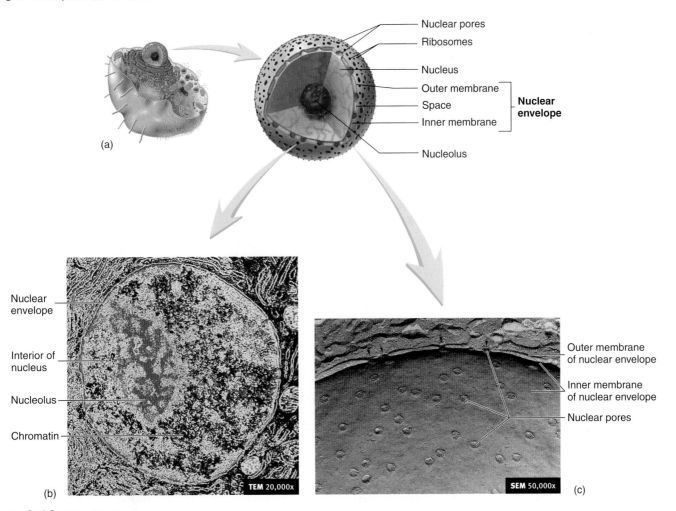

Nuclear pores
Ribosomes
Nucleus
Outer membrane
Space
Inner membrane
Nuclear envelope
Nucleolus

(a)

Nuclear envelope

Interior of nucleus

Nucleolus

Chromatin

(b)   TEM 20,000x

Outer membrane of nuclear envelope

Inner membrane of nuclear envelope

Nuclear pores

SEM 50,000x   (c)

Figure 3.12   The Nucleus

(a) The nuclear envelope consists of inner and outer membranes that become fused at the nuclear pores. The nucleolus is a condensed region of the nucleus not bounded by a membrane and consisting mostly of RNA and protein. (b) Transmission electron micrograph of the nucleus. (c) Scanning electron micrograph showing the membranes of the nuclear envelope and the nuclear pores.

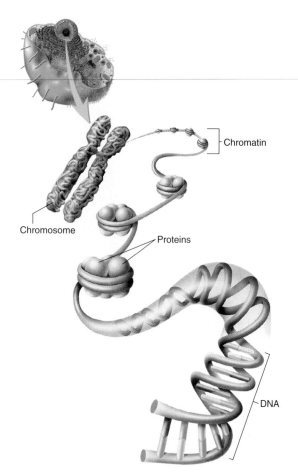

Chromatin

Chromosome

Proteins

DNA

Figure 3.13   Structure of a Chromosome

nuclear pores, through which materials can pass into or out of the nucleus.

The nucleus contains loosely coiled fibers called **chromatin** consisting of deoxyribonucleic acid (DNA) and proteins (figure 3.13 and see figure 3.12*b*). During cell division, the chromatin fibers become more tightly coiled to form the 23 pairs of **chromosomes** (krō′mō-sōmz, *chroma,* color + *soma,* body) characteristic of human cells (see the section on Cell Division on p. 62). The genes that influence the structural and functional features of every individual are portions of DNA molecules. These sections of DNA molecules determine the structure of proteins. By determining the structure of proteins, genes direct cell structure and function.

Nucleoli and Ribosomes

Nucleoli (noo-klē′ō-lī; sing. nucleolus, little nucleus) number from one to four per nucleus. They are rounded, dense, well-defined nuclear bodies with no surrounding membrane (see figure 3.12). The subunits of ribosomes are formed within a nucleolus. Proteins produced in the cytoplasm move through the nuclear pores into the nucleus and to the nucleolus. These proteins are joined to **ribosomal ribonucleic** (rī′bō-noo-klē′ik) **acid (rRNA),** produced within the nucleolus, to form large and small ribosomal subunits (figure 3.14). The ribosomal subunits then move from the nucleus through the nuclear pores into the cytoplasm, where one large and one small subunit join to form a ribosome.

Ribosomes (rī′bō-sōmz, *ribos,* a specific sugar) are the organelles where proteins are produced (see section on Protein

1. Ribosomal proteins, produced in the cytoplasm, are transported through nuclear pores into the nucleolus.

2. rRNA, most of which is produced in the nucleolus, is assembled with ribosomal proteins to form small and large ribosomal subunits.

3. The small and large ribosomal subunits leave the nucleolus and the nucleus through nuclear pores.

4. The small and large subunits, now in the cytoplasm, combine with each other and with mRNA.

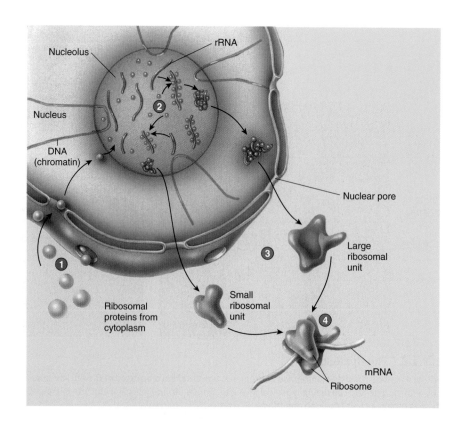

Process Figure 3.14   Production of Ribosomes

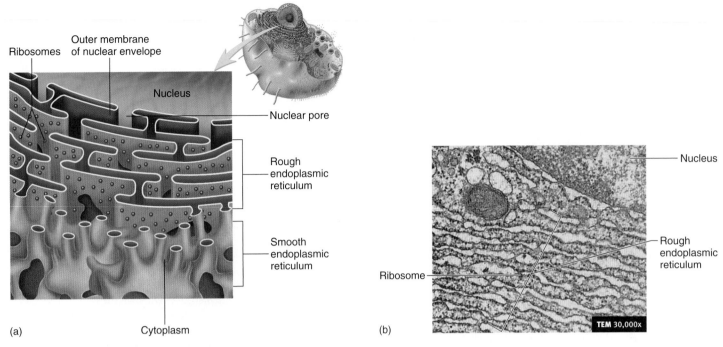

Figure 3.15 The Endoplasmic Reticulum

(*a*) The endoplasmic reticulum is continuous with the nuclear envelope and can exist as either rough endoplasmic reticulum (with ribosomes) or smooth endoplasmic reticulum (without ribosomes). (*b*) Transmission electron micrograph of the rough endoplasmic reticulum.

Synthesis on p. 60). Free ribosomes are not attached to any other organelles in the cytoplasm, whereas other ribosomes are attached to a network of membranes called the endoplasmic reticulum.

Rough and Smooth Endoplasmic Reticulum

The **endoplasmic reticulum** (en′dō-plas′mik re-tik′ū-lŭm, *rete,* a network) **(ER)** is a series of membranes forming sacs and tubules that extends from the outer nuclear membrane into the cytoplasm (figure 3.15). **Rough ER** is ER with ribosomes attached to it. A large amount of rough ER in a cell indicates that it is synthesizing large amounts of protein for export from the cell. On the other hand, ER without ribosomes is called **smooth ER.** Smooth ER is a site for lipid synthesis and also participates in detoxification of chemicals within cells. In skeletal muscle cells, the smooth ER stores calcium ions.

The Golgi Apparatus

The **Golgi** (gol′jē) **apparatus** (named for Camillo Golgi [1843–1926], an Italian histologist) consists of closely packed stacks of curved, membrane-bound sacs (figure 3.16). It collects, modifies, packages, and distributes proteins and lipids manufactured by the ER. For example, proteins produced at the ribosomes enter the Golgi apparatus from the ER. In some cases, the Golgi apparatus chemically modifies the proteins by attaching carbohydrate or lipid molecules to them. The proteins then are packaged into membrane sacs that pinch off

from the margins of the Golgi apparatus (see the following section on Secretory Vesicles). The Golgi apparatus is present in larger numbers and is most highly developed in cells that secrete protein, such as the cells of the salivary glands or the pancreas.

Secretory Vesicles

A **vesicle** (ves′i-kl, a bladder) is a small, membrane-bound sac that transports or stores materials within cells. **Secretory vesicles** pinch off from the Golgi apparatus and move to the surface of the cell (see figure 3.16). Their membranes then fuse with the cell membrane, and the contents of the vesicles are released to the exterior of the cell. In many cells, secretory vesicles accumulate in the cytoplasm and are released to the exterior when the cell receives a signal. For example, nerve cells release substances called neurotransmitters from secretory vesicles to communicate with other cells. Also, secretory vesicles containing the hormone insulin remain in the cytoplasm of pancreatic cells until rising blood glucose levels stimulate their release.

Lysosomes and Peroxisomes

Lysosomes (lī′sō-sōmz, *lysis,* a loosening + *soma,* body) (figure 3.17) are membrane-bound vesicles formed from the Golgi apparatus. They contain a variety of enzymes that function as intracellular digestive systems. Vesicles formed by endocytosis may fuse with lysosomes. The enzymes within the lysosomes break down the materials in the endocytotic vesicle. For example, white blood cells phagocytize bacteria. Enzymes within

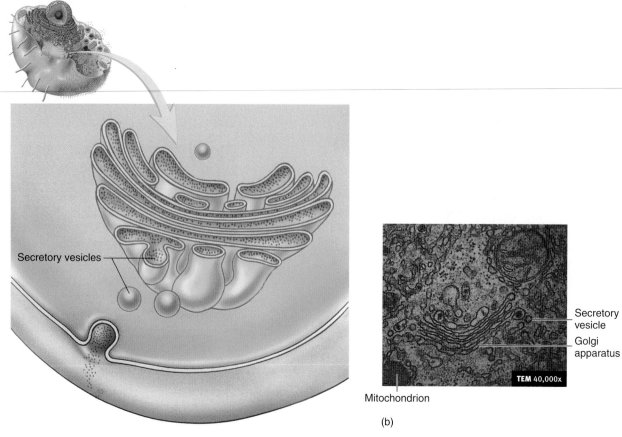

(a)   (b)

Figure 3.16   Golgi Apparatus

(*a*) The Golgi apparatus is composed of flattened membranous sacs and resembles a stack of dinner plates or pancakes. (*b*) Transmission electron micrograph of the Golgi apparatus.

1. A vesicle forms around material outside the cell.

2. The vesicle is pinched off from the cell membrane and becomes a separate vesicle inside the cell.

3. A lysosome is pinched off the Golgi apparatus.

4. The lysosome fuses with the vesicle.

5. The enzymes from the lysosome mix with the material in the vesicle, and the enzymes digest the material.

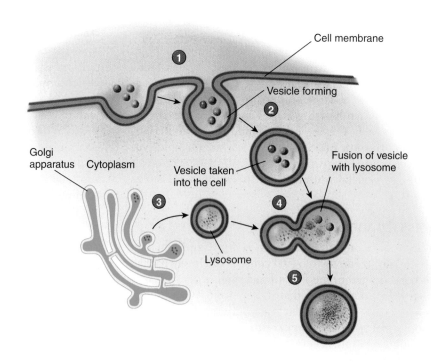

Process Figure 3.17   Action of Lysosomes

lysosomes destroy the bacteria. Also, when tissues are damaged, ruptured lysosomes within the damaged cells release their enzymes and digest both healthy and damaged cells. The released enzymes are responsible for part of the resulting inflammation (see chapter 4).

Carbohydrate and Lipid Disorders

Some diseases result from nonfunctional lysosomal enzymes. For example, **Pompe's disease** results from the inability of lysosomal enzymes to break down the carbohydrate glycogen produced in certain cells. Glycogen accumulates in large amounts in the heart, liver, and skeletal muscle cells. Glycogen accumulation in heart muscle cells often leads to heart failure. **Lipid storage disorders** are often hereditary and are characterized by the accumulation of large amounts of lipid in phagocytic cells. These cells take up the lipid by phagocytosis, but they lack the enzymes required to break down the lipid droplets. Symptoms include enlargement of the spleen and liver and replacement of bone marrow by lipid-filled phagocytes.

Peroxisomes (per-ok′si-sōmz) are small, membrane-bound vesicles containing enzymes that break down fatty acids, amino acids, and hydrogen peroxide (H_2O_2). Hydrogen peroxide is a by-product of fatty acid and amino acid breakdown and can be toxic to the cell. The enzymes in peroxisomes break down hydrogen peroxide to water and oxygen. Cells active in detoxification, such as liver and kidney cells, have many peroxisomes.

Mitochondria

Mitochondria (mī′tō-kon′drē-ă; sing. mitochondrion, *mitos,* thread + *chondros,* granule) are small, bean-shaped, rod-shaped, or long, threadlike organelles with inner and outer membranes separated by a space (figure 3.18 and see figure 3.1). The outer membranes have a smooth contour, but the inner membranes have numerous folds called **cristae** (kris′tē), which project like shelves into the interior of the mitochondria.

Mitochondria are the major sites of adenosine triphosphate (ATP) production within cells. ATP is the major energy source for most chemical reactions within the cell, and cells with a large energy requirement have more mitochondria than cells that require less energy. Mitochondria carry out aerobic respiration (discussed in greater detail in the section Cell Metabolism on p. 60) in which oxygen is required to allow the

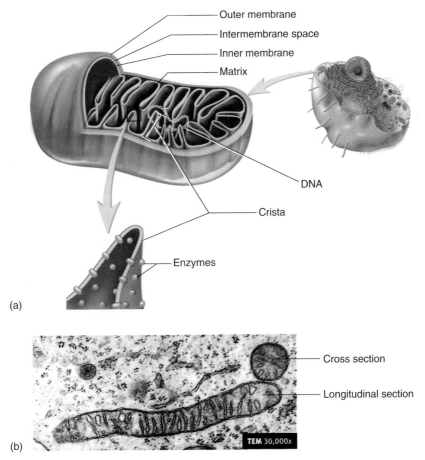

(a)

(b)

Figure 3.18 Mitochondrion

(*a*) Typical mitochondrion structure. (*b*) Transmission electron micrograph of mitochondria in longitudinal and cross section.

reactions that produce ATP to proceed. Cells that carry out extensive active transport, which is described on p. 50, contain many mitochondria. When muscles enlarge as a result of exercise, the mitochondria increase in number within the muscle cells and provide the additional ATP required for muscle contraction.

Increases in the number of mitochondria result from the division of preexisting mitochondria. The information for making some mitochondrial proteins and for mitochondrial division is contained in a unique type of DNA within the mitochondria. This DNA is more like bacterial DNA than that of the cell's nucleus.

Cytoskeleton

The **cytoskeleton** (sī-tō-skel′ĕ-ton) consists of proteins that support the cell, hold organelles in place, and enable the cell to change shape. The cytoskeleton consists of microtubules, microfilaments, and intermediate filaments (figure 3.19).

Microtubules are hollow structures formed from protein subunits. They perform a variety of roles, such as helping to provide support to the cytoplasm of cells, assisting in the process of cell division, and forming essential components of certain organelles such as cilia and flagella.

Microfilaments are small fibrils formed from protein subunits that structurally support the cytoplasm. Some microfilaments are involved with cell movements. For example, microfilaments in muscle cells enable the cells to shorten or contract.

Intermediate filaments are fibrils formed from protein subunits that are smaller in diameter than microtubules but larger in diameter than microfilaments. They provide mechanical support to the cell.

Centrioles

The **centrosome** (sen′trō-sōm) is a specialized zone of cytoplasm close to the nucleus that is the center of microtubule formation. It contains two **centrioles** (sen′trē-ōlz), normally oriented perpendicular to each other. Each centriole is a small, cylindrical organelle composed of nine triplets, each consisting of three parallel microtubules joined together

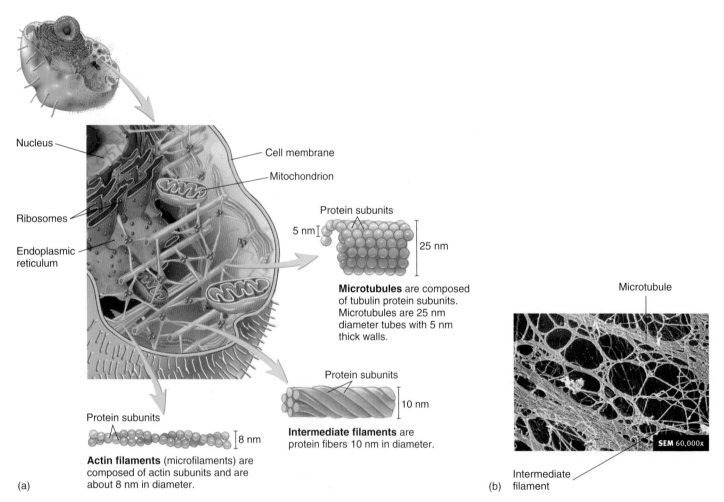

Nucleus

Ribosomes

Endoplasmic reticulum

Cell membrane

Mitochondrion

Protein subunits

5 nm

25 nm

Microtubules are composed of tubulin protein subunits. Microtubules are 25 nm diameter tubes with 5 nm thick walls.

Protein subunits

10 nm

Intermediate filaments are protein fibers 10 nm in diameter.

Protein subunits

8 nm

Actin filaments (microfilaments) are composed of actin subunits and are about 8 nm in diameter.

Microtubule

SEM 60,000x

Intermediate filament

(a)   (b)

Figure 3.19   Cytoskeleton

(*a*) Diagram of the cytoskeleton. (*b*) Scanning electron micrograph of the cytoskeleton.

Each cell is well adapted for the functions it performs, and the abundance of organelles in each cell reflects the function of the cell. For example, epithelial cells lining the larger-diameter respiratory passages secrete mucus and transport it toward the throat, where it is either swallowed or expelled from the body by coughing. Particles of dust and other debris suspended in the air become trapped in the mucus. The production and transport of mucus from the respiratory passages function to keep these passages clean. Cells of the respiratory system have abundant rough ER, Golgi apparatuses, secretory vesicles, and cilia. The ribosomes on the rough ER are the sites where proteins, a major component of mucus, are produced. The Golgi apparatuses package the proteins and other components of mucus into secretory vesicles, which move to the surface of the epithelial cells. The contents of the secretory vesicles are released onto the surface of the epithelial cells. Cilia on the cell surface then propel the mucus toward the throat.

In people who smoke, prolonged exposure of the respiratory epithelium to the irritation of tobacco smoke causes the respiratory epithelial cells to change in structure and function. The cells flatten and form several layers of epithelial cells. These flattened epithelial cells no longer contain abundant rough ER, Golgi apparatuses, secretory vesicles, or cilia. The altered respiratory epithelium is adapted to protect the underlying cells from irritation, but it can no longer secrete mucus nor does it have cilia to move the mucus toward the throat to clean the respiratory passages. Extensive replacement of normal epithelial cells in respiratory passages is associated with chronic inflammation of the respiratory passages (bronchitis), which is common in people who smoke heavily. Coughing is a common manifestation of inflammation and mucus accumulation in the respiratory passages.

(figure 3.20). During cell division, additional microtubules, extending from the area of the centrioles, facilitate the movement of chromosomes toward the centrosomes.

Cilia, Flagella, and Microvilli

Cilia (sĭl′ē-ă; sing. cilium, an eyelash) project from the surface of cells, are capable of moving (see figure 3.1), and vary in number from none to thousands per cell. Cilia have a cylindrical shape, contain specialized microtubules similar to the orientation in centrioles, and are enclosed by the cell membrane. Cilia are numerous on surface cells that line the respiratory tract. Their coordinated movement moves mucus, in which dust particles are embedded, upward and away from the lungs. This action helps keep the lungs clear of debris.

Flagella (flă-jel′ă; sing. flagellum, a whip) have a structure similar to that of cilia but are much longer, and usually occur only one per cell. Sperm cells each have one flagellum, which functions to propel the sperm cell.

Microvilli (mī′krō-vil′ī, *mikros*, small + *villus*, shaggy hair) are specialized extensions of the cell membrane that are supported by microfilaments (see figure 3.1), but they do not actively move like cilia and flagella. Microvilli are numerous on cells that have them and function to increase the surface area of those cells. They are abundant on the surface of cells that line the intestine, kidney, and other areas in which absorption is an important function.

P R E D I C T ❸
List the organelles that are common in cells that (a) synthesize and secrete proteins, (b) actively transport substances into cells, and (c) ingest foreign substances by endocytosis. Explain the function of each organelle you list.

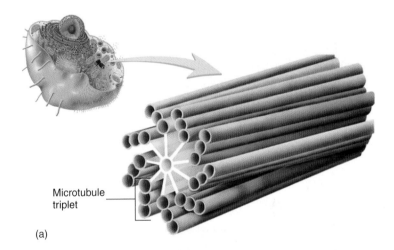

Microtubule triplet

(a)

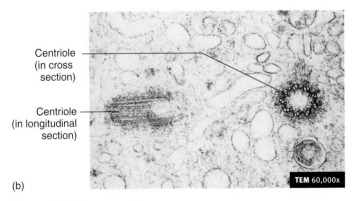

Centriole (in cross section)

Centriole (in longitudinal section)

(b)

TEM 60,000x

Figure 3.20 Centriole

(*a*) Structure of a centriole, which comprises nine triplets of microtubules. Each triplet contains one complete microtubule fused to two incomplete microtubules. (*b*) Transmission electron micrograph of a pair of centrioles, which are normally located together near the nucleus. One is shown in cross section and one in longitudinal section.

Whole-Cell Activity

Interactions between organelles must be considered to understand how a cell functions. For example, the transport of many food molecules into the cell requires ATP and cell membrane proteins. Most ATP is produced by mitochondria. ATP is required to transport amino acids across the cell membrane. Amino acids are assembled to synthesize proteins, including the transport proteins of the cell membrane and mitochondrial proteins. Information contained in DNA within the nucleus determines which amino acids are combined at ribosomes to form proteins. The mutual interdependence of cellular organelles is coordinated to maintain homeostasis within the cell and the entire body. The following sections, Cell Metabolism, Protein Synthesis, and Cell Division, illustrate the interactions of organelles that result in a functioning cell.

Cell Metabolism

Cell metabolism is the sum of all the chemical reactions in the cell (figure 3.21). The breakdown of food molecules releases energy that is used to synthesize ATP (see chapter 17). When ATP is broken down, energy is released that can be used to drive other chemical reactions or processes such as active transport. The breakdown of the sugar glucose, such as the sugar from a candy bar, by a series of reactions within the cytoplasm of a cell is called **glycolysis** (glī-kol′i-sis). Glucose is converted to pyruvic acid, which can enter alternative biochemical pathways, depending on oxygen availability.

Aerobic (ār-ō′bik, living in air) **respiration** occurs when oxygen is available. Pyruvic acid molecules enter mitochondria and, through a series of chemical reactions called the citric acid

cycle and the electron-transport chain, are converted to carbon dioxide and water. Aerobic respiration can yield up to 38 ATP molecules from each glucose molecule, depending on the cells involved (brain and muscle cells produce 36 ATPs; liver, kidney, and heart cells produce 38 ATPs). Aerobic respiration requires oxygen because the last reaction in the series is the combination of oxygen with hydrogen to form water. If this reaction does not take place, the reactions immediately preceding it do not occur either. This explains why breathing oxygen is necessary for animal life: without oxygen, aerobic respiration is inhibited, and the cells do not produce enough ATP to sustain life. During aerobic respiration, glucose molecules from the food are broken down and the carbon atoms released from the glucose are combined with oxygen to produce carbon dioxide. Thus, the carbon in the carbon dioxide humans breathe out comes from the food they eat.

Anaerobic respiration occurs without oxygen and includes the conversion of pyruvic acid to lactic acid. There is a net production of two ATP molecules for each glucose molecule. Anaerobic respiration does not produce as much ATP as aerobic respiration, but it allows cells to function for short periods when oxygen levels are too low for aerobic respiration to provide all the needed ATP. For example, during intense exercise, when the demand for ATP exceeds the rate at which aerobic metabolism can produce it, anaerobic respiration can provide additional ATP.

Protein Synthesis

DNA contains the information that directs protein synthesis. The proteins produced in a cell are structural components inside the cell, structural proteins secreted to the outside of the cell, and enzymes that regulate chemical reactions in the cell. DNA influences the structural and functional characteristics of the entire organism because it directs protein synthesis. Whether an individual has blue eyes, brown hair, or other inherited traits is determined ultimately by DNA.

A DNA molecule consists of nucleotides joined together to form two nucleotide strands (see figures 2.19 and 3.23). The two strands are connected and resemble a ladder that is twisted around its long axis. The nucleotides function as chemical "letters" that form chemical "words." A **gene** is a sequence of nucleotides (making a word) providing a chemical set of instructions for making a specific protein. Each DNA molecule contains many different genes.

Recall from chapter 2 that proteins consist of amino acids. The unique structural and functional characteristics of different proteins are determined by the kinds, numbers, and arrangement of their amino acids. The nucleotide sequence of a gene determines the amino acid sequence of a specific protein.

DNA directs the production of proteins in two steps—transcription and translation—which can be illustrated with an analogy. Suppose a chef wants a recipe that is found only in a reference book in the library. Because the book cannot be checked out, the chef makes a copy, or **transcription,** of the recipe. Later, in the kitchen the information contained in the copied recipe is used to prepare a meal. The changing of something from one form to another (from recipe to meal) is called **translation.**

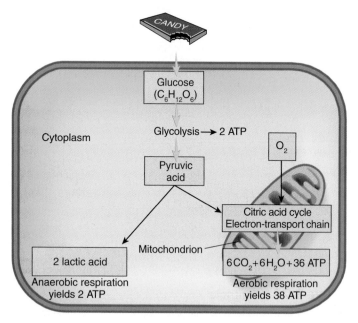

Figure 3.21 **Overview of Cell Metabolism**

Aerobic respiration requires oxygen and produces more ATP per glucose molecule than does anaerobic respiration.

In terms of this analogy, DNA (the reference book) contains many genes (recipes) for making different proteins (meals). DNA, however, is too large a molecule to pass through the nuclear pores to go to the ribosomes (kitchen) where the proteins (the meal) are prepared. Just as the reference book stays in the library, DNA remains in the nucleus. Through transcription therefore the cell makes a copy of the information in DNA necessary to make a particular protein. The copy, which is called **messenger RNA (mRNA),** travels from the nucleus to the ribosomes in the cytoplasm, where the information in the copy is used to construct a protein by means of translation. Of course, the actual ingredients are needed to turn a recipe into a meal. The ingredients necessary to synthesize a protein are amino acids. Specialized molecules, called **transfer RNA (tRNA),** carry the amino acids to the ribosome (figure 3.22).

In summary, the synthesis of proteins involves transcription—making a copy of part of the information in DNA (a gene), and translation—converting that copied information into a protein. The details of transcription and translation are considered next.

Transcription

The events leading to protein synthesis begin in the nucleus. DNA determines the structure of mRNA through transcription.

The double strands of a DNA segment separate, and DNA nucleotides pair with RNA nucleotides (figure 3.23). Each nucleotide of DNA contains one of the following organic bases: thymine, adenine, cytosine, or guanine; and each nucleotide of mRNA contains uracil, adenine, cytosine, or guanine. The number and sequence of nucleotides in the DNA serve as a template to determine the number and sequence of nucleotides in the mRNA. DNA nucleotides only pair with specific RNA nucleotides: DNA's thymine with RNA's adenine, DNA's adenine with RNA's uracil, DNA's cytosine with RNA's guanine, and DNA's guanine with RNA's cytosine.

After the DNA nucleotides pair up with the RNA nucleotides, an enzyme catalyzes reactions that form chemical bonds between the RNA nucleotides to form a long mRNA segment. Once the mRNA segment has been transcribed, portions of the mRNA molecule can be removed, or two or more mRNA molecules can be combined.

Translation

Translation, the synthesis of proteins based on the information in mRNA, occurs at ribosomes. The mRNA molecules produced by transcription pass through the nuclear pores to the ribosomes. Ribosomes consist of small and large subunits, which combine with each other and with mRNA. The information in mRNA is

1. DNA contains the information necessary to produce proteins.

2. Transcription of one DNA strand results in mRNA, which is a copy of the information in the DNA strand needed to make a protein.

3. The mRNA leaves the nucleus and goes to a ribosome.

4. Amino acids, the building blocks of proteins, are carried to the ribosome by tRNAs.

5. In the process of translation, the information contained in mRNA is used to determine the number, kinds, and arrangement of amino acids in the protein (see fig. 3.24).

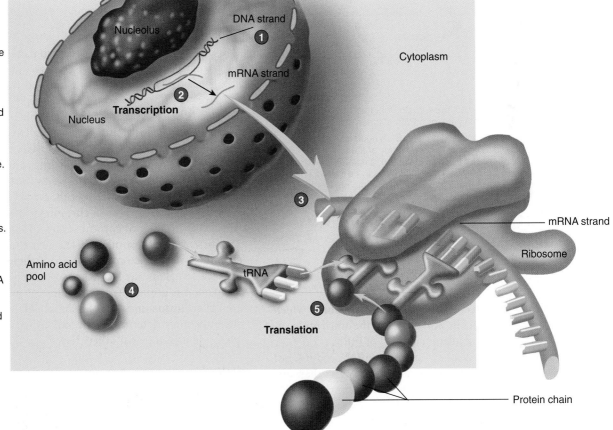

Process Figure 3.22 Overview of Protein Synthesis

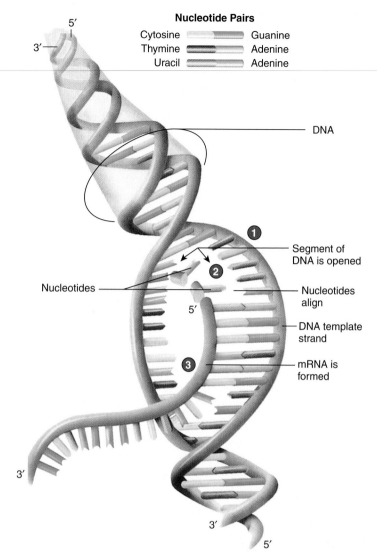

Nucleotide Pairs

Cytosine	Guanine
Thymine	Adenine
Uracil	Adenine

DNA

① Segment of DNA is opened

② Nucleotides align

Nucleotides

DNA template strand

mRNA is formed

③

1. The strands of the DNA molecule separate from each other. One DNA strand serves as a template for mRNA synthesis.

2. Nucleotides that will form mRNA pair with DNA nucleotides according to the base-pair combinations shown in the key at the top of the figure. Thus the sequence of nucleotides in the template DNA strand (*purple*) determines the sequence of nucleotides in mRNA (*gray*). RNA polymerase (the enzyme is not shown) joins the nucleotides of mRNA together.

3. As nucleotides are added, an mRNA molecule is formed.

Process Figure 3.23 Formation of mRNA by Transcription of DNA

carried in groups of three nucleotides called **codons,** which code for specific amino acids. For example, the nucleotide sequence uracil, cytosine, and adenine (UCA) of mRNA codes for the amino acid serine. There are 64 possible mRNA codons, but only 20 amino acids are in proteins. As a result, more than one codon can code for the same amino acid. For example, CGA, CGG, CGT, and CGC code for the amino acid alanine, and UUU and UAC code for phenylalanine. Some codons do not code for amino acids but perform other functions. For example, UAA

codes for no amino acid. It is called a stop codon, which acts as a signal for stopping the addition of amino acids to a protein.

Protein synthesis requires two types of RNA in addition to mRNA: tRNA and **ribosomal RNA (rRNA).** There is one type of tRNA for each mRNA codon. A series of three nucleotides of each tRNA molecule, the **anticodon,** pairs with the codon of the mRNA. Another part of each tRNA molecule binds to a specific amino acid. For example, the tRNA that pairs with the UUU codon of mRNA has the anticodon AAA and binds only to the amino acid phenylalanine.

The ribosomes, which consist of ribosomal RNA and proteins, align mRNA with tRNA molecules so that the anticodons of tRNAs pair with the codons of mRNA while the mRNA is attached to a ribosome (figure 3.24). The amino acids bound to the tRNAs are then joined to one another by an enzyme associated with the ribosome. The enzyme causes the formation of a chemical bond, called a **peptide bond,** between the adjacent amino acids to form a **polypeptide chain,** consisting of many amino acids bound together by peptide bonds. The polypeptide chain then becomes folded to form the three-dimensional structure of the protein molecule (see figure 2.16). A protein can consist of a single polypeptide chain or two or more polypeptide chains that are joined after each chain is produced on separate ribosomes.

P R E D I C T ④
Explain how changing one nucleotide within a DNA molecule of a cell could change the structure of a protein produced by the cell.

Cell Division

Cell division is the formation of two daughter cells from a single parent cell. The new cells necessary for growth and tissue repair are formed through mitosis, and the sex cells necessary for reproduction are formed through meiosis (discussed in chapter 20).

During mitosis and meiosis the DNA within the parent cell is distributed to the daughter cells. The DNA is found within chromosomes. Each cell of the human body, except for sex cells, contains 46 chromosomes. Sex cells have half the number of chromosomes as other cells (see the section on Meiosis on p. 595). The 46 chromosomes are called a **diploid** (dip′loyd) number of chromosomes and are organized to form 23 pairs of chromosomes. Of the 23 pairs, one pair is the sex chromosomes, which consist of two **X chromosomes** if the person is a female or an X chromosome and a **Y chromosome** if the person is a male. The remaining 22 pairs of chromosomes are called **autosomes** (aw′tō-sōmz). The combination of sex chromosomes determines the individual's sex, and the autosomes determine most other characteristics.

Mitosis

All cells of the body, except those that give rise to sex cells, divide by **mitosis** (mī-tō′sis, *mitos,* thread; threadlike microtubules form during mitosis). Mitosis involves two steps: (1) the genetic material within a cell is **replicated,** or duplicated, and (2) the cell

1. To start protein synthesis a ribosome binds to mRNA. The ribosome also has two binding sites for tRNA, one of which is occupied by a tRNA with its amino acid. Note that the codon of mRNA and the anticodon of tRNA are aligned and joined. The other tRNA binding site is open.

2. By occupying the open tRNA binding site, the next tRNA is properly aligned with mRNA and with the other tRNA.

3. An enzyme within the ribosome catalyzes a synthesis reaction to form a peptide bond between the amino acids. Note that the amino acids are now associated with only one of the tRNAs.

4. The ribosome shifts position by three nucleotides. The tRNA without the amino acid is released from the ribosome, and the tRNA with the amino acids takes its position. A tRNA binding site is left open by the shift. Additional amino acids can be added by repeating steps *2* through *4*. Eventually a stop codon in the mRNA ends the addition of amino acids to the protein (polypeptide), which is released from the ribosome.

5. Multiple ribosomes attach to a single mRNA. As the ribosomes move down the mRNA, proteins attached to the ribosomes lengthen and eventually detach from the mRNA.

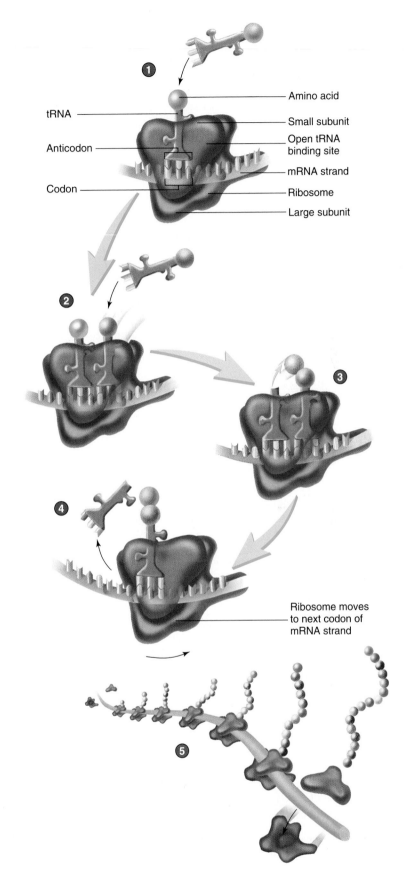

Process Figure 3.24 Translation of mRNA to Produce a Protein

divides to form two daughter cells with the same amount and type of DNA as the parent cell. Because DNA determines the structure and function of cells, the daughter cells, which have the same DNA as the parent cell, can have the same structure and perform the same functions as the parent cell.

The period between active cell divisions is called **interphase,** during which DNA is replicated. The two strands of DNA separate from each other, and each strand serves as a template for the production of a new strand of DNA (figure 3.25). Nucleotides in the DNA of an old strand, forming a template,

pair with nucleotides that are subsequently joined by enzymes to form a new strand of DNA. The sequence of nucleotides in the DNA template determines the sequence of nucleotides in the new strand of DNA because adenine pairs with thymine, and cytosine pairs with guanine. The two new strands of DNA combine with the two template strands to form two double strands of DNA.

At the end of interphase, each cell has two complete sets of genetic material. The DNA is dispersed throughout the nucleus as thin threads called **chromatin** (krō′mă-tin) (figure 3.26, step 1; see also figures 3.12b and 3.13).

Mitosis follows interphase. For convenience, mitosis is divided into four stages. Although each stage represents major events, the process of mitosis is continuous. Learning each of the stages is helpful, but the most important concept to understand is how each of the two cells produced by mitosis obtains the same number and type of chromosomes as the parent cell. There are four stages in mitosis:

1. *Prophase.* During **prophase** (*prophasis,* to foreshadow) (figure 3.26, step 2), the chromatin condenses to form visible chromosomes. After interphase, each chromosome is made up of two separate but genetically identical strands of chromatin, called **chromatids** (krō′mă-tidz), which are linked at one point by a specialized region called the **centromere** (sen′trō-mēr, *kentron,* center + *meros,* part). Replication of the genetic material during interphase results in the two identical chromatids of each chromosome. Also during prophase, microtubules called spindle fibers extend from the centrioles to the centromeres (see figures 3.1, 3.20, and 3.26, step 2). The centrioles divide and migrate to each pole of the cell. In late prophase, the nucleolus and nuclear envelope disappear.
2. *Metaphase.* In **metaphase** (*meta,* between + *phasis,* an appearance) (figure 3.26, step 3), the chromosomes align near the center of the cell.
3. *Anaphase.* At the beginning of **anaphase** (*ana,* up + *phasis,* an appearance) (figure 3.26, step 4), the centromeres separate. When this happens, each chromatid is then referred to as a chromosome. Thus, when the centromeres divide, the chromosome number doubles to form two identical sets of 46 chromosomes. Each of the two sets of 46 chromosomes is moved by the spindle fibers toward the centriole at one of the poles of the cell. At the end of anaphase, each set of chromosomes has reached an opposite pole of the cell, and the cytoplasm begins to divide.
4. *Telophase.* During **telophase** (tel′ō-fāz, *telas,* end + *phasis,* an appearance) (figure 3.26, step 5), the chromosomes in each of the daughter cells become organized to form two separate nuclei. The chromosomes begin to unravel and resemble the genetic material during interphase.

Following telophase, the cytoplasm of the two cells completes division, and two separate daughter cells are produced (figure 3.26, step 6).

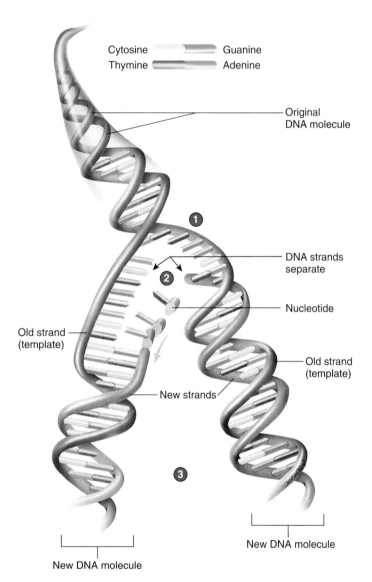

1. The strands of the DNA molecule separate from each other.
2. Each old strand (*dark purple*) functions as a template on which a new strand (*light purple*) is formed. The base-pairing relationship between nucleotides determines the sequence of nucleotides in the newly formed strands.
3. Two identical DNA molecules are produced.

Process Figure 3.25   Replication of DNA

Replication of DNA during interphase produces two identical molecules of DNA.

1. **Interphase** is the time between cell divisions. DNA is found as thin threads of chromatin in the nucleus. DNA replication occurs during interphase.

2. In **prophase**, the chromatin condenses into chromosomes. Each chromosome consists of two chromatids joined at the centromere. The centrioles move to the opposite ends of the cell, and the nucleolus and the nuclear envelope disappear.

3. In **metaphase**, the chromosomes align in the center of the cell in association with the spindle fibers.

4. In **anaphase**, the chromatids separate to form two sets of identical chromosomes. The chromosomes, assisted by the spindle fibers, move toward the centrioles at each end of the cell.

5. In **telophase**, the chromosomes disperse, the nuclear envelopes and the nucleoli form, and the cytoplasm begins to divide to form two cells.

6. Mitosis is complete, and a new interphase begins. The chromosomes have unraveled to become chromatin. Cell division has produced two daughter cells, each with DNA that is identical to the DNA of the parent cell.

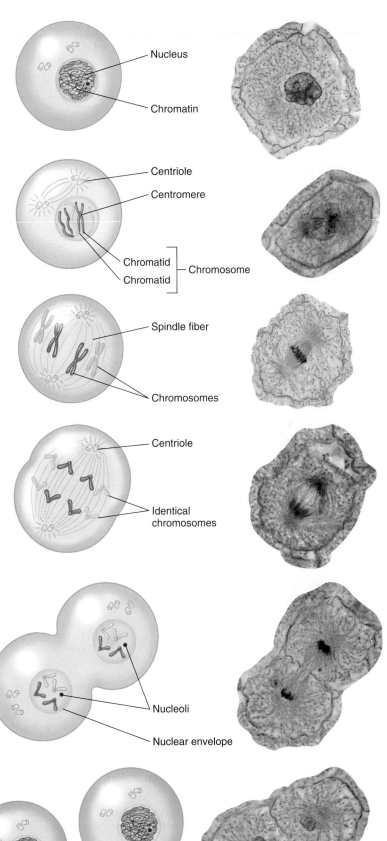

Process Figure 3.26 Mitosis

Tim C. is a 5-year-old white male. He is small for his age and has had frequent bouts of pulmonary infections all his life. Tim always seemed to have a "runny nose." None of the infections were very serious, mostly just irritating. This time, however, his congestion became so extreme that he was unable to breathe and was rushed to the hospital. There, a series of tests demonstrated that Tim suffers from cystic fibrosis.

Cystic fibrosis is a genetic disorder that occurs at a rate of approximately one per 2000 births and currently affects 33,000 people in the United States. It is the most common lethal genetic disorder among whites. The diagnosis is based on the existence of recurrent respiratory disease, increased Na^+ in the sweat, and high levels of unabsorbed fats in the stool. Approximately 98% of all cases of cystic fibrosis are diagnosed before the patient is 18 years old.

At the molecular level, cystic fibrosis results from an abnormality in Cl^- channels. There are three types of cystic fibrosis: (1) In about 70% of cases, a defective channel protein fails to reach the cell membrane from its site of production inside the cell. (2) In the second group, the channel protein is incorporated into the cell membrane but fails to bind ATP. (3) In the final category, the channel protein is incorporated into the cell membrane and ATP is bound to the channel protein, but the channel does not open. The result of any of these defects is that Cl^- do not exit cells at a normal rate.

Normally, as Cl^- move out of cells lining tubes, such as ducts or respiratory passages in the body, water follows by osmosis. In cystic fibrosis, Cl^- do not exit these cells at normal rates and, therefore, less water moves into the tubes. With less water present, the mucus produced by cells lining those tubes is thick and cannot be readily moved over the surface of the cells by their cilia. As a result, the tubes become clogged with mucus, and much of their normal function is lost.

The most critical effects of cystic fibrosis, accounting for 90% of the deaths, are on the respiratory system. Cystic fibrosis also affects the secretory cells lining ducts of the pancreas, sweat glands, and salivary glands.

In normal lungs, a thin fluid layer of mucus is moved by ciliated cells. In people with cystic fibrosis, the viscous mucus resists movement by cilia and accumulates in the lung passages. The mucus accumulation obstructs the passageways and increases the likelihood of infections. The infections are frequent and may result in chronic pneumonia. Chronic coughing occurs as the affected person attempts to remove the mucus.

Cystic fibrosis was once fatal during early childhood, but many patients are now surviving into young adulthood because of modern medical treatment. Currently, approximately 80% of people with cystic fibrosis live past age 20. Pulmonary therapy consists of supporting and enhancing existing respiratory functions, and infections are treated with antibiotics.

The buildup of thick mucus in the pancreatic and hepatic ducts blocks them so that pancreatic digestive enzymes and bile salts are prevented from reaching the small intestine. As a result, fats and fat-soluble vitamins, which require bile salts for absorption and which cannot be adequately digested without pancreatic enzymes, are not taken up by intestinal cells in normal amounts. The patient suffers

Differentiation

A sperm cell and oocyte unite to form a single cell, and a new individual begins. The trillions of cells that ultimately make up the body of an adult stem from that single cell. Therefore, all the cells in an individual's body contain the same amount and type of DNA that contains all of the genetic information for that individual. Not all cells look and function alike, even though the genetic information contained in them is identical. Bone cells, for example, do not look like or function as muscle cells, nerve cells, or red blood cells (figure 3.27).

The process by which cells develop with specialized structures and functions is called **differentiation.** The single cell formed during fertilization divides by mitosis to form two cells, which divide to form four cells, and so on (see chapter 20). The cells continue to divide until there are thousands of cells, which differentiate and give rise to the different cell types.

During differentiation of a cell, some portions of DNA are active, but others are inactive. The active and inactive sections of DNA differ with each cell type. The portion of DNA responsible for the structure and function of a bone cell is different from that responsible for the structure and function of a muscle cell. Differentiation, then, results from the selective activation and inactivation of segments of DNA. The mechanisms that determine which portions of DNA are active in any one cell type are not fully understood, but the resulting differentiation produces the many cell types that function together to make a person. Eventually, as cells differentiate and mature, the rate at which they divide slows or even stops.

System	Interactions
Integumentary	Cystic fibrosis is characterized by increased perspiration with abnormally high quantities of Na^+ in the sweat, which can lead to decreased blood Na^+ levels. A number of skin rashes and other disorders can develop as a result of the abnormal perspiration.
Nervous	Night blindness can develop as a result of vitamin A deficiency caused by insufficient absorption of the vitamin in the digestive tract.
Endocrine	Diabetes mellitus resulting from decreased production of the hormone insulin may develop because blockage of the pancreatic duct by mucus results in pancreatic digestive enzymes, retained within the pancreas, destroying the pancreatic tissues (pancreatic islets) that produce insulin.
Cardiovascular	Fragile blood vessels can develop, resulting in excessive bleeding. Decreased blood clotting results from insufficient vitamin K absorption from the digestive tract. Red blood cell membranes become fragile because of inadequate vitamin E absorption.
Respiratory	The respiratory passages become clogged with viscous mucus, which blocks the airways and inhibits respiration. Recurrent respiratory infections also occur. Decreased airflow into and out of the lungs results in reduced oxygen flow to the tissues. Respiratory complications account for most deaths.
Digestive	Pancreatic ducts and ducts from the liver and salivary glands are blocked with thick mucus. Fats and the fat-soluble vitamins, A, D, E, and K, are poorly absorbed. Deficiencies in fat-soluble vitamins result that affect many other systems. The intestine can become impacted with dehydrated stool. Gallstones can form in the gallbladder or liver ducts.
Reproductive	Reproductive ability is greatly decreased. In 95% of males with cystic fibrosis, there is an absence of living sperm cells in the semen. Viscous secretions in the male or female reproductive tracts decrease fertility.

from deficiencies of vitamins A, D, E, and K, which result in conditions such as night blindness, skin disorders, rickets, and excessive bleeding. Therapy includes administering the missing vitamins to the patient and reducing dietary fat intake.

Future treatments could include the development of drugs that correct or assist Cl^- transport. Alternatively, cystic fibrosis may some day be cured through gene therapy; that is, inserting a functional copy of the defective gene into the cells of people with the disease.

P R E D I C T **5**

Predict the effect of cystic fibrosis on the concentrations of Cl^- inside and outside the cell. In normal muscle and nerve cells at rest, many K^+ channels are open and K^+ tend to flow out of the cell down their concentration gradient. How is this flow of K^+ affected in cells of people with cystic fibrosis?

Cloning

Through the process of differentiation, cells become specialized to certain functions and are no longer capable of producing an entire organism if isolated. Over 30 years ago, however, it was demonstrated in frogs that if the nucleus is removed from a differentiated cell and is transferred to an oocyte with the nucleus removed, a complete normal frog can develop from that oocyte. This process, called **cloning,** demonstrated that during differentiation, genetic information is not irrevocably lost. Because mammalian oocytes are considerably smaller than frog oocytes, cloning of mammalian cells has been technically much more difficult. Dr. Ian Wilmut and his colleagues at the Roslin Institute in Edinburgh, Scotland, overcame those technical difficulties in 1996, when they successfully cloned the first mammal, a sheep. Since that time, many other mammalian species have been cloned.

Apoptosis

Apoptosis (ăp′op-tō′sis), or **programmed cell death,** is a normal process by which cell numbers within various tissues are adjusted and controlled. During development, extra tissue is removed by apoptosis, such as cells between the developing fingers and toes, to fine-tune the contours of the developing fetus. The number of cells in most adult tissues is maintained at a specific level. Apoptosis eliminates excess cells within some adult tissues to maintain a constant number of cells within the tissue. Damaged or potentially dangerous cells, virus-infected cells, and potential cancer cells are also eliminated by apoptosis.

Apoptosis is regulated by specific genes. The proteins coded for by those genes initiate events within the cell that ultimately lead to the cell's death. As apoptosis begins, the chromatin within the nucleus condenses and fragments. This is followed by fragmentation of the nucleus and finally by death and fragmentation of the cell. Specialized cells called macrophages phagocytize the cell fragments.

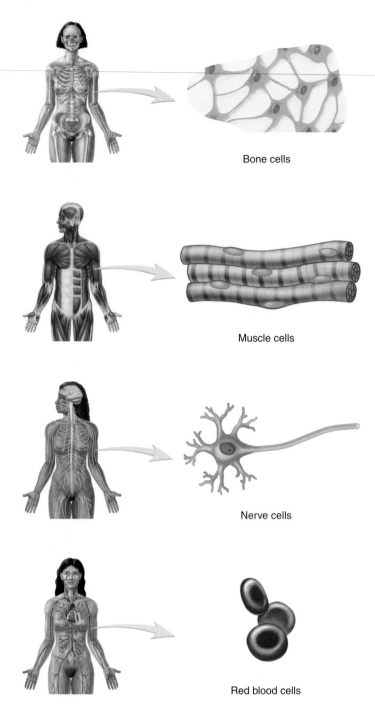

Bone cells

Muscle cells

Nerve cells

Red blood cells

Figure 3.27   **Diversity of Cell Types**

Cellular Aspects of Aging

A number of cellular structures or events appear to be involved in the process of aging. The major theories of aging concentrate on molecules within the cell, such as lipids, proteins, and nucleic acids. It is estimated that at least 35% of the factors affecting aging are genetic.

1. *Cellular clock.* One theory of aging suggests that there is a cellular clock, which, after a certain passage of time or a certain number of cell divisions, results in death of a given cell line.

2. *Death genes.* Another theory suggests that there are "death genes," which turn on late in life, or sometimes prematurely, causing cells to deteriorate and die.

3. *DNA damage.* Other theories suggest that through time, DNA is damaged, resulting in cell degeneration and death.

4. *Free radicals.* DNA is also susceptible to direct damage, resulting in mutations, which may result in cellular dysfunction and, ultimately, cell death. One of the major sources of DNA damage is apparently from **free radicals,** which are atoms or molecules with an unpaired electron.

5. *Mitochondrial damage.* It may be that mitochondrial DNA is more sensitive to free-radical damage than is nuclear DNA. Mitochondrial DNA damage may result in loss of proteins critical to mitochondrial function. Because the mitochondria are the source of ATP, loss of mitochondrial function could result in the loss of energy critical to cell function and, ultimately, to cell death. One proposal suggests that reduced caloric intake may reduce free radical damage to mitochondria.

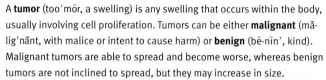

Cancer

A **tumor** (too′mōr, a swelling) is any swelling that occurs within the body, usually involving cell proliferation. Tumors can be either **malignant** (mă-lig′nănt, with malice or intent to cause harm) or **benign** (bē-nīn′, kind). Malignant tumors are able to spread and become worse, whereas benign tumors are not inclined to spread, but they may increase in size.

Cancer (kan′ser, a crab, suggesting crab-like movement) refers to a malignant, spreading tumor and the illness that results from it. Benign tumors are usually less dangerous than malignant ones, but they can also cause problems. As a benign tumor enlarges, it can compress surrounding tissues and impair their functions. In some cases (e.g., brain tumors), the results can be fatal.

Malignant tumors can spread by local growth and expansion or by **metastasis** (mĕ-tas′tă-sis, moving to another place), which results from tumor cells separating from the main neoplasm and being carried by the lymphatic or circulatory system to a new site, where a second tumor forms.

Cancers lack the normal growth control that is exhibited by most other adult tissues. Cancer results when a cell or group of cells, for some reason, breaks away from the normal control of growth and differentiation. This breaking loose involves the genetic machinery and can be induced by viruses, environmental toxins, and other causes.

The illness associated with cancer usually occurs as the tumor invades and destroys the healthy surrounding tissues, eliminating their functions.

Promising anticancer therapies are being developed in which cells responsible for immune responses can be stimulated to recognize tumor cells and destroy them. A major advantage in such anticancer treatments is that the cells of the immune system can specifically attack the tumor cells and not other, healthy tissues. Other therapies currently under investigation include techniques to starve a tumor to death by cutting off its blood supply. Drugs that can inhibit blood vessel development are currently in use (bevacizumab, Avastin) or are under investigation (thalidomide).

P R E D I C T   6
Cancer cells divide continuously. The normal mechanisms that regulate whether cell division occurs or ceases do not function properly in cancer cells. Cancer cells, such as breast cancer cells, do not look like normal, mature cells. Explain.

S U M M A R Y

Cell Structure (p. 43)

1. Cells are highly organized units containing organelles, which perform specific functions.
2. The nucleus contains genetic material, and cytoplasm is living material between the nucleus and cell membrane.

Functions of the Cell (p. 44)

1. Cells are the basic unit of life.
2. They protect and support the body, as well as provide for movement, communication, metabolism, and inheritance.

Cell Membrane (p. 44)

1. The cell membrane forms the outer boundary of the cell. It determines what enters and leaves the cell.
2. The cell membrane is composed of a double layer of phospholipid molecules in which proteins float. The proteins function as membrane channels, carrier molecules, receptor molecules, enzymes, and structural components of the membrane.

Movement Through the Cell Membrane (p. 45)

1. Lipid-soluble molecules pass through the cell membrane readily by dissolving in the lipid portion of the membrane.
2. Small molecules and ions can pass through membrane channels.
3. Large molecules that are not lipid-soluble can be transported through the membrane by carrier molecules.
4. Large molecules that are not lipid-soluble, particles, and cells can be transported across the membrane by vesicles.

Diffusion

1. Diffusion is the movement of a solute from an area of higher concentration to an area of lower concentration within a solvent. At equilibrium, there is a uniform distribution of molecules.
2. A concentration gradient is the concentration of a solute at one point in a solution minus the concentration of that solute at another point in the solution divided by the distance between the points.

Osmosis

1. Osmosis is the diffusion of a solvent (water) point across a selectively permeable membrane.
2. Osmotic pressure is the force required to prevent movement of water across a selectively permeable membrane.
3. In a hypotonic solution, cells swell (and can undergo lysis); in an isotonic solution, cells neither swell nor shrink; and in a hypertonic solution, cells shrink and undergo crenation.

Filtration

Filtration is the passage of a solution through a partition in response to a pressure difference. Some materials, usually larger particles, in the solution do not pass through the partition.

Mediated Transport Mechanisms

1. Mediated transport is the movement of a substance across a membrane by means of a carrier molecule. The substances transported tend to be large, water-soluble molecules or ions.
2. Facilitated diffusion moves substances from a higher to a lower concentration and does not require energy in the form of ATP.
3. Active transport can move substances from a lower to a higher concentration and requires ATP.
4. Secondary active transport uses the energy of one substance moving down its concentration gradient to move another substance across the cell membrane. In cotransport, both substances move in the same direction; in countertransport, they move in opposite directions.

Endocytosis and Exocytosis

1. Endocytosis is the movement of materials into cells by the formation of a vesicle. Phagocytosis is the movement of solid material into cells by the formation of a vesicle. Pinocytosis is similar to phagocytosis, except that the material ingested is much smaller and is in solution. Receptor-mediated endocytosis involves cell receptors attaching to molecules that are then phagocytized by the cell.
2. Exocytosis is the secretion of materials from cells by vesicle formation.

Organelles (p. 52)

Nucleus

1. The nuclear envelope consists of two separate membranes with nuclear pores.
2. DNA and associated proteins are found inside the nucleus as chromatin. DNA is the hereditary material of the cell and controls the activities of the cell.

Nucleoli and Ribosomes

1. Nucleoli consist of RNA and proteins and are the sites of ribosomal subunit assembly.
2. Ribosomes are the sites of protein synthesis.

Rough and Smooth Endoplasmic Reticulum

1. Rough ER is ER with ribosomes attached. It is a major site of protein synthesis.
2. Smooth ER does not have ribosomes attached and is a major site of lipid synthesis.

The Golgi Apparatus

The Golgi apparatus is a series of closely packed membrane sacs that function to collect, modify, package, and distribute proteins and lipids produced by the ER.

Secretory Vesicles

Secretory vesicles are membrane-bound sacs that carry substances from the Golgi apparatus to the cell membrane, where the vesicle contents are released.

Lysosomes and Peroxisomes

Membrane-bound sacs containing enzymes include lysosomes and peroxisomes. Within the cell the lysosomes break down phagocytized material. Peroxisomes break down fatty acids, amino acids, and hydrogen peroxide.

Mitochondria

Mitochondria are the major sites of ATP production, which cells use as an energy source. Mitochondria carry out aerobic respiration (requires oxygen).

Cytoskeleton

1. The cytoskeleton supports the cytoplasm and organelles and is involved with cell movements.
2. The cytoskeleton is composed of microtubules, microfilaments, and intermediate filaments.

Centrioles

Centrioles, located in the centrosome, are made of microtubules and facilitate chromosome movement during cell division.

Cilia, Flagella, and Microvilli

1. Cilia move substances over the surface of cells.
2. Flagella are much longer than cilia and propel sperm cells.
3. Microvilli increase the surface area of cells and aid in absorption.

Whole-Cell Activity (p. 60)

The interactions between organelles must be considered for cell function to be fully understood. That function is reflected in the quantity and distribution of organelles.

Cell Metabolism

1. Aerobic respiration requires oxygen and produces carbon dioxide, water, and up to 38 ATP molecules from a molecule of glucose.
2. Anaerobic respiration does not require oxygen and produces lactic acid and two ATP molecules from a molecule of glucose.

Protein Synthesis

1. Cell activity is regulated by enzymes (proteins), and DNA controls enzyme production.
2. During transcription, the sequence of nucleotides in DNA (a gene) determines the sequence of nucleotides in mRNA; the mRNA moves through the nuclear pores to ribosomes.
3. During translation the sequence of codons in mRNA is used at ribosomes to produce proteins. Anticodons of tRNA bind to the codons of mRNA, and the amino acids carried by tRNA are joined to form a protein.

Cell Division

1. Cell division that occurs by mitosis produces new cells for growth and tissue repair.
2. DNA replicates during interphase, the time between cell division.
3. Mitosis is divided into four stages:
   Prophase—Each chromosome consists of two chromatids joined at the centromere.
   Metaphase—Chromosomes align at the center of the cell.
   Anaphase—Chromatids separate at the centromere and migrate to opposite poles.
   Telophase—The two new nuclei assume their normal structure, and cell division is completed, producing two new daughter cells.

Differentiation

Differentiation, the process by which cells develop specialized structures and functions, results from the selective activation and inactivation of DNA sections.

Cellular Aspects of Aging (p. 68)

Aging may result from the presence of "cellular clocks," from the function of "death genes," from DNA damage, from mitochondrial damage, or by some combination of these effects.

R E V I E W   A N D   C O M P R E H E N S I O N

1. Define cytoplasm and cell organelle.
2. List the functions of the cell.
3. Describe the structure of the cell membrane. What functions does it perform?
4. How do lipid-soluble molecules, small molecules that are not lipid-soluble, and large molecules that are not lipid-soluble cross the cell membrane?
5. Define solution, solute, solvent, diffusion, and concentration gradient.
6. Define osmosis and osmotic pressure.
7. What happens to cells that are placed in isotonic solutions? In hypertonic or hypotonic solutions? What are crenation and lysis?
8. Define filtration.
9. What is mediated transport? How are facilitated diffusion and active transport similar, and how are they different?
10. How does secondary active transport work? Define cotransport and countertransport.
11. Describe receptor-mediated endocytosis, phagocytosis, pinocytosis, and exocytosis. What do they accomplish?
12. Describe the structure of the nucleus and nuclear envelope. Name the organelles found in the nucleus, and give their functions.
13. Where are ribosomes assembled, and what kinds of molecules are found in them?

14. What is endoplasmic reticulum? Compare the functions of rough and smooth endoplasmic reticulum.
15. Describe the Golgi apparatus, and state its function.
16. Where are secretory vesicles produced? What are their contents, and how are they released?
17. What is the function of lysosomes and peroxisomes?
18. Describe the structure and function of mitochondria.
19. Name the components of the cytoskeleton, and give their functions.
20. Describe the structure and function of centrioles.
21. Describe the structure and function of cilia, flagella, and microvilli.
22. Describe how proteins are synthesized and how the structure of DNA determines the structure of proteins.
23. Define autosome, sex chromosome, and diploid number.
24. How do the sex chromosomes of males and females differ?
25. Describe what happens during interphase and each phase of mitosis. What kinds of tissues undergo mitosis?
26. Define differentiation. In general terms, how does differentiation occur?
27. List the principle theories of aging.

C R I T I C A L   T H I N K I N G

1. The body of a man was found floating in the salt water of Grand Pacific Bay, which has a concentration that is slightly greater than body fluids. When seen during an autopsy, the cells in his lung tissues were clearly swollen. Choose the most logical conclusion.
   a. He probably drowned in the bay.
   b. He may have been murdered elsewhere.
   c. He did not drown.
2. Patients with kidney failure can be kept alive by dialysis, which removes toxic waste products from the blood. In a dialysis machine, blood flows past one side of a selectively permeable dialysis membrane,

and dialysis fluid flows on the other side of the membrane. Small substances, such as ions, glucose, and urea, can pass through the dialysis membrane, but larger substances, such as proteins, cannot. If you wanted to use a dialysis machine to remove only the toxic waste product urea from blood, what could you use for the dialysis fluid?
   a. A solution that is isotonic and contains only protein
   b. A solution that is isotonic and contains the same concentration of substances as blood, except for having no urea in it
   c. Distilled water
   d. Blood

3. Secretory vesicles fuse with the cell membrane to release their contents to the outside of the cell. In this process the membrane of the secretory vesicle becomes part of the cell membrane. Because small pieces of membrane are continually added to the cell membrane, one would expect the cell membrane to become larger and larger as secretion continues. The cell membrane stays the same size, however. Explain how this happens.

4. Suppose that a cell has the following characteristics: many mitochondria, well-developed rough ER, well-developed Golgi apparatuses, and numerous vesicles. Predict the major function of the cell. Explain how each characteristic supports your prediction.

5. The proteins (hemoglobin) in red blood cells normally organize relative to one another forming "stacks" of proteins, which are, in part, responsible for the normal shape of red blood cells. In sickle cell anemia, proteins inside red blood cells do not stack normally. Consequently, the red blood cells become sickle-shaped and plug up small blood vessels. It is known that sickle cell anemia is hereditary and results from changing one nucleotide for a different nucleotide within the gene that is responsible for producing the protein. Explain how this change results in an abnormally functioning protein.

Answers in Appendix D

Visit this textbook's website at www.mhhe.com/seeleyess6 for practice quizzes, animations, interactive learning exercises, and other study tools.

4

Tissues, Glands, and Membranes

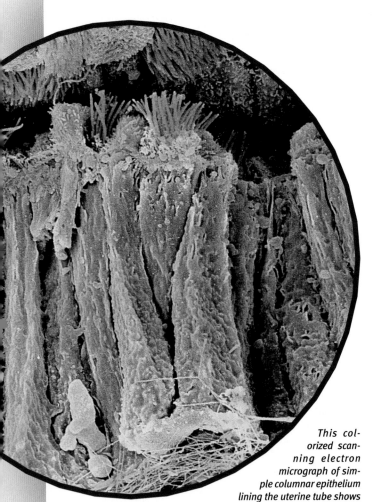

This colorized scanning electron micrograph of simple columnar epithelium lining the uterine tube shows the columnar epithelial cells (blue) resting on a weblike basement membrane. The cells are taller than they are wide, and all of them extend from the basement membrane to the free surface of the epithelial layer. Organelles of the epithelial cells synthesize a small volume of mucus and secrete it on their free surface. Cilia, the hairlike structures seen at the free surface of some of the epithelial cells, move. They carry the mucus over the surface of the epithelial cells. The epithelial cells are bound tightly to one another along their lateral surfaces, thus forming a continuous layer of columnar epithelial cells. The layer of epithelial cells is selectively permeable so that only certain substances can pass across it.

Chapter Outline and Objectives

Epithelial Tissue (p. 73)

1. List the characteristics of epithelial tissue.
2. Classify and give an example of the major types of epithelial tissue.
3. Explain the function in epithelial tissue of the following: cell layers, cell shapes, free cell surfaces, and connections between cells.
4. Define and categorize glands.

Connective Tissue (p. 81)

5. Describe the classification of connective tissues, and give examples of each major type.

Muscle Tissue (p. 87)

6. Name the three types of muscle tissue, and list their functions.

Nervous Tissue (p. 89)

7. State the functions of nervous tissue, and describe a neuron.

Membranes (p. 89)

8. List the structural and functional characteristics of mucous and serous membranes.

Inflammation (p. 90)

9. Describe the process of inflammation, and explain why inflammation protects the body.

Tissue Repair (p. 92)

10. Describe the major events involved in tissue repair.

Tissues and Aging (p. 93)

11. Describe the age-related changes that occur in cells and in extracellular matrix.

In some ways, the human body is like a complex machine. For example, all parts of an automobile cannot be made from a single type of material. Metal, capable of withstanding the heat of the engine, cannot be used for windows or tires. Similarly, the many parts of the human body are made of collections of specialized cells and the materials surrounding them. Muscle cells that contract to produce body movements have a different structure and function from epithelial cells that protect, secrete, or absorb.

A **tissue** (tish´ū, to weave) is a group of cells with similar structure and function, as well as similar extracellular substances located between the cells. The microscopic study of tissue structure is called **histology** (his-tol´ō-jē, *histo-*, tissue + *-ology*, study). Knowledge of tissue structure and function is important in understanding how individual cells are organized to form tissues and how tissues are organized to form organs, organ systems, and the complete organism. There is a relationship between the structure of each tissue type and its function and between the tissues in an organ and the organ's function.

Development, growth, aging, trauma, and diseases result from changes in tissues.

For example, enlargement of skeletal muscles occurs because skeletal muscle cells increase in size in response to exercise. Reduced elasticity of blood vessel walls in aging people results from slowly developing changes in connective tissue. Many tissue abnormalities, including cancer, result from changes in tissues that can be identified by microscopic examination.

The four basic tissue types are epithelial, connective, muscle, and nervous tissue. This chapter emphasizes epithelial and connective tissues. Muscle and nervous tissues are considered in more detail in later chapters.

Epithelial Tissue

Epithelium (ep-i-thē´lē-ŭm; pl. epithelia, ep-i-thē´lē-ă, *epi*, on + *thele*, covering or lining), or **epithelial** (ep-i-thē´lē-ăl) **tissue,** is found throughout the body where it covers internal and external surfaces. It also forms most glands. Surfaces of the body include the outer layer of the skin and the lining of cavities such as the digestive tract, respiratory passages, and blood vessels. Epithelium consists almost entirely of cells with very little extracellular material between them. Although there are some exceptions, most epithelia have a **free surface,** which is not in contact with other cells, and a surface adjacent to a basement membrane, which attaches the epithelial cells to underlying tissues (figure 4.1). Epithelium may consist of a single layer of cells or multiple layers of epithelial cells between the free surface and the basement membrane.

The basement membrane is secreted partly by epithelial cells and partly by the cells of the underlying tissues. It consists of a meshwork of protein molecules with other molecules bound to them. Substances that cross the epithelium must also cross the basement membrane. It can function as a filter and as a barrier to the movement of cells. For example, if some epithelial cells are converted to cancer cells, the basement membrane can, for some time, help prevent the spread of the cancer into the underlying tissues.

Blood vessels do not extend from the underlying tissues into epithelium, so gases and nutrients that reach the epithelium must diffuse across the basement membrane from the underlying tissues, where blood vessels are abundant. Waste products produced by the epithelial cells diffuse across the basement membrane to blood vessels.

Functions of Epithelia

The major functions of epithelia include:

1. *Protecting underlying structures.* Examples include the outer layer of the skin and the epithelium of the oral cavity, which protect the underlying structures from abrasion.
2. *Acting as barriers.* Epithelium prevents the movement of many substances through the epithelial layer. For example, the epithelium of the skin acts as a barrier to water and reduces water loss from the body. The epithelium of the skin is also a barrier that prevents the entry of many toxic molecules and microorganisms into the body.
3. *Permitting the passage of substances.* Epithelium allows the movement of many substances through the epithelial layer. For example, oxygen and carbon dioxide are exchanged between the air and blood by diffusion through the epithelium in the lungs.
4. *Secreting substances.* Examples include the sweat glands, mucous glands, and the enzyme-secreting portion of the pancreas.
5. *Absorbing substances.* The cell membranes of certain epithelial tissues contain carrier molecules (see chapter 3) that regulate the absorption of materials. For example, the epithelial cells of the intestine absorb digested food molecules, vitamins, and ions.

Classification of Epithelia

Epithelia are classified according to the number of cell layers and the shape of the cells (table 4.1). **Simple epithelium** consists of a single layer of cells. **Stratified** (layers) **epithelium** consists of more than one layer of epithelial cells, with some cells sitting on top of other cells. Categories of epithelium based on cell shape are **squamous** (skwā´mŭs, relating to scales), **cuboidal** (cubelike), and **columnar** (tall and thin). In most cases, each epithelium is given two names. Examples include simple squamous, simple columnar, and stratified squamous epithelia. When epithelium is stratified, it is named according to the shape of the cells at the free surface.

Simple squamous epithelium is a single layer of thin, flat cells (table 4.2*a*). It is often found where diffusion or filtration take place. Some substances easily pass through this thin layer of cells and other substances do not. For example, the respiratory passages end as small sacs called **alveoli** (al-vē´o-lī, sing. alveolus, hollow sac). The alveoli consist of simple squamous epithelium that allows oxygen from the air to diffuse into the body and carbon dioxide to diffuse out of the body into the air. Simple squamous epithelial tissue in the filtration membranes of the

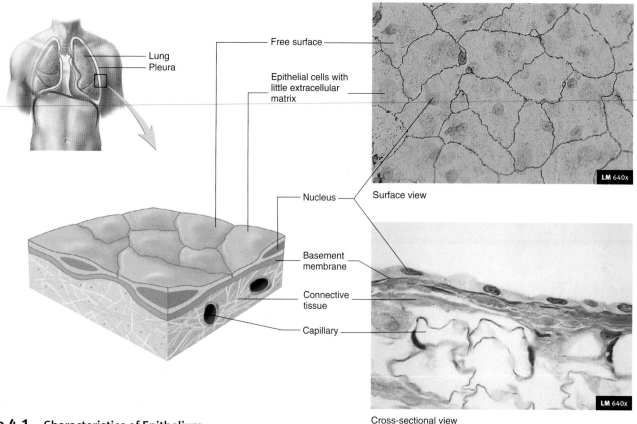

Figure 4.1   Characteristics of Epithelium

Surface and cross-sectional views of epithelium illustrate the following characteristics: little extracellular material between cells, a free surface, and a basement membrane attaching epithelial cells to underlying tissues. Capillaries in connective tissue do not penetrate the basement membrane. Nutrients, oxygen, and waste products diffuse across the basement membrane between the capillaries and the epithelial cells.

Table 4.1	Classification of Epithelia
Number of Layers	**Cell Shape**
Simple (one layer)	Squamous Cuboidal Columnar
Pseudostratified (a modified form of simple epithelium)	Columnar

Stratified (more than one layer)	Squamous
	  Keratinized
	  Nonkeratinized (moist)

Transitional (a type of stratified epithelium)	Roughly cuboidal to columnar when not stretched and squamouslike when stretched

kidneys forms thin barriers through which small molecules, but not large ones, can pass. Small molecules including water from blood are filtered through these barriers as a major step in urine formation. Large molecules, such as proteins and blood cells, remain in the blood vessels of the kidneys.

Simple squamous epithelium also functions to prevent abrasion between organs in the pericardial, pleural, and peritoneal cavities (see chapter 1). The outer surfaces of organs are covered with simple squamous epithelium that secretes a slippery fluid. The fluid lubricates the surfaces between the organs, preventing damage from friction when the organs rub against one another.

Simple cuboidal epithelium is a single layer of cubelike cells (table 4.2*b*) that carry out active transport, facilitated diffusion, or secretion. Epithelial cells that secrete molecules such as proteins contain organelles that synthesize them. These cells have a greater volume than simple squamous epithelial cells and contain more cell organelles. The organelles of simple cuboidal cells that actively transport molecules into and out of the cells include mitochondria, which produce ATP, and organelles needed to synthesize the carrier molecules. Transport of molecules across a layer of simple cuboidal epithelium can be regulated by the amount of ATP produced or by the type of carrier molecules synthesized. The many kidney tubules have large portions of their walls composed of simple cuboidal epithelium. These cuboidal epithelial cells secrete waste products into the tubules and reabsorb useful materials from the tubules as urine is formed. Some cuboidal epithelial cells have cilia that move mucus over the free surface or microvilli that increase the surface area for secretion and absorption.

Simple columnar epithelium is a single layer of tall, thin cells (table 4.2*c*). These large cells contain organelles that enable them to perform complex functions. For example, the simple columnar epithelium of the small intestine produces and secretes mucus and digestive enzymes. The mucus protects the

Table 4.2   Simple Epithelium

(a) Simple Squamous Epithelium

Structure: Single layer of flat, often hexagonal cells. The nuclei appear as bumps when viewed as a cross section because the cells are so flat.

Function: Diffusion, filtration, some secretion, and some protection against friction.

Location: Lining of blood vessels and the heart, lymphatic vessels, alveoli of the lungs, portions of the kidney tubules, lining of serous membranes of body cavities (pleural, pericardial, peritoneal).

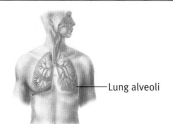

Lung alveoli

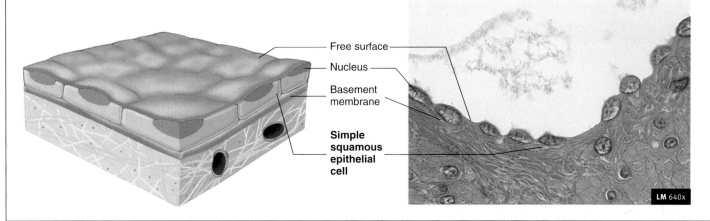

Free surface
Nucleus
Basement membrane
Simple squamous epithelial cell

LM 640x

(b) Simple Cuboidal Epithelium

Structure: Single layer of cube-shaped cells; some cells have microvilli (kidney tubules) or cilia (terminal bronchioles of the lungs).

Function: Active transport and facilitated diffusion result in secretion and absorption by cells of the kidney tubules; secretion by cells of glands and choroid plexuses; movement of particles embedded in mucus out of the terminal bronchioles by ciliated cells.

Location: Kidney tubules, glands and their ducts, choroid plexuses of the brain, lining of terminal bronchioles of the lungs, and surfaces of the ovaries.

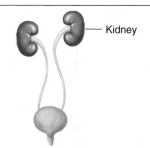

Kidney

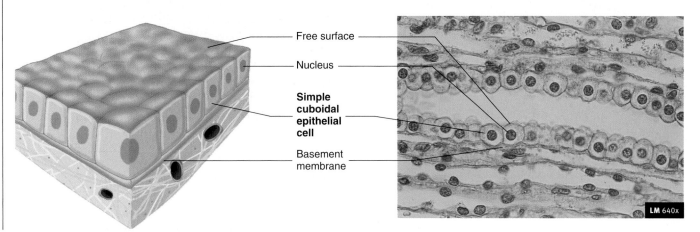

Free surface
Nucleus
Simple cuboidal epithelial cell
Basement membrane

LM 640x

continued

Table 4.2   *continued*

(c) Simple Columnar Epithelium

Structure: Single layer of tall, narrow cells. Some cells have cilia (bronchioles of lungs, auditory tubes, uterine tubes, and uterus) or microvilli (intestines).

Function: Movement of particles out of the bronchioles of the lungs by ciliated cells; partially responsible for the movement of the oocytes through the uterine tubes by ciliated cells. Secretion by cells of the glands, the stomach, and the intestine. Absorption by cells of the intestine.

Location: Glands and some ducts, bronchioles of lungs, auditory tubes, uterus, uterine tubes, stomach, intestines, gallbladder, bile ducts, and ventricles of the brain.

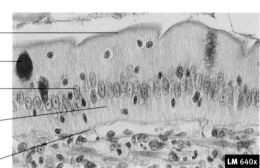

Lining of stomach and intestines

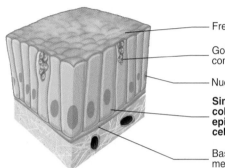

Free surface

Goblet cell containing mucus

Nucleus

Simple columnar epithelial cell

Basement membrane

LM 640x

(d) Pseudostratified Columnar Epithelium

Structure: Single layer of cells; some cells are tall and thin and reach the free surface, and others do not; the nuclei of these cells are at different levels and appear stratified; the cells are almost always ciliated and are associated with goblet cells that secrete mucus onto the free surface.

Function: Synthesize and secrete mucus onto the free surface and move mucus (or fluid) that contains foreign particles over the surface of the free surface and from passages.

Location: Lining of nasal cavity, nasal sinuses, auditory tubes, pharynx, trachea, and bronchi of lungs.

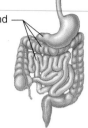

Trachea

Bronchus

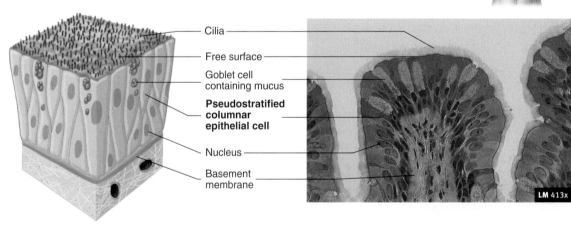

Cilia

Free surface

Goblet cell containing mucus

Pseudostratified columnar epithelial cell

Nucleus

Basement membrane

LM 413x

lining of the intestine, and the digestive enzymes complete the process of digesting food. The columnar cells then absorb the digested foods by active transport, facilitated diffusion, or simple diffusion.

Pseudostratified (*pseudo,* false) **columnar epithelium** is a special type of simple epithelium (table 4.2*d*). The prefix pseudo-means false, so this type of epithelium appears to be stratified but is not. It consists of one layer of cells, with all the cells attached to the basement membrane. There is an appearance of two or more layers of cells because some of the cells are tall and reach the free surface, whereas others are short and do not reach the free surface. Pseudostratified columnar epithelium is found lining some glands and ducts, auditory tubes, some of the respiratory passages, such as the nasal cavity, nasal sinuses, pharynx, trachea, and bronchi. Pseudostratified columnar epithelium secretes mucus, which covers its free surface. Cilia located on the free surface move the mucus and the debris that accumulates in it over the surfaces. For example, cilia of the respiratory passages move mucus toward the throat, where it is swallowed.

Stratified squamous epithelium forms a thick epithelium because it consists of several layers of cells (table 4.3*a*). The deepest cells are cuboidal or columnar and are capable of dividing and producing new cells. As these newly formed cells are pushed to the surface, they become flat and thin. If cells at the surface are damaged or rubbed away, they are replaced by cells formed in the deeper layers. One type of stratified squamous

epithelium forms the outer layer of the skin and is called keratinized stratified squamous epithelium (see chapter 5). The outer layers consist of dead stratified epithelial cells tightly bound to one another. As the cells flatten, the cytoplasm of the epithelial cells is replaced by a protein called keratin, and the cells die. The dead cells provide protection against abrasion, form a barrier that prevents microorganisms and toxic chemicals from entering the body, and reduces the loss of water from the body. In contrast, stratified squamous epithelium of the mouth is composed of living cells with a moist surface. This nonkeratinized (moist) stratified squamous epithelium also provides protection against abrasion and acts as a mechanical barrier, preventing the entry of microorganisms into the body. Water, however, can move across it more readily than across the skin.

Stratified cuboidal epithelium consists of more than one layer of cuboidal epithelial cells. This epithelial type is relatively rare and is found in sweat gland ducts, ovarian follicular cells, and the salivary glands. It functions in absorption, secretion, and protection.

Stratified columnar epithelium consists of more than one layer of epithelial cells, but only the surface cells are columnar in shape. The deeper layers are irregular in shape or cuboidal. Like stratified cuboidal epithelium, stratified columnar epithelium is relatively rare. It is found in locations such as the mammary gland ducts, the larynx, and a portion of the male urethra. This epithelium carries out secretion, protection, and some absorption.

Table 4.3 Stratified Epithelium

(a) Stratified Squamous Epithelium

Structure: Several layers of cells that are cuboidal in the basal layer and progressively flattened toward the surface. The epithelium can be nonkeratinized (moist) or keratinized. In nonkeratinized stratified squamous epithelium, the surface cells retain a nucleus and cytoplasm. In keratinized stratified epithelium, the cytoplasm of cells at the surface is replaced by a protein called keratin, and the cells are dead.

Function: Protection against abrasion, barrier against infection, and reduces loss of water from the body.

Location: Keratinized—outer layer of the skin. Nonkeratinized—mouth, throat, larynx, esophagus, anus, vagina, inferior urethra, and corneas.

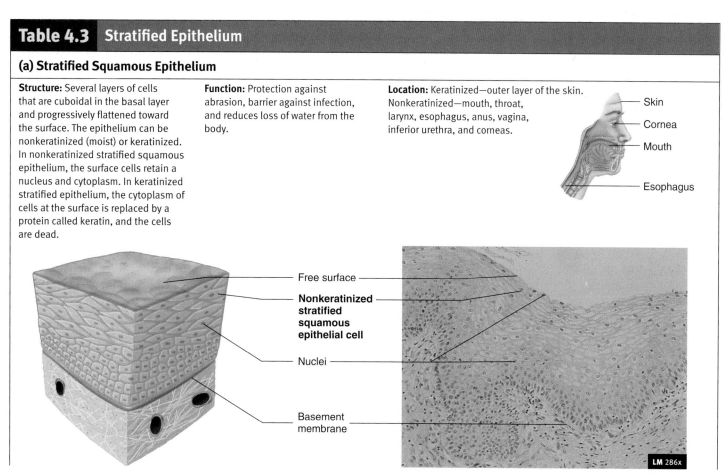

continued

Table 4.3 *continued*

(b) Transitional Epithelium

Structure: Stratified cells that appear cuboidal when the organ or tube is not stretched and squamous when the organ or tube is stretched by fluid.

Function: Accommodates fluctuations in the volume of fluid in an organ or tube; protection against the caustic effects of urine.

Location: Lining of urinary bladder, ureters, and superior urethra.

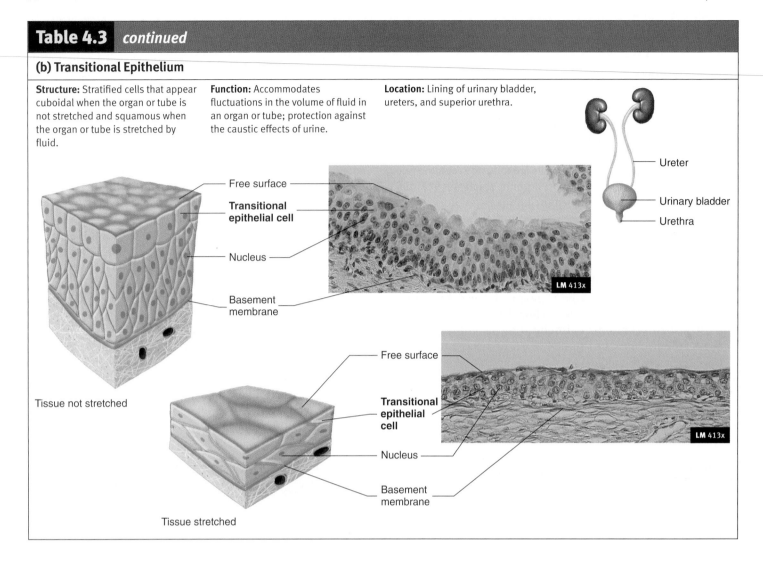

Free surface
Transitional epithelial cell
Nucleus
Basement membrane
Tissue not stretched
LM 413x

Free surface
Transitional epithelial cell
Nucleus
Basement membrane
Tissue stretched
LM 413x

Ureter
Urinary bladder
Urethra

Transitional epithelium is a special type of stratified epithelium that can be greatly stretched (table 4.3*b*). In the unstretched state, transitional epithelium consists of five or more layers of cuboidal or columnar cells that often are dome-shaped at the free surface. As transitional epithelium is stretched, the cells change shape to a low cuboidal or squamous shape, and the number of cell layers decreases. Transitional epithelium is found lining cavities that can expand greatly, such as the urinary bladder. It also protects underlying structures from the caustic effects of urine.

Structural and Functional Relationships

Cell Layers and Cell Shapes

The number of cell layers and the shape of the cells in a specific type of epithelium reflect the function the epithelium performs. Two important functions are controlling the passage of materials through the epithelium and protecting the underlying tissues. Simple epithelium, with its single layer of cells, is found in organs in which the principal function is the movement of materials. Examples include diffusion of gases across the wall of the alveoli of the lungs, filtration of fluid across the filtration membranes in the

kidneys, secretion in glands, and nutrient absorption in the intestines. The movement of materials through a stratified epithelium is hindered by its many layers. Stratified epithelium is well adapted for its protective function. As the outer cell layers are damaged, they are replaced by cells from deeper layers. Stratified squamous epithelium is found in areas of the body where abrasion can occur, such as in the skin, anal canal, and vagina.

Differences in function are also reflected in cell shape. Cells are normally flat and thin when the function is diffusion, such as in the alveoli of the lungs, or filtration, such as in kidney tubules. Cells with the major function of secretion or absorption are usually cuboidal or columnar. They are larger because they contain more organelles, which are responsible for the function of the cell. The stomach, for example, is lined with simple columnar epithelium. These cells contain many **secretory vesicles** (ves′i-klz) filled with **mucus** (mū′kŭs), which is a clear, viscous material. The large amounts of mucus produced by the simple columnar epithelium protect the stomach lining against the digestive enzymes and acid produced in the stomach. An ulcer, or irritation in the epithelium and underlying tissue, can develop if this protective mechanism fails. Simple cuboidal epithelial cells that secrete or absorb molecules, such as in the kidney tubules, contain many mitochondria, which produce the ATP required for active transport.

P R E D I C T   1
Explain the consequences of having (a) nonkeratinized stratified epithelium rather than simple columnar epithelium lining the digestive tract, and (b) nonkeratinized stratified squamous epithelium rather than keratinized stratified squamous epithelium in the skin.

The shape and number of layers of epithelial cells can change if they are subjected to long-term irritation or to other abnormal conditions. People who smoke cigarettes eventually experience changes in the epithelium of the larger respiratory passages. The delicate pseudostratified columnar epithelium, which performs a cleaning function by moving mucus and debris from the passageways, is replaced by stratified squamous epithelium that is more resistant to irritation, but does not perform a cleaning function. Also, lung cancer most often results from changes in epithelial cells of the lung passageways of smokers. The changes in the structure of the cells are used to identify the cancer.

A CASE IN POINT | Detecting Cancer

Wanna Wonder was wary about having a Pap test. She insisted on knowing why she had to endure such a procedure. Her physician explained that it was a test that could detect cancer of the uterus. Cells are collected, using a swab, from the cervix of the uterus. The cells are stained and examined with a microscope. Compared to normal cells, cancer cells have variable sizes and shapes, with large and unusually shaped nuclei. If uterine cancer is detected early, it can be treated before it spreads to other areas of the body.

Free Cell Surfaces

Most epithelia have a free surface that is not in contact with other cells and faces away from underlying tissues. The characteristics of the free surface reflect the functions it performs. The free surface can be smooth, or it can have microvilli or cilia. **Smooth surfaces** reduce friction. For example, the lining of blood vessels is simple squamous epithelium with a smooth surface, which reduces friction as blood flows through the vessels. **Microvilli** are cylindrical extensions of the cell membrane that function to increase the cell surface area (see chapter 3). Normally many microvilli cover the free surface of each cell involved in absorption or secretion, such as the cells lining the small intestine. **Cilia** (see chapter 3) propel materials along the surface of cells. The nasal cavity and trachea are lined with pseudostratified columnar ciliated epithelium. Intermixed with the ciliated cells are specialized mucus-producing cells called **goblet** (glass with a stem and foot) **cells** (see table 4.2d). Dust and other materials are trapped in the mucus that covers the epithelium, and movement of the cilia

propels the mucus with its entrapped particles to the back of the throat, where it is swallowed or coughed up. The constant movement of mucus helps to keep the respiratory passages clean.

Cell Connections

Epithelial cells are connected to one another in several ways (figure 4.2). **Tight junctions** bind adjacent cells together and form permeability barriers. Tight junctions prevent the passage of materials between epithelial cells because they completely surround each cell, similar to the way a belt surrounds the waist. Materials that pass through the epithelial layer must pass through the cells, which can regulate what materials cross the epithelial layer. Tight junctions are found in the lining of the intestines and most other simple epithelia. **Desmosomes** (dez′mō-sōmz, *desmos*, a band + *soma*, body) are mechanical links that function to bind cells together. Modified desmosomes, called **hemidesmosomes** (hem-ē-dez′mō-sōmz, *hemi*, one half), also anchor cells to the basement membrane. Many desmosomes are found in epithelia subjected to stress, such as the stratified squamous epithelium of the skin. **Gap junctions** are small channels that allow small molecules and ions to pass from one epithelial cell to an adjacent one. Most epithelial cells are connected to one another by gap junctions, and it is believed

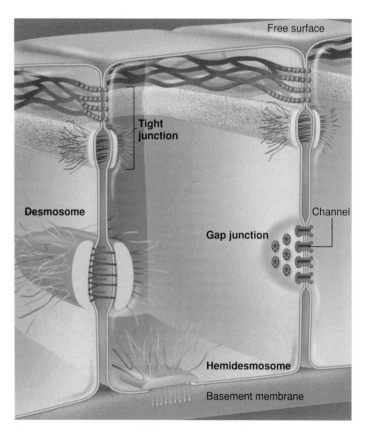

Figure 4.2 Cell Connections
Desmosomes anchor cells to one another, and hemidesmosomes anchor cells to the basement membrane. Tight junctions bind adjacent cells together. Gap junctions have channels that allow adjacent cells to communicate with each other. Few cells have all of these different connections.

that molecules or ions moving through the gap junctions act as communication signals to coordinate the activities of the cells.

Glands

A **gland** is a multicellular structure that secretes substances onto a surface, into a cavity, or into the blood. Most glands are composed primarily of epithelium. Sometimes single goblet cells are classified as unicellular glands because they secrete mucus onto epithelial surfaces. Glands with ducts are called **exocrine** (ek′sō-krin, *exo,* outside + *krino,* to separate) glands (figure 4.3). The exocrine glands can be **simple,** with ducts that have no branches, or **compound,** with ducts that have many branches. The end of a duct can be **tubular.** Some tubular glands are straight and others have coiled tubules. Some glands have ends that are expanded into a saclike structure called an **acinus** (as′i-nŭs, grapelike), or **alveolus** (al-vē′ō-lŭs, small cavity). Some compound glands have acini and tubules (tubuloacinar or tubuloalveolar glands) that secrete substances. Secretions from exocrine glands pass through the ducts onto a surface or into an organ. For example, sweat from sweat glands and oil from sebaceous glands flow onto the skin surface.

Endocrine (en′dō-krin, *endo,* within) glands have no ducts and empty their secretions into the blood. These secretions, called **hormones** (hōr′mōnz), are carried by the blood to

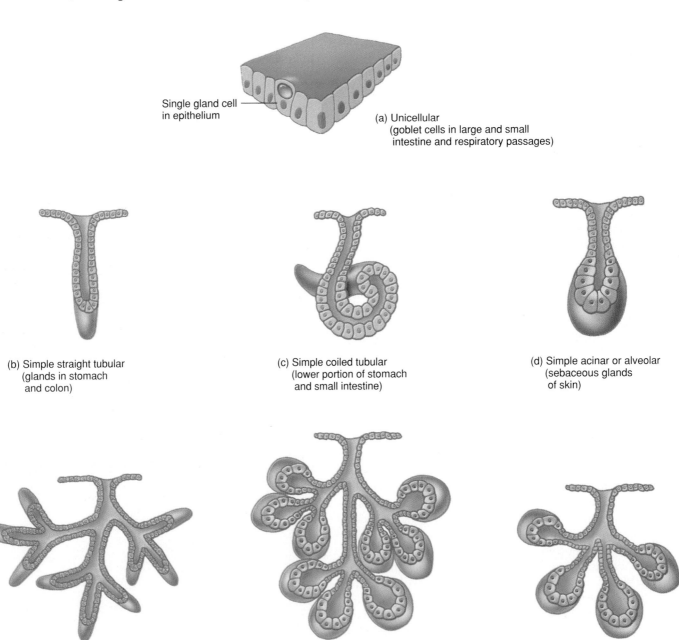

Single gland cell in epithelium

(a) Unicellular
(goblet cells in large and small
intestine and respiratory passages)

(b) Simple straight tubular
(glands in stomach
and colon)

(c) Simple coiled tubular
(lower portion of stomach
and small intestine)

(d) Simple acinar or alveolar
(sebaceous glands
of skin)

(e) Compound tubular
(mucous glands of duodenum)

(f) Compound acinar or alveolar
(mammary glands)

(g) Compound tubuloacinar or
tubuloalveolar (pancreas)

Figure 4.3 **Structure of Exocrine Glands**

The names of exocrine glands are based on the shapes of their secretory units and their ducts.

other parts of the body. Endocrine glands include the thyroid gland and the insulin-secreting portions of the pancreas. Endocrine glands are discussed more fully in chapter 10.

Connective Tissue

Connective tissue structure is usually characterized by large amounts of extracellular materials that separate cells from one another. The extracellular material, or **extracellular matrix** (mā′triks), has three major components: (1) protein fibers, (2) ground substance consisting of nonfibrous protein and other molecules, and (3) fluid.

Three types of protein fibers help to form most connective tissues. **Collagen** (kol′lă-jen, glue producing) **fibers,** which resemble microscopic ropes, are flexible but resist stretching. **Reticular** (rē-tik′ū-lăr) **fibers** are very fine, short collagen fibers that branch to form a supporting network. **Elastic fibers** have a structure similar to coiled metal bed springs. After being stretched, elastic fibers have the ability to recoil to their original shape.

Ground substance is the shapeless background against which cells and collagen fibers are seen in the light microscope. Although ground substance appears shapeless, the molecules within the ground substance are highly structured. **Proteoglycans** (prō′tē-ō-glī′kanz, *proteo,* protein + *glycan,* polysaccharide) resemble the limbs of pine trees, with proteins forming the branches and polysaccharides forming the pine needles. This structure enables proteoglycans to trap large quantities of water between the polysaccharides.

Connective tissue cells are named according to their functions. **Blast** (germ) cells produce the matrix, **cyte** (cell) cells maintain it, and **clast** (break) cells break it down for remodeling. For example, **fibroblasts** (fī′bro-blast, *fibra,* fiber) are cells that form fibers and ground substance in the extracellular matrix of fibrous connective tissue and fibroyctes are cells that maintain it. **Osteoblasts** (os′tē-ō-blasts, *osteo,* bone) form bone, **osteocytes** (os′tē-ō-sītz) maintain bone, and **osteoclasts** (os′tē-ō-klasts, broken) break down bone. Cells associated with the immune system are also found in connective tissue. **Macrophages** (mak′rō-fāj-ez, *makros,* large + *phago,* to eat) are large cells that are capable of moving about and ingesting foreign substances, including microorganisms that are found in the connective tissue. **Mast cells** are nonmotile cells that release chemicals, such as histamine, that promote inflammation.

Functions of Connective Tissue

Connective tissues perform the following major categories of functions:

1. *Enclosing and separating.* Sheets of connective tissues form capsules around organs such as the liver and kidneys. Connective tissue also forms layers that separate tissues and organs. For example, connective tissues separate muscles, arteries, veins, and nerves from one another.
2. *Connecting tissues to one another.* For example, tendons are strong cables, or bands, of connective tissue that

attach muscles to bone, and ligaments are connective tissue bands that hold bones together.
3. *Supporting and moving.* Bones of the skeletal system provide rigid support for the body, and semirigid cartilage supports structures such as the nose, ears, and surfaces of joints. Joints between bones allow one part of the body to move relative to other parts.
4. *Storing.* Adipose tissue (fat) stores high-energy molecules, and bones store minerals such as calcium and phosphate.
5. *Cushioning and insulating.* Adipose tissue (fat) cushions and protects the tissues it surrounds and provides an insulating layer beneath the skin that helps conserve heat.
6. *Transporting.* Blood transports substances throughout the body, such as gases, nutrients, enzymes, hormones, and cells of the immune system.
7. *Protecting.* Cells of the immune system and blood provide protection against toxins and tissue injury, as well as from microorganisms. Bones protect underlying structures from injury.

Classification of Connective Tissue

During embryonic development, embryonic connective tissue gives rise to six major categories of connective tissue (table 4.4). The extracellular matrix predominates in nearly all types of connective tissue and is largely responsible for the functional characteristics of each connective tissue type.

Loose, or **areolar** (a-re′ō-lar, small areas), **connective tissue** has extracellular matrix consisting mostly of collagen and a few elastic fibers. These fibers are widely separated from one another (table 4.5). The most common cells found in loose connective tissue are the **fibroblasts** (fī′brō-blasts), which are responsible for the production of the matrix. Loose connective tissue is widely distributed throughout the body and is the loose packing material of the body that fills the spaces between glands, muscles, and nerves. The basement membranes of

Table 4.4	Classification of Connective Tissues
Loose (areolar) connective tissue	
Adipose tissue	
Dense connective tissue Dense collagenous connective tissue Collagen fibers arranged in the same direction Collagen fibers arranged in many directions Dense elastic connective tissue Elastic fibers arranged in the same direction Elastic fibers arranged in many directions	
Cartilage Hyaline cartilage Fibrocartilage Elastic cartilage	
Bone	
Blood	

Table 4.5   Loose Connective Tissue

Structure: A fine network of fibers (mostly collagen fibers with a few elastic fibers) with spaces between the fibers. Fibroblasts, macrophages, and lymphocytes are located in the spaces.

Function: Loose packing, support, and nourishment for the structures with which it is associated.

Location: Widely distributed throughout the body; substance on which epithelial basement membranes rest; packing between glands, muscles, and nerves. Attaches the skin to underlying tissues.

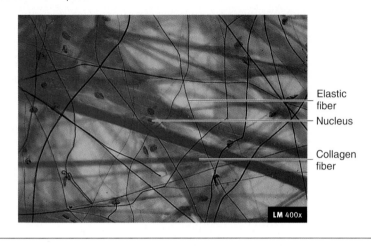
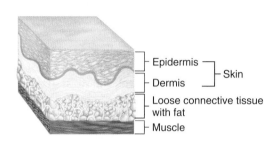

epithelia often rest on loose connective tissue, and it attaches the skin to underlying tissues.

Adipose (ad′i-pōs, fat) **tissue** has an extracellular matrix with collagen and elastic fibers but it is not a typical connective tissue because it has very little extracellular matrix. The individual cells are large and closely packed together (table 4.6). Adipose cells are filled with lipids and function to store energy. Adipose tissue also pads and protects parts of the body and acts as a thermal insulator.

Dense connective tissue has extracellular matrix that consists of densely packed fibers produced by fibroblasts. There are two major subcategories of dense connective tissue. In one,

Table 4.6   Connective Tissue with Special Properties

Adipose Tissue

Structure: Little extracellular matrix surrounding cells. The adipocytes, or fat cells, are so full of lipid that the cytoplasm is pushed to the periphery of the cell.

Function: Packing material, thermal insulator, energy storage, and protection of organs against injury from being bumped or jarred.

Location: Predominantly in subcutaneous areas, mesenteries, renal pelves, around kidneys, attached to the surface of the colon, mammary glands, and in loose connective tissue that penetrates into spaces and crevices.

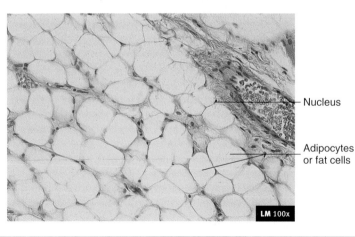
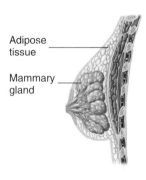

collagen fibers predominate, and in the other, elastic fibers predominate. **Dense collagenous connective tissue** has extracellular matrix consisting mostly of collagen fibers (table 4.7*a*). Structures made up of dense collagenous connective tissue include tendons, which attach muscle to bone; many ligaments, which attach bones to other bones; and much of the dermis, which is the connective tissue of the skin. Dense collagenous connective tissue also forms many capsules that surround organs such as the liver and kidneys. The collagen fibers are oriented in the same direction in tendons and ligaments. They are oriented in many different directions in the dermis of the skin and in organ capsules.

PREDICT ②
In tendons, collagen fibers are oriented parallel to the length of the tendon. In the skin, collagen fibers are oriented in many directions. What are the functional advantages of the fiber arrangements in tendons and in the skin?

Dense elastic connective tissue has abundant elastic fibers among collagen fibers. The elastic fibers allow the tissue to stretch and recoil. Examples include the dense elastic connective tissue of the vocal cords (table 4.7*b*), the walls of large arteries and elastic ligaments. The elastic fibers are oriented in the same direction in elastic ligaments and in the vocal cords, and they are oriented in many different directions in the walls of arteries.

A CASE IN POINT | Marfan Syndrome

Izzy Taller is a 6-foot, 10-inch-tall high school student. The basketball coach convinced Izzy to try out for the basketball team, which required a physical exam. The physician determined that Izzy's limbs, fingers, and toes are disproportionately long in relation to the rest of his body; he has an abnormal heart sound, indicating a problem with a heart valve; and he has poor vision because the lenses of his eyes are positioned abnormally. The physician diagnosed Izzy's condition as Marfan syndrome, an autosomal dominant (see chapter 20) genetic disorder that results in the production of an abnormal protein called fibrillin-1. This protein is necessary for the normal formation and maintenance of elastic fibers and the regulation of a growth factor called TGFß. Because elastic fibers are found in connective tissue throughout the body, several systems of the body can be affected. The major cause of death in Marfan syndrome is rupture of the aorta due to weakening of the connective tissue in its wall. Izzy Taller should not join the basketball team.

Table 4.7	**Dense Connective Tissue**

(a) Dense Collagenous Connective Tissue

Structure: Matrix composed of collagen fibers running in somewhat the same direction in tendons and ligaments. Collagen fibers run in several directions in the dermis of the skin and in organ capsules.	**Function:** Ability to withstand great pulling forces exerted in the direction of fiber orientation, great tensile strength, and stretch resistance.	**Location:** Found in tendons (attach muscle to bone) and ligaments (attach bones to each other). Also found in the dermis of the skin, organ capsules, and the outer layer of many blood vessels.

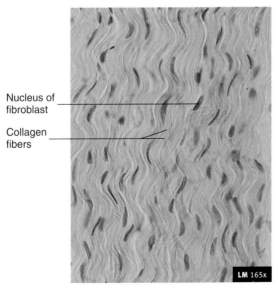

Nucleus of fibroblast

Collagen fibers

LM 165x

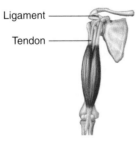

Ligament

Tendon

continued

Table 4.7   *continued*

(b) Dense Elastic Connective Tissue

Structure: Matrix composed of collagen fibers and elastin fibers running in somewhat the same direction in elastic ligaments. Elastic fibers run in several directions in the elastic connective tissue of blood vessel walls.

Function: Capable of stretching and recoiling like a rubber band with strength in the direction of fiber orientation.

Location: Elastic ligaments between the vertebrae and along the dorsal aspect of the neck (nucha) and in the vocal cords. Also found in elastic connective tissue of blood vessel walls.

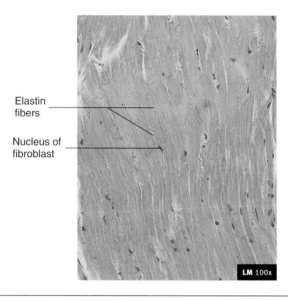

Elastin fibers

Nucleus of fibroblast

LM 100x

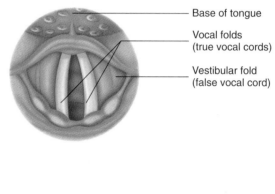

Base of tongue

Vocal folds (true vocal cords)

Vestibular fold (false vocal cord)

Table 4.8   Connective Tissue: Cartilage

(a) Hyaline Cartilage

Structure: Collagen fibers are small and evenly dispersed in the matrix, making the matrix appear transparent. The cartilage cells, or chondrocytes, are found in spaces, or lacunae, within the firm but flexible matrix.

Function: Allows growth of long bones. Provides rigidity with some flexibility in the trachea, bronchi, ribs, and nose. Forms rugged, smooth, yet somewhat flexible articulating surfaces. Forms the embryonic skeleton.

Location: Growing long bones, cartilage rings of the respiratory system, costal cartilage of ribs, nasal cartilages, articulating surface of bones, and the embryonic skeleton.

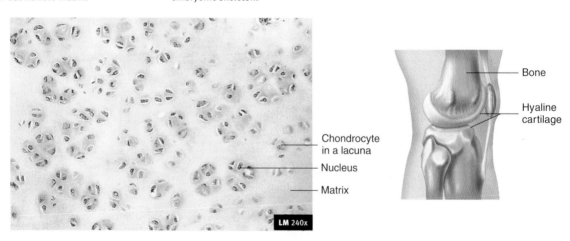

Chondrocyte in a lacuna

Nucleus

Matrix

LM 240x

Bone

Hyaline cartilage

(b) Fibrocartilage

Structure: Collagenous fibers similar to those in hyaline cartilage. The fibers are more numerous than in other cartilages and are arranged in thick bundles.

Function: Somewhat flexible and capable of withstanding considerable pressure. Connects structures subjected to great pressure.

Location: Intervertebral disks, symphysis pubis, articular disks (e.g., knees and temporomandibular [jaw] joints).

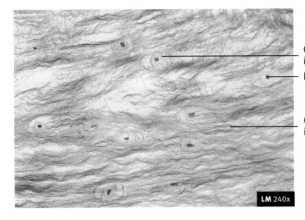

Chondrocyte in lacuna
Nucleus
Collagen fibers in matrix

LM 240x

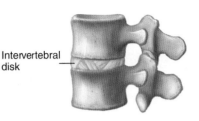

Intervertebral disk

(c) Elastic Cartilage

Structure: Similar to hyaline cartilage, but matrix also contains elastin fibers.

Function: Provides rigidity with even more flexibility than hyaline cartilage because elastic fibers return to their original shape after being stretched.

Location: External ears, epiglottis, and auditory tubes.

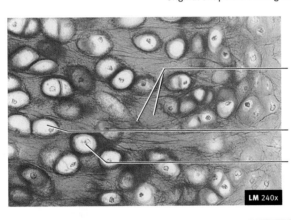

Elastic fibers in matrix

Chondrocytes in lacunae

Nucleus

LM 240x

Cartilage (kar′ti-lij, gristle) is composed of cartilage cells, or **chondrocytes** (kon′drō-sītz, cartilage cell), located in spaces called **lacunae** (lă-koo′nē, a small space) within an extensive matrix (table 4.8a). Collagen in the matrix gives cartilage flexibility and strength. Cartilage is resilient because the proteoglycans of the matrix trap water, which makes the cartilage relatively rigid and enables it to spring back after being compressed. Cartilage provides support, but, if it is bent or slightly compressed, it resumes its original shape. Cartilage heals slowly after an injury because blood vessels do not penetrate it. Thus cells and nutrients necessary for tissue repair do not easily reach the damaged area.

Hyaline (hī′ă-lin, clear or glassy) **cartilage** (see table 4.8a) is the most abundant type of cartilage and has many functions. It covers the ends of bones where bones come together to form joints. In joints, hyaline cartilage forms smooth, resilient surfaces that can withstand repeated compression. Hyaline cartilage also forms the costal cartilages, which attach the ribs to the sternum (breastbone), the cartilage rings of the respiratory tract, and nasal cartilages.

Fibrocartilage (table 4.8b) has more collagen than does hyaline cartilage, and bundles of collagen fibers can be seen in the matrix. In addition to withstanding compression, it is able to resist pulling or tearing forces. It is found in the disks between vertebrae (bones of the back), and in some joints such as the knee and temporomandibular (jaw) joints, for example.

Elastic cartilage (table 4.8c) contains elastic fibers in addition to collagen and proteoglycans. The elastic fibers appear as coiled fibers among bundles of collagen fibers. Elastic cartilage is able to recoil to its original shape when bent. The external ear, epiglottis, and auditory tube contain elastic cartilage.

Scars consist of dense connective tissue made of collagen fibers. Vitamin C is required for collagen synthesis. Predict the effect of scurvy, which is a nutritional disease caused by vitamin C deficiency, on wound healing.

Bone is a hard connective tissue that consists of living cells and a mineralized matrix (table 4.9). Bone cells, or **osteocytes**

(*osteo,* bone), are located within spaces in the matrix called lacunae. The strength and rigidity of the mineralized matrix enables bones to support and protect other tissues and organs of the body. The two types of bone, **compact** and **cancellous** (kan′sĕ-lŭs, a lattice work), are considered in greater detail in chapter 6.

Blood is unique because the matrix is liquid, enabling blood cells to move through blood vessels (table 4.10). Some blood cells even leave the blood and wander into other tissues.

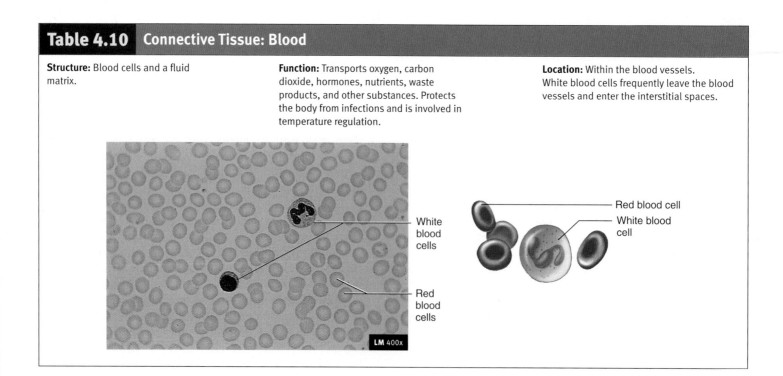

Table 4.9	Bone

Structure: Hard, bony matrix predominates. Many osteocytes (not seen in this bone preparation) are located within lacunae; the matrix is organized into layers called lamellae.

Function: Provides great strength and support and protects internal organs such as the brain. Bone also provides attachment sites for muscles and ligaments. The joints of bones allow movements.

Location: All bones of the body.

Lacuna

Central canal

Matrix organized into lamellae

LM 240x

Cancellous bone

Compact bone

Table 4.10	Connective Tissue: Blood

Structure: Blood cells and a fluid matrix.

Function: Transports oxygen, carbon dioxide, hormones, nutrients, waste products, and other substances. Protects the body from infections and is involved in temperature regulation.

Location: Within the blood vessels. White blood cells frequently leave the blood vessels and enter the interstitial spaces.

White blood cells

Red blood cells

LM 400x

Red blood cell

White blood cell

The liquid matrix enables blood to flow rapidly through the body, carrying food, oxygen, waste products, and other materials. Blood is discussed more fully in chapter 11.

Muscle Tissue

The main characteristic of **muscle tissue** is its ability to contract, or shorten, making movement possible. Muscle contraction results from contractile proteins located within the muscle cells (see chapter 7). The length of muscle cells is greater than the diameter. Muscle cells are sometimes called **muscle fibers** because they often resemble tiny threads.

The three types of muscle tissue are skeletal, cardiac, and smooth muscle. **Skeletal muscle** is what normally is thought of as "muscle" (table 4.11*a*). It is the meat of animals and constitutes about 40% of a person's body weight. As the name implies, skeletal muscle attaches to the skeleton and enables body movement. Skeletal muscle is described as being under voluntary (conscious) control because one can purposefully cause skeletal muscle contraction to achieve specific body movements. However, the nervous system can cause skeletal muscles to contract without conscious involvement such as during reflex movements and maintenance of muscle tone. Skeletal muscle cells tend to be long, cylindrical cells with several nuclei per cell. The nuclei of these cells are located near the periphery of the cell. Some skeletal muscle cells extend the length of an entire muscle. Skeletal muscle

cells are **striated** (strī′āt-ed), or banded, because of the arrangement of contractile proteins within the cells (see chapter 7).

Cardiac muscle is the muscle of the heart and is responsible for pumping blood (table 4.11*b*). It is under involuntary (unconscious) control although one can learn to influence the heart rate by using techniques such as meditation and biofeedback. Cardiac muscle cells are cylindrical in shape but much shorter in length than skeletal muscle cells. Cardiac muscle cells are striated and usually have one nucleus per cell. They often are branched and connected to one another by **intercalated** (in-ter′kǎ-lā-ted, inserted between) **disks.** The intercalated disks, which contain specialized gap junctions, are important in coordinating the contractions of the cardiac muscle cells (see chapter 12).

Smooth muscle forms the walls of hollow organs (except the heart) and also is found in the skin and the eyes (table 4.11*c*). It is responsible for a number of functions, such as movement of food through the digestive tract and emptying of the urinary bladder. Like cardiac muscle, smooth muscle is controlled involuntarily. Smooth muscle cells are tapered at each end, have a single nucleus, and are not striated.

P R E D I C T
Make a table that summarizes the characteristics of the three major muscle types. The muscle types should form a column at the left side of the table and the characteristics of muscle should form a row at the top of the table.

Table 4.11 **Muscle Tissue**

(a) Skeletal Muscle

Structure: Skeletal muscle cells or fibers appear striated (banded). Cells are large, long, and cylindrical, with many nuclei located at the periphery.

Function: Movement of the body; under voluntary control.

Location: Attaches to bone or other connective tissue.

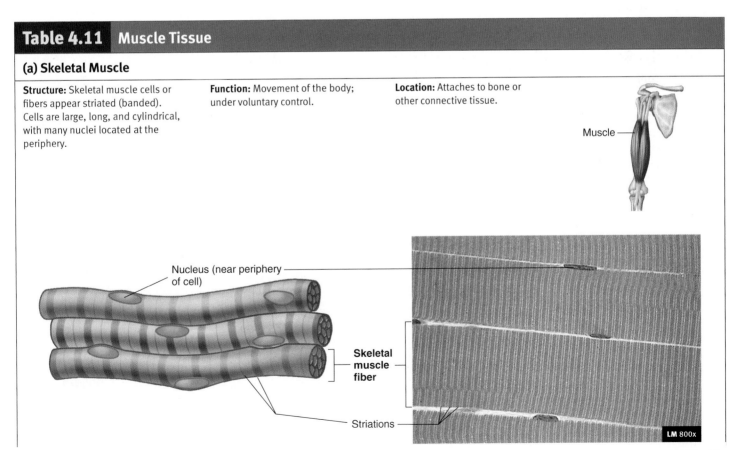

Muscle

Nucleus (near periphery of cell)

Skeletal muscle fiber

Striations

LM 800x

continued

Table 4.11   *continued*

(b) Cardiac Muscle

Structure: Cardiac muscle cells are cylindrical and striated and have a single, centrally located nucleus. They are branched and connected to one another by intercalated disks, which contain gap junctions.

Function: Pumps the blood; under involuntary control.

Location: Cardiac muscle is in the heart.

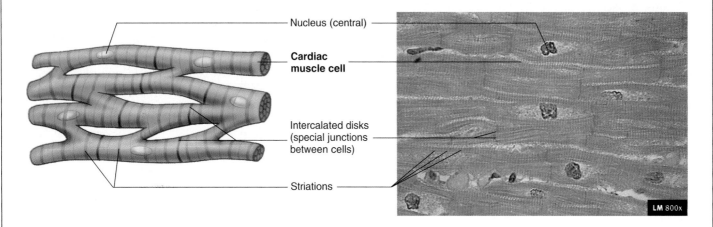

Nucleus (central)

Cardiac muscle cell

Intercalated disks (special junctions between cells)

Striations

LM 800x

(c) Smooth Muscle

Structure: Smooth muscle cells are tapered at each end, are not striated, and have a single nucleus.

Function: Regulates the size of organs, forces fluid through tubes, controls the amount of light entering the eye, and produces "goose flesh" in the skin; under involuntary control.

Location: Smooth muscle is in hollow organs such as the stomach and intestine.

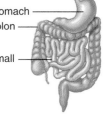

Wall of stomach

Wall of colon

Wall of small intestine

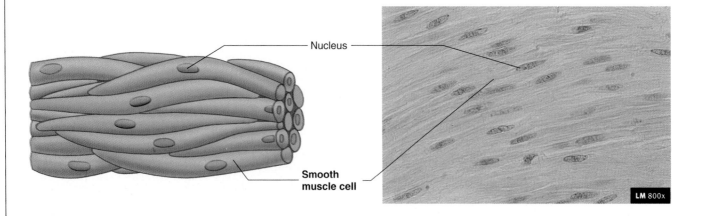

Nucleus

Smooth muscle cell

LM 800x

Nervous Tissue

Nervous tissue forms the brain, spinal cord, and nerves. It is responsible for coordinating and controlling many bodily activities. For example, the conscious control of skeletal muscles and the unconscious regulation of cardiac muscle are accomplished by nervous tissue. Awareness of ourselves and the external environment, emotions, reasoning skills, and memory are other functions performed by nervous tissue. Many of these functions depend on the ability of nervous tissue cells to communicate with one another and with the cells of other tissues by electrical signals called **action potentials.**

Nervous tissue consists of neurons and support cells. The **neuron** (noor'on), or **nerve cell,** is responsible for the conduction of action potentials. It is composed of three parts (table 4.12). The **cell body** contains the nucleus and is the site of general cell functions. **Dendrites** (den'drītz, relating to a tree) and axons (ak'sonz) are nerve cell processes (extensions). Dendrites usually receive stimuli that lead to electrical changes that either increase or decrease action potentials in the neuron's axon. There is usually one axon per neuron. Action potentials usually originate at the base of an axon where it joins the cell body and travel to the end of the axon. **Neuroglia** (noo-rog'lē-ă, *glia,* glue) are the support cells of the nervous system, and they function to nourish, protect, and insulate the neurons. Nervous tissue is considered in greater detail in chapter 8.

Membranes

A **membrane** is a thin sheet or layer of tissue that covers a structure or lines a cavity. Most membranes consist of epithelium and the connective tissue on which the epithelium rests. The two major categories of membranes are mucous membranes and serous membranes.

Mucous Membranes

Mucous (mū'kŭs) **membranes** consist of various kinds of epithelium resting on a thick layer of loose connective tissue. They line cavities that open to the outside of the body, such as the digestive, respiratory, excretory, and reproductive tracts (figure 4.4). Many, but not all, mucous membranes have mucous glands, which secrete mucus. The functions of mucous membranes vary, depending on their location, and include protection, absorption, and secretion. For example, the stratified squamous epithelium of the oral cavity (mouth) performs a protective function, whereas the simple columnar epithelium of the intestine absorbs nutrients and secretes digestive enzymes and mucus. Inflammation of the mucous membrane of the nasal passages caused by the common cold or allergies is called **rhinitis** (rī-nī'tis, *rhin-* refers to the nose).

Serous Membranes

Serous (sēr'ŭs, produces watery secretion) membranes consist of simple squamous epithelium resting on a delicate layer of loose

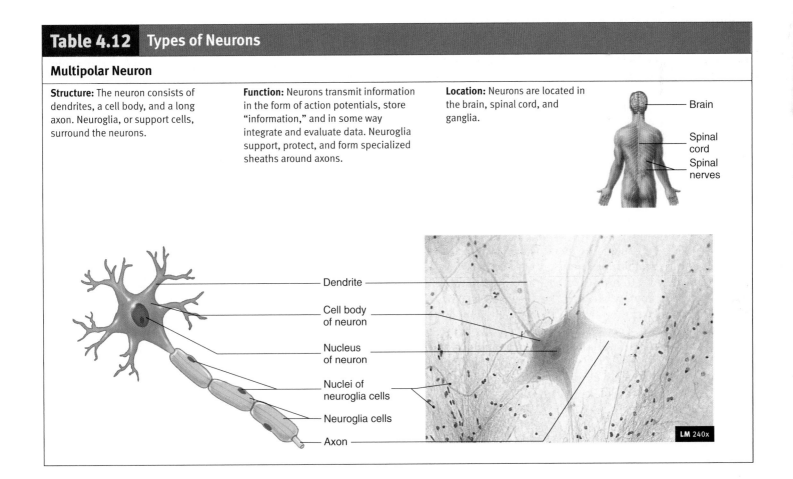

Table 4.12   Types of Neurons

Multipolar Neuron

Structure: The neuron consists of dendrites, a cell body, and a long axon. Neuroglia, or support cells, surround the neurons.

Function: Neurons transmit information in the form of action potentials, store "information," and in some way integrate and evaluate data. Neuroglia support, protect, and form specialized sheaths around axons.

Location: Neurons are located in the brain, spinal cord, and ganglia.

Brain

Spinal cord

Spinal nerves

Dendrite

Cell body of neuron

Nucleus of neuron

Nuclei of neuroglia cells

Neuroglia cells

Axon

LM 240x

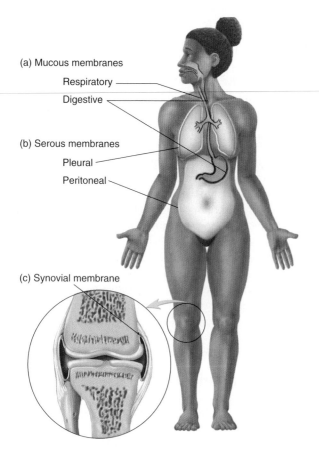

(a) Mucous membranes
Respiratory
Digestive

(b) Serous membranes
Pleural
Peritoneal

(c) Synovial membrane

Figure 4.4   **Membranes**

(*a*) Mucous membranes line cavities that open to the outside and often contain mucous glands, which secrete mucus. (*b*) Serous membranes line cavities that do not open to the exterior, and do not contain glands, but do secrete serous fluid. (*c*) Synovial membranes line cavities that surround synovial joints.

connective tissue. Serous membranes line the trunk cavities and cover the organs located within these cavities (see figure 4.4). The serous membranes secrete serous fluid, which covers the surface of the membranes. The smooth surface of the epithelial cells of the serous membranes, together with the lubricating qualities of the serous fluid, combine to prevent damage from abrasion when organs in the thoracic or abdominopelvic cavities rub against one another. The serous membranes are named according to their location: the **pleural** (ploor′ăl, a rib or the side) **membranes** are associated with the lungs, the **pericardial** (per-i-kar′dē-ăl, around the heart) **membranes** are associated with the heart, and the **peritoneal** (per′i-tō-nē′ăl, to stretch over) **membranes** are located in the abdominopelvic cavity (see figure 1.14). When the suffix "-itis" is added to the name of a structure, it means that the structure is inflamed. **Pericarditis** (per′i-kar-dī′tis) and **peritonitis** (per′i-tō-nī′tis) are inflammations of the pericardial membranes and peritoneal membranes, respectively. **Pleurisy** (ploor′i-sē) is inflammation of the pleural membranes (see section on Inflammation for a description).

Other Membranes

In addition to mucous and serous membranes, there are several other membranes in the body. The **skin,** or **cutaneous** (kū-tā′nē-ŭs, skin) **membrane,** is stratified squamous epithelium and dense connective tissue (see chapter 5). Other membranes are made up of only connective tissue. **Synovial** (si-nō′vē-ăl) **membranes** line the inside of joint cavities (the space where bones come together within a movable joint) (see figure 4.4), and the **periosteum** (per-ē-os′tē-ŭm, around bone) surrounds bone. These connective tissue membranes are discussed in chapter 6.

Inflammation

The **inflammatory response,** or **inflammation** (*flamma,* flame), occurs when tissues are damaged (figure 4.5). For example, viruses infect epithelial cells of the upper respiratory tract to produce inflammation and the symptoms of the common cold. Also, inflammation results from the immediate and painful events that follow trauma such as closing your finger in a car door or cutting yourself with a knife. Inflammation mobilizes the body's defenses and isolates and destroys microorganisms, foreign materials, and damaged cells so that tissue repair can proceed. Inflammation produces five major symptoms: redness, heat, swelling, pain, and disturbance of function. Although unpleasant, the processes producing the symptoms are usually beneficial.

Following an injury, chemical substances called **mediators of inflammation** are released or activated in the injured tissues and adjacent blood vessels. The mediators include **histamine** (his′tă-mēn), **kinins** (kī′ninz, to move), **prostaglandins** (pros-tă-glan′dinz), **leukotrienes** (lū-kō-trī′ēnz), and others. Some mediators cause dilation of blood vessels, which produces the symptoms of redness and heat, similar to what occurs when a person blushes. Dilation of blood vessels is beneficial because it increases the speed with which blood cells and other substances important for fighting infections and repairing the injury are brought to the injury site.

Mediators of inflammation also increase the permeability of blood vessels, allowing materials and blood cells to move out of the vessels and into the tissue, where they can deal directly with the injury. **Edema** (e-dē′mă), or swelling, of the tissues results when water, proteins, and other substances from the blood move into the tissues. One of the proteins, fibrin, forms a fibrous network that "walls off" the site of injury from the rest of the body. This mechanism can help prevent the spread of infectious agents. One type of blood cell that enters the tissues is the **neutrophil** (noo′trō-fil), a phagocytic white blood cell that fights infections by ingesting bacteria. Macrophages ingest tissue debris, clearing the area for tissue repair. Neutrophils die after ingesting a small number of bacteria; the mixture of dead neutrophils, other cells, and fluid that can accumulate is called **pus.**

Pain associated with inflammation is produced in several ways. Nerve cell endings are stimulated by direct damage and by some mediators of inflammation to produce pain sensations. In addition, the increased pressure in the tissue caused by edema and accumulation of pus can cause pain.

Pain, limitation of movement resulting from edema, and tissue destruction all contribute to the disturbance of function. This disturbance of function can be adaptive because it warns the person to protect the injured area from further damage.

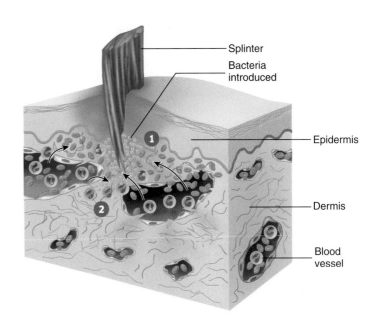

1. A splinter in the skin causes damage and introduces bacteria. Mediators of inflammation are released or activated in injured tissues and adjacent blood vessels. Some blood vessels are ruptured, causing bleeding.

2. Mediators of inflammation cause capillaries to dilate, causing the skin to become red. Mediators of inflammation also increase capillary permeability, and fluid leaves the capillaries, producing swelling *(arrows)*.

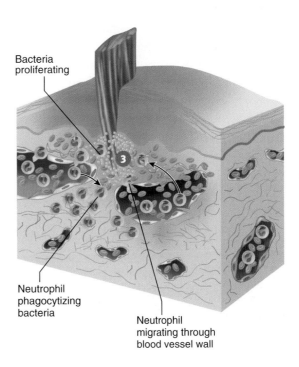

3. White blood cells (e.g., neutrophils and macrophages) leave the dilated blood vessels and move to the site of bacterial infection, where they begin to phagocytize bacteria and other debris.

Process Figure 4.5   Inflammation

Reducing Inflammation

Sometimes the inflammatory response lasts longer or is more intense than is desirable, and drugs are used to suppress the symptoms by inhibiting the synthesis, release, or actions of the mediators of inflammation. For example, the effects of histamine released in people with hay fever are suppressed by antihistamines. Aspirin and related drugs such as ibuprofen and naproxen are effective anti-inflammatory agents that relieve pain by preventing the synthesis of prostaglands and related substances.

Chronic Inflammation

Chronic, or prolonged, inflammation results when the agent responsible for an injury is not removed or something else interferes with the process of healing. Infections of the lungs or kidneys usually result in a brief period of inflammation followed by repair. However, prolonged infections, or prolonged exposure to irritants, can result in chronic inflammation. Chronic inflammation caused by irritants, such as silica in the lungs, or abnormal immune responses can result in the replacement of normal tissue with fibrous connective tissue.

Chronic inflammation of the stomach or small intestine may result in ulcers. The loss of normal tissue leads to the loss of normal organ functions. Consequently, chronic inflammation of organs such as the lungs, liver, or kidneys, can lead to death.

PREDICT

In some injuries, tissues are so severely damaged that areas exist where cells are killed and blood vessels are destroyed. For injuries such as these, where do the signs of inflammation, such as redness, heat, edema, and pain, occur?

Tissue Repair

Tissue repair (figure 4.6), the substitution of viable cells for dead cells, can occur by regeneration or replacement. In **regeneration,** the new cells are the same type as those that were destroyed, and normal function is usually restored. In **replacement,** a new type of tissue develops that eventually causes scar production and the loss of some tissue function. The tissues involved and the severity of the wound determine the type of tissue repair that dominates.

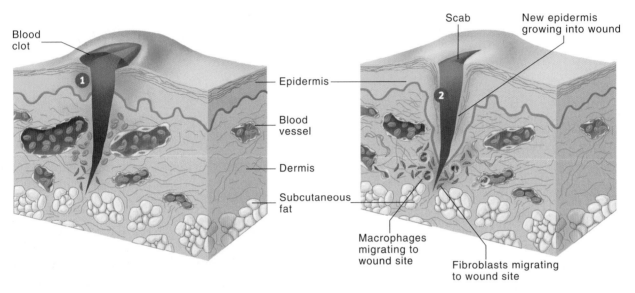

1. Fresh wound cuts through the epithelium (epidermis) and underlying connective tissue (dermis), and a clot forms.

2. Approximately 1 week after the injury, a scab is present, and epithelium (new epidermis) is growing into the wound.

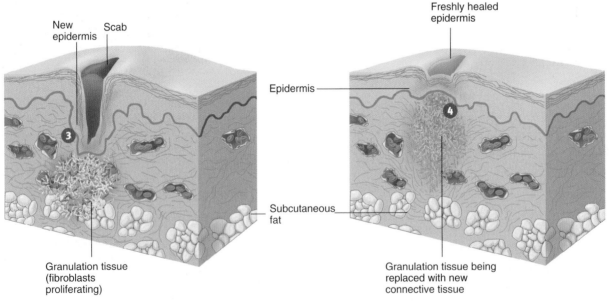

3. Approximately 2 weeks after the injury, the epithelium has grown completely into the wound, and fibroblasts have formed granulation tissue.

4. Approximately 1 month after the injury, the wound has completely closed, the scab has been sloughed, and the granulation tissue is being replaced by new connective tissue.

Process Figure 4.6 Tissue Repair

Cells can be classified into three groups on the basis of their ability to divide and produce new cells. **Labile** (not fixed) **cells** continue to divide throughout life. Damage to such labile cells as the cells of the skin and mucous membranes can be repaired completely by regeneration. **Stable cells** do not actively divide after growth ceases, but they do retain the ability to divide after an injury. For example, connective tissue and glands, including the liver and pancreas, are capable of regeneration. **Permanent cells** have little or no ability to divide. Neurons and skeletal muscle cells are examples of permanent cells. If they are killed, they are usually replaced by connective tissue. Permanent cells, however, can usually recover from a limited amount of damage. For example, if the axon of a neuron is damaged, the neuron can grow a new axon. If the cell body is sufficiently damaged, however, the neuron dies and is replaced by connective tissue.

Stem Cells

Stem cells are precursor cells that are not fully differentiated into mature cells. For example, a small population of stem cells in the brain that can divide and form new neurons has been discovered. Although mature neurons do not form additional neurons, it may be possible to develop treatments for some brain injuries that stimulate the stem cells. A class of chemicals called growth factors has been identified that stimulate stem cells to divide and make injured neurons recover more rapidly. The new neurons may be incorporated with other functional neurons of the central nervous system.

In addition to the type of cells involved, the severity of an injury can influence whether repair is by regeneration or replacement. Generally, the more severe the injury, the greater the likelihood that repair involves replacement.

Repair of the skin is an illustration of tissue repair (see figure 4.6). When the edges of a wound are close together, the wound fills with blood, and a clot forms (see chapter 11). The **clot** contains a threadlike protein, fibrin, which binds the edges of the wound together and stops the bleeding. The surface of the clot dries to form a **scab,** which seals the wound and helps to prevent infection.

An inflammatory response is activated to fight infectious agents in the wound and to help the repair process. Dilation of blood vessels brings blood cells and other substances to the injury area, and increased blood vessel permeability allows them to enter the tissue. The area is "walled off" by the fibrin, and neutrophils enter the tissue from the blood.

The epithelium at the edge of the wound undergoes regeneration and migrates under the scab while the inflammatory response proceeds. Eventually the epithelial cells from the edges meet, and the epithelium is restored. After the epithelium is repaired, the scab is sloughed off (shed).

A second type of phagocytic cell, called a **macrophage,** removes the dead neutrophils, cellular debris, and the decomposing clot. Fibroblasts from the surrounding connective tissue migrate into the area, producing collagen and other extracellular matrix components. Capillaries grow from blood vessels at the edge of the wound and revascularize the area. The result is the replacement of the clot by a delicate connective tissue called **granulation** (appears granular) **tissue,** which consists of

fibroblasts, collagen, and capillaries. Eventually normal connective tissue replaces the granulation tissue. Sometimes a large amount of granulation tissue persists as a scar, which at first is bright red because of the vascularization of the tissue. The scar turns from red to white as collagen accumulates and the blood vessels decrease in number.

When the wound edges are far apart, the clot may not completely close the gap, and it takes much longer for the epithelial cells to regenerate and cover the wound. With increased tissue damage, the degree of the inflammatory response is greater, there is more cell debris for the phagocytes to remove, and the risk of infection is greater. Much more granulation tissue forms, and **wound contracture,** a result of the contraction of fibroblasts in the granulation tissue, pulls the edges of the wound closer together. Although wound contracture reduces the size of the wound and speeds healing, it can lead to disfiguring and debilitating scars.

P R E D I C T   6

Explain why it is advisable to suture large wounds.

Tissues and Aging

The consequences of some age-related changes in tissues are obvious, whereas others are subtle. It is clear that the appearance of skin changes as people age. There is a clear improvement in athletic performance as children mature, and a well-documented decline after approximately 30–35 years. With advanced age there is ultimately a substantial decrease in the number of neurons and muscle cells. Reduced visual acuity, smell, taste, and touch occur, and there is a decline in the functional capacities of the respiratory and cardiovascular systems.

At the tissue level, age-related changes affect cells and the extracellular matrix produced by them. In general, cells divide more slowly in older than in younger people. The rate of red blood cell synthesis declines in the elderly. Injuries in the very young heal more rapidly and more completely than in older people, in part, because of the more rapid cell division. For example, a fracture in the femur of an infant is likely to heal quickly and eventually leave no evidence of the fracture in the bone. A similar fracture in an adult heals more slowly and a scar, seen in radiographs of the bone, is likely to persist throughout life.

The consequences of changes in the extracellular matrix are important. Collagen fibers become more irregular in structure, even though they may increase in number. As a consequence, connective tissues with abundant collagen, such as tendons and ligaments, become less flexible and more fragile. Elastic fibers fragment, bind to Ca^{2+}, and become less elastic. Consequently, elastic connective tissues, such as elastic ligaments, become less elastic. Reduced flexibility and elasticity of connective tissue is responsible for increased wrinkling of skin as well as the increased tendency for bones to break in older people.

The walls of arteries become less elastic because of changes in collagen and elastic fibers. Atherosclerosis results as plaques form in the walls of blood vessels, which contain collagen fibers, lipids, and calcium deposits. These changes result in reduced blood supply to tissues and increased susceptibility to blockage and rupture of arteries.

Systems Pathology
Cancer: Malignant Melanoma

Ms. B. was proud of the tan she worked so hard to perfect each summer. When she was in college she noticed one of the small dark moles on her shoulder seemed to be getting larger. It was tender, appeared to be ulcerated in the center, and had developed irregular edges. Ms. B's friend, who was taking anatomy and physiology, insisted that she have a physician examine the mole. The physician quickly arranged to take a **biopsy** (bī′op-sē, *bios,* life + *opsis,* vision) of the mole and sent it to a pathology laboratory for analysis. The results confirmed that the mole was a **malignant melanoma** (mă-lig′nănt mel′ă-nō′mă). Ms. B's physician explained how tumor cells differ from normal cells, explained the major difference between benign and malignant tumors, and described the effect of malignant tumor cells on normal tissues. He then explained that the tumor must be surgically removed as soon as possible.

A **tumor** (too′mŏr) is any swelling, but modern usage has limited the term to swellings that involve neoplastic tissue. A **neoplasm** (nē′ō-plazm, "new growth") refers to abnormal tissue growth resulting from cellular divisions that continue after normal cell division of the tissue has stopped or slowed considerably. A neoplasm can be **benign** (bē-nīn′) or **malignant** (mă-lig′nănt). Benign tumors do not invade and destroy surrounding healthy tissue. They may have an irregular shape, but are generally surrounded by a connective tissue capsule. Although benign tumors are less dangerous than malignant tumors, they can cause major problems when they enlarge and compress surrounding tissues. For example, benign brain tumors can compress nervous tissue, resulting in loss of brain function and even death.

The term **cancer** (kan′ser) refers to a malignant, spreading tumor and the illness that results from such a tumor. Malignant tumors normally do not have capsules and they can spread by local growth and tissue destruction into healthy surrounding tissues. The margins of malignant tumors are very irregular and inflammation is normally evident between the

tumor and the normal surrounding tissues. Malignant tumors also spread to distant sites by **metastasis** (mě-tas′tă-sis), which occurs when tumor cells separate from the main mass and are carried by the lymphatic or circulatory system to a new site, where a second neoplasm is formed. The illness associated with cancer usually occurs as tumors invade and destroy healthy surrounding tissue, eliminating its function.

Malignant neoplasms lack the normal growth control that most other adult tissues have. This lack of normal control involves the genetic machinery and can be induced by viruses, environmental toxins, and other causes. Cancer therapy concentrates on trying to confine and then kill the malignant cells by killing the tissue with radiation or lasers, by removing the tumor surgically, by treating the patient with drugs that selectively kill cells undergoing division, or by stimulating the patient's immune system to destroy the tumor. **Oncology** (ong-kol′ō-jē, tumor study) is the study of cancer and its associated problems.

Malignant melanomas are malignant tumors of the skin. The malignant cells originate from pigment-producing cells of the skin called **melanocytes** (mel′ă-nō-sītz). Some factors that may contribute to the formation of malignant melanoma are genetic predisposition, solar radiation, and steroid hormone activity. A malignant melanoma normally develops in a **nevus** (nē′vŭs, pl. **nevi,** nē′vī), or mole, which is a benign neoplasm consisting of an aggregation of melanocytes. Characteristics consistent with the formation of malignant melanomas in nevi include change in color, change in size, an irregular notched margin, itching, bleeding or oozing, nodular features, scab formation, and ulceration.

Treatment of malignant melanomas with no evidence of metastasis involves surgical excision of both the primary lesion site and regional lymph nodes. Malignant melanomas are dangerous because of their ability to metastasize. They invade deeper layers of the dermis and the subcutaneous tissues. Once malignant cells enter the

---

S U M M A R Y

1. A tissue is a group of cells with similar structure and function, as well as the extracellular substances located between the cells.
2. Histology is the study of tissues.

Epithelial Tissue   (p. 73)

Epithelial tissue covers surfaces, usually has a basement membrane, has little extracellular material, and has no blood vessels.

Functions of Epithelia

General functions of epithelia include protection, permitting the passage of substances, secreting substances, and absorption of substances.

Classification of Epithelia

1. Epithelia are classified according to the number of cell layers and the shape of the cells.
2. Simple epithelium has one layer of cells, whereas stratified epithelium has more than one.
3. Pseudostratified columnar epithelium is simple epithelium that appears to have two or more cell layers.
4. Transitional epithelium is stratified epithelium that can be greatly stretched.

System	Interactions
Muscle	Metastasis to skeletal muscle is not the most important feature. Severe atrophy of skeletal muscle and other tissues, however, can occur in advanced stages of the disease.
Circulatory	Provides a route for metastasis because cells can enter the capillaries or venules and pass to distant sites, such as the lungs and brain.
Lymphatic	Provides a route for metastasis because cells can enter lymphatic vessels and spread to lymph nodes where they initiate tumors. The lymph nodes closest to the original tumor are affected first and those more distant are affected later.
Nervous	Metastasis can result in tumors that affect the function of the nervous system. Pain, paralysis, and loss of sensations result as tumors compress and destroy nervous tissue in the brain or spinal cord. Death results when a tumor destroys or compresses an essential part of the central nervous system.
Respiratory	The lungs are common sites for metastasis. Malignant tumors destroy lung tissue and block air passageways, resulting in reduced gas exchange.
Digestive	The digestive system is a potential site for metastasis. Damage caused by malignant tumors in the liver, for example, can be life-threatening.
Urinary	The urinary system is a potential site for metastasis. Damage caused by metastatic tumors in the kidneys can be life-threatening.
Endocrine	Endocrine organs are potential sites for metastasis and may result in the destruction of these tissues.
Reproductive	The reproductive system is a potential site for metastasis and may result in the destruction of these tissues.
Skeletal	Metastasis can result in tumors that can invade and destroy the bone marrow that produces cells of the circulatory system, such as red blood cells (erythrocytes), white blood cells (leukocytes), and platelets.

lymphatic and circulatory systems, they can rapidly spread to distant sites. Only 20–40% of patients with metastatic malignant melanoma are alive and cured 5 years after diagnosis. Different forms of malignant melanoma are classified on the basis of their structure, which, in part, determines their tendency to invade the deeper layers of the skin and metastasize to other parts of the body. Malignant melanomas that develop on the extremities have the best, head and neck lesions have the next best, and trunk lesions have the poorest rate of cure.

Most melanomas occur in the 40- to 70-year-old age group, but the frequency of malignant melanoma is increasing among those between the ages of 20 and 40 years because of increased exposure of the skin to ultraviolet light.

P R E D I C T
A person who had some time ago been diagnosed as having malignant melanoma began to experience severe headaches, which increased in intensity as time passed. She also began to develop shortness of breath and a constant urge to cough. Explain how malignant melanoma could be responsible for these manifestations.

Structural and Functional Relationships

1. Simple epithelium is involved with diffusion, secretion, or absorption. Stratified epithelium serves a protective role. Squamous cells function in diffusion or filtration. Cuboidal or columnar cells, which contain more organelles, secrete or absorb.
2. A smooth, free surface reduces friction. Microvilli increase surface area, and cilia move materials over the cell surface.
3. Tight junctions bind adjacent cells together and form a permeability barrier.
4. Desmosomes mechanically bind cells together and hemidesmosomes mechanically bind cells to the basement membrane.
5. Gap junctions allow intercellular communication.

Glands

1. A gland is a single cell or a multicellular structure that secretes.
2. Exocrine glands have ducts, and endocrine glands do not.

Connective Tissue (p. 81)

1. Connective tissue holds cells and tissues together.
2. Connective tissue has an extracellular matrix consisting of protein fibers, ground substance, and fluid.
3. Collagen fibers are flexible but resist stretching, reticular fibers form a fiber network, and elastic fibers recoil.
4. Blast cells form the matrix, cyte cells maintain it, and clast cells break it down.

Functions of Connective Tissue

Connective tissues enclose and separate; connect tissues to one another; play a role in support and movement; store, cushion, insulate, transport, and protect.

Classification of Connective Tissue

1. Embryonic connective tissue gives rise to six major categories of connective tissue.
2. Loose, or areolar, connective tissue is the "loose packing" material of the body, which fills the spaces between organs and holds them in place.
3. Adipose, or fat, tissue functions to store energy. Adipose tissue also pads and protects parts of the body and acts as a thermal insulator.
4. Dense connective tissue consists of matrix containing densely packed collagen fibers (tendons, ligaments, and dermis of the skin) or of matrix containing densely packed elastic fibers (elastic ligaments and in the walls of arteries).
5. Cartilage provides support and is found in structures such as the costal cartilages, disks between vertebrae, and the external ear.
6. Bone has a mineralized matrix and forms most of the skeleton of the body.
7. Blood has a liquid matrix and is found in blood vessels.

Muscle Tissue   (p. 87)

1. Muscle tissue is specialized to shorten, or contract.
2. The three types of muscle tissue are skeletal, cardiac, and smooth muscle.

Nervous Tissue   (p. 89)

1. Nervous tissue is specialized to conduct action potentials (electrical signals).
2. Neurons conduct action potentials, and neuroglia support the neurons.

Membranes   (p. 89)

Mucous Membranes

Mucous membranes line cavities that open to the outside of the body (digestive, respiratory, excretory, and reproductive tracts). They contain glands and secrete mucus.

Serous Membranes

Serous membranes line trunk cavities that do not open to the outside of the body (pleural, pericardial, and peritoneal cavities). They do not contain glands but do secrete serous fluid.

Other Membranes

Other membranes include the cutaneous membrane (skin), synovial membranes (line joint cavities), and periosteum (around bone).

Inflammation   (p. 90)

1. The function of the inflammatory response is to isolate and destroy harmful agents.
2. The inflammatory response produces five symptoms: redness, heat, swelling, pain, and disturbance of function.

Chronic Inflammation

Chronic inflammation results when the agent causing injury is not removed or something else interferes with the healing process.

Tissue Repair   (p. 92)

1. Tissue repair is the substitution of viable cells for dead cells. Labile cells divide throughout life and can undergo regeneration. Stable cells do not ordinarily divide but can regenerate if necessary. Permanent cells have little or no ability to divide. If killed, repair is by replacement.
2. Tissue repair involves clot formation, inflammation, formation of granulation tissue, and the regeneration or replacement of tissues. In severe wounds, wound contracture can occur.

Tissues and Aging   (p. 93)

1. Cells divide more slowly as people age. Injuries heal more slowly.
2. Extracellular matrix containing collagen and elastic fibers become less flexible and less elastic. Consequently, skin wrinkles, elasticity in arteries is reduced, and bones break more easily.

R E V I E W   A N D   C O M P R E H E N S I O N

1. Define tissue and histology.
2. In what areas of the body is epithelium located? What are four characteristics of epithelial tissue?
3. Explain how epithelial tissue is classified according to the number of cell layers and cell shape. What is pseudostratified columnar and transitional epithelium?
4. What kinds of functions does a single layer of epithelium perform? A stratified layer? Give an example of each.
5. Contrast the functions performed by squamous cells with those of cuboidal or columnar cells. Give an example of each.
6. What is the function of an epithelial free surface that is smooth, one that has microvilli, and one that has cilia?
7. Name the ways in which epithelial cells are connected to one another, and give the function for each way.
8. Define gland. Distinguish between an exocrine and an endocrine gland.
9. Explain the difference between connective tissue cells that are termed blast, cyte, and clast cells.
10. What are the functions of connective tissues?
11. What are the six major connective tissue types? How are they used to classify connective tissue?
12. Describe loose or areolar connective tissue and give an example.
13. How is adipose tissue different from other connective tissues? List the functions of adipose tissue.
14. Describe dense collagenous connective tissue and dense elastic connective tissue, and give two examples.
15. Describe the components of cartilage. Give an example of hyaline cartilage, fibrocartilage, and elastic cartilage.
16. Describe the components of bone.
17. Describe the cell types and the matrix of blood and list its functions.
18. Functionally, what is unique about muscle? Which of the muscle types is under voluntary control? What tasks does each type perform?
19. Functionally, what is unique about nervous tissue? What do neurons and neuroglia accomplish? What is the difference between an axon and a dendrite?

20. Compare mucous and serous membranes according to the type of cavity they line and their secretions. Name the serous membranes associated with the lungs, heart, and abdominopelvic organs.

21. What is the function of the inflammatory response? Name the five symptoms of inflammation, and explain how each is produced.

22. Define tissue repair. What is the difference between repair by regeneration and by replacement?

23. Differentiate between labile cells, stable cells, and permanent cells. Give an example of each type. What is the significance of these cell types to tissue repair?

24. Describe the process of tissue repair when the edges of a wound are close together versus when they are far apart.

25. Describe the effect of aging on cell division and the formation of extracellular matrix.

C R I T I C A L   T H I N K I N G

1. What types of epithelium are likely to be found lining the trachea of a heavy smoker; predict the changes that are likely to occur after he or she stops smoking for 1 or 2 years.

2. The blood–brain barrier is a specialized epithelium in capillaries that prevents many materials from passing from the blood into the brain. What kind of cell connections would be expected in the blood–brain barrier?

3. One of the functions of the pancreas is to secrete digestive enzymes that are carried by ducts to the small intestine. How many cell layers and what cell shape, cell surface, and type of cell-to-cell connections would be expected in the epithelium that is responsible for producing the digestive enzymes?

4. Explain the consequences:
   a. if simple columnar epithelium replaced nonkeratinized stratified squamous epithelium that lines the mouth.
   b. if tendons were dense elastic connective tissue instead of dense collagenous connective tissue.
   c. if bones were made entirely of elastic cartilage.

5. Some dense connective tissue has elastic fibers in addition to collagen fibers. This enables a structure to stretch and then recoil to its original shape. Examples are certain ligaments that hold together the vertebrae (bones of the back). When the back is bent (flexed), the ligaments are stretched. How does the elastic nature of these ligaments help the back to function? How are the fibers in the ligaments organized?

6. The aorta is a large blood vessel that is attached to the heart. When the heart beats, blood is ejected into the aorta, which expands to accept the blood. The wall of the aorta is constructed with dense connective tissue that has elastic fibers. How are the fibers arranged?

7. Antihistamines block the effect of a chemical mediator, histamine, that is released during the inflammatory response. Give an example of when it could be harmful to use an antihistamine and an example of when it could be beneficial.

8. Granulation tissue and scars consist of dense irregular collagenous connective tissue. Vitamin C is required for collagen synthesis. Predict the effect of scurvy, which is a nutritional disease caused by vitamin C deficiency, on wound healing.

Answers in Appendix D

Visit this textbook's website at www.mhhe.com/seeleyess6 for practice quizzes, animations, interactive learning exercises, and other study tools.

McGraw-Hill offers a study CD that features interactive cadaver dissection. *Anatomy & Physiology Revealed* includes cadaver photos that allow you to peel away layers of the human body to reveal structures beneath the surface. This program also includes animations, radiologic imaging, audio pronunciations, and practice quizzing.

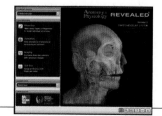

5

Integumentary System

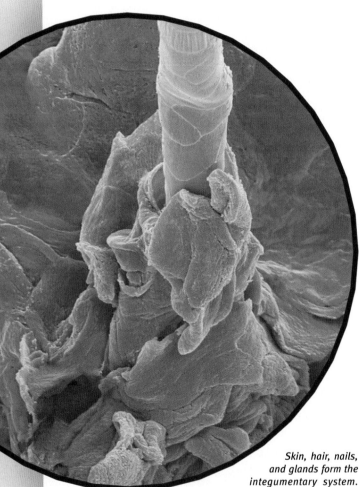

Skin, hair, nails, and glands form the integumentary system. This colorized scanning electron micrograph shows the shaft of a hair (yellow) protruding through the surface of the skin. Tightly bound epithelial cells form the hair shaft, and flat, scalelike epithelial cells form the skin's surface. Like the shingles on the roof of a house, the skin's tough cells protect underlying structures. Dandruff is excessive flaking of these epithelial cells from the scalp.

Chapter Outline and Objectives

the **integumentary** (in-teg-ū-men′tă-rē) **system** consists of the skin and accessory structures, such as hair, nails, and glands. Integument means covering, and the integumentary system is familiar to most people because it covers the outside of the body and is easily observed. In addition, humans are concerned with the appearance of the integumentary system. Skin without blemishes is considered attractive, whereas acne (ak′nē) is a source of embarrassment for many teenagers. The development of wrinkles and the graying or loss of hair is a sign of aging that some people find unattractive. Because of these feelings, much time, effort, and money is spent on changing the appearance of the integumentary system. For example, people apply lotion to their skin, color their hair, and trim their nails. They also try to prevent sweating with antiperspirants and reduce or mask body odor with washing, deodorants, and perfumes.

The appearance of the integumentary system can indicate physiological imbalances in the body. Some disorders affect just the integumentary system, such as acne or warts.

Disorders of other parts of the body can be reflected in the integumentary system and thus are useful for diagnosis. For example, reduced blood flow through the skin during a heart attack can cause a pale appearance, whereas increased blood flow as a result of fever can cause a flushed appearance. Also, the rashes of some diseases are characteristic, such as the rashes of measles, chickenpox, and allergic reactions. In addition, the integumentary system and the other systems often interact in complex ways in both health and disease (see Systems Pathology: Burns on p. 110).

Functions of the Integumentary System

Although we are often concerned with how the integumentary system looks, it has many important functions that go beyond appearance. Major functions of the integumentary system include:

1. *Protection.* The skin provides protection against abrasion and ultraviolet light. It also prevents the entry of microorganisms and dehydration by reducing water loss from the body.
2. *Sensation.* The integumentary system has sensory receptors that can detect heat, cold, touch, pressure, and pain.
3. *Vitamin D production.* When exposed to ultraviolet light, the skin produces a molecule that can be transformed into vitamin D.
4. *Temperature regulation.* Body temperature is regulated by controlling blood flow through the skin and the activity of sweat glands.
5. *Excretion.* Small amounts of waste products are lost through the skin and in gland secretions.

Hypodermis

Just as a house rests on a foundation, the skin rests on the **hypodermis** (hī-pō-der′mis, under the dermis), which attaches it to underlying bone and muscle and supplies it with blood vessels and nerves (figure 5.1). The hypodermis, which is not part of the skin, is sometimes called **subcutaneous** (sŭb-koo-tā′nē-ŭs, under the skin) **tissue.** The hypodermis is loose connective tissue that contains about half the body's stored fat, although the amount and location vary with age, sex, and diet. Fat in the hypodermis functions as padding and insulation, and it is responsible for some of the differences in appearance between men and women and between individuals of the same sex.

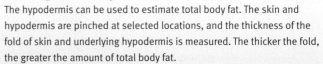

Estimating Total Body Fat

The hypodermis can be used to estimate total body fat. The skin and hypodermis are pinched at selected locations, and the thickness of the fold of skin and underlying hypodermis is measured. The thicker the fold, the greater the amount of total body fat.

Skin

The skin is made up of two major tissue layers. The **dermis** (der′mis, skin) is a layer of dense connective tissue, and the **epidermis** (ep-i-der′mis, upon the dermis) is a layer of epithelial tissue that rests on the dermis (see figure 5.1). The thickness of the dermis and epidermis varies depending on location, but on the average, the dermis is 10 to 20 times thicker than the epidermis and is responsible for most of the structural strength of skin. The epidermis prevents water loss and resists abrasion. If the hypodermis is the foundation on which the house rests, the dermis forms most of the house, and the epidermis is its roof.

Dermis

The dense collagenous connective tissue that makes up the dermis contains fibroblasts, fat cells, and macrophages. Compared with the hypodermis, the dermis has fewer fat cells and blood vessels. Nerves, hair follicles, smooth muscles, glands, and lymphatic vessels extend into the dermis (see figure 5.1).

Collagen and elastic fibers are responsible for the structural strength of the dermis. In fact, the dermis is that part of an animal hide from which leather is made. The epidermis is removed, and the dermis is preserved by tanning. The collagen fibers of the dermis are oriented in many different directions and can resist stretch. More collagen fibers are oriented in some directions than in others, however. This produces **cleavage,** or **tension, lines** in the skin, and the skin is most resistant to

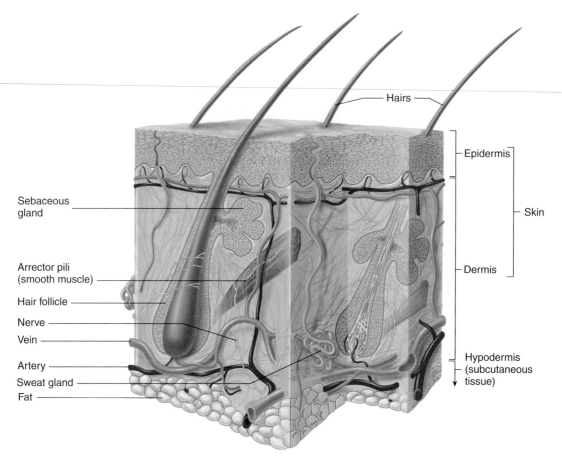

Figure 5.1 **Skin and Hypodermis**

The skin, consisting of the epidermis and the dermis, is connected by the hypodermis to underlying structures. Note the accessory structures (hairs, glands, and arrector pili), some of which project into the hypodermis, and the large amount of fat in the hypodermis.

stretch along these lines (figure 5.2). It is important for surgeons to be aware of cleavage lines. An incision made across the cleavage lines is likely to gap and produce considerable scar tissue, but an incision made parallel with the lines tends to gap less and produce less scar tissue (see chapter 4). If the skin is overstretched for any reason, the dermis can be damaged, leaving lines that are visible through the epidermis. These lines, called **striae** (strī′ē, furrow), or **stretch marks,** can develop on the abdomen and breasts of a woman during pregnancy.

The upper part of the dermis has projections called **dermal papillae** (pă-pil′ē, nipple), which extend toward the epidermis (figure 5.3a). The dermal papillae contain many blood vessels that supply the overlying epidermis with nutrients, remove waste products, and aid in regulating body temperature. The dermal papillae in the palms of the hands, the soles of the feet, and the tips of the digits are in parallel, curving ridges that shape the overlying epidermis into fingerprints and footprints. The ridges increase friction and improve the grip of the hands and feet.

A CASE IN POINT | Injections

Howey Stickum, a student nurse, learns three ways to give injections. An **intradermal injection,** such as the tuberculin skin test, is an injection into the dermis that is administered by drawing the skin taut and inserting a small needle at a shallow angle into the skin. A **subcutaneous injection** into the fatty tissue of the hypodermis, such as an insulin injection, is achieved by pinching the skin to form a "tent" into which a short needle is inserted. An **intramuscular injection** into a muscle deep to the hypodermis is accomplished by inserting a long needle at a 90-degree angle to the skin. Intramuscular injections are used for most vaccines and certain antibiotics.

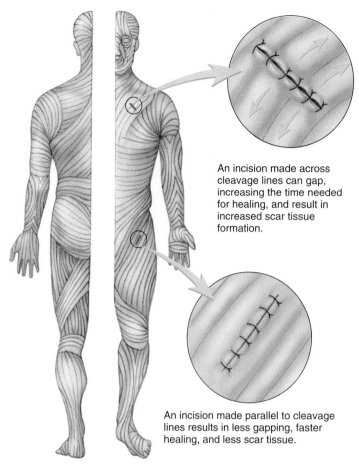

An incision made across cleavage lines can gap, increasing the time needed for healing, and result in increased scar tissue formation.

An incision made parallel to cleavage lines results in less gapping, faster healing, and less scar tissue.

Figure 5.2   Cleavage Lines

The orientation of collagen fibers produces cleavage, or tension, lines in the skin.

Epidermis

The epidermis is stratified squamous epithelium; in its deepest layers, cells are produced by mitosis. As new cells are formed, they push older cells to the surface, where they slough, or flake off. The outermost cells protect the cells underneath, and the deeper replicating cells replace cells lost from the surface. During their movement the cells change shape and chemical composition. This process is called **keratinization** (ker′ă-tin-i-zā′shŭn) because the cells become filled with the protein **keratin** (ker′ă-tin), which makes the cells hard. As keratinization proceeds, epithelial cells eventually die and produce an outer layer of dead, hard cells that resists abrasion and forms a permeability barrier.

Although keratinization is a continuous process, distinct layers called **strata** (stra′tă, layer) are recognized (figure 5.3b). The deepest stratum, the **stratum basale** (bā′să-lē, a base), consists of cuboidal or columnar cells that undergo mitotic divisions about every 19 days. One daughter cell becomes a new stratum basale cell and can divide again. The other cell is pushed toward the surface, a journey that takes about 40–56 days. As cells move to the surface, changes in the cells produce intermediate strata.

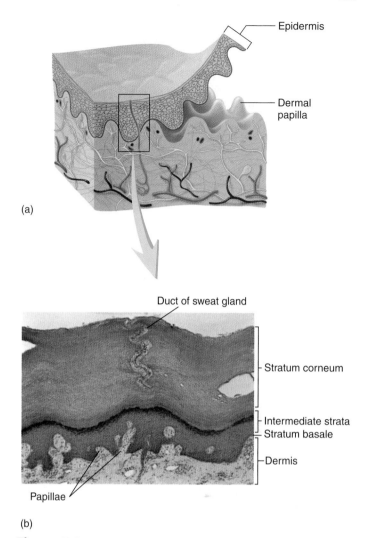

Figure 5.3   Epidermis and Dermis

(*a*) The epidermis rests on the dermis. Dermal papillae project toward the epidermis. (*b*) Photomicrograph of the epidermis resting on the dermis. Note the strata of the epidermis and the papillae of the dermis.

The **stratum corneum** (kōr′nē-ŭm, horny) is the most superficial stratum of the epidermis. It consists of dead, squamous cells filled with the hard protein keratin. Keratin gives the stratum corneum its structural strength. The stratum corneum cells are also coated and surrounded by lipids, which help prevent fluid loss through the skin.

P R E D I C T   ❶
What kinds of substances could easily pass through the skin by diffusion? What kinds would have difficulty?

The stratum corneum is composed of 25 or more layers of dead squamous cells joined by desmosomes (see chapter 4). Eventually the desmosomes break apart, and the cells are sloughed from the skin. Excessive stratum corneum cells sloughed from the surface of the scalp is called dandruff. In skin subjected to friction, the number of layers in the stratum corneum greatly increases, producing a thickened area called a **callus** (kal′ŭs, hard skin). Over a bony prominence,

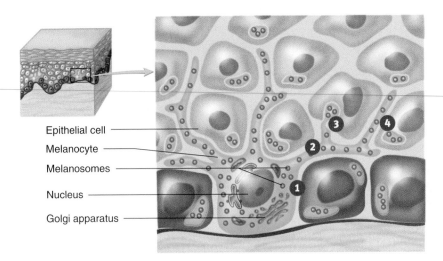

1. Melanosomes are produced by the Golgi apparatus of the melanocyte.

2. Melanosomes move into melanocyte cell processes.

3. Epithelial cells phagocytize the tips of the melanocyte cell processes.

4. The melanosomes, which were produced inside the melanocytes, have been trasferred to epithelial cells and are now inside them.

Epithelial cell

Melanocyte

Melanosomes

Nucleus

Golgi apparatus

Process Figure 5.4   Melanin Transfer from Melanocyte to Epithelial Cells

Melanocytes make melanin, which is packaged into melanosomes and transferred to many epithelial cells.

the stratum corneum can thicken to form a cone-shaped structure called a **corn.**

Skin Color

Pigments in the skin, blood circulating through the skin, and the thickness of the stratum corneum determine skin color. **Melanin** (mel′ă-nin, black) is the term used to describe a group of pigments responsible for skin, hair, and eye color. Most melanin molecules are brown to black pigments, but some are yellowish or reddish. Melanin provides protection against ultraviolet light from the sun.

Melanin is produced by **melanocytes** (mel′ă-nō-sītz, *melano,* black + *kytos,* cell), which are irregularly shaped cells with many long processes that extend between the epithelial cells of the deep part of the epidermis (figure 5.4). The Golgi apparatuses of the melanocytes package melanin into vesicles called **melanosomes** (mel′ă-nō-sōmz, *melano,* black + *soma,* body), which move into the cell processes of the melanocytes. Epithelial cells phagocytize the tips of the melanocyte cell processes, thereby acquiring melanosomes. Although all the epithelial cells of the epidermis can contain melanin, only the melanocytes produce it.

Large amounts of melanin occur in some regions of the skin, such as freckles; moles; the genitalia; the nipples; and the pigmented, circular areas around the nipples. Other areas, such as the lips, palms of the hands, and soles of the feet, have less melanin. Racial variations in skin color are determined by the amount, kind, and distribution of melanin. All races have about the same number of melanocytes.

Melanin production is determined by genetic factors, exposure to light, and hormones. Genetic factors are responsible for the amounts of melanin produced in different races. Although many genes are responsible for skin color, a single mutation can prevent the manufacture of melanin. For example, **albinism** (al′bi-nizm) is a recessive genetic trait that causes a deficiency or absence of melanin. Albinos have fair skin, white hair, and unpigmented irises in the eyes.

Exposure to ultraviolet light, for example in sunlight, stimulates melanocytes to increase melanin production. The result is a suntan.

Certain hormones, such as estrogen and melanocyte-stimulating hormone, cause an increase in melanin production during pregnancy in the mother, darkening the nipples, the pigmented circular areas around the nipples, and the genitalia. The cheekbones, forehead, and chest can also darken, resulting in "the mask of pregnancy," and a dark line of pigmentation can appear on the midline of the abdomen.

Blood flowing through the skin imparts a reddish hue, and when blood flow increases the red color intensifies. Examples include blushing, anger, and the inflammatory response. A decrease in blood flow such as occurs in shock can make the skin appear pale. A decrease in the blood oxygen content produces a bluish color called **cyanosis** (sī-ă-nō′sis, dark blue color). Birthmarks are congenital (present at birth) disorders of the blood vessels (capillaries) in the dermis.

Carotene (kar′ō-tēn) is a yellow pigment found in plants such as squash and carrots. Humans normally ingest carotene and use it as a source of vitamin A. Carotene is lipid-soluble, and, when consumed, it accumulates in the lipids of the stratum corneum and in the fat cells of the dermis and hypodermis. This gives the skin a slight yellowish tint. If large amounts of carotene are consumed, the skin can become quite yellowish.

The location of pigments and other substances in the skin affects the color produced. If a dark pigment is located in the dermis or hypodermis, light reflected off the dark pigment can be scattered by collagen fibers of the dermis to produce a blue color. The same effect produces the blue color of the sky as light is reflected from dust particles in the air. The deeper within the dermis or hypodermis any dark pigment is located, the bluer the pigment appears because of the light-scattering effect of the overlying tissue. This effect causes the blue color of tattoos, bruises, and some superficial blood vessels.

Adaptive Advantages of Skin Color

The evolution of skin color in humans is intriguing because it helps to explain certain modern-day health problems. During human evolution, the skeletal system of our ancestors changed, resulting in an upright posture and the ability to walk and run greater distances. Excess heat is produced as a result of increased physical activity, however, which can cause overheating. An increase in the number of sweat glands in the skin and a reduction in the amount of hair (fur) covering the skin helps to eliminate the excess heat.

Prolonged exposure of the skin to sunlight, however, can be harmful in two ways. It promotes the development of skin cancer (see Skin Cancer on p. 109) by damaging DNA and decreases the levels of the B vitamin folate in the blood by breaking it down through a photochemical reaction. Low folate levels are known to increase the risk of abnormal development of the fetal nervous system (see Neural Tube Defects on p. 581). On the other hand, exposure to ultraviolet light from the sun stimulates the production of vitamin D (see Vitamin D Production on p. 105). Vitamin D promotes the uptake of calcium from the small intestine, which is important for the normal development of the skeletal system in the fetus and in children. Inadequate quantities of vitamin D can result in rickets, in which the bones are soft, weak, and easily broken. Thus, increased skin pigmentation protects against skin cancer and abnormal development of the nervous system but impairs skeletal system development.

The best amount of melanin in the skin should be large enough to protect against the harmful effects of ultraviolet light, but small enough to allow ultraviolet light to stimulate vitamin D production. Ultraviolet light intensity is high in the tropics, but diminishes toward the poles. The skin color of populations is a genetic adaptation to their different exposure to ultraviolet light. Dark-skinned people in the tropics have more melanin, which provides protection against ultraviolet light, but can still produce vitamin D year-round. Light-skinned people at higher latitudes have less melanin, which increases the ability to produce vitamin D while providing ultraviolet light protection.

When people migrate from the regions where their ancestors evolved, cultural adaptations have helped them adjust to their changed ultraviolet light environment. For example, clothing and portable shade, such as tents or umbrellas, provides protection against ultraviolet light; and eating vitamin D-rich foods, such as fish, provides vitamin D. Still, light-skinned people who move to southern climates are more likely to develop skin cancers, and dark-skinned people who move to northern climates are more likely to develop rickets.

P R E D I C T 2
Explain the differences in skin color between: (a) the palms of the hands and the lips, (b) the palms of the hands of a person who does heavy manual labor and one who does not, and (c) the anterior and posterior surfaces of the forearm.

Accessory Skin Structures

Accessory skin structures are hair; smooth muscles called the arrector pili, glands, and nails.

Hair

The presence of **hair** is one of the characteristics common to all mammals. If the hair is thick and covers most of the body

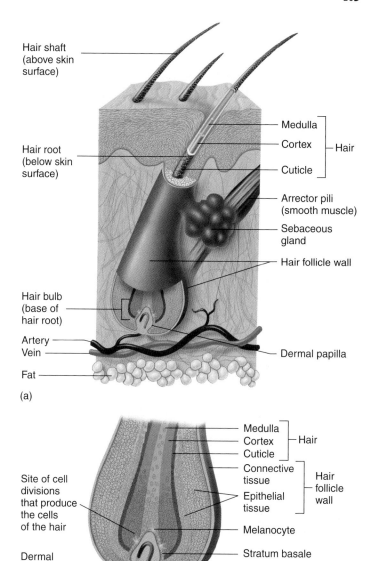

Figure 5.5 Hair Follicle

(*a*) Hair within a hair follicle. (*b*) Enlargement of the hair bulb and hair follicle wall.

surface, it is called fur. In humans, some hair is found everywhere in the skin except the palms, soles, lips, nipples, parts of the genitalia, and the distal segments of the fingers and toes.

The **shaft** of the hair protrudes above the surface of the skin, whereas the **root** and **hair bulb** are below the surface (figure 5.5*a*). A hair has a hard **cortex** (bark), which surrounds a softer center, the **medulla** (me-dool′ă). The cortex is covered by the **cuticle** (kū′ti-kl, skin), a single layer of overlapping cells that holds the hair in the hair follicle. The **hair follicle** is an extension of the epidermis deep into the dermis, and it can play an important role in tissue repair. If the surface epidermis is damaged, the epithelial cells within the hair follicle can divide and serve as a source of new epithelial cells.

Hair is produced in the hair bulb, which rests on a dermal papilla (figure 5.5b). Blood vessels within the papilla supply the hair bulb with the nourishment needed to produce the hair. Hair is produced in cycles. During the growth stage, it is formed by epithelial cells within the hair bulb. These cells, like the cells of the stratum basale in the skin, divide and undergo keratinization. The hair grows longer as these cells are added to the base of the hair within the hair bulb. Thus, the hair root and shaft consist of columns of dead keratinized epithelial cells. During the resting stage, growth stops, and the hair is held in the hair follicle. When the next growth stage begins, a new hair is formed, and the old hair falls out. The duration of each stage depends on the individual hair. Eyelashes grow for about 30 days and rest for 105 days, whereas scalp hairs grow for 3 years and rest for 1–2 years. The loss of hair normally means that the hair is being replaced, because the old hair falls out of the hair follicle when the new hair begins to grow. In some men, however, a permanent loss of hair results in "pattern baldness." Although many of the hair follicles are lost, some remain and produce a very short, transparent hair, which for practical purposes is invisible. These changes occur when male sex hormones act on the hair follicles of men who have the genetic predisposition for "pattern baldness."

Hair color is determined by varying amounts and types of melanin. The production and distribution of melanin by melanocytes occurs in the hair bulb by the same method as in the skin. With age, the amount of melanin in hair can decrease, causing the hair to become faded in color, or the hair can have no melanin and be white. Gray hair is usually a mixture of unfaded, faded, and white hairs.

P R E D I C T ❸
Marie Antoinette's hair supposedly turned white overnight after she heard she would be sent to the guillotine. Explain why you believe or disbelieve this story.

Muscles

Associated with each hair follicle are smooth muscle cells, the **arrector** (ă-rek′tōr, that which raises) **pili** (pī′lī, hair) (see figure 5.5a). Contraction of the arrector pili causes the hair to become more perpendicular to the skin's surface, or to "stand on end," and also produces a raised area of skin called "goose flesh." In animals with fur, contraction of the arrector pili is beneficial because it increases the thickness of the fur by raising the hairs. In the cold, the thicker layer of fur traps air and becomes a better insulator. The thickened fur can also make the animal appear larger and more ferocious, which might deter an attacker. It is unlikely that humans, with their sparse amount of hair, derive any important benefit from contraction of their arrector pili.

Glands

The major glands of the skin are the **sebaceous** (sē-bā′shŭs) **glands** and the **sweat glands** (figure 5.6). Sebaceous glands are simple, branched acinar glands. Most are connected by a duct to

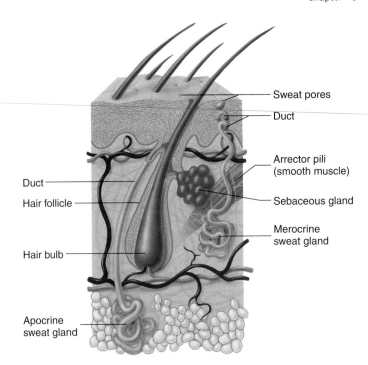

Figure 5.6   Glands of the Skin
Sebaceous and apocrine sweat glands empty into the hair follicle. Merocrine sweat glands empty onto the surface of the skin.

the superficial part of a hair follicle. They produce **sebum,** an oily, white substance rich in lipids. The sebum lubricates the hair and the surface of the skin, which prevents drying and protects against some bacteria.

There are two kinds of sweat glands. **Merocrine** (mer′ō-krin) **sweat glands** are simple, coiled tubular glands located in almost every part of the skin and are most numerous in the palms and soles. They produce a secretion that is mostly water with a few salts. Merocrine sweat glands have ducts that open onto the surface of the skin through sweat pores. When the body temperature starts to rise above normal levels, the sweat glands produce sweat, which evaporates and cools the body. Sweat can also be released in the palms, soles, axillae (armpits), and other places because of emotional stress.

Sweat and Lie Detection
Emotional sweating is used in lie detector (polygraph) tests because sweat gland activity usually increases when a person tells a lie. Even small amounts of sweat can be detected because the salt solution conducts electricity and lowers the electrical resistance of the skin.

Apocrine (ap′ō-krin) **sweat glands** are simple, coiled tubular glands that produce a thick secretion rich in organic substances. They open into hair follicles, but only in the axillae and genitalia. Apocrine sweat glands become active at puberty because of the influence of sex hormones. The organic secretion, which is essentially odorless when released, is quickly broken down by bacteria into substances responsible for what is commonly known as body odor.

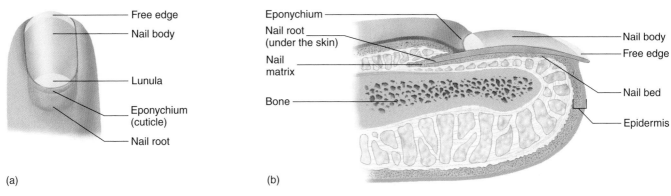

Figure 5.7 Nail

(*a*) Dorsal view. (*b*) Lateral view of a sagittal section through the nail. Most of the epidermis is absent from the nail bed.

Nails

The distal ends of the digits of humans and other primates have nails, whereas reptiles, birds, and most mammals have claws or hooves. The **nail** is a thin plate, consisting of layers of dead stratum corneum cells that contain a very hard type of keratin. The visible part of the nail is the **nail body,** and the part of the nail covered by skin is the **nail root** (figure 5.7). The **eponychium** (ep-ō-nik′ē-ŭm, *epi,* upon + *onyx,* nail), or **cuticle,** is stratum corneum that extends onto the nail body. The nail root extends distally from the **nail matrix.** The nail also attaches to the underlying **nail bed,** which is located distal to the nail matrix. The nail matrix and bed are epithelial tissue with a stratum basale that gives rise to the cells that form the nail. The nail matrix is thicker than the nail bed and produces most of the nail. A small part of the nail matrix, the **lunula** (loo′noo-lă, moon), can be seen through the nail body as a whitish, crescent-shaped area at the base of the nail. The production of cells within the nail matrix results in growth of the nail. Unlike hair, nails grow continuously and do not have a resting stage.

Physiology of the Integumentary System

Protection

The integumentary system performs many protective functions.

1. The intact skin plays an important role in preventing water loss because its lipids act as a barrier to the diffusion of water.
2. The skin prevents the entry of microorganisms and other foreign substances into the body. Secretions from skin glands also produce an environment unsuitable for some microorganisms.
3. The stratified squamous epithelium of the skin protects underlying structures against abrasion.
4. Melanin absorbs ultraviolet light and protects underlying structures from its damaging effects.

5. Hair provides protection in several ways: the hair on the head acts as a heat insulator, eyebrows keep sweat out of the eyes, eyelashes protect the eyes from foreign objects, and hair in the nose and ears prevents the entry of dust and other materials.
6. The nails protect the ends of the digits from damage and can be used in defense.

Sensation

The skin has receptors in the epidermis and dermis that can detect pain, heat, cold, and pressure (see chapter 9). Although hair does not have a nerve supply, movement of the hair can be detected by sensory receptors around the hair follicle.

Vitamin D Production

When the skin is exposed to ultraviolet light, a precursor molecule of **vitamin D** is formed. The precursor is carried by the blood to the liver, where it is modified, and then to the kidneys, where the precursor is modified further to form active vitamin D. If exposed to enough ultraviolet light, humans can produce all the vitamin D they need. Many people need to ingest vitamin D, however, because clothing and indoor living reduce their exposure to ultraviolet light. Fatty fish (and fish oils) and vitamin D-fortified milk are the best sources of vitamin D. Eggs, butter, and liver contain small amounts of vitamin D, but are not considered significant sources because too large a serving size is necessary to meet daily vitamin D requirements. Adequate levels of vitamin D are necessary because vitamin D stimulates calcium and phosphate uptake in the intestines. These substances are necessary for normal bone metabolism (see chapter 6) and normal muscle function (see chapter 7).

Temperature Regulation

Body temperature normally is maintained at about 37°C (98.6°F). Regulation of body temperature is important because the rate of chemical reactions within the body can be increased or decreased by changes in body temperature. Even slight changes in temperature can make enzymes operate less

Clinical Focus Diseases of the Skin

Bacterial Infections

Acne (ak'nē) is an inflammation of the hair follicles and sebaceous glands. Four factors are believed to be involved: hormones, sebum, abnormal keratinization, and the bacterium *Propionibacterium acnes*. The lesions of acne begin with the overproduction of epidermal cells in the hair follicle. These cells are shed from the wall of the hair follicle, and they stick to one another to form a mass of cells mixed with sebum that blocks the hair follicle. During puberty, hormones, especially testosterone, stimulate the sebaceous glands, and sebum production increases. Because both the testes and the ovaries produce testosterone, the effect is seen in males and females. An accumulation of sebum behind the blockage produces a whitehead. A blackhead develops when the accumulating mass of cells and sebum pushes through the opening of the hair follicle. Although there is general agreement that dirt is not responsible for the black color, the exact cause of the black color in blackheads is disputed. A pimple results if the wall of the hair follicle ruptures, forming an entry into the surrounding tissue. *P. acnes* and other bacteria stimulate an inflammatory response that results in the formation of a red pimple filled with pus. If tissue damage is extensive, scarring occurs.

Impetigo (im-pe-tī'gō, a scabby eruption) is a skin disease caused by *Staphylococcus aureus*. It usually affects children, producing small blisters containing pus that easily rupture to form a thick, yellowish crust. The bacteria are transmitted by direct contact (touching) and enter the skin through abrasions or small breaks in the skin.

Decubitis (dē-kū'bi-tŭs, to lie down) **ulcers,** also known as **bedsores** or **pressure sores,** can develop in people who are bedridden or confined to a wheelchair. The weight of the body, especially in areas over bony projections such as the hipbones and heels, compresses tissue and reduces circulation. The lack of blood flow results in the destruction of the hypodermis and the skin. After the skin dies, bacteria gain entry to produce an infected ulcer.

Viral Infections

Interestingly, many of the viruses that cause skin diseases do not enter the body through the skin. Instead, the viruses enter through the respiratory system, where they reside and multiply for about 2 weeks. Then they are carried by the blood to the skin where they cause lesions. Examples are rubeola, rubella, and chickenpox. **Rubeola** (rū-bē'ō-lă, measles) can become dangerous because it can develop into pneumonia, or the virus can invade the brain and cause damage. **Rubella** (rŭ-bel'ă, German measles) is a mild disease but can prove dangerous if contracted during pregnancy. The virus can cross the placenta and damage the fetus, resulting in deafness, cataracts, heart defects, mental retardation, or death. **Chickenpox** is a mild disease if contracted during childhood. **Herpes** (her'pēz, a spreading skin eruption) **zoster** (zos'ter, a girdle), or **shingles,** is a disease caused by the chickenpox virus that occurs after the childhood infection. The virus remains dormant within nerve cells. Trauma, stress, or another illness somehow activates the virus, which moves through the nerve to the skin, where it causes very painful lesions along the nerve's pathway.

Cold sores, or **fever blisters,** are caused by the herpes simplex I virus, which is related to the chickenpox virus. The initial infection usually does not produce symptoms. Dormant viruses can become active, however, and produce lesions in the skin around the mouth and in the mucous membrane of the mouth. The virus is transmitted by oral or respiratory routes. The herpes simplex II virus is transmitted by sexual contact and produces genital lesions, referred to as **genital herpes.**

Warts are uncontrolled growths of the epidermis caused by the human *Papillomavirus*. Usually the growths are benign and disappear spontaneously, or they can be removed by a variety of techniques. The viruses are transmitted to the skin by direct contact with contaminated objects or an infected person. They can also be spread by scratching.

Ringworm

Ringworm is a fungal infection that produces patchy scaling and an inflammatory response in the skin. The lesions are often circular with a raised edge and in ancient times were thought to be caused by worms. Several species of fungus cause ringworm in humans, and they usually are described by their location on the body. Ringworm in the scalp is called ringworm, ringworm of the groin is called jock itch, and ringworm of the feet is called athlete's foot.

Eczema and Dermatitis

Eczema (ek'zĕ-mă, eg-zē'mă, to boil over) and **dermatitis** (der-mă-tī'tis) are general terms used for inflammatory conditions of the skin. Causes of the inflammation can be allergy; infection; poor circulation; or exposure to physical factors, such as chemicals, heat, cold, or sunlight.

Psoriasis

Psoriasis (sō-rī'ă-sis, the itch) is characterized by increased cell division in the stratum basale, abnormal keratin production, and elongation of the dermal papillae toward the skin surface. The result is a thicker-than-normal stratum corneum that sloughs to produce large, silvery scales. If the scales are scraped away, bleeding occurs from the blood vessels at the top of the dermal papillae. Evidence suggests that the disease has a genetic component and that the immune system stimulates the increased cell divisions. Psoriasis is a chronic disease that can be controlled with drugs and phototherapy (UV light), but as yet has no cure.

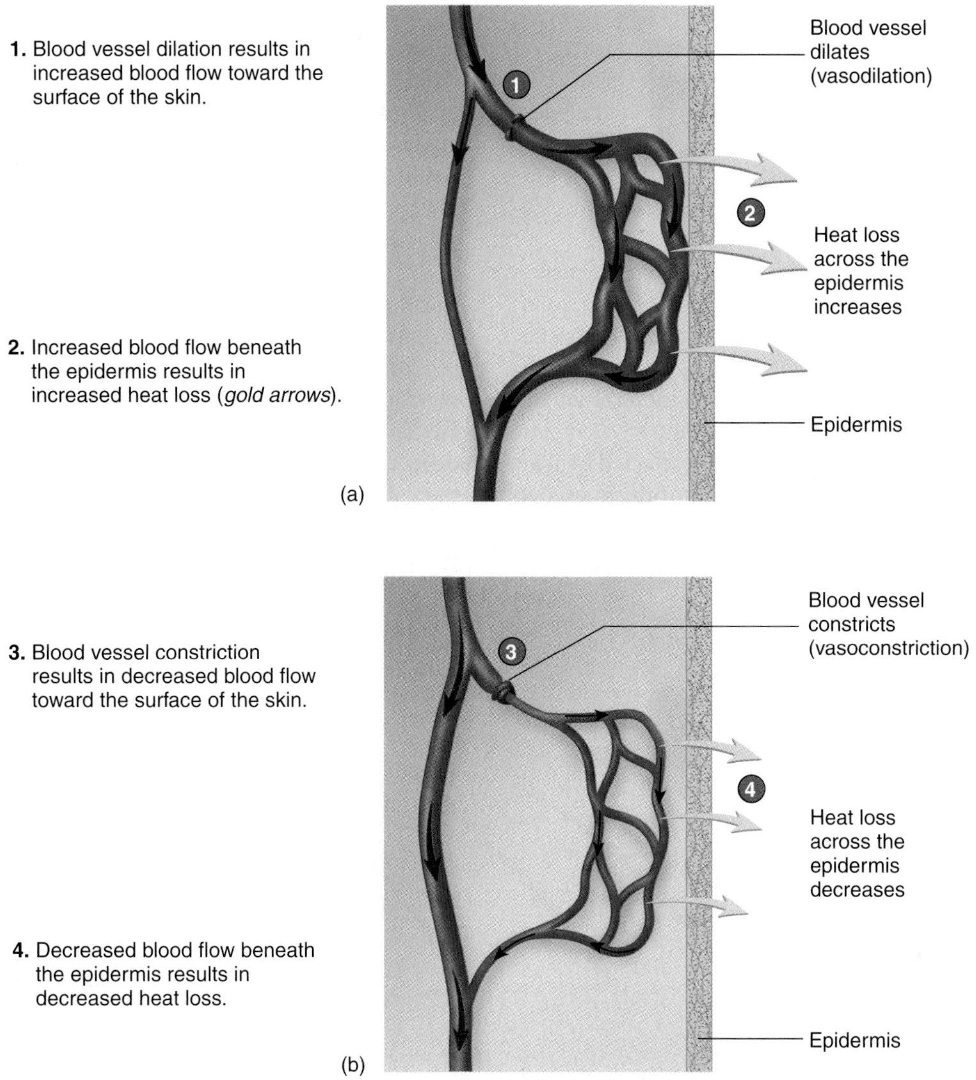

1. Blood vessel dilation results in increased blood flow toward the surface of the skin.

Blood vessel dilates (vasodilation)

2

Heat loss across the epidermis increases

2. Increased blood flow beneath the epidermis results in increased heat loss (*gold arrows*).

Epidermis

(a)

3. Blood vessel constriction results in decreased blood flow toward the surface of the skin.

Blood vessel constricts (vasoconstriction)

4

Heat loss across the epidermis decreases

4. Decreased blood flow beneath the epidermis results in decreased heat loss.

Epidermis

(b)

Figure 5.8   Heat Exchange in the Skin

efficiently and disrupt the normal rates of chemical changes in the body.

Exercise, fever, or an increase in environmental temperature tend to raise body temperature. Homeostasis requires the loss of excess heat. Blood vessels (arterioles) in the dermis dilate and enable more blood to flow through the skin, thus transferring heat from deeper tissues to the skin (figure 5.8a), where the heat is lost by radiation (infrared energy), convection (air movement), or conduction (direct contact with an object). Sweat that spreads over the surface of the skin and evaporates also carries away heat and reduces body temperature.

If body temperature begins to drop below normal, heat can be conserved by constriction of dermal blood vessels, which reduces blood flow to the skin (figure 5.8b). Thus, less heat is transferred from deeper structures to the skin, and heat loss is reduced. With smaller amounts of warm blood flowing through the skin, however, the skin temperature decreases. If the skin temperature drops below about 15°C (59°F), blood vessels dilate.

P R E D I C T

You may have noticed that on cold winter days, people's noses and ears turn red. Can you explain the advantage of this response?

Excretion

Excretion is the removal of waste products from the body. In addition to water and salts, sweat contains a small amount of waste products such as urea, uric acid, and ammonia. Even though large amounts of sweat can be lost from the body, the sweat glands do not play a significant role in the excretion of waste products.

The Integumentary System as a Diagnostic Aid

The integumentary system is useful in diagnosis because it is observed easily and often reflects events occurring in other parts of the body. For example, **cyanosis** (sī-ă-nō′sis), a bluish color caused by decreased blood oxygen content, is an indication of impaired circulatory or respiratory function. A yellowish skin color, **jaundice** (jawn′dis), can occur when the liver is damaged by a disease such as viral hepatitis. Normally the liver secretes bile pigments, which are products of the breakdown of worn-out red blood cells, into the intestine. Bile pigments are yellow, and their buildup in the blood and tissues can indicate an impairment of liver function.

Rashes and lesions in the skin can be symptoms of problems elsewhere in the body. For example, scarlet fever results from a bacterial infection in the throat. The bacteria in the throat release a toxin into the blood that causes a pink–red rash in the skin. The development of a rash can also indicate an allergic reaction to foods or drugs such as penicillin.

The condition of the skin, hair, and nails is affected by nutritional status. In vitamin A deficiency the skin produces excess keratin and assumes a characteristic sandpaper texture, whereas in iron-deficiency anemia the nails lose their normal contour and become flat or concave (spoon-shaped).

Hair concentrates many substances that can be detected by laboratory analysis, and comparison of a patient's hair to a "normal" hair can be useful in certain diagnoses. For examples, lead poisoning results in high levels of lead in the hair. The use of hair analysis as a screening test to determine the general health or nutritional status of an individual is unreliable, however.

Burns

Burns are classified according to the depth of the burn (figure 5.9). In **partial-thickness burns** some part of the stratum basale remains viable, and regeneration of the epidermis occurs from within the burn area as well as from the edges of the burn. Partial-thickness burns are divided into first- and second-degree burns.

First-degree burns involve only the epidermis and are red and painful, and slight **edema** (e-dē′mă), or swelling, can be present. They can be caused by sunburn or brief exposure to hot or cold objects, and they heal without scarring in about a week.

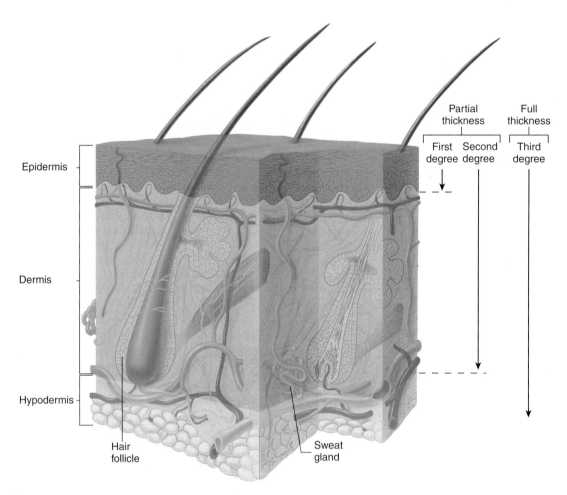

Figure 5.9 Burns

Parts of the skin damaged by different types of burns are shown. Partial-thickness burns are subdivided into first-degree burns (damage to only the epidermis) and second-degree burns (damage to the epidermis and part of the dermis). Full-thickness, or third-degree, burns destroy the epidermis, dermis, and sometimes deeper tissues.

Second-degree burns damage the epidermis and the dermis. If there is minimal dermal damage, symptoms include redness, pain, edema, and blisters. Healing takes about 2 weeks, and there is no scarring. If the burn goes deep into the dermis, however, the wound appears red, tan, or white; can take several months to heal; and might scar. In all second-degree burns, the epidermis regenerates from epithelial tissue in hair follicles and sweat glands, as well as from the edges of the wound.

In **full-thickness,** or **third-degree, burns** the epidermis and the dermis are completely destroyed, and recovery occurs from the edges of the burn wound. Third-degree burns often are surrounded by areas of first- and second-degree burns. Although the first- and second-degree burn areas are painful, the region of third-degree burn is usually painless because sensory receptors in the epidermis and dermis have been destroyed. Third-degree burns appear white, tan, brown, black, or deep cherry red.

Deep partial-thickness and full-thickness burns take a long time to heal, and they form scar tissue with disfiguring and debilitating wound contracture (see chapter 4). To prevent these complications and to speed healing, skin grafts are often performed. In a split skin graft the epidermis and part of the dermis are removed from another part of the body and placed over the burn. Interstitial fluid from the burn nourishes the graft until blood vessels grow into the graft and supply it with nourishment. Meanwhile, the donor tissue produces new epidermis from epithelial tissue in the hair follicles and sweat glands in the same manner as in superficial second-degree burns.

Replacing Skin

When it is not possible or practical to move skin from one part of the body to a burn site, artificial skin or grafts from human cadavers or from pigs are used. These techniques are often unsatisfactory because the body's immune system recognizes the graft as a foreign substance and rejects it. A solution to this problem is laboratory-grown skin. A piece of healthy skin from the burn victim is removed and placed in a flask with nutrients and hormones that stimulate rapid growth. The skin that is produced consists only of epidermis and does not contain glands or hair.

Skin Cancer

Skin cancer is the most common type of cancer. Although chemicals and radiation (x-rays) are known to induce cancer, the development of skin cancer most often is associated with exposure to ultraviolet (UV) light from the sun. Consequently, most skin cancers develop on the face, neck, or hands. The group of people most likely to have skin cancer are fair-skinned (i.e., they have less protection from the sun) or are older than 50 (i.e., they have had long exposure to the sun).

Basal cell carcinoma (kar-si-nō′mă), the most frequent skin cancer, begins with cells in the stratum basale and extends into the dermis to produce an open ulcer (figure 5.10*a*). Surgical removal or radiation therapy cures this type of cancer. Fortunately there is little danger that this type of cancer will spread, or **metastasize** (mĕ-tas′tă-sīz), to other areas of the body. **Squamous cell carcinoma** develops from cells immediately superficial to the stratum basale. Normally these cells undergo little or no cell division. In squamous cell carcinoma, however, the cells continue to divide as they produce keratin. Typically the result is a nodular, keratinized tumor confined to the epidermis (figure 5.10*b*). If untreated, the tumor can invade the dermis, metastasize, and cause death. **Malignant melanoma** (mel′ă-nō′mă) is a rare form of skin cancer that arises from melanocytes, usually in a preexisting mole. A mole is an aggregation, or "nest," of melanocytes. The melanoma can appear as a large, flat, spreading lesion or as a deeply pigmented nodule (figure 5.10*c*). Metastasis is common, and unless diagnosed and treated early in development, this cancer is often fatal.

Limiting exposure to the sun and using sunscreens that block ultraviolet light can reduce the likelihood of developing skin cancer. Ultraviolet light is classified into two types based on their wavelengths: UVA has a longer wavelength than UVB. Exposure to UVA causes most tanning of the skin but is associated with the development of malignant melanoma. Exposure to UVB causes most burning of the skin and is associated with the development of basal cell and squamous cell carcinomas. It is advisable to use sunscreens that effectively block both UVA and UVB.

Someday, protection against ultraviolet light may be achieved by stimulating tanning. Melanotan I is being tested. It is a synthetic version of melanocyte-stimulating hormone, which stimulates increased melanin production by melanocytes.

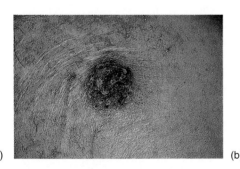
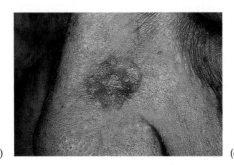
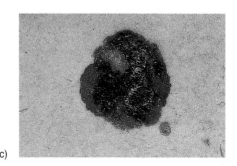

(a)   (b)   (c)

Figure 5.10   **Cancer of the Skin**
(*a*) Basal cell carcinoma. (*b*) Squamous cell carcinoma. (*c*) Malignant melanoma.

Systems Pathology
Burns

Mr. S is a 23-year-old man who had difficulty falling asleep at night. He often stayed up late watching TV or reading until he fell asleep. Mr. S was also a chain smoker. One night he took several sleeping pills. Unfortunately, he fell asleep before putting out his cigarette, which started a fire. As a result, Mr. S was severely burned, receiving full-thickness and partial-thickness burns (figure Aa). He was rushed to the emergency room and was eventually transferred to a burn unit.

For the first day after his accident, his condition was critical because he went into shock. Administration of large volumes of intravenous fluid stabilized his condition. As part of his treatment, Mr. S was also given a high-protein, high-calorie diet.

A week later, dead tissue was removed from the most serious burns (figure Ab), and skin grafts were performed. Despite the use of topical antimicrobial drugs and sterile bandages, however, some of the burns became infected. An additional complication was the development of a venous thrombosis (blood clot) in his leg.

Although the burns were painful and the treatment was prolonged, Mr. S made a full recovery. He no longer smokes.

Background Information

When large areas of skin are severely burned, systemic effects are produced that can be life-threatening. Within minutes of a major burn injury, there is increased permeability of capillaries, which are the small blood vessels in which fluid, gases, nutrients, and waste products are normally exchanged between the blood and tissues. This increased permeability occurs at the burn site and throughout the body. As a result, fluid and ions are lost from the burn wound and into tissue spaces. The loss of fluid decreases blood volume, which decreases the ability of the heart to pump blood. The resulting decrease in blood delivery to tissues can cause tissue damage, shock, and even death. Treatment consists of administering intravenous fluid at a faster rate than it leaks out of the capillaries. Although this can reverse the shock and prevent death, fluid continues to leak into tissue spaces, causing pronounced edema of the tissues.

Typically, after 24 hours, capillary permeability returns to normal, and the amount of intravenous fluid administered can be greatly decreased. How burns result in capillary permeability changes is not well understood. It is clear that following a burn, immunological and metabolic changes occur that affect not only capillaries, but the rest of the body as well. For example, mediators of inflammation (see chapter 4), which are released in response to the tissue damage, contribute to changes in capillary permeability throughout the body.

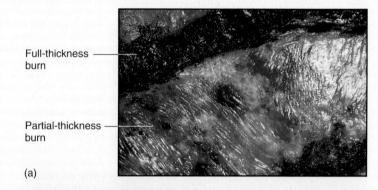

Full-thickness burn

Partial-thickness burn

(a)

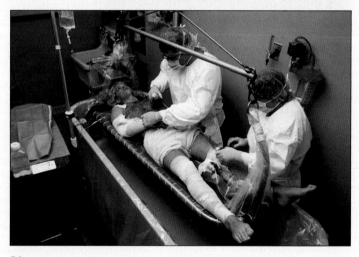

(b)

Figure A   Burn Victim
(a) Partial- and full-thickness burns. (b) Patient in a burn unit.

Substances released from the burn may also play a role in causing cells to function abnormally. Burn injuries result in an almost immediate hypermetabolic state that persists until wound closure. Two other factors contributing to the increased metabolism are (1) a resetting of the temperature control center in the brain to a higher temperature and (2) hormones released by the endocrine system (e.g., epinephrine and norepinephrine from the adrenal glands increase cell metabolism). Compared with a normal body temperature of approximately 37°C (98.6°F), a body temperature of 38.5°C

System Interactions — Effects of Burns on Other Systems

System	Interactions
Skeletal	Red bone marrow replaces red blood cells destroyed in the burnt skin.
Muscular	Loss of muscle mass resulting from the hypermetabolic state caused by the burn.
Nervous	Pain is sensed in the partial-thickness burns. The temperature-regulatory center in the brain is set to a higher temperature, contributing to increased body temperature. Abnormal K^+ concentrations disturb normal nervous system activity: elevated levels are caused by release of K^+ from damaged cells; low levels can be caused by rapid loss of these ions in fluid from the burn.
Endocrine	Increased secretion of epinephrine and norepinephrine from the adrenal glands in response to the injury contributes to increased body temperature by increasing metabolism.
Cardiovascular	Increased capillary permeability causes decreased blood volume, resulting in decreased blood delivery to tissues, edema, and shock. The pumping effectiveness of the heart is impaired by ion imbalance and substances released from the burn. Increased blood clotting causes venous thrombosis. Preferential delivery of blood to the injury promotes healing.
Lymphatic and immune	Increased inflammation in response to tissue damage. Later, depression of the immune system can result in infection.
Respiratory	Airway obstruction is caused by edema. Increased respiration rate results from increased metabolism and lactic acid buildup.
Digestive	Decreased blood delivery as a result of the burn causes degeneration of the intestinal lining and liver. Bacteria from the intestine can cause systemic infections. The liver releases blood-clotting factors in response to the injury. Increased nutrients necessary to support increased metabolism and for repair of the integumentary system are absorbed.
Urinary	The kidneys compensate for the increased fluid loss caused by the burn by greatly reducing or even stopping urine production. Decreased blood volume causes decreased blood flow to the kidneys, which reduces urine output, but can cause kidney tissue damage. Hemoglobin, released from red blood cells damaged in the burnt skin, can decrease urine production by blocking fluid movement from the blood into the kidney.

(101.3°F) is typical in burn patients, despite the higher loss of water by evaporation from the burn.

In severe burns, the increased metabolic rate can result in weight loss as great as 30–40% of the patient's preburn weight. To help compensate, caloric intake may double or even triple. In addition, the need for protein, which is necessary for tissue repair, is greater.

Normal skin maintains homeostasis by preventing the entry of microorganisms. Because burns damage and even completely destroy the skin, microorganisms can cause infections. For this reason, burn patients are maintained in an aseptic environment, which attempts to prevent the entry of microorganisms into the wound. They are also given antimicrobial drugs, which kill microorganisms or suppress their growth. **Debridement** (dā-brēd-mon′), the removal of dead tissue from the burn, helps to prevent infections by cleaning the wound and removing tissue in which infections could develop. Skin grafts, performed within a week of the injury, also prevent infections by closing the wound and preventing the entry of microorganisms.

Despite these efforts, however, infections are still the major cause of death of burn victims. Depression of the immune system during the first or second week after the injury contributes to the high infection rate. The thermally altered tissue is recognized as a foreign substance that can stimulate the immune system. As a result, the immune system is overwhelmed as immune system cells become less effective and production of the chemicals that normally provide resistance to infections decreases (see chapter 14). The greater the magnitude of the burn, the greater the depression of the immune system, and the greater the risk of infection.

Venous thrombosis (throm-bō′sis), the development of a clot in a vein, is also a complication of burns. Blood normally forms a clot when exposed to damaged tissue, such as at a burn site, but clotting can also occur elsewhere, such as in veins. Clots can block blood flow, resulting in tissue destruction. The concentration of chemicals in the blood that cause clotting increases for two reasons: loss of fluid from the burn patient concentrates the chemicals and there is an increased release of clotting factors from the liver.

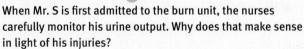

P R E D I C T   5
When Mr. S is first admitted to the burn unit, the nurses carefully monitor his urine output. Why does that make sense in light of his injuries?

Effects of Aging on the Integumentary System

As the body ages, the skin is more easily damaged because the epidermis thins and the amount of collagen in the dermis decreases. Skin infections are more likely and repair of the skin occurs more slowly. A decrease in the number of elastic fibers in the dermis and loss of fat from the hypodermis cause the skin to sag and wrinkle. A decrease in the activity of sweat glands and a decrease in the blood supply to the dermis result in a poor ability to regulate body temperature. The skin becomes drier as sebaceous gland activity decreases. The number of melanocytes generally decreases, but in some areas, the number of melanocytes increases to produce **age spots.** Note that age spots are different from **freckles,** which are caused by increased melanin production. Gray or white hair also results because of a decrease in or a lack of melanin production. Skin that is exposed to sunlight shows signs of aging more rapidly than nonexposed skin, so avoiding overexposure to sunlight and using sun blockers is advisable.

SUMMARY

The integumentary system consists of the skin, hair, glands, and nails.

Functions of the Integumentary System   (p. 99)

The integumentary system protects us from the external environment. Other functions include sensation, vitamin D production, temperature regulation, and excretion of small amounts of waste products.

Hypodermis   (p. 99)

1. The hypodermis, which is not part of the skin, is loose connective tissue that attaches the skin to underlying tissues.
2. About half of the body's fat is stored in the hypodermis.

Skin   (p. 99)

Dermis

1. The dermis is dense connective tissue.
2. Collagen and elastic fibers provide structural strength, and the blood vessels of the papillae supply the epidermis with nutrients.

Epidermis

1. The epidermis is stratified squamous epithelium divided into strata.
   - Cells are produced in the stratum basale.
   - The stratum corneum is many layers of dead, squamous cells containing keratin. The most superficial layers are sloughed.
2. Keratinization is the transformation of stratum basale cells into stratum corneum cells.
   - Structural strength results from keratin inside the cells and from desmosomes, which hold the cells together.
   - Permeability characteristics result from lipids surrounding the cells.

Skin Color

1. Melanocytes produce melanin, which is responsible for different racial skin colors. Melanin production is determined genetically but can be modified by hormones and ultraviolet light (tanning).
2. Carotene, a plant pigment ingested as a source of vitamin A, can cause the skin to appear yellowish.
3. Scattering of light by collagen produces a bluish color.
4. Increased blood flow produces a red skin color, whereas a decreased blood flow causes a pale skin color. Decreased blood oxygen results in the blue color of cyanosis.

Accessory Skin Structures   (p. 103)

Hair

1. Hairs are columns of dead, keratinized epithelial cells. Each hair consists of a shaft (above the skin), root (below the skin), and hair bulb (site of hair cell formation).
2. Hairs have a growth phase and a resting phase.

Muscles

Contraction of the arrector pili, which are smooth muscles, causes hair to "stand on end" and produces "goose flesh."

Glands

1. Sebaceous glands produce sebum, which oils the hair and the surface of the skin.
2. Merocrine sweat glands produce sweat, which cools the body.
3. Apocrine sweat glands produce an organic secretion that can be broken down by bacteria to cause body odor.

Nails

1. The nail consists of the nail body and nail root.
2. The nail matrix produces the nail, which is stratum corneum containing hard keratin.

Physiology of the Integumentary System   (p. 105)

Protection

The skin prevents the entry of microorganisms, acts as a permeability barrier, and provides protection against abrasion and ultraviolet light.

Sensation

The skin contains sensory receptors for pain, heat, cold, and pressure.

Vitamin D Production

1. Ultraviolet light stimulates the production of a precursor molecule in the skin that is modified by the liver and kidneys into vitamin D.
2. Vitamin D increases calcium uptake in the intestines.

Temperature Regulation

1. Through dilation and constriction of blood vessels, the skin controls heat loss from the body.
2. Evaporation of sweat cools the body.

Excretion

Skin glands remove small amounts of waste products but are not important in excretion.

The Integumentary System as a Diagnostic Aid   (p. 107)

The integumentary system is easily observed and often reflects events occurring in other parts of the body (e.g., cyanosis, jaundice, rashes).

Burns   (p. 108)

1. Partial-thickness burns damage only the epidermis (first-degree burn) or the epidermis and the dermis (second-degree burn).
2. Full-thickness burns (third-degree burns) destroy the epidermis, dermis, and usually underlying tissues.

Skin Cancer   (p. 108)

1. Basal cell carcinoma involves the cells of the stratum basale and is readily treatable.

2. Squamous cell carcinoma involves the cells immediately superficial to the stratum basale and can metastasize.
3. Malignant melanoma involves melanocytes, can metastasize, and is often fatal.

Effects of Aging on the Integumentary System   (p. 112)

1. Blood flow to the skin is reduced, the skin becomes thinner, and elasticity is lost.
2. Sweat and sebaceous glands are less active, and the number of melanocytes decreases.

R E V I E W   A N D   C O M P R E H E N S I O N

1. Name the components of the integumentary system.
2. What type of tissue is the hypodermis, and what are its functions?
3. What type of tissue is the dermis? What is responsible for its structural strength? How does the dermis supply the epidermis with blood?
4. What kind of tissue is the epidermis? In which stratum of the epidermis are new cells formed? From which stratum are they sloughed?
5. Define keratinization. What structural changes does keratinization produce to make the skin resistant to abrasion and water loss?
6. Name the cells that produce melanin. What happens to the melanin after it is produced? What is the function of melanin?
7. Describe the factors that determine the amount of melanin produced in the skin.
8. How do melanin, blood, carotene, and collagen affect skin color?
9. Define the root, shaft, and hair bulb of a hair. What kind of cells are found in a hair?
10. What is a hair follicle? Why is it important in the repair of skin?
11. What part of a hair is the site of hair growth? What are the stages of hair growth?
12. What happens when the arrector pili of the skin contract?
13. What secretion is produced by the sebaceous glands? What is the function of the secretion?

14. Which glands of the skin are responsible for cooling the body? Which glands are involved in producing body odor?
15. Name the parts of a nail. Where are the cells that make up the nail produced, and what kind of cells make up a nail? What is the lunula? Describe nail growth.
16. How does the integumentary system provide protection?
17. List the types of sensations detected by receptors in the skin.
18. Describe the production of vitamin D by the body. What is the function of vitamin D?
19. How does the integumentary system assist in the regulation of body temperature?
20. Name the substances excreted by skin glands. Is the skin an important site of excretion?
21. Why is the skin a useful diagnostic aid? Give three examples of how the skin functions as a diagnostic aid.
22. Define the different categories of burns. How is repair accomplished after each type?
23. What is the most common cause of skin cancer? Describe three types of skin cancer and the risks of each type.
24. What changes occur in the skin as a result of aging?

C R I T I C A L   T H I N K I N G

1. A woman has stretch marks on her abdomen, yet she states that she has never been pregnant. Is this possible?
2. The rate of water loss from the skin of the hand was measured. Following the measurement the hand was soaked in alcohol for 15 minutes. After all the alcohol was removed from the hand, the rate of water loss was again measured. Compared with the rate of water loss before soaking the hand in alcohol, what difference, if any, would you expect in the rate of water loss after soaking the hand in alcohol?
3. In has been several weeks since Goodboy Player has competed in a tennis match. After the match he discovers that a blister has formed beneath an old callus on his foot and the callus has fallen off. When he examines the callus he discovers that it appears yellow. Can you explain why?
4. The lips are muscular folds forming the anterior boundary of the oral cavity. A mucous membrane covers the lips internally and the skin of the face covers them externally. The vermillion border, which is the red part of the lips, is covered by keratinized epithelium that is a transition

between the epithelium of the mucous membrane and the facial skin. The vermillion border can become chapped (dry and cracked), whereas the mucous membrane and the facial skin do not. Propose as many reasons as you can to explain why the vermillion border is more prone to drying than the mucous membrane or facial skin.

5. Pulling on hair can be quite painful, yet cutting hair is not painful. Explain.
6. Given what you know about the cause of acne, propose some ways to prevent or treat the disorder.
7. Consider the following statement: Dark-skinned children are more susceptible to rickets (insufficient calcium in the bones) than fair-skinned children. Defend or refute this statement.
8. Harry Fastfeet, a white man, jogs on a cool day. What color would you expect his skin to be (a) after going outside and just before starting to run, (b) during the run, and (c) 5 minutes after the run?

Answers in Appendix D

Visit this textbook's website at www.mhhe.com/seeleyess6 for practice quizzes, animations, interactive learning exercises, and other study tools.

Skeletal System: Bones and Joints

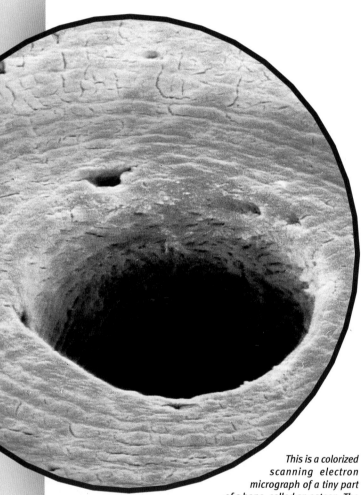

This is a colorized scanning electron micrograph of a tiny part of a bone, called an osteon. The large opening is the space through which blood vessels bring blood to the bone. The surrounding bone matrix is organized into circular layers.

Chapter Outline and Objectives

itting, standing, walking, picking up a pencil, and taking a breath all involve the skeletal system (figure 6.1). Without the skeletal system, we would have no rigid framework to support the soft tissues of the body and no system of joints and levers so critical for movement. The skeletal system consists of bones and their associated connective tissues, including cartilage, tendons, and ligaments. The term *skeleton* is derived from a Greek word meaning dried. Despite this concept of the skeleton as dry, and nonliving, the skeletal system actually consists of dynamic, living tissues that are capable of growth, detect pain stimuli, adapt to stress, and undergo repair after injury.

A **joint,** or **articulation,** is a place where two bones come together. Many joints are movable, but not all. Many joints allow only limited movement, and others allow no apparent movement. The structure of a given joint is directly correlated with its degree of movement.

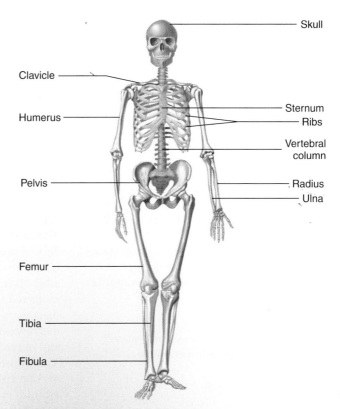

Figure 6.1   The Skeletal System

Functions of the Skeletal System

The skeletal system provides support and protection, allows body movements, stores minerals and fats, and is the site of blood cell production.

1. *Support.* Rigid, strong bone is well suited for bearing weight and is the major supporting tissue of the body. Cartilage provides a firm yet flexible support within certain structures, such as the nose, external ear, thoracic cage, and trachea. Ligaments are strong bands of fibrous connective tissue that attach to bones and hold them together.
2. *Protection.* Bone is hard and protects the organs it surrounds. For example, the skull encloses and protects the brain, and the vertebrae surround the spinal cord. The rib cage protects the heart, lungs, and other organs of the thorax.
3. *Movement.* Skeletal muscles attach to bones by tendons, which are strong bands of connective tissue. Contraction of the skeletal muscles moves the bones, producing body movements. Joints, which are formed where two or more bones come together, allow movement between bones. Smooth cartilage covers the ends of bones within some joints, allowing the bones to move freely. Ligaments allow some movement between bones but prevent excessive movements.

4. *Storage.* Some minerals in the blood are taken into bone and stored. Should blood levels of these minerals decrease, the minerals are released from bone into the blood. The principal minerals stored are calcium and phosphorus. Fat (adipose tissue) is also stored within bone cavities. If needed, the fats are released into the blood and used by other tissues as a source of energy.
5. *Blood cell production.* Many bones contain cavities filled with bone marrow that gives rise to blood cells and platelets (see chapter 11).

Connective Tissue

Bone, cartilage, tendons, and ligaments are connective tissues. In connective tissues the extracellular matrix is largely responsible for their characteristics. The types and relative quantities of molecules in the extracellular matrix determine its characteristics. The matrix contains collagen, proteoglycan (ground substance), and other organic molecules, as well as water and minerals. **Collagen** (kol′lă-jen; *koila,* glue + *-gen,* producing) is a tough, ropelike protein. **Proteoglycans** (prō′tē-ō-glī′kanz; *proteo,* protein + *glycan,* polysaccharide) are large molecules consisting of polysaccharides attached to core proteins, much like the needles of a pine tree are attached to the tree's branches. The proteoglycans form large aggregates, much like pine branches combine to form a whole tree. Proteoglycans can attract and retain large amounts of water between their polysaccharide "needles."

The extracellular matrix of **tendons** and **ligaments** contains large amounts of collagen fibers, making these structures very tough, like ropes or cables. The extracellular matrix of **cartilage** (kar′ti-lij, gristle) contains collagen and proteoglycans. Collagen makes cartilage tough, whereas the water-filled proteoglycans make it smooth and resilient. As a result, cartilage is relatively rigid, but springs back to its original shape if it is bent or slightly compressed. It is an excellent shock absorber.

The extracellular matrix of bone contains collagen and minerals, including calcium and phosphate. The matrix resembles reinforced concrete. The ropelike collagen fibers, like the reinforcing steel bars of reinforced concrete, lend flexible strength to the bone. The mineral component, like the concrete part of reinforced concrete, gives the bone compression (weight-bearing) strength. Most of the mineral in bone is in the form of calcium phosphate crystals called **hydroxyapatite** (hī-drok′sē-ap-ă-tīt).

PREDICT 1

What would a bone be like if all of the mineral was removed? What would it be like if all of the collagen was removed?

A CASE IN POINT | Osteogenesis Imperfecta

May Trix is a 10-year-old girl who has a history of numerous broken bones. At first, it was suspected that she was a victim of child abuse, but eventually it was determined that she has **osteogenesis imperfecta,** which literally means imperfect bone formation. May is short for her age, and her limbs are short and bowed. Her vertebral column is also abnormally curved. Osteogenesis imperfecta is a rare disorder caused by any one of a number of faulty genes that results in either too little collagen formation or a poor quality of collagen. As a result, bone matrix has decreased flexibility and is more easily broken than normal bone. Osteogenesis imperfecta is also known as the "brittle bone" disorder.

General Features of Bone

There are four types of bone, based on their shape: long, short, flat, and irregular. **Long bones** are longer than they are wide. Most of the bones of the upper and lower limbs are long bones. **Short bones** are approximately as broad as they are long, such as the bones of the wrist and ankle. **Flat bones** have a relatively thin, flattened shape. Examples of flat bones are certain skull bones, ribs, scapulae (shoulder blades), and the sternum. **Irregular bones** include the vertebrae and facial bones, with shapes that do not fit readily into the other three categories.

Each long bone consists of a central shaft, called the **diaphysis** (dī-af′i-sis, growing between), and two ends, each called an **epiphysis** (e-pif′i-sis, growing upon) (figure 6.2a and b). A thin layer of **articular** (ar-tik′ū-lăr, joint) **cartilage** covers the ends of the epiphyses where the bone articulates with other bones. A long bone that is still growing has an **epiphyseal plate,** or **growth plate,** composed of cartilage, between each epiphysis and the diaphysis (see figure 6.2a). The epiphyseal plate is the site of growth in bone length. When bone growth stops, the cartilage of each epiphyseal plate is replaced by bone and is called an **epiphyseal line** (see figure 6.2b).

Bones contain cavities such as the large **medullary cavity** in the diaphysis (see figure 6.2), as well as smaller cavities in the epiphyses of long bones and in the interior of other bones. These spaces are filled with either yellow or red marrow. **Marrow** is the soft tissue in the medullary cavities of the bone. **Yellow marrow** consists mostly of fat. **Red marrow** consists of blood-forming cells and is the only site of blood formation in adults (see chapter 11). Children's bones have proportionately more red marrow than do adult bones. As a person ages, red marrow is mostly replaced by yellow marrow. In adults, red marrow is confined to the bones in the central axis of the body and in the most proximal epiphyses of the limbs.

Most of the outer surface of bone is covered by dense connective tissue called the **periosteum** (per-ē-os′tē-ŭm, *peri,* around + *osteon,* bone), which contains blood vessels and nerves (figure 6.2c). The surface of the medullary cavity is lined with a thinner connective tissue membrane, the **endosteum** (en-dos′tē-ŭm, *endo,* inside). The periosteum and endosteum contain **osteoblasts** (os′tē-ō-blastz, bone-forming cells), which function in the formation of bone, as well as in the repair and remodeling of bone. When osteoblasts become surrounded by matrix, they are referred to as **osteocytes** (os′tē-ō-sītz, bone cells).

Bone is formed in thin sheets of extracellular matrix called **lamellae** (lă-mel′ē, plate), with osteocytes, located between the lamellae (figure 6.3). The osteocytes are located within spaces called **lacunae** (lă-koo′nē, a hollow). Cell processes extend from the osteocytes across the extracellular matrix of the lamellae within tiny canals called **canaliculi** (kan-ă-lik′ū-lī, sing. canaliculus, little canal).

There are two major types of bone, based on their histological structure. **Compact bone** is mostly solid matrix and cells. **Cancellous** (kan′sĕ-lŭs) **bone** consists of a lacy network of bone with many small, marrow-filled spaces.

Compact Bone

Compact bone (figure 6.3a) forms most of the diaphysis of long bones and the thinner surfaces of all other bones. Most of the lamellae of compact bone are organized into sets of concentric rings, with each set surrounding a **central,** or **haversian** (ha-ver′shan, named for Clopton Havers, anatomist, 1650–1702), **canal.** Blood vessels that run parallel to the long axis of the bone are contained within the central canals. Each central canal, with the lamellae and osteocytes surrounding it, is called an **osteon** (os′tē-on), or **haversian system.** Each osteon, seen in cross section, looks like a microscopic target, with the central canal as the "bull's eye" (figure 6.3b). Osteocytes, located in lacunae, are connected to one another by cell processes in

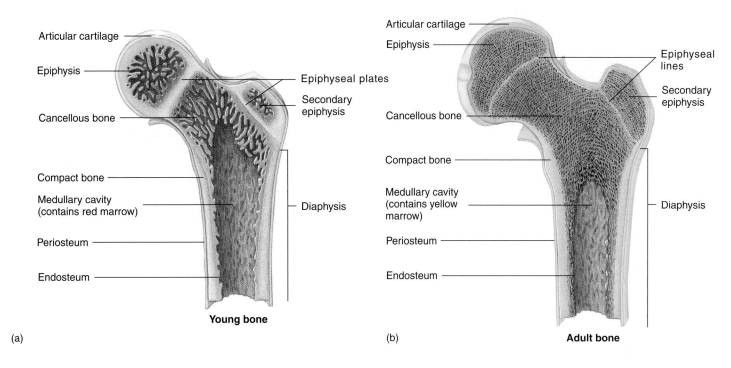

Young bone

(a)

Adult bone

(b)

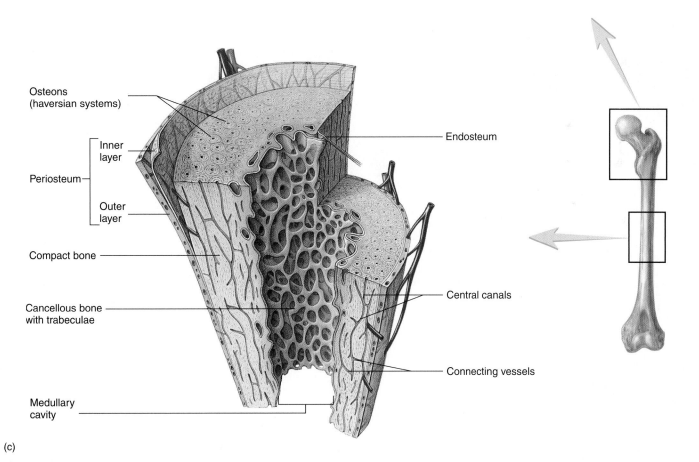

(c)

Figure 6.2 Long Bone

(a) Young long bone (the femur) showing epiphysis, epiphyseal plates, and diaphysis. (b) Adult long bone with epiphyseal lines. (c) Internal features of a portion of the diaphysis in (a).

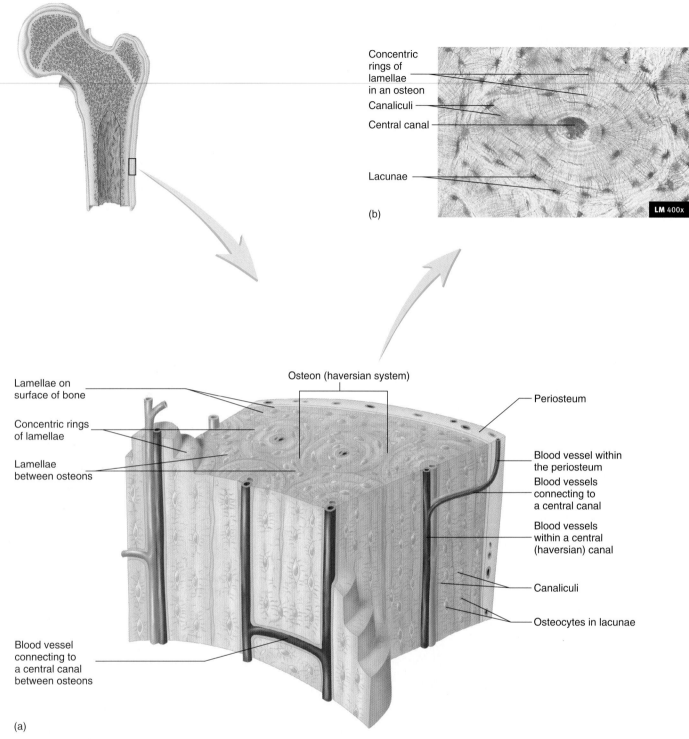

Concentric rings of lamellae in an osteon

Canaliculi

Central canal

Lacunae

(b)

LM 400x

Osteon (haversian system)

Lamellae on surface of bone

Concentric rings of lamellae

Lamellae between osteons

Blood vessel connecting to a central canal between osteons

(a)

Periosteum

Blood vessel within the periosteum

Blood vessels connecting to a central canal

Blood vessels within a central (haversian) canal

Canaliculi

Osteocytes in lacunae

Figure 6.3   Compact Bone

(*a*) Fine structure of compact bone. (*b*) Photomicrograph of compact bone.

canaliculi. The canaliculi give the osteon the appearance of having tiny cracks in the lamellae.

Nutrients leave the blood vessels of the central canals and diffuse to the osteocytes through the canaliculi. Waste products diffuse in the opposite direction. The blood vessels in the central canals, in turn, are connected to blood vessels in the periosteum and endosteum.

Cancellous Bone

Cancellous bone (see figures 6.2*b* and 6.4), also called spongy bone because of its appearance, is located mainly in the epiphyses of long bones, and it forms the interior of all other bones. It consists of delicate interconnecting rods or plates of bone called **trabeculae** (tră-bek′ū-lē, beams), which resemble the beams or

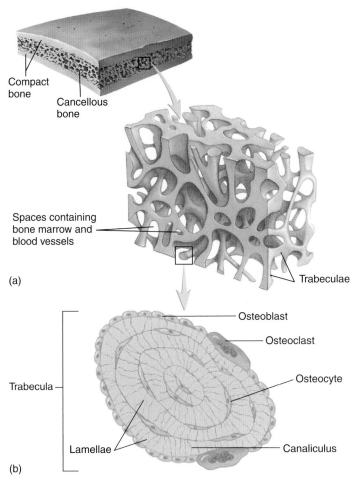

Figure 6.4 Cancellous Bone

(a) Beams of bone, the trabeculae, surround spaces in the bone. In life, the spaces are filled with red or yellow bone marrow and with blood vessels. (b) Transverse section of a trabecula.

scaffolding of a building (figure 6.4a). Like scaffolding, the trabeculae add strength to a bone without the added weight that would be present if the bone were solid mineralized matrix. The spaces between the trabeculae are filled with marrow. Each trabecula consists of several lamellae with osteocytes between the lamellae (figure 6.4b). Usually no blood vessels penetrate the trabeculae, and the trabeculae have no central canals. Nutrients exit vessels in the marrow and pass by diffusion through canaliculi to the osteocytes of the trabeculae.

Bone Ossification

Ossification (os′i-fi-kā′shŭn, *os,* bone + *facio,* to make) is the formation of bone by osteoblasts. It involves the synthesis of an organic matrix containing collagen and proteoglycans and the addition of hydroxyapatite crystals to the matrix. After an osteoblast becomes completely surrounded by bone matrix, it becomes a mature bone cell, or osteocyte. Bones develop in the fetus by two processes, each involving the formation of bone matrix on preexisting connective tissue (figure 6.5). Bone formation that occurs within connective tissue membranes is called intramembranous ossification, and bone formation that

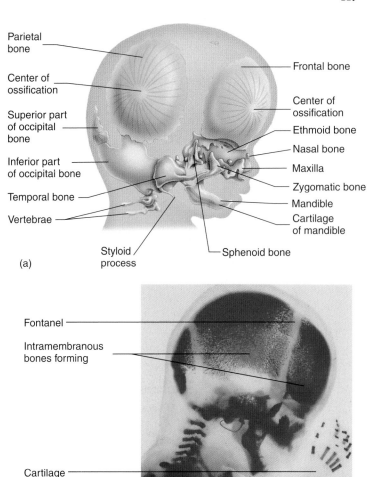

Figure 6.5 Bone Formation

(a) Intramembranous ossification occurs in a 12-week-old fetus at ossification centers in the flat bones of the skull (*yellow*). Endochondral ossification occurs in the bones forming the inferior part of the skull (*blue*). (b) Radiograph of an 18-week-old fetus, showing intramembranous and endochondral ossification. Intramembranous ossification occurs at centers of ossification in the flat bones of the skull. Endochondral ossification has formed bones in the diaphyses of long bones. The epiphyses are still cartilage at this stage of development.

occurs inside cartilage is called endochondral ossification. Both types of bone formation result in compact and cancellous bone.

Intramembranous (in′tră-mem′brā-nŭs, between membranes) **ossification** occurs when osteoblasts begin to produce bone in connective tissue membranes. This occurs primarily in the bones of the skull. Osteoblasts line up on the surface of connective tissue fibers and begin depositing bone matrix to form trabeculae. The process begins in areas called **ossification centers** (see figure 6.5a), and the trabeculae radiate out from the centers. Usually two or more ossification centers exist in each flat skull bone, and the skull bones result from fusion of these centers as they enlarge. The trabeculae are constantly

1. A cartilage model, with the general shape of the mature bone, is produced by chondrocytes. A perichondrium surrounds most of the cartilage model.

2. A bone collar is produced and the perichondrium of the diaphysis becomes the periosteum. The chondrocytes hypertrophy, and cartilage is calcified.

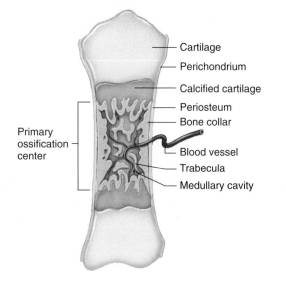

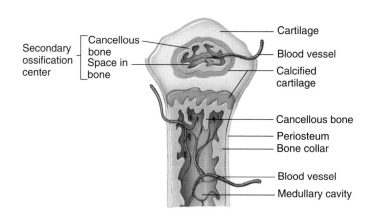

3. A primary ossification center forms as blood vessels and osteoblasts invade the calcified cartilage. The osteoblasts lay down bone matrix, forming trabeculae.

4. Secondary ossification centers form in the epiphyses of long bones.

Process Figure 6.6   Endochondral Ossification of a Long Bone

remodeled after their initial formation, and they may enlarge or be replaced by compact bone.

The bones at the base of the skull and most of the remaining skeletal system develop through the process of **endochondral ossification** from cartilage models. The cartilage models have the general shape of the mature bone (figure 6.6, step 1). During endochondral ossification, cartilage cells, called **chondrocytes,** increase in number, hypertrophy (enlarge), and die. Then the cartilage matrix becomes calcified (figure 6.6, step 2). As this process is occurring in the center of the cartilage model, blood vessels accumulate in the perichondrium. The presence of blood vessels in the outer surface of future bone causes some of the unspecified connective tissue cells on the surface to become osteoblasts. These osteoblasts then produce a collar of bone around part of the outer surface of the diaphysis, and

the perichondrium becomes periosteum in that area (see figure 6.6, step 2). Blood vessels also grow into the center of the diaphyses, bringing in osteoblasts and stimulating ossification to occur. The center part of the diaphysis, where bone first begins to appear, is called the **primary ossification center** (figure 6.6, step 3). Osteoblasts invade spaces in the center of the bone left by the dying cartilage cells. Some of the calcified cartilage matrix is removed by cells called **osteoclasts** (os′tē-ō-klastz, bone-eating cells; these cells have several nuclei), and the osteoblasts line up on the remaining calcified matrix and begin to form bone trabeculae. As the bone develops it is constantly remodeled. A medullary cavity forms in the center of the diaphysis as osteoclasts remove bone and calcified cartilage, which are replaced by bone marrow. Later, **secondary ossification centers** form in the epiphyses (see figure 6.6, steps 3 and 4).

Bone Growth

Bone growth occurs by the deposition of new bone lamellae onto existing bone or other connective tissue. This process is called appositional growth. As osteoblasts deposit new bone matrix on the surface of bones between the periosteum and the existing bone matrix, the bone increases in width or diameter. Growth in the length of a bone, which is the major source of increased height in the individual, occurs in the epiphyseal plate (figure 6.7). Chondrocytes increase in number on the epiphyseal side of the epiphyseal plate. They line up in columns

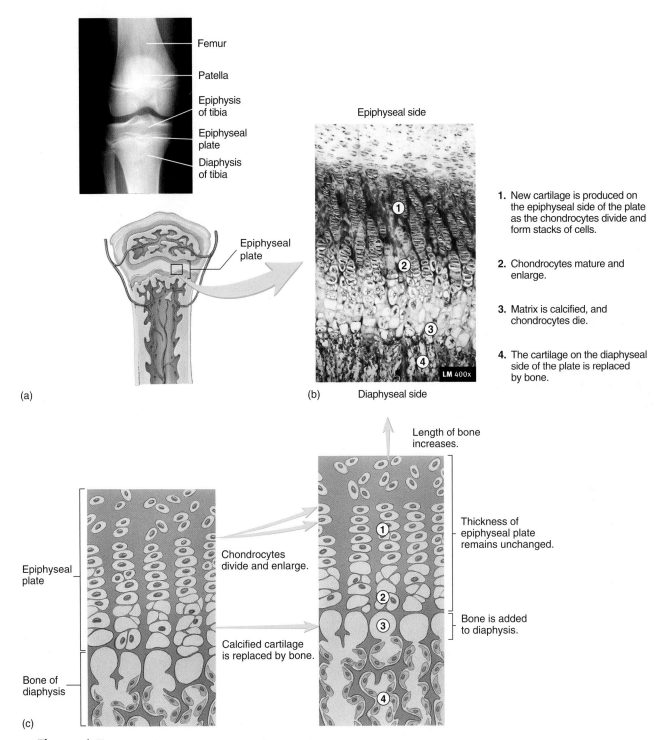

1. New cartilage is produced on the epiphyseal side of the plate as the chondrocytes divide and form stacks of cells.

2. Chondrocytes mature and enlarge.

3. Matrix is calcified, and chondrocytes die.

4. The cartilage on the diaphyseal side of the plate is replaced by bone.

Process Figure 6.7 Endochondral Bone Growth

(*a*) Location of the epiphyseal plate in a long bone. (*b*) Photomicrograph of an epiphyseal plate, demonstrating proliferation and hypertrophy, and the areas of calcification and ossification. (*c*) As the chondrocytes of the epiphyseal plate divide and align in columns, the cartilage expands toward the epiphysis, and the bone elongates. At the same time, the older cartilage is calcified and then replaced by bone, which is, in turn, remodeled, resulting in expansion of the medullary cavity of the diaphysis. The net result is an epiphyseal plate that remains uniform in thickness through time but that is constantly moving toward the epiphysis, resulting in elongation of the bone.

parallel to the long axis of the bone, causing elongation of the bone, and then hypertrophy and die. The cartilage matrix is calcified. Much of the cartilage that forms around the hypertrophied cells is removed by osteoclasts, and the dying chondrocytes are replaced by osteoblasts. The osteoblasts start forming bone by depositing bone lamellae on the surface of the calcified cartilage. This process produces bone on the diaphyseal side of the epiphyseal plate.

PREDICT ❷
Describe the appearance of an adult if cartilage growth did not occur in the long bones during childhood.

Bone Remodeling

Bone remodeling involves the removal of existing bone by osteoclasts and the deposition of new bone by osteoblasts. Bone remodeling occurs in all bone. Remodeling of newly formed cancellous bone in the epiphyseal plate to form compact bone is involved in bone growth. Remodeling is responsible for changes in bone shape, the adjustment of bone to stress, bone repair, and calcium ion regulation in the body fluids. A long bone increases in length and diameter by deposition of new bone on the outer surface and by growth at the epiphyseal plate. At the same time, bone is removed from the inner, medullary surface of the bone. As the bone diameter increases, the thickness of the compact bone relative to the medullary cavity tends to remain fairly constant. If the size of the medullary cavity did not also increase as bone size increased, the compact bone of the diaphysis would become thick and very heavy.

Bone is the major storage site for calcium in the body. Blood calcium levels must be maintained within narrow limits for many functions to occur normally. Calcium is removed from bones when blood calcium levels decrease, and it is deposited when dietary calcium is adequate. This removal and deposition is under hormonal control (see section on Bone and Calcium Homeostasis p. 123).

If too much bone is deposited, the bones become thick or have abnormal spurs or lumps that can interfere with normal function. Too little bone formation or too much bone removal weakens the bones and makes them susceptible to fracture.

Bone Repair

When a bone is broken, blood vessels in the bone are also damaged. The vessels bleed, and a clot forms in the damaged area (figure 6.8, step 1). Two to three days after the injury, blood vessels and cells from surrounding tissues begin to invade the clot. Some of these cells produce a fibrous network of connective tissue between the broken bones, which holds the bone fragments together and fills the gap between the fragments. Other cells produce islets of cartilage in the fibrous network. The zone of tissue repair between the two bone fragments is called a **callus** (figure 6.8, step 2).

Osteoblasts enter the callus and begin forming cancellous bone (figure 6.8, step 3). Cancellous bone formation in the callus is usually complete 4–6 weeks after the injury. Immobilization of the bone is critical up to this time because movement can refracture the delicate new matrix. Subsequently, the cancellous bone is slowly remodeled to form compact and cancellous

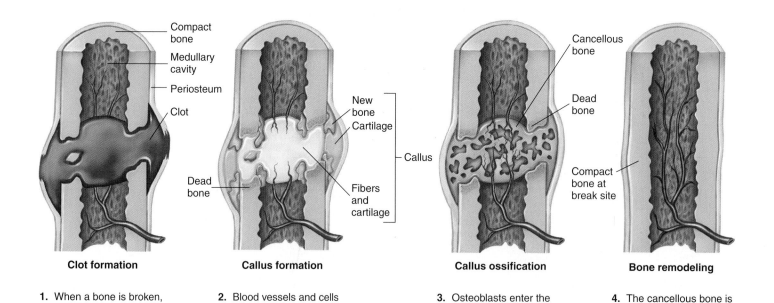

Clot formation | **Callus formation** | **Callus ossification** | **Bone remodeling**

1. When a bone is broken, a clot forms in the damaged area.
2. Blood vessels and cells invade the clot and produce a fibrous network and cartilage between the broken bones, called a callus.
3. Osteoblasts enter the callus and form cancellous bone.
4. The cancellous bone is slowly remodeled to form compact bone and the repair is complete.

Process Figure 6.8   Bone Repair

bone, and the repair is complete (figure 6.8, step 4). Total healing of the fracture may require several months. If bone healing occurs properly, the healed region can be even stronger than the adjacent bone.

Bone and Calcium Homeostasis

Bone is the major storage site for calcium in the body, and movement of calcium into and out of bone helps to determine blood calcium levels, which is critical for normal muscle and nervous system function. Calcium moves into bone as osteoblasts build new bone and out of bone as osteoclasts break down bone (figure 6.9). When osteoblast and osteoclast activity is balanced, the movement of calcium into and out of a bone is equal.

When blood calcium levels are too low, osteoclast activity increases, calcium is released by osteoclasts from bone into the blood, and blood calcium levels increase. Conversely, if blood calcium levels are too high, osteoclast activity decreases, calcium is taken from the blood by osteoblasts to produce new bone, and blood calcium levels decrease.

Parathyroid hormone (PTH) from the parathyroid glands stimulates increased bone breakdown and increased blood calcium levels by indirectly stimulating osteoclast activity. PTH also increases calcium reabsorption from the urine in the kidney. PTH also stimulates the kidneys to form active vitamin D, which increases calcium absorption from the small intestine. Decreasing blood calcium levels stimulate PTH secretion.

Calcitonin (kal-si-tō′nin), secreted from the thyroid gland, decreases osteoclast activity and thus decreases blood calcium levels. Increasing blood calcium levels stimulate calcitonin secretion. PTH and calcitonin are described more fully in chapter 10.

Immobilization of Joints

Although immobilization at a fracture point is critical during the early stages of bone healing, complete immobilization is not good for the bone, muscles, or joints. Not long ago, it was common practice to completely immobilize a bone for as long as 10 weeks. It is now known that, if a bone is immobilized for as little as 2 weeks, the muscles associated with that bone may lose as much as half their strength. Furthermore, if a bone is completely immobilized, it is not subjected to the normal mechanical stresses that help it to form. Bone matrix is reabsorbed, and the strength of the bone decreases. In experimental animals, complete immobilization of the back for 1 month resulted in up to a threefold decrease in vertebral compression strength. Modern therapy attempts to balance bone immobilization with enough exercise to minimize muscle and bone atrophy and to maintain joint mobility. These goals are accomplished by limiting the amount of time a cast is left on the patient, and by employing "walking casts," which allow some stress on the bone and some movement.

1. Osteoclasts break down bone and release Ca²⁺ into the blood, and osteoblasts remove Ca²⁺ from the blood to make bone (*blue arrows* represent the movement of Ca²⁺). Parathyroid hormone (PTH) regulates blood Ca²⁺ levels by indirectly stimulating osteoclast activity, resulting in increased Ca²⁺ release into the blood. Calcitonin plays a minor role in Ca²⁺ maintenance by inhibiting osteoclast activity.

2. In the kidneys, PTH increases Ca²⁺ reabsorption from the urine.

3. In the kidneys, PTH also promotes the formation of active vitamin D (*green arrows*), which increases Ca²⁺ absorption from the small intestine.

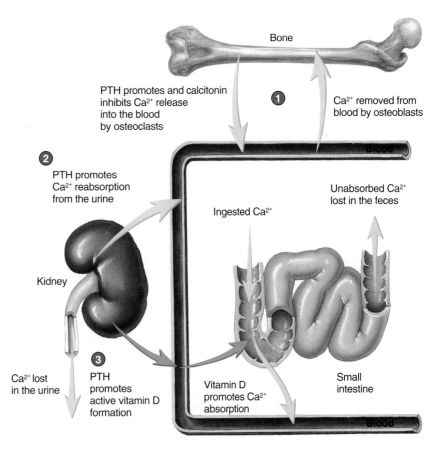

Process Figure 6.9   Calcium Homeostasis

Growth and Developmental Disorders

Giantism (jī'an-tizm) is a condition of abnormally increased size that usually involves excessive endochondral growth at the epiphyseal plates of long bones. **Dwarfism,** the condition in which a person is abnormally small, may result from improper growth in the epiphyseal plates (figure A; see also chapter 10).

Osteogenesis imperfecta (os'tē-ō-jen'ĕ-sis im-per-fek'tă, *osteo,* bone + *genesis,* production; imperfect), a group of genetic disorders producing very brittle bones that are easily fractured, occurs because insufficient collagen or abnormal collagen is formed. Collagen normally strengthens bones and makes them flexible. In severe cases, prenatal fractures of the limbs often occur in the fetus. These fractures usually heal in poor alignment, causing the limbs to appear bent and shortened. In less severe cases, the disease first becomes apparent during childhood.

Rickets (rik'ets, *wrick,* to twist; bones become twisted in the disease) is a condition involving growth retardation resulting from nutritional deficiencies either in minerals (calcium and phosphate) necessary for normal ossification or in vitamin D, which is necessary for calcium and phosphate absorption. The condition results in bones that are soft, weak, and easily broken.

Rickets most often occurs in children who receive inadequate amounts of sunlight (necessary for vitamin D production by the body) and whose diets are deficient in vitamin D.

Bacterial Infections

Osteomyelitis (os'tē-ō-mī-e-lī'tis, *osteo,* bone + *myelos,* marrow + *itis,* inflammation) is bone inflammation that often results from bacterial infection, and it can lead to complete destruction of the bone. **Staphylococcus** (staf'i-lō-kok'ŭs, *staphyle,* a bunch of grapes + *kokkos,* a berry; these terms describe the organization and shape of the bacterium) (staph) infections, introduced into the body through wounds, are the most common cause of osteomyelitis. Tuberculosis is primarily a lung disease, but it can also affect bones. Because of milk pasteurization and other improvements in hygiene, tuberculosis became rare in the United States. Because tuberculosis can be a complication in AIDS and because a drug-resistant form of tuberculosis has emerged, tuberculosis has once more become a clinical problem in the United States.

Tumors

There are many types of bone **tumors** with a wide range of resultant bone defects. Tumors may be benign or malignant. Malignant bone tumors may metastasize (spread) to other parts of the body or may result from metastasizing tumors elsewhere.

Decalcification

Osteomalacia (os'tē-ō-mă-lā'shē-ă, *osteo,* bone + *malakia,* softness), or the softening of bones, results from calcium depletion from bones. If the body has an unusual need for calcium (e.g., during pregnancy when fetal growth requires large amounts of calcium), it may be removed from the mother's bones, which consequently soften and weaken. Osteomalacia is sometimes called adult rickets and can result from vitamin D deficiency.

Osteoporosis (os'tē-ō-pō-rō'sis, *osteo,* bone + *poros,* pore + *osis,* condition), or porous bone, results from reduction in the overall quantity of bone tissue (see Systems Pathology: Osteoporosis on p. 152).

Figure A Giant and Dwarf (both are adults)

General Considerations of Bone Anatomy

It is traditional to list 206 bones in the average adult skeleton (table 6.1 and figure 6.10), although the actual number varies from person to person and decreases with age as some bones become fused.

Several common terms are used to describe the features of bones (table 6.2). For example, a hole in a bone is called a **foramen** (fō-rā'men, pl. foramina, fō-rā'min-ă, *foro,* to pierce). A foramen usually exists in a bone because some structure, such as a nerve or blood vessel, passes through the bone at that point. If the hole is elongated into a tunnellike passage through the bone, it is called a **canal** or a **meatus** (mē-ā'tus, a passage). A depression in a bone is called a **fossa** (fos'ă). A lump on a bone is called a **tubercle** (too'ber-kl, a knob), or **tuberosity** (too'ber-os'i-tē), and a projection from a bone is called a **process.** Most tubercles and processes are sites of muscle attachment on the bone. Increased muscle pull, such as

when a person lifts weights to build up muscle mass, can increase the size of some tubercles. The smooth, rounded end of a bone, where it forms an articulation (a joint) with another bone, is called a **condyle** (kon'dīl, knuckle).

Axial Skeleton

The axial skeleton is divided into the skull, the vertebral column, and the thoracic cage.

Skull

The 22 bones of the skull (table 6.1) are divided into two groups: those of the braincase and those of the face. The **braincase,** which encloses the cranial cavity, consists of 8 bones that immediately surround and protect the brain; the 14 **facial bones** form the structure of the face. Thirteen of the facial bones are rather solidly connected to form the bulk of the face. The mandible, however, forms a freely movable articulation

Clinical Focus   Bone Fractures

Bone fractures (figure B) can be classified as **open,** or **compound,** if the bone protrudes through the skin, and **closed,** or **simple,** if the skin is not perforated. If the fracture totally separates the two bone fragments, it is called **complete;** if it doesn't, it is called **incomplete.** An incomplete fracture that occurs on the convex side of the curve of a bone is called a **greenstick fracture.** A **comminuted** (kom'i-nū-ted; broken into small pieces) fracture is one in which the bone breaks into more than two fragments. An **impacted** fracture occurs when one of the fragments of one part of the bone is driven into the cancellous bone of another fragment. Fractures can also be classified according to the direction of the fracture line as **linear** (parallel to the long axis), **transverse** (at right angles to the long axis), or **oblique** and **spiral** (at an angle other than a right angle to the long axis).

Figure B   Bone Fractures

(*a*) Complete and incomplete. (*b*) Transverse and comminuted. (*c*) Impacted. (*d*) Spiral and oblique.

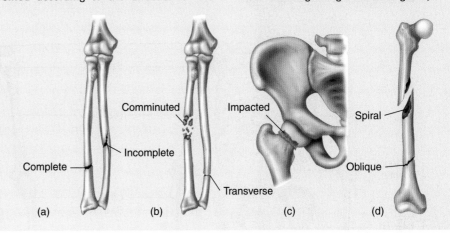

Comminuted
Incomplete
Complete
Transverse
Impacted
Spiral
Oblique

(a)   (b)   (c)   (d)

Table 6.1   Number of Named Bones Listed by Category

Bones			Number	Bones	Number
Axial Skeleton				*Thoracic Cage*	
Skull				Ribs	24
Braincase				Sternum (3 parts, sometimes considered 3 bones)	1
Paired		Parietal	2	*TOTAL THORACIC CAGE*	25
		Temporal	2	*TOTAL AXIAL SKELETON*	80
Unpaired		Frontal	1		
		Occipital	1	**Appendicular Skeleton**	
		Sphenoid	1	*Pectoral Girdle*	
		Ethmoid	1	Scapula	2
Face				Clavicle	2
Paired		Maxilla	2	*Upper Limb*	
		Zygomatic	2	Humerus	2
		Palatine	2	Ulna	2
		Nasal	2	Radius	2
		Lacrimal	2	Carpals	16
		Inferior nasal concha	2	Metacarpals	10
Unpaired		Mandible	1	Phalanges	28
		Vomer	1	*TOTAL GIRDLE AND UPPER LIMB*	64
		TOTAL SKULL	22	*Pelvic Girdle*	
Auditory ossicles		Malleus	2	Coxa	2
		Incus	2	*Lower Limb*	
		Stapes	2	Femur	2
		TOTAL	6	Tibia	2
Hyoid			1	Fibula	2
				Patella	2
Vertebral Column				Tarsals	14
Cervical vertebrae			7	Metatarsals	10
Thoracic vertebrae			12	Phalanges	28
Lumbar vertebrae			5	*TOTAL GIRDLE AND LOWER LIMB*	62
Sacrum			1	*TOTAL APPENDICULAR SKELETON*	126
Coccyx			1	*TOTAL BONES*	206
		TOTAL VERTEBRAL COLUMN	26		

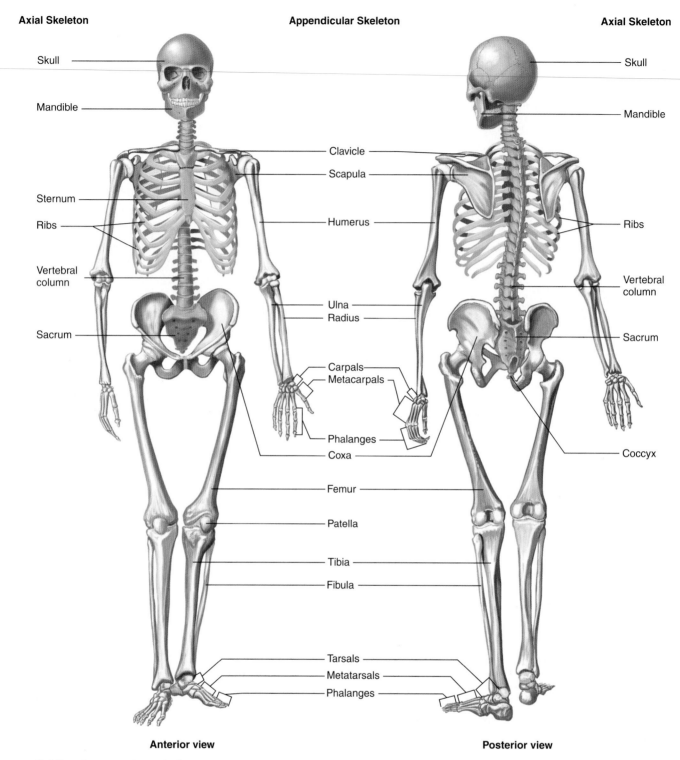

Axial Skeleton

Skull

Mandible

Sternum

Ribs

Vertebral
column

Sacrum

Appendicular Skeleton

Clavicle

Scapula

Humerus

Ulna
Radius

Carpals
Metacarpals

Phalanges
Coxa

Femur

Patella

Tibia

Fibula

Tarsals
Metatarsals
Phalanges

Axial Skeleton

Skull

Mandible

Ribs

Vertebral
column

Sacrum

Coccyx

Anterior view

Posterior view

Figure 6.10   The Complete Skeleton

(The skeleton is not shown in the anatomic position.)

with the rest of the skull. There are also three auditory ossicles (os′i-klz) in each middle ear (six total).

Many students studying anatomy never see the separate, individual bones of the skull. Even if they do, it makes more sense from a functional, or clinical, perspective to study most of the bones as they appear together in the intact skull. Many of the anatomical features of the skull cannot be fully

appreciated by examining the separate bones. For example, several ridges on the skull cross more than one bone, and several foramina are located between bones rather than within a single bone. For these reasons, it is more relevant to think of the skull, excluding the mandible, as a single unit. The major features of the intact skull are therefore described from four views.

Table 6.2 General Anatomical Terms for Various Features of Bones

Term	Description
Major Features	
Body, shaft	Main portion
Head	Enlarged (often rounded) end
Neck	Constricted area between head and body
Condyle	Smooth, rounded articular surface
Facet	Small, flattened articular surface
Crest	Prominent ridge
Process	Prominent projection
Tubercle or tuberosity	Knob or enlargement
Trochanter	Large tuberosity found only on the proximal femur
Epicondyle	Enlargement near or above a condyle
Openings or Depressions	
Foramen	Hole
Canal, meatus	Tunnel
Fissure	Cleft
Sinus	Cavity
Fossa	Depression

Lateral View

The **parietal** (pă-rī′ĕ-tăl, wall) and **temporal** (tem′pŏ-răl, refers to time; the hairs of the temples turn white, indicating the passage of time) **bones** form a large portion of the side of the head (figure 6.11). These two bones join each other on the side of the head at the **squamous** (skwā′mŭs, scalelike) **suture.** A suture is a joint uniting bones of the skull. Anteriorly, the parietal bone is joined to the **frontal** (forehead) **bone** by the **coronal** (kōr′ŏ-năl, *corona,* crown) **suture,** and posteriorly it is joined to the **occipital** (ok-sip′i-tăl, back of the head) bone by the **lambdoid** (lam′doyd, shaped like the Greek letter lambda, λ) **suture.** A prominent feature of the temporal bone is a large opening, the **external acoustic meatus** (mē-ā′tus), a canal that enables sound waves to reach the eardrum. The **mastoid** (mas′toyd, resembling a breast) **process** of the temporal bone can be seen and felt as a prominent lump just posterior to the ear. Important neck muscles involved in rotation of the head attach to the mastoid process.

Part of the **sphenoid** (sfē′noyd, wedge-shaped) **bone** can be seen immediately anterior to the temporal bone. Although it appears to be two small, paired bones on either side of the skull, the sphenoid bone is actually a single bone that extends completely across the skull. It resembles a butterfly, with its body in the center of the skull and its wings extending to the sides

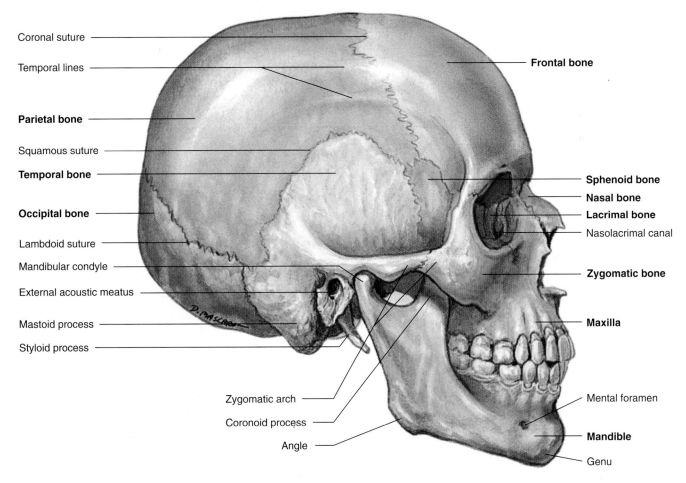

Coronal suture
Temporal lines
Parietal bone
Squamous suture
Temporal bone
Occipital bone
Lambdoid suture
Mandibular condyle
External acoustic meatus
Mastoid process
Styloid process
Zygomatic arch
Coronoid process
Angle

Frontal bone
Sphenoid bone
Nasal bone
Lacrimal bone
Nasolacrimal canal
Zygomatic bone
Maxilla
Mental foramen
Mandible
Genu

Figure 6.11   The Skull as Seen from the Right Side   **Lateral view**

of the skull. Anterior to the sphenoid bone is the **zygomatic** (zī-gō-mat′ik, yoke) **bone,** or cheekbone, which can be easily felt. The **zygomatic arch,** which consists of joined processes of the temporal and zygomatic bones, forms a bridge across the side of the face and provides a major attachment site for a muscle moving the mandible.

The **maxilla** (mak-sil′ă, jawbone) forms the upper jaw, and the **mandible** (man′di-bl, a jaw) forms the lower jaw. The maxilla articulates by sutures to the temporal bone. The maxilla contains the superior set of teeth, and the mandible contains the inferior teeth.

Frontal View

The major structures seen from the frontal view are the frontal bone, the zygomatic bones, the maxillae, and the mandible (figure 6.12a). The teeth are very prominent in this

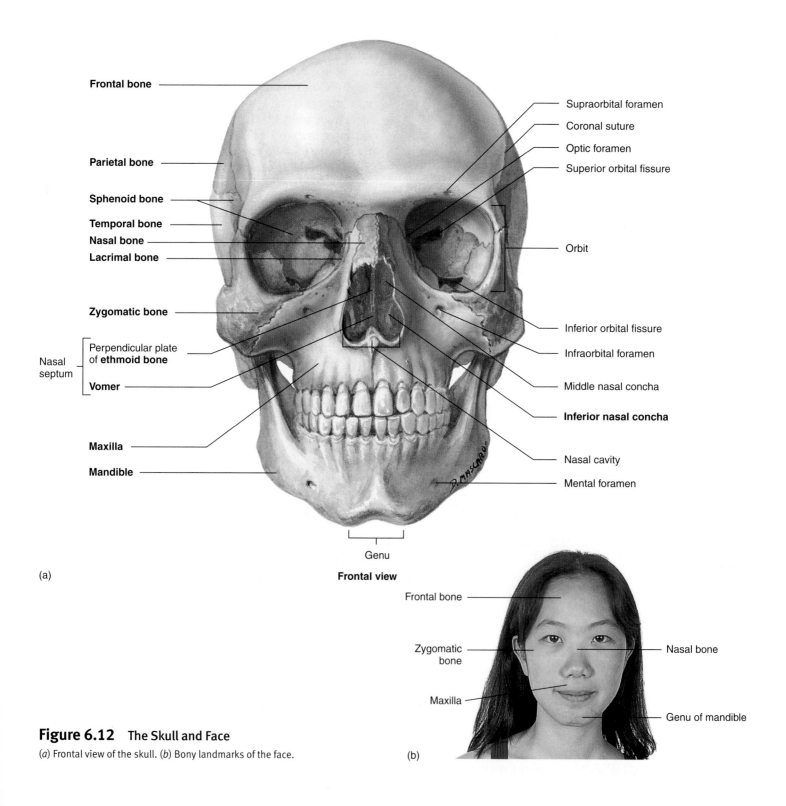

(a)

Frontal view

Figure 6.12 The Skull and Face

(a) Frontal view of the skull. (b) Bony landmarks of the face.

(b)

view. Many bones of the face can be easily felt through the skin (figure 6.12*b*).

From this view, the most prominent openings into the skull are the **orbits** (ōr′bitz, eye sockets) and the **nasal cavity.** The orbits are cone-shaped fossae, so named because of the rotation of the eyes within them. The bones of the orbits provide both protection for the eyes and attachment points for the muscles that move the eyes. The orbit is a good example of why it is valuable to study the skull as an intact structure. No fewer than seven bones come together to form the orbit, and, for the most part, the contribution of each bone to the orbit cannot be appreciated when the bones are examined individually.

Each orbit (see figure 6.12*a*) has several openings through which structures communicate with other cavities. The largest of these are the **superior** and **inferior orbital fissures.** They provide openings through which nerves and blood vessels communicate with the orbit or pass to the face. The optic nerve, for the sense of vision, passes from the eye through the **optic foramen** and enters the cranial cavity. The **nasolacrimal** (nā-zō-lak′ri-măl, *nasus,* nose + *lacrima,* tear) **canal** (see figure 6.11) passes from the orbit into the nasal cavity. It contains a duct that carries tears from the eyes to the nasal cavity. A small **lacrimal** (lak′ri-măl, tear) bone can be seen in the orbit just above the opening of this canal (see figure 6.11).

P R E D I C T   3
Why does your nose run when you cry?

The nasal cavity is divided into right and left halves by a **nasal septum** (sep′tŭm, wall) (see figure 6.12*a*). The bony part of the nasal septum consists primarily of the **vomer** (vō′mer, shaped like a plowshare) inferiorly and the **perpendicular plate** of the **ethmoid** (eth′moyd, sieve-shaped) **bone** superiorly. The anterior part of the nasal septum is formed by cartilage.

The external part of the nose is formed mostly of cartilage. The bridge of the nose is formed by the **nasal bones.**

Each of the lateral walls of the nasal cavity has three bony shelves, the **nasal conchae** (kon′kē, resembling a conch shell). The inferior nasal concha is a separate bone, and the middle and superior conchae are projections from the ethmoid bone. The conchae function to increase the surface area in the nasal cavity. The increased surface area of the overlying epithelium facilitates moistening and warming of the air inhaled through the nose (see chapter 15).

Several of the bones associated with the nasal cavity have large cavities within them, called the **paranasal** (par-ă-nā′săl, *para,* alongside) **sinuses** (figure 6.13), which open into the nasal cavity. The sinuses decrease the weight of the skull and act as resonating chambers during voice production. Compare the normal voice to the voice of a person who has a cold and whose sinuses are "stopped up." The sinuses are named for the bones where they are located and include the frontal, maxillary, ethmoidal, and sphenoidal sinuses.

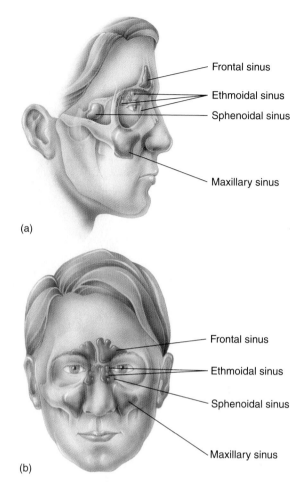

(a)

(b)

Figure 6.13   The Paranasal Sinuses
(*a*) Lateral view. (*b*) Anterior view.

The skull has additional sinuses, the **mastoid air cells,** which are located inside the mastoid processes of the temporal bone. These air cells open into the middle ear instead of into the nasal cavity. An auditory tube connects the middle ear to the nasopharynx (upper part of throat).

Interior of the Cranial Cavity

When the floor of the cranial cavity is viewed from above with the roof cut away (figure 6.14), it can be divided roughly into three cranial fossae (anterior, middle, and posterior), which are formed as the developing skull conforms to the shape of the brain. The bones forming the floor of the cranial cavity, from anterior to posterior, are the frontal, ethmoid, sphenoid, temporal, and occipital. Several foramina can be seen in the floor of the middle fossa (see figure 6.14). These allow passage of nerves and blood vessels through the skull. For example, the foramen rotundum and foramen ovale transmit important nerves to the face. A major artery to the meninges (the membranes around the brain) passes through the foramen spinosum. The internal carotid artery passes through the carotid canal and the internal jugular vein passes through the jugular foramen (see chapter 13).

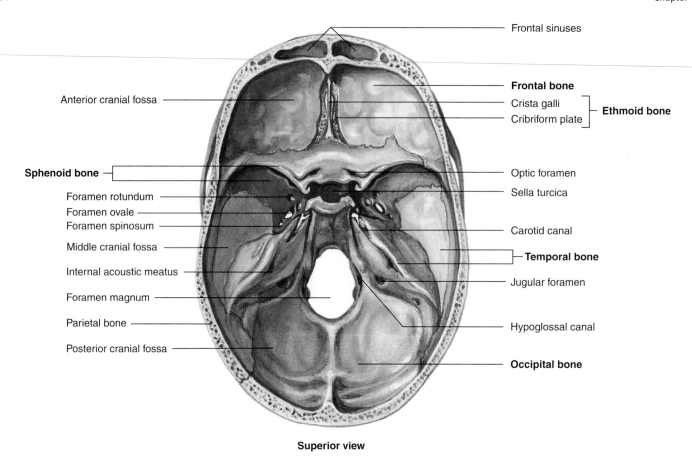

Superior view

Figure 6.14   Floor of the Cranial Cavity

The roof of the skull has been removed, and the floor is viewed from above.

The large **foramen magnum,** through which the spinal cord joins the brain, is located in the posterior fossa. The central region of the sphenoid bone is modified into a structure resembling a saddle, the **sella turcica** (sel′ă tŭr′sĭ-kă, Turkish saddle), which contains the pituitary gland.

Base of Skull Seen from Below

Many of the same foramina that can be seen in the interior of the skull also can be seen in the base of the skull, when seen from below, with the mandible removed (figure 6.15). Other specialized structures, such as processes for muscle attachments, can also be seen. The foramen magnum is located in the occipital bone near the center of the skull base. **Occipital condyles** (ok-sip′i-tăl kon′dīlz), the smooth points of articulation between the skull and the vertebral column, are located beside the foramen magnum.

Two long, pointed **styloid** (stī′loyd, stylus or penshaped) **processes** project from the inferior surface of the temporal bone. Muscles involved in movement of the tongue, the hyoid bone, and the pharynx (throat) originate from this process. The **mandibular fossa,** where the mandible articulates with the temporal bone, is anterior to the mastoid process.

The **hard palate** (pal′ăt) forms the floor of the nasal cavity and the roof of the mouth. The anterior two-thirds are formed by the maxillae, and the posterior one-third by the **palatine** (pal′ă-tīn) **bones.** The connective tissue and muscles that make up the **soft palate** extend posteriorly from the hard or bony palate. The hard and soft palates function to separate the nasal cavity and nasopharynx from the mouth, enabling us to chew and breathe at the same time.

Hyoid

The **hyoid bone** (figure 6.16) is an unpaired, U-shaped bone. It is not part of the skull (see table 6.1) and has no direct bony attachment to the skull. Muscles and ligaments attach it to the skull. The hyoid bone provides an attachment for some tongue muscles, and it's also an attachment point for important neck muscles that elevate the larynx during speech or swallowing.

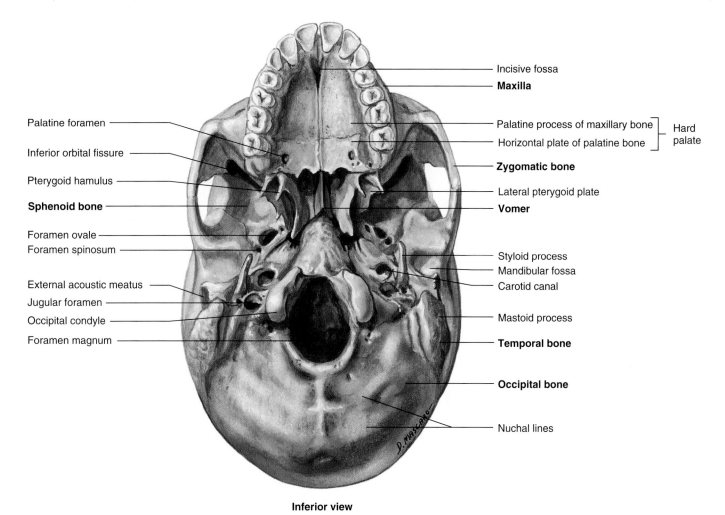

Inferior view

Figure 6.15   Base of the Skull as Seen from Below

Viewed from below, mandible removed.

Vertebral Column

The **vertebral column,** or backbone, is the central axis of the skeleton, extending from the base of the skull to slightly past the end of the pelvis. It usually consists of 26 individual bones, grouped into five regions (figure 6.17; see table 6.1): 7 **cervical** (ser′vĭ-kal, neck) **vertebrae** (ver′tĕ-brē, *verto,* to turn), 12 **thoracic** (thō-ras′ik) **vertebrae,** 5 **lumbar** (lŭm′bar) **vertebrae,** 1 **sacral** (sā′krăl) **bone,** and 1 **coccygeal** (kok-sij′ē-ăl) **bone.** For convenience, each of the five regions is identified by a letter and the vertebrae within each region by numbers: C1–C7, T1–T12, L1–L5, S, and CO.

The adult vertebral column has four major curvatures (see figure 6.17). The cervical region curves anteriorly, the thoracic region curves posteriorly, the lumbar region curves anteriorly, and the sacral and coccygeal regions together curve posteriorly.

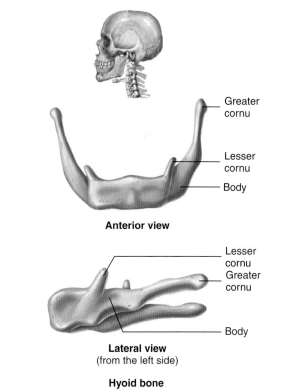

Anterior view

Lateral view
(from the left side)

Hyoid bone

Figure 6.16   Hyoid Bone

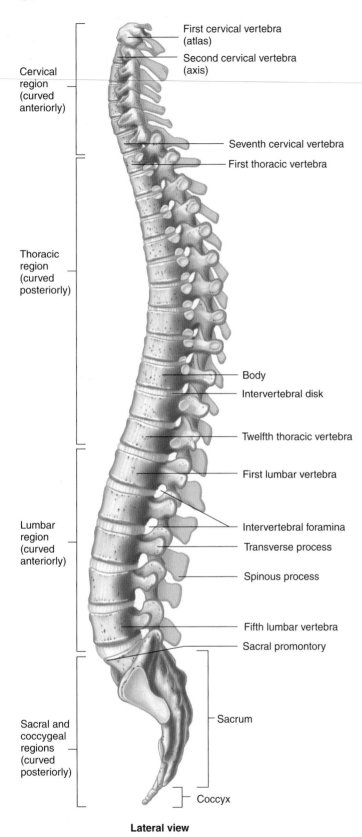

Figure 6.17 Vertebral Column

Complete column viewed from the left side.

Cervical region (curved anteriorly)

First cervical vertebra (atlas)

Second cervical vertebra (axis)

Seventh cervical vertebra

First thoracic vertebra

Thoracic region (curved posteriorly)

Body

Intervertebral disk

Twelfth thoracic vertebra

First lumbar vertebra

Intervertebral foramina

Transverse process

Spinous process

Lumbar region (curved anteriorly)

Fifth lumbar vertebra

Sacral promontory

Sacral and coccygeal regions (curved posteriorly)

Sacrum

Coccyx

Lateral view

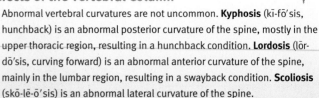

Defects of the Vertebral Column

Abnormal vertebral curvatures are not uncommon. **Kyphosis** (kĭ-fō′sis, hunchback) is an abnormal posterior curvature of the spine, mostly in the upper thoracic region, resulting in a hunchback condition. **Lordosis** (lōr-dō′sis, curving forward) is an abnormal anterior curvature of the spine, mainly in the lumbar region, resulting in a swayback condition. **Scoliosis** (skō-lē-ō′sis) is an abnormal lateral curvature of the spine.

The vertebral column performs the following five major functions: (1) supports the weight of the head and trunk; (2) protects the spinal cord; (3) allows spinal nerves to exit the spinal cord; (4) provides a site for muscle attachment; and (5) permits movement of the head and trunk.

General Plan of the Vertebrae

Each vertebra consists of a body, an arch, and various processes (figure 6.18). The weight-bearing portion of each vertebra is the **body.** The vertebral bodies are separated by **intervertebral disks** (see figure 6.17), which are formed by dense fibrous connective tissue. The **vertebral arch** surrounds a large opening called the **vertebral foramen.** The vertebral foramina of all the vertebrae form the **vertebral canal,** where the spinal cord is located. The vertebral canal protects the spinal cord from injury. Each vertebral arch consists of two **pedicles** (ped′ĭ-klz, feet), which extend from the body to the transverse process of each vertebra, and two **laminae** (lam′i-nē, thin plates), which extend from the transverse processes to the spinous process. A **transverse process** extends laterally from each side of the arch, between the pedicle and lamina, and a single **spinous process** projects dorsally from where the two laminae meet. The spinous processes can be seen and felt as a series of lumps down the midline of the back (the spinous processes can be seen in figure 6.24). The transverse and spinous processes provide attachment sites for muscles that move the vertebral column. Spinal nerves exit the spinal cord through the **intervertebral foramina,**

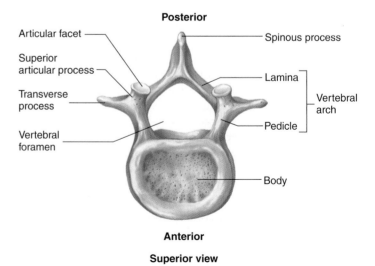

Posterior

Articular facet

Superior articular process

Transverse process

Vertebral foramen

Spinous process

Lamina

Vertebral arch

Pedicle

Body

Anterior

Superior view

Figure 6.18 Vertebra

which are formed by notches in the pedicles of adjacent vertebrae (see figure 6.17). Each vertebra has a superior and inferior **articular process** where the vertebrae articulate with each other. Each articular process has a smooth "little face" called an **articular facet** (fas′et).

Regional Differences in Vertebrae

The **cervical vertebrae** (figure 6.19a–c) have very small bodies, except for the atlas, which has no body. Each of the transverse processes has a transverse foramen through which the vertebral arteries pass toward the brain. Several of the cervical vertebrae also have partly split spinous processes. The first cervical vertebra (figure 6.19a) is called the **atlas** because it holds up the head, as Atlas in classical mythology held up the world. Movement between the atlas and the occipital bone is responsible for a "yes" motion of the head. It also allows a slight tilting of the head from side to side. The second cervical vertebra (figure 6.19b) is called the **axis**

because a considerable amount of rotation occurs at this vertebra, as in shaking the head "no." This rotation occurs around a process called the **dens** (denz), which extends superiorly from the axis.

The **thoracic vertebrae** (figure 6.19d) possess long, thin spinous processes that are directed inferiorly. The thoracic vertebrae also have extra articular facets on their lateral surfaces that articulate with the ribs.

The **lumbar vertebrae** (figure 6.19e) have large, thick bodies and heavy, rectangular transverse and spinous processes. The superior articular facets of the lumbar vertebrae face medially, whereas the inferior articular facets face laterally. This arrangement tends to "lock" adjacent lumbar vertebrae together, giving the lumbar part of the vertebral column more strength. The articular facets in other regions of the vertebral column have a more "open" position, allowing for more rotational movement but less stability than in the lumbar region.

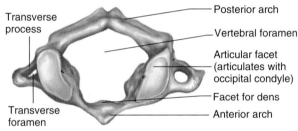

(a) **Atlas (first cervical vertebra), superior view**

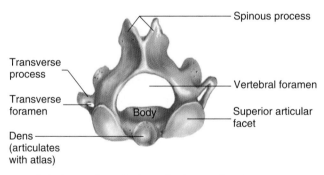

(b) **Axis (second cervical vertebra), superior view**

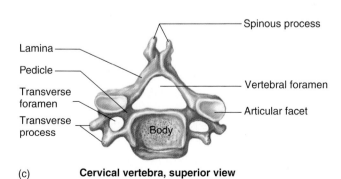

(c) **Cervical vertebra, superior view**

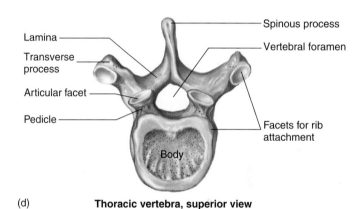

(d) **Thoracic vertebra, superior view**

(e) **Lumbar vertebra, superior view**

Figure 6.19 Regional Differences in Vertebrae

Posterior is shown at the top of each illustration.

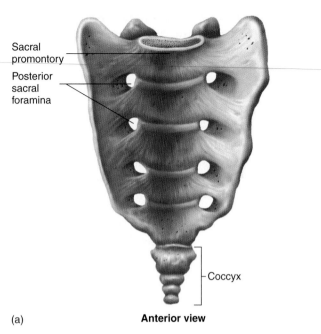

Sacral promontory

Posterior sacral foramina

Coccyx

(a)   **Anterior view**

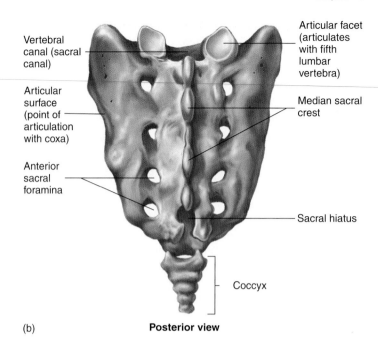

Vertebral canal (sacral canal)

Articular surface (point of articulation with coxa)

Anterior sacral foramina

Articular facet (articulates with fifth lumbar vertebra)

Median sacral crest

Sacral hiatus

Coccyx

(b)   **Posterior view**

Figure 6.20   Sacrum

Vertebral Column Damage

Because the cervical vertebrae are relatively delicate and have small bodies, dislocations and fractures are more common in this area than in other regions of the vertebral column. Because the lumbar vertebrae have massive bodies and carry a large amount of weight, ruptured intervertebral disks are more common in this area than in other regions of the column. Each intervertebral disk is made up of a ring of fibrous connective tissue with a softer center of semifluid tissue. The weight of the body may compress the disk, causing the fibrous ring to bulge or even break. This causes compression of the nerves exiting the intervertebral foramina (figure C). The coccyx is easily broken in falls during which a person sits down hard on a solid surface. Also, a mother's coccyx may be fractured during childbirth.

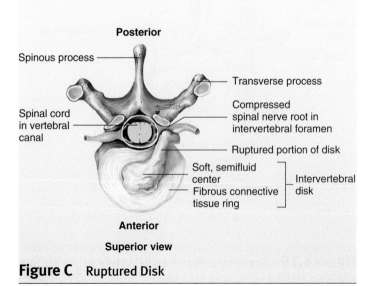

Posterior

Spinous process

Spinal cord in vertebral canal

Transverse process

Compressed spinal nerve root in intervertebral foramen

Ruptured portion of disk

Soft, semifluid center

Fibrous connective tissue ring

Intervertebral disk

Anterior

Superior view

Figure C   Ruptured Disk

The five sacral vertebrae are fused into a single bone called the **sacrum** (figure 6.20). The spinous processes of the first four sacral vertebrae form the **median sacral crest.** The spinous process of the fifth vertebra does not form, leaving a **sacral hiatus** (hī-ā-tŭs) at the inferior end of the sacrum, which is often the site of "caudal" anesthetic injections given just before childbirth. The anterior edge of the body of the first sacral vertebra bulges to form the **sacral promontory** (prom'on-tō-rē) (see figure 6.17), a landmark that can be felt during a vaginal examination. It is used as a reference point during measurement to determine if the pelvic openings are large enough to allow for normal vaginal delivery of a baby.

The **coccyx** (kok'siks, shaped like a cuckoo's bill), or tail-bone, usually consists of four more-or-less fused vertebrae. The vertebrae of the coccyx do not have the typical structure of most other vertebrae. They consist of extremely reduced vertebral bodies, without the foramina or processes, usually fused into a single bone.

Thoracic Cage

The **thoracic cage,** or **rib cage,** protects the vital organs within the thorax and prevents the collapse of the thorax during respiration. It consists of the thoracic vertebrae, the ribs with their associated cartilages, and the sternum.

Ribs and Costal Cartilages

The 12 pairs of ribs (figure 6.21) can be divided into true and false ribs. The superior seven pairs, called the **true ribs,** attach directly to the sternum by means of costal cartilages. The inferior five pairs, called **false ribs,** do not attach directly to the sternum. Three pairs, ribs 8 through 10, attach to the sternum by a common cartilage; two pairs, ribs 11 and 12, called the **floating ribs,** do not attach to the sternum.

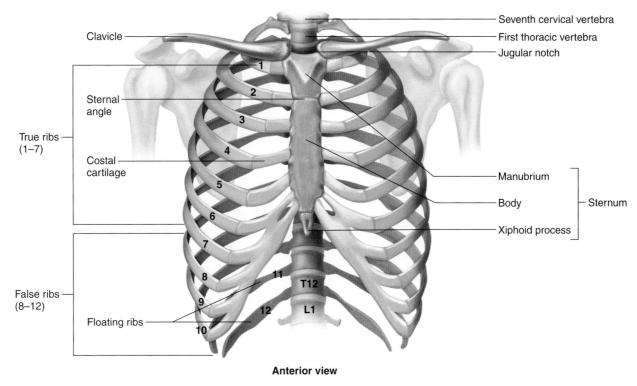

Figure 6.21   Thoracic Cage

Han D. Mann's ladder fell as he was working on his roof and he landed on the ladder with his chest. Three ribs were fractured on his right side. It was difficult for Han to cough, laugh, or even breathe without severe pain in the right side of his chest. The middle ribs are those most commonly fractured and the portion of each rib that forms the lateral wall of the thorax is the weakest and most commonly broken. The pain from rib fractures occurs because the broken ends move during respiration and other chest movements, stimulating pain receptors. Broken rib ends can damage internal organs such as the lungs, spleen, liver, and diaphragm. Fractured ribs often are not dislocated, but dislocated ribs may have to be set for proper healing to occur. Binding the chest to limit movement can facilitate healing and lessen pain.

Sternum

The **sternum** (ster′nŭm), or the breastbone (see figure 6.21), is divided into three parts: the **manubrium** (mă-nū′brē-ŭm, handle), the **body,** and the **xiphoid** (zif′oyd, or zī′foyd, sword) **process.** The sternum resembles a sword, with the manubrium forming the handle, the body forming the blade,

and the xiphoid process forming the tip. At the superior end of the sternum, a depression, called the **jugular notch,** is located between the ends of the clavicles where they articulate with the sternum. A slight elevation, called the **sternal angle,** can be felt at the junction of the manubrium and the body of the sternum. This junction is an important landmark because it identifies the location of the second rib. This identification allows the ribs to be counted and, for example, allows location of the apex of the heart, which is located between the fifth and sixth ribs.

The xiphoid process is another important landmark of the sternum. During cardiopulmonary resuscitation (CPR), it is very important to place the hands over the body of the sternum rather than over the xiphoid process. If the hands are placed over the xiphoid process, the pressure applied during CPR could break the xiphoid process and drive it into an underlying abdominal organ such as the liver, causing internal bleeding.

Appendicular Skeleton

The **appendicular** (ap′en-dik′ū-lăr, *ap-pendo,* to hang something on) skeleton consists of the bones of the upper and lower limbs, as well as the girdles, which attach the limbs to the axial skeleton (figure 6.22).

Pectoral Girdle

The **pectoral** (pek′tŏ-răl), or **shoulder, girdle** consists of four bones, two scapulae and two clavicles, which attach the upper limb to the body: the **scapula** (skap′ū-lă), or **shoulder blade**

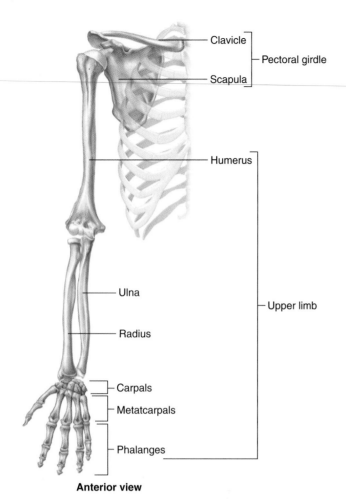

Anterior view

Figure 6.22   Bones of the Pectoral Girdle and Right Upper Limb

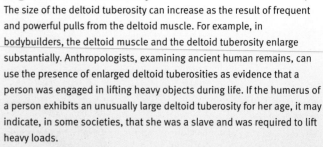

Upper Limb

The upper limb consists of the bones of the arm, forearm, wrist, and hand (see figure 6.22).

Arm

The **arm** is the region between the shoulder and the elbow and contains the **humerus** (hū′mer-ŭs, shoulder) (figure 6.25). The proximal end of the humerus has a smooth, rounded **head,** which attaches the humerus to the scapula at the glenoid cavity. Around the edge of the humeral head is the anatomical neck. This neck is not easily accessible if the proximal end of the humerus must be removed and replaced. A more accessible site for surgical removal is at the surgical neck, located at the proximal end of the humeral shaft. Lateral to the head are two tubercles, a **greater tubercle** and a **lesser tubercle.** Muscles originating on the scapula attach to the greater and lesser tubercles and hold the humerus to the scapula. Approximately one-third of the way down the shaft of the humerus, on the lateral surface, is the **deltoid tuberosity,** where the deltoid muscle attaches. The distal end of the humerus is modified into specialized condyles that connect the humerus to the forearm bones. **Epicondyles** (ep′i-kon′dīlz, *epi,* upon + *kondylos,* knuckle) on the distal end of the humerus, just lateral to the condyles, provide attachment sites for forearm muscles.

Forearm

The **forearm** has two bones: the **ulna** (ŭl′nă) on the medial (little finger) side of the forearm and the **radius** on the lateral (thumb) side (figure 6.26). The proximal end of the ulna forms a **semilunar notch** that fits tightly over the end of the humerus, forming most of the elbow joint. Just proximal to the semilunar notch is an extension of the ulna, called the **olecranon** (ō-lek′ră-non, elbow) **process,** which can be felt as the point of the elbow (the olecranon process is shown in figure 6.28). Just distal to the semilunar notch is a **coronoid** (kōr′ŏ-noyd, crow-like) **process,** which helps complete the "grip" of the ulna on the distal end of the humerus. The distal end of the ulna forms a head, which articulates with the bones of the wrist, and a **styloid process** is located on its medial side. The ulnar head can be seen as a prominent lump on the posterior ulnar side of the wrist. The proximal end of the radius has a head by which the radius articulates with both the humerus and the ulna. The radius does not attach as firmly to the humerus as does the ulna. The radial

(figures 6.23 and 6.24; also see figure 6.10), and the **clavicle** (klav′i-kl, key), or **collarbone** (figure 6.23*c* and *d*). The scapula is a flat, triangular bone with three large fossae, where muscles extending to the arm are attached. A fourth fossa, the **glenoid** (glen′oyd) **cavity,** is where the head of the humerus connects to the scapula. A ridge, called the **spine,** runs across the posterior surface of the scapula. A projection, called the **acromion** (ă-krō′mē-on, *akron,* tip + *omos,* shoulder) **process,** extends from the scapular spine to form the point of the shoulder. The clavicle articulates with the scapula at the acromion process. The proximal end of the clavicle is attached to the sternum, providing the only bony attachment of the scapula to the remainder of the skeleton. The **coracoid** (kōr′ă-koyd, crow's beak) **process** curves below the clavicle and provides attachment for arm and chest muscles.

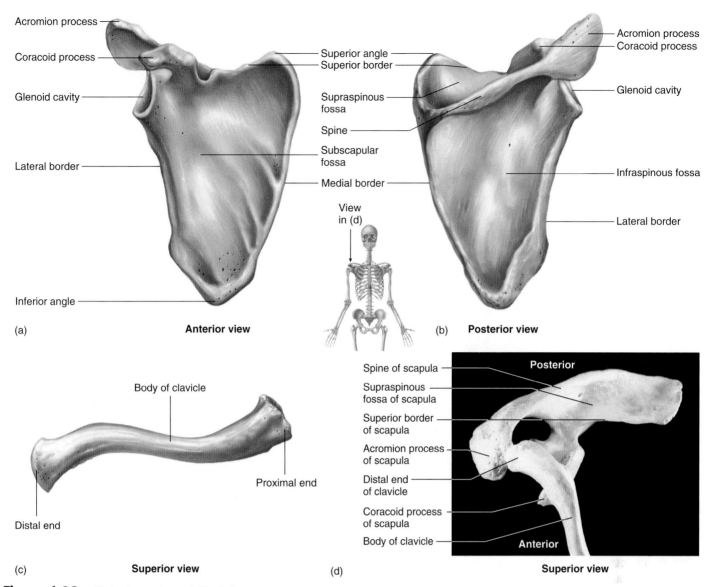

Figure 6.23 Right Scapula and Clavicle

(*a*) Right scapula, anterior view. (*b*) Right scapula, posterior view. (*c*) Right clavicle, superior view. (*d*) Photograph of the right scapula and clavicle from a superior view. showing the relationship between the distal end of the clavicle and the acromion process of the scapula.

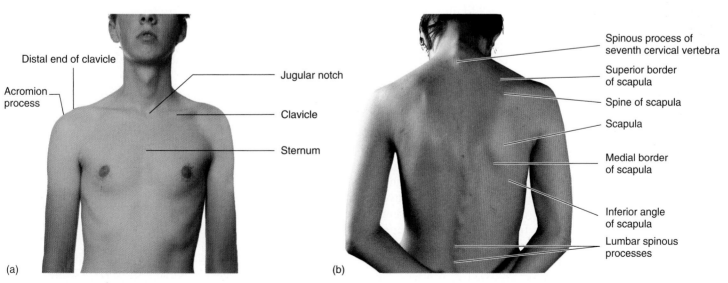

Figure 6.24 Bones of the Thorax

(*a*) Surface anatomy showing bones of the anterior thorax. (*b*) Surface anatomy showing bones of the posterior vertebral column and scapula.

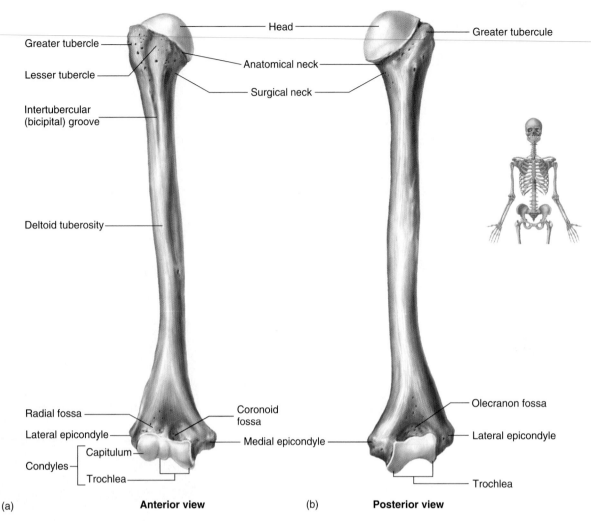

Head

Greater tubercle
Lesser tubercle

Anatomical neck
Surgical neck

Greater tubercule

Intertubercular
(bicipital) groove

Deltoid tuberosity

Radial fossa
Lateral epicondyle
Capitulum
Condyles
Trochlea

Coronoid
fossa

Medial epicondyle

Olecranon fossa

Lateral epicondyle

Trochlea

(a)   **Anterior view**   (b)   **Posterior view**

Figure 6.25   Right Humerus

head rotates against the humerus and ulna. Just distal to the radial head is a **radial tuberosity,** where one of the arm muscles,
the biceps brachii, attaches. The distal end of the radius articulates with the wrist bones. A styloid process is located on the lateral side of the distal end of the radius. The radial and ulnar
styloid processes provide attachments for ligaments of the
wrist.

Wrist

The **wrist** is a relatively short region between the forearm and
hand and is composed of eight **carpal** (kar′păl, wrist) bones
(figure 6.27). These eight bones are the scaphoid (skaf′oyd,
boatlike), lunate (lū′nāt, moon), triquetrum (trī-kwē′trŭm,
three-cornered), pisiform (pis′i-fōrm, pea-shaped), trapezium (tra-pē′zē-ŭm, table), trapezoid (trap′ĕ-zoyd, resembling a table), capitate (kap′i-tāt, head), and hamate (ha′māt,
hook). The eight carpal bones are arranged in two rows of
four bones each and form a slight curvature that is concave
anteriorly and convex posteriorly. A number of mnemonics

have been developed to help students remember the carpal
bones. This one allows students to remember them in order
from lateral to medial for the proximal row (top) and from
medial to lateral (by the thumb) for the distal row: **S**o **L**ong
Top **P**art, **H**ere **C**omes **T**he **T**humb—that is, **S**caphoid,
Lunate, **T**riquetrum, **P**isiform, **H**amate, **C**apitate, **T**rapezoid,
and **T**rapezium.

Carpal Tunnel Syndrome

The bones and ligaments on the anterior side of the wrist form a **carpal
tunnel,** which does not have much "give." Tendons and nerves pass
from the forearm through the carpal tunnel to the hand. Fluid and
connective tissue can accumulate in the carpal tunnel as a result of
inflammation associated with overuse or trauma. The inflammation can
also cause the tendons in the carpal tunnel to enlarge. The accumulated
fluid and enlarged tendons can apply pressure to a major nerve passing
through the tunnel. The pressure on this nerve causes **carpal tunnel
syndrome,** the symptoms of which are tingling, burning, and numbness
in the hand.

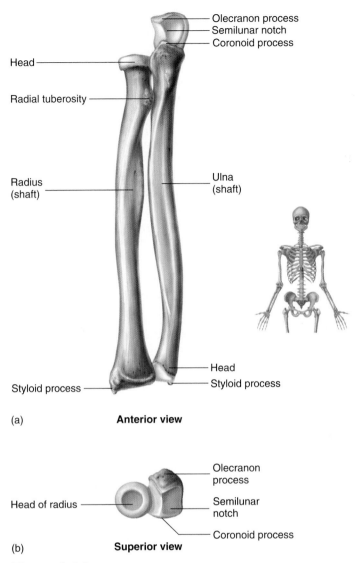

Olecranon process
Semilunar notch
Coronoid process

Head

Radial tuberosity

Radius
(shaft)

Ulna
(shaft)

Head

Styloid process

Styloid process

(a)   **Anterior view**

Olecranon
process

Head of radius

Semilunar
notch

Coronoid process

(b)   **Superior view**

Figure 6.26   Right Ulna and Radius

(*a*) Anterior view of right ulna and radius. (*b*) Proximal ends of the right ulna and radius.

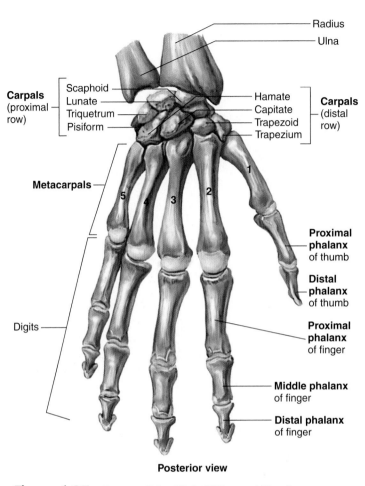

Radius
Ulna

Carpals
(proximal
row)

Scaphoid
Lunate
Triquetrum
Pisiform

Hamate
Capitate
Trapezoid
Trapezium

Carpals
(distal
row)

Metacarpals

Digits

**Proximal
phalanx**
of thumb

**Distal
phalanx**
of thumb

**Proximal
phalanx**
of finger

Middle phalanx
of finger

Distal phalanx
of finger

Posterior view

Figure 6.27   Bones of the Right Wrist and Hand

Hand

Five **metacarpals** (met′ă-kar′pălz, after the carpals) are attached to the carpal bones and form the bony framework of the hand (see figure 6.27). The metacarpals are aligned with the five **digits:** the thumb and fingers. They are numbered 1 to 5 from the thumb to the little finger. The ends, or heads, of the five metacarpals, associated with the thumb and fingers, form the knuckles (figure 6.28). Each finger consists of three small bones called **phalanges** (fă-lan′jēz, sing. phalanx, fā′langks; the Greek phalanx is a wedge of soldiers holding their spears, tips outward, in front of them). The phalanges of each finger are called proximal, middle, and distal, according to their position in the digit. The thumb has two phalanges, proximal and distal. The digits are also numbered 1 to 5, starting from the thumb.

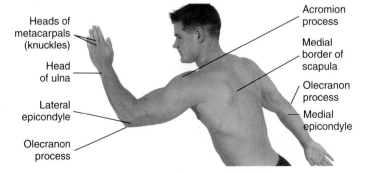

Heads of
metacarpals
(knuckles)

Head
of ulna

Lateral
epicondyle

Olecranon
process

Acromion
process

Medial
border of
scapula

Olecranon
process

Medial
epicondyle

Figure 6.28   Surface Anatomy Showing Bones of the Pectoral Girdle and Upper Limb

Pelvic Girdle

The **pelvic girdle** is the place where the lower limbs attach to the body (figure 6.29). The right and left **coxae** (kok'sē, Latin for hip), or hip bones, join each other anteriorly and the **sacrum** posteriorly to form a ring of bone called the **pelvic girdle**. The **pelvis** (pel'vis, basin) (figure 6.30) includes the pelvic girdle and the coccyx. The sacrum and coccyx form part of the pelvis but are also part of the axial skeleton. Each coxa (figure 6.31) is formed by three bones fused to one another to form a single bone. The **ilium** (il'ē-ŭm, groin) is the most superior, the **ischium** (is'kē-ŭm, Greek for hip) is inferior and posterior, and the **pubis** (pū'bis, refers to the genital hair) is inferior and anterior. An **iliac crest** can be seen along the superior margin of each ilium, and an **anterior superior iliac spine,** an important hip landmark, is located at the anterior end of the iliac crest. The coxae join each other anteriorly at the **pubic** (pū'bik) **symphysis** and join the sacrum posteriorly at the **sacroiliac** (sā-krō-il'ē-ak) **joints** (see figure 6.30). The **acetabulum** (as-ĕ-tab'ū-lŭm, vinegar cup) is the socket of the hip joint. The **obturator** (ob'too-rā-tŏr, to stop up) **foramen** is the large hole in each coxa that is closed off by muscles and other structures.

The male pelvis can be distinguished from the female pelvis because it is usually larger and more massive, but the female pelvis tends to be broader (figure 6.32 and table 6.3). Both

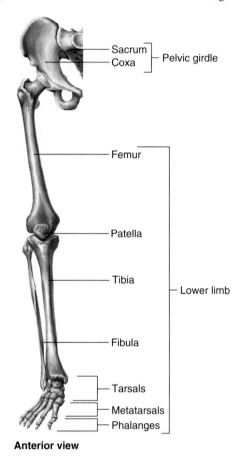

Anterior view

Figure 6.29   **Bones of the Pelvic Girdle and Right Lower Limb**

Table 6.3	Differences Between Male and Female Pelvic Girdles
Area	**Description of Difference**
General	Female pelvis somewhat lighter in weight and wider laterally, but shorter superiorly to inferiorly and less funnel-shaped; less obvious muscle attachment points in female than in male
Sacrum	Broader in female, with the inferior portion directed more posteriorly; the sacral promontory projects less anteriorly in female
Pelvic inlet	Heart-shaped in male; oval in female
Pelvic outlet	Broader and more shallow in female
Subpubic angle	Less than 90 degrees in male; 90 degrees or more in female
Ilium	More shallow and flared laterally in female
Ischial spines	Farther apart in female
Ischial tuberosities	Turned laterally in female and medially in male (not shown in figure 6.31)

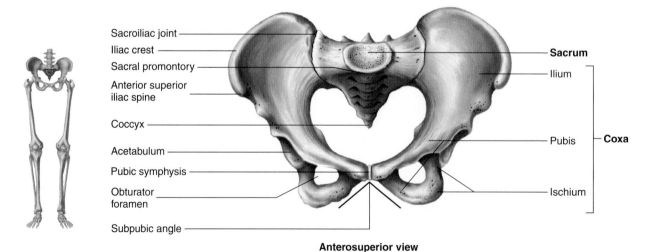

Anterosuperior view

Figure 6.30   **Anterior View of the Pelvis**

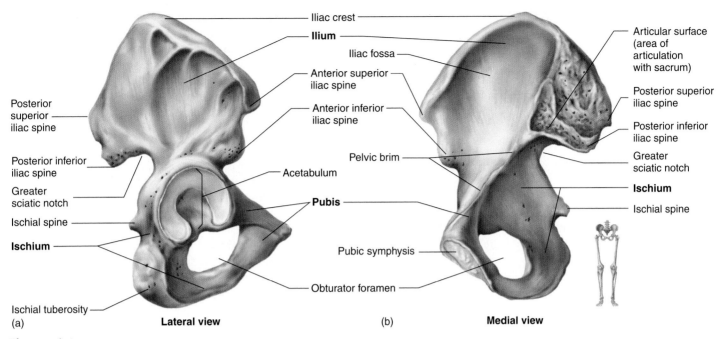

Figure 6.31 Right Coxa

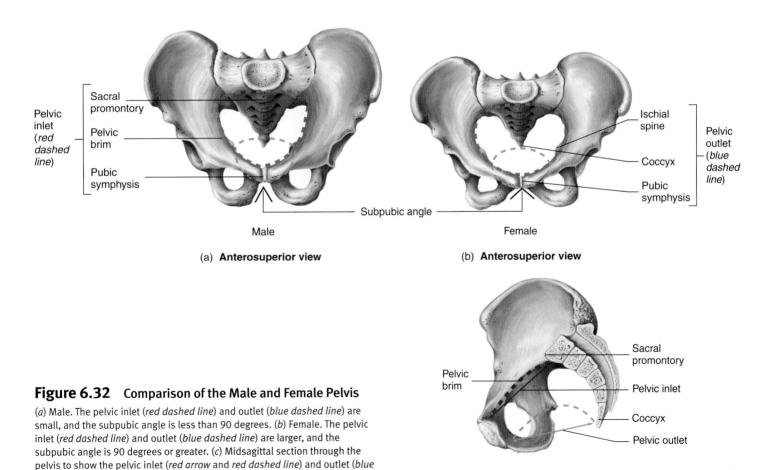

Figure 6.32 Comparison of the Male and Female Pelvis

(*a*) Male. The pelvic inlet (*red dashed line*) and outlet (*blue dashed line*) are small, and the subpubic angle is less than 90 degrees. (*b*) Female. The pelvic inlet (*red dashed line*) and outlet (*blue dashed line*) are larger, and the subpubic angle is 90 degrees or greater. (*c*) Midsagittal section through the pelvis to show the pelvic inlet (*red arrow* and *red dashed line*) and outlet (*blue arrow* and *blue dashed line*).

the inlet and outlet of the female pelvis are larger than those of the male pelvis, and the subpubic angle is greater in the female (figure 6.32a and b). The increased size of these openings helps accommodate the fetus during childbirth. The **pelvic inlet** is formed by the pelvic brim and the sacral promontory. The **pelvic outlet** is bounded by the ischial spines, the pubic symphysis, and the coccyx (figure 6.32c).

Lower Limb

The lower limb consists of the bones of the thigh, leg, ankle, and foot (see figure 6.29).

Thigh

The **thigh** (figure 6.33a) is the region between the hip and the knee. It contains a single bone called the **femur.** The **head** of the femur articulates with the acetabulum of the coxa; and the **condyles,** at the distal end of the femur, articulate with the tibia. **Epicondyles,** located medial and lateral to the condyles,

are points of ligament attachment. The femur can be distinguished from the humerus by its long neck located between the head and the **trochanters** (trō′kan-terz, runners). The trochanters are points of muscle attachment. The **patella** (pa-tel′ă), or kneecap (figure 6.33b), is located within the major tendon of the anterior thigh muscles and enables the tendon to turn the corner over the knee.

Leg

The **leg** (figure 6.34) is the region between the knee and the ankle. It contains two bones, called the **tibia** (tib′ē-ă, shinbone) and the **fibula** (fib′ū-lă, resembling a clasp or buckle). The tibia is the larger of the two and is the major weight-bearing bone of the leg. The rounded condyles of the femur rest on the flat **condyles** on the proximal end of the tibia. Just distal to the condyles of the tibia, on its anterior surface, is the **tibial tuberosity,** where the muscles of the anterior thigh attach. The fibula does not articulate with the femur but its

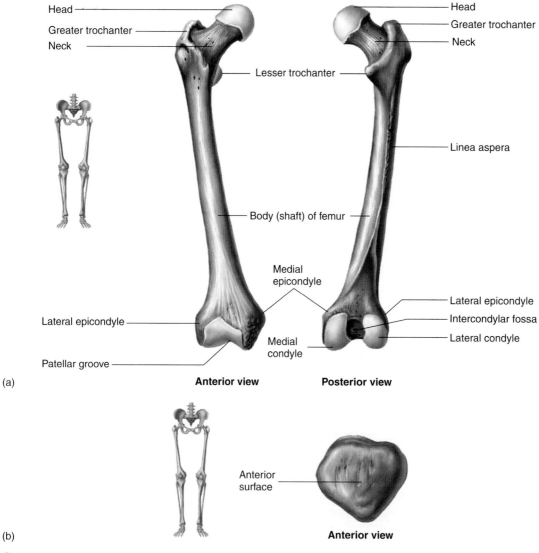

(a)

Head

Greater trochanter

Neck

Lesser trochanter

Head

Greater trochanter

Neck

Linea aspera

Body (shaft) of femur

Medial epicondyle

Lateral epicondyle

Lateral epicondyle

Intercondylar fossa

Lateral condyle

Medial condyle

Patellar groove

Anterior view   **Posterior view**

(b)

Anterior surface

Anterior view

Figure 6.33   **Right Femur and Patella**

(a) Right femur, (b) Patella.

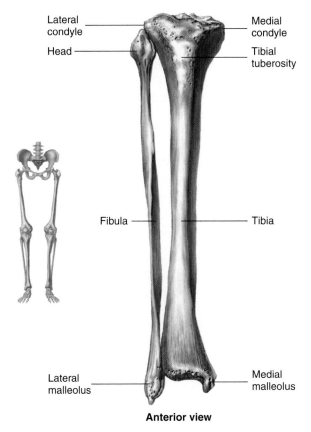

Figure 6.34   Right Tibia and Fibula

head is attached to the proximal end of the tibia. The distal ends of the tibia and fibula form a partial socket that articulates with a bone of the ankle (the talus). A prominence can be seen on each side of the ankle (these are shown in figure 6.34). These are the medial **malleolus** (mal-ē′ō-lŭs, a hammer or mallet) of the tibia and the lateral malleolus of the fibula.

Ankle

The **ankle** consists of seven **tarsal** (tar′săl, the sole of the foot) bones (figure 6.35). The tarsal bones are the **talus** (tā′lŭs, ankle bone), **calcaneus** (kal-kā′nē-ŭs, heel), **cuboid** (kū′boyd, cube-shaped), and **navicular** (nă-vik′yū-lăr, boat-shaped); and the medial, intermediate, and lateral **cuneiforms** (kū′nē-i-fōrmz, wedge-shaped). The talus articulates with the tibia and fibula to form the ankle joint, and the calcaneus forms the heel (figure 6.36). A mnemonic for the distal row is: **MILC**—that is, **M**edial, **I**ntermediate, and **L**ateral cuneiforms, and the **C**uboid. That for the proximal three bones is: **No Thanks Cow**—that is, **N**avicular, **T**alus, and **C**alcaneus.

Broken Hip

A "broken hip" usually is a break of the femoral neck. A broken hip is difficult to repair and often requires pinning to hold the femoral head to the shaft. A major complication can occur if the blood vessels between the femoral head and the acetabulum are damaged. If this occurs, the femoral head may degenerate from lack of nourishment.

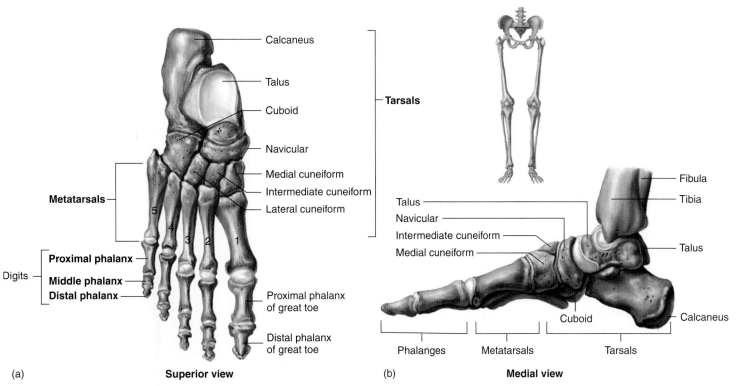

Figure 6.35   Bones of the Right Foot

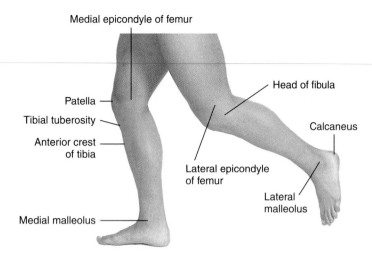

Figure 6.36 Surface Anatomy Showing Bones of the Lower Limb

Foot

The **metatarsals** (met′ă-tar′sălz, after the tarsals) and **phalanges** of the foot are arranged and numbered in a manner very similar to the metacarpals and phalanges of the hand (see figure 6.35). The metatarsals are somewhat longer than the metacarpals, whereas the phalanges of the foot are considerably shorter than those of the hand.

There are three primary **arches** in the foot, formed by the positions of the tarsals and the metatarsals, and held in place by ligaments. Two longitudinal arches extend from the heel to the ball of the foot, and a transverse arch extends across the foot. The arches function similarly to the springs of a car, allowing the foot to give and spring back.

Articulations

An **articulation,** or **joint,** is a place where two bones come together. A joint is usually considered movable, but that is not always the case. Many joints exhibit limited movement, and others are completely, or almost completely, immovable.

One method of classifying joints is a functional classification, based on the degree of motion at each joint and includes the terms **synarthrosis** (sin′ar-thrō′sis, nonmovable joint), **amphiarthrosis** (am′fi-ar-thrō′sis, slightly movable joint), and **diarthrosis** (dī-ar-thrō′sis, freely movable joint). Functional classification is somewhat restrictive and is not used in this text. Another method of classifying joints is a structural classification. Joints are classified according to the major connective tissue type that binds the bones together and according to whether there is a fluid-filled joint capsule. The three major structural classes of joints are fibrous, cartilaginous, and synovial. The structural classification with its various subclasses allows for a more precise classification and thus is used in this text.

Fibrous Joints

Fibrous joints consist of two bones that are united by fibrous tissue and that exhibit little or no movement. Joints in this group are further classified on the basis of structure as sutures, syndesmoses, or gomphoses. **Sutures** (soo′choorz) are fibrous joints between the bones of the skull (see figure 6.11). In a newborn, some parts of the sutures are quite wide and are called **fontanels** (fon′tă-nelz′), or soft spots (figure 6.37). They allow flexibility in the skull during the birth process, as well as growth of the head after birth. **Syndesmoses** (sin′dez-mō′sēz) are fibrous joints in which the bones are separated by some distance and are held together by ligaments. An example is the fibrous membrane connecting most of the distal parts of the radius and ulna. **Gomphoses** (gom-fō′sēz) consist of pegs fitted into sockets and held in place by ligaments. The joint between a tooth and its socket is a gomphosis.

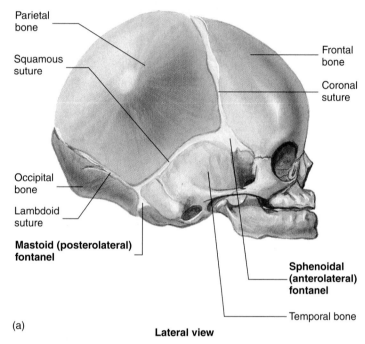

(a)   **Lateral view**

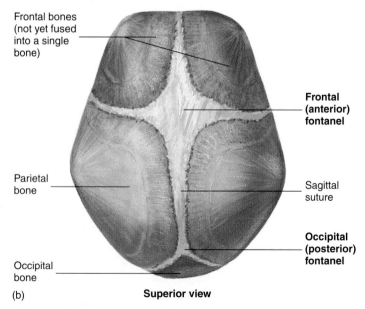

(b)   **Superior view**

Figure 6.37 Fetal Skull Showing Fontanels and Sutures

Cartilaginous Joints

Cartilaginous joints unite two bones by means of cartilage. Only slight movement can occur at these joints. Examples are the cartilage in the epiphyseal plates of growing long bones and the cartilages between the ribs and sternum. The cartilage of some cartilaginous joints, where much strain is placed on the joint, may be reinforced by the presence of additional collagen fibers. This type of cartilage, called **fibrocartilage** (see chapter 4), forms joints such as the intervertebral disks.

Synovial Joints

Synovial (si-nō′vē-ăl; Greek *syn,* together + *oon,* egg) **joints** are freely movable joints that contain synovial fluid in a cavity surrounding the ends of articulating bones. Most joints that unite the bones of the appendicular skeleton are large, synovial joints, whereas many of the joints that unite the bones of the axial skeleton are not. This pattern reflects the greater mobility of the appendicular skeleton compared with the axial skeleton.

Several features of synovial joints are important to their function (figure 6.38). The articular surfaces of bones within synovial joints are covered with a thin layer of **articular cartilage,** which provides a smooth surface where the bones meet. The **joint cavity** is filled with synovial fluid. The cavity is enclosed by a **joint capsule,** which helps hold the bones together and, at the same time, allows for movement. Portions of the fibrous part of the joint capsule may be thickened to form ligaments. In addition, ligaments and tendons outside the joint capsule contribute to the strength of the joint.

A **synovial membrane** lines the joint cavity everywhere except over the articular cartilage. The membrane produces **synovial fluid,** which is a complex mixture of polysaccharides, proteins, fat, and cells. Synovial fluid forms a thin lubricating film covering the surfaces of the joint. In certain synovial joints, the synovial membrane may extend as a pocket or sac, called a **bursa** (ber′să, pocket). Bursae are located between structures that rub together, such as where a tendon crosses a bone; they function to reduce friction, which could damage the structures involved. Inflammation of a bursa, often resulting from abrasion, is called a **bursitis.** A synovial membrane may extend as a **tendon sheath** along some tendons associated with joints (see figure 6.38).

Types of Synovial Joints

Synovial joints are classified according to the shape of the adjoining articular surfaces (figure 6.39). **Plane,** or **gliding, joints** consist of two opposed flat surfaces that glide over each other. Examples of these joints are the articular facets between vertebrae. **Saddle joints** consist of two saddle-shaped articulating surfaces oriented at right angles to each other. Movement in

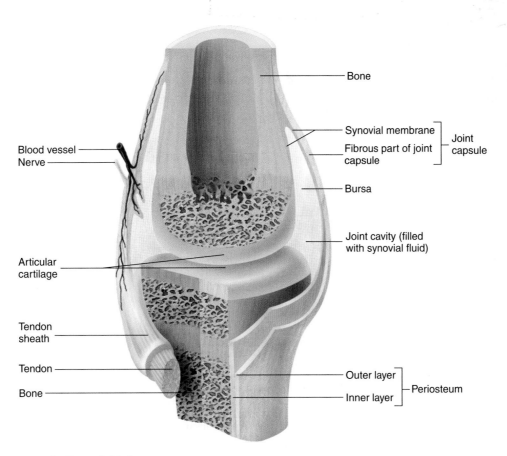

Figure 6.38 Structure of a Synovial Joint

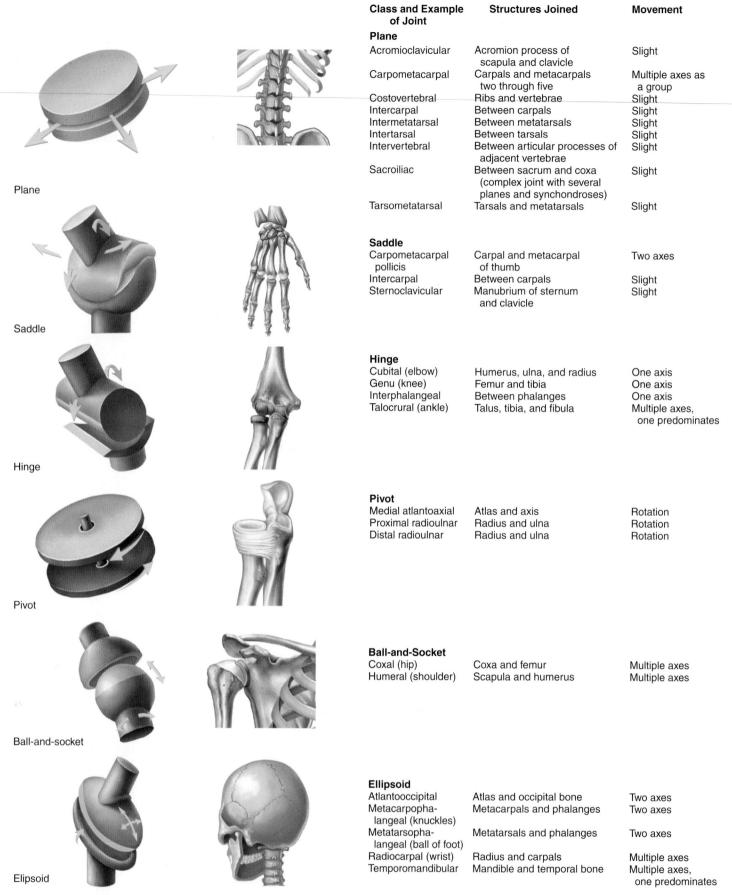

Class and Example of Joint	Structures Joined	Movement
Plane		
Acromioclavicular	Acromion process of scapula and clavicle	Slight
Carpometacarpal	Carpals and metacarpals two through five	Multiple axes as a group
Costovertebral	Ribs and vertebrae	Slight
Intercarpal	Between carpals	Slight
Intermetatarsal	Between metatarsals	Slight
Intertarsal	Between tarsals	Slight
Intervertebral	Between articular processes of adjacent vertebrae	Slight
Sacroiliac	Between sacrum and coxa (complex joint with several planes and synchondroses)	Slight
Tarsometatarsal	Tarsals and metatarsals	Slight
Saddle		
Carpometacarpal pollicis	Carpal and metacarpal of thumb	Two axes
Intercarpal	Between carpals	Slight
Sternoclavicular	Manubrium of sternum and clavicle	Slight
Hinge		
Cubital (elbow)	Humerus, ulna, and radius	One axis
Genu (knee)	Femur and tibia	One axis
Interphalangeal	Between phalanges	One axis
Talocrural (ankle)	Talus, tibia, and fibula	Multiple axes, one predominates
Pivot		
Medial atlantoaxial	Atlas and axis	Rotation
Proximal radioulnar	Radius and ulna	Rotation
Distal radioulnar	Radius and ulna	Rotation
Ball-and-Socket		
Coxal (hip)	Coxa and femur	Multiple axes
Humeral (shoulder)	Scapula and humerus	Multiple axes
Ellipsoid		
Atlantooccipital	Atlas and occipital bone	Two axes
Metacarpophalangeal (knuckles)	Metacarpals and phalanges	Two axes
Metatarsophalangeal (ball of foot)	Metatarsals and phalanges	Two axes
Radiocarpal (wrist)	Radius and carpals	Multiple axes
Temporomandibular	Mandible and temporal bone	Multiple axes, one predominates

Plane

Saddle

Hinge

Pivot

Ball-and-socket

Elipsoid

Figure 6.39   Types of Synovial Joints

these joints can occur in two planes. The joint between the metatarsal and carpal (trapezium) of the thumb is a saddle joint. **Hinge joints** permit movement in one plane only. They consist of a convex cylinder of one bone applied to a corresponding concavity of the other bone. Examples are the elbow, knee, and finger joints (figure 6.40*a* and *b*). The flat condylar surface of the knee joint is modified into a concave surface by shock-absorbing fibrocartilage pads, called **menisci** (mĕ-nis'sī, crescent). **Pivot joints** restrict movement to rotation around a single axis. Each pivot joint consists of a cylindrical bony process that rotates within a ring composed partly of bone and partly of ligament. The rotation that occurs between the axis and atlas when shaking the head "no" is an example. The

articulation between the proximal ends of the ulna and radius is also a pivot joint.

Ball-and-socket joints consist of a ball (head) at the end of one bone and a socket in an adjacent bone into which a portion of the ball fits. This type of joint allows a wide range of movement in almost any direction. Examples are the shoulder and hip joints (figure 6.40*c* and *d*). **Ellipsoid** (ē-lip′soyd), or **condyloid** (kon′di-loyd), **joints** are elongated ball-and-socket joints. The shape of the joint limits its range of movement nearly to a hinge motion, but in two planes. The joint between the occipital condyles of the skull and the atlas of the vertebral column and the joints between the metacarpals and phalanges are examples of ellipsoid joints.

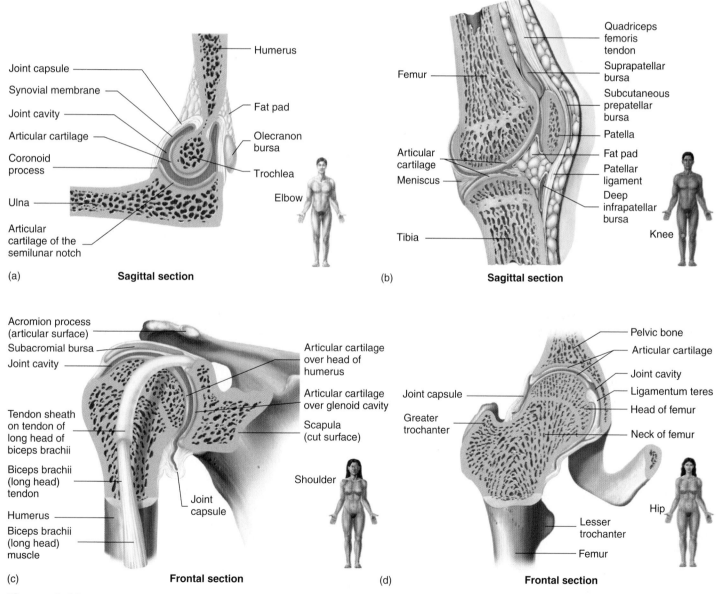

Figure 6.40 Examples of Synovial Joints

(*a*) Elbow. (*b*) Knee. (*c*) Shoulder. (*d*) Hip.

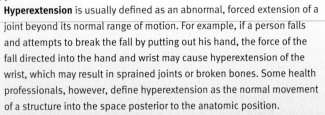

The shoulder joint is the most commonly dislocated joint in the body. Loosh Holder dislocated his shoulder joint while playing basketball. As a result of a "charging" foul, Loosh was knocked backward and fell. As he broke his fall with his extended right upper limb, the head of the right humerus was forced out of the glenoid cavity. When he was helped up from the floor, he felt severe pain in his shoulder, his right arm sagged, and he could not move his arm at the shoulder. Most dislocations result in stretching of the joint capsule and the movement of the humeral head to the inferior, anterior side of the glenoid cavity. The dislocated humeral head is moved back to its normal position by carefully pulling it laterally over the inferior lip of the glenoid cavity and then superiorly into the glenoid cavity. Once the shoulder joint capsule has been stretched by a shoulder dislocation, the shoulder joint may be predisposed to future dislocations. Some individuals have hereditary "loose" joints and are more likely to experience a dislocated shoulder.

Types of Movement

The types of movement occurring at a given joint are related to the structure of that joint. Some joints are limited to only one type of movement, whereas others permit movement in several directions. All the movements are described relative to the anatomic position. Because most movements are accompanied by movements in the opposite direction, they are listed in pairs (figure 6.41).

There are a number of ways to define **flexion** and **extension,** but in each case there are exceptions to the definition. The literal definition is to bend and straighten, respectively. We have chosen to use a definition with more utility and fewer exceptions. Flexion moves a part of the body in the anterior or ventral to the coronal plane. Extension moves a part in a posterior or dorsal to the coronal plane (figure 6.41*a* and *b*). An exception is the knee, in which flexion moves the leg in a posterior direction and extension moves it in an anterior direction.

Movement of the foot toward the plantar surface (sole of the foot), such as when standing on the toes, is commonly called **plantar flexion.** Movement of the foot toward the shin, such as when walking on the heels, is called **dorsiflexion.**

Joint Damage

A sprain results when the bones of a joint are forcefully pulled apart and the ligaments around the joint are pulled or torn. A separation exists when the bones remain apart after an injury to a joint. A dislocation is when the end of one bone is pulled out of the socket in a ball-and-socket, ellipsoid, or pivot joint.

Hyperextension

Hyperextension is usually defined as an abnormal, forced extension of a joint beyond its normal range of motion. For example, if a person falls and attempts to break the fall by putting out his hand, the force of the fall directed into the hand and wrist may cause hyperextension of the wrist, which may result in sprained joints or broken bones. Some health professionals, however, define hyperextension as the normal movement of a structure into the space posterior to the anatomic position.

Abduction (ab-dŭk′shun, to take away) is movement away from the median or midsagittal plane; **adduction** (to bring together) is movement toward the median plane (figure 6.41*c*). Moving the legs away from the midline of the body, as in the outward movement of "jumping jacks," is abduction, and bringing the legs back together is adduction.

Pronation (prō-nā′shŭn) and **supination** (soo′pi-nā′shun) are best described with the elbow flexed at a 90-degree angle. When the elbow is flexed, pronation is rotation of the forearm so that the palm is down, and supination is rotation of the forearm so that the palm faces up (figure 6.41*d*).

Eversion (ē-ver′zhŭn) is turning the foot so that the plantar surface (bottom of the foot) faces laterally; **inversion** (in-ver′zhŭn) is turning the foot so that the plantar surface faces medially.

Rotation is the turning of a structure around its long axis, as in shaking the head "no." Rotation of the arm can best be demonstrated with the elbow flexed (figure 6.41*e*), so that rotation is not confused with supination and pronation of the forearm. With the elbow flexed, medial rotation of the arm brings the forearm against the anterior surface of the abdomen, and lateral rotation moves it away from the body.

Protraction (prō-trak′shŭn) is a movement in which a structure, such as the mandible, glides anteriorly. In **retraction** (rē-trak′shŭn), the structure glides posteriorly.

Elevation is movement of a structure in a superior direction. Closing the mouth involves elevation of the mandible. **Depression** is movement of a structure in an inferior direction. Opening the mouth involves depression of the mandible.

Excursion is the movement of a structure to one side or the other, such as in moving the mandible from side to side.

Opposition is a movement unique to the thumb and little finger. It occurs when the tips of the thumb and little finger are brought toward each other across the palm of the hand. The thumb can also oppose the other digits. **Reposition** returns the digits to the anatomic position.

Circumduction (ser-kŭm-dŭk′shŭn) occurs at freely movable joints such as the shoulder. In circumduction, the arm moves so that it describes a cone with the shoulder joint at the apex (figure 6.41*f*).

Most movements that occur in the course of normal activities are combinations of movements. A complex movement can be described by naming the individual movements involved.

P R E D I C T   4

What combination of movements is required at the shoulder and elbow joints for a person to perform a crawl stroke in swimming?

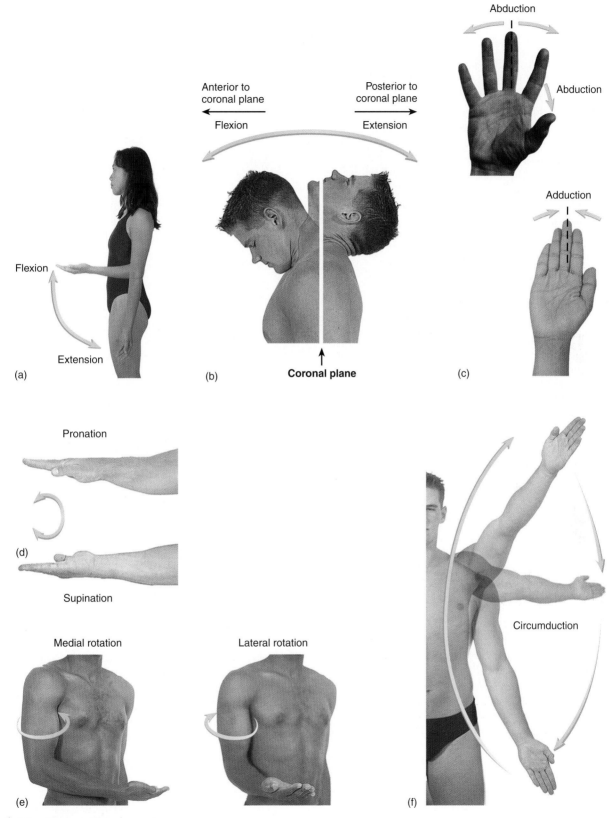

Figure 6.41 Types of Movement

(a) Flexion and extension of the elbow. (b) Flexion and extension of the neck. (c) Abduction and adduction of the fingers. (d) Pronation and supination of the hand. (e) Medial and lateral rotation of the arm. (f) Circumduction of the arm.

Arthritis

Arthritis (ar-thrī'tis, *arthron*, joint + *itis*, inflammation) (figure D), the inflammation of a joint, is the most common and best known of the joint disorders, affecting 10% of the world's population. There are more than 100 different types of arthritis, which differ in their cause and progress. Causes include infectious agents, metabolic disorders, trauma, and immune disorders.

Rheumatoid (rū'mă-toyd, *rheuma*, flux + *eidos*, resemblance) **arthritis** affects about 3% of all women and about 1% of all men in the United States. It is a general connective tissue disorder that affects the skin, vessels, lungs, and other organs, but it is most pronounced in the joints. It is severely disabling and most commonly destroys small joints such as those in the hands and feet. The initial cause is unknown but may involve a transient infection. It appears to be an autoimmune disease, which is an immune reaction against one's own tissues. There may be a genetic predisposition to this disease. In rheumatoid arthritis, the synovial membrane and associated connective tissue cells proliferate, forming a pannus (cloth-like layer) in the joint capsule, which can grow into the articulating surfaces of the bones, destroying the articular cartilage. In advanced stages, the bones forming the joint may become fused.

Degenerative Joint Disease

Degenerative joint disease (DJD), also called **osteoarthritis** (os'tē-ō-ar-thrī'tis), results from the gradual "wear and tear" of a joint that occurs with advancing age. Slowed metabolic rates with increased age seem to contribute to DJD. It is very common in older individuals and affects 85% of all people in the United States over the age of 70. It tends to occur in the weight-bearing joints such as the knees and hips and is more common in overweight individuals. Mild exercise retards joint degeneration and enhances mobility.

Gout

Gout (gowt) is caused by an increase in uric acid in the body. Uric acid is a waste product, which can accumulate as crystals in various tissues, including the kidneys and joint capsules. Gout is more common in males than in females.

Frequently, only one or two joints are affected. The most commonly affected joints (85% of the cases) are the base of the great toe and other foot and leg joints. Any joint may ultimately be involved, and damage to the kidneys from crystal formation occurs in almost all advanced cases.

Bursitis and Bunions

Bursitis (ber-sī'tis) is the inflammation of a bursa. The bursae around the shoulders and elbows are common sites of bursitis. A **bunion** (bun'yun) is a bursitis that develops over the joint at the base of the great toe. Bunions are frequently irritated by shoes that are too tight and that rub on them.

Joint Replacement

As a result of recent advancements in biomedical technology, many joints of the body can now be replaced by artificial joints. **Joint replacement,** or **arthroplasty** (ar'thrō-plas-tē) was developed in the late 1950s. It is used in patients with joint disorders to eliminate unbearable pain and to increase joint mobility. Degenerative joint disease is the leading disease requiring

Effects of Aging on the Skeletal System and Joints

The most significant age-related changes in the skeletal system affect the joints as well as the quality and quantity of bone matrix. The bone matrix in an older bone is more brittle than in a younger bone because decreased collagen production results in matrix that has relatively more mineral and less collagen fibers. With aging, the amount of matrix also decreases because the rate of matrix formation by osteoblasts becomes slower than the rate of matrix breakdown by osteoclasts.

Bone mass is at its highest around age 30, and men generally have denser bones than women because of the effects of testosterone and greater body weight. Race also affects bone mass. African-Americans and Hispanics have higher bone masses than Caucasians and Asians. After age 35, both men and women have an age-related loss of bone of 0.3–0.5% a year. This loss can increase 10-fold in women after menopause, and they can have a bone loss of 3–5% a year for approximately 5–7 years (see "Systems Pathology: Osteoporosis").

Significant loss of bone increases the likelihood of having bone fractures. For example, loss of trabeculae greatly increases the risk of fractures of the vertebrae. In addition, loss of bone

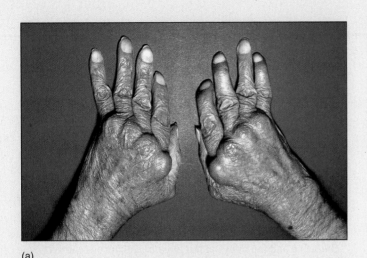

(a)

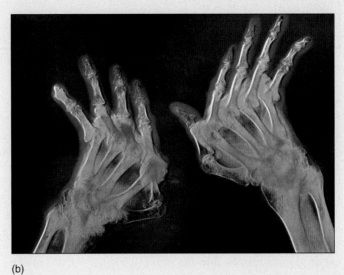

(b)

Figure D The Effects of Rheumatoid Arthritis on the Hands

(*a*) Photograph of hands with rheumatoid arthritis. (*b*) Radiograph of the same hands.

joint replacement, accounting for two-thirds of the patients. Rheumatoid arthritis accounts for more than half the remaining cases.

Artificial joints usually are composed of metal (e.g., stainless steel, titanium alloys, or cobalt-chrome alloys) in combination with modern plastics (e.g., high-density polyethylene, silicone rubber, or elastomer). The bone of the articular area is removed on one side (hemireplacement) or both sides (total replacement) of the joint, and the artificial articular structures are attached to the bone. The smooth metal surface rubbing against the smooth plastic surface provides a low-friction contact with a range of movement that depends on the design.

and the resulting fractures can cause deformity, loss of height, pain, and stiffness. Loss of bone from the jaws can also lead to tooth loss.

A number of changes occur within many joints as a person ages. Those that occur in synovial joints have the greatest effect and often present major problems for elderly people. With use, the cartilage covering articular surfaces can wear down. When a person is young, production of new, resilient matrix compensates for the wear. As a person ages, the rate of replacement declines and the matrix becomes more rigid, thus

adding to its rate of wear. The production rate of lubricating synovial fluid also declines with age, further contributing to the wear of the articular cartilage. Many people also experience arthritis, an inflammatory degeneration of joints, with advancing age. In addition, the ligaments and tendons surrounding a joint shorten and become less flexible with age, resulting in a decrease in the range of motion of the joint. Furthermore, older people often experience a general decrease in activity, which causes the joints to become less flexible and their range of motion to decrease.

Systems Pathologies
Osteoporosis

Mrs. B is a 65-year-old woman. She has smoked heavily for 50 years. She does not exercise, seldom goes outside, has a poor diet, and is slightly underweight. While attending a family picnic, Mrs. B tripped on a lawn sprinkler and fell. She was unable to stand because of severe hip pain, so she was rushed to the hospital where a radiograph revealed a fracture in the neck of her femur (figure E*a*) and that she had osteoporosis (figure E*b*).

Because of the location of the fracture and the osteoporosis, Mrs. B's orthopedic surgeon suggested that she should undergo hip replacement surgery. Mrs. B had a successful hip transplant and then began physical therapy. She was also placed on a diet and exercise program designed to improve the strength of her bones. In addition she was prescribed alendronate (see following discussion) therapy and was advised to quit smoking.

Background Information

Osteoporosis, or porous bone, results from reduction in the overall quantity of bone matrix. It occurs when the rate of bone reabsorption exceeds the rate of bone formation. The loss of bone mass makes bones so porous and weakened that they become deformed and prone to fracture. The occurrence of osteoporosis increases with age. In both men and women, bone mass starts to decrease at about age 40, and continually decreases thereafter. Women can eventually lose approximately one-half, and men one-quarter, of their cancellous bone. Osteoporosis is 2.5 times more common in women than in men.

In postmenopausal women, the decreased production of the female sex hormone, estrogen, can cause osteoporosis. The degeneration occurs mostly in cancellous bone, especially in the vertebrae of the spine and the bones of the forearm. Collapse of the vertebrae can cause a decrease in height or, in more severe cases, produce kyphosis in the upper back.

Conditions that result in decreased estrogen other than menopause can also cause osteoporosis. Examples include removal of the ovaries before menopause, extreme exercise to the point of amenorrhea (lack of menstrual flow), anorexia nervosa (self starvation), and cigarette smoking.

In males, reduction in testosterone levels can cause loss of bone tissue. Decreasing testosterone levels are usually less of a

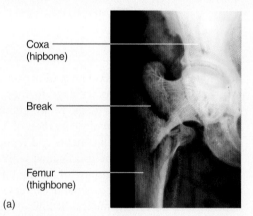

Coxa (hipbone)

Break

Femur (thighbone)

(a)

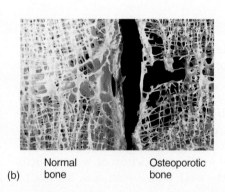

Normal bone

Osteoporotic bone

(b)

Figure E   Osteoporosis

(*a*) Radiograph of a broken hip. A "broken hip" is usually a break in the femur in the hip region (in this case, the break is at the base of the neck).
(*b*) Photomicrograph of normal bone and osteoporotic bone.

System Interactions — Effects of Osteoporosis on Other Systems

System	Interactions with Skeletal System
Integumentary	Mrs. B has little exposure to the sun because of her indoor lifestyle, resulting in little vitamin D production and decreased calcium absorption.
Muscular	Mrs. B's muscle atrophy and weakness made it difficult for her to maintain her balance. Thus she was more likely to fall and injure herself. Following the surgery, her physical therapy placed stress on her bones and improved her muscular strength.
Nervous	Pain sensations following the injury and during her rehabilitation help to prevent further injury.
Endocrine	Calcitonin is being used to treat osteoporosis.
Cardiovascular	Blood clotting following the injury starts the process of tissue repair. Blood cells are carried to the injury site to fight infections and remove cell debris. Blood vessels grow into the recovering tissue, providing nutrients and removing waste products.
Lymphatic and Immune	Immune cells resist possible infection following surgery, such as hip-replacement surgery, and release chemicals that promote tissue repair.
Respiratory	Excessive smoking lowers estrogen levels, which increases bone loss.
Digestive	Inadequate calcium and vitamin D in the diet can contribute to inadequate calcium absorption by the digestive system and osteoporosis.
Reproductive	Decreased estrogen following menopause contributed to her osteoporosis.

problem for men than decreasing estrogen levels are for women for two reasons. First, because males have denser bones than females, loss of some bone tissue has less of an effect. Second, testosterone levels generally don't decrease significantly until after age 65, and even then the rate of decrease is often slow.

Inadequate dietary intake or absorption of calcium can contribute to osteoporosis. Absorption of calcium from the small intestine decreases with age, and individuals with osteoporosis often have insufficient intake of calcium, vitamin D, and vitamin C. Drugs that interfere with calcium uptake or use can also increase the risk of osteoporosis.

Finally, osteoporosis can result from inadequate exercise or disuse caused by fractures or paralysis. Significant amounts of bone are lost after 8 weeks of immobilization.

Treatments for osteoporosis are designed to reduce bone loss or increase bone formation (or both). Increased dietary calcium and vitamin D can increase calcium uptake and promote bone formation. Daily doses of 1000–1500 mg of calcium and 800 IU (20 µg) of vitamin D are recommended. Exercise, such as walking or using light weights, also appears to be effective not only in reducing bone loss, but in increasing bone mass.

In postmenopausal women, estrogen replacement therapy decreases osteoclast activity, which reduces bone loss, but does not result in an increase in bone mass because osteoclast activity still exceeds osteoblast activity. Estrogen therapy also apparently increases the risk of breast cancer, strokes, heart attacks, and Alzheimer's disease and is no longer recommended for treating osteoporosis. Calcitonin (Miacalcin), which inhibits osteoclast activity, is now available as a nasal spray. Calcitonin can be used to treat osteoporosis in men and women and has been shown to produce a slight increase in bone mass. Alendronate (Fosamax) belongs to a class of drugs called bisphosphonates. Bisphosphonates bind to hydroxyapatite and inhibit bone breakdown by osteoclasts. Alendronate increases bone mass and reduces fracture rates even more effectively than calcitonin. Slow-releasing sodium fluoride (Slow Fluoride) in combination with calcium citrate (Citracal) also appear to increase bone mass.

Early diagnosis of osteoporosis can lead to more preventative treatments. Instruments that measure the absorption of photons (particles of light) by bone are currently used, of which dual-energy x-ray absorptiometry (DEXA) is considered the best.

P R E D I C T
What advice should Mrs. B give to her granddaughter so that the granddaughter will be less likely to develop osteoporosis when she is Mrs. B's age?

The skeletal system consists of bone, cartilage, tendons, and ligaments.

Functions of the Skeletal System   (p. 115)

1. Cartilage provides a model for bone formation and growth, provides a smooth cushion between adjacent bones, and provides firm, flexible support.
2. Tendons attach muscles to bones, and ligaments attach bones to bones.
3. The skeletal system provides the major support for the body.
4. Bone protects internal organs.
5. Joints allow movement between bones.
6. Bones store and release minerals as needed by the body.
7. Bone marrow gives rise to blood cells and platelets.

Connective Tissue   (p. 115)

1. Connective tissue consists of matrix and the cells that produce matrix.
2. Varying amounts of collagen, proteoglycan, and mineral in the matrix determine the characteristics of the connective tissue.

General Features of Bone   (p. 116)

Long bones consist of a diaphysis (shaft), epiphyses (ends), and epiphyseal (growth) plates. The diaphysis contains a medullary cavity, which is filled with marrow, and the end of the epiphysis is covered by articular cartilage.

Compact Bone

1. Compact bone tissue consists of osteons.
2. Osteons consist of osteocytes organized into lamellae surrounding central canals.

Cancellous Bone

Cancellous bone tissue consists of trabeculae without central canals.

Bone Ossification

1. Bone ossification is either intramembranous or endochondral.
2. Intramembranous ossification occurs within connective tissue membranes.
3. Endochondral ossification occurs within cartilage.

Bone Growth

Bone growth occurs by apposition. Bone elongation occurs at the epiphyseal plate as chondrocytes proliferate, hypertrophy, die, and are replaced by bone.

Bone Remodeling

Bone remodeling consists of removal of existing bone by osteoclasts and deposition of new bone by osteoblasts.

Bone Repair

During bone repair, cells move into the damaged area and form a callus, which is replaced by bone.

Bone and Calcium Homeostasis   (p. 123)

1. Osteoclasts remove calcium from bone, causing blood calcium levels to increase.
2. Osteoblasts deposit calcium into bone, causing blood calcium levels to decrease.
3. Parathyroid hormone increases bone breakdown, whereas calcitonin decreases bone breakdown.

General Considerations of Bone Anatomy   (p. 124)

There are 206 bones.

Axial Skeleton   (p. 124)

The axial skeleton includes the skull, vertebral column, and thoracic cage.

Skull

1. The skull consists of 22 bones: 8 forming the braincase, and 14 facial bones. The hyoid bone and 6 auditory ossicles are associated with the skull.
2. From a lateral view, the parietal, temporal, and sphenoid bones can be seen.
3. From a frontal view, the orbits and nasal cavity can be seen, as well as associated bones and structures, such as the frontal bone, zygomatic bone, maxilla, and mandible.
4. The interior of the cranial cavity contains three fossae with several foramina.
5. Seen from below, the base of the skull reveals numerous foramina and other structures, such as processes for muscle attachment.

Vertebral Column

1. The vertebral column contains 7 cervical, 12 thoracic, and 5 lumbar vertebrae, plus 1 sacral and 1 coccygeal bone.
2. Each vertebra consists of a body, an arch, and processes.
3. Regional differences in vertebrae are as follows: cervical vertebrae have transverse foramina; thoracic vertebrae have long spinous processes and attachment sites for the ribs; lumbar vertebrae have rectangular transverse and spinous processes, and the position of their facets limit rotation; the sacrum is a single, fused bone; the coccyx is four or fewer fused vertebrae.

Thoracic Cage

1. The thoracic cage consists of thoracic vertebrae, ribs, and sternum.
2. There are 12 pairs of ribs: 7 true and 5 false (two of the false ribs are also called floating ribs).
3. The sternum consists of the manubrium, body, and xiphoid process.

Appendicular Skeleton   (p. 135)

The appendicular skeleton consists of the bones of the upper and lower limbs and their girdles.

Pectoral Girdle

The pectoral girdle includes the scapulae and clavicles.

Upper Limb

The upper limb consists of the arm (humerus), forearm (ulna and radius), wrist (eight carpal bones), and hand (five metacarpals, three phalanges in each finger, and two phalanges in the thumb).

Pelvic Girdle

The pelvic girdle is made up of the two coxae. Each coxa consists of an ilium, ischium, and pubis. The coxae, sacrum, and coccyx form the pelvis.

Lower Limb

The lower limb includes the thigh (femur), leg (tibia and fibula), ankle (seven tarsals), and foot (metatarsals and phalanges, similar to the bones in the hand).

Articulations   (p. 144)

An articulation is a place where bones come together.

Fibrous Joints

Fibrous joints consist of bones united by fibrous connective tissue. They allow little or no movement.

Cartilaginous Joints

Cartilaginous joints consist of bones united by cartilage, and they exhibit slight movement.

Synovial Joints

1. Synovial joints consist of articular cartilage over the uniting bones, a joint cavity lined by a synovial membrane and containing synovial fluid, and a joint capsule. They are highly movable joints.
2. Synovial joints can be classified as plane, saddle, hinge, pivot, ball-and-socket, or ellipsoid.

Types of Movement

The major types of movement include flexion/extension, abduction/adduction, pronation/supination, eversion/inversion, rotation, protraction/retraction, elevation/depression, excursion, opposition/reposition, and circumduction.

Effects of Aging on the Skeletal System and Joints   (p. 150)

1. Bone matrix becomes more brittle and decreases in total amount during aging.
2. Joints lose articular cartilage and become less flexible with age.

R E V I E W   A N D   C O M P R E H E N S I O N

1. What are the primary functions of the skeletal system?
2. Name the major types of fibers and molecules found in the extracellular matrix of the skeletal system. How do they contribute to the functions of tendons, ligaments, cartilage, and bones?
3. Define the terms diaphysis, epiphysis, epiphyseal plate, medullary cavity, articular cartilage, periosteum, and endosteum.
4. Describe the structure of compact bone. How do nutrients reach the osteocytes in compact bone?
5. Describe the structure of cancellous bone. What are trabeculae? How do nutrients reach osteocytes in trabeculae?
6. Define and describe intramembranous and endochondral ossification.
7. How do bones grow in diameter? How do long bones grow in length?
8. What is accomplished by bone remodeling? How does bone repair occur?
9. Define the axial skeleton and the appendicular skeleton.
10. Name the bones of the braincase and the face.
11. Give the locations of the paranasal sinuses. What are their functions?
12. What is the function of the hard palate?
13. Through what foramen does the brain connect to the spinal cord?
14. How do the vertebrae protect the spinal cord? Where do spinal nerves exit the vertebral column?
15. Name and give the number of each type of vertebra. Describe the characteristics that distinguish the different types of vertebrae from one another.
16. What is the function of the thoracic cage? Name the parts of the sternum. Distinguish true, false, and floating ribs.
17. Name the bones that make up the pectoral girdle, arm, forearm, wrist, and hand. How many phalanges are in each finger and in the thumb?
18. Define the pelvic girdle. What bones fuse to form each coxa? Where and with what bones do the coxae articulate?
19. Name the bones of the thigh, leg, ankle, and foot.
20. Define the term articulation, or joint. Name and describe the differences between the three major classes of joints.
21. Describe the structure of a synovial joint. How do the different parts of the joint function to permit joint movement?
22. On what basis are synovial joints classified? Describe the different types of synovial joints, and give examples of each. What movements do each type of joint allow?
23. Describe and give examples of flexion/extension, abduction/adduction, and supination/pronation.

C R I T I C A L   T H I N K I N G

1. A 12-year-old boy fell while playing basketball. The physician explained that the head (epiphysis) of the femur was separated from the shaft (diaphysis). Although the bone was set properly, by the time the boy was 16 it was apparent that the injured lower limb was shorter than the normal one. Explain why this difference occurred.
2. Justin Time leaped from his hotel room to avoid burning to death in a fire. If he landed on his heels, what bone was likely to fracture? Unfortunately for Justin, a 240-pound fireman, Hefty Stomper, ran by and stepped heavily on the distal part of Justin's foot (not the toes). What bones now could be broken?
3. One day while shopping, Ms. Wantta Bargain picked up her 3-year-old son, Somm, by his right wrist and lifted him into a shopping cart. She heard a clicking sound and Somm immediately began to cry and hold his elbow. Given that lifting the child caused a separation at the elbow, which is more likely: separation of the radius and humerus or separation of the ulna and humerus?
4. Why are women knock-kneed more commonly than men?
5. A skeleton was discovered in a remote mountain area. The coroner determined that not only was the skeleton human but that it was female. Explain.

Answers in Appendix D

Visit this textbook's website at www.mhhe.com/seeleyess6 for practice quizzes, animations, interactive learning exercises, and other study tools.

Muscular System

The unique feature of the muscular system is that it consists of cells that contract. This photograph shows a color-enhanced scanning electron micrograph of six skeletal muscle fibers. Skeletal muscle cells are very large, multinucleated cells that are long and uniform in diameter. The nuclei show up as egg-shaped bumps on the surface of the cell. Two nuclei have broken out of the middle cell. One is gone and the other is partially dislodged from the cell. The curly material in the upper right corner is connective tissue associated with the muscle cells.

as a runner rounds the last corner of the track and sprints for the finish line, her arms and legs are pumping as she tries to reach her maximum speed. Her heart is beating rapidly and her breathing is rapid, deep, and regular. Blood is shunted away from digestive organs, and a greater volume is delivered to skeletal muscles to maximize the oxygen supply to them. These actions are accomplished by muscle tissue, the most abundant tissue of the body, and one of the most adaptable.

You don't have to be running for the muscular system to be at work. Even when you aren't "moving," postural muscles keep you sitting or standing upright, respiratory muscles keep you breathing, the heart continually pumps blood to all parts of the body, and blood vessels constrict or relax to direct blood to organs where it is needed.

Functions of the Muscular System

Movement within the body is accomplished by cilia or flagella on the surface of certain cells, by the force of gravity, or by the contraction of muscles. Most of the body's movement results from muscle contraction. As described in chapter 4, there are three types of muscle tissue: skeletal, cardiac, and smooth. This chapter deals primarily with the structure and function of skeletal muscle; cardiac and smooth muscle are described briefly. The major functions of the muscular system are:

1. *Body movement.* Contraction of skeletal muscles is responsible for the overall movements of the body, such as walking, running, or manipulating objects with the hands.
2. *Maintenance of posture.* Skeletal muscles constantly maintain tone, which keeps us sitting or standing erect.
3. *Respiration.* Muscles of the thorax are responsible for the movements necessary for respiration.
4. *Production of body heat.* When skeletal muscles contract, heat is given off as a by-product. This released heat is critical to the maintenance of body temperature.
5. *Communication.* Skeletal muscles are involved in all aspects of communication, such as speaking, writing, typing, gesturing, and facial expression.
6. *Constriction of organs and vessels.* The contraction of smooth muscle within the walls of internal organs and vessels causes constriction of those structures. This constriction can help propel and mix food and water in the digestive tract, propel secretions from organs, and regulate blood flow through vessels.
7. *Heart beat.* The contraction of cardiac muscle causes the heart to beat, propelling blood to all parts of the body.

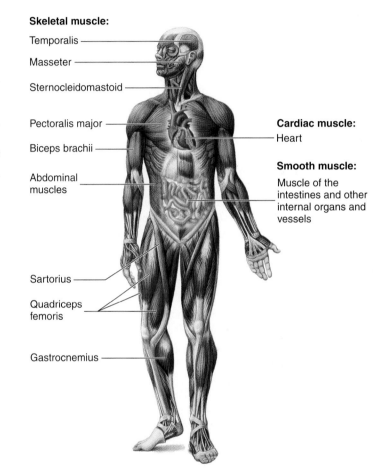

Skeletal muscle:
Temporalis
Masseter
Sternocleidomastoid
Pectoralis major
Biceps brachii
Abdominal muscles
Sartorius
Quadriceps femoris
Gastrocnemius

Cardiac muscle:
Heart

Smooth muscle:
Muscle of the intestines and other internal organs and vessels

Figure 7.1   The Muscular System

Characteristics of Skeletal Muscle

Skeletal muscle (figure 7.1), with its associated connective tissue, constitutes approximately 40% of body weight. Skeletal muscle is so named because most skeletal muscles are attached to the skeletal system. It is also called **striated muscle** because of the transverse bands or striations that can be seen in the muscle under the microscope.

Skeletal muscle has four major functional characteristics: contractility, excitability, extensibility, and elasticity.

1. **Contractility** (kon-trak-til′i-tē) is the ability of skeletal muscle to shorten with force. When skeletal muscles contract, they cause movement of the structures to which they are attached. Skeletal muscles shorten forcefully during contraction, but they lengthen passively. Either gravity or the contraction of an opposing muscle produces a force that pulls on the shortened muscle, causing it to lengthen.

2. **Excitability** (ek-sī′tă-bil′i-tē) is the capacity of skeletal muscle to respond to a stimulus. Normally skeletal muscle contracts as a result of stimulation by nerves.
3. **Extensibility** (eks-ten′sĭ-bil′i-tē) means that skeletal muscles can be stretched. After a contraction, skeletal muscles can be stretched to their normal resting length and beyond to a limited degree.
4. **Elasticity** (ĕ-las-tis′i-tē) is the ability of skeletal muscles to recoil to their original resting length after they have been stretched.

Structure

Each skeletal muscle is surrounded by a connective tissue sheath called the **epimysium** (ep-i-mis′ē-ŭm, *epi*, upon + *mys*, muscle), or **fascia** (fash′ē-ă, fillet) (figure 7.2*a*). A muscle is composed of numerous visible bundles called **muscle fasciculi** (fă-sik′ū-lī, *fascis*, bundle), which are surrounded by loose connective tissue called the **perimysium** (per′i-mis′ē-ŭm). A fasciculus is composed of several **muscle cells** or **muscle fibers** (figure 7.2*b*). Each muscle fiber is surrounded by loose connective tissue, called **endomysium** (en′dō-mis′ē-ŭm). A muscle fiber is a single cylindrical cell that contains several nuclei located at the periphery of the muscle fiber. The largest, longest human muscle cells are up to 30 cm long and 0.15 mm in diameter. Such giant cells may contain several thousand nuclei.

Muscle contraction is much easier to understand when we understand the structure of a muscle cell. The cytoplasm of each muscle fiber, called the **sarcoplasm** (sar′kō-plazm, *sarco*, flesh + *plasma*, formed), contains numerous myofibrils (figure 7.2*c*). Each **myofibril** (mī-ō-fī′bril, *myo*, muscle) is a threadlike structure that extends from one end of the muscle fiber to the other. Myofibrils consist of two major kinds of protein fibers: **actin** (ak′tin) and **myosin** (mī′ō-sin) **myofilaments** (mī-ō-fil′ă-ments) (figure 7.2*d*).

The actin and myosin myofilaments are arranged into highly ordered repeating units along the myofibril called **sarcomeres** (sar′kō-mērz) (see figures 7.2*c* and *d*). The sarcomere is the basic structural and functional unit of skeletal muscle because it is the smallest portion of skeletal muscle capable of contracting. Each sarcomere extends from one Z disk to another Z disk. Each **Z disk** is a network of protein fibers forming an attachment site for actin myofilaments.

Actin myofilaments, or thin myofilaments, resemble two minute strands of pearls twisted together (figure 7.2*e*). **Troponin** (trō′pō-nin, *trope*, a turning) molecules are attached at specific intervals along the actin myofilaments and provide calcium-binding sites on the actin myofilament. **Tropomyosin** (trō-pō-mī′ō-sin) filaments are located along the groove between the twisted strands of actin myofilament subunits. These filaments expose attachment sites on the actin myofilament when calcium is bound to troponin, and they cover attachment sites on the actin myofilament when calcium is not bound to troponin.

Myosin myofilaments, or thick myofilaments, resemble bundles of minute golf clubs (figure 7.2*f*). The part of the myosin molecule that resembles golf club heads can bind to the exposed attachment sites on the actin myofilaments.

The cell membrane of the muscle fiber is called the **sarcolemma** (sar′kō-lem′ă) (see figure 7.2*b*). The multiple nuclei of the muscle fiber are located just deep to the sarcolemma. The sarcolemma has along its surface many tubelike invaginations called **transverse,** or **T, tubules,** which are located at regular intervals along the muscle fiber and wrap around sarcomeres where the actin and myosin myofilaments overlap. The T tubules are associated with a highly organized, smooth endoplasmic reticulum called the **sarcoplasmic reticulum** (re-tik′ū-lŭm). T tubules connect the sarcolemma to the sarcoplasmic reticulum. The sarcoplasmic reticulum has a relatively high concentration of calcium ions, which play a major role in muscle contractions.

The arrangement of the actin and myosin myofilaments in sarcomeres gives the myofibril a banded appearance (figure 7.3). A light **I band,** which consists only of actin myofilaments, spans each Z disk and ends at the myosin myofilaments. A darker, central region in each sarcomere, called an **A band,** extends the length of the myosin myofilaments. The actin and myosin myofilaments overlap for some distance at both ends of the A band. In the center of each sarcomere is a second light zone, called the **H zone,** which consists only of myosin myofilaments. The myosin myofilaments are anchored in the center of the sarcomere at a dark-staining band, called the **M line.** The alternating I bands and A bands of the sarcomeres are responsible for the striations seen in skeletal muscle fibers observed through the microscope (see table 4.11*a*).

Membrane Potentials

Muscle fibers, like other cells of the body, have electrical properties. This section describes the electrical properties of skeletal muscle fibers and later sections illustrate their role in contraction.

The outside of most cell membranes is positively charged compared with the inside of the cell membrane (figure 7.4, step 1). The charge difference, called the **resting membrane potential,** develops for two reasons: (1) the concentration of K^+ inside the cell membrane is higher than that outside the cell membrane, and (2) the cell membrane is more permeable to K^+ than it is to other ions, including negatively charged molecules, such as proteins, located inside the cell. This occurs because some K^+ channels are open, whereas other ion channels, such as those for Na^+, are closed.

A few K^+ are able to diffuse down their concentration gradient from inside to just outside the cell membrane. Negatively charged ions cannot diffuse through the cell membrane with the K^+ because the cell membrane is less permeable to the negatively charged ions. Because K^+ are positively charged, their movement from inside the cell to the outside causes the outside of the cell membrane to become positively charged compared with the inside of the cell membrane.

Potassium ions only diffuse down their concentration gradient until the charge difference across the cell membrane is great enough to prevent any additional diffusion of K^+ out of the cell. The resting membrane potential is an equilibrium in which the tendency for K^+ to diffuse out of the cell is opposed

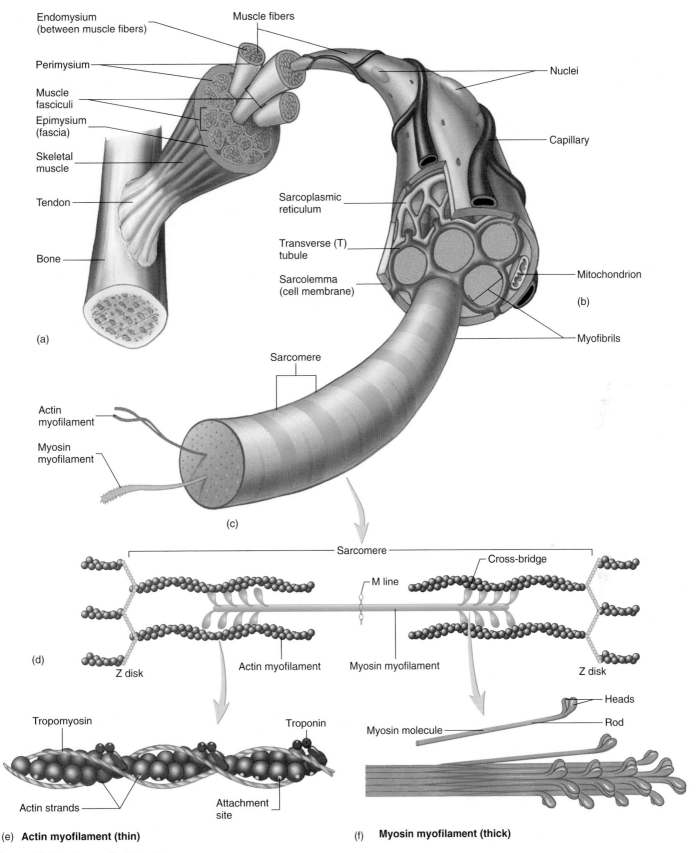

Endomysium (between muscle fibers)

Muscle fibers

Perimysium

Nuclei

Muscle fasciculi

Epimysium (fascia)

Skeletal muscle

Tendon

Capillary

Sarcoplasmic reticulum

Transverse (T) tubule

Sarcolemma (cell membrane)

Bone

Mitochondrion

(b)

Myofibrils

(a)

Sarcomere

Actin myofilament

Myosin myofilament

(c)

Sarcomere

Cross-bridge

M line

Actin myofilament

Myosin myofilament

(d)

Z disk

Z disk

Tropomyosin

Troponin

Heads

Rod

Myosin molecule

Actin strands

Attachment site

(e) **Actin myofilament (thin)**

(f) **Myosin myofilament (thick)**

Figure 7.2 Parts of a Muscle

(a) Part of a muscle attached by a tendon to a bone. A muscle is composed of muscle fasciculi, each surrounded by perimysium. The fasciculi are composed of bundles of individual muscle fibers (muscle cells), each surrounded by endomysium. The entire muscle is surrounded by a connective tissue sheath called epimysium or fascia. (b) Enlargement of one muscle fiber. The muscle fiber contains several myofibrils. (c) A myofibril extended out the end of the muscle fiber. The banding patterns of the sarcomeres are shown in the myofibril. (d) A single sarcomere of a myofibril is composed mainly of actin myofilaments and myosin myofilaments. The Z disks anchor the actin myofilaments, and the myosin myofilaments are held in place by the M line. (e) Part of an actin myofilament is enlarged. (f) Part of a myosin myofilament is enlarged.

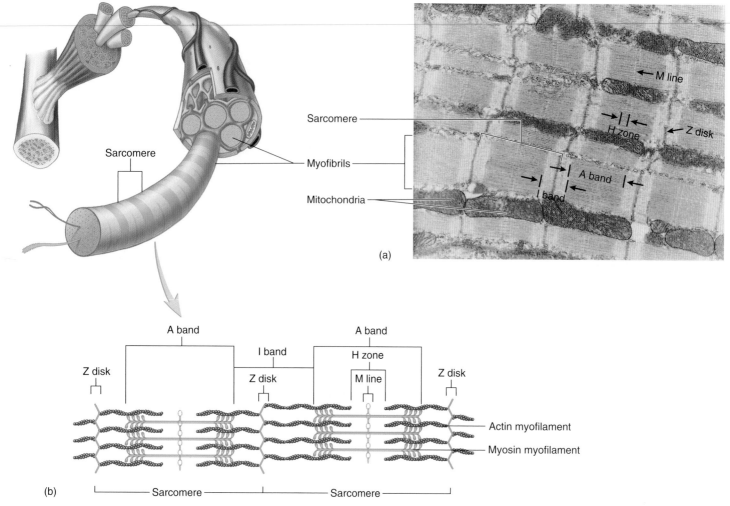

Figure 7.3 Skeletal Muscle

(a) Electron micrograph of skeletal muscle. Several sarcomeres are shown in a muscle fiber. (b) Diagram of two adjacent sarcomeres, depicting the structures responsible for the banding pattern.

by the negative charges inside the cell, which tend to attract the positively charged K^+ into the cell.

When a muscle cell or nerve cell is stimulated, Na^+ channels quickly open, and the membrane becomes very permeable to Na^+ for a brief time (figure 7.4, step 2). The Na^+ concentration is much greater outside the cell than inside, and, therefore, a few positively charged Na^+ quickly diffuse down their concentration gradient into the cell, causing the inside of the cell membrane to become more positive than the outside of the cell. This change is called **depolarization.** Near the end of depolarization, Na^+ channels close and additional K^+ channels open (figure 7.4, step 3). Consequently, the tendency for Na^+ to enter the cell is decreased, and the tendency for K^+ to leave the cell is increased. These changes cause the inside of the cell membrane to become more negative than the outside once again. Additional K^+ channels then close as the charge across the cell membrane returns to its resting condition (return to figure 7.4, step 1). The change back to the resting membrane potential is called **repolarization.** The rapid depolarization

and repolarization of the cell membrane is called an **action potential.** In a muscle fiber, an action potential results in muscle contraction. The resting membrane potential and action potential are described in more detail in chapter 8 (p. 203).

The sodium–potassium exchange pump transports K^+ from outside the cell to the inside and transports Na^+ from inside the cell to the outside (see chapter 3). The active transport of K^+ and Na^+ maintains the normal, resting concentrations of ions on either side of the cell membrane.

Nerve Supply

Skeletal muscle fibers do not contract unless they are stimulated by motor neurons. This section describes how neurons stimulate skeletal muscle fibers.

Motor neurons are nerve cells along which action potentials travel to skeletal muscle fibers. Axons of these neurons enter muscles and send out branches to several muscle fibers. Each branch forms a junction with a muscle fiber, called a

1. Resting membrane potential.
Na$^+$ channels (*pink*) and some,
but not all, K$^+$ channels (*purple*)
are closed. A few K$^+$ diffuse down
their concentration gradient
through the open K$^+$ channels,
making the outside of the cell
membrane positively charged
compared to the inside.

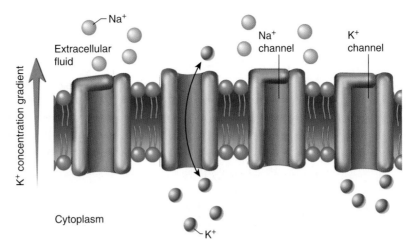

2. Depolarization. Na$^+$ channels are
open. A few Na$^+$ diffuse down their
concentration gradient through the
open Na$^+$ channels, making the
inside of the cell membrane
positively charged compared to
the outside.

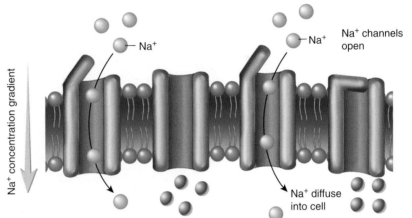

3. Repolarization. Na$^+$ channels are
closed, and Na$^+$ movement into
the cells stops. More K$^+$ channels
open. K$^+$ movement out of the cell
increases, making the outside of
the cell membrane positively
charged compared to the inside.

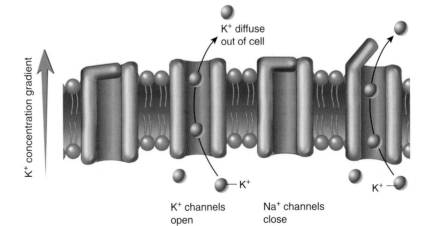

Process Figure 7.4 Ion Channels and the Action Potential

Step 1 illustrates the status of Na$^+$ and K$^+$ channels in a resting cell. Steps 2 and 3 show how the channels open and close to produce an action potential. Next to each step, the charge difference across the plasma membrane is illustrated.

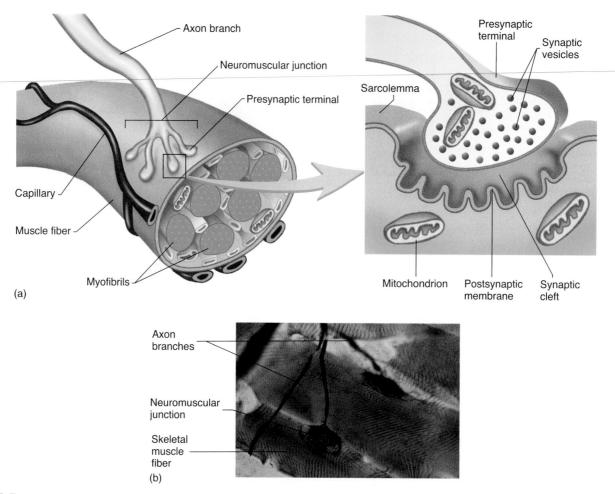

Figure 7.5 Neuromuscular Junction

(a) Diagram showing the neuromuscular junction. Several branches of an axon from the neuromuscular junction with a single muscle fiber. (b) Photomicrograph of neuromuscular junctions.

neuromuscular junction, or **synapse** (sin′aps) (figure 7.5a). The term synapse is a more general term; it refers to the cell-to-cell junction between a nerve cell and either another nerve cell or an effector cell, such as a muscle or gland cell. Neuromuscular junctions are located near the center of a muscle fiber. A single motor neuron and all the skeletal muscle fibers it innervates constitute a **motor unit.** A motor unit in a small, precisely controlled muscle may have only one or a few muscle fibers per unit, whereas the motor units of large thigh muscles may have as many as 1000 muscle fibers per motor unit. Many motor units constitute a single muscle.

A neuromuscular junction is formed by a cluster of enlarged axon terminals resting in indentations of the muscle fiber's cell membrane (figure 7.5b). An enlarged axon terminal is the **presynaptic terminal,** the space between the presynaptic terminal and the muscle fiber membrane is the **synaptic cleft,** and the muscle fiber membrane is the **postsynaptic membrane.** Each presynaptic terminal contains many small vesicles, called **synaptic vesicles.** These vesicles contain **acetylcholine** (as-e-til-kō′lēn, ACh), which functions as a **neurotransmitter,**

a molecule released by a presynaptic nerve cell that stimulates or inhibits a postsynaptic cell.

When an action potential reaches the presynaptic terminal, it causes Ca^{2+} channels to open. Ca^{2+} enter the presynaptic terminal and cause several synaptic vesicles to release acetylcholine into the synaptic cleft by exocytosis (figure 7.6). The acetylcholine diffuses across the synaptic cleft and binds to acetylcholine receptor sites on the Na^+ channels in the muscle fiber cell membrane. The combination of acetylcholine with its receptor opens Na^+ channels and therefore causes an increase in the permeability of the cell membrane to Na^+. The resulting movement of Na^+ into the muscle fiber initiates an action potential in the muscle fiber, which travels along the length of the muscle fiber and causes it to contract. The acetylcholine released into the synaptic cleft between the neuron and muscle fiber is rapidly broken down by an enzyme, **acetylcholinesterase** (as′e-til-kō-lin-es′ter-ās). This enzymatic breakdown ensures that one action potential in the neuron yields only one action potential in the skeletal muscle fibers of that motor unit, and only one contraction of each muscle fiber.

Acetylcholine Antagonists

Anything that affects the production, release, or degradation of acetylcholine or its ability to bind to its receptor on the muscle cell membrane can affect the transmission of action potentials across the neuromuscular junction. Some insecticides bind to and inhibit acetylcholinesterase. Consequently, acetylcholine accumulates in the synaptic cleft and acts as a constant stimulus to the muscle fiber. The insects die, partly because their respiratory muscles contract and cannot relax or because they keep contracting until they fatigue and can no longer contract. Other poisons, such as **curare** (koo-rǎ′rē), the poison originally used by South American Indians in poison arrows, bind to the acetylcholine receptors on the muscle cell membrane and prevent acetylcholine from binding to them. The muscle fibers cannot therefore be stimulated by acetylcholine and do not contract, resulting in paralysis.

A CASE IN POINT | Myasthenia Gravis

Les Strong was 45 years old when he first noticed muscle weakness in his face. While shaving, he noticed that he could not tighten the left side of his face. Within a few days he noticed the same problem in the right side of his face. He also experienced ptosis (drooping) of the left eyelid and slightly slurred speech. These changes concerned him and he scheduled an appointment with his physician. After a thorough examination, Les's physician referred him to a neurologist, who was unable to make a clear diagnosis by a simple examination. The neurologist considered the possibility that myasthenia gravis was the cause of Les's muscular problems and ordered a blood test. The blood test revealed the presence of acetylcholine receptor antibodies.

Myasthenia gravis is an autoimmune disorder in which antibodies are formed against acetylcholine receptors. As a result, acetylcholine receptors in the postsynaptic membranes of skeletal muscles are destroyed. With fewer receptors, acetylcholine is less likely to stimulate muscle contraction, resulting in muscle weakness and fatigue. Les was prescribed the anticholinesterase drug neostigmine (nē′ō-stig′min), which greatly reduced his symptoms and allowed him to continue a relatively normal life. Anticholinesterase drugs inhibit acetylcholinesterase activity, preventing the breakdown of acetylcholine. Consequently, acetylcholine levels in the synapse remain elevated, which increases the stimulation of the reduced number of functional acetylcholine receptors.

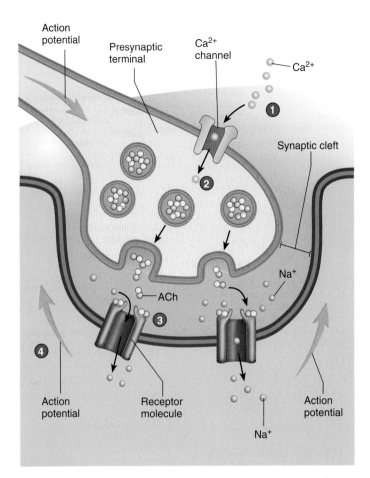

1. An action potential arrives at the presynaptic terminal causing Ca^{2+} channels to open, increasing the Ca^{2+} permeability of the presynaptic terminal.
2. Calcium ions enter the presynaptic terminal and initiate the release of a neurotransmitter, acetylcholine (ACh), from synaptic vesicles into the presynaptic cleft.
3. Diffusion of ACh across the synaptic cleft and binding of ACh to ACh receptors on the postsynaptic muscle fiber membrane opens Na^+ channels and increases the permeability of the postsynaptic membrane to Na^+.
4. The increase in Na^+ permeability results in depolarization of the postsynaptic membrane; once threshold has been reached a postsynaptic action potential results.

Process Figure 7.6 Function of the Neuromuscular Junction

Release of ACh in response to an action potential at the neuromuscular junction.

Muscle Contraction

Contraction of skeletal muscle tissue occurs as actin and myosin myofilaments slide past one another, causing the sarcomeres to shorten. Many sarcomeres joined end to end form myofibrils. Shortening of the sarcomeres causes myofibrils to shorten. Shortening of myofibrils causes the entire muscle to shorten.

The sliding of actin myofilaments past myosin myofilaments during contraction is called the **sliding filament model** of muscle contraction. During contraction, neither the actin nor myosin fibers shorten. The H zones and I bands shorten during contraction, but the A bands do not change in length (figure 7.7).

1. Actin and myosin myofilaments in a relaxed muscle (*right*) and a contracted muscle (*#4 below*) are the same length. Myofilaments do not change length during muscle contraction.

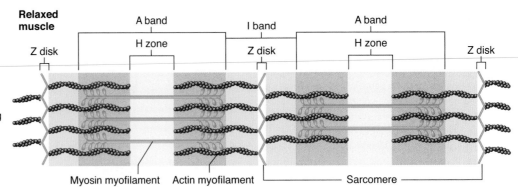

Myosin myofilament   Actin myofilament   Sarcomere

2. During contraction, actin myofilaments at each end of the sarcomere slide past the myosin myofilaments toward each other. As a result, the Z disks are brought closer together, and the sarcomere shortens.

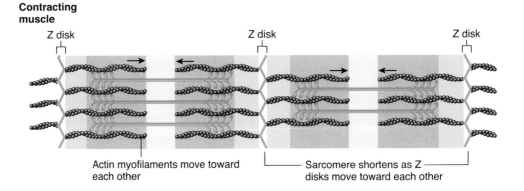

Actin myofilaments move toward each other   Sarcomere shortens as Z disks move toward each other

3. As the actin myofilaments slide over the myosin myofilaments, the H zones (*yellow*) and the I bands (*blue*) narrow. The A bands, which are equal to the length of the myosin myofilaments, do not narrow, because the length of the myosin myofilaments does not change.

H zone narrows   I band narrows   A band does not narrow

4. In a fully contracted muscle, the ends of the actin myofilaments overlap and the H zone disappears.

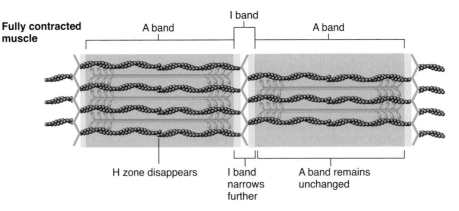

H zone disappears   I band narrows further   A band remains unchanged

Process Figure 7.7   Sarcomere Shortening

Table 7.1	Summary of Skeletal Muscle Contraction (see figures 7.7 and 7.8)

1. An action potential travels along an axon membrane to a neuromuscular junction.

2. Ca^{2+} channels open and Ca^{2+} enter the presynaptic terminal.

3. Acetylcholine is released from the synaptic vesicles in the presynaptic terminal of the neuron.

4. Acetylcholine diffuses across the synaptic cleft and binds to receptor sites on Na^+ channels in the muscle cell postsynaptic membrane causing Na^+ channels to open.

5. Sodium ions diffuse into the muscle cell, initiating an action potential in the muscle cell. The action potentials travel along the sarcolemma and T tubule membranes.

6. Action potentials in the T tubules cause the sarcoplasmic reticulum to release Ca^{2+}.

7. Calcium ions bind to troponin molecules, associated with actin myofilaments. The binding causes tropomyosin molecules to move into grooves along the actin myofilament, which exposes myosin attachment sites.

8. Muscle contraction requires energy. ATP molecules, bound to the myosin heads, are broken down to ADP and P, releasing energy, which is briefly stored in the myosin heads.

9. Some of the stored energy is used to supply the energy for movement of the myosin heads. The heads of myosin myofilaments bind to exposed attachment sites on the actin myofilaments, forming cross-bridges. The P are released from the myosin heads.

10. The heads of the myosin myofilaments bend, causing the actin myofilaments to slide over the surface of the myosin myofilaments.

11. ADP molecules are released from the myosin head.

12. ATP molecules bind to the myosin heads and are broken down to ADP and P. Energy is stored in the myosin heads. The actin myofilament is released and the myosin heads bend back to their resting position.

13. As long as Ca^{2+} remain attached to troponin, and as long as ATP remains available, steps 8 through 11 are repeated and the muscle continues to contract .

Sarcomeres lengthen during muscle relaxation. This lengthening requires some opposing force, such as that produced by other muscles or gravity.

Action potentials produced in skeletal muscle fibers at the neuromuscular junction travel along the sarcolemma and the T tubule membranes (table 7.1). The action potentials cause the membranes of the sarcoplasmic reticulum adjacent to the T tubules to become more permeable to Ca^{2+}, and Ca^{2+} diffuse into the sarcoplasm. The Ca^{2+} bind to troponin molecules attached to the actin myofilaments (figure 7.8). This binding causes tropomyosin molecules to move into a groove along the actin molecule, exposing myosin attachment sites on the actin myofilament. The exposed attachment sites on the actin myofilament bind to the heads of the myosin myofilaments to form **cross-bridges** between the actin and myosin myofilaments.

Energy for muscle contraction is supplied to the muscles in the form of adenosine triphosphate (ATP), a high-energy molecule produced from the energy that is released during the metabolism of food (see chapters 3 and 17). The energy in ATP is released as it is broken down to adenosine diphosphate (ADP) and phosphate (P). During muscle contraction, the energy released from ATP is briefly stored in the myosin head. This energy is used to move the heads of the myosin myofilaments toward the center of the sarcomere, causing the actin myofilaments to slide past the myosin myofilaments. In the process, ADP and P are released from the myosin heads.

As a new ATP molecule attaches to the head of the myosin molecule, the cross-bridge is released, the ATP is broken down to ADP and P, which both remain bound to the myosin head, and the myosin head is restored to its original position, where it can attach to the next attachment site. As long as Ca^{2+} remain attached to troponin, and as long as ATP remains available, the cycle of cross-bridge formation, movement, and release is repeated (see table 7.1). A new ATP must bind to myosin before the cross-bridge can be released. When ATP is not available after a person dies, the cross-bridges that have formed are not released, causing muscles to become rigid. This condition is called **rigor mortis** (rig′er mōr′tis, stiffness + death).

Part of the energy from ATP involved in muscle contraction is required for the formation and movement of the cross-bridges, and part is released as heat. The heat released during muscle contraction increases body temperature, and a person becomes warmer during exercise. Shivering, a type of generalized muscle contraction, is one of the body's mechanisms for dealing with cold. The muscle movement involved in shivering produces heat, which raises the body temperature.

Muscle relaxation occurs as calcium ions are actively transported back into the sarcoplasmic reticulum (a process that requires ATP). As a consequence, the attachment sites on the actin molecules are once again covered by tropomyosin so that cross-bridges cannot reform.

P R E D I C T   1

Predict the consequences of having the following conditions develop in a muscle in response to a stimulus: (a) inadequate ATP is present in the muscle fiber before a stimulus is applied; (b) adequate ATP is present within the muscle fiber, but action potentials occur at a frequency so great that calcium ions are not transported back into the sarcoplasmic reticulum between individual action potentials.

Muscle Twitch, Summation, Tetanus, and Recruitment

A **muscle twitch** is the contraction of a muscle fiber in response to a stimulus. Because most muscle fibers are grouped into motor units, a muscle twitch is usually the contraction of all muscle fibers in a motor unit in response to a stimulus.

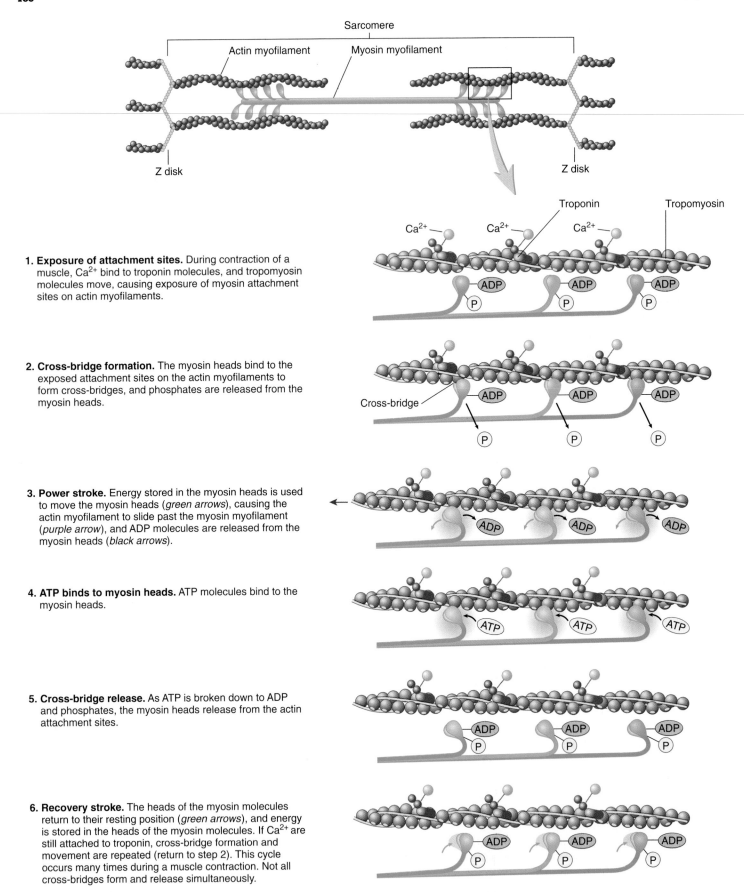

1. **Exposure of attachment sites.** During contraction of a muscle, Ca^{2+} bind to troponin molecules, and tropomyosin molecules move, causing exposure of myosin attachment sites on actin myofilaments.

2. **Cross-bridge formation.** The myosin heads bind to the exposed attachment sites on the actin myofilaments to form cross-bridges, and phosphates are released from the myosin heads.

3. **Power stroke.** Energy stored in the myosin heads is used to move the myosin heads (*green arrows*), causing the actin myofilament to slide past the myosin myofilament (*purple arrow*), and ADP molecules are released from the myosin heads (*black arrows*).

4. **ATP binds to myosin heads.** ATP molecules bind to the myosin heads.

5. **Cross-bridge release.** As ATP is broken down to ADP and phosphates, the myosin heads release from the actin attachment sites.

6. **Recovery stroke.** The heads of the myosin molecules return to their resting position (*green arrows*), and energy is stored in the heads of the myosin molecules. If Ca^{2+} are still attached to troponin, cross-bridge formation and movement are repeated (return to step 2). This cycle occurs many times during a muscle contraction. Not all cross-bridges form and release simultaneously.

Process Figure 7.8 Breakdown of ATP and Cross-Bridge Movement During Muscle Contraction

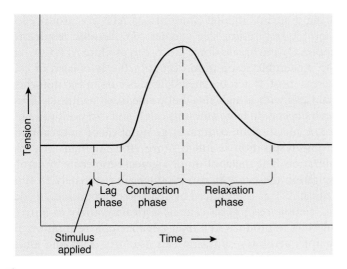

Figure 7.9 Phases of a Muscle Twitch

Hypothetical muscle twitch in a single muscle fiber. There is a short lag phase after stimulus application, followed by a contraction phase and a relaxation phase.

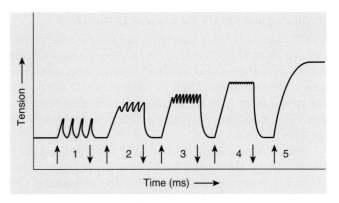

Figure 7.10 Multiple-Wave Summation

Stimuli 1–5 are stimuli of increasing frequency. For each stimulus, the up arrow indicates the start of stimulation, and the down arrow indicates the end of stimulation. Stimulus frequency 1 produces successive muscle twitches with complete relaxation between stimuli. Stimuli frequencies 2–4 do not allow complete relaxation between stimuli. Stimulus frequency 5 causes tetanus—no relaxation between stimuli.

A muscle twitch has three phases (figure 7.9). The **lag phase** is the time between the application of a stimulus and the beginning of contraction. The **contraction phase** is the time of contraction, and the **relaxation phase** is the time during which the muscle relaxes.

During the lag phase, action potentials are produced in one or more motor neurons. An action potential is conducted along the axon of a motor neuron to a neuromuscular junction (see figure 7.5). At the neuromuscular junction, acetylcholine molecules released from the presynaptic terminal diffuse across the synaptic cleft and bind to receptor sites on Na^+ channels in the muscle cell postsynaptic membrane, resulting in the production of an action potential (see figure 7.6). The action potential travels along the sarcolemma and T tubule membranes (see figure 7.2). In response to action potentials in the T tubules, Ca^{2+} are released from the sarcoplasmic reticulum and diffuse into myofibrils. Within the myofibrils, Ca^{2+} bind to troponin causing tropomyosin to move, attachment sites to be exposed, and cross-bridges to be formed (see figure 7.8, steps 1–2).

The contraction phase results from cross-bridge movement and cycling (see figure 7.8), which increases the tension produced by the muscle fibers (see figure 7.9).

During the relaxation phase, Ca^{2+} are actively transported back into the sarcoplasmic reticulum. As Ca^{2+} diffuse away from the troponin molecules, tropomyosin molecules once again block the attachment sites. Cross-bridge formation is prevented, and the tension produced by the muscle fibers decreases (see figure 7.9).

The strength of muscle contractions can vary from weak to strong. For example, the force generated by muscles to lift a feather is much less than the force required to lift a 25-pound weight. The force of contraction produced by a muscle is increased in two ways: 1. **summation,** which involves increasing the force of contraction of the muscle fibers within the muscle and 2. **recruitment,** which involves increasing the number of

muscle fibers contracting. In summation, the force of contraction of individual muscle fibers is increased by rapidly stimulating them. When stimulus frequency, which is the number of times a motor neuron is stimulated per second, is low, there is time for complete relaxation of muscle fibers between muscle twitches (figure 7.10, stimulus frequency 1). As stimulus frequency increases (figure 7.10, stimuli frequencies 2–4), there is not enough time between contractions for muscle fibers to completely relax. Thus, one contraction summates, or is added onto, a previous contraction. As a result, the overall force of contraction increases. **Tetanus** (tet′a-nus, convulsive tension) is a sustained contraction that occurs when the frequency of stimulation is so rapid that there is no relaxation (figure 7.10, stimulus frequency 5). The increased force of contraction produced in summation and tetanus occurs because of a buildup of Ca^{2+} in myofibrils, which promotes cross-bridge formation and cycling. The buildup of Ca^{2+} occurs because the rapid production of action potentials in muscle fibers causes Ca^{2+} to be released from the sarcoplasmic reticulum faster than they are actively transported back into the sarcoplasmic reticulum.

In **recruitment,** the strength of contraction of a muscle is increased by increasing the number of motor units stimulated. When only a few motor units are stimulated, a small force of contraction is produced, because only a small number of muscle fibers are contracting. As the number of motor units stimulated increases, more muscle fibers are stimulated to contract, and the force of contraction increases. Maximum force of contraction is produced in a given muscle when all the motor units of that muscle are stimulated (recruited).

If all the motor units in a muscle could be stimulated simultaneously, a quick, jerking motion would occur. Because the motor units are recruited gradually so that some are stimulated and held in tetanus while additional motor units are recruited, slow, smooth, sustained contractions occur. Smooth relaxation of muscle occurs because some motor units are held in tetanus while other motor units relax.

Energy Requirements for Muscle Contraction

The ATP required to provide energy for muscle contraction is produced primarily by aerobic (in air) respiration in numerous mitochondria located within the muscle fiber sarcoplasm between the myofibrils. Even in resting muscle fibers, fairly large amounts of ATP are required for cell maintenance. It is therefore necessary for muscle fibers to constantly produce ATP.

Muscle fibers cannot stockpile ATP in preparation for periods of activity. The muscle fibers, however, can store another high-energy molecule, **creatine** (krē′ă-tēn) **phosphate.** Creatine phosphate provides a means of storing energy that can be used rapidly to help maintain an adequate amount of ATP in the contracting muscle fiber. During periods of inactivity, as excess ATP is produced in the muscle fiber, the energy contained in the ATP is used to synthesize creatine phosphate. During periods of activity, the small reserves of ATP existing in the cell are used first. Then the energy stored in creatine phosphate is accessed quickly to produce ATP, which is used in muscle contraction and to restore ATP reserves (figure 7.11).

ATP is produced by both anaerobic and aerobic cellular respiration (see chapter 17 for details). **Anaerobic respiration,** which does not require O_2, results in the breakdown of glucose to yield ATP and lactic acid. During intense exercise, such as during a sprint, smaller amounts of ATP are produced by anaerobic respiration (see chapter 17). **Aerobic respiration** requires O_2 and breaks down glucose to produce ATP, CO_2, and H_2O. Anaerobic respiration occurs in the cytoplasm of cells, whereas most of aerobic metabolism occurs in the mitochondria. Cells with a high metabolic rate, such as muscle fibers, which depend on large amounts of O_2 and carry out primarily aerobic metabolism, contain large numbers of mitochondria. Aerobic respiration is much more efficient than anaerobic respiration. The metabolism of a glucose molecule by aerobic respiration theoretically can produce approximately 18 times as much ATP as is produced by anaerobic respiration. In addition, aerobic respiration can use a greater variety of nutrient molecules to produce ATP than can anaerobic respiration. For example, aerobic respiration can use fatty acids and amino acids to generate ATP.

Although aerobic respiration produces more ATP molecules than anaerobic respiration for each glucose molecule metabolized, anaerobic respiration can occur faster and is important when O_2 availability limits aerobic respiration. By using many glucose molecules, anaerobic respiration can rapidly produce much ATP, but anaerobic respiration can proceed for only a short time. Lactic acid is an end product of anaerobic respiration. Some of the lactic acid can diffuse out of the muscle

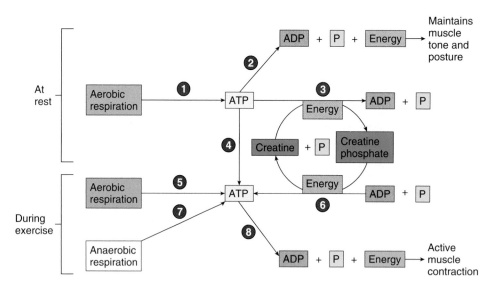

1. At rest, ATP is produced by aerobic respiration.

2. Small amounts of ATP are used in muscle contractions that maintain muscle tone and posture.

3. Excess ATP is used to produce creatine phosphate, an energy-storage molecule.

4. As exercise begins, ATP already in the cell is used first.

5. During moderate (aerobic) exercise, aerobic respiration provides most of the ATP necessary for muscle contraction.

6. Energy stored in creatine phosphate can also be used to produce ATP.

7. During times of extreme (anaerobic) exercise, anaerobic respiration provides small amounts of ATP that can sustain muscle contraction for brief periods.

8. Throughout the time of exercise, ATP from all of these sources (4–7) provides energy for active muscle contraction.

Figure 7.11 Fate of ATP in Resting and Exercising Muscle

fiber into the blood, but some remains in the muscle fibers. Lactic acid can irritate muscle fibers, causing short-term pain. Muscle pain that lasts for several days, however, indicates mechanical injury to the muscle.

Resting muscles or muscles undergoing long-term exercise, such as during long-distance running, depend primarily on aerobic respiration for ATP synthesis. Although some glucose is used as an energy source, fatty acids are a more important energy source during both sustained exercise and resting conditions. During short periods of intense exercise such as in sprinting, ATP is used up more quickly than it can be produced by aerobic respiration. Once the ATP reserves begin to decrease, the energy stored in creatine phosphate is used to maintain ATP levels in the contracting muscle fiber. Once the creatine phosphate stores are depleted, anaerobic respiration becomes important. Anaerobic respiration rapidly metabolizes available glucose to provide enough ATP to support intense muscle contraction for up to 2–3 minutes. During intense exercise, glycogen stored in muscle fibers can be broken down to glucose, which can then be used to produce more ATP. Anaerobic metabolism is ultimately limited by depletion of glucose and a buildup of lactic acid within the muscle fiber.

After intense exercise, the respiration rate and volume remain elevated for a time, even though the muscles are no longer actively contracting. This increased respiration provides the O_2 to pay back the oxygen debt. The *oxygen debt,* or *excess postexercise O_2 consumption,* is the amount of O_2 needed in chemical reactions that occur to (1) convert lactic acid to glucose, (2) replenish the depleted ATP and creatine phosphate stores in muscle fibers, and (3) replenish O_2 stores in the lungs, blood, and muscles. After the lactic acid produced by anaerobic respiration is converted to glucose and creatine phosphate levels are restored, respiration rate returns to normal. The magnitude of the oxygen debt depends on the intensity of the exercise, the length of time it was sustained, and the physical condition of the individual. The metabolic capacity of an individual in poor physical condition is much lower than that of a well-trained athlete. With exercise and training, a person's ability to carry out both aerobic and anaerobic activities is enhanced.

P R E D I C T
After a 10-mile run with a sprint at the end, a runner continues to breathe heavily for a time. Indicate the type of metabolism that is producing energy during the run, during the sprint, and after the run.

Fatigue

The most common type of fatigue, **psychological fatigue,** involves the central nervous system rather than the muscles themselves. The muscles are still capable of contracting, but the individual "perceives" that continued muscle contraction is impossible. A determined burst of activity in a tired athlete, such as a runner, in response to pressure from a competitor is an example of how psychological fatigue can be overcome.

Muscle fatigue results when ATP is used during muscle contraction faster than it can be produced in the muscle fibers and lactic acid builds up faster than it can be removed. As a consequence, ATP falls to levels too low for muscle fibers to produce their maximum force of contraction, and contractions become weaker and weaker.

An example of muscle fatigue occurs when a runner collapses on the track and must be helped off. The runner's muscles can no longer function regardless of how determined the runner may be. Under conditions of extreme muscular fatigue, muscles may become incapable of either contracting or relaxing. This condition, called **physiological contracture,** occurs when there is too little ATP to bind to myosin myofilaments. Because binding of ATP to the myosin heads is necessary for crossbridge release between the actin and myosin, the cross-bridges between the actin and myosin myofilaments cannot be broken, and the muscle cannot relax.

Types of Muscle Contractions

Muscle contractions are classified as either isometric or isotonic. In **isometric** (equal distance) **contractions,** the amount of tension increases during the contraction process, but the length of the muscle does not change. Isometric contractions are responsible for the constant length of the postural muscles of the body, such as the muscles of the back. On the other hand, in **isotonic** (equal tension) **contractions,** the amount of tension produced by the muscle is constant during contraction, but the length of the muscle decreases. Movements of the arms or fingers are predominantly isotonic contractions. Most muscle contractions are a combination of isometric and isotonic contractions in which the muscles shorten some distance and the degree of tension increases.

Concentric (kon-sen′trik) **contractions** are isotonic contractions in which muscle tension increases as the muscle shortens. Most movements performed by muscle contractions are concentric contractions. **Eccentric** (ek-sen′trik) **contractions** are isotonic contractions in which tension is maintained as the muscle lengthens. Eccentric contractions are used when a person lets a heavy weight down slowly. Substantial force is produced in muscles during eccentric contractions, and muscle injuries can occur from repetitive eccentric contractions, such as in the hamstring muscles when a person runs downhill.

Muscle Tone

Muscle tone refers to the constant tension produced by muscles of the body over long periods of time. Muscle tone is responsible for keeping the back and legs straight, the head held in an upright position, and the abdomen from bulging. Muscle tone depends on a small percentage of all the motor units in a muscle being stimulated at any point in time, causing their muscle fibers to contract tetanically and out of phase with one another.

Slow and Fast Fibers

Muscle fibers are sometimes classified as either fast-twitch or slow-twitch muscle fibers. This classification is based on differences in the rod portion of the myosin myofilament (see figure 7.2). Slow-twitch fibers contain type I myosin as the predominant or even exclusive type. Fast-twitch fibers contain either type IIa or IIx myosin myofilaments. Each of these three myosin types is the product of a different myosin gene. Fast-twitch muscle fibers, also called type II muscle fibers, contract quickly and fatigue quickly, whereas slow-twitch muscle fibers, also called type I muscle fibers, contract more slowly and are more resistant to fatigue. Type IIx fibers can contract 10 times faster than type I fibers, and type IIa fibers contract at an intermediate speed. Type IIa fibers are more fatigue-resistant than type IIx fibers. Fast-twitch muscles have larger stores of glycogen and are well adapted to performing anaerobic metabolism, whereas slow-twitch muscles contain more mitochondria and are better suited for aerobic metabolism.

The white meat of a chicken's breast is composed mainly of fast-twitch fibers. The muscles are adapted to contract rapidly for a short time but fatigue quickly. Chickens normally do not fly long distances. They spend most of their time walking. Ducks, on the other hand, fly for much longer periods and over greater distances. The red, or dark, meat of a chicken's leg or a duck's breast is composed of slow-twitch fibers. The darker appearance is due partly to a richer blood supply and partly to the presence of **myoglobin,** which stores oxygen temporarily. Myoglobin can continue to release oxygen in a muscle, even when a sustained contraction has interrupted the continuous flow of blood.

Humans exhibit no clear separation of slow-twitch and fast-twitch muscle fibers in individual muscles. Most muscles have both types of fibers, although the number of each type varies in a given muscle. The large postural muscles contain more slow-twitch fibers, whereas muscles of the upper limb contain more fast-twitch fibers. People who are good sprinters have a greater percentage of fast-twitch, type II muscle fibers in their lower limbs, whereas good long-distance runners have a higher percentage of slow-twitch, type I fibers. Athletes who are able to perform a variety of anaerobic and aerobic exercises tend to have a more balanced mixture of fast-twitch and slow-twitch muscle fibers.

Effects of Exercise on Muscles

Average, healthy, active adults have roughly equal numbers of slow- and fast-twitch fibers in their muscles and over three times as many type IIa as type IIx fibers. A world-class sprinter may have over 80% type II fibers, with type IIa slightly predominating. A world-class endurance athlete, on the other hand, may have 95% type I fibers. The ratio of muscle fiber types in world-class sprinters or endurance athletes apparently has a large hereditary component but can be considerably influenced by training. Exercise increases the vascularity of muscles, increases the number of mitochondria per muscle fiber, and causes enlargement of muscle fibers by increasing the number of myofibrils and myofilaments. With weight training, type IIx myosin myofilaments can be replaced by type IIa myosin myofilaments as muscles enlarge. Muscle nuclei quit expressing type IIx genes and begin expressing type IIa, which are more resistant to fatigue. If the exercise stops, the type IIa genes turn off and the type IIx genes turn back on. Vigorous exercise programs can cause a limited number of type I myofilaments to be replaced by type IIa myofilaments.

In addition, weight training and other exercises can cause muscles to **hypertrophy** (enlarge). This occurs as more myofilaments and thus myofibrils are produced inside a myofiber, causing them to enlarge. The number of mitochondria in the myofibers also increases, causing additional enlargement of the myofiber. Exercise to produce muscle hypertrophy is often referred to as bodybuilding (see box on Bodybuilding p. 172).

The number of cells in a skeletal muscle remains relatively constant following birth. Enlargement of muscles after birth is therefore primarily the result of an increase in the size of the existing muscle fibers. As people age, the number of muscle fibers actually decreases, and new ones cannot be added.

Smooth Muscle and Cardiac Muscle

Smooth muscle cells are small and spindle-shaped, usually with one nucleus per cell (table 7.2). They contain less actin and myosin than do skeletal muscle cells, and the myofilaments are not organized into sarcomeres. As a result, smooth muscle cells are not striated. Smooth muscle cells contract more slowly than skeletal muscle cells when stimulated by neurotransmitters from the nervous system and do not develop an oxygen debt. The resting membrane potential of some smooth muscle cells fluctuates between slow depolarization and repolarization phases. As a result, smooth muscle cells can periodically and spontaneously generate action potentials that cause the smooth muscle cells to contract. The resulting periodic spontaneous contraction of smooth muscle is called **autorhythmicity.** Smooth muscle is under involuntary control, whereas skeletal muscle is under voluntary motor control. Some hormones, such as those that regulate the digestive system, can stimulate smooth muscle to contract.

Smooth muscle cells are organized to form layers. Most of those cells have gap junctions, specialized cell-to-cell contacts (see chapter 4), that allow action potentials to spread to all the smooth muscle cells in a given tissue. Thus, all the smooth muscle cells tend to function as a unit and contract at the same time.

Cardiac muscle shares some characteristics with both smooth and skeletal muscle (see table 7.2). Cardiac muscle cells are long, striated, and branching, with usually only one nucleus per cell. The actin and myosin myofilaments are organized into sarcomeres, but the distribution of myofilaments is not as uniform as in skeletal muscle. As a result, cardiac muscle cells are striated, but not as distinctly striated as skeletal muscle. When stimulated by neurotransmitters, the rate of cardiac muscle contraction is between that of smooth and skeletal muscle. Cardiac muscle contraction is autorhythmic. Cardiac muscle exhibits limited anaerobic metabolism. Instead, it continues to contract at a level that can be sustained by aerobic metabolism and consequently does not fatigue.

Cardiac muscle cells are connected to one another by **intercalated** (in-ter′kă-lā-ted) **disks.** Intercalated disks are specialized structures that include tight junctions and gap junctions and that facilitate action potential conduction between the cells. This cell-to-cell connection allows cardiac muscle cells to function as a unit. As a result, an action potential in one cardiac muscle cell can stimulate action potentials in adjacent cells, causing all to contract together. As with smooth muscle, cardiac muscle is under involuntary control and is influenced by hormones, such as epinephrine.

Clinical Focus Disorders and Other Conditions of Muscle Tissue

Cramps

Cramps are painful, spastic contractions of muscle that are usually the result of an irritation within a muscle. Local inflammation from buildup of lactic acid or connective tissue inflammation can cause contraction of muscle fibers surrounding the irritated region. Muscle pulls are described in A Case in Point on p. 190.

Fibromyalgia

Fibromyalgia (fī-brō-mī-al'ja), or chronic muscle pain syndrome, has muscle pain as its main symptom. Fibromyalgia has no known cure, but it is not progressive, crippling, or life-threatening. The pain is chronic and widespread in muscles and muscle-tendon junctions.

Hypertrophy and Atrophy

Exercise causes muscular **hypertrophy** (hī-per'trō-fē), which is an enlargement of a muscle resulting from an increase in the number of myofibrils within muscle fibers. Muscle hypertrophy is greater in males than in females, mainly because of greater concentrations of the male sex hormone, testosterone, in males. Disuse of muscle results in

muscular **atrophy** (at'rō-fē), which is a decrease in muscle size because of a decrease in myofilaments within muscle fibers. Severe atrophy involves the permanent loss of skeletal muscle fibers and the replacement of those fibers by connective tissue. Immobility resulting from damage to the nervous system or casting a broken limb leads to muscular atrophy. If the nerve supply to a muscle is severed, the muscle becomes flaccid (having no tone) and atrophies.

Muscular Dystrophy

Muscular dystrophy (dis'trō-fē, *dys,* bad + *trophe,* nourishment) refers to a group of inherited muscle disorders in which skeletal muscle, cardiac muscle, and smooth muscle tissue degenerates and the person experiences progressive weakness and other symptoms, including heart problems. The disorders are characterized by the progressive degeneration of muscle fibers leading to atrophy and their eventual replacement by fat and other connective tissue.

Duchenne's (dū-shān', Guillaume Duchenne, neurologist, 1806–1875) **muscular dystrophy** is described in the Systems Pathology on p. 192.

Myotonic (mī-ō-ton'ik) **muscular dystrophy** is characterized by the failure of muscles to relax following a forceful contraction, as well as by muscular weakness. The disorder is inherited as a dominant trait in both males and females and occurs in about 1 in every 20,000 births. The disorder progresses slowly, usually affecting the face and neck muscles first and affecting the hands most severely.

Myasthenia Gravis

Myasthenia (mī-as-thē'nē-ă, *mys,* muscle + *astheneia,* weakness) **gravis** was described in A Case in Point on p. 163. It is a chronic, progressive autoimmune disease resulting from the destruction of acetylcholine receptors in the neuromuscular junction. Treatment includes anticholinesterase drugs and/or immunosuppresive drugs.

Tendinitis

As the name implies, **tendinitis** (ten-di-nī'tis) is an inflammation of a tendon or its attachment point. It usually occurs in athletes who overtax the muscle to which the tendon is attached.

Table 7.2	Comparison of Muscle Types		
Feature	**Skeletal Muscle**	**Cardiac Muscle**	**Smooth Muscle**
Location	Attached to bone	Heart	Wall of hollow organs, blood vessels, and glands
Appearance			
Cell shape	Long, cylindrical	Branched	Spindle-shaped
Nucleus	Multiple, peripheral	Usually single, central	Single, central
Special features		Intercalated disks	Cell–cell attachments
Striations	Yes	Yes	No
Autorhythmic	No	Yes	Yes
Control	Voluntary	Involuntary	Involuntary
Function	Move the whole body	Heart contraction to propel blood through the body	Compression of organs, ducts, tubes, etc.

Skeletal Muscle Anatomy
General Principles

Most muscles extend from one bone to another and cross at least one joint. Muscle contraction causes most body movements by pulling one of the bones toward the other across the movable joint. Some muscles are not attached to bone at both ends. For example, some facial muscles attach to the skin, which moves as the muscles contract.

The two points of attachment of each muscle are its origin and insertion. At these attachment points, the muscle is connected to the bone by a **tendon.** Some broad, sheetlike tendons are called **aponeuroses** (ap′ō-noo-rō′sēz, *apo,* from + *neuron,* sinew). A **retinaculum** (ret-i-nak′ū-lum, bracelet) is a band of connective tissue that holds down the tendons at each wrist and ankle. The **origin,** also called the **head,** is the most stationary end of the muscle. The **insertion** is the end of the muscle attached to the bone undergoing the greatest movement. Origins are usually, but not always, proximal or medial to the insertion of a given muscle. The part of the muscle between the origin and the insertion is the **belly** (figure 7.12). Some muscles, such as the biceps brachii with two heads and the triceps brachii with three heads, have multiple origins, or heads.

Muscles are typically grouped so that the action of one muscle or group of muscles is opposed by that of another muscle or group of muscles. For example, the biceps brachii flexes the elbow and the triceps brachii extends the elbow. A muscle that accomplishes a certain movement, such as flexion, is called the **agonist** (ag′ō-nist, *agon,* a contest). A muscle acting in opposition to an agonist is called an **antagonist** (an-tag′ō-nist). The biceps brachii is the agonist in elbow flexion, whereas the triceps brachii is the antagonist, which extends the elbow.

Muscles also tend to function in groups to accomplish specific movements. For example, the deltoid, biceps brachii, and pectoralis major all help flex the shoulder. Furthermore, many muscles are members of more than one group, depending on the type of movement being considered. For example, the anterior part of the deltoid muscle functions with the flexors of the shoulder, whereas the posterior part functions with the extensors of the shoulder. Members of a group of muscles working together to produce a movement are called **synergists** (sin′er-jistz). For example, the biceps brachii and brachialis are synergists in elbow flexion. Among a group of synergists, if one muscle plays the major role in accomplishing the desired movement, it is the **prime mover.** The brachialis is the prime mover in flexing the elbow. **Fixators** are muscles that hold one bone in place relative to the body while a usually more distal bone is moved. The muscles of the scapula act as fixators to hold the scapula in place while other muscles contract to move the humerus.

Nomenclature

Most muscles have names that are descriptive (figure 7.13). Some muscles are named according to their location, such as the pectoralis (chest) muscles in the chest. Other muscles are named according to their origin and insertion, such as the brachioradialis (*brachio,* arm) muscle, which extends from the arm to the radius. Some muscles are named according to the number of heads, such as the biceps (*bi,* two + *ceps,* head) brachii, which has two heads; and some according to their function, such as the flexor digitorum, which flexes the digits (fingers). Other muscles are named according to their size, such as vastus, which means large; their shape, such as deltoid, which means triangular; or the orientation of their fasciculi, such as rectus, which means straight. Recognizing the descriptive nature of muscle names makes learning those names much easier. The most superficial muscles are shown in figure 7.13. Examining surface anatomy can be a great advantage to the anatomy student in gaining a better understanding of muscle anatomy. We have pointed out some of the muscles of the upper and lower limbs that can be seen on the surface of the body. Some muscles are especially well developed in bodybuilders (figure 7.14).

Bodybuilding

Bodybuilding is a popular sport enjoyed by thousands of men and women. Participants in this sport combine specific weight training and diet to develop maximum muscle mass and minimum body fat. Their major goal is to develop a well-balanced, complete physique. Bodybuilding requires knowledge of exercises that develop all muscles and necessitates consistent and rigorous training. Exercising the appropriate muscles to the proper degree is required to develop a well-proportioned body.

Vigorous weight training can double or triple the size of a muscle. The cardiorespiratory fitness of bodybuilders is often similar to that of other well-trained athletes. This fitness has not always been present in bodybuilders, but it can now be attributed to modern bodybuilding techniques that include aerobic exercise and running, along with "pumping iron."

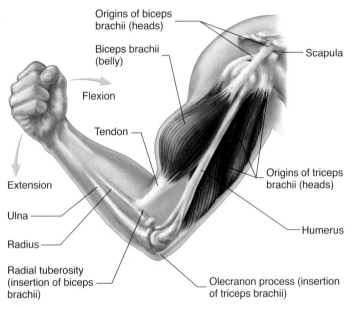

Figure 7.12   Muscle Attachment

Muscles are attached to bones by tendons. The biceps brachii has two heads that originate on the scapula. The triceps brachii has three heads that originate on the scapula and humerus. The biceps tendon inserts onto the radial tuberosity and onto nearby connective tissue. The triceps brachii inserts onto the olecranon of the ulna.

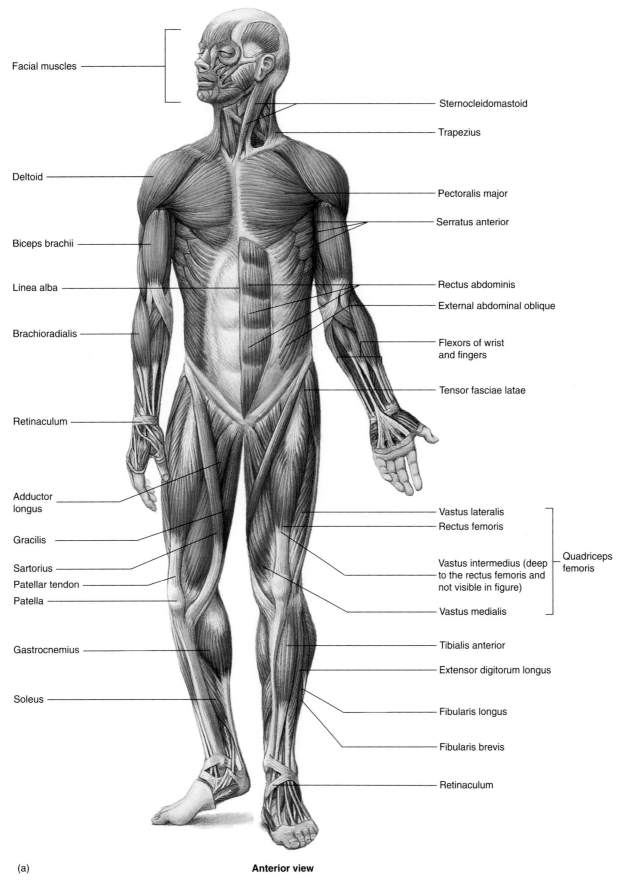

Facial muscles

Deltoid

Biceps brachii

Linea alba

Brachioradialis

Retinaculum

Adductor longus

Gracilis

Sartorius

Patellar tendon

Patella

Gastrocnemius

Soleus

Sternocleidomastoid

Trapezius

Pectoralis major

Serratus anterior

Rectus abdominis

External abdominal oblique

Flexors of wrist and fingers

Tensor fasciae latae

Vastus lateralis
Rectus femoris

Vastus intermedius (deep to the rectus femoris and not visible in figure)

Vastus medialis

Quadriceps femoris

Tibialis anterior

Extensor digitorum longus

Fibularis longus

Fibularis brevis

Retinaculum

(a)   **Anterior view**

Figure 7.13   General Overview of the Superficial Body Musculature

Red is muscle; white is connective tissue, such as tendons, aponeuroses, and retinacula.

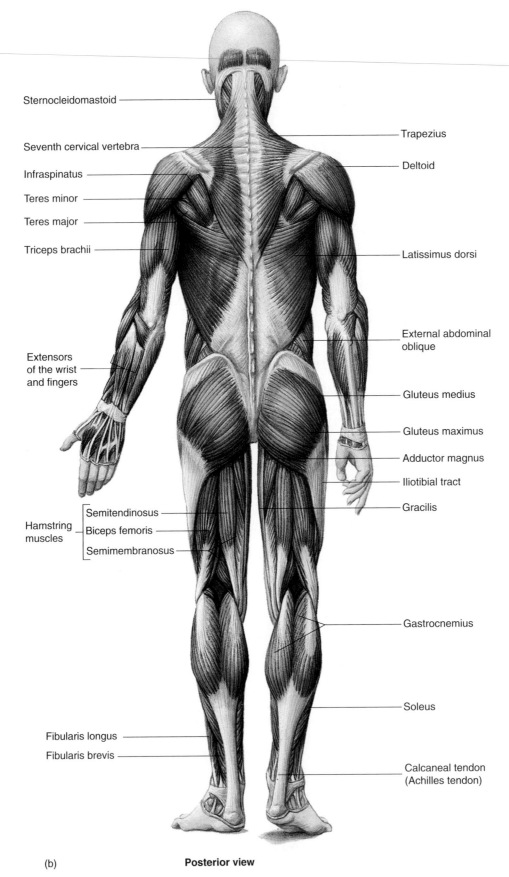

Sternocleidomastoid

Seventh cervical vertebra

Infraspinatus

Teres minor

Teres major

Triceps brachii

Extensors
of the wrist
and fingers

Hamstring
muscles

Semitendinosus

Biceps femoris

Semimembranosus

Fibularis longus

Fibularis brevis

Trapezius

Deltoid

Latissimus dorsi

External abdominal
oblique

Gluteus medius

Gluteus maximus

Adductor magnus

Iliotibial tract

Gracilis

Gastrocnemius

Soleus

Calcaneal tendon
(Achilles tendon)

(b)

Posterior view

Figure 7.13   *continued*

Anterior view

Lateral view

Figure 7.14   Bodybuilders

Name as many muscles as you can from the photos. Compare these photos with the labeled muscles in figure 7.13.

Muscles of the Head and Neck

The muscles of the head and neck include those involved in facial expression, mastication (chewing), movement of the tongue, swallowing, voice production, eye movements, and movements of the head and neck.

Facial Expression

Several muscles act on the skin around the eyes and eyebrows (figure 7.15 and table 7.3). The **occipitofrontalis** (ok-sip′i-tō-frŭn-tā′lis) raises the eyebrows. The occipital and frontal portions of the muscle are connected by the epicranial (galea) aponeurosis. The **orbicularis oculi** (ōr-bik′ū-la′ris, circular + ok′ū-lī, eye) encircle the eyes, tightly close the eyelids, and cause "crow's feet" wrinkles in the skin at the lateral corners of the eyes.

Several other muscles function in moving the lips and the skin surrounding the mouth (see figure 7.15). The **orbicularis oris** (ōr′is, mouth), which encircles the mouth, and the **buccinator** (buk′sĭ-nā′tōr, *bucca*, cheek) are sometimes called the kissing muscles because they pucker the mouth. The buccinator also flattens the cheeks as in whistling or blowing a trumpet and is therefore sometimes called the trumpeter's muscle. Smiling is accomplished primarily by the **zygomaticus** (zī′gō-mat′i-kŭs, *zygon*, yoke) muscles, which elevate the upper lip and corner of the mouth. Sneering is accomplished by the **levator labii superioris** (le-vā′ter lā′bē-ī soo-pēr′ē-ōr′is, *labium*, lip) because the muscle elevates one side of the upper lip, and frowning or pouting largely by the **depressor anguli oris** (dē-pres′ŏr an′gū-lī ōr′ŭs), which depresses the corner of the mouth.

> **P R E D I C T**   **3**
> Harry Wolf, a notorious flirt, on seeing Sally Gorgeous, raises his eyebrows, winks, whistles, and smiles. Name the facial muscles he uses to carry out this communication. Sally, thoroughly displeased with this exhibition, frowns and sneers in disgust. What muscles does she use?

Mastication

The four pairs of muscles of chewing, or **mastication** (mas-ti-kā′shŭn), are some of the strongest muscles of the body (table 7.4). The **temporalis** (tem′pŏ-rā′lis) and **masseter** (mă-sē′ter, chewer) muscles (see figure 7.15) can be easily seen and felt on the side of the head during mastication. The **pterygoid** (ter′ĭ-goyd, wing-shaped) muscles, consisting of two pairs, are deep to the mandible.

Tongue and Swallowing Muscles

The tongue is very important in mastication and speech. It moves food around in the mouth, and with the buccinator muscle, holds the food in place while the teeth grind the food. The tongue pushes food up to the palate and back toward the pharynx to initiate swallowing. The tongue consists of a mass of **intrinsic muscles,** which are located entirely within the tongue and function to change its shape. The **extrinsic muscles** are

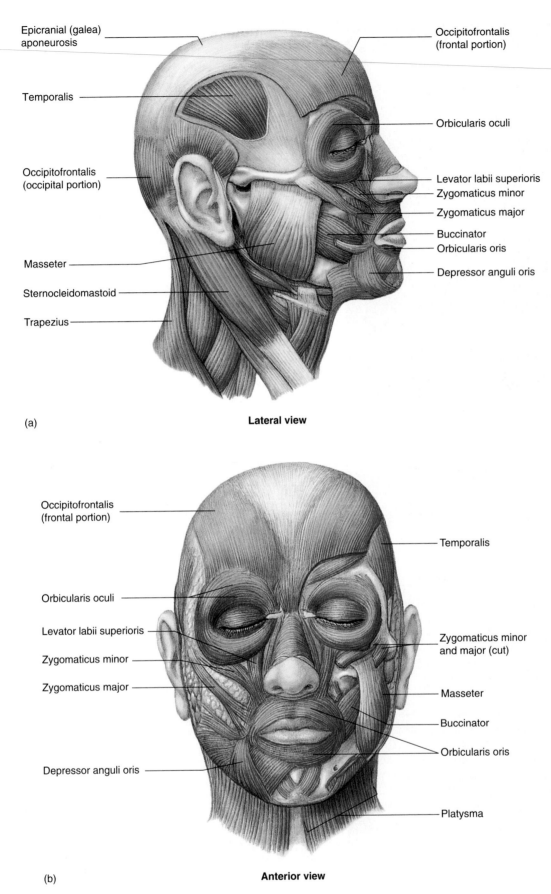

Epicranial (galea) aponeurosis

Occipitofrontalis (frontal portion)

Temporalis

Orbicularis oculi

Occipitofrontalis (occipital portion)

Levator labii superioris

Zygomaticus minor

Zygomaticus major

Buccinator

Orbicularis oris

Masseter

Depressor anguli oris

Sternocleidomastoid

Trapezius

(a)   **Lateral view**

Occipitofrontalis (frontal portion)

Temporalis

Orbicularis oculi

Levator labii superioris

Zygomaticus minor and major (cut)

Zygomaticus minor

Zygomaticus major

Masseter

Buccinator

Orbicularis oris

Depressor anguli oris

Platysma

(b)   **Anterior view**

Figure 7.15   Muscles of Facial Expression

Table 7.3 Muscles of Facial Expression (see figure 7.15)

Muscle	Origin	Insertion	Action
Buccinator (buk′sĭ-nā′tŏr)	Maxilla and mandible	Orbicularis oris at angle of mouth	Retracts angle of mouth; flattens cheek
Depressor anguli oris (dĕ-pres′ŏr an′gŭ-lī ōr′ŭs)	Lower border of mandible	Lip near angle of mouth	Depresses angle of mouth
Levator labii superioris (le-vā′ter lā′bē-ī soo-pēr′ē-ōr′is)	Maxilla	Skin and orbicularis oris of upper lip	Elevates upper lip
Occipitofrontalis (ok-sip′i-tō-frŭn′tā′lis)	Occipital bone	Skin of eyebrow and nose	Moves scalp; elevates eyebrows
Orbicularis oculi (ōr-bik′ū-lā′ris ok′ū-lī)	Maxilla and frontal bones	Circles orbit and inserts near origin	Closes eye
Orbicularis oris (ōr-bik′ū-lā′ris ōr′is)	Nasal septum, maxilla, and mandible	Fascia and other muscles of lips	Closes lip
Zygomaticus major (zī′gō-mat′i-kŭs)	Zygomatic bone	Angle of mouth	Elevates and abducts upper lip
Zygomaticus minor (zī′gō-mat′i-kŭs)	Zygomatic bone	Orbicularis oris of upper lip	Elevates and abducts upper lip

Table 7.4 Muscles of Mastication (see figure 7.15)

Muscle	Origin	Insertion	Action
Temporalis (tem′pŏ-rā′lis)	Temporal fossa	Anterior portion of mandibular ramus and coronoid process	Elevates and retracts mandible; involved in excursion
Masseter (mă-sē′ter)	Zygomatic arch	Lateral side of mandibular ramus	Elevates and protracts mandible; involved in excursion
Lateral pterygoid (ter′i-goyd) (not shown in illustration)	Lateral pterygoid plate and greater wing of sphenoid	Condylar process of mandible and articular disk	Protracts and depresses mandible; involved in excursion
Medial pterygoid (not shown in illustration)	Lateral pterygoid plate of sphenoid and tuberosity of maxilla	Medial surface of mandible	Protracts and elevates mandible; involved in excursion

located outside the tongue but are attached to and move the tongue (figure 7.16 and table 7.5).

Swallowing involves a number of structures and their associated muscles, including the hyoid muscles, soft palate, pharynx (throat), and larynx (voice box). The **hyoid** (hī′oyd, U-shaped) **muscles** are divided into a suprahyoid group (superior to the hyoid bone) and an infrahyoid group (inferior to the hyoid) (see figure 7.16 and table 7.5). When the suprahyoid muscles hold the hyoid bone in place from above, the infrahyoid muscles can elevate the larynx. To observe this effect, place your hand on your larynx (Adam's apple) and swallow.

The muscles of the soft palate close the posterior opening to the nasal cavity during swallowing, preventing food and liquid from entering the nasal cavity. Swallowing is accomplished by elevation of the pharynx and larynx, followed by constriction of the pharynx. The **pharyngeal** (fă-rin′jē-ăl) **elevators** elevate the pharynx, and the **pharyngeal constrictors** constrict the pharynx from superior to inferior, forcing the food into the esophagus. Pharyngeal muscles also open the auditory tube, which connects the middle ear with the pharynx. Opening the auditory tube equalizes the pressure between the middle ear and the atmosphere. This is why it is sometimes helpful to chew gum or swallow when ascending or descending a mountain in a car or changing altitude in an airplane.

Neck Muscles

The deep neck muscles (figure 7.17 and table 7.6) include neck flexors, located along the anterior surfaces of the vertebral bodies, and neck extensors, which are located posteriorly. Rotation and lateral flexion of the head are accomplished by lateral and posterior neck muscles. The **sternocleidomastoid** (ster′nō-klī′dō-mas′toyd) muscle (see figure 7.15a), the prime mover of the lateral muscle group, is easily seen on the anterior and lateral sides of the neck. Contraction of only one sternocleidomastoid muscle causes rotation of the head. Contraction of both sternocleidomastoids results in flexion of the neck or extension of the head, depending on what other neck muscles are doing. **Torticollis** (tōr′ti-kol′is, a twisted neck), or wry neck, may result from injury to one of the sternocleidomastoid muscles. It is sometimes caused by damage to a baby's neck muscles during a difficult birth and usually can be corrected by exercising the muscle.

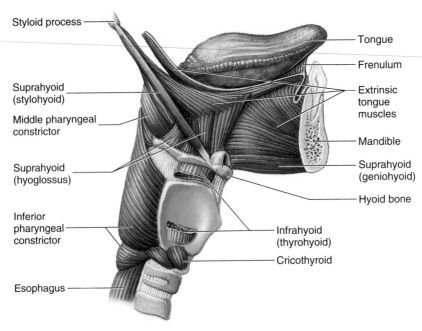

Lateral view

Figure 7.16   **Tongue and Swallowing Muscles**

Muscles of the tongue, hyoid, pharynx, and larynx as seen from the right.

Table 7.5	**Tongue and Swallowing Muscles (see figure 7.16)**		
Muscle	**Origin**	**Insertion**	**Action**
Tongue muscles			
Intrinsic (not shown)	Inside tongue	Inside tongue	Changes shape of tongue
Extrinsic	Bones around oral cavity or soft palate	Onto tongue	Moves the tongue
Hyoid muscles			
Suprahyoid	Base of skull, mandible	Hyoid bone	Elevates or stabilizes hyoid
(e.g., geniohyoid, stylohyoid,			
and hyoglossus)			
Infrahyoid (e.g., thyrohyoid)	Sternum, larynx	Hyoid bone	Depresses or stabilizes hyoid
Soft palate (not shown)	Skull or soft palate	Palate, tongue, or pharynx	Moves soft palate, tongue, or pharynx
Pharyngeal muscles			
Elevators (not shown)	Soft palate and auditory tube	Pharynx	Elevate pharynx
Constrictors	Larynx and hyoid	Pharynx	Constrict pharynx
Superior (not shown)			
Middle			
Inferior			

P R E D I C T
Shortening of the right sternocleidomastoid muscle rotates the head in which direction?

Trunk Muscles

Trunk muscles include those that move the vertebral column, those of the thorax and abdominal wall, and those of the pelvic floor.

Muscles Moving the Vertebral Column

In humans, the back muscles are very strong to maintain erect posture. The **erector spinae** (ē-rek′tŏr spī′nē) group of muscles on each side of the back are the muscles primarily responsible

for keeping the back straight and the body erect (table 7.7 and see figure 7.17). **Deep back muscles,** located between the spinous and transverse processes of adjacent vertebrae, are responsible for several movements of the vertebral column such as extension, lateral flexion, and rotation.

Low Back Pain

Muscle strains and sprains of lumbar vertebral ligaments are the most common causes of low back pain. Low back muscle strain occurs when the deep back muscles are stretched abnormally or torn. Treatments include anti-inflammatory medication and RICE: *rest, ice, compression,* and *elevation.* Low back exercises can also help the problem.

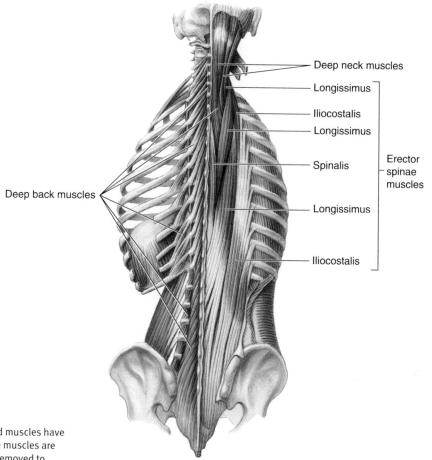

Figure 7.17 Deep Back Muscles

The upper limb, pectoral girdle, and associated muscles have been removed. On the right, the erector spinae muscles are demonstrated. On the left, these muscles are removed to reveal the deeper back muscles.

Posterior view

Table 7.6	Neck Muscles (see figures 7.13, 7.15, 7.17, and 7.21)		
Muscle	**Origin**	**Insertion**	**Action**
Deep neck muscles			
Flexors (not shown)	Anterior side of vertebrae	Base of skull	Flex head and neck
Extensors	Posterior side of vertebrae	Base of skull	Extend head and neck
Sternocleidomastoid (ster′nō-klī′dō-mas′toyd)	Manubrium of sternum and medial part of clavicle	Mastoid process and nuchal line of skull	Individually rotate head, together flex neck or extend head
Trapezius (tra-pē′zē-ŭs)	Posterior surface of skull and upper vertebral column (C7–T12)	Clavicle, acromion process and scapular spine	Extends head and neck

Table 7.7	Muscles Acting on the Vertebral Column (see figure 7.17)		
Muscle	**Origin**	**Insertion**	**Action**
Superficial			
Erector spinae (ĕ-rek′tōr spī′nē) divides into three columns:	Sacrum, ilium, vertebrae, and ribs	Ribs, vertebrae, and skull	Extends vertebral column
Ilicostalis (il′ē-ō-kos-tā′lis)			
Longissimus (lon-gis′i-mŭs)			
Spinalis (spī-nā′lis)			
Deep back muscles	Vertebrae	Vertebrae	Extend vertebral column and help bend vertebral column laterally

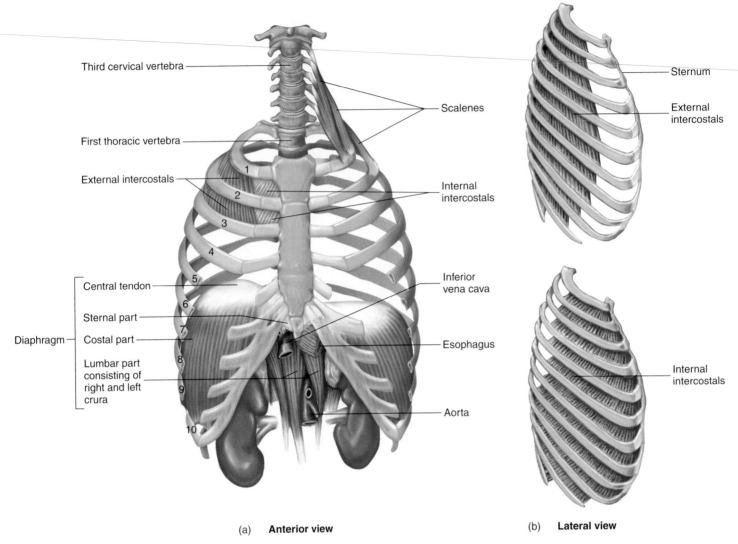

(a)   **Anterior view**   (b)   **Lateral view**

Figure 7.18   Muscles of the Thorax

(*a*) Anterior view. A few selected intercostal muscles and the diaphragm are demonstrated. (*b*) Lateral view.

Table 7.8	**Muscles of the Thorax (see figures 7.18 and 7.21)**		
Muscle	**Origin**	**Insertion**	**Action**
Scalenes (skā′lēnz)	Cervical vertebrae	First and second ribs	Inspiration; elevate ribs
External intercostals (in′ter-kos′tŭlz)	Inferior edge of each rib	Superior edge of next rib below origin	Inspiration; elevates ribs
Internal intercostals (in′ter-kos′tŭlz)	Superior edge of each rib	Inferior edge of next rib above origin	Forced expiration; depresses ribs
Diaphragm (dī′ă-fram)	Interior ribs, sternum, and lumbar vertebrae	Central tendon of diaphragm	Inspiration; depresses floor of thorax

Thoracic Muscles

The muscles of the thorax (figure 7.18 and table 7.8) are involved almost entirely in the process of breathing. The **external intercostals** (in′ter-kos′tŭlz, between ribs) elevate the ribs during inspiration. The **internal intercostals** contract during forced expiration, depressing the ribs.

The major movement produced in the thorax during quiet breathing, however, is accomplished by the dome-shaped **diaphragm** (dī′ă-fram). When it contracts, the dome is flattened, causing the volume of the thoracic cavity to increase and resulting in inspiration.

Abdominal Wall Muscles

The muscles of the anterior abdominal wall (figure 7.19 and table 7.9) flex and rotate the vertebral column, compress the abdominal cavity, and hold in and protect the abdominal

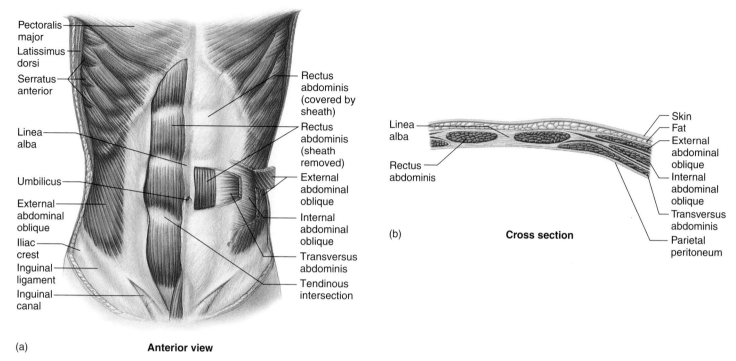

Figure 7.19 Muscles of the Anterior Abdominal Wall

(*a*) Anterior view. (Windows in the side reveal the various muscle layers.) (*b*) Cross section.

Table 7.9	Muscles of the Abdominal Wall (see figures 7.13, 7.19, 7.21)		
Muscle	**Origin**	**Insertion**	**Action**
Rectus abdominis (rek'tŭs ab-dom'i-nis)	Pubic crest and symphysis pubis	Xiphoid process and inferior ribs	Flexes vertebral column; compresses abdomen
External abdominal oblique	Ribs 5 to 12	Iliac crest, inguinal ligament, and fascia of rectus abdominis	Compresses abdomen; flexes and rotates vertebral column
Internal abdominal oblique	Iliac crest, inguinal ligament, and lumbar fascia	Ribs 10–12 and fascia of rectus abdominis	Compresses abdomen; flexes and rotates vertebral column
Transversus abdominis (trans-ver'sŭs ab-dom'in-is)	Seventh to twelfth costal cartilages, lumbar fascia, iliac crest, and inguinal ligament	Xiphoid process, fascia of rectus abdominis, and pubic tubercle	Compresses abdomen

organs. In a relatively muscular person with little fat, a vertical indentation, extending from the sternum through the navel to the pubis, is visible. This tendinous area of the abdominal wall, called the **linea alba** (lin'ē-ă al'bă, white line), consists of white connective tissue rather than muscle. On each side of the linea alba is the **rectus abdominis** (rek'tŭs ab-dom'ĭ-nis, *rectus,* straight) muscle. **Tendinous intersections** cross the rectus abdominis at three or more locations, causing the abdominal wall of a well-muscled lean person to appear segmented. Lateral to the rectus abdominis are three layers of muscle. From superficial to deep, these muscles are the **external abdominal oblique, internal abdominal oblique,** and **transversus abdominis** (trans-ver'sŭs ab-dom'in-is)

muscles. The fasciculi of these three muscle layers are oriented in different directions to one another. When these muscles contract, they flex and rotate the vertebral column or compress the abdominal contents.

Pelvic Floor and Perineal Muscles

The pelvis is a ring of bone with an inferior opening that is closed by a muscular floor through which the anus and the openings of the urinary tract and reproductive tract penetrate. Most of the **pelvic floor,** also referred to as the **pelvic diaphragm,** is formed by the **levator ani** (le-vā'ter ā'nī) muscle. The area inferior to the pelvic floor is the **perineum** (per'i-nē'ŭm), which contains a number of muscles associated with

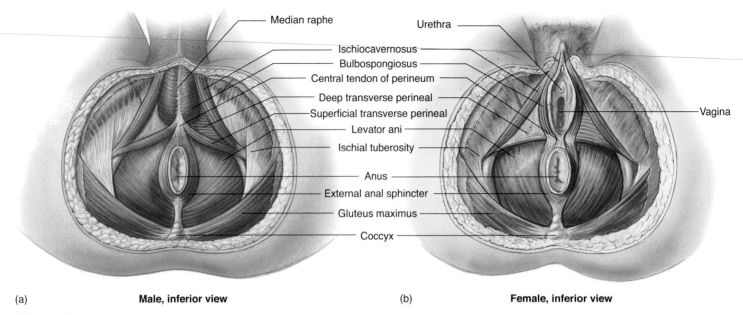

Figure 7.20 Muscles of the Pelvic Floor and Perineum

Table 7.10	Muscles of the Pelvic Floor and Perineum (see figure 7.20)		
Muscle	**Origin**	**Insertion**	**Action**
Pelvic Floor			
Levator ani (le′vā-ter ā′nī)	Posterior pubis and ischial spine	Sacrum and coccyx	Elevates anus; supports pelvic viscera
Perineum			
Bulbospongiosus (bul′bō-spŭn′jē-ō′sŭs)	Male—central tendon of perineum and median raphe of penis Female—central tendon of perineum	Dorsal surface of penis and bulb of penis Base of clitoris	Constricts urethra; erects penis Erects clitoris
Ischiocavernosus (ish′ē-ō-kav′er-nō′sŭs)	Ischial ramus	Corpus cavernosum	Compresses base of penis or clitoris
External anal sphincter	Coccyx	Central tendon of perineum	Keeps orifice of anal canal closed
Transverse perinei (pĕr′i-nē′ī) Deep Superficial	Ischial ramus Ischial ramus	Midline connective tissue Central tendon of perineum	Supports pelvic floor Fixes central tendon

the male or female reproductive structures (figure 7.20 and table 7.10). Several of these muscles help regulate urination and defecation.

Upper Limb Muscles

The muscles of the upper limb include those that attach the limb and pectoral girdle to the body and those that are in the arm, forearm, and hand.

Scapular Movements

The connection of the upper limb to the body is accomplished primarily by muscles. The muscles that attach the scapula to the thorax and move the scapula include the **trapezius** (tra-pē′zē-ŭs, table), **levator scapulae** (le-vā′ter skap′ū-lē), **rhomboids** (rom′boydz, rhombohedrin-shaped), **serratus** (ser-ā′tŭs, serrated) **anterior,** and **pectoralis** (pek′tō-ra′lis) **minor** (figure 7.21 and table 7.11). These muscles act as fixators to hold the scapula firmly in position when the muscles of the arm contract. The scapular muscles also move the scapula into different positions, thereby increasing the range of movement of the upper limb. The trapezius forms the upper line from each shoulder to the neck, and the origin of the serratus anterior from the first eight or nine ribs can be seen along the lateral thorax.

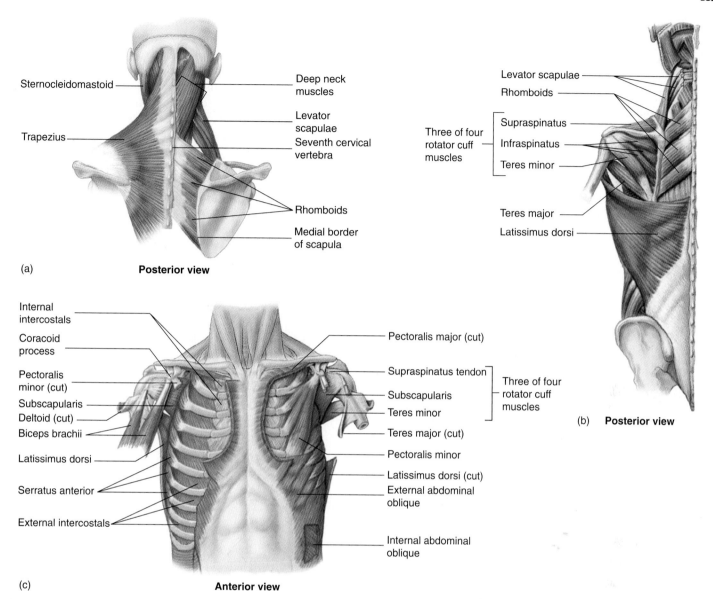

Figure 7.21 Muscles of the Shoulder

(a) Posterior view of the neck and upper shoulder. The left side shows the superficial muscles. On the right, the superficial muscles are removed to show the deep muscles. (b) Posterior view of the thoracic region, with the trapezius and deltoid muscles removed. (c) Anterior view of the thoracic region.

Table 7.11   Muscles Acting on the Scapula (see figures 7.13, 7.21, and 7.22)

Muscle	Origin	Insertion	Action
Levator scapulae (le-vā′ter skap′ū-lē)	C1–C4	Superior angle of scapula	Elevates, retracts, and rotates scapula; laterally flexes neck
Pectoralis minor (pek′tō-ra′lis)	Third to fifth ribs	Coracoid process of scapula	Depresses scapula or elevates ribs
Rhomboids (rom′boydz)			
Major	T1–T4	Medial border of scapula	Retracts, rotates, and fixes scapula
Minor	T1–T4	Medial border of scapula	Retracts, slightly elevates, rotates, and fixes scapula
Serratus anterior (ser-ā′tŭs)	First to ninth ribs	Medial border of scapula	Rotates and protracts scapula; elevates ribs
Trapezius (tra-pē′zē-ŭs)	Posterior surface of skull and C7–T12	Clavicle, acromion process, and scapular spine	Elevates, depresses, retracts, rotates, and fixes scapula; extends head and neck

Arm Movements

The arm is attached to the thorax by the **pectoralis major** and **latissimus dorsi** (lă-tis′i-mŭs dōr′sī, wide back) muscles (figure 7.22a and table 7.12; see figure 7.21c). The pectoralis major adducts the arm and flexes the shoulder. It can also extend the shoulder from a flexed position. The latissimus dorsi medially rotates and adducts the arm, and powerfully extends the shoulder. Because a swimmer uses these three motions during the power stroke of the crawl, the latissimus dorsi is often called the swimmer's muscle.

Another group of four muscles, called the **rotator cuff muscles,** attaches the humerus to the scapula and forms a cuff or cap over the proximal humerus (see table 7.12 and figure 7.21b and c). These muscles function to stabilize the joint by holding the head of the humerus in the glenoid cavity during arm movements, especially abduction. A rotator cuff injury involves damage to one or more of these muscles or their tendons. The **deltoid** (del′toyd, triangular) muscle attaches the humerus to the scapula and clavicle and is the major abductor of the upper limb. The pectoralis major forms the upper chest,

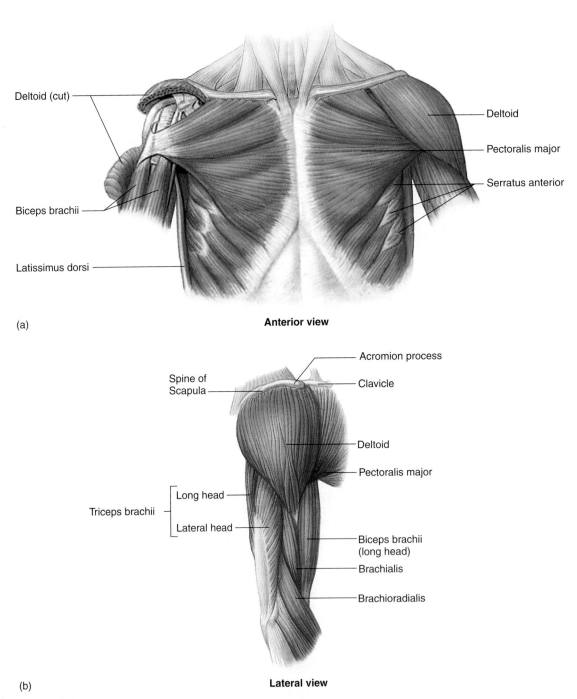

Anterior view

(a)

Lateral view

(b)

Figure 7.22 Arm Muscles

Table 7.12 Arm Movements (see figures 7.13, 7.21, 7.22, and 7.24)

Muscle	Origin	Insertion	Action
Deltoid (del′toyd)	Clavicle, acromion process, and scapular spine	Deltoid tuberosity	Flexes and extends shoulder; abducts and medially and laterally rotates arm
Latissimus dorsi (lă-tis′i-mŭs dŏr′sī)	T7–L5, sacrum and iliac crest	Medial crest of intertubercular groove	Extends shoulder; adducts and medially rotates arm
Pectoralis major (pek′tō-ră′lis)	Clavicle, sternum, and abdominal muscles	Lateral crest of intertubercular groove	Flexes shoulder; extends shoulder from flexed position; adducts and medially rotates arm
Teres major (ter′ēz)	Lateral border of scapula	Medial crest of intertubercular groove	Extends shoulder; adducts and medially rotates arm
Rotator cuff			
Infraspinatus (in′fră-spī-nā′tŭs)	Infraspinous fossa of scapula	Greater tubercle of humerus	Stabilizes and extends shoulder and laterally rotates arm
Subscapularis (sŭb′skap-ū-lăr′is)	Subscapular fossa of scapula	Lesser tubercle of humerus	Stabilizes and extends shoulder and medially rotates arm
Supraspinatus (sŭ′pră-spī-nā′tŭs)	Supraspinous fossa of scapula	Greater tubercle of humerus	Stabilizes shoulder and abducts arm
Teres minor (te′rēz)	Lateral border of scapula	Greater tubercle of humerus	Stabilizes and extends shoulder; adducts and laterally rotates arm

Table 7.13 Arm Muscles (see figures 7.13, 7.21, 7.22, and 7.24)

Muscle	Origin	Insertion	Action
Arm			
Biceps brachii (bī′seps brā′kē-ī)	Long head—supraglenoid tubercle; Short head—coracoid process	Radial tuberosity	Flexes elbow; supinates forearm; flexes shoulder
Brachialis (brā′kē-al′is)	Humerus	Coronoid process of ulna	Flexes elbow
Triceps brachii (trī′seps brā′kē-ī)	Long head—lateral border of scapula; Lateral head—lateral and posterior surface of humerus; Medial head—posterior humerus	Olecranon process of ulna	Extends elbow; extends shoulder; adducts arm

and the deltoid forms the rounded mass of the shoulder (see figure 7.24). The deltoid is a common site for administering injections.

Forearm Movements

The arm can be divided into anterior and posterior compartments. The **triceps brachii** (trī′seps brā′kē-ī, three heads, arm), the primary extensor of the elbow, occupies the posterior compartment (figure 7.22b and table 7.13). The anterior compartment is occupied mostly by the **biceps** (bī′seps) **brachii** and the **brachialis** (brā′kē-ăl-is), the primary flexors of the elbow. The **brachioradialis** (brā′kē-ō-rā′dē-al′is), which is actually a posterior forearm muscle, helps flex the elbow.

Supination and Pronation

Supination of the forearm, or turning the flexed forearm so that the palm is up, is accomplished by the **supinator** (soo′pi-nā-ter) (see table 7.13; figure 7.23 and table 7.14) and the biceps brachii, which tends to supinate the forearm while flexing the elbow. Pronation, turning the forearm so that the palm is down, is a function of two **pronator** (prō-nā′ter) muscles.

Wrist and Finger Movements

The 20 muscles of the forearm can also be divided into anterior and posterior groups. Only a few of these muscles, the most superficial, are listed in table 7.14 and are illustrated in figure 7.23. Most of the anterior forearm muscles are responsible for flexion of the wrist and fingers, whereas most of the posterior forearm muscles cause extension. A strong band of fibrous connective tissue, the **retinaculum** (ret-i-nak′ū-lŭm, bracelet) (see figure 7.23b), covers the flexor and extensor tendons and holds them in place around the wrist so that they do not "bowstring" during muscle contraction. Because the retinaculum does not stretch as a result of pressure, this characteristic is a contributing factor in carpal tunnel syndrome (see discussion in chapter 6).

The **flexor carpi** (kar′pī, wrist) muscles flex the wrist, and the **extensor carpi** muscles extend the wrist. The tendons of the wrist extensors are visible on the posterior surface of the forearm (figure 7.24). The tendon of the flexor carpi radialis is used as a landmark for locating the radial pulse. Flexion of the fingers is the function of the **flexor digitorum** (dij′i-tōr′ŭm, flexor of the digits, or fingers). Extension of the fingers is accomplished

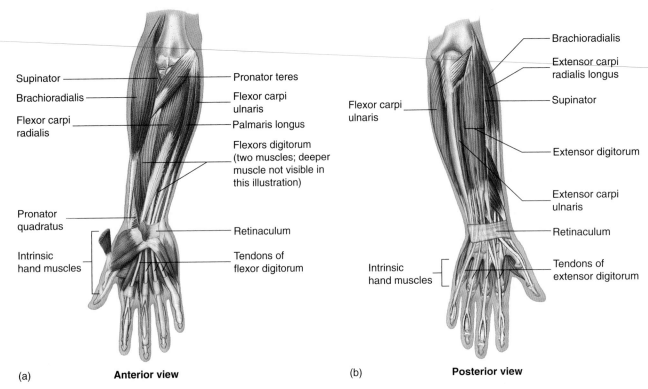

Figure 7.23   Muscles of the Forearm

(*a*) Anterior view. The flexor retinaculum has been removed. (*b*) Posterior view.

Table 7.14	Forearm Muscles (see figure 7.23)		
Muscle	**Origin**	**Insertion**	**Action**
Anterior Forearm			
Palmaris longus (pawl-mār′is lon′gus)	Medial epicondyle of humerus	Aponeurosis over palm	Tighten skin of palm
Flexor carpi radialis (kar′pī rā′dē-a-lĭs)	Medial epicondyle of humerus	Second and third metacarpals	Flexes and abducts wrist
Flexor carpi ulnaris (kar′pī ŭl-nār′is)	Medial epicondyle of humerus and ulna	Pisiform	Flexes and abducts wrist
Flexor digitorum profundus (dij′i-tōr′ŭm prō-fŭn′dŭs) (not shown)	Ulna	Distal phalanges of digits 2 through 5	Flexes fingers and wrist
Flexor digitorum superficialis (sū′per-fish′ē-a′lis)	Medial epicondyle of humerus, coronoid process, and radius	Middle phalanges of digits 2 through 5	Flexes fingers and wrist
Pronator Quadratus (prō′nā-tōr kwah-drā′tŭs)	Distal ulna	Distal radius	Pronates forearm
Teres (prō′nā-tōr te′rēz)	Medial epicondyle of humerus and coronoid process of ulna	Radius	Pronates forearm
Posterior Forearm			
Brachioradialis (brā′kē-ō-rā′dē-a′lis)	Lateral supracondylar ridge of humerus	Styloid process of radius	Flexes forearm
Extensor carpi radialis brevis (kar′pī rā′dē-a′lis brev′is) (not shown)	Lateral epicondyle of humerus	Base of third metacarpal	Extends and abducts wrist
Extensor carpi radialis longus (lon′gus)	Lateral supracondylar ridge of humerus	Base of second metacarpal	Extends and abducts wrist
Extensor carpi ulnaris (kar′pī ŭl-nār′is)	Lateral epicondyle of humerus and ulna	Base of fith metacarpal	Extends and adducts wrist
Extensor digitorum (dij′i-tōr′ŭm)	Lateral epicondyle of humerus	Bases of phalanges of digits 2 through 5	Extends fingers and wrist
Supinator (sū′pi-nā′tōr)	Lateral epicondyle of humerus and ulna	Radius	Supinates forearm

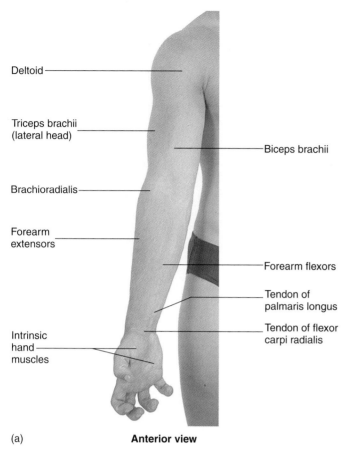

(a) **Anterior view**

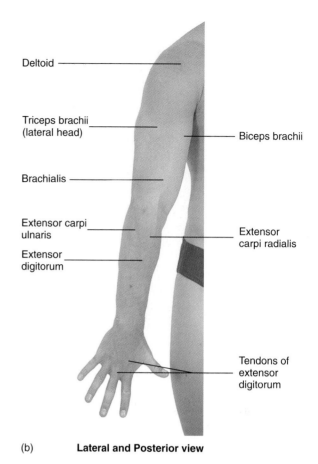

(b) **Lateral and Posterior view**

Figure 7.24 Surface Anatomy, Muscles of the Upper Limb

(*a*) Anterior view. (*b*) Lateral and posterior view.

by the **extensor digitorum.** The tendons of this muscle are very visible on the dorsal surface of the hand (see figure 7.24). The thumb has its own set of flexors, extensors, adductors, and abductors. The little finger also has some similar muscles.

Tennis Elbow

Forceful, repeated contraction of the wrist extensor muscles, such as occurs in a tennis backhand, may result in inflammation and pain where the extensor muscles attach to the lateral humeral epicondyle. This condition is sometimes referred to as "tennis elbow."

Nineteen muscles, called **intrinsic hand muscles,** are located within the hand. **Interossei** (in′ter-os′ē-ī, between bones) muscles, located between the metacarpals, are responsible for abduction and adduction of the fingers. Other intrinsic hand muscles are responsible for many other movements of the thumb and fingers. These muscles account for the fleshy masses at the base of the thumb and little finger and the fleshy region between the metacarpals of the thumb and index finger.

Lower Limb Muscles

The muscles of the lower limb include those located in the hip, thigh, leg, and foot.

Thigh Movements

Several hip muscles originate on the coxa and insert onto the femur (figure 7.25 and table 7.15). The anterior muscle, the **iliopsoas** (il′ē-ō-sō′ŭs, *psoa,* muscle of the loin), flexes the hip (figure 7.25*a*). The posterior and lateral hip muscles consist of the **gluteal muscles** and the tensor fascia latae (figure 7.25*a* and *b*). The **tensor fascia latae** (ten′sōr fa′shē-ă la′tē) is so named because it tenses a thick band of fascia on the lateral side of the thigh called the iliotibial tract. By so doing, it helps steady the femur on the tibia when a person is standing. The **gluteus** (glū′tē-us, buttock) **maximus,** which extends the hip and abducts and laterally rotates the thigh, contributes most of the mass that can be seen as the buttocks. The **gluteus medius,** which abducts and medially rotates the thigh, creates a smaller mass just superior and lateral to the maximus (figure 7.26*b*). The gluteus maximus functions optimally to extend the hip when the thigh is flexed at a 45-degree angle. The gluteus medius is a common site for injections in the buttocks because the sciatic nerve lies deep to the gluteus maximus and could be damaged during an injection.

In addition to the hip muscles, some of the muscles located in the thigh also attach to the coxa and can cause movement of the thigh. There are three groups of thigh muscles: the anterior thigh muscles, which flex the hip; the posterior thigh

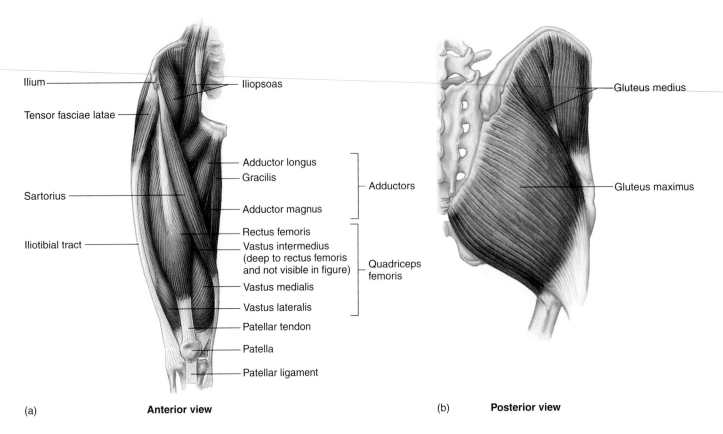

Ilium

Tensor fasciae latae

Sartorius

Iliotibial tract

Iliopsoas

Adductor longus

Gracilis

Adductor magnus

Rectus femoris

Vastus intermedius
(deep to rectus femoris
and not visible in figure)

Vastus medialis

Vastus lateralis

Patellar tendon

Patella

Patellar ligament

Adductors

Quadriceps
femoris

(a) **Anterior view**

Gluteus medius

Gluteus maximus

(b) **Posterior view**

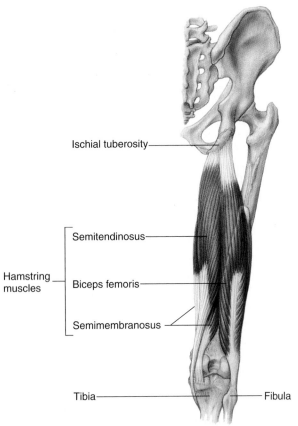

Ischial tuberosity

Hamstring
muscles

Semitendinosus

Biceps femoris

Semimembranosus

Tibia

Fibula

(c) **Posterior view**

Figure 7.25 Muscles of the Hip and Thigh

(*a*) Anterior view. The vastus intermedius is labeled on the figure to allow for a complete listing of the quadriceps femoris muscles, but the muscle lies deep to the rectus femoris and cannot be seen in the figure. (*b*) Posterior view, hip muscles. (*c*) Posterior view, thigh muscles.

Table 7.15 Muscles Moving the Thigh (see figure 7.25)

Muscle	Origin	Insertion	Action
Iliopsoas (il′ē-ō-sō′ŭs)	Iliac fossa and vertebrae T12–L5	Lesser trochanter of femur and hip capsule	Flexes hip
Gluteus maximus (glū′tē-ŭs mak′si-mŭs)	Ilium, sacrum, and coccyx	Gluteal tuberosity of femur and lateral fascia of thigh	Extends, hip; abducts and laterally rotates thigh
Gluteus medius (glū′tē-ŭs mē′dē-ŭs)	Ilium	Greater trochanter of femur	Abducts and medially rotates thigh
Gluteus minimus (glū′tē-ŭs min′i-mŭs) (not shown)	Ilium	Greater trochanter of femur	Abducts and medially rotates thigh
Tensor fasciae latae (ten′sōr fa′shē-ē lā′tē)	Anterior superior iliac spine	Through lateral fascia of thigh to lateral condyle of tibia	Steadies femur on tibia through the iliotibial tract when standing; flexes hip; medially rotates and abducts thigh

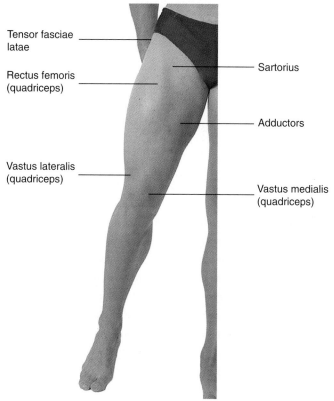

Tensor fasciae latae

Rectus femoris (quadriceps)

Vastus lateralis (quadriceps)

Sartorius

Adductors

Vastus medialis (quadriceps)

(a)   **Anterior view**

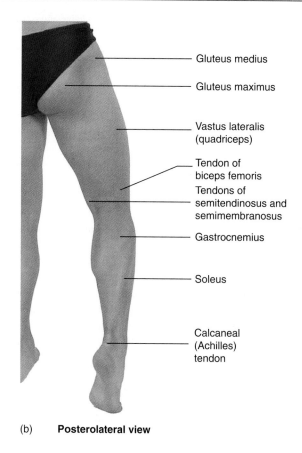

Gluteus medius

Gluteus maximus

Vastus lateralis (quadriceps)

Tendon of biceps femoris

Tendons of semitendinosus and semimembranosus

Gastrocnemius

Soleus

Calcaneal (Achilles) tendon

(b)   **Posterolateral view**

Figure 7.26 Surface Anatomy of the Lower Limb

(*a*) Anterior view. (*b*) Posterolateral view.

muscles, which extend the hip; and the medial thigh muscles, which adduct the thigh.

PREDICT 5

Consider the sprinter's stance and the bicyclist's racing posture, and explain why these postures are used by these athletes.

Leg Movements

The anterior thigh muscles are the **quadriceps femoris** (kwah′dri-seps fe-mōr′is, four muscles) and the **sartorius** (sar-tōr′ē-ŭs, tailor, refers to the way a tailor sits with legs crossed)

(table 7.16; see figures 7.25*a* and 7.26*a*). The quadriceps femoris muscles are the primary extensors of the knee. They have a common insertion, the patellar tendon, on and around the patella. The patellar ligament is an extension of the patellar tendon onto the tibial tuberosity. The patellar ligament is tapped with a rubber hammer when testing the knee-jerk reflex in a physical examination (see figure 8.19). One of the quadriceps muscles, the vastus lateralis, is often used as an intermuscular injection site. The sartorius, the longest muscle in the body, is called the "tailor's muscle" because it flexes the hip and knee and rotates the thigh laterally for sitting cross-legged, as tailors used to sit while sewing.

Table 7.16 Leg Movements (see figures 7.13, 7.25, and 7.27)

Muscle	Origin	Insertion	Action
Anterior Compartment			
Quadriceps femoris (kwah′dri-seps fem′ō-ris)			
Rectus femoris (rek′tŭs fem′ō-ris)	Ilium	Tibial tuberosity via patellar ligament	Extends knee; flexes hip
Vastus lateralis (vas′tus lat-er-ā′lis)	Femur	Tibial tuberosity via patellar ligament	Extends knee
Vastus medialis (vas′tus mē′dē-ā′lis)	Femur	Tibial tuberosity via patellar ligament	Extends knee
Vastus intermedius (vas′tŭs in′ter-mē′dē-ŭs)	Femur	Tibial tuberosity via patellar ligament	Extends knee
Sartorius (sar-tōr′ē-ŭs)	Anterior superior iliac spine	Medial side of tibial tuberosity	Flexes hip and knee; laterally rotates thigh
Medial Compartment			
Adductor longus (a′dŭk-ter lon′gŭs)	Pubis	Femur	Adducts and laterally rotates thigh; flexes hip
Adductor magnus (a′dŭk-ter mag′nŭs)	Pubis and ischium	Femur	Adducts and laterally rotates thigh; extends knee
Gracilis (gras′i-lis)	Pubis near symphysis	Tibia	Adducts thigh; flexes knee
Posterior Compartment (Hamstring Muscles)			
Biceps femoris (bi′seps fem′ō-ris)	Long head—ischial tuberosity; Short head—femur	Head of fibula	Flexes knee; laterally rotates leg; extends hip
Semimembranosus (se′mē-mem′bră-nō′sŭs)	Ischial tuberosity	Medial condyle of tibia and collateral ligament	Flexes knee; medially rotates leg; extends hip
Semitendinosus (se′mē-ten′di-nō′sŭs)	Ischial tuberosity	Tibia	Flexes knee; medially rotates leg; extends hip

The posterior thigh muscles are called **hamstring muscles,** and they are responsible for flexing the knee (see table 7.16 and figures 7.13*b* and 7.25*c*). Their tendons are easily felt and seen on the medial and lateral posterior aspect of a slightly bent knee (see figure 7.26*b*). The hamstrings were so named because these tendons in hogs or pigs could be used to suspend hams during curing. Animals such as wolves often bring down their prey by biting through the hamstrings, thus preventing the prey animal from running. "To hamstring" someone is therefore to render him or her helpless. A "pulled hamstring" consists of tearing one or more of these muscles or their tendons, usually where the tendons attach to the coxa.

The medial thigh muscles, the **adductor** (a′dŭk-ter) **muscles,** are involved, as the name implies, primarily in adduction of the thigh (see table 7.16).

A CASE IN POINT | Groin Pull

Lowe Kikker was playing soccer on a very cold October day. He raced up to kick the ball, just as a player for the other team kicked the ball from the opposite direction. The ball did not move initially. Lowe felt a sudden, sharp pain in the medial side of his right thigh and fell to the ground, holding the injured area. His coach and team trainer ran out onto the field. As the trainer palpated the injured area, Lowe felt severe pain as the trainer pressed on the center part of his medial thigh. He was able to stand and was helped off the field. On the sideline, ice was applied to the adductor region of Lowe's right thigh. He was placed on rest for the next few days, along with icing, compression of the muscles, and elevation of the thigh (RICE). He was also given anti-inflammitory drugs. After a couple days of rest, Lowe was referred for massage therapy. He missed the next game, but was able to play in the finals. Lowe's injury was considered a Grade 1 injury as there was pain and tenderness in the medial thigh muscles but no swelling. Grade 2 injury, which involves partial muscle tear, is more painful and involves swelling of the area. Grade 3 injury, which involves complete muscle tear, involves a great amount of pain and swelling, and the injured person cannot even walk. A groin pull involves one or more of the adductor muscles, the adductor longus being the most commonly damaged. The damage usually occurs at the musculotendon junction, near its insertion.

Ankle and Toe Movements

The 13 muscles in the leg, with tendons extending into the foot, can be divided into three groups: anterior, posterior, and lateral. As with the forearm, only the most superficial muscles are illustrated in figure 7.27 and are listed in table 7.17. The anterior

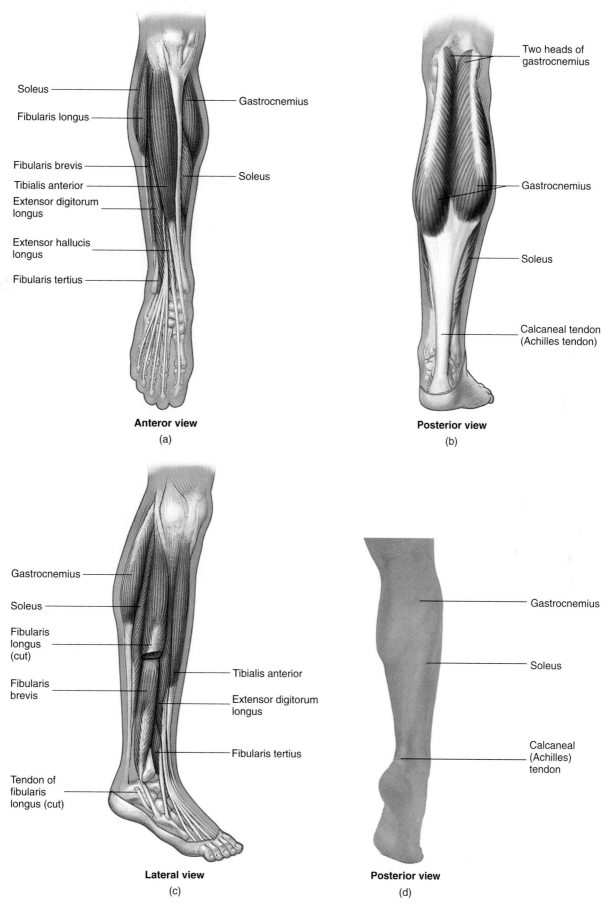

Anteror view
(a)

Soleus
Fibularis longus
Fibularis brevis
Tibialis anterior
Extensor digitorum longus
Extensor hallucis longus
Fibularis tertius
Gastrocnemius
Soleus

Posterior view
(b)

Two heads of gastrocnemius
Gastrocnemius
Soleus
Calcaneal tendon (Achilles tendon)

Lateral view
(c)

Gastrocnemius
Soleus
Fibularis longus (cut)
Fibularis brevis
Tibialis anterior
Extensor digitorum longus
Fibularis tertius
Tendon of fibularis longus (cut)

Posterior view
(d)

Gastrocnemius
Soleus
Calcaneal (Achilles) tendon

Figure 7.27 Superficial Muscles of the Leg

(*a*) Anterior view. (*b*) Posterior view. (*c*) Lateral view. (*d*) Surface anatomy.

Systems Pathology
Duchenne's Muscular Dystrophy

A couple became concerned about their 3-year-old boy when they noticed that he was much weaker than other boys his age and the differences appeared to become more obvious as time passed. He had difficulty sitting, standing, and walking. He seemed clumsy and he fell often. He had difficulty climbing stairs, and he often got from a sitting position on the floor to a standing position by using his hands and arms to climb up his legs. His muscles appeared to be poorly developed. The couple took their son to a physician to have him examined. After several tests, they were informed that their son had Duchenne muscular dystrophy.

Background Information

Duchenne's muscular dystrophy (DMD) is usually identified in children at around 3 years of age when the parents notice slow motor development with progressive weakness and muscle wasting (figure A). Typically, muscular weakness begins in the pelvic girdle, causing a waddling gait. Rising from the floor by "climbing up the legs" is characteristic and is caused by weakness of the lumbar and gluteal muscles. Within 3 to 5 years, muscles of the shoulder girdle become involved. Wasting of the muscles and their replacement with connective tissue contribute to muscular atrophy and deformity of the skeleton.

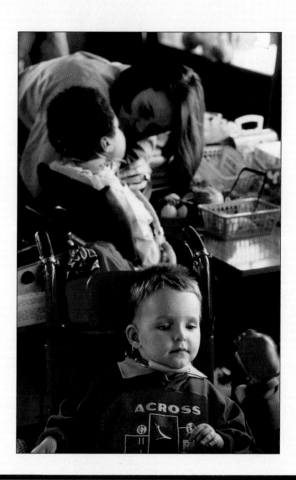

Figure A Child with Duchenne's Muscular Dystrophy

Table 7.17	Muscles of the Leg Acting on the Leg, Ankle, and Foot (see figures 7.13, 7.26, and 7.27)		
Muscle	**Origin**	**Insertion**	**Action**
Anterior Compartment			
Extensor digitorum longus (dij′i-tōr′ŭm lon′gŭs)	Lateral condyle of tibia and fibula	Four tendons to phalanges of four lateral toes	Extends four lateral toes; dorsiflexes and everts foot
Extensor hallicus longus (hal′i-sis lon′gŭs)	Middle fibula and interosseous membrane	Distal phalanx of great toe	Extends great toe; dorsiflexes and inverts foot
Tibialis anterior (tib′ē-a′lis)	Tibia and interosseous membrane	Medical cuneiform and first metatarsal	Dorsiflexes and inverts foot
Fibularis tertius (peroneus tertius) (per′ō-nē′ŭs ter′shē-ŭs)	Fibula and interosseous membrane	Fifth metatarsal	Dorsiflexes and everts foot
Posterior Compartment			
Superficial			
Gastrocnemius (gas′trok-nē′mē-ŭs)	Medial and lateral condyles of femur	Through calcaneal (Achilles) tendon to calcaneus	Plantar flexes foot; flexes leg
Soleus (sō′lē-ŭs)	Fibula and tibia	Through calcaneal tendon to calcaneus	Plantar flexes foot

System Interactions — Effects of Duchenne's Muscular Dystrophy on Other Systems

System	Interactions with the Muscular System
Skeletal	Replacement of muscles by connective tissue results in shortened, inflexible muscles that result in severe deformities of the skeletal system. Curvature of the spinal column can be so severe that normal respiratory movements cannot be carried out. Deformities of the limbs also occur as a result of shortened muscles.
Nervous	Some degree of mental retardation occurs in a large percentage of cases.
Cardiovascular	Cardiac muscle is affected by DMD. Consequently, heart failure occurs in a large number of people with this condition in its advanced stages. Death caused by cardiac failure usually occurs before age 20. Cardiac involvement becomes serious in as high as 95% of cases.
Lymphatic	There are no obvious direct effects on the lymphatic system, but phagocytosis of muscle fibers is accomplished mainly by macrophages.
Respiratory	Deformity of the thorax and increasing weakness of the respiratory muscles results in inadequate respiratory movements and an increase in respiratory infections such as pneumonia. Inadequate respiratory movements because of weak respiratory muscles is a major factor in many deaths.
Digestive	Smooth muscle tissue is also influenced by muscular dystrophy. The reduced ability of smooth muscle to contract can result in abnormalities of the digestive system such as an enlarged colon diameter, a twisting of the intestine resulting in increased intestinal obstruction, cramping, and reduced absorption of nutrients.
Urinary	Reduced smooth muscle function and being confined to a wheelchair increase the frequency of urinary tract infections.

People with DMD are usually unable to walk by 10 to 12 years of age, and few live beyond 20 years of age. There is no effective treatment to prevent the progressive deterioration of muscles in DMD.

Duchenne's muscular dystrophy results from an abnormal gene located on the X chromosome and is therefore a sex-linked (X-linked) condition. The disease affects 1 in 3000 boys. Although the gene is carried by females, DMD affects males almost exclusively. The DMD gene is responsible for producing a protein called **dystrophin,** which plays a role in attaching myofibrils to other proteins in the cell membrane and regulating their activity. Dystrophin is thought to protect muscle cells against mechanical stress in the normal individual. In DMD, part of the gene is missing, and the protein it produces malfunctions, resulting in progressive muscular weakness.

P R E D I C T 6

A 15-year-old boy with DMD developed pulmonary edema and then pneumonia. His physician diagnosed the condition in the following way: the pulmonary edema was the result of heart failure and the increased fluid in the lungs acted as a site where bacteria invaded and grew. The fact that the boy could not breathe deeply or cough effectively made the condition worse. Explain why a 15-year-old with this condition might develop heart failure and ineffective respiratory movements.

Table 7.17 *continued*

Posterior Compartment (Cont.)

Deep

Flexor digitorum longus (dij'i-tōr'ŭm lon'gŭs) (not shown)	Tibia	Four tendons to distal phalanges of four lateral toes	Flexes four lateral toes; plantar flexes and inverts foot
Flexor hallucis longus (hal'i-sis lon'gŭs) (not shown)	Fibula	Distal phalanx of great toe	Flexes great toe; plantar flexes and inverts foot
Tibialis posterior (tib'ē-a'lis) (not shown)	Tibia, interosseous membrane, and fibula	Navicular, cuneiforms, cuboid, and second through fourth metatarsals	Plantar flexes and inverts foot

Lateral Compartment

Fibularis brevis (peroneus brevis) (fib-ū-lā'ris brev'is)	Fibula	Fifth metatarsal	Everts and plantar flexes foot
Fibularis longus (peroneus longus) (fib-ū-lā'ris lon'gŭs)	Fibula	Medial cuneiform and first metatarsal	Everts and plantar flexes foot

muscles (figure 7.27a) are extensor muscles involved in dorsi-flexion (extension) of the foot and extension of the toes.

The superficial muscles of the posterior compartment of the leg (figure 7.27b), the **gastrocnemius** (gas′trok-nē′mē-us, gastro, stomach + kneme, leg) and **soleus** (sō′lē-ŭs, solea, sandal or flatfish), form the bulge of the calf (posterior leg; figure 7.27b). They join to form the common **calcaneal** (kal-kā′nē-ăl, heel), or **Achilles, tendon.** These muscles are flexors and are involved in plantar flexion (flexion) of the foot. The deep muscles of the posterior compartment plantar flex and invert the foot and flex the toes.

Achilles Tendon

The Achilles tendon derives its name from a hero of Greek mythology. As a baby, Achilles was dipped into magic water that made him invulnerable to harm everywhere it touched his skin. His mother, however, holding him by the back of his heel, overlooked submerging his heel under the water. Consequently, his heel was vulnerable and proved to be his undoing; he was shot in the heel with an arrow at the battle of Troy and died. Thus, saying that someone has an "Achilles heel" means that the person has a weak spot that can be attacked.

The lateral muscles of the leg, called the **fibularis** (fib-ū-lā′ris, buckle) **muscles** (figure 7.27c), are primarily everters (turning the lateral side of the foot outward) of the foot, but they also aid in plantar flexion.

The 20 muscles located within the foot, called the **intrinsic foot muscles,** flex, extend, abduct, and adduct the toes. They are arranged in a manner similar to the intrinsic muscles of the hand.

Effects of Aging on Skeletal Muscle

Several changes occur in aging skeletal muscle that reduce muscle mass, increase the time that muscle takes to contract in response to nervous stimuli, reduce stamina, and increase recovery time. There is a loss of muscle fibers as aging occurs. The loss begins as early as 25 years of age and by 80 years of age, the muscle mass is reduced by approximately 50%. Weight lifting exercises help slow the loss of muscle mass, but it doesn't prevent the loss of muscle fibers. In addition, fast-twitch muscle fibers decrease in number more rapidly than slow-twitch fibers. Most of the loss of strength and speed is due to the loss of muscle fibers and the loss of fast-twitch muscle fibers. The surface area of the neuromuscular junction decreases and, as a result, action potentials in neurons stimulate action potential production in muscle cells more slowly and fewer action potentials are produced in muscle fibers. The number of motor neurons also decreases. The remaining neurons innervate more muscle fibers. This makes motor units in skeletal muscle fewer in number, with a greater number of muscle fibers for each neuron. This may result in less precise control of muscles. Aging is associated with a decrease in the density of capillaries in skeletal muscles, and after exercise a longer period of time is required to recover.

Many of the age-related changes in skeletal muscle can be dramatically slowed if people remain physically active. As people age, they often assume a sedentary life-style. It has been demonstrated that elderly people who are sedentary can become stronger and more mobile in response to exercise.

S U M M A R Y

Functions of the Muscular System   (p. 157)

The muscular system functions to produce body movement, maintain posture, cause respiration, produce body heat, produce movements involved in communication, constrict organs and vessels, and pump blood.

Characteristics of Skeletal Muscle   (p. 157)

Skeletal muscle has contractility, excitability, extensibility, and elasticity.

Structure

1. Muscle fibers are organized into fasciculi, and fasciculi are organized into muscles by associated connective tissue.
2. Each skeletal muscle fiber is a single cell containing numerous myofibrils.
3. Myofibrils are composed of actin and myosin myofilaments.
4. Sarcomeres are joined end to end to form myofibrils.

Membrane Potentials

1. Cell membranes have a positive charge on the outside relative to a negative charge inside. This is called the resting membrane potential.
2. Action potentials are a brief reversal of the membrane charge. They are carried rapidly along the cell membrane.
3. Sodium ions move into cells during depolarization and K^+ move out of cells during repolarization.

Nerve Supply

1. Motor neurons carry action potentials to skeletal muscles, where the neuron and muscle fibers form neuromuscular junctions.
2. Neurons release acetylcholine, which binds to receptors on muscle cell membranes, stimulates an action potential in the muscle cell, and causes the muscle to contract.

Muscle Contraction

1. Action potentials are carried along T tubules to the sarcoplasmic reticulum, where they cause the release of calcium ions.
2. Calcium ions, released from the sarcoplasmic reticulum, bind to the actin myofilaments, exposing attachment sites.
3. Myosin forms cross-bridges with the exposed actin attachment sites.
4. The myosin molecules bend, causing the actin molecules to slide past; this is the sliding filament model. The H and I bands shorten, the A bands do not.
5. This process requires ATP breakdown.
6. A muscle twitch is the contraction of a muscle fiber in response to a stimulus; it consists of a lag phase, contraction phase, and relaxation phase.
7. Tetanus occurs when stimuli occur so rapidly that a muscle does not relax between twitches.
8. Small contraction forces are generated when small numbers of motor units are recruited, and greater contraction forces are generated when large numbers of motor units are recruited.

9. Energy is produced by anaerobic (without oxygen) and aerobic (with oxygen) respiration.
10. After intense exercise, the rate of aerobic metabolism remains elevated to repay the oxygen debt.
11. Muscle fatigue occurs as ATP is depleted during muscle contraction. Physiological contracture occurs in extreme fatigue when a muscle can neither contract nor relax.
12. Muscles contract either isometrically (tension increases, but muscle length stays the same) or isotonically (tension remains the same, but muscle length decreases).
13. Muscle tone consists of a small percentage of muscle fibers contracting tetanically and is responsible for posture.
14. Muscles contain a combination of slow-twitch and fast-twitch fibers.
15. Slow-twitch fibers are better suited for aerobic metabolism, and fast-twitch fibers are adapted for anaerobic metabolism.
16. Sprinters have more fast-twitch fibers, whereas distance runners have more slow-twitch fibers.

Smooth Muscle and Cardiac Muscle   (p. 170)

1. Smooth muscle is not striated, has one nucleus per cell, contracts more slowly than skeletal muscle, can be autorhythmic, and is under involuntary control.
2. Cardiac muscle is striated, usually has one nucleus per cell, has intercalated disks, is autorhythmic, and is under involuntary control.

Skeletal Muscle Anatomy   (p. 172)
General Principles

1. Most muscles have an origin on one bone and an insertion onto another and cross at least one joint.
2. A muscle causing a specific movement is an agonist. A muscle causing the opposite movement is an antagonist.
3. Muscles working together are synergists.
4. A prime mover is the one muscle of a synergistic group that is primarily responsible for the movement.

Nomenclature

Muscles are named according to location, origin and insertion, number of heads, or function.

Muscles of the Head and Neck

1. Muscles of facial expression are associated primarily with the mouth and eyes.
2. Four pairs of muscles are involved in mastication.
3. Tongue movements involve intrinsic and extrinsic muscles.
4. Swallowing involves suprahyoid and infrahyoid muscles, plus muscles of the soft palate, pharynx, and larynx.
5. Neck muscles move the head.

Trunk Muscles

1. Erector spinae muscles hold the body erect.
2. Intercostal muscles and the diaphragm are involved in breathing.
3. Muscles of the abdominal wall flex and rotate the vertebral column, compress the abdominal cavity, and hold in and protect the abdominal organs.
4. Muscles form the floor of the pelvis.

Upper Limb Muscles

1. The upper limb is attached to the body primarily by muscles.
2. Arm movements are accomplished by pectoral, rotator cuff, and deltoid muscles.
3. The elbow is flexed and extended by anterior and posterior arm muscles, respectively.
4. Supination and pronation of the forearm are accomplished by supinators and pronators in the forearm.
5. Movements of the wrist and fingers are accomplished by most of the 20 forearm muscles and 19 intrinsic muscles in the hand.

Lower Limb Muscles

1. Hip muscles flex and extend the hip and abduct the thigh.
2. Thigh muscles flex and extend the hip and adduct the thigh. They also flex and extend the knee.
3. Muscles of the leg and foot are similar to those of the forearm and hand.

Effects of Aging on Skeletal Muscle   (p. 194)

Aging is associated with a decrease in muscle mass, slower reaction time, reduced stamina, and increased recovery time.

R E V I E W   A N D   C O M P R E H E N S I O N

1. List the seven major functions of the muscular system.
2. Define contractility, excitability, extensibility, and elasticity.
3. List the connective tissue layers associated with muscles.
4. What are fasciculi?
5. What is a muscle fiber?
6. Describe the composition of a myofibril.
7. What is a sarcomere?
8. Describe the structure of actin and myosin myofilaments.
9. Describe the resting membrane potential and how it is produced.
10. Describe the production of an action potential.
11. What is a neuromuscular junction? What happens there?
12. Describe the sliding filament model of muscle contraction.
13. Explain how an action potential results in a muscle contraction.
14. Define muscle twitch, tetanus, and recruitment.
15. Describe the two ways energy is produced in skeletal muscle.
16. Compare isometric, isotonic, concentric, and eccentric contraction.
17. Explain fatigue.
18. What is muscle tone?
19. Compare slow-twitch and fast-twitch muscle fibers.
20. How do smooth muscles and cardiac muscles differ from skeletal muscles?
21. Define origin, insertion, agonist, antagonist, synergist, prime mover, and fixator.
22. Describe the muscles of facial expression.
23. What is mastication? What muscles are involved?
24. What are intrinsic and extrinsic tongue muscles?
25. What muscles are involved in swallowing?
26. What muscles are involved in respiration?
27. Describe the functions of the muscles of the anterior abdominal wall.
28. What is primarily responsible for attaching the upper limb to the body?
29. Describe, by muscle groups, movements of the arm, forearm, and hand.
30. Describe, by muscle groups, movements of the thigh, leg, and foot.

C R I T I C A L   T H I N K I N G

1. Bob Canner improperly canned some home-grown vegetables. As a result, he contracted botulism poisoning after eating the vegetables. Botulism results from a toxin, produced by bacteria, which prevents skeletal muscles from contracting. Symptoms include difficulty in swallowing and breathing. Eventually he died of respiratory failure because his respiratory muscles relaxed and would not contract. Assuming that botulism toxin affects the neuromuscular junction, propose as many ways as you can how botulism toxin could produce the observed symptoms.

2. Harvey Leche milked cows by hand each morning before school. One morning he overslept and had to hurry to get to school on time. As he was milking the cows as fast as he could, his hands became very tired, and then for a short time he could neither release his grip nor squeeze harder. Explain what happened.

3. A researcher was investigating the fast-twitch versus slow-twitch composition of muscle tissue in the gastrocnemius muscle (in the calf of the leg) of athletes. Describe the general differences this researcher would see when comparing the muscles from athletes who were outstanding in the following events: 100-m dash, weight lifting, the 10,000-m run.

4. Describe an exercise routine that would build up each of the following groups of muscles: anterior arm, posterior arm, anterior forearm, anterior thigh, posterior leg, and abdomen.

5. Sherri Speedster started a 100-m dash but fell to the ground in pain. Examination of her right lower limb revealed the following symptoms: the knee was held in a slightly flexed position, but she could not flex it voluntarily; she could extend the knee with difficulty, but this caused her considerable pain; and there was considerable pain and bulging of the muscles in the posterior thigh. Explain the nature of her injury.

Answers in Appendix D

Visit this textbook's website at www.mhhe.com/seeleyess6 for practice quizzes, animations, interactive learning exercises, and other study tools.

Chapter Outline and Objectives

Nervous System

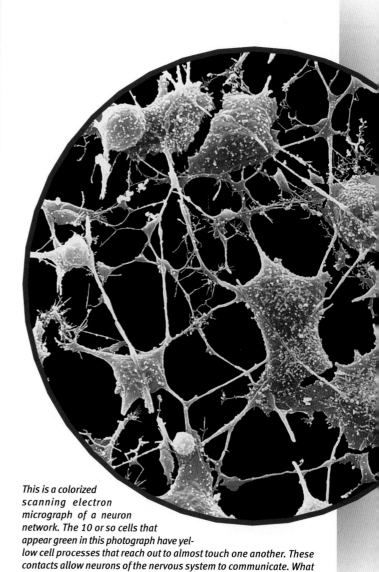

This is a colorized scanning electron micrograph of a neuron network. The 10 or so cells that appear green in this photograph have yellow cell processes that reach out to almost touch one another. These contacts allow neurons of the nervous system to communicate. What is seen here is only a minute portion of the total connections among nerve cells. All the tiny processes branching off the larger processes contact processes of other cells.

a man is walking down a street on a beautiful autumn day, enjoying the warmth of the sun. He passes a produce stand. The apples look good to him and he decides that a big, juicy one would taste wonderful. He selects just the right one and picks it up. At the register, he reaches into his pocket, feels the coins there, and pulls out a few. He counts out the correct change, pays for the apple, and puts the rest of the change back into his pocket. While walking down the street, he takes a big bite of the apple. He feels the apple's juicy crispness in his mouth and savors its tart-sweet flavor. Life is wonderful. All these sensations, actions, and emotions are made possible by the nervous system, which consists of the brain, spinal cord, nerves, and sensory receptors.

Functions of the Nervous System

The nervous system is involved in some way in nearly every body function. Some major functions of the nervous system are:

1. *Sensory input.* Sensory receptors monitor numerous external and internal stimuli that may be interpreted as touch, temperature, taste, smell, sound, blood pressure, and body position. Action potentials from the sensory receptors travel along nerves to the spinal cord and brain, where they are interpreted.
2. *Integration.* The brain and spinal cord are the major organs for processing sensory input and initiating responses. The input may produce an immediate response, may be stored as memory, or may be ignored.
3. *Homeostasis.* The nervous system plays an important role in the maintenance of homeostasis. This function depends on the ability of the nervous system to detect, interpret, and respond to changes in internal and external conditions. In response, the nervous system can stimulate or inhibit the activities of other systems to help maintain a constant internal environment.
4. *Mental activity.* The brain is the center of mental activity, including consciousness, memory, and thinking.
5. *Control of muscles and glands.* Skeletal muscles normally contract only when stimulated by the nervous system. Thus, through the control of skeletal muscle, the nervous system controls the major movements of the body. The nervous system also participates in controlling cardiac muscle, smooth muscle, and many glands.

Divisions of the Nervous System

The nervous system can be divided into the central and the peripheral nervous systems (figure 8.1). The **central nervous system (CNS)** consists of the brain and spinal cord. The

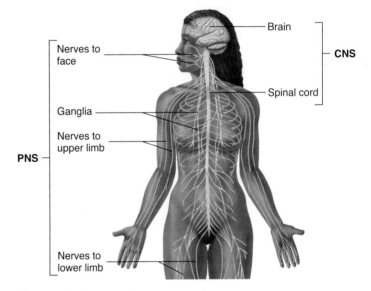

Figure 8.1   The Nervous System

The central nervous system (CNS) consists of the brain and spinal cord. The peripheral nervous system (PNS) consists of nerves and ganglia.

peripheral nervous system (PNS) lies outside the CNS and consists of nerves and ganglia.

The PNS has two subdivisions: The **sensory,** or **afferent division** conducts action potentials from sensory receptors to the CNS (figure 8.2). The neurons that transmit action potentials from the periphery to the CNS are **sensory neurons.** The **motor,** or **efferent division** conducts action potentials from the CNS to effector organs such as muscles and glands. The neurons that transmit action potentials from the CNS toward the periphery are **motor neurons.**

The motor division can be further subdivided into the **somatic** (sō-mat′ik, bodily) **motor nervous system,** which transmits action potentials from the CNS to skeletal muscles;

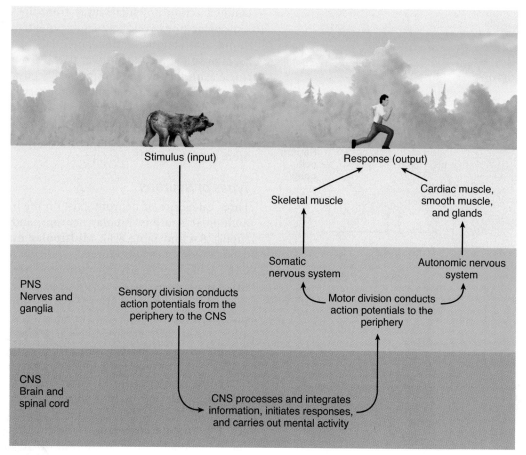

Figure 8.2 Organization of the Nervous System

The sensory division of the peripheral nervous system (PNS) detects stimuli and conducts action potentials to the central nervous system (CNS). The CNS interprets incoming action potentials and initiates action potentials that are conducted through the motor division to produce a response. The motor division is divided into the *somatic nervous system* and the *autonomic nervous system.*

and the **autonomic** (aw-tō-nom′ik) **nervous system (ANS),** which transmits action potentials from the CNS to cardiac muscle, smooth muscle, and glands. The autonomic nervous system, in turn, is divided into sympathetic and parasympathetic divisions, and the **enteric nervous system, (ENS),** which is associated with the digestive tract.

Cells of the Nervous System

Cells of the nervous system are neurons and neuroglia.

Neurons

Neurons (noor′onz, nerve), or **nerve cells** (figure 8.3), receive stimuli and transmit action potentials to other neurons or to effector organs. Each neuron consists of a cell body and two types of processes: dendrites and axons (see table 4.12).

Each neuron cell body contains a single nucleus. As with any other cell, the nucleus of the neuron is the source of information for protein synthesis. If an axon, which is one of the neuron cell processes, is separated from the cell body, it dies because it has no connection to the nucleus, and no protein

synthesis occurs in the axon. Extensive rough endoplasmic reticulum (rough ER), Golgi apparatus, and mitochondria surround the nucleus. Large numbers of neurofilaments (intermediate filaments) and microtubules course through the cytoplasm in all directions and separate the rough ER into distinct areas in the cell body. The areas of rough ER concentration, when stained with a specific dye, appear as microscopic granules called **Nissl** (nis′l, Franz Nissl, German neurologist [1860–1919]) **bodies.**

Dendrites (den′drītz, trees) are short, often highly branching cytoplasmic extensions that are tapered from their bases at the neuron cell body to their tips. Most dendrites are extensions of the neuron cell body, but dendritelike structures also project from the peripheral ends of some sensory axons. Dendrites usually function to receive information from other neurons or from sensory receptors and transmit the information toward the neuron cell body.

An **axon** is a long cell process extending from the neuron cell body. Where the axon leaves the neuron cell body is an area called the **axon hillock,** which is devoid of Nissl bodies. Each axon has a constant diameter and may vary in length from a few millimeters to more than a meter. Axons of motor neurons

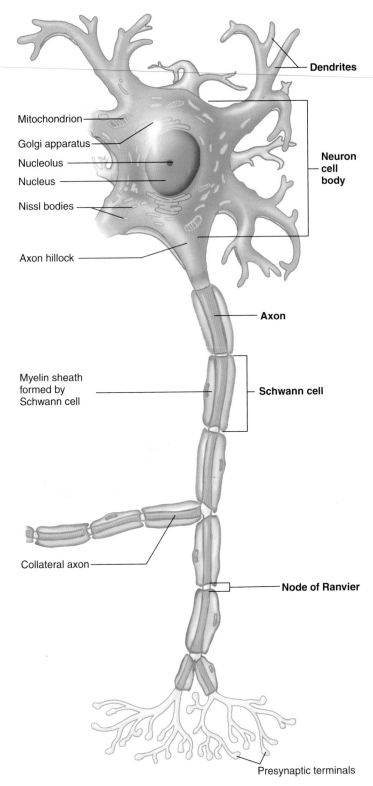

Mitochondrion

Golgi apparatus

Nucleolus

Nucleus

Nissl bodies

Axon hillock

Dendrites

Neuron
cell
body

Axon

Myelin sheath
formed by
Schwann cell

Schwann cell

Collateral axon

Node of Ranvier

Presynaptic terminals

Figure 8.3 Neuron

Structural features of a neuron include a cell body and two types of cell
processes: dendrites and an axon.

conduct action potentials away from the CNS and axons of
sensory neurons conduct action potentials toward the CNS.
Axons also conduct action potentials from one part of the
brain or spinal cord to another part. Each motor neuron has a
single axon that extends from the CNS toward a target tissue.
An axon may remain unbranched or may branch to form **collateral** (ko-lat′er-ăl) **axons.** Axons are surrounded by neuroglia called **Schwann cells,** which form a highly specialized
insulating layer of cells called the myelin sheath (described in
more detail on p. 201).

Types of Neurons

Three categories of neurons exist on the basis of their shape:
multipolar neurons, bipolar neurons, and unipolar neurons
(figure 8.4 and table 8.1). **Multipolar neurons** have many
dendrites and a single axon. Most of the neurons within the
CNS, including nearly all motor neurons, are multipolar.
Bipolar neurons have two processes: one dendrite and one
axon. Bipolar neurons are located in some sensory organs,
such as in the retina of the eye and in the nasal cavity. Most
other sensory neurons are unipolar. **Unipolar neurons** have a
single process extending from the cell body. This process divides into two processes a short distance from the cell body.
One process extends to the periphery, and the other process
extends to the CNS. The two extensions function as a single
axon with small dendritelike sensory receptors at the periphery. The axon receives sensory information at the periphery
and transmits that information in the form of action potentials to the CNS.

Neuroglia

Neuroglia (noo-rog′lē-ă, nerve glue), or **glial** (glī′ăl or glē′ăl)
cells, are the nonneuronal cells of the CNS and PNS. Neuroglia are far more numerous than neurons. Most neuroglia
retain the ability to divide, whereas most neurons do not.
There are five types of neuroglia. **Astrocytes** (as′trō-sītz,
astron, star + *kytos,* hollow-cell) serve as the major supporting tissue in the CNS and participate with the blood vessel
endothelium to form a permeability barrier, called the
blood–brain barrier, between the blood and the neurons.
Ependymal (ep-en′di-măl, *ependyma,* an upper garment)
cells line the fluid-filled cavities (ventricles and canals)
within the CNS. Some ependymal cells produce cerebrospinal
fluid, and others, with cilia on the surface, help move the
cerebrospinal fluid through the CNS. **Microglia** (mī-krog′lē-ă)
help remove bacteria and cell debris from the CNS. **Oligodendrocytes** (ol′i-gō-den′drō-sītz; cells with many dendritic
processes) in the CNS and **Schwann cells** in the PNS surround axons (figure 8.5; see table 8.1). Schwann cells are also
referred to as **neurolemmocytes** (noor-ō-lem′ahsītz, *neuro-,*
nerve + *lemma,* husk), or **neurolemma cells.**

Myelin Sheaths

Axons are surrounded by the cell processes of oligodendrocytes
in the CNS and Schwann cells in the PNS (see figure 8.5).

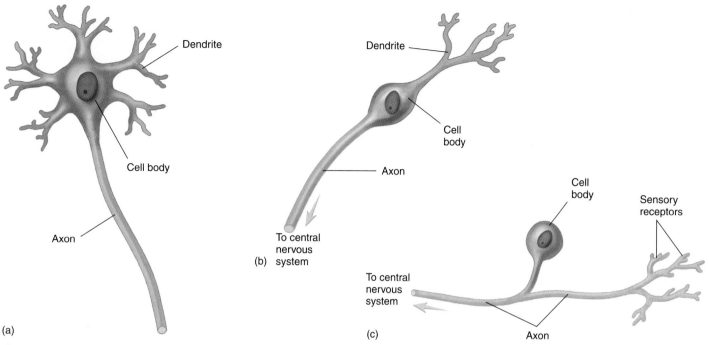

Figure 8.4 Types of Neurons

(a) A multipolar neuron has many dendrites and one axon. (b) A bipolar neuron has a dendrite and an axon. (c) A unipolar neuron has an axon and no dendrites.

Table 8.1	Cells of the Nervous System (see figures 8.4 and 8.5)	
Cell Type	**Description**	**Function**
Neuron		
Multipolar	Several dendrites and one axon	Most motor neurons and most CNS neurons
Bipolar	One dendrite and one axon	Found in special sense organs such as the eye and the nose
Unipolar	A neuron with a single axon	Most sensory neurons
Neuroglia		
Astrocytes	Star-shaped	Provide structural support; form a layer around blood vessels, contribute to blood–brain barrier
Ependymal cells	Squamous epithelial-like	Line ventricles of brain, circulate cerebrospinal fluid (CSF); some form choroid plexuses, which produce CSF
Microglia	Small mobile cells	Protect CNS from infection; become phagocytic in response to inflammation
Oligodendrocytes	Cell with processes that can surround several axons	Cell processes form myelin sheaths around axons, or enclose unmyelinated axons in the CNS
Schwann cells	Single cells surrounding axons	Form myelin sheaths around axons, or enclose unmyelinated axons in the PNS

Unmyelinated axons rest in indentations of the oligodendrocytes in the CNS and the Schwann cells in the PNS (figure 8.6a). **Myelinated axons** have specialized sheaths, called **myelin sheaths,** wrapped around them. Each oligodendrocyte process or Schwann cell repeatedly wraps around a segment of an axon to form a series of tightly wrapped cell membranes. A typical small nerve usually contains more unmyelinated than myelinated axons. Myelin is an excellent insulator, which prevents almost all electrical current flow through the cell membrane. Gaps in the myelin sheath, called **nodes of Ranvier** (ron′vē-ā, Louis Ranvier, French pathologist [1835–1922]), can be seen about every millimeter between the oligodendrocyte segments or between individual Schwann cells (figure 8.6b). At the nodes of Ranvier, current flows easily between the extracellular fluid and the axon, and action potentials can develop.

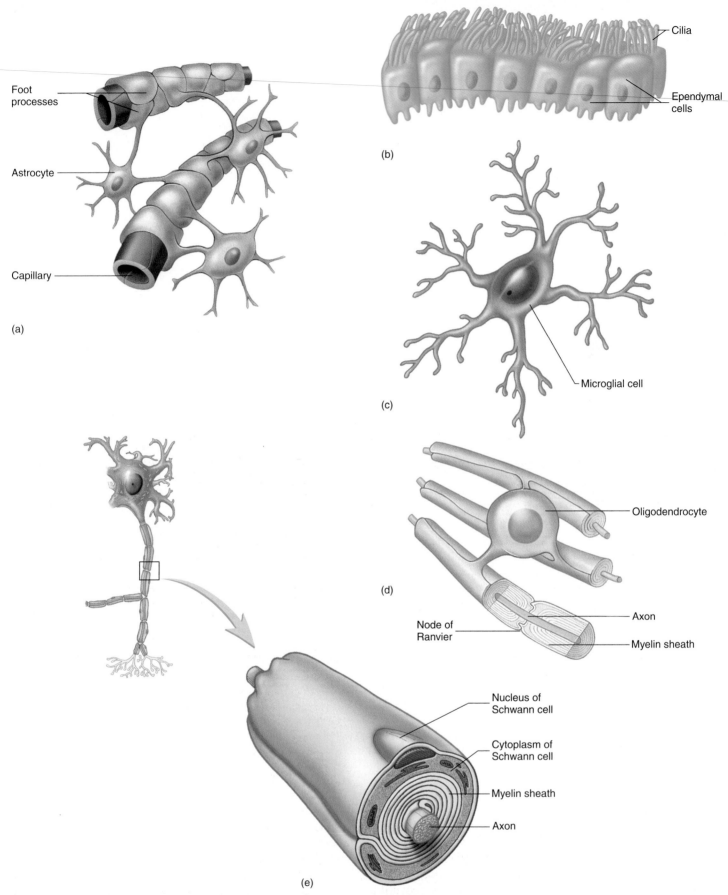

Foot
processes

Astrocyte

Capillary

(a)

Cilia

Ependymal
cells

(b)

Microglial cell

(c)

Oligodendrocyte

Axon

Node of
Ranvier

Myelin sheath

Nucleus of
Schwann cell

Cytoplasm of
Schwann cell

Myelin sheath

Axon

(d)

(e)

Figure 8.5 Types of Neuroglia

(*a*) Astrocytes, with foot processes surrounding a blood capillary. (*b*) Ependymal cells, with cilia extending from the surfaces. (*c*) Microglia. (*d*) Oligodendrocyte, forming a myelin sheath around parts of three axons within the CNS. (*e*) Schwann cell forming part of the myelin sheath of an axon in the PNS.

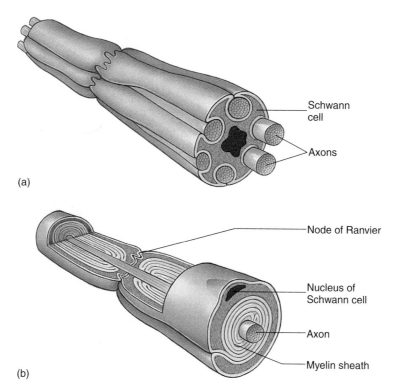

Figure 8.6 Comparison of Myelinated and Unmyelinated Axons

(a) Unmyelinated axons with two Schwann cells surrounding several axons in parallel formation. Each Schwann cell surrounds part of several axons. (b) Myelinated axon with two Schwann cells forming the myelin sheath around a single axon. Each Schwann cell surrounds part of one axon.

Organization of Nervous Tissue

Groups of neuron cell bodies and their dendrites, where there is very little myelin, form **gray matter.** Gray matter on the surface of the brain is called the **cortex,** and clusters of gray matter located deeper within the brain are called **nuclei.** In the PNS, a cluster of neuron cell bodies is called a **ganglion** (gang′glē-on, pl. ganglia, a swelling or knot). Bundles of parallel axons with their myelin sheaths are whitish in color and are called **white matter.** White matter of the CNS forms conduction **pathways,** or nerve **tracts,** which propagate action potentials from one area in the CNS to another. In the PNS, bundles of axons and their connective tissue sheaths are called **nerves.**

Electric Signals and Neural Pathways

The Resting Membrane Potential

All cells exhibit electrical properties. The outside of most cell membranes is positively charged compared with the inside of the cell membrane, which is negatively charged (as discussed in chapter 7). This charge difference across the membrane of an unstimulated cell is called the **resting membrane potential.** The cell is said to be **polarized.** The outside of the cell membrane can be thought of as the positive pole of a battery, and the inside as the negative pole. Thus, a small voltage difference, or potential, can be measured across the resting cell membrane.

The resting membrane potential results from differences in the concentration of ions across the cell membrane and the permeability characteristics of the cell membrane. There is a higher concentration of sodium ions (Na^+) immediately outside the cell membrane than inside and a higher concentration of potassium ions (K^+) immediately inside the cell membrane than outside (figure 8.7a). The concentration of Na^+ outside the cell membrane and of K^+ inside is maintained by the **Na^+–K^+ pump,** which actively transports K^+ into and Na^+ out of the cell (figure 8.7b).

When a cell is at rest, some K^+ **channels** are open and Na^+ channels are not. The cell membrane is therefore more permeable to K^+ than to Na^+ (figure 8.8). This allows a few K^+ to diffuse down their concentration gradient out of the cell, carrying their positive charges with them. In addition, the cell membrane has more K^+ channels than Na^+ channels. Larger molecules, such as proteins, which are negatively charged, are too large to diffuse out of the cell. As positive K^+ leave the cell, the charge inside the cell becomes more negative. The molecules inside the cell with negative charges tend to attract the positive K^+ back into the cell. A point of equilibrium is reached at which the tendency for K^+ to move down their concentration gradient out of the cell is balanced by the negative charge within the cell, which tends to attract the K^+ back into the cell. This point of K^+ equilibrium is the point at which the resting membrane potential is established

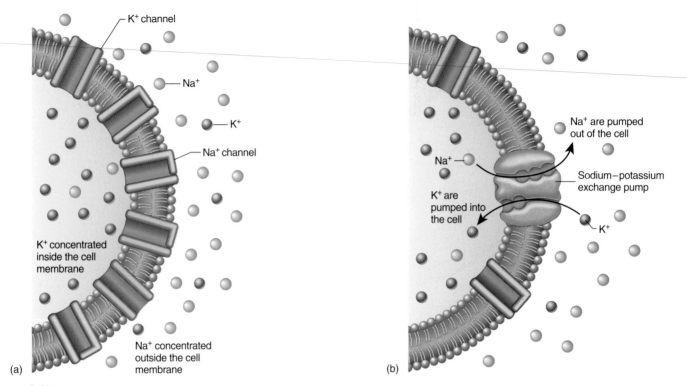

Figure 8.7 **Major Ion Concentration Differences Across a Cell Membrane**

(a) There is a higher concentration of K$^+$ inside the cell and a higher concentration of Na$^+$ outside the cell. (b) The Na$^+$–K$^+$ pump moves Na$^+$ out of the cell in exchange for K$^+$ that move into the cell.

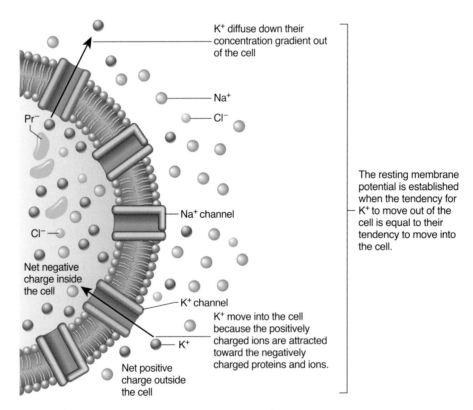

Figure 8.8 **Ion Channels and Ion Concentrations Across a Cell Membrane**

Concentrations of Na$^+$, K$^+$, Cl$^-$, and negatively charged proteins across the cell membrane are represented. Some K$^+$ channels are open but Na$^+$ channels are not. As a result, K$^+$ diffuse out of the cell down their concentration gradient. The membrane is not permeable to the negatively charged proteins inside the cell. The tendency for the K$^+$ to diffuse to the outside of the cell down their concentration gradient is opposed by the tendency for the positively charged K$^+$ to be attracted back into the cell by the negatively charged proteins inside the cell.

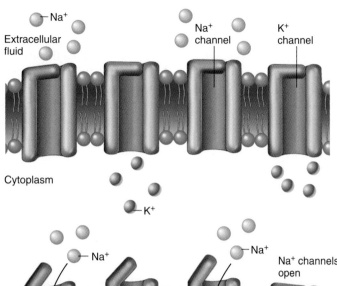

1. Resting membrane potential. Na⁺ channels (*pink*) and most, but not all, K⁺ channels (*purple*) are closed. The outside of the plasma membrane is positively charged compared to the inside.

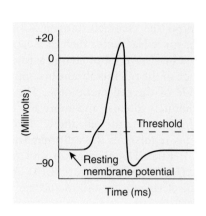

2. Depolarization. Na⁺ channels open. K⁺ channels begin to open. Depolarization results because the inward movement of Na⁺ makes the inside of the membrane more positive.

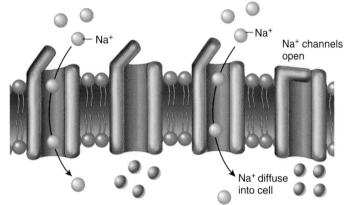

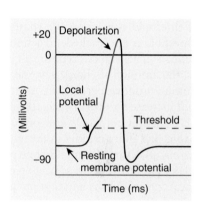

3. Repolarization. Na⁺ channels close and additional K⁺ channels open. Na⁺ movement into the cell stops and K⁺ movement out of the cell increases, causing repolarization.

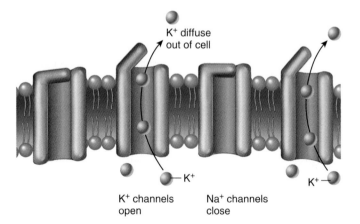

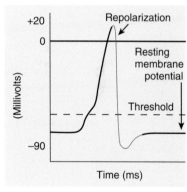

Process Figure 8.9   Ion Channels and the Action Potential

and there is no more net K⁺ movement (figure 8.9, step 1). At equilibrium, there is a net positive charge outside the cell and a net negative charge inside the cell.

Action Potentials

Muscle and nerve cells are **excitable.** When a stimulus is applied to a muscle cell or nerve cell, some Na⁺ channels open for a very brief time, and Na⁺ diffuse quickly into the cell (figure 8.9, step 2). The movement of Na⁺ into a cell is called a **local current.** The positively charged Na⁺ entering the cell cause the inside of the cell membrane to become more positive, a change called **depolarization.** This depolarization results in a **local potential.** If threshold is not reached, the Na⁺ channels close again, and the local potential disappears without being conducted along the nerve cell membrane. If enough Na⁺ enter the cell so that the local potential reaches a **threshold** value, this threshold depolarization causes many more Na$^+$ channels to open. K$^+$ channels also begin to open. As more Na$^+$ enter the cell, depolarization occurs until there is a brief reversal of charge across the membrane, and the inside of the cell membrane becomes positive relative to the outside of the cell membrane. The charge reversal causes Na$^+$ channels to close and more K$^+$ channels to open. Na$^+$ then stop entering the cell, and K$^+$ leave the cell (figure 8.9, step 3). This repolarizes the cell membrane to its resting membrane potential. Depolarization and repolarization constitute an **action potential** (figure 8.10). At the end of repolarization, the charge on the cell membrane briefly becomes more negative than the resting membrane potential; this condition is called hyperpolarization. The elevated permeability to K$^+$ lasts only a very short time.

Action potentials occur in an **all-or-none** fashion; that is, if threshold is reached, the charge reversal is complete; if the threshold is not reached, no action potential occurs. Action potentials in a given cell type are all of the same magnitude—that is, the amount of charge reversal is always the same. Stronger stimuli produce a greater frequency of action potentials but do not increase the size of each action potential.

Action potentials are conducted slowly in unmyelinated axons and more rapidly in myelinated axons. In unmyelinated axons, an action potential in one part of a cell membrane stimulates local currents in adjacent parts of the cell membrane. The local currents in the adjacent membrane produce an action potential. By this means, the action potential is conducted along the entire axon cell membrane (figure 8.11). In myelinated axons, an action potential at one node of Ranvier causes a local current to flow through the surrounding extracellular fluid and

through the cytoplasm of the axon to the next node, stimulating an action potential at that node of Ranvier. By this means, action potentials "jump" from one node of Ranvier to the next along the length of the axon (figure 8.12). This type of action potential conduction is called **saltatory** (sal'tă-tōr-ē, to leap) **conduction.** Saltatory conduction greatly increases the conduction velocity because the nodes of Ranvier make it unnecessary for action potentials to travel along the entire cell membrane. Action potential conduction in a myelinated fiber is like a grasshopper jumping, whereas in an unmyelinated axon it is like a grasshopper walking.

Medium-diameter, lightly myelinated axons, characteristic of autonomic neurons, conduct action potentials at the rate of about 3–15 m/s, whereas large-diameter, heavily myelinated axons conduct action potentials at the rate of 15–120 m/s. These rapidly conducted action potentials, carried by sensory and motor neurons, allow for rapid responses to changes in the external environment. In addition, several hundred times fewer ions cross the cell membrane during conduction in myelinated cells than in unmyelinated cells. Much less energy is therefore required for the sodium–potassium exchange pump to maintain the ion distribution.

The Synapse

A **synapse** (sin'aps) is a junction where the axon of one neuron interacts with another neuron or an effector organ such as a muscle or gland (figure 8.13). The end of the axon forms a **presynaptic terminal.** The membrane of the dendrite or effector cell is the **postsynaptic membrane,** and the space separating the presynaptic and postsynaptic membranes is the **synaptic cleft.** Chemical substances called **neurotransmitters** (noor'ō-trans-mit'ers, *neuro-*, nerve + *transmitto*, to send across) are stored in **synaptic vesicles** in the presynaptic terminal. When an action potential reaches the presynaptic terminal, Ca^{2+} channels open and Ca^{2+} move into the cell. This influx of Ca^{2+} causes the release of neurotransmitters by exocytosis from the presynaptic terminal. The neurotransmitters diffuse across the synaptic cleft and bind to specific receptor molecules on the postsynaptic membrane. The binding of neurotransmitters to these membrane receptors causes channels for Na$^+$, K$^+$, or Cl$^-$ to open or close in the postsynaptic membrane, depending on the type of neurotransmitter in the presynaptic terminal and the type of receptors on the postsynaptic membrane. The response may be either stimulation or an inhibition of an action potential in the postsynaptic cell. For example, if Na$^+$ channels open, the postsynaptic cell becomes depolarized, and an action potential will result if threshold is reached. If K$^+$ or Cl$^-$ channels open, the inside of the postsynaptic cell tends to become more negative, or **hyperpolarized** (hī'per-pō'lăr-ī-zed, *hyper*, above + *polaris*, polar), and an action potential is inhibited from occurring.

Of the many neurotransmitter substances or suspected neurotransmitter substances, the best known are **acetylcholine** (as'e-til-kō'lēn) and **norepinephrine** (nōr'ep-i-nef'rin). Other neurotransmitters include serotonin (sēr-ō-tō'nin), dopamine (dō'pă-men), γ (gamma)-aminobutyric (gam'ă ă-mē'nō-bū-tēr'ik) acid (GABA), glycine, and endorphins (en'dōr-finz,

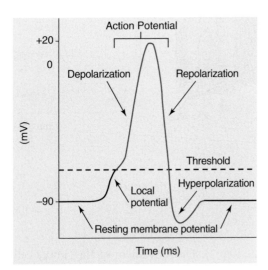

Figure 8.10   The Action Potential

Once a local depolarization reaches threshold, an all-or-none action potential is started. During the depolarization phase, the voltage across the cell membrane changes from approximately −90 mV to approximately +20 mV. During the repolarization phase, the voltage across the cell membrane returns to −90 mV. There is a brief period of hyperpolarization at the end of repolarization before the membrane returns to its resting membrane potential. The entire process lasts 1 or 2 milliseconds (ms).

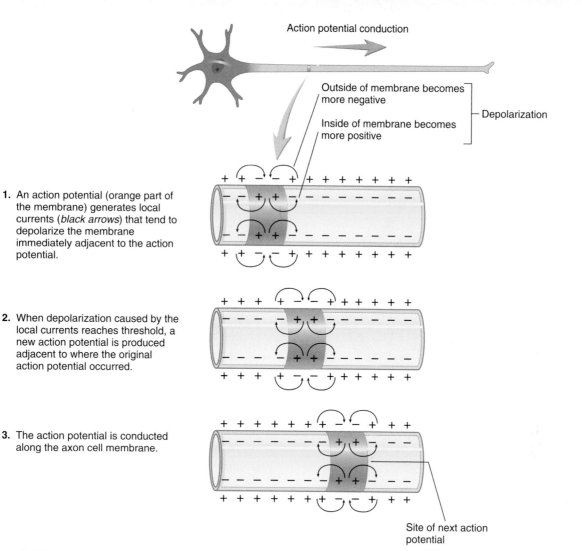

Action potential conduction

Outside of membrane becomes more negative
Inside of membrane becomes more positive
} Depolarization

1. An action potential (orange part of the membrane) generates local currents (*black arrows*) that tend to depolarize the membrane immediately adjacent to the action potential.

2. When depolarization caused by the local currents reaches threshold, a new action potential is produced adjacent to where the original action potential occurred.

3. The action potential is conducted along the axon cell membrane.

Site of next action potential

Process Figure 8.11 Action Potential Conduction in an Unmyelinated Axon

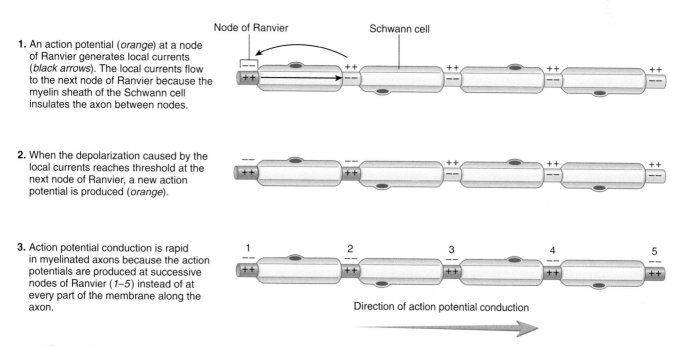

Node of Ranvier   Schwann cell

1. An action potential (*orange*) at a node of Ranvier generates local currents (*black arrows*). The local currents flow to the next node of Ranvier because the myelin sheath of the Schwann cell insulates the axon between nodes.

2. When the depolarization caused by the local currents reaches threshold at the next node of Ranvier, a new action potential is produced (*orange*).

3. Action potential conduction is rapid in myelinated axons because the action potentials are produced at successive nodes of Ranvier (*1–5*) instead of at every part of the membrane along the axon.

Direction of action potential conduction

Process Figure 8.12 Saltatory Conduction: Action Potential Conduction in a Myelinated Axon

the neuromuscular junction (see chapter 7), an enzyme called **acetylcholinesterase** (as′e-til-kō-lin-es′-ter-ās) breaks down the acetylcholine. The breakdown products are then returned to the presynaptic terminal for reuse. Norepinephrine is either actively transported back into the presynaptic terminal or it is broken down by enzymes. The release and breakdown or removal of neurotransmitters occurs so rapidly that a postsynaptic cell can be stimulated many times a second.

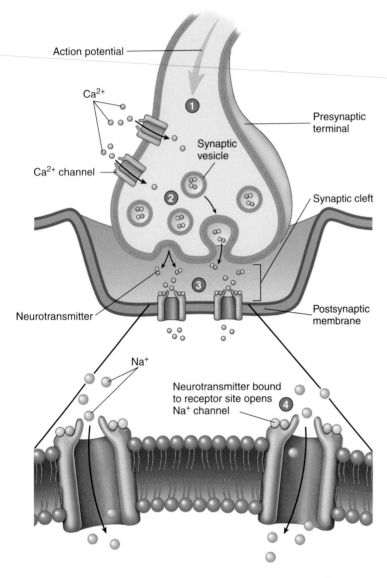

1. Action potentials arriving at the presynaptic terminal cause Ca²⁺ channels to open.
2. Ca²⁺ diffuse into the cell and cause synaptic vesicles to release neurotransmitter molecules.
3. Neurotransmitter molecules diffuse from the presynaptic terminal across the synaptic cleft.
4. Neurotransmitter molecules combine with their receptor sites and cause Na⁺ channels to open. Na⁺ diffuse into the cell (*shown in illustration*) or out of the cell (*not shown*) and cause a change in membrane potential.

Process Figure 8.13 The Synapse

A synapse consists of the end of a neuron (presynaptic terminal), a small space (synaptic cleft), and the postsynaptic membrane of another neuron or an effector cell such as a muscle or gland cell.

A CASE IN POINT | Botulism

Ima Kannar removed a 5-year-old bottle of green beans from the shelf in her food storage room. She dusted off the lid and noticed that the lid was a bit bowed out; but the beans inside looked okay and smelled okay when she opened the lid. She ate some of the beans for dinner, watched some television, and then went to bed. The next afternoon, as she was working in her garden, she felt that her throat was dry. She started to experience double vision and she was finding it difficult to breathe. She dialed 911. By the time the paramedics arrived, 20 minutes later, Ima was not breathing. She was put on a respirator and rushed to the hospital. She recovered very slowly and was in the hospital for several months, much of that time on a respirator. Botulism is caused by the toxin of the bacterium *Clostridium botulinium,* which grows in anaerobic (without air) environments such as improperly processed canned food. The toxin blocks the release of acetylcholine in synapses. As a result, muscles cannot contract, and the person may die from respiratory failure.

Cocaine and Amphetamines

Cocaine and amphetamines are strong CNS stimulants that cause elation, euphoria, psychosis, hallucinations, and paranoia. Cocaine and amphetamines also function in the PNS to increase the release and block the reuptake of norepinephrine, resulting in overstimulation of postsynaptic neurons. Symptoms include dilation of the pupils, restlessness, exaggerated reflexes, muscle spasms, tachycardia, vasoconstriction, hypertension, nausea, and vomiting. Death may occur from heart failure or respiratory failure.

endogenous morphine) (table 8.2). Neurotransmitter substances are rapidly broken down by enzymes within the synaptic cleft or are transported back into the presynaptic terminal. Consequently, they are removed from the synaptic cleft so their effects on the postsynaptic membrane are very short term. In synapses where acetylcholine is the neurotransmitter, such as in

Reflexes

A **reflex** is an involuntary reaction in response to a stimulus applied to the periphery and transmitted to the CNS. Reflexes allow a person to react to stimuli more quickly than is possible if conscious thought is involved. A **reflex arc** is the neuronal pathway by

Table 8.2 Neurotransmitters

Substance	Location	Effect	Clinical Example
Acetylcholine	Many nuclei scattered throughout the brain and spinal cord. Nerve tracts from the nuclei extend to many areas of the brain and spinal cord. Also found in the neuromuscular junction of skeletal muscle and many ANS synapses.	Excitatory or inhibitory	Alzheimer's disease (a type of senile dementia) is associated with a decrease in acetylcholine-secreting neurons. Myasthenia gravis (weakness of skeletal muscles) results from a reduction in acetylcholine receptors.
Norepinephrine	A small number of small-sized nuclei in the brainstem. Nerve tracts extend from the nuclei to many areas of the brain and spinal cord. Also in some ANS synapses.	Excitatory or inhibitory	Cocaine and amphetamines increase the release and block the reuptake of norepinephrine, resulting in overstimulation of postsynaptic neurons.
Serotonin	A small number of small-sized nuclei in the brainstem. Nerve tracts extend from the nuclei to many areas of the brain and spinal cord.	Generally inhibitory	Involved with mood, anxiety, and sleep induction. Levels of serotonin are elevated in schizophrenia (delusions, hallucinations, and withdrawal).
Dopamine	Confined to a small number of nuclei and nerve tracts. Distribution is more restricted than that of norepinephrine or serotonin. Also found in some ANS synapses.	Generally excitatory	Parkinson's disease (depression of voluntary motor control) results from destruction of dopamine-secreting neurons. Drugs used to increase dopamine production induce vomiting and schizophrenia.
Gamma-aminobutyric acid (GABA)	GABA-secreting neurons mostly control activities in their own area and are not usually involved with transmission from one part of the CNS to another. Most neurons of the CNS have GABA receptors.	Generally inhibitory	Drugs that increase GABA function have been used to treat epilepsy (excessive discharge of neurons).
Glycine	Spinal cord and brain. Like GABA, glycine predominantly produces local effects.	Generally inhibitory	Glycine receptors are inhibited by the poison strychnine. Strychnine increases the excitability of certain neurons by blocking their inhibition. Strychnine poisoning results in powerful muscle contractions and convulsions. Tetanus of respiratory muscles can cause death.
Endorphins	Widely distributed in the CNS and PNS.	Generally inhibitory	The opiates morphine and heroin bind to endorphin receptors on presynaptic neurons and reduce pain by blocking the release of a neurotransmitter.

which a reflex occurs. The reflex arc (figure 8.14) is the basic functional unit of the nervous system because it is the smallest, simplest pathway capable of receiving a stimulus and yielding a response. A reflex arc has five basic components: (1) a **sensory receptor;** (2) a **sensory neuron;** (3) **interneurons,** which are neurons located between and communicating with two other neurons; (4) a **motor neuron;** and (5) an **effector organ.** Most reflexes occur in the spinal cord or brainstem and not the higher brain centers.

The result of a reflex can be seen when a person's finger touches a hot stove. Pain receptors in the skin are stimulated by the hot stove, and action potentials are produced. Sensory neurons conduct the action potentials to the spinal cord, where they synapse with interneurons. The interneurons, in turn, synapse with motor neurons in the spinal cord that conduct action potentials along their axons to flexor muscles in the upper limb. These muscles contract

and pull the finger away from the stove. No conscious thought is required for this reflex, and withdrawal of the finger from the stimulus begins before the person is consciously aware of any pain.

Neuronal Pathways

Neurons are organized within the CNS to form pathways ranging from relatively simple to extremely complex. The two simplest pathways are converging and diverging pathways. **Converging pathways** have two or more neurons that synapse with (converge on) the same neuron (figure 8.15a). This allows information transmitted in more than one neuronal pathway to converge into a single pathway. In **diverging pathways,** the axon from one neuron divides (diverges) and synapses with more than one other neuron (figure 8.15b). This allows information transmitted in one neuronal pathway to diverge into two or more pathways.

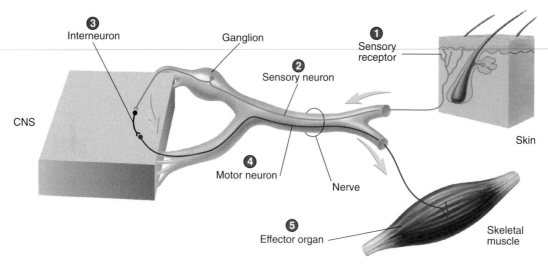

Figure 8.14   Reflex Arc

The parts of a reflex arc are labeled in the order in which action potentials pass through them. The five components are the (1) sensory receptor, (2) sensory neuron, (3) interneuron, (4) motor neuron, and (5) effector organ.

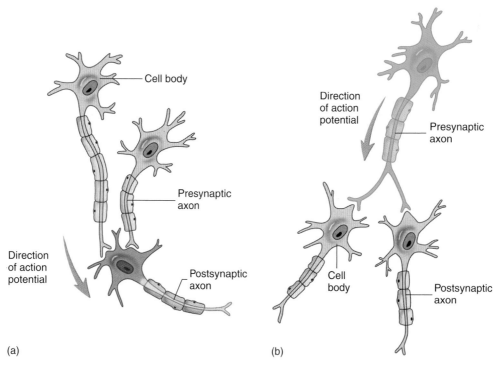

(a)   (b)

Figure 8.15   Convergent and Divergent Pathways

(a) General model of a convergent pathway, showing the neurons converging onto one neuron. (b) General model of a divergent pathway, showing one neuron diverging onto two neurons.

Central and Peripheral Nervous Systems

The **central nervous system (CNS)** consists of the brain and spinal cord. The brain is that part of the CNS housed within the braincase. The spinal cord is in the vertebral column.

The **peripheral nervous system (PNS)** consists of all the nerves and ganglia located outside the brain and spinal cord. The PNS collects information from numerous sources both inside and on the surface of the body and relays it by way of sensory fibers to the CNS, where the information is ignored, triggers a reflex, or is evaluated more extensively. Motor fibers in the PNS relay information from the CNS to muscles and glands in various parts of the body, regulating activity in those structures. Nerves of the PNS can be divided into two groups: 12 pairs of cranial nerves and 31 pairs of spinal nerves.

Spinal Cord

The **spinal cord** extends from the foramen magnum at the base of the skull to the second lumbar vertebra (figure 8.16). Spinal nerves communicate between the spinal cord and the body. The inferior end of the spinal cord and the spinal nerves exiting there resemble a horse's tail and are called the **cauda equina** (kaw′dă; tail; ē-kwī′nă; horse).

A cross section of the spinal cord reveals that the cord consists of a peripheral white matter portion and a central gray matter portion (figure 8.17a). The white matter consists of myelinated axons, and the gray matter is mainly a collection of neuron cell bodies. The white matter in each half of the spinal cord is organized into three **columns** called the **dorsal** (posterior), **ventral** (anterior), and **lateral columns.** Each column of the spinal cord contains ascending and descending tracts. **Ascending tracts** consist of axons that conduct action potentials toward the brain, and **descending tracts** consist of axons that conduct action potentials away from the brain. Ascending tracts

are discussed more fully on page 220. Descending tracts are discussed on page 222.

The gray matter of the spinal cord is shaped like the letter H, with **posterior** and **anterior horns.** Small **lateral horns** exist in levels of the cord associated with the autonomic nervous system. The **central canal** is a fluid-filled space in the center of the cord.

Spinal nerves arise from numerous rootlets along the dorsal and ventral surfaces of the spinal cord (see figure 8.17a). The ventral rootlets combine to form a **ventral root** on the ventral (anterior) side of the spinal cord, and the dorsal rootlets combine to form a **dorsal root** on the dorsal (posterior) side of the cord at each segment. The ventral and dorsal roots unite just lateral to the spinal cord to form a spinal nerve. The dorsal root contains a ganglion, called the **dorsal root ganglion** (gang′glē-on, a swelling or knot).

The cell bodies of unipolar sensory neurons are in the dorsal root ganglia (see figure 8.17b). The axons of these neurons originate in the periphery of the body. They pass through spinal nerves and the dorsal roots to the posterior horn of the spinal cord gray matter. In the posterior horn, the axons either synapse with interneurons or pass into the white matter and ascend or descend in the spinal cord.

The cell bodies of motor neurons, which supply muscles and glands, are located in the anterior and lateral horns of the spinal cord gray matter. Somatic motor neurons are in the anterior horn and autonomic neurons are in the lateral horns. Axons from the motor neurons form the ventral roots and pass into the spinal nerves. Thus, the dorsal root contains sensory axons, the ventral root motor axons, and the spinal nerve has both sensory and motor axons.

P R E D I C T   1
Explain why the dorsal root ganglia are larger in diameter than the spinal nerves.

An example of a converging pathway is one involving a motor neuron in the spinal cord, which stimulates muscle contraction (figure 8.18a). Sensory fibers from pain receptors carry action potentials to the spinal cord and synapse with interneurons, which in turn synapse with a motor neuron. Neurons in the cerebral cortex, controlling conscious movement, also synapse with the same motor neuron by way of axons in descending tracts. Both the interneurons and the neurons in the cerebral cortex have axons that converge onto the motor neuron, which can therefore be stimulated either through the reflex arc or by conscious thought.

An example of a diverging pathway involves sensory neurons within the spinal cord (figure 8.18b). The axon of a sensory neuron carrying action potentials from pain receptors branches within the spinal cord. One branch produces a reflex response by synapsing with an interneuron. The interneuron, in turn, synapses with a motor neuron, which stimulates a muscle to withdraw the injured region of the body from the source of the pain. The other branch synapses with an ascending neuron that carries action potentials through a nerve tract to the brain, where the stimulation is interpreted as pain.

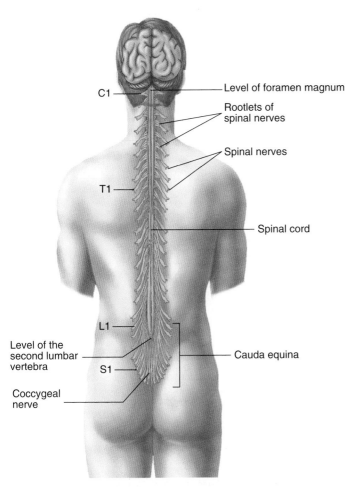

C1 — Level of foramen magnum

Rootlets of spinal nerves

Spinal nerves

T1

Spinal cord

L1

Level of the second lumbar vertebra

S1 — Cauda equina

Coccygeal nerve

Figure 8.16   Spinal Cord and Spinal Nerve Roots

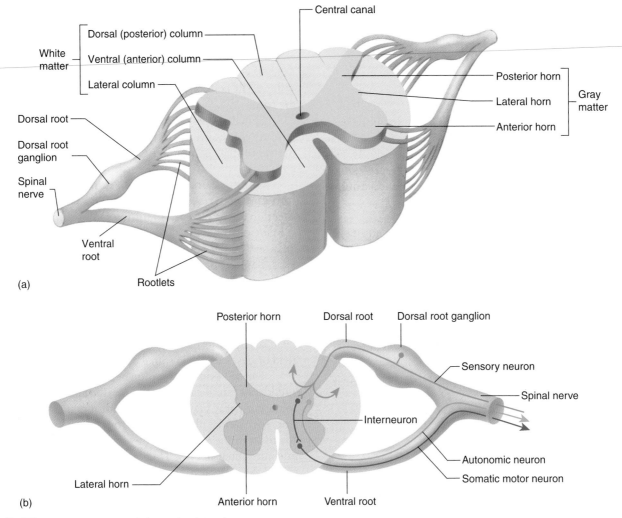

Figure 8.17 Cross Section of the Spinal Cord

(a) A 3-D drawing of a segment of the spinal cord showing one dorsal and one ventral root on each side and the rootlets that form them. (b) Relationship of sensory and motor neurons to the spinal cord.

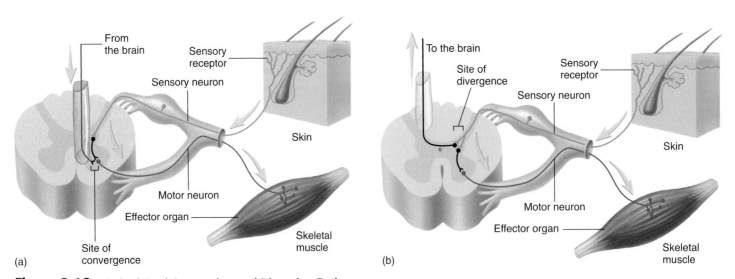

Figure 8.18 Spinal Cord Converging and Diverging Pathways

(a) Example of converging circuit in the spinal cord. A sensory neuron from the periphery and a descending neuron from the brain converge on a single motor neuron. (b) Example of a diverging circuit in the spinal cord. A sensory neuron from the periphery diverges and sends information to a motor neuron by way of an interneuron and sends information to the brain.

Spinal Cord Reflexes

Knee-Jerk Reflex

The simplest reflex is the **stretch reflex,** a reflex in which muscles contract in response to a stretching force applied to them. The **knee-jerk reflex,** or **patellar reflex** (figure 8.19), is a classic example of the stretch reflex. When the patellar ligament is tapped, the quadriceps femoris muscle tendon and the muscles themselves are stretched. Sensory receptors within these muscles are also stretched, and the stretch reflex is activated. Consequently, contraction of the muscles extends the leg, producing the characteristic knee-jerk response.

Descending neurons within the spinal cord synapse with the neurons of the stretch reflex and modulate their activity. This activity is important in maintaining posture and in coordinating muscular activity. Clinicians use the knee-jerk reflex to determine if the higher CNS centers that normally influence this reflex are functional. All spinal reflexes are lost below the level of injury for a few weeks after a severe spinal cord injury. By about 2 weeks after injury, the knee-jerk reflex returns but it is often exaggerated. When the stretch reflex is absent or greatly exaggerated, it indicates that the neurons within the brain or spinal cord that modify this reflex have been damaged.

Withdrawal Reflex

The function of the **withdrawal,** or **flexor reflex,** is to remove a limb or other body part from a painful stimulus. The sensory receptors are pain receptors (see chapter 9). Action potentials resulting from painful stimuli are conducted by sensory neurons through the dorsal root to the spinal cord, where they synapse with interneurons, which in turn synapse with motor neurons (figure 8.20). These neurons stimulate muscles, usually flexor muscles, that remove the limb from the source of the painful stimulus.

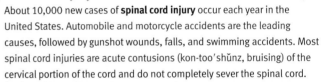

Spinal Cord Injury

About 10,000 new cases of **spinal cord injury** occur each year in the United States. Automobile and motorcycle accidents are the leading causes, followed by gunshot wounds, falls, and swimming accidents. Most spinal cord injuries are acute contusions (kon-too′shŭnz, bruising) of the cervical portion of the cord and do not completely sever the spinal cord.

Spinal cord injuries can interrupt ascending and/or descending tracts. Reflexes can still function below the level of injury but sensations and/or motor functions and reflex modulation may be disrupted.

At the time of spinal cord injury, two types of tissue damage occur: (1) primary mechanical damage, and (2) secondary tissue damage extending into a much larger region of the cord than the primary damage. The only treatment for primary damage is prevention, such as wearing seat belts when riding in automobiles and not diving in shallow water. Once an accident occurs, however, little can be done at present about the primary damage. On the other hand, it is now known that much of the secondary damage can be prevented or reversed. Secondary spinal cord damage, which begins within minutes of the primary damage, is caused by ischemia (lack of blood supply), edema (fluid accumulation), ion imbalances, the release of "excitotoxins" such as glutamate, and inflammatory cell invasion.

With quick treatment, directed at the mechanisms of secondary tissue damage, much of the total damage to the spinal cord can be prevented. Treatment of the damaged spinal cord with large doses of methylprednisolone, a synthetic antiinflammatory steroid, within 8 hours of the injury, can dramatically reduce the secondary damage to the cord. Current treatment includes anatomic realignment and stabilization of the vertebral column, decompression of the spinal cord, and administration of methylprednisolone. Rehabilitation is based on retraining the patient to use whatever residual connections exist across the site of damage.

It had long been thought that the spinal cord is incapable of regeneration following severe damage. It is now known that following injury, most neurons of the adult spinal cord survive and begin to regenerate, growing about 1 mm into the site of damage, but then they regress to an inactive, atrophic state. The major block to adult spinal cord regeneration is the formation of a scar, consisting mainly of astrocytes, at the site of injury. Myelin in the scar is apparently the primary inhibitor of regeneration. Implantation of peripheral nerves, Schwann cells, or fetal CNS tissue can bridge the scar and stimulate some regeneration. Certain growth factors can also stimulate some regeneration. Current research continues to look for the right combination of chemicals and other factors to stimulate regeneration of the spinal cord following injury.

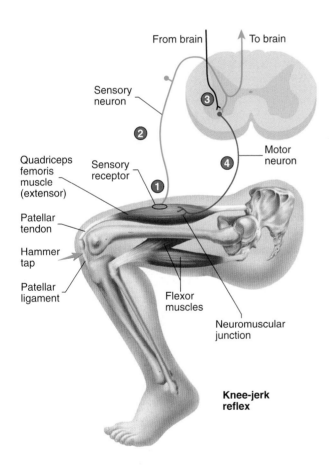

1. Sensory receptors in the muscle detect stretch of the muscle.

2. Sensory neurons conduct action potentials to the spinal cord.

3. Sensory neurons synapse with motor neurons. Descending neurons (*black*) within the spinal cord also synapse with the neurons of the stretch reflex and modulate their activity.

4. Stimulation of the motor neurons causes the muscle to contract and resist being stretched.

Process Figure 8.19 Knee-Jerk Reflex

1. Pain receptors detect a painful stimulus.

2. Sensory neurons conduct action
   potentials to the spinal cord.

3. Sensory neurons synapse with interneurons
   that synapse with motor neurons.

4. Excitation of the motor neurons
   results in contraction of the flexor
   muscles and withdrawal of the limb
   from the painful stimulus.

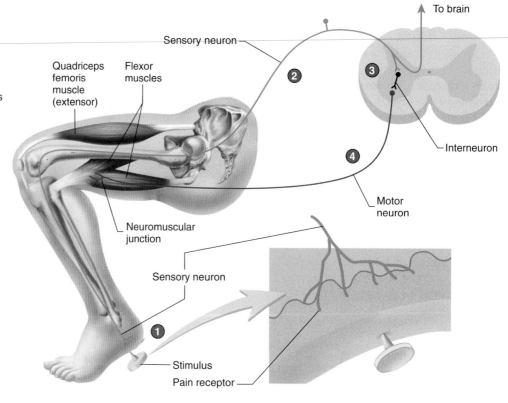

Withdrawal reflex

Process Figure 8.20   Withdrawal Reflex

Spinal Nerves

The **spinal nerves** arise along the spinal cord from the union of the dorsal roots and ventral roots (see figures 8.19 and 8.20). All the spinal nerves are mixed nerves because they contain axons of both sensory and somatic motor neurons. Some spinal nerves also contain parasympathetic or sympathetic axons. Most of the spinal nerves exit the vertebral column between adjacent vertebrae. Spinal nerves are categorized by the region of the vertebral column from which they emerge—cervical (C), thoracic (T), lumbar (L), sacral (S), and coccygeal (Co). The spinal nerves are also numbered (starting superiorly) according to their order within that region. The 31 pairs of spinal nerves are therefore C1 through C8, T1 through T12, L1 through L5, S1 through S5, and Co.

Most of the spinal nerves are organized into three **plexuses** (plek′sŭs-ēz, braids) where nerves come together and then separate: the cervical plexus, the brachial plexus, and the lumbosacral plexus (table 8.3 and figure 8.21). The major nerves of the neck and limbs are branches of these plexuses. Spinal nerves T2 through T11 do not join a plexus. Instead, these nerves extend around the thorax between the ribs, giving off branches to muscles and skin. Motor nerve fibers derived from plexuses innervate skeletal muscles, and sensory nerve fibers in those plexuses supply sensory innervation to the skin overlying those muscles (see table 8.3).

Cervical Plexus

The **cervical plexus** originates from spinal nerves C1 to C4. Branches from this plexus innervate several of the muscles attached to the hyoid bone, as well as the skin of the neck and posterior portion of the head. One of the most important branches of the cervical plexus is the **phrenic nerve,** which innervates the diaphragm. Contraction of the diaphragm is largely responsible for the ability to breathe (see chapter 15).

> **P R E D I C T   2**
> The phrenic nerve may be damaged where it descends along the neck or during open thorax or open heart surgery. Explain how damage to the right phrenic nerve affects the diaphragm. Describe the effect on the diaphragm of completely severing the spinal cord in the thoracic region versus in the upper cervical region.

Brachial Plexus

The **brachial plexus** originates from spinal nerves C5 to T1. Five major nerves emerge from the brachial plexus to supply the upper limb and shoulder. The **axillary nerve** innervates two shoulder muscles and the skin over part of the shoulder. The **radial nerve** innervates all the muscles located in the posterior arm and forearm. It innervates the skin over the posterior surface of the arm, forearm, and hand. The **musculocutaneous** (mŭs′kū-lō-kū-tā′nē-ŭs, muscle + skin) **nerve** innervates the

Table 8.3 Plexuses of the Spinal Nerves (see figure 8.21)

Plexus	Origin	Major Nerves	Muscles Innervated	Skin Innervated
Cervical	C1–C4		Several neck muscles	Neck and posterior head
		Phrenic	Diaphragm	
Brachial	C5–T1	Axillary	Two shoulder muscles	Part of shoulder
		Radial	Posterior arm and forearm muscles (extensors)	Posterior arm, forearm, and hand
		Musculocutaneous	Anterior arm muscles (flexors)	Radial surface of forearm
		Ulnar	Two anterior forearm muscles (flexors), most intrinsic hand muscles	Ulnar side of hand
		Median	Most anterior forearm muscles (flexors), some intrinsic hand muscles	Radial side of hand
Lumbosacral	L1–S4	Obturator	Medial thigh muscles (adductors)	Medial thigh
		Femoral	Anterior thigh muscles (extensors)	Anterior thigh, medial leg and foot
		Ischiadic (sciatic) Tibial	Posterior thigh muscles (flexors), anterior and posterior leg muscles, most foot muscles	Posterior leg and sole of foot
		Common fibular	Lateral thigh and leg, some foot muscles	Anterior and lateral leg, and dorsal (top) part of foot

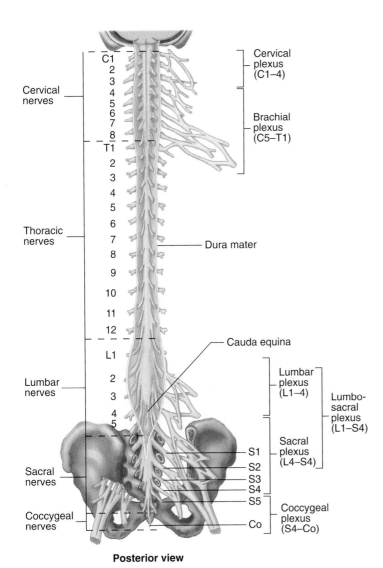

Posterior view

Figure 8.21 Plexuses

anterior muscles of the arm and the skin over the radial surface of the forearm. The **ulnar nerve** innervates two anterior forearm muscles and most of the intrinsic hand muscles. It also innervates the skin over the ulnar side of the hand. The ulnar nerve can be easily damaged where it passes posterior to the medial side of the elbow. The ulnar nerve at this location is called the "funny bone." The **median nerve** innervates most of the anterior forearm muscles and some of the intrinsic hand muscles. It also innervates the skin over the radial side of the hand.

Radial Nerve Damage

The radial nerve lies very close to the medial side of the humerus in the proximal part of the arm and is susceptible to damage in that area. If a person uses crutches improperly, so that the weight of the body is carried in the axilla and upper arm rather than by the hands, the top of the crutch can compress the radial nerve against the humerus. This compression can cause dysfunction of the radial nerve, resulting in paralysis of the posterior arm and forearm muscles, referred to as "crutch paralysis." There is also a loss of sensation over the back of the forearm and hand. The condition is usually temporary as long as the patient begins to use the crutches correctly. The radial nerve can be permanently damaged by a fracture of the humerus in the proximal part of the arm. A sharp edge of the broken bone may cut the nerve, resulting in permanent paralysis unless the nerve is surgically repaired. Because of potential damage to the radial nerve, a broken humerus should be treated very carefully.

Lumbosacral Plexus

The **lumbosacral** (lŭm′bō-sā′krăl, *lumbus,* loin + *sacrum,* sacred) **plexus** originates from spinal nerves L1 to S4. Four major nerves exit the plexus to supply the lower limb. The **obturator** (ob′tū-rā-tŏr, *obturo,* to stop up) **nerve** innervates the muscles of the medial thigh and the skin over the same region. The **femoral nerve** innervates the anterior thigh muscles and the skin over the anterior thigh and medial side of the leg. The **tibial nerve** innervates the posterior thigh muscles, the anterior and posterior leg muscles, and most of the intrinsic foot muscles. It also innervates the skin over the sole of the foot. The **common fibular** (fib′ū-lăr, *fibula,* a clasp or buckle) **nerve** innervates the muscles of the lateral thigh and leg and some intrinsic foot muscles. It innervates the skin over the anterior and lateral leg and the dorsal surface (top) of the foot. The tibial and common fibular nerves are bound together within a connective tissue sheath and together are called the **sciatic** (sī-at′ik, *ischion,* the hip joint) **nerve.**

Brain

The major regions of the brain are the brainstem, the diencephalon, the cerebrum, and the cerebellum (figure 8.22).

Brainstem

The **brainstem** (figure 8.23) connects the spinal cord to the remainder of the brain. It consists of the medulla oblongata, pons, and midbrain and contains several nuclei involved in vital body functions such as the control of heart rate, blood pressure, and breathing. Damage to small areas of the brainstem can cause death, whereas damage to relatively large areas of the cerebrum or cerebellum often do not cause death. Nuclei for all but the first two cranial nerves are also located in the brainstem.

Medulla Oblongata

The **medulla oblongata** (ob′long-gă′tă, *oblongus,* rather long) is the most inferior portion of the brainstem (see figure 8.23) and is continuous with the spinal cord. It extends from the level of the foramen magnum to the pons. In addition to ascending and descending nerve tracts, the medulla oblongata contains discrete nuclei with specific functions such as regulation of heart rate and blood vessel diameter, breathing, swallowing, vomiting, coughing, sneezing, balance, and coordination.

On the anterior surface, two prominent enlargements called **pyramids** extend the length of the medulla oblongata (see figure 8.23). The pyramids consist of descending nerve tracts, which transmit action potentials from the brain to motor neurons of the spinal cord and are involved in the conscious control of skeletal muscles.

P R E D I C T ❸
A large tumor or **hematoma** (hē-mă-tō′mă), a mass of blood that occurs as the result of bleeding into the tissues, can cause increased pressure within the skull. This pressure can force the medulla oblongata downward toward the foramen magnum of the skull. The displacement can compress the medulla oblongata and lead to death. Give two likely causes of death, and explain why they would occur.

Pons

Immediately superior to the medulla oblongata is the **pons.** It contains ascending and descending nerve tracts, as well as several nuclei. Some of the nuclei in the pons relay information between the cerebrum and the cerebellum. The term *pons* means bridge, and it describes both the structure and function of the pons. Not only is the pons a functional bridge between the cerebrum and cerebellum, but on the anterior surface, it resembles an arched footbridge (see figure 8.23*a*). Several nuclei of the medulla oblongata, described earlier, extend into the lower part of the pons, so that functions such as breathing, swallowing, and balance are controlled in the lower pons, as well as in the

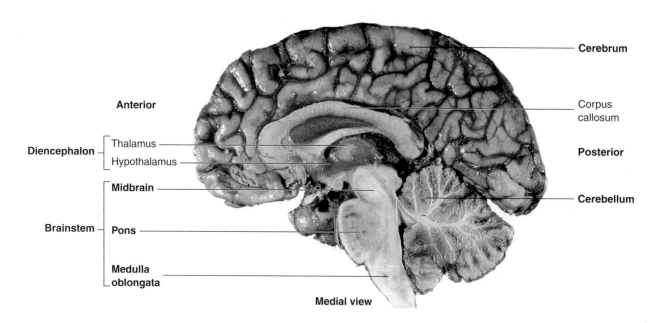

Medial view

Figure 8.22   Regions of the Right Half of the Brain
(as seen in a median section)

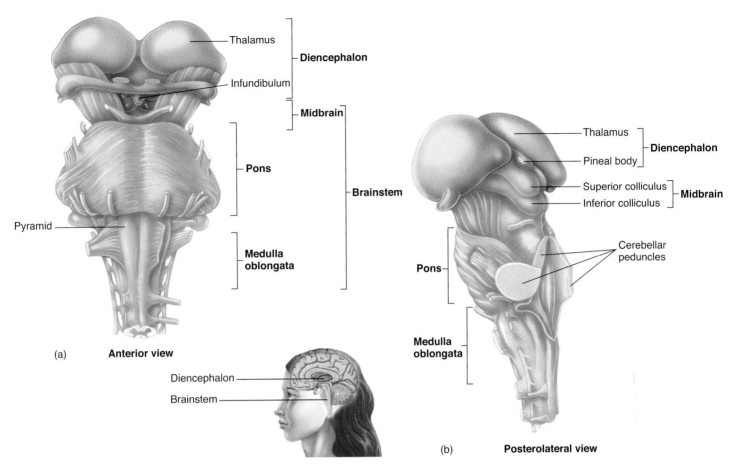

Figure 8.23   Brainstem and Diencephalon

medulla oblongata. Other nuclei in the pons control functions such as chewing and salivation.

Midbrain

The **midbrain,** just superior to the pons, is the smallest region of the brainstem (see figure 8.23*b*). The dorsal part of the midbrain consists of four mounds called the **colliculi** (colliculus, sing., ko-lik′ū-lī, hills). The two inferior colliculi are major relay centers for the auditory nerve pathways in the CNS. The two superior colliculi are involved in visual reflexes. Turning the head toward a tap on the shoulder, a sudden loud noise, or a bright flash of light are reflexes controlled in the superior colliculi. The midbrain contains nuclei involved in the coordination of eye movements and in the control of pupil diameter and lens shape. The midbrain also contains a black nuclear mass, called the **substantia nigra** (sŭb-stan′shē-ă nī′gră, black substance), which is part of the basal nuclei (see the section on Basal Nuclei on p. 223) and is involved in the regulation of general body movements. The rest of the midbrain consists largely of ascending tracts from the spinal cord to the cerebrum and descending tracts from the cerebrum to the spinal cord or cerebellum.

Reticular Formation

Scattered throughout the brainstem is a group of nuclei collectively called the **reticular formation.** The reticular formation plays important regulatory functions in the brain. It is particularly involved in regulating cyclical motor functions such as respiration, walking, and chewing. The reticular formation is a major component of the **reticular activating system,** which plays an important role in arousing and maintaining consciousness and in regulating the sleep–wake cycle. Stimuli such as an alarm clock ringing, sudden bright lights, smelling salts, or cold water being splashed on the face can arouse consciousness. Conversely, removal of visual or auditory stimuli may lead to drowsiness or sleep. General anesthetics function by suppressing the reticular activating system. Damage to cells of the reticular formation can result in coma.

Cerebellum

Cerebellum (ser-e-bel′ŭm) means little brain (see figure 8.22). The cerebellum is attached to the brainstem by several large connections called **cerebellar peduncles** (pe-dŭng′kl, *pes*, foot). These connections provide routes of communication between the cerebellum and other parts of the CNS. The structure and function of the cerebellum are discussed on page 224.

Diencephalon

The **diencephalon** (dī′en-sef′ă-lon, *dia*, through + *enkephalos*, brain) (figure 8.24) is the part of the brain between the brainstem and the cerebrum. Its main components are the thalamus, epithalamus, and hypothalamus.

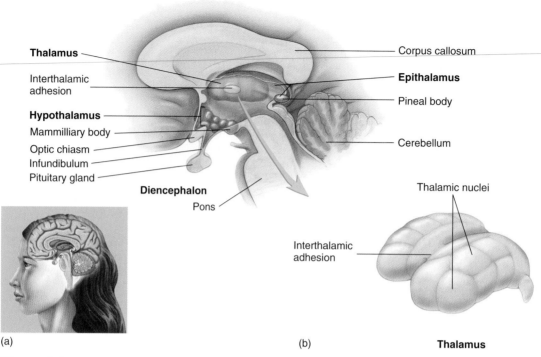

Figure 8.24   Diencephalon

(*a*) Median section of the diencephalon showing the thalamus, epithalamus, and hypothalamus. (*b*) Both halves of the thalamus as seen from an anterior, dorsolateral view with the separations between nuclei depicted by indentations on the surface.

Thalamus

The **thalamus** (thal′ a-mŭs, bedroom) is by far the largest part of the diencephalon. It consists of a cluster of nuclei and is shaped somewhat like a yo-yo, with two large, lateral parts connected in the center by a small **interthalamic adhesion** (see figure 8.24). Most sensory input that ascends through the spinal cord and brainstem projects to the thalamus, where ascending neurons synapse with thalamic neurons. Thalamic neurons, in turn, send their axons to the cerebral cortex. The thalamus also influences mood and registers an unlocalized, uncomfortable perception of pain.

Epithalamus

The **epithalamus** (ep′i-thal′ă-mŭs, *epi*, upon + thalamus) is a small area superior and posterior to the thalamus (see figure 8.24). It consists of a few small nuclei that are involved in the emotional and visceral response to odors, and the pineal body. The **pineal body** (pin′ē-ăl, pinecone-shaped) is an endocrine gland that may influence the onset of puberty. It also may play a role in controlling some long-term cycles that are influenced by the light–dark cycle. The pineal body is known to influence annual behaviors such as migration in birds, as well as changes in fur color and density in some mammals (see chapter 10).

Hypothalamus

The **hypothalamus** is the most inferior part of the diencephalon and contains several small nuclei (see figure 8.24), which are very important in maintaining homeostasis. The hypothalamus plays a central role in the control of body temperature, hunger, and thirst. Sensations such as sexual pleasure, feeling relaxed and "good" after a meal, rage, and fear are related to hypothalamic functions. Emotional responses, which seem to be inappropriate to the circumstances, such as "nervous perspiration" in response to stress or feeling hungry as a result of depression, also involve the hypothalamus. A funnel-shaped stalk, the **infundibulum** (in-fŭn-dib′ū-lŭm, a funnel), extends from the floor of the hypothalamus to the pituitary gland. The hypothalamus plays a major role in controlling the secretion of hormones from the pituitary gland (see chapter 10). The **mamillary** (mam′i-lār-ē, *mamilla*, nipple) **bodies** form externally visible swellings on the posterior portion of the hypothalamus and are involved in emotional responses to odors and in memory.

Cerebrum

The **cerebrum** (ser′ĕ-brŭm, sĕ-rē′brŭm, brain) is the largest part of the brain (figure 8.25). It is divided into left and right hemispheres by a **longitudinal fissure.** The most conspicuous features on the surface of each hemisphere are numerous folds called **gyri** (jī′rī, sing. gyrus, *gyros*, circle), which greatly increase the surface area of the cortex, and intervening grooves called **sulci** (soo′kī, sing. sulcus, a furrow or ditch).

Each cerebral hemisphere is divided into lobes (see figure 8.25), named for the skull bones overlying them. The **frontal lobe** is important in the control of voluntary motor functions, motivation, aggression, mood, and olfactory (smell) reception. The **parietal lobe** is the principal center for the reception and conscious perception of most sensory information, such as touch,

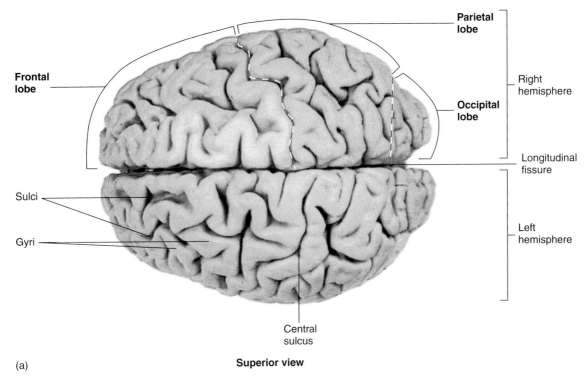

(a)   **Superior view**

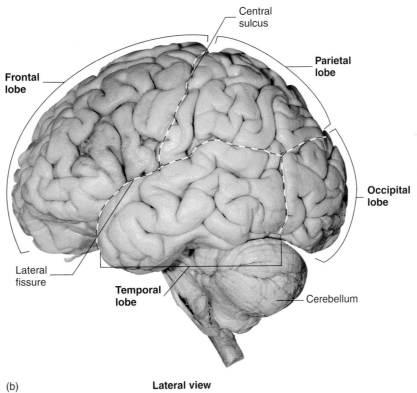

(b)   **Lateral view**

Figure 8.25   The Brain

(*a*) Superior view. (*b*) Lateral view.

pain, temperature, balance, and taste. The frontal and parietal lobes are separated by a prominent sulcus called the **central sulcus.** The **occipital lobe** functions in the reception and perception of visual input and is not distinctly separate from the other lobes. The **temporal lobe** (figure 8.25*b*) is involved in olfactory (smell) and auditory (hearing) sensations and plays an important role in memory. Its anterior and inferior portions are referred to as the "psychic cortex," and they are associated with functions such as abstract thought and judgment. Most of the temporal lobe is separated from the rest of the cerebrum by the **lateral fissure.**

Sensory Functions

The CNS constantly receives large amounts of sensory input in response to a variety of stimuli originating both inside and outside the body. We are unaware of a large part of this sensory input, but it is vital to our survival and normal functions. Sensory input to the brainstem and diencephalon helps maintain homeostasis. Input to the cerebrum and cerebellum keeps us informed of our environment and allows the CNS to control motor functions. A small portion of the sensory input results in **sensation,** the conscious awareness of stimuli (sensation is discussed in chapter 9).

Ascending Tracts

The spinal cord and brainstem contain a number of **ascending** (sensory) **tracts,** or pathways, that transmit action potentials from the periphery to various parts of the brain (table 8.4 and figure 8.26). Each tract is involved with a limited type of sensory input, such as pain, temperature, touch, position, or pressure, because each tract contains axons from specific

sensory receptors specialized to detect a particular type of stimulus (see chapter 9).

Tracts are usually given composite names that indicate their origin and termination. The names of ascending tracts usually begin with the prefix *spino-,* indicating that they begin in the spinal cord. For example, the spinothalamic tract is one that begins in the spinal cord and terminates in the thalamus.

Most ascending tracts consist of two or three neurons in sequence from the periphery to the brain. Almost all neurons relaying information to the cerebrum terminate in the thalamus. Another neuron then relays the information from the thalamus to the cerebral cortex. The **lateral spinothalamic tract,** which transmits action potentials dealing with pain and temperature to the thalamus and on to the cerebral cortex, is an example of an ascending tract. The **dorsal column,** which transmits action potentials dealing with touch, position, and pressure, is another example (figure 8.27).

Table 8.4	Ascending Tracts (see figures 8.26 and 8.27)
Pathway	**Function**
Spinothalamic	Pain, temperature, light touch, pressure, tickle, and itch sensations
Dorsal column	Proprioception, touch, deep pressure, and vibration
Spinocerebellar	Proprioception to cerebellum

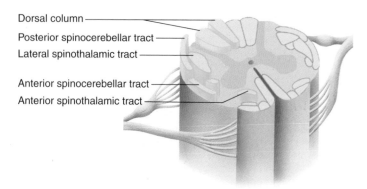

Dorsal column
Posterior spinocerebellar tract
Lateral spinothalamic tract
Anterior spinocerebellar tract
Anterior spinothalamic tract

Figure 8.26 Ascending Tracts of the Spinal Cord

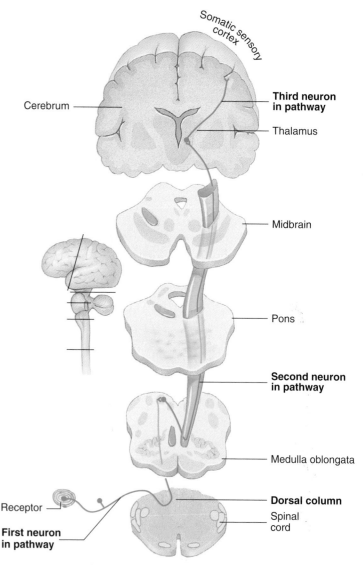

Somatic sensory cortex

Cerebrum

Third neuron in pathway

Thalamus

Midbrain

Pons

Second neuron in pathway

Medulla oblongata

Dorsal column

Spinal cord

Receptor

First neuron in pathway

Figure 8.27 Dorsal Column

The dorsal column transmits action potentials dealing with touch, position, and pressure. Lines on the inset indicate levels of section

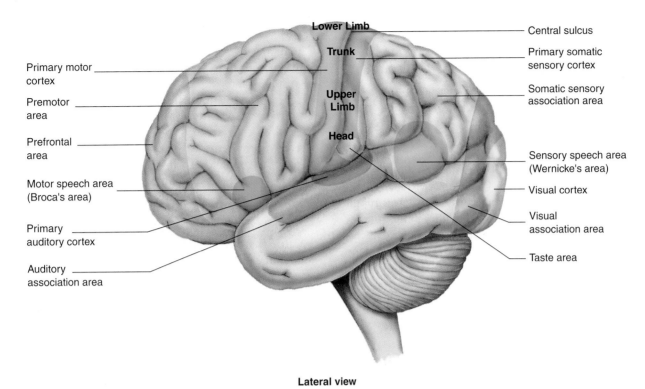

Lateral view

Figure 8.28 Functional Regions of the Lateral Side of the Left Cerebral Cortex

Sensory tracts typically cross from one side of the body in the spinal cord or brainstem to the other side of the body. Thus, the left side of the brain receives sensory input from the right side of the body, and vice versa.

Ascending tracts also terminate in the brainstem or cerebellum. The anterior and posterior **spinocerebellar tracts,** for example, transmit information about body position to the cerebellum (see the Cerebellum section on p. 224).

Sensory Areas of the Cerebral Cortex

Figure 8.28 depicts a lateral view of the left cerebral cortex with some of the sensory and motor areas indicated. The terms area and cortex are often used interchangeably for these regions of the cerebral cortex. Ascending sensory tracts project to specific regions of the cerebral cortex, called **primary sensory areas,** where sensations are perceived. The **primary somatic sensory cortex,** or **general sensory area,** is located in the parietal lobe posterior to the central sulcus. Sensory fibers carrying general sensory input, such as pain, pressure, and temperature, synapse in the thalamus, and thalamic neurons relay the information to the primary somatic sensory cortex. Sensory fibers from specific parts of the body project to specific regions of the primary somatic sensory cortex so that a topographic map of the body, with the head most inferior, exists in this part of the cerebral cortex (see figure 8.28). Other primary sensory areas include the visual cortex in the occipital lobe, the primary auditory cortex in the temporal lobe, and the taste area in the parietal lobe.

Cortical areas immediately adjacent to the primary sensory areas, called **association areas,** are involved in the process of recognition. For example, sensory action potentials originating in the retina of the eye reach the visual cortex, where the image is perceived. Action potentials then pass from the visual cortex to the visual association area, where the present visual information is compared with past visual experience ("Have I seen this before?"). On the basis of this comparison, the visual association area "decides" whether or not the visual input is recognized and judges whether the input is significant. For example, if you pass a man walking down a street, you usually pay less attention to him if you've never seen him before than if you know him, unless some unique characteristic of the unknown person draws your attention. Other examples of association areas include the auditory association area, adjacent to the primary auditory cortex, and the somatic sensory association area, adjacent to the primary somatic sensory cortex.

Motor Functions

The motor system of the brain and spinal cord is responsible for maintaining the body's posture and balance; as well as moving the trunk, head, limbs, tongue, and eyes; and communicating through facial expressions and speech. Reflexes mediated through the spinal cord and brainstem are responsible for some body movements. They occur without conscious thought. **Voluntary movements,** on the other hand, are movements consciously activated to achieve a specific goal, such as walking or typing. Although consciously activated, the details of most voluntary movements occur automatically. After walking begins, it is not necessary to think about the moment-to-moment control of every muscle because neural circuits in the reticular formation automatically control the limbs. After learning how to perform complex tasks, such as typing, they can be performed relatively automatically.

Voluntary movements result from the stimulation of upper and lower motor neurons. **Upper motor neurons** have cell bodies in the cerebral cortex. The axons of upper motor neurons form descending tracts that connect to lower motor neurons. **Lower motor neurons** have cell bodies in the anterior horn of the spinal cord gray matter or in cranial nerve nuclei. Their axons leave the central nervous system and extend through spinal or cranial nerves to skeletal muscles. Lower motor neurons are the neurons forming the motor units described in chapter 7.

Motor Areas of the Cerebral Cortex

The **primary motor cortex** is located in the posterior portion of the frontal lobe, directly anterior to the central sulcus (see figure 8.28). Action potentials initiated in this region control voluntary movements of skeletal muscles. Upper motor neuron axons project from specific regions of this cortex to specific parts of the body, so that a topographic map of the body exists in the primary motor cortex, with the head inferior, analogous to the topographic map of the primary somatic sensory cortex (see figure 8.28). The **premotor area** of the frontal lobe is the staging area where motor functions are organized before they are actually initiated in the primary motor cortex. For example, if a person decides to take a step, the neurons of the premotor area are first stimulated, and the determination is made there as to which muscles must contract, in what order, and to what degree. Action potentials are then passed to the upper motor neurons of the primary motor cortex, which initiate each planned movement.

The motivation and the foresight to plan and initiate movements occur in the anterior portion of the frontal lobes, the **prefrontal area.** This is a region of association cortex that is well developed only in primates, especially in humans. It is involved in motivation and regulation of emotional behavior and mood. The large size of this area in humans may account for our relatively well-developed forethought and motivation and for our emotional complexity.

Frontal Lobotomy

In relation to its involvement in motivation, the prefrontal area is also thought to be the functional center for aggression. Beginning in 1935, one method used to eliminate uncontrollable aggression or anxiety in mental patients was to surgically remove or destroy the prefrontal regions of the brain, a procedure called a **prefrontal, or frontal, lobotomy.** This operation appeared to be successful in eliminating aggression, but most patients developed epilepsy or abnormal personality changes, such as lack of inhibition or a lack of initiative and drive. Later studies failed to confirm the usefulness of lobotomies, and the practice was largely discontinued in the late 1950s.

Descending Tracts

The names of the descending tracts are based on their origin and termination (table 8.5 and figure 8.29). The corticospinal tracts are so named because they begin in the cerebral cortex and terminate in the spinal cord. The corticospinal tracts are referred to as **direct** tracts because they extend directly from upper motor neurons in the cerebral cortex to lower motor neurons in the spinal cord (a similar direct tract extends to lower motor neurons in the brainstem). Other tracts are named after the part of the brainstem from which they originate. Although they originate in the brainstem, these tracts are controlled by the cerebral cortex, basal nuclei, and cerebellum. These tracts are called **indirect** because there is not a direct connection between cortical and spinal neurons.

The descending tracts in the lateral columns (see figures 8.17 and 8.29) are most important for the control of goal-directed limb movements such as reaching and manipulating. The **lateral corticospinal tracts** are especially important in controlling the speed and precision of skilled movements of the hands. The descending tracts in the ventral columns are most important for maintaining posture, balance, and limb position through their control of neck, trunk, and proximal limb muscles.

Table 8.5	**Descending Tracts (see figures 8.29 and 8.30)**
Pathway	**Function**
Direct	
Lateral corticospinal	Muscle tone and skilled movements, especially of the hands
Anterior corticospinal	Muscle tone and movement of trunk muscles
Indirect	
Rubrospinal	Movement coordination
Reticulospinal	Posture adjustment, especially during movement
Vestibulospinal	Posture, balance
Tectospinal	Movement in response to visual reflexes

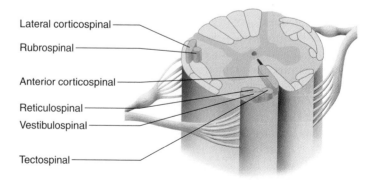

Lateral corticospinal

Rubrospinal

Anterior corticospinal

Reticulospinal

Vestibulospinal

Tectospinal

Figure 8.29   **Descending Tracts of the Spinal Cord**

The lateral corticospinal tract (figure 8.30) begins in the cerebral cortex and descends into the brainstem. At the inferior end of the pyramids of the medulla oblongata, the axons cross over to the opposite side of the body and continue into the spinal cord. Cross over of axons in the brainstem or spinal cord to the opposite side of the body is typical of descending pathways. Thus, the left side of the brain controls skeletal muscles on the right side of the body, and vice versa.

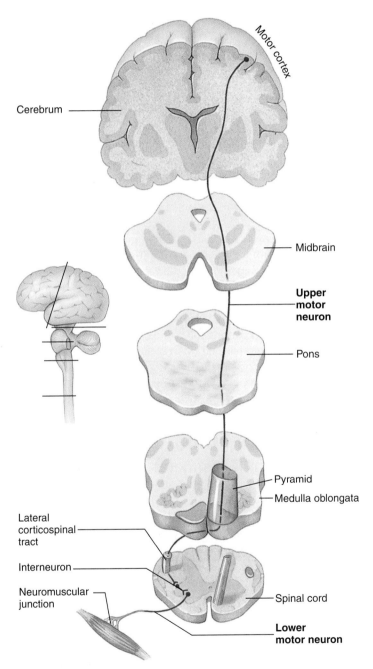

Figure 8.30 **Example of a Direct Tract**

Lateral corticospinal tract, which is responsible for movement below the head. Lines on the inset indicate levels of section.

A CASE IN POINT | Spinal Cord Injury

Harley Chopper was taking a curve on his motorcycle when he hit a patch of loose gravel on the side of the road. His motorcycle slid onto its side. Harley fell off and slid, head-first, as he was spun around, into a retaining wall. In spite of the fact that Harley was wearing a helmet, the compression of hitting the wall with great force injured his neck. A passing motorist stopped at the wreck and saw that Harley was still conscious and breathing. The motorist (wisely) did not move Harley but dialed 911. The paramedics placed Harley on a back-support bed and rushed him to the hospital. Examination there revealed that Harley had suffered a fracture of the fifth and sixth cervical vertebrae. Methylprednisone was administered to reduce inflammation around the spinal cord. His neck vertebrae were realigned and stabilized. Initially, Harley had no feeling or movement in his lower limbs and some feeling but little movement in his upper limbs. He was hospitalized for several weeks. Sensation and motor function gradually returned in all four limbs, but he required many months of physical therapy. Three years later, Harley had still not fully regained his ability to walk. Initially, both ascending and descending tracts experienced a lack of function. This is not uncommon with a spinal cord injury, as the entire cord experiences the "shock" of the trauma. Without rapid, modern treatment, that initial shock can become permanent loss. Sometimes, even with treatment, the damage is so severe as to cause permanent loss of function. In Harley's case, all ascending tracts regained normal function but there was still some functional deficit in the lateral corticospinal tract to the lower limbs after 3 years.

Basal Nuclei

The **basal nuclei** are a group of functionally related nuclei (figure 8.31). Two primary nuclei are the **corpus striatum** (kōr′-pŭs strī-ā′tŭm; striped body), located deep within the cerebrum, and the **substantia nigra,** a group of darkly pigmented cells located in the midbrain.

The **basal nuclei** are important in planning, organizing, and coordinating motor movements and posture. Complex neural circuits link the basal nuclei with each other, with the thalamus, and with the cerebral cortex. These connections form several feedback loops, some of which are stimulatory and others inhibitory. The stimulatory circuits facilitate muscle activity, especially at the beginning of a voluntary movement like rising

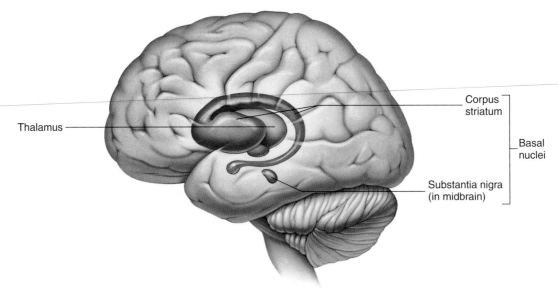

Thalamus

Corpus striatum

Basal nuclei

Substantia nigra (in midbrain)

Lateral view

Figure 8.31 Basal Nuclei

from a sitting position or beginning to walk. The inhibitory circuits facilitate the actions of the stimulatory circuits by inhibiting muscle activity in antagonist muscles. Inhibitory circuits also decrease muscle tone when the body, limbs, and head are at rest. Disorders of the basal nuclei result in difficulty in rising from a sitting position and difficulty in initiating walking. People with basal nuclei disorders exhibit increased muscle tone and exaggerated, uncontrolled movements when they are at rest. A specific feature of some basal nuclei disorders is a "resting tremor," a slight shaking of the hands when a person is not performing a task. Parkinson's disease and cerebral palsy are basal nuclei disorders. They are discussed in the Clinical Focus on "Central Nervous System Disorders"(p. 232).

Cerebellum

The **cerebellum** is attached by cerebellar peduncles to the brainstem (see figure 8.23). The cerebellar cortex is composed of gray matter and has gyri and sulci, but the gyri are much smaller than those of the cerebrum. Internally, the cerebellum consists of gray nuclei and white nerve tracts. The cerebellum is involved in balance, maintenance of muscle tone, and coordination of fine motor movement. If the cerebellum is damaged, muscle tone decreases, and fine motor movements become very clumsy.

A major function of the cerebellum is that of a **comparator** (figure 8.32). Action potentials from the cerebral motor cortex descend into the spinal cord to initiate voluntary movements. Collateral branches are also sent from the motor cortex to the cerebellum, giving information representing the intended movement. Simultaneously, action potentials from proprioceptive neurons reach the cerebellum. **Proprioceptive** (prō-prē-ō-sep′tiv, *proprius,* one's own + *capio,* to take) **neurons** innervate joints, tendons, and muscles, providing information about the position of body parts. The cerebellum compares information about the intended movement from

the motor cortex with sensory information from the moving structures. If a difference is detected, the cerebellum sends action potentials to motor neurons in the motor cortex and the spinal cord to correct the discrepancy. The result is smooth and coordinated movements. For example, if you close your eyes, the cerebellar comparator function allows you to touch your nose smoothly and easily with your finger. If the cerebellum is not functioning, your finger tends to overshoot the target. One effect of alcohol is to inhibit the function of the cerebellum.

Another function of the cerebellum involves learning motor skills such as playing the piano or riding a bicycle. When such a skill is being learned, the cerebellum participates with the cerebrum in learning these highly specialized movements. Once the cerebrum and cerebellum "learn" these skills, the movements can be accomplished smoothly and automatically.

Other Brain Functions
Right and Left Hemispheres

The right cerebral hemisphere receives sensory input from and controls muscular activity in the left half of the body. The left cerebral hemisphere receives input from and controls muscles in the right half of the body. Sensory information received by one hemisphere is shared with the other through connections between the two hemispheres called **commissures** (kom′ĭ-shūrz, a joining together). The largest of these commissures is the corpus callosum (kōr′pus kă-lō′sŭm, callous body), a broad band of nerve tracts at the base of the longitudinal fissure (see figures 8.22 and 8.24).

Language and perhaps other functions, such as artistic activities, are not shared equally between the two hemispheres. The left hemisphere is thought to be the more analytical hemisphere, emphasizing such skills as mathematics and speech. The right hemisphere is thought to be involved more in functions such as three-dimensional or spatial perception and musical ability.

1. The motor area of the cerebral cortex sends action potentials to lower motor neurons in the spinal cord.

2. Action potentials from the motor cortex inform the cerebellum of the intended movement.

3. Lower motor neurons in the spinal cord send action potentials to skeletal muscles, causing them to contract.

4. Proprioceptive signals from the skeletal muscles and joints to the cerebellum convey information concerning the status of the muscles and the structure being moved during contraction.

5. The cerebellum compares the information from the motor cortex to the proprioceptive information from the skeletal muscles and joints.

6. Action potentials from the cerebellum to the spinal cord modify the stimulation from the motor cortex to the lower motor neurons.

7. Action potentials from the cerebellum are sent to the motor cortex, which modify its motor activity.

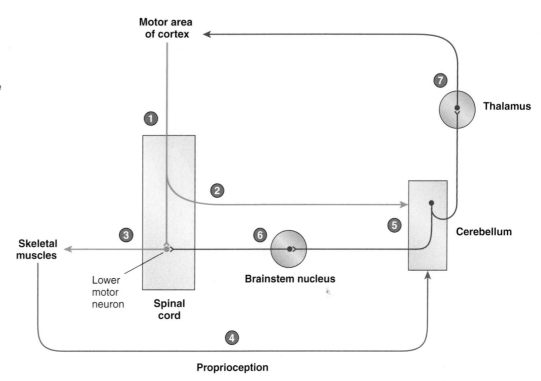

Process Figure 8.32 Cerebellar Comparator Function

Speech

In most people, the speech area is in the left cerebral cortex. Two major cortical areas are involved in speech: the **sensory speech area** (Wernicke's area), a portion of the parietal lobe, and the **motor speech area** (Broca's area) in the inferior portion of the frontal lobe (see figure 8.28). Damage to these parts of the brain or to associated brain regions may result in **aphasia** (ă-fā′zē-ă, *a-*, without + *phasis*, speech), absent or defective speech or language comprehension.

To repeat a word that one hears requires the functional integrity of the following primary pathway. Action potentials from the ear reach the primary auditory cortex, where the word is perceived; the word is recognized in the auditory association area and is comprehended in portions of the sensory speech area. Action potentials representing the word are then conducted through nerve tracts that connect the sensory and motor speech areas. In the motor speech area, the word is formulated as it is to be repeated; action potentials then go to the premotor area, where the movements are programmed, and finally to the primary motor cortex, where specific movements are triggered.

Speaking a written word is somewhat similar. The information enters the visual cortex; passes to the visual association area, where it is recognized; and continues to the sensory speech area, where it is understood and formulated as it is to be spoken. From the sensory speech area it follows the same route for repeating words that one hears: through nerve tracts to the motor speech area, to the premotor area, and then to the primary motor cortex.

> **P R E D I C T 4**
> Describe a neural pathway that allows a blindfolded person to name an object placed in her hand.

Brain Waves

Electrodes placed on a person's scalp and attached to a recording device can record the brain's electrical activity, producing an **electroencephalogram** (ē-lek′trō-en-sef′ă-lō-gram, *elektron*, amber; which can generate static electricity + *enkephalos*, brain + *grapho*, to write) (**EEG**) (figure 8.33). These electrodes are not positioned so that they can detect individual action potentials, but they can detect the simultaneous action potentials in large numbers of neurons. As a result, the EEG displays wavelike patterns known as **brain waves.** This electrical activity is constant, but the intensity and frequency of electrical discharge differ from time to time based on the state of brain activity. Distinct EEG patterns occur with specific brain disorders such as epileptic seizures. Neurologists use these patterns to diagnose and determine the treatment for the disorders.

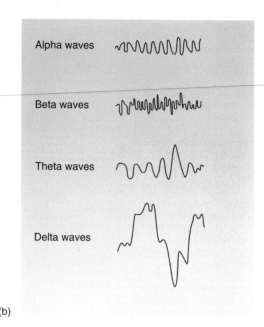

(a)   (b)

Figure 8.33   Electroencephalogram

(*a*) Patient with electrodes attached to his head. (*b*) Four electroencephalographic tracings.
1. Alpha waves, often seen in a relaxed individual with eyes closed.
2. Beta waves, typical of an alert individual.
3. Theta waves, seen in the first stage of sleep.
4. Delta waves, characteristic of deep sleep.

Memory

Memory consists of four processes: *encoding, consolidation, storage,* and *retrieval.* Encoding consists of the brief retention of sensory input received by the brain while something is scanned, evaluated, and acted on. This process occurs in the temporal lobe and lasts less than a second.

Once a given piece of data is encoded, it can be consolidated within the temporal lobe into **short-term memory,** or **working memory,** where information is retained for a few seconds to a few minutes. This memory is limited primarily by the number of bits of information (about seven) that can be stored at any one time. When new information is presented, old information, previously stored in short-term memory, is eliminated. What happens to a telephone number you just looked up if you are distracted? If the temporal lobe is damaged, the transition from encoding to short-term memory may not occur, and the person always lives only in the present and in the more remote past, with memory already stored before the injury. This person is unable to add new memory.

Certain pieces of information are transferred for storage from short-term to **long-term memory,** some of which may last for only a few minutes, and some of which may become permanent. The length of time that memory is stored may depend on how often it is retrieved and used. Long-term memory may involve a physical change in neuron shape. A whole series of neurons, called **memory engrams** or **memory traces,** are probably involved in the long-term retention of a given piece of information, thought, or idea. Rehearsal of information assists in the transfer of information from short-term to long-term memory.

Limbic System

The olfactory cortex and certain deep cortical regions and nuclei of the cerebrum and diencephalon are grouped together under the title **limbic** (lim′bik, a boundary) **system** (figure 8.34). The limbic system responds to olfactory stimulation by initiating responses necessary for survival, such as hunger and thirst. It influences memory, emotions, visceral responses to emotions, motivation, and mood. The limbic system is connected to, and functionally associated with, the hypothalamus. Lesions in the limbic system can result in voracious appetite, increased (often perverse) sexual activity, and docility (including loss of normal fear and anger responses).

Meninges and Cerebrospinal Fluid

Meninges

Three connective tissue membranes, the **meninges** (mĕ-nin′jēz *meninx*, membrane) (figure 8.35), surround and protect the brain and spinal cord. The most superficial and thickest of the meninges is the **dura mater** (doo′rǎ mā′ter, tough mother). Folds of dura mater extend into the longitudinal fissure between the two cerebral hemispheres and between the cerebrum and cerebellum. Within these folds, the dura mater contains spaces called **dural venous sinuses,** which collect blood from the small veins of the brain (see figure 8.34*a*). The dural venous sinuses empty into the internal jugular veins, which exit the skull.

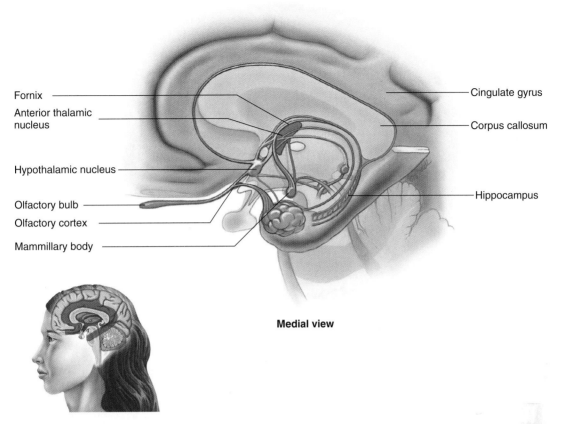

Medial view

Figure 8.34 The Limbic System

The limbic system includes the olfactory cortex, the cingulate gyrus (an area of the cerebral cortex on the medial side of each hemisphere), nuclei such as those of the hypothalamus and thalamus, the hippocampus (a mass of neuron cell bodies in the temporal lobe), and connecting nerve tracts such as the fornix.

The dura mater around the brain is tightly attached to the periosteum of the skull to form a single functional layer. The dura mater of the spinal cord is surrounded by an **epidural space** between the dura mater and the periosteum of the vertebrae (see figure 8.35*b*). **Epidural anesthesia** of the spinal nerves is induced by injecting anesthetics into the epidural space.

The second meningeal membrane is the very thin, wispy **arachnoid** (ă-rak′noyd, spiderlike, i.e.; cobwebs) **mater.** The space between the dura mater and arachnoid mater is the **subdural space,** which is normally only a potential space containing a very small amount of serous fluid.

A CASE IN POINT | Subdural Hematoma

May Fall is a 70-year-old woman who fell in her living room and struck her head on her coffee table. Some days later she complained to her son, Earl E., that she was experiencing headaches and that her left hand seemed weak. Earl also noticed that his mother, who was always very happy and positive, seemed irritable and short-tempered. Earl convinced his mother to see a physician.

Based on May's age and history, her physician ordered a CT scan. The CT scan revealed an image consistent with a subdural hematoma over the right hemisphere. May was scheduled for surgery. After making a large surgical flap, the dura was opened, and the hematoma was removed by suction. After the surgery, May required several months of recovery and physical therapy, but she did recover. Damage to the veins crossing between the cerebral cortex and the dural venous sinuses can cause bleeding into the subdural space, resulting in a subdural hematoma, which can cause pressure on the brain. The pressure can result in decreased brain function in the affected area. The primary motor cortex is located in the posterior portion of the frontal lobe. Pressure on the portion of the right primary motor cortex involved in hand movements can cause decreased function in the left hand. The frontal lobe is also involved with mood, and pressure in that area can result in mood changes, such as a normally happy person becoming irritable. Subdural hematoma is more common in people over age 60 because their veins are less resilient and are more easily damaged.

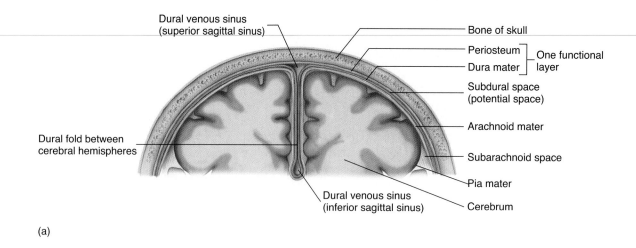

(a)

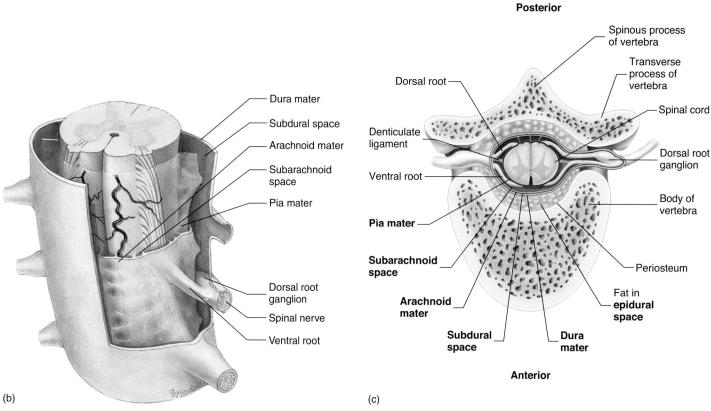

(b)   (c)

Figure 8.35   Meninges

(*a*) Frontal section of head to show the meninges. (*b*) Meningeal membranes surrounding the spinal cord. (*c*) Cross section of spinal cord and vertebra.

The third meningeal membrane, the **pia mater** (pī′ă, or pē′ă, affectionate mother), is very tightly bound to the surface of the brain and spinal cord. Between the arachnoid mater and the pia mater is the **subarachnoid space,** which is filled with cerebrospinal fluid and contains blood vessels.

Subarachnoid Space Around Spinal Cord

The spinal cord extends only to approximately the level of the second lumbar vertebra. Spinal nerves surrounded by meninges extend to the end of the vertebral column. Because there is no spinal cord in the inferior portion of the vertebral canal, a needle can be introduced into the subarachnoid space inferior to the end of the spinal cord to induce **spinal anesthesia** (spinal block), by injecting anesthesia into the area, or to take a sample of cerebrospinal fluid in a **spinal tap,** without damaging the spinal cord. The cerebrospinal fluid can be examined for infectious agents (meningitis) or for blood (hemorrhage). A radiopaque substance can be injected into this area, and a **myelograph** (x-ray film of the spinal cord) can be taken to visualize spinal cord defects or damage.

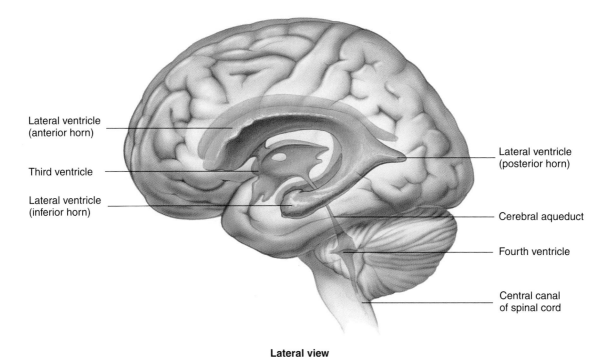

Lateral ventricle
(anterior horn)

Third ventricle

Lateral ventricle
(inferior horn)

Lateral ventricle
(posterior horn)

Cerebral aqueduct

Fourth ventricle

Central canal
of spinal cord

Lateral view

Figure 8.36   Ventricles of the Brain Viewed from the Left

Ventricles

The CNS contains fluid-filled cavities, called **ventricles,** that may be quite small in some areas and large in others (figure 8.36). Each cerebral hemisphere contains a relatively large cavity, called the **lateral ventricle.** The **third ventricle** is a smaller midline cavity located in the center of the diencephalon between the two halves of the thalamus and is connected by foramina (holes) to the lateral ventricles. The **fourth ventricle** is located at the base of the cerebellum and is connected to the third ventricle by a narrow canal, called the **cerebral aqueduct.** The fourth ventricle is continuous with the **central canal** of the spinal cord. The fourth ventricle also opens into the subarachnoid space through foramina in its walls and roof.

Cerebrospinal Fluid

Cerebrospinal fluid (CSF) bathes the brain and spinal cord, providing a protective cushion around the CNS. It is produced by the **choroid** (kō′royd, lacy) **plexuses,** specialized structures made of ependymal cells, which are located in the ventricles (figure 8.37). CSF fills the brain ventricles, the central canal of the spinal cord, and the subarachnoid space. The CSF flows from the lateral ventricles into the third ventricle and then through the cerebral aqueduct into the fourth ventricle. A small amount of CSF enters the central canal of the spinal cord. The CSF exits from the fourth ventricle through small openings in its walls and roof and enters the subarachnoid space. Masses of arachnoid tissue, called **arachnoid granulations,** penetrate into the superior sagittal sinus, a dural venous sinus in the longitudinal fissure, and CSF passes from the subarachnoid space into the blood through these granulations.

Hydrocephalus

Blockage of the openings in the fourth ventricle or of the cerebral aqueduct can result in accumulation of CSF in the ventricles, a condition known as **hydrocephalus.** The accumulation of fluid inside the brain ventricles causes increased pressure that dilates the ventricles and compresses the brain tissue. Compression of the nervous tissue usually results in irreversible brain damage. If the skull bones are not completely ossified when the hydrocephalus occurs, such as in a fetus or newborn, the pressure can also cause severe enlargement of the head. Hydrocephalus is treated by placing a drainage tube (shunt) from the ventricles to the abdominal cavity to eliminate the high internal pressures.

Cranial Nerves

The 12 **cranial nerves** (figure 8.38) are listed in table 8.6. They are designated by Roman numerals from I to XII. There are two general categories of cranial nerve function: sensory and motor. The motor functions of the cranial nerves are further subdivided into somatic motor and parasympathetic. Sensory functions can be divided into the special senses such as vision and the more general senses such as touch and pain in the face. Somatic motor cranial nerves innervate skeletal muscles in the head and neck. Parasympathetic cranial nerves innervate glands, smooth muscle, and cardiac muscle. Some cranial nerves are only sensory, and some are only somatic motor, whereas other cranial nerves are mixed nerves with sensory, somatic motor, and parasympathetic functions.

Three cranial nerves—the olfactory(I), optic(II), and vestibulocochlear (VIII) nerves—are sensory only. Four other cranial nerves—the trochlear (IV), abducent (VI), accessory (XI),

1. Cerebrospinal fluid (CSF) is produced by the choroid plexuses of each of the four ventricles (*inset*).

2. CSF from the lateral ventricles flows to the third ventricle.

3. CSF flows from the third ventricle through the cerebral aqueduct to the fourth ventricle.

4. CSF exits the fourth ventricle through openings in the wall of the fourth ventricle and enters the subarachnoid space. Some CSF enters the central canal of the spinal cord.

5. CSF flows through the subarachnoid space to the arachnoid granulations in the superior sagittal sinus, where it enters the venous circulation (*inset*).

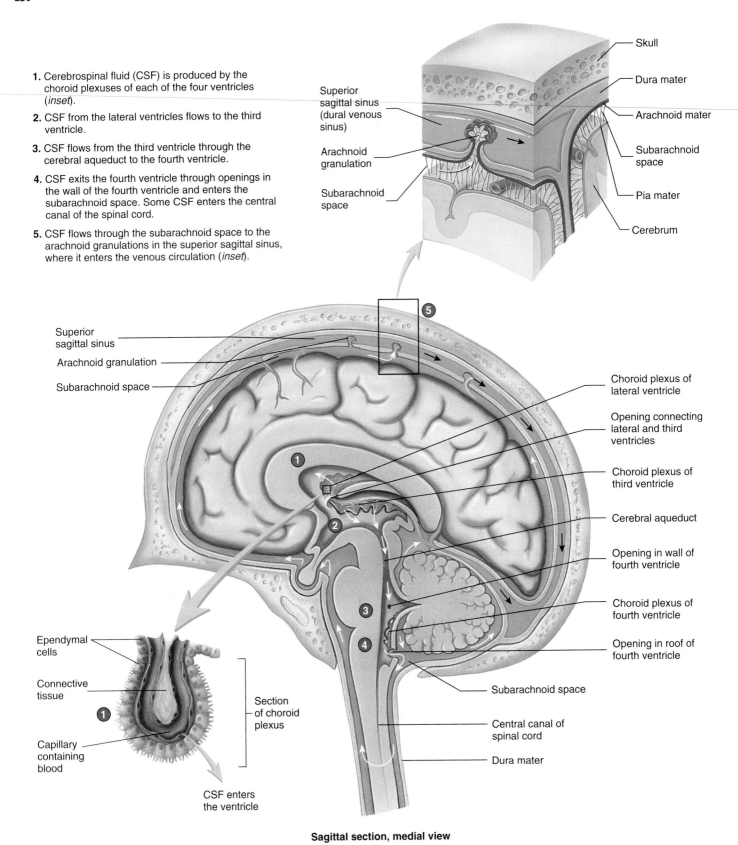

Sagittal section, medial view

Process Figure 8.37   Flow of CSF

CSF flow through the ventricles and subarachnoid space is shown by *white arrows*. Those going through the foramina in the wall and roof of the fourth ventricle depict the CSF entering the subarachnoid space. CSF passes back into the blood through the arachnoid granulations (*white and black arrow*), which penetrate the dural venous sinus. The *black arrows* show the direction of blood flow in the sinuses.

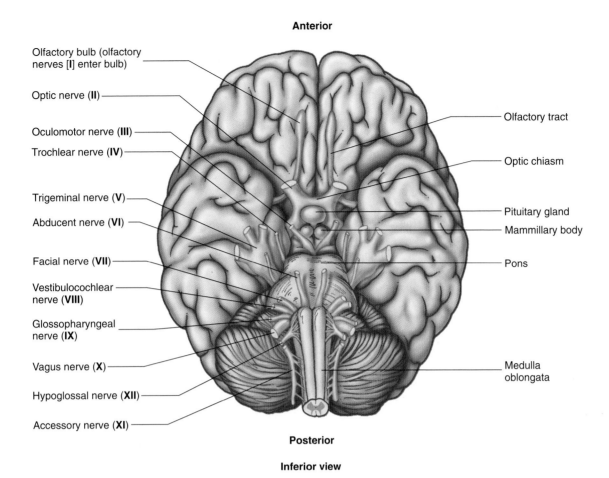

Anterior

Olfactory bulb (olfactory nerves [I] enter bulb)

Optic nerve (**II**)

Oculomotor nerve (**III**)

Trochlear nerve (**IV**)

Trigeminal nerve (**V**)

Abducent nerve (**VI**)

Facial nerve (**VII**)

Vestibulocochlear nerve (**VIII**)

Glossopharyngeal nerve (**IX**)

Vagus nerve (**X**)

Hypoglossal nerve (**XII**)

Accessory nerve (**XI**)

Olfactory tract

Optic chiasm

Pituitary gland

Mammillary body

Pons

Medulla oblongata

Posterior

Inferior view

Figure 8.38   Inferior Surface of the Brain Showing the Origin of the Cranial Nerves

Table 8.6	Cranial Nerves and Their Functions (see figure 8.38)		
Number	**Name**	**General Function***	**Specific Function**
I	Olfactory	S	Smell
II	Optic	S	Vision
III	Oculomotor	M, P	Motor to four of six eye extrinsic muscles and upper eyelid; parasympathetic: constricts pupil; thickens lens
IV	Trochlear	M	Motor to one extrinsic eye muscle
V	Trigeminal	S, M	Sensory to face and teeth; motor to muscles of mastication (chewing)
VI	Abducent	M	Motor to one extrinsic eye muscle
VII	Facial	S, M, P	Sensory: taste; motor to muscles of facial expression; parasympathetic to salivary and tear glands
VIII	Vestibulocochlear	S	Hearing and balance
IX	Glossopharyngeal	S, M, P	Sensory: taste and touch to back of tongue; motor to pharyngeal muscles; parasympathetic to salivary glands
X	Vagus	S, M, P	Sensory to pharynx, larynx, and viscera; motor to palate, pharynx, and larynx; parasympathetic to viscera of thorax and abdomen
XI	Accessory	M	Motor to two neck and upper back muscles
XII	Hypoglossal	M	Motor to tongue muscles

*S, sensory; M, somatic motor; P, parasympathetic.

Infections

Encephalitis (en-sef-ă-lī′tis) is an inflammation of the brain, most often caused by a virus and less often by bacteria or other agents. A large variety of symptoms result, including fever, coma, and convulsions. Encephalitis can also result in death.

Meningitis (men-in-jī′tis) is an inflammation of the meninges. It can be caused by either a viral or a bacterial infection. Symptoms usually include stiffness in the neck, headache, and fever. In severe cases, meningitis can also cause paralysis, coma, or death.

Rabies (rā′bēz) is a viral disease transmitted by the bite of an infected animal such as a skunk, bat, or dog. The rabies virus infects the brain, salivary glands, muscles, and connective tissue. When the patient attempts to swallow, the effort can produce pharyngeal muscle spasms. Sometimes even the thought of swallowing water or the sight of water can induce the spasms. Thus, the term "hydrophobia," which means the fear of water, is applied to the disease. The brain infection results in abnormal excitability, aggression, and, in the later stages, paralysis and death.

Tabes dorsalis (tā′bēz dōr-sā′lis) is a progressive disorder occurring as a result of a syphilis infection. It can occur many years after the initial infection. Tabes means a wasting away, and dorsalis refers to the dorsal roots and dorsal regions of the spinal cord. The symptoms are anesthesia and ataxia. Anesthesia is a loss of sensation resulting from damage of the dorsal roots, and ataxia is the inability to coordinate voluntary muscle activity such as walking. The inability to walk results from a loss of proprioceptive function, which is carried in the dorsal column of the cord. Eventually paralysis develops as the infection spreads.

Tetanus (tet′ă-nŭs) is a disease caused by bacteria (*Clostridium tetani*) found in soil contaminated with animal wastes. It is often introduced into the body through an open wound. The bacteria produce a potent neurotoxin that affects lower motor neurons in the spinal cord and brainstem, as well as interneurons synapsing with those neurons. It binds to the lower motor neurons and causes them to initiate action potentials. It also blocks release of inhibitory neurotransmitters normally released from the interneurons synapsing with the lower motor neurons. As a result, the toxin causes muscle contraction and prevents muscle relaxation, so that the body becomes rigid. The jaw muscles are affected early in the disease, locking the jaw in a closed position. For this reason, tetanus is sometimes referred to as "lockjaw." Death results from spasms in the diaphragm and other respiratory muscles.

Multiple sclerosis (sklĕ-rō′sis) **(MS)** is an autoimmune condition that may be initiated by a viral infection. The disease results in inflammation in areas of the brain and spinal cord. The inflammation, promoted by the immune response, results in localized brain lesions and demyelination of axons in the brain and spinal cord. The myelin sheaths around axons become sclerotic, or hard, resulting in poor conduction of axons in action potentials. Multiple sclerosis exhibits symptomatic periods that are separated by periods of apparent remission. With each recurrence of a symptomatic period, however, many neurons are permanently damaged. Progressive symptoms of the disease include exaggerated reflexes, tremor, nystagmus (tremorous movement of the eyes), and speech defects.

Movement Disorders

Movement disorders may result from disorders of the basal nuclei, cerebellum, or other CNS structures. **Dyskinesias** (dis-ki-nē′zē-ăz), are a group of disorders involving the basal nuclei that result in a resting tremor, as well as in brisk, jerky, purposeless movements, resembling fragments of voluntary movements. **Sydenham's chorea** (St. Vitus' dance) is a disease usually associated with a toxic or infectious disorder that apparently causes temporary dysfunction of the basal nuclei in children. **Huntington's chorea** (kōr-ē′ă) is a dominant hereditary disorder that begins in middle life and causes progressive degeneration of the basal nuclei in affected persons.

Cerebral palsy (pawl′zē) is a general term referring to defects in motor functions or coordination resulting from several types of brain damage, which may be caused by abnormal brain development or birth-related injury. Some symptoms of cerebral palsy are related to basal nuclei dysfunction, such as increased muscle tone and resting tremors. One of the features of cerebral palsy is the presence of slow, writhing, aimless movements. When the face, neck, and tongue muscles are involved, characteristics are grimacing, protrusion and writhing of the tongue, and difficulty in speaking and swallowing.

Parkinson's (par′kin-sonz) **disease,** characterized by muscular rigidity, resting tremor, a slow, shuffling gait, and general lack of movement, is caused by a lesion in another part of the basal nuclei. A resting tremor called "pill-rolling" is characteristic of Parkinson's disease and consists of circular movement of the opposed thumb and index finger tip. The increased muscular rigidity in Parkinson's disease results from defective inhibition of muscle tone by some of the basal nuclei. In this disease, dopamine, an inhibitory neurotransmitter substance, is deficient. Because dopamine cannot cross the barrier, Parkinson's disease is treated (to a limited extent) with L-dopa, a precursor to dopamine that crosses the blood–brain barrier from the capillaries of the brain into the brain tissue. However, there are long-term, negative side effects with L-dopa. As a result, other drugs, such as ropinirole and pramipexole, which have fewer side effects, are now being used to treat the symptoms. Parkinson's disease is also being treated by transplanting fetal cells capable of producing dopamine into the Parkinson patient. Removal of a portion of the corpus striatum, or implantation of an electrical pulse generator to stimulate specific basal nuclei, are now being used to effectively treat Parkinson's disease.

Cerebellar lesions (lĕ'zhŭnz) result in a spectrum of characteristic functional disorders that are essentially opposite of those seen in basilar nuclei dysfunctions. There is a decrease in muscle tone and a tendency to point past a mark that one tries to touch with the finger. A cerebellar tremor is an "intention tremor"; that is, the more carefully one tries to control a given movement, the greater the tremor becomes. For example, when a person with a cerebellar tremor tries to drink a glass of water, the closer the glass comes to the mouth, the shakier the movement becomes. This type of tremor is in direct contrast to basal nuclei tremors described previously, in which the resting tremor largely or completely disappears during purposeful movement.

Other Disorders

Nearly all brain **tumors** (too'mŏrz) develop from neuroglia and not from neurons. Symptoms vary widely, depending on the location of the tumor, but may include headaches, neuralgia (pain along the distribution of a peripheral nerve), paralysis, seizures, coma, and death.

Stroke is a term meaning a sudden blow, suggesting the speed with which this type of defect can occur. It is also referred to clinically as a **cerebrovascular** (ser'ĕ-brō-vas'kŭ-lăr or sĕ-rē'brō-vas'kŭ-lăr) **accident (CVA).** A CVA may be caused by a **hemorrhage** (hem'ŏ-rij), bleeding into the tissue; by a clot, called a **thrombus** (throm'bŭs), in a blood vessel; by a piece of a clot, called an **embolus** (em'bō-lŭs) that has broken loose and floats through the circulation until it reaches and blocks a small vessel; or by **vasospasm** (vā'sō-spazm), constricting the cerebral blood vessels. A hemorrhage, thrombus, embolism, or vasospasm can result in a local area of cell death, called an **infarct** (in'farkt), caused by a lack of blood supply, surrounded by an area of cells that are secondarily affected. Symptoms depend on the location of the stroke and the size of the infarct, but they may include anesthesia (a lack of feeling) or paralysis on the side of the body opposite the cerebral infarct.

Cyclin-dependent kinases (CDK) normally stimulate cell division, especially at sites of injury. In mature neurons, however, CDK causes cell death. Studies have shown that CDK is produced in the cells surrounding a cerebral infarct and that CDK inhibitors can prevent the death of these secondarily effected cells. Aspirin or warfarin treatment can help prevent strokes by preventing blood clotting.

Alzheimer's (ălz'hī-merz) **disease** is a severe type of mental deterioration, or dementia, usually affecting older people, but occasionally affecting people younger than 60. It accounts for half of all dementias; the other half result from drug and alcohol abuse, infections, or strokes. Alzheimer's disease is estimated to affect 10% of all people older than 65 and nearly half of those older than 85.

Alzheimer's disease involves a general decrease in brain size that results from loss of neurons in the cerebral cortex. The gyri become more narrow, and the sulci widen. The frontal lobes and specific regions of the temporal lobes are affected most severely. Symptoms include general intellectual deficiency, memory loss, short attention span, moodiness, disorientation, and irritability.

Localized axonal enlargements, called **amyloid** (am'i-loyd) **plaques,** containing large amounts of β(beta)-amyloid protein, form in the cortex of patients with Alzheimer's disease. There is some evidence that Alzheimer's disease may have characteristics of a chronic inflammatory disease, and antiinflammatory drugs have some effect in treating the disease. Estrogen, which affects some brain functions such as emotion, memory, and cognition may be involved in the disease.

The gene for β-amyloid protein has been mapped to chromosome 21 (see chapter 20), but this gene accounts for only a small portion of the cases. It is noteworthy that people with Down syndrome, which results from having three copies of chromosome 21 (trisomy 21), exhibit the cortical and other changes associated with Alzheimer's disease.

The more common, late-onset form of the disease maps to chromosome 19. A protein, **apolipoprotein E** (ap'ō-lip-ō-prō'tēn; **apo E**), which binds β-amyloid protein has also been associated with Alzheimer's disease. This protein maps to the same part of chromosome 19 as the late-onset form of Alzheimer's. Apo E may also be involved in the regulation of yet another protein, called τ (tau), which is involved in microtubule formation inside neurons. If τ does not function properly, microtubules do not form normally, and the τ proteins become tangled within the neurons, decreasing their function. Nitric oxide production, which stimulates cerebral blood flow and memory in the brain, may help protect cerebral blood vessels and brain tissue from the toxic effects of β-amyloid protein.

Tay-Sachs (tā saks) **disease** is a hereditary lipid-storage disorder of infants that primarily affects neurons of the CNS and results in severe brain dysfunction. Symptoms include paralysis, blindness, and death, usually before age 5.

Epilepsy (ep'i-lep'sē) is actually a group of brain disorders that have seizure episodes in common. The seizure, a sudden massive neuronal discharge, can be either partial or complete, depending on the amount of brain involved and whether or not consciousness is impaired. The neuronal discharges may stimulate muscles innervated by the nerves involved, resulting in involuntary muscle contractions (i.e., convulsions).

Headaches have a variety of causes that can be grouped into two basic classes: extracranial and intracranial. Extracranial headaches can be caused by inflammation of the paranasal sinuses, dental irritations, eye disorders, or tension in the muscles moving the head and neck. Intracranial headaches may result from inflammation of the brain or meninges, ventricular enlargement, vascular changes, mechanical damage, or tumors.

Anesthesia, Neuritis, and Neuralgias

Anesthesia (an'es-thē'zē-ă, *a-*, without + *aistesis,* sensation) is the loss of sensation. It may be a pathological condition, or it may be induced temporarily to facilitate surgery or some other medical action. **Neuritis** (noo-rī'tis, *neuri,* nerve + *-itis,* inflammation) is an inflammation of a nerve resulting from any one of a number of causes, including injury or infection. In motor nerves, neuritis can result in the loss of motor function. In sensory nerves, neuritis can result in anesthesia or neuralgia. **Neuralgia** is a general term meaning nerve pain. It involves severe spasms of throbbing or stabbing pain along the pathway of a nerve. Neuralgia can result from inflammation or nerve damage, or it may be of unknown cause. **Trigeminal neuralgia** involves the trigeminal nerve and consists of sharp bursts of pain in the face. **Facial palsy** involves the facial nerve and results in unilateral paralysis of the facial muscles. The affected side of the face droops because of the absence of muscle tone. **Ischiadica** (is-kē'ad-ik-ă, sciatica) is a neuralgia of the ischiadic nerve, with pain radiating down the back of the thigh and leg. The most common cause of ischiadica is a herniated lumbar disk putting pressure on the spinal nerves forming the lumbosacral plexus.

Infections

Leprosy (lep'rō-sē, *lepros,* scaly), or Hansen's disease, is a bacterial disease that kills the skin and other tissues, such as cells of the peripheral nervous system. Disfiguring nodules form on the body, and tissue necrosis occurs. Even though the disease can be treated with sulfone drugs, millions of people in Asia and Africa are still affected by it. Leprosy is not highly contagious and is usually transmitted from lesions or clothing of an infected person through cuts or abrasions on the skin of another. The time from infection to the appearance of symptoms may be several years. Leprosy itself is usually not fatal, but patients may die from complications.

Herpes (her'pēz, *herpo,* to creep) is a family of viral diseases characterized by skin lesions. The viruses apparently reside in the ganglia of sensory nerves and cause lesions along the course of the nerves. Herpes simplex I causes lesions in the area of the lips and nose. The lesions are prone to occur during times of decreased resistance, such as during a cold. For this reason they are called cold sores or fever blisters. Herpes simplex II (genital herpes) is responsible for a sexually transmitted disease causing lesions on the external genitalia. Another herpes-type virus causes **chickenpox** in children and **shingles,** also called **herpes zoster,** in older adults.

Poliomyelitis (pō'lē-ō-mī' ĕlī'tis, polio, *polios,* gray) is a viral infection of the CNS, but it damages the somatic motor neurons, which extend into the PNS. Many somatic motor neurons degenerate, leaving muscles without innervation. Without stimulation from the CNS, the muscles are paralyzed, and they atrophy, or waste away.

Other Disorders

Myotonic dystrophy (mī-ō-tō'nik dis'trō-fē, *myo-,* muscle + *tonos,* tension; *dys-,* bad + *trophe,* nourishment) is a dominant hereditary disease characterized by muscle weakness, dysfunction, and atrophy and by visual impairment as a result of nerve degeneration.

Neurofibromatosis (noor'ō-fī-brō-mă-tō'sis) is also a genetic disorder, in which neurofibromas (benign tumors along peripheral nerve tracts) occur in early childhood and result in large skin growths, which can result in substantial disfigurement.

Myasthenia gravis (mī-as-thē'nē-ă gra'vis, *mys,* muscle + *astheneia,* weakness; *gravis,* heavy) is an autoimmune disorder in which the immune system attacks neuromuscular junctions. The receptors for acetylcholine are destroyed, which makes the neuromuscular junction less functional, resulting in muscle weakness and increased fatigability. Myasthenia gravis can eventually lead to complete muscle paralysis.

---

and hypoglossal (XII) nerves—are considered to be somatic motor only (although these motor nerves have position function).

The trigeminal nerve (V) has sensory and somatic motor functions. It has the greatest general sensory distribution of all the cranial nerves and is the only cranial nerve supplying sensory information to the brain from the skin of the face. Sensory information from the skin over all the rest of the body is carried to the CNS by spinal nerves. Injections of anesthetic administered by a dentist are designed to block sensory transmission carried through branches of the trigeminal nerve from the teeth. These dental branches of the trigeminal nerve are probably anesthetized more often than any other nerves in the body.

The oculomotor nerve (III) is somatic motor and parasympathetic. The facial (VII), glossopharyngeal (IX), and vagus (X) nerves have all three functions: sensory, somatic motor, and parasympathetic (see table 8.6). The vagus nerve is probably the most important parasympathetic nerve in the body. It helps regulate the functions of the thoracic and abdominal organs such as heart rate, respiration rate, and digestion.

As is the case for the spinal nerves, the sensory and motor functions of many cranial nerves are crossed between the face and the cerebral cortex. Sensory input from the right side of the face via the trigeminal nerve (V), for example, crosses in the brainstem and projects to the left cerebral cortex. Motor output from the left cerebral cortex crosses in the brainstem to the right side of the face via the facial nerve (VII).

Autonomic Nervous System

Somatic motor neuron cell bodies are located in the CNS, and their axons extend from there to skeletal muscles. Axons from autonomic motor neurons, on the other hand, do not extend all the way from the CNS to target tissues. Instead, two neurons in series extend from the CNS to the target organs. The first is called the **preganglionic neuron,** and the second is called the **postganglionic neuron.** The neurons are so named because preganglionic neurons synapse with postganglionic neurons in **autonomic ganglia** outside the CNS. An exception is the preganglionic neuron that extends to the adrenal gland. There

Table 8.7	Sympathetic and Parasympathetic Divisions of the Autonomic Nervous System (see figure 8.39)		
Division	**Location of Preganglionic Cell Body**	**Location of Postganglionic Cell Body**	**Function**
Sympathetic	T1–L2	Sympathetic chain ganglia or collateral ganglia	"Fight-or-flight"; prepares the body for physical activity
Parasympathetic	Cranial nerves III, VII, IX, X; S2–S4 spinal nerves	Terminal ganglia near or embedded in the walls of target organs	Stimulates vegetative activities; slows heart and respiration rates; constricts pupil; thickens lens

the postganglionic neurons are actually the hermone-secreting cells of the adrenal medulla. Autonomic motor neurons innervate smooth muscle, cardiac muscle, and glands. Autonomic functions are largely controlled unconsciously.

The autonomic nervous system is composed of sympathetic and parasympathetic divisions (table 8.7 and figure 8.39) and enteric nervous system. Increased activity in sympathetic neurons generally prepares the individual for physical activity, whereas parasympathetic stimulation generally activates "vegetative" functions, such as digestion, normally associated with the body at rest.

Most organs that receive autonomic motor neurons are innervated by both the parasympathetic and sympathetic divisions. Sweat glands and blood vessels, however, are innervated by sympathetic neurons almost exclusively, whereas the smooth muscles associated with the lens of the eye are innervated primarily by parasympathetic neurons. In most cases, the influence of the two autonomic divisions is opposite on structures that receive dual innervation. For example, sympathetic stimulation of the heart causes an increase in heart rate, whereas parasympathetic stimulation causes a decrease in heart rate.

Sympathetic Division

The **sympathetic** (sim-pă-thet′ik; *sympatheo,* to feel with, + *pathos,* suffering) **division** of the autonomic nervous system prepares a person for physical activity by increasing heart rate and blood pressure, by dilating respiratory passageways, and by stimulating perspiration. The sympathetic division also stimulates the release of glucose from the liver for energy. At the same time, it inhibits digestive activities. The sympathetic division is sometimes referred to as the "fight-or-flight" system because, as it prepares the body for physical activity, it prepares the person to either stand and face a threat or leave as quickly as possible.

Cell bodies of sympathetic preganglionic neurons are in the lateral horn of the spinal cord gray matter (see figure 8.17) between the first thoracic (T1) and the second lumbar (L2) segments. The axons of the preganglionic neurons exit through ventral roots and project to either sympathetic chain ganglia or collateral ganglia. The **sympathetic chain ganglia** are connected to one another and form a chain along both sides of the spinal cord. **Collateral ganglia** are located nearer target organs and consist of the celiac, superior mesenteric, and inferior mesenteric ganglia (see figure 8.39). Some preganglionic neurons synapse in the sympathetic chain ganglia, but others pass through these ganglia and synapse in the collateral ganglia.

Postganglionic neurons arise in the sympathetic chain ganglia or in the collateral ganglia and project to target tissues.

Parasympathetic Division

The **parasympathetic** (par-ă-sim-pa-thet′ik; para, alongside of) **division** of the autonomic nervous system stimulates vegetative activities, such as digestion, defecation, and urination. At the same time, it slows the heart rate and respiration. It also causes the pupil of the eye to constrict and the lens to thicken.

> **PREDICT 5**
> List some of the responses stimulated by the autonomic nervous system in (a) a person who is extremely angry and (b) a person who has just finished eating and is now relaxing.

Preganglionic cell bodies of the parasympathetic division are located either within brainstem nuclei of the oculomotor nerve (III), facial nerve (VII), glossopharyngeal nerve (IX), or vagus nerve (X), or within the lateral part of the central gray matter of the spinal cord in the regions giving rise to spinal nerves S2 through S4.

Axons of the preganglionic neurons extend through spinal nerves or cranial nerves to **terminal ganglia** located either near target organs in the head (see figure 8.39) or embedded in the walls of target organs in the thorax, abdomen, and pelvis. The axons of the postganglionic neurons extend a relatively short distance from the terminal ganglia to the target organ. Most of the thoracic and abdominal organs are supplied by preganglionic neurons of the vagus nerve, extending from the brainstem.

Enteric Nervous System

The enteric nervous system consists of plexuses within the wall of the digestive tract (see figure 16.2). The plexuses include: (1) sensory neurons that connect the digestive tract to the CNS; (2) sympathetic and parasympathetic motor neurons that connect the CNS to the digestive tract; and (3) enteric neurons, located entirely within the enteric plexuses. Enteric neurons are capable of monitoring and controlling the digestive tract independently of the CNS through local reflexes. For example, stretch of the digestive tract is detected by enteric sensory neurons, which stimulate enteric interneurons. The enteric interneurons stimulate enteric motor neurons, which stimulate glands to secrete. Although the enteric nervous system is capable of controlling the activities of the digestive tract completely independently of the CNS, normally the two systems work together.

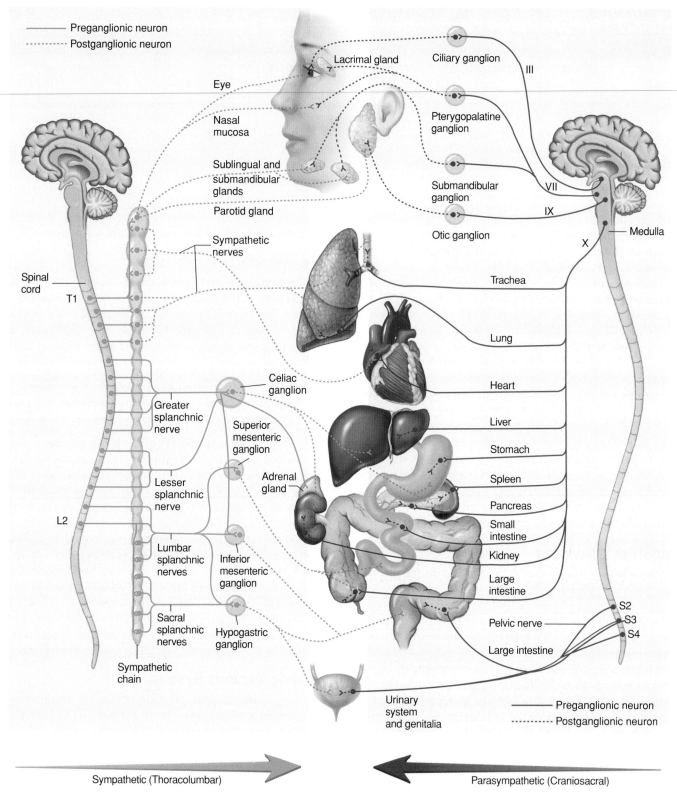

Figure 8.39 Innervation of Organs by the ANS

Preganglionic fibers are indicated by *solid lines*, and postganglionic fibers are indicated by *dashed lines*.

Biofeedback takes advantage of electronic instruments or other techniques to monitor subconscious activities, many of which are regulated by the autonomic nervous system. Skin temperature, heart rate, and brain waves are monitored electronically. By watching the monitor and using biofeedback techniques, a person can learn to consciously reduce his heart rate and blood pressure or reduce the severity of migraine headaches.

Some people use biofeedback methods to relax by learning to reduce the heart rate or change the pattern of brain waves. The severity of peptic (stomach) ulcers, high blood pressure, anxiety, or depression can be reduced by using biofeedback techniques.

Meditation is another technique that influences autonomic functions. It is also claimed that meditation can improve one's spiritual well-being, consciousness, and holistic view of the universe. Some people find meditation techniques to be useful in reducing heart rate, blood pressure, the severity of ulcers, and other symptoms that are frequently associated with stress.

Autonomic Neurotransmitter Substances

All preganglionic neurons of both the sympathetic and parasympathetic divisions and all postganglionic neurons of the parasympathetic division secrete acetylcholine as a neurotransmitter. Most postganglionic neurons of the sympathetic division secrete norepinephrine, but some secrete acetylcholine. Many body functions can be stimulated or inhibited by drugs that either mimic these neurotransmitters or prevent the neurotransmitters from activating their target tissues.

Autonomic Dysfunctions

Raynaud's (rā-nōz′) **disease** involves the spasmodic contraction of blood vessels in the periphery, especially in the digits, and results in pale, cold hands that are prone to ulcerations and gangrene as a result of poor circulation. This condition may be caused by exaggerated sensitivity of the blood vessels to sympathetic stimulation. Cutting the preganglionic neurons is occasionally performed to alleviate the condition.

Dysautonomia (dis′aw-tō-nō′mē-ă) is an inherited condition involving reduced tear secretion, poor vasomotor control, dry mouth and throat, and other symptoms. It is the result of poorly controlled autonomic reflexes.

Effects of Aging on the Nervous System

As a person ages, there's a gradual decline in sensory function because the number of sensory neurons decreases, the function of remaining neurons decreases, and CNS processing decreases. As a result of decreases in the number of skin receptors, elderly people are less conscious of something touching or pressing on the skin and have a more difficult time identifying objects by touch. These changes leave elderly people more prone to skin injuries.

A decreased sense of position of the limbs and in the joints can affect balance and coordination. Information on the position, tension, and length of tendons and muscles also decreases, resulting in additional reduction in the senses of movement, posture, and position, as well as reduced control and coordination of movement.

Other sensory neurons with reduced function include those that monitor blood pressure, thirst, objects in the throat, the amount of urine in the urinary bladder, and the amount of feces in the rectum. As a result, elderly people are more prone to high blood pressure, dehydration, swallowing and choking problems, urinary incontinence, and constipation or bowel incontinence.

There's also a general decline in the number of motor neurons. As many as 50% of the lower motor neurons in the lumbar region of the spinal cord may be lost by age 60. Muscle fibers innervated by the lost motor neurons are also lost, resulting in a general decline in muscle mass. Loss of motor units leads to more rapid fatigue as the remaining units must perform compensatory work.

Reflexes slow as people age because both the generation and conduction of action potentials and synaptic functions slow. The number of neurotransmitters and receptors declines. As reflexes slow, older people are less able to react automatically, quickly, and accurately to changes in internal and external conditions.

The size and weight of the brain decrease as a person ages. At least part of these changes result from the loss of neurons within the cerebrum. The remaining neurons can apparently compensate for much of this loss. Structural changes also occur in neurons. Neuron plasma membranes become more rigid, the endoplasmic reticulum becomes more irregular in structure, neurofibrillar tangles develop in the cells, and amyloid plaques form in synapses. All of these changes decrease the ability of neurons to function.

Short-term memory is decreased in most older people. This change varies greatly among individuals, but, in general, such changes are slow until about age 60 and then become more rapid, especially after age 70. However, the total amount of memory loss is normally not great for most people. The most difficult information for older people to assimilate is that which is unfamiliar and presented verbally and rapidly. Long-term memory appears to be unaffected or even improved in older people.

As with short-term memory, thinking, which includes problem solving, planning, and intelligence, in general, declines slowly to age 60 but more rapidly thereafter. These changes, however, are slight and quite variable. Many older people show no change and about 10% show an increase in thinking ability. Many of these changes are affected by a person's background, education, health, motivation, and experience.

Among older people, more time is required to fall asleep, there are more periods of waking during the night, and the wakeful periods are of greater duration. Factors that can affect sleep include pain, indigestion, rhythmic leg movements, sleep apnea, decreased urinary bladder capacity, and poor peripheral circulation.

Systems Pathology

Stroke

Mr. S is approaching middle age, is somewhat overweight, and has high blood pressure. He was seated on the edge of his couch watching TV, at least most of the time. However, he often jumped to his feet and shouted at the referees for making bad calls. He was surrounded by empty pizza boxes, bowls of chips and salsa, empty beer cans, and full ash trays. As he cheered on his favorite team in a hotly contested big game, which they would be winning easily if it weren't for the lousy officiating, he noticed that he felt drowsy and that the television screen seemed blurry. He began to feel dizzy. As he tried to stand up, he suddenly vomited and collapsed to the floor, unconscious.

Mr. S was rushed to the local hospital where the following signs and symptoms were observed. He exhibited weakness in his limbs, especially on the right side, and ataxia, the inability to walk. He had loss of pain and temperature sensation in his right lower limb and the left side of his face. The dizziness persisted, and he appeared disoriented and lacked attentiveness. He also exhibited hoarseness and dysphagia, the inability to swallow. He had nystagmus, which is a rhythmic oscillation of the eyes. His pupils were slightly dilated, his respiration was short and shallow, and his pulse rate and blood pressure were elevated.

Background Information

Mr. S suffered a "stroke," also referred to as a **cerebrovascular accident (CVA)**. The term **stroke** describes a heterogeneous group of conditions involving death of brain tissue resulting from disruption of its vascular supply. There are two types of stroke: **hemorrhagic stroke,** results from bleeding of arteries supplying brain tissue, and **ischemic** (is-kē′mik) **stroke,** results from blockage of arteries supplying brain tissue (figure A). The blockage in ischemic stroke can result from a thrombus, which is a clot that develops in place within an artery, or an embolism, which is a plug, composed of a detached thrombus or other foreign body, such as a fat globule or gas bubble, which becomes lodged in an artery, blocking it. Mr. S was at high risk for developing a stroke. He was approaching middle age, was overweight, did not exercise enough, smoked, was under stress, and had a poor diet.

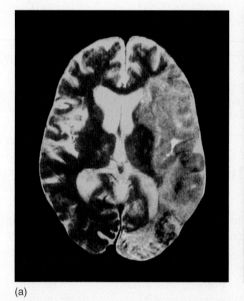 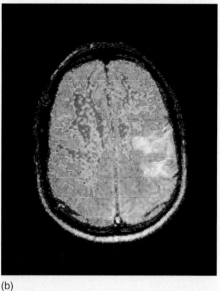

Figure A

(a) MRI of a massive stroke in the left side (the viewer's right) of the brain. (b) Colorized NMR showing disruption of blood flow to the left side (the viewer's right) of the brain (*yellow*). This disruption could cause a stroke.

(a)   (b)

S U M M A R Y

Functions of the Nervous System (p. 198)

The functions of the nervous system include the reception, interpretation, and integration of sensory stimuli; the control of homeostasis; mental activity; and the control of muscles and glands.

Divisions of the Nervous System (p. 198)

1. The central nervous system (CNS) consists of the brain and spinal cord, whereas the peripheral nervous system (PNS) consists of nerves and ganglia.

System Interactions Effects of Stroke on Other Systems

System	Interactions with the Nervous System
Integumentary	Decubitus (dē-kū′bi-tŭs) ulcers (bed sores) result from immobility because of the loss of motor function. Immobility causes compression of blood vessels in areas of the skin, which leads to death of the skin in areas supplied by those vessels.
Skeletal	Loss of bone mass, if muscles are dysfunctional for a prolonged time; in the absence of muscular activity, the bones to which those muscles are attached begin to be resorbed by osteoclasts.
Muscular	Absence of innervation because of damaged pathways or neurons leads to decreased motor function and may result in muscle atrophy.
Endocrine	Strokes in other parts of the brain could involve the hypothalamus, pineal body, or pituitary gland functions.
Cardiovascular	Risks: Phlebothrombosis (fleb′-ō-throm-bō′sis, blood clot in a vein) can occur from inactivity. Edema around the brain can increase the pressure inside the skull. This increased pressure activates reflexes that increase the blood pressure. If the cardioregulatory center in the brain is damaged, death may occur rapidly because of a dramatically decreased blood pressure. Bleeding may result from the use of anticoagulants, and hypotension may result from use of antihypertensives.
Respiratory	Pneumonia may result from aspiration (fluid from the digestive tract entering the lungs when the patient vomits). Breathing may slow or stop completely if the respiratory center is damaged.
Digestive	Vomiting or dysphagia (dis-fā′jē-ă, difficulty swallowing) may occur if the brain centers controlling these functions are damaged. Hypovolemia (decreased blood volume) results from decreased fluid intake because of dysphagia; there may be a loss of bowel control.
Urinary	Motor innervation of the bladder is often affected. Urinary tract infection may result from catheter implantation or from urine retention.
Reproductive	Loss of libido (sex drive); innervation of the reproductive organs is often affected.

The combination of motor loss, which was seen as weakness in his limbs, and sensory loss, seen as loss of pain and temperature sensation in his left lower limb and loss of all sensation in the right side of his face, along with the ataxia, dizziness, nystagmus, and hoarseness suggest that the stroke affected the brainstem and cerebellum. Blockage of the vertebral artery, a major artery supplying the brain, or its branches can result in what is called a lateral medullary infarction (an area of dead tissue resulting from a loss of blood supply to an area). Damage to the descending motor pathways in that area, above the medullary crossover point, results in muscle weakness. Damage to ascending pathways can result in loss of pain and temperature sensation, or other sensory modalities depending on the affected tract. Damage to cranial nerve nuclei results in the loss of pain and temperature sensation in the face, dizziness, blurred vision, nystagmus, vomiting, and hoarseness. These signs and symptoms are not observed unless the lesion is in the brainstem, where these nuclei are located. Some damage to the cerebellum, also supplied by branches of the vertebral artery, can account for the ataxia.

Drowsiness, disorientation, inattentiveness, loss of consciousness, and even seizures are generalized responses to neurological damage. Depression from neurological damage or from discouragement is also common. Slight dilation of the pupils; short, shallow respiration; and increased pulse rate and blood pressure are all signs of Mr. S's anxiety, not about the outcome of the game, but about his current condition and his immediate future. With a loss of consciousness, Mr. S would not remember the last few minutes of what he saw in the game he was watching. People in these circumstances are often worried about how they are going to deal with work tomorrow. They often have no idea that they may be permanently debilitated because the motor and sensory losses may be permanent, or that they will have a long stretch of therapy and rehabilitation ahead.

P R E D I C T

Given that Mr. S exhibited weakness in his right limbs and the loss of pain and the temperature sensation in his right lower limb and the left side of his face, state which side of the brainstem was most severely affected by the stroke. Explain your answer.

---

2. The sensory division of the PNS transmits action potentials to the CNS; the motor division carries action potentials away from the CNS.
3. The somatic motor nervous system innervates skeletal muscle and is mostly under voluntary control. The autonomic nervous system innervates cardiac muscle, smooth muscle, and glands, and it is mostly under involuntary control.
4. The autonomic nervous system is divided into sympathetic, parasympathetic, and enteric portions.

Cells of the Nervous System   (p. 199)
Neurons

1. Neurons receive stimuli and transmit action potentials. A neuron consists of a cell body, dendrites, and an axon.
2. Neurons are multipolar, bipolar, or unipolar.

Neuroglia

Neuroglia are the support cells of the nervous system. They include astrocytes, microglia, ependymal cells, oligodendrocytes, and Schwann cells.

Myelin Sheaths

Axons are either unmyelinated or myelinated.

Organization of Nervous Tissue

Nervous tissue consists of white matter and gray matter. Gray matter forms the cortex and nuclei in the brain and ganglia in the PNS. White matter forms nerve tracts in the CNS and nerves in the PNS.

Electric Signals and Neural Pathways   (p.203)

The Resting Membrane Potential

A resting membrane potential results from the charge difference that exists across the membrane of cells.

Action Potentials

An action potential occurs when the charge across the cell membrane is briefly reversed.

The Synapse

1. The synapse is the point of contact between two neurons or between a neuron and some other cell, such as a muscle or gland cell.
2. An action potential arriving at the synapse causes the release of a neurotransmitter from the presynaptic terminal, which diffuses across the synaptic cleft and binds to the receptors of the postsynaptic membrane.

Reflexes

1. Reflexes are the functional units of the nervous system.
2. A reflex arc consists of a sensory receptor, a sensory neuron, interneurons, a motor neuron, and an effector organ.

Neuronal Pathways

Neuronal pathways are either diverging or converging.

Central and Peripheral Nervous Systems   (p. 210)

The CNS consists of the brain and spinal cord. The PNS consists of nerves and ganglia outside the CNS.

Spinal Cord   (p. 211)

1. The spinal cord extends from the foramen magnum to the second lumbar vertebra below which is the cauda equina.
2. The spinal cord has a central gray part organized into horns and a peripheral white part forming nerve tracts.
3. Roots of spinal nerves extend out of the cord.

Spinal Cord Reflexes

1. The knee-jerk reflex occurs when the quadriceps femoris muscle is stretched.
2. The withdrawal reflex removes a body part from a painful stimulus.

Spinal Nerves   (p. 214)

1. The spinal nerves exit from the cervical, thoracic, lumbar, and sacral regions.
2. The nerves are grouped into plexuses.
3. The phrenic nerve, which supplies the diaphragm, is the most important branch of the cervical plexus.
4. The brachial plexus supplies nerves to the upper limb.
5. The lumbosacral plexus supplies nerves to the lower limb.

Brain   (p. 216)

Brainstem

1. The brainstem contains several nuclei, as well as ascending and descending tracts.
2. The medulla oblongata contains nuclei that control such activities as heart rate, breathing, swallowing, and balance.
3. The pons contains relay nuclei between the cerebrum and cerebellum.
4. The midbrain is involved in hearing and in visual reflexes.
5. The reticular formation is scattered throughout the brainstem and is important in regulating cyclical motor functions. It is involved in maintaining consciousness and in the sleep-wake cycle.

Cerebellum

The cerebellum is attached to the brainstem.

Diencephalon

The diencephalon consists of the thalamus (main sensory relay center), epithalamus (the pineal body may play a role in sexual maturation), and hypothalamus (important in maintaining homeostasis).

Cerebrum

The cerebrum has two hemispheres divided into lobes. The lobes are the frontal, parietal, occipital, and temporal.

Sensory Functions   (p. 220)

1. The CNS constantly receives sensory input.
2. We are unaware of much of the input, but it is vital to survival.
3. Some sensory input results in sensation.

Ascending Tracts

1. Ascending tracts transmit action potentials from the periphery to the brain.
2. Each tract carries a specific type of sensory information.

Sensory Areas of the Cerebral Cortex

1. Ascending tracts project to primary sensory areas of the cerebral cortex.
2. Association areas are involved in recognition of the sensory input.

Motor Functions   (p. 221)

1. Motor functions include involuntary and voluntary movements.
2. Upper motor neurons in the cerebral cortex connect to lower motor neurons in the spinal cord or cranial nerve nuclei.

Motor Areas of the Cerebral Cortex

1. Upper motor neurons are located in the primary motor cortex.
2. The premotor and prefrontal areas regulate movements.

Descending Tracts

Descending tracts project directly from upper motor neurons in the cerebral cortex to lower motor neurons in the spinal cord and brainstem, or project indirectly through basal nuclei or cerebellum.

Basal Nuclei

1. Basal nuclei help plan, organize, and coordinate motor movements and posture.
2. People with basal nuclei disorders exhibit increased muscle tone and exaggerated, uncontrolled movements when at rest.

Cerebellum

1. The cerebellum is involved in balance and muscle coordination. An important function is to compare the intended action with what is occurring and modify the action to eliminate differences.
2. If the cerebellum is damaged, muscle tone decreases, and fine motor movements become very clumsy.

Other Brain Functions   (p. 224)

Right and Left Hemispheres

1. Each hemisphere controls the opposite half of the body.
2. Commissures connect the two hemispheres.
3. The left hemisphere is thought to be the dominant analytical hemisphere, and the right hemisphere is thought to be dominant for spatial perception and musical ability.

Speech

Speech involves the sensory speech area, the motor speech area, and the interactions between them and other cortical areas.

Brain Waves

An EEG monitors brain waves, which are a summation of the electrical activity of the brain.

Memory

Memory consists of sensory (less than 1 second), short-term (lasting a few minutes), and long-term (permanent) memory.

Limbic System

The limbic system includes the olfactory cortex, deep cortical regions, and nuclei. It responds to olfactory stimulation and is involved with memory, motivation, mood, and other visceral functions.

Meninges and Cerebrospinal Fluid   (p. 226)

Meninges

Three connective tissue meninges cover the CNS: the dura mater, arachnoid mater, and pia mater.

Ventricles

1. The brain and spinal cord contain fluid-filled cavities: the lateral ventricles in the cerebral hemispheres, a third ventricle in the diencephalon, a cerebral aqueduct in the midbrain, a fourth ventricle at the base of the cerebellum, and a central canal in the spinal cord.
2. The fourth ventricle has openings into the subarachnoid space.

Cerebrospinal Fluid

Cerebrospinal fluid is formed in the choroid plexuses in the ventricles, it exits through the fourth ventricle, and it reenters the blood through arachnoid granulations in the superior sagittal sinus.

Cranial Nerves   (p. 229)

1. There are 12 cranial nerves: 3 with only sensory function (S), 4 with only somatic motor function (M), 1 with motor and sensory function, 1 with somatic motor and parasympathetic (P) function, and 3 with all three functions. Four of the cranial nerves have parasympathetic function.
2. The cranial nerves are: olfactory (I; S), optic (II; S), oculomotor (III; M, P), trochlear (IV; M), trigeminal (V; S, M), abducens (VI; M), facial (VII; S, M, P), vestibulocochlear (VIII; S), glossopharyngeal (IX; S, M, P), vagus (X; S, M, P), accessory (XI; M), and hypoglossal (XII; M).

Autonomic Nervous System   (p. 234)

1. The autonomic nervous system contains preganglionic and postganglionic neurons.
2. The autonomic nervous system has sympathetic and parasympathetic divisions, and the enteric nervous system.

Sympathetic Division

1. The sympathetic division is involved in preparing the person for action by increasing heart rate, blood pressure, and respiration rate.
2. Preganglionic cell bodies of the sympathetic division lie in the thoracic and upper lumbar regions of the spinal cord.
3. Postganglionic cell bodies are located in the sympathetic chain ganglia or in collateral ganglia.

Parasympathetic Division

1. The parasympathetic division is involved in vegetative activities, such as the digestion of food, defecation, and urination.
2. Preganglionic cell bodies of the parasympathetic division are associated with some of the cranial and sacral nerves.
3. Postganglionic cell bodies are located in terminal ganglia, located either near or within target organs.

Enteric Nervous System

1. The enteric nervous system forms plexuses in the digestic tract wall.
2. Enteric neurons are sensory, motor, or interneurons, and receive CNS input.

Autonomic Neurotransmitter Substances

1. All autonomic preganglionic and parasympathetic postganglionic neurons secrete acetylcholine.
2. Most sympathetic postganglionic neurons secrete norepinephrine.

Effects of Aging on the Nervous System   (p. 237)

1. There is a general decline in sensory and motor functions with age.
2. Mental functions, including memory, may decline with age, but this varies from person to person.

R E V I E W   A N D   C O M P R E H E N S I O N

1. List the functions of the nervous system.
2. Describe the CNS and PNS.
3. Define the sensory and motor divisions of the PNS, and the somatic motor and autonomic nervous systems.
4. What are the functions of neurons? Name the three parts of a neuron.
5. List the three types of neurons based on their shapes.
6. Define neuroglia. Name and describe the functions of the different neuroglia.
7. What are the differences between unmyelinated and myelinated axons? Which conducts action potentials more rapidly? Why?
8. For nerve tracts, nerves, cortex, nuclei, and ganglia, name the cells or parts of cells found in each, state if they are white or gray matter, and name the part (CNS or PNS) of the nervous system in which they are found.
9. Explain the resting membrane potential and how an action potential is generated.

10. Describe the operation of the synapse, starting with an action potential in the presynaptic neuron and ending with the generation of an action potential in the postsynaptic neuron.

11. Define a reflex, and name the five components of a reflex arc and explain its operation.

12. Describe the two major types of neuronal pathways.

13. Describe the spinal cord gray matter. Where are sensory and motor neurons located in the gray matter?

14. Differentiate among dorsal root, ventral root, and spinal nerve. Which contain sensory fibers, and which contain motor fibers?

15. Describe the knee-jerk and withdrawal reflexes.

16. List the spinal nerves by name and number.

17. Name the main plexuses and the major nerves derived from each.

18. Name the four parts of the brainstem, and describe the general functions of each.

19. Name the three main components of the diencephalon, describing their functions.

20. Name the four lobes of the cerebrum, and describe the location and function of each.

21. List the ascending tracts and state their functions.

22. Describe the locations in the cerebral cortex of the primary sensory areas and of their association areas. How do the association areas interact with the primary areas?

23. Distinguish between upper and lower motor neurons.

24. Distinguish the functions of the primary motor cortex, premotor area, and prefrontal area.

25. List the descending tracts and state their functions.

26. Describe the function of the basal nuclei.

27. Describe the comparator activities of the cerebellum.

28. What are the differences in function between right and left cerebral hemispheres?

29. Describe the process required to speak a word that is seen or heard.

30. Name the three types of memory, and describe the processes that result in long-term memory.

31. What is the function of the limbic system?

32. Name and describe the three meninges that surround the CNS.

33. Describe the production and circulation of the cerebrospinal fluid. Where does the cerebrospinal fluid return to the blood?

34. What are the three principal functional categories of the cranial nerves? List a specific function for each cranial nerve.

35. Define the terms preganglionic and postganglionic neuron.

36. Compare the structure of the somatic motor nervous system and the autonomic nervous system in terms of the number of neurons between the CNS and the effector organs and the types of effector organs.

37. Contrast the functions of the sympathetic and parasympathetic divisions of the autonomic nervous system.

38. What kinds of neurons (sympathetic or parasympathetic, preganglionic or postganglionic) are found in the following:
    a. Cranial nerve nuclei
    b. Lateral horn of the thoracic spinal cord gray matter
    c. Lateral portion of the sacral spinal cord gray matter
    d. Chain ganglia
    e. Ganglia in the wall of an organ

39. List the three parts of the enteric nervous system.

40. List all the reasons why an elderly person may have a difficult time walking from the living room to the bedroom.

C R I T I C A L   T H I N K I N G

1. Given two series of neurons, explain why action potentials could be propagated along one series more rapidly than the other series.

2. The left lung of a cancer patient was removed. To reduce the empty space left in the thorax after the lung was removed, the diaphragm on the left side was paralyzed to allow the abdominal viscera to push the diaphragm upward into the space. What nerve should be cut to paralyze the left half of the diaphragm?

3. Name the nerve that, if damaged, produces the following symptoms:
   a. The elbow and wrist on one side are held in a flexed position and cannot be extended.
   b. The patient is unable to extend the right hip and knee (as in kicking a ball).

4. A patient suffered brain damage in an automobile accident. It was suspected that the cerebellum was the part of the brain affected. On the basis of what you know about cerebellar function, how could

you determine that the cerebellum was involved? What symptoms would you expect to see?

5. Louis Ville was accidentally struck in the head with a baseball bat. He fell to the ground unconscious. Later, when he had regained consciousness, he was unable to remember any of the events that happened during the 10 minutes before the accident. Explain.

6. Name the cranial nerve that, if damaged, produces the following symptoms:
   a. The patient is unable to move the tongue.
   b. The patient is unable to see out of one eye.
   c. The patient is unable to feel one side of the face.
   d. The patient is unable to move the facial muscles on one side.
   e. The pupil of one eye is dilated and does not constrict.

7. Why doesn't injury to the spinal cord at the level of C6 significantly interfere with nervous system control of the digestive system?

Visit this textbook's website at www.mhhe.com/seeleyess6 for practice quizzes, animations, interactive learning exercises, and other study tools.

Chapter Outline and Objectives

General Senses (p. 244)

1. Define sensation; list the types of sensation and the receptor type associated with each.
2. Define and describe pain and referred pain.

Special Senses (p. 246)

3. List the special senses.

Olfaction (p. 246)

4. Describe olfactory neurons and explain what is known about how airborne molecules can stimulate action potentials in the olfactory nerves.

Taste (p. 247)

5. Outline the structure and function of a taste bud.

Vision (p. 248)

6. List the accessory structures of the eye and explain their functions.
7. Name the tunics of the eye, list the parts of each tunic, and give the functions of each part.
8. Explain the differences in function between the rods and cones.
9. Describe the chambers of the eye and the fluids they contain.
10. Explain how images are focused on the retina.

Hearing and Balance (p. 258)

11. Describe the structures of the outer and middle ear, and state the function of each.
12. Describe the anatomy of the cochlea and explain how sounds are detected.
13. Explain how the structures of the vestibule and semicircular canals function in static and kinetic equilibrium.

Effects of Aging on the Special Senses (p. 265)

14. Describe changes that occur in the special senses with aging.

Senses

This is a photomicrograph of an isolated cochlea from the inner ear. This remarkable structure is responsible for detecting the initial stimuli that result in the complexity of sounds we hear. The coiled structure allows receptors that detect a wide range of tones to occupy a very small space.

Sense is the ability to perceive stimuli. The senses are the means by which the brain receives information about the environment and the body. **Sensation,** or **perception,** is the conscious awareness of stimuli received by sensory receptors. The brain constantly receives a wide variety of stimuli from both inside and outside the body. Stimulation of sensory receptors does not immediately result in sensation. Sensory receptors respond to stimuli by generating action potentials that are propagated to the spinal cord and brain. Sensations result when action potentials reach the cerebral cortex. Some other parts of the brain are involved in sensation. For example, the thalamus is involved in the sensation of pain.

Historically, five senses were recognized: smell, taste, sight, hearing, and touch. Today we recognize many more senses. They are divided into two basic groups: general and special senses. The **general senses** are those with receptors distributed over a large part of the body. They are divided into two groups: the somatic and visceral senses. The **somatic senses** provide sensory information about the body and the environment. The **visceral senses,** which provide information about various internal organs, consist primarily of pain and pressure.

Special senses are more specialized in structure and are localized to specific parts of the body. The special senses are smell, taste, sight, hearing, and balance.

General Senses

The **general senses** are widely distributed throughout the body and include the senses of touch, pressure, pain, temperature, vibration, itch, and proprioception (prō-prē-ō-sep′shun), which is the sense of movement and position of the body and limbs.

Receptors (rē-sep′tŏrz) are sensory nerve endings or specialized cells capable of responding to stimuli by developing action potentials. There are several types of receptors associated with both the special and general senses, each responding to a different type of stimulus. **Mechanoreceptors** (mek′ă-nō-rē-sep′tŏrz) respond to mechanical stimuli such as the bending or stretching of receptors, **chemoreceptors** (kem′ō-rē-sep′tŏrz) respond to chemicals such as odor molecules, **photoreceptors** (fō′tō-rē-sep′tŏrz) respond to light, **thermoreceptors** (ther′mō-rē-sep′tŏrz) respond to temperature changes, and **nociceptors** (nō′si-sep′tŏrz, *noceo,* to injure) respond to stimuli that result in the sensation of pain.

Many of the receptors for the general senses are associated with the skin (figure 9.1); others are associated with deeper structures, such as tendons, ligaments, and muscles. Structurally, the simplest and most common type of receptor nerve endings are **free nerve endings,** which are relatively unspecialized neuronal branches similar to dendrites. Free nerve endings are distributed throughout almost all parts of the body. Some free nerve endings respond to painful stimuli, some to temperature, some to itch, and some to movement. Receptors for temperature are either **cold receptors** or **hot receptors.** Cold receptors respond to decreasing temperatures but stop responding at temperatures below 12°C (54°F). Hot receptors respond to increasing temperatures, but stop responding at temperatures above 47°C (117°F). It is sometimes difficult to distinguish very cold from very warm

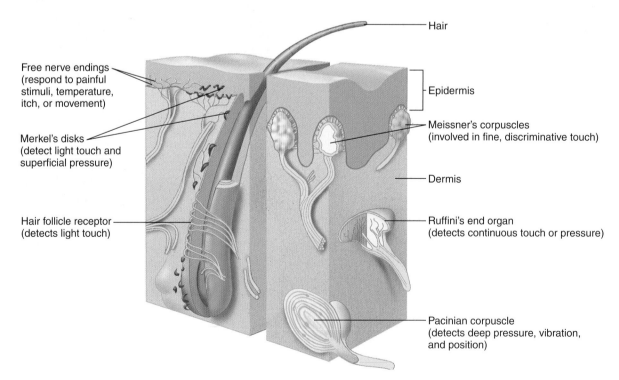

Figure 9.1   **Sensory Nerve Endings in the Skin**

objects touching the skin because only pain receptors are stimulated at temperatures below 12°C or above 47°C.

Touch receptors are structurally more complex than free nerve endings, and many of them are enclosed by capsules (see figure 9.1). **Merkel's disks** are small, superficial nerve endings involved in detecting light touch and superficial pressure. **Hair follicle receptors,** associated with hairs, are also involved in detecting light touch. Light touch receptors are very sensitive but are not very discriminative, meaning that the point being touched cannot be precisely located. Receptors for fine, discriminative touch, called **Meissner's corpuscles,** are located just deep to the epidermis. These receptors are very specific in localizing tactile sensations. Deeper tactile receptors, called **Ruffini's end organs,** play an important role in detecting continuous pressure in the skin. The deepest receptors are the receptors associated with tendons and joints and are called **pacinian corpuscles.** These receptors relay information concerning deep pressure, vibration, and position (proprioception).

Pain

Pain is a sensation characterized by a group of unpleasant perceptual and emotional experiences. There are two types of pain sensation: (1) sharp, well-localized, pricking, or cutting pain resulting from rapidly conducted action potentials and (2) diffuse, burning, or aching pain resulting from action potentials that are propagated more slowly.

Superficial pain sensations in the skin are highly localized as a result of the simultaneous stimulation of pain receptors and tactile receptors, which help to localize the source of the pain stimuli. Deep or visceral pain sensations are not highly localized because of the absence of tactile receptors in the deeper structures. Visceral pain stimuli are normally perceived as diffuse pain.

Action potentials from pain receptors in local areas of the body can be suppressed by chemical anesthetics injected near a sensory receptor or nerve and result in reduced pain sensation. This treatment is called **local anesthesia.** Pain sensations can also be suppressed if loss of consciousness is produced. This is usually accomplished by chemical anesthetics that affect the reticular formation. This treatment is called **general anesthesia.**

Pain sensations can also be influenced by inherent control systems. Sensory axons from tactile receptors in the skin have collateral branches that synapse with neurons in the dorsal horn of the spinal cord. Those neurons, in turn, synapse with and inhibit neurons in the dorsal horn that give rise to the lateral spinothalamic tract. Rubbing the skin in the area of an injury stimulates the tactile receptors, which send action potentials along the sensory axons to the spinal cord. According to the **gate control theory,** these action potentials "close the gate" and inhibit action potentials carried to the brain by the lateral spinothalamic tract. Action potentials carried by the lateral spinothalamic tract can also be inhibited by action potentials carried by descending neurons of the dorsal column system (see chapter 8). These neurons are stimulated by mental or physical activity, especially involving movement of the limbs. The descending neurons synapse with and inhibit neurons in the dorsal horn that give rise to the lateral spinothalamic tract. Vigorous mental or physical activity increases the rate of action potentials in neurons of the dorsal column and can reduce the sensation of pain.

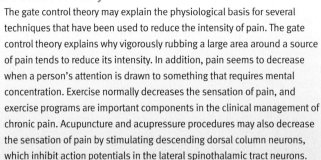

Gate Control Theory

The gate control theory may explain the physiological basis for several techniques that have been used to reduce the intensity of pain. The gate control theory explains why vigorously rubbing a large area around a source of pain tends to reduce its intensity. In addition, pain seems to decrease when a person's attention is drawn to something that requires mental concentration. Exercise normally decreases the sensation of pain, and exercise programs are important components in the clinical management of chronic pain. Acupuncture and acupressure procedures may also decrease the sensation of pain by stimulating descending dorsal column neurons, which inhibit action potentials in the lateral spinothalamic tract neurons.

Referred Pain

Referred pain is a painful sensation perceived to originate in a region of the body that is not the source of the pain stimulus. Most commonly, referred pain is sensed in the skin or other superficial structures when deeper structures such as internal organs are damaged or inflamed (figure 9.2). This occurs because sensory neurons from the superficial area to which the pain is referred and the neurons from the deeper, visceral area where the pain stimulation originates converge onto the same ascending neurons in the spinal cord. The brain cannot distinguish between the two sources of pain stimuli, and the painful sensation is referred to the most superficial structures innervated, such as the skin.

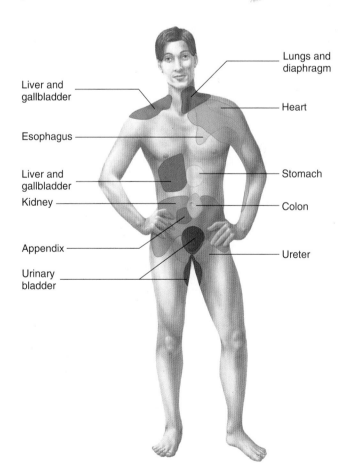

Figure 9.2   **Areas of Referred Pain on the Body Surface**

Pain from the indicated internal organs is referred to the surface areas shown.

Referred pain is clinically useful in diagnosing the actual cause of the painful stimulus. For example, during a heart attack, pain receptors in the heart are stimulated when blood flow is blocked to some of the heart muscle. Heart attack victims, however, often do not feel the pain in the heart, but instead feel what they perceive as cutaneous pain radiating from the left shoulder down the arm (see figure 9.2).

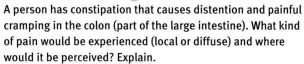

P R E D I C T   1

A person has constipation that causes distention and painful cramping in the colon (part of the large intestine). What kind of pain would be experienced (local or diffuse) and where would it be perceived? Explain.

Phantom Pain

Phantom pain occurs in people who have had appendages amputated. These people may, at times, perceive intense pain in the amputated structure as if it were still there. If a sensory pathway is stimulated at any point, action potentials are initiated and propagated toward the central nervous system. Integration in the cerebral cortex results in the perception of pain that is projected to the site of the sensory receptors for that pathway, even if those sensory receptors are no longer present. A similar phenomenon can be easily demonstrated by bumping the ulnar nerve as it crosses the elbow (the funny bone). A sensation of pain is felt in the fourth and fifth digits, even though the neurons are stimulated at the elbow.

Special Senses

The senses of smell, taste, sight, hearing, and balance are associated with very specialized, localized sensory receptors. The sensations of smell and taste are closely related, both structurally and functionally, and are both initiated by the interaction of chemicals with sensory receptors. The sense of vision is initiated by the interaction of light with sensory receptors. Both hearing and balance function in response to the interaction of mechanical stimuli with sensory receptors. Hearing occurs in response to sound waves, and balance occurs in response to gravity or motion.

Olfaction

The sense of smell, called **olfaction** (ol-fak′shŭn), occurs in response to airborne molecules called **odorants** that enter the nasal cavity (figure 9.3). **Olfactory neurons** are bipolar neurons within the olfactory epithelium lining the superior part of the nasal cavity. The dendrites of the olfactory neurons extend to the epithelial surface of the nasal cavity, and their ends are modified into bulbous enlargements. These enlargements possess long specialized cilia, which lie in a thin mucous film on the epithelial surface. The mucus keeps the nasal epithelium moist, traps and dissolves airborne molecules, and facilitates removal of molecules and particles from the nasal epithelium.

Airborne odorants become dissolved in the mucus on the surface of the epithelium and bind to receptor molecules on the membranes of the specialized cilia. The odorants must first be dissolved in fluid in order to reach the olfactory receptors. The exact nature and site of the interaction is not fully understood, but in some way the combination of odorant with receptors

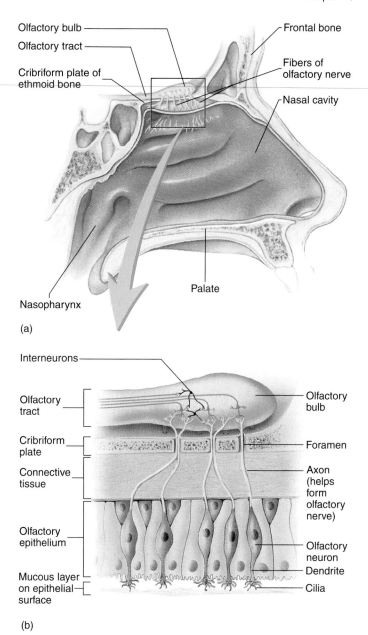

Figure 9.3   Olfactory Epithelium and Bulb

(a) The lateral wall of the nasal cavity (cut in sagittal section), showing the olfactory nerves, olfactory bulb, and olfactory tract. (b) The olfactory neurons within the olfactory epithelium are shown. The axons of olfactory neurons passing through the cribriform plate and the fine structure of the olfactory bulb are also shown.

causes the olfactory neurons to depolarize. The threshold for the detection of odors is very low, so very few odorants bound to an olfactory neuron can initiate an action potential. Once an odorant has become bound to its receptor, however, that receptor does not respond to another odor molecule for some time. It is unlikely that there is a different type of receptor for each of the thousands of detectable odors. It has been proposed that a wide variety of detectable odors are actually combinations of a smaller number of (perhaps as few as seven) primary odors interacting with a limited number of receptor types.

Neuronal Pathways for Olfaction

Axons from olfactory neurons form the olfactory nerves (cranial nerve I), which pass through foramina of the cribriform plate and enter the **olfactory bulb.** There they synapse with interneurons that relay action potentials to the brain through the **olfactory tracts.** Each olfactory tract terminates in an area of the brain called the **olfactory cortex,** located within the temporal and frontal lobes. Within the olfactory bulb and olfactory cortex, feedback loops occur that tend to inhibit transmission of action potentials resulting from prolonged exposure to a given odorant. This feedback, plus the temporary decreased sensitivity at the level of the receptors, results in adaptation to a given odor. For example, if you enter a room that has an odor, you are aware of the odor, but you adapt to the odor and cannot smell it as well after the first few minutes. If you leave the room for some time and then reenter the room, the odor again seems more intense.

Taste

The sensory structures that detect **taste** stimuli are the **taste buds** (figure 9.4). Taste buds are oval structures located on the surface of certain **papillae** (pă-pil′ē, a nipple), which are enlargements on the surface of the tongue. Taste buds are also distributed throughout other areas of the mouth and pharynx, such as on the palate, root of the tongue, and epiglottis. Each taste bud consists of two types of cells. Specialized epithelial cells form the exterior supporting capsule of the taste bud, and the interior of each bud consists of about 40 **taste cells.** Each taste cell contains hairlike processes, called **taste hairs,** that extend into a tiny opening in the surrounding stratified epithelium, called a **taste pore.** Dissolved molecules or ions bind to receptors on the taste hairs and initiate action potentials that are carried by sensory neurons to the parietal lobe of the cerebral cortex.

A CASE IN POINT | Loss of Taste

T. Burnes boiled a pot of water to make some tea. She poured the water over the tea bags in her tea cup and then, absent-mindedly took a large gulp of hot tea, severely burning her tongue. She noticed that for the next several days, sensation in her tongue, including her sense of touch and sense of taste, were severely reduced. Heat damage of the tongue epithelial tissue can cause injury to or even death of epithelial cells, including taste cells in the taste buds. If the epithelial cells are only damaged, taste sensation returns within a few hours to a few days. If the cells die, it takes about two weeks for the epithelial cells to be replaced, including the replacement of taste cells by the supporting cells.

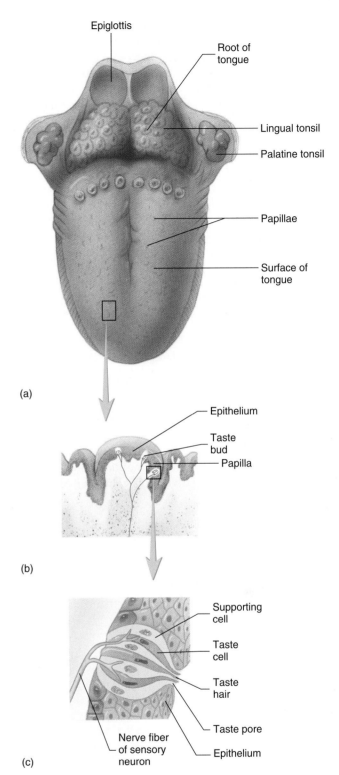

(a)

(b)

(c)

Figure 9.4 **The Tongue**

(*a*) Dorsal surface of the tongue. (*b*) Section through a papilla showing the location of taste buds. (*c*) Enlarged view of a section through a taste bud.

Taste sensations can be divided into five basic types: sour, salty, bitter, sweet, and umami (ū-ma′mē, savory). Even though there are only five primary taste sensations, a fairly large number of different tastes can be perceived, presumably by combining the five basic taste sensations. Although all taste buds are able to detect all five of the basic taste sensations, each taste bud is usually most sensitive to one class of taste stimuli.

Many taste sensations are strongly influenced by olfactory sensations. This influence can be demonstrated by comparing the taste of some food before and after pinching your nose. It is easy to detect that the sense of taste is reduced while the nose is pinched.

Neuronal Pathways for Taste

Taste sensations from the anterior two-thirds of the tongue are carried by the facial nerve (cranial nerve VII). Taste sensations from the posterior third of the tongue are carried by the glossopharyngeal nerve (cranial nerve IX). In addition, the vagus nerve (cranial nerve X) carries some taste sensations from the root of the tongue. Axons from these three cranial nerves synapse in the gustatory (taste) portion of brainstem nuclei. Axons of neurons in these brainstem nuclei synapse in the thalamus and axons from neurons in the thalamus project to the taste area in the parietal lobe of the cerebral cortex (see figure 8.28).

P R E D I C T   2
Why does food not taste as good when a person has a cold?

Vision

The visual system includes the eyes, the accessory structures, and the sensory neurons that project to the cerebral cortex where action potentials conveying visual information are interpreted. Much of the information we obtain about the world around us is detected by the visual system. Our education is largely based on visual input and depends on our ability to read words and numbers. Visual input includes information about light and dark, movement, color, and hue.

Accessory Structures

Accessory structures (figures 9.5 and 9.6) protect, lubricate, and move the eye. They include the eyebrows, eyelids, conjunctiva, lacrimal apparatus, and extrinsic eye muscles.

Eyebrows

The **eyebrows** protect the eyes by preventing perspiration, which can irritate the eyes, from running down the forehead and into them (see figures 9.5 and 9.6). They also help shade the eyes from direct sunlight.

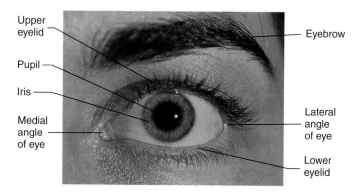

Figure 9.5   The Left Eye and Its Accessory Structures

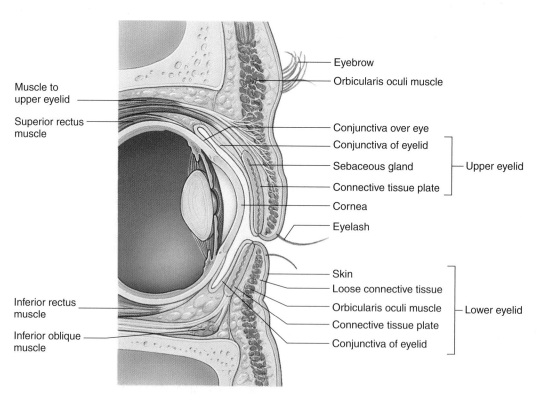

Figure 9.6   Sagittal Section Through the Eye Showing Its Accessory Structures

Eyelids

The **eyelids,** with their associated lashes, protect the eyes from foreign objects (see figures 9.5 and 9.6). If an object suddenly approaches the eye, the eyelids protect the eye by closing and then opening quite rapidly (blink reflex). Blinking, which normally occurs about 20 times per minute, also helps to keep the eyes lubricated by spreading tears over the surface of the eye.

Conjunctiva

The **conjunctiva** (kon-jŭnk-tī′vă, to bind together) is a thin, transparent mucous membrane covering the inner surface of the eyelids and the anterior surface of the eye. Conjunctivitis is an inflammation of the conjunctiva (see the Clinical Focus: Eye Disorders on p. 256).

Lacrimal Apparatus

The **lacrimal** (lak′ri-măl, tear) **apparatus** (figure 9.7) consists of a lacrimal gland situated in the superior lateral corner

of the orbit and a nasolacrimal duct and associated structures in the inferior medial corner of the orbit. The **lacrimal gland** produces tears, which pass over the anterior surface of the eye. Most of the fluid produced by the lacrimal glands evaporates from the surface of the eye, but excess tears are collected in the medial angle of the eye by small ducts called **lacrimal canaliculi** (kan-ă-lik′ū-lī, little canal). These canaliculi open into a **lacrimal sac,** an enlargement of the **nasolacrimal** (nă-zō-lak′ri-măl) **duct,** which opens into the nasal cavity. Tears serve to lubricate the eye and cleanse it. In addition, tears contain an enzyme that helps combat eye infections.

P R E D I C T ❸
Explain why it is often possible to smell (or "taste") medications, such as eyedrops, that have been placed into the eyes.

Extrinsic Eye Muscles

Movement of each eyeball is accomplished by six skeletal muscles called the **extrinsic eye muscles** (figure 9.8). Four of these

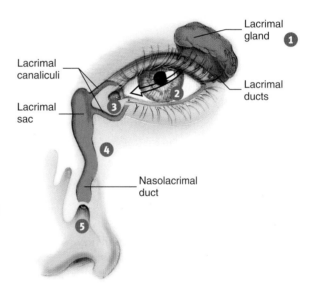

1. Tears are produced in the lacrimal gland and pass through several ducts to the surface of the eye.

2. The tears pass over the surface of the eye.

3. Tears enter the lacrimal canaliculi.

4. Tears are carried through the lacrimal sac and nasolacrimal duct.

5. Tears enter the nasal cavity from the nasolacrimal duct.

Process Figure 9.7 The Lacrimal Apparatus

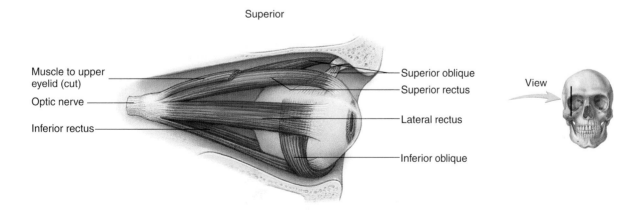

Figure 9.8 Muscles of the Eye
Extrinsic muscles of the right eye as seen from a lateral view with the lateral wall of the orbit removed. The medial rectus muscle cannot be seen from this view.

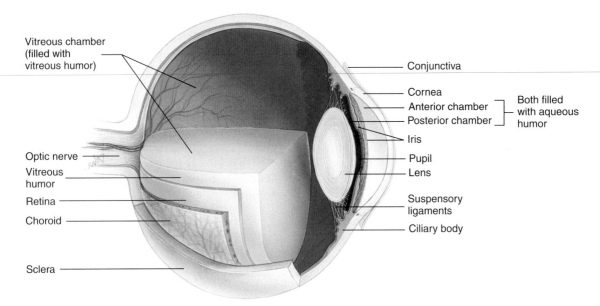

Figure 9.9   Sagittal Section of the Eye Demonstrating Its Layers

muscles run more or less straight from their origins in the posterior portion of the orbit to the eye, to attach to the four quadrants of the eyeball. They are the superior, inferior, medial, and lateral **rectus muscles.** Two muscles, the superior and inferior **oblique muscles,** are located at an angle to the long axis of the eyeball.

Anatomy of the Eye

The eyeball is a hollow, fluid-filled sphere. The sphere has a larger, posterior compartment, which makes up about five-sixths of the eye, and a much smaller anterior compartment, which makes up about one-sixth of the eye. The chambers within the eye and the fluid within each chamber are discussed on pp. 253 and 254.

The wall of the eye consists of three layers, or **tunics** (figure 9.9). The outer, or **fibrous, tunic** consists of the sclera and cornea. The middle, or **vascular, tunic** consists of the choroid, ciliary body, and iris. The inner, or **nervous, tunic** consists of the retina.

Fibrous Tunic

The **sclera** (sklēr′ă, hard) is the firm, white, outer connective tissue layer of the posterior five-sixths of the fibrous tunic. The sclera helps maintain the shape of the eye, protects the internal structure, and provides attachment sites for the extrinsic eye muscles. A small portion of the sclera can be seen as the "white of the eye."

The **cornea** (kōr′nē-ă, hornlike) is the transparent, anterior sixth of the eye that permits light to enter the eye. As part of the focusing system of the fibrous tunic, it also bends, or refracts, the entering light.

Cornea Transplants

The cornea was one of the first organs to be successfully transplanted. Several characteristics make it relatively easy to transplant: it is easily accessible and relatively easily removed from the donor and grafted to the recipient; it does not have blood vessels and therefore does not require the growth of an extensive circulation into the tissue after grafting; and it is less likely to stimulate the immune system and is therefore less likely to be rejected than other tissues.

Vascular Tunic

The middle tunic of the eye is called the **vascular tunic** because it is the layer containing most of the blood vessels of the eye. The posterior portion of the vascular tunic, associated with the sclera, is the **choroid** (kō′royd, like a membrane). This is a very thin structure consisting of a vascular network and many melanin-containing pigment cells, so that it appears black in color. The black color absorbs light so that it is not reflected inside the eye. If light was reflected inside the eye, the reflection would interfere with vision. The interiors of cameras are black for the same reason.

Anteriorly the vascular tunic consists of the ciliary body and iris. The **ciliary** (sil′ē-ar-ē, like an eyelash) **body** is continuous with the anterior margin of the choroid. The ciliary body contains smooth muscles called **ciliary muscles,** which attach to the perimeter of the lens by **suspensory ligaments** (figure 9.10). The **lens** is a flexible, biconvex, transparent disc (see figure 9.9).

The **iris** is the colored part of the eye. It is attached to the anterior margin of the ciliary body, anterior to the lens. The iris is a contractile structure consisting mainly of smooth muscle that surrounds an opening called the **pupil.** Light passes

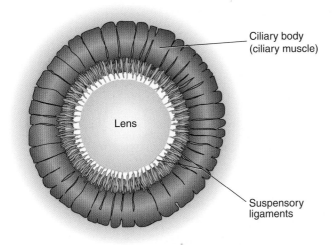

Figure 9.10   The Lens and Ciliary Body

through the pupil, and the iris regulates the diameter of the pupil, which controls the amount of light entering the eye. Parasympathetic stimulation from the oculomotor nerve (cranial nerve III) causes the circular smooth muscles of the iris to contract, resulting in pupillary constriction, whereas sympathetic stimulation causes radial smooth muscles of the iris to contract, resulting in pupillary dilation (figure 9.11). As light intensity increases, the pupil constricts; as light intensity decreases, the pupil dilates.

Nervous Tunic

The **retina,** or **nervous tunic,** is the innermost tunic and it covers the posterior five-sixths of the eye. It consists of an outer **pigmented retina** and an inner **sensory retina** (figure 9.12a). The pigmented retina, with the choroid, keeps light from reflecting back into the eye. The sensory retina contains photoreceptor cells, called **rods** and **cones,** which respond to light. The sensory retina also contains numerous interneurons, some of

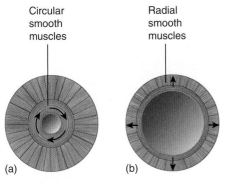

Figure 9.11   The Iris

(a) Circular smooth muscles of the iris constrict the pupil. (b) Radial smooth muscles of the iris dilate the pupil.

which are named in figure 9.12. Over most of the retina, rods are 20 times more common than cones. Rods are very sensitive to light and can function in very dim light, but they do not provide color vision. Cones require much more light, and they do provide color vision. There are three types of cones, each sensitive to a different color: blue, green, or red. The many colors that we can see result from stimulation of combinations of these three types of cones.

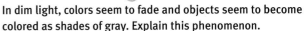

P R E D I C T   4

In dim light, colors seem to fade and objects seem to become colored as shades of gray. Explain this phenomenon.

The outer segments of rod and cone cells are modified by numerous foldings of the cell membrane to form discs (figure 9.12b–d). Rod cells contain a photosensitive pigment called **rhodopsin** (rō-dop′sin, purple pigment), which is made up of the colorless protein **opsin** (op′sin) in loose chemical combination with a yellow pigment called **retinal** (ret′i-năl) (figures 9.12e and 9.13). When light strikes a rod cell, retinal changes shape. This causes opsin to change shape and retinal loses its attachment to the opsin molecule. The change in rhodopsin's shape stimulates a response in the rod cell that results in vision. Retinal then completely detaches from opsin. Energy (ATP) is required to reattach retinal to opsin and, at the same time, to return rhodopsin to the shape that it had before being stimulated by light (see figure 9.13).

Manufacture of retinal in rod cells takes time and requires vitamin A. In bright light, much of the rhodopsin in rod cells is dissociated. When a person goes into a dark building on a bright day, several seconds are required to adjust to the dark as opsin and retinal reassociate to form rhodopsin in the rod cells, which can then react to the dim light. A person with a vitamin A deficiency may have a condition called **night blindness,** which is difficulty in seeing, especially in dim light. Retinal degeneration or detachment may also cause night blindness. Because retinal detachment (see Clinical Focus, p. 256) affects the periphery of the retina, where the rods are located, more than the center of the retina, where the cones are located, night blindness is a major result of retinal detachment.

Cone cells contain slightly different photosensitive pigments, which are sensitive to colored light. Each color results from stimulation by a certain wavelength of light. Three major types of color-sensitive opsin exist: that sensitive to blue, that sensitive to red, and that sensitive to green. Color vision, over a wide range of colors, results from combined stimulation of cone cells containing blue, red, and green opsin.

The rod and cone cells synapse with bipolar cells of the sensory retina (see figure 9.12). These and the horizontal cells of the retina modify the output of the rod and cone cells. For example, this modification is involved in the sharpening of borders between objects of contrasting brightness. The bipolar and horizontal cells synapse with ganglion cells, whose axons converge at the posterior of the eye to form the **optic nerve** (cranial nerve II; see figures 9.9 and 9.12a).

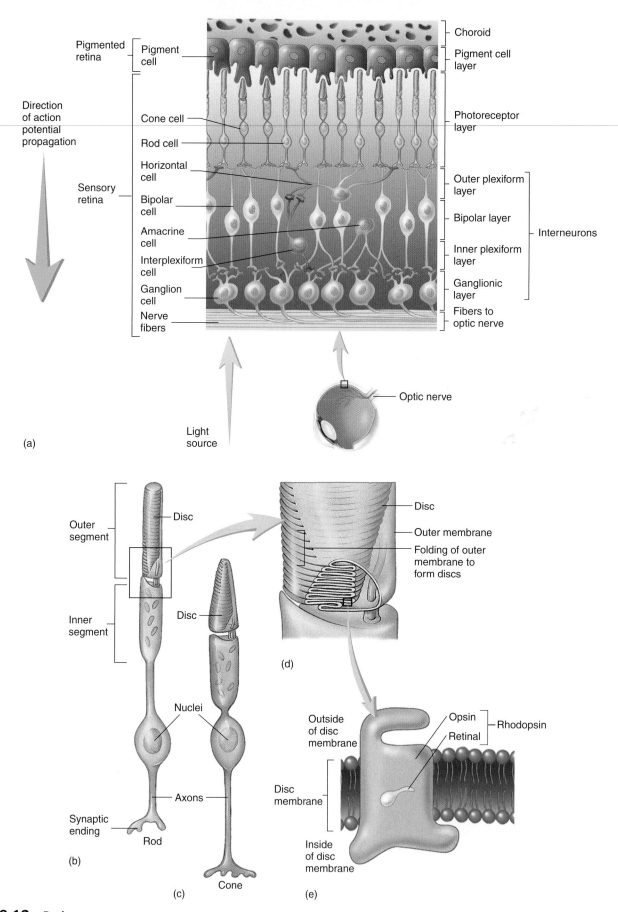

(a)

Pigmented retina

Pigment cell

Direction of action potential propagation

Sensory retina

Cone cell

Rod cell

Horizontal cell

Bipolar cell

Amacrine cell

Interplexiform cell

Ganglion cell

Nerve fibers

Choroid

Pigment cell layer

Photoreceptor layer

Outer plexiform layer

Bipolar layer

Inner plexiform layer

Ganglionic layer

Fibers to optic nerve

Interneurons

Light source

Optic nerve

(b)

Outer segment

Inner segment

Disc

Nuclei

Axons

Synaptic ending

Rod

(c)

Disc

Cone

(d)

Disc

Outer membrane

Folding of outer membrane to form discs

(e)

Outside of disc membrane

Disc membrane

Inside of disc membrane

Opsin

Retinal

Rhodopsin

Figure 9.12 Retina

(*a*) Enlarged section through the retina with its major layers labeled. (*b*) Rod cell. (*c*) Cone cell. (*d*) An enlargement of the discs in the outer segment.
(*e*) An enlargement of one of the discs, showing the relation of rhodopsin to the membrane.

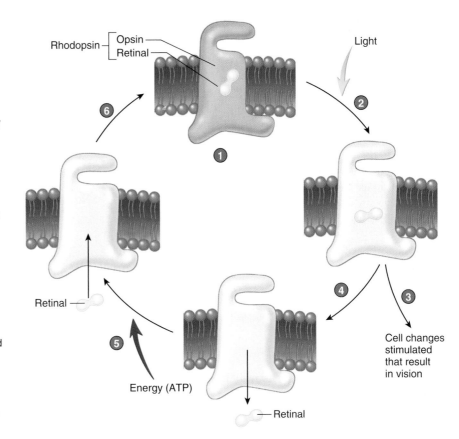

1. Retinal is attached inside opsin to make rhodopsin.

2. Light activates rhodopsin by causing retinal to change shape, which causes opsin to change shape.

3. Activated rhodopsin stimulates cell changes that result in vision.

4. Following rhodopsin activation, retinal detaches from opsin.

5. Energy from ATP is required to bring retinal back to its original form.

6. Retinal attaches to opsin to form rhodopsin (return to step 1).

Process Figure 9.13 Rhodopsin Cycle

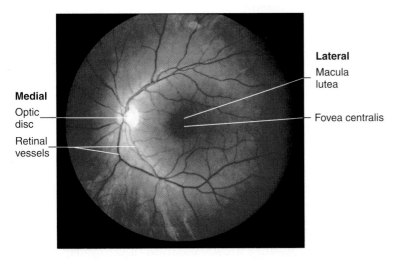

Figure 9.14 Ophthalmoscopic View of the Retina

This view shows the posterior wall of the left eye as seen through the pupil. Notice the vessels entering the eye through the optic disc. The macula lutea, with the fovea centralis in the center, is located lateral to the optic disc.

When the posterior region of the retina is examined with an ophthalmoscope (of-thal′mō-skōp), two major features can be observed: the macula lutea and the optic disc (figure 9.14). Near the center of the posterior retina is a small yellow spot called the **macula lutea** (mak′ū-lă lü′tē-ă,

spot + yellow). In the center of the macula lutea is a small pit, the **fovea** (fō′vē-ă, pit) **centralis.** The fovea centralis is the part of the retina where light is normally most focused when the eye is looking directly at an object. The fovea centralis contains only cone cells.

Just medial to the macula lutea is a white spot, the **optic disc,** through which a number of blood vessels enter the eye and spread over the surface of the retina. This is also the spot at which axons from the retina meet, pass through the outer two tunics, and exit the eye as the optic nerve. The optic disc contains no photoreceptor cells and does not respond to light; it is therefore called the **blind spot** of the eye. A small image projected onto the blind spot cannot be seen. You can demonstrate this by drawing two small dots about 2 inches apart on a card, closing one eye, and holding the card about 1 foot in front of your open eye. As you move the card toward you, focusing on one dot, the other dot seems to disappear.

Chambers of the Eye

The interior of the eye is divided into three chambers: **anterior chamber, posterior chamber,** and **vitreous** (vit′rē-ŭs, glassy), or postremal, **chamber** (see figure 9.9). The anterior and posterior chambers are located between the cornea and the lens. The iris separates the anterior chamber from the posterior chamber, which are continuous with each other through the pupil. The much larger vitreous chamber is posterior to the lens.

The anterior and posterior chambers are filled with **aqueous humor** (watery fluid), which helps maintain pressure within the eye, refracts (bends) light, and provides nutrients to the inner surface of the eye. Aqueous humor is produced by the ciliary body as a blood filtrate and is returned to the circulation through a venous ring that surrounds the cornea. The presence of aqueous humor keeps the eye inflated, much like the air in a basketball. If flow of the aqueous humor from the eye through the venous ring is blocked, the pressure in the eye increases, resulting in a condition called **glaucoma** (see the Clinical Focus: Eye Disorders on p. 256). Glaucoma can eventually lead to blindness because the fluid compresses the retina, thereby restricting blood flow through it.

The vitreous chamber of the eye is filled with a transparent, jellylike substance called the **vitreous humor.** The vitreous humor helps maintain pressure within the eye and holds the lens and the retina in place. It also functions to refract the light. Unlike the aqueous humor, the vitreous humor does not circulate.

Functions of the Complete Eye

The eye functions much like a camera. The iris allows light into the eye, which is focused by the cornea, lens, and humors onto the retina. The light striking the retina produces action potentials that are relayed to the brain.

Light Refraction

An important characteristic of light is that it can be refracted (bent). As light passes from air to some other, denser transparent substance, the light rays are refracted. If the surface of a lens is concave, the light rays are bent so that they diverge as they pass through the lens; if the surface is convex, they converge. As the light rays converge, they finally reach a point at which they cross. The crossing point is called the **focal point** (figure 9.15), and causing light to converge is called **focusing.** The focal point in the eye occurs just anterior to the retina, and the tiny image that is focused on the retina is inverted compared with the actual object.

Focusing of Images on the Retina

The cornea is a convex structure, and as light rays pass from the air through the cornea, they converge (see figure 9.15). Additional convergence occurs as light passes through the aqueous humor, lens, and vitreous humor. The greatest contrast in media density is between the air and the cornea. The greatest amount of convergence therefore occurs at that point. The shape of the cornea and its distance from the retina are fixed, however, so that no adjustment in focus can be made by the cornea. Fine adjustments in focus are accomplished by changing the shape of the lens.

When the ciliary muscles are relaxed, the suspensory ligaments of the ciliary body maintain elastic pressure on the perimeter of the lens, keeping it relatively flat and allowing for distant vision (see figure 9.15a). When an object is brought closer than 20 feet (about $6\frac{1}{2}$ m) to the eye, the ciliary muscles contract as a result of parasympathetic stimulation,

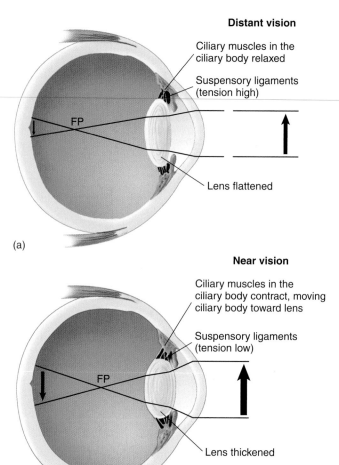

Distant vision

Ciliary muscles in the ciliary body relaxed

Suspensory ligaments (tension high)

FP

Lens flattened

(a)

Near vision

Ciliary muscles in the ciliary body contract, moving ciliary body toward lens

Suspensory ligaments (tension low)

FP

Lens thickened

(b)

Figure 9.15   Focus and Accommodation by the Eye

The focal point (FP) is where light rays cross. (a) Distant image. The lens is flattened, and the image is focused on the retina. (b) Accommodation for near vision. The lens is more rounded, and the image is focused on the retina.

pulling the ciliary body toward the lens. This reduces the tension on the suspensory ligaments of the lens and allows the lens to assume a more spherical form because of its own internal elastic nature (figure 9.15b). The spherical lens then has a more convex surface, causing greater refraction of light. This process is called **accommodation** (ă-kom′ō-dā′shŭn), and it enables the eye to focus on objects closer than 20 feet on the retina.

Vision Tests

When a person's vision is tested, a chart is placed 20 feet from the eye, and the person is asked to read a line that has been standardized for normal vision. If the person can read the line, he has 20/20 vision, which means that he can see at 20 feet what people with normal vision see at 20 feet. On the other hand, if the person can read only letters at 20 feet that people with normal vision see at 40 feet, the person's eyesight is 20/40.

PREDICT 5
What changes occur, as you are driving a car, when you look down at the speedometer and then back up at the road?

Neuronal Pathways for Vision

The **optic nerve** (figure 9.16) leaves the eye at the optic disc and exits the orbit through the optic foramen to enter the cranial cavity. Just inside the cranial cavity, the two optic nerves connect to each other at the **optic chiasm** (kī′azm, crossing). Axons from the nasal (medial) part of each retina cross through the optic chiasm and project to the opposite side of the brain. Axons from the temporal (lateral) part of each retina pass through the optic nerves and project to the brain on the same side of the body without crossing.

Beyond the optic chiasm, the route of the ganglionic axons is through the two **optic tracts** (see figure 9.16). Most of the optic tract axons terminate in the thalamus. Some axons do

1. Each visual field is divided into a temporal and nasal half.

2. After passing through the lens, light from each half of a visual field projects to the opposite side of the retina.

3. An optic nerve consists of axons extending from the retina to the optic chiasm.

4. In the optic chiasm, axons from the nasal part of the retina cross and project to the opposite side of the brain. Axons from the temporal part of the retina do not cross.

5. An optic tract consists of axons that have passed through the optic chiasm (with or without crossing) to the thalamus.

6. The axons synapse in the thalamus. Collateral branches of the axons in the optic tracts synapse in the superior colliculi.

7. An optic radiation consists of axons from thalamic neurons that project to the visual cortex.

8. The right part of each visual field (*dark green* and *light blue*) projects to the left side of the brain, and the left part of each visual field (*light green* and *dark blue*) projects to the right side of the brain.

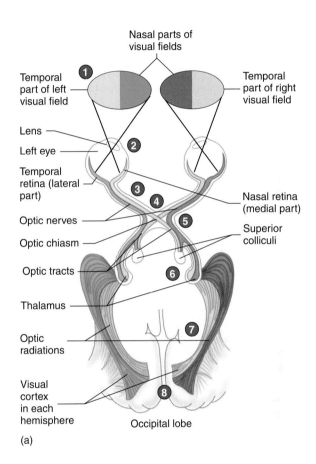

(a)

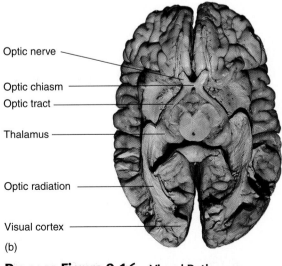

(b)

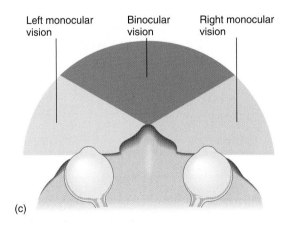

(c)

Process Figure 9.16 Visual Pathways

(a) Pathways for both eyes (*superior view*). (b) Photograph of the visual nerves, tracts, and pathways (*inferior view*). (c) Overlap of the fields of vision (*superior view*).

Clinical Focus Eye Disorders

Infections

Conjunctivitis (kon-jŭnk-ti-vī′tis) is an inflammation of the conjunctiva, usually resulting from a bacterial infection. **Contagious conjunctivitis** (pinkeye) occurs primarily in children. It can be transmitted by hand contact, flies, or contaminated water, such as in swimming pools. **Neonatal gonorrheal ophthalmia** (nē-ō-nā′tăl gon-ō-rē′ăl of-thal′mē-ă, eye inflammation) is a severe form of conjunctivitis that is contracted by an infant passing through the birth canal of a mother with gonorrhea. This infection carries a high risk of blindness. The treatment of newborn eyes with silver nitrate is effective in preventing the disease. **Chlamydial** (kla-mid′ē-ăl) **conjunctivitis** is contracted as an infant passes through the birth canal of a mother with a *Chlamydia* infection. This infection is not affected by silver nitrate, so in many places, newborns are treated with antibiotics against both *Chlamydia* and gonorrhea. **Trachoma** (tră-kō′mă) is the greatest single cause of blindness in the world today. The disease, also caused by *Chlamydia,* is transmitted by hand contact, flies, or objects such as towels. It is a conjunctivitis that leads to scarring of the cornea and blindness. It is most common in arid parts of Africa and Asia.

A **chalazion** (ka-lā′zē-on, a sty) is a cyst caused by infection of the sebaceous glands along the edge of the eyelid. A **stye** (stī) is an infection of an eyelash hair follicle.

Defects of Focus

Myopia (mī-ō′pē-ă), or nearsightedness, is the ability to see close objects but not distant ones. It is a defect of the eye in which the refractive power of the cornea and lens is too great for the relative length of the eye. As a result, the focal point is too near to the lens and the image is focused in front of the retina when looking at distant objects (figure A*a*). Myopia usually occurs because the eye is abnormally long but may also result from corneal or lens defects. Myopia is corrected by a concave lens that spreads out the light rays coming to the eye so that when the light is focused by the eye, it is focused on the retina (figure A*b*).

Another technique for correcting myopia is **radial keratotomy** (ker′ă-tot′ō-mē), which consists of making a series of radiating cuts in the cornea. The cuts are intended to slightly weaken the dome of the cornea so that it becomes more flattened and eliminates the myopia. One problem with the technique is that it is difficult to predict exactly how much flattening will occur and the surgery may overcorrect or undercorrect the problem. Another problem is that some patients are bothered by glare following radial keratotomy because the slits apparently don't heal evenly.

An alternative procedure is **laser corneal sculpturing,** in which a thin portion of the cornea is etched away to make the cornea less convex. The advantage of this procedure

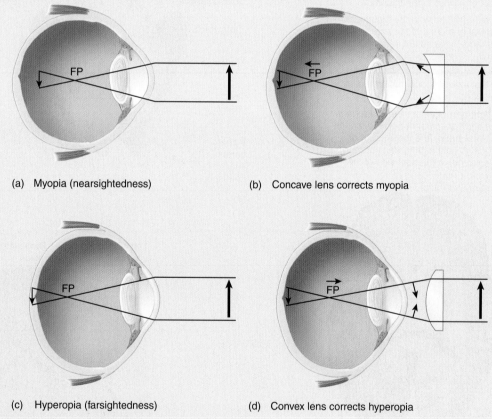

(a) Myopia (nearsightedness)

(b) Concave lens corrects myopia

(c) Hyperopia (farsightedness)

(d) Convex lens corrects hyperopia

Figure A Visual Disorders and Their Correction by Various Lenses

FP is the focal point. (*a*) Myopia (nearsightedness). (*b*) Correction of myopia with a concave lens. (*c*) Hyperopia (farsightedness). (*d*) Correction of hyperopia with a convex lens.

is that the results can be more accurately predicted than those from radial keratotomy.

Hyperopia (hī-per-ō′pē-ă), or farsightedness, is the ability to see distant objects but not close ones. It is a disorder in which the relative length of the eye is too short, the cornea is too flat, or the lens has too little refractive power. As a result, the focal point is too far from the lens, and the image is focused "behind" the retina when looking at a close object (figure A*c*). In hyperopia, the lens must thicken and accommodate to bring somewhat distant objects into focus, which would not be necessary for a normal eye. Closer objects cannot be brought into focus because the lens cannot thicken enough to focus the image on the retina. Hyperopia is corrected by a convex lens that causes light rays to converge as they approach the eye and to focus on the retina (figure A*d*).

Presbyopia (prez-bē-ō′pē-ă) is the decrease in the ability of the eye to accommodate for near vision. This occurs as a normal part of aging and the lens becomes less flexible. The average age of onset of presbyopia is the midforties. Presbyopia can be corrected by the use of "reading glasses" or by bifocals, which have different lenses in the top and bottom. The bottom half of a bifocal lens is more convex for a person with hyperopia to allow for near vision when the person reads, and the top half is less convex for distant vision. For a person with myopia, the bottom half of the lens is less concave to allow for near vision when the person reads, and the top half is more concave for distant vision.

Astigmatism (ă-stig′mă-tizm) is a defect in which the cornea or lens is not uniformly curved and the image is not sharply focused. Glasses may be made to adjust for the abnormal curvature as long as the curvature is not too irregular. If the curvature of the cornea or lens is too irregular, the condition is difficult to correct.

Strabismus, Diplopia, and Nystagmus

Strabismus (stra-biz′mŭs) is a condition in which one eye or both eyes are directed medially or laterally. The condition can result from abnormally weak eye muscles.

Diplopia (di-plō′pē-ă) is double vision.

Nystagmus (nī-stag′mus) is an involuntary, rythmic, repeated oscillation of one or both eyes. The cause is unknown and there is no known treatment. Eye muscle surgery helps in some cases.

Color Blindness

Color blindness is the absence of perception of one or more colors (figure B). There may be a complete loss of color perception or only a decrease in perception. The loss may involve perception of all three colors or of one or two colors. Most forms of color blindness occur more frequently in males and are X-linked genetic traits (see chapter 20). In western Europe, about 8% of all males have some form of color blindness, whereas only about 1% of the females are color blind.

Blindness

Cataract (kat′ă-rakt, a waterfall) is the most common cause of blindness in the United States. It is a condition in which clouding of the lens occurs as the result of advancing age, infection, or trauma. Excess exposure to ultraviolet radiation may be a factor in causing cataracts, so wearing sunglasses that reduce exposure to ultraviolet radiation in bright sunshine is recommended. Cataracts occur to some degree in 95% of people over age 65. Surgery performed to remove a cataract involves removal of the lens. More than 400,000 cataract lenses are removed in the United States each year. In almost all cases, an artificial lens is implanted in place of the natural one. Clear vision is restored for distant vision, but the ability to accommodate for near vision is lost and glasses are required for near vision.

Macular degeneration is common in older people. It does not cause total blindness but results in the loss of acute vision. This degeneration has a variety of causes, including hereditary disorders, infections, trauma, tumor, or most often, poorly understood degeneration associated with aging. Because no satisfactory medical treatment has been developed, optical aids, such as magnifying glasses, are used to improve visual function.

Glaucoma (glaw-kō′mă, *glaukos,* bluegreen) is a condition involving excessive pressure buildup in the aqueous humor. Glaucoma results from an interference with normal reentry of aqueous humor into the blood or from an overproduction of aqueous humor. The increased pressure within

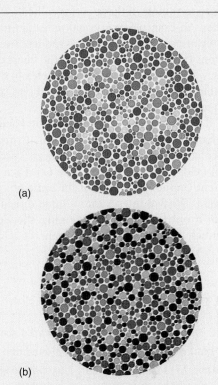

(a)

(b)

Figure B Color Blindness Charts

(*a*) A person with normal vision can see the number 74, whereas a person with red–green color blindness sees the number 21. (*b*) A person with normal vision can see the number 42. A person with red color blindness sees the number 2, and a person with green color blindness sees the number 4.

The above has been reproduced from Ishihara's Tests for Colour Deficiency published by Kanehara Trading, Inc., Tokyo, Japan. Tests for color deficiency cannot be conducted with this material. For accurate testing, the original plates should be used.

the eye can close off the blood vessels entering the eye and may destroy the retina or optic nerve, resulting in blindness.

Diabetes mellitus (dī-ă-bē′tēz me-lī′tus) is a major cause of blindness (diabetic retinopathy) in the United States. Diabetes can result in optic nerve degeneration, cataracts, and retinal detachment. These defects are often caused by blood vessel degeneration and hemorrhage, which are common in diabetic patients.

Retinal detachment, the separation of the sensory retina from the pigmented retina, is a relatively common problem. If a hole or tear occurs in the retina, fluid can accumulate between the sensory and pigmented retina. As a result, the sensory retina may become detached from the pigmented retina and degenerate, resulting in loss of vision.

not terminate in the thalamus but separate from the optic tracts to terminate in the superior colliculi, the center for visual reflexes. An example of a visual reflex is turning the head and eyes toward a stimulus such as a sudden noise or flash of light. Neurons from the thalamus form the fibers of the **optic radiations,** which project to the **visual cortex** in the occipital lobe of the brain (see figure 9.16). The visual cortex is the area of the brain where vision is perceived.

The image seen by each eye is the **visual field** of that eye (see figure 9.16*a*). Depth perception (three-dimensional, or binocular, vision) requires both eyes and occurs where the two visual fields overlap (see figure 9.16*c*). Each eye sees a slightly different (monocular) view of the same object. The brain then processes the two images into a three-dimensional view of the object. If only one eye is functioning, the view of the object is flat, much like viewing a photograph.

A CASE IN POINT | Double Vision

I.C. Double awoke one morning and discovered that her vision was blurred. When she closed one eye or the other, her vision was clear. She tried to ignore the problem and went about her work, but the double vision persisted. Eventually she learned from an optometrist that her blurred vision resulted from double vision, or **diplopia.** The most common cause of diplopia is misalignment of the two eyes (binocular diplopia). This often results from weakness of the muscles moving the eyes. Most adults cannot easily ignore the double image. Wearing a patch over one eye to eliminate one image may be the only option for such people. In children with diplopia, the brain may compensate for the two discordant images by ignoring one of the images, and the problem appears to go away. Double vision should not be ignored because it may be a symptom of a serious neurological problem, such as an expanding brain tumor. A tumor may compress the nerves to the eye muscles, decreasing their function. A physician should be consulted as soon as possible.

Hearing and Balance

The organs of hearing and balance are divided into three parts: external, middle, and inner ear (figure 9.17). The external ear is the part extending from the outside of the head to the eardrum. The middle ear is an air-filled chamber medial to the eardrum. The inner ear is a set of fluid-filled chambers medial to the middle ear. The external and middle ears are involved in hearing only, whereas the inner ear functions in both hearing and balance.

The Ear and Its Functions
External Ear

The **auricle** (aw′ri-kl, ear) is the fleshy part of the external ear on the outside of the head. The auricle opens into the **external acoustic meatus** (mē-ā′tŭs, passage), a passageway that leads to the eardrum. The auricle collects sound waves, and directs them toward the external acoustic meatus, which transmits them to the eardrum. The meatus is lined with hairs and **ceruminous** (sĕ-roo′mi-nŭs, *cera*, wax) **glands,** which produce **cerumen** (sĕ-roo′men), a modified sebum commonly called earwax. The hairs and cerumen help prevent foreign objects from reaching the delicate eardrum.

The **tympanic** (tim-pan′ik, drumlike) **membrane,** or **eardrum,** is a thin membrane that separates the external ear from the middle ear. It consists of a thin layer of connective tissue sandwiched between two epithelial layers. Sound waves reaching the tympanic membrane cause it to vibrate.

Middle Ear

Medial to the tympanic membrane is the air-filled cavity of the middle ear. Two covered openings, the **oval window** and the **round window** on the medial side of the middle ear, connect the middle ear with the inner ear. The middle ear contains three **auditory ossicles** (os′i-klz, ear bones): the **malleus** (mal′ē-ŭs, hammer), **incus** (ing′kŭs, anvil), and **stapes** (stā′pēz, stirrup). These bones transmit vibrations from the tympanic membrane to the oval window. The malleus is attached to the medial surface of the tympanic membrane. The incus connects the malleus to the stapes. The base of the stapes is seated in the oval window and is surrounded by a flexible ligament. As the vibrations are transmitted from the malleus to the stapes, the force of the vibrations is amplified about 20-fold because the area of the tympanic membrane is about 20 times that of the oval window. Two small muscles in the middle ear, one attached to the malleus and the other to the stapes, help dampen vibrations caused by loud noises, thus protecting the delicate inner ear structures.

There are two unblocked openings into the middle ear. One opens into the mastoid air cells in the mastoid process of the temporal bone. The other, called the **auditory tube,** or **eustachian** (ū-stā′shŭn) **tube,** opens into the pharynx and enables air pressure to be equalized between the outside air and the middle ear cavity. Unequal pressure between the middle ear and the outside environment can distort the tympanic membrane, dampen its vibrations, and make hearing difficult. Distortion of the tympanic membrane also stimulates pain fibers associated with that structure. That distortion is why, as a person changes altitude, sounds seem muffled and the tympanic membrane may become painful. These symptoms can be relieved by opening the auditory tube, allowing air to enter or exit the middle ear. Swallowing, yawning, chewing, and holding the nose and mouth shut while gently trying to force air out of the lungs are methods that can be used to open the auditory tube.

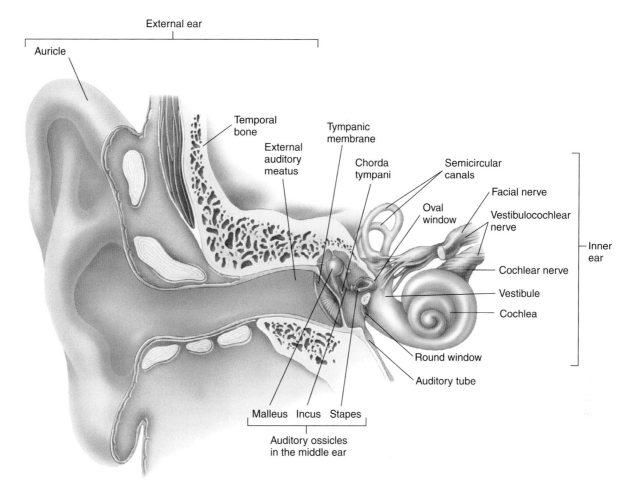

Figure 9.17   External, Middle, and Inner Ear

Inner Ear

The inner ear consists of interconnecting tunnels and chambers within the temporal bone, called the **bony labyrinth** (lab′i-rinth, maze) (figure 9.18a). Inside the bony labyrinth is a similarly shaped but smaller set of membranous tunnels and chambers called the **membranous labyrinth** (figure 9.18b). The membranous labyrinth is filled with a clear fluid called **endolymph** (en′dō-limf, *lymph,* clear spring water), and the space between the membranous and bony labyrinth is filled with a fluid called **perilymph** (per′i-limf). The bony labyrinth can be divided into three regions: the cochlea, vestibule, and semicircular canals. The cochlea is involved in hearing, and the vestibule and semicircular canals are involved primarily in balance.

Hearing

The **cochlea** (kok′lē-ă, snail shell) (see figure 9.18a) is shaped like a snail shell and contains a bony core shaped like a screw. The threads of this screw are called the **spiral lamina.** A Y-shaped membranous complex divides the cochlea into three portions (see figure 9.18b). The base of the Y is the spiral lamina. One branch of the Y is the **vestibular** (ves-tib′ū-lăr, entrance hall) **membrane,** and the other branch is the **basilar membrane.** The space between these membranes is called the **cochlear duct.** This complex is the membranous labyrinth, and it is filled with endolymph. If the Y is viewed lying on its right side, as in figure 9.18b, the space above the Y is called the **scala vestibuli** (skā′lă ves-tib′ū-lī, *scala,* stairway), and the space below the Y is called the **scala tympani** (tim-pa′nē). These two spaces are filled with perilymph. The scala vestibuli extends from the oval window to the apex of the cochlea, and the scala tympani extends from the apex to the round window. The two scalae are continuous with each other at the apex of the cochlea.

Inside the cochlear duct is a specialized structure called the **spiral organ,** or **organ of Corti** (see figure 9.18c). The spiral organ contains specialized sensory cells called **hair cells,** which have hairlike microvilli on their surfaces (see figure 9.18c and d). The microvilli are stiffened by actin filaments. The hair tips are embedded within an acellular gelatinous shelf called the **tectorial** (tek-tōr′ē-ăl, a covering) **membrane,** which is attached to the spiral lamina (see figure 9.18b and c).

Hair cells have no axons of their own, but each hair cell is associated with axon terminals of sensory neurons, the cell bodies of which are located within the **spiral ganglion.** Axons of the sensory neurons join to form the cochlear nerve. This nerve joins the vestibular nerve to become the **vestibulocochlear nerve** (cranial nerve VIII), which carries action potentials to the brain.

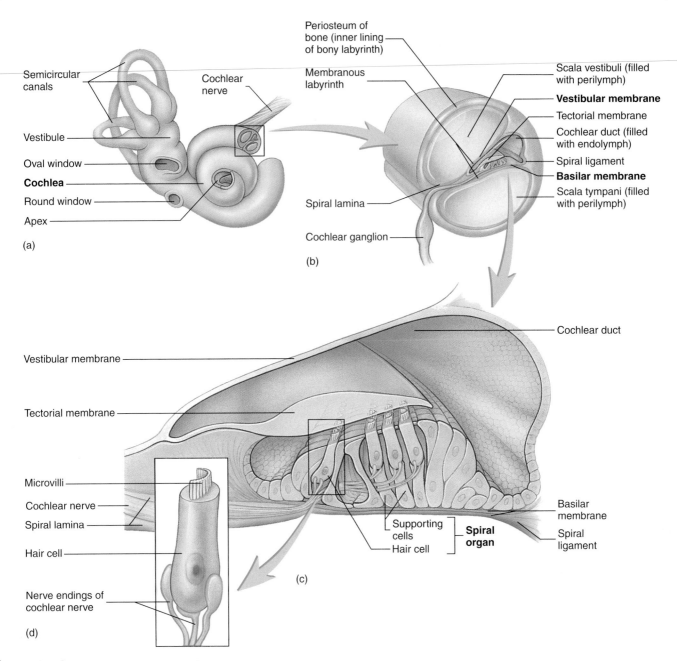

Figure 9.18   Structure of the Cochlea

(*a*) **The inner ear.** The outer surface (*gray*) is the periosteum lining the inner surface of the bony labyrinth. (*b*) **A cross section of the cochlea.** The outer layer is the periosteum lining the inner surface of the bony labyrinth. The membranous labyrinth is very small in the cochlea and consists of the vestibular and basilar membranes. The space between the membranous and bony labyrinth consists of two parallel tunnels: the scala vestibuli and scala tympani. (*c*) An enlarged section of the cochlear duct (membranous labyrinth). (*d*) A greatly enlarged individual sensory hair cell.

Sound waves are collected by the auricle and are conducted through the external acoustic meatus toward the tympanic membrane. Sound waves strike the tympanic membrane and cause it to vibrate. This vibration causes vibration of the three ossicles of the middle ear, and by this mechanical linkage the force of vibration is amplified and transferred to the oval window (figure 9.19).

P R E D I C T   6
Why is it that when you hear a faint sound, you turn your head toward it?

Vibrations of the base of the stapes, seated in the oval window, produce waves in the perilymph of the cochlea. The two scalae can be thought of as a continuous U-shaped tube, with the oval window at one end of the scala vestibuli and the round window at the other end of the scala tympani (see figure 9.19). The vibrations of the stapes in the oval window cause movement of the perilymph, which pushes against the membrane covering the round window. This phenomenon is similar to pushing against a rubber diaphragm on one end of a fluid-filled glass tube. If the tube has a rubber diaphragm on each end, the

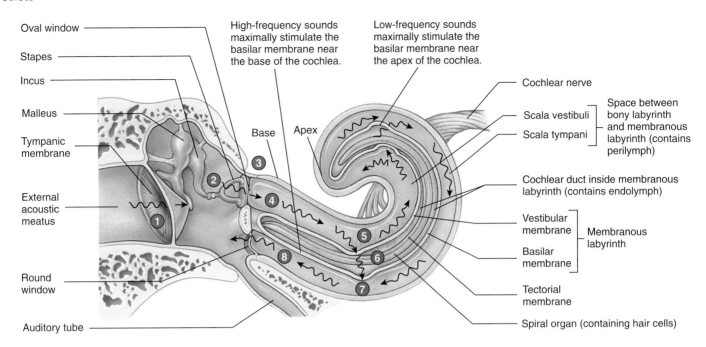

1. Sound waves strike the tympanic membrane and cause it to vibrate.

2. Vibration of the tympanic membrane causes the three bones of the middle ear to vibrate.

3. The stapes vibrates in the oval window.

4. Vibration of the stapes causes the perilymph in the scala vestibuli to vibrate.

5. Vibration of the perilymph passes through the vestibular membrane and causes vibration of endolymph in the cochlear duct.

6. Vibration of the endolymph causes displacement of the basilar membrane. High-pitch sound waves cause displacement of the basilar

membrane near the base of the cochlea (near the oval window), and low-pitch sound waves cause displacement of the basilar membrane near the apex of the cochlea. Movement of the basilar membrane is detected by the hair cells of the spiral organ, which are attached to the basilar membrane and the tectorial membrane. Bending of hair cell microvilli induces action potentials in the cochlear nerves. These action potentials are conducted to the CNS, where they are perceived as sound.

7. Vibrations of the perilymph in the scala vestibuli and of the endolymph in the cochlear duct are transferred to the perilymph of the scala tympani.

8. Vibrations in the perilymph of the scala tympani are transferred to the round window, which is flexible and allows movement of the entire fluid column of perilymph.

Process Figure 9.19 Effect of Sound Waves on Middle and Inner Ear Structures

fluid can move. If one end of the glass tube or of the cochlear tubes were solid, no fluid movement would occur.

The waves produced in the perilymph pass through the vestibular membrane and cause vibrations of the endolymph. Waves in the endolymph, within the cochlear duct, cause displacement of the basilar membrane. As the basilar membrane is displaced, the hair cells, seated on the basilar membrane, move with the movements of the membrane. The microvilli of the hair cells are embedded into the tectorial membrane, which is a rigid shelf that does not move. Because one end of the microvilli move with the hair cells and their other ends are embedded into the nonmoving tectorial membrane, the microvilli bend. The bending of the microvilli causes stimulation of the hair cells, which induces action potentials in the cochlear nerves.

The basilar membrane is not uniform throughout its length. The membrane is narrower and denser near the oval window and wider and less dense near the tip of the cochlea. The various regions of the membrane can be compared to the strings in a piano (i.e., some are short and thick, and others are longer and thinner). As a result of this organization, sounds with higher **pitches** cause the basilar membrane

nearer the oval window to distort maximally, whereas sounds with lower pitches cause the basilar membrane nearer the apex of the cochlea to distort maximally. Different hair cells are stimulated in each case, and, because of the differences in which hair cells are maximally stimulated, a person is able to detect variations in pitch. Sound **volume** is a function of sound wave amplitude, which causes the basilar membrane to distort more intensely and the hair cells to be stimulated more strongly.

Loud sounds can damage the delicate microvilli of the hair cells, resulting in permanent hearing loss. Avoiding loud music and wearing ear protection around loud noises are recommended.

Neuronal Pathways for Hearing

The cochlear nerves, whose cell bodies are located in the cochlear ganglion, send axons to the **cochlear nucleus** in the brainstem. Neurons in the cochlear nucleus project to other areas of the brainstem and to the **inferior colliculus** in the midbrain. From the inferior colliculus, fibers project to the thalamus, and from there to the auditory cortex of the cerebrum.

Clinical Focus Ear Disorders

Hearing Impairment

The term **hearing impairment** refers to any type or degree of hearing loss. Hearing impairment can have many causes. In general, there are two categories of hearing impairment: conduction deafness and sensorineural hearing loss. **Conduction deafness** involves a mechanical deficiency in transmission of sound waves from the outer ear to the spiral organ. Hearing aids may help people with such hearing deficiencies by boosting the sound volume that reaches the ear. Some conduction deafness can be corrected surgically. For example, **otosclerosis** (ō′tō-sklē-rō′sis, *oto*, ear + *sklerosis*, hardening) is an ear disorder in which bone grows over the oval window and immobilizes the stapes. This disorder can be surgically corrected by breaking away the bony growth and the stapes. The base of the stapes, located in the oval window, is replaced by a fat pad or synthetic membrane, and the rest of the stapes is replaced by a small metal rod connected to the oval window at one end and to the incus at the other end.

    Sensorineural hearing loss involves the spiral organ or nerve pathways and is more difficult to correct. One approach involves the direct stimulation of the cochlear nerve by action potentials. The mechanism consists of a microphone for picking up sound waves; a microelectronic processor for converting the sound into electrical signals; a transmission system for relaying the signals to the inner ear, and a long, slender electrode that is threaded into the cochlea. This electrode delivers electrical signals directly to the cochlear nerve.

Tinnitus

Tinnitus (ti-nī′tus) consists of phantom sound sensations, such as roaring, buzzing, or ringing in the ears. This is a common problem, affecting 17% of the world's population. It is treated primarily by training people to ignore the sounds.

Ear Infections

Infections of the middle ear, called **otitis media** (ō-tī′tis mē′dē-ă) are quite common in young children. These infections usually result from the spread of infection from the mucous membrane of the pharynx through the auditory tube. The symptoms of low-grade fever, lethargy, and irritability, and pulling at the ear are not often recognized by the parent as signs of middle ear infection. The infection also can cause a temporary decrease or loss of hearing because fluid buildup can dampen the tympanic membrane. In extreme cases, the infection can damage or rupture the tympanic membrane.

    Chronic middle ear infections increase the chances of inner ear infections. Inner ear infections can decrease the inner ear's detection of sound and maintenance of equilibrium.

Motion Sickness

Motion sickness consists of nausea and weakness caused by continuous stimulation of the semicircular canals because of the motion occurring in a boat, automobile, or airplane. The brain compares input from the semicircular canals to that from the eyes and position sensors in the back and lower limbs. Conflicting information from these inputs may result in motion sickness. About 90% of the population experiences motion sickness at some time. Often the cases are mild. Severe cases may be treated with an antihistamine, such as meclizine or dimenhydrinate. Long-term patches containing scopolamine may be placed behind the ear for up to three days. Both antihistamines and scopolamine have anticholinergic actions, decreasing the rate of action potentials through the vestibular nerve to the brain.

Space Sickness

Space sickness is a balance disorder occurring in zero gravity and resulting from unfamiliar sensory input to the brain. The brain must adjust to these unusual signals, or severe symptoms such as headache and dizziness can result. Space sickness is unlike motion sickness in that motion sickness results from excessive stimulation to the brain, and space sickness results from too little stimulation as a result of weightlessness.

Meniere's Disease

Meniere's disease is the most common disease involving dizziness from the inner ear. Its cause is unknown but it appears to involve a fluid abnormality in one (usually) or both ears. Symptoms include vertigo, hearing loss, tinnitus, and a feeling of "fullness" in the affected ear. Treatment includes a low-salt diet and diuretics (water pills). Symptoms may also be treated with medications for motion sickness.

Equilibrium

The sense of equilibrium, or balance, has two components: static equilibrium and kinetic equilibrium. **Static equilibrium** is associated with the vestibule and is involved in evaluating the position of the head relative to gravity. **Kinetic equilibrium** is associated with the semicircular canals and is involved in evaluating changes in the direction and rate of head movements.

    The **vestibule** (ves′ti-bool) can be divided into two chambers: the **utricle** (ū′tri-kl, leather bag) and the **saccule** (sak′ūl, sac) (figure 9.20*a*). Each chamber contains specialized patches of epithelium called the **maculae** (mak′ū-lē, a spot), which are surrounded by endolymph (figure 9.20*b*). The maculae, like the spiral organ, contain hair cells. The tips of the microvilli of these cells are embedded in a gelatinous mass weighted by **otoliths** (ō′tō-liths, ear stones). Otoliths are particles composed of protein and calcium carbonate (figure 9.20*c*). The weighted gelatinous mass moves in response to gravity, bending the hair cell microvilli and initiating action potentials in the associated neurons. The action potentials from these neurons are carried by axons of the vestibular portion of the vestibulocochlear nerve (cranial nerve VIII) to the brain, where they are interpreted as a change in position of the head. For example, when a person bends over, the maculae are displaced by gravity, and the resultant action potentials provide information to the brain concerning the position of the head relative to gravity (figure 9.21).

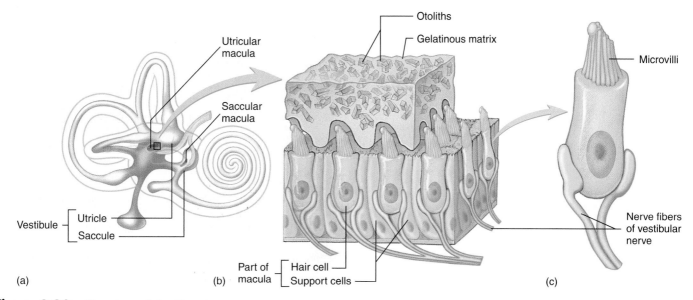

Figure 9.20 **Structure of the Macula**

(*a*) Vestibule showing the location of the utricular and saccular maculae. (*b*) Enlargement of the utricular macula, showing hair cells and otoliths in the macula. (*c*) An enlarged hair cell, showing the microvilli.

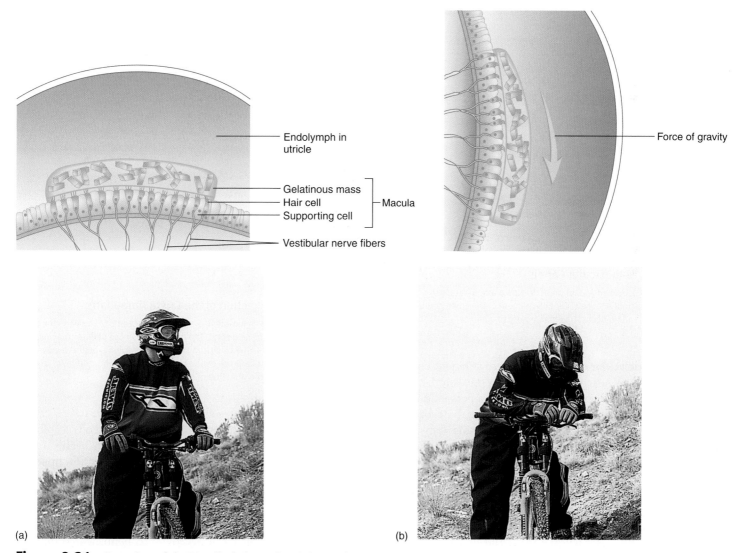

Figure 9.21 **Function of the Vestibule in Maintaining Balance**

(*a*) In an upright position, the maculae don't move. (*b*) As the position of the head changes, such as when a person bends over, the maculae respond to changes in position of the head relative to gravity by moving in the direction of gravity.

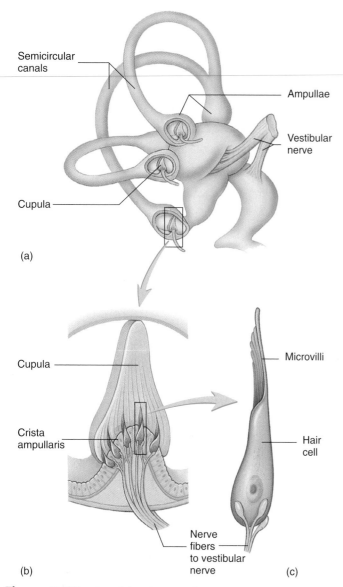

(a)

(a)

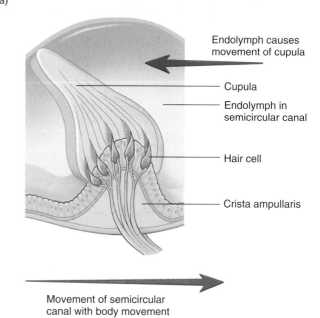

(b)

(b)   (c)

Figure 9.22   Semicircular Canals

(a) Semicircular canals showing location of the crista ampullaris in the ampullae of the semicircular canals. (b) Enlargement of the crista ampullaris, showing the cupula and hair cells. (c) Enlargement of a hair cell.

Figure 9.23   Function of the Crista Ampullaris

(a) As a person begins to tumble, the semicircular canals (b) move in the same direction as the body (*blue arrow*). The endolymph in the semicircular canals tends to stay in place as the body and the crista ampullaris begin to move. As a result, the cupula is displaced by the moving endolymph (*red arrow*) in a direction opposite to the direction of movement.

Three **semicircular canals** are involved in kinetic equilibrium and placed at nearly right angles to one another. The placement of the semicircular canals enables a person to detect movements in essentially any direction. The base of each semicircular canal is expanded into an **ampulla** (am-pul′ă, two-handled bottle). Within each ampulla the epithelium is specialized to form a **crista ampullaris** (kris′tă am-pūl′ar′is) (figure 9.22a). Each crista consists of a ridge of epithelium with a curved gelatinous mass, the **cupula** (koo′poo-lă, a tub), suspended over the crest (figure 9.22b). The cupula is structurally and functionally very similar to the maculae, except that it contains no otoliths. The hairlike microvilli of the crista hair cells (figure 9.22c) are embedded in the cupula. The cupula functions as a float that is displaced by endolymph movement within the semicircular canals (figure 9.23). As the head begins

to move in a given direction, the endolymph tends to remain stationary, while the cupula moves with the head. This difference causes the cupula to be displaced in a direction opposite to that of the movement of the head. As movement continues, the fluid "catches up." When movement of the head and the cupula stops, the fluid tends to continue to move, displacing the cupula in the direction of the movement. Movement of the cupula causes the hair cell microvilli to bend, which initiates depolarization in the hair cells. This depolarization initiates action potentials in the vestibular nerves, which join the cochlear nerves to form the vestibulocochlear nerves.

Neuronal Pathways for Equilibrium

Axons forming the vestibular portion of the vestibulocochlear nerve project to the vestibular nucleus in the brainstem, where they synapse. Equilibrium is a complex sensation involving several brainstem areas, as well as the cerebellum and the cerebral cortex. Sensory input for equilibrium comes not only from the inner ear, but from the eyes and limbs (proprioception) as well.

A CASE IN POINT | Sea Sickness

Earl E. Fisher booked his first trip on a charter fishing boat. Crossing over "the bar" was refreshing and invigorating, and Earl was surprised that all those warnings about passengers becoming sea sick did not seem to apply to him. At last the boat arrived at the fishing site, the engine was cut, and the sea anchor was set. As the boat drifted, it began to roll and pitch. Earl noticed, for the first time, that the bait smelled oily and unpleasant. The smell mixed poorly with the diesel fumes. Earl felt a little light-headed and a bit drowsy. He then noticed that he was definitely nauseous and felt pale. "I'm sea sick," he realized. Sea sickness is a form of motion sickness, which is discussed on p. 262. The cause of the problem is conflicting information reaching the brain from sensory sources, such as the inner ear and eyes. The brain reacts with a feeling of vertigo (dizziness) and nausea. Trying to fish seemed to worsen his condition. Eventually his nausea intensified, he leaned over the boat rail and vomited into the ocean. "Improves the fishing," the ship owner shouted to him cheerily. Earl noticed that he felt somewhat better. His condition continued to improve. He found that looking at the horizon rather than at the water helped his condition by decreasing some of the sensory input to the brain. He enjoyed the rest of the trip and even caught a couple of very nice fish.

Effects of Aging on the Special Senses

Elderly people experience only a slight loss in the ability to detect odors. However, the ability to correctly identify specific odors is decreased, especially in men over age 70.

In general, the sense of taste decreases as people age. The number of sensory receptors decreases and the ability of the brain to interpret taste sensations declines.

The lenses of the eyes lose flexibility as a person ages because the connective tissue of the lenses becomes more rigid. Consequently there is first a reduction and then an eventual loss in the ability of the lenses to change shape. This condition, called **presbyopia,** is the most common age-related change in the eyes. It is discussed more fully in the Clinical Focus on "Eye Disorders" earlier in the chapter.

The most common visual problem in older people requiring medical treatment, such as surgery, is the development of cataracts. Macular degeneration is the second most common defect, glaucoma is third, and diabetic retinopathy is fourth. These defects are also described more fully in the Clinical Focus on "Eye Disorders" on p. 257.

The number of cones decreases, especially in the fovea centralis. These changes cause a gradual decline in visual acuity and color perception.

As people age, the number of hair cells in the cochlea decreases, resulting in age-related hearing loss called **presbyacusis.** This decline doesn't occur equally in both ears. As a result, because direction is determined by comparing sounds coming into each ear, elderly people may experience a decreased ability to localize the origin of certain sounds. In some people, this may lead to a general sense of disorientation. In addition, CNS defects in the auditory pathways can result in difficulty understanding sounds with echoes or background noise. Such deficit makes it difficult for elderly people to understand rapid or broken speech.

With age, the number of hair cells in the saccule, utricle, and ampullae decreases. The number of otoliths also declines. As a result, elderly people experience a decreased sensitivity to gravity, acceleration, and rotation. Because of these decreases, elderly people experience dizziness (instability) and vertigo (a feeling of spinning). They often feel that they can't maintain posture and are prone to fall.

S U M M A R Y

Sensations result only from those stimuli that reach the cerebral cortex and are consciously perceived. Senses can be defined as special or general.

General Senses   (p. 244)

Receptors for general senses, such as pain, temperature, touch, pressure, and proprioception, are scattered throughout the body.

Pain

1. Pain is an unpleasant sensation and is either sharp or diffuse.
2. Pain can be "gated," referred, or phantom.

Special Senses   (p. 246)

Smell and taste respond to chemical stimulation, vision to light stimulation, and hearing and balance to mechanical stimulation.

Olfaction   (p. 246)

1. Olfactory neurons have enlarged distal ends with long cilia. The cilia have receptors that respond to dissolved substances in the nasal mucus.
2. The wide range of detectable odors may result from combinations of receptor responses stimulated by only a few primary odors.

Neuronal Pathways for Olfaction

Axons of the olfactory neurons form the olfactory nerves, which enter the olfactory bulb. Olfactory tracts carry action potentials from the olfactory bulbs to the olfactory cortex of the brain.

Taste　(p. 247)

1. Taste buds contain taste cells with hairs that extend into taste pores. Receptors on the hairs detect dissolved substances.
2. There are five basic types of taste: sour, salty, bitter, sweet, and umami.

Neuronal Pathways for Taste

The facial nerves carry taste from the anterior two-thirds of the tongue, the glossopharyngeal from the posterior one-third of the tongue, and the vagus from the root of the tongue.

Vision　(p. 248)

Accessory Structures

1. The eyebrows prevent perspiration from entering the eyes.
2. The eyelids protect the eyes from foreign objects.
3. The conjunctiva covers the inner eyelids and the anterior surface of the eye.
4. Lacrimal glands produce tears that flow across the surface of the eye. Tears lubricate and protect the eye. Excess tears pass through the nasolacrimal duct into the nasal cavity.
5. The extrinsic eye muscles move the eyeball.

Anatomy of the Eye

1. The fibrous tunic is the outer layer of the eye. It consists of the sclera and cornea.
2. The vascular tunic is the middle layer of the eye. It consists of the choroid, ciliary body, and iris.
3. The lens is held in place by the suspensory ligaments, which are attached to the smooth muscles of the ciliary body.
4. The retina (nervous tunic) is the inner layer of the eye and contains neurons sensitive to light.
5. Rods are responsible for vision in low illumination (night vision).
6. Cones are responsible for color vision.
7. Light causes retinal to change shape, causing opsin to change shape, causing cellular changes that result in vision.
8. The fovea centralis in the center of the macula lutea has the highest concentration of cones and is the area in which images are detected most clearly.
9. The optic disc, or blind spot, is where the optic nerve exits the eye and blood vessels enter.
10. The anterior and posterior chambers of the eye are anterior to the lens and are filled with aqueous humor. The vitreous chamber is filled with vitreous humor. The humors keep the eye inflated, refract light, and provide nutrients to the inner surface of the eye.

Functions of the Complete Eye

1. Light passing through a concave surface diverges. Light passing through a convex surface converges.
2. Converging light rays cross at the focal point and are said to be focused.
3. The cornea, aqueous humor, lens, and vitreous humor all refract light. The cornea is responsible for most of the convergence, whereas the lens can adjust the focus by changing shape (accommodation).

Neuronal Pathways for Vision

1. Axons pass through the optic nerves to the optic chiasm, where some cross. Axons from the nasal retina cross and those from the temporal retina do not.
2. Optic tracts from the chiasm lead to the thalamus.
3. Optic radiations extend from the thalamus to the visual cortex in the occipital lobe.

Hearing and Balance　(p. 258)

The Ear and Its Functions

1. The external ear consists of the auricle and the external acoustic meatus.
2. The middle ear connects the external and inner ear.
3. The tympanic membrane (eardrum) is stretched across the external auditory meatus.
4. The malleus, incus, and stapes connect the tympanic membrane to the oval window of the inner ear.
5. The auditory, or eustachian, tube connects the middle ear to the pharynx and equalizes pressure. The middle ear is also connected to the mastoid air cells.
6. The inner ear has three parts: the semicircular canals, the vestibule, and the cochlea.

Hearing

1. The cochlea is a canal shaped like a snail's shell.
2. The cochlea is divided into three compartments by the vestibular and basilar membranes.
3. The spiral organ consists of hair cells that attach to the basilar and tectorial membranes.
4. Sound waves are funneled by the auricle down the external acoustic meatus, causing the tympanic membrane to vibrate.
5. The tympanic membrane vibrations are passed along the ossicles to the oval window of the inner ear.
6. Movement of the stapes in the oval window causes the perilymph to move the vestibular membrane, which causes the endolymph to move the basilar membrane. Movement of the basilar membrane causes movement of the hair cells in the spiral organ and generation of action potentials, which travel along the vestibulocochlear nerve.

Neuronal Pathways for Hearing

Action potentials travel along the cochlear portion of the vestibulo-cochlear nerve to the cochlear nucleus and on to the cerebral cortex.

Equilibrium

1. Static equilibrium evaluates the position of the head relative to gravity.
2. Maculae, located in the vestibule, consist of hair cells with the microvilli embedded in a gelatinous mass that contains otoliths. The gelatinous mass moves in response to gravity.
3. Kinetic equilibrium evaluates movements of the head.
4. There are three semicircular canals in the inner ear, arranged perpendicular to each other. The ampulla of each semicircular canal contains a crista ampullaris, which has hair cells with microvilli embedded in a gelatinous mass, the cupula.

Neuronal Pathways for Equilibrium

Axons in the vestibular portion of the vestibulocochlear nerve project to the vestibular nucleus and on to the cerebral cortex.

Effects of Aging on the Special Senses　(p. 265)

There is a general decline in taste, vision, hearing, and balance in elderly people.

R E V I E W   A N D   C O M P R E H E N S I O N

1. Define stimulus and sensation.
2. List and describe the receptors associated with the general senses.
3. Explain how pain occurs and how it can be modified.
4. Define and explain referred pain and phantom pain.
5. Describe the process by which airborne molecules produce the sensation of smell.
6. How is the sense of taste related to the sense of smell?
7. What are the five primary tastes? How do they produce many different kinds of taste sensations?
8. Describe the following structures and state their functions: eyebrows, eyelids, conjunctiva, lacrimal apparatus, and extrinsic eye muscles.
9. Name the three layers (tunics) of the eye. Describe the structures composing each layer, and explain the functions of these structures.
10. Describe the three chambers of the eye, the substances that fill each, and the function of the substances.
11. Describe the lens of the eye and how it is held in place.
12. Describe the arrangement of cones and rods in the fovea centralis and in the periphery of the eye.
13. What is the blind spot of the eye, and what causes it?
14. What causes the pupil to constrict and dilate?
15. What causes light to refract? What is a focal point?
16. Define accommodation. What does accommodation accomplish?
17. Name the three regions of the ear, name the structures found in each region, and state the functions of each structure.
18. Describe the relationship between the tympanic membrane, the ear ossicles, and the oval window of the inner ear.
19. Describe the structure of the cochlea.
20. Starting with the auricle, trace sound into the inner ear to the point at which action potentials are generated in the vestibulocochlear nerve.
21. Describe the maculae and their function.
22. What is the function of the semicircular canals? Describe the crista ampullaris and its mode of operation.

C R I T I C A L   T H I N K I N G

1. An elderly male with normal vision developed cataracts. He was treated surgically by removing the lenses of his eyes. What kind of glasses do you recommend to compensate for the removal of his lenses?
2. On a camping trip, Starr Gazer was admiring all the stars that could be seen in the night sky. She noticed a little cluster of dim stars at the edge of her vision. When she looked directly at that part of the sky, however, she could not see the cluster. When she looked toward the stars but not directly at them, she could see them. Explain what was happening.
3. Skin divers are subject to increased pressure as they descend toward the bottom of the ocean. Sometimes this pressure can lead to damage to the ear and loss of hearing. Describe the normal mechanisms that adjust for changes in pressure. Explain how increased pressure might cause reduced hearing and suggest at least one other common condition that might interfere with this pressure adjustment.
4. If a vibrating tuning fork is placed against the mastoid process of the temporal bone, the vibrations are perceived as sound, even if the external acoustic meatus is plugged. Explain how this happens.
5. The main way that people "catch" colds is through their hands. After touching an object contaminated with the cold virus, the person transfers the virus to the nasal cavity where it causes an infection. Other than the obvious entry of the virus through the nose, how could the virus get into the nasal cavity?

Visit this textbook's website at www.mhhe.com/seeleyess6 for practice quizzes, animations, interactive learning exercises, and other study tools.

Endocrine System

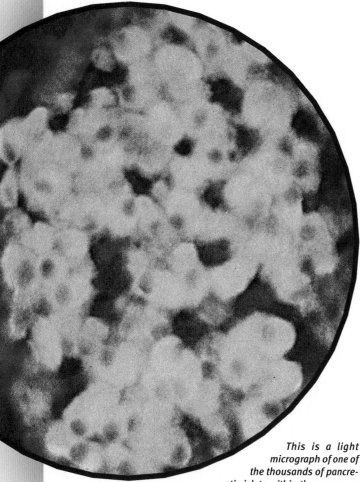

This is a light micrograph of one of the thousands of pancreatic islets within the pancreas. Insulin-secreting cells are stained green, and glucagon-secreting cells are stained red. Insulin binds to receptors on most cells of the body and increases their uptake of glucose. Glucagon binds to receptors, mainly on liver cells, causing them to release glucose into the blood.

homeostasis depends on the precise regulation of the organs and organ systems of the body. The nervous and endocrine systems are the two major systems responsible for that regulation. Together they regulate and coordinate the activity of nearly all other body structures. When these systems fail to function properly, homeostasis is not maintained. Failure of some component of the endocrine system to function can result in diseases such as diabetes mellitus or Addison's disease. Early in the 1900s there were no effective treatments for most endocrine diseases and people with them often died. Now, more is known about the endocrine system, and treatments for many endocrine diseases are available.

The regulatory functions of the nervous and endocrine systems are similar in some respects, but they differ in other important ways. The nervous system controls the activity of tissues by sending action potentials along axons, which release chemical signals at their ends, near the cells they control. The endocrine system releases chemical signals into the circulatory system, which carries them to all parts of the body. The cells that can detect those chemical signals produce responses.

The nervous system usually acts more quickly and has short-term effects, whereas the endocrine system usually responds more slowly and has longer-lasting effects. In general, each nervous stimulus controls a specific tissue or organ, whereas each endocrine stimulus controls several tissues or organs.

Functions of the Endocrine System

The main regulatory functions of the endocrine system include:

1. *Water balance.* The endocrine system regulates water balance by controlling the solute concentration of the blood.
2. *Uterine contractions and milk release.* The endocrine system regulates uterine contractions during delivery of the newborn and stimulates milk release from the breasts in lactating females.
3. *Growth, metabolism, and tissue maturation.* The endocrine system regulates the growth of many tissues, such as bone and muscle, and the rate of metabolism of many tissues, which helps maintain a normal body temperature and normal mental functions. Maturation of tissues, which results in the development of adult features and adult behavior, are also influenced by the endocrine system.
4. *Ion regulation.* The endocrine system regulates Na^+, K^+, and Ca^{2+} concentrations in the blood.
5. *Heart rate and blood pressure regulation.* The endocrine system helps regulate the heart rate and blood pressure and helps prepare the body for physical activity.
6. *Blood glucose control.* The endocrine system regulates blood glucose levels and other nutrient levels in the blood.
7. *Immune system regulation.* The endocrine system helps control the production and functions of immune cells.
8. *Reproductive functions control.* The endocrine system controls the development and the functions of the reproductive systems in males and females.

Chemical Signals

Major categories of chemical signals are presented here to allow the comparison of hormones to other chemical signals. **Chemical signals,** or **ligands** (lī′gandz, to bind), are molecules released from one location that move to another location to produce a response. **Intracellular chemical signals** are produced in one part of a cell, such as the cell membrane, and travel to another part of the same cell and bind to receptors, either in the cytoplasm or in the nucleus of the cell. **Intercellular chemical signals** are released from one cell, are carried in the intercellular fluid, and bind to their receptors, which are found in some cells, but usually not in all cells of the body.

Intercellular chemical signals can be placed into functional categories on the basis of the tissues from which they are secreted and the tissues they regulate (table 10.1).

Autocrine (aw′tō-krin, *autos,* self + *krinō,* to separate) **chemical signals** are released by cells and have a local effect on the same cell type from which the chemical signals are released. Examples include chemicals, such as the eicosanoids (*eicosa,* refers to 20-carbon atoms), that are released from smooth muscle cells and from platelets in response to inflammation. These chemicals cause the relaxation of blood vessel smooth muscle cells and the aggregation of platelets. As a result the blood vessels dilate and blood clots.

Paracrine (par′ă-krin, *para,* alongside + crine) **chemical signals** are released by cells that have effects on other cell types near the cells from which they are released, without being transported in blood. For example, a peptide called somatostatin is released by cells in the pancreas and functions locally to inhibit the secretion of insulin from other cells of the pancreas.

Hormones and neurohormones are intercellular chemical signals secreted into the circulatory system (see section, Hormones, on p. 275). They are carried in the blood to the organs they control, where they bind to receptor molecules and produce a response.

Neuromodulators and **neurotransmitters** are intercellular chemical signals, secreted by nerve cells, which play important roles in the function of the nervous system (see chapter 8).

Pheromones (fer′ō-mōnz, *pherō,* to carry + *hormaō,* to excite) are chemical signals secreted into the environment that modify the behavior and the physiology of other individuals. For example, pheromones released in the urine of cats and dogs at certain times are olfactory signals that indicate fertility. Pheromones also appear to be produced by humans, although their importance is unclear. For example, pheromones

Table 10.1   Functional Classification of Intercellular Chemical Signals

Intercellular Chemical Signal	Description	Example	
Autocrine	Secreted by cells in a local area and influences the activity of the same cell type from which it was secreted	Eicosanoids (prostaglandins, thromboxanes, prostacylins, and leukotrienes)	Autocrine chemical signal
Paracrine	Produced by a wide variety of tissues and secreted into tissue spaces; usually has a localized effect on other tissues	Somatostatin histamine, eicosanoids	Paracrine chemical signal
Hormone	Secreted into the blood by specialized cells; travels some distance to target tissues; influences specific activities	Thyroid hormones, growth hormone, insulin, epinephrine, estrogen, progesterone, testosterone	Hormone
Neurohormone, (often simply referred to as hormones)	Produced by neurons and functions like hormones	Oxytocin, antidiuretic hormone, hypothalamic-releasing and inhibiting hormones	Neuron Neurohormone
Neurotransmitter or neuromodulator	Produced by neurons and secreted into extracellular spaces by presynaptic nerve terminals; travels short distances; influences postsynaptic cells	Acetylcholine, epinephrine	Neurotransmitter Neuron
Pheromone	Secreted into the environment; modifies physiology and behavior of other individuals	Sex pheromones are released by humans and many other animals. They are released in the urine of animals, such as dogs and cats. Pheromones produced by women can influence the timing of the menstrual cycles of other women.	Pheromone

produced by women can influence the timing of the menstrual cycle of other women.

Many intercellular chemical signals consistently fit one specific definition, although others do not. For example, norepinephrine functions both as a neurotransmitter and as a hormone; and prostaglandins function as neurotransmitters, neuromodulators, paracrine chemical signals, and autocrine chemical signals. The schemes used to classify chemicals on the basis of their functions are useful, but they do not indicate that a specific molecule always acts as the same type of chemical signal. For that reason, the study of endocrinology often includes the study of autocrine and paracrine chemical signals in addition to hormones and neurohormones.

Receptors

Chemical signals bind to proteins or glycoprotiens, called **receptor** (*recipio*, to receive) **molecules,** to produce a response. The portion of each receptor molecule where a chemical signal binds is a **receptor site.** The shape and other characteristics of each receptor site allows only a specific chemical signal to bind to it (figure 10.1). The tendency for each receptor site to bind to a specific chemical signal and not to others is called **specificity.**

Receptor Types for Intercellular Chemical Signals

There are two major types of receptor molecules that respond to an intercellular chemical signal (figure 10.2):

1. **Membrane-bound receptors** extend through the cell membrane, with their receptor sites on the outer surface of the cell membrane. These receptors respond to intercellular chemical signals that are large, water-soluble molecules that do not diffuse through the cell membrane. When a chemical signal binds to the receptor site of a

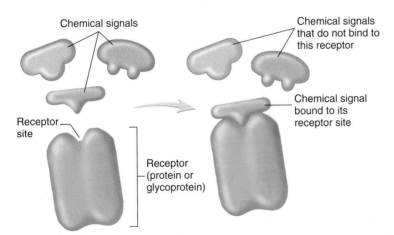

Figure 10.1 **Receptors and Specificity of Receptor Sites**

Chemical signals bind to receptor molecules. The shape and chemical characteristics of each receptor site allow certain chemical signals that have a compatible shape and compatible chemical characteristics to bind to it, not others. This relationship is called specificity.

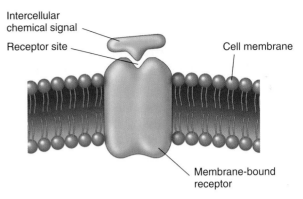

(a) Intercellular chemical signals that are large and water-soluble interact with membrane-bound receptors that extend across the cell membrane and have receptor sites exposed to the outer surface of the cell membrane. The portion of the receptor inside the cell produces the response.

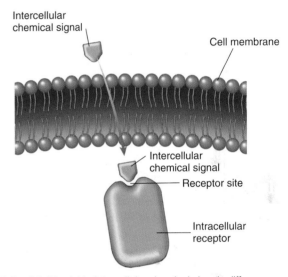

(b) Small lipid-soluble intercellular chemical signals diffuse through the cell membrane and combine with the receptor sites of intracellular receptors. The combination of intercellular chemical signals and receptors produces a response.

Figure 10.2 **Intracellular and Membrane-Bound Receptors**

membrane-bound receptor, the part of the receptor that extends to the inside of the cell produces a response (see figure 10.2a).

2. **Intracellular receptors** are located in either the cytoplasm or nucleus of the cell. Intercellular chemical signals diffuse through the cell membrane and into the cytoplasm of the cell. They then bind to receptor sites on intracellular receptors (see figure 10.2b) either in the cytoplasm or nucleus of the cell.

Receptor Responses

Once intercellular signals bind to their receptors, the combinations produce responses.

Membrane-Bound Receptor Responses

There are many membrane-bound receptors. When intercellular chemical signals bind to their receptors, there are three major mechanisms by which they can produce responses:

1. *Receptors that directly alter membrane permeability.* Some intercellular chemical signals bind to receptor sites, causing ion channels in the cell membrane to open or close. The resulting change in membrane permeability alters the movement of ions across the cell membrane, which is responsible for the cell's response. For example, acetylcholine from presynaptic terminals binds to receptors that are part of the membrane channels for Na^+ in skeletal muscle cell membranes. As a consequence, the Na^+ channels open. Sodium ions diffuse through the open Na^+ channels into the cell to produce an action potential, which stimulates contraction of the skeletal muscle fiber (table 10.2a and figure 10.3).

2. *Receptors and G proteins.* Some intercellular chemical signals bind to membrane-bound receptor sites, and the result is the activation of a complex of proteins at the inner surface of the cell membrane called G proteins (figure 10.4 and table 10.2b). The inactive G protein complex has alpha (α), beta (β), and gamma (γ) subunits bound together. Guanosine diphosphate (GDP) is bound to the α subunit. The α, β, and γ subunit complex can bind to the intercellular portion of the receptor molecule. When an intercellular chemical signal binds to the membrane-bound receptor, the α subunit separates from the β and γ subunits. Guanosine triphosphate (GTP) then replaces guanosine diphosphate (GDP) on the α subunit. This is the activated form of the G protein complex. The α subunit with GTP bound to it can either (1) open or close membrane channels, to produce a response, (2) activate enzymes that produce intracellular chemical signals, or (3) affect gene expression.

Several types of intracellular chemical signals are produced in response to activated G proteins. The type depends on the receptor molecule, the specific function of the G proteins, and the specific enzyme activated in the target tissue. Examples of intracellular chemical signals include **cyclic adenosine monophosphate (cAMP), diacylglycerol** (dī′as-il-glis′er-ol), **(DAG),** and **inosotol** (in-ō′si-tōl) **triphosphate (IP3).** These intracellular chemical signals bind to receptor molecules in the cell and alter their activity to produce responses.

3. *Receptors that directly alter the activity of enzymes.* Some intercellular chemical signals bind to receptor sites and directly increase or decrease the activity of enzymes

Table 10.2	Some Major Intercellular Chemical Signals, Their Source, Target Tissues, and Effects	
Signal	**Source**	**Target Tissue and Effect**
a. Intercellular Chemical Signals That Bind to Membrane-Bound Receptors and Directly Alter Membrane Permeability		
Acetylcholine	Nerve endings	Skeletal muscle, smooth muscle, glands, and neurons; usually opens Na^+ channels
Serotonin	Nerve endings	Neurons; opens Na^+ channels
GABA*	Nerve endings	Neurons; opens Cl^- channels
b. Intercellular Chemical Signals That Bind to Membrane-Bound Receptors and Activate G Proteins		
Luteinizing hormone	Anterior Pituitary	Ovaries; estrogen secretion and ovulation: Testes; testosterone secretion
Thyroid-stimulating hormone	Anterior Pituitary	Thyroid gland; secretion of thyroid hormones
Oxytocin	Posterior Pituitary	Uterus contraction and expulsion of milk from mammary glands
Vasopressin or Antidiuretic hormone	Posterior Pituitary	Kidney; conserves water
c. Intercellular Chemical Signals That Bind to Membrane-Bound Receptors and Alter the Activity of Enzymes		
Insulin	Pancreatic islets	Most cells; increases glucose and amino acid uptake
Growth hormone	Anterior pituitary gland	Most cells; increases protein synthesis and resists protein breakdown
Growth factors	Various tissues	Stimulate cell division and growth in most tissue
d. Chemical Signals That Combine with Intracellular Receptors		
Testosterone	Testis	Responsible for development of the reproductive structures and development of male secondary sexual characteristics
Estrogen	Ovary	Causes increased cell division in the lining of the uterus
Cortisol	Adrenal cortex	Increases breakdown of proteins and fats and increases blood levels of glucose
Triiodothyronine (T_3)	Thyroid gland	Regulates development and metabolism

Abbreviations: GABA: gamma-aminobutyric acid.

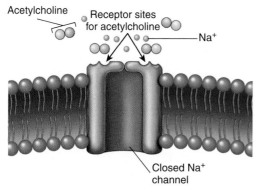

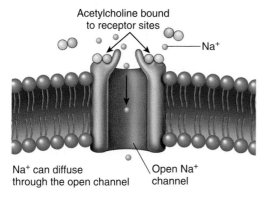

1. The sodium ion (Na⁺) channel has two receptor sites for acetylcholine. The receptor sites are part of the Na⁺ channel. When acetylcholine is not bound to the receptor sites, the Na⁺ channel is closed.

2. When acetylcholine binds to its receptor sites, the Na⁺ channel opens. Na⁺ diffuse into the cell to produce an action potential.

Process Figure 10.3 Intercellular Chemical Signals That Directly Cause Ion Channels in the Cell Membrane to Open

(figure 10.5 and table 10.2c). In some cases the enzymes either increase or decrease the synthesis of an intracellular chemical signal such as cyclic **guanosine monophosphate (cGMP)** (figure 10.6). In some cases the enzyme adds phosphate groups to certain proteins inside the cells. The proteins with phosphates attached to them produce the response of the cell to the chemical signal (figure 10.7 and table 10.2c).

Intracellular Receptor Responses

Intercellular chemical signals that diffuse across cell membranes and bind to intracellular receptors are relatively small and are soluble in lipids. Some intracellular receptors are enzymes. Intercellular chemical signals bind to these kinds of receptors and change the activity of the enzymes. Other intracellular receptors bind to DNA in the nucleus. Intercellular chemical signals bind to these kinds of receptors, which then bind to DNA in the nucleus and increase specific messenger RNA in the nucleus of the cell. The messenger RNA then moves to the ribosomes in the cytoplasm where new proteins are produced (figure 10.8, table 10.2d, and chapter 3).

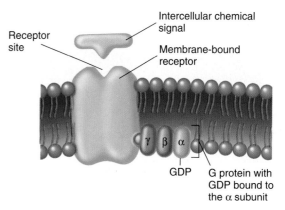

1. The membrane-bound receptor can associate with a G protein at the inner surface of the cell membrane. When the receptor site at the outside of the cell membrane is not occupied, the G protein (α, β, and γ subunits) remains as a unit and inactive.

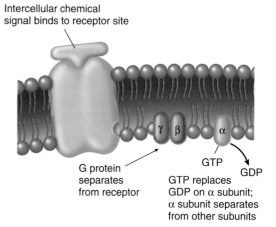

2. The intercellular chemical signal binds to its receptor site on the outside of the cell membrane. The combination causes a G protein to separate from the receptor and the α subunit to separate from the γ and the β subunits. GTP then replaces GDP on the α subunit. The α subunit, with GTP bound to it, is active and can open membrane channels or activate enzymes.

Process Figure 10.4 Membrane-Bound Receptors That Activate G Proteins

Comparison of Receptor Responses

There are some important differences in the responses produced by membrane-bound and intracellular receptors. Because of the time required for the synthesis of new mRNA and proteins, several hours can pass between the time the chemical signals bind to their intracellular receptors and the response. In contrast, chemical signals that bind to membrane-bound receptors often produce rapid responses. A few intercellular chemical signal molecules can bind to their membrane-bound receptors and each activated receptor can produce many intracellular chemical signal molecules. The intracellular chemical signal molecules, in turn, rapidly activate many specific enzymes inside the cell. This pattern of response is called a **cascade** (*cascare,* to fall) **effect** because a few intercellular chemical signals can produce many

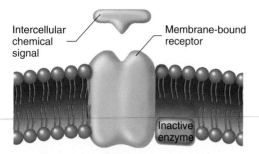

1. The membrane-bound receptor is associated with an enzyme at the inner surface of the cell membrane. When the receptor site at the outside of the cell membrane is not occupied by an intercellular chemical signal, the enzyme remains inactive.

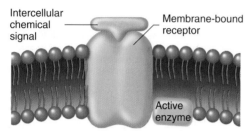

2. The intercellular chemical signal binds to its receptor site on the outside of the cell membrane and causes the enzyme at the inner surface of the cell membrane to become active.

Process Figure 10.5 **Membrane-Bound Receptors That Control Enzyme Activity**

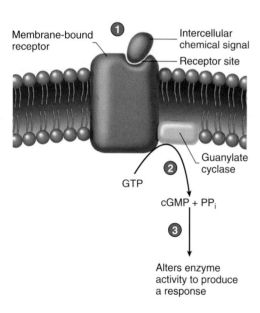

1. An intercellular chemical signal combines with its receptor site on the membrane-bound receptor.

2. The combination activates the enzyme guanylate cyclase at the inner surface of the cell membrane. Guanylate cyclase converts GTP to cyclic GMP plus 2 inorganic phosphate groups (PPᵢ).

3. cGMP is an intracellular chemical signal, and it functions to alter the activity of other intracellular enzymes to produce a response.

Process Figure 10.6 **Membrane-Bound Receptors Activating Intracellular Enzymes That Increase cGMP Synthesis**

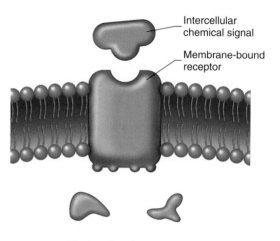

Unphosphorylated receptor and proteins

1. The membrane-bound receptor and other proteins have sites that can be phosphorylated at the inner surface of the cell membrane. When the receptor site at the outside of the cell membrane is not occupied by an intercellular chemical signal, these sites remain unphosphorylated.

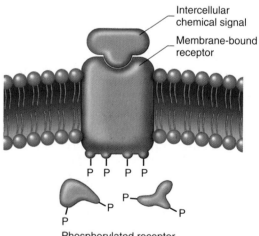

Phosphorylated receptor and proteins result in the cell response

2. The intercellular chemical signal binds to its receptor site on the outside of the cell membrane, activating an enzyme at the inner surface of the cell membrane that phosphorylates the sites on the receptor and associated proteins. These phosphorylated proteins produce a response inside the cell.

Process Figure 10.7 **Membrane-Bound Receptors That Phosphorylate Intracellular Proteins**

1. The lipid-soluble intercellular chemical signal diffuses through the cell membrane and enters the cytoplasm of the cell.

2. The intercellular chemical signal combines with a receptor in the cytoplasm (or in the nucleus).

3. The receptor with the chemical signal bound to it interacts with DNA and increases the synthesis of specific messenger RNA (mRNA) molecules.

4. The mRNA molecule passes from the nucleus to the cytoplasm.

5. In the cytoplasm of the cell, the mRNA molecule combines with ribosomes, and new protein molecules are synthesized.

6. The new proteins produce the response of the cell to the chemical signal.

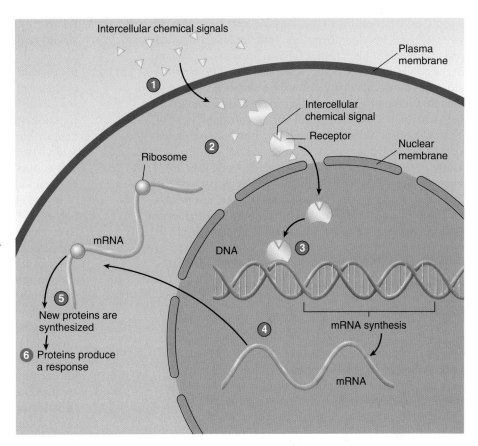

Process Figure 10.8   Intercellular Chemical Signals That Diffuse into Cells and Bind to Intracellular Receptors

intracellular chemical signals, and the intracellular chemical signals can, in turn, quickly activate many proteins inside the cell.

> **P R E D I C T   1**
> A drug binds to a receptor and prevents the response of a target tissue to a chemical signal. It is known that the drug is lipid-soluble and it prevents the synthesis of messenger RNA. Explain how the chemical signal produces a response in its target tissue.

Hormones

The term **endocrine** (en′dō-krin) is derived from the Greek words *endo* and *krino* meaning "within" and "to separate." The word implies that intercellular chemical signals are produced within and secreted from **endocrine glands,** but the chemical signals have effects at locations that are away from, or separate from, the endocrine glands that secrete them. The intercellular chemical signals are transported in the blood to tissues some distance from the glands. Examples of endocrine glands are the thyroid and the adrenal glands. In contrast, **exocrine** (ek′sō-krin, *exō,* outside + crine) **glands** secrete their products into ducts, which exit the glands and carry the secretory products to an external or internal surface, such as the skin or digestive tract. Examples of exocrine glands are the sweat and the salivary glands. The **endocrine system** consists of the endocrine glands of the body (figure 10.9).

The intercellular chemical signals secreted by endocrine glands are called **hormones** (hōr′mōnz), a term derived from the

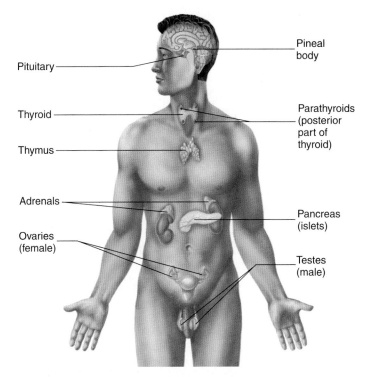

Figure 10.9   The Major Endocrine Glands and Their Locations in the Human Body

Greek word *hormon* meaning to "set into motion," because hormones set responses by cells into motion. Traditionally a hormone is defined as an intercellular chemical signal, produced in minute amounts by a collection of cells, that is secreted into the interstitial fluid and enters the circulatory system to be transported some distance, and that acts on tissues at another site in the body to influence their activity in a specific way. All hormones exhibit most of these characteristics. **Neurohormones** are hormones secreted from cells of the nervous system, although neurohormones are commonly referred to as hormones (see table 10.1 and page 269).

Hormones are distributed in the blood to all parts of the body, but only certain tissues, called **target tissues,** respond to each type of hormone. A target tissue for a hormone is made up of cells that have receptor molecules for the hormone. Each hormone can bind only to its receptor molecules and cannot influence the function of cells that do not have receptor molecules for the hormone (figure 10.10).

Chemistry

Hormones fall into the following chemical categories:

1. *Proteins, peptides, and amino acid derivatives.* Hormones in this category bind to membrane-bound receptors, with the exception of peptide hormones secreted by the thyroid gland, which diffuse through membranes and bind to intracellular receptors.

a. Some hormones are proteins, consisting of many amino acids bound together by peptide bonds. Carbohydrate molecules are bound to some of the protein hormones. Most hormones of the anterior pituitary gland, including those that control functions such as growth, metabolism, and reproductive functions, are examples of protein hormones.

b. Peptide hormones consist of short chains of amino acids. The hormones of the posterior pituitary gland that control functions such as milk "let-down" in the breast and urine volume in the kidneys are examples of peptide hormones.

c. Some hormones consist of single amino acids that have been chemically modified. Hormones secreted by the adrenal medulla, which help a person prepare for physical activity or respond to emergency conditions, are examples.

2. *Lipid hormones.*

a. Steroid hormones are lipids, all of which are derived from cholesterol. The steroids all have a structure that varies only slightly among the different types. The small differences, however, give each type of steroid unique functional characteristics. Steroid hormones are produced mostly by the adrenal cortex and the gonads (testes and ovaries). Hormones in this category diffuse across the cell membrane and bind to intracellular receptor molecules.

b. The **eicosanoids** (ī′kō-să-noydz) make up a class of chemicals derived from the fatty acid arachidonic acid. The eicosanoids include the prostaglandins, thromboxanes, prostacyclins, and leukotrienes. A major group of eicosanoids is the **prostaglandins** (pros′tă-glan′dinz, named after the prostate gland and other accessory glands where it was first found), which are produced by many tissues and generally have local effects. They play important roles in regulating smooth muscle contractions and inflammation. They bind to membrane-bound receptors that are associated with G proteins to produce a response.

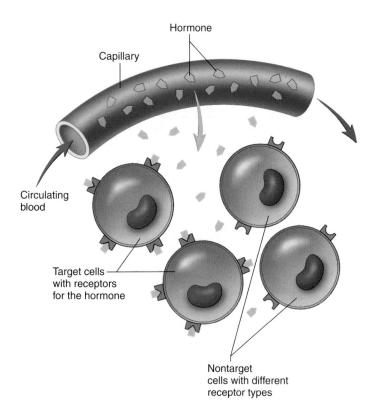

Figure 10.10   **Target Cell Response to Hormones**

Hormones are secreted into the blood and distributed throughout the body, where they diffuse from the blood into the interstitial fluid (*yellow arrow*). Only target cells have receptors to which hormones can bind; therefore, even though a hormone is distributed throughout the body, only target cells for that hormone can respond to it.

Lipid- and Water-Soluble Hormones in Medicine

Specific hormones are given as treatments for certain illnesses. Hormones that are soluble in lipids, such as steroids, can be taken orally because they can diffuse across the wall of the stomach and intestine into the circulatory system. Examples include the synthetic estrogen and progesterone-like hormones in birth control pills and steroids that reduce the severity of inflammation, such as prednisone (pred′ni-sōn). In contrast to lipid-soluble hormones, protein hormones cannot diffuse across the wall of the intestine because they are not lipid-soluble. Furthermore, protein hormones are broken down to individual amino acids before they are transported across the wall of the digestive system. The normal structure of a protein hormone is therefore destroyed, and its physiological activity is lost. Consequently, protein hormones must be injected rather than taken orally. The most commonly administered protein hormone is insulin, which is prescribed for the treatment of diabetes mellitus.

Regulation of Hormone Secretion

The secretion of hormones is controlled by negative-feedback mechanisms (see chapter 1). Negative-feedback mechanisms keep the body functioning within a narrow range of values consistent with life. For example, insulin is a hormone that regulates the concentration of blood glucose, or blood sugar. When blood glucose levels increase after a meal, insulin is secreted. Insulin acts on several target tissues and causes them to take up glucose, causing blood glucose levels to decline. As blood glucose levels begin to fall, however, the rate at which insulin is secreted falls also. As insulin levels fall, the rate at which glucose is taken up by the tissues decreases, keeping the blood glucose levels from declining too much. This negative-feedback mechanism counteracts increases and decreases in blood glucose levels to maintain homeostasis.

Hormone secretion is regulated in three ways. The secretion of some hormones is regulated by one of these methods, whereas the secretion other hormones can be regulated by two or even all of these methods:

1. *Blood levels of chemicals.* The secretion of some hormones is directly controlled by the blood levels of certain chemicals. For example, insulin secretion is controlled by blood glucose levels, and secretion of parathyroid hormone is controlled by blood calcium levels.

2. *Hormones.* The secretion of some hormones is controlled by other hormones. For example, hormones from the pituitary gland act on the ovaries and the testes, causing those organs to secrete sex hormones.

3. *Nervous system.* The secretion of some hormones is controlled by the nervous system. An example is epinephrine, which is released from the adrenal medulla as a result of nervous system stimulation.

Endocrine Glands and Their Hormones

The endocrine system consists of ductless glands, that secrete hormones into the interstitial fluid. The hormones then enter the circulatory system (table 10.3; see figure 10.9). The endocrine glands are supplied by an extensive network of blood vessels; organs with the richest blood supply include endocrine glands such as the adrenal and thyroid glands.

Some glands of the endocrine system perform functions in addition to hormone secretion. For example, the endocrine part of the pancreas has cells that secrete hormones. The much larger exocrine portion of the pancreas secretes digestive enzymes. Portions of the ovaries and testes secrete hormones. Other parts of the ovaries and testes produce oocytes (female sex cells) or sperm cells (male sex cells), respectively.

Table 10.3	Endocrine Glands, Hormones, and Their Target Tissues		
Gland	**Hormone**	**Target Tissue**	**Response**
Pituitary gland Anterior	Growth hormone	Most tissues	Increases protein synthesis, breakdown of lipids, and release of fatty acids from cells; increases blood glucose levels
	Thyroid-stimulating hormone (TSH)	Thyroid gland	Increases thyroid hormone secretion (thyroxine and triiodothyronine)
	Adrenocorticotropic hormone (ACTH)	Adrenal cortex	Increases secretion of glucocorticoid hormones such as cortisol; increases skin pigmentation at high concentrations
	Melanocyte-stimulating hormone (MSH)	Melanocytes in skin	Increases melanin production in melanocytes to make the skin darker in color
	Luteinizing hormone (LH) or interstitial cell-stimulating hormone (ICSH)	Ovary in females, testis in males	Promotes ovulation and progesterone production in the ovary; testosterone synthesis and support for sperm cell production in testis
	Follicle-stimulating hormone (FSH)	Follicles in ovary in females, seminiferous tubules in males	Promotes follicle maturation and estrogen secretion in ovary; sperm cell production in testis
	Prolactin	Ovary and mammary gland in females, testis in males	Stimulates milk production and prolongs progesterone secretion following ovulation and during pregnancy in women; increases sensitivity to LH in males
Posterior	Antidiuretic hormone (ADH)	Kidney	Increases water reabsorption (less H_2O is lost as urine)
	Oxytocin	Uterus	Increases uterine contractions
		Mammary gland	Increases milk "let-down" from mammary glands

continued

Table 10.3 *continued*

Gland	Hormone	Target Tissue	Response
Thyroid gland	Thyroid hormones (thyroxine and triiodothyronine)	Most cells of the body	Increase metabolic rates, essential for normal process of growth and maturation
	Calcitonin	Primarily bone	Decreases rate of bone breakdown; prevents large increase in blood Ca^{2+} levels following a meal
Parathyroid glands	Parathyroid hormone	Bone, kidney	Increases rate of bone breakdown by osteoclasts; increases vitamin D synthesis, essential for maintenance of normal blood calcium levels
Adrenal medulla	Epinephrine mostly, some norepinephrine	Heart, blood vessels, liver, fat cells	Increases cardiac output; increases blood flow to skeletal muscles and heart; increases release of glucose and fatty acids into blood; in general, prepares the body for physical activity
Adrenal cortex	Mineralocorticoids (aldosterone)	Kidneys; to lesser degree, intestine and sweat glands	Increase rate of sodium transport into body; increase rate of potassium excretion; secondarily favor water retention
	Glucocorticoids (cortisol)	Most tissues (e.g., liver, fat, skeletal muscle, immune tissues)	Increase fat and protein breakdown; increase glucose synthesis from amino acids; increase blood nutrient levels; inhibit inflammation and immune response
	Adrenal androgens	Most tissues	Insignificant in males; increase female sexual drive, pubic hair and axillary hair growth.
Pancreas	Insulin	Especially liver, skeletal muscle, adipose tissue	Increases uptake and use of glucose and amino acids
	Glucagon	Primarily liver	Increases breakdown of glycogen and release of glucose into the circulatory system
Reproductive organs Testes	Testosterone	Most tissues	Aids in sperm cell production, maintenance of functional reproductive organs, secondary sexual characteristics, and sexual behavior
Ovaries	Estrogens and progesterone	Most tissues	Aid in uterine and mammary gland development and function, external genitalia structure, secondary sexual characteristics, sexual behavior, and menstrual cycle
Uterus, ovaries, inflamed tissues	Prostaglandins	Most tissues	Mediate inflammatory responses; increase uterine contractions, and ovulation
Thymus gland	Thymosin	Immune tissues	Promotes immune system development and function
Pineal body	Melatonin	At least the hypothalamus	Inhibits secretion of gonadotropin-releasing hormone, thereby inhibiting reproduction

Pituitary and Hypothalamus

The **pituitary** (pi-too′i-tār-rē, *pituita*, phlegm or a thick mucous secretion) **gland** is also called the **hypophysis** (hī-pof′i-sis, *hypo*, under + *physis*, growth). It is a small gland about the size of a pea (figure 10.11). It rests in a depression of the sphenoid bone inferior to the hypothalamus of the brain. The **hypothalamus** (hī′pō-thal′ă-mŭs, *hypo*, under + *thalamos*, bedroom) is an important autonomic nervous system and endocrine control center of the brain located inferior to the thalamus. The pituitary gland is located posterior to the optic chiasma and is connected to the hypothalamus by a stalk called

the **infundibulum** (in-fŭn-dib′ū-lŭm, a funnel). The pituitary gland is divided into two parts: the **anterior pituitary** is made up of epithelial cells derived from the embryonic oral cavity; the **posterior pituitary** is an extension of the brain and is made up of nerve cells. The hormones secreted from each lobe of the pituitary gland are listed in table 10.3.

Hormones from the pituitary gland control the functions of many other glands in the body, such as the ovaries, testes, thyroid gland, and adrenal cortex (see figure 10.11). The pituitary gland also secretes hormones that influence growth, kidney function, birth, and milk production by the breast. The

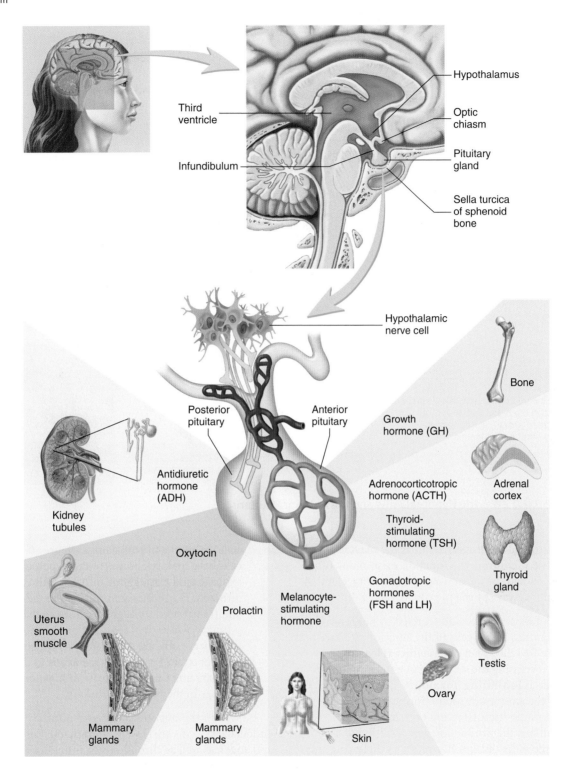

Figure 10.11 The Pituitary Gland Hormones and Their Target Tissues

Overview of the pituitary gland, hormones secreted by the pituitary gland, and their target tissues. Hormones secreted by the posterior pituitary are indicated by the *blue* shading and hormones secreted by the anterior pituitary by *peach* shading. The inset shows the location of the pituitary gland and its relationship to the hypothalamus of the brain. The infundibulum connects the hypothalamus and the pituitary gland.

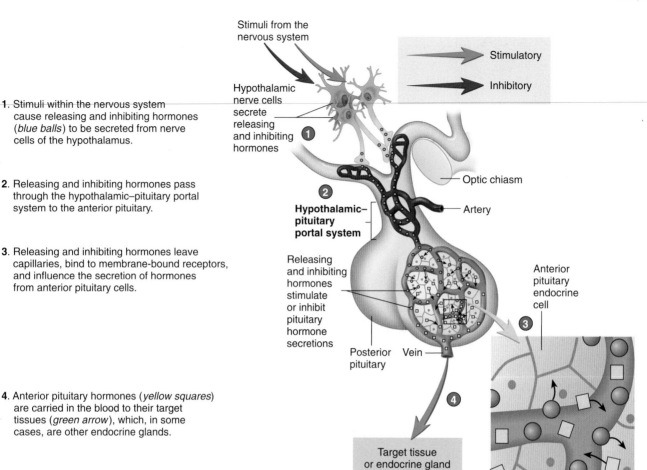

1. Stimuli within the nervous system cause releasing and inhibiting hormones (*blue balls*) to be secreted from nerve cells of the hypothalamus.

2. Releasing and inhibiting hormones pass through the hypothalamic–pituitary portal system to the anterior pituitary.

3. Releasing and inhibiting hormones leave capillaries, bind to membrane-bound receptors, and influence the secretion of hormones from anterior pituitary cells.

4. Anterior pituitary hormones (*yellow squares*) are carried in the blood to their target tissues (*green arrow*), which, in some cases, are other endocrine glands.

Process Figure 10.12a   The Hypothalamus and the Anterior Pituitary

pituitary gland historically was referred to as the **master gland** of the body because it controls the function of so many other glands. It is now known, however, that the pituitary gland is itself controlled in two ways by the hypothalamus of the brain:

1. Neurohormones, produced and secreted by neurons of the hypothalamus, act on cells of the anterior pituitary gland (figure 10.12.*a*). They act as either **releasing hormones,** increasing the secretion of anterior pituitary hormones, or as **inhibiting hormones,** decreasing the secretion of the anterior pituitary hormones. Each releasing hormone stimulates and each inhibiting hormone inhibits the production and secretion of a specific hormone by the anterior pituitary. Releasing and inhibiting hormones enter a capillary bed in the hypothalamus and are transported through veins to a second capillary bed in the anterior pituitary. There they leave the blood and bind to membrane-bound receptors involved with the regulation of anterior pituitary hormone secretion. The capillary beds and veins that transport the releasing and inhibiting hormones are called the **hypothalamic–pituitary** (hī′pō-thal′ă-mik–pi-too′i-tār-ē) **portal** (*porta,* gate) **system.**

2. Secretion of hormones from the posterior pituitary is controlled by nervous system stimulation of nerve

cells within the hypothalamus (figure 10.12*b*). These nerve cells have their cell bodies in the hypothalamus. Their axons extend through the infundibulum to the posterior pituitary. Hormones are produced in the nerve cell bodies and transported through the axons to the posterior pituitary, where they are stored in the axon endings. When these nerve cells are stimulated, action potentials from the hypothalamus travel along the axons to the posterior pituitary and cause the release of hormones from the axon endings in the posterior pituitary.

Within the hypothalamus and pituitary, the nervous and endocrine systems are closely interrelated. Emotions such as joy and anger, as well as chronic stress, influence the endocrine system through the hypothalamus. Conversely, hormones of the endocrine system can influence the functions of the hypothalamus and other parts of the brain.

Hormones of the Anterior Pituitary

Growth hormone (GH) stimulates the growth of bones, muscles, and other organs by increasing protein synthesis. It also resists protein breakdown during periods of food deprivation and favors fat breakdown. Too little growth hormone secretion can be the result of abnormal development of the pituitary gland. A young person suffering from a deficiency of growth hormone remains a small,

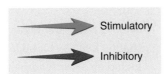

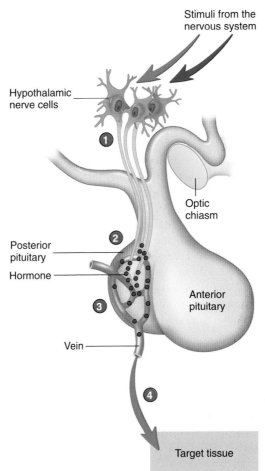

1. Stimuli within the nervous system stimulate hypothalamic nerve cells to produce action potentials.

2. Action potentials are carried by axons of nerve cells to the posterior pituitary. The axons of nerve cells store hormones in the posterior pituitary.

3. In the posterior pituitary gland, action potentials cause the release of hormones (*red balls*) from the axons into the circulatory system.

4. The hormones pass through the circulatory system and influence the activity of their target tissues (*green arrow*).

Process Figure 10.12b   The Hypothalamus and the Posterior Pituitary

though normally proportioned, person called a **pituitary dwarf** (dwōrf). This condition can be treated by administering growth hormone. Excess growth hormone secretion can result from hormone-secreting tumors of the pituitary gland. If excess growth hormone is present before bones complete their growth in length, exaggerated bone growth occurs. The result is **giantism** (jī′an-tizm), and the person becomes abnormally tall. If excess hormone is secreted after growth in bone length is complete, growth in bone diameter, but not in length, continues. As a result, the facial features and hands become abnormally large, a condition called **acromegaly** (ak-rō-meg′ă-lē, *akron*, limb + *megas*, large).

The secretion of growth hormone is controlled by two hormones from the hypothalamus. A releasing hormone stimulates growth hormone secretion, and an inhibiting hormone inhibits its secretion. Most people have a rhythm of growth hormone secretion, with daily peak levels occurring during deep sleep. Growth hormone secretion also increases during periods of fasting and exercise. Blood growth hormone levels do not become greatly elevated during periods of rapid growth, although children tend to have somewhat higher blood levels of growth hormone than do adults. In addition to growth hormone, genetics, nutrition, and sex hormones influence growth.

Part of the effect of growth hormone is influenced by a group of protein chemical signals called **insulin-like growth factors (IGFs)** or **somatomedins** (sō′mă-tō-mē′dinz, *sōma*, body + mediator). Growth hormone increases IGF secretion from tissues such as the liver, and the IGF molecules bind to receptors on cells of tissues such as bone and cartilage and stimulate growth. The IGFs's are similar in structure to insulin and can bind, to some degree, to insulin receptors. Also, insulin, at high concentrations, can bind to IGF receptors.

GH and Genetic Engineering

Because GH is a protein, it is difficult to produce artificially using conventional techniques. However, human genes for GH have been successfully introduced into bacteria using genetic engineering techniques. The gene in the bacteria causes GH synthesis, and the GH can be extracted from the medium in which the bacteria are grown. Thus modern genetic engineering has provided a source of human GH for people who produce inadequate quantities.

P R E D I C T ②

Mr. Hoops has a son who wants to be a basketball player. Mr. Hoops knows something about GH. He asked his son's doctor if he would prescribe some GH for his son so he can grow taller. What do you think the doctor tells Mr. Hoops?

Thyroid-stimulating hormone (**TSH,** thī′royd, refers to shield-shaped) binds to membrane-bound receptors on cells of the thyroid gland and cause the cells to secrete thyroid hormone. When too much TSH is secreted, it causes the thyroid gland to enlarge and secrete too much thyroid hormone. When too little TSH is secreted, the thyroid gland decreases in size, and too little thyroid hormone is secreted. The rate of TSH secretion is increased by a releasing hormone from the hypothalamus.

Adrenocorticotropic (a-drē′nō-kōr′ti-kō-trō′pik, *ad,* near + *ren,* kidney + *cortico,* cortex + *trophrē,* nurture) **hormone (ACTH)** binds to membrane-bound receptors on cells in the cortex of the adrenal glands. ACTH increases the secretion of a hormone from the adrenal cortex called **cortisol** (kōr′ti-sol), also called hydrocortisone, and ACTH is required to keep the adrenal cortex from degenerating. ACTH molecules also bind to melanocytes in the skin and increase skin pigmentation (see chapter 5). One symptom of too much ACTH secretion is a darkening of the skin. The rate of ACTH secretion is increased by a releasing hormone from the hypothalamus.

Gonadotropins (gō′nad-ō-trō′pinz, *gonē,* seed + *trope,* a turning) are hormones that bind to membrane-bound receptors on the cells of the gonads (ovaries and testes). They regulate the growth, development, and functions of the gonads. In females, **luteinizing** (loo′tē-ĭ-nīz-ing, refers to the corpus luteum of the ovary) **hormone (LH)** causes the ovulation of oocytes and the secretion of the sex hormones estrogen and progesterone from the ovaries. In males, LH stimulates the secretion of the sex hormone testosterone from the testes. LH is sometimes referred to as **interstitial** (in-ter-stish′ăl, spaces in a structure) **cell-stimulating hormone (ICSH)** in males because it stimulates interstitial cells of the testes to secrete testosterone. **Follicle-stimulating hormone (FSH)** stimulates the development of follicles in the ovaries and sperm cells in the testes. Without LH and FSH, the ovaries and testes decrease in size, no longer produce oocytes or sperm cells, and no longer secrete hormones. A single releasing hormone from the hypothalamus increases the secretion of both LH and FSH.

Prolactin (prō-lak′tin, *pro,* precursor + *lact,* milk) binds to membrane-bound receptors in cells of the breast and helps promote development of the breast during pregnancy and stimulates the production of milk in the breast following pregnancy. The regulation of prolactin secretion is complex, and several substances released from the hypothalamus may regulate its secretion. There are two main releasing hormones: one increases prolactin secretion and one decreases it.

Melanocyte-stimulating (mel′ă-nō-sīt, *melas,* black + *kytos,* cell) **hormone (MSH)** binds to membrane-bound receptors on melanocytes and causes them to synthesize melanin. Oversecretion of MSH causes the skin to darken. The structure of MSH is similar to that of ACTH, and both hormones cause the skin to darken. Regulation of MSH is not well understood, but there appears to be two releasing hormones from the hypothalamus: one increases MSH secretion and one decreases it.

Hormones of the Posterior Pituitary

Antidiuretic (an′tē-dī-ū-ret′ik, *anti,* against + *uresis,* urine volume) **hormone (ADH)** binds to membrane-bound receptors and increases water reabsorption by kidney tubules. This results in less water lost as urine. ADH can also cause blood vessels to constrict when released in large amounts. Consequently, it is sometimes called **vasopressin** (vā-sō-pres′in, *vaso,* blood vessel + *pressum,* to press down). Reduced ADH release from the posterior pituitary results in the formation of large amounts of dilute urine.

A lack of ADH secretion causes diabetes insipidus, which is the production of a large amount of dilute urine. The consequences of diabetes insipidus are not obvious until the condition becomes severe. When the condition is severe, the volume of urine produced is many liters each day. The result is an increase in the osmolality of the body fluids and the loss of important electrolytes such as Ca^{2+}, Na^+, and K^+ in the large urine volume.

Oxytocin (ok′sĭ-tō′sin, swift birth) binds to membrane-bound receptors and causes contraction of the smooth muscle cells of the uterus and milk ejection, or milk "let-down," from the breasts in lactating women. Commercial preparations of oxytocin are given under certain conditions to assist in childbirth and to constrict uterine blood vessels following childbirth.

Willy Leek was recovering from an emergency appendectomy. Three days after the surgery he developed a reduced sense of taste, a poor appetite, fatigue, weakness, abdominal cramps, and vomiting. Willy's kidneys retained a large volume of water. His urine volume was low, his urine was very concentrated, and his blood Na^+ levels were reduced (hyponatremia). Willy produced a small volume of concentrated urine because of elevated ADH secretion. Many of the manifestations he exhibited were caused by the effect of hyponatremia on the nervous system. Fortunately, the condition was temporary and lasted only a few days. During that time, Willy's water intake was restricted and a hypertonic saline solution was administered. This kept his blood Na^+ levels within normal levels. This condition occurs in some cases following surgery as the result of abnormally increased ADH secretion. It is usually temporary.

Thyroid Gland

The **thyroid** (thī'royd) **gland** is made up of two lobes connected by a narrow band called the **isthmus** (is'mŭs, a constriction). The lobes are located on either side of the trachea, just inferior to the larynx (figure 10.13a and b). The thyroid

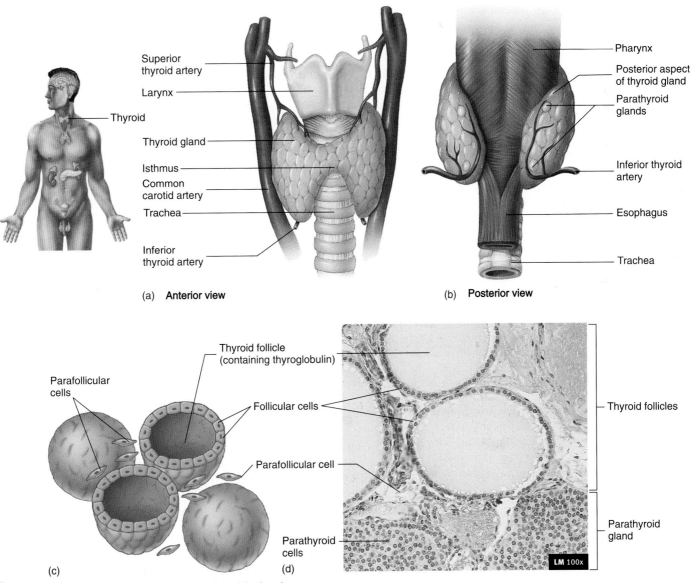

(a) **Anterior view**

(b) **Posterior view**

(c)

(d)

Figure 10.13   The Thyroid and Parathyroid Glands

(a) Anterior view of the thyroid gland. (b) Posterior view of the thyroid gland with the four small parathyroid glands embedded in the posterior surface of the thyroid gland. (c) Three-dimensional interpretive drawing of thyroid follicles and parafollicular cells. (d) Light micrograph of thyroid and parathyroid tissue.

gland is one of the largest endocrine glands. It appears more red than surrounding tissues because it is highly vascular. It is surrounded by a connective tissue capsule. The thyroid gland contains numerous **thyroid follicles,** which are small spheres with walls that consist of simple cuboidal epithelium (figure 10.13c and d). Each thyroid follicle is filled with proteins to which thyroid hormones are attached. The cells of the thyroid follicles synthesize thyroid hormones, which are stored in the follicles. Between the follicles is a network of loose connective tissue that contains capillaries and scattered **parafollicular** (par-ă-fo-lik′ū-lăr, refers to cells beside the thyroid follicle) **cells** (figure 10.13d).

The main function of the thyroid gland is to secrete **thyroid hormones,** which bind to intracellular receptors in cells and regulate the rate of metabolism in the body (see table 10.3). Also, growth and development cannot proceed normally without a normal rate of thyroid hormone secretion. A lack of thyroid hormones is called **hypothyroidism** (hī′pō-thī′royd-izm). In infants, hypothyroidism can result in **cretinism** (krē′tin-izm), a condition in which the person is mentally retarded and has a short stature with abnormally formed skeletal structures. In adults, the lack of thyroid hormones results in a reduced rate of metabolism, sluggishness, a reduced ability to perform routine tasks, and **myxedema** (mik-se-dē′mă), which is the accumulation of fluid and other molecules in the subcutaneous tissue. An elevated rate of thyroid hormone secretion is known as **hyperthyroidism** (hī-per-thī′royd-izm). It results in an elevated rate of metabolism, extreme nervousness, and chronic fatigue. **Graves' disease** is a type of hyperthyroidism resulting from the production of abnormal proteins by the immune system that are similar in structure and function to TSH. Graves' disease is often accompanied by bulging of the eyes, a condition, called **exophthalmia** (ek-sof-thal′mē-ă, *ex*, out + *ophthalmos,* eye) (see Systems Pathology "Graves' Disease" on p. 296).

The thyroid gland requires iodine to synthesize thyroid hormones. Iodine is taken up by the thyroid follicles in which hormone synthesis occurs. One thyroid hormone, called **thyroxine** (thī-rok′sin) or **tetraiodothyronine** (tet′ră-ī′ō-dō-thī′rō-nēn), contains four iodine atoms and is abbreviated T_4. The other thyroid hormone, called **triiodothyronine** (trī-ī′ō-dō-thī′rō-nēn), contains three iodine atoms and is abbreviated T_3. If the quantity of iodine present is not sufficient, the production and secretion of the thyroid hormones decrease. Thyroid hormones are stored in combination with a protein called thyroglobulin (thī-rō-glob′ū-lin) within the thyroid follicles.

A lack of iodine in the diet results in reduced T_3 and T_4 synthesis. A deficiency of iodine is not as common as it used to be. Table salt with iodine added to it (iodized salt) is available in most grocery stores, and vegetables are shipped throughout the country so the vegetables grown in soil rich in iodine are available in most places.

Thyroid hormone secretion is regulated by thyroid-stimulating hormone (TSH) from the anterior pituitary (figure 10.14). Small fluctuations occur in blood TSH levels on a daily basis, with a small increase at night. Increasing blood levels of TSH increase the synthesis of thyroid hormones and the release of thyroid hormones from thyroglobulin. Decreasing blood levels of TSH decrease the synthesis and release of thyroid hormones.

The thyroid hormones have a negative-feedback effect on the hypothalamus and pituitary so that increasing levels of thyroid hormones inhibit the secretion of TSH-releasing hormone from the hypothalamus and inhibit TSH secretion from the anterior pituitary gland. Decreasing thyroid hormone levels allow additional releasing hormone and TSH to be secreted. Because of the negative-feedback effect, the thyroid hormones fluctuate within a narrow concentration range in the blood.

When thyroid hormone secretion is reduced below normal levels, the secretion of TSH-releasing hormone from the hypothalamus and TSH secretion from the anterior pituitary gland increase substantially. Excess TSH causes the thyroid gland to enlarge, a condition called a **goiter** (goy′ter). Iodine deficiency goiters develop if iodine in the diet is too low. Less T_3 and T_4 is synthesized and secreted. Consequently, TSH-releasing hormone and TSH secretion increase above normal levels and cause dramatic enlargement of the thyroid gland.

P R E D I C T   3

In people with Graves' disease (hyperthyroidism), the immune system produces a large amount of a protein that is so much like TSH that it binds to cells of the thyroid gland and acts like TSH. Unlike TSH, however, the secretion of this protein does not respond to negative feedback. Predict the effect of this abnormal protein on the structure and function of the thyroid gland and the release of hormones from the hypothalamus and anterior pituitary gland.

In addition to secreting thyroid hormones, the thyroid gland secretes a hormone called **calcitonin** (kal-si-tō′nin) from the parafollicular cells (see figure 10.13c). Calcitonin is secreted if the blood concentration of Ca^{2+} becomes too high, and it causes Ca^{2+} levels to decrease to their normal range (figure 10.15). Calcitonin binds to membrane-bound receptors of osteoclasts and reduces the rate of Ca^{2+} resorption from bone by inhibiting them. Calcitonin may function to prevent blood Ca^{2+} levels from becoming overly elevated following a meal that contains a high concentration of Ca^{2+}.

Calcitonin helps prevent elevated blood Ca^{2+} levels, but a lack of calcitonin secretion does not result in a prolonged increase in those levels. Other mechanisms controlling blood Ca^{2+} levels are able to compensate for the lack of calcitonin secretion.

Parathyroid Glands

Four tiny **parathyroid** (par-ă-thī′royd) **glands** are embedded in the posterior wall of the thyroid gland (see figure 10.13b and d). The parathyroid glands secrete a hormone called **parathyroid hormone (PTH),** which is essential for the

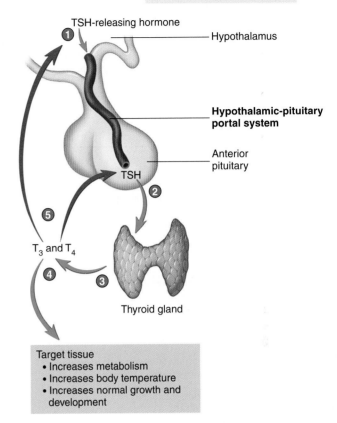

1. TSH-releasing hormone is released from neurons within the hypothalamus into the blood. It passes through the hypothalamic-pituitary portal system to the anterior pituitary.

2. TSH-releasing hormone causes cells of the anterior pituitary to secrete (TSH), which passes through the general circulation to the thyroid gland.

3. TSH causes increased release of thyroid hormones (T_3 and T_4) into the general circulation.

4. T_3 and T_4 act on target tissues to produce a response.

5. T_3 and T_4 also have an inhibitory effect on the secretion of TSH-releasing hormone from the hypothalamus and TSH from the anterior pituitary.

Process Figure 10.14   Regulation of Thyroid Hormone (T_3 and T_4) Secretion

regulation of blood calcium levels (see table 10.3). PTH is more important than calcitonin in regulating blood levels of Ca^{2+}. PTH binds to membrane-bound receptors of renal tubule cells, which increases active vitamin D formation. Vitamin D causes the epithelial cells of the intestine to increase Ca^{2+} absorption. PTH also binds to receptors on osteoblasts. Substances released by the osteoblasts increase osteoclast activity and cause resorption (breakdown) of bone tissue to release Ca^{2+} into the circulatory system. PTH binds to receptors on cells of the renal tubules and decreases the rate at which Ca^{2+} are lost in the urine. PTH acts on its target tissues to raise blood Ca^{2+} levels to normal.

Vitamin D is produced from precursors in the skin that are modified by the liver and kidneys. Ultraviolet light acting on the skin is required for the first stage of vitamin D synthesis, and the final stage of synthesis in the kidney is stimulated by PTH. Vitamin D can also be supplied in the diet.

P R E D I C T   **4**
Explain why a lack of vitamin D results in bones that are softer than normal.

Decreasing blood Ca^{2+} levels stimulate an increase in PTH secretion (see figure 10.15). For example, if too little Ca^{2+} is consumed in the diet or if a person suffers from a prolonged lack of vitamin D, blood Ca^{2+} levels decrease, and PTH secretion increases. The increased PTH increases the rate of bone resorption. Blood Ca^{2+} levels can be maintained within a normal range, but prolonged resorption of bone results in reduced bone density. The reduced bone density causes soft, flexible bones that are easily deformed in young people and porous, fragile bones in older people.

Increasing blood Ca^{2+} levels cause a decrease in PTH secretion (see figure 10.15). The decreased PTH secretion results in a reduction in blood Ca^{2+} levels. In addition, increasing blood Ca^{2+} levels stimulate calcitonin secretion, which also causes blood Ca^{2+} levels to decline (see figure 10.15).

An abnormally high rate of PTH secretion is called **hyperparathyroidism.** It can result from a tumor of a parathyroid gland. The elevated blood levels of PTH increase bone resorption and elevate blood Ca^{2+} levels. As a result, bones can become soft, deformed, and easily fractured. In addition,

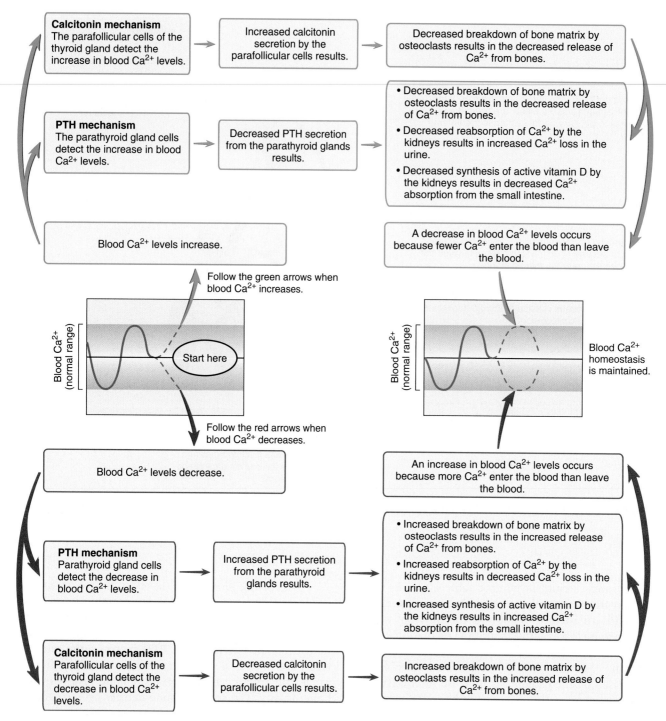

Homeostasis Figure 10.15   Regulation of Blood Calcium Ion Levels

the elevated blood Ca^{2+} levels make nerve and muscle less excitable, resulting in fatigue and muscle weakness. The excess Ca^{2+} can be deposited in soft tissues of the body, and kidney stones can result. The Ca^{2+} deposits in soft tissues cause inflammation.

An abnormally low rate of PTH secretion is called **hypoparathyroidism.** It can be the result of injury or the surgical removal of the thyroid and parathyroid glands. The low blood levels of PTH result in a reduced rate of bone resorption and reduced vitamin D formation. As a result, blood Ca^{2+} levels decrease. In response to low blood Ca^{2+} levels, nerves and muscles become excitable and produce spontaneous action potentials. The result is frequent muscle cramps or tetanus. Severe tetanus can affect the respiratory muscles; breathing stops, resulting in death.

Adrenal Glands

The **adrenal** (ă-drē′năl, near or on the kidneys) **glands,** or **suprarenal glands,** are two small glands that are located superior to each kidney (see figure 10.16 and table 10.3). Each adrenal gland has an inner part, called the **adrenal medulla** (marrow or middle), and an outer part, called the **adrenal cortex** (bark or outer). The adrenal medulla and the adrenal cortex function as separate endocrine glands.

Adrenal Medulla

The principal hormone released from the adrenal medulla is **epinephrine** (ep′i-nef′rin, *epi*, upon + *nephros*, kidney), or **adrenalin** (ă-dren′ă-lin, from the adrenal gland), but small amounts of **norepinephrine** (nōr′ep-i-nef′rin) are also released.

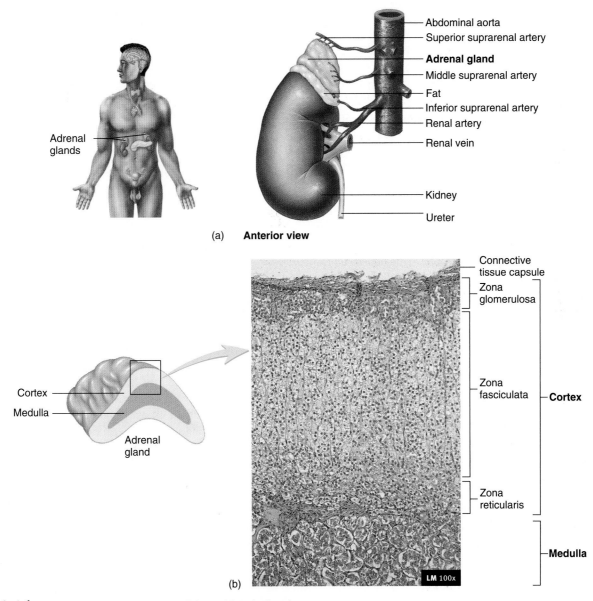

(a)   **Anterior view**

(b)

Figure 10.16   Anatomy and Histology of the Adrenal Gland

(*a*) An adrenal gland is at the superior pole of each kidney. (*b*) The adrenal glands have an outer cortex and an inner medulla. The cortex is surrounded by a connective tissue capsule.

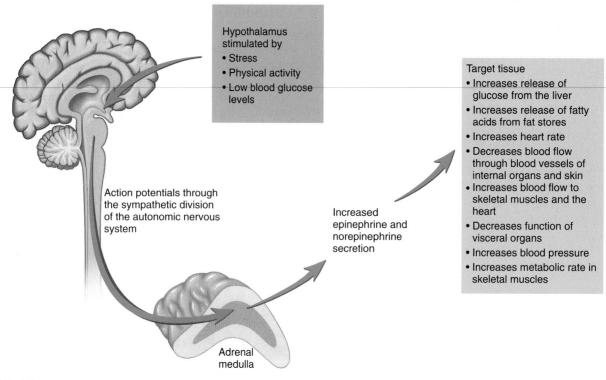

Figure 10.17 **Regulation of Adrenal Medullary Secretions**

Stimulation of the hypothalamus by stress, physical activity, or low blood glucose levels causes action potentials to travel through the sympathetic nervous system to the adrenal medulla. In response, the adrenal medulla releases epinephrine and smaller amounts of norepinephrine into the general circulation. These hormones have several effects on the body to prepare it for physical activity.

Epinephrine and norepinephrine are released in response to stimulation by the sympathetic nervous system, which becomes most active when a person is excited or physically active (figure 10.17). These hormones bind to membrane-bound receptors in their target tissues. Stress and low blood glucose levels can also result in increased sympathetic stimulation of the adrenal medulla. Epinephrine and norepinephrine are referred to as the **fight-or-flight** hormones because of their role in preparing the body for vigorous physical activity. Some of the major effects of the hormones released from the adrenal medulla are:

1. Increases in the breakdown of glycogen to glucose in the liver, the release of the glucose into the blood, and the release of fatty acids from fat cells. The glucose and fatty acids are used as energy sources to maintain the body's increased rate of metabolism.
2. Increase in heart rate, which causes blood pressure to increase.
3. Stimulation of smooth muscle in the walls of arteries supplying the internal organs and the skin, but not those supplying skeletal muscle. Blood flow to internal organs and the skin decreases, and functions of the internal organs decrease. Blood flow through skeletal muscles increases.
4. Increase in blood pressure because of smooth muscle contraction in the walls of blood vessels in the internal organs and the skin.

5. Increase in the metabolic rate of several tissues, especially skeletal muscle, cardiac muscle, and nervous tissue.

Responses to hormones from the adrenal medulla reinforce the effect of the sympathetic division of the autonomic nervous system. Thus the adrenal medulla and the sympathetic division function together to prepare the body for physical activity, to produce the "fight-or-flight" response, and to produce many of the responses to stress.

Pheochromocytoma and Neuroblastoma

The two major disorders of the adrenal medulla are both tumors: pheochromocytoma (fē′ō-krō′mō-sī-tō′mă), a benign tumor, and neuroblastoma (noor′ō-blas-tō′mă), a malignant tumor. Symptoms result from the release of large amounts of epinephrine and norepinephrine and include hypertension (high blood pressure), sweating, nervousness, pallor, and tachycardia (rapid heart rate). The high blood pressure results from the effect of these hormones on the heart and blood vessels and is correlated with an increased chance of heart disease and stroke.

Adrenal Cortex

Three classes of steroid hormones are secreted from the adrenal cortex. The molecules of all three classes of steroid hormones enter their target cells and bind to intracellular receptor molecules. The hormones and the receptors of each

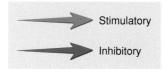

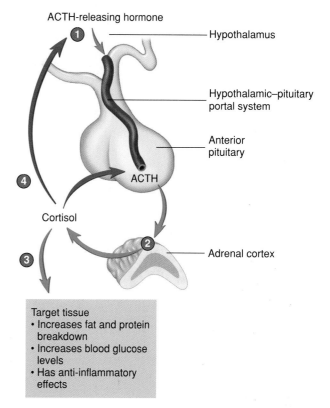

1. In response to stress or low blood glucose, ACTH-releasing hormone passes from the hypothalamus through the hypothalamic–pituitary portal system to the anterior pituitary. The releasing hormone binds to and stimulates cells that secrete ACTH into the general circulation.

2. ACTH acts on the adrenal cortex and stimulates the secretion of cortisol into the general circulation.

3. Cortisol acts on its target tissues to increase protein breakdown and increase blood glucose.

4. Cortisol acts on the hypothalamus and anterior pituitary to decrease ACTH secretion.

Target tissue
• Increases fat and protein breakdown
• Increases blood glucose levels
• Has anti-inflammatory effects

Process Figure 10.18   Regulation of Cortisol Secretion from the Adrenal Cortex

class, however, have unique structural and functional characteristics.

The **glucocorticoids** (gloo-kō-kōr′ti-koydz) help regulate blood nutrient levels in the body. The major glucocorticoid hormone is **cortisol** (kōr′ti-sol), which increases the breakdown of protein and fat and increases their conversion to forms that can be used as energy sources by the body. For example, cortisol acts on the liver, causing it to convert amino acids to glucose, and it acts on adipose tissue, causing fat stored in fat cells to be broken down to fatty acids. The glucose and fatty acids are released into the circulatory system, taken up by tissues, used as a source of energy. Cortisol also causes proteins to be broken down to amino acids, which are then released into the circulatory system (figure 10.18).

Cortisol reduces the inflammatory and immune responses. **Cortisone** (kōr′ti-sōn), a steroid closely related to cortisol, or closely related drugs, are often given as a medication to reduce inflammation that occurs in response to injuries. It is also given to reduce the immune and inflammatory responses that occur as a result of allergic reactions or diseases resulting from abnormal immune responses, such as rheumatoid arthritis or asthma. In response to stressful conditions, cortisol is secreted in larger than normal amounts. It aids the body in responding to stressful conditions by providing energy sources for tissues. However, if stressful conditions are prolonged, the immune system can be suppressed enough to make the body susceptible to stress-related conditions (see Clinical Focus, page 295).

When blood glucose levels decline, cortisol secretion increases. The low blood glucose acts on the hypothalamus to increase the secretion of the ACTH-releasing hormone, which, in turn, stimulates ACTH secretion from the anterior pituitary. ACTH, in turn, stimulates cortisol secretion.

Adrenocorticotropic hormone (ACTH) molecules from the anterior pituitary bind to membrane-bound receptors and regulate the secretion of cortisol from the adrenal cortex. Without ACTH, the adrenal cortex atrophies and loses its ability to secrete cortisol.

P R E D I C T   5
Cortisone, a drug similar to cortisol, is sometimes given to people who have severe allergies or extensive inflammation, or to people who suffer from autoimmune diseases. Taking this substance for long periods of time can damage the adrenal cortex. Explain how this damage can occur.

The second class of hormones secreted by the adrenal cortex, the **mineralocorticoids** (min′er-al-ō-kōr′ti-koydz), help regulate blood volume and blood levels of K^+ and Na^+. **Aldosterone** (al-dos′ter-ōn) is the major hormone of this class (figure 10.19). Aldosterone binds to receptor molecules primarily in the kidney, but it also affects the intestine, sweat glands, and salivary glands. Aldosterone causes Na^+ and H_2O to be retained in the body and increases the rate at which K^+ are eliminated.

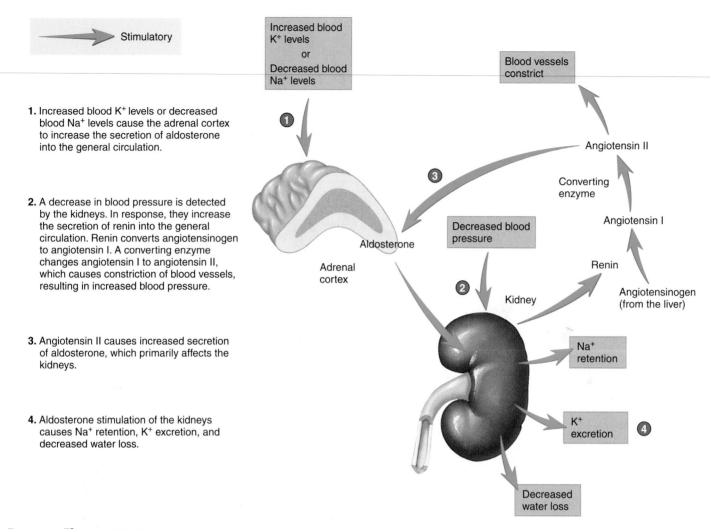

1. Increased blood K$^+$ levels or decreased blood Na$^+$ levels cause the adrenal cortex to increase the secretion of aldosterone into the general circulation.

2. A decrease in blood pressure is detected by the kidneys. In response, they increase the secretion of renin into the general circulation. Renin converts angiotensinogen to angiotensin I. A converting enzyme changes angiotensin I to angiotensin II, which causes constriction of blood vessels, resulting in increased blood pressure.

3. Angiotensin II causes increased secretion of aldosterone, which primarily affects the kidneys.

4. Aldosterone stimulation of the kidneys causes Na$^+$ retention, K$^+$ excretion, and decreased water loss.

Process Figure 10.19   Regulation of Aldosterone Secretion from the Adrenal Cortex

Blood levels of K$^+$ and Na$^+$ directly affect the adrenal cortex to influence aldosterone secretion. The adrenal gland is much more sensitive to changes in blood K$^+$ levels than to changes in blood Na$^+$ levels. The rate of aldosterone secretion increases when blood K$^+$ levels increase or when blood Na$^+$ levels decrease.

Changes in blood pressure indirectly affect the rate of aldosterone secretion. Low blood pressure causes the release of a protein molecule called **renin** (rē′nin, from kidneys) from the kidney. Renin, which acts as an enzyme, causes a blood protein called **angiotensinogen** (an′jē-ō-ten-sin′ō-jen, *angio*, blood vessel) to be converted to **angiotensin I** (an-jē-ō-ten′sin). Then, a protein called **angiotensin-converting enzyme** causes angiotensin I to be converted to **angiotensin II.** Angiotensin II causes smooth muscle in blood vessels to constrict, and angiotensin II acts on the adrenal cortex to increase aldosterone secretion. Aldosterone causes retention of sodium and water, which causes an increase in blood volume (see figure 10.19). Both blood vessel constriction and increased blood volume help raise blood pressure.

P R E D I C T   6
Predict the effects of reduced aldosterone secretion on blood levels of Na$^+$, K$^+$, and blood pressure.

The third class of hormones secreted by the adrenal cortex is the **androgens** (an′drō-jenz, *andros,* male). They are named for their ability to stimulate the development of male sexual characteristics. Small amounts of androgens are secreted from the adrenal cortex in both males and females. In adult males, most androgens are secreted by the testes. In adult females, the adrenal androgens influence the female sex drive. If the secretion of sex hormones from the adrenal cortex is abnormally high, exaggerated male characteristics develop in both males and females. This condition is most apparent in females and in males before puberty, when the effects are not masked by the secretion of androgens by the testes.

Pancreas, Insulin, and Diabetes

The endocrine part of the **pancreas** (pan′krē-as) consists of **pancreatic islets** (small islands; islets of Langerhans)

Figure 10.20   Structure of the Pancreas

The endocrine portion of the pancreas is made up of scattered pancreatic islets. Alpha cells secrete glucagon, and beta cells secrete insulin. The exocrine portion of the pancreas surrounds the pancreatic islets and produces digestive enzymes that are carried through a system of ducts to the small intestine. The stain used for the light micrograph does not distinguish between alpha and beta cells.

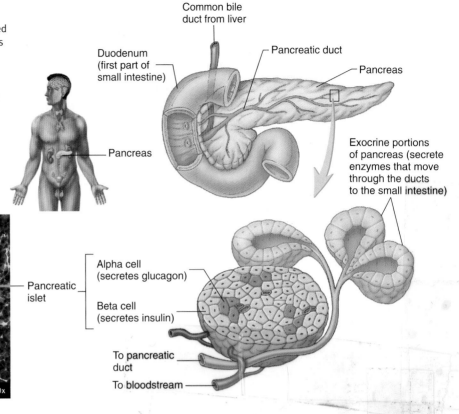

dispersed among the exocrine portion of the pancreas (figure 10.20). The islets secrete two hormones—insulin and glucagon—which function to help regulate blood nutrient levels, especially blood glucose (table 10.4). **Alpha cells** of the pancreatic islets secrete glucagon, and **beta cells** of the pancreatic islets secrete insulin.

It is very important to maintain blood glucose levels within a normal range of values (figure 10.21). A decline in the blood glucose level below its normal range causes the nervous system to malfunction because glucose is the nervous system's main source of energy. When blood glucose decreases, fats and proteins are broken down rapidly by other tissues to provide an alternative energy source. As fats are broken down, some of the fatty acids are converted by the liver to acidic **ketones** (kē′tōnz), which are released into the circulatory system. When blood glucose levels are very low, the breakdown of fats can cause the release of enough fatty acids and ketones to cause the pH of the body fluids to decrease below normal, a condition called **acidosis** (as-i-dō′sis). The amino acids of proteins are broken down and used to synthesize glucose by the liver.

If blood glucose levels are too high, the kidneys produce large volumes of urine containing substantial amounts of glucose. Because of the rapid loss of water in the form of urine, dehydration can result.

Insulin (in′sŭ-lin, *insula,* island) is released from the beta cells primarily in response to the elevated blood glucose levels and increased parasympathetic stimulation that is associated with digestion of a meal. Increased blood levels of certain amino acids also stimulate insulin secretion. Decreased insulin secretion results from decreasing blood glucose levels and from stimulation by the sympathetic division of the nervous system. Sympathetic stimulation of the pancreas occurs during physical activity. Decreased insulin levels allow blood glucose to be conserved to provide the brain with adequate glucose and to allow other tissues to metabolize fatty acids and glycogen stored in the cells.

The major target tissues for insulin are the liver, adipose tissue, muscles, and the area of the hypothalamus that controls appetite, called the **satiety** (sa-tī′-ĕ-tē, fulfillment of hunger) **center.** Insulin binds to membrane-bound receptors and, either directly or indirectly, increases the rate of glucose and amino acid uptake in these tissues. Glucose is converted to glycogen or fat, and the amino acids are used to synthesize protein. The effects of insulin on target tissues are summarized in table 10.4.

Diabetes mellitus (dī-ă-bē′tēz′ me-lī′tŭs, much urine + honey or sweetened) can result from any of the following: Type 1 diabetes mellitus is caused by the secretion of too little insulin from the pancreas and type 2 diabetes mellitus is caused by insufficient numbers of insulin receptors on target cells, or defective receptors that do not respond normally to insulin.

In people who have type 1 diabetes mellitus, tissues cannot take up glucose effectively, causing blood glucose levels to become very high, a condition called **hyperglycemia** (hī′per-glī-sē′mē-ă, *hyper,* above + *glycemia,* blood glucose). Because glucose cannot enter cells of the satiety center of the brain without insulin, the satiety center responds as if there were very little blood glucose, resulting in an exaggerated appetite. The excess glucose in the blood is excreted in the urine, causing the urine volume to be much greater than normal. Because of

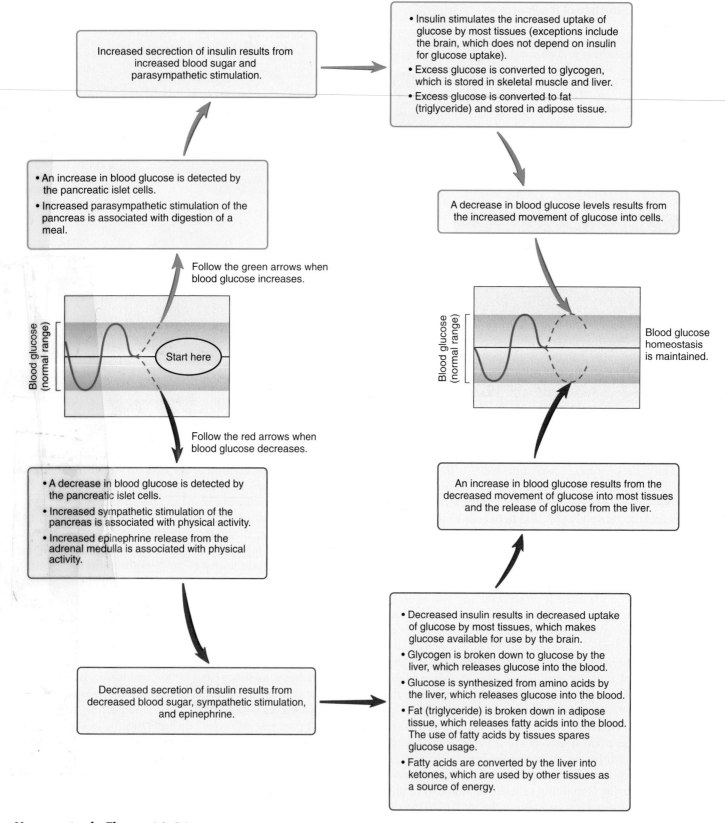

Homeostasis Figure 10.21   Regulation of Blood Glucose Levels

Table 10.4 Effects of Insulin and Glucagon on Target Tissues		
Target Tissue	**Insulin Responses**	**Glucagon Responses**
Skeletal muscle, cardiac muscle, cartilage, bone fibroblasts, blood cells, and mammary glands	Increases glucose uptake and glycogen synthesis; increases uptake of amino acids	Has little effect
Liver	Increases glycogen synthesis; increases use of glucose for energy	Causes rapid increase in the breakdown of glycogen to glucose and release of glucose into the blood; increases the formation of glucose from amino acids and, to some degree, from fats; increases metabolism of fatty acids
Adipose cells	Increases glucose uptake, glycogen synthesis, fat synthesis	High concentrations cause breakdown of fats; probably unimportant under most conditions
Nervous system	Has little effect except to increase glucose uptake in the satiety center	Has no effect

excessive urine production, the person has a tendency to become dehydrated and thirsty. Even though blood glucose levels are high, fats and proteins are broken down to provide an energy source for metabolism, resulting in the wasting away of body tissues, acidosis, and ketosis. People with this condition also exhibit a lack of energy.

Insulin Shock

When too much insulin is present, such as occurs when a diabetic is injected with too much insulin or has not eaten after an insulin injection, blood glucose levels become very low. The brain, which depends primarily on glucose for an energy source, malfunctions. This condition, called insulin shock, can include symptoms of disorientation and convulsions, and may result in loss of consciousness.

Glucagon (gloo′kă-gon, glucose + *agō*, to lead) is released from the alpha cells when blood glucose levels are low. Glucagon binds to membrane-bound receptors primarily in the liver and cause the conversion of glycogen stored in the liver to glucose. The glucose is then released into the blood to increase blood glucose levels. After a meal, when blood glucose levels are elevated, glucagon secretion is reduced.

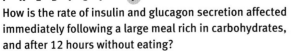

P R E D I C T **7**

How is the rate of insulin and glucagon secretion affected immediately following a large meal rich in carbohydrates, and after 12 hours without eating?

Insulin and glucagon function together to regulate blood glucose levels (see figure 10.21). When blood glucose levels increase, insulin secretion increases, and glucagon secretion decreases. When blood glucose levels decrease, the rate of insulin secretion declines, and the rate of glucagon secretion increases. Other hormones, such as epinephrine,

A CASE IN POINT | Type 2 Diabetes Mellitus

Kandy Barr is 60 years old. She is overweight and has not been feeling well. She is lethargic, feels weak, and she has had two urinary tract infections in the past 6 months. She visited her optometrist because she could not read the handwriting on checks at the bank where she works. Her optometrist recommended that Kandy see her physician, who ordered a blood test. The results indicated that she had high blood glucose and high blood lipid levels. After additional tests, Kandy's physician told her that she has type II diabetes. Type I diabetes is caused by reduced insulin secretion. Type II diabetes results from reduced sensitivity of tissues to the effects of insulin because of abnormal insulin receptors or abnormal responses to the insulin receptors. Consequently, insulin is less able to facilitate the entry of glucose into the liver, skeletal muscles, and adipose tissues. Kandy's physician recommended that she try to control her diabetes by restricting food intake, especially carbohydrates and fats, and increasing exercise. Kandy is fortunate because her symptoms were reduced as a result of faithful adherence to her physician's recommendations. People who are obese are approximately 10 times more likely to suffer from type 2 diabetes than people of normal weight. Frequent infections, changes in vision and fatigue are associated with high blood glucose levels.

cortisol, and growth hormone, also function to maintain blood levels of nutrients. When blood glucose levels decrease, these hormones are secreted at a greater rate. Epinephrine and cortisol cause the breakdown of protein and fat and the synthesis of glucose to help increase blood levels of nutrients. Growth hormone slows protein breakdown and favors fat breakdown.

Testes and Ovaries

The testes of the male and the ovaries of the female secrete sex hormones, in addition to producing sperm cells or oocytes. The hormones produced by these organs play important roles in the development of sexual characteristics. Structural and functional differences between males and females and the ability to reproduce all depend on the sex hormones (see figure 10.9 and table 10.3).

The main sex hormone produced in the male is **testosterone** (tes′tos′tĕ-rōn), which is secreted by the testes. It is responsible for the growth and development of the male reproductive structures, muscle enlargement, growth of body hair, voice changes, and the male sexual drive.

In the female two main classes of sex hormones, secreted by the ovaries, affect sexual characteristics: **estrogen** (es′trō-jen) and **progesterone** (prō-jes′ter-ōn). Together these hormones contribute to the development and function of female reproductive structures and other female sexual characteristics. These characteristics include enlargement of the breasts and distribution of fat, which influences the shape of the hips, breasts, and thighs. The female menstrual cycle is controlled by the cyclical release of estrogen and progesterone from the ovaries.

LH and FSH stimulate the secretion of hormones from the ovaries and testes. Releasing hormone from the hypothalamus controls the rate of LH and FSH secretion in males and females. LH and FSH, in turn, control the secretion of hormones from the ovaries and testes. Hormones secreted by the ovaries and testes also have a negative-feedback effect on the hypothalamus and anterior pituitary. The control of hormones that regulate reproductive functions is discussed in greater detail in chapter 19.

Thymus Gland

The **thymus** gland lies in the upper part of the thoracic cavity (see figure 10.9 and table 10.3). It is important in the function of the immune system. Part of the function of the thymus gland is to secrete a hormone called **thymosin** (thī′mō-sin), which helps in the development of certain white blood cells, called T cells. T cells help protect the body against infection by foreign organisms. The thymus gland is most important early in life, becoming smaller in older adults. If an infant is born without a thymus gland, the immune system does not develop normally, and the body is less capable of fighting infections (see chapter 14).

Pineal Body

The **pineal** (pin′ē-ăl, pinecone) **body** is a small, pinecone-shaped structure located superior and posterior to the thalamus of the brain (see chapter 8). The pineal body produces a hormone called **melatonin** (mel-ă-tōn′in), which is thought to decrease the secretion of LH and FSH by decreasing the release of hypothalamic-releasing hormones (see table 10.3). Thus, melatonin acts to inhibit the functions of the reproductive system. Animal studies have demonstrated that the amount of light available controls the rate of melatonin secretion. In many animals, short day length causes an increase in melatonin secretion, whereas longer day length causes a decrease in melatonin secretion. Some evidence suggests that melatonin may play an important role in the onset of puberty in humans. Tumors may develop in the pineal body, which, in some cases, can increase pineal secretions and in others decrease them.

P R E D I C T   **8**
Predict the effect on a young person's reproductive system of a tumor that destroys the ability of the pineal body to secrete melatonin.

Other Hormones

Cells in the lining of the stomach and small intestine secrete hormones that stimulate the production of digestive juices from the stomach, pancreas, and liver. These hormones aid the process of digestion by causing secretion of digestive juices when food is present in the digestive system but not at other times. Hormones secreted from the small intestine also help regulate the rate at which food passes from the stomach into the small intestine, so that food enters the small intestine at an optimal rate (see chapter 16).

Prostaglandins are widely distributed in tissues of the body and function as intercellular signals. Unlike most hormones, they are usually not transported long distances in the circulatory system but function mainly as autocrine or paracrine chemical signals (see table 10.1). Thus, their effects occur in the tissues where they are produced. Some prostaglandins cause relaxation of smooth muscle, such as dilation of blood vessels. Others cause contraction of smooth muscle, such as contraction of the uterus during the delivery of a baby. Because of their action on the uterus, prostaglandins have been used medically to initiate abortion. Prostaglandins also play a role in inflammation. They are released by damaged tissues and cause blood vessel dilation, localized swelling, and pain. Prostaglandins produced by platelets appear to be necessary for blood clotting to occur normally. The ability of aspirin and related substances to reduce pain and inflammation, help prevent the painful cramping of uterine smooth muscle, and treat headache is a result of their inhibitory effect on prostaglandin synthesis.

Stress, in the form of disease, physical injury, or emotional anxiety initiates a specific response from the body that involves the nervous and endocrine systems. The stressful condition influences the hypothalamus, and through the hypothalamus the sympathetic division of the autonomic nervous system is activated. The sympathetic division prepares the body for physical activity. It increases heart rate and blood pressure, shunts blood from the gut and other visceral structures to skeletal muscles, and increases the rate of metabolism in several tissues, especially in skeletal muscle. Part of the response of the sympathetic division is due to the release of epinephrine from the adrenal medulla.

In addition to sympathetic responses, stress causes the release of ACTH from the pituitary. ACTH acts on the adrenal cortex to cause the release of glucocorticoids. These hormones increase blood glucose levels and break down protein and fat, making nutrients more readily available to tissues.

Although the ability to respond to stress is adaptive for short periods of time, responses triggered by stressful conditions are harmful if they occur for long periods. Prolonged stress can lead to hypertension (elevated blood pressure); heart disease; ulcers; inhibited immune system function; changes in mood, including depression; and other conditions. Humans are frequently exposed to prolonged psychological stress from high-pressure jobs, the

inability to meet monetary obligations, or social expectations. Although responses to stress prepare a person for physical activity, often increased physical activity is not an appropriate response to the situation causing the stress. Long-term exposure to stress under conditions in which physical activity and emotions must be constrained may be harmful. Techniques that effectively reduce responses to stressful conditions are of substantial advantage to people who are exposed to chronic stress. Biofeedback, meditation, or other relaxation exercises are useful. Getting adequate rest, relaxation, and regular physical exercise are important in maintaining good health and reducing unhealthy responses to stressful situations.

The kidneys secrete a hormone in response to reduced oxygen levels in the kidney. The hormone is called **erythropoietin** (ĕ-rith′rō-poy′ĕ-tin, *erythro* refers to red blood cells). It acts on bone marrow to increase the production of red blood cells (see chapter 11).

In pregnant women, the placenta is an important source of hormones that function to maintain pregnancy and stimulate breast development. These hormones include estrogen, progesterone, and **human chorionic gonadotropin** (gō′nad-ō-trō′pin), which is similar in structure and function to LH. These hormones are essential to the maintenance of pregnancy (see chapter 20).

Effects of Aging on the Endocrine System

Age-related changes of the endocrine system include a gradual decrease in the secretion of some, but not all, endocrine glands. Some of the decreases in secretion may be due to a decrease in physical activity as people age.

GH secretion decreases as people age, and the decrease is greatest in people who do not exercise. It may not occur in older people who exercise regularly. Decreasing GH levels may explain some of the gradual decrease in bone and muscle mass and some of the increase in adipose tissue in many elderly people. Administering GH to slow or prevent the consequences of

aging has not yet been established to be effective, and unwanted side effects are possible.

A decrease in melatonin secretion may influence age-related changes in sleep patterns and decrease in the secretion of some hormones, such as GH and testosterone.

The secretion of thyroid hormones decreases slightly with age. Age-related damage to the thyroid gland by the immune system can occur. This occurs in women more than in men. Approximately 10% of elderly women have some reduction in thyroid hormone secretion.

The kidneys of the elderly secrete less renin, resulting in a reduced ability to respond to decreases in blood pressure.

Reproductive hormone secretion gradually declines in elderly men, and women experience menopause (see chapter 19).

Thymosin from the thymus gland decreases with age. Fewer functional lymphocytes are produced, and the immune system becomes less effective in protecting the body against infections and cancer.

Parathyroid hormone secretion increases to maintain blood calcium levels if dietary Ca^{2+} and vitamin D levels decrease, as they often do in the elderly. Consequently, a substantial decrease in bone matrix may occur.

There is no age-related decrease in the ability to regulate blood glucose in most people. There is an age-related tendency to develop type II diabetes mellitus for those who have a familial tendency to do so, and it is correlated with age-related increases in body weight.

Systems Pathology
Graves' Disease (Hyperthyroidism)

Mrs. G. owns a business, has several employees, and works hard to manage her business and make time for her husband and two children. Over several months, she slowly recognized that she felt warm when others did not, she would sweat excessively and her skin was often flushed. She often felt as if her heart was pounding, she was much more nervous than usual, and it was difficult for her to concentrate. She began to feel weak and lose weight even though her appetite was greater than normal. Her family recognized some of these changes and that her eyes seemed larger than usual. They encouraged her to see her physician. After an examination and some blood tests, it was concluded that she had Graves' disease, a type of hyperthyroidism.

Background Information

Graves' disease is an example of altered regulation of hormone secretion. It is characterized by the elevated secretion of thyroid hormones from the thyroid gland. In approximately 95% of Graves' disease cases, an unusual antibody type is produced by the immune system, which binds to receptors on the cells of the thyroid follicle and stimulates them to secrete increased amounts of thyroid hormone. The secretion of the releasing hormone and thyroid-stimulating hormone is inhibited by elevated thyroid hormones. The antibody is produced in large amounts, however, and is not inhibited by thyroid hormones. A very elevated rate of thyroid hormone secretion is therefore maintained. In addition, the size of the thyroid gland increases and connective tissue components are deposited behind the eyes, causing them to bulge (figure A). The thyroid gland can enlarge somewhat or it can become very large. Enlargement of the thyroid gland is called a goiter.

Mrs. G. was treated with radioactive iodine (^{131}I) atoms that were actively transported into thyroid cells where they destroyed a substantial

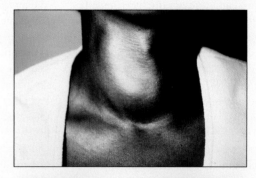

Figure A　Person with Hyperthyroidism

portion of the thyroid gland. Data indicate that this treatment has few side effects and is effective in treating most cases of Graves' disease. Other options include drugs that inhibit the synthesis and secretion of thyroid hormones and surgery to remove part of the thyroid gland.

P R E D I C T　⑨
Explain why removal of part of the thyroid gland is an effective treatment for Graves' disease.

S U M M A R Y

The nervous and endocrine systems are the two major regulatory systems in the body.

1. The nervous system controls structures by sending action potentials along axons, which release chemicals signals at their ends near the cells they control, whereas the endocrine system releases chemicals into the circulatory system, which carries the chemicals to the cells they control.
2. The endocrine system has a more general effect, acts more slowly, and has a longer lasting effect than the nervous system.

Functions of the Endocrine System　(p. 269)

The main functions of the endocrine system include regulation of water balance; uterine contractions during delivery; milk release from the breasts; growth; metabolism; tissue maturation; Na^+, K^+, and Ca^{2+} concentration of blood; heart rate; blood pressure; preparation for physical activity; blood glucose concentration; immune cell production; and reproductive functions in males and females.

Chemical Signals　(p. 269)

Chemical signals bind to receptor sites on receptor molecules.

1. Intracellular chemical signals are produced in one part of a cell and travel to another part of the same cell and bind to receptors.
2. Intercellular chemical signals are released from one cell, are carried in the intercellular fluid, and bind to receptors in other cells.
3. Intercellular chemical signals can be classified as autocrine, paracrine, hormone, neurohormone, neuromodulator, neurotransmitter, or pheromone chemical signals.

Systems Interactions — The Effect of Graves' Disease (Hyperthyroidism) on Other Systems

System	Interactions
Integumentary	Excessive sweating, flushing, and warm skin result from the elevated body temperature caused by the increase rate of metabolism. Sweating and dilation of blood vessels in the skin are mechanisms by which the skin increases heat loss. Fine, soft, straight hair, along with hair loss, result from the reduced protein synthesis. The elevated rate of metabolism makes amino acids unavailable for protein synthesis.
Skeletal	Some increased bone resorption, which can decrease bone density, and increased blood Ca^{2+} levels can occur in severe cases.
Muscular	Muscle atrophy and muscle weakness are the result of increased metabolism. Elevated metabolism results in the breakdown of muscle and the increased use of muscle proteins as energy sources.
Nervous	Enlargement of the extrinsic eye muscles, edema in the area of the orbits, and the accumulation of fibrous connective tissue results in protrusion of the eyes in 50–70% of individuals with Graves' disease. Damage to the retina and optic nerve and paralysis of the extraocular muscles can occur. The effects on the eyes may be influenced directly by the abnormal antibodies, and they may progress even after the hyperthyroid condition is successfully treated. Restlessness, short attention span, compulsive movement, tremor, insomnia, and increased emotional responses are consistent with hyperactivity of the nervous system.
Cardiovascular	Cardiovascular responses are consistent with an increased rate of metabolism and increased stimulation by the sympathetic division of the autonomic nervous system. There is an increased amount of blood pumped by the heart and increased blood flow through the tissues, including the skin. The heart rate is greater than normal, heart sounds are louder than normal, and the heartbeats may be out of rhythm periodically.
Lymphatic, Immune	Antibodies that bind to receptors for thyroid-stimulating hormone on the cells of the thyroid gland have been found in nearly all people who suffer from Graves' disease. The condition, therefore, is classified as an autoimmune disease in which antibodies produced by one's lymphatic system results in abnormal functions.
Respiratory	Breathing may be labored, and the volume of air taken in with each breath may be decreased. Weak contractions of muscles of inspiration contribute.
Digestive	Weight loss occurs with an associated increase in appetite. Increased peristalsis in the intestines leads to frequent stools or diarrhea. Nausea, vomiting, and abdominal pain also result. There is an increased use of hepatic glycogen stores and of adipose and protein stores for energy, and there is a decrease in serum lipid levels (including triglycerides, phospholipids, and cholesterol) and increased tendency to develop vitamin deficiencies. There is a reduced ability of the body to absorb nutrients from the intestine and increased metabolism. This results in an increased use of body stores of fat for energy and increased breakdown of proteins for energy.
Reproductive	Reduced regularity of menstruation or lack of menstruation may occur in women because of the elevated metabolism. In men the primary effect is a loss of sex drive.

Receptors (p. 271)

Chemical signals bind to receptor sites on receptor molecules to produce a response.

Receptor Types for Intercellular Chemical Signals

Membrane-bound receptors extend through the cell membrane and intracellular receptors are located in the cytoplasm or nucleus of the cell.

Receptor Responses

1. Membrane-bound receptors can produce a response by:
   a. Altering membrane permeability by directly opening or closing membrane channels.
   b. Activating G proteins, which, in turn, can open or close membrane channels, or activate enzymes that produce intracellular chemical signals.
   c. Alter the activity of enzymes directly. The enzymes increase or decrease the synthesis of intracellular chemical signals or add phosphate groups to proteins in the cell.

2. Intracellular receptors are located in the cytoplasm or nuclei and can regulate enzyme activity or regulate the synthesis of specific messenger RNA.

Hormones (p. 275)

1. Endocrine glands produce hormones that are released into the circulatory system and travel some distance, where they act on target tissues to produce a response.
2. A target tissue for a given hormone has receptor molecules for that hormone.

Chemistry

1. Hormones are basically proteins, peptides, or lipids.
2. Protein, and most peptide hormones bind to receptors on the cell membrane and cause permeability changes or the production of intracellular chemical signals inside the cell. Eicosanoids also bind to receptors on the cell membrane. Lipid-soluble hormones, such as

the steroids and thyroid hormones, enter the cell and bind to
receptors inside the cell.

3. The combining of hormones with their receptors results in a
   response.

Regulation of Hormone Secretion

1. The secretion of hormones is controlled by negative-feedback
   mechanisms.
2. Secretion of hormones from a specific gland is controlled by
   blood levels of some chemical, another hormone, or the nervous
   system.
3. The endocrine system consists of ductless glands.
4. Some glands of the endocrine system perform more than one function.

Endocrine Glands and Their Hormones   (p. 277)

Pituitary and Hypothalamus

1. The pituitary is connected to the hypothalamus of the brain by
   the infundibulum. It is divided into the anterior and posterior
   pituitary.
2. Secretions from the anterior pituitary are controlled by hormones
   that pass through the hypothalamic–pituitary portal system from
   the hypothalamus.
3. Hormones secreted from the posterior pituitary are controlled by
   action potentials carried by axons that pass from the hypothalamus
   to the posterior pituitary.
4. The hormones released from the anterior pituitary are growth
   hormone (GH), thyroid-stimulating hormone (TSH),
   adrenocorticotropic hormone (ACTH), luteinizing hormone (LH),
   follicle-stimulating hormone (FSH), prolactin, and melanocyte-
   stimulating hormone (MSH).
5. Hormones released from the posterior pituitary include antidiuretic
   hormone (ADH) and oxytocin.

Thyroid Gland

The thyroid gland secretes thyroid hormones, which control the metabolic
rate of tissues, and calcitonin, which helps regulate blood Ca^{2+} levels.

Parathyroid Glands

The parathyroid glands secrete parathyroid hormone, which helps regulate blood Ca^{2+} levels. Active vitamin D also helps regulate blood Ca^{2+}
levels.

Adrenal Glands

1. The adrenal medulla secretes primarily epinephrine and some
   norepinephrine. These hormones help prepare the body for physical
   activity.
2. The adrenal cortex secretes three classes of hormones.
3. Glucocorticoids (cortisol) reduce inflammation and break down fat
   and protein, making them available as energy sources to other
   tissues.

4. Mineralocorticoids (aldosterone) help regulate blood Na^+ and K^+
   levels and water volume in the body. Renin, secreted by the kidneys,
   helps regulate blood pressure by increasing angiotensin II and
   aldosterone production. These hormones cause blood vessels to
   constrict and enhance Na^+ and water retention by the kidney.
5. Adrenal androgens increase female sexual drive; normally adrenal
   androgens have little effect in males.

Pancreas, Insulin, and Diabetes

1. The pancreas secretes insulin in response to elevated levels of blood
   glucose and amino acids. Insulin increases the rate at which many
   tissues, including adipose tissue, liver, and skeletal muscles, take up
   glucose and amino acids.
2. The pancreas secretes glucagon in response to reduced blood glucose
   and increases the rate at which the liver releases glucose into the blood.

Testes and Ovaries

1. The testes secrete testosterone, and the ovaries secrete estrogen
   and progesterone. These hormones help control reproductive
   processes.
2. LH and FSH from the pituitary gland control hormone secretion
   from the ovaries and testes.

Thymus Gland

The thymus gland secretes thymosin, which enhances the ability of the
immune system to function.

Pineal Body

The pineal body secretes melatonin, which may help regulate the onset of
puberty by acting on the hypothalamus.

Other Hormones   (p. 294)

1. Hormones secreted by cells in the stomach and intestine help
   regulate stomach, pancreatic, and liver secretions.
2. The prostaglandins are hormones that have a local effect, produce
   numerous effects on the body, and play a role in inflammation.
3. Erythropoietin from the kidney stimulates red blood cell production.
4. The placenta secretes human chorionic gonadotropin, estrogen,
   and progesterone, which are essential to the maintenance of
   pregnancy.

Effects of Aging on the Endocrine System   (p. 295)

1. Age-related changes include a gradual decrease in:
   a. GH in people who do not exercise
   b. Melatonin
   c. Thyroid hormones (slight decrease)
   d. Reproductive hormones
   e. Thymus hormones
2. Parathyroid hormones increase if vitamin D and Ca^{2+} levels
   decrease.
3. There is an increase in type II diabetes.

R E V I E W   A N D   C O M P R E H E N S I O N

1. What are the major functional differences between the endocrine
   and the nervous systems?
2. List the functions of the endocrine system.
3. List the major differences between intracellular and intercellular
   chemical signals.
4. List the intercellular chemical signals that are classified on the basis
   of the location of the cells from which they are secreted and the
   location of their target cells.
5. Explain the relationship between a chemical signal and its receptor.

6. Describe the mechanisms by which membrane-bound receptors
   produce responses in their target tissues.
7. Describe the mechanisms by which intracellular receptors produce
   responses in their target tissues.
8. Compare the means by which hormones that can and cannot cross
   the cell membrane produce a response.
9. Define endocrine gland and hormone.
10. What makes one tissue a target tissue and another not a target tissue
    for a hormone?

11. Into what chemical categories can hormones be classified?

12. Name three ways that hormone secretion is regulated.

13. Describe how secretions of the anterior and posterior pituitary hormones are controlled.

14. What are the functions of growth hormone? What happens when too little or too much growth hormone is secreted?

15. Describe the effect of gonadotropins on the ovary and testis.

16. What are the functions of the thyroid hormones, and how is their secretion controlled? What happens when too little or too much of the thyroid hormones are secreted?

17. Explain how calcitonin, parathyroid hormone, and vitamin D are involved in maintaining blood Ca^{2+} levels. What happens when too little or too much parathyroid hormone is secreted?

18. List the hormones secreted from the adrenal gland, give their functions, and compare the means by which the secretion rate of each is controlled.

19. What are the major functions of insulin and glucagon? How is their secretion regulated? What is the effect if too little insulin is secreted or the target tissues are not responsive to insulin?

20. List the effects of testosterone, progesterone, and estrogen.

21. What hormones are produced by the thymus gland and pineal body? Name the effects of these hormones.

22. List the effects of prostaglandins. How is aspirin able to reduce the severity of the inflammatory response?

23. List the hormones secreted by the placenta.

24. List the major age-related changes in the endocrine system.

C R I T I C A L   T H I N K I N G

1. A hormone is known to bind to a membrane-bound receptor. A drug that inhibits the breakdown of cyclic-AMP causes an increased response to the hormone. A drug that inhibits the binding of GTP to proteins reduces the response to this drug. Based on these observations, describe the mechanism by which the membrane-bound receptor is most likely to produce a response to the hormone.

2. Aldosterone and antidiuretic hormone play important roles in the regulation of blood volume and concentration of blood. The response to one of these hormones is evident within minutes and the response to the other requires several hours. Explain the difference in response time for these two hormones.

3. Biceps Benny figured that if a small amount of a vitamin was good, a lot should do more good. He therefore began to take vitamins that included a large amount of vitamin D. Predict the effect of vitamin D on Biceps' blood Ca^{2+} levels and on the rate of secretion of hormones responsible for the regulation of blood Ca^{2+} levels.

4. What would be the consequences if the adrenal cortex degenerated and no longer secreted hormones?

5. Predict the consequences of elevated aldosterone secretion from the adrenal cortex.

6. Explain how the blood levels of glucocorticoids, epinephrine, insulin, and glucagon change after a person has gone without food for 24 hours.

7. Stetha Scope wanted to go to medical school to become a physician. While attending college she knew her grades had to be excellent. Stetha worked very hard and worried constantly. By the end of each school year she had a cold and suffered from stomach pains. Explain why she might be susceptible to these symptoms.

Answers in Appendix D

Visit this textbook's website at www.mhhe.com/seeleyess6 for practice quizzes, animations, interactive learning exercises, and other study tools.

11

Blood

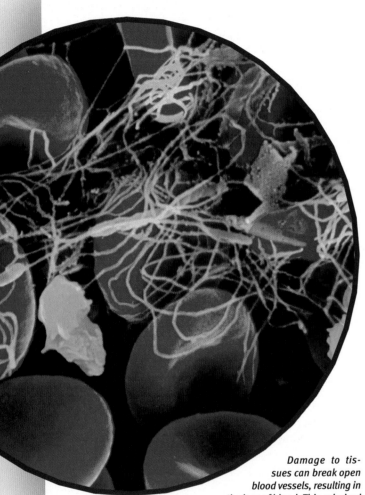

Damage to tissues can break open blood vessels, resulting in the loss of blood. This colorized scanning electron micrograph shows a blood clot, which functions to prevent blood loss. The blue particles are platelets, which are activated when tissues are damaged. Activated platelets promote the formation of protein fibers called fibrin, which are seen here as yellow strands. The fibrin forms a network that captures red blood cells (red discs) and prevents their loss from the body.

Chapter Outline and Objectives

Functions of Blood (p. 301)
1. State the functions of blood.

Composition of Blood (p. 301)
2. List the components of blood.

Plasma (p. 301)
3. Name the components of plasma and give their functions.

Formed Elements (p. 303)
4. Describe the origin and production of the formed elements.
5. Describe the structure, function, and life history of red blood cells.
6. Compare the structures and functions of the five different types of white blood cells.
7. Describe the origin and structure of platelets.

Preventing Blood Loss (p. 308)
8. Explain the formation and function of platelet plugs and clots.
9. Describe the regulation of clot formation and how clots are removed.

Blood Grouping (p. 311)
10. Explain the basis of ABO and Rh incompatibilities.

Diagnostic Blood Tests (p. 315)
11. Describe diagnostic blood tests and the normal values for the tests, and give examples of disorders that produce abnormal test values.

blood has always fascinated humans, and historically there has been much speculation about its function. Blood was considered the "essence of life" because the uncontrolled loss of it can result in death. Many cultures around the world, both ancient and modern, share beliefs in the magical qualities of blood. Blood was also thought to define our character and emotions. People of a noble bloodline were described as "blue bloods," whereas criminals were considered to have "bad" blood. It was said that anger caused the blood to "boil," and fear makes the blood "curdle." The scientific study of blood reveals characteristics as fascinating as any of these fantasies. Blood performs many functions essential to life and can reveal much about our health.

Functions of Blood

Blood is pumped by the heart through blood vessels, which extend throughout the body. Blood helps to maintain homeostasis in several ways:

1. *Transport of gases, nutrients, and waste products.* Oxygen enters blood in the lungs and is carried to cells. Carbon dioxide, produced by cells, is carried in the blood to the lungs, from which it is expelled. Ingested nutrients, ions, and water are transported by the blood from the digestive tract to cells, and the waste products of the cells are transported to the kidneys for elimination.
2. *Transport of processed molecules.* Many substances are produced in one part of the body and transported in the blood to another part, where they are modified. For example, the precursor to vitamin D is produced in the skin (see chapter 5) and transported by the blood to the liver and then to the kidneys for processing into active vitamin D. The active vitamin D is transported to the small intestines, where it promotes the uptake of calcium. Another example is lactic acid produced by skeletal muscles during anaerobic respiration (see chapter 7). The lactic acid is carried to the liver, where it is converted into glucose.
3. *Transport of regulatory molecules.* Many of the hormones and enzymes that regulate body processes are carried from one part of the body to another within the blood.
4. *Regulation of pH and osmosis.* Buffers (see chapter 2), which help keep the blood's pH within its normal limits of 7.35–7.45, are found in the blood. The osmotic composition of blood is also critical for maintaining normal fluid and ion balance.
5. *Maintenance of body temperature.* Blood is involved with body temperature regulation because warm blood is transported from the interior to the surface of the body, where heat is released from the blood.
6. *Protection against foreign substances.* Cells and chemicals of the blood constitute an important part of the immune system, protecting against foreign substances such as microorganisms and toxins.
7. *Clot formation.* Blood clotting provides protection against excessive blood loss when blood vessels are damaged. When tissues are damaged, the blood clot that forms is also the first step in tissue repair and the restoration of normal function (see chapter 4).

Composition of Blood

Blood is a type of connective tissue that consists of cells and cell fragments surrounded by a liquid matrix. The cells and cell fragments are the **formed elements,** and the liquid is the **plasma** (plaz′mă, something formed) (figure 11.1). The formed elements account for slightly less than half and plasma accounts for slightly more than half the total blood volume. The total blood volume in the average adult is about 4–5 liters (L) in females and 5–6 L in males. Blood makes up about 8% of total body weight.

Plasma

Plasma is a pale yellow fluid that consists of about 91% water; 7% proteins; and 2% other substances, such as ions, nutrients, gases, and waste products (see figure 11.1 and table 11.1). Plasma proteins include albumin, globulins, and fibrinogen. **Albumin** (al-bū′min, the white of egg) makes up 58% of the plasma proteins. Although the osmotic pressure (see chapter 3) of blood results primarily from sodium chloride, albumin makes an important contribution. The water balance between blood and tissues is determined by the movement of water into and out of the blood by osmosis. **Globulins** (glob′ū-linz, globule) account for 38% of the plasma proteins. Some globulins, such as antibodies and complement, are part of the immune system (see chapter 14). Other globulins and albumin function as transport molecules because they bind to molecules such as hormones (see chapter 10) and carry them in the blood throughout the body. Some globulins are clotting factors, which are necessary for the formation of blood clots. **Fibrinogen** (fi-brin′ō-jen, *fibra,* fiber + *gen,* produce) is a clotting factor that constitutes 4% of plasma proteins. Activation of clotting factors results in the conversion of fibrinogen into **fibrin** (fi′brin), a threadlike protein that forms blood clots (see discussion on p. 310). **Serum** (ser′um, whey) is plasma without the clotting factors.

Plasma volume remains relatively constant. Normally water intake through the digestive tract closely matches water loss through the kidneys, lungs, digestive tract, and skin. Oxygen enters blood in the lungs, and carbon dioxide enters blood from tissues. Other suspended or dissolved substances in the blood come from the liver, kidneys, intestines, endocrine glands, and immune tissues such as the lymph nodes and spleen. The concentration of these substances in the blood is also regulated and maintained within narrow limits.

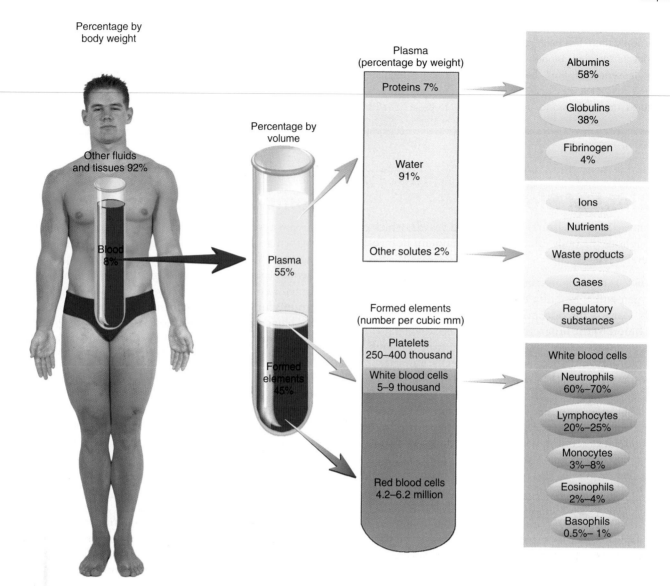

Figure 11.1   Composition of Blood

Approximate values for the components of blood in a normal adult.

Table 11.1	Composition of Plasma
Plasma Components	**Functions and Examples**
Water	Acts as a solvent and suspending medium for blood components
Proteins	Maintain osmotic pressure (albumin), destroy foreign substances (antibodies and complement), transport molecules (albumin, globulins), and form clots (fibrinogen)
Ions	Involved in osmotic pressure (sodium and chloride ions), membrane potentials (sodium and potassium ions), and acid–base balance (hydrogen, hydroxide, and bicarbonate ions)
Nutrients	Source of energy and "building blocks" of more complex molecules (glucose, amino acids, triacylglycerides)
Gases	Involved in aerobic respiration (oxygen and carbon dioxide)
Waste products	Breakdown products of protein metabolism (urea and ammonia salts), erythrocytes (bilirubin), and anaerobic respiration (lactic acid)
Regulatory substances	Catalyze chemical reactions (enzymes) and stimulate or inhibit many body functions (hormones)

Formed Elements

About 95% of the volume of the formed elements consists of **red blood cells (RBCs)**, or **erythrocytes** (ĕ-rith′rō-sītz, erythro-, red + *kytos*, cell). The remaining 5% of the volume of the formed elements consists of **white blood cells (WBCs)**, or **leukocytes** (loo′kō-sītz, leuko-, white + *kytos*, cell), and cell fragments called **platelets** (plāt′letz), or **thrombocytes** (throm′bō-sītz, thrombo-, clot + *kytos*, cell). Red blood cells are 700 times more numerous than white blood cells and 17 times more numerous than platelets. The formed elements of the blood are outlined and illustrated in table 11.2.

Production of Formed Elements

The process of blood cell production is called **hematopoiesis** (hē′mă-tō-poy-ē′sis, hemato-, blood + *poiēsis*, a making). In the fetus, hematopoiesis occurs in several tissues such as the liver, thymus gland, spleen, lymph nodes, and red bone marrow. After birth, hematopoiesis is confined primarily to red bone marrow, but some white blood cells are produced in lymphatic tissues (see chapter 14).

All the formed elements of blood are derived from a single population of cells called **stem cells** or **hemocytoblasts**.

These stem cells differentiate to give rise to different cell lines, each of which ends with the formation of a particular type of formed element (figure 11.2). The development of each cell line is regulated by specific growth factors. That is, the types of formed element derived from the stem cells and how many formed elements are produced are determined by the growth factors.

Stem Cells and Cancer Therapy

Many cancer therapies attack dividing cells such as those found in tumors. An undesirable side effect, however, can be the destruction of nontumor cells that divide rapidly, such as the stem cells and their derivatives in red bone marrow. After treatment for cancer, growth factors are used to stimulate the rapid regeneration of the red bone marrow. Although not a treatment for the cancer itself, the use of growth factors can speed recovery from the side effects of cancer therapy.

Some types of leukemia and genetic immune deficiency diseases can be treated with a bone marrow transplant containing blood stem cells. To avoid problems of tissue rejection, families with a history of these disorders can freeze the umbilical cord blood of their newborn children. The cord blood contains many stem cells and can be used instead of a bone marrow transplant.

Table 11.2	Formed Elements of the Blood		
Cell Type	**Illustration**	**Description**	**Function**
Red blood cell		Biconcave disk; no nucleus; contains hemoglobin, which colors the cell red; 6.5–8.5μm in diameter	Transports oxygen and carbon dioxide
White blood cells		Spherical cells with a nucleus	Five types of white blood cells, each with specific functions
Granulocytes Neutrophil		Nucleus with two to four lobes connected by thin filaments; cytoplasmic granules stain a light pink or reddish purple; 10–12 μm in diameter	Phagocytizes microorganisms and other substances
Basophil		Nucleus with two indistinct lobes; cytoplasmic granules stain blue-purple; 10–12 μm in diameter	Releases histamine, which promotes inflammation, and heparin, which prevents clot formation
Eosinophil		Nucleus often bilobed; cytoplasmic granules stain orange-red or bright red; 11–14 μm in diameter	Releases chemicals that reduce inflammation; attacks certain worm parasites
Agranulocytes Lymphocyte		Round nucleus; cytoplasm forms a thin ring around the nucleus; 6–14 μm in diameter	Produces antibodies and other chemicals responsible for destroying microorganisms; contributes to allergic reactions, graft rejection, tumor control, and regulation of the immune system
Monocyte		Nucleus round, kidney, or horseshoe-shaped; contains more cytoplasm than does lymphocyte; 12–20 μm in diameter	Phagocytic cell in the blood; leaves the blood and becomes a macrophage, which phagocytizes bacteria, dead cells, cell fragments, and other debris within tissues
Platelet		Cell fragment surrounded by a plasma membrane and containing granules; 2–4 μm in diameter	Forms platelet plugs; releases chemicals necessary for blood clotting

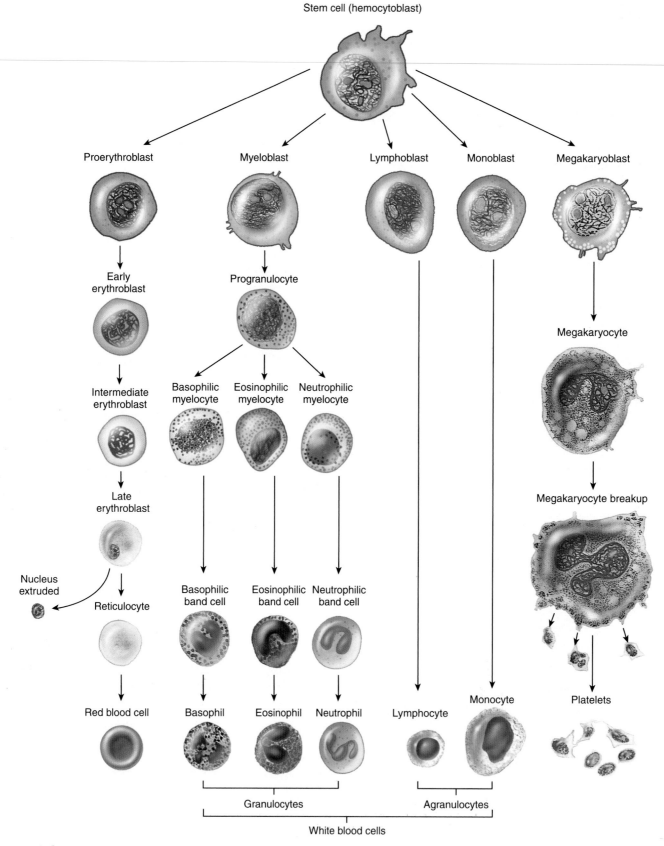

Figure 11.2   **Hematopoiesis**

Stem cells give rise to the cell lines that produce the formed elements.

which is necessary for the normal function of hemoglobin. Each iron in a heme molecule can reversibly associate with an oxygen molecule. Hemoglobin picks up oxygen in the lungs and releases oxygen in other tissues (see chapter 15). Hemoglobin that is bound to oxygen is bright red in color, whereas hemoglobin without bound oxygen is a darker red color. Hemoglobin is responsible for 98.5% of the oxygen transported in blood. The remaining 1.5% is transported dissolved in plasma.

Hemoglobin-Based Oxygen Carriers

Hemoglobin-based oxygen carriers (HBOCs) are being studied as an alternative to blood transfusions for people who have inadequate delivery of oxygen to tissues as a result of blood loss, anemia, etc. The preparation of HBOCs involves removing and purifying the hemoglobin in red blood cells. These hemoglobin molecules are chemically cross-linked and placed in an isotonic salt solution. HBOCs have been produced using human hemoglobin (PolyHeme and HemoLink) and cow hemoglobin (HemoPure). The use of HBOCs for blood transfusions has several benefits compared with using blood. They have a longer shelf life than blood and can be used when blood is not available. The free oxygen-carrying hemoglobin molecules are smaller than red blood cells, allowing them to flow past partially blocked arteries. There are no transfusion reactions because HBOCs do not have red blood cell membranes, which are necessary for transfusion reactions (see Blood Grouping on p. 311). HBOCs prepared from cow blood can be used by those who oppose transfusions of human blood on religious grounds.

Because iron is necessary for oxygen transport, it is not surprising that two-thirds of the body's iron is found in hemoglobin. Small amounts of iron are required in the diet to replace the small amounts lost in the urine and feces. Women need more dietary iron than men do because women lose iron as a result of menstruation.

Carbon Monoxide and Oxygen Transport

Carbon monoxide is a gas produced by the incomplete combustion of hydrocarbons such as gasoline. It binds to the iron in hemoglobin about 210 times as readily as does oxygen and does not tend to unbind. As a result, the hemoglobin bound to carbon monoxide no longer transports oxygen. Nausea, headache, unconsciousness, and death are possible consequences of prolonged exposure to carbon monoxide.

Carbon dioxide is produced in tissues and transported in the blood to the lungs, where it is removed from the blood (see chapter 15). Carbon dioxide transport involves bicarbonate ions, hemoglobin, and plasma. Approximately 70% of the carbon dioxide in blood is transported in the form of bicarbonate ions. The enzyme **carbonic anhydrase** (kar-bon′ik an-hī′drās), found primarily inside red blood cells, catalyzes a reaction that converts carbon dioxide (CO_2) and water (H_2O) into a hydrogen ion (H^+) and a bicarbonate ion (HCO_3^-) (see chapter 15 for more details).

$$CO_2 + H_2O \rightleftarrows H^+ + HCO_3^-$$

Carbon dioxide can bind reversibly to the globin part of hemoglobin. About 23% of the CO_2 in blood is transported

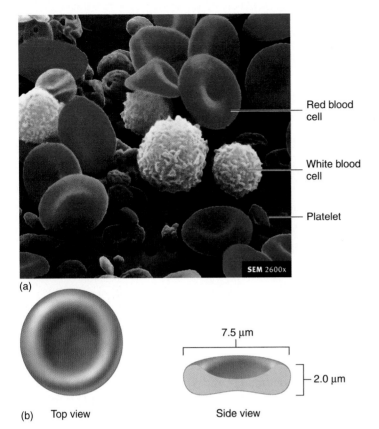

Figure 11.3 Red Blood Cells and White Blood Cells

(*a*) Scanning electron micrograph of formed elements: red blood cells (*red doughnut shapes*) and white blood cells (*yellow*). (*b*) Shape and dimensions of a red blood cell.

Red Blood Cells

Normal red blood cells are disk-shaped cells with edges that are thicker than the center of the cell (figure 11.3). The biconcave shape increases the surface area of the red blood cell compared with a flat disk of the same size. The greater surface area makes it easier for gases to move into and out of the red blood cell. In addition, the red blood cell can bend or fold around its thin center, decreasing its size and enabling it to pass more easily through small blood vessels.

During their development, red blood cells lose their nuclei and most of their organelles. Consequently, they are unable to divide. Red blood cells live for about 120 days in males and 110 days in females. The main component of a red blood cell is the pigmented protein **hemoglobin** (hē-mō-glō′bin, hemo-, blood + *globus,* a ball), which accounts for about a third of the cell's volume and is responsible for its red color.

Function

The primary functions of red blood cells are to transport oxygen from the lungs to the various tissues of the body and to assist in the transport of carbon dioxide from the tissues to the lungs. Oxygen transport is accomplished by hemoglobin, which consists of four protein chains and four heme groups. Each protein, called a **globin** (glō′bin), is bound to one **heme** (hēm), a red-pigmented molecule. Each heme contains one iron atom,

bound to hemoglobin or other blood proteins. The remaining 7% of CO_2 is transported dissolved in plasma.

Life History of Red Blood Cells

Under normal conditions, about 2.5 million red blood cells are destroyed every second. Fortunately, new red blood cells are produced as rapidly as old red blood cells are destroyed. Stem cells form **proerythroblasts** (prō-ĕ-rith′rō-blastz, pro-, before + erythro-, red + *blastos,* germ), which give rise to the red blood cell line (see figure 11.2). Red blood cells are the final cells produced from a series of cell divisions. After each cell division, the newly formed cells change and become more like a mature red blood cell. For example, following one of these cell divisions, the newly formed cells manufacture large amounts of hemoglobin. After the final cell division, the nucleus is lost from the cell, and a completely mature red blood cell is formed.

The process of cell division requires the B vitamins folate and B_{12}, which are necessary for the synthesis of DNA (see chapter 3). Iron is required for the production of hemoglobin. Consequently, lack of folate, vitamin B_{12}, or iron can interfere with normal red blood cell production.

Red blood cell production is stimulated by low blood oxygen levels. Typical causes of low blood oxygen are decreased numbers of red blood cells, decreased or defective hemoglobin, diseases of the lungs, high altitude, inability of the cardiovascular system to deliver blood to tissues, and increased tissue demands for oxygen such as during endurance exercises.

Low blood oxygen levels increase red blood cell production by increasing the formation of the glycoprotein **erythropoietin** (ĕ-rith-rō-poy′ĕ-tin, erythrocyte + *poiēsis,* a making) by the kidneys (figure 11.4). Erythropoietin stimulates red bone marrow to produce more red blood cells. Thus, when oxygen levels in the blood decrease, the production of erythropoietin increases, which increases red blood cell production. The increased number of red blood cells increases the ability of the blood to transport oxygen. This mechanism returns blood oxygen levels to normal and maintains homeostasis by increasing the delivery of oxygen to tissues. Conversely, if blood oxygen levels increase, less erythropoietin is released, and red blood cell production decreases.

> **P R E D I C T**   ①
>
> Cigarette smoke produces carbon monoxide. If a nonsmoker smoked a pack of cigarettes a day for a few weeks, what would happen to the number of red blood cells in the person's blood? Explain.

Old, abnormal, or damaged red blood cells are removed from the blood by macrophages located in the spleen and liver (figure 11.5). Within the macrophage the globin part of the molecule is broken down into amino acids that are reused to produce other proteins. The iron released from heme is transported in the blood to the red bone marrow and is used to produce new hemoglobin. Only small amounts of iron are required in the daily diet because the iron is recycled. The heme molecules are converted to **bilirubin** (bil-i-roo′bin, bili-, bile + *ruber,* red), a yellow pigment molecule. Bilirubin normally is taken up by the liver and released into the small intestine as part of the bile (see chapter 16). If the liver is not functioning

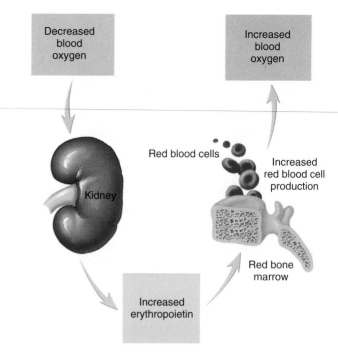

Figure 11.4   **Red Blood Cell Production**

In response to decreased blood oxygen, the kidneys release erythropoietin into the general circulation. The increased erythropoietin stimulates red blood cell production in the red bone marrow. This process increases blood oxygen levels.

normally, or if the flow of bile from the liver to the small intestine is hindered, bilirubin builds up in the circulation and produces **jaundice** (jawn′dis, *jaune,* yellow), a yellowish color of the skin. After it enters the intestine, bilirubin is converted by bacteria into other pigments. Some of these pigments give feces their brown color, whereas others are absorbed from the intestine into the blood, modified by the kidneys, and excreted in the urine, contributing to the characteristic yellow color of urine.

White Blood Cells

White blood cells or **leukocytes,** are spherical cells that lack hemoglobin. White blood cells form a thin, white layer of cells between plasma and red blood cells when the components of blood are separated from each other (see figure 11.1). They are larger than red blood cells, and each has a nucleus (see table 11.2). Although white blood cells are components of the blood, the blood serves primarily as a means to transport these cells to other tissues of the body. White blood cells can leave the blood and move by **ameboid** (ă-mē′boyd, like an ameba) **movement** through the tissues. In this process, the cell projects a cytoplasmic extension that attaches to an object. Then the rest of the cell's cytoplasm flows into the extension. Two functions of white blood cells are (1) to protect the body against invading microorganisms and (2) to remove dead cells and debris from the tissues by phagocytosis.

Each white blood cell type is named according to its appearance in stained preparations. Those containing large cytoplasmic granules are **granulocytes** (gran′ū-lō-sītz, granulo-, granular + *kytos,* cell), and those with very small granules that cannot be easily seen with the light microscope are **agranulocytes** (ă-gran′ū-lō-sītz, a-, without).

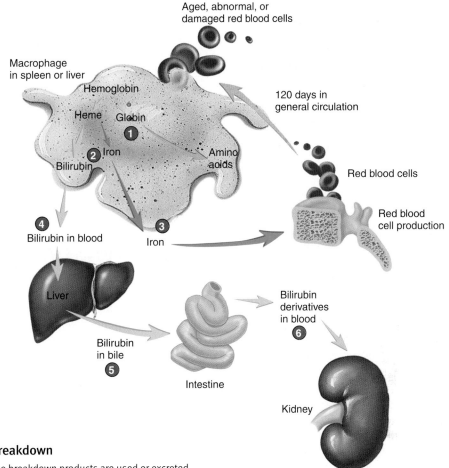

1. The globin chains of hemoglobin are broken down to individual amino acids (*pink arrow*) and are metabolized or used to build new proteins.

2. Iron is released from the heme of hemoglobin. The heme is converted into bilirubin.

3. Iron is transported in the blood to the red bone marrow and used in the production of new hemoglobin (*green arrows*).

4. Bilirubin (*blue arrow*) is transported in the blood to the liver.

5. Bilirubin is excreted as part of the bile into the small intestine.

6. Bilirubin derivatives contribute to the color of feces or are reabsorbed from the intestine into the blood and excreted from the kidneys in the urine.

Process Figure 11.5 Hemoglobin Breakdown

Hemoglobin is broken down in macrophages, and the breakdown products are used or excreted.

There are three kinds of granulocytes: neutrophils, basophils, and eosinophils. **Neutrophils** (noo′trō-filz, neutro-, neutral + *philos*, loving), the most common type of white blood cells, have small cytoplasmic granules that stain with both acidic and basic dyes (figure 11.6). Their nuclei are commonly lobed, with the number of lobes varying from two to four. Neutrophils usually remain in the blood for a short time (10–12 hours), move into other tissues, and phagocytize microorganisms and other foreign substances. Dead neutrophils, cell debris, and fluid can accumulate as **pus** at sites of infections.

Basophils (bā′sō-filz, baso-, base + *philos*, loving), the least common of all white blood cells, contain large cytoplasmic granules that stain blue or purple with basic dyes (see table 11.2). Basophils release histamine and other chemicals that promote inflammation (see chapters 4 and 14). They also release heparin, which prevents the formation of clots.

Eosinophils (ē-ō-sin′o-filz, eosin, an acidic dye + *philos*, loving) contain cytoplasmic granules that stain bright red with eosin, an acidic stain. They often have a two-lobed nucleus (see table 11.2). Eosinophils release chemicals that reduce inflammation. In addition, chemicals from eosinophils are involved with the destruction of certain worm parasites.

There are two kinds of agranulocytes: lymphocytes and monocytes. **Lymphocytes** (lim′fō-sītz, lympho-, lymph +

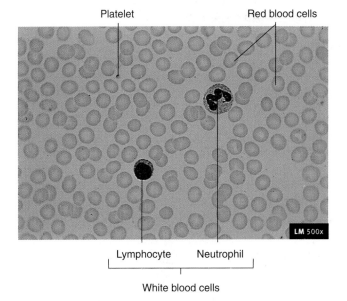

Figure 11.6 Standard Blood Smear

A thin film of blood is spread on a microscope slide and stained. The white blood cells have pink-colored cytoplasm and purple-colored nuclei. The red blood cells do not have nuclei. The center of a red blood cell appears whitish because light more readily shines through the thin center of the disk than through the thicker edges. The platelets are purple-colored cell fragments.

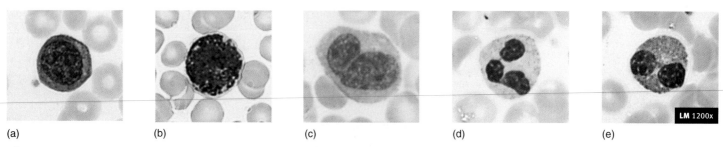

(a)   (b)   (c)   (d)   (e)

Figure 11.7   Identification of White Blood Cells

See predict question 2 below.

kytos, cell) are the smallest of the white blood cells (see figure 11.6). The lymphocytic cytoplasm consists of only a thin, sometimes imperceptible ring around the nucleus. There are several types of lymphocytes, and they play an important role in the body's immune response. Their diverse activities involve the production of antibodies and other chemicals that destroy microorganisms, contribute to allergic reactions, reject grafts, control tumors, and regulate the immune system. Chapter 14 considers these cells in more detail.

   Monocytes (mon′ō-sītz, mono-, one + *kytos,* cell) are the largest of the white blood cells (see table 11.2). After they leave the blood and enter tissues, monocytes enlarge and become **macrophages** (mak′rō-fā-jez, macro-, large + *phagō,* to eat), which phagocytize bacteria, dead cells, cell fragments, and any other debris within the tissues. In addition, macrophages can break down phagocytized foreign substances and present the processed substances to lymphocytes, which results in activation of the lymphocytes (see chapter 14).

P R E D I C T   (2)

Based on their morphology, identify each of the white blood cells shown in figure 11.7

Platelets

Platelets (plăt′letz), or **thrombocytes** (throm′bō-sītz), are minute fragments of cells, each consisting of a small amount of cytoplasm surrounded by a cell membrane (see figure 11.6). They are produced in the red bone marrow from **megakaryocytes** (meg-ă-kar′ē-ō-sītz, mega-, large + *karyon,* nucleus + *kytos,* cell), which are large cells (see figure 11.2). Small fragments of these cells break off and enter the blood as platelets, which play an important role in preventing blood loss. This prevention is accomplished in two ways: (1) the formation of platelet plugs, which seal holes in small vessels, and (2) the formation of clots, which help seal off larger wounds in the vessels.

Preventing Blood Loss

When a blood vessel is damaged, blood can leak into other tissues and interfere with normal tissue function, or blood can be lost from the body. A small amount of blood loss from the body can be tolerated, and new blood is produced to replace it. If a large amount of blood is lost, death can occur. Fortunately, when a blood vessel is damaged, vascular spasm, platelet plug formation, and blood clotting minimize the loss of blood.

Vascular Spasm

Vascular spasm is an immediate but temporary constriction of a blood vessel resulting from contraction of smooth muscle within the wall of the vessel. This constriction can close small vessels completely and stop the flow of blood through them. Damage to blood vessels can activate nervous system reflexes that cause vascular spasms. Chemicals also produce vascular spasms. For example, platelets release **thromboxanes** (throm′bok-zānz), which are derived from certain prostaglandins, and endothelial (epithelial) cells lining blood vessels release the peptide **endothelin** (en-dō′thē-lin).

Platelet Plugs

A **platelet plug** is an accumulation of platelets that can seal up a small break in a blood vessel. Platelet plug formation is very important in maintaining the integrity of the circulatory system because small tears occur in the smaller vessels and capillaries many times each day, and platelet plug formation quickly closes them. People who lack the normal number of platelets tend to develop numerous small hemorrhages in their skin and internal organs.

   The formation of a platelet plug can be described as a series of steps, but in actuality many of these steps occur at the same time (figure 11.8). **Platelet adhesion** results in platelets sticking to collagen exposed by blood vessel damage. Most platelet adhesion is mediated through **von Willebrand factor,** which is a protein produced and secreted by blood vessel endothelial cells. Von Willebrand factor forms a bridge between collagen and platelets by binding to platelet surface receptors and collagen. After platelets adhere to collagen, they become activated, change shape, and release chemicals. In the **platelet release reaction,** platelets release chemicals, such as ADP and thromboxane. ADP and thromboxane bind to their respective receptors on the surfaces of platelets, resulting in the activation of the platelets. These activated platelets also release ADP and thromboxane, which activates more platelets. Thus, a cascade of chemical release activates many platelets. As platelets become activated they express surface receptors called **fibrinogen receptors,** which can bind to fibrinogen, a plasma protein. In **platelet aggregation,** fibrinogen forms bridges between the fibrinogen receptors of numerous platelets, resulting in the formation of a platelet plug.

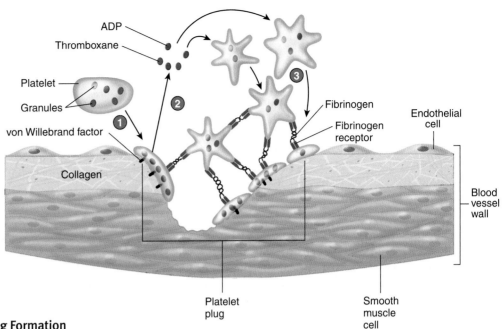

1. Platelet adhesion occurs when von Willebrand factor connects collagen and platelets.

2. The platelet release reaction is the release of ADP, thromboxanes, and other chemicals that activate other platelets.

3. Platelet aggregation occurs when fibrinogen receptors on activated platelets bind to fibrinogen, connecting the platelets to one another. A platelet plug is formed by the accumulating mass of platelets.

Process Figure 11.8 Platelet Plug Formation

Clinical Importance of Activating Platelets

Platelet activation results in platelet plug formation and the production of chemicals, such as phospholipids, that are important for blood clotting. Thus, inhibition of platelet activation reduces the formation of blood clots (see the next section, "Blood Clotting").

Thromboxanes, which activate platelets, are derived from certain prostaglandins. Aspirin inhibits prostaglandin synthesis and, therefore, thromboxane synthesis, which results in reduced platelet activation. If an expectant mother ingests aspirin near the end of pregnancy, thromboxane synthesis is inhibited, with several effects. Two of these effects are (1) the mother experiences excessive bleeding after delivery because of decreased platelet function, and (2) the baby can exhibit numerous localized hemorrhages over the surface of its body as a result of decreased platelet function. If the quantity of ingested aspirin is large, the infant, mother, or both may die as a result of bleeding.

Platelet plug and clot formation can cause blockage of blood vessels, producing heart attacks and strokes. Suspected heart attack victims are routinely given aspirin enroute to the emergency room as part of their treatment. The United States Preventive Services Task Force (USPSTF) and the American Heart Association recommend low-dose aspirin therapy (75–160 mg/day) for the prevention of cardiovascular disease for all men and women at high risk for cardiovascular disease. Determining high risk involves analyzing many risk factors and should be done in consultation with a physician. Decreased risk of cardiovascular disease from aspirin therapy must be weighed against the increased risk of hemorrhagic stroke and gastrointestinal bleeding. Risk factors include age (men over 40 and postmenopausal women), high cholesterol levels, high blood pressure, a history of smoking, diabetes, a family history of cardiovascular disease, or a previous clotting event, such as a heart attack, transient ischemic attack, or occlusive stroke.

Plavix (clopidogrel bisulfate) reduces the activation of platelets by blocking the ADP receptors on the surface of platelets. It is used to prevent clotting, and, along with other anticlotting drugs, to treat heart attacks.

A CASE IN POINT | Idiopathic Thrombocytopenic Purpura

Bruz Moore noticed that he developed a number of small bruises following a weekend touch football game. Although the bruising was unusual for him, he dismissed it. A few days later, Bruz suddenly developed a rash on his legs, which he also ignored. But when he noticed bleeding from his gums while brushing his teeth, he went to see his doctor. The rash, which was caused by many pinhead-sized hemorrhages in the skin, the bruising, and the bleeding from the gums all indicated a below-normal ability to stop bleeding. A blood test confirmed that Bruz had a lower than normal platelet count. In the absence of any evidence of toxic exposure to chemicals or diseases associated with a low platelet count, the doctor concluded that Bruz has idiopathic thrombocytopenic purpura (id′ē-ō-path′ik throm′ bō-sī-tō-pē′nē-ik pŭr′poo-ra) (ITP). In this disorder, the immune system makes antibodies (see chapter 14) that bind to platelets. As a result, the "tagged" platelets are removed from the blood by phagocytic cells in the spleen faster than they are produced in the red bone marrow. Consequently, platelet numbers decrease, resulting in bleeding problems. For more information on thrombocytopenia, see "Platelet Count" on p. 317.

Blood Clotting

Blood vessel constriction and platelet plugs alone are not sufficient to close large tears or cuts in blood vessels. When a blood vessel is severely damaged, **blood clotting,** or **coagulation** (kō-ag-ū-lā′shŭn), results in the formation of a clot. A clot is a network of threadlike protein fibers, called **fibrin** (fī′brin), that traps blood cells, platelets, and fluid.

The formation of a blood clot depends on a number of proteins found within plasma called **clotting factors.** Normally the clotting factors are inactive and do not cause clotting. Following injury, however, the clotting factors are activated to produce a clot. This is a complex process involving many chemical reactions, but it can be summarized in three main stages (figure 11.9).

1. The chemical reactions can be started in two ways:
   (a) the contact of inactive clotting factors with exposed connective tissue can result in their activation;
   (b) chemicals, such as **thromboplastin,** released from injured tissues can cause activation of clotting factors. After the initial clotting factors are activated, they in turn activate other clotting factors. A series of reactions results in which each clotting factor activates the next in the series until the clotting factor **prothrombinase** (prō-throm′bi-nās) is formed.
2. Prothrombinase acts on an inactive clotting factor called **prothrombin** (prō-throm′bin) to convert it to its active form called **thrombin** (throm′bin).
3. Thrombin converts the inactive clotting factor **fibrinogen** (fī-brin′ō-jen) into its active form, fibrin, a threadlike protein. A clot is a network of fibrin that traps blood cells, platelets, and fluid.

At each step of the clotting process, each clotting factor activates many additional clotting factors. Consequently, a large quantity of clotting factors is activated, resulting in the formation of the clot.

Most clotting factors are manufactured in the liver, and many of them require vitamin K for their synthesis. In addition, many of the chemical reactions of clot formation require Ca^{2+} and the chemicals released from platelets. Low levels of vitamin K, low levels of Ca^{2+}, low numbers of platelets, or reduced synthesis of clotting factors because of liver dysfunction can seriously impair the blood-clotting process.

Sources of Vitamin K

Humans rely on two sources of vitamin K. About half comes from the diet, and half from bacteria within the large intestine. Antibiotics taken to fight bacterial infections sometimes kill these intestinal bacteria, reducing vitamin K levels and resulting in bleeding problems. Vitamin K supplements may be necessary for patients on prolonged antibiotic therapy. Newborns lack these intestinal bacteria, and a vitamin K injection is routinely given to infants at birth. Infants can also obtain vitamin K from food such as milk. Because cow's milk contains more vitamin K than does human milk, breast-fed infants are more susceptible to bleeding than bottle-fed infants.

Control of Clot Formation

Without control, clotting would spread from the point of its initiation throughout the entire circulatory system. The blood contains several **anticoagulants** (an′tē-kō-ag′ū-lantz), which prevent clotting factors from forming clots. **Antithrombin** (an-tē-throm′bin) and **heparin** (hep′ă-rin), for example, inactivate thrombin. Without thrombin, fibrinogen is not converted to fibrin, and no clot

1. Stage 1. Inactive clotting factors are activated by exposure to connective tissue or by chemicals released from tissues.

2. Stage 1. Through a series of reactions, the activated clotting factors form prothrombinase.

3. Stage 2. Prothrombin is converted to thrombin by prothrombinase.

4. Stage 3. Fibrinogen is converted to fibrin (the clot) by thrombin.

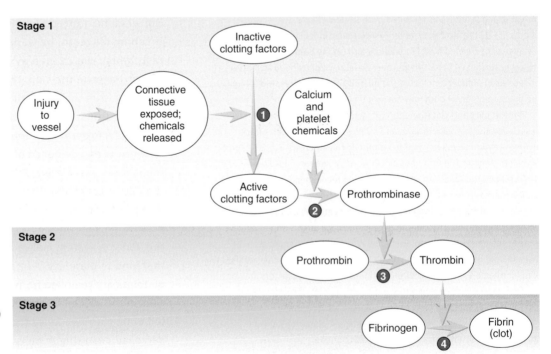

Process Figure 11.9   Clot Formation

Clot formation has three stages.

forms. Normally there are enough anticoagulants in the blood to prevent clot formation. At an injury site, however, the activation of clotting factors is very rapid. Enough clotting factors are activated so that the anticoagulants can no longer prevent a clot from forming. Away from the injury site there are enough anticoagulants to prevent clot formation from spreading.

The Danger of Unwanted Clots

When platelets encounter damaged or diseased areas of blood vessels or heart walls, an attached clot, called a **thrombus** (throm′bus), can form. A thrombus that breaks loose and begins to float through the circulation is called an **embolus** (em′bō-lŭs). Both thrombi and emboli can result in death if they block vessels that supply blood to essential organs such as the heart, brain, or lungs. Abnormal coagulation can be prevented or hindered by the injection of anticoagulants such as heparin, which acts rapidly. Warfarin (war′fă-rin), also called COUMADIN (koo′mă-din), acts more slowly than heparin. Warfarin prevents clot formation by suppressing the production of vitamin K-dependent clotting factors by the liver.

Clot Retraction and Fibrinolysis

After a clot has formed, it begins to condense into a more compact structure by a process known as **clot retraction.** Platelets contain the contractile proteins, actin and myosin, which operate in a similar fashion to the actin and myosin in muscle (see chapter 7). Platelets form small extensions that attach to fibrin through surface receptors. Contraction of the extensions pulls on the fibrin and is responsible for clot retraction. **Serum** (sēr′ŭm), which is plasma without the clotting factors, is squeezed out of the clot during clot retraction.

Retraction of the clot pulls the edges of the damaged blood vessel together, helping to stop the flow of blood, reducing the probability of infection, and enhancing healing. The damaged vessel is repaired by the movement of fibroblasts into the damaged area and the formation of new connective tissue. In addition, epithelial cells around the wound divide and fill in the torn area (see chapter 4).

Clots are dissolved by a process called **fibrinolysis** (fī-bri-nol′-i-sis, fibrino-, fiber + *lysis,* dissolution) (figure 11.10). An inactive plasma protein called **plasminogen** (plaz-min′ō-jen, plasmin + *gen,* produce) is converted to its active form, **plasmin** (plaz′min). Thrombin, other clotting factors activated during clot formation, and **tissue plasminogen activator (t-PA)** released from surrounding tissues can stimulate the conversion of plasminogen to plasmin. Over a period of a few days, plasmin slowly breaks down the fibrin.

Dissolving Clots

A heart attack can result from blockage by a clot of blood vessels that supply blood to the heart. One treatment for a heart attack is to inject into the blood chemicals that activate plasmin. Unlike aspirin and anticoagulant therapies, which are used to prevent heart attacks, the strategy in using plasmin activators is to quickly dissolve the clot and restore blood flow to cardiac muscle, thus reducing damage to tissues. **Streptokinase** (strep-tō-kin′ăs), a bacterial enzyme, and t-PA, produced through genetic engineering, have been used successfully to dissolve clots.

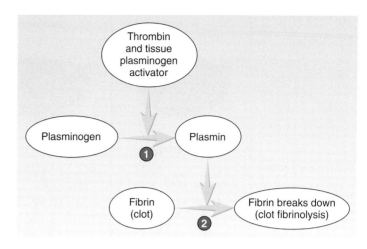

1. Thrombin and tissue plasminogen activator convert inactive plasminogen into plasmin.

2. Plasmin breaks down the fibrin in a blood clot, resulting in clot fibrinolysis.

Process Figure 11.10 Fibrinolysis

Blood Grouping

If large quantities of blood are lost during surgery or in an accident, the patient can go into shock and die unless a transfusion or infusion is performed. A **transfusion** is the transfer of blood or blood components from one individual to another. When large quantities of blood are lost, red blood cells must be replaced so that the oxygen-carrying capacity of the blood is restored. An **infusion** is the introduction of a fluid other than blood, such as a saline or glucose solution, into the blood. In many cases, the return of blood volume to normal levels is all that is necessary to prevent shock. Eventually, the body produces red blood cells to replace those that were lost.

Early attempts to transfuse blood were often unsuccessful because they resulted in **transfusion reactions,** which included clumping of blood cells, rupture of blood cells, and clotting within blood vessels. It is now known that transfusion reactions are caused by interactions between antigens and antibodies (see chapter 14). In brief, the surfaces of red blood cells have molecules called **antigens** (an′ti-jenz, anti (body) + -*gen,* producing), and the plasma includes proteins called **antibodies** (an′te-bod-ēz, anti-, against + body, a thing). Antibodies are very specific, meaning that each antibody can combine only with a certain antigen. When the antibodies in the plasma bind to the antigens on the surface of the red blood cells, they form molecular bridges that connect the red blood cells together. As a result, **agglutination** (ă-gloo-ti-nā′shŭn, *ad,* to + *gluten,* glue), or clumping of the cells, occurs. The combination of the antibodies with the antigens also can initiate reactions that cause **hemolysis** (hē-mol′i-sis, hemo-, blood + *lysis,* destruction), or rupture of the red blood cells. The debris formed from the ruptured red blood cells can trigger clotting within small blood vessels. As a result of these changes, tissue damage and death may occur.

The antigens on the surface of red blood cells have been categorized into **blood groups.** Although many blood groups

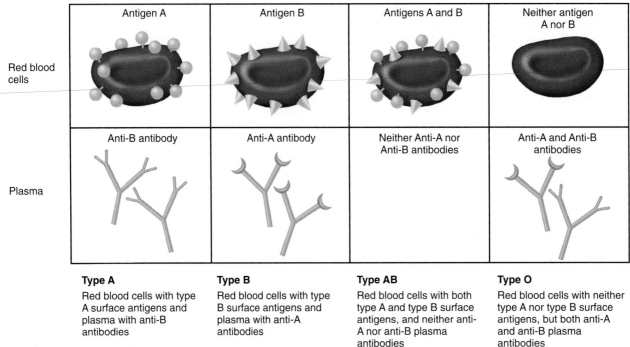

Figure 11.11   ABO Blood Groups

are recognized, the ABO and Rh blood groups are the most important for transfusion reactions.

ABO Blood Group

The **ABO blood group** system is used to categorize human blood. ABO antigens appear on the surface of the red blood cells. Type A blood has type A antigens, type B blood has type B antigens, type AB blood has both types of antigens, and type O blood has neither A nor B antigens (figure 11.11). In addition, plasma from type A blood contains anti-B antibodies, which act against type B antigens; whereas plasma from type B blood contains anti-A antibodies, which act against type A antigens. Type AB blood plasma has neither type of antibody, and type O blood plasma has both anti-A and anti-B antibodies.

The ABO blood types are not found in equal numbers. In white people in the United States the distribution is type O, 47%; type A, 41%; type B, 9%; and type AB, 3%. Among African-Americans the distribution is type O, 46%; type A, 27%; type B, 20%; and type AB, 7%.

Antibodies do not normally develop against an antigen unless the body is exposed to that antigen. One possible explanation for the production of anti-A and/or anti-B antibodies is that type A or B antigens on bacteria or food in the digestive tract stimulate the formation of antibodies against antigens that are different from one's own antigens. In support of this explanation is the observation that anti-A and anti-B antibodies are not found in the blood until about 2 months after birth. For example, an infant with type A blood would produce anti-B antibodies against the B antigens on bacteria or food. An infant with A antigens would not produce antibodies against the A antigen on bacteria or food because mechanisms exist in the body to prevent the production of antibodies that would react with the body's own antigens (see chapter 14).

A **donor** is a person who gives blood, and a **recipient** is a person who receives blood. Usually a recipient can receive blood from a donor if they both have the same blood type. For example, a person with type A blood can receive blood from a person with type A blood. There would be no ABO transfusion reaction because the recipient has no anti-A antibodies against the type A antigen. On the other hand, if type A blood were donated to a person with type B blood, a transfusion reaction would occur because the person with type B blood has anti-A antibodies against the type A antigen, and agglutination would result (figure 11.12).

Historically, people with type O blood have been called universal donors because they usually can give blood to the other ABO blood types without causing an ABO transfusion reaction. Their red blood cells have no ABO surface antigens and therefore do not react with the recipient's anti-A or anti-B antibodies. For example, if type O blood is given to a person with type A blood, the type O red blood cells do not react with the anti-B antibodies in the recipient's blood. In a similar fashion, if type O blood is given to a person with type B blood, there would be no reaction with the recipient's anti-A antibodies.

It should be noted, however, that the term "universal donor" is misleading. There are two ways in which transfusion of type O blood can produce a transfusion reaction. First, mismatching blood groups other than the ABO blood group can cause a transfusion reaction. To reduce the likelihood of a transfusion reaction, all the blood groups must be correctly matched. Second, antibodies in the blood of the donor can react with antigens on the red blood cells in the blood of the recipient. For example, type O blood has anti-A and anti-B antibodies. If type O blood is transfused into a person with type A blood, the anti-A antibodies (in the type O blood) react against the A antigens (on the red blood cells in the type A blood). Usually such reactions are not serious because the antibodies in the

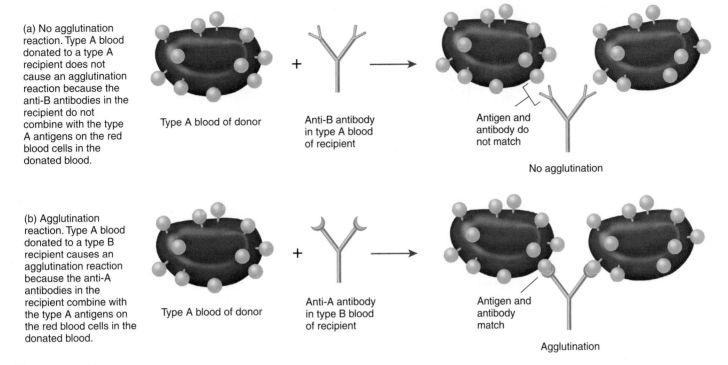

(a) No agglutination reaction. Type A blood donated to a type A recipient does not cause an agglutination reaction because the anti-B antibodies in the recipient do not combine with the type A antigens on the red blood cells in the donated blood.

Type A blood of donor

Anti-B antibody in type A blood of recipient

Antigen and antibody do not match

No agglutination

(b) Agglutination reaction. Type A blood donated to a type B recipient causes an agglutination reaction because the anti-A antibodies in the recipient combine with the type A antigens on the red blood cells in the donated blood.

Type A blood of donor

Anti-A antibody in type B blood of recipient

Antigen and antibody match

Agglutination

Figure 11.12   **Agglutination Reaction**

donor's blood are diluted in the large volume of the recipient's blood, and few reactions take place. Type O blood is given to a person with another blood type only in life-or-death conditions because it sometimes can cause a transfusion reaction.

> **P R E D I C T   ❸**
> Historically, people with type AB blood were called universal recipients. What is the rationale for this term? Explain why the term is misleading.

Rh Blood Group

Another important blood group is the **Rh blood group,** so named because it was first studied in the rhesus monkey. People are Rh-positive if they have certain Rh antigens on the surface of their red blood cells, and they are Rh-negative if they do not have these Rh antigens. About 85% of whites and 95% of African-Americans are Rh-positive. The ABO blood type and the Rh blood type usually are designated together. For example, a person designated as A positive is type A in the ABO blood group and Rh-positive. The rarest combination in the United States is AB negative, which occurs in less than 1% of all Americans.

Antibodies against the Rh antigens do not develop unless an Rh-negative person is exposed to Rh-positive red blood cells. This can occur through a transfusion or by the transfer of blood across the placenta to a mother from her fetus. When an Rh-negative person receives a transfusion of Rh-positive blood, the recipient becomes sensitized to the Rh antigens and produces anti-Rh antibodies. If the Rh-negative person is unfortunate enough to receive a second transfusion of Rh-positive blood after becoming sensitized, a transfusion reaction results.

Rh incompatibility can pose a major problem in some pregnancies, when the mother is Rh-negative and the fetus is

Rh-positive (figure 11.13). If fetal blood leaks through the placenta and mixes with the mother's blood, the mother becomes sensitized to the Rh antigen. The mother produces anti-Rh antibodies that cross the placenta and cause agglutination and hemolysis of fetal red blood cells. This disorder is called **hemolytic** (hē-mō-lit′ik) **disease of the newborn (HDN)** or **erythroblastosis fetalis** (ĕ-rith′rō-blas-tō′sis fē-ta′lis), and it can be fatal to the fetus. In the mother's first pregnancy, there is often no problem. The leakage of fetal blood is usually the result of a tear in the placenta that takes place either late in the pregnancy or during delivery. Thus, there is not enough time for the mother to produce enough anti-Rh antibodies to harm the fetus. In later pregnancies, however, a problem can arise because the mother has been sensitized to the Rh antigen. Consequently, if the fetus is Rh-positive and if any fetal blood leaks into the mother's blood, she rapidly produces large amounts of anti-Rh antibodies, which can cross the placenta to the fetus, and HDN develops. Therefore, the levels of anti-Rh antibodies in the mother's blood should be monitored. If they increase to unacceptable levels, the fetus should be tested to determine the severity of the HDN. In severe cases, a transfusion to replace lost red blood cells can be performed through the umbilical cord, or the baby can be delivered if mature enough.

Prevention of HDN is often possible if the Rh-negative mother is given an injection of a specific type of antibody preparation called Rho(D) immune globulin (RhoGAM). The injection can be given during the pregnancy, before delivery, or immediately after each delivery, miscarriage, or abortion. The injection contains antibodies against Rh antigens. The injected antibodies bind to the Rh antigens of any fetal red blood cells that may have entered the mother's blood. This treatment inactivates the fetal Rh antigens and prevents sensitization of the mother.

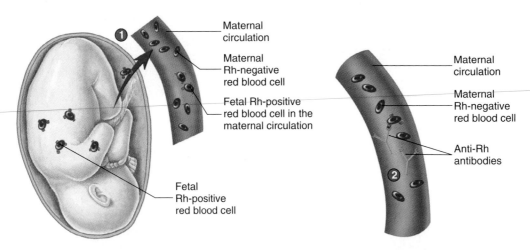

1. Before or during delivery, Rh-positive red blood cells from the fetus enter the blood of an Rh-negative woman through a tear in the placenta.

2. The mother is sensitized to the Rh antigen and produces anti-Rh antibodies. Because this usually happens after delivery, there is no effect on the fetus in the first pregnancy.

3. During a subsequent pregnancy with an Rh-positive fetus, Rh-positive red blood cells cross the placenta, enter the maternal circulation, and stimulate the mother to produce antibodies against the Rh antigen. Antibody production is rapid because the mother has been sensitized to the Rh antigen.

4. The anti-Rh antibodies from the mother cross the placenta, causing agglutination and hemolysis of fetal red blood cells, and hemolytic disease of the newborn (HDN) develops.

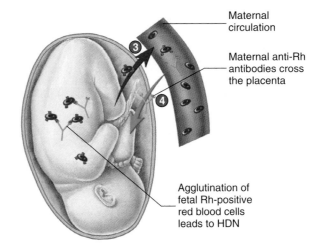

Process Figure 11.13   Hemolytic Disease of the Newborn (HDN)

A CASE IN POINT | Treatment of HDN

Billy Rubin was born with HDN. He was treated with phototherapy, exchange transfusion, and erythropoietin. Phototherapy, in which blood passing through the skin is exposed to blue or white lights, results in the breakdown of bilirubin to less toxic compounds that are removed by the newborn's liver. During fetal development, the increased rate of red blood cell destruction caused by the mother's anti-Rh antibodies results in the increased production of bilirubin. Although high levels of bilirubin can damage the brain by killing nerve cells, this is not usually a problem in the fetus because the bilirubin is removed by the placenta. Following birth, bilirubin levels can increase because of the continued lysis of red blood cells and the inability of the newborn's liver to handle the large bilirubin load.

An exchange transfusion replaces Billy Rubin's blood with donor blood, resulting in a decrease in bilirubin and anti-Rh antibody levels. Fewer anti-Rh antibodies decreases the agglutination and lysis of red blood cells. In HDN, red blood cells are destroyed, resulting in low numbers of red blood cells, a condition called **anemia** (ă-nē′mē-ă). Exchange transfusion and administering erythropoietin can be used to treat anemia. The exchange transfusion replaces the newborn's blood with blood that has more red blood cells. Erythropoietin stimulates the newborn to increase the production of red blood cells.

PREDICT
When treating HDN with an exchange transfusion, should the donor's blood be Rh-positive or negative? Explain.

Diagnostic Blood Tests

Type and Crossmatch

To prevent transfusion reactions the blood is typed, and a crossmatch is made. **Blood typing** determines the ABO and Rh blood groups of the blood sample. Typically, the cells are separated from the serum. The cells are tested with known antibodies to determine the type of antigen on the cell surface. For example, if a patient's blood cells agglutinate when mixed with type A antibodies but do not agglutinate when mixed with type B antibodies, it is concluded that the cells have type A antigen. In a similar fashion, the serum is mixed with known cell types (antigens) to determine the type of antibodies in the serum.

Normally, donor blood must match the ABO and Rh type of the recipient. Because other blood groups can cause a transfusion reaction, however, a crossmatch is performed. In a **crossmatch,** the donor's blood cells are mixed with the recipient's serum, and the donor's serum is mixed with the recipient's cells. The donor's blood is considered safe for transfusion only if no agglutination occurs in either match.

Complete Blood Count

The **complete blood count (CBC)** is an analysis of the blood that provides much information. It consists of a red blood cell count, hemoglobin and hematocrit measurements, and a white blood cell count.

Red Blood Count

Blood cell counts are usually done electronically with a machine, but they can be done manually with a microscope. A normal **red blood count (RBC)** for a male is 4.6–6.2 million red blood cells per microliter (μL) of blood, and for a female it is 4.2–5.4 million μL of blood. A microliter is equivalent to one cubic millimeter (mm^3) or 10^{-6} L, and one drop of blood is approximately 50 μL. Erythrocytosis (ĕ-rith′rō-sī-tō′sis) is an overabundance of red blood cells (see Clinical Focus "Some Disorders of the Blood" on p. 316).

Hemoglobin Measurement

The hemoglobin measurement determines the amount of hemoglobin in a given volume of blood, usually expressed as grams of hemoglobin per 100 mL of blood. The normal hemoglobin measurement for a male is 14–18 grams (g)/100 mL of blood, and for a female it is 12 to 16 g/100 mL of blood. An abnormally low hemoglobin measurement is an indication of **anemia** (ă-nē′mē-ă), which is either a reduced number of red blood cells or a reduced amount of hemoglobin in each red blood cell (see Clinical Focus "Some Disorders of the Blood" on p. 316).

Hematocrit Measurement

The percentage of total blood volume composed of red blood cells is the **hematocrit** (hē′mă-tō-krit, hem′a-tō-krit). One way to determine hematocrit is to place blood in a tube and spin the tube in a centrifuge. The formed elements are heavier than the plasma and are forced to one end of the tube. White blood cells and platelets form a thin, whitish layer, called the **buffy coat,** between the plasma and the red blood cells (figure 11.14). The red blood cells account for 40–52% of the total blood volume in males and 38–48% in females. The hematocrit measurement is affected by the number and size of red blood cells because it is based on volume. For example, a decreased hematocrit can result from a decreased number of normal-sized red blood cells or a normal number of small-sized red blood cells. The average size of a red blood cell is calculated by dividing the hematocrit by the red blood cell count. A number of disorders cause red blood cells to be smaller or larger than normal. For example, inadequate iron in the diet can impair hemoglobin production. Consequently, red blood cells do not fill up with hemoglobin during their formation, and they remain smaller than normal.

White Blood Count

A **white blood count (WBC)** measures the total number of white blood cells in the blood. There are normally 5000–9000 white blood cells per microliter of blood. **Leukopenia** (loo-kō-pē′nē-ă) is a lower than normal WBC and often indicates decreased

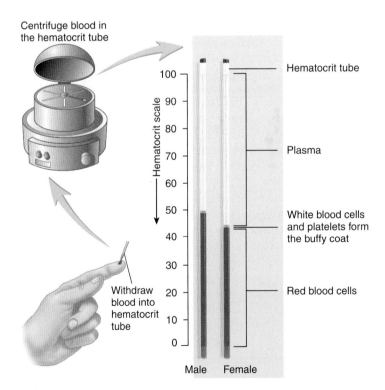

Figure 11.14 Hematocrit

Blood is withdrawn into a capillary tube and placed in a centrifuge. The blood is separated into plasma, red blood cells, and a small amount of white blood cells and platelets, which rest on the red blood cells. The hematocrit measurement is the percent of the blood volume that is red blood cells. It doesn't measure the white blood cells and platelets. Normal hematocrits for a male and a female are shown.

Clinical Focus Some Disorders of the Blood

Erythrocytosis

Erythrocytosis (ĕ-rith'rō-sī-tō'sis) is an overabundance of red blood cells, resulting in increased blood viscosity, reduced flow rates, and, if severe, plugging of the capillaries. **Relative erythrocytosis** results from decreased plasma volume, such as that caused by dehydration, diuretics, and burns. **Primary erythrocytosis,** often called **polycythemia vera** (pol'ē-sī-thē'mē-ă ve'ra), is a stem cell defect of unknown cause that results in the overproduction of red blood cells, granulocytes, and platelets. Erythropoietin levels are low and the spleen can be enlarged. **Secondary erythrocytosis (polycythemia)** results from a decreased oxygen supply, such as that which occurs at high altitudes, in chronic obstructive pulmonary disease, or in congestive heart failure. The resulting decrease in oxygen delivery to the kidneys stimulates erythropoietin secretion and causes an increase in red blood cell production. In primary and secondary erythrocytosis, the increased number of red blood cells increases blood viscosity and blood volume. There can be clogging of capillaries and the development of hypertension.

Anemia

Anemia (ă-nē'mē-ă) is a deficiency of normal hemoglobin in the blood, resulting from a decreased number of red blood cells, a decreased amount of hemoglobin in each red blood cell, or both. Anemia can also be the result of abnormal hemoglobin production.

Anemia reduces the ability of the blood to transport oxygen. People with anemia suffer from a lack of energy and feel excessively tired and listless. They may appear pale and quickly become short of breath with only slight exertion.

One general cause of anemia is insufficient production of red blood cells. **Aplastic** (ă-plas'tik) **anemia** is caused by an inability of the red bone marrow to produce red blood cells and, often, white blood cells and platelets. It is usually acquired as a result of damage to the stem cells in red marrow by chemicals such as benzene, drugs such as certain antibiotics and sedatives, or radiation.

Red blood cell production also can be lower than normal as a result of nutritional deficiencies. **Iron-deficiency anemia** results from a deficient intake or absorption of iron or from excessive iron loss. Consequently not enough hemoglobin is produced, the number of red blood cells decreases, and the red blood cells that are manufactured are smaller than normal.

Folate deficiency can also cause anemia. Inadequate amounts of folate in the diet is the usual cause of folate deficiency, with the disorder developing most often in the poor, in pregnant women, and in chronic alcoholics. Because folate helps in the synthesis of DNA, a folate deficiency results in fewer cell divisions and, therefore, decreased red blood cell production. A deficiency in folate during pregnancy is also associated with birth disorders called neural tube defects, such as spina bifida.

Another type of nutritional anemia is **pernicious** (per-nish'ŭs) **anemia,** which is caused by inadequate vitamin B_{12}. A 2- to 3-year supply of vitamin B_{12} can be stored in the liver. Because vitamin B_{12} is important for folate synthesis, inadequate amounts of vitamin B_{12} can also result in decreased red blood cell production. Although inadequate levels of vitamin B_{12} in the diet can cause pernicious anemia, the usual cause is insufficient absorption of the vitamin. Normally the stomach produces **intrinsic factor,** a protein that binds to vitamin B_{12}. The combined molecules pass into the lower intestine, where intrinsic factor facilitates the absorption of the vitamin. Without adequate levels of intrinsic factor, insufficient vitamin B_{12} is absorbed, and pernicious anemia develops. Most cases of pernicious anemia probably result from an autoimmune disease in which the body's immune system damages the cells in the stomach that produce intrinsic factor.

Another general cause of anemia is loss or destruction of red blood cells. **Hemorrhagic** (hem-ŏ-raj'ik) **anemia** results from a loss of blood such as can result from trauma, ulcers, or excessive menstrual bleeding. Chronic blood loss, in which small amounts of blood are lost over a period of time, can result in iron-deficiency anemia. **Hemolytic** (hē-mō-lit'ik) **anemia** is a disorder in which red blood cells rupture or are destroyed at an excessive rate. It can be caused by inherited defects in the red blood cells. For example, one kind of inherited hemolytic anemia results from a defect in the cell membrane that causes red blood cells to rupture easily. Many kinds of hemolytic anemia result from unusual damage to the red blood cells by drugs, snake venom, artificial heart valves, autoimmune disease, or hemolytic disease of the newborn.

Anemia can result from a reduced rate of synthesis of the globin chains in hemoglobin. **Thalassemia** (thal-ă-sē'mē-ă) is a hereditary disease found in people of Mediterranean, Asian, and African ancestry. If hemoglobin production is severely depressed, death usually occurs before age 20. In less severe cases, thalassemia produces a mild anemia.

Some anemias are caused by defective hemoglobin production. **Sickle cell anemia** is a hereditary disease found mostly in people of African descent that results in the formation of an abnormal hemoglobin. The red blood cells assume a rigid, sickle shape and plug up small blood vessels. They are also

production or destruction of the red marrow. Radiation, drugs, tumors, viral infections, or a deficiency of the vitamins folate or B_{12} can cause leukopenia. **Leukocytosis** (loo'kō-sī-tō'sis) is an abnormally high WBC. Bacterial infections often cause leukocytosis by stimulating neutrophils to increase in number. **Leukemia** (loo-kē'mē-ă), a cancerous tumor of the red marrow, can cause leukocytosis, but the white blood cells do not function normally.

Differential White Blood Count

A **differential white blood count** determines the percentage of each of the five kinds of white blood cells in the white blood cell count. Normally neutrophils account for 60–70%, lymphocytes 20–25%, monocytes 3–8%, eosinophils 2–4%, and basophils 0.5–1% of all white blood cells. Much insight about a patient's condition can be obtained from a differential

more fragile than normal. In severe cases, there is so much abnormal hemoglobin production that the disease is usually fatal before age 30. In many cases, however, the production of normal hemoglobin compensates for the abnormal hemoglobin, and the person exhibits no symptoms.

Leukemia

Leukemia (loo-kē'mē-ă) is a cancer in which abnormal production of one or more of the white blood cell types occurs. Because these cells are usually immature or abnormal and lack normal immunological functions, people with leukemia are very susceptible to infections. The excess production of white blood cells in the red marrow can also interfere with red blood cell and platelet formation and thus lead to anemia and bleeding.

Disseminated Intravascular Coagulation

Disseminated intravascular coagulation (DIC) is a complex disorder involving clotting throughout the vascular system followed by bleeding. Normally excessive clotting is prevented by anticoagulants. DIC can develop when these control mechanisms are overwhelmed. Many conditions can cause DIC by overstimulating blood clotting. Examples include massive tissue damage, such as burns, or alteration of the lining of blood vessels caused by infections or snake bites. If DIC occurs slowly, the predominant effect is thrombosis and blockage of blood vessels. If DIC occurs rapidly, massive clot formation occurs, quickly using up available blood clotting factors and platelets. The result is continual bleeding around wounds, intravenous lines, catheters, as well as internal bleeding. The best therapy for DIC is to treat and stop whatever condition is stimulating blood clotting.

Von Willebrand Disease

Von Willebrand disease is the most common inherited bleeding disorder, occurring as frequently as 1 in 1000 individuals. Von Willebrand factor helps platelets adhere to collagen and become activated. In von Willebrand disease, platelet plug formation and the contribution of activated platelets to blood clotting is impaired. Treatments for von Willebrand disease include injections of von Willebrand factor or the administration of drugs that increase von Willebrand factor levels in the blood.

Hemophilia

Hemophilia (hē-mō-fil'ē-ă) is a genetic disorder in which clotting is abnormal or absent. It is most often found in people from northern Europe and their descendants. Hemophilia is most often a sex-linked trait, and it occurs almost exclusively in males (see chapter 20). There are several types of hemophilia, each the result of a deficiency or dysfunction of a clotting factor. Treatment of hemophilia involves injection of the missing clotting factor taken from donated blood or produced by genetic engineering.

Infectious Diseases of the Blood

After entering the body, many microorganisms are transported by the blood to the tissues they infect. For example, the poliovirus enters through the small intestine and is carried to nervous tissue. After microorganisms are established at a site of infection, some of them can enter the blood. These microorganisms can be transported to other locations in the body, multiply within the blood, or be eliminated by the body's immune system.

Septicemia (sep-ti-sē'mē-ă), or blood poisoning, is the spread of microorganisms and their toxins by the blood. Often septicemia results from the introduction of microorganisms by a medical procedure such as the insertion of an intravenous tube into a blood vessel. The release of toxins by bacteria can cause **septic shock,** which can produce a decrease in blood pressure that can result in death.

There are a few diseases in which microorganisms actually multiply within blood cells. **Malaria** (mă-lār'ē-ă) is caused by a protozoan that is introduced into the blood by the bite of the *Anopheles* mosquito. Part of the protozoan's development occurs inside red blood cells. The symptoms of chills and fever are produced by toxins released when the protozoan causes the red blood cells to rupture. **Infectious mononucleosis** (mon'ō-noo-klē-ō'sis) is caused by a virus (Epstein-Barr virus) that infects the salivary glands and lymphocytes. The lymphocytes are altered by the virus, and the immune system attacks and destroys the lymphocytes. The immune system response is believed to produce the symptoms of fever, sore throat, and swollen lymph nodes. The human immunodeficiency virus (HIV) also infects lymphocytes and causes immune system suppression, resulting in **acquired immunodeficiency syndrome (AIDS)** (see chapter 14).

The presence of microorganisms in blood is a concern when transfusions are made, because it is possible to infect the blood recipient. Blood is routinely tested in an effort to eliminate this risk, especially for AIDS and hepatitis. One cause of **hepatitis** (hep-ă-tī'tis) is an infection of the liver by viruses. After recovering, hepatitis victims can become carriers. Although they show no signs of the disease, they release the virus into their blood or bile. To prevent infection of others, anyone who has had hepatitis is asked not to donate blood products.

white blood count. For example, in bacterial infections, the neutrophil count is often greatly increased, whereas in allergic reactions the eosinophil and basophil counts are elevated.

Clotting

Two measurements that test the ability of the blood to clot are the platelet count and the prothrombin time.

Platelet Count

A normal **platelet count** is 250,000–400,000 platelets per microliter of blood. **Thrombocytopenia** (throm'bō-sī-tō-pē'nē-ă) is a condition in which the platelet count is greatly reduced, resulting in chronic bleeding through small vessels and capillaries. It can be caused by decreased platelet production as a result of hereditary disorders, lack of vitamin B_{12} (pernicious anemia), drug therapy, or radiation therapy.

Prothrombin Time Measurement

Prothrombin time is a measure of how long it takes for the blood to start clotting, which is normally 9–12 seconds. Prothrombin time is determined by adding thromboplastin to whole plasma. Thromboplastin is a chemical released from injured tissues that starts the process of clotting (see figure 11.9). Prothrombin time is officially reported as the International Normalized Ratio (INR), which standardizes the time it takes to clot on the basis of the slightly different thromboplastins used by different labs. Because many clotting factors have to be activated to form fibrin, a deficiency of any one of them can cause an abnormal prothrombin time. Vitamin K deficiency, certain liver diseases, and drug therapy can cause an increased prothrombin time.

Blood Chemistry

The composition of materials dissolved or suspended in the plasma can be used to assess the functioning of many of the body's systems. For example, high blood glucose levels can indicate that the pancreas is not producing enough insulin, high blood urea nitrogen (BUN) is a sign of reduced kidney function, increased bilirubin can indicate liver dysfunction, and high cholesterol levels can indicate an increased risk of developing cardiovascular disease. A number of blood chemistry tests are routinely done when a blood sample is taken, and additional tests are available.

> **P R E D I C T** **5**
> When a patient complains of acute pain in the abdomen, the physician suspects appendicitis, which is a bacterial infection of the appendix. What blood test could provide supporting evidence for the diagnosis?

S U M M A R Y

Functions of Blood   (p. 301)

1. Blood transports gases, nutrients, waste products, processed molecules, and regulatory molecules.
2. Blood regulates pH, fluid, and ion balance.
3. Blood is involved with temperature regulation and protects against foreign substances such as microorganisms and toxins.
4. Blood clotting prevents fluid and cell loss and is part of tissue repair.

Composition of Blood   (p. 301)

1. Blood is a connective tissue consisting of plasma and formed elements.
2. Total blood volume is approximately 5 L.

Plasma   (p. 301)

1. Plasma is 91% water and 9% suspended or dissolved substances.
2. Plasma maintains osmotic pressure, is involved in immunity, prevents blood loss, and transports molecules.

Formed Elements   (p. 303)

The formed elements are cells (red blood cells and white blood cells) and cell fragments (platelets).

Production of Formed Elements

Formed elements arise (hematopoiesis) in red bone marrow from stem cells.

Red Blood Cells

1. Red blood cells are disk-shaped cells containing hemoglobin, which transports oxygen and carbon dioxide. Red blood cells also contain carbonic anhydrase, which is involved with carbon dioxide transport.
2. In response to low blood oxygen levels, the kidneys produce erythropoietin, which stimulates red blood cell production in red bone marrow.
3. Worn-out red blood cells are phagocytized by macrophages in the spleen or liver. Hemoglobin is broken down, iron and amino acids are reused, and heme becomes bilirubin that is secreted in bile.

White Blood Cells

1. White blood cells protect the body against microorganisms and remove dead cells and debris.
2. Granulocytes contain cytoplasmic granules, and there are three types of granulocytes: neutrophils are small phagocytic cells, basophils promote inflammation, and eosinophils reduce inflammation.
3. Agranulocytes have very small granules and are of two types: lymphocytes are involved in antibody production and other immune system responses; monocytes become macrophages that ingest microorganisms and cellular debris.

Platelets

Platelets are cell fragments involved with preventing blood loss.

Preventing Blood Loss   (p. 308)
Vascular Spasm

Blood vessels constrict in response to injury, resulting in decreased blood flow.

Platelet Plugs

1. Minor damage to blood vessels is repaired by platelet plugs.
2. Platelets adhere to collagen, release chemicals (ADP and thromboxanes) that activate other platelets, and connect to one another with fibrinogen to form platelet plugs.

Blood Clotting

1. Blood clotting, or coagulation, is formation of a clot (a network of protein fibers called fibrin).
2. There are three steps in the clotting process: Activation of clotting factors by connective tissue and chemicals, resulting in the formation of prothrombinase. Conversion of prothrombin to thrombin by prothrombinase. Conversion of fibrinogen to fibrin by thrombin.

Control of Clot Formation

Anticoagulants in the blood, such as antithrombin and heparin, prevent clot formation.

Clot Retraction and Fibrinolysis

1. Clot retraction condenses the clot, pulling the edges of damaged tissue closer together.
2. Serum is plasma without clotting factors.
3. Fibrinolysis (clot breakdown) is accomplished by plasmin.

Blood Grouping (p. 311)

1. Blood groups are determined by antigens on the surface of red blood cells.
2. In transfusion reactions, antibodies can bind to red blood cell antigens, resulting in agglutination or hemolysis of red blood cells.

ABO Blood Group

1. Type A blood has A antigens, type B blood has B antigens, type AB blood has A and B antigens, and type O blood has neither A or B antigens.
2. Type A blood has anti-B antibodies, type B blood has anti-A antibodies, type AB blood has neither A or B antibodies, and type O blood has both anti-A and anti-B antibodies.
3. Mismatching the ABO blood group can result in transfusion reactions.

Rh Blood Group

1. Rh-positive blood has Rh antigens, whereas Rh-negative blood does not.
2. Antibodies against the Rh antigen are produced when an Rh-negative person is exposed to Rh-positive blood.
3. The Rh blood group is responsible for hemolytic disease of the newborn, which can occur when the fetus is Rh-positive and the mother is Rh-negative.

Diagnostic Blood Tests (p. 315)

Type and Crossmatch

1. Blood typing determines the ABO and Rh blood groups of a blood sample.
2. A crossmatch tests for agglutination reactions between donor and recipient blood.

Complete Blood Count

The complete blood count consists of the following: red blood count (million/μL), hemoglobin measurement (grams of hemoglobin per 100 mL of blood), hematocrit measurement (percent volume of red blood cells), and white blood count (million/μL).

Differential White Blood Count

The differential white blood count determines the percentage of each type of white blood cell.

Clotting

Platelet count and prothrombin time measure the ability of the blood to clot.

Blood Chemistry

The composition of materials dissolved or suspended in plasma (e.g., glucose, urea nitrogen, bilirubin, and cholesterol) can be used to assess the functioning and status of the body's systems.

REVIEW AND COMPREHENSION

1. Describe the functions of blood.
2. Define plasma. List the functions of plasma.
3. Define the formed elements, and name the different types of formed elements. Explain how and where the formed elements arise through hematopoiesis.
4. Describe the two basic parts of a hemoglobin molecule. Which part is associated with iron? What gases are transported by each part?
5. What is the role of carbonic anhydrase in gas transport?
6. Why are the vitamins folate and B_{12} important in red blood cell production?
7. Explain how low blood oxygen levels result in increased red blood cell production.
8. Where are red blood cells broken down? What happens to the breakdown products?
9. Give two functions of white blood cells.
10. Name the five types of white blood cells, and state a function for each type.
11. What are platelets, and how are they formed?
12. Describe the role of blood vessel constriction and platelet plugs in preventing bleeding. Describe the three steps of platelet plug formation.
13. What are clotting factors? Describe the three steps of activation that result in the formation of a clot.
14. Explain the function of anticoagulants in the blood, and give an example of an anticoagulant.
15. What is clot retraction, and what does it accomplish?
16. Define fibrinolysis, and name the chemicals responsible for this process.
17. What are blood groups, and how do they cause transfusion reactions? List the four ABO blood types. Why is type O blood considered a universal donor?
18. What is meant by the term Rh-positive? How can Rh incompatibility affect a pregnancy?
19. For each of the following tests, define the test and give an example of a disorder that would cause an abnormal test result:
    a. Type and crossmatch
    b. Red blood count
    c. Hemoglobin measurement
    d. Hematocrit measurement
    e. White blood count
    f. Differential white blood count
    g. Platelet count
    h. Prothrombin time
    i. Blood chemistry tests

C R I T I C A L   T H I N K I N G

1. Red Packer, a physical education major, wanted to improve his performance in an upcoming marathon race. About 6 weeks before the race, 1 L of blood was removed from his body, and the formed elements were separated from the plasma. The formed elements were frozen, and the plasma was reinfused into his body. Just before the race, the formed elements were thawed and injected into his body. Explain why this procedure, called blood doping or blood boosting, would help Red's performance. Can you suggest any possible bad effects?

2. Chemicals such as benzene can destroy red bone marrow, causing aplastic anemia. What symptoms would you expect to develop as a result of the lack of (a)red blood cells, (b)platelets, and (c)white blood cells?

3. E. Z. Goen habitually used barbiturates to ease feelings of anxiety. Because barbiturates depress the respiratory centers in the brain, they cause hypoventilation (i.e., slower than normal rate of breathing). What happens to the red blood cell count of a habitual user of barbiturates? Explain.

4. What blood problems would you expect to observe in a patient after total gastrectomy (removal of the stomach)?

5. According to the old saying, "Good food makes good blood." Name three substances in the diet that are essential for "good blood." What blood disorders develop if these substances are absent from the diet?

6. Why do anemic patients often have gray-colored feces? (*Hint:* The feces is lacking its normal coloration.)

7. Reddie Popper has a cell membrane defect in her red blood cells that makes them more susceptible to rupturing. Her red blood cells are destroyed faster than they can be replaced. Would her RBC, hemoglobin, hematocrit, and bilirubin levels be below normal, normal, or above normal? Explain.

Answers in Appendix D

Visit this textbook's website at www.mhhe.com/seeleyess6 for practice quizzes, animations, interactive learning exercises, and other study tools.

McGraw-Hill offers a study CD that features interactive cadaver dissection. *Anatomy & Physiology Revealed* includes cadaver photos that allow you to peel away layers of the human body to reveal structures beneath the surface. This program also includes animations, radiologic imaging, audio pronunciations, and practice quizzing.

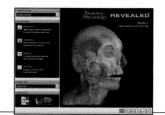

Heart

Chapter Outline and Objectives

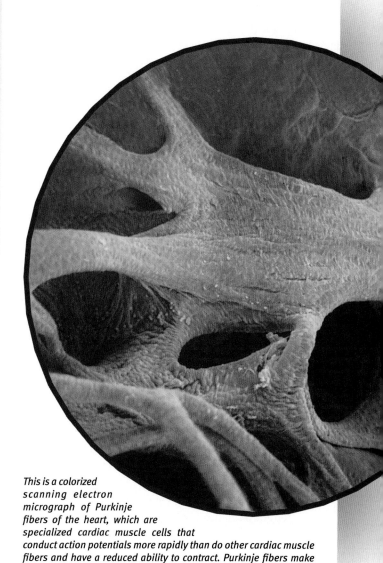

This is a colorized scanning electron micrograph of Purkinje fibers of the heart, which are specialized cardiac muscle cells that conduct action potentials more rapidly than do other cardiac muscle fibers and have a reduced ability to contract. Purkinje fibers make up much of the conducting system of the heart.

People often refer to the heart as if it were the seat of certain strong emotions. A very determined person may be described as having "a lot of heart," and a person who has been disappointed romantically can be described as having a "broken heart." A popular holiday in February not only dramatically distorts the heart's anatomy, it also attaches romantic emotions to it. The heart is a muscular organ that is essential for life because it pumps blood through the body. Emotions are a product of brain function, not heart function.

Fluids flow through a pipe only if they are forced to do so. The force is commonly produced by a pump, which increases the pressure of the liquid at the pump above the pressure in the pipe. Thus, the liquid flows from the pump through the pipe from an area of higher pressure to an area of lower pressure. If the pressure produced by the pump increases, flow of liquid through the pipe increases. If the pressure produced by the pump decreases, flow of liquid through the pipe decreases.

Like a pump that forces water to flow through a pipe, the heart contracts forcefully to pump blood through the blood vessels of the body (figure 12.1). The heart of a healthy adult, at rest, pumps approximately 5 liters (L) of blood per minute. For most people, the heart continues to pump at approximately that rate for more than 75 years; and, during short periods of vigorous exercise, the amount of blood pumped per minute increases several fold. If the heart loses its pumping ability for even a few minutes, however, blood flow

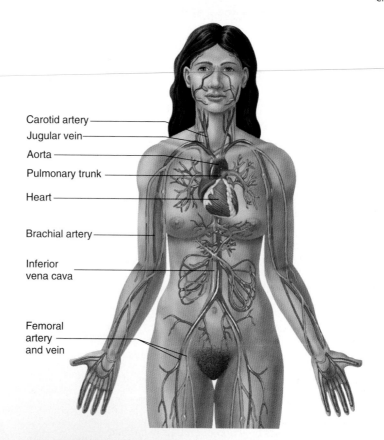

Carotid artery
Jugular vein
Aorta
Pulmonary trunk
Heart
Brachial artery
Inferior vena cava
Femoral artery and vein

Figure 12.1   The Cardiovascular System
The heart, blood, and blood vessels are the major components of the cardiovascular system.

through the blood vessels stops, and the life of the individual is in danger.

The heart is actually two pumps in one. The right side of the heart pumps blood to the lungs and back to the left side of the heart through vessels of the **pulmonary circulation** (figure 12.2). The left side of the heart pumps blood to all other tissues of the body and back to the right side of the heart through vessels of the **systemic circulation.**

Functions of the Heart

Functions of the heart include:

1. *Generating blood pressure.* Contractions of the heart generate blood pressure, which is required for blood flow through the blood vessels.
2. *Routing blood.* The heart separates the pulmonary and systemic circulations, which ensures the flow of oxygenated blood to tissues.
3. *Ensuring one-way blood flow.* The valves of the heart ensure a one-way flow of blood through the heart and blood vessels.
4. *Regulating blood supply.* Changes in the rate and force of heart contraction match blood flow to the changing metabolic needs of the tissues during rest, exercise, and changes in body position.

Size, Form, and Location of the Heart

The adult heart is shaped like a blunt cone and is approximately the size of a closed fist. It is larger in physically active adults than in less active but otherwise healthy adults, and it generally decreases in size after approximately age 65, especially in those who are not physically active. The blunt, rounded point of the cone is the **apex;** and the larger, flat part at the opposite end of the cone is the **base.**

The heart is located in the thoracic cavity between the two pleural cavities, which surround the lungs. The heart, trachea, esophagus, and associated structures form a midline partition, the **mediastinum** (me′dē-as-tī′nŭm; see figure 1.12). The heart is surrounded by its own cavity, the **pericardial cavity** (*peri,* around + *cardio,* heart) (see chapter 1).

It is important for clinical reasons to know the location and shape of the heart in the thoracic cavity. This knowledge

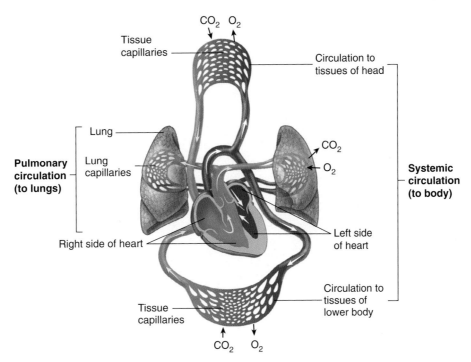

Figure 12.2   Overview of the Circulatory System

The circulatory system consists of the pulmonary and systemic circulations. The right side of the heart pumps blood through vessels to the lungs and back to the left side of the heart through the pulmonary circulation. The left side of the heart pumps blood through vessels to the tissues of the body and back to the right side of the heart through the systemic circulation.

allows a person to accurately place a stethoscope to hear the heart sounds, place chest leads to record an electrocardiogram (ē-lek-trō-kar′dē-ō-gram; **ECG** or **EKG**), or administer effective cardiopulmonary resuscitation (kar′dē-ō-pŭl′mo-nār-ē-rē-sŭs′i-tā-shun; **CPR**).

The heart lies obliquely in the mediastinum, with its base directed posteriorly and slightly superiorly and the apex directed anteriorly and slightly inferiorly. The apex is also directed to the left so that approximately two-thirds of the heart's mass lies to the left of the midline of the sternum (figure 12.3). The base of the heart is located deep to the sternum and extends to the level of the second intercostal space. The apex is located deep to the left fifth intercostal space, approximately 7–9 centimeters (cm) to the left of the sternum near the midclavicular line, which is a perpendicular line that extends down from the middle of the clavicle (see figure 12.3).

Anatomy of the Heart
Pericardium

The heart is surrounded by the pericardial cavity. The pericardial cavity is formed by the **pericardium** (per-i-kar′dē-ŭm), or **pericardial sac,** which surrounds the heart and anchors it within the mediastinum (see figures 12.3 and 12.4). The pericardium consists of two layers. The tough, fibrous connective tissue outer layer is called the **fibrous pericardium** and the inner layer of flat epithelial cells, with a thin layer of connective tissue, is called the **serous pericardium.** The portion of the serous pericardium lining the fibrous pericardium is the **parietal pericardium,** whereas the portion covering the heart surface is the **visceral pericardium,** or **epicardium** (ep-i-kar′dē-ŭm, upon the heart). The parietal and visceral pericardia are continuous with each other where the great vessels enter or leave the heart. The pericardial cavity, located between the visceral and parietal pericardia, is filled with a thin layer of **pericardial fluid** produced by the serous pericardium. The pericardial fluid helps reduce friction as the heart moves within the pericardial sac.

Disorders of the Pericardium

Pericarditis (per′i-kar-dī′tis) is an inflammation of the serous pericardium. The cause is frequently unknown, but it can result from infection, diseases of connective tissue, or damage due to radiation treatment for cancer. It can be extremely painful, with sensations of pain referred to the back and to the chest, which can be confused with the pain of a myocardial infarction (heart attack). Pericarditis can result in a small amount of fluid accumulation within the pericardial sac.

Cardiac tamponade (tam-pŏ-nād′, a pack or plug) is a potentially fatal condition in which fluid or blood accumulates in the pericardial sac. The fluid compresses the heart from the outside. The heart is a powerful muscle, but it relaxes passively. When it is compressed by fluid within the pericardial sac, it cannot dilate when the cardiac muscle relaxes. Consequently, the heart cannot fill with blood during relaxation, which makes it impossible for it to pump. Cardiac tamponade can cause a person to die quickly unless the fluid is removed. Causes of cardiac tamponade include rupture of the heart wall following a myocardial infarction, rupture of blood vessels in the pericardium after a malignant tumor invades the area, damage to the pericardium resulting from radiation therapy, and trauma such as occurs in a traffic accident.

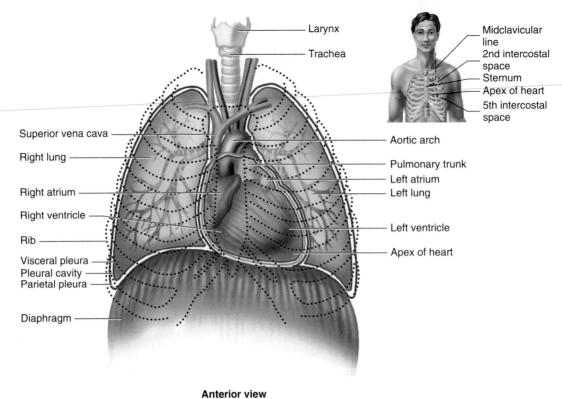

Anterior view

Figure 12.3 Location of the Heart in the Thorax

The heart is located in the thoracic cavity between the lungs, deep and slightly to the left of the sternum. The base of the heart, located deep to the sternum, extends superiorly to the second intercostal space, and the apex of the heart is located deep to the fifth intercostal space, approximately 7–9 cm to the left of the sternum where the midclavicular line intersects with the fifth intercostal space (see inset).

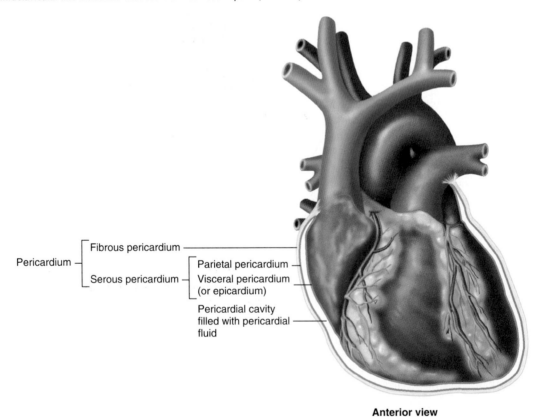

Anterior view

Figure 12.4 The Heart in the Pericardium

The heart is located in the pericardium, which consists of an outer fibrous pericardium and an inner serous pericardium. The serous pericardium has two parts: the parietal pericardium lines the fibrous pericardium, and the visceral pericardium (epicardium) covers the surface of the heart. The pericardial cavity, between the parietal and visceral pericardium, is filled with a small amount of pericardial fluid.

External Anatomy

The right and left **atria** (a′trē-ă, entrance chambers; sing. atrium) are located at the base of the heart, and the right and left **ventricles** (ven′tri-klz, a cavity) extend from the base of the heart toward the apex (figure 12.5). A **coronary** (kōr′o-nār-ē, circling like a crown) **sulcus** (sul′kus, ditch) extends around the heart, separating the atria from the ventricles. In addition, two grooves, or sulci, which indicate the division between the right and left ventricles, extend inferiorly from the coronary sulcus. The **anterior interventricular sulcus** extends inferiorly from the coronary sulcus on the anterior surface of the heart, and the **posterior interventricular sulcus** extends inferiorly from the coronary sulcus on the posterior surface of the heart (see figure 12.5).

Six large veins carry blood to the heart (see figure 12.5*a* and *c*): the **superior vena cava** and **inferior vena cava** carry

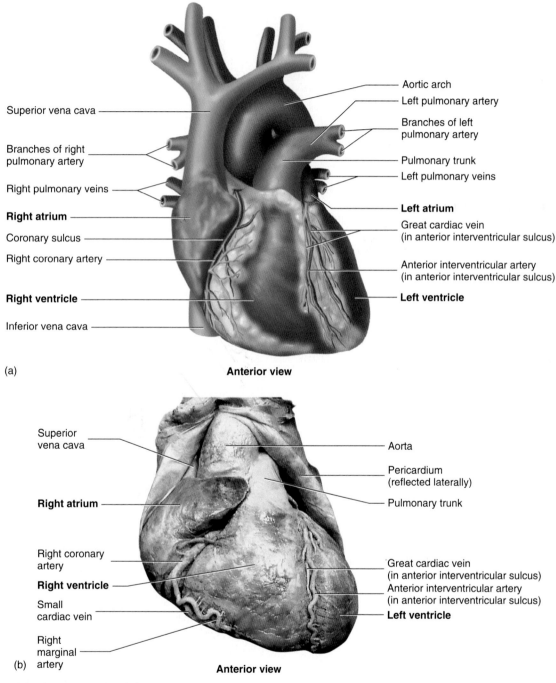

(a) **Anterior view**

- Superior vena cava
- Branches of right pulmonary artery
- Right pulmonary veins
- **Right atrium**
- Coronary sulcus
- Right coronary artery
- **Right ventricle**
- Inferior vena cava

- Aortic arch
- Left pulmonary artery
- Branches of left pulmonary artery
- Pulmonary trunk
- Left pulmonary veins
- **Left atrium**
- Great cardiac vein (in anterior interventricular sulcus)
- Anterior interventricular artery (in anterior interventricular sulcus)
- **Left ventricle**

(b) **Anterior view**

- Superior vena cava
- **Right atrium**
- Right coronary artery
- **Right ventricle**
- Small cardiac vein
- Right marginal artery

- Aorta
- Pericardium (reflected laterally)
- Pulmonary trunk
- Great cardiac vein (in anterior interventricular sulcus)
- Anterior interventricular artery (in anterior interventricular sulcus)
- **Left ventricle**

Figure 12.5   Anterior Surface View of the Heart

(*a*) The anterior view of the heart. The two atria (right and left) are located superiorly, and the two ventricles (right and left) are located inferiorly. The superior and inferior venae cava enter the right atrium. The pulmonary veins enter the left atrium. The pulmonary trunk exits the right ventricle, and the aorta exits the left ventricle. (*b*) Photograph of the anterior surface of the heart.

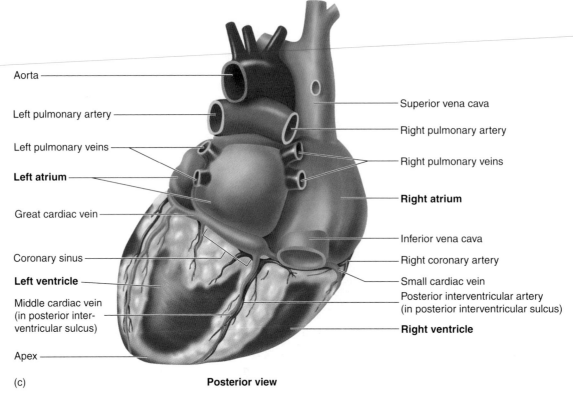

Aorta

Left pulmonary artery

Left pulmonary veins

Left atrium

Great cardiac vein

Coronary sinus

Left ventricle

Middle cardiac vein
(in posterior inter-
ventricular sulcus)

Apex

Superior vena cava

Right pulmonary artery

Right pulmonary veins

Right atrium

Inferior vena cava

Right coronary artery

Small cardiac vein

Posterior interventricular artery
(in posterior interventricular sulcus)

Right ventricle

(c)   **Posterior view**

Figure 12.5   *continued*

(*c*) The posterior surface view of the heart. The two atria (right and left) are located superiorly, and the two ventricles (right and left) are located inferiorly. The superior and inferior venae cava enter the right atrium, and the four pulmonary veins enter the left atrium.

blood from the body to the right atrium, and four **pulmonary** (pŭl′mō-nār-ē, lung) **veins** carry blood from the lungs to the left atrium. Two arteries, the **pulmonary trunk** and the **aorta** (ā-ōr′tă), exit the heart. The pulmonary trunk, arising from the right ventricle, splits into the right and left **pulmonary arteries,** which carry blood to the lungs. The aorta, arising from the left ventricle, carries blood to the rest of the body.

Heart Chambers and Internal Anatomy

The heart is a muscular pump consisting of four chambers: two atria and two ventricles (figure 12.6).

Right and Left Atria

The atria of the heart receive blood from veins. The atria function primarily as reservoirs, where blood returning from veins collects before it enters the ventricles. Contraction of the atria forces blood into the ventricles to complete ventricular filling. The right atrium receives blood through three major openings. The superior vena cava and the inferior vena cava drain blood from most of the body (see figure 12.6), and the smaller coronary sinus drains blood from most of the heart muscle. The left atrium receives blood through the four pulmonary veins (see figure 12.6), which drain blood from the lungs. The two atria are separated from each other by a partition called the **interatrial** (between the atria) **septum.**

Right and Left Ventricles

The ventricles of the heart are its major pumping chambers. They eject blood into the arteries and force it to flow through the circulatory system. The atria open into the ventricles, and each ventricle has one large outflow route located superiorly near the midline of the heart. The right ventricle pumps blood into the pulmonary trunk, and the left ventricle pumps blood into the aorta. The two ventricles are separated from each other by the muscular **interventricular** (between the ventricles) **septum** (see figure 12.6).

The wall of the left ventricle is thicker than the wall of the right ventricle, and the wall of the left ventricle contracts more forcefully and generates a greater blood pressure than the wall of the right ventricle. When the left ventricle contracts, the pressure increases to approximately 120 mm Hg. When the right ventricle contracts, the pressure increases to approximately one-fifth of the pressure in the left ventricle. However, the left and right ventricles pump nearly the same volume of blood. The higher pressure generated by the left ventricle moves blood through the larger systemic circulation, whereas the lower pressure generated by the right ventricle moves blood through the smaller pulmonary circulation (see figure 12.2).

Heart Valves

The **atrioventricular (AV) valves** are located between the right atrium and the right ventricle and between the left atrium and left ventricle. The AV valve between the right

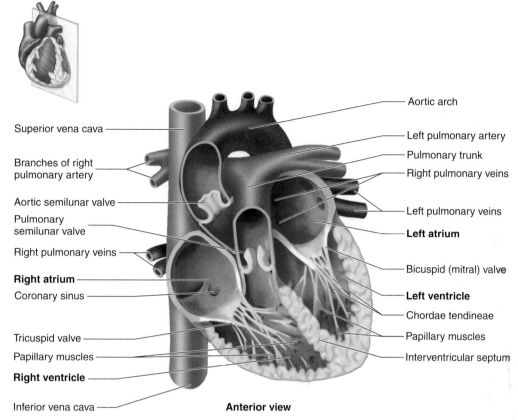

Anterior view

Figure 12.6 Internal Anatomy of the Heart

The heart is cut in a frontal plane to show the internal anatomy.

atrium and the right ventricle has three cusps and is called the **tricuspid valve** (see figures 12.6 and 12.7*a*). The AV valve between the left atrium and left ventricle has two cusps and is called the **bicuspid,** or **mitral** (resembling a bishop's miter, a two-pointed hat) **valve** (see figures 12.6 and 12.7*b*). These valves allow blood to flow from the atria into the ventricles but prevent it from flowing back into the atria. When the ventricles relax, the higher pressure in the atria forces the AV

valves to open and blood flows from the atria into the ventricles (figure 12.8*a*). In contrast, when the ventricles contract, blood flows toward the atria and causes the AV valves to close (figure 12.8*b*).

Each ventricle contains cone-shaped muscular pillars called **papillary** (pap′ĭ-lār-ē, nipple- or pimple-shaped) **muscles.** These muscles are attached by thin, strong connective tissue strings called **chordae tendineae** (kōr′dē ten′di-nē-ē, heart strings) to the

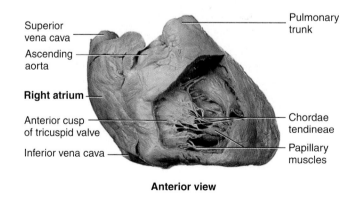

Anterior view

(a) View of the tricuspid valve, the chordae tendineae, and the papillary muscles.

Superior view

(b) A superior view of the heart valves. Note the three cusps of each semilunar valve meeting to prevent the backflow of blood.

Figure 12.7 Heart Valves

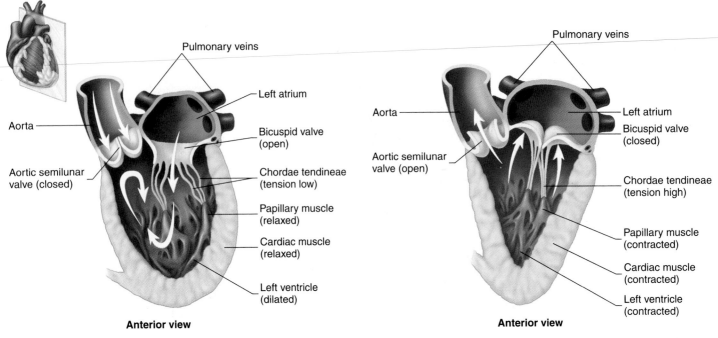

(a) When the bicuspid valve is open, the cusps of the valve are pushed by blood into the ventricle. Papillary muscles are relaxed and tension on the chordae tendineae is low. Blood flows from the left atrium into the left ventricle. When the aortic semilunar valve is closed, the cusps of the valve overlap as they are pushed by the blood in the aorta toward the ventricle. There is no blood flow from the aorta into the ventricle.

(b) When the bicuspid valve is closed, the cusps of the valves overlap as they are pushed by the blood toward the left atrium. There is no blood flow from the ventricle into the atrium. Papillary muscles are contracted and tension on the chordae tendineae is increased, which prevents the bicuspid valve from opening into the left atrium. When the aortic semilunar valve is open, the cusps of the valve are pushed by the blood toward the aorta. Blood then flows from the left ventricle into the aorta.

Figure 12.8   Function of the Heart Valves

free margins of the cusps of the atrioventricular valves. When the ventricles contract, the papillary muscles contract and prevent the valves from opening into the atria by pulling on the chordae tendineae attached to the valve cusps (see figures 12.7a and 12.8).

The aorta and pulmonary trunk possess **aortic** and **pulmonary semilunar** (halfmoon-shaped) **valves,** respectively (see figure 12.6). Each valve consists of three pocketlike semilunar cusps (see figures 12.7b and 12.8). When the ventricles contract, the increasing pressure within the ventricles forces the semilunar valves to open (see figure 12.8b). When the ventricles relax, the pressure in the aorta is higher than in the ventricles and pulmonary trunk. Blood flows back from the aorta or pulmonary trunk toward the ventricles, and enters the pockets of the cusps, causing them to bulge toward and meet in the center of the aorta or pulmonary trunk, thus closing the vessels and blocking blood flow back into the ventricles (see figure 12.8a).

A plate of fibrous connective tissue, sometimes called the **cardiac skeleton,** consisting mainly of fibrous rings around the atrioventricular and semilunar valves, provides a solid support for the valves (figure 12.9). This connective tissue plate also serves as electrical insulation between the atria and the ventricles and provides a rigid site of attachment for cardiac muscle.

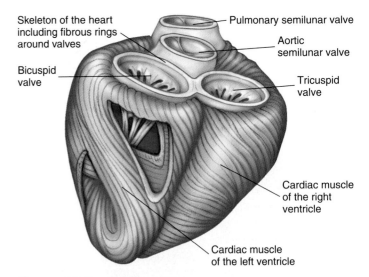

Figure 12.9   Cardiac Skeleton

The cardiac skeleton consists of fibrous connective tissue rings that surround the heart valves and separate the atria from the ventricles. Cardiac muscle attaches to the fibrous connective tissue. The muscle fibers are arranged so that when the ventricles contract, a wringing motion is produced and the distance between the apex and base of the heart shortens.

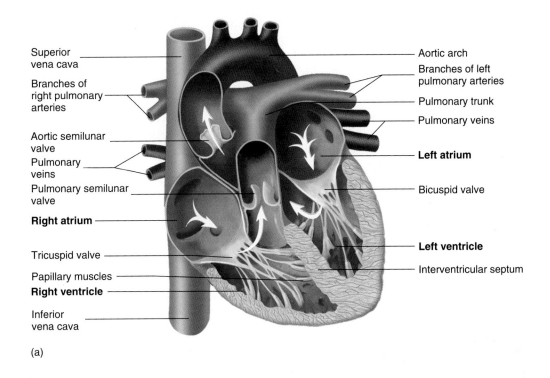

(a)

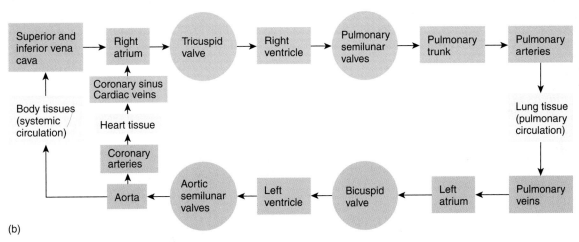

(b)

Figure 12.10 Blood Flow Through the Heart

(a) Frontal section of the heart revealing the four chambers and the direction of blood flow through the heart. (b) Diagram listing in order the structures through which blood flows in the systemic and pulmonary circulations. The heart valves are indicated by *circles;* deoxygenated blood (*blue*); oxygenated blood (*red*).

Route of Blood Flow Through the Heart

The route of blood flow through the heart is depicted in figure 12.10. Even though blood flow through the heart is described for the right and then the left side of the heart, it is important to understand that both atria contract at the same time, and both ventricles contract at the same time. This concept is most important when the electrical activity, pressure changes, and heart sounds are considered.

Blood enters the right atrium from the systemic circulation through the superior and inferior venae cava, and from heart muscle through the coronary sinus (see figure 12.10*a* and *b*). Most of the blood flowing into the right atrium flows into the right ventricle while the right ventricle relaxes following the previous contraction. Before the end of ventricular relaxation, the right atrium contracts, and enough blood is pushed from the right atrium into the right ventricle to complete right ventricular filling.

Following right atrial contraction, the right ventricle begins to contract. Contraction of the right ventricle pushes blood against the tricuspid valve, forcing it closed. After pressure within the right ventricle increases, the pulmonary semilunar valve is forced open, and blood flows into the pulmonary trunk. As the right ventricle relaxes, its pressure falls rapidly, and pressure in the pulmonary trunk becomes greater than in the right ventricle. The back-flow of blood forces the pulmonary semilunar valve to close.

The pulmonary trunk branches to form the right and left pulmonary arteries, which carry blood to the lungs, where carbon dioxide is released and oxygen is picked up. Blood returning from the lungs enters the left atrium through the four pulmonary veins (see figure 12.10a and b). Most of the blood flowing into the left atrium passes into the left ventricle while the left ventricle relaxes following the previous contraction. Before the end of ventricular relaxation, the left atrium contracts, and enough blood is pushed from the left atrium into the left ventricle to complete left ventricular filling.

Following left atrial contraction, the left ventricle begins to contract. Contraction of the left ventricle pushes blood against the bicuspid valve, forcing it closed. After pressure within the left ventricle increases, the aortic semilunar valve is forced open, and blood flows into the aorta (see figure 12.10a and b). Blood flowing through the aorta is distributed to all parts of the body, except to that part of the lung supplied by the pulmonary blood vessels. As the left ventricle relaxes, its pressure falls rapidly, and pressure in the aorta becomes greater than in the left ventricle. The back-flow of blood forces the aortic semilunar valve to close.

Blood Supply to the Heart

Coronary Arteries

Cardiac muscle in the wall of the heart is thick and metabolically very active. Two coronary arteries supply blood to the wall of the heart (figure 12.11a). The **coronary arteries** originate

from the base of the aorta, just above the aortic semilunar valves. The **left coronary artery** originates on the left side of the aorta. It has three major branches: The **anterior interventricular artery** lies in the anterior interventricular sulcus, the **circumflex artery** extends around the coronary sulcus on the left to the posterior surface of the heart, and the **left marginal artery** extends inferiorly along the lateral wall of the left ventricle from the circumflex artery. The branches of the left coronary artery supply much of the anterior wall of the heart and most of the left ventricle. The **right coronary artery** originates on the right side of the aorta. It extends around the coronary sulcus on the right to the posterior surface of the heart and gives rise to the **posterior interventricular artery,** which lies in the posterior interventricular sulcus. The **right marginal artery** extends inferiorly along the lateral wall of the right ventricle. The right coronary artery and its branches supply most of the wall of the right ventricle.

In a resting person, blood flowing through the coronary arteries of the heart gives up approximately 70% of its oxygen. In comparison, blood flowing through arteries to skeletal muscle gives up only about 25% of its oxygen. The percentage of oxygen the blood releases to skeletal muscle increases to 70% or more during exercise. The percentage of oxygen the blood releases to cardiac muscle cannot increase substantially during exercise. Cardiac muscle is therefore very dependent on an increased rate of blood flow through the coronary arteries above its resting level to provide an adequate oxygen supply during exercise. Blood flow into the coronary circulation is greatest during

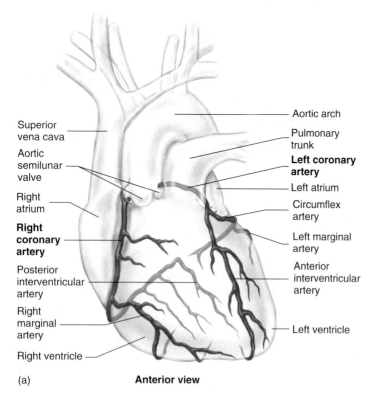

(a) **Anterior view**

(b) **Anterior view**

Figure 12.11 **Blood Supply to the Heart**

The vessels of the anterior surface of the heart are seen directly and are a darker color, whereas the vessels of the posterior surface are seen through the heart and are a lighter color. (a) Coronary arteries supply blood to the wall of the heart. (b) Cardiac veins carry blood from the wall of the heart back to the right atrium.

relaxation of the ventricles of the heart when contraction of the cardiac muscle does not compress the coronary arteries. Blood flow into other arteries of the body is highest during contraction of the ventricles.

> **P R E D I C T**
>
> Predict the effect on the heart if blood flow through the anterior interventricular artery is restricted or completely blocked (*Hint:* See figure 12.11*a*).

Cardiac Veins

The **cardiac veins** drain blood from the cardiac muscle. Their pathways are nearly parallel to the coronary arteries and most drain blood into the **coronary sinus,** a large vein located within the coronary sulcus on the posterior aspect of the heart. Blood flows from the coronary sinus into the right atrium (see figure 12.11*b*). Some small cardiac veins drain directly into the right atrium.

Disorders of Coronary Arteries

When a blood clot, or **thrombus** (throm′bŭs, a clot), suddenly blocks a coronary blood vessel, a **heart attack,** or **coronary thrombosis** (throm′bō-sis), occurs. The area that has been cut off from its blood supply suffers from a lack of oxygen and nutrients and dies if the blood supply is not quickly reestablished. The region of dead heart tissue is called an **infarct** (in′farkt), or **myocardial infarction.** If the infarct is large enough, the heart may be unable to pump enough blood to keep the person alive. People who are at risk for coronary thromboses can reduce the likelihood of heart attack by taking small amounts of aspirin daily, which inhibits thrombus formation (see chapter 11).

Aspirin is also administered to many people who are exhibiting clear symptoms of a heart attack. In some cases, it is possible to treat heart attacks with enzymes such as **streptokinase** (strep-tō-kī′nās) or **tissue plasminogen** (plaz-min′o-jen) **activator (t-PA),** which break down blood clots. One of the enzymes is injected into the circulatory system of a heart attack patient, where it reduces or removes the blockage in the coronary artery. If the clot is broken down quickly, the blood supply to cardiac muscle is reestablished, and the heart may suffer little permanent damage.

Coronary arteries can become blocked more gradually by **atherosclerotic** (ath′er-ō-skler-ot′ik, *athero,* pasty material + *sclerosis,* hardness) **lesions.** These thickenings in the walls of arteries can contain deposits that are high in cholesterol and other lipids. The lesions narrow the lumen (opening) of the arteries, thus restricting blood flow. The ability of cardiac muscle to function is reduced when it is deprived of an adequate blood supply. The person suffers from fatigue and often pain in the area of the chest and usually in the left arm with the slightest exertion. The pain is called **angina pectoris** (an-jī′nă, pain; pek′tō-ris, in the chest).

Angioplasty (an′jē-ō-plas-tē) is a surgical procedure in which a small balloon is threaded through the aorta and into a coronary artery. After the balloon has entered a partially blocked coronary artery, it is inflated, flattening the atherosclerotic deposits against the vessel wall and opening the blocked blood vessel. This technique improves the function of cardiac muscle in patients suffering from inadequate blood flow to the cardiac muscle through the coronary arteries. Some controversy exists about its effectiveness, at least in some patients, because dilation of the coronary arteries can be reversed within a few weeks or months and because blood clots can form in coronary arteries following angioplasty. Small rotating blades and lasers are also used to remove lesions from coronary vessels, or a small coil device, called a **stent,** is placed in the vessels to hold them open following angioplasty.

A **coronary bypass** is a surgical procedure that relieves the effects of obstructions in the coronary arteries. The technique involves taking healthy segments of blood vessels from other parts of the patient's body and using them to bypass, or create an alternative path around, obstructions in the coronary arteries. The technique is common for those who suffer from severe blockage of parts of the coronary arteries.

Histology of the Heart
Heart Wall

The heart wall is composed of three layers of tissue: the epicardium, the myocardium, and the endocardium (figure 12.12). The **epicardium** (ep-i-kar′dē-ŭm), also called the **visceral pericardium,** is a thin serous membrane forming the smooth outer surface of the heart. It consists of simple squamous

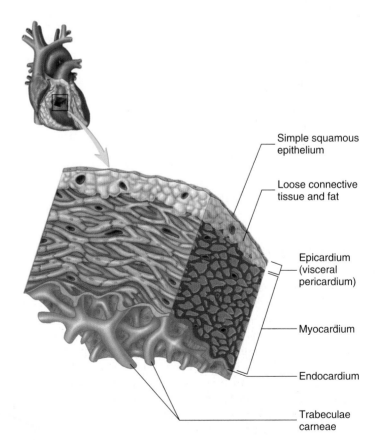

Figure 12.12 **Heart Wall**

Part of the wall of the heart has been removed, enlarged, and rotated so that the inner surface is visible. The enlarged section illustrates the epicardium (visceral pericardium), the myocardium, and the endocardium.

Labels in figure:
- Simple squamous epithelium
- Loose connective tissue and fat
- Epicardium (visceral pericardium)
- Myocardium
- Endocardium
- Trabeculae carneae

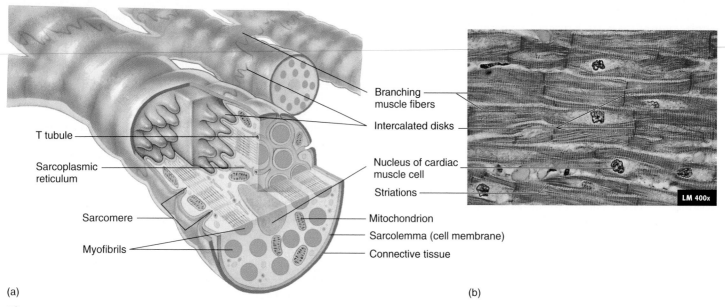

Figure 12.13   Cardiac Muscle Cells

(a) Cardiac muscle cells are branching cells with centrally located nuclei. As in skeletal muscle, sarcomeres join end-to-end to form myofibrils, and mitochondria provide ATP for contraction. The cells are joined to one another by intercalated disks, which allow action potentials to pass from one cardiac muscle cell to the next. Sarcoplasmic reticulum and T tubules are visible but are not as numerous as they are in skeletal muscle. (b) A light micrograph of cardiac muscle tissue. The cardiac muscle fibers appear to be striated because of the arrangement of the individual myofilaments.

epithelium overlying a layer of loose connective tissue and fat. The thick middle layer of the heart, the **myocardium** (mī-ō-kar′dē-ŭm), is composed of cardiac muscle cells and is responsible for contractions of the heart chambers. The smooth inner surface of the heart chambers is the **endocardium** (en-dō-kar′dē-ŭm), which consists of simple squamous epithelium over a layer of connective tissue. The endocardium allows blood to move easily through the heart. The heart valves are formed by folds of endocardium that include a thick layer of connective tissue.

The surfaces of the interior walls of the ventricles are modified by ridges and columns of cardiac muscle called **trabeculae carnea.** Smaller muscular ridges are also found in portions of the atria.

Cardiac Muscle

Cardiac muscle cells are elongated, branching cells that contain one, or occasionally two, centrally located nuclei (figure 12.13). The cardiac muscle cells contain actin and myosin myofilaments organized to form sarcomeres, which are joined end-to-end to form myofibrils (see chapter 7). The actin and myosin myofilaments are responsible for muscle contraction, and their organization gives cardiac muscle a striated (banded) appearance much like that of skeletal muscle. However, the striations are less regularly arranged and less numerous than is the case in skeletal muscle. Ca^{2+} enter cardiac muscle cells in response to action potentials and activate the process of contraction much like they do in skeletal muscle (see chapter 7).

Adenosine triphosphate (ATP) provides the energy for cardiac muscle contraction and ATP production depends on oxygen availability. Cardiac muscle cells are rich in

mitochondria, which produce ATP at a rate rapid enough to sustain the normal energy requirements of cardiac muscle. An extensive capillary network provides an adequate oxygen supply to the cardiac muscle cells. Unlike skeletal muscle, cardiac muscle cannot develop a significant oxygen debt. Development of a large oxygen debt could result in muscular fatigue and cessation of cardiac muscle contraction.

Cardiac muscle cells are organized into spiral bundles or sheets (see figure 12.9). When cardiac muscle fibers contract, not only do the muscle fibers shorten, the spiral bundles twist to compress the contents of the heart chambers. Cardiac muscle cells are bound end-to-end and laterally to adjacent cells by specialized cell-to-cell contacts called **intercalated** (in-ter′kă-lā-ted, insertion between two others) **disks** (see figure 12.13). The membranes of the intercalated disks are highly folded, and the adjacent cells fit together, greatly increasing contact between them and preventing cells from pulling apart. Specialized cell membrane structures in the intercalated disks called **gap junctions** (see chapter 4) reduce electrical resistance between the cells, allowing action potentials to pass easily from one cell to adjacent cells. The cardiac muscle cells of the atria or ventricles, therefore, contract at nearly the same time. The highly coordinated pumping action of the heart depends on this characteristic.

Electrical Activity of the Heart
Action Potentials in Cardiac Muscle

Like action potentials in skeletal muscle and neurons, those in cardiac muscle exhibit depolarization followed by repolarization. In cardiac muscle, however, a period of slow repolarization greatly prolongs the action potential (figure 12.14). In contrast to action potentials in skeletal muscle, which take less than 2 milliseconds (ms) to complete, action potentials in cardiac muscle take approximately 200 to 500 ms to complete.

Unlike in skeletal muscle, action potentials in cardiac muscle are conducted from cell to cell. The action potentials take longer, and their rate of conduction in cardiac muscle from cell to cell is slower than the rate of conduction of action potentials in single skeletal muscle cells and neurons.

In cardiac muscle, each action potential consists of a **depolarization phase** followed by a rapid, but partial **early repolarization phase.** This is followed by a longer period of slow repolarization, called the **plateau phase.** At the end of the plateau phase, a more rapid **final repolarization phase** takes place. During the final repolarization phase, the membrane potential achieves its maximum degree of repolarization (see figure 12.14).

Opening and closing of membrane channels are responsible for the changes in the permeability of the cell membrane that produce action potentials. The depolarization phase of the action potential results from three permeability changes. **Na⁺ channels** open, increasing the permeability of the cell membrane to Na^+. Sodium ions then diffuse into the cell, causing depolarization. This causes K^+ channels to close quickly, decreasing the permeability of the cell membrane to K^+. The decreased diffusion of K^+ out of the cell also causes depolarization. **Ca²⁺ channels** slowly open, increasing the permeability of the cell membrane to Ca^{2+}. Calcium ions then diffuse into the cell and cause depolarization. It is not until the plateau phase that most of the Ca^{2+} channels open.

Early repolarization occurs when the Na^+ channels close and a small number of **K⁺ channels** open. Diffusion of Na^+ into the cell stops, and there is some movement of K^+ out of the cell. These changes in ion movement result in an early, but small repolarization.

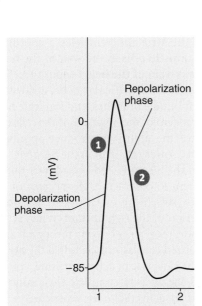

(a) Permeability changes due to voltage-gated channels opening and closing during an action potential in skeletal muscle:

1. **Depolarization phase**
   - Na⁺ channels open.
   - K⁺ channels begin to open.

2. **Repolarization phase**
   - Na⁺ channels close.
   - K⁺ channels continue to open causing repolarization.
   - K⁺ channels close at the end of repolarization and return the membrane potential to its resting value.

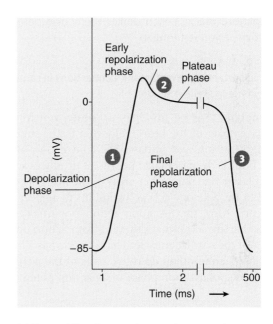

(a) Permeability changes due to voltage-gated channels opening and closing during an action potential in cardiac muscle:

1. **Depolarization phase**
   - Na⁺ channels open.
   - K⁺ channels close.
   - Ca²⁺ channels begin to open.

2. **Early repolarization and plateau phases**
   - Na⁺ channels close.
   - Some K⁺ channels open, causing early repolarization.
   - Ca²⁺ channels are open, producing the plateau by slowing further repolarization.

3. **Final repolarization phase**
   - Ca²⁺ channels close.
   - Many K⁺ channels open.

Process Figure 12.14   Comparison of Action Potentials in Skeletal and Cardiac Muscle

(a) An action potential in skeletal muscle consists of depolarization and repolarization phases. (b) An action potential in cardiac muscle consists of depolarization, early repolarization, plateau, and final repolarization phases. Cardiac muscle does not repolarize as rapidly as skeletal muscle (*indicated by the break in the curve*) because of the plateau phase.

The plateau phase occurs as Ca^{2+} channels continue to open, and the diffusion of Ca^{2+} into the cell counteracts the potential change produced by the diffusion of K^+ out of the cell. The plateau phase ends and final repolarization begins as the Ca^{2+} channels close, and many K^+ channels open. Diffusion of Ca^{2+} into the cell decreases and diffusion of K^+ out of the cell increases. These changes cause the membrane potential to repolarize during the final repolarization phase.

Action potentials in cardiac muscle exhibit a **refractory period,** like that of action potentials in skeletal muscle and in neurons. The refractory period lasts about the same length of time as the prolonged action potential in cardiac muscle. The prolonged action potential and refractory period allow cardiac muscle to contract and almost complete relaxation to take place before another action potential can be produced. Also, the long refractory period in cardiac muscle prevents tetanic contractions from occurring, thus ensuring a rhythm of contraction and relaxation for cardiac muscle. Therefore, action potentials in cardiac muscle are different from those in skeletal muscle because of the plateau phase, which makes the action potential and its refractory period last longer.

PREDICT ②
Why is it important to prevent tetanic contractions in cardiac muscle but not in skeletal muscle?

The **sinoatrial (SA)** (sī′nō-a′trē′-ăl) **node,** which functions as the pacemaker of the heart, is located in the superior wall of the right atrium and initiates the contraction of the heart. The SA node is the pacemaker because it produces action potentials at a faster rate than other areas of the heart. The action potential of the SA node acts as a stimulus to adjacent areas of the heart. Also, the SA node action potentials have characteristics that are somewhat different from action potentials in the rest of the cardiac muscle. The SA node has a larger number of Ca^{2+} channels than do other areas of the heart. As soon as the final depolarization phase of an action potential is

completed, some Na^+ enter the cell through nongated channels. Also, the permeability of the membrane to K^+ decreases and some of the Ca^{2+} channels open. As they open, Ca^{2+} and some Na^+ begin to diffuse into the cell and cause depolarization. The depolarization stimulates additional Ca^{2+} channels to open. Once threshold is reached, a large number of Ca^{2+} channels open. Ca^{2+} diffuse into the cell and quickly cause depolarization. At the peak of the action potential, Ca^{2+} channels close and K^+ channels open once again. The outward movement of K^+ causes repolarization. The cycle repeats itself when the K^+ channels begin to close once again. Ca^{2+} channel blocking agents are drugs that slow the heart by decreasing the rate of action potential production in the SA node. Ca^{2+} channel blockers decrease the rate at which Ca^{2+} move through Ca^{2+} channels. As a result, it takes longer for depolarization to reach threshold and the intervals between action potential production increases.

Conduction System of the Heart
Contraction of the atria and ventricles is coordinated by specialized cardiac muscle cells in the wall of the heart that form the **conduction system of the heart** (figure 12.15).

The SA node, atrioventricular node, atrioventricular bundle, right and left bundle branches, and Purkinje fibers comprise the conduction system of the heart. All of the cells of the conduction system have the ability to produce spontaneous action potentials, but at a lower rate than in the SA node. Action potentials originate in the SA node and spread over the right and left atria, causing them to contract. A second area of the heart, the **atrioventricular (AV)** (ā-trē-ō-ven′trik′-ū′lăr) **node,** is located in the lower portion of the right atrium. When action potentials reach the AV node, they spread slowly through it and then into a bundle of specialized cardiac muscle called the **atrioventricular (AV) bundle.** The slow rate of action potential conduction in the AV node allows the atria to complete their contraction before action potentials are delivered to the ventricles.

1. Action potentials originate in the sinoatrial (SA) node and travel across the wall of the atrium (*arrows*) from the SA node to the atrioventricular (AV) node.

2. Action potentials pass through the AV node and along the atrioventricular (AV) bundle, which extends from the AV node, through the fibrous skeleton, into the interventricular septum.

3. The AV bundle divides into right and left bundle branches, and action potentials descend to the apex of each ventricle along the bundle branches.

4. Action potentials are carried by the Purkinje fibers from the bundle branches to the ventricular walls.

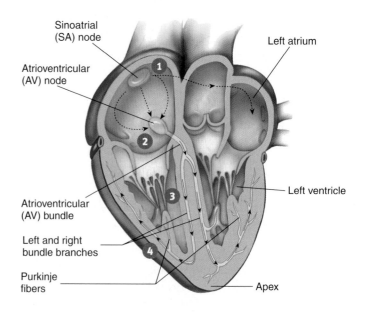

Process Figure 12.15   Conduction System of the Heart

After action potentials pass through the AV node, they are rapidly transmitted through the AV bundle, which projects through the fibrous connective tissue plate that separates the atria from the ventricles (see figure 12.9). The AV bundle then divides into two branches of conducting tissue called the **left** and **right bundle branches** (see figure 12.15). At the tips of the left and right bundle branches, the conducting tissue forms many small bundles of **Purkinje** (pŭr-kĭn′jē, Johannes von Purkinje, Bohemian anatomist, 1787–1869) **fibers.** These Purkinje fibers pass to the apex of the heart and then extend to the cardiac muscle of the ventricle walls. The AV bundle, the bundle branches, and the Purkinje fibers are composed of specialized cardiac muscle fibers that conduct action potentials more rapidly than do other cardiac muscle fibers. Consequently, action potentials are rapidly delivered to all the cardiac muscle of the ventricles. The coordinated contraction of the ventricles depends on the conduction of action potentials by the conduction system.

PREDICT 3

If blood supply is reduced in a small area of the heart through which the left bundle branch passed, predict the effect on ventricular contractions.

Following their contraction, the ventricles begin to relax. After the ventricles have completely relaxed, another action potential originates in the SA node to begin the next cycle of contractions.

The SA node is the pacemaker of the heart, but other cells of the conduction system also are capable of producing action potentials spontaneously. For example, if the SA node is unable to function, another area, such as the AV node, becomes the pacemaker. The resulting heart rate is much slower than normal. When action potentials originate in an area of the heart other than the SA node, the result is called an **ectopic** (ek-top′ik) **beat.**

Fibrillation of the Heart

Cardiac muscle can also act as if there are thousands of pacemakers, each making a very small portion of the heart contract rapidly and independently of all other areas. This condition is called **fibrillation** (fĭ-bri-lā′shŭn), and it reduces the output of the heart to only a few milliliters of blood per minute when it occurs in the ventricles. Death of the individual results in a few minutes unless fibrillation of the ventricles is stopped.

To stop the process of fibrillation, defibrillation is used, in which a strong electrical shock is applied to the chest region. The electrical shock causes simultaneous depolarization of all cardiac muscle fibers. Following depolarization, the SA node can recover and produce action potentials before any other area of the heart. Consequently, the normal pattern of action potential generation and the normal rhythm of contraction can be reestablished.

Defibrillator machines have changed considerably over the years. Portable models are now available that can be included in emergency equipment in workplaces and even in the home, and nonprofessionals can be trained to use them. Consequently, the time required to respond to an emergency requiring a defibrillator can be greatly shorted.

Fibrillation of the heart is more likely to occur when action potentials originate at ectopic sites in the heart. For example, people who have ectopic beats that originate from one of their ventricles are more likely to develop fibrillation of the heart than people who have normal heart beats.

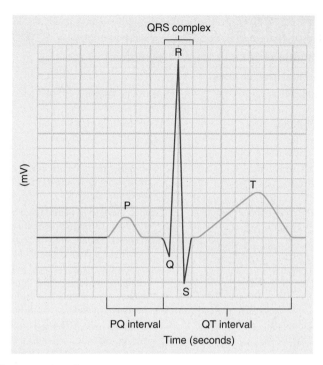

Figure 12.16 Electrocardiogram

The major waves and intervals of an electrocardiogram are labeled. Each thin horizontal line on the ECG recording represents 1 mV, and each thin vertical line represents 0.04 second.

Electrocardiogram

Action potentials conducted through the heart during the cardiac cycle produce electrical currents that can be measured at the surface of the body. Electrodes placed on the surface of the body and attached to a recording device can detect the small electrical changes resulting from the action potentials in all of the cardiac muscle cells. The record of these electrical events is an **electrocardiogram** (**ECG** or **EKG**) (figure 12.16).

The normal ECG consists of a P wave, a QRS complex, and a T wave. The **P wave** results from depolarization of the atrial myocardium, and the beginning of the P wave precedes the onset of atrial contraction. The **QRS complex** consists of three individual waves: the Q, R, and S waves. The QRS complex results from depolarization of the ventricles, and the beginning of the QRS complex precedes ventricular contraction. The **T wave** represents repolarization of the ventricles, and the beginning of the T wave precedes ventricular relaxation. A wave representing repolarization of the atria cannot be seen because it occurs during the QRS complex.

The time between the beginning of the P wave and the beginning of the QRS complex is the **PQ interval,** commonly called the **PR interval** because the Q wave is very small. During the PQ interval the atria contract and begin to relax. At the end of the PQ interval the ventricles begin to depolarize.

The **QT interval** extends from the beginning of the QRS complex to the end of the T wave and represents the length of time required for ventricular depolarization and repolarization. Table 12.1 describes several conditions associated with abnormal heart rhythms.

Table 12.1   Major Cardiac Arrhythmias

Condition	Symptoms	Possible Causes
Abnormal Heart Rhythms		
Tachycardia	Heart rate in excess of 100 bpm	Elevated body temperature, excessive sympathetic stimulation, toxic conditions
Bradycardia	Heart rate less than 60 bpm	Increased stroke volume in athletes, excessive vagus nerve stimulation, nonfunctional SA node, carotid sinus syndrome
Sinus arrhythmia	Heart rate varies as much as 5% during respiratory cycle and up to 30% during deep respiration	Cause not always known; occasionally caused by ischemia, inflammation, or cardiac failure
Paroxysmal atrial tachycardia	Sudden increase in heart rate to 95–150 bpm for a few seconds or even for several hours; P waves precede every QRS complex; P wave inverted and superimposed on T wave	Excessive sympathetic stimulation, abnormally elevated permeability of cardiac muscle to Ca^{2+}
Atrial flutter	As many as 300 P waves/min and 125 QRS complexes/min; resulting in two or three P waves (atrial contractions) for every QRS complex (ventricular contraction)	Ectopic beats in the atria
Atrial fibrillation	No P waves, normal QRS and T waves, irregular timing, ventricles are constantly stimulated by atria, reduced ventricle filling; increased chance of fibrillation	Ectopic beats in the atria
Ventricular tachycardia	Frequently causes fibrillation	Often associated with damage to AV node or ventricular muscle
Heart Blocks		
SA node block	No P waves, low heart rate resulting from AV node acting as the pacemaker, normal QRS complexes and T waves	Ischemia, tissue damage resulting from infarction; cause sometimes is unknown
AV node blocks		
First-degree	PQ interval greater than 0.2 s	Inflammation of AV bundle
Second-degree	PQ interval 0.25–0.45 s; some P waves trigger QRS complexes and others do not; examples of 2:1, 3:1, and 3:2 P wave/QRS complex ratios	Excessive vagus nerve stimulation, AV node damage
Complete heart block	P wave dissociated from QRS complex, atrial rhythm about 100 bpm, ventricular rhythm less than 40 bpm	Ischemia of AV node or compression of AV bundle
Premature Contractions		
Premature atrial contractions	Occasional shortened intervals between one contraction and the succeeding contraction; frequently occurs in healthy people	Excessive smoking, lack of sleep, or too much caffeine
Premature ventricular contractions (PVCs)	Prolonged QRS complex, exaggerated voltage because only one ventricle may depolarize, possible inverted T wave, increased probability of fibrillation	Ectopic beat in ventricles, lack of sleep, too much coffee, irritability; occasionally occurs with coronary thrombosis

P R E D I C T **4**
Explain how the ECGs appear for a person who has a damaged left bundle branch (see Predict 3) and for a person who has many ectopic beats originating from her atria.

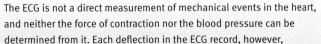

The ECG as a Diagnostic Tool

The ECG is not a direct measurement of mechanical events in the heart, and neither the force of contraction nor the blood pressure can be determined from it. Each deflection in the ECG record, however, indicates an electrical event within the heart and correlates with a subsequent mechanical event. Consequently, it is an extremely valuable diagnostic tool in identifying a number of cardiac abnormalities, particularly because it is painless, easy to record, and does not require surgical procedures. Abnormal heart rates or rhythms, abnormal conduction pathways such as blockages in the conduction pathways, hypertrophy or atrophy of portions of the heart, and the approximate location of damaged cardiac muscle can be determined from analysis of an ECG.

Willie May Kitt is a 65-year-old man. While walking up a short flight of stairs to his office where he works as a bank manager, he experienced a crushing pain in his chest and he exhibited substantial pallor. Willie fell to the floor, lost consciousness, and then stopped breathing. A coworker noticed the pallor, saw Willie fall, and ran to his aid. He could detect no pulse and decided to administer cardiopulmonary resuscitation (CPR). Another coworker called 911 and then assisted the first coworker. One coworker pushed down firmly on Willie's sternum at a rate of approximately 100 compressions per minute. After every 15 compressions, he paused. The other coworker forced air into Willie's lungs by tipping Willie's head back slightly, placing her mouth over Willie's mouth, and blowing air forcefully into his mouth two times while holding his nasal passages closed. Pushing down on the sternum compresses the ventricles of the heart and forces blood to flow into the aorta and pulmonary trunk. Between compressions, blood flows into the ventricles from the atria.

Fortunately, a fire station was only a few blocks away and it took only about 5 minutes for emergency medical technicians to arrive. They confirmed the lack of a pulse and used portable equipment to record an electrocardiogram, which indicated that the heart was fibrillating (see Fibrillation of the Heart, p. 335). They quickly used a portable defibrillator to apply a strong electrical shock to Willie's chest. Fortunately, Willie's heart responded by beginning to beat rhythmically.

Willie's heart may have first developed arrhythmia and then ventricular fibrillation developed. Willie was very fortunate. Most people who suffer from sudden cessation of the pumping activity of the heart do not survive. In Willie's case, CPR was administered quickly and effectively, and emergency help arrived in a very short period of time.

Willie was transported to a hospital. His condition could be due to a myocardial infarction (see Myocardial Infarction, p. 348) or to some other condition. It is important to identify the underlying cause of the condition and treat it.

Cardiac Cycle

The heart can be viewed as two separate pumps represented by the right and left halves of the heart. Each pump consists of a primer pump—the atrium—and a power pump—the ventricle. The atria act as primer pumps because they complete the filling of the ventricles with blood, and the ventricles act as power pumps because they produce the major force that causes blood to flow through the pulmonary and systemic circulations. The term **cardiac cycle** refers to the repetitive pumping process that begins with the onset of cardiac muscle contraction and ends with the beginning of the next contraction (figure 12.17). Pressure changes produced within the heart chambers as a result of cardiac muscle contraction are responsible for blood movement because blood moves from areas of higher pressure to areas of lower pressure.

Atrial systole (sis′tō-lē, a contracting) refers to contraction of the two atria. **Ventricular systole** refers to contraction of the two ventricles. **Atrial diastole** (dī-as′tō-lē, dilation) refers to relaxation of the two atria, and **ventricular diastole** refers to relaxation of the two ventricles. When the terms **systole** and **diastole** are used without reference to the atria or ventricles, they refer to ventricular contraction or relaxation. The ventricles contain more cardiac muscle than the atria and produce far greater pressures, which force blood to circulate throughout the vessels of the body.

The major events of the cardiac cycle are:

1. *Systole*—At the beginning of systole, contraction of the ventricles pushes blood toward the atria, causing the AV valves to close as the pressure begins to increase (see figure 12.17, step 1). As systole continues, the increasing pressure in the ventricles exceeds the pressure in the pulmonary trunk and aorta, the semilunar valves are forced open, and blood is ejected into the pulmonary trunk and aorta (see figure 12.17, step 2).
2. *Diastole*—At the beginning of ventricular diastole, the pressure in the ventricles decreases below the pressure in the aorta and pulmonary trunk. The semilunar valves close and prevent blood from flowing back into the ventricles (see figure 12.17, step 3).

   As diastole continues, the pressure continues to decline in the ventricles until atrial pressures are greater than ventricular pressures. Then the AV valves open and blood flows directly from the atria into the relaxed ventricles. During the previous ventricular systole, the atria were relaxed and blood collected in them. When the ventricles relax and the AV valves open, blood flows into the ventricles (see figure 12.17, step 4) and fills them to approximately 70% of their volume.
3. At the end of ventricular diastole, the atria contract and then relax. Atrial systole forces additional blood to flow into the ventricles to complete their filling (see figure 12.17, step 5). The semilunar valves remain closed.

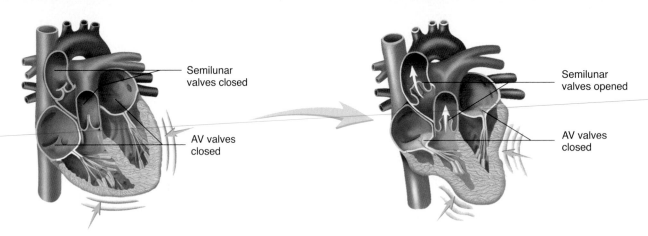

Semilunar
valves closed

AV valves
closed

Semilunar
valves opened

AV valves
closed

1.
 Contraction of the ventricles causes
 pressure in the ventricle to increase.
 Almost immediately the AV valves
 close (the first heart sound). The
 pressure in the ventricle continues
 to increase.

2.
 Continued ventricular contraction
 causes the pressure in the ventricle
 to exceed the pressure in the pulmonary
 trunk and aorta. As a result, the
 semilunar valves are forced open
 and blood is ejected into the
 pulmonary trunk and aorta.

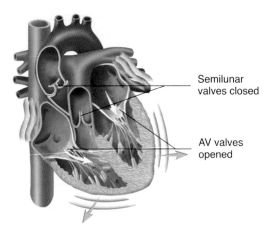

Semilunar
valves closed

AV valves
opened

Semilunar
valves closed

AV valves
closed

 The atria contract and complete
 ventricular filling.

3.
 At the beginning of ventricular
 diastole the ventricles relax, and the
 semilunar valves close (the second
 heart sound).

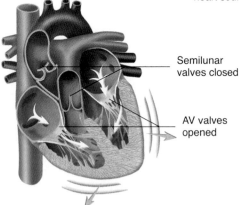

Semilunar
valves closed

AV valves
opened

4.
 The AV valves open and blood flows into
 the ventricle and the ventricles fill
 to approximately 70% of their
 volume.

Process Figure 12.17 The Cardiac Cycle

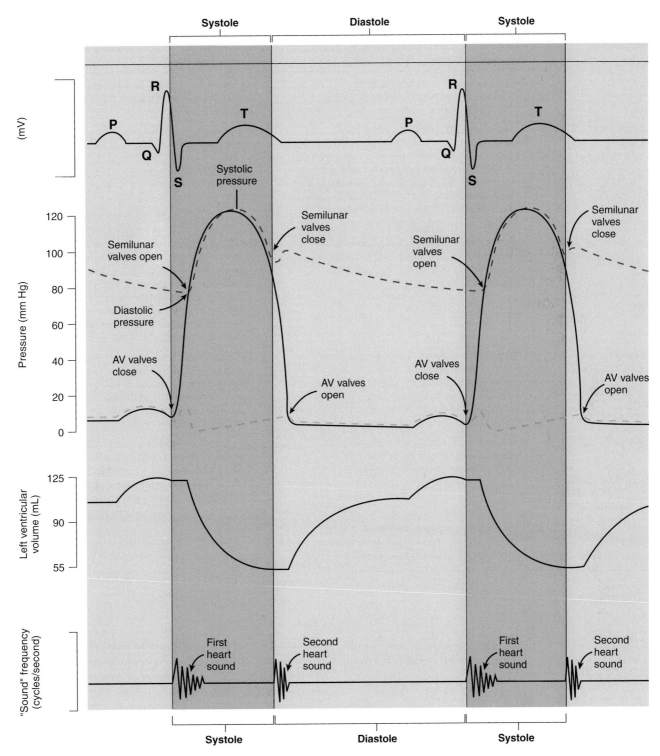

Figure 12.18 Events Occurring During the Cardiac Cycle (see Table 12.2)

The cardiac cycle is divided into systole and diastole (*see top of figure*). Within systole and diastole, four graphs are presented. From *top* to *bottom*, the electrocardiogram; pressure changes for the left atrium (*blue line*), left ventricle (*black line*), and aorta (*red line*); left ventricular volume curve; and heart sounds are illustrated.

Figure 12.18 displays the main events of the cardiac cycle in graphic form and should be examined from top to bottom for each period of the cardiac cycle. The ECG indicates the electrical events that cause contraction and relaxation of the atria and ventricles. The pressure graph shows the pressure changes within the left atrium, left ventricle, and aorta resulting from atrial and ventricular contraction and relaxation. The pressure changes on the right side of the heart are not shown here, but they are similar to those in the left side, only lower. The volume graph presents the changes in left ventricular volume as blood flows into and out of the left ventricle as a result of the pressure changes. The sound graph records the closing of valves caused

Table 12.2 | Summary of Events of the Cardiac Cycle for the Left Atrium and Ventricle (see figure 12.18)

	Ventricular Systole	Ventricular Diastole
ECG	The QRS complex is completed and the ventricles are stimulated to contract. The T wave begins.	The T wave is completed and the ventricles relax. Then the P wave stimulates the atria to contract, after which they relax.
Ventricular pressure curve (*black*)	Pressure increases rapidly as a result of left ventricular contraction. When left ventricular pressure exceeds aortic pressure, blood pushes the aortic semilunar valve open. Continued contraction increases ventricular pressure to a peak value of 120 mm Hg. Ventricular pressure then decreases as blood flows out of the left ventricle into the aorta.	Ventricular pressure decreases rapidly to nearly zero as the left ventricle relaxes.
Aortic pressure curve (*red*)	As ventricular contraction forces blood into the aorta, pressure in the aorta increases to its highest value (120 mm Hg), called the systolic pressure.	Ventricular pressure decreases below aortic pressure. Blood flows back toward the left ventricle and the aortic semilunar valve closes. As blood flows out of the aorta toward the body, elastic recoil of the aorta prevents a sudden decrease in pressure. Just before the aortic semilunar valve opens, pressure in the aorta decreases to its lowest value (80 mm Hg), called the diastolic pressure.
Atrial pressure curve (*blue*)	Atrial pressure increases slightly as contraction of the left ventricle pushes blood through the aorta and toward the left atrium. After closure of the bicuspid valve, pressure drops in the left atrium as it relaxes, then increases as blood flows into the left atrium from the four pulmonary veins.	After the bicuspid valve opens, pressure decreases slightly as blood flows into the left ventricle. At the end of ventricular diastole, contraction of the left atrium increases the pressure slightly.
Volume graph	Blood pushes the aortic semilunar valve open, blood is ejected from the left ventricle, and ventricular volume decreases.	Blood flows from the left atrium into the left ventricle, accounting for 70% of ventricular filling. Near the end of ventricular diastole, contraction of the left atrium pushes blood into the left ventricle, completing ventricular filling.
Sound graph	As contraction of the ventricles pushes blood toward the atria, the AV valves close, preventing the flow of blood into the atria and producing the first heart sound.	As blood flows back toward the heart, the semilunar valves close, preventing the flow of blood into the ventricles and producing the second heart sound.

by blood flow. See figure 12.17 for illustration of the valves and blood flow and table 12.2 for a summary of the events occurring during each period.

PREDICT 5
Predict the effect of a leaky (incompetent) aortic semilunar valve on the volume of blood in the left ventricle just before ventricular contraction. Predict the effect of a severely narrowed opening through the aortic semilunar valves on the amount of work the heart must do to pump the normal volume of blood into the aorta during each beat of the heart.

The Consequences of an Incompetent Bicuspid Valve

Incompetent valves do not close completely and therefore they leak when they are supposed to be closed. Incompetent valves allow blood to flow in the reverse direction. For example, an incompetent bicuspid valve allows blood to flow from the left ventrical to the left atrium during ventricular systole. This reduces the amount of blood pumped into the

aorta. It also dramatically increases the blood pressure in the left atrium and in the pulmonary veins during ventricular systole. During diastole, the excess blood pumped into the atrium once again flows into the ventricle along with the blood that normally flows from the lungs to the left atrium. Therefore, the volume of blood entering the left ventricle is greater than normal. The increased filling of the left ventricle gradually causes it to hypertrophy and can lead to heart failure. The increased pressure in the pulmonary veins can cause edema in the lungs.

Heart Sounds

A **stethoscope** (steth′ō-skōp, *stetho,* the chest) was originally developed to listen to the sounds of the lungs and heart and is now used to listen to other sounds of the body (figure 12.19). There are two main heart sounds. The **first heart sound** can be represented by the syllable **lubb,** and the **second heart sound** can be represented by **dupp.** The first heart sound has a lower pitch than

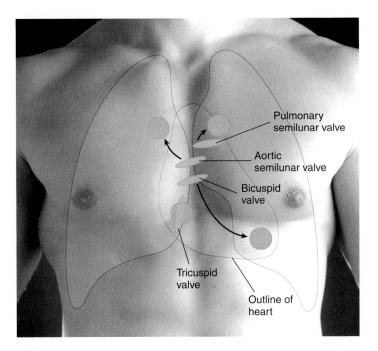

Figure 12.19 Location of the Heart Valves in the Thorax

Surface markings of the heart in the male. The positions of the four heart valves are indicated by *blue ellipses,* and the sites where the sounds of the valves are best heard with the stethoscope are indicated by *pink circles.*

the second. The first heart sound occurs at the beginning of ventricular systole and results from closure of the AV valves (see figure 12.17, step 1 and figure 12.18). The second heart sound occurs at the beginning of ventricular diastole and results from closure of the semilunar valves (see figure 12.17, step 3 and figure 12.18). The valves usually do not make sounds when they open.

Clinically, ventricular systole occurs between the first and second heart sounds. Ventricular diastole occurs between the second heart sound and the first heart sound of the next beat. Because ventricular diastole lasts longer than ventricular systole, there is less time between the first and second heart sounds than between the second heart sound and the first heart sound of the next beat.

P R E D I C T **6**
Compare the rate of blood flow out of the ventricles between the first and second heart sounds of the same beat with the rate of blood flow out of the ventricles between the second heart sound of one beat and the first heart sound of the next beat.

Abnormal heart sounds called **murmurs** are usually a result of faulty valves. For example, an **incompetent valve** fails to close tightly and blood leaks through the valve when it is closed. A murmur caused by an incompetent valve makes a swishing sound immediately after closure of the valve. For example, an incompetent bicuspid valve results in a swishing sound immediately after the first heart sound.

When the opening of a valve is narrowed, or **stenosed** (sten′ozd, a narrowing), a swishing sound precedes closure of the stenosed valve. For example, when the bicuspid valve is stenosed, a swishing sound precedes the first heart sound.

P R E D I C T
If normal heart sounds are represented by lubb–dupp, lubb–dupp, what does a heart sound represented by lubb–duppshhh, lubb–duppshhh represent? What does lubb–shhhdupp, lubb–shhhdupp represent (assume that shhh represents a swishing sound)?

A CASE IN POINT | Paroxysmal Atrial Tachycardia

Speedy Beat is a 70-year-old man. He and his daughter Normal were getting out of the car at one of his favorite restaurants where they planned to have dinner. Before Speedy could get completely out of the car he became dizzy. He exhibited substantial pallor and experienced chest pains. Normal checked her father's pulse, which was close to 180–200 bpm and irregular. She helped him back into the car and drove to the emergency room at a nearby hospital. There, it was determined that Speedy's blood pressure was low even though his heart rate was rapid. Speedy was previously diagnosed as suffering from paroxysmal atrial tachycardia and he regularly takes a calcium channel blocking agent to control it.

The term *paroxysmal* means that the cause of Speedy's tachycardia is not known, but it results from rapid and ectopic beats that originate in the atria. The heart rate is irregular because the rate at which action potentials responsible for atrial contractions occur is greater than the rate at which the ventricles can contract. Consequently, not every atrial contraction is followed by a ventricular contraction. Speedy's blood pressure is low because the heart rate is so fast that there is little time for blood to fill the rapidly contracting chambers of the heart between the contractions. Consequently, Speedy's stroke volume and cardiac output are low. Chest pains result because the heart muscle is working hard, but blood flow to cardiac muscle through coronary vessels is reduced so that the heart muscle suffers from an inadequate supply of oxygen (ischemia).

Once Speedy's heart rate and rhythm were stabilized, the amount of calcium channel blocking agent prescribed for him was adjusted and he was released from the hospital the next day. Another drug used to treat this condition along with calcium channel blocking agent is digoxin, which has the overall effect of increasing the force and slowing the rate of cardiac muscle contraction.

Clinical Focus Conditions and Diseases Affecting the Heart

Heart Diseases

Inflammation of Heart Tissues

Endocarditis (en′dō-kar-dī′tis) is inflammation of the endocardium. It affects the valves more severely than other areas of the heart and may lead to deposition of scar tissue, causing valves to become stenosed or incompetent.

Pericarditis (per′i-kar-dī′tis) is inflammation of the pericardium. Pericarditis can result from bacterial or viral infections and can be extremely painful.

Cardiomyopathy (kar′dē-ō-mī-op′ă-thē) is a disease of the myocardium of unknown cause or occurring secondarily to other diseases. It may develop secondarily to infections or from exposure to toxins including alcohol, drugs, or environmental pollutants. It may also result from nutritional disorders or other systemic diseases such as leukemia. In most cases the cause is unknown. The most common form of cardiomyopathy results in weakened cardiac muscle, which causes all of the chambers of the heart to dilate or enlarge. This can eventually lead to congestive heart failure.

Rheumatic (roo-mat′ik) **heart disease** can result from a **streptococcal** (strep′tō-kok′ăl) **infection** in young people. Toxin produced by the bacteria can cause an immune reaction called rheumatic fever approximately 2–4 weeks after the infection. The immune reaction can cause inflammation of the endocardium, called **rheumatic endocarditis.** The inflamed valves, especially the bicuspid valve, can become stenosed or incompetent. The effective treatment of streptococcal infections with antibiotics has reduced the frequency of rheumatic heart disease.

Reduced Blood Flow to Cardiac Muscle

Coronary heart disease reduces the amount of blood that the coronary arteries are able to deliver to the myocardium. The reduction in blood flow can damage the myocardium. The degree of damage depends on the size of the arteries involved, whether occlusion (blockage) is partial or complete,

and whether occlusion is gradual or sudden. As the walls of the arteries thicken and harden with age, the volume of blood they can supply to the heart muscle declines, and the ability of the heart to pump blood decreases. Inadequate blood flow to the heart muscle can result in **angina pectoris,** which is a poorly localized sensation of pain in the region of the chest, left arm, and left shoulder.

Degenerative changes in the artery wall can cause the inside surface of the artery to become roughened. The chance of platelet aggregation increases at the rough surface, which increases the chance of **coronary thrombosis** (throm-bō′sis), the formation of a blood clot in a coronary vessel. Inadequate blood flow can cause an **infarct** (in′farkt), an area of damaged cardiac tissue. A heart attack is often referred to as a coronary thrombosis or a **myocardial infarction.** The outcome of coronary thrombosis depends on the extent of the damage to heart muscle caused by inadequate blood flow and whether other blood vessels can supply enough blood to maintain the function of the heart. Death can occur swiftly if the infarct is large; if the infarct is small, the heart can continue to function. In some cases, the infarct weakens the wall of the heart, and the wall ruptures; but in most cases scar tissue replaces damaged cardiac muscle in the area of the infarct.

People who survive infarctions often lead fairly normal lives if they take precautions. Most cases call for moderate exercise, adequate rest, a disciplined diet, and reduced stress. Small doses of aspirin and treatments, including drugs, to reduce elevated blood pressure appear to provide protection against the development of myocardial infarcts.

Congenital Conditions Affecting the Heart

Congenital (occuring at birth) **heart disease** is heart disease present at birth and is the result of abnormal development of the heart. The following are common congenital defects:

A **septal defect** is a hole in a septum between the left and right sides of the heart. The hole may be in the interatrial or interventricular septum. These defects allow blood to flow from one side of the heart to the other and, as a consequence, greatly reduce the pumping effectiveness of the heart.

Patent (to lie open) **ductus arteriosus** (dŭk′tŭs artēr′ē-ō-sŭs) results when a blood vessel called the **ductus arteriosus,** which is present in the fetus, fails to close after birth. The ductus arteriosus extends between the pulmonary trunk and the aorta. It allows blood to pass from the pulmonary trunk to the aorta, thus bypassing the lungs. This is normal before birth because the lungs are not functioning. If the ductus arteriosus fails to close after birth, however, blood flows in the opposite direction, from the aorta to the pulmonary trunk. As a consequence, blood flows through the lungs under a higher pressure and damages them. In addition, the amount of work required of the left ventricle to maintain an adequate systemic blood pressure increases.

Stenosis (ste-nō′sis) **of the heart valves** is a narrowed opening through one or more of the heart valves. In aortic or pulmonary semilunar valve stenosis, the workload of the heart is increased because the ventricles must contract with a much greater force to pump blood from the ventricles. Stenosis of the bicuspid valve prevents the flow of blood into the left ventricle, causing blood to back up in the left atrium and the lungs, resulting in edema in the lungs. Stenosis of the tricuspid valve causes blood to back up in the right atrium and systemic veins, causing edema in the periphery.

Cyanosis (sī-ă-nō′sis, *cyan,* blue + *osis,* condition of) is a symptom of inadequate heart function in babies suffering from congenital heart disease. The term "blue baby" is sometimes used to refer to infants with cyanosis. The blueness of the skin is caused by low oxygen levels in the blood in peripheral blood vessels.

Regulation of Heart Function

Cardiac output (CO) is the volume of blood pumped by either ventricle of the heart each minute. Cardiac output can be calculated by multiplying the stroke volume times the heart rate. **Stroke volume (SV)** is the volume of blood pumped per ventricle each time the heart contracts, and the **heart rate (HR)** is the number of times the heart contracts each minute.

$$CO \quad = \quad SV \quad \times \quad HR$$
$$\text{(mL/min)} \qquad \text{(mL/beat)} \qquad \text{(beats/min)}$$

Under resting conditions, the heart rate is approximately 72 beats/min (or bpm) and the stroke volume is approximately 70 mL/beat. Consequently, the cardiac output is slightly more than 5 L/min:

$$CO = SV \times HR$$
$$= 70 \text{ mL/beat} \times 72 \text{ bpm}$$
$$= 5040 \text{ mL/min (approximately 5 L/min)}$$

The heart rate and the stroke volume vary considerably among people. Athletes tend to have a larger stroke volume and lower heart rate at rest because exercise has increased the size of their hearts. Nonathletes are more likely to have a higher heart rate and lower stroke volume. During exercise the heart in a nonathlete can increase to 190 bpm and the stroke volume can increase to 115 mL/beat. Therefore, the cardiac output increases to approximately 22 L/min:

$$CO = SV \times HR$$
$$= 115 \text{ mL/beat} \times 190 \text{ bpm}$$
$$= 21,850 \text{ mL/min (approximately 22 L/min)}$$

This produces a cardiac output that is several times greater than the cardiac output under resting conditions. Athletes can increase their cardiac output to a greater degree than nonathletes.

The control mechanisms that modify the stroke volume and the heart rate are classified as intrinsic and extrinsic mechanisms.

Intrinsic Regulation of the Heart

Intrinsic regulation of the heart refers to mechanisms contained within the heart itself. The force of contraction produced by cardiac muscle is related to the degree of stretch of cardiac muscle fibers. The amount of blood in the ventricles at the end of ventricular diastole determines the degree to which cardiac muscle fibers are stretched. **Venous return** is the amount of blood that returns to the heart, and the degree to which the ventricular walls are stretched at the end of diastole is called **preload.** If venous return increases, the heart fills to a greater volume and stretches the cardiac muscle fibers, producing an increased preload. In response to the increased preload, cardiac muscle fibers contract with a greater force. The greater force of contraction causes an increased volume of blood to be ejected from the heart, resulting in an increased stroke volume. As venous return increases, resulting in an increased preload, cardiac output increases. Conversely, if venous return decreases, resulting in a decreased

preload, the cardiac output decreases. The relationship between preload and stroke volume is called **Starling's law of the heart.**

Because venous return is influenced by many conditions, Starling's law of the heart has a major influence on cardiac output. For example, muscular activity during exercise causes increased venous return, resulting in an increased preload, stroke volume, and cardiac output. This is beneficial because an increased cardiac output is needed during exercise to supply oxygen to exercising skeletal muscles.

Afterload refers to the pressure against which the ventricles must pump blood. People suffering from hypertension have an increased afterload because they have an elevated aortic pressure during contraction of the ventricles. The heart must do more work to pump blood from the left ventricle into the aorta, which increases the workload on the heart and can eventually lead to heart failure. A reduced afterload decreases the work the heart must do. People who have a lower blood pressure have a reduced afterload and develop heart failure less often than people who have hypertension. The afterload, however, influences cardiac output less than preload influences it. The afterload must increase substantially before it decreases the volume of blood pumped by a healthy heart.

Consequences of Heart Failure

Although heart failure can occur in young people, it usually results from a progressive weakening of the heart muscle in elderly people. A failing heart gradually enlarges and eventually fails. In heart failure, the heart is not capable of pumping all the blood that is returned to it because further stretching of the cardiac muscle fibers does not increase the stroke volume of the heart. Consequently, blood backs up in the veins. For example, heart failure that affects the right ventricle is called **right heart failure** and causes blood to back up in the veins that return blood from systemic vessels to the heart. Filling of the veins with blood causes edema, especially in the legs and feet. Edema results from the accumulation of fluid in tissues outside of blood vessels. Heart failure that affects the left ventricle is called **left heart failure** and causes blood to back up in the veins that return blood from the lungs to the heart. Filling of these veins causes edema in the lungs, which makes breathing difficult.

Extrinsic Regulation of the Heart

Extrinsic regulation refers to mechanisms external to the heart, such as either hormonal or nervous regulation (figure 12.20). Nervous influences are carried through the autonomic nervous system. Both sympathetic and parasympathetic nerve fibers innervate the heart, and have a major effect on the SA node. Stimulation by sympathetic nerve fibers causes the heart rate and the stroke volume to increase, whereas stimulation by parasympathetic nerve fibers causes the heart rate to decrease.

The **baroreceptor** (bar′ō-rē-sep′ter, *baro*, pressure) **reflex** plays an important role in regulating the function of the heart. **Baroreceptors** are stretch receptors that monitor blood pressure in the aorta and in the wall of the internal carotid arteries, which

1. Sensory (*green*) neurons carry action potentials from baroreceptors to the cardioregulatory center. Chemoreceptors in the medulla oblongata influence the cardioregulatory center.

2. The cardioregulatory center controls the frequency of action potentials in the parasympathetic (*red*) neurons extending to the heart. The parasympathetic neurons decrease the heart rate.

3. The cardioregulatory center controls the frequency of action potential in the sympathetic (*blue*) neurons extending to the heart. The sympathetic neurons increase the heart rate and the stroke volume.

4. The cardioregulatory center influences the frequency of action potentials in the sympathetic (*blue*) neurons extending to the adrenal medulla. The sympathetic neurons increase the secretion of epinephrine and some norepinephrine into the general circulation. Epinephrine and norepinephrine increase the heart rate and stroke volume.

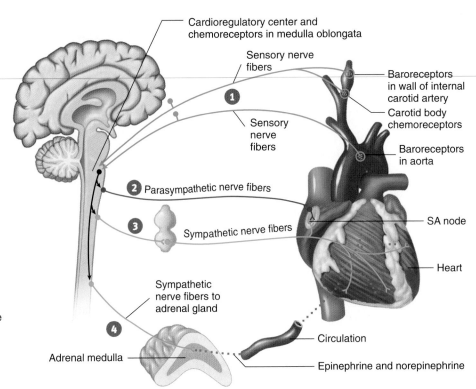

Process Figure 12.20   **Baroreceptor and Chemoreceptor Reflexes**

Sensory (*green*) nerves carry action potentials from sensory receptors to the medulla oblongata. Sympathetic (*blue*) and parasympathetic (*red*) nerves exit the spinal cord or medulla oblongata and extend to the heart to regulate its function. Epinephrine and norepinephrine from the adrenal gland also help regulate the heart's action. (SA = sinoatrial)

carry blood to the brain (see figure 12.20). Changes in blood pressure result in changes in the stretch of the walls of these blood vessels. Thus, changes in blood pressure cause changes in the frequency of action potentials produced by the baroreceptors. The action potentials are transmitted along nerve fibers from the stretch receptors to the medulla oblongata of the brain.

Within the medulla oblongata of the brain is a **cardioregulatory center,** which receives and integrates action potentials from the baroreceptors. The cardioregulatory center controls the action potential frequency in sympathetic and parasympathetic nerve fibers that extend from the brain and spinal cord to the heart. The cardioregulatory center also influences sympathetic stimulation of the adrenal gland (see figure 12.20). Epinephrine and norepinephrine, released from the adrenal gland, increase the stroke volume and heart rate.

When the blood pressure increases, the baroreceptors are stimulated. There is increased frequency of action potentials, sent along the nerve fibers to the medulla oblongata of the brain. This prompts the cardioregulatory center to increase parasympathetic stimulation and to decrease sympathetic stimulation of the heart. As a result, the heart rate and stroke volume decrease, causing blood pressure to decline (figure 12.21).

When the blood pressure decreases, there is less stimulation of the baroreceptors. A lower frequency of action potentials is sent to the medulla oblongata of the brain and this triggers a response in the cardioregulatory center. The cardioregulatory

center responds by increasing sympathetic stimulation of the heart and decreasing parasympathetic stimulation. Consequently, the heart rate and stroke volume increase. If the decrease in blood pressure is large, sympathetic stimulation of the adrenal medulla also increases. The epinephrine and norepinephrine secreted by the adrenal medulla increase the heart rate and stroke volume, also causing the blood pressure to increase toward its normal value (see figure 12.21).

P R E D I C T   **8**

In response to a severe hemorrhage, blood pressure lowers, the heart rate increases dramatically, and the stroke volume lowers. If low blood pressure activates a reflex that increases sympathetic stimulation of the heart, why is the stroke volume low?

Emotions integrated in the cerebrum of the brain can influence the heart. Excitement, anxiety, or anger can affect the cardioregulatory center, resulting in increased sympathetic stimulation of the heart and an increased cardiac output. Depression, on the other hand, can increase parasympathetic stimulation of the heart, causing a slight reduction in cardiac output.

Epinephrine and small amounts of norepinephrine released from the adrenal medulla in response to exercise, emotional excitement, or stress also influence the heart's function (see figures 12.20 and 12.21). Epinephrine and norepinephrine bind to receptor molecules on cardiac muscle and cause increased heart rate and stroke volume.

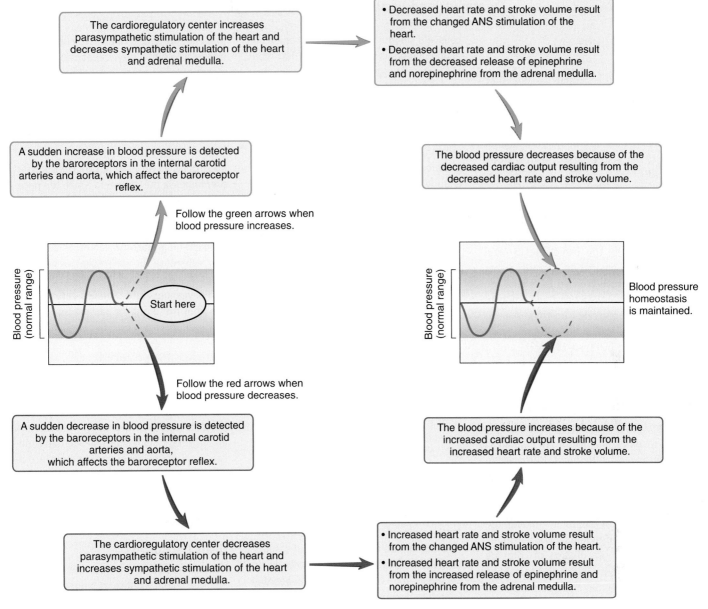

Homeostasis Figure 12.21   Baroreceptor Reflex

The baroreceptor reflex maintains homeostasis in response to changes in blood pressure. (ANS = autonomic nervous system)

Exercise and Cardiac Output

During exercise, the cardiac output of the heart increases, resulting in increased delivery of blood to skeletal muscles. Cardiac output increases because of the increased heart rate and stroke volume that result from increased sympathetic stimulation of the heart and from the effects of epinephrine and norepinephrine on cardiac muscle. Starling's law of the heart also contributes to the increase in stroke volume during exercise. Blood vessels in exercising skeletal muscles dilate, which increases blood flow to the muscle tissue. The dilation of the blood vessels also increases venous return to the heart because the rate of blood flow from exercising skeletal muscle through the veins is greatly increased. As venous return increases, preload increases, and cardiac muscle is stretched. Consequently, the muscle contracts more forcefully (Starling's law) and stroke volume increases.

The medulla oblongata of the brain also contains chemoreceptors that are sensitive to changes in pH and carbon dioxide levels (see figure 12.20). A decrease in pH, often caused by an increase in carbon dioxide, results in sympathetic stimulation of the heart (figure 12.22).

Changes in the extracellular concentration of K^+, Ca^{2+}, and Na^+, which influence other electrically excitable tissues, also affect cardiac muscle function. Excess extracellular K^+ cause the heart rate and stroke volume to decrease. If the extracellular K^+ concentration increases further, normal conduction of action potentials through cardiac muscle is blocked, and death can result. An excess of extracellular Ca^{2+} causes the heart to contract arrhythmically. Reduced extracellular Ca^{2+} cause both the heart rate and stroke volume to decrease.

Clinical Focus **Treatment and Prevention of Heart Disease**

Heart Medications

Digitalis

Digitalis (dij'i-tal'is, refers to fingerlike flowers), an extract of foxglove plants, slows and strengthens contractions of the heart muscle. This drug is frequently given to people who suffer from heart failure, although it can also be used to treat atrial tachycardia.

Nitroglycerin

During exercise, when the heart rate and stroke volume are increased, dilation of blood vessels in the exercising skeletal muscles and constriction in most other blood vessels results in an increased venous return to the heart and an increased preload. **Nitroglycerin** (nī-trō-glis'er-in) causes dilation of all of the veins and arteries without an increase in heart rate or stroke volume. When all blood vessels dilate, a greater volume of blood pools in the dilated blood vessels, causing a decrease in the venous return to the heart. The reduced preload causes cardiac output to decrease, resulting in a decreased amount of work performed by the heart. Nitroglycerin is frequently given to people who suffer from coronary artery disease, which restricts coronary blood flow. The decreased work performed by the heart reduces the amount of oxygen required by the cardiac muscle. In addition, dilation of coronary arteries can increase blood flow to cardiac muscle. Consequently, the heart does not suffer from a lack of oxygen, and angina pectoris does not develop.

Beta-Adrenergic Blocking Agents

Beta-adrenergic (bā-tă ad-rĕ-ner'jik) **blocking agents** reduce the rate and strength of cardiac muscle contractions, thus reducing the oxygen demand of the heart. They bind to receptors for norepinephrine and epinephrine and prevent these substances from having their normal effects. Beta-adrenergic blocking agents are often used to treat people who suffer from rapid heart rates, certain types of arrhythmias, and hypertension.

Calcium Channel Blockers

Calcium channel blockers reduce the rate at which Ca^{2+} diffuse into cardiac muscle cells and smooth muscle cells. Because the action potentials that produce cardiac muscle contractions depend in part on the flow of Ca^{2+} into the cardiac muscle cells, the Ca^{2+} channel blockers can be used to control the force of heart contractions and reduce arrhythmia, tachycardia, and hypertension. Because entry of Ca^{2+} to smooth muscle cells causes contraction, Ca^{2+} channel blockers cause dilation of blood vessels. They dilate coronary blood vessels and increase blood flow to cardiac muscle. Consequently, they can be used to treat angina pectoris.

Antihypertensive Agents

Several drugs are used specifically to treat hypertension. These drugs reduce blood pressure and therefore reduce the work required by the heart to pump blood. In addition, the reduction of blood pressure reduces the risk of heart attacks and strokes. Medications used to treat hypertension include drugs that reduce the activity of the sympathetic division, those that dilate arteries and veins, those that increase urine production (diuretics), and those that block the conversion of angiotensin I to angiotensin II (see chapter 13).

Anticoagulants

Anticoagulants (an'tē-kō-ag'ū-lants) prevent clot formation in persons with damage to heart valves or blood vessels or in persons who have had a myocardial infarction. Aspirin functions as a weak anticoagulant by inhibiting the synthesis of prostaglandins in platelets, which in turn reduces clot formation. Some data suggest that taking a small dose of aspirin regularly reduces the chance of a heart attack. Approximately one baby aspirin each day may benefit those who are likely to experience a coronary thrombosis.

Instruments

Artificial Pacemaker

An **artificial pacemaker** is an instrument placed beneath the skin that is equipped with an electrode that extends to the heart. An artificial pacemaker provides an electrical stimulus to the heart at a set frequency. Artificial pacemakers are used in patients in whom the natural pacemaker of the heart does not produce a heart rate high enough to sustain normal physical activity. Modern electronics has made it possible to design artificial pacemakers that can increase the heart rate as physical activity increases. In addition, special artificial pacemakers can defibrillate the heart if it becomes arrhythmic. It is likely that rapid development of electronics for artificial pacemakers will further increase the degree to which the pacemakers can regulate the heart.

Heart–Lung Machine

A **heart–lung machine** serves as a temporary substitute for the patient's heart and lungs. It pumps blood throughout the body and oxygenates and removes carbon dioxide from the blood. It has made possible many surgeries on the heart and lungs.

Surgical Procedures

Heart Valve Replacement or Repair

Heart valve replacement or repair is a surgical procedure performed on those who have diseased valves that are so deformed and scarred from conditions such as endocarditis that the valves are severely incompetent or stenosed. Substitute valves made of synthetic materials such as plastic or Dacron are effective; valves transplanted from pigs are also used.

Heart Transplants

Heart transplants are possible when the immune characteristics of a donor and the recipient are closely matched. The heart of a recently deceased donor is transplanted to the recipient, and the diseased heart of the recipient is removed. People who have received heart transplants must remain on drugs that suppress their immune responses for the rest of their lives. Unless they do so, their immune system rejects the transplanted heart.

Artificial Hearts

Artificial hearts have been used on an experimental basis to extend the lives of individuals until an acceptable transplant can be found or to replace the heart permanently. The technology currently available for artificial hearts has not yet reached the point at which a high quality of life can be achieved with a permanent artificial heart.

Prevention of Heart Disease

Proper nutrition is important in reducing the risk of heart disease. A recommended diet is low in fats, especially saturated fats and cholesterol, and low in refined sugar. Diets should be high in fiber, whole grains, fruits, and vegetables. Total food intake should be limited to avoid obesity, and sodium chloride intake should be reduced.

Tobacco and excessive use of alcohol should be avoided. Smoking increases the risk of heart disease by at least 10-fold, and excessive use of alcohol also substantially increases the risk of heart disease. However, moderate use of alcohol may reduce the risk of heart disease.

Chronic stress, frequent emotional upsets, and a lack of physical exercise can increase the risk of cardiovascular disease. Remedies include relaxation techniques and aerobic exercise programs involving gradual increases in duration and difficulty in activities such as swimming, walking, jogging, or aerobic dancing.

Hypertension is an abnormally high systemic blood pressure. Hypertension affects approximately one-fifth of the population. Regular blood pressure measurements are important because hypertension does not produce obvious symptoms. If hypertension cannot be controlled by diet and exercise, it is important to treat the condition with prescribed drugs. The cause of hypertension in the majority of cases is unknown.

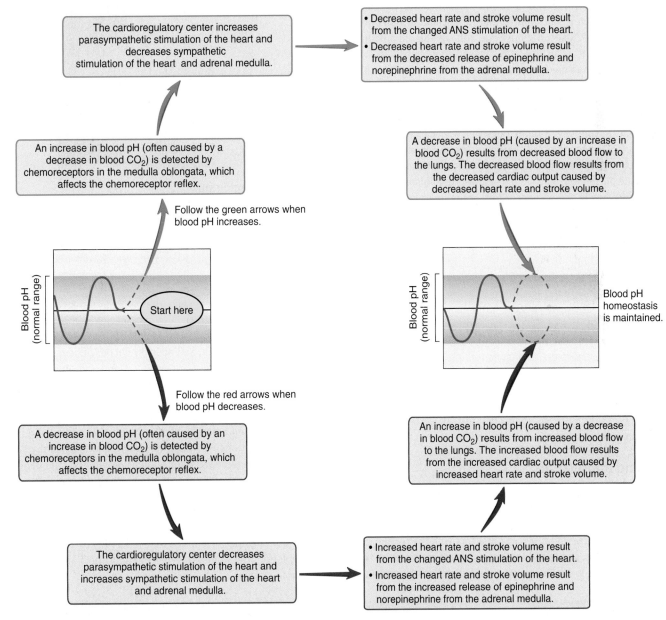

Homeostasis Figure 12.22 Chemoreceptor Reflex–pH

The chemoreceptor reflex maintains homeostasis in response to changes in blood concentrations of CO_2 and H^+ (or pH). (ANS = autonomic nervous system)

Systems Pathology
Myocardial Infarction

Mr. P. was an overweight, out-of-shape executive who regularly consumed food with a high fat content and smoked. He viewed his job as frustrating because he was frequently confronted with stressful deadlines. He had not had a physical examination for several years so he was not aware that his blood pressure was high. One evening, Mr. P. was walking to his car after work when he began to feel pain in his chest that also radiated down his left arm. Shortly after the onset of pain he became out of breath, developed marked pallor, became dizzy, and had to lie down on the sidewalk. The pain in his chest and arm was poorly localized, but intense and he became anxious and then disoriented. Mr. P. lost consciousness, although he did not stop breathing. After a short delay, one of his coworkers noticed him and called for help. When paramedics arrived, they determined that Mr. P's blood pressure was low and he exhibited arrhythmia and tachycardia. The paramedics transmitted the electrocardiogram they took to a physician by way of their electronic communication system, and they discussed Mr. P.'s symptoms with the physician who was at the hospital. The paramedics were directed to administer oxygen and medication to control arrythmias and to transport him to the hospital. At the hospital, tissue plasminogen activator (t-PA) was administered, which improved blood flow to the damaged area of the heart by activating plasminogen, which dissolves blood clots. Blood levels of enzymes such as creatine phosphokinase increased in Mr. P.'s blood over the next few days, which confirmed that damage to cardiac muscle resulted from an infarction.

In the hospital, Mr. P. began to experience shortness of breath because of pulmonary edema, and after a few days in the hospital he developed pneumonia. He was treated for pneumonia and gradually improved over the next few weeks. An **angiogram** (an′jē-ō-gram, *angeion*, a vessel + *gramma,* a writing) (figure A), which is an imaging technique used to visualize the coronary arteries, was performed several days after Mr. P.'s infarction. The angiogram indicated that Mr. P. suffered from a permanent reduction in a significant part of the lateral wall of his left ventricle and that neither angioplasty nor bypass surgery was necessary. Mr. P. has some other restrictions to blood flow in his coronary arteries.

Background Information

Mr. P. experienced a myocardial infarction. A thrombosis in one of the branches of the left coronary artery reduced blood supply to the lateral wall of the left ventricle, resulting in ischemia of the left ventricle wall. That t-PA was an effective treatment is consistent with the conclusion that the infarction was due to a thrombosis. An ischemic area of the heart wall was not able to contract normally and, therefore, the pumping

Figure A Angiogram

An angiogram is a picture of a blood vessel. It is usually obtained by placing a catheter into a blood vessel and injecting a dye that can be detected with x-rays. Note the partially occluded (blocked) coronary blood vessel in this angiogram, which has been computer-enhanced to show colors.

effectiveness of the heart was dramatically reduced. The reduced pumping capacity of the heart was responsible for the low blood pressure, which caused the blood flow to his brain to decrease resulting in confusion, disorientation, and possibly unconsciousness.

Low blood pressure, increasing blood carbon dioxide levels, pain, and anxiousness increased sympathetic stimulation of the heart and adrenal glands. Increased sympathetic stimulation of the adrenal medulla resulted in the release of epinephrine from the adrenal medulla. Increased parasympathetic stimulation of the heart resulted from pain sensations. The heart rate was periodically arrhythmic because of the combined effects of parasympathetic stimulation, epinephrine and norepinephrine from the adrenal gland, and sympathetic stimulation. In addition, ectopic beats were produced by the ischemic areas of the left ventricle.

Pulmonary edema resulted from the increased pressure in the pulmonary veins because of the reduced ability of the left ventricle to pump blood. The edema allowed bacteria to infect the lungs and cause pneumonia.

Systems Interactions: Effects of Myocardial Infarctions on Other Systems

System	Interactions
Integumentary	Pallor of the skin resulted from intense vasoconstriction of peripheral blood vessels, including those in the skin.
Muscular	Reduced skeletal muscle activity required for activities such as walking result because of the effect of a lack of blood flow to the brain and because blood is shunted from blood vessels that supply skeletal muscles to those that supply the heart and brain.
Nervous	Decreased blood flow to the brain, decreased blood pressure, and pain because of ischemia of heart muscle result in increased sympathetic and parasympathetic stimulation of the heart. Loss of consciousness occurs when the blood flow to the brain decreases enough to result in too little oxygen to maintain normal brain function, especially in the reticular activating system.
Endocrine	When blood pressure decreases to low values, antidiuretic hormone (ADH) is released from the posterior pituitary gland and renin, released from the kidney, activates the renin–angiotensin–aldosterone mechanism. ADH, secreted in large amounts, and angiotensin II cause vasoconstriction of peripheral blood vessels. ADH and aldosterone act on the kidneys to retain water and ions. An increased blood volume increases venous return, which results in an increased stroke volume of the heart and an increase in blood pressure unless damage to the heart is very severe.
Lymphatic or Immune	White blood cells, including macrophages, move to the area of cardiac muscle damage and phagocytize any dead cardiac muscle cells.
Respiratory	Decreased blood pressure results in a decreased blood flow to the lungs. The decrease in gas exchange results in increased blood carbon dioxide levels, acidosis, and decreased blood oxygen levels. Initially, respiration becomes deep and labored because of the elevated carbon dioxide levels, decreased blood pH, and depressed oxygen levels. If the blood oxygen levels decrease too much the person loses consciousness. Pulmonary edema can result when the pumping effectiveness of the left ventricle is substantially reduced.
Digestive	Intense sympathetic stimulation decreases blood flow to the digestive system to very low levels, which often results in increased nausea and vomiting.
Urinary	Blood flow to the kidney decreases dramatically in response to sympathetic stimulation. If the kidney becomes ischemic, damage to the kidney tubules can occur, resulting in the development of acute renal failure. Acute renal failure results in reduced urine production. Increased blood urea nitrogen, increased blood levels of K^+, and edema are indications that the kidneys cannot eliminate waste products and excess water. If damage is not too great, the period of reduced urine production may last up to 3 weeks, and then the rate of urine production slowly returns to normal as the kidney tubules heal.

The heart began to beat rhythmically in response to medication because the infarction did not damage the conducting system of the heart, which is an indication that there were no permanent arrhythmias. Permanent arrhythmias are indications of damage done to cardiac muscle cells specialized to conduct action potentials in the heart.

Analysis of the electrocardiogram, blood pressure measurements, and the angiogram indicated that the infarction, in this case, was located on the left side of Mr P.'s heart.

Mr. P.'s physician made it very clear to him that he was lucky to have survived a myocardial infarction and recommended a weight-loss program and a low-sodium and low-fat diet. Mr. P.'s physician also recommended that Mr. P. regularly take a small amount of aspirin and stop smoking. He explained that Mr. P. would have to take medication for high blood pressure if his blood pressure did not decrease in response to the recommended changes. After a period of recovery, the physician recommended an aerobic exercise program and suggested that Mr. P. seek ways to reduce the stress associated with his job. Mr. P. followed the doctor's recommendations, and after several months he began to feel better than he had in years and his blood pressure was normal.

PREDICT 9
Severe ischemia in the wall of a ventricle can result in the death of cardiac muscle cells. Inflammation around the necrotic tissue results and macrophages invade the necrotic tissue and phagocytize dead cells. At the same time, blood vessels and connective tissue grow into the necrotic area and begin to deposit connective tissue to replace the necrotic tissue. A person who entered the hospital at about the same time with a very similar condition to Mr. P.'s was recovering. After about a week, his blood pressure suddenly decreased to very low levels and he died within a very short time. Upon autopsy, a large amount of blood was found in the pericardial sac, and the wall of the left ventricle was ruptured. Explain.

Body temperature affects metabolism in the heart like it affects other tissues. Elevated body temperature increases the heart rate, and reduced body temperature slows the heart rate. For example, during fever the heart rate is usually elevated. During heart surgery the body temperature is sometimes intentionally lowered to slow the heart rate and metabolism.

Effects of Aging on the Heart

Gradual changes in the function of the heart are associated with aging. These changes are minor under resting conditions, but become more obvious during exercise and in response to age-related diseases.

By age 70 cardiac output often decreases by approximately one-third. Because of the decrease in the reserve strength of the heart, many elderly people are limited in their ability to respond to emergencies, infections, blood loss, or stress.

Hypertrophy (enlargement) of the left ventricle is a common age-related change. This appears to result from a gradual increase in the pressure in the aorta (afterload) against which the left ventricle must pump. The increased aortic pressure results from a gradual decrease in the elasticity of the aorta, and there is an increased stiffness of the cardiac muscle. The enlarged left ventricle has a reduced ability to pump blood out of the left ventricle. This can cause an increase in left atrial pressure, which can result in increased pulmonary edema.

Consequently, there is an increased tendency for people to feel out of breath when they exercise strenuously.

Aging cardiac muscle requires a greater amount of time to contract and relax. Thus, there is a decrease in the maximum heart rate. Both the resting and maximum cardiac output slowly decrease as people age and, by 85 years of age, the cardiac output is decreased by 30–60%.

Age-related changes in the connective tissue of the heart valves occur. The connective tissue becomes less flexible, and calcium deposits develop in the valves. As a result, there is an increased tendency for the aortic semilunar valve to become stenosed or incompetent.

There is an age-related increase in cardiac arrhythmias as a consequence of a decrease in the number of cardiac cells in the SA node and because of the replacement of cells of the AV bundle.

The development of coronary artery disease and heart failure also are age-related. Approximately 10% of elderly people over age 80 have heart failure, and a major contributing factor is coronary heart disease. Advanced age, malnutrition, chronic infections, toxins, severe anemias, hyperthyroidism, and hereditary factors can lead to heart failure.

Exercise has many beneficial effects on the heart. Regular aerobic exercise improves the functional capacity of the heart at all ages, providing there are no conditions that cause the increased workload of the heart to be harmful.

S U M M A R Y

Functions of the Heart   (p. 322)

The heart functions include
1. The heart generates blood pressure.
2. The heart routes blood through the systemic and pulmonary circulation.
3. The pumping action of the heart and the valves of the heart ensure a one-way flow of blood through the heart and blood vessels.
4. The heart helps regulate blood supply to tissues.

Size, Form, and Location of the Heart   (p. 322)

The heart is approximately the size of a fist and is located in the pericardial cavity.

Anatomy of the Heart   (p. 323)
Pericardium

1. The pericardial sac consists of a fibrous and serous pericardium. The fibrous pericardium is lined by the parietal pericardium.
2. The outer surface of the heart is lined by the visceral pericardium (epicardium).
3. Between the visceral and parietal pericardium is the pericardial cavity, which is filled with pericardial fluid.

External Anatomy

1. Atria are separated externally from the ventricles by the coronary sulcus. The right and left ventricles are separated externally by the interventricular sulci.
2. The inferior and superior venae cava enter the right atrium. The four pulmonary veins enter the left atrium.
3. The pulmonary trunk exits the right ventricle, and the aorta exits the left ventricle.

Heart Chambers and Internal Anatomy

1. There are four chambers in the heart. The left and right atria receive blood from veins and function mainly as reservoirs. Contraction of the atria completes ventricular filling.
2. The atria are separated internally from each other by the interatrial septum.
3. The ventricles are the main pumping chambers of the heart. The right ventricle pumps blood into the pulmonary trunk and the left ventricle, which has a thicker wall, and pumps blood into the aorta.
4. The ventricles are separated internally by the interventricular septum.

Heart Valves

1. The heart valves ensure one-way flow of blood.
2. The tricuspid valve (three cusps) separates the right atrium and right ventricle, and the bicuspid valve (two cusps) separates the left atrium and left ventricle.
3. The papillary muscles attach by the chordae tendineae to the cusps of the tricuspid and bicuspid valves and adjust tension on the valves.
4. The aorta and pulmonary trunk are separated from the ventricles by the semilunar valves.
5. The skeleton of the heart is a plate of fibrous connective tissue that separates the atria from the ventricles, acts as an electrical barrier between the atria and ventricles, and supports the valves of the heart.

Route of Blood Flow Through the Heart

1. The left and right sides of the heart can be considered separate pumps.
2. Blood flows from the systemic vessels to the right atrium and from the right atrium to the right ventricle. From the right ventricle blood flows to the pulmonary trunk and from the pulmonary trunk

to the lungs. From the lungs blood flows through the pulmonary veins to the left atrium, and from the left atrium blood flows to the left ventricle. From the left ventricle blood flows into the aorta and then through the systemic vessels.

Blood Supply to the Heart

1. The left and right coronary arteries originate from the base of the aorta and supply the heart.
2. The left coronary artery has three major branches: the anterior interventricular, the circumflex, and the left marginal arteries.
3. The right coronary artery has two major branches: the posterior interventricular and the right marginal arteries.
4. Blood returns from heart tissue through cardiac veins to the coronary sinus and into the right atrium. Small cardiac veins also return blood directly to the right atrium.

Histology of the Heart   (p. 331)
Heart Wall

The heart wall consists of the outer epicardium, the middle myocardium, and the inner endocardium.

Cardiac Muscle

1. Cardiac muscle is striated and depends on ATP for energy. It depends on aerobic metabolism.
2. Cardiac muscle cells are joined by intercalated disks that allow action potentials to be propagated throughout the heart.

Electrical Activity of the Heart   (p. 332)
Action Potentials in Cardiac Muscle

1. Action potentials in cardiac muscle are prolonged compared with those in skeletal muscle and have a depolarization phase, an early repolarization phase, a plateau phase, and a final repolarization phase.
2. The depolarization is due mainly to opening of the voltage-gated Na^+ channels, and the plateau phase is due to opened voltage-gated Ca^{2+} channels. Repolarization at the end of the plateau phase is due to the opening of K^+ channels for a brief period.
3. The prolonged action potential in cardiac muscle ensures that contraction and relaxation occurs and prevents tetany in cardiac muscle.
4. The SA node located in the upper wall of the right atrium is the normal pacemaker of the heart and cells of the SA node have more voltage-gated Ca^{2+} channels than other areas of the heart.

Conduction System of the Heart

1. The conduction system of the heart is made up of specialized cardiac muscle cells.
2. The SA node produces action potentials that are propagated over the atria to the AV node.
3. The AV node and atrioventricular bundle conduct action potentials to the ventricles.
4. The right and left bundle branches conduct action potentials from the atrioventricular bundle through Purkinje fibers to the ventricular muscle.
5. An ectopic beat results from an action potential that originates in an area of the heart other than the SA node.

Electrocardiogram

1. The ECG is a record of electrical events within the heart.
2. The ECG can be used to detect abnormal heart rates or rhythms, conduction pathways, hypertrophy or atrophy of the heart, and the approximate location of damaged cardiac muscle.
3. The normal ECG consists of a P wave (atrial depolarization), a QRS complex (ventricular depolarization), and a T wave (ventricular repolarization).
4. Atrial contraction occurs during the PQ interval, and the ventricles contract and relax during the QT interval.

Cardiac Cycle   (p. 337)

1. Atrial systole is contraction of the atria, and ventricular systole is contraction of the ventricles. Atrial diastole is relaxation of the atria, and ventricular diastole is relaxation of the ventricles.
2. During ventricular systole, the AV valves close, pressure increases in the ventricles, the semilunar valves are forced to open, and blood flows into the aorta and pulmonary trunk.
3. At the beginning of ventricular diastole, pressure in the ventricles decreases. The semilunar valves close to prevent backflow of blood from the aorta and pulmonary trunk into the ventricles.
4. When the pressure in the ventricles is low enough, the AV valves open and blood flows from the atria into the ventricles.
5. During atrial systole, the atria contract and complete filling of the ventricles.

Heart Sounds   (p. 340)

1. The first heart sound results from closure of the AV valves. The second heart sound results from closure of the semilunar valves.
2. Abnormal heart sounds are called murmurs. They can result from incompetent (leaky) valves or stenosed (narrowed) valves.

Regulation of Heart Function   (p. 343)

Cardiac output (volume of blood pumped per ventricle per minute) is equal to the stroke volume (volume of blood ejected per beat) times the heart rate (beats per minute).

Intrinsic Regulation of the Heart

1. Intrinsic regulation refers to regulation that is contained in the heart.
2. As venous return to the heart increases, the heart wall is stretched, and the increased stretch of the ventricular walls is called preload.
3. An increase in preload causes the cardiac output to increase because stroke volume increases (Starling's law of the heart).
4. Afterload is the pressure against which the ventricles must pump blood.

Extrinsic Regulation of the Heart

1. Extrinsic regulation refers to nervous and hormonal mechanisms.
2. Sympathetic stimulation increases stroke volume and heart rate; parasympathetic stimulation decreases heart rate.
3. The baroreceptor reflex detects changes in blood pressure and causes a decrease in heart rate and stroke volume in response to a sudden increase in blood pressure or an increase in heart rate and stroke volume in response to a sudden decrease in blood pressure.
4. Emotions influence heart function by increasing sympathetic stimulation of the heart in response to exercise, excitement, anxiety, or anger and by increasing parasympathetic stimulation in response to depression.
5. Alterations in body fluid levels of carbon dioxide, pH, and ion concentrations, as well as changes in body temperature, influence heart function.

Effects of Aging on the Heart   (p. 350)

The following age-related changes are common
1. By age 70 cardiac output often decreases by one-third.
2. Hypertrophy of the left ventricle can cause pulmonary edema.
3. Decrease in the maximum heart rate by 30–60% by age 85 leads to a decrease in cardiac output.
4. Aortic semilunar valves can become stenotic or incompetent.
5. Coronary artery disease and congestive heart failure can develop.
6. Aerobic exercise improves the functional capacity of the heart at all ages.

R E V I E W   A N D   C O M P R E H E N S I O N

1. Describe the size and location of the heart, including its base and apex.

2. Describe the structure and function of the pericardium.

3. What chambers make up the left and right side of the heart? What are their functions?

4. Describe the structure and location of the tricuspid, bicuspid, and semilunar valves. What is the function of these valves?

5. What are the functions of the atria and ventricles?

6. Starting in the right atrium, describe the flow of blood through the heart.

7. Describe the vessels that supply blood to the cardiac muscle.

8. Define coronary thrombosis and infarct. How do atherosclerotic lesions affect the heart?

9. Describe the three layers of the heart. Which of the three layers is most important in causing contractions of the heart?

10. Describe the structure of cardiac muscle cells, including the structure and function of intercalated disks.

11. Describe the events that result in an action potential in cardiac muscle.

12. Explain how cardiac muscle cells in the SA node produce action potentials spontaneously and why the SA node is the pacemaker of the heart.

13. What is the function of the conduction system of the heart? Starting with the SA node, describe the route taken by an action potential as it goes through the conduction system of the heart.

14. Explain the electrical events that generate each portion of the electrocardiogram. How do they relate to contraction events?

15. What contraction and relaxation events occur during the P-Q interval and the Q-T interval of the electrocardiogram?

16. Define cardiac cycle, systole, and diastole.

17. Describe blood flow and the opening and closing of heart valves during the cardiac cycle.

18. Describe the pressure changes that occur in the left atrium, left ventricle, and aorta during ventricular systole and diastole (see figure 12.18).

19. What events cause the first and second heart sounds?

20. Define murmur. Describe how either an incompetent or a stenosed valve can cause a murmur.

21. Define cardiac output, stroke volume, and heart rate.

22. What is Starling's law of the heart? What effect does an increase or a decrease in venous return have on cardiac output?

23. Describe the effect of parasympathetic and sympathetic stimulation on heart rate and stroke volume.

24. How does the nervous system detect and respond to the following:
    a. A decrease in blood pressure
    b. An increase in blood pressure

25. What is the effect of epinephrine on the heart rate and stroke volume?

26. Explain how emotions affect heart function.

27. What effects do the following have on cardiac output:
    a. Decrease in blood pH
    b. Increase in blood carbon dioxide

28. How do changes in body temperature influence the heart rate?

29. List the common age-related heart diseases that develop in elderly people.

C R I T I C A L   T H I N K I N G

1. A friend tells you that her son had an ECG, and it revealed that he has a slight heart murmur. Should you be convinced that he has a heart murmur? Explain.

2. Predict the effect on Starling's law of the heart if the parasympathetic (vagus) nerves to the heart are cut.

3. Predict the effect on heart rate if the sensory nerve fibers from the baroreceptors are cut.

4. An experiment is performed on a dog in which the arterial blood pressure in the aorta is monitored before and after the common carotid arteries are clamped (at time A). Explain the change in arterial blood pressure (*Hint:* Baroreceptors are located in the internal carotid arteries, which are superior to the site of clamping of the common carotid arteries).

5. Predict the consequences on the heart if a person took a large dose of a drug that blocks Ca^{2+} channels.

6. What happens to cardiac output following the ingestion of a large amount of fluid?

7. At rest, the cardiac output of athletes and nonathletes can be equal, but the heart rate of athletes is lower than that of nonathletes. At maximum exertion, the maximum heart rate of athletes and nonathletes can be equal, but the cardiac output of athletes is greater than that of nonathletes. Explain.

8. Explain why the walls of the ventricles are thicker than those of the atria.

9. Predict the effect of having an incompetent aortic semilunar valve on ventricular and aortic pressure during ventricular systole and diastole.

Answers in Appendix D

Visit this textbook's website at www.mhhe.com/seeleyess6 for practice quizzes, animations, interactive learning exercises, and other study tools.

Blood Vessels and Circulation

Chapter Outline and Objectives

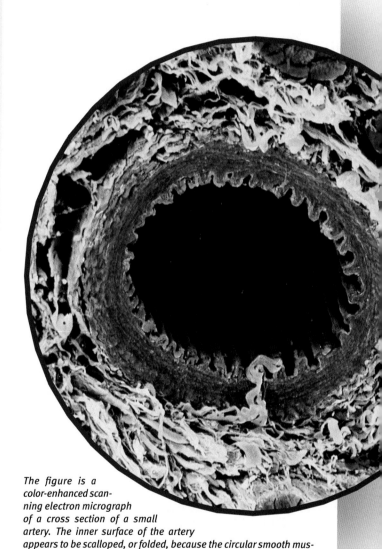

The figure is a color-enhanced scanning electron micrograph of a cross section of a small artery. The inner surface of the artery appears to be scalloped, or folded, because the circular smooth muscle cells in the wall of the artery partially contract when tissues are prepared for microscopic examination. Arteries carry blood from the heart to all tissues of the body. The tissues of the body cannot survive without being able to exchange waste products for nutrients with blood.

Complex urban water systems are simple in their design compared with the blood vessels of the body. The blood vessels carry blood to within two or three cell diameters of nearly all of the trillions of cells that make up the body. Blood flow through them is regulated so that cells are provided with adequate nutrients and waste products are removed. Blood vessels remain functional, in most cases, in excess of 70 years, and when they are damaged, they repair themselves.

Blood vessels outside of the heart are divided into two classes: (1) the **systemic vessels,** which transport blood through all parts of the body from the left ventricle and back to the right atrium, and (2) the **pulmonary vessels,** which transport blood from the right ventricle through the lungs and back to the left atrium (see chapter 12 and figure 12.2). The systemic vessels and the pulmonary vessels together constitute the **peripheral circulation.**

Functions of the Peripheral Circulation

The heart provides the major force that causes blood to circulate, and the peripheral circulation functions to:

1. *Carry blood.* Blood vessels carry blood from the heart to all tissues of the body and back to the heart.
2. *Exchange nutrients, waste products, and gases.* Nutrients and oxygen diffuse from blood vessels to cells in essentially all areas of the body. Waste products and carbon dioxide diffuse from the cells, where they are produced, to blood vessels.
3. *Transport.* Hormones, components of the immune system, molecules required for coagulation, enzymes, nutrients, gases, waste products, and other substances are transported in the blood to all areas of the body.
4. *Regulate blood pressure.* The peripheral circulatory system and the heart work together to regulate blood pressure within a normal range of values.
5. *Direct blood flow.* The peripheral circulatory system directs blood to tissues when increased blood flow is required to maintain homeostasis.

General Features of Blood Vessel Structure

Arteries (ar'ter-ēz, resembling a windpipe) are blood vessels that carry blood away from the heart. Blood is pumped from the ventricles of the heart into large elastic arteries, which branch repeatedly to form progressively smaller arteries. As they become smaller, the arteries undergo a gradual transition from having walls containing more elastic tissue than smooth muscle to having walls with more smooth muscle than elastic tissue (figure 13.1). The arteries are normally classified as (1) elastic arteries, (2) muscular arteries, or (3) arterioles, although they form a continuum from the largest to the smallest branches.

Blood flows from arterioles into **capillaries** (kap'i-lār-ēz, resembling fine hair), where exchange occurs between the blood and tissue fluid. Capillaries have thinner walls. Blood flows through them more slowly, and there are far more of them than any other blood vessel type.

From the capillaries, blood flows into veins. **Veins** (vānz) are blood vessels that carry blood toward the heart. Compared with arteries, the walls of veins are thinner and contain less elastic tissue and fewer smooth muscle cells. Going from capillaries toward the heart, small-diameter veins come together to form larger-diameter veins, which are fewer in number. Veins increase in diameter and decrease in number as they project toward the heart, and their walls increase in thickness. Veins are classified as (1) venules, (2) small veins, (3) medium-sized veins, or (4) large veins (see figure 13.1).

Blood vessel walls consist of three layers, except in capillaries and venules. The relative thickness and composition of each layer varies with the type and diameter of the blood vessel. From the inner to the outer wall of the blood vessels, the layers, or **tunics** (too'niks), are (1) the tunica intima, (2) the tunica media, and (3) the tunica adventitia, or tunica externa (see figure 13.1 and figure 13.2).

The **tunica intima** (too'ni-kǎ in'ti-mǎ, a coat + *intima,* innermost) consists of an endothelium composed of simple squamous epithelial cells, a basement membrane, and a small amount of connective tissue. In muscular arteries, the tunica intima also contains a layer of thin elastic connective tissue. The **tunica media,** or middle layer, consists of smooth muscle cells arranged circularly around the blood vessel. It also contains variable amounts of elastic and collagen fibers, depending on the size and type of the vessel. In muscular arteries, there is a layer of elastic connective tissue at the outer margin of the tunica media. The **tunica adventitia** (ad-ven-tish'ǎ, a coat + *adventicius,* to come from abroad) is composed of connective tissue. It is a denser connective tissue adjacent to the tunica media that becomes loose connective tissue toward the outer portion of the blood vessel wall.

Arteries

Elastic arteries are the largest-diameter arteries and have the thickest walls (see figure 13.1*a*). A greater proportion of their walls is elastic tissue, and a smaller proportion is smooth muscle compared with other arteries. Elastic arteries are stretched when the ventricles of the heart pump blood into them. The elastic recoil of the elastic arteries prevents blood pressure from falling rapidly and maintains blood flow while the ventricles are relaxed.

The **muscular arteries** include medium-sized and small-diameter arteries. The walls of medium-sized arteries are relatively thick compared with their diameter. Most of the thickness of the wall results from smooth muscle cells of the tunica media (see figure 13.1*b*). Medium-sized arteries are frequently called **distributing arteries** because the smooth

(a) **Elastic Arteries.** The tunica media is mostly elastic connective tissue. Elastic arteries recoil when stretched, which prevents blood pressure from falling rapidly.

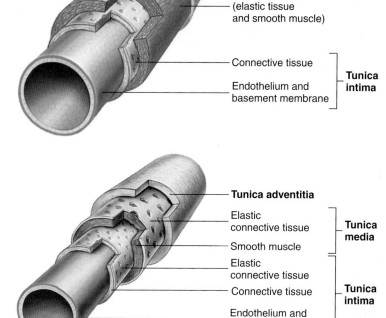

Tunica adventitia

Tunica media (elastic tissue and smooth muscle)

Connective tissue ⎤
Endothelium and basement membrane ⎦ **Tunica intima**

(b) **Muscular Arteries.** The tunica media is a thick layer of smooth muscle. Muscular arteries regulate blood flow to different regions of the body.

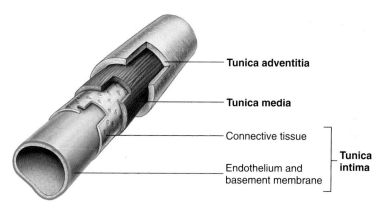

Tunica adventitia

Elastic connective tissue ⎤ **Tunica media**
Smooth muscle ⎦

Elastic connective tissue ⎤
Connective tissue ⎥ **Tunica intima**
Endothelium and basement membrane ⎦

(c) **Medium and Large Veins.** All three tunics are present. The tunica media is thin, but can regulate vessel diameter because blood pressure in the venous system is low. The predominant layer is the tunica adventitia.

Tunica adventitia

Tunica media

Connective tissue ⎤
Endothelium and basement membrane ⎦ **Tunica intima**

(d) Folds in the endothelium form the valves of veins, which allow blood to flow toward the heart but not in the opposite direction.

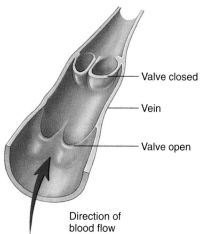

Valve closed

Vein

Valve open

Direction of blood flow

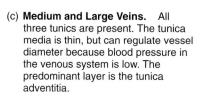

Figure 13.1 Blood Vessel Structure

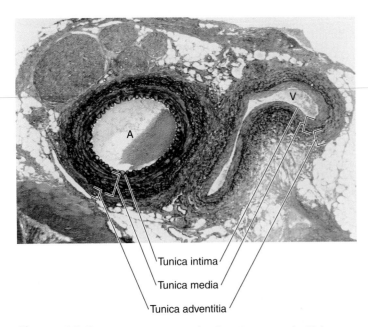

Figure 13.2 Photomicrograph of an Artery and a Vein

The typical structure of a medium-sized artery (A) and vein (V). Note that the artery has a thicker wall than the vein. The tunica intima, tunica media, and tunica adventitia make up the walls of the blood vessels. The predominant layer in the wall of the artery is the tunica media with its circular layers of smooth muscle. The predominant layer in the wall of the vein is the tunica adventitia, and the tunica media is thinner.

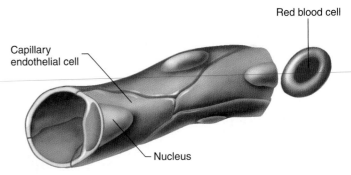

Figure 13.3 Capillary

The capillary wall is a thin layer of simple squamous epithelium, which facilitates exchange of gases, nutrients, and waste products between the blood and tissues.

Blood flows from arterioles into capillaries, which branch to form networks (figure 13.4). Red blood cells flow through most capillaries in single file and are frequently folded as they pass through the smaller diameter capillaries. Blood flow through capillaries is regulated by smooth muscle cells called **precapillary sphincters** located at the origin of the branches. As blood flows through capillaries, blood gives up oxygen and nutrients to the tissue spaces and takes up carbon dioxide and other by-products of metabolism. Capillary networks are more numerous and more extensive in the lungs and in highly metabolic tissues, such as the liver, kidneys, skeletal muscle, and cardiac muscle, than in other tissue types.

muscle tissue enables these vessels to control blood flow to different regions of the body. Contraction of the smooth muscle in blood vessels, which is called **vasoconstriction** (vā′sō-kon-strik′shŭn), decreases blood vessel diameter and blood flow. Relaxation of the smooth muscle in blood vessels, which is called **vasodilation** (vā′sō-dī-lā′shŭn), increases blood vessel diameter and blood flow.

Medium-sized arteries supply blood to small arteries. **Small arteries** have about the same structure as the medium-sized arteries, except that small arteries have a smaller diameter and their walls are thinner. The smallest of the small arteries have only three or four layers of smooth muscle in their walls.

Arterioles (ar′ter-ē′ōlz) transport blood from small arteries to capillaries and are the smallest arteries in which the three tunics can be identified. The tunica media consists of only one or two layers of circular smooth muscle cells. Small arteries and arterioles are adapted for vasodilation and vasoconstriction.

Capillaries

Capillary walls consist of **endothelium** (en-dō-thē′lē-ŭm, *endon* within + *thēlē*, nipple), which is a layer of simple squamous epithelium surrounded by a delicate loose connective tissue. The thin walls of capillaries facilitate diffusion between the capillaries and surrounding cells. Each capillary is 0.5–1 millimeter (mm) long. Capillaries branch without changing their diameter, which is approximately the same as the diameter of a red blood cell (7.5 μm) (figure 13.3).

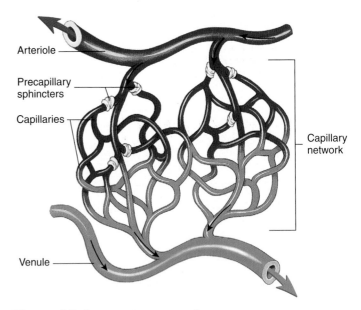

Figure 13.4 Capillary Network

An arteriole giving rise to a capillary network. The network forms numerous branches. Blood flows from capillaries into venules. Smooth muscle cells, called precapillary sphincters, regulate blood flow through the capillaries. Blood flow decreases when the precapillary sphincters constrict and increases when they dilate.

Veins

Blood flows from capillaries into venules and from venules into small veins. **Venules** (ven'oolz) are tubes with a diameter slightly larger than that of capillaries and are composed of endothelium resting on a delicate connective tissue layer. The structure of venules, except for their diameter, is very similar to that of capillaries. **Small veins** are slightly larger in diameter than venules. All three tunics are present in small veins. The tunica media contains a continuous layer of smooth muscle cells, and the connective tissue of the tunica adventitia surrounds the tunica media.

Medium-sized veins collect blood from small veins and deliver it to large veins. The three thin but distinctive tunics make up the wall of the medium-sized and large veins. The tunica media contains some circular smooth muscle and sparsely scattered elastic fibers. The predominant layer is the outer tunica adventitia, which consists primarily of dense collagen fibers (see figures 13.1c and 13.2). Consequently, veins are more distensible than arteries. The connective tissue of the tunica adventitia determines the degree to which they can distend.

Veins having diameters greater than 2 mm contain **valves,** which allow blood to flow toward the heart but not in the opposite direction. Each valve consists of folds in the tunica intima that form two flaps, which are shaped like and function like the semilunar valves of the heart. There are many valves in medium-sized veins (see figure 13.1d). There are more valves in veins of the lower limbs than in veins of the upper limbs. This prevents the flow of blood toward the feet in response to the pull of gravity.

Varicose Veins

Varicose (văr'ĭ-kōs) **veins** (dilated veins) result when the veins of the lower limbs become so dilated that the cusps of the valves no longer overlap to prevent the backflow of blood. As a consequence, venous pressure is greater than normal in the veins of the lower limbs and can result in edema. Some people have a genetic tendency for the development of varicose veins. The development of varicose veins is encouraged by conditions that increase the pressure in veins, causing them to stretch. Examples include standing in place for prolonged periods and pregnancy. Standing in place allows the pressure of the blood to stretch the veins, and pregnancy allows compression of the veins in the pelvis by the enlarged uterus, resulting in increased venous pressure in the veins that drain the lower limbs. Blood in the veins can become so stagnant that the blood clots. The clots, called **thromboses** (throm-bō'sēz), can result in inflammation of the veins, a condition called **phlebitis** (fle-bī'tis, *phlebo*, vein). If the condition becomes severe enough, the blocked veins can prevent blood flow through capillaries that are drained by the veins. The lack of blood flow can lead to tissue death, or **necrosis** (nĕ-krō'sis, death), and infection of the tissue with anaerobic bacteria, a condition called **gangrene** (gang'grēn, an eating sore). In addition, fragments of the clots can dislodge and travel through the veins to the lungs where they can cause severe damage. Fragments of thromboses that dislodge and float in the blood are called **emboli** (em'bō-lī, a plug).

Blood Vessels of the Pulmonary Circulation

Blood from the right ventricle is pumped into the **pulmonary** (pŭl'mō-nār-ē, relating to the lungs) **trunk** (figures 13.5 and 13.6). This short vessel branches into the **right** and **left pulmonary arteries,** which extend to the right and left lungs, respectively. Poorly oxygenated blood is carried by these arteries to the pulmonary capillaries in the lungs, where oxygen is taken up by the blood and carbon dioxide is released. Blood rich in oxygen flows from the lungs to the left atrium. Four **pulmonary veins** (two from each lung) exit the lungs and carry the oxygenated blood to the left atrium.

Blood Vessels of the Systemic Circulation: Arteries

The **systemic circulation** is the flow of blood through the system of blood vessels that carries blood from the left ventricle of the heart to the tissues of the body and back to the right atrium. Oxygenated blood from the pulmonary veins passes from the left atrium into the left ventricle and from the left ventricle into the aorta. Blood is distributed from the aorta to all portions of the body (see figures 13.5 and 13.6).

Aorta

All arteries of the systemic circulation branch directly or indirectly from the **aorta** (ā-ōr'tă, *aortē*, to lift up). The aorta is usually considered in three parts: the ascending aorta, the aortic arch, and the descending aorta. The descending aorta is further divided into the thoracic aorta and the abdominal aorta (figure 13.7a).

The part of the aorta that passes superiorly from the left ventricle is called the **ascending aorta.** The right and left **coronary arteries** arise from the base of the ascending aorta and supply blood to the heart (see chapter 12).

The aorta arches posteriorly and to the left as the **aortic arch.** Three major arteries, which carry blood to the head and upper limbs, originate from the aortic arch. They are the brachiocephalic artery, the left common carotid artery, and the left subclavian artery (figure 13.7b).

The **descending aorta** is the longest part of the aorta. It extends through the thorax and abdomen to the upper margin of the pelvis. The part of the descending aorta that extends through the thorax to the diaphragm is called the **thoracic** (thō-ras'ik) **aorta** (see figure 13.7b). The part of the descending aorta that extends from the diaphragm to the point at which it divides into the two **common iliac** (il'ē-ak, relating to the flank area) **arteries** is called the **abdominal** (ab-dom'i-năl) **aorta** (see figure 13.7a and figure 13.7c).

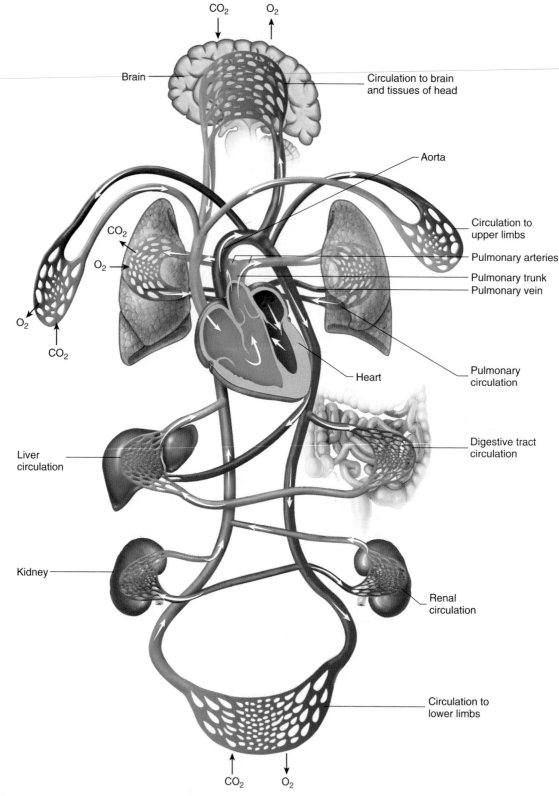

Figure 13.5   Blood Flow Through the Circulatory System

Blood is returned from the body to the right atrium. After passing from the right atrium to the right ventricle, blood is pumped into the pulmonary trunk. The pulmonary trunk divides into the right and left pulmonary arteries, which carry oxygen-poor blood to the lungs. In the lung capillaries, carbon dioxide is given off, and oxygen is picked up by the blood. Blood, now rich in oxygen, flows from each lung to the left atrium. Blood then passes from the left atrium to the left ventricle. The left ventricle then pumps the blood into the aorta, which distributes the blood through its branches to all of the body.

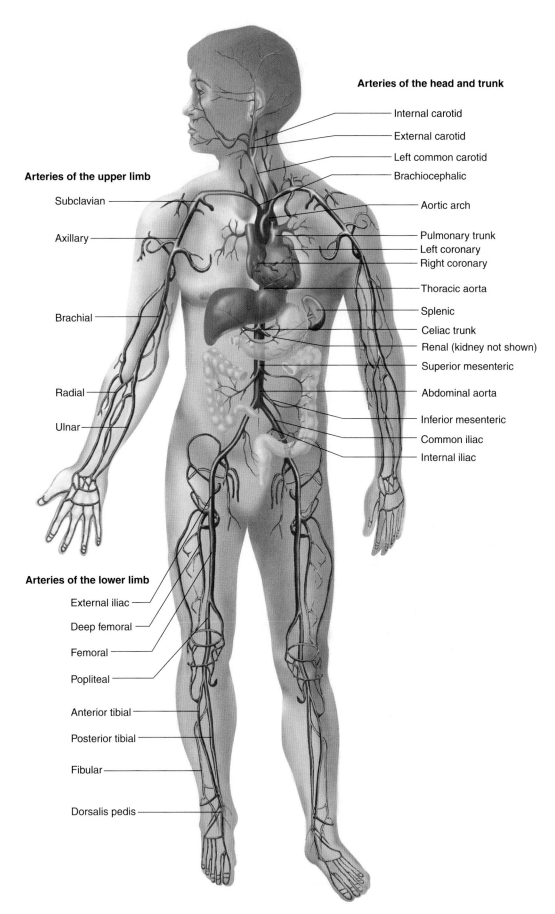

Arteries of the head and trunk

- Internal carotid
- External carotid
- Left common carotid
- Brachiocephalic
- Aortic arch
- Pulmonary trunk
- Left coronary
- Right coronary
- Thoracic aorta
- Splenic
- Celiac trunk
- Renal (kidney not shown)
- Superior mesenteric
- Abdominal aorta
- Inferior mesenteric
- Common iliac
- Internal iliac

Arteries of the upper limb

- Subclavian
- Axillary
- Brachial
- Radial
- Ulnar

Arteries of the lower limb

- External iliac
- Deep femoral
- Femoral
- Popliteal
- Anterior tibial
- Posterior tibial
- Fibular
- Dorsalis pedis

Figure 13.6 The Major Arteries

The major arteries that carry blood from the left ventricle of the heart to the tissues of the body.

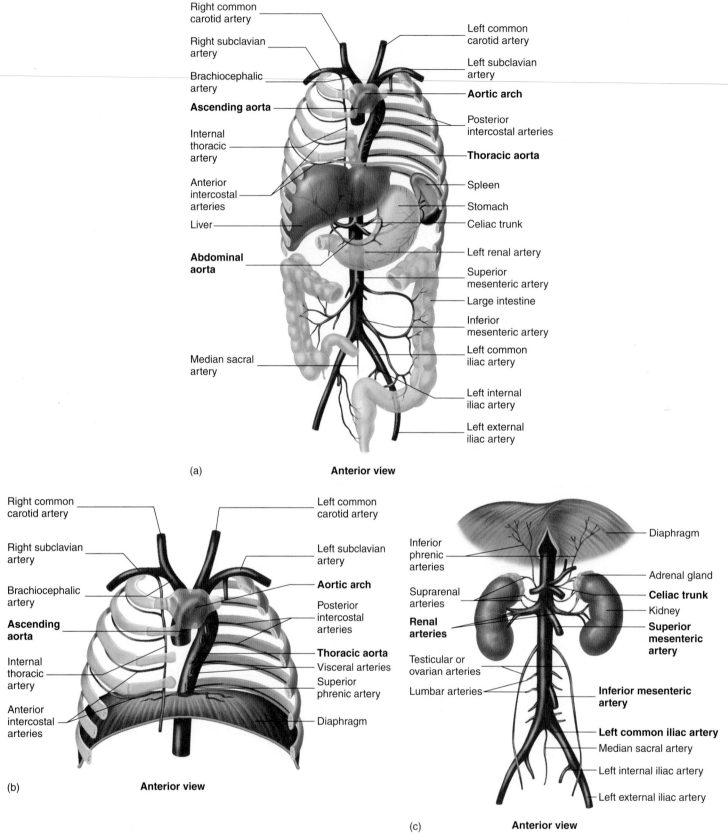

Figure 13.7 Branches of the Aorta

(a) The aorta is considered in three portions: the ascending aorta, the aortic arch, and the descending aorta. The descending aorta consists of the thoracic and abdominal aorta. (b) The thoracic aorta. (c) The abdominal aorta.

Aneurysms

An **arterial aneurysm** (an′ū-rizm, a dilation) is a localized dilation of an artery that usually develops in response to trauma or a congenital (existing at birth) weakness of the artery wall. Ruptures of aneurysms are serious. Rupture of a large aneurysm of the aorta is almost always fatal, and rupture of an aneurysm of an artery in the brain causes massive damage to brain tissue and even death. If aneurysms are discovered they often can be surgically corrected. For example, large aneurysms of the aorta that leak blood slowly can often be surgically repaired.

Arteries of the Head and Neck

The first vessel to branch from the aortic arch is the **brachiocephalic** (brā′kē-ō-se-fal′ik, vessel to the arm and head) **artery.** It is a short artery, and it branches at the level of the clavicle to form the **right common carotid** (ka-rot′id, to put to sleep) **artery,** which transports blood to the right side of the head and neck, and the **right subclavian** (sŭb-klā′vē-an, beneath the clavicle) **artery,** which transports blood to the right upper limb (see figures 13.7*b*, 13.8, and 13.9).

There is no brachiocephalic artery on the left side of the body. Instead, both the left common carotid and the left subclavian arteries branch directly off the aortic arch (see figures 13.6, 13.7*a* and *b*, and 13.9). They are the second and third branches of the aortic arch. The **left common carotid artery** transports blood to the left side of the head and neck, and the **left subclavian artery** transports blood to the left upper limb.

The common carotid arteries extend superiorly along each side of the neck to the angle of the mandible, where they branch into **internal** and **external carotid arteries** (see figures 13.8 and 13.9). The base of each internal carotid artery is slightly dilated to form a **carotid sinus,** which contains structures important in monitoring blood pressure (baroreceptors). The external carotid arteries have several branches that supply the structures of the neck, face, nose, and mouth. The internal carotid arteries pass through the carotid canals and contribute to the **cerebral arterial circle** (circle of Willis) at the base of the brain. The blood

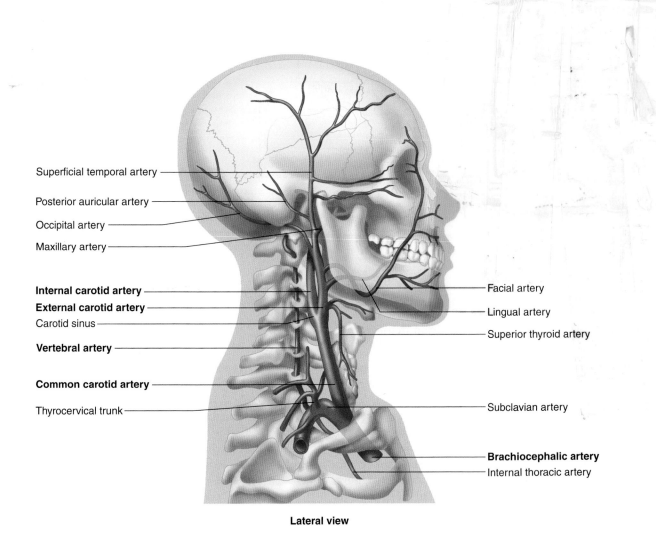

Superficial temporal artery

Posterior auricular artery

Occipital artery

Maxillary artery

Internal carotid artery

External carotid artery

Carotid sinus

Vertebral artery

Common carotid artery

Thyrocervical trunk

Facial artery

Lingual artery

Superior thyroid artery

Subclavian artery

Brachiocephalic artery

Internal thoracic artery

Lateral view

Figure 13.8   Arteries of the Head and Neck

The brachiocephalic artery, the right common carotid artery, and the right vertebral artery are major arteries that supply the head and neck. The right common carotid artery branches from the brachiocephalic artery, and the vertebral artery branches from the subclavian artery.

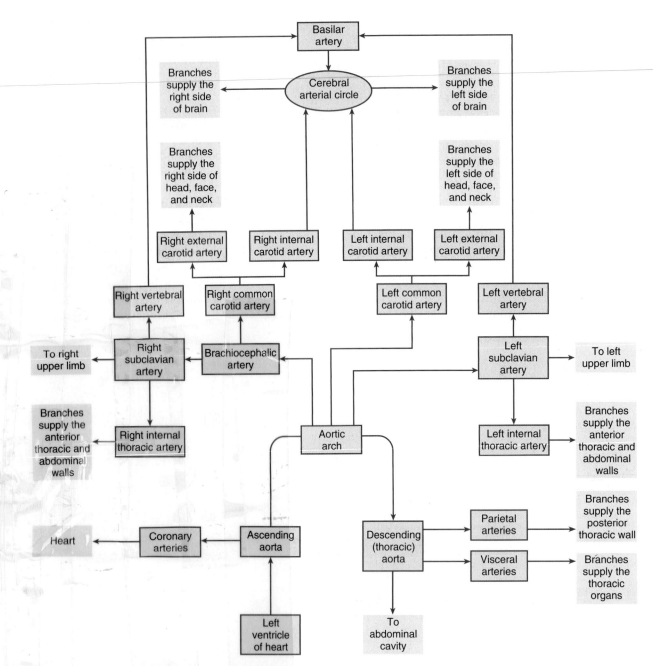

Figure 13.9 Major Arteries of the Head and Thorax

vessels of the cerebral arterial circle are illustrated and named in figure 13.10. The blood vessels that supply blood to most of the brain branch from the cerebral arterial circle.

Some of the blood to the brain is supplied by the **vertebral** (ver′tĕ-brăl) **arteries,** which branch from the subclavian arteries (see figures 13.8 and 13.9) and pass to the head through the transverse foramina of the cervical vertebrae. The vertebral arteries then pass into the cranial cavity through the foramen magnum. Branches of the vertebral arteries supply blood to the spinal cord, as well as to the vertebrae, muscles, and ligaments in the neck.

Within the cranial cavity, the vertebral arteries unite to form a single **basilar** (bas′i-lăr, relating to the base of the brain) **artery** located along the anterior, inferior surface of the brainstem (see figures 13.9 and 13.10). The basilar artery gives

off branches that supply blood to the pons, cerebellum, and midbrain. It also forms right and left branches that contribute to the cerebral arterial circle. Most of the blood supply to the brain is through the internal carotid arteries; however, not enough blood is supplied to the brain to maintain life if either the carotid arteries or the vertebral arteries are blocked.

P R E D I C T   1

The term "carotid" means to put to sleep, implying that, if the carotid arteries are blocked for several seconds, the patient can lose consciousness. Interruption of the blood supply for a few minutes can result in permanent brain damage. What is the physiological significance of atherosclerosis in the carotid arteries?

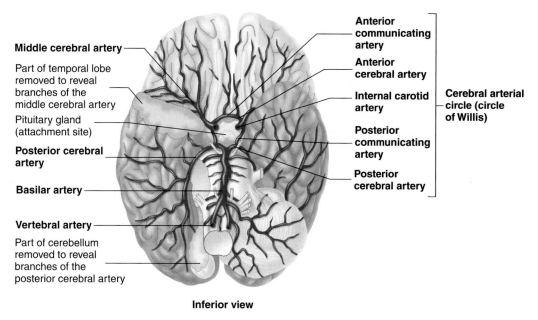

Inferior view

Figure 13.10 The Cerebral Arterial Circle (Circle of Willis)

The blood supply to the brain is carried by the internal carotid and vertebral arteries. The vertebral arteries join to form the basilar artery. Branches of the internal carotid arteries and basilar artery supply blood to the brain and complete a circle of arteries around the pituitary gland and the base of the brain called the cerebral arterial circle (circle of Willis).

Arteries of the Upper Limbs

The arteries of the upper limbs are named differently as they pass into different body regions, even though no major branching occurs. The subclavian artery, located deep to the clavicle, becomes the **axillary** (ak′sil-ār-ē, refers to the axillary area) **artery,** in the axilla (armpit). The **brachial** (brā′kē-ăl, relating to the arm) **artery,** located in the arm, is a continuation of the axillary artery (see figures 13.6 and 13.11). Blood pressure measurements are normally taken from the brachial artery. The brachial artery branches at the elbow to form the **ulnar** (ul′năr) **artery** and the **radial** (rā′dē-ăl) **artery,** which supply blood to the forearm and hand. The radial artery is the artery most commonly used for taking a pulse. The pulse can be detected conveniently on the thumb (radial) side of the anterior surface of the wrist.

The Thoracic Aorta and Its Branches

The branches of the thoracic aorta can be divided into two groups: the **visceral** (vis′er-ăl, refers to internal organs) **arteries** supply the thoracic organs, and the **parietal arteries** (pă-rī′ĕ-tăl, wall) supply the thoracic wall. The visceral branches of the thoracic aorta supply the esophagus, trachea, parietal pericardium, and part of the lung. The major parietal arteries are the **posterior intercostal** (in-ter-kos′tăl, *inter* + *costa*, between the ribs) **arteries,** which arise from the thoracic aorta and extend between the ribs (see figure 13.7). They supply intercostal muscles, the vertebrae, the spinal cord, and deep muscles of the back (see figure 13.7a and b). The **superior phrenic** (fren′ik, diaphragm) **arteries** supply the diaphragm.

The **internal thoracic arteries** are branches of the subclavian arteries. They descend along the internal surface of the anterior thoracic wall and give rise to branches called the

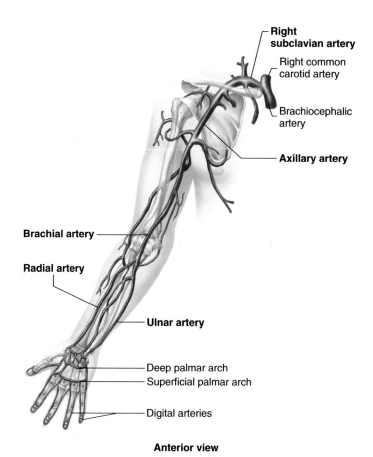

Anterior view

Figure 13.11 The Arteries of the Upper Limb

Anterior view of the arteries of the right upper limb and their branches.

anterior intercostal arteries, which extend between the ribs to supply the anterior chest wall (see figure 13.7*a* and *b*).

The Abdominal Aorta and Its Branches

The branches of the abdominal aorta, like those of the thoracic aorta, can be divided into visceral and parietal groups. The visceral arteries are divided into paired and unpaired branches. There are three major unpaired branches: the **celiac** (sē′lē-ak, belly) **trunk, superior mesenteric** (mez-en-ter′ik, relating to membranes attached to the intestine) **artery,** and the **inferior mesenteric artery.** The celiac trunk supplies blood to the stomach, pancreas, spleen, upper duodenum, and liver. The superior mesenteric artery supplies blood to the small intestine and the upper portion of the large intestine, and the inferior mesenteric artery supplies blood to the remainder of the large intestine (see figure 13.7*a* and *c* and figure 13.12).

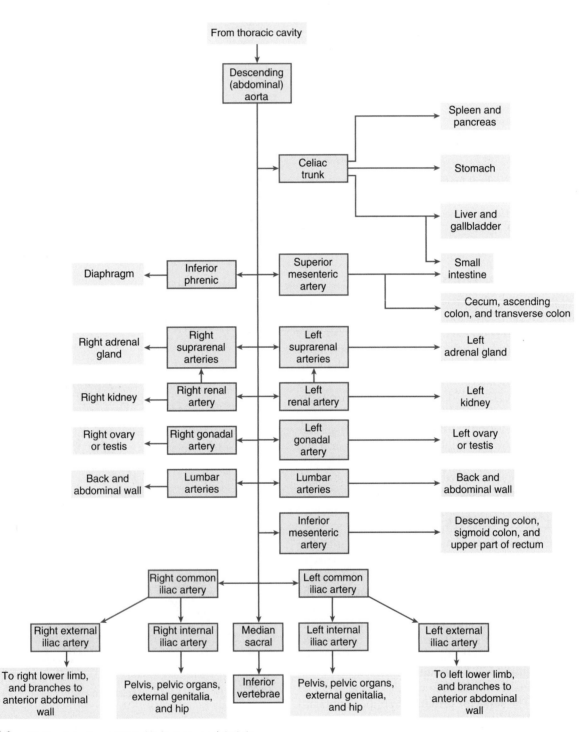

Figure 13.12   **Major Arteries of the Abdomen and Pelvis**

Visceral branches include those that are unpaired (celiac trunk, superior mesenteric, inferior mesenteric) and those that are paired (renal, suprarenal, testicular or ovarian). Parietal branches include inferior phrenic, lumbar, and median sacral.

There are three paired visceral branches of the abdominal aorta. The **renal** (rē′nal, kidney) **arteries** supply the kidneys and the **suprarenal** (sū′pră-rē′năl, superior to the kidney) **arteries** supply the adrenal glands. The **testicular arteries** supply the testes in males or the **ovarian arteries** supply the ovaries in females.

The parietal branches of the abdominal aorta supply the diaphragm and abdominal wall. The **inferior phrenic arteries** supply the diaphragm, the **lumbar** (between the ribs and pelvis) **arteries** supply the lumbar vertebrae and back muscles, and the **median sacral** (area of sacrum) **artery** supplies the inferior vertebrae.

Arteries of the Pelvis

The abdominal aorta divides at the level of the fifth lumbar vertebra into two **common iliac** (relating to the ilium) **arteries.** Each common iliac artery divides to form an **external iliac artery,** which enters a lower limb, and an **internal iliac artery,** which supplies the pelvic area (see figures 13.7c and 13.12). Visceral branches of the internal iliac artery supply organs such as the urinary bladder, rectum, uterus, and vagina. Parietal branches supply blood to the walls and floor of the pelvis; the lumbar, gluteal, and proximal thigh muscles; and the external genitalia.

Arteries of the Lower Limbs

Like the arteries of the upper limbs, arteries of the lower limbs are named differently as they pass into different body regions, even though there are no major branches. The external iliac artery, in the pelvis, becomes the **femoral** (fem′ŏ-răl, relating to the thigh) **artery** in the thigh, and it becomes the **popliteal** (pop-lit′ē-ăl) **artery** in the popliteal space, which is the posterior region of the knee. The popliteal artery branches slightly inferior to the knee to give off the **anterior tibial artery** and the **posterior tibial artery.** The anterior and posterior tibial arteries give rise to arteries that supply blood to the leg and feet (figure 13.13). The anterior tibial artery becomes the **dorsalis pedis** (dōr-sāl′lis pē′dis, *pes*, foot) **artery** at the ankle. The posterior tibial artery gives rise to the **fibular,** or peroneal, artery, which supplies the lateral leg and foot.

The Femoral Triangle

The femoral triangle is found in the superior and medial area of the thigh. Its margins are formed by the inguinal ligament, the medial margin of the sartorius muscle, and the lateral margin of the adductor longus muscle (see figures 7.13a and 7.25a). The femoral artery, vein, and nerve pass though the femoral triangle. A pulse in the femoral artery can be detected in the area of the femoral triangle, and it is an area that is susceptible to serious traumatic injuries that can result in hemorrhage and nerve damage. In addition, pressure can be applied to this area to help prevent bleeding from wounds in more inferior areas of the lower limb. The femoral triangle is also an important access point for certain medical procedures.

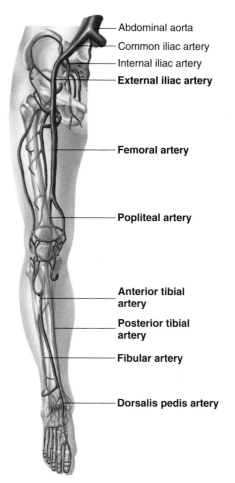

Abdominal aorta
Common iliac artery
Internal iliac artery
External iliac artery
Femoral artery
Popliteal artery
Anterior tibial artery
Posterior tibial artery
Fibular artery
Dorsalis pedis artery

Anterior view

Figure 13.13   The Arteries of the Lower Limb

Anterior view of the arteries of the right lower limb and their branches.

Blood Vessels of the Systemic Circulation: Veins

The **superior vena cava** (vē′nă kā′vă, venous cave) returns blood from the head, neck, thorax, and upper limbs to the right atrium of the heart, and the **inferior vena cava** returns blood from the abdomen, pelvis, and lower limbs to the right atrium (figure 13.14).

Veins of the Head and Neck

The two pairs of major veins that drain blood from the head and neck are the **external** and **internal jugular** (jŭg′ū-lar, neck) **veins** (see figures 13.14 and 13.15). The external jugular veins are the more superficial of the two sets, and they drain blood from the posterior head and neck. The external jugular veins empty primarily into the subclavian veins. The internal jugular veins are much larger and deeper. They drain blood from the brain and the anterior head, face, and neck. The internal jugular veins join the **subclavian veins** on each side of the body to form

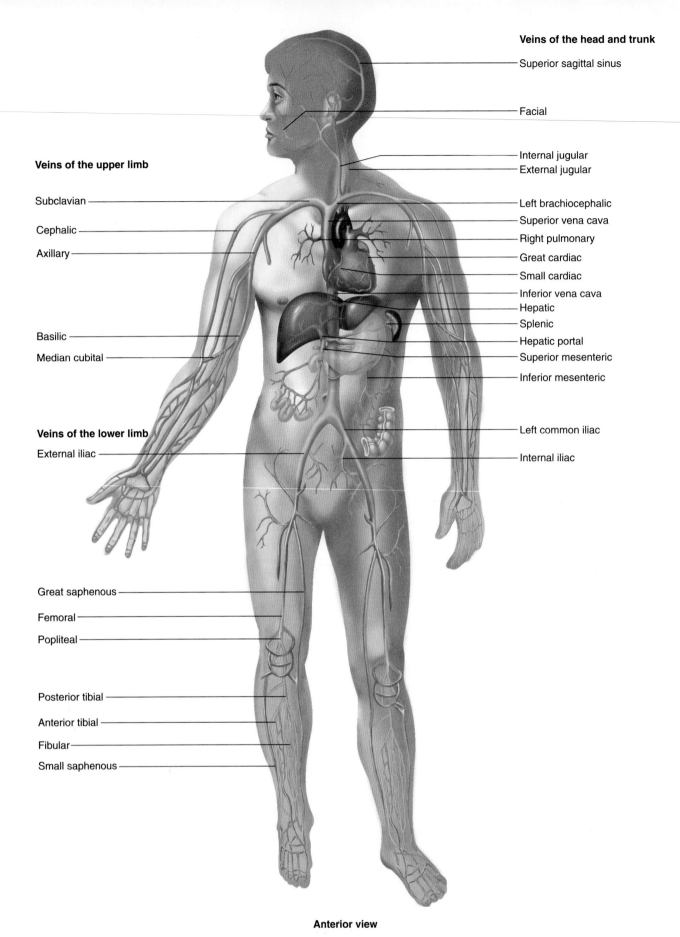

Veins of the head and trunk

Superior sagittal sinus

Facial

Internal jugular

External jugular

Left brachiocephalic

Superior vena cava

Right pulmonary

Great cardiac

Small cardiac

Inferior vena cava

Hepatic

Splenic

Hepatic portal

Superior mesenteric

Inferior mesenteric

Left common iliac

Internal iliac

Veins of the upper limb

Subclavian

Cephalic

Axillary

Basilic

Median cubital

Veins of the lower limb

External iliac

Great saphenous

Femoral

Popliteal

Posterior tibial

Anterior tibial

Fibular

Small saphenous

Anterior view

Figure 13.14   The Major Veins

The major veins carry blood from the tissues of the body and return it to the right atrium.

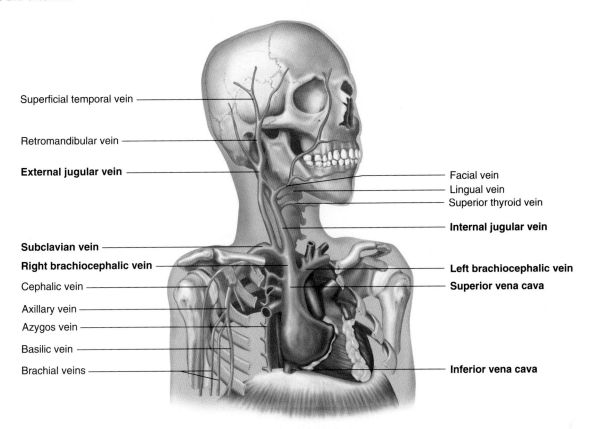

Anterior view

Figure 13.15 **Veins of the Head and Neck**

The external and internal jugular veins drain blood from the head and neck. The internal jugular veins join the subclavian veins on each side of the body to form the brachiocephalic veins. The external jugular veins drain into the subclavian veins.

the **brachiocephalic veins.** The brachiocephalic veins join to form the superior vena cava.

Veins of the Upper Limbs

The veins of the upper limbs (figure 13.16) can be divided into deep and superficial groups. The deep veins, which drain the deep structures of the upper limbs, follow the same course as the arteries and are named for the arteries they accompany. The only noteworthy deep veins are the **brachial veins,** which accompany the brachial artery and empty into the axillary vein. The superficial veins drain the superficial structures of the upper limbs and then empty into the deep veins. The **cephalic** (sĕ′fal′ik, toward the head) **vein,** which empties into the **axillary vein** and the **basilic** (ba-sil′ik, toward the base of the arm) **vein,** which becomes the axillary vein, are the major superficial veins (see figure 13.16). Many of their tributaries in the forearm and hand can be seen through the skin. The **median cubital** (kū′bi-tal, elbow) **vein** usually connects the cephalic vein or its tributaries with the basilic vein. Although this vein varies in size among people, it is usually quite prominent on the anterior surface of the upper limb at the level of the elbow, an area called the **cubital fossa,** and is often used as a site for drawing blood.

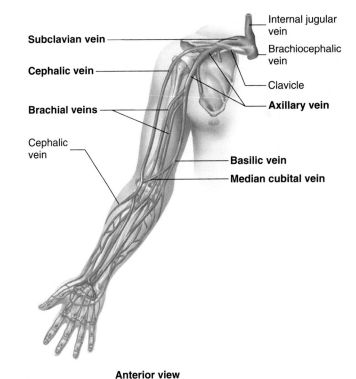

Anterior view

Figure 13.16 **Veins of the Upper Limb**

Anterior view of the major veins of the upper right limb and their branches.

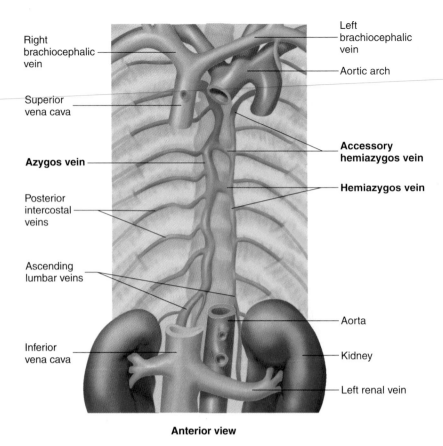

Right
brachiocephalic
vein

Left
brachiocephalic
vein

Aortic arch

Superior
vena cava

**Accessory
hemiazygos vein**

Azygos vein

Hemiazygos vein

Posterior
intercostal
veins

Ascending
lumbar veins

Aorta

Inferior
vena cava

Kidney

Left renal vein

Anterior view

Figure 13.17   Veins of the Thoracic Wall
The azygous and hemiazygos veins and their tributaries.

Veins of the Thorax

Three major veins return blood from the thorax to the superior vena cava: the **right** and **left brachiocephalic veins** and the **azygos** (az-ī′gos or az′i-gos, unpaired) **vein** (figure 13.17). Blood drains from the anterior thoracic wall by way of the **anterior intercostal veins.** These veins empty into the **internal thoracic veins,** which empty into the brachiocephalic veins. Blood from the posterior thoracic wall is collected by **posterior intercostal veins,** which drain into the azygos vein or right brachiocephalic vein, on the right and the **hemiazygos vein** or **accessory hemiazygos vein,** or left brachiocephalic vein, on the left. The hemiazygos and accessory hemiazygos veins empty into the azygos vein, which drains into the superior vena cava (see figure 13.15).

Veins of the Abdomen and Pelvis

Blood from the posterior abdominal wall drains through **ascending lumbar veins** into the azygos vein. Blood from the rest of the abdomen and from the pelvis and lower limbs returns to the heart through the inferior vena cava. The gonads (testes or ovaries), kidneys, adrenal glands, and liver are the only abdominal organs outside the pelvis from which blood drains directly into the inferior vena cava. The **internal iliac veins** drain the pelvis and join the **external iliac veins** (figure 13.18) from the lower limbs to form the **common iliac veins.** The common iliac veins combine to form the inferior vena cava (see figures 13.14 and 13.18).

Blood from the capillaries within most of the abdominal viscera, such as the stomach, intestines, pancreas, and spleen, drains through a specialized system of blood vessels to the liver. The liver is a major processing center for substances absorbed by the intestinal tract. A **portal** (pōr′tăl, *porta,* gate) **system** is a vascular system that begins and ends with capillary beds and has no pumping mechanism such as the heart between them. The **hepatic** (he-pa′tik, relating to the liver) **portal system** (see figures 13.18 and 13.19) begins with capillaries in the viscera and ends with capillaries in the liver. The major tributaries of the hepatic portal system are the **splenic** (splen′ik) **vein** and the **superior mesenteric vein.** The **inferior mesenteric vein** empties into the splenic vein. The splenic vein carries blood from the spleen and pancreas. The superior and inferior mesenteric veins carry blood from the intestines. The splenic vein and the superior mesenteric vein join to form the **hepatic portal vein,** which enters the liver.

Blood from the liver flows into **hepatic veins,** which join the inferior vena cava. Blood entering the liver through the hepatic portal vein is rich with nutrients collected from the intestines, but it may also contain a number of toxic substances harmful to the tissues of the body. Within the liver, nutrients are taken up and stored or modified so they can be used by other cells of the body. Toxic substances are converted to nontoxic substances and are removed from the blood or are carried by the blood to the kidneys (the liver is discussed more fully in chapter 16).

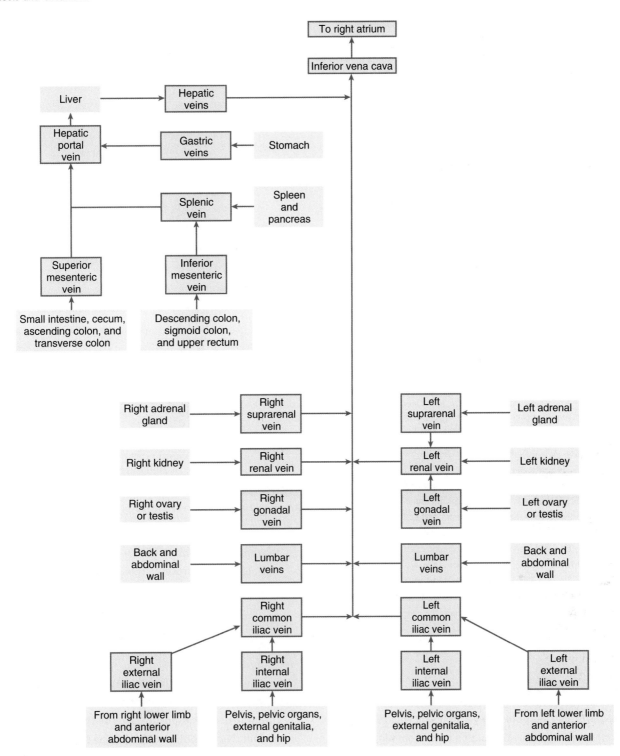

Figure 13.18 Major Veins of the Abdomen and Pelvis

Veins of the Lower Limbs

The veins of the lower limbs (figure 13.20), like those of the upper limbs, consist of deep and superficial groups. The deep veins follow the same path as the arteries and are named for the arteries they accompany. The superficial veins consist of the great and small **saphenous** (să-fē′nŭs or sa′fĕ-nŭs, visible) **veins.** The **great saphenous vein** originates over the dorsal and medial side of the foot and ascends along the medial side of the leg and thigh to empty into the femoral vein (see figure 13.20). The **small saphenous vein** begins over the lateral side of the foot and joins the **popliteal vein** which, in turn, becomes the femoral vein. The femoral vein empties into the external iliac vein.

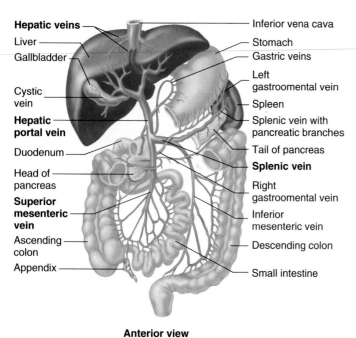

Figure 13.19   Veins of the Hepatic Portal System

The hepatic portal system begins as capillary beds in the stomach, pancreas, spleen, small intestine, and large intestine. The veins of the hepatic portal system converge on the hepatic portal vein, which carries blood to a series of capillaries in the liver. Hepatic veins carry blood from capillaries in the liver to the hepatic veins, which then enter the inferior vena cava.

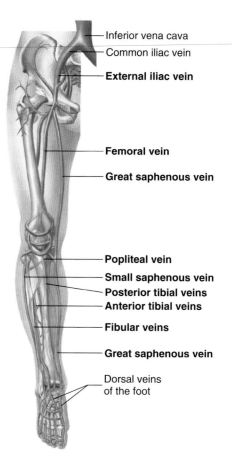

Figure 13.20   Veins of the Lower Limb

Anterior view of the major veins of the lower right limb and their branches.

Blood Vessels Used for Coronary Bypass Surgery

The great saphenous vein often is removed surgically and used in coronary bypass surgery. Portions of the saphenous vein are grafted to create a route of blood flow that bypasses blocked portions of the coronary arteries. The circulation interrupted by the removal of the saphenous vein flows through other veins of the lower limb. The internal thoracic artery is also used for coronary bypasses. The distal end of the artery is freed and attached to a coronary artery at a point that bypasses the blocked portion of the coronary artery. This technique appears to be better because the internal thoracic artery does not become blocked as fast as the saphenous vein.

A CASE IN POINT | A Venous Thrombosis

Harry Leggs is a 55-year-old college professor who teaches a night class in a small town about 50 miles from his home. As he walked to his car after teaching the class, Harry was aware of a slight pain in his right leg. When he reached home about 90 minutes later, the calf of his right leg was swollen. When he extended his knee and plantar flexed his foot, the pain in his right leg increased. Harry knew this was a serious condition, so he drove to the emergency room.

In the emergency room, a Doppler test, which monitors the flow of blood through blood vessels, was performed on his right leg. The test confirmed that a thrombus had formed in one of the deep veins of his right leg. Small proteins released from fibrinogen during clot formation cause the activation of kinins that stimulate pain receptors, (see chapter 14). Also, the blocked vein causes edema and can lead to tissue ischemia, which results in tissue damage and pain. Edema results because of the increase in venous pressure inferior to the blocked vein (see figure 13.24).

A major danger associated with a deep venous thrombosis is that portions of the thrombus will form emboli. The emboli are likely to be transported in the blood to the lungs, where they form pulmonary emboli. Pulmonary emboli are life-threatening because they can be large enough or numerous enough to block blood flow to large areas of the lungs.

Harry was admitted to the hospital for a few days. His movements were restricted and he was treated with an anticoagulant. Subsequently, he was released from the hospital and an anticoagulant was prescribed for a few weeks until the clot dissolved.

P R E D I C T ❷
If a thrombus in the posterior tibial vein gave rise to an embolus, name in order the parts of the circulatory system the embolus would pass through before lodging in a blood vessel in the lungs. Why are the lungs the most likely place the embolus will lodge?

The Physiology of Circulation

The function of the circulatory system is to maintain adequate blood flow to all tissues. An adequate blood flow is required to provide nutrients and oxygen to tissues and to remove waste products of metabolism from the tissues. Blood flows through the arterial system primarily as a result of the pressure produced by contractions of the heart ventricles.

Blood Pressure

Blood pressure is a measure of the force blood exerts against the blood vessel walls. In arteries, blood pressure values exhibit a cycle dependent on the rhythmic contractions of the heart. When the ventricles contract, blood is forced into the arteries, and the pressure reaches a maximum called the **systolic pressure.**

When the ventricles relax, blood pressure in the arteries falls to a minimum value called the **diastolic pressure.** The standard unit for measuring blood pressure is millimeters of mercury (mm Hg). If the blood pressure is 100 mm Hg, the pressure is great enough to lift a column of mercury 100 mm.

The **auscultatory** (aws-kŭl′tă-tō-rē, to listen) **method** of determining blood pressure is used under most clinical conditions (figure 13.21). A blood pressure cuff connected to a **sphygmomanometer** (sfĭg′mō-mă-nom′ĕ-ter, *sphygmic,* relating to the pulse + manometer, instrument for measuring pressure) is placed around the patient's arm, and a **stethoscope** (steth′ō-skōp, *stēthos,* chest + *skopeō,* to view) is placed over the brachial artery. The blood pressure cuff is then inflated until the brachial artery is completely blocked. Because no blood flows through the constricted area, no sounds can be heard through the stethoscope at this point. The pressure in the cuff is then gradually lowered. As soon as the pressure in the cuff declines below the systolic pressure, blood flows through the constricted area each time the left ventricle contracts. The blood flow is turbulent immediately downstream from the constricted area. This turbulence produces vibrations in the blood and surrounding tissues that can be heard through the stethoscope. These sounds are

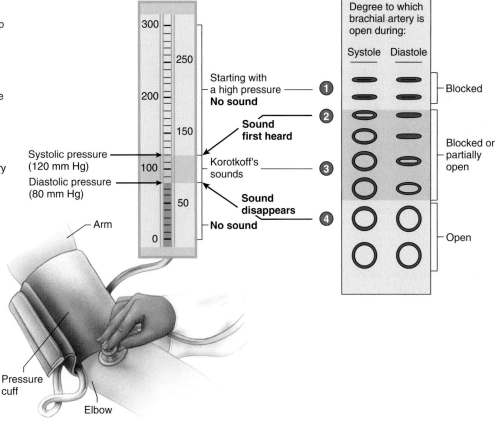

1. There is no blood flow and no sound is heard when the cuff pressure is high enough to keep the brachial artery closed.

2. **Systolic pressure** is the pressure at which a sound is first heard. When cuff pressure decreases and is no longer able to keep the brachial artery closed, blood is pushed through the partially opened brachial artery during systole producing turbulent blood flow and a sound. The brachial artery remains closed during diastole.

3. As cuff pressure continues to decrease, the brachial artery opens even more during systole. At first, the artery is closed during diastole, but as cuff pressure continues to decrease, the brachial artery partially opens during diastole. Turbulent blood flow during systole produces Korotkoff's sounds, although the pitch of the sounds change as the artery becomes more open.

4. **Diastolic pressure** is the pressure at which the sound disappears. Eventually cuff pressure decreases below the pressure in the brachial artery and it remains open during systole and diastole. Nonturbulent flow is reestablished and no sounds are heard.

Process Figure 13.21   **Blood Pressure Measurement**

called **Korotkoff** (Kō-rot′kof, Nikolai Korotkoff, Russian physician [1874–1920]) **sounds,** and the pressure at which the first Korotkoff sound is heard is the systolic pressure.

As the pressure in the blood pressure cuff is lowered still more, the Korotkoff sounds change tone and loudness. When the pressure has dropped until the brachial artery is no longer constricted and blood flow is no longer turbulent, the sound disappears completely. The pressure at which the Korotkoff sounds disappear is the diastolic pressure. The brachial artery remains open during systole and diastole, and continuous blood flow is reestablished.

The systolic pressure is the maximum pressure produced in the large arteries. It is also a good measure of the maximum pressure within the left ventricle. The diastolic pressure is close to the lowest pressure within the large arteries. During relaxation of the left ventricle, the aortic semilunar valve closes, trapping the blood that was ejected during ventricular contraction in the aorta. The pressure in the ventricles falls to 0 mm Hg during ventricular relaxation. The blood trapped in the elastic arteries is compressed by the recoil of the elastic arteries, however, and the pressure falls more slowly, reaching the diastolic pressure (see figure 12.18).

Pressure and Resistance

The values for systolic and diastolic pressure vary among healthy people, making the range of normal values quite broad. In addition, the values for blood pressure in a normal person are affected by factors such as physical activity and emotions. A standard blood pressure for a resting young adult male is 120 mm Hg for the systolic pressure and 80 mm Hg for the diastolic pressure, which is reported as 120/80.

Blood pressure falls progressively as blood flows from arteries through the capillaries and veins to about 0 mm Hg or even slightly lower by the time blood is returned to the right atrium. In addition, the pressure is damped, in that the difference between the systolic and diastolic pressures is decreased in the small-diameter vessels. By the time blood reaches the capillaries, there is no variation in blood pressure, and only a steady pressure of about 30 mm Hg remains (figure 13.22).

Hypertension

Hypertension, or **high blood pressure,** affects at least 20% of all people at some time in their lives.

Guidelines* categorize blood pressure according to the following for adults:

- *Normal:* Less than 120 mm Hg systolic and 80 mm Hg diastolic
- *Prehypertension:* From 120 mm Hg systolic and 80 mm Hg diastolic to 139 mm Hg systolic to 89 mm Hg diastolic
- *Stage 1 hypertension:* From 140 mm Hg systolic and 90 mm Hg diastolic to 159 mm Hg systolic and 99 mm Hg diastolic
- *Stage 2 hypertension:* Greater than 160 mm Hg systolic and 100 mm Hg diastolic

---

* Joint National Committee on Prevention, Detection, Evaluation, and Treatment of High Blood Pressure.

Individuals with prehypertension pressure should monitor their blood pressure for changes and consider life-style changes that can reduce blood pressure.

Hypertension requires the heart to perform a greater-than-normal amount of work because of the increased afterload on the heart (see chapter 12). The extra work leads to hypertrophy of cardiac muscle, especially in the left ventricle, and can lead to heart failure. Hypertension also increases the rate of arteriosclerosis development. Arteriosclerosis, in turn, increases the chance that blood clots will form and that blood vessels will rupture. Common conditions associated with hypertension are cerebral hemorrhage, coronary infarction, hemorrhage of renal blood vessels, and poor vision resulting from burst blood vessels in the retina.

Treatments that dilate blood vessels, increase the rate of urine production, or decrease cardiac output are normally used for hypertension. Low-salt diets also are normally recommended to reduce the amount of sodium chloride and water absorbed from the intestine into the bloodstream.

The greater the resistance in a blood vessel, the more rapidly the pressure decreases as blood flows through it. The most rapid decline in blood pressure occurs in the arterioles and capillaries, because their small diameters increase the resistance to blood flow. Blood pressure declines slowly as blood flows from large to medium-sized arteries because their diameters are larger and the resistance to blood flow is not great.

Resistance to blood flow in veins is low because of their larger diameters. Also, the valves that prevent backflow of blood in the veins, as well as skeletal muscle movements that

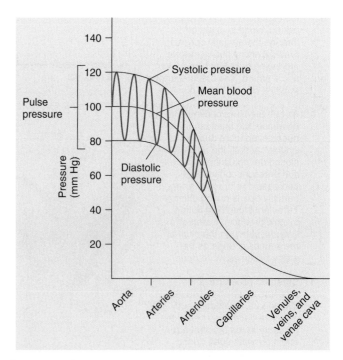

Figure 13.22   Blood Pressure in Major Blood Vessel Types

Blood pressure fluctuations between systole and diastole are damped, so the fluctuation become smaller, in small arteries and arterioles. There are no large fluctuations in blood pressure in capillaries and veins.

periodically compress veins, force blood to flow toward the heart. Consequently, blood flows through veins, even though the pressure in them is low.

The muscular arteries, arterioles, and precapillary sphincters are capable of constricting (vasoconstriction) and dilating (vasodilation). If constriction occurs, resistance to blood flow increases, and the volume of blood flow through the vessels declines. Because they are able to constrict and dilate, muscular arteries help control the amount of blood flowing to each region of the body. In contrast, arterioles and precapillary sphincters regulate blood flow through local tissues.

A CASE IN POINT | Atherosclerosis

Mori Payne and Leslie Payne were preparing for their 40th wedding anniversary celebration. While they were at the shopping mall, Mori's legs began to ache, especially his left leg. He had noticed before that when he walked briskly or very far, his legs hurt. Mori sat down to rest while Leslie shopped at a nearby store. After about 5 minutes or so, Mori experienced excruciating pain in his left leg. The pain was not relieved by massage or rest. Leslie asked someone to help her get Mori to their car and she drove him to the hospital.

In the emergency room, Mori's left leg was observed to be mottled and cyanotic distal to the knee, and it was cool to touch. His right leg was pink and warm. A Doppler test, which monitors the flow of blood through blood vessels, was performed on his left leg. It revealed decreased pulses with faint bruits (broo-ez) in his distal left poplitial artery. Normal pulses and no obvious bruits were observed in other arteries of his thigh and leg, including the right poplitial artery. His temperature was slightly elevated and his blood pressure was 165/95 mm Hg. His pulse was 96 beats per minute and regular. His respiratory rate was 20 respirations per minute. The physician explained to Leslie that the reduced pulses indicate reduced blood flow, and the bruits are sounds that indicate turbulent blood flow. Both of these observations and the sudden onset of the pain are consistent with the formation of an arterial thrombus (see p. 331 and 342) that is partially blocking Mori's distal left poplitial artery. He was quickly treated with an enzyme that breaks down the fibrin of blood clots (see p. 331). This enzyme successfully dissolved the thrombus. Pain and cyanosis decreased as blood flow to his left leg increased.

An angiogram on Mori's left leg revealed a severe stenosis (narrowing) in his left distal poplitial artery due to an atherosclerotic plaque. At rest, blood delivery past the partially blocked artery is adequate to maintain tissue homeostasis. During exercise, however, blood flow is not adequate (see also angina pectoris on p. 342). Other smaller atherosclerotic lesions are likely to exist in both of his legs because both of his legs hurt in response to brisk walking. The lack of adequate oxygen resulted in anaerobic metabolism and the production of acidic byproducts of anaerobic metabolism such as lactic acid.

The sudden dramatic increase in pain that Mori experienced resulted from the formation of a thrombus on a large atherosclerotic plaque that nearly blocked the blood flow through the left poplitial artery. Angioplasty was performed to further increase blood flow through the left poplitial artery after the thrombus was dissolved.

It was later determined that Mori has high blood cholesterol levels and high blood glucose levels. It is important for Mori to reduce his blood pressure, blood cholesterol levels, and blood glucose levels. He may be suffering from type II diabetes mellitus (see chapter 10). These conditions are associated with the increased development of atherosclerotic plaques in arteries.

Pulse Pressure

The difference between the systolic and diastolic pressure is called the **pulse pressure.** If a person has a systolic pressure of 120 mm Hg and a diastolic pressure of 80 mm Hg, the pulse pressure is 40 mm Hg. When the stroke volume increases, the systolic pressure increases more than the diastolic pressure, causing the pulse pressure to increase. During periods of exercise, the stroke volume and pulse pressure are increased substantially.

In those who suffer from arteriosclerosis, the arteries are less elastic than normal. In these people, arterial pressure increases rapidly and falls rapidly. The systolic pressure increases substantially, and the diastolic pressure may be somewhat lower than normal or slightly increased, resulting in a large pulse pressure. The same amount of blood ejected into a less elastic artery results in a higher systolic pressure than that in a more elastic artery. In people who suffer from arteriosclerosis, the pulse pressure is greater than normal, even though the same amount of blood is ejected into the aorta as in a normal person. Arteriosclerosis increases the amount of work performed by the heart because the left ventricle must produce a greater pressure to eject the same amount of blood into a less elastic artery. In severe cases, the increased workload on the heart leads to heart failure.

Ejection of blood from the left ventricle into the aorta produces a pressure wave, or **pulse,** which travels rapidly along the arteries. A pulse can be felt at locations where large arteries are close to the surface of the body (figure 13.23). It is helpful to know the major locations where the pulse can be detected because monitoring the pulse is important clinically. The heart rate, heart rhythm, and other characteristics can be determined by feeling the pulse. For example, a weak pulse usually indicates a decreased stroke volume or increased constriction of the arteries.

P R E D I C T 3
A weak pulse occurs in response to premature beats of the heart and during cardiovascular shock that is due to hemorrhage. Stronger than normal pulses occur in a healthy person during exercise. Explain the causes for the changes in the pulse under these conditions.

Capillary Exchange

There are about 10 billion capillaries in the body. Nutrients diffuse across the capillary walls into the interstitial spaces, and waste products diffuse in the opposite direction. In addition, a small amount of fluid is forced out of the capillaries into the interstitial space at their arteriolar ends. Most of that fluid, but not all, reenters the capillaries at their venous ends.

Two major forces are responsible for the movement of fluid through the capillary wall (figure 13.24). Blood pressure forces fluid out of the capillary, and osmosis moves fluid into it. Fluid moves by osmosis from the interstitial space into the capillary because blood has a greater osmotic pressure than does the interstitial fluid. The greater the concentration of molecules dissolved in a fluid, the greater the osmotic pressure of the fluid (see chapter 3). The greater osmotic pressure of blood is caused by the large concentration of blood proteins (see chapter 11) that are unable to cross the capillary wall. The concentration of proteins in

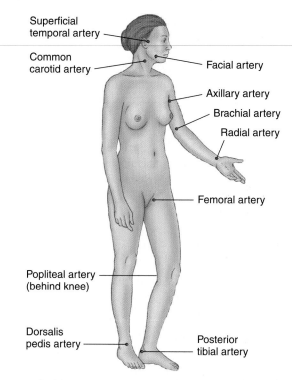

Figure 13.23 **Location of Major Points at Which the Pulse Can Be Monitored**

Each pulse point is named after the artery on which it occurs.

1. At the arterial end of the capillary, the movement of fluid out of the capillary due to blood pressure is greater than the movement of fluid into the capillary due to osmosis.

2. At the venous end of the capillary, the movement of fluid into the capillary due to osmosis is greater than the movement of fluid out of the capillary due to blood pressure.

3. Approximately nine-tenths of the fluid that leaves the capillary at its arterial end reenters the capillary at its venous end. About one-tenth of the fluid passes into the lymphatic capillaries.

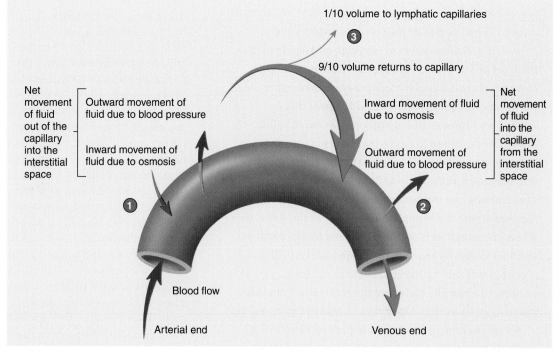

Process Figure 13.24 **Capillary Exchange**

the interstitial space is much lower than that in the blood. The capillary wall acts as a selectively permeable membrane, which prevents proteins from moving from the capillary into the interstitial space but allows fluid to move across the wall of the capillary.

At the arteriole end of the capillary, the movement of fluid out of the capillary resulting from blood pressure is greater than the movement of fluid into the capillary as a result of osmosis. Consequently, there is a net movement of fluid out of the capillary into the interstitial space (see figure 13.24).

At the venous end of the capillary, blood pressure is lower than at the arteriolar end because of the resistance to blood flow through the capillary (see figure 13.24). Consequently, the movement of fluid out of the capillary resulting from blood pressure is less than the movement of fluid into the capillary resulting from osmosis, and there is a net movement of fluid from the interstitial space into the capillary (see figure 13.24).

Approximately nine-tenths of the fluid that leaves the capillary at the arteriolar end reenters the capillary at its venous end. The remaining one-tenth of the fluid enters the lymphatic capillaries and is eventually returned to the general circulation (see chapter 14).

Edema

Edema, or swelling, results from a disruption in the normal inwardly and outwardly directed pressures across the capillary walls. For example, inflammation results in an increase in the permeability of capillaries. Proteins, mainly albumen, leak out of the capillaries into the interstitial spaces. The proteins increase the osmotic pressure in the interstitial fluid. Consequently, fluid passes from the arteriolar end of capillaries into the interstitial spaces at a greater rate, and fluid passes from the interstitial spaces into the venous ends of capillaries at a slower rate. The lymphatic capillaries cannot carry all of the fluid away. Thus, fluid accumulates in the interstitial spaces, resulting in edema.

PREDICT 4

Explain edema (a) in response to a decrease in plasma protein concentration, and (b) as a result of increased blood pressure within a capillary.

Local Control of Blood Vessels

Local control of blood flow is achieved by periodic contraction and relaxation of the precapillary sphincters. Blood flow through the capillaries is cyclical because of this contraction and relaxation. The precapillary sphincters are controlled by the metabolic needs of the tissues. Blood flow increases when oxygen levels decrease or, to a lesser degree, when glucose, amino acids, fatty acids, and other nutrients decrease. Blood flow also increases when by-products of metabolism build up in tissue spaces. An increase in carbon dioxide or a decrease in pH causes the precapillary sphincters to relax. For example, during exercise, the metabolic needs of skeletal muscle increase dramatically, and the by-products of metabolism are produced at a more rapid rate. The precapillary sphincters therefore dilate, and blood flow through capillaries in exercising muscle increases dramatically (figure 13.25 and table 13.1).

PREDICT 5

A student has been sitting for a short time with her legs crossed. After getting up to walk out of class she noticed a red blotch on the back of one of her legs. On the basis of what you know about the local control of blood flow, explain why this happens.

In addition to the control of blood flow through existing capillaries, if the metabolic activity of a tissue increases often, additional capillaries gradually grow into the area. The additional capillaries allow local blood flow to be increased to a level that matches the metabolic demand of the tissue. For example, the density of capillaries in the well-trained skeletal muscles of athletes is greater than that in poorly trained skeletal muscles (see table 13.1).

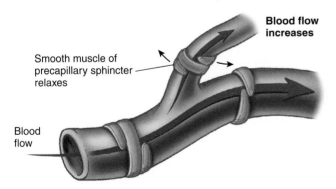

(a) Precapillary sphincters relax as the tissue concentration of nutrients such as O_2, glucose, amino acids, and fatty acids decreases.

Precapillary sphincters relax as the concentration of tissue metabolic by-products increases as a result of an increase in CO_2 and lactic acid and a decrease in pH.

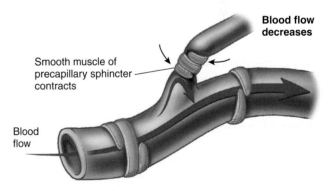

(b) Precapillary sphincters contract as the tissue concentration of nutrients such as O_2, glucose, amino acids, and fatty acids increases.

Precapillary sphincters contract as the tissue concentration of metabolic by-products decreases as a result of a decrease in CO_2 and lactic acid and an increase in pH.

Process Figure 13.25 *Control of Local Blood Flow Through Capillary Beds*
(*a*) Dilation of precapillary sphincters. (*b*) Constriction of precapillary sphincters.

Table 13.1 — Homeostasis: Local Control of Blood Flow*

Stimulus	Response
Regulation by Metabolic Need of Tissues	
Increased carbon dioxide and decreased pH or decreased nutrients, such as oxygen, glucose, amino acids and fatty acids, as a result of increased metabolism	Relaxation of precapillary sphincters and subsequent increase in blood flow through capillaries
Decreased carbon dioxide and increased pH or increased nutrients, such as oxygen, glucose, amino acids, and fatty acids	Contraction of precapillary sphincters and subsequent decrease in blood flow through capillaries
Regulation by Nervous Mechanisms	
Increased physical activity or increased sympathetic activity	Constriction of blood vessels in skin and viscera
Increased body temperature detected by neurons of the hypothalamus	Dilation of blood vessels in skin (see chapter 5)
Decreased body temperature detected by neurons of the hypothalamus	Constriction of blood vessels in skin (see chapter 5)
Decrease in skin temperature below a critical value	Dilation of blood vessels in skin (protects skin from extreme cold)
Anger or embarrassment	Dilation of blood vessels in skin of face and upper thorax
Regulation by Hormonal Mechanisms	
(reinforces increased activity of the sympathetic division)	
Increased physical activity and increased sympathetic activity, causing release of epinephrine and small amounts of norepinephrine from the adrenal medulla	Constriction of blood vessels in skin and viscera; dilation of blood vessels in skeletal and cardiac muscle
Long-Term Local Blood Flow	
Increased metabolic activity of tissues over a long period such as in athletes who train regularly	Increased number of capillaries
Decreased metabolic activity of tissues over a long period such as during periods of reduced physical activity	Decreased number of capillaries

* The mechanisms operate when the systemic blood pressure is maintained within a normal range of values.

Nervous Control of Blood Vessels

Nervous control of blood vessels is carried out primarily through the sympathetic division of the autonomic nervous system. Sympathetic vasoconstrictor fibers innervate most blood vessels of the body, except the capillaries and precapillary sphincters, which have no nerve supply (figure 13.26).

An area of the lower pons and upper medulla oblongata, called the **vasomotor center,** continually transmits a low frequency of action potentials to the sympathetic vasoconstrictor fibers. As a consequence, the peripheral blood vessels are continually in a partially constricted state, a condition called **vasomotor** (vā-sō-mō′ter) **tone.** An increase in vasomotor tone causes blood vessels to constrict further and blood pressure to increase. A decrease in vasomotor tone causes blood vessels to dilate and blood pressure to decrease. Nervous control of blood vessel diameter is an important way blood pressure is regulated.

Nervous control of blood vessels also causes blood to be shunted from one large area of the body to another. For example, nervous control of blood vessels during exercise increases vasomotor tone in the viscera and skin and reduces vasomotor tone in exercising skeletal muscles. As a result, blood flow to the viscera and skin decreases, and blood flow to skeletal muscle increases. Nervous control of blood vessels during exercise

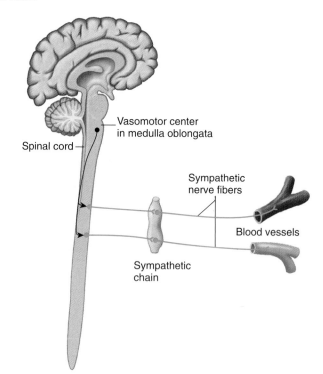

Figure 13.26 **Nervous Regulation of Blood Vessels**

Most arteries and veins are innervated by sympathetic nerve fibers. The vasomotor center within the medulla oblongata regulates the frequency of action potentials in nerve fibers that innervate blood vessels. Increased action potential frequencies cause vasoconstriction and decreased action potential frequencies cause vasodilation in most blood vessels.

and dilation of precapillary sphincters as muscle activity increases function together to increase blood flow through exercising skeletal muscle by severalfold (see table 13.1).

P R E D I C T
Raynaud's syndrome is a condition in which blood vessels, primarily in the digits, undergo exaggerated vasoconstriction in response to exposure to cold or emotions. Although treatments are available for Raynaud's syndrome patients, predict the consequences for the digits of a person who suffers from severe Raynaud's syndrome. Explain why the consequences occur.

Regulation of Arterial Pressure

An adequate blood pressure is required to maintain blood flow through the blood vessels of the body, and several regulatory mechanisms ensure that an adequate blood pressure is maintained. The **mean arterial blood pressure (MAP)** is slightly less than the average of the systolic and diastolic pressures in the aorta because diastole lasts longer than systole. The mean arterial pressure is about 70 mm Hg at birth, is maintained at about 95 mm Hg from adolescence to middle age, and may reach 110 mm Hg in a healthy older person.

The MAP in the body is equal to the **cardiac output (CO)** times the **peripheral resistance (PR)**, which is the resistance to blood flow in all the blood vessels.

$$MAP = CO \times PR$$

Because the cardiac output is equal to the **heart rate (HR)** times the **stroke volume (SV)**, the mean arterial pressure is equal to the heart rate times the stroke volume times the peripheral resistance.

$$MAP = HR \times SV \times PR$$

Thus, the MAP increases in response to increases in HR, SV, or PR and MAP decreases in response to decreases in HR, SV, or PR.

The MAP is controlled on a minute-to-minute basis by changes in the heart rate, stroke volume, and peripheral resistance. For example, when blood pressure suddenly drops because of hemorrhage or some other cause, control systems attempt to reestablish blood pressure by increasing the HR, SV, and PR so that blood pressure is maintained at a value consistent with life. Mechanisms are also activated to increase the blood volume to its normal value.

Baroreceptor Reflexes

Baroreceptor reflexes activate responses that keep the blood pressure within its normal range of values.

Baroreceptors respond to stretch in arteries caused by increased pressure. They are scattered along the walls of most of the large arteries of the neck and thorax, and there are many in the carotid sinus at the base of the internal carotid artery and in the walls of the aortic arch (figure 13.27). Action potentials are transmitted from the baroreceptors to the medulla oblongata along sensory nerve fibers (see figure 13.27).

A sudden increase in blood pressure stretches the artery walls and increases action potential frequency in the baroreceptors.

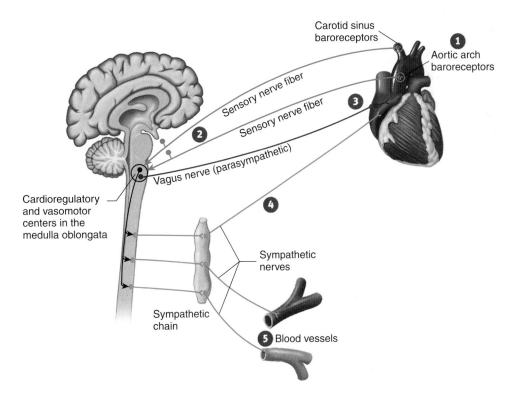

1. Baroreceptors in the carotid sinus and aortic arch monitor blood pressure.

2. Action potentials are conducted by sensory nerves to the cardioregulatory and vasomotor centers in the medulla oblongata.

3. Increased parasympathetic stimulation of the heart decreases the heart rate.

4. Increased sympathetic stimulation of the heart increases the heart rate and stroke volume.

5. Increased sympathetic stimulation of blood vessels increases vasoconstriction.

Process Figure 13.27 The Baroreceptor Reflex Mechanisms
Baroreceptor reflex control of blood pressure.

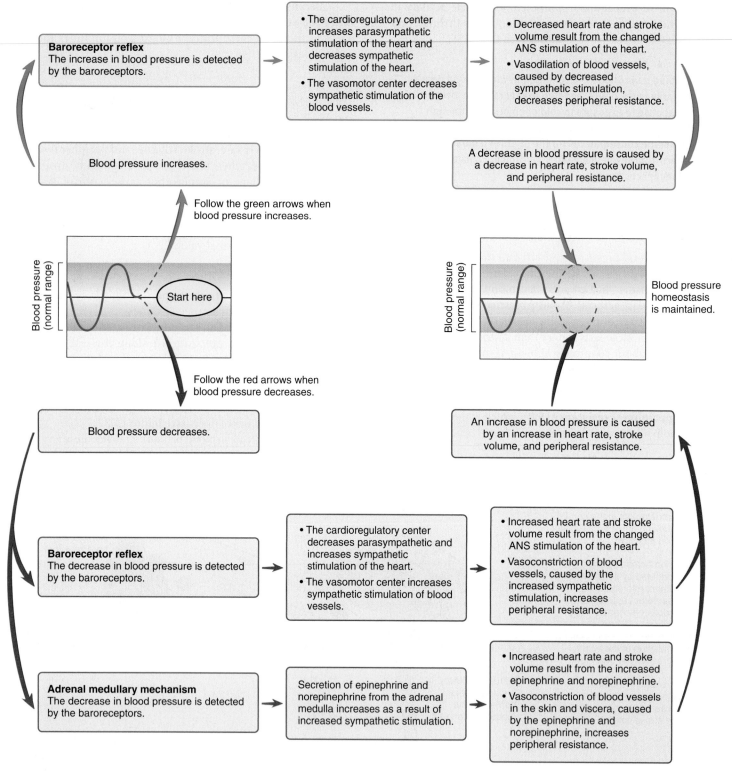

Homeostasis Figure 13.28   Baroreceptor Effects on Blood Pressure

The increased action potential frequency delivered to the vasomotor and cardioregulatory centers in the medulla oblongata causes responses that lower the blood pressure. One major response is a decrease in vasomotor tone, resulting in vasodilation of blood vessels and a decrease in peripheral resistance. Other responses, controlled by the cardioregulatory center, are an increase in the parasympathetic stimulation of the heart, which decreases the heart rate, and a decrease in sympathetic stimulation of the heart, which reduces the stroke volume. The decreased heart rate, stroke volume, and peripheral resistance lower the blood pressure toward its normal value (figure 13.28).

Circulatory shock is defined as inadequate blood flow throughout the body. As a consequence, tissues suffer damage resulting from a lack of oxygen. Severe shock may damage vital body tissues and lead to death.

There are several causes of circulatory shock, but hemorrhagic shock resulting from excessive blood loss can be used to illustrate the general characteristics of shock. If hemorrhagic shock is not severe, blood pressure decreases only a moderate amount. Under these conditions, the mechanisms that normally regulate blood pressure function to reestablish normal pressure and blood flow. The baroreceptor reflexes produce strong sympathetic responses, resulting in intense vasoconstriction and increased heart rate.

As a result of the reduced blood flow through the kidneys, increased amounts of renin are released. The elevated renin level results in a greater rate of angiotensin II formation, causing vasoconstriction and increased aldosterone release from the adrenal cortex. Aldosterone, in turn, promotes water and salt retention by the kidneys. In response to reduced blood pressure, antidiuretic hormone (ADH) is released from the posterior pituitary gland, and ADH also enhances the retention of water by the kidneys. An intense sensation

of thirst leads to increased water intake, which helps restore the normal blood volume.

In mild cases of shock, the baroreceptor reflexes can be adequate to compensate for blood loss until the blood volume is restored, but in more severe cases, all mechanisms are required to sustain life.

In even more severe cases of shock, the regulatory mechanisms are not adequate to compensate for the effects of shock. As a consequence, a positive-feedback cycle begins to develop in which the blood pressure regulatory mechanisms lose their ability to control the blood pressure, and shock worsens. As shock becomes worse, the effectiveness of the regulatory mechanisms deteriorates even further. The positive-feedback cycle proceeds until death occurs or until treatment, such as a transfusion, terminates the cycle. Several types of shock are classified by the cause of the condition:

1. **Hypovolemic shock** is the result of reduced blood volume. **Hemorrhagic shock** is caused by internal or external bleeding. **Plasma loss shock** results from a loss of plasma. An example is the loss of plasma from severely burned areas. **Interstitial fluid loss shock** is reduced blood volume

resulting from a loss of interstitial fluid. Examples include diarrhea, vomiting, or dehydration.

2. **Neurogenic shock** is caused by vasodilation in response to emotional upset or anesthesia.

3. **Anaphylactic shock** is caused by an allergic response, resulting in the release of inflammatory substances that cause vasodilation and an increase in capillary permeability. Large amounts of fluid then move from capillaries into the interstitial spaces.

4. **Septic shock,** or **"blood poisoning,"** is caused by infections that result in the release of toxic substances into the circulatory system, which depress the activity of the heart and lead to vasodilation and increased capillary permeability.

5. **Cardiogenic shock** results from a decrease in cardiac output caused by events that decrease the ability of the heart to function. Heart attack (myocardial infarction) is a common cause of cardiogenic shock. Fibrillation of the heart, which can be initiated by stimuli such as cardiac arrhythmias or exposure to electrical shocks, also results in cardiogenic shock.

A sudden decrease in blood pressure results in a decreased action potential frequency in the baroreceptors. The decreased action potential frequency delivered to the vasomotor and cardioregulatory centers in the medulla oblongata produces responses that raise blood pressure. There is an increase in sympathetic stimulation of the heart, which increases the heart rate and stroke volume, and an increase in vasomotor tone, which increases peripheral resistance. The increased heart rate, stroke volume, and peripheral resistance raise the blood pressure toward its normal value (see figures 13.27 and 13.28).

These **baroreceptor reflexes** regulate blood pressure on a moment-to-moment basis. When a person rises rapidly from a sitting or lying position to a standing position, blood pressure in the neck and thoracic regions drops dramatically as a result of the pull of gravity on the blood. This reduction in blood pressure can be so great that blood flow to the brain is reduced enough to cause dizziness or even loss of consciousness. The falling blood pressure activates the baroreceptor reflexes, which reestablish normal blood pressure within a few seconds. In a healthy person, a temporary sensation of dizziness is all that may be experienced.

Chemoreceptor Reflexes

Carotid bodies are small structures that lie near the carotid sinuses, and **aortic bodies** are structures near the aortic arch. These structures contain sensory receptors that respond to changes in blood oxygen concentration, carbon dioxide concentration, and pH. Because they are sensitive to chemical changes in the blood, they are called **chemoreceptors.** They send action potentials along sensory nerve fibers to the medulla oblongata (figure 13.29). There are also chemoreceptors in the medulla oblongata.

When oxygen levels decrease, carbon dioxide levels increase, or pH decreases, the chemoreceptors respond with an increased frequency of action potentials and activate the **chemoreceptor reflexes.** In response, the vasomotor and cardiovascular centers decrease parasympathetic stimulation of the heart, which increases the heart rate. The vasomotor and cardioregulatory centers also increase sympathetic stimulation of the heart, which increases the heart rate and stroke volume and increases the vasomotor tone. All of these changes result in an increased blood pressure. This increased blood pressure causes a greater rate of blood flow to the lungs, which helps increase

1. Chemoreceptors in the carotid and aortic bodies monitor blood O_2, CO_2, and pH.

2. Chemoreceptors in the medulla oblongata monitor blood CO_2 and pH.

3. Decreased blood O_2, increased CO_2, and decreased pH decrease parasympathetic stimulation of the heart, which increases the heart rate.

4. Decreased blood O_2, increased CO_2, and decreased pH increase sympathetic stimulation of the heart, which increases the heart rate and stroke volume.

5. Decreased blood O_2, increased CO_2, and decreased pH increase sympathetic stimulation of blood vessels, which increases vasoconstriction.

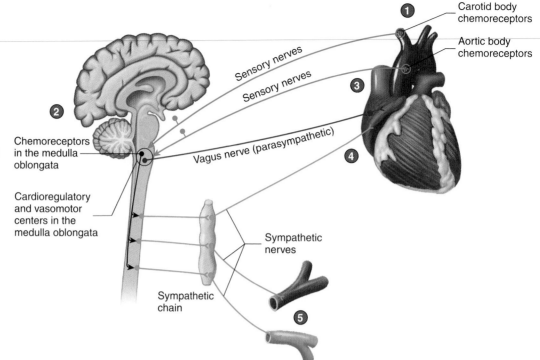

Process Figure 13.29 Chemoreceptor Reflex Control of Blood Pressure

blood oxygen levels and reduce blood carbon dioxide levels. The chemoreceptor reflexes function under emergency conditions and usually do not play an important role in the regulation of the cardiovascular system. They respond strongly only when the oxygen levels in the blood fall to very low levels or when carbon dioxide levels become substantially elevated.

Hormonal Mechanisms

In addition to the rapidly acting baroreceptor and chemoreceptor reflexes, there are important hormonal mechanisms that help control blood pressure.

Adrenal Medullary Mechanism

Stimuli that result in increased sympathetic stimulation of the heart and blood vessels also cause increased stimulation of the adrenal medulla. The adrenal medulla responds by releasing epinephrine into the blood (figure 13.30). Epinephrine increases heart rate and stroke volume and causes vasoconstriction, especially of blood vessels in the skin and viscera. Epinephrine also causes vasodilation of blood vessels in skeletal muscle and cardiac muscle. Epinephrine therefore increases the supply of blood flowing to skeletal and cardiac muscle, and this prepares one for physical activity.

Renin–Angiotensin–Aldosterone Mechanism

In response to reduced blood flow, the kidneys release an enzyme called **renin** (rē′nin, *rena,* kidney) into the circulatory system (figure 13.31). Renin acts on the blood protein **angiotensinogen** (an′jē-ō-ten-sin′ō-jen) to produce **angiotensin I** (an-jē-ō-ten′sin, *angio,* blood vessel + *tensus,* to stretch). Another enzyme,

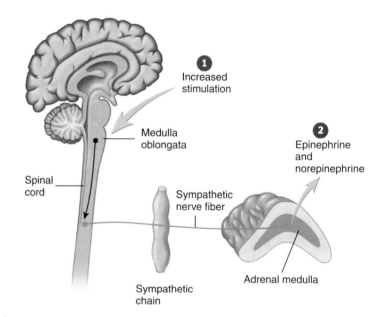

1. The same stimuli that increase sympathetic stimulation of the heart and blood vessels cause action potentials to be carried to the medulla oblongata.

2. Descending pathways from the medulla oblongata to the spinal cord increase sympathetic stimulation of the adrenal medulla, resulting in increased epinephrine and some norepinephrine secretion.

Process Figure 13.30 Hormonal Regulation: The Adrenal Medullary Mechanism

Stimuli that increase sympathetic stimulation of the heart and blood vessels also result in increased sympathetic stimulation of the adrenal medulla and result in epinephrine and some norepinephrine secretion.

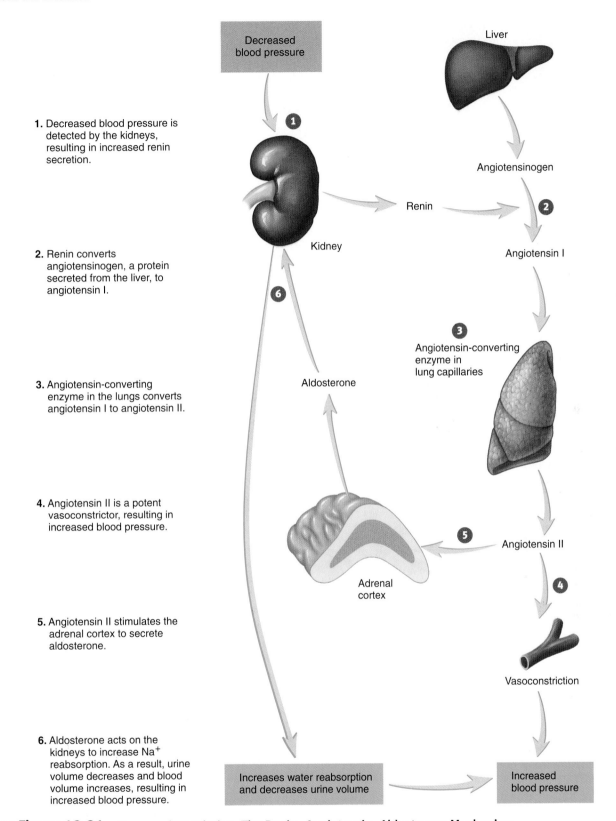

1. Decreased blood pressure is detected by the kidneys, resulting in increased renin secretion.

2. Renin converts angiotensinogen, a protein secreted from the liver, to angiotensin I.

3. Angiotensin-converting enzyme in the lungs converts angiotensin I to angiotensin II.

4. Angiotensin II is a potent vasoconstrictor, resulting in increased blood pressure.

5. Angiotensin II stimulates the adrenal cortex to secrete aldosterone.

6. Aldosterone acts on the kidneys to increase Na^+ reabsorption. As a result, urine volume decreases and blood volume increases, resulting in increased blood pressure.

Process Figure 13.31   Hormonal Regulation: The Renin–Angiotensin–Aldosterone Mechanism

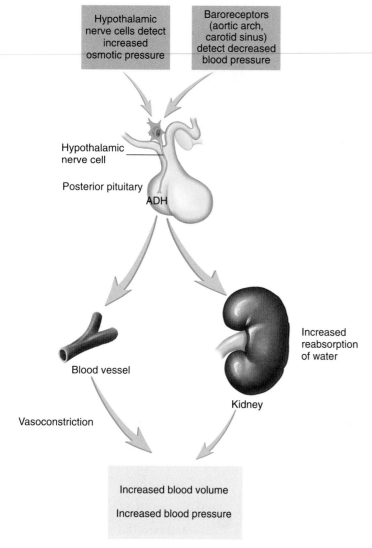

Figure 13.32   Hormonal Regulation: The Vasopressin
(ADH) Mechanism

Increases in osmolality of blood or decreases in blood pressure result in ADH
secretion. ADH increases water reabsorption by the kidney, and large amounts
of ADH result in vasoconstriction. These changes function to maintain blood
pressure.

called **angiotensin-converting enzyme,** found in large amounts
in organs such as the lungs, acts on angiotensin I to convert it to
its most active form, called **angiotensin II.** Angiotensin II is a potent vasoconstrictor substance. Thus, in response to a reduced
blood pressure, the release of renin by the kidney acts to increase
the blood pressure toward its normal value.

Angiotensin II also acts on the adrenal cortex to increase the secretion of **aldosterone** (al-dos′ter-ōn). Aldosterone acts on the kidneys causing them to conserve Na^+
and water. As a result, the volume of water lost from the blood
into the urine is reduced. The decrease in urine volume results in less fluid loss from the body, which functions to
maintain blood volume (see figure 13.31). An adequate blood
volume is essential for the maintenance of normal venous return to the heart and therefore for the maintenance of blood
pressure (see chapter 12).

Vasopressin Mechanism

When the concentration of solutes in the plasma increases, or
blood pressure decreases substantially, nerve cells in the hypothalamus respond by causing the release of **antidiuretic**
(an′tē-dī-ū-ret′ik, to decrease urine production) **hormone
(ADH),** also called **vasopressin** (vā-sō-pres′in, to cause vasoconstriction), from the posterior pituitary gland (figure 13.32).
ADH acts on the kidneys and causes a greater reabsorption of
water by the kidneys, thereby decreasing urine volume. This response helps maintain blood volume and blood pressure. The
release of large amounts of ADH causes vasoconstriction of
blood vessels, which causes blood pressure to increase.

Atrial Natriuretic Mechanism

A peptide hormone called **atrial natriuretic** (ā′trē-ăl nā′trē-ū-
ret′ik, to excrete sodium in the urine) **hormone** is released
primarily from specialized cells of the right atrium in response
to elevated blood pressure. Atrial natriuretic hormone acts on
the kidneys and causes them to promote the loss of Na^+ in the
urine and to increase urine volume. Loss of water in the urine
causes blood volume to decrease, thus decreasing the blood
pressure (figure 13.33).

Short-Term and Long-Term Regulation

Baroreceptor mechanisms are most important in controlling blood
pressure on a short-term basis (see figures 13.27 and 13.28). They
are sensitive to sudden changes in blood pressure, and they respond
quickly. The chemoreceptor and adrenal medullary reflexes are also
sensitive to sudden changes in blood pressure and respond quickly,
but they respond to large changes in blood pressure. The renin–
angiotensin–aldosterone vasopressin, and atrial natriuretic mechanisms, are more important in the maintenance of blood pressure
on a long-term basis. They are influenced by small changes in
blood pressure or concentration and respond by gradually bringing the blood pressure back to its normal range (see figure 13.33).

P R E D I C T   7

Explain the differences in blood pressure regulation mechanisms
in response to hemorrhage that results in a rapid loss of a large
volume of blood compared to hemorrhage that results in the loss
of the same volume of blood but over a period of several hours.

Effects of Aging on the Blood Vessels

The walls of all arteries undergo changes as they age. Some arteries change more rapidly than others and some individuals are
more susceptible to change than others. The most significant effects of aging occur in the large elastic arteries such as the aorta,
large arteries carrying blood to the brain, and coronary arteries.

Changes in arteries that make them less elastic are referred
to as **arteriosclerosis** (ar-tēr′ē-ō-skler-ō′sis, hardening of the arteries). These changes occur in nearly every individual, and they
become more severe with advancing age. A type of arteriosclerosis
called **atherosclerosis** (ath′er-ō-skler-ō′sis) results from the deposition of material in the walls of arteries to form plaques

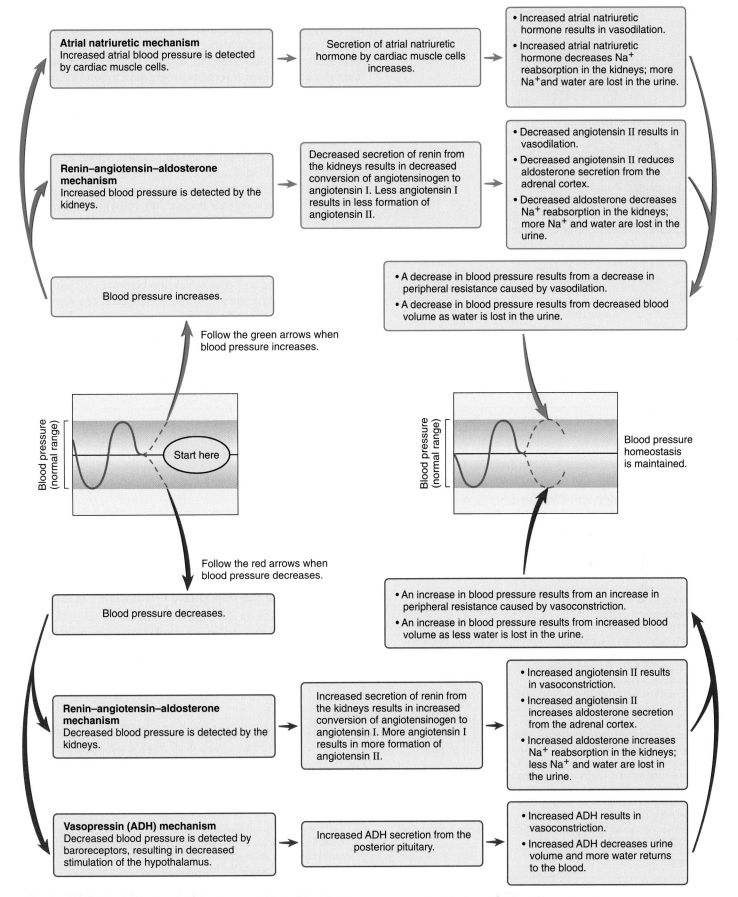

Homeostasis Figure 13.33   Control of Blood Pressure Long-Term (Slow-Acting) Mechanisms

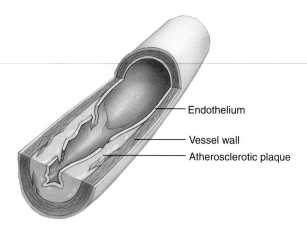

- Endothelium
- Vessel wall
- Atherosclerotic plaque

Figure 13.34　Atherosclerotic Plaque in an Artery

Atherosclerotic plaques develop within the tissue of the artery wall.

(figure 13.34). The material includes a fatlike substance containing cholesterol. The fatty material can eventually be dominated by the deposition of dense connective tissue and calcium salts.

The development of atherosclerosis is influenced by several factors. Lack of exercise, smoking, obesity, and a diet high in cholesterol and fats appear to increase the severity and the rate at which atherosclerosis develops. Severe atherosclerosis is more prevalent in some families than in others, which suggests a genetic influence. Some evidence suggests that a low-fat diet, mild exercise, and relaxation exercises slow the progression of atherosclerosis and may even reverse its progression to some degree.

Atherosclerosis greatly increases resistance to blood flow because the deposits reduce the inside diameter of the arteries. The increased resistance hampers normal circulation to tissues and greatly increases the work performed by the heart. The rough atherosclerotic plaques attract platelets, which adhere to them and increase the chance of thrombus formation.

Capillaries narrow and become more irregular in shape with age. Their walls become thicker and, consequently, there is a decrease in the efficiency of capillary exchange.

Veins tend to develop patchy thickenings in their walls, resulting in narrowing in these areas. There is an increased tendency to develop varicose veins (see varicose veins, page 357). There is age-related tendency to develop hemorrhoids (varicose veins of the rectum or anus) because the diameter of some veins increases due to weakening of the connective tissue in the walls of veins. There is a related increase in the tendency for thrombi and emboli to develop, especially in veins that are dilated and in which blood flow is sluggish.

S U M M A R Y

The peripheral circulatory system can be divided into the systemic and the pulmonary vessels. The peripheral circulatory system and the heart are regulated to maintain sufficient blood flow to tissues.

Functions of the Peripheral Circulation　(p. 354)

The peripheral circulation functions to carry blood, exchange nutrients and gases, transport hormones, regulate blood pressure, and direct blood flow.

General Features of Blood Vessel Structure　(p. 354)

1. Blood is pumped from the heart through elastic arteries, muscular arteries, and arterioles to the capillaries.
2. Blood returns to the heart from the capillaries through venules, small veins, and large veins.
3. Except for capillaries and venules, blood vessels have three layers:
   a. The tunica intima consists of endothelium, basement membrane, and connective tissue.
   b. The tunica media, the middle layer, contains circular smooth muscle and elastic fibers.
   c. The outer tunica adventitia is connective tissue.

Arteries

1. Large elastic arteries have many elastic fibers but little smooth muscle in their walls and carry blood from the heart to smaller arteries with little decrease in pressure.
2. Muscular arteries have much smooth muscle and some elastic fibers and undergo vasodilation and vasoconstriction to control blood flow to different regions of the body.
3. Arterioles are the smallest arteries and have smooth muscle cells and a few elastic fibers and undergo vasodilation and vasoconstriction to control blood flow to local areas.

Capillaries

1. Capillaries consist of only endothelium and are surrounded by a basement membrane and loose connective tissue.
2. Nutrient and waste product exchange is the principal function of capillaries.

3. Blood is supplied to capillaries by arterioles. Precapillary sphincters regulate blood flow through capillary networks.

Veins

1. Venules are endothelium surrounded by a basement membrane.
2. Small veins are venules covered with a layer of smooth muscle and a layer of connective tissue.
3. Medium-sized and large veins contain less smooth muscle and elastic fibers than arteries of the same size.
4. Valves prevent the backflow of blood in the veins.

Blood Vessels of the Pulmonary Circulation　(p. 357)

The pulmonary circulation moves blood to and from the lungs. The pulmonary trunk carries oxygen-poor blood from the heart to the lungs, and pulmonary veins carry oxygen-rich blood from the lungs to the left atrium of the heart.

Blood Vessels of the Systemic Circulation: Arteries　(p. 357)

Aorta

The aorta leaves the left ventricle to form the ascending aorta, aortic arch, and descending aorta, which consists of the thoracic and abdominal aorta.

Arteries of the Head and Neck

1. The brachiocephalic, left common carotid, and left subclavian arteries branch from the aortic arch to supply the head and the upper limbs.
2. The common carotid arteries and the vertebral arteries supply the head. The common carotid arteries divide to form the external carotids (which supply the face and mouth) and the internal carotids (which supply the brain).

Arteries of the Upper Limbs

The subclavian artery continues as the axillary artery and then as the brachial artery, which branches to form the radial and ulnar arteries.

The Thoracic Aorta and Its Branches

The thoracic aorta has visceral branches, which supply the thoracic organs, and parietal branches, which supply the thoracic wall.

The Abdominal Aorta and Its Branches

The abdominal aorta has visceral branches, which supply the abdominal organs, and parietal branches, which supply the abdominal wall.

Arteries of the Pelvis

Branches of the internal iliac arteries supply the pelvis.

Arteries of the Lower Limbs

The common iliac arteries give rise to the external iliac arteries, and the external iliac artery continues as the femoral artery and then as the popliteal artery in the leg. The popliteal artery divides to form the anterior and posterior tibial arteries.

Blood Vessels of the Systemic Circulation: Veins (p. 365)

The superior vena cava drains the head, neck, thorax, and upper limbs. The inferior vena cava drains the abdomen, pelvis, and lower limbs.

Veins of the Head and Neck

1. The internal jugular veins drain the brain, anterior head, and anterior neck.
2. The external jugular veins drain the posterior head and posterior neck.

Veins of the Upper Limbs

The deep veins are the brachial, axillary, and subclavian; the superficial veins are the basilic, cephalic, and median cubital.

Veins of the Thorax

The left and right brachiocephalic veins and the azygos veins return blood to the superior vena cava.

Veins of the Abdomen and Pelvis

1. Posterior abdominal wall veins join the azygos veins.
2. Veins from the kidneys, adrenal glands, and gonads directly enter the inferior vena cava.
3. Veins from the stomach, intestines, spleen, and pancreas connect with the hepatic portal vein, which transports blood to the liver for processing. The hepatic veins from the liver join the inferior vena cava.

Veins of the Lower Limbs

1. The deep veins course with the deep arteries and have similar names.
2. The superficial veins are the small and great saphenous veins.

The Physiology of Circulation (p. 371)

Blood Pressure

1. Blood pressure is a measure of the force exerted by blood against the blood vessel wall. Blood pressure moves blood through vessels.
2. Blood pressure can be measured by listening for Korotkoff sounds produced as blood flows through arteries partially constricted by a blood pressure cuff.

Pressure and Resistance

Blood pressure fluctuates between 120 mm Hg (systolic) and 80 mm Hg (diastolic) in the aorta. If constriction of blood vessels occurs, resistance to blood flow increases, and blood flow decreases.

Pulse Pressure

1. Pulse pressure is the difference between systolic and diastolic pressure. Pulse pressure increases when stroke volume increases.
2. A pulse can be detected when large arteries are near the surface of the body.

Capillary Exchange

1. Most exchange across the wall of the capillary is by diffusion.
2. Blood pressure, capillary permeability, and osmosis affect movement of fluid across the wall of the capillaries. There is a net movement of fluid from the blood into the tissues. The fluid gained by the tissues is removed by the lymphatic system.

Local Control of Blood Vessels (p. 375)

Blood flow through a tissue is usually proportional to the metabolic needs of the tissue and is controlled by the precapillary sphincters.

Nervous Control of Blood Vessels (p. 376)

1. The vasomotor center (sympathetic division) controls blood vessel diameter. Other brain areas can excite or inhibit the vasomotor center.
2. Vasomotor tone is a state of partial contraction of blood vessels.
3. The nervous system is responsible for routing the flow of blood, except in the capillaries and precapillary sphincters, and is responsible for maintaining blood pressure.

Regulation of Arterial Pressure (p. 377)

Mean arterial pressure (MAP) is proportional to cardiac output times the peripheral resistance.

Baroreceptor Reflexes

1. Baroreceptors are sensory receptors that are sensitive to stretch.
2. Baroreceptors are located in the carotid sinuses and the aortic arch.
3. The baroreceptor reflex changes peripheral resistance, heart rate, and stroke volume in response to changes in blood pressure.

Chemoreceptor Reflexes

1. Chemoreceptors are sensitive to changes in blood oxygen, carbon dioxide, and pH.
2. Chemoreceptors are located in the carotid bodies and the aortic bodies.
3. The chemoreceptor reflex increases peripheral resistance in response to low oxygen levels, high carbon dioxide levels, and reduced blood pH.

Hormonal Mechanisms

1. Epinephrine released from the adrenal medulla as a result of sympathetic stimulation increases heart rate, stroke volume, and vasoconstriction.
2. Renin is released by the kidneys in response to low blood pressure. Renin promotes the production of angiotensin II, which causes vasoconstriction and an increase in aldosterone secretion. Aldosterone reduces urine output. Angiotensin II can also cause vasoconstriction.
3. ADH released from the posterior pituitary causes vasoconstriction and reduces urine output.
4. Atrial natriuretic hormone is released from the heart when atrial blood pressure increases. It stimulates an increase in urine production, causing a decrease in blood volume and blood pressure.

Short-Term and Long-Term Regulation

1. The baroreceptor mechanisms are most important in short-term regulation of blood pressure.
2. Hormonal mechanisms, such as the renin–angiotensin–aldosterone system and atrial natriuretic hormone, are more important in long-term regulation of blood pressure.

Age-Related Changes in Blood Vessels (p. 382)

1. Reduced elasticity and thickening of arterial walls result in hypertension and reduced ability to respond to changes in blood pressure.
2. Atherosclerosis is an age-related condition.
3. There is a decrease in the efficiency of capillary exchange.
4. Walls of veins thicken in some areas and dilate in others. Thrombi, emboli, varicose veins, and hemorrhoids are age-related conditions.

R E V I E W   A N D   C O M P R E H E N S I O N

1. Name, in order, all the types of blood vessels, starting at the heart, going to the tissues, and returning to the heart.

2. Name the three layers of a blood vessel. What kinds of tissue are in each layer?

3. Relate the structure of the different types of arteries to their functions.

4. Describe the structure of capillaries, and explain their major function.

5. Describe the structure of a capillary network. Name the structure that regulates blood flow through the capillary network.

6. Describe the structure of veins.

7. What is the function of valves in blood vessels and in which blood vessels are valves found?

8. List the different parts of the aorta. Name the major arteries that branch from the aorta and deliver blood to the vessels that supply the heart, the head and upper limbs, and the lower limbs.

9. Name the arteries that supply the major areas of the head, upper limbs, thorax, abdomen, and lower limbs. Describe the area each artery supplies.

10. Name the major vessels that return blood to the heart. What area of the body does each drain?

11. List the veins that drain blood from the thorax, abdomen, and pelvis. What specific area of the body does each drain? Describe the hepatic portal system.

12. List the major veins that drain the upper and lower limbs.

13. Define blood pressure, and describe how it is normally measured.

14. Describe the changes in blood pressure starting in the aorta, moving through the vascular system, and returning to the right atrium.

15. Define pulse pressure, and explain what information can be determined from monitoring the pulse.

16. Explain how blood pressure and osmosis affect the movement of fluid between capillaries and tissues. What happens to excess fluid that enters the tissues?

17. Explain what is meant by the local control of blood flow through tissues, and describe what carries out local control.

18. Describe nervous control of blood vessels. Define vasomotor tone.

19. Define mean arterial pressure. How is it related to heart rate, stroke volume, and peripheral resistance?

20. Where are baroreceptors located? Describe the baroreceptor reflex when blood pressure increases and when it decreases.

21. Where are the chemoreceptors for carbon dioxide and pH located? Describe what happens when oxygen levels in the blood decrease.

22. For each of the following hormones—epinephrine, renin, angiotensin II, aldosterone, ADH, and atrial natriuretic hormone—state where each is produced, what stimulus causes an increased hormone production, and what effect the hormone has on the circulatory system.

23. Describe the changes that occur in arteries as they age.

C R I T I C A L   T H I N K I N G

1. For each of the following destinations, name all the arteries that a red blood cell encounters if it starts its journey in the left ventricle:
   a. The brain
   b. External part of the skull
   c. The left hand
   d. Anterior portion of the right leg

2. For each of the following starting places, name all the veins that a red blood cell encounters on its way back to the right atrium:
   a. The left side of the brain
   b. External part of the right side of the skull
   c. The left hand
   d. Medial portion of right leg
   e. Kidney
   f. Small intestine

3. In angioplasty, a catheter is threaded through blood vessels to a blocked coronary artery. The tip of the catheter can expand, stretching the coronary artery and unblocking it, or the tip of the catheter can be equipped with tiny blades that can remove the blockage. Typically, the catheter is first inserted into a large blood vessel in the superior, medial part of the thigh. Starting with this blood vessel, name all the blood vessels that the catheter passes through to reach the anterior interventricular artery.

4. A 55-year-old man has a colonoscopy to check for colon cancer. A large tumor is found in his colon. His doctor orders a liver scan to determine if the cancer has spread from the colon to the liver. Based on your knowledge of blood vessels, explain why it makes sense that cancer cells from the colon could end up in the liver.

5. High blood pressure can be caused by advanced atherosclerosis of the renal arteries, even though there appears to be enough blood flow to allow a normal volume of urine to be produced. Explain how atherosclerosis of the renal arteries can result in high blood pressure.

6. Hugo Faster ran a race. During the race his stroke volume and heart rate increased. Vasoconstriction occurred in his viscera, and his blood pressure increased, but not dramatically. Explain these changes in his circulatory system.

7. Nitroglycerin is a drug often given to people who suffer from angina pains. This drug causes vasodilation of arteries and veins, which results in a reduced amount of work performed by the heart, and increases blood flow through coronary arteries. Explain why dilation of arteries and veins reduces the amount of work performed by the heart.

Answers in Appendix D

Visit this textbook's website at www.mhhe.com/seeleyess6 for practice quizzes, animations, interactive learning exercises, and other study tools.

Chapter Outline and Objectives

The Lymphatic System and Immunity

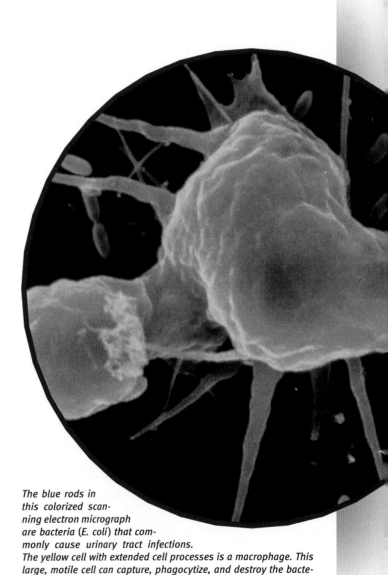

The blue rods in this colorized scanning electron micrograph are bacteria (E. coli) that commonly cause urinary tract infections.
The yellow cell with extended cell processes is a macrophage. This large, motile cell can capture, phagocytize, and destroy the bacteria. The immune system functions to provide protection against microorganisms.

One of the basic tenets of life is that many organisms consume or use other organisms in order to survive. Some microorganisms, such as certain bacteria or viruses, use humans as a source of nutrients and as an environment where they can survive and reproduce. As a result, some of these microorganisms can damage the body, causing disease or even death. Not surprisingly, the body has ways to resist or destroy harmful microorganisms. This chapter considers how the lymphatic system and the components of other systems, such as white blood cells and phagocytes, continually provide protection against invading microorganisms.

Lymphatic System

The **lymphatic** (lim-fat'ik) **system** includes lymph, lymphocytes, lymphatic vessels, lymph nodes, tonsils, the spleen, and the thymus gland (figure 14.1).

Functions of the Lymphatic System

The lymphatic system is part of the body's defense system against microorganisms and other harmful substances. In addition, it helps to maintain fluid balance in tissues and to absorb fats from the digestive tract.

1. *Fluid balance.* About 30 liters (L) of fluid pass from the blood capillaries into the interstitial spaces each day, whereas only 27 L pass from the interstitial spaces back into the blood capillaries (see chapter 13). If the extra 3 L of interstitial fluid remained in the interstitial spaces, edema would result, causing tissue damage and eventually death. The 3 L of fluid enters the lymphatic capillaries, where the fluid is called **lymph** (limf, meaning clear spring water), and it passes through the lymphatic vessels to return to the blood. In addition to water, lymph contains solutes derived from two sources: (a) substances in plasma, such as ions, nutrients, gases, and some proteins, pass from blood capillaries into the interstitial spaces and become part of the lymph; and (b) substances, such as hormones, enzymes, and waste products, derived from cells within the tissues are also part of the lymph.

2. *Fat absorption.* The lymphatic system absorbs fats and other substances from the digestive tract (see figure 16.14).

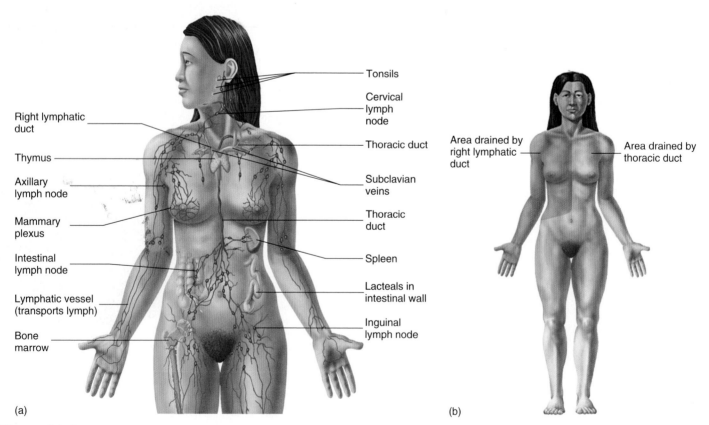

Labels (a): Right lymphatic duct, Thymus, Axillary lymph node, Mammary plexus, Intestinal lymph node, Lymphatic vessel (transports lymph), Bone marrow, Tonsils, Cervical lymph node, Thoracic duct, Subclavian veins, Thoracic duct, Spleen, Lacteals in intestinal wall, Inguinal lymph node

Labels (b): Area drained by right lymphatic duct, Area drained by thoracic duct

(a)

(b)

Figure 14.1   **Lymphatic System and Lymph Drainage**

(*a*) The major lymphatic organs and vessels are shown. (*b*) Lymph from the uncolored areas drains through the thoracic duct. Lymph from the brown areas drains through the right lymphatic duct.

Special lymphatic vessels called **lacteals** (lak′tē-ălz, relating to milk) are located in the lining of the small intestine. Fats enter the lacteals and pass through the lymphatic vessels to the venous circulation. The lymph passing through these lymphatic vessels has a milky appearance because of its fat content, and it is called **chyle** (kīl, juice).

3. *Defense.* Microorganisms and other foreign substances are filtered from lymph by lymph nodes and from blood by the spleen. In addition, lymphocytes and other cells are capable of destroying microorganisms and foreign substances.

Lymphatic Capillaries and Vessels

The lymphatic system, unlike the circulatory system, does not circulate fluid to and from tissues. Instead, the lymphatic system carries fluid in one direction, from tissues to the circulatory system. Fluid moves from blood capillaries into tissue spaces (see figure 13.24). Most of the fluid returns to the blood, but some of the fluid moves from the tissue spaces into lymphatic capillaries to become lymph (figure 14.2*a*). The **lymphatic capillaries** are tiny, closed-ended vessels consisting of simple squamous epithelium. The lymphatic capillaries are more permeable than blood capillaries because they lack a basement membrane, and fluid moves easily into the lymphatic capillaries. Overlapping squamous cells of the lymphatic capillary walls act as valves that prevent the back-flow of fluid (figure 14.2*b*). After fluid enters lymphatic capillaries, it flows through them.

Lymphatic capillaries are in most tissues of the body. Exceptions are the central nervous system, bone marrow, and tissues without blood vessels such as the epidermis and cartilage. A superficial group of lymphatic capillaries drains the dermis and hypodermis, and a deep group drains muscle, viscera, and other deep structures.

The lymphatic capillaries join to form larger **lymphatic vessels,** which resemble small veins (see figure 14.2*b*). Small lymphatic vessels have a beaded appearance because of one-way valves that are similar to the valves of veins. When a lymphatic vessel is compressed, backward movement of lymph is prevented by the valves. Consequently, compression of lymphatic vessels causes lymph to move forward through them. Three factors cause compression of the lymphatic vessels: (1) contraction of surrounding skeletal muscle during activity, (2) periodic contraction of smooth muscle in the lymphatic vessel wall, and (3) pressure changes in the thorax during respiration.

The lymphatic vessels converge and eventually empty into the blood at two locations in the body. Lymphatic vessels from the upper right limb and the right half of the head, neck, and chest form the **right lymphatic duct,** which empties into the right subclavian vein. Lymphatic vessels from the rest of the body enter the **thoracic duct,** which empties into the left subclavian vein (see figure 14.1).

Lymphatic Organs

Lymphatic organs include the tonsils, lymph nodes, the spleen, and the thymus gland. **Lymphatic tissue,** which consists of

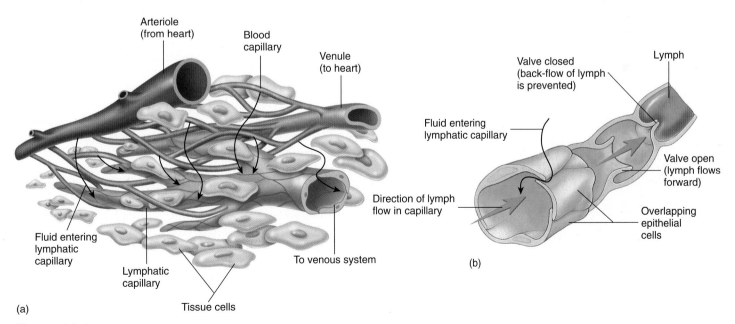

(a)

(b)

Figure 14.2   Lymph Formation and Movement

(*a*) Movement of fluid from blood capillaries into tissues and from tissues into lymphatic capillaries to form lymph. (*b*) The overlap of epithelial cells of the lymphatic capillary allows easy entry of fluid but prevents movement back into the tissue. Valves, located farther along in lymphatic vessels, also ensure one-way flow of lymph.

many lymphocytes and other cells such as macrophages, is found within lymphatic organs. The lymphocytes originate from red bone marrow (see chapter 11) and are carried by the blood to lymphatic organs. These lymphocytes divide and increase in number when the body is exposed to microorganisms or foreign substances. The increased number of lymphocytes is part of the immune response that causes the destruction of microorganisms and foreign substances. In addition to cells, lymphatic tissue has very fine reticular fibers (see chapter 4). These fibers form an interlaced network that holds the lymphocytes and other cells in place. When lymph or blood filters through lymphatic organs, the fiber network also traps microorganisms and other items in the fluid.

Tonsils

There are three groups of **tonsils** (figure 14.3 and figure 15.2). The **palatine** (pal′ă-tīn, palate) **tonsils** are located on each side of the posterior opening of the oral cavity. They usually are referred to as "the tonsils." The **pharyngeal** (fă-rin′jē-ăl) **tonsil,** is located near the internal opening of the nasal cavity. When the pharyngeal tonsil is enlarged, it is commonly referred to as the **adenoid** (ad′ē-noid, glandlike) or **adenoids.** An enlarged pharyngeal tonsil can interfere with normal breathing. The **lingual** (ling′gwăl, tongue) **tonsil** is on the posterior surface of the tongue.

The tonsils form a protective ring of lymphatic tissue around the openings between the nasal and oral cavities and the pharynx. They provide protection against pathogens and other potentially harmful material entering from the nose and mouth. Sometimes the palatine or pharyngeal tonsils become chronically infected and must be removed. The lingual tonsil becomes infected less often than the other tonsils and is more difficult to remove. In adults, the tonsils decrease in size and may eventually disappear.

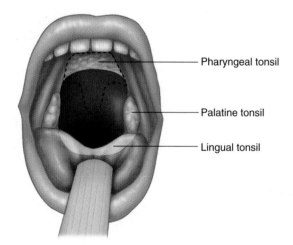

Figure 14.3

Anterior view of the oral cavity showing the tonsils. Part of the palate is removed (*dotted line*) to show the pharyngeal tonsil.

- Pharyngeal tonsil
- Palatine tonsil
- Lingual tonsil

A CASE IN POINT | Tonsillectomy and Adenoidectomy

Audie Tory is a 4-year-old child who has some hearing loss and an associated delay in speech development. She has a history of frequent sore throats and middle ear infections, which have been treated with antibiotics. Recently, she has experienced difficulty in swallowing and snores and sleeps with her mouth open. Audie's physician tells her parents she has enlarged and chronically infected (palatine) tonsils and adenoid. Her enlarged tonsils make swallowing difficult and infected tonsils stimulate inflammation that causes her throat to hurt. Her adenoid is restricting air flow, causing her to snore and sleep with her mouth open. In addition, an enlarged and/or infected adenoid is associated with chronic middle ear infections because the openings to the auditory tubes are located next to the adenoid. Chronic middle ear infections are associated with loss of hearing, which, in turn, affects speech development. The doctor recommends a procedure called a T&A—that is, a tonsillectomy (ton′si-lek′tō-mē), which is removal of the palatine tonsils, and an adenoidectomy (ad′ě-noy-dek′tō-mē), which is removal of the adenoid.

Lymph Nodes

Lymph nodes are rounded structures, varying in size from that of small seeds to that of shelled almonds. Lymph nodes are distributed along the various lymphatic vessels (see figure 14.1), and most lymph passes through at least one lymph node before entering the blood. Although lymph nodes are found throughout the body, there are three superficial aggregations of lymph nodes on each side of the body: inguinal nodes in the groin, axillary nodes in the axilla (armpit), and cervical nodes in the neck.

A dense connective tissue **capsule** surrounds each lymph node (figure 14.4). Extensions of the capsule, called **trabeculae,** subdivide lymph nodes into compartments containing lymphatic tissue and lymphatic sinuses. The lymphatic tissue consists of lymphocytes and other cells that can form dense aggregations of tissue called **lymph nodules. Lymphatic sinuses** are spaces between lymphatic tissue which contain macrophages on a network of fibers. Lymph enters the lymph node through afferent vessels, passes through the lymphatic tissue and sinuses, and exits through efferent vessels.

As lymph moves through the lymph nodes, two functions are performed. One function is activation of the immune system. Microorganisms or other foreign substances in the lymph can stimulate lymphocytes in the lymphatic tissue to divide. The lymph nodules containing the rapidly dividing lymphocytes are called **germinal centers.** The newly produced lymphocytes are released into the lymph and eventually reach the blood, where they circulate and enter other lymphatic tissues. The lymphocytes

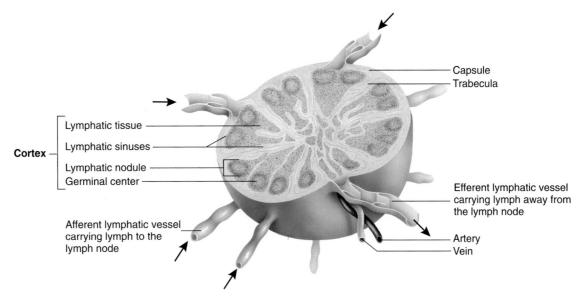

Capsule
Trabecula

Cortex

Lymphatic tissue
Lymphatic sinuses
Lymphatic nodule
Germinal center

Efferent lymphatic vessel carrying lymph away from the lymph node

Afferent lymphatic vessel carrying lymph to the lymph node

Artery
Vein

Figure 14.4 **Lymph Node**

Arrows indicate direction of lymph flow. As lymph moves through the sinuses, phagocytic cells remove foreign substances. The germinal centers are sites of lymphocyte production.

are part of the adaptive immune response (see p. 397) that destroys microorganisms and foreign substances. Another function of the lymph nodes is the removal of microorganisms and foreign substances from the lymph by macrophages.

PREDICT 1

Cancer cells can spread from a tumor site to other areas of the body through the lymphatic system. At first, however, as the cancer cells pass through the lymphatic system they are trapped in the lymph nodes, which filter the lymph. During radical cancer surgery, malignant (cancerous) lymph nodes are removed, and their vessels are cut and tied off to prevent the spread of the cancer. Predict the consequences of tying off the lymphatic vessels.

Spleen

The **spleen** (splēn) is roughly the size of a clenched fist, and it is located in the left, superior corner of the abdominal cavity (figure 14.5). The spleen has an outer **capsule** of dense connective tissue and a small amount of smooth muscle. **Trabeculae** from the capsule divide the spleen into small, interconnected compartments containing two specialized types of lymphatic tissue. **White pulp** is lymphatic tissue surrounding the arteries within the spleen. **Red pulp** is associated with the veins. It consists of a fibrous network, filled with macrophages and red blood cells, and enlarged capillaries that connect to the veins.

The spleen filters blood instead of lymph. Cells within the spleen detect and respond to foreign substances in the blood

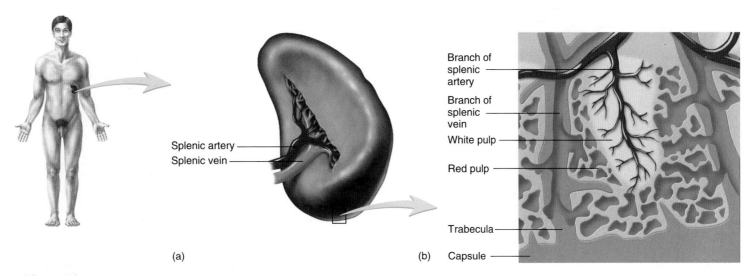

Splenic artery
Splenic vein

(a)

Branch of splenic artery
Branch of splenic vein
White pulp
Red pulp
Trabecula
Capsule

(b)

Figure 14.5 **Spleen**

(*a*) Inferior view of the spleen. (*b*) Section showing the arrangement of arteries, veins, white pulp, and red pulp. White pulp is associated with arteries, and red pulp is associated with veins.

and destroy worn-out red blood cells. Lymphocytes in the white pulp can be stimulated in the same manner as in lymph nodes. Before blood leaves the spleen through veins, it passes through the red pulp. Macrophages in the red pulp remove foreign substances and worn-out red blood cells through phagocytosis.

The spleen also functions as a blood reservoir, holding a small volume of blood. In emergency situations such as hemorrhage, smooth muscle in splenic blood vessels and in the splenic capsule can contract. The result is the movement of a small amount of blood out of the spleen into the general circulation.

The Ruptured Spleen

Although the spleen is protected by the ribs, it is often ruptured in traumatic abdominal injuries. A ruptured spleen can cause severe bleeding, shock, and possibly death. Surgical intervention may stop the bleeding. Cracks in the spleen are repaired using sutures and blood-clotting agents. Mesh wrapped around the spleen can hold it together. A **splenectomy** (splē-nek′tō-mē), removal of the spleen, may be necessary if these techniques do not stop the bleeding. Other lymphatic organs and the liver compensate for the loss of the spleen's functions.

Thymus

The **thymus** (thī′mŭs, sweetbread) is a bilobed gland roughly triangular in shape (figure 14.6). It is located in the superior mediastinum, the partition dividing the thoracic cavity into left and right parts. It was once thought that the thymus increases in size until puberty after which it dramatically decreases in size. It is now believed that the thymus increases in size until the first year of life, after which it remains approximately the same size, even though the size of the individual increases. After 60 years of age, it decreases in size, and in older adults, the thymus may be so small that it is difficult to find during dissection. Although the size of the thymus is fairly constant throughout much of life, by 40 years of age much of the thymus has been replaced with adipose tissue.

Each lobe of the thymus is surrounded by a thin connective tissue **capsule. Trabeculae** from the capsule divide each lobe into **lobules.** Near the capsule and trabeculae, the lymphocytes are numerous and form dark-staining areas called the **cortex.** A lighter staining central portion of the lobules, called the **medulla,** has fewer lymphocytes.

The thymus functions as a site for the production and maturation of lymphocytes. Large numbers of lymphocytes are produced in the thymus, but for unknown reasons, most degenerate. While in the thymus, lymphocytes do not respond to foreign substances. After thymic lymphocytes have matured, however, they enter the blood and travel to other lymphatic tissues, where they help to protect against microorganisms and other foreign substances.

Overview of the Lymphatic System

Figure 14.7 summarizes the parts of the lymphatic system and their functions. Lymphatic capillaries and vessels remove fluid from tissues and absorb fats from the small intestines. Lymph nodes filter lymph and the spleen filters blood.

Figure 14.7 also illustrates two types of lymphocytes called B and T cells. B cells originate and mature in red bone marrow. Pre-T cells are produced in red bone marrow and migrate to the thymus, where they mature to become T cells. B cells from red bone marrow and T cells from the thymus circulate to, and populate, other lymphatic tissues.

B and T cells are responsible for much of immunity. In response to infections, B and T cells increase in number and circulate to lymphatic and other tissues. How B and T cells protect us is discussed in the section on Adaptive Immunity on p. 397.

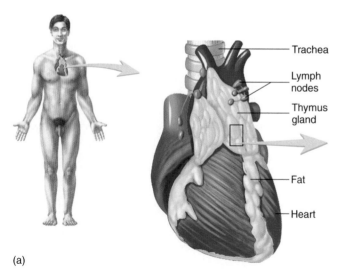

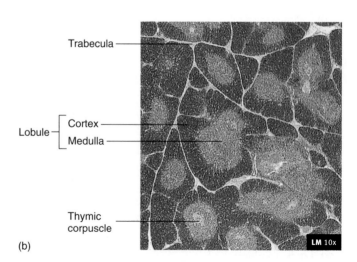

Figure 14.6

(*a*) Location and shape of the thymus. (*b*) Histology of thymic lobules, showing outer cortex and inner medulla.

(a) Lymphatic capillaries remove fluid from tissues. The fluid becomes lymph (see figure 14.2a).

(b) Lymph flows through lymphatic vessels, which have valves that prevent the back-flow of lymph (see figure 14.2b).

(c) Lymph nodes filter lymph (see figure 14.4) and are sites where lymphocytes respond to infections, etc.

(d) Lymph enters the thoracic duct or the right lymphatic duct.

(e) Lymph enters the blood.

(f) Lacteals in the small intestine (see figure 16.14) absorb fats, which enter the thoracic duct.

(g) Chyle, which is lymph containing fats, enters the blood.

(h) The spleen (see figure 14.5) filters blood and is a site where lymphocytes respond to infections, etc.

(i) Lymphocytes (pre-B and pre-T cells) originate from stem cells in the red bone marrow (see figure 14.9). The pre-B cells become mature B cells in the red bone marrow and are released into the blood. The pre-T cells enter the blood and migrate to the thymus.

(j) The thymus (see figure 14.6) is where pre-T cells derived from red bone marrow increase in number and become mature T cells that are released into the blood (see figure 14.9).

(k) B and T cells from the blood enter and populate all lymphatic tissues. These lymphocytes can remain in tissues or pass through them and return to the blood. B and T cells can also respond to infections, etc. by dividing and increasing in number. Some of the newly formed cells enter the blood and circulate to other tissues.

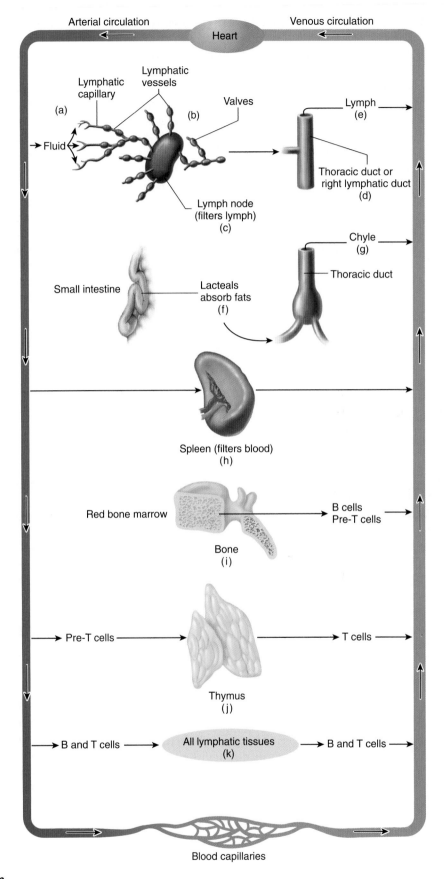

Figure 14.7 Overview of the Lymphatic System

It is not surprising that many infectious diseases produce symptoms associated with the lymphatic system, because the lymphatic system is involved with the production of lymphocytes that fight infectious diseases, as well as filtering blood and lymph to remove microorganisms. **Lymphadenitis** (lim-fad′ĕ-nī′tis) is an inflammation of the lymph nodes, causing them to become enlarged and tender. It is an indication that microorganisms are being trapped and destroyed within the lymph nodes. **Lymphangitis** (lim-fan-jī′tis) is an inflammation of the lymphatic vessels. This often results in visible red streaks in the skin that extend away from the site of infection. If microorganisms pass through the lymphatic vessels and nodes to reach the blood, **septicemia** (sep-ti-sē′mē-ă), or blood poisoning, can result (see chapter 11).

A **lymphoma** (lim-fō′mă) is a neoplasm (tumor) of lymphatic tissue that is almost always malignant. Lymphomas are usually divided into two groups: **Hodgkin's disease** and all other lymphomas, which are called **non-Hodgkin's lymphomas.** The different types of lymphomas are diagnosed based on their histological appearance and cell of origin.

Typically, a lymphoma begins as an enlarged, painless mass of lymph nodes. Enlargement of the lymph nodes can compress surrounding structures and produce complications. The immune system is depressed, and the patient has an increased susceptibility to infections. Fortunately, treatment with drugs and radiation is effective for many people who suffer from lymphomas.

Bubonic plague and elephantiasis are diseases of the lymphatic system. **Bubonic** (bū-bon′ik) **plague** is caused by bacteria that are transferred to humans from rats by the bite of the rat flea. The bacteria localize in the lymph nodes, causing them to enlarge. The term bubonic is derived from a Greek word referring to the groin, because the disease often causes the inguinal lymph nodes of the groin to swell. Without treatment, septicemia followed rapidly by death occurs in 70–90% of those infected. In the sixth, fourteenth, and nineteenth centuries the bubonic plague killed large numbers of people in Europe. Fortunately, there are relatively few cases today.

Elephantiasis (el-ĕ-fan-tī′ă-sis) is caused by long, slender roundworms. The adult worms lodge in the lymphatic vessels and can cause blockage of lymph flow. Consequently, edema develops and a limb can become permanently swollen and enlarged. The resemblance of an affected limb to that of an elephant's leg is the basis for the name of the disease. The offspring of the adult worms pass through the lymphatic system into the blood. They can be transferred from an infected person to other humans by mosquitoes.

Immunity

Immunity (i-mū′ni-tē) is the ability to resist damage from foreign substances, such as microorganisms, and harmful chemicals, such as toxins released by microorganisms. Immunity is categorized as **innate** (i′nāt, i-nāt′) **immunity** (also called nonspecific resistance) or **adaptive immunity** (also called specific immunity). In innate immunity, the body recognizes and destroys certain foreign substances, but the response to them is the same each time the body is exposed to them. In adaptive immunity, the body recognizes and destroys foreign substances, but the response to them improves each time the foreign substance is encountered.

Specificity and memory are characteristics of adaptive immunity but not innate immunity. **Specificity** is the ability of adaptive immunity to recognize a particular substance. For example, innate immunity can act against bacteria in general, whereas adaptive immunity can distinguish among different kinds of bacteria. **Memory** is the ability of adaptive immunity to "remember" previous encounters with a particular substance. As a result, the response is faster, stronger, and lasts longer.

In innate immunity, each time the body is exposed to a substance, the response is the same because specificity and memory of previous encounters are not present. For example, each time a bacterial cell is introduced into the body, it is phagocytized with the same speed and efficiency. In adaptive immunity, the response during the second exposure is faster and stronger than the response to the first exposure because the immune system exhibits memory for the bacteria from the first exposure. For example, following the first exposure to the bacteria, the body can take many days to destroy them. During this time the bacteria damage tissues, producing the symptoms of disease. Following the second exposure to the same bacteria, the response is rapid and effective. Bacteria are destroyed before any symptoms develop, and the person is said to be **immune.**

Innate Immunity

Innate immunity is accomplished by mechanical mechanisms, chemical mediators, cells, and the inflammatory response.

Mechanical Mechanisms

Mechanical mechanisms prevent the entry of microorganisms and chemicals into the body in two ways: (1) the skin and mucous membranes form barriers that prevent their entry, and (2) tears, saliva, and urine act to wash them from the surfaces of the body. Microorganisms cannot cause a disease if they cannot get into the body.

Chemical Mediators

Chemical mediators are molecules responsible for many aspects of innate immunity. Some chemicals that are found on the surface of cells kill microorganisms or prevent their entry into the cells. Lysozyme in tears and saliva kills certain bacteria, and mucus on the mucous membranes prevents the entry of some microorganisms. Other chemical mediators, such as histamine (his′tă-mēn), complement, prostaglandins (pros-tă-glan′dinz), and leukotrienes (loo-kō-trī′ēnz), promote inflammation by causing vasodilation, increasing vascular permeability, and stimulating phagocytosis. In addition, interferons protect cells against viral infections.

Complement

Complement (kom′plĕ-ment) is a group of approximately 20 proteins found in plasma. The operation of complement proteins is similar to that of clotting proteins (see chapter 11). Normally, complement proteins circulate in the blood in an inactive form. Certain complement proteins can be activated by combining with foreign substances, such as parts of a bacterial cell, or by combining with antibodies (see the discussion, Effects of Antibodies, on p. 401). Once activation begins, a series of reactions results, in which each complement protein activates the next. Once activated, certain complement proteins promote inflammation and phagocytosis and can directly lyse (rupture) bacterial cells.

Interferons

Interferons (in-ter-fēr′onz) are proteins that protect the body against viral infections. When a virus infects a cell, the infected cell produces viral nucleic acids and proteins, which are assembled into new viruses. The new viruses are then released to infect other cells. Because infected cells usually stop their normal functions or die during viral replication, viral infections are clearly harmful to the body. Fortunately, viruses often stimulate infected cells to produce interferons. Interferons do not protect the cell that produces them. Instead, interferons bind to the surface of neighboring cells where they stimulate those cells to produce antiviral proteins. These antiviral proteins inhibit viral reproduction by preventing the production of new viral nucleic acids and proteins.

Some interferons play a role in the activation of immune cells such as macrophages and natural killer cells (see the following Cells section).

Treating Viral Infections and Cancer with Interferons

Because some cancers are induced by viruses, interferons may play a role in controlling cancers. Interferons activate macrophages and natural killer cells (a type of lymphocyte, see following Cells section), which attack tumor cells. Through genetic engineering, interferons currently are produced in sufficient quantities for clinical use and, along with other therapies, have been effective in treating certain viral infections and cancers. For example, interferons are used to treat hepatitis C, a viral disorder that can cause cirrhosis and cancer of the liver, and to treat genital warts, caused by the herpes virus. Interferons are also approved for the treatment of Kaposi's sarcoma, a cancer that can develop in AIDS patients.

Cells

White blood cells and the cells derived from white blood cells (see chapter 11) are the most important cellular components of immunity. White blood cells are produced in red bone marrow and lymphatic tissue and are released into the blood. Chemicals released from microorganisms or damaged tissues attract the white blood cells, and they leave the blood and enter affected tissues. Important chemicals known to attract white blood cells include complement, leukotrienes, kinins (kī′ ninz), and histamine. The movement of white blood cells toward these chemicals is called **chemotaxis** (kem-ō-tak′sis, kē-mō-tak′sis).

Phagocytic Cells

Phagocytosis (fag′ō-sī-tō′sis) is the ingestion and destruction of particles by cells called **phagocytes** (fag′ō-sītz) (see chapter 3). The particles can be microorganisms or their parts, foreign substances, or dead cells from the individual's body. The most important phagocytes are neutrophils and macrophages, although other white blood cells also have limited phagocytic ability.

Neutrophils (noo′trō-filz) are small phagocytic cells that are usually the first cells to enter infected tissues from the blood in large numbers; however, neutrophils often die after phagocytizing a single microorganism. **Pus** is an accumulation of fluid, dead neutrophils, and other cells at a site of infection.

Macrophages (mak′rō-fā′jes) are monocytes that leave the blood, enter tissues, and enlarge about fivefold. Monocytes and macrophages form the **mononuclear phagocytic system** because they are phagocytes with a single (mono), unlobed nucleus. Sometimes macrophages are given specific names such as dust cells in the lungs, Kupffer cells in the liver, and microglia in the central nervous system. Macrophages can ingest more and larger items than can neutrophils. Macrophages usually appear in infected tissues after neutrophils and are responsible for most of the phagocytic activity in the late stages of an infection, including the cleanup of dead neutrophils and other cellular debris.

In addition to leaving the blood in response to an infection, macrophages are also found in uninfected tissues. If microorganisms enter uninfected tissue, the macrophages may phagocytize the microorganisms before they can replicate or cause damage. For example, macrophages are located at potential points of entry for microorganisms into the body, such as beneath the skin and mucous membranes, and around blood and lymphatic vessels. They also protect lymph in lymph nodes and blood in the spleen and liver.

Cells of Inflammation

Basophils, which are derived from red bone marrow, are motile white blood cells that can leave the blood and enter infected tissues. **Mast cells,** which are also derived from red bone marrow, are nonmotile cells in connective tissue, especially near capillaries. Like macrophages, mast cells are located at potential points of entry for microorganisms into the body such as the skin, lungs, gastrointestinal tract, and urogenital tract.

Basophils and mast cells can be activated through innate immunity (e.g., by complement) or through adaptive immunity (see the Antibodies section, on p. 401). When activated, they release chemicals such as histamine and leukotrienes that produce an inflammatory response or activate other mechanisms such as smooth muscle contraction in the lungs.

Eosinophils are produced in red bone marrow, enter the blood, and within a few minutes enter tissues. Enzymes released by eosinophils break down chemicals released by basophils and mast cells. Thus at the same time that inflammation is initiated, mechanisms are activated that contain and reduce the inflammatory response.

Inflammation is beneficial in the fight against microorganisms, but too much inflammation can be harmful, resulting in the unnecessary destruction of healthy tissues as well as the destruction of the microorganisms.

Natural Killer Cells

Natural killer (NK) cells are a type of lymphocyte produced in red bone marrow, and they account for up to 15% of lymphocytes. NK cells recognize classes of cells, such as tumor cells or virus-infected cells in general, rather than specific tumor cells or cells infected by a specific virus. For this reason, and because NK cells do not exhibit a memory response, NK cells are classified as part of innate immunity. NK cells use a variety of methods to kill their target cells, including the release of chemicals that damage cell membranes, causing the cells to lyse.

Inflammatory Response

The **inflammatory response** to injury involves many of the chemicals and cells previously discussed. Most inflammatory responses are very similar, although some details can vary depending on the intensity of the response and the type of injury. A bacterial infection is used here to illustrate an inflammatory response (figure 14.8). The bacteria, or damage to tissues, cause the release or activation of chemical mediators, such as histamine, prostaglandins, leukotrienes, complement, and kinins. The chemicals produce several effects: (1) vasodilation, which increases blood flow and brings phagocytes and other white blood cells to the area; (2) chemotactic attraction of phagocytes, which leave the blood and enter the tissue; and (3) increased vascular permeability, allowing fibrinogen and complement to enter the tissue from the blood. Fibrinogen is converted to fibrin (see chapter 11), which isolates the infection by walling off the infected area. Complement further enhances the inflammatory response and attracts additional phagocytes. This process of releasing chemical mediators and attracting phagocytes and other white blood cells continues until the bacteria are destroyed. Phagocytes remove microorganisms and dead tissue, and the damaged tissues are repaired.

Inflammation can be localized or systemic. **Local inflammation** is an inflammatory response confined to a specific area of the body. Symptoms of local inflammation include redness, heat, swelling, pain, and loss of function. Redness, heat, and swelling result from increased blood flow and increased vascular permeability. Pain is caused by swelling and by chemical

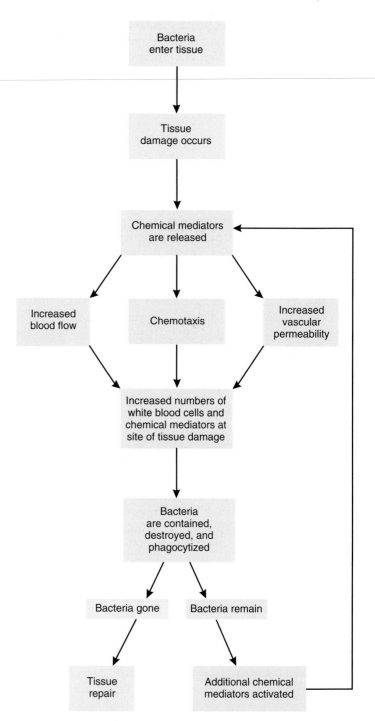

Figure 14.8   Inflammatory Response

Bacteria cause tissue damage and the release of chemical mediators that initiate inflammation and phagocytosis, resulting in the destruction of the bacteria. If any bacteria remain, additional chemical mediators are activated. After all the bacteria are destroyed, the tissue is repaired.

mediators acting on pain receptors. Loss of function results from tissue destruction, swelling, and pain. See the Inflammation section in chapter 4 for more details on the symptoms of inflammation

Systemic inflammation is an inflammatory response that is generally distributed throughout the body. In addition to the local symptoms at the sites of inflammation, three additional features can be present:

1. Red bone marrow produces and releases large numbers of neutrophils, which promote phagocytosis.
2. **Pyrogens** (pī′rō-jenz, fever producing), chemicals released by microorganisms, neutrophils, and other cells, stimulate fever production. Pyrogens affect the body temperature-regulating mechanism in the hypothalamus of the brain. As a consequence, heat production and conservation increase, and body temperature increases. Fever promotes the activities of the immune system, such as phagocytosis, and inhibits the growth of some microorganisms.
3. In severe cases of systemic inflammation, vascular permeability can increase so much that large amounts of fluid are lost from the blood into the tissues. The decreased blood volume can cause shock and death.

Adaptive Immunity

Adaptive immunity exhibits specificity and memory. Specificity is the ability to recognize a particular substance, and memory is the ability to respond with increasing effectiveness to successive exposures to the antigen. Substances that stimulate adaptive immune responses are called **antigens** (an′ti-jenz, anti(body)+ -*gen*, producing). Antigens can be divided into two groups: foreign antigens and self-antigens. **Foreign antigens** are introduced from outside the body. Microorganisms, such as bacteria and viruses, cause diseases, and components of microorganism and chemicals released by microorganisms are examples of foreign antigens. Pollen, animal hairs, foods, and drugs can cause an **allergic reaction** because they are foreign antigens that produce an overreaction of the immune system. Transplanted tissues and organs contain foreign antigens, and the response to these antigens can result in the rejection of the transplant.

Self-antigens are molecules produced by the person's body that stimulate an immune system response. The response to self-antigens can be beneficial. For example, the recognition of tumor antigens can result in destruction of the tumor. The response to self-antigens can also be harmful. **Autoimmune disease** results when self-antigens stimulate unwanted destruction of normal tissue. An example is rheumatoid arthritis, which results in the destruction of tissue within joints.

The adaptive immune system response to antigens was historically divided into two parts: **humoral immunity** and **cell-mediated immunity.** Early investigators of the immune system found that when plasma from an immune animal was injected into the blood of a nonimmune animal, the nonimmune animal became immune. Because this process involved body fluids (humors), it was called humoral immunity. It was also discovered that blood cells alone could be responsible for immunity, and this process was called cell-mediated immunity.

It is now known that both types of immunity involve the activities of lymphocytes. There are two types of lymphocytes: B cells and T cells. **B cells** give rise to cells that produce proteins called **antibodies** (an′tĕ-bod-ēz), which are found in the plasma. The antibodies are responsible for humoral immunity, which is now called **antibody-mediated immunity.**

T cells are responsible for cell-mediated immunity. Several subpopulations of T cells exist. For example, **cytotoxic** (sī-tō-tok′sik, destructive to cells) **T cells** produce the effects of cell-mediated immunity and **helper T** cells can promote or inhibit the activities of both antibody-mediated immunity and cell-mediated immunity.

Table 14.1 summarizes and contrasts the main features of innate immunity and adaptive immunity (i.e., antibody-mediated immunity and cell-mediated immunity).

Origin and Development of Lymphocytes

To understand how lymphocytes are responsible for antibody-mediated and cell-mediated immunity, it is important to know how lymphocytes originate and become specialized immune cells. **Stem cells** in red bone marrow are cells that are capable of giving rise to all the blood cells (see figure 11.2). Some stem cells give rise to pre-T cells, which migrate through the blood to the thymus gland, where they divide and are processed into T cells (figure 14.9). Other stem cells produce pre-B cells, which are processed in the red bone marrow into B cells.

B cells are released from red bone marrow, and T cells are released from the thymus. Both types of cells move through the blood to lymphatic tissues (see figure 14.7). These lymphocytes live for a few months to many years and continually circulate between the blood and the lymphatic tissues. Normally there are about five T cells for every B cell in the blood. When stimulated by an antigen, B cells and T cells divide, producing cells that are responsible for the destruction of antigens.

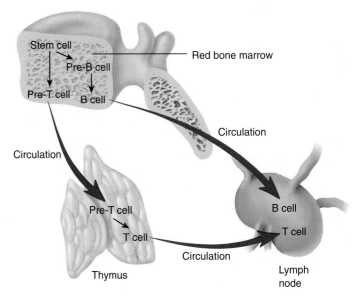

Figure 14.9   Origin and Processing of B and T Cells

Both B cells and T cells originate from stem cells in red bone marrow. B cells are processed from pre-B cells in the red marrow, whereas T cells are processed from pre-T cells in the thymus. Both B and T cells circulate to other lymphatic tissues, such as lymph nodes.

Table 14.1 Comparison of Innate and Adaptive Immunity

Primary Cells	Origin of Cells	Site of Maturation
Innate Immunity		
Neutrophils, eosinophils, basophils, mast cells, monocytes, and macrophages	Red bone marrow	Red bone marrow (neutrophils, eosinophils, basophils, and monocytes) and tissues (mast cells and macrophages)
Adaptive Immunity		
Antibody-mediated immunity B cells	Red bone marrow	Red bone marrow
Cell-mediated immunity T cells	Red bone marrow	Thymus

Evidence suggests that small groups of identical B or T cells, called **clones,** are formed during embryonic development. Each clone is derived from a single, unique B or T cell. Each clone can respond only to a particular antigen. However, there is such a large variety of clones that the immune system can react to most antigens. Among the antigens to which the clones can respond, however, are self-antigens. Because this response could destroy self-cells, clones acting against self-antigens are normally eliminated or suppressed. Most of this process occurs during prenatal development, but it also continues after birth and throughout a person's lifetime.

Activation and Multiplication of Lymphocytes

The specialized B or T cell clones can respond to antigens and produce an adaptive immune response. For the adaptive immune response to be effective, two events must occur: (1) antigen recognition by lymphocytes and (2) proliferation of the lymphocytes recognizing the antigen.

Antigen Recognition

Lymphocytes have proteins, called **antigen receptors,** on their surfaces. The antigen receptors on B cells are called **B-cell receptors** and those on T cells are called **T-cell receptors.** Each receptor binds with only a specific antigen. Each clone consists of lymphocytes that have identical antigen receptors on their surfaces. When antigens combine with the antigen receptors of a clone, the lymphocytes in that clone can be activated and the adaptive immune response begins.

B and T cells typically recognize antigens after large molecules have been processed or broken down into smaller fragments. Antigen-presenting cells, such as macrophages, present antigens to B and T cells. The antigens are taken into macrophages by phagocytosis and are broken down into smaller antigen fragments. The processed antigen fragments are bound to major histocompatibility complex molecules, transported to the surface of the macrophages, and presented (figure 14.10, step 1).

Major histocompatibility complex (MHC) molecules are glycoproteins, and they have binding sites for antigens. Different MHC molecules have different binding sites, that is, they are specific for certain antigens. The MHC molecules function as "serving trays" that hold and present a processed antigen on the outer surface of the cell membrane. The combined MHC molecule and processed antigen can then bind to the antigen receptor on a B or T cell and stimulate them. For example, a special type of T cell, called a **helper T cell,** can be stimulated (figure 14.10, step 2)

The MHC molecule/antigen combination is usually only the first signal necessary to produce a response from a B or T cell. In many cases, **costimulation** by a second signal is also required. Costimulation can be achieved by **cytokines** (sī′tō-kīnz, cell movement), which are proteins or peptides secreted by one cell as a regulator of neighboring cells. For example, **interleukin-1** (in-ter-loo′kin, between white blood cells) is a cytokine released by macrophages, which can stimulate helper T cells (figure 14.10, step 3).

Lymphocytes have other surface molecules besides MHC molecules that help to bind cells together and stimulate a response. For example, helper T cells have a glycoprotein called CD4, which helps to connect helper T cells to the macrophage by binding to MHC molecules. For this reason, helper T cells are sometimes referred to as **CD4 or T4, cells.** In a similar fashion, cytotoxic T cells are sometimes called **CD8,** or **T8, cells** because they have a glycoprotein called CD8, which helps to connect cytotoxic T cells to cells displaying MHC molecules. The CD designation stands for "cluster of differentiation," which is a system used to classify many surface molecules.

Lymphocyte Proliferation

Before exposure to a particular antigen, the number of helper T cells that can respond to that antigen is too small to produce an effective response against it. After the antigen is processed and presented to a helper T cell by a macrophage, the helper T cell

Location of Mature Cells	Primary Secretory Product	Primary Actions	Allergic Reactions
Blood, connective tissue, and lymphatic tissue	Histamine, complement, prostaglandins, leukotrienes, kinins, and interferons	Inflammatory response and phagocytosis	None
Blood and lymphatic tissue	Antibodies	Protection against extracellular antigens (bacteria, toxins, and viruses outside of cells)	Immediate hypersensitivity
Blood and lymphatic tissue	Cytokines	Protection against intracellular antigens (viruses and intracellular bacteria) and tumors; responsible for graft rejection	Delayed hypersensitivity

1. Antigen-presenting cells such as macrophages phagocytize, process, and display antigens on the cell's surface.

2. The antigens are bound to major histocompatibility complex (MHC) molecules, which function to present the processed antigen to the T-cell receptor of the helper T cell.

3. Costimulation results from interleukin-1, secreted by the macrophage, and the CD4 glycoprotein of the helper T cell.

4. Interleukin-1 stimulates the helper T cell to secrete interleukin-2 and to produce interleukin-2 receptors.

5. The helper T cell stimulates itself to divide when interleukin-2 binds to interleukin-2 receptors.

6. The "daughter" helper T cells resulting from this division can be stimulated to divide again if they are exposed to the same antigen that stimulated the "parent" helper T cell. This greatly increases the number of helper T cells.

7. The increased number of helper T cells can facilitate the activation of B cells or effector T cells.

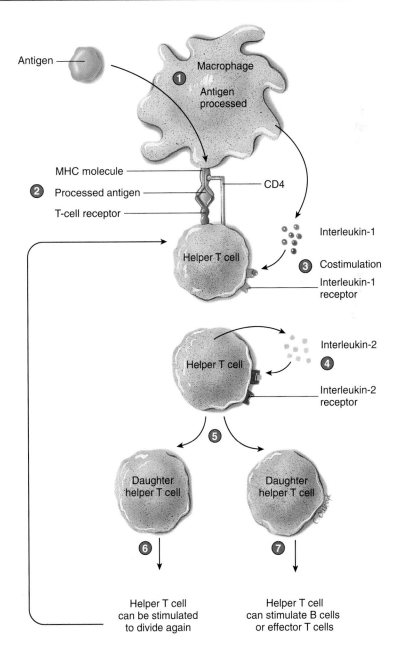

Process Figure 14.10 Proliferation of Helper T Cells

An antigen-presenting cell (macrophage) stimulates helper T cells to divide.

responds by producing **interleukin-2 receptors** and **interleukin-2** (figure 14.10, step 4). Interleukin-2 binds to the receptors and stimulates the helper T cell to divide (figure 14.10, step 5). The helper T cells produced by this division can again be presented with antigen by macrophages and again be stimulated to divide. Thus, the number of helper T cells is greatly increased (figure 14.10, step 6)

Inhibiting and Stimulating Immunity

Decreasing the production or activity of cytokines can suppress the immune system. For example, cyclosporine, a drug used to prevent the rejection of transplanted organs, inhibits the production of interleukin-2. Conversely, genetically engineered interleukins can be used to stimulate the immune system. Administering interleukin-2 has promoted the destruction of cancer cells in some cases by increasing the activities of T cells.

It is important for the number of helper T cells to increase because helper T cells are necessary for the activation of most B or T cells (figure 14.10, step 7). For example, B cells have receptors that can recognize antigens. Most B cells, however, do not respond to antigens without stimulation from helper T cells. This process begins when a B cell takes in the same kind of antigen that stimulated the helper T cell (figure 14.11, step 1). The antigen is processed by the B cell and presented on the B cell surface by a MHC molecule (figure 14.11, step 2). A helper T cell is stimulated when it binds to the MHC molecule/antigen complex (figure 14.11, step 3). There is also costimulation involving CD4 and interleukins (figure 14.11, steps 4 and 5). The result is the division of the B cell into two cells (figure 14.11, step 6). The division process continues, eventually producing many cells that are capable of producing antibodies (figure 14.11, step 7). Thus, many cells producing antibodies results in sufficient antibodies to destroy all of the antigen.

P R E D I C T ②

How does elimination of the antigen stop the production of antibodies?

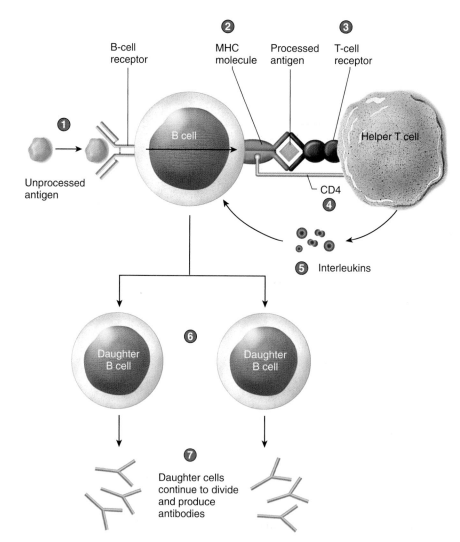

1. Before a B cell can be activated by a helper T cell, the B cell must phagocytize and process the same antigen that activated the helper T cell. The antigen binds to a B-cell receptor, and both the receptor and antigen are taken into the cell by endocytosis.

2. The B cell uses a MHC molecule to present the processed antigen to the helper T cell.

3. The T-cell receptor binds to the MHC molecule/antigen complex.

4. There is costimulation of the B cell by CD4 and other surface molecules.

5. There is costimulation by interleukins (cytokines) released from the helper T cell.

6. The B cell divides, and the resulting daughter cells divide, and so on, eventually producing many cells (only two are shown here).

7. The increased number of cells produce antibodies, which are part of the antibody-mediated immune system response that eliminates the antigen.

B-cell receptor

MHC molecule

Processed antigen

T-cell receptor

Unprocessed antigen

B cell

Helper T cell

CD4 ④

Interleukins ⑤

Daughter B cell

Daughter B cell

Daughter cells continue to divide and produce antibodies

Process Figure 14.11   Proliferation of B Cells

A helper T cell stimulates a B cell to divide.

Antibody-Mediated Immunity

Exposure of the body to an antigen can result in the activation of B cells and the production of antibodies. The antibodies bind to the antigens, and through several different mechanisms, the antigens can be destroyed. Because antibodies are in body fluids, antibody-mediated immunity is effective against extracellular antigens, such as bacteria, viruses (when they are outside cells), and toxins. Antibody-mediated immunity is also involved with certain allergic reactions.

Antibodies

Antibodies are proteins produced in response to an antigen. They are Y-shaped molecules consisting of four polypeptide chains: two identical heavy chains and two identical light chains (figure 14.12). The end of each "arm" of the antibody is the **variable region,** which is the part of the antibody that combines with the antigen. The variable region of a particular antibody can only join with a particular antigen. This is similar to the lock-and-key model of enzymes (see chapter 2). The rest of the antibody is the **constant region,** which has several functions. For example, the constant region can activate complement, or it can attach the antibody to cells such as macrophages, basophils, and mast cells.

Antibodies make up a large portion of the proteins in plasma. Most plasma proteins can be separated into albumin and alpha, beta, and gamma globulin portions. Antibodies are called **gamma globulins** (glob′ū-linz, globule) because they are found mostly in the gamma globulin part of plasma. Antibodies are also called **immunoglobulins (Ig)** because they are globulin proteins involved in immunity. The five general classes of immunoglobulins are denoted IgG, IgM, IgA, IgE, and IgD (table 14.2).

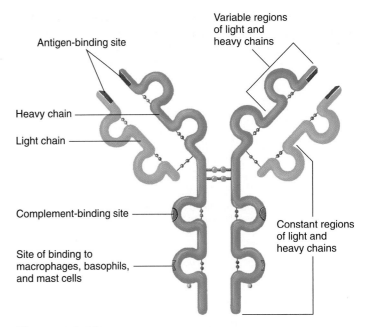

Figure 14.12 **Structure of an Antibody**

The Y-shaped antibody has two "arms." Each arm has a variable region that functions as an antigen-binding site. The constant region can activate complement or bind to other immune system cells, such as macrophages, basophils, or mast cells.

Effects of Antibodies

Antibodies can affect antigens either directly or indirectly. Direct effects occur when a single antibody binds to an antigen and inactivates the antigen, or when many antigens are bound together and are inactivated by many antibodies (figure 14.13a and b). The ability of antibodies to join antigens together is the basis for many clinical tests, such as blood typing, because when enough antigens are bound together, they form visible clumps.

Table 14.2	Classes of Antibodies and Their Functions		
Antibody	**Total Serum Antibody (%)**	**Structure**	**Description**
IgG	80–85		Activates complement and functions to increase phagocytosis; can cross the placenta and provide immune protection to the fetus and newborn; responsible for Rh reactions such as hemolytic disease of the newborn
IgM	5–10		Activates complement and acts as an antigen-binding receptor on the surface of B cells; responsible for transfusion reactions in the ABO blood system; often the first antibody produced in response to an antigen
IgA	15		Secreted into saliva, tears, and onto mucous membranes to provide protection on body surfaces; found in colostrum and milk to provide immune protection to the newborn
IgE	0.002		Binds to mast cells and basophils and stimulates the inflammatory response
IgD	0.2		Functions as antigen-binding receptors on B cells

Antigen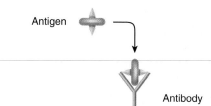

(a) **Inactivates the antigen.** An antibody binds to an antigen and inactivates it.

Antibody

(b) **Binds antigens together.** Antibodies bind several antigens together.

(c) **Activates the complement cascade.** An antigen binds to an antibody. As a result, the antibody can activate complement proteins, which can produce inflammation, chemotaxis, and lysis.

Complement cascade activated → Inflammation Chemotaxis Lysis

(d) **Initiates the release of inflammatory chemicals.** An antibody binds to a mast cell or basophil. When an antigen binds to the antibody, it triggers a release of chemicals that cause inflammation.

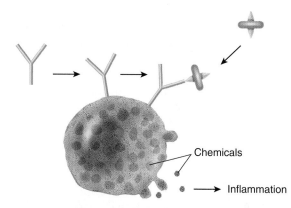

Chemicals

Inflammation

Mast cell or basophil

(e) **Facilitates phagocytosis.** An antibody binds to an antigen and then to a macrophage, which phagocytizes the antibody and antigen.

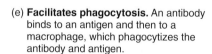

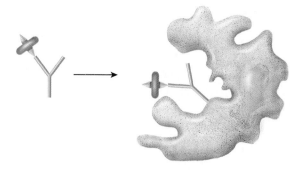

Macrophage

Figure 14.13 Effects of Antibodies

Antibodies directly affect antigens by inactivating the antigens or binding the antigens together. Antibodies indirectly affect antigens by activating other mechanisms through the constant region of the antibody. Indirect mechanisms include activation of complement, increased inflammation resulting from the release of inflammatory chemicals from mast cells or basophils, and increased phagocytosis resulting from antibody attachment to macrophages.

Use of Monoclonal Antibodies

Each type of **monoclonal antibody** is a pure antibody preparation that is specific for only one antigen. When the antigen is injected into a laboratory animal, it activates a B-cell clone against the antigen. These B cells are removed from the animal and fused with tumor cells. The resulting cells have two ideal characteristics: (1) they produce only one (mono) specific antibody because they are derived from one B-cell clone, and (2) they divide rapidly because they are derived from tumor cells. The result is many cells producing a specific antibody.

Monoclonal antibodies are used for determining pregnancy and for diagnosing diseases such as gonorrhea, syphilis, hepatitis, rabies, and cancer. These tests are specific and rapid because the monoclonal antibodies bind only to the antigen being tested. Monoclonal antibodies may some day be used to effectively treat cancer by delivering drugs to cancer cells (see the Immunotherapy section on p. 407).

Most of the effectiveness of antibodies results from indirect effects (figure 14.13*c* to *e*). After an antibody has attached by its variable region to an antigen, the constant region of the antibody can activate other mechanisms that destroy the antigen. For example, the constant region of antibodies can activate complement, which stimulates inflammation, attracts white blood cells through chemotaxis, and lyses bacteria. When an antigen combines with the antibody, the constant region triggers a release of inflammatory chemicals from mast cells and basophils. Finally, macrophages can attach to the constant region of the antibody and phagocytize both the antibody and antigen.

Antibody Production

The production of antibodies after the first exposure to an antigen is different from that following a second or subsequent exposure. The **primary response** results from the first exposure of a B cell to an antigen (figure 14.14, step 1). When the antigen binds to the antigen-binding receptor on the B cell, the B cell undergoes several divisions to form plasma cells and memory B cells. **Plasma cells** produce antibodies. The primary response normally takes 3–14 days to produce enough antibodies to be effective against the antigen. In the meantime, the individual usually develops disease symptoms because the antigen has had time to cause tissue damage.

Memory B cells are responsible for the **secondary,** or **memory, response,** which occurs when the immune system is exposed to an antigen against which it has already produced a

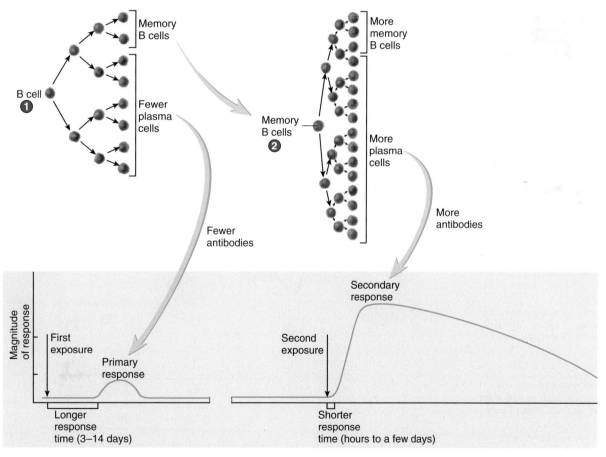

1. **Primary response.** The primary response occurs when a B cell is first activated by an antigen. The B cell proliferates to form plasma cells and memory cells. The plasma cells produce antibodies.

2. **Secondary response.** The secondary response occurs when another exposure to the same antigen causes the memory cells to rapidly form plasma cells and additional memory cells. The secondary response is faster and produces more antibodies than the primary response.

Process Figure 14.14   **Antibody Production**

primary response (figure 14.14, step 2). When exposed to the antigen, the B memory cells quickly divide to form plasma cells, which rapidly produce antibodies. The secondary response provides better protection than the primary response for two reasons: (1) the time required to start producing antibodies is less (hours to a few days), and (2) more plasma cells and antibodies are produced. As a consequence, the antigen is quickly destroyed, no disease symptoms develop, and the person is immune.

The memory response also includes the formation of new memory cells, which provide protection against additional exposures to a specific antigen. Memory cells are the basis of adaptive immunity. After destruction of the antigen, plasma cells die, the antibodies they released are degraded, and antibody levels decline to the point where they can no longer provide adequate protection. Memory cells persist for many years, however; probably for life in some cases. If memory cell production is not stimulated, or if the memory cells produced are short-lived, it is possible to have repeated infections of the same disease. For example, the same cold virus can cause the common cold more than once in the same person.

P R E D I C T   3

One theory for long-lasting immunity assumes that humans are continually exposed to disease-causing agents. Explain how this exposure could produce lifelong immunity.

Cell-Mediated Immunity

Cell-mediated immunity is a function of cytotoxic T cells and is most effective against microorganisms that live inside the cells of the body. Viruses and some bacteria are examples of intracellular microorganisms. Cell-mediated immunity is also involved with some allergic reactions, control of tumors, and graft rejections.

Cell-mediated immunity is essential for fighting viral infections. When viruses infect cells, they direct the cells to make new viruses, which are then released and infect other cells. Thus, cells are turned into virus manufacturing plants. While inside the cell, viruses have a safe haven from antibody-mediated immunity because antibodies cannot cross the cell membrane. Cell-mediated immunity fights viral infections by destroying virally infected cells. When viruses infect cells, some viral proteins are broken down and become processed antigens that are combined with MHC molecules and displayed on the surface of the infected cell (figure 14.15, step 1). Cytotoxic T cells can distinguish between virally infected cells and noninfected cells because the T-cell receptor can bind to the MHC molecule/viral antigen complex, which is not present on uninfected cells.

The T-cell receptor binding with the MHC molecule/antigen complex is a signal for activating cytotoxic T cells (figure 14.15, step 2). Costimulation by other surface molecules such as CD8 also occurs (figure 14.15, step 3). Helper T cells provide costimulation by releasing cytokines, such as interleukin-2, which stimulates activation and cell division of cytotoxic T cells (figure 14.15, step 4). Unlike their interactions with macrophages and B cells, however, helper T cells do not connect to cytotoxic T cells through MHC molecule/antigen complexes or other surface molecules.

Increasing the number of helper T cells (figure 14.15, step 5) results in greater stimulation of cytotoxic T cells. In cell-mediated responses, helper T cells are activated and stimulated to divide in the same fashion as in antibody-mediated responses (see figure 14.10).

1. A MHC molecule displays an antigen, such as a viral protein, on the surface of a target cell.

2. Activation of a cytotoxic T cell begins when the T-cell receptor binds to the MHC molecule/antigen complex.

3. There is costimulation of the cytotoxic T cell by CD8 and other surface molecules.

4. There is costimulation by cytokines, such as interleukin-2, released from helper T cells.

5. The activated cytotoxic T cell divides, and the resulting daughter cells divide, and so on, eventually producing many cytotoxic T cells (only two are shown here).

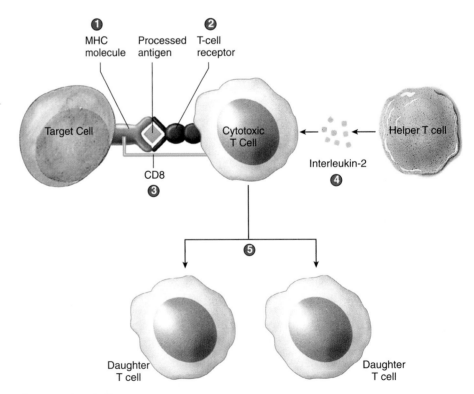

Process Figure 14.15   Proliferation of Cytotoxic T Cells

After cytotoxic T cells are activated by antigen on the surface of a target cell, they undergo a series of divisions to produce additional cytotoxic T cells and memory T cells (figure 14.16). The cytotoxic T cells are responsible for the cell-mediated immune response, and the **memory T cells** provide a secondary response and long-lasting immunity in the same fashion as memory B cells.

Cytotoxic T cells have two main effects:

1. They release cytokines that activate additional components of the immune system. For example, some cytokines attract innate immune cells, especially macrophages. These cells are then responsible for phagocytosis of the antigen and the production of an inflammatory response. Cytokines also activate additional cytotoxic T cells, which increases the effectiveness of the cell-mediated response.
2. Cytotoxic T cells can come into contact with other cells and kill them. Virus-infected cells have viral antigens, tumor cells have tumor antigens, and tissue transplants have foreign antigens that can stimulate cytotoxic T-cell activity. The cytotoxic T cells bind to the antigens on the surfaces of these cells and cause the cells to lyse.

Immune Interactions

Although the immune system can be described in terms of innate, antibody-mediated, and cell-mediated immunity, there is really only one immune system. These categories are an artificial division that is used to emphasize particular aspects of immunity. Actually, immune system responses often involve components of more than one type of immunity (figure 14.17). For example, although adaptive immunity can recognize and remember specific antigens, once recognition has occurred, many of the events that lead to the destruction of the antigen are innate immunity activities such as inflammation and phagocytosis (see table 14.1).

A CASE IN POINT | Sjogren Syndrome

Ima Akin has a toothache, so she goes to her dentist, Dr. Dekay, for help. Dr. Dekay finds an abscessed tooth, several dental cavities, and enlarged parotid salivary glands, but little saliva production. Upon questioning, Ima tells the dentist she has been experiencing a dry mouth and dry eyes and she feels fatigued all of the time. Dr. Dekay refers Ima to her family physician, Dr. Hurtt, for further testing. Eventually it is determined that Ima Akin has **Sjogren** (show'grin) **syndrome,** which is a systemic, autoimmune inflammatory disorder affecting glands and mucous membranes. Innate, antibody-mediated, and cell-mediated immunity normally work together to protect us against foreign antigens. In autoimmune disorders, self-antigens activate immune responses, resulting in the destruction of healthy tissues. In Sjogren syndrome, damage to salivary glands results in decreased saliva production and a dry mouth, which increases the likelihood of developing cavities. Damage to lacrimal glands results in decreased tear production and dry eyes, which damages the conjunctiva. Sjogren syndrome is one of the most common autoimmune disorders. About 50% of the time it occurs alone and about 50% of the time with other autoimmune diseases such as rheumatoid arthritis, systemic lupus erythematosus, and scleroderma. Nine out of ten people with Sjogren syndrome are women.

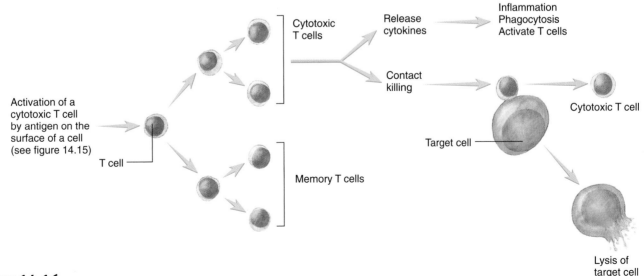

Figure 14.16

When activated, cytotoxic T cells form many cytotoxic T cells and memory T cells. The cytotoxic T cells release cytokines that promote the destruction of the antigen or cause the lysis of target cells, such as virus-infected cells, tumor cells, or transplanted cells. The memory T cells are responsible for the secondary response.

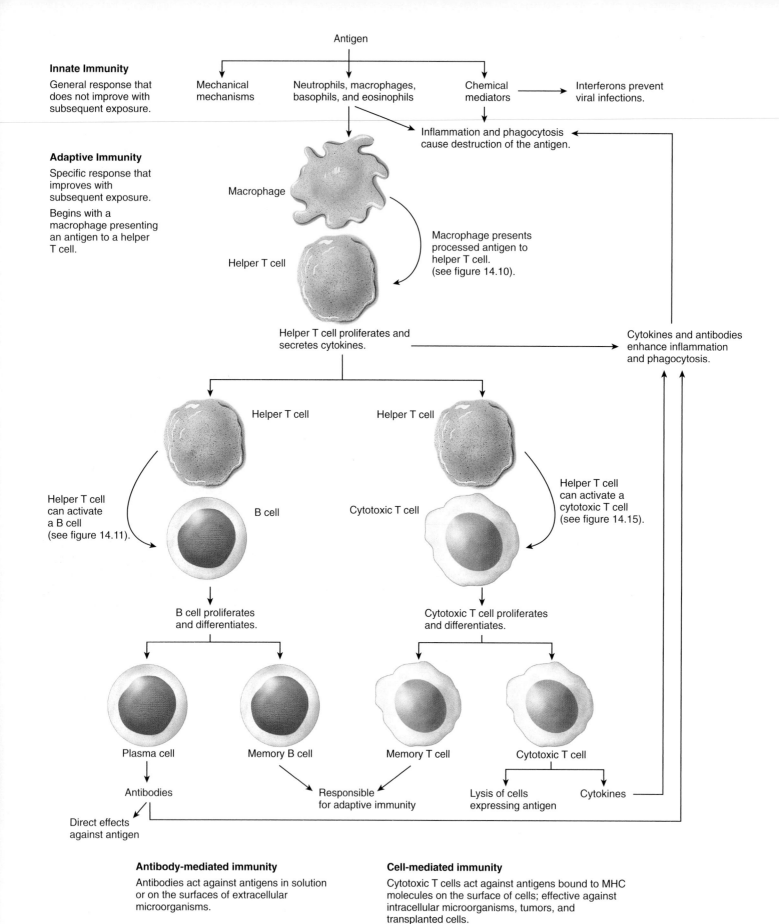

Innate Immunity

General response that does not improve with subsequent exposure.

Adaptive Immunity

Specific response that improves with subsequent exposure.

Begins with a macrophage presenting an antigen to a helper T cell.

Antigen

Mechanical mechanisms

Neutrophils, macrophages, basophils, and eosinophils

Chemical mediators → Interferons prevent viral infections.

Inflammation and phagocytosis cause destruction of the antigen.

Macrophage

Helper T cell

Macrophage presents processed antigen to helper T cell. (see figure 14.10).

Helper T cell proliferates and secretes cytokines.

Cytokines and antibodies enhance inflammation and phagocytosis.

Helper T cell

Helper T cell

Helper T cell can activate a B cell (see figure 14.11).

B cell

Cytotoxic T cell

Helper T cell can activate a cytotoxic T cell (see figure 14.15).

B cell proliferates and differentiates.

Cytotoxic T cell proliferates and differentiates.

Plasma cell

Memory B cell

Memory T cell

Cytotoxic T cell

Antibodies

Responsible for adaptive immunity

Lysis of cells expressing antigen

Cytokines

Direct effects against antigen

Antibody-mediated immunity

Antibodies act against antigens in solution or on the surfaces of extracellular microorganisms.

Cell-mediated immunity

Cytotoxic T cells act against antigens bound to MHC molecules on the surface of cells; effective against intracellular microorganisms, tumors, and transplanted cells.

Figure 14.17 Immune Interactions

The major interactions and responses of innate and adaptive immunity to an antigen.

Immunotherapy

Knowledge of the basic ways that the immune system operates has produced two fundamental benefits: (1) an understanding of the cause and progression of many diseases and (2) the development or proposed development of effective methods to prevent, stop, or even reverse diseases.

Immunotherapy treats disease by altering immune system function or by directly attacking harmful cells. Some approaches attempt to boost immune system function in general. For example, administering cytokines or other agents can promote inflammation and the activation of immune cells, which can help in the destruction of tumor cells. On the other hand, sometimes inhibiting the immune system is helpful. For example, multiple sclerosis is an autoimmune disease in which the immune system treats self-antigens as foreign antigens, destroying the myelin that covers axons. Interferon beta, which is a cytokine, blocks the expression of MHC molecules that display self-antigens and is now being used to treat multiple sclerosis.

Some immunotherapy takes a more specific approach. For example, vaccination can prevent many diseases (see section, Acquired Immunity, following). The ability to produce monoclonal antibodies may result in therapies that are effective for treating tumors. If an antigen unique to tumor cells can be found, then monoclonal antibodies could be used to deliver radioactive isotopes, drugs, toxins, enzymes, or cytokines that kill the tumor cell or activate the immune system to kill the cell. Unfortunately, no antigen on tumor cells has been found that is not also found on normal cells. Nonetheless, this approach may be useful if damage to normal cells is minimal.

The use of monoclonal antibodies to treat tumors is mostly in the research stage of development, but a few clinical trials are now yielding promising results. For example, monoclonal antibodies with radioactive iodine (^{131}I) have been found to cause the regression of B-cell lymphomas, while producing few side effects. Herceptin is a monoclonal antibody that binds to a growth factor receptor that is overexpressed in 25–30% of primary breast cancers. The antibody serves to "tag" cancer cells, which are then lysed by natural killer cells. Herceptin slows disease progression and increases survival time, but is not a cure for breast cancer.

Many other approaches for immunotherapy are being studied, and the development of treatments that use the immune system are certain to increase in the future. Your knowledge of the immune system will enable you to understand and appreciate these therapies.

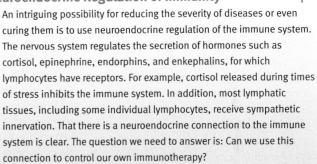

Neuroendocrine Regulation of Immunity

An intriguing possibility for reducing the severity of diseases or even curing them is to use neuroendocrine regulation of the immune system. The nervous system regulates the secretion of hormones such as cortisol, epinephrine, endorphins, and enkephalins, for which lymphocytes have receptors. For example, cortisol released during times of stress inhibits the immune system. In addition, most lymphatic tissues, including some individual lymphocytes, receive sympathetic innervation. That there is a neuroendocrine connection to the immune system is clear. The question we need to answer is: Can we use this connection to control our own immunotherapy?

Acquired Immunity

There are four ways to acquire adaptive immunity: active natural, active artificial, passive natural, and passive artificial (figure 14.18). "Natural" and "artificial" refer to the method of exposure. Natural exposure implies that contact with the antigen occurs as part of everyday living and is not deliberate.

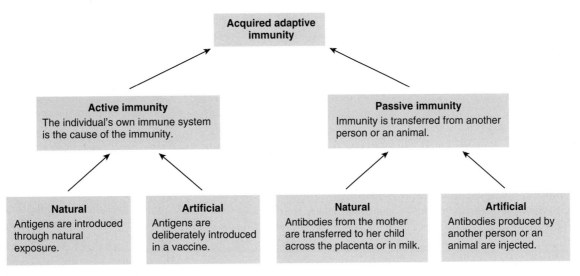

Figure 14.18 Ways to Acquire Adaptive Immunity

Allergy

An **allergy,** or **hypersensitivity reaction,** is a harmful response to an antigen that does not stimulate an adaptive immune response in most people. Immune and allergic reactions involve the same mechanisms, and the differences between them are unclear. Both require exposure to an antigen and stimulation of antibody-mediated or cell-mediated immunity. If immunity to the antigen is established, later exposure to the antigen results in an immune response that eliminates the antigen, and no symptoms appear. In allergic reactions, the antigen is called the **allergen** (al′er-jen, allergy + -gen, producing), and later exposure to the allergen stimulates much the same processes that occur during a normal immune response. The processes that eliminate the allergen, however, also produce undesirable side effects such as a strong inflammatory reaction, which can be more harmful than beneficial.

Immediate hypersensitivities produce symptoms within a few minutes of exposure to the allergen and are caused by antibodies. The reaction takes place rapidly because the antibodies are already present because of prior exposure to the allergen. For example, in people with **hay fever,** the allergens, usually plant pollens, are inhaled and absorbed through the respiratory mucous membrane. The combination of the allergen with antibodies stimulates mast cells to release inflammatory chemicals such as histamine (see figure 14.13). The resulting localized inflammatory response produces swelling and excess mucus production. In **asthma** (az′mă), resulting from an allergic reaction, the allergen combines with antibodies on mast cells or basophils in the lungs. As a result, these cells release inflammatory chemicals such as leukotrienes and histamine. The chemicals cause constriction of smooth muscle in the walls of the tubes that transport air throughout the lungs. Consequently, less air flows into and out of the lungs, and the patient has difficulty breathing. **Urticaria** (er′ti-kar′i-ă, to burn), or **hives,** is a skin rash or localized swelling that can be caused by an ingested allergen. **Anaphylaxis** (an′ă-fī-lak′sis, an′ă-fī-lak′sis, *ana,* away from + *phylaxis,* protection) is a systemic allergic reaction, often resulting from insect stings or drugs such as penicillin. The chemicals released from mast cells and basophils cause systemic vasodilation, increased vascular permeability, a drop in blood pressure, and possibly death. Transfusion reactions and hemolytic disease of the newborn (see chapter 11) are also examples of immediate hypersensitivity reactions.

Delayed hypersensitivities take hours to days to develop and are caused by T cells. It takes some time for this reaction to develop because it takes time for the T cells to move by chemotaxis to the allergen. It also takes time for the T cells to release cytokines that attract other immune system cells involved with producing inflammation. The most common type of delayed hypersensitivity reactions result from contact of an allergen with the skin or mucous membranes. For example, poison ivy, poison oak, soaps, cosmetics, and drugs can cause a delayed hypersensitivity reaction. The allergen is absorbed by epithelial cells, which are then destroyed by T cells, causing inflammation and tissue destruction. Although itching can be intense, scratching is harmful because it damages tissues and causes additional inflammation.

Autoimmune Disease

In **autoimmune disease,** the immune system incorrectly treats self-antigens as foreign antigens. Autoimmune disease operates through the same mechanisms as hypersensitivity reactions except that the reaction is stimulated by self-antigens. Examples of autoimmune diseases include thrombocytopenia, lupus erythematosus, rheumatoid arthritis, rheumatic fever, diabetes mellitus (type I), and myasthenia gravis.

Immunodeficiency

Immunodeficiency is a failure of some part of the immune system to function properly. It can be congenital (present at birth) or acquired. Congenital immunodeficiencies usually involve failure of the fetus to form adequate numbers of B cells, T cells, or both. **Severe combined immunodeficiency (SCID),** in which both B cells and T cells fail to form, is probably the best known. Unless the person suffering from SCID is kept in a sterile environment or is provided with a compatible bone marrow transplant, death from infection results.

Acquired immunodeficiency can result from many different causes. For example, inadequate protein in the diet inhibits protein synthesis and, therefore, antibody levels decrease. Immunity can be depressed as a result of stress, illness, or drugs such as those used to prevent graft rejection. Diseases such as leukemia cause an overproduction of lymphocytes that do not function properly.

Acquired immunodeficiency syndrome (AIDS) is a life-threatening disease caused by the **human immunodeficiency virus (HIV).** Two strains of HIV are recognized: HIV-1 is responsible for most cases of AIDS, whereas HIV-2 is increasingly being found in West Africa. AIDS was first reported in 1981 in the United States. Since then over one million cases have been reported to the Centers for Disease Control (CDC). The United Nations Program on AIDS (UNAIDS) estimates that 70 million people have been infected by HIV worldwide, and 20 million have died. The course of HIV infection varies. After contracting an HIV infection, some people die within a year; most, however, survive for 10 or 11 years, and some have survived beyond 20 years.

HIV is transmitted from an infected to a noninfected person in body fluids such as blood, semen, or vaginal secretions. The major methods of transmission are intimate sexual contact, contaminated needles used by intravenous drug users, and tainted blood products. Present evidence indicates household, school, or work contacts do not result in transmission.

In the United States, most cases of AIDS have appeared in homosexual or bisexual men and in intravenous drug users. A small percentage of cases have resulted from transfusions or contaminated clotting factors used by hemophiliacs. Children can be infected before birth, during delivery, or after birth from breast-feeding. A few cases of AIDS have occurred in health care workers accidentally exposed to HIV-infected blood or body fluids, and even fewer cases of health care workers infecting patients has been documented. The most rapidly increasing group of AIDS patients in the United States is heterosexual women or men who have had sexual contact with an infected person. UNAIDS estimates that over 90% of all HIV infections globally are transmitted heterosexually.

Preventing transmission of HIV is presently the only way to prevent AIDS. The risk of transmission can be reduced by such safe sexual practices as reducing the number of one's sexual partners, avoiding anal intercourse, and using a condom. Public education also includes warnings to intravenous drug users of the dangers of using contaminated needles. Ensuring the safety of the blood supply is another important preventive measure. In 1985, a test for HIV antibodies in blood became available. Heat treatment of clotting factors taken from blood has also been effective in preventing transmission of HIV to hemophiliacs.

HIV infection begins when the virus binds to a CD4 surface molecule. The CD4 molecule is found primarily on helper T cells but also on certain monocytes, macrophages, neurons, and neuroglial cells. Once attached to a CD4 molecule, the virus injects its genetic material (RNA) and enzymes into the cell. The viral RNA and enzymes produce DNA that can direct the formation of new HIV ribonucleic acid and proteins, that is, additional viruses that can infect other cells. Most of the manifestations of AIDS can be explained by the loss of helper T cell functions or the infection of other cells with CD4 molecules. Without helper T cells, cytotoxic T cell and B cell activation is impaired, and adaptive resistance is suppressed.

Following infection by the HIV, within 3 weeks to 3 months, some patients develop mononucleosislike symptoms such as fever, sweats, fatigue, muscle and joint aches, headache, sore throat, diarrhea, rash, and swollen lymph nodes. Within 1–3 weeks, these symptoms disappear as the immune system responds to the virus by producing antibodies and activating cytotoxic T cells that kill HIV-infected cells. The immune system is not able to completely eliminate HIV, however, and the virus continues to replicate at a low, but steady, rate. This chronic stage of infection lasts, on the average, for 8–10 years, and the infected person exhibits few, if any, symptoms.

Over a period of years the HIV numbers gradually increase and helper T cell numbers decrease. Normally approximately 1200 helper T cells are present per microliter of blood. An HIV-infected person is considered to have AIDS when one or more of the following conditions appear: the helper T cell count falls below 200 cells/mL, an opportunistic infection occurs, or Kaposi's sarcoma develops.

Opportunistic infections involve organisms that normally do not cause disease but can do so when adaptive resistance is suppressed. Examples of opportunistic infections include pneumocystis (noo-mō-sis′tis) pneumonia (caused by an intracellular fungus, *Pneumocystis carinii*), tuberculosis (caused by an intracellular bacterium, *Mycobacterium tuberculosis*), syphilis (caused by a sexually transmitted bacterium, *Treponema pallidum*), candidiasis (kan-di-dī′ă-sis; a yeast infection of the mouth or vagina caused by *Candida albicans*), and protozoans that cause severe, persistent diarrhea. Kaposi's sarcoma is a type of cancer that produces lesions in the skin, lymph nodes, and visceral organs. Also associated with AIDS are symptoms resulting from the effects of HIV on the nervous system, including motor retardation, behavioral changes, progressive dementia, and possibly psychosis.

There currently is no cure for AIDS. Management of AIDS can be divided into two

categories: (1) management of secondary infections or malignancies associated with AIDS and (2) treatment of HIV infection itself. In order for HIV to replicate, the viral RNA is used to make viral DNA, which is inserted into the host cell's DNA. The inserted viral DNA directs the production of new viral RNA and proteins, which are assembled to form new HIV. Key steps in the replication of HIV require viral enzymes. **Reverse transcriptase** promotes the formation of viral DNA from viral RNA, and **integrase** (in′te-grās) inserts the viral DNA into the host cell's DNA. A viral **protease** (prō′tē-ās) breaks large viral proteins into smaller proteins, which are incorporated into the new HIV.

Blocking the activity of HIV enzymes can inhibit replication of HIV. The first effective treatment of AIDS was the drug azidothymidine (AZT) (az′i-dō-thī′mi-dēn), also called zidovudine (zī-dō′voo-dēn). AZT is a **reverse transcriptase inhibitor,** which prevents HIV RNA from producing viral DNA. AZT can delay the onset of AIDS but doesn't appear to increase the survival time of AIDS patients. However, the number of babies who contract AIDS from their HIV-infected mothers can be dramatically reduced by giving AZT to the mothers during pregnancy and to the babies following birth. AZT can produce serious side effects such as anemia or even total bone marrow failure.

Often after 6–18 months of treatment with AZT, viral mutations result in HIV that is resistant to AZT. Other drugs that inhibit viral nucleic acid replication, such as dideoxyinosine (DDI) (dī′dē-oks-ē-ī′nō-sēn), have been developed. These drugs have been used for patients who are resistant to, or do not respond to, AZT.

Protease inhibitors are drugs that interfere with viral proteases. Examples of protease inhibitors are ritonavir and indinavir.

The current treatment for suppressing HIV replication is **highly active antiretroviral therapy (HAART).** This therapy involves the use of drugs from at least two distinct

continued

continued

classes of antivirals. Treatment may involve a combination of three drugs, such as two reverse transcriptase inhibitors and one protease inhibitor. It's less likely that HIV will develop resistance to all three drugs. This strategy has proven very effective in reducing the death rate from AIDS and partially restoring health in some individuals.

Still in clinical trials are **integrase inhibitors,** which prevent the insertion of viral DNA into the host cell's DNA. Perhaps someday integrase inhibitors will be part of a combination drug therapy for AIDS.

Another advance in AIDS treatment is a test for measuring viral load, which is the quantity of HIV in a milliliter of blood. It has been learned that viral load is a good predictor of how soon a person will develop AIDS. If viral load is high, the onset of AIDS is much sooner than if viral load is low.

New drug therapies and the ability to monitor viral load can be combined to produce an effective treatment of AIDS by matching drug types and doses with viral load. If viral load can be kept low, then HIV infections may not develop into full-blown AIDS.

An effective treatment for AIDS is not a cure. Even if viral load decreases to the point that the virus is not detected in the blood, the virus is still hidden in cells throughout the body. In addition, the long-term effects of these drug therapies is unknown.

Although research to develop vaccines, as well as clinical trials to test them, is occurring around the world, an effective vaccine has not yet been demonstrated.

Tumor Control

According to the concept of **immune surveillance,** the immune system detects tumor cells and destroys them before a tumor can form. T cells, NK cells, and macrophages are involved in the destruction of tumor cells. Immune surveillance may exist for some forms of cancer caused by viruses. The immune response appears to be directed more against the viruses, however, than against tumors in general. Only a few cancers are known to be caused by viruses in humans. For most tumors the response of the immune system may be ineffective and too late.

Transplantation

The genes that code for the production of the MHC molecules are generally called the **major histocompatibility complex genes.** Histocompatibility refers to the ability of tissues (*histo*) to get along (compatibility) when tissues are transplanted from one individual to another. In humans, the MHC genes are often referred to as **human leukocyte antigen (HLA) genes** because they were first identified in leukocytes. There are millions of possible combinations of the HLA genes, and it is very rare for two individuals (except identical twins)

to have the same set of HLA genes. Because they are genetically determined, however, the closer the relationship between two individuals, the greater the likelihood of sharing the same HLA genes.

The immune system can distinguish between self and foreign cells because self-cells have self-HLAs, whereas foreign cells have foreign HLAs. Rejection of a graft is caused by a normal immune response to foreign HLAs.

Graft rejection can occur in two different directions. In **host-versus-graft rejection,** the recipient's immune system recognizes the donor tissue as foreign and rejects the transplant. In a **graft-versus-host rejection,** the donor tissue (e.g., bone marrow) recognizes the recipient's tissue as foreign, and the transplant rejects the recipient, causing destruction of the recipient's tissue and possibly death.

To reduce graft rejection, a tissue match is performed. Only tissue with HLAs similar to the recipient's have a chance of being accepted. An exact match is possible only for a graft from one part to another part of the same person, or between identical twins. For all other graft situations, drugs such as cyclosporine (sī-klō-spōr'in) that suppress the immune system must be administered throughout the patient's life to prevent graft rejection. Unfortunately, the person then has a drug-produced immunodeficiency and is more susceptible to infections.

Artificial exposure is a deliberate introduction of an antigen or antibody into the body.

Active immunity results when an individual is exposed to an antigen (either naturally or artificially) and the response of the individual's own immune system is the cause of the immunity. Passive immunity occurs when another person or an animal develops immunity and the immunity is transferred to a nonimmune individual.

Active Natural Immunity

Active natural immunity results from natural exposure to an antigen such as a disease-causing microorganism that stimulates an individual's immune system to respond against the antigen. Because the individual is not immune during the first exposure, he usually develops the symptoms of the disease.

Active Artificial Immunity

In **active artificial immunity,** an antigen is deliberately introduced into an individual to stimulate her immune system. This process is called **vaccination** (vak′si-nā-shŭn), and the introduced antigen is a **vaccine** (vak′sēn, vak-sēn′, *vaccinus*, relating to a cow). Injection of the vaccine is the usual mode of administration. Examples of injected vaccinations are the DTP injection against diphtheria, tetanus, and pertussis (whooping cough); and the MMR injection against mumps, measles, and rubella (German measles). Sometimes the vaccine is ingested, as in the oral poliomyelitis vaccine (OPV).

The vaccine usually consists of some part of a microorganism, a dead microorganism, or a live, altered microorganism. The antigen has been changed so that it will stimulate an immune response but will not cause the symptoms of disease. Because active artificial immunity produces long-lasting immunity without disease symptoms, it is the preferred method of acquiring adaptive immunity.

> **P R E D I C T**
> In some cases, a "booster" shot is used as part of a vaccination procedure. A booster shot is another dose of the original vaccine given some time after the original dose was administered. Why are booster shots given?

Passive Natural Immunity

Passive natural immunity results from the transfer of antibodies from a mother to her child across the placenta before birth. During her life, the mother has been exposed to many antigens, either naturally or artificially, and she has antibodies against many of these antigens. These antibodies protect the mother and the developing fetus against disease. Some of the antibodies (IgG) can cross the placenta and enter the fetal blood. Following birth, the antibodies provide protection for the first few months of the baby's life. Eventually the antibodies are broken down, and the baby must rely on his own immune system. If the mother nurses her baby, antibodies (IgA) in the mother's milk may also provide some protection for the baby.

Passive Artificial Immunity

Achieving **passive artificial immunity** begins with vaccinating an animal such as a horse. After the animal's immune system responds to the antigen, antibodies are removed from the animal and are injected into the individual requiring immunity. Alternatively, a human who has developed immunity through natural exposure or vaccination can be used as a source of antibodies. Passive artificial immunity provides immediate protection because the antibodies either directly or indirectly destroy the antigen. Passive artificial immunity is therefore the preferred treatment when not enough time is available for the individual to develop her own active immunity. The technique provides only temporary immunity, however, because the antibodies are used or eliminated by the recipient.

Antiserum is the general term used for antibodies that provide passive artificial immunity because the antibodies are found in serum, which is plasma minus the clotting factors. Antisera are available against microorganisms that cause disease such as rabies, hepatitis, and measles; bacterial toxins such as tetanus, diphtheria, and botulism; and venoms from poisonous snakes and spiders.

Effects of Aging on the Lymphatic System and Immunity

Aging appears to have little effect on the ability of the lymphatic system to remove fluid from tissues, absorb fats from the digestive tract, or remove defective red blood cells from the blood.

Aging also seems to have little direct effect on the ability of B cells to respond to antigens, and the number of circulating B cells remains stable in most individuals. With age, thymic tissue is replaced with adipose tissue, and the ability to produce new, mature T cells in the thymus is eventually lost. Nonetheless, the number of T cells remains stable in most individuals due to the replication (not maturation) of T cells in lymphatic tissues. In many individuals, however, there is a decreased ability of helper T cells to proliferate in response to antigens. Thus, antigen exposure produces fewer helper T cells, which results in less stimulation of B cells and cytotoxic T cells. Consequently, both antibody-mediated immunity and cell-mediated immunity responses to antigens decrease.

Primary and secondary antibody responses decrease with age. More antigen is required to produce a response, the response is slower, less antibody is produced, and fewer memory cells result. Thus, the ability to resist infections and develop immunity decreases.

The ability of cell-mediated immunity to resist intracellular pathogens decreases with age. For example, the elderly are more susceptible to influenza (flu) and should be vaccinated every year. Some pathogens cause disease but are not eliminated from the body. With age, a decrease in immunity can result in reactivation of the pathogen. For example, the virus that causes chickenpox in children can remain latent within nerve cells even though the disease seems to have disappeared. Later in life, the virus can leave the nerve cells and infect skin cells, causing painful lesions known as herpes zoster or shingles.

Autoimmune disease occurs when immune responses destroy otherwise healthy tissue (see "Autoimmune Disease" on p. 408). There is very little increase in the number of new-onset autoimmune diseases in the elderly. However, the chronic inflammation and immune responses that began earlier in life have a cumulative, damaging effect. The increased incidence of cancer in the elderly is assumed to be related to a decrease in the immune response.

Systems Pathology
Systemic Lupus Erythematosus

Mrs. L. is a 30-year-old divorced woman with two children. Despite the fact that she had to work to support herself and the children, she entered college, determined to become a nurse and provide a better life for her family. Mrs. L. was an excellent student, but her class attendance and her performance on tests was somewhat erratic. Sometimes she seemed very energetic and earned high grades, but other times she seemed depressed and did not do as well. Toward the end of the course, she developed a rash on her face (figure A), a large red lesion on her arm, and was obviously not feeling well.

Mrs. L. went to the instructor to ask if she could take an incomplete grade and take the last exam at a later time. She explained that she has had lupus since she was 25 years old. Normally, medication helps to control her symptoms, but the stress of being a single parent combined with the challenges of school seemed to be making her condition worse. She further explained that the symptoms of lupus come and go, and bedrest was often helpful. Mrs. L. finished the course requirements later that summer. She went on to complete her education and now has a full-time job as a nurse at a local hospital.

Background Information

Systemic lupus erythematosus (SLE) (lū′pŭs er-i-thē′mă-tō-sŭs) is a disease of unknown cause in which tissues and cells are damaged by the immune system. The name describes some of the characteristics of the disease. The term lupus literally means wolf and was originally used to refer to eroded (as if gnawed by a wolf) lesions of the skin. Erythematosus refers to a redness of the skin resulting from inflammation. Unfortunately, as the term systemic implies, the disorder is not confined to the skin but can affect tissues and cells throughout the body. Another systemic effect is the presence of low-grade fever in most cases of active SLE.

SLE is an autoimmune disorder in which a large variety of antibodies are produced that recognize self-antigens, such as nucleic acids, phospholipids, coagulation factors, red blood cells, and platelets. The combination of the antibodies with self-antigens forms immune complexes that circulate throughout the body to be deposited

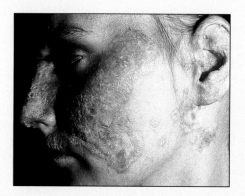

Figure A Systemic Lupus Erythematosus

The butterfly rash resulting from inflammation in the skin caused by systemic lupus erythematosus.

in various tissues, in which they stimulate inflammation and tissue destruction. Thus, SLE is a disease that can affect many systems of the body. For example, the most common antibodies act against DNA that is released from damaged cells. Normally the liver removes the DNA, but when DNA and antibodies form immune complexes, they tend to be deposited in the kidneys and other tissues. Approximately 40–50% of individuals with SLE develop renal disease. In some cases, the antibodies can bind to antigens on cells, resulting in lysis of the cells. For example, antibodies binding to red blood cells cause hemolysis and the development of anemia.

The cause of SLE is unknown. The most popular hypothesis is that a viral infection disrupts the function of T cells that normally prevent an immune response to self-antigens. Genetic factors probably contribute to the development of the disease. The likelihood of developing SLE is much higher if a family member also has it.

Approximately 1 out of 2000 individuals in the United States has SLE. The first symptoms of SLE usually appear between 15 and 25 years of age, affecting women approximately nine times as often as

---

S U M M A R Y

Lymphatic System   (p. 388)

The lymphatic system consists of lymph, lymphocytes, lymphatic vessels, lymph nodes, tonsils, the spleen, and the thymus gland.

Functions of the Lymphatic System

The lymphatic system maintains fluid balance in tissues, absorbs fats from the small intestine, and defends against microorganisms and foreign substances.

Lymphatic Capillaries and Vessels

1. Lymphatic vessels carry lymph away from tissues. Valves in the vessels ensure the one-way flow of lymph.
2. Skeletal muscle contraction, contraction of lymphatic vessel smooth muscle, and thoracic pressure changes move the lymph through the vessels.
3. The thoracic duct and right lymphatic duct empty lymph into the blood.

System	Interactions
Integumentary	Skin lesions frequently occur and are made worse by exposure to the sun. There are three forms: (1) an inflammatory redness that can take the form of the butterfly-shaped rash, which extends from the bridge of the nose to the cheeks; (2) small, localized, pimplelike eruptions accompanied by scaling of the skin; (3) areas of atrophied, depigmented skin with borders of increased pigmentation. Hair loss results in diffuse thinning of the hair.
Skeletal	Arthritis, tendonitis, and death of bone tissue can develop.
Muscular	Destruction of muscle tissue and muscular weakness.
Nervous	Memory loss, intellectual deterioration, disorientation, psychosis, reactive depression, headache, seizures, nausea, and loss of appetite can occur. Stroke is a major cause of dysfunction and death. Cranial nerve involvement results in facial muscle weakness, drooping of the eyelid, and double vision. Central nervous system lesions can cause paralysis.
Endocrine	Sex hormones may play a role in SLE because 90% of the cases occur in females, and females with SLE have reduced levels of androgens.
Cardiovascular	Inflammation of the pericardium (pericarditis) with chest pain can develop. Damage to heart valves, inflammation of cardiac tissue, tachycardia, arrhythmias, angina, and myocardial infarction also occurs. Hemolytic anemia and leukopenia can be present (see chapter 11). Antiphospholipid antibody syndrome, through an unknown mechanism, increases coagulation and thrombus formation, which increases the risk of stroke and heart attack.
Respiratory	Chest pain caused by inflammation of the pleural membranes; fever, shortness of breath, and hypoxemia caused by inflammation of the lungs; and alveolar hemorrhage can develop.
Digestive	Ulcers develop in the oral cavity and pharynx. Abdominal pain and vomiting are common, but no cause can be found. Inflammation of the pancreas and occasionally enlargement of the liver and minor abnormalities in liver function tests occur.
Urinary	Renal lesions and glomerulonephritis can result in progressive failure of kidney functions. Excess proteins are lost in the urine, resulting in lower than normal blood proteins, which can produce edema.

men. The progress of the disease is unpredictable, with flare-ups of symptoms followed by periods of remission. The survival after diagnosis is greater than 90% after 10 years. The most frequent causes of death involve kidney failure, CNS dysfunction, infections, and cardiovascular disease.

There is no cure for SLE, nor is there one standard of treatment because the course of the disease is highly variable and there are many differences between patients with SLE. Treatment usually begins with mild medications and proceeds to more and more potent therapies as conditions warrant. Aspirin and nonsteroidal antiinflammatory drugs are used to suppress inflammation. Antimalarial drugs are used to treat skin rash and arthritis in SLE, but the mechanism of action is unknown. Patients who do not respond to these drugs or those with severe SLE are helped by steroids. Although steroids effectively suppress inflammation, they can produce undesirable side effects including suppression of normal adrenal gland functions. In patients with life-threatening SLE, very high doses of steroids are used.

P R E D I C T   5
The red lesion Mrs. L. developed on her arm is called purpura (pŭr′pū-ră), which is caused by bleeding into the skin. The lesions gradually change color and disappear in 2 or 3 weeks. Explain how SLE produces purpura.

Lymphatic Organs

1. Lymphatic tissue produces lymphocytes, when exposed to foreign substances, and it filters lymph and blood.
2. The tonsils protect the openings between the nasal and oral cavities and the pharynx.
3. Lymph nodes, located along lymphatic vessels, filter lymph.
4. The white pulp of the spleen responds to foreign substances in the blood, whereas the red pulp phagocytizes foreign substances and worn out red blood cells. The spleen also functions as a reservoir for blood.
5. The thymus processes lymphocytes that move to other lymphatic tissue to respond to foreign substances.

Overview of the Lymphatic System

The lymphatic system removes fluid from tissues, absorbs fats from the small intestine, and produces B and T cells, which are responsible for much of immunity.

Immunity   (p. 394)

Immunity is the ability to resist the harmful effects of microorganisms and other foreign substances.

Innate Immunity   (p. 394)

Mechanical Mechanisms

1. The skin and mucous membranes are barriers that prevent the entry of microorganisms into the body.
2. Tears, saliva, and urine act to wash away microorganisms.

Chemical Mediators

1. Chemical mediators kill microorganisms, promote phagocytosis, and increase inflammation.
2. Lysozyme in tears and complement in plasma are examples of chemicals involved in innate immunity.
3. Interferons prevent the replication of viruses.

Cells

1. Chemotaxis is the ability of cells to move toward microorganisms or sites of tissue damage.
2. Neutrophils are the first phagocytic cells to respond to microorganisms.
3. Macrophages are large phagocytic cells that are active in the latter part of an infection. Macrophages are positioned at sites of potential entry of microorganisms into tissues.
4. Basophils and mast cells promote inflammation, whereas eosinophils inhibit inflammation.
5. Natural killer cells lyse tumor cells and virus-infected cells.

Inflammatory Response

1. Chemical mediators cause vasodilation and increase vascular permeability, allowing the entry of chemicals into damaged tissues. Chemicals also attract phagocytes.
2. The amount of chemical mediators and phagocytes increases until the cause of the inflammation is destroyed. Then the tissues undergo repair.
3. Local inflammation produces the symptoms of redness, heat, swelling, pain, and loss of function. Symptoms of systemic inflammation include an increase in neutrophil numbers, fever, and shock.

Adaptive Immunity   (p. 397)

1. Antigens are molecules that stimulate adaptive immunity.
2. B cells are responsible for humoral, or antibody-mediated, immunity. T cells are involved with cell-mediated immunity.

Origin and Development of Lymphocytes

1. B and T cells originate in red bone marrow. T cells are processed in the thymus and B cells are processed in red bone marrow.
2. B and T cells move to lymphatic tissue from their processing sites. They continually circulate from one lymphatic tissue to another.

Activation and Multiplication of Lymphocytes

1. B cells and T cells have antigen receptors on their surfaces. Clones are lymphocytes with the same antigen receptor.
2. Major histocompatibility complex (MHC) molecules present processed antigens to B or T cells.
3. Costimulation by cytokines, such as interleukins, and surface molecules, such as CD4, are required in addition to MHC molecules.

4. Macrophages present processed antigens to helper T cells, which divide and increase in number.
5. Helper T cells stimulate B cells to divide and differentiate into cells that produce antibodies.

Antibody-Mediated Immunity

1. Antibodies are proteins. The variable region combines with antigens and is responsible for antibody specificity. The constant region activates complement or attaches the antibody to cells. The five classes of antibodies are IgG, IgM, IgA, IgE, and IgD.
2. Antibodies directly inactivate antigens or cause them to clump together. Antibodies indirectly destroy antigens by promoting phagocytosis and inflammation.
3. The primary response results from the first exposure to an antigen. B cells form plasma cells, which produce antibodies, and memory B cells.
4. The secondary (memory) response results from exposure to an antigen after a primary response. Memory B cells quickly form plasma cells and new memory B cells.

Cell-Mediated Immunity

1. Exposure to an antigen activates cytotoxic T cells and produces memory T cells.
2. Cytotoxic T cells lyse virus-infected cells, tumor cells, and tissue transplants. Cytotoxic T cells produce cytokines, which promote inflammation and phagocytosis.

Immune Interactions   (p. 405)

Innate immunity, antibody-mediated immunity, and cell-mediated immunity can function together to eliminate an antigen.

Immunotherapy   (p. 407)

Immunotherapy stimulates or inhibits the immune system to treat diseases.

Acquired Immunity   (p. 407)

1. Active natural immunity results from everyday exposure to an antigen against which the person's own immune system mounts a response.
2. Active artificial immunity results from deliberate exposure to an antigen (vaccine) to which the person's own immune system responds.
3. Passive natural immunity is the transfer of antibodies from a mother to her fetus during gestation or baby during breastfeeding.
4. Passive artificial immunity is the transfer of antibodies from an animal or another person to a person requiring immunity.

Effects of Aging on the Lymphatic System and Immunity   (p. 411)

1. Aging has little effect on the ability of the lymphatic system to remove fluid from tissues, absorb fats from the digestive tract, or remove defective red blood cells from the blood.
2. Decreased helper T cell proliferation results in decreased antibody-mediated immunity and cell-mediated immunity responses to antigens.
3. The primary and secondary antibody responses decrease with age.
4. The ability to resist intracellular pathogens decreases with age.

R E V I E W   A N D   C O M P R E H E N S I O N

1. List the parts of the lymphatic system, and describe the three main functions of the lymphatic system.
2. What is the function of the valves in lymphatic vessels? What causes lymph to move through lymphatic vessels?
3. Which parts of the body are drained by the right lymphatic duct and which by the thoracic duct?
4. Describe the cells and fibers of lymphatic tissue, and explain the functions of lymphatic tissue.

5. Name the three groups of tonsils. What is their function?
6. Where are lymph nodes found? What is the function of the germinal centers within lymph nodes?
7. Where is the spleen located? What is the function of white pulp and red pulp within the spleen? What other function does the spleen perform?
8. Where is the thymus located, and what function does it perform?
9. What is the difference between innate immunity and adaptive immunity?

10. How do mechanical mechanisms and chemical mediators provide protection against microorganisms? Describe the effects of complement and interferons.

11. Describe the functions of the two major phagocytic cell types of the body. What is the mononuclear phagocytic system?

12. Name the cells involved in promoting and inhibiting inflammation.

13. What protective function is performed by natural killer cells?

14. Describe the effects that take place during an inflammatory response. What are the symptoms of local and systemic inflammation?

15. Define antigen. What is the difference between a self-antigen and a foreign antigen?

16. Which cells are responsible for antibody-mediated and for cell-mediated immunity?

17. Describe the origin and development of B and T cells.

18. What is the function of antigen receptors and major histocompatibility proteins?

19. What is costimulation? Give an example.

20. Describe the process by which an antigen can cause an increase in helper T-cell numbers.

21. Describe the process by which helper T cells can stimulate B cells to divide, differentiate, and produce antibodies.

22. What are the functions of the variable and constant regions of an antibody?

23. Describe the direct and indirect ways that antibodies function to destroy antigens.

24. What are the functions of plasma cells and B memory cells?

25. Define the primary and memory (secondary) response. How do they differ from each other in regard to speed of response and amount of antibody produced?

26. Explain how cytotoxic T cells are activated.

27. What are the functions of cytotoxic T cells and T memory cells?

28. Define active natural, active artificial, passive natural, and passive artificial immunity. Give an example of each.

29. What effect does aging have on the major functions of the lymphatic system?

30. Describe the effects of aging on B cells and T cells. Give examples of how this affects antibody-mediated immunity and cell-mediated immunity responses.

C R I T I C A L   T H I N K I N G

1. A patient is suffering from edema in the lower right limb. Explain why elevation and massage of the limb helps to remove the excess fluid.

2. If the thymus of an experimental animal is removed immediately following birth, the animal exhibits the following characteristics:
   a. It is more susceptible to infections.
   b. It has decreased numbers of lymphocytes.
   c. Its ability to reject grafts is greatly decreased.

   Explain these observations.

3. Adjuvants are substances that slow, but do not stop, the release of an antigen from an injection site into the blood. Suppose injection A of a given amount of antigen is given without an adjuvant and injection B of the same amount of antigen is given with an adjuvant that caused the release of antigen over a period of 2 to 3 weeks. Does injection A or B result in the greater amount of antibody production? Explain.

4. Compare how long active immunity and passive immunity last. Explain the difference between the two types of immunity. In what situations is one type preferred over the other?

5. Tetanus is caused by bacteria (*Clostridium tetani*) that enter the body through wounds in the skin. The bacteria produce a toxin that causes spastic muscle contractions. Death often results from failure of the respiratory muscles. A patient comes to the emergency room after stepping on a nail. If the patient has been vaccinated against tetanus, he is given a tetanus booster shot, which consists of the toxin altered so that it is harmless. If the patient has never been vaccinated against tetanus, he is given an antiserum shot against tetanus. Explain the rationale for this treatment strategy. Sometimes both a booster and an antiserum shot are given, but at different locations of the body. Explain why this is done, and why the shots are given in different locations.

6. A child appears to be healthy until he reaches approximately 9 months of age. Then he develops severe bacterial infections, one after another. Fortunately, the infections are successfully treated with antibiotics. When infected with measles and other viral diseases, the child recovers without difficulty. Explain.

7. A patient had many allergic reactions (see Clinical Focus: Immune System Problems of Clinical Significance on p. 408). As part of the treatment scheme, it was decided to try to identify the allergens that stimulated the allergic reaction. A series of solutions, each containing an allergen that commonly causes a reaction, was composed. Each solution was then injected into the skin at different locations on the patient's back. The following results were obtained:
   a. At one location, within a few minutes the injection site became red and swollen.
   b. At another injection site, swelling and redness did not appear until 2 days later.
   c. No redness or swelling developed at the other sites. Explain what happened for each observation and what caused the redness and swelling.

8. Ivy Hurtt developed a poison ivy rash after a camping trip. Her doctor prescribed a cortisone ointment to relieve the inflammation. A few weeks later Ivy scraped her elbow, which became inflamed. Because she had some of the cortisone ointment left over, she applied it to the scrape. Explain why the ointment was or was not a good idea for the poison ivy and for the scrape.

9. Suzy Withitt has just had her ears pierced. To her dismay, she finds that, when she wears inexpensive (but tasteful) jewelry, by the end of the day there is an inflammatory (allergic) reaction to the metal in the jewelry. Is this because of antibodies or cytokines?

Answers in Appendix D

Visit this textbook's website at www.mhhe.com/seeleyess6 for practice quizzes, animations, interactive learning exercises, and other study tools.

15

Respiratory System

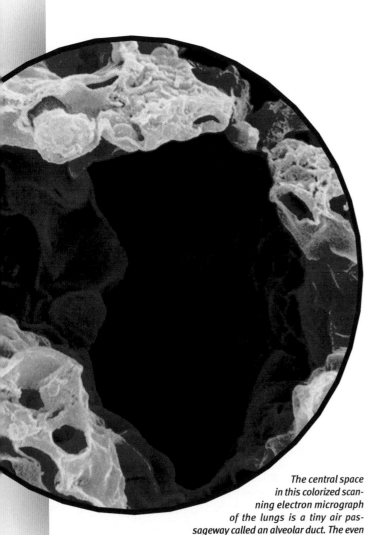

The central space in this colorized scanning electron micrograph of the lungs is a tiny air passageway called an alveolar duct. The even smaller spaces around the alveolar duct are alveoli, which are small chambers where gas exchange takes place between the air and the blood. Air enters the nose or mouth, passes through the throat (pharynx) and then the larynx (voice box) to reach the windpipe (trachea). Arising from the trachea, a series of branching tubes carries air into the lungs. After each branching, the tubes become progressively smaller until they become alveolar ducts, which connect to the alveoli.

Chapter Outline and Objectives

Functions of the Respiratory System   (p. 417)
1. Describe the functions of the respiratory system.

Anatomy of the Respiratory System   (p. 417)
2. Describe the anatomy of the respiratory passages, beginning at the nose and ending with the alveoli.
3. Describe the anatomy of the lungs, and define the respiratory membrane.

Ventilation and Lung Volumes   (p. 426)
4. Explain how contraction of the muscles of respiration causes changes in thoracic volume during quiet breathing and during labored breathing.
5. Describe the changes in alveolar pressure that are responsible for the movement of air into and out of the lungs.
6. Explain how surfactant and pleural pressure prevent the collapse of the lungs and how changes in pleural pressure cause changes in alveolar volume.
7. List the pulmonary volumes and capacities and define each of them.

Gas Exchange   (p. 432)
8. Explain the factors that affect gas movement through the respiratory membrane.
9. Describe the partial pressure gradients for oxygen and carbon dioxide.

Gas Transport in the Blood   (p. 434)
10. Explain how oxygen and carbon dioxide are transported in the blood.

Rhythmic Ventilation   (p. 435)
11. Describe the respiratory areas of the brainstem and how they produce a rhythmic pattern of ventilation.

Modification of Ventilation   (p. 436)
12. Name the neural mechanisms that can modify the normal rhythmic pattern of ventilation.
13. Explain how alterations in blood pH, carbon dioxide, and oxygen levels affect ventilation.

Respiratory Adaptations to Exercise   (p. 440)
14. Describe the regulation of ventilation during exercise and describe the changes in the respiratory system that result from exercise training.

Effects of Aging on the Respiratory System   (p. 442)
15. Describe the effects of aging on the respiratory system.

Studying, sleeping, talking, eating, and exercising all involve breathing. From our first breath at birth, the rate and depth of our breathing is unconsciously matched to our activities. Although we can voluntarily stop breathing, within a few minutes we must breathe again. Breathing is so characteristic of life that, along with the pulse, it is one of the first things we check for to determine if an unconscious person is alive.

Respiration includes the following processes: (1) ventilation, or breathing, which is the movement of air into and out of the lungs; (2) exchange of oxygen and carbon dioxide between the air in the lungs and the blood; (3) transport of oxygen and carbon dioxide in the blood; and (4) exchange of oxygen and carbon dioxide between the blood and the tissues. The term respiration is also used in reference to cell metabolism. In aerobic respiration, for example, cells use oxygen and produce carbon dioxide. Cellular respiration is considered in chapter 17.

Functions of the Respiratory System

Respiration is necessary because all living cells of the body require oxygen and produce carbon dioxide. The respiratory system assists in gas exchange and performs other functions as well.

1. *Gas exchange.* The respiratory system allows oxygen from the air to enter the blood and carbon dioxide to leave the blood and enter the air. The cardiovascular system transports oxygen from the lungs to the cells of the body and carbon dioxide from the cells of the body to the lungs. Thus the respiratory and cardiovascular systems work together to supply oxygen to all cells and to remove carbon dioxide. Without healthy respiratory and cardiovascular systems, the capacity to carry out normal activity is reduced, and without adequate respiratory and cardiovascular system functions, life itself is impossible.
2. *Regulation of blood pH.* The respiratory system can alter blood pH by changing blood carbon dioxide levels.
3. *Voice production.* Air movement past the vocal cords makes sound and speech possible.
4. *Olfaction.* The sensation of smell occurs when airborne molecules are drawn into the nasal cavity.
5. *Innate immunity* (see chapter 14). The respiratory system provides protection against some microorganisms by preventing their entry into the body and by removing them from respiratory surfaces.

Anatomy of the Respiratory System

The **respiratory system** consists of the external nose, the nasal cavity, the pharynx, the larynx, the trachea, the bronchi, and the lungs (figure 15.1). Although air frequently passes through the oral cavity, it is considered to be part of the digestive system instead of the respiratory system. The **upper respiratory tract** refers to the external nose, nasal cavity, pharynx, and associated structures; and the **lower respiratory tract** includes the larynx, trachea, bronchi, and lungs. These terms are not official

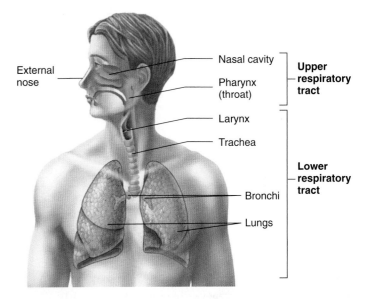

Figure 15.1   The Respiratory System

The upper respiratory tract consists of the external nose, nasal cavity, and pharynx (throat). The lower respiratory tract consists of the larynx, trachea, bronchi, and lungs.

anatomical terms, however, and there are several alternative definitions. For example, one alternative places the larynx in the upper respiratory tract.

Nose

The **nose** consists of the external nose and the nasal cavity. The **external nose** is the visible structure that forms a prominent feature of the face. Most of the external nose is composed of hyaline cartilage, although the bridge of the external nose consists of bone (see figure 6.12). The bone and cartilage are covered by connective tissue and skin.

The **nasal** (nā′zăl) **cavity** extends from the nares to the choane (figure 15.2). The **nares** (nā′rēs; sing. nā′ris), or **nostrils,** are the external openings of the nose and the **choane** (kō′an-ē, funnel) are the openings into the pharynx. The **nasal septum** is a partition dividing the nasal cavity into right and left parts. A **deviated nasal septum** occurs when the septum bulges to one side or the other. The **hard palate** (pal′ăt) forms the floor of the nasal cavity, separating the

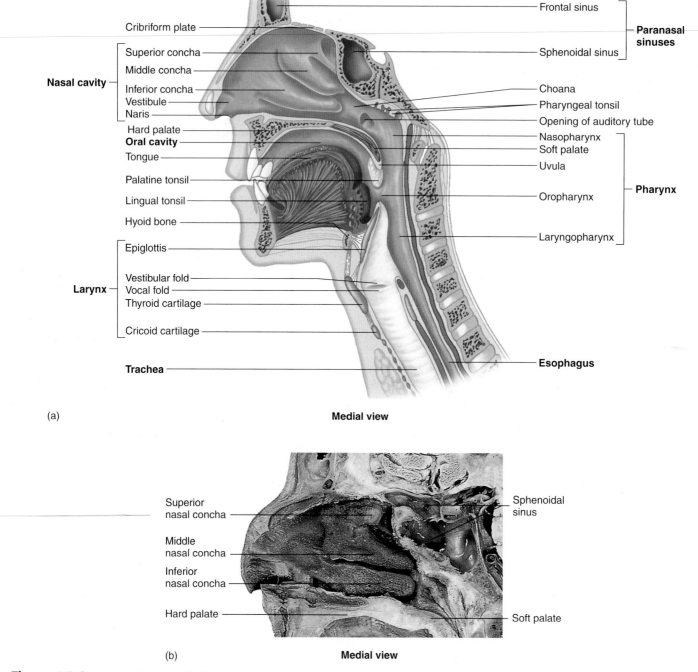

Figure 15.2   **Nasal Cavity and Pharynx**

(*a*) Sagittal section through the nasal cavity and pharynx. (*b*) Photograph of sagittal section of the head.

nasal cavity from the oral cavity. Air can flow through the nasal cavity when the mouth is closed or when the oral cavity is full of food.

Three prominent bony ridges called **conchae** (kon′kē, resembling a conch shell) are present on the lateral walls on each side of the nasal cavity. The conchae increase the surface area of the nasal cavity.

Paranasal (par-ă-nā′săl) **sinuses** are air-filled spaces within bone. The maxillary, frontal, ethmoidal, and sphenoidal sinuses are named after the bones in which they are located. The paranasal sinuses open into the nasal cavity and are lined with a mucous membrane. They reduce the weight of the skull, produce mucus, and influence the quality of the voice by acting as resonating chambers.

A CASE IN POINT | Sinusitis

Cy Ness has a cold. He feels tired and achy and he is sneezing. Although Cy recovers from the cold, for several weeks he has a stuffy nose and recurrent headaches. Cy's doctor listens to his history of symptoms, examines his throat and nose, and orders a sinus x-ray. It is concluded that Cy Ness has sinusitis. **Sinusitis** (sī-nŭ-sī′tis) is an inflammation of the mucous membrane of any sinus, especially of one or more paranasal sinuses. Viral infections, such as the common cold, can cause mucous membranes to become inflamed, swell, and produce excess mucus. As a result, the sinus opening into the nasal cavity can be partially or completely blocked. When mucus accumulates within the sinus, it can promote the development of a bacterial infection. The buildup of mucus and the inflammation of mucous membranes caused by infection produce pain. Treatment consists of taking antibiotics to kill the bacteria and decongestants to promote sinus drainage; drinking fluids to maintain hydration; and inhaling steam. Decongestants reduce tissue swelling or edema. When mucous membranes are less swollen, breathing is easier and the movement of mucus out of the paranasal sinuses and nasal cavity is increased. Decongestants, such as pseudoephedrine hydrochloride (e.g., Sudafed), reduce swelling by causing the release of norepinephrine from sympathetic neurons supplying blood vessels. Increased vasoconstriction of blood vessels in mucous membranes reduces blood flow and the movement of fluid from the blood into tissues. Sinusitis can also result from inflammation caused by allergies or from benign growths, called polyps, that obstruct a sinus opening into the nasal cavity.

The **nasolacrimal** (nā-zō-lak′ri-măl) **ducts,** which carry tears from the eyes, also open into the nasal cavity. Sensory receptors for the sense of smell are found in the superior part of the nasal cavity (see chapter 9).

Air enters the nasal cavity through the nares. Just inside the nares the epithelial lining is composed of stratified squamous epithelium containing coarse hairs. The hairs trap some of the large particles of dust suspended in the air. The rest of the nasal cavity is lined with pseudostratified columnar epithelial cells containing cilia and many mucus-producing goblet cells (see chapter 4). Mucus produced by the goblet cells also traps debris in the air. The cilia sweep the mucus posteriorly to the pharynx, where it is swallowed. As air flows through the nasal cavities, it is humidified by moisture from the mucous epithelium and is warmed by blood flowing through the superficial capillary networks underlying the mucous epithelium.

PREDICT 1

Explain what happens to your throat when you sleep with your mouth open, especially when your nasal passages are plugged as a result of having a cold. Explain what may happen to your lungs when you run a long way in very cold weather while breathing rapidly through your mouth.

Pharynx

The **pharynx** (far′ingks, throat) is the common passageway of both the respiratory and digestive systems. It receives air from the nasal cavity and air, food, and water from the mouth. Inferiorly, the pharynx leads to the rest of the respiratory system through the opening into the larynx and to the digestive system through the esophagus. The pharynx can be divided into three regions: the nasopharynx, the oropharynx, and the laryngopharynx (see figure 15.2*a*).

The **nasopharynx** (nā′zō-far′ingks) is the superior part of the pharynx. It is located posterior to the choanae and superior to the **soft palate,** which is an incomplete muscle and connective tissue partition separating the nasopharynx from the oropharynx. The **uvula** (ū′vū-lă, a little grape) is the posterior extension of the soft palate. The soft palate forms the floor of the nasopharynx. The nasopharynx is lined with pseudostratified ciliated columnar epithelium that is continuous with the nasal cavity. The auditory tubes extend from the middle ears and open into the nasopharynx. The posterior part of the nasopharynx contains the **pharyngeal** (fă-rin′jē-ăl) **tonsil,** which aids in defending the body against infection (see chapter 14). The soft palate is elevated during swallowing; this movement results in the closure of the nasopharynx, which prevents food from passing from the oral cavity into the nasopharynx.

The Sneeze Reflex

The **sneeze reflex** functions to dislodge foreign substances from the nasal cavity. Sensory receptors detect the foreign substances, and action potentials are conducted along the trigeminal nerves to the medulla oblongata in which the reflex is triggered. During the sneeze reflex, the uvula and the soft palate are depressed so that rapidly flowing air from the lungs is directed primarily through the nasal passages, although a considerable amount passes through the oral cavity.

Some people have a photic sneeze reflex in which exposure to bright light, such as the sun, can stimulate a sneeze reflex. The pupillary reflex causes the pupils to constrict in response to bright light. It is speculated that the complicated "wiring" of the pupillary and sneeze reflexes is intermixed in some people so that when bright light activates a pupillary reflex, it also activates a sneeze reflex.

The **oropharynx** (ōr′ō-far′ingks) extends from the uvula to the epiglottis, and the oral cavity opens into the oropharynx. Thus food, drink, and air all pass through the oropharynx. The oropharynx is lined with stratified squamous epithelium, which protects against abrasion. Two sets of tonsils, the palatine tonsil and the lingual tonsils, are located near the opening between the mouth and the oropharynx. The **palatine** (pal′ă-tīn) **tonsils** are located in the lateral walls near the border of the oral cavity and the oropharynx. The **lingual tonsil** is located on the surface of the posterior part of the tongue.

The **laryngopharynx** (lă-ring′gō-far-ingks) passes posterior to the larynx and extends from the tip of the epiglottis to the esophagus. Food and drink pass through the laryngopharynx to the esophagus. A small amount of air is usually swallowed with the food and drink. Swallowing too much air can cause excess gas in the stomach and may result in belching. The laryngopharynx is lined with stratified squamous epithelium and ciliated columnar epithelium.

Larynx

The **larynx** (lar′ingks) is located in the anterior throat, and it is continuous superiorly with the pharynx and inferiorly with the trachea. The larynx consists of an outer casing of nine cartilages that are connected to one another by muscles and ligaments (figure 15.3). Three of the nine cartilages are unpaired, and six of them form three pairs. The largest cartilage is the unpaired

thyroid (thī′royd, shield-shaped) **cartilage,** or **Adam's apple.** The thyroid cartilage is attached superiorly to the hyoid bone. The most inferior cartilage of the larynx is the unpaired **cricoid** (krī′koyd, ring-shaped) **cartilage,** which forms the base of the larynx on which the other cartilages rest. The thyroid and cricoid cartilages maintain an open passageway for air movement.

The third unpaired cartilage is the **epiglottis** (ep-i-glot′is, on the glottis). It differs from the other cartilages in that it consists of elastic cartilage rather than hyaline cartilage. Its inferior margin is attached to the thyroid cartilage anteriorly, and the superior part of the epiglottis projects as a free flap toward the tongue. The epiglottis helps prevent swallowed materials from entering the larynx. As the larynx elevates during swallowing, the epiglottis tips posteriorly to cover the opening of the larynx.

The six paired cartilages consist of three cartilages on either side of the posterior part of the larynx (see figure 15.3b). The top cartilage on each side is the **cuneiform** (kū′nē-i-fŏrm, wedge-shaped) **cartilage,** the middle cartilage is the **corniculate** (kŏr-nik′ū-lāt, horn-shaped) **cartilage,** and the bottom cartilage is the **arytenoid** (ar-i-tē′noyd, ladle-shaped) **cartilage.** The arytenoid cartilages articulate with the cricoid cartilage inferiorly. The paired cartilages form an attachment site for the vocal folds.

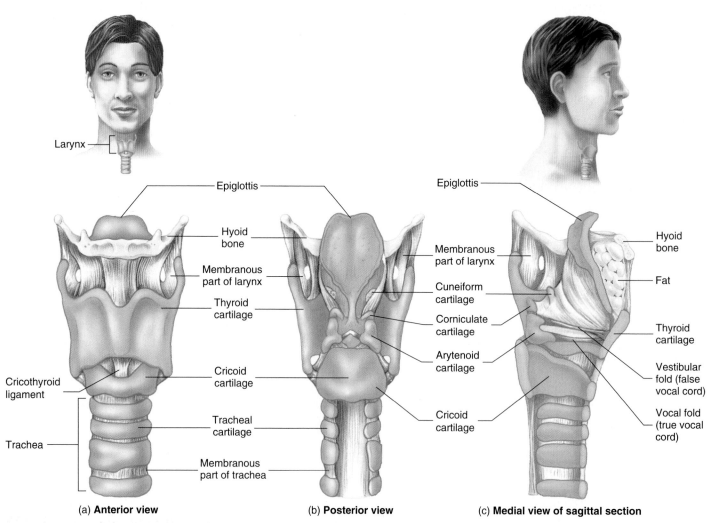

(a) **Anterior view**   (b) **Posterior view**   (c) **Medial view of sagittal section**

Figure 15.3   Anatomy of the Larynx

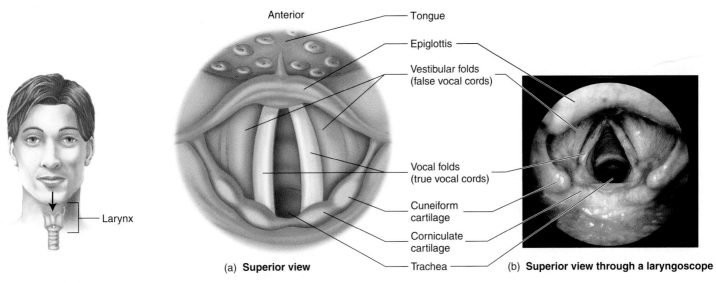

(a) Superior view

(b) Superior view through a laryngoscope

Figure 15.4 Vestibular and Vocal Folds

Arrow shows the direction of viewing the vestibular and vocal folds. (*a*) The relationship of the vestibular folds to the vocal folds and the laryngeal cartilages.
(*b*) Laryngoscopic view of the vestibular and vocal folds.

Two pairs of ligaments extend from the posterior surface of the thyroid cartilage to the paired cartilages. The superior pair forms the **vestibular** (ves-tib′ū-lăr) **folds,** or **false vocal cords,** and the inferior pair composes the **vocal folds,** or **true vocal cords** (figure 15.4). When the vestibular folds come together, they prevent air from leaving the lungs, such as when a person holds his breath. Along with the epiglottis, the vestibular folds also prevent food and liquids from entering the larynx.

The vocal folds are the primary source of voice production. Air moving past the vocal folds causes them to vibrate, producing sound. Muscles control the length and tension of the vocal folds. The force of air moving past the vocal folds controls the loudness, and the tension of the vocal folds controls the pitch of the voice. An inflammation of the mucous epithelium of the vocal folds is called **laryngitis** (lar-in-jī′tis). Swelling of the vocal folds during laryngitis inhibits voice production.

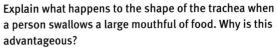

The Cough Reflex

The function of the **cough reflex** is to dislodge foreign substances from the trachea. Sensory receptors detect the foreign substances, and action potentials are conducted along the vagus nerves to the medulla oblongata in which the cough reflex is triggered. During coughing, contraction of smooth muscle decreases the diameter of the trachea. As a result, air moves rapidly through the trachea, which helps to expel mucus and foreign substances. Also, the uvula and soft palate are elevated so that air primarily passes through the oral cavity.

Trachea

The **trachea** (trā′kē-ă), or windpipe, is a membranous tube that consists of connective tissue and smooth muscle, reinforced with 16–20 C-shaped pieces of cartilage (see figure

15.3). The adult trachea is about 1.4–1.6 centimeters (cm) in diameter and about 10–11 cm long. It begins immediately inferior to the cricoid cartilage, which is the most inferior cartilage of the larynx. The trachea projects through the mediastinum, and divides into the right and left primary bronchi at the level of the fifth thoracic vertebra (figure 15.5). The esophagus lies immediately posterior to the trachea (see figure 15.2*a*).

C-shaped cartilages form the anterior and lateral sides of the trachea. The cartilages protect the trachea and maintain an open passageway for air. The posterior wall of the trachea has no cartilage and consists of a ligamentous membrane and smooth muscle (see figure 15.3*b*). The smooth muscle can alter the diameter of the trachea.

P R E D I C T   2
Explain what happens to the shape of the trachea when a person swallows a large mouthful of food. Why is this advantageous?

The trachea is lined with pseudostratified columnar epithelium, which contains numerous cilia and goblet cells. The cilia propel mucus produced by the goblet cells, as well as foreign particles embedded in the mucus, out of the trachea, through the larynx, and into the pharynx, from which they are swallowed.

Constant irritation of the trachea by cigarette smoke can cause the tracheal epithelium to change to stratified squamous epithelium. The stratified squamous epithelium has no cilia and therefore lacks the ability to clear the airway of mucus and debris. The accumulations of mucus provide a place for microorganisms to grow, resulting in respiratory infections. Constant irritation and inflammation of the respiratory passages stimulate the cough reflex, resulting in "smoker's cough."

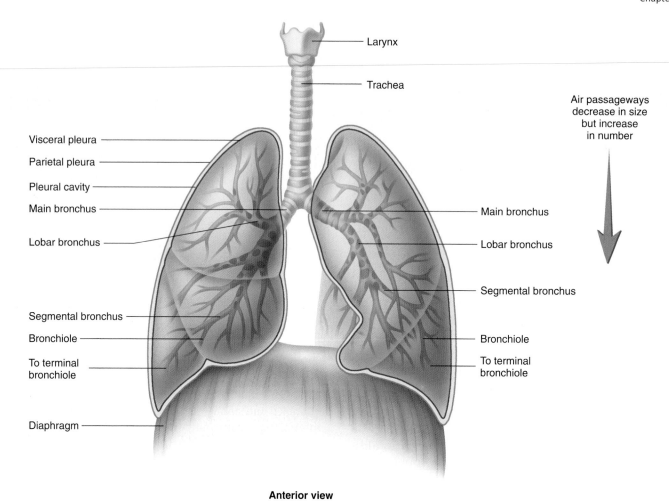

Air passageways
decrease in size
but increase
in number

Anterior view

Figure 15.5   Anatomy of the Trachea and Lungs

Drawing of the trachea and lungs, showing the branching of the bronchi. Each lung is surrounded by a pleural cavity, formed by the visceral and parietal pleurae.

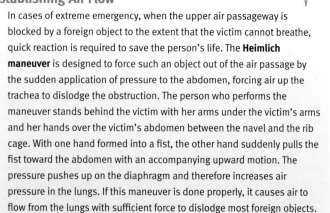

Establishing Air Flow

In cases of extreme emergency, when the upper air passageway is blocked by a foreign object to the extent that the victim cannot breathe, quick reaction is required to save the person's life. The **Heimlich maneuver** is designed to force such an object out of the air passage by the sudden application of pressure to the abdomen, forcing air up the trachea to dislodge the obstruction. The person who performs the maneuver stands behind the victim with her arms under the victim's arms and her hands over the victim's abdomen between the navel and the rib cage. With one hand formed into a fist, the other hand suddenly pulls the fist toward the abdomen with an accompanying upward motion. The pressure pushes up on the diaphragm and therefore increases air pressure in the lungs. If this maneuver is done properly, it causes air to flow from the lungs with sufficient force to dislodge most foreign objects.

There are other ways to establish air flow, but they should be performed only by trained medical personnel. **Intubation** is the insertion of a tube into an opening, canal, or hollow organ. A tube can be passed through the mouth or nose into the pharynx and then through the larynx to the trachea.

Sometimes it is necessary to make an opening through which to pass the tube. The preferred point of entry in an emergency is through the membrane between the cricoid and thyroid cartilages (see figure 15.3*a*). This procedure is referred to as a **cricothyrotomy** (krī′kō-thī-rot′ō-mē). A tube is inserted into the opening to facilitate the passage of air.

A **tracheostomy** (trā′kē-os′tō-mē, *tracheo- + stoma*, mouth) is an operation to make an opening into the trachea. Usually the opening is intended to be permanent, and a tube is inserted into the trachea to allow airflow and provide a way to remove secretions. The term **tracheotomy** (trā-kē-ot′ō-mē, *tracheo- + tome*, incision) refers to the actual cutting into the trachea. Sometimes the terms *tracheostomy* and *tracheotomy* are used interchangeably. It is not advisable to enter the air passageway through the trachea in emergency cases because arteries, nerves, and the thyroid gland overlie the anterior surface of the trachea.

Bronchi

The trachea divides into the left and right **main (primary) bronchi,** (brong′kī; sing., bronchus, brong′kŭs; windpipe), each of which connects to a lung. The left main bronchus is more horizontal than the right main bronchus because it is displaced by the heart (see figure 15.5). Foreign objects that enter the trachea usually lodge in the right main bronchus, because it is more vertical than the left main bronchus and therefore more in direct line with the trachea. The main bronchi extend from the trachea to the lungs. Like the trachea, the main bronchi are lined with pseudostratified ciliated columnar epithelium and are supported by C-shaped pieces of cartilage.

Lungs

The **lungs** are the principal organs of respiration. Each lung is cone-shaped, with its base resting on the diaphragm and its apex extending superiorly to a point about 2.5 cm above the clavicle (see figure 15.5). The right lung has three **lobes** called the superior, middle, and inferior lobes. The left lung has two lobes called the superior and inferior lobes (figure 15.6). The lobes of the lungs are separated by deep, prominent fissures on the surface of the lung. Each lobe is divided into **bronchopulmonary segments** separated from one another by connective tissue septa, but these separations are not visible as surface fissures. Individual diseased bronchopulmonary segments can be surgically removed, leaving the rest of the lung relatively intact, because major blood vessels and bronchi do not cross the septa. There are 9 bronchopulmonary segments in the left lung and 10 in the right lung.

The main bronchi branch many times to form the **tracheobronchial tree** (see figure 15.5). Each main bronchus divides into lobar bronchi as they enter their respective lungs (see figure 15.6). The **lobar (secondary) bronchi,** two in the left lung and three in the right lung, conduct air to each lobe. The lobar bronchi in turn give rise to **segmental (tertiary) bronchi,** which extend to the bronchopulmonary segments of the lungs. The bronchi continue to branch many times, finally giving rise to **bronchioles** (brong′kē-ōlz). The bronchioles also subdivide numerous times to give rise to **terminal bronchioles,** which then subdivide into **respiratory bronchioles** (figure 15.7). Each respiratory bronchiole subdivides to form **alveolar** (al-vē′ō-lăr) **ducts,** which are like long, branching hallways with many open doorways. The doorways open into **alveoli** (al-vē′ō-lī, hollow sacs), which are small air sacs. The alveoli become so numerous that the alveolar duct wall is little more than a succession of alveoli. The alveolar ducts end as two or three **alveolar sacs,** which are chambers connected to two or more alveoli. There are about 300 million alveoli in the lungs.

As the air passageways of the lungs become smaller, the structure of their walls changes. The amount of cartilage decreases and the amount of smooth muscle increases, until at the terminal bronchioles, the walls have a prominent smooth muscle layer, but no cartilage. Relaxation and contraction of the

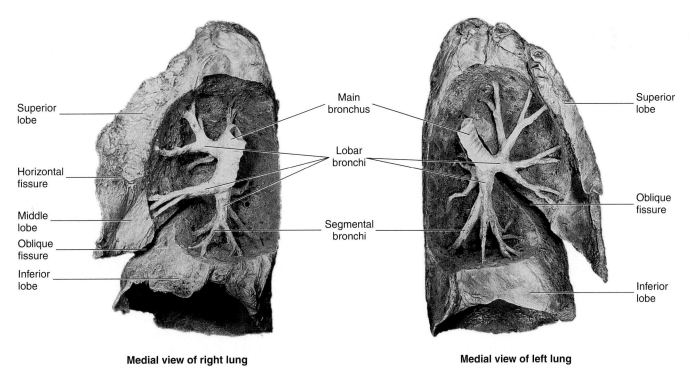

Medial view of right lung   **Medial view of left lung**

Figure 15.6 Lungs, Lung Lobes, and Bronchi

The right lung is divided into three lobes by the horizontal and oblique fissures. The left lung is divided into two lobes by the oblique fissure. A main bronchus supplies each lung, a lobar bronchus supplies each lung lobe, and segmental bronchi supply the bronchopulmonary segments (not visible).

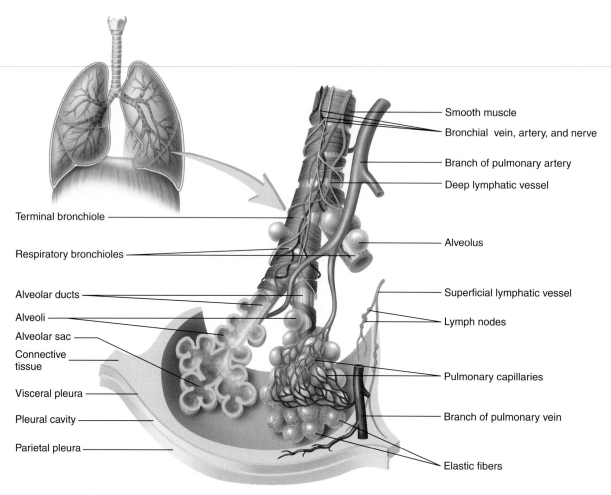

Figure 15.7 **Bronchioles and Alveoli**

A terminal bronchiole branches to form respiratory bronchioles, which give rise to alveolar ducts. Alveoli connect to the alveolar ducts and respiratory bronchioles. The alveolar ducts end as two or three alveolar sacs.

smooth muscle within the bronchi and bronchioles can change the diameter of the air passageways. For example, during exercise the diameter can increase, thus increasing the volume of air moved. During an **asthma attack,** however, contraction of the smooth muscle in the terminal bronchioles can result in greatly reduced air flow. In severe cases, air movement can be so restricted that death results.

As the air passageways of the lungs become smaller, the lining of their walls also changes. The trachea and bronchi have pseudostratified ciliated columnar epithelium, the bronchioles have ciliated simple columnar epithelium, and the terminal bronchioles have ciliated simple cuboidal epithelium. The ciliated epithelium of the air passageways functions as a mucus-cilia escalator, which traps debris in the air and removes it from the respiratory system.

As the air passageways beyond the terminal bronchioles become smaller, their walls become thinner. The walls of the respiratory bronchioles are cuboidal epithelium and those of the alveolar ducts and alveoli are simple squamous epithelium. The **respiratory membrane** of the lungs is where gas exchange between the air and blood takes place. It is mainly formed by the

walls of the alveoli and surrounding capillaries (figure 15.8), but there's some contribution by the alveolar ducts and respiratory bronchioles. The respiratory membrane is very thin to facilitate the diffusion of gases. It consists of:

1. A thin layer of fluid lining the alveolus
2. The alveolar epithelium composed of simple squamous epithelium
3. The basement membrane of the alveolar epithelium
4. A thin interstitial space
5. The basement membrane of the capillary endothelium
6. The capillary endothelium composed of simple squamous epithelium

The elastic fibers surrounding the alveoli (see figure 15.8) allow them to expand during inspiration and recoil during expiration. The lungs are very elastic, and when inflated, they are capable of expelling the air and returning to their original, uninflated state. Specialized secretory cells within the walls of the alveoli (see figure 15.8) secrete a chemical called surfactant that reduces the tendency of alveoli to recoil (see Lung Recoil on p. 429).

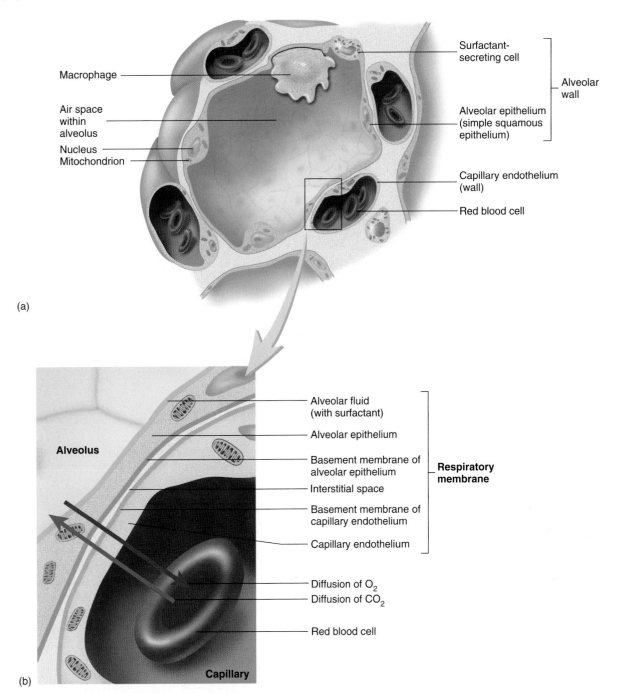

Figure 15.8   Alveolus and the Respiratory Membrane

(*a*) Section of an alveolus showing the air-filled interior and thin walls composed of simple squamous epithelium. The alveolus is surrounded by elastic connective tissue and blood capillaries. (*b*) Diffusion of O_2 and CO_2 across the six thin layers of the respiratory membrane.

Pleural Cavities

The lungs are contained within the thoracic cavity. In addition, each lung is surrounded by a separate **pleural** (ploor'ăl, relating to the ribs) **cavity.** Each pleural cavity is lined with a serous membrane called the **pleura.** The pleura consists of a parietal and visceral part. The **parietal pleura,** which lines the walls of the thorax, diaphragm, and mediastinum, is continuous with

the **visceral pleura,** which covers the surface of the lung (see figures 15.5 and 15.9).

The pleural cavity, between the parietal and visceral pleurae, is filled with a small volume of pleural fluid produced by the pleural membranes. The pleural fluid performs two functions: (1) it acts as a lubricant, allowing the visceral and parietal pleurae to slide past each other as the lungs and thorax change shape during

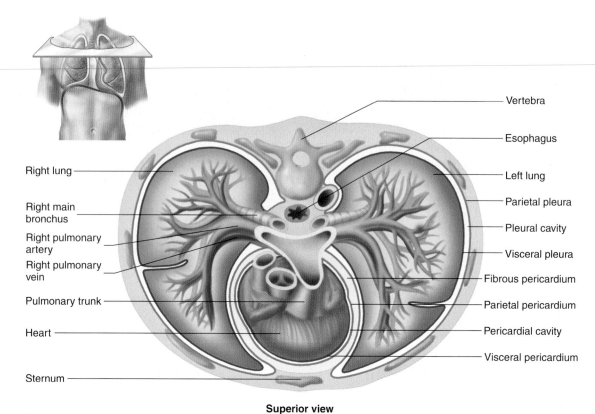

Superior view

Figure 15.9 Pleural Cavities and Membranes

Transverse section of the thorax showing the relationship of the pleural cavities to the thoracic organs. Each lung is surrounded by a pleural cavity. The parietal pleura lines the wall of each pleural cavity, and the visceral pleura covers the surface of the lungs. The space between the parietal and visceral pleurae is small and filled with pleural fluid.

respiration, and (2) it helps hold the pleural membranes together. The pleural fluid acts like a thin film of water between two sheets of glass (the visceral and parietal pleurae); the glass sheets can slide over each other easily, but it is difficult to separate them.

P R E D I C T   3
Pleurisy is an inflammation of the pleural membranes. Explain why this condition is so painful, especially when a person takes deep breaths.

Lymphatic Supply

The lungs have two lymphatic supplies (see figure 15.7). The **superficial lymphatic vessels** are deep to the visceral pleura and function to drain lymph from the superficial lung tissue and the visceral pleura. The **deep lymphatic vessels** follow the bronchi and function to drain lymph from the bronchi and associated connective tissues. No lymphatic vessels are located in the walls of the alveoli. Both the superficial and deep lymphatic vessels exit the lungs at the main bronchi.

Phagocytic cells within the lungs phagocytize carbon particles and other debris from inspired air and move them to the lymphatic vessels. In older people, the surface of the lungs can appear gray to black because of the accumulation of these particles, especially if the person smoked or lived most of his life in a city with air pollution. Cancer cells from the lungs can also spread to other parts of the body through the lymphatic vessels.

Ventilation and Lung Volumes

Ventilation, or **breathing,** is the process of moving air into and out of the lungs. There are two phases of ventilation: (1) **inspiration,** or **inhalation,** is the movement of air into the lungs; (2) **expiration,** or **exhalation,** is the movement of air out of the lungs. Changes in thoracic volume, which produce changes in air pressure within the lungs, are responsible for ventilation.

Changing Thoracic Volume

Muscles associated with the ribs are responsible for ventilation (figure 15.10). The **muscles of inspiration** include the diaphragm and muscles that elevate the ribs and sternum, such as the external intercostals. The **diaphragm** (dī′a-fram, partition) is a large dome of skeletal muscle that separates the thoracic cavity from the abdominal cavity (see figure 7.18). The **muscles of expiration,** such as the internal intercostals, depress the ribs and sternum.

At the end of a normal, quiet expiration, the respiratory muscles are relaxed (see figure 15.10a). During quiet inspiration, contraction of the diaphragm causes the top of the dome to move inferiorly, which increases the volume of the thoracic cavity. The largest change in thoracic volume results from movement of the diaphragm. Contraction of the external intercostals also elevates the ribs and sternum (see figure 15.10b), which increases thoracic volume by increasing the diameter of the thoracic cage.

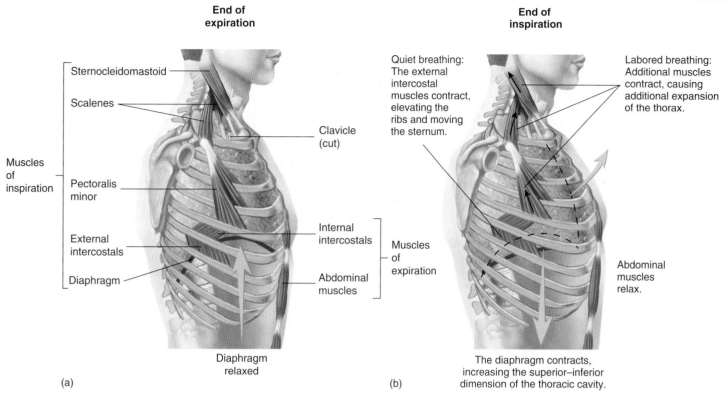

End of expiration

Sternocleidomastoid

Scalenes

Clavicle (cut)

Muscles of inspiration

Pectoralis minor

External intercostals

Diaphragm

Internal intercostals

Muscles of expiration

Abdominal muscles

Diaphragm relaxed

(a)

End of inspiration

Quiet breathing: The external intercostal muscles contract, elevating the ribs and moving the sternum.

Labored breathing: Additional muscles contract, causing additional expansion of the thorax.

Abdominal muscles relax.

The diaphragm contracts, increasing the superior–inferior dimension of the thoracic cavity.

(b)

Figure 15.10   Effect of the Muscles of Respiration on Thoracic Volume

(*a*) Muscles of respiration at the end of expiration. (*b*) Muscles of respiration at the end of inspiration.

P R E D I C T   4
During inspiration, the abdominal muscles relax. How is this advantageous?

Expiration during quiet breathing occurs when the diaphragm and external intercostals relax and the elastic properties of the thorax and lungs cause a passive decrease in thoracic volume.

There are several differences between normal, quiet breathing and labored breathing. During labored breathing, all of the inspiratory muscles are active and they contract more forcefully than during quiet breathing, causing a greater increase in thoracic volume (see figure 15.10*b*). During labored breathing, forceful contraction of the internal intercostals and the abdominal muscles produces a more rapid and greater decrease in thoracic volume than would be produced by the passive recoil of the thorax and lungs.

Pressure Changes and Airflow

The flow of air into and out of the lungs is governed by two physical principles:

1. *Changes in volume result in changes in pressure.* As the volume of a container increases, the pressure within the container decreases. As the volume of a container decreases, the pressure within the container increases. The muscles of respiration change thoracic volume and therefore pressure within the thoracic cavity.

2. *Air flows from areas of higher to lower pressure.* If the pressure is higher at one end of a tube than at the other, air or fluid (see chapter 13) flows from the area of higher pressure toward the area of lower pressure. The greater the pressure difference, the greater the rate of airflow. Air flows through the respiratory passages because of pressure differences between the outside of the body and the alveoli inside the body. These pressure differences are produced by changes in thoracic volume.

The volume and pressure changes responsible for one cycle of inspiration and expiration can be described as follows:

1. At the end of expiration, **alveolar pressure,** which is the air pressure within the alveoli, is equal to **atmospheric pressure,** which is the air pressure outside the body. There is no movement of air into or out of the lungs because alveolar pressure and atmospheric pressure are equal (figure 15.11, step 1).

2. During inspiration, contraction of the muscles of inspiration increases the volume of the thoracic cavity. The increased thoracic volume causes the lungs to expand, resulting in an increase in alveolar volume (see the section, Changing Alveolar Volume, on p. 430). As the alveolar volume increases, alveolar pressure becomes less than atmospheric pressure, and air flows from outside the body through the respiratory passages to the alveoli (figure 15.11, step 2).

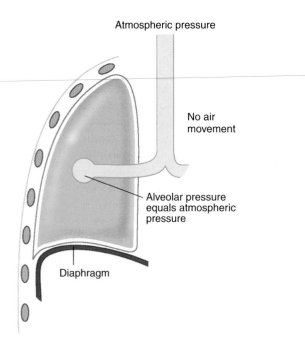

Atmospheric pressure

No air
movement

Alveolar pressure
equals atmospheric
pressure

Diaphragm

1. At the end of expiration, alveolar pressure is
   equal to atmospheric pressure. Therefore,
   there is no air movement.

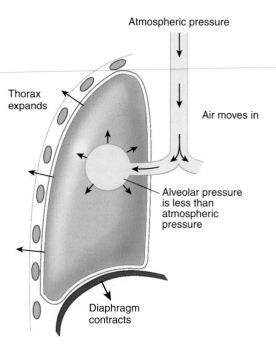

Atmospheric pressure

Thorax
expands

Air moves in

Alveolar pressure
is less than
atmospheric
pressure

Diaphragm
contracts

2. During inspiration, increased thoracic
   volume results in decreased pressure inside
   the alveoli. Therefore, air moves into the
   lungs.

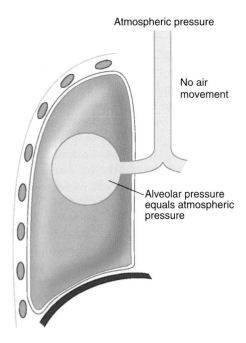

Atmospheric pressure

No air
movement

Alveolar pressure
equals atmospheric
pressure

3. At the end of inspiration, alveolar pressure is
   equal to atmospheric pressure. Therefore,
   there is no air movement.

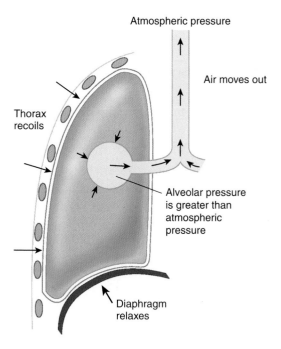

Atmospheric pressure

Air moves out

Thorax
recoils

Alveolar pressure
is greater than
atmospheric
pressure

Diaphragm
relaxes

4. During expiration, decreased thoracic
   volume results in increased pressure
   inside the alveoli. Therefore, air moves out
   of the lungs.

Process Figure 15.11 Alveolar Pressure Changes During Inspiration and Expiration

The combined space of all the alveoli is represented by a large "bubble." The alveoli are actually microscopic in size and cannot be seen in the illustration.

3. At the end of inspiration, the thorax and alveoli stop expanding. When the alveolar pressure and atmospheric pressure become equal, airflow stops (figure 15.11, step 3).

4. During expiration, the thoracic volume decreases, producing a decrease in alveolar volume. Consequently, alveolar pressure increases above the air pressure outside the body, and air flows from the alveoli through the respiratory passages to the outside (figure 15.11, step 4).

As expiration ends, the decrease in thoracic volume stops and the process repeats beginning at step 1.

Lung Recoil

During quiet expiration, thoracic volume and lung volume decrease because of passive recoil of the thoracic wall and lungs. The recoil of the thoracic wall results from the elastic properties of the thoracic wall tissues. **Lung recoil** is the tendency for an expanded lung to decrease in size. It occurs for two reasons: (1) the elastic fibers in the connective tissue of the lungs and (2) surface tension of the film of fluid that lines the alveoli. **Surface tension** exists because the oppositely charged ends of water molecules attract each other (see chapter 2). As the water molecules pull together, they also pull on the alveolar walls, causing the alveoli to recoil and become smaller.

Two factors keep the lungs from collapsing: (1) surfactant, and (2) pressure in the pleural cavity.

Surfactant

Surfactant (ser-fak′tănt, <u>surf</u>ace <u>act</u>ing <u>ag</u>ent) is a mixture of lipoprotein molecules produced by secretory cells of the alveolar epithelium. The surfactant molecules form a single layer on the surface of the thin fluid layer lining the alveoli, reducing surface tension. Without surfactant, the surface tension causing the alveoli to recoil can be 10 times greater than when surfactant is present. Thus, surfactant greatly reduces the tendency of the lungs to collapse.

Infant Respiratory Distress Syndrome

Surfactant is not produced in adequate quantities until about the seventh month of gestation. Thereafter, the amount produced increases as the fetus matures. In premature infants, **infant respiratory distress syndrome (IRDS)**, or **hyaline** (hī′ă-lin, glass) **membrane disease**, is caused by too little surfactant. It is common, especially for infants delivered before the seventh month of pregnancy. Cortisol can be given to pregnant women who are likely to deliver prematurely, because it crosses the placenta into the fetus and stimulates surfactant synthesis.

If too little surfactant has been produced by the time of birth, the lungs tend to collapse, and a great deal of energy must be exerted by the muscles of respiration to keep the lungs inflated; even then, inadequate ventilation occurs. Without specialized treatment, most babies with this condition die soon after birth as a result of inadequate ventilation of the lungs and fatigue of the respiratory muscles. Treatment strategies include forcing enough oxygen-rich air into the lungs to inflate them and administering surfactant.

A CASE IN POINT | Infant Respiratory Distress Syndrome

Tu Soun was born 3 months prematurely. She presented with a respiration rate of 68 breaths per minute; blue lips, tongue, and nail beds; nasal flaring during inspiration; inward movement of the thoracic cage and outward movement of the abdomen during inspiration; an expiratory grunt; and a negative shake test. Tu Soun has infant respiratory distress syndrome.

The normal respiration rate for a newborn is around 40 breaths per minute. Tu's high respiratory rate is stimulated by high blood carbon dioxide levels and low blood oxygen levels (see Chemical Control of Ventilation on p. 437). Her blue lips, tongue, and nail beds are signs of cyanosis caused by deoxygenated blood. The nasal flaring is expansion of the nares to maximize air intake.

During quiet inspiration, thoracic volume increases as the thoracic cage and the abdomen expand (see Changing Thoracic Volume on p. 426). Inferior movement of the contracting diaphragm is responsible for most of the increase in thoracic volume, and therefore, most of the decrease in pressure within the thoracic cavity that results in the movement of air into the alveoli. During labored inspiration in newborns, the thoracic cage moves inward as the abdomen expands. The thoracic cage in newborns is very pliable. During labored inspiration, the increased inferior movement of the diaphragm causes such a decrease in thoracic cavity pressure that the thoracic cage is pulled inward. The more labored the breathing, the more exaggerated the expansion of the abdomen and the inward movement of the thoracic cage.

An expiratory grunt is a gruff, throaty sound made during expiration. It is caused by partial closure of the vestibular and vocal folds. Expiratory grunting increases airway pressure and helps to prevent alveolar collapse.

The shake test determines the presence of surfactant in lung fluid. Fetal lung fluid is either swallowed by the fetus or passes out the mouth into the amniotic fluid. A sample of gastric fluid collected within 30 minutes after delivery contains swallowed lung fluid and amniotic fluid. The gastric sample, saline, and alcohol are placed in a tube and shaken. A positive shake test produces bubbles and a negative shake test does not, which indicates there is very little surfactant present because the bubbles collapse.

Pleural Pressure

When **pleural pressure,** which is the pressure in the pleural cavity, is less than alveolar pressure, the alveoli tend to expand. This principle can be understood by considering a balloon. The balloon expands when the pressure outside it is less than the pressure inside. This pressure difference is normally achieved by increasing the pressure inside the balloon when a person forcefully blows into it. This pressure difference, however, can also be achieved by decreasing the pressure outside the balloon. For example, if the balloon is placed in a chamber from which air is removed, the pressure around the balloon becomes lower than atmospheric pressure, and the balloon expands. The lower the pressure outside the balloon, the greater the tendency for the higher pressure inside the balloon to cause it to expand. In a similar fashion, decreasing pleural pressure can result in expansion of the alveoli.

Normally, the alveoli are expanded because pleural pressure is lower than alveolar pressure. Pleural pressure is lower than alveolar pressure because of a "suction effect" caused by fluid removal by the lymphatic system (see p. 389) and by lung recoil. As the lungs recoil, the visceral and parietal pleurae tend to be pulled apart. Normally the lungs do not pull away from the thoracic wall because pleural fluid holds the visceral and parietal pleurae together. Nonetheless, this pull decreases pressure in the pleural cavity, an effect that can be appreciated by putting water on the palms of the hands and putting them together. A sensation of negative pressure is felt as the hands are gently pulled apart.

When pleural pressure is lower than alveolar pressure, the alveoli tend to expand. This expansion is opposed by the tendency of the lungs to recoil. Therefore, the alveoli expand when the pleural pressure is low enough that lung recoil is overcome. If the pleural pressure is not low enough to overcome lung recoil, then the alveoli collapse.

PREDICT 5

Treatment of a pneumothorax involves closing the opening into the pleural cavity that caused the pneumothorax. Then a tube is placed into the pleural cavity. In order to inflate the lung, should this tube pump in air under pressure (as in blowing up a balloon) or should the tube apply suction? Explain.

Pneumothorax

A **pneumothorax** (noo-mō-thōr′aks) is the introduction of air into the pleural cavity. Air can enter by an external route when a sharp object, such as a bullet or broken rib, penetrates the thoracic wall; or air can enter the pleural cavity by an internal route if alveoli at the lung surface rupture, such as can occur in a patient with emphysema. When the pleural cavity is connected to the outside by such openings, the pressure in the pleural cavity increases and becomes equal to the air pressure outside the body. Thus, pleural pressure is also equal to alveolar pressure because pressure in the alveoli at the end of expiration is equal to air pressure outside the body. When pleural pressure and alveolar pressure are equal, there is no tendency for the alveoli to expand, lung recoil is unopposed, and the lungs collapse. A pneumothorax can occur in one lung while the lung on the opposite side remains inflated because the two pleural cavities are separated by the mediastinum.

Changing Alveolar Volume

Changes in alveolar volume result in the changes in alveolar pressure that are responsible for the movement of air into and out of the lungs (see figure 15.11). Alveolar volume changes result from changes in pleural pressure. For example, during inspiration, pleural pressure decreases and the alveoli expand. The decrease in pleural pressure occurs for two reasons:

1. Increasing the volume of the thoracic cavity results in a decrease in pleural pressure because of the effect of changing volume on pressure.
2. As the lungs expand, lung recoil increases, resulting in an increased suction effect and a lowering of pleural pressure. The increased lung recoil of the stretched lung is similar to the increased force generated in a stretched rubber band.

The events of inspiration and expiration can be summarized as follows:

1. During inspiration, pleural pressure decreases because of increased thoracic volume and increased lung recoil. As pleural pressure decreases, alveolar volume increases, alveolar pressure decreases, and air flows into the lungs.
2. During expiration, pleural pressure increases because of decreased thoracic volume and decreased lung recoil. As pleural pressure increases, alveolar volume decreases, alveolar pressure increases, and air flows out of the lungs.

Pulmonary Volumes and Capacities

Spirometry (spī-rom′ĕ-trē, *spiro*, to breathe + *metron*, to measure) is the process of measuring volumes of air that move into and out of the respiratory system, and the **spirometer** (spī-rom′ĕ-ter) is the device that is used to measure these pulmonary volumes (figure 15.12*a*). Measurements of the respiratory volumes can provide information about the health of the lungs. The four **pulmonary volumes** and their normal values (figure 15.12*b*) for a young adult male are as follows:

1. **Tidal volume** is the volume of air inspired or expired with each breath. At rest, quiet breathing results in a tidal volume of about 500 milliliters (mL).
2. **Inspiratory reserve volume** is the amount of air that can be inspired forcefully after inspiration of the resting tidal volume (about 3000 mL).
3. **Expiratory reserve volume** is the amount of air that can be expired forcefully after expiration of the resting tidal volume (about 1100 mL).
4. **Residual volume** is the volume of air still remaining in the respiratory passages and lungs after a maximum expiration (about 1200 mL).

The tidal volume increases when a person is more active. Because the maximum volume of the respiratory system does not change from moment to moment, the increase in the tidal volume causes a decrease in the inspiratory and expiratory reserve volumes.

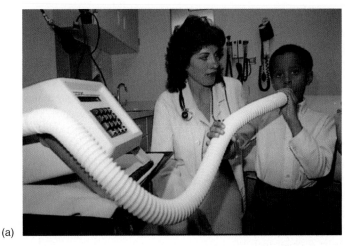

(a)

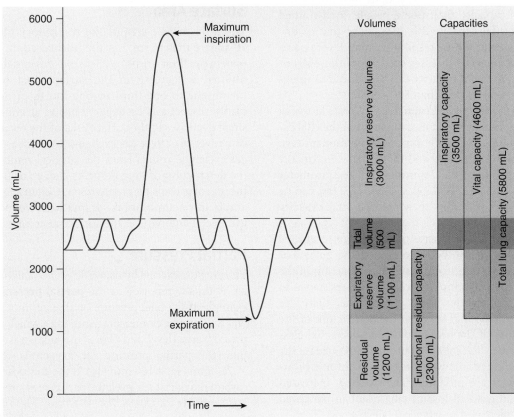

(b)

Figure 15.12 Spirometer, Lung Volumes, and Lung Capacities

(a) A spirometer used to measure lung volumes and capacities. (b) Lung volumes and capacities. The tidal volume in the figure is the tidal volume during resting conditions.

P R E D I C T   6

The **minute ventilation** is the total amount of air moved into and out of the respiratory system each minute, and it is equal to the tidal volume times the respiratory rate. The **respiratory rate** is the number of breaths taken per minute. Calculate the minute ventilation of a resting person, who has a tidal volume of 500 mL and a respiratory rate of 12 respirations/min, and an exercising person, who has a tidal volume of 4000 mL and a respiratory rate of 24 respirations/min.

A **pulmonary capacity** is the sum of two or more pulmonary volumes (see figure 15.12b). For example:

1. **Functional residual capacity** is the expiratory reserve volume plus the residual volume, which is the amount of air remaining in the lungs at the end of a normal expiration (about 2300 mL at rest).
2. **Inspiratory capacity** is the tidal volume plus the inspiratory reserve volume, which is the amount of air that a person can inspire maximally after a normal expiration (about 3500 mL at rest).

3. **Vital capacity** is the sum of the inspiratory reserve volume, the tidal volume, and the expiratory reserve volume; it is the maximum volume of air that a person can expel from his respiratory tract after a maximum inspiration (about 4600 mL).

4. **Total lung capacity** is the sum of the inspiratory and expiratory reserves and the tidal and residual volumes (about 5800 mL). The total lung capacity is also equal to the vital capacity plus the residual volume.

Factors such as sex, age, and body size influence the respiratory volumes and capacities. For example, the vital capacity of adult females is usually 20–25% less than that of adult males. The vital capacity reaches its maximum amount in the young adult and gradually decreases in the elderly. Tall people usually have a greater vital capacity than short people, and thin people have a greater vital capacity than obese people. Well-trained athletes can have a vital capacity 30–40% above that of untrained people. In patients whose respiratory muscles are paralyzed by spinal cord injury or diseases such as poliomyelitis or muscular dystrophy, the vital capacity can be reduced to values not consistent with survival (less than 500–1000 mL).

The **forced expiratory vital capacity** is the rate at which lung volume changes during direct measurement of the vital capacity. It is a simple and clinically important pulmonary test. The individual inspires maximally and then exhales maximally and as rapidly as possible into a spirometer. The spirometer records the volume of air expired per second. This test can be used to help identify conditions in which the vital capacity might not be affected, but in which the expiratory flow rate is reduced. Abnormalities that increase the resistance to airflow slow the rate at which air can be forced out of the lungs. For example, in people who suffer from asthma, contraction of the smooth muscle in the bronchioles increases the resistance to airflow. In people who suffer from emphysema, there are changes in the lung tissue that result in the destruction of the alveolar walls, collapse of the bronchioles, and decreased elasticity of the lung tissue. The collapsed bronchioles increase the resistance to airflow. In people who suffer from chronic bronchitis, the air passages are inflamed. The swelling, increased mucus secretion, and gradual loss of cilia result in narrowed bronchioles and an increased resistance to airflow.

Gas Exchange

Ventilation supplies atmospheric air to the alveoli. The next step in the process of respiration is the diffusion of gases between the alveoli and the blood in the pulmonary capillaries. The **respiratory membrane** is all of the areas in which gas exchange between air and blood occurs (see figure 15.8). The major area of gas exchange is in the alveoli, although some occurs in the respiratory bronchioles and alveolar ducts. Gas exchange between blood and air does not occur in such other areas of the respiratory passageways as the bronchioles, bronchi, and trachea. The volume of these passageways is therefore called **dead space.**

The exchange of gases across the respiratory membrane is influenced by the thickness of the membrane, the total surface area of the membrane, and the partial pressure of gases across the membrane.

Respiratory Membrane Thickness

Increasing the thickness of the respiratory membrane decreases the rate of diffusion across it. The thickness of the respiratory membrane increases during certain respiratory diseases. For example, in patients with pulmonary edema, fluid accumulates in the alveoli, and gases must diffuse through a thicker than normal layer of fluid. If the thickness of the respiratory membrane is doubled or tripled, the rate of gas exchange is markedly decreased. Oxygen exchange is affected before carbon dioxide exchange because oxygen diffuses through the respiratory membrane about 20 times less easily than does carbon dioxide.

Surface Area

The total surface area of the respiratory membrane is about 70 square meters (m^2) in the normal adult, which is approximately the floor area of a 25- $\times$ 30-ft room. Under resting conditions, a decrease in the surface area of the respiratory membrane to one-third or one-fourth of normal can significantly restrict gas exchange. During strenuous exercise, even small decreases in the surface area of the respiratory membrane can adversely affect gas exchange. A decreased surface area for gas exchange results from the surgical removal of lung tissue, the destruction of lung tissue by cancer, or the degeneration of the alveolar walls by emphysema. Collapse of the lung, for example in pneumothorax, dramatically reduces the volume of the alveoli and reduces the surface area for gas exchange.

Partial Pressure

The pressure exerted by a specific gas in a mixture of gases, such as air, is usually reported as the **partial pressure** of that gas. For example, if the total pressure of all gases in a mixture of gases is 760 millimeters of mercury (mm Hg), which is the atmospheric pressure at sea level, and 21% of the mixture is made up of oxygen, then the partial pressure for oxygen is 160 mm Hg (0.21 $\times$ 760 mm Hg = 160 mm Hg). If the composition of air is 0.04% carbon dioxide at sea level, the partial pressure for carbon dioxide is 0.3 mm Hg (0.0004 $\times$ 760 = 0.3 mm Hg) (table 15.1). It is traditional to designate the partial pressure of individual gases in a mixture with a capital P followed by the symbol for the gas. Thus the partial pressure of oxygen is P_{O_2}, and carbon dioxide is P_{CO_2}.

When air is in contact with a liquid such as water, gases such as carbon dioxide and oxygen in the air dissolve in the liquid. The gases dissolve in the liquid until the partial pressure of each gas in the liquid is equal to the partial pressure of that gas in the air. Gases in a liquid, like gases in air, diffuse from areas of higher partial pressure toward areas of lower partial pressure, until the partial pressures of the gases are equal throughout the liquid.

Diffusion of Gases in the Lungs

The cells of the body use oxygen and produce carbon dioxide. Thus, blood returning from tissues and entering the lungs has a decreased P_{O_2} and increased P_{CO_2} compared with alveolar air (figure 15.13). Oxygen diffuses from the alveoli into the

Table 15.1 Partial Pressures of Gases at Sea Level

Gases	Dry Air mm HG	Dry Air %	Humidified Air mm HG	Humidified Air %	Alveolar Air mm HG	Alveolar Air %	Expired Air mm HG	Expired Air %
Nitrogen	600.2	78.98	563.4	74.09	569.0	74.9	566.0	74.5
Oxygen	159.5	20.98	149.3	19.67	104.0	13.6	120.0	15.7
Carbon dioxide	0.3	0.04	0.3	0.04	40.0	5.3	27.0	3.6
Water vapor	0.0	0.0	47.0	6.20	47.0	6.2	47.0	6.2

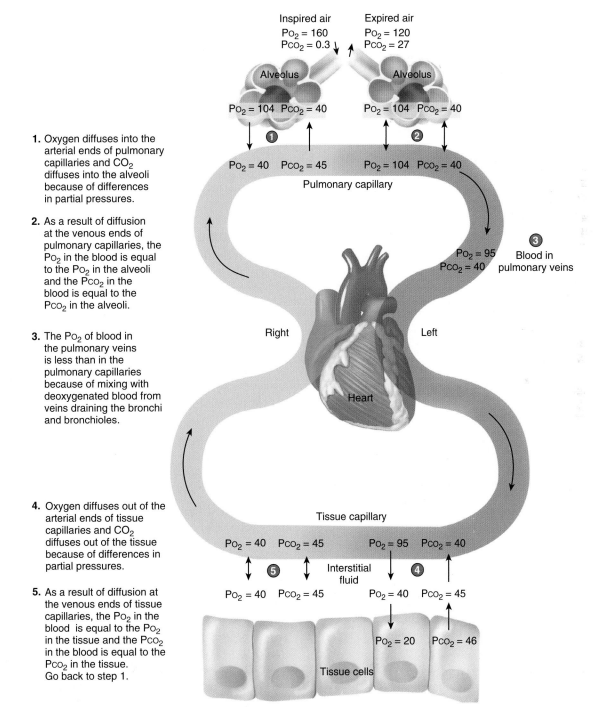

1. Oxygen diffuses into the arterial ends of pulmonary capillaries and CO_2 diffuses into the alveoli because of differences in partial pressures.

2. As a result of diffusion at the venous ends of pulmonary capillaries, the P_{O_2} in the blood is equal to the P_{O_2} in the alveoli and the P_{CO_2} in the blood is equal to the P_{CO_2} in the alveoli.

3. The P_{O_2} of blood in the pulmonary veins is less than in the pulmonary capillaries because of mixing with deoxygenated blood from veins draining the bronchi and bronchioles.

4. Oxygen diffuses out of the arterial ends of tissue capillaries and CO_2 diffuses out of the tissue because of differences in partial pressures.

5. As a result of diffusion at the venous ends of tissue capillaries, the P_{O_2} in the blood is equal to the P_{O_2} in the tissue and the P_{CO_2} in the blood is equal to the P_{CO_2} in the tissue. Go back to step 1.

Inspired air
$P_{O_2} = 160$
$P_{CO_2} = 0.3$

Expired air
$P_{O_2} = 120$
$P_{CO_2} = 27$

Alveolus
Alveolus

$P_{O_2} = 104$ $P_{CO_2} = 40$
$P_{O_2} = 104$ $P_{CO_2} = 40$

$P_{O_2} = 40$ $P_{CO_2} = 45$
$P_{O_2} = 104$ $P_{CO_2} = 40$
Pulmonary capillary

$P_{O_2} = 95$
$P_{CO_2} = 40$
Blood in pulmonary veins

Right
Left

Heart

Tissue capillary

$P_{O_2} = 40$ $P_{CO_2} = 45$
$P_{O_2} = 95$ $P_{CO_2} = 40$

Interstitial fluid

$P_{O_2} = 40$ $P_{CO_2} = 45$
$P_{O_2} = 40$ $P_{CO_2} = 45$

$P_{O_2} = 20$ $P_{CO_2} = 46$

Tissue cells

Process Figure 15.13 Gas Exchange

Oxygen and carbon dioxide partial pressure diffusion gradients between the alveoli and the pulmonary capillaries and between the tissues and the tissue capillaries are responsible for gas exchange.

pulmonary capillaries because the P_{O_2} in the alveoli is greater than that in the pulmonary capillaries. In contrast, carbon dioxide diffuses from the pulmonary capillaries into the alveoli because the P_{CO_2} is greater in the pulmonary capillaries than in the alveoli.

When blood enters a pulmonary capillary, the P_{O_2} and P_{CO_2} in the capillary are different from the P_{O_2} and P_{CO_2} in the alveolus. By the time blood flows through the first third of the pulmonary capillary, an equilibrium is achieved, and the P_{O_2} and P_{CO_2} in the capillary are the same as in the alveolus. Thus, the blood gains oxygen and loses carbon dioxide in the lungs.

During breathing, atmospheric air is mixed with alveolar air. The air entering and leaving the alveoli keeps the P_{O_2} higher in the alveoli than in the pulmonary capillaries. Increasing the rate of ventilation makes the P_{O_2} even higher in the alveoli than during slow breathing. During heavy breathing, the rate of oxygen diffusion into the pulmonary capillaries increases because the difference in partial pressure between the alveoli and the pulmonary capillaries is increased.

Increasing the rate of ventilation also makes the P_{CO_2} lower in the alveoli than during slow breathing. Because the alveolar P_{CO_2} decreases, the difference in partial pressure between the alveoli and the pulmonary capillaries increases, which increases the rate of carbon dioxide diffusion from the pulmonary capillaries into the alveoli.

On the other hand, inadequate ventilation causes a smaller difference in the P_{O_2} and P_{CO_2} across the respiratory membrane. The rate of oxygen and carbon dioxide diffusion across the membrane therefore decreases, causing oxygen levels in the blood to decrease and carbon dioxide levels to increase.

Diffusion of Gases in the Tissues

Blood flows from the lungs through the left side of the heart to the tissue capillaries. Figure 15.13 illustrates the partial pressure differences for oxygen and carbon dioxide across the wall of a tissue capillary. Oxygen diffuses from the capillary into the interstitial fluid because the P_{O_2} in the interstitial fluid is lower than in the capillary. Oxygen diffuses from the interstitial fluid into cells, in which the P_{O_2} is less than in the interstitial fluid. Within the cells, oxygen is used in aerobic metabolism. There is a constant difference in P_{O_2} from the tissue capillaries to the cells because oxygen is continuously used by cells. There is also a constant diffusion gradient for carbon dioxide from the cells to the tissue capillaries because carbon dioxide is continuously produced by cells. Carbon dioxide therefore diffuses from cells into the interstitial fluid and from the interstitial fluid into the tissue capillaries.

P R E D I C T (7)
During exercise, the movement of oxygen into skeletal muscle cells and the movement of carbon dioxide out of skeletal muscle cells increases. Explain how this happens.

Gas Transport in the Blood
Oxygen Transport

After oxygen diffuses across the respiratory membrane into the blood, about 98.5% of the oxygen transported in the blood combines reversibly with the iron-containing heme groups of hemoglobin (see chapter 11). About 1.5% of the oxygen remains dissolved in the plasma. Hemoglobin with oxygen bound to its heme groups is called **oxyhemoglobin** (ok′sē-hē-mō-glō′bin).

The ability of hemoglobin to bind to oxygen depends on the P_{O_2}. At high P_{O_2}, hemoglobin binds to oxygen and at low P_{O_2} hemoglobin releases oxygen. In the lungs, P_{O_2} normally is sufficiently high that hemoglobin holds as much oxygen as it can. In the tissues, P_{O_2} is lower because the tissues are using oxygen. Consequently, hemoglobin releases oxygen in the tissues. Oxygen then diffuses into cells and is used by the cells in aerobic metabolism. At rest, approximately 23% of the oxygen picked up by hemoglobin in the lungs is released to the tissues.

The amount of oxygen released from oxyhemoglobin is influenced by several factors. More oxygen is released from hemoglobin if (1) the P_{O_2} is low, (2) the P_{CO_2} is high, (3) the pH is low, and (4) the temperature is high. Increased muscular activity results in a decreased P_{O_2}, an increased P_{CO_2}, a reduced pH, and an increased temperature. Consequently, during periods of physical exercise, as much as 73% of the oxygen picked up by hemoglobin in the lungs is released in skeletal muscles.

Carbon Dioxide Transport and Blood pH

Carbon dioxide diffuses from cells, where it is produced, into the tissue capillaries. After carbon dioxide enters the blood, it is transported in three ways: (1) About 7% is transported as carbon dioxide dissolved in the plasma, (2) 23% is transported in combination with blood proteins, primarily hemoglobin, and (3) 70% is transported in the form of bicarbonate ions.

Carbon dioxide (CO_2) reacts with water to form carbonic acid (H_2CO_3), which then dissociates to form H^+ and bicarbonate ions (HCO_3^-).

$$\underset{\substack{\text{carbon}\\\text{dioxide}}}{CO_2} + \underset{\text{water}}{H_2O} \overset{\substack{\text{Carbonic}\\\text{anhydrase}}}{\leftrightarrows} \underset{\substack{\text{carbonic}\\\text{acid}}}{H_2CO_3} \leftrightarrows \underset{\substack{\text{hydrogen}\\\text{ion}}}{H^+} + \underset{\substack{\text{bicarbonate}\\\text{ion}}}{HCO_3^-}$$

An enzyme called **carbonic anhydrase** (kar-bon′ik an-hī′drās) is found inside red blood cells and on the surface of capillary epithelial cells. Carbonic anhydrase increases the rate at which CO_2 reacts with water to form H^+ and HCO_3^- in the tissue capillaries (figure 15.14). Thus, carbonic anhydrase promotes the uptake of CO_2 by red blood cells.

In the capillaries of the lungs, the process is reversed so that the HCO_3^- and H^+ combine to produce H_2CO_3, which then forms CO_2 and H_2O. The CO_2 diffuses into the alveoli and is expired.

Carbon dioxide has an important effect on the pH of blood. As CO_2 levels increase, the blood pH decreases (becomes more acidic) because CO_2 reacts with H_2O to form H_2CO_3. The H^+ that result from the dissociation of H_2CO_3 are responsible for the decrease in pH. Conversely, as blood levels of CO_2 decline, the blood pH increases (becomes less acidic or more basic).

(a) In the tissue capillaries, CO_2 released from tissue cells diffuses into red blood cells and combines with H_2O to form carbonic acid, a reaction catalyzed by the enzyme carbonic anhydrase. Carbonic acid then dissociates to form H and HCO_3^-. This process promotes the uptake and transport of CO_2 by red blood cells.

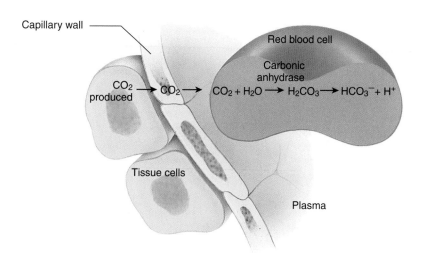

(b) In the pulmonary capillaries, CO_2 diffuses out of red blood cells into the aveoli. The loss of CO_2 promotes the formation of additional CO_2 from carbonic acid, a reaction catalyzed by carbonic anhydrase. H and HCO_3^- then combine to replace the carbonic acid. This process promotes the formation and release of CO_2 by red blood cells.

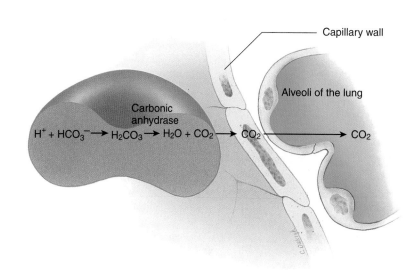

Figure 15.14 Carbon Dioxide Transport

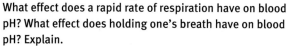

PREDICT 8
What effect does a rapid rate of respiration have on blood pH? What effect does holding one's breath have on blood pH? Explain.

Rhythmic Ventilation

The normal rate of respiration in adults is between 12 and 20 respirations per minute. In children, the rates are higher and may vary from 20 to 40 per minute. The generation of the basic rhythm of ventilation is controlled by neurons within the medulla oblongata that stimulate the muscles of respiration. An increased depth of respiration results from stronger contractions of the respiratory muscles caused by recruitment of muscle fibers and increased frequency of stimulation of muscle fibers. The rate of respiration is determined by the number of times respiratory muscles are stimulated.

Respiratory Areas in the Brainstem

In the classic view of respiratory areas, distinct inspiratory and expiratory centers were thought to be located in the brainstem. This view is now known to be too simplistic. Although neurons involved with respiration are aggregated in certain parts of the brainstem, neurons that are active during inspiration are intermingled with neurons active during expiration.

The **medullary respiratory center** consists of **two dorsal respiratory groups,** each forming a longitudinal column of cells located bilaterally in the dorsal part of the medulla oblongata, and two **ventral respiratory groups,** each forming a longitudinal column of cells located bilaterally in the ventral part of the medulla oblongata (figure 15.15). The dorsal respiratory groups are primarily responsible for stimulating contraction of the diaphragm. The ventral respiratory groups are primarily responsible for stimulating

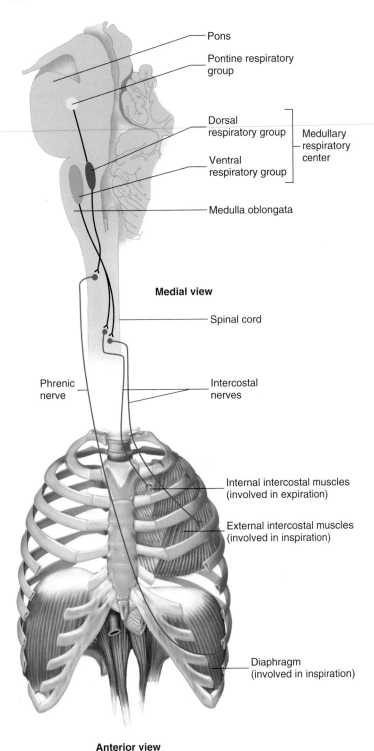

- Pons
- Pontine respiratory group
- Dorsal respiratory group ⎫ Medullary respiratory center
- Ventral respiratory group ⎭
- Medulla oblongata

Medial view

- Spinal cord

- Phrenic nerve
- Intercostal nerves

- Internal intercostal muscles (involved in expiration)
- External intercostal muscles (involved in inspiration)

- Diaphragm (involved in inspiration)

Anterior view

Figure 15.15   Respiratory Structures in the Brainstem

The relationship of respiratory structures to each other and to the nerves innervating the muscles of respiration.

the external intercostal, internal intercostal, and abdominal muscles.

The **pontine respiratory group** is a collection of neurons in the pons (see figure 15.15). It has connections with the medullary respiratory center and appears to play a role in the switching between inspiration and expiration.

Generation of Rhythmic Ventilation

The medullary respiratory center generates the basic pattern of spontaneous, rhythmic ventilation. The precise mechanism by which it is able to do so is not well understood. The generation of rhythmic ventilation involves the integration of stimuli that start and stop inspiration.

1. *Starting inspiration.* Certain neurons in the medullary respiratory center that promote inspiration are continuously active. The medullary respiratory center constantly receives stimulation from many sources, such as receptors that monitor blood gas levels and the movements of muscles and joints. In addition, stimulation from parts of the brain concerned with voluntary respiratory movements and emotions can occur. When the inputs from all these sources reach a threshold level, neurons that stimulate respiratory muscles produce action potentials and inspiration starts.
2. *Increasing inspiration.* Once inspiration begins, more and more neurons are activated. The result is progressively stronger stimulation of the respiratory muscles that lasts for approximately 2 seconds (s).
3. *Stopping inspiration.* The neurons stimulating the muscles of respiration also stimulate other neurons in the medullary respiratory center that are responsible for stopping inspiration. The neurons responsible for stopping inspiration also receive input from the pontine respiratory neurons, stretch receptors in the lungs, and probably other sources. When the inputs to these neurons exceeds a threshold level, they cause the neurons stimulating respiratory muscles to be inhibited. Relaxation of respiratory muscles results in expiration, which lasts approximately 3 s. For the next inspiration, go back to step 1.

Modification of Ventilation

Although the medullary neurons establish the basic rate and depth of breathing, their activities can be influenced by input from other parts of the brain and by input from peripherally located receptors.

Nervous Control of Ventilation

Higher brain centers can modify the activity of the respiratory center (figure 15.16). For example, controlling air movements out of the lungs makes speech possible, and emotions can make us sob or gasp. There is conscious control over respiration. It is possible to breathe voluntarily or to stop respiratory movements voluntarily. Some people can hold their breath until they lose consciousness as a result of the lack of oxygen in the brain. Some children have used this strategy to encourage parents to give them what they want. As soon as conscious control of respiration is lost, however, the automatic control of respiration resumes, and the person starts to breathe again.

Several reflexes, such as sneeze and cough reflexes, can modify ventilation. The **Hering-Breuer** (her′ing broy′er)

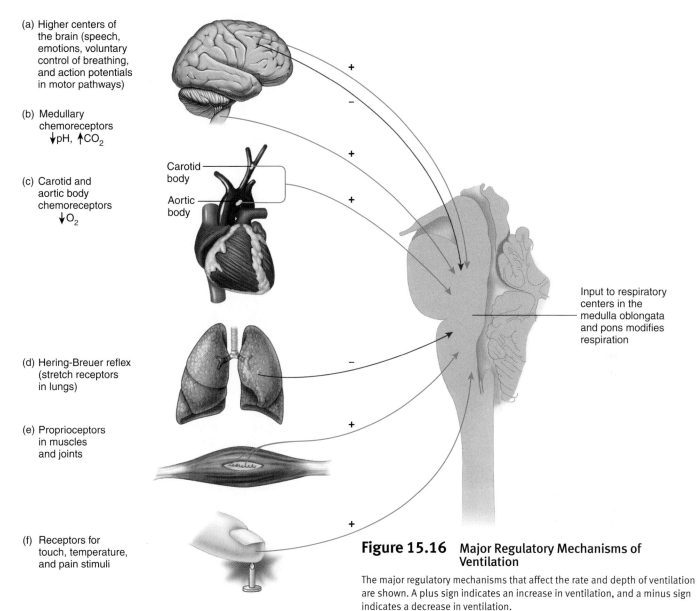

(a) Higher centers of the brain (speech, emotions, voluntary control of breathing, and action potentials in motor pathways)

(b) Medullary chemoreceptors
↓pH, ↑CO_2

(c) Carotid and aortic body chemoreceptors
↓O_2

Carotid body

Aortic body

(d) Hering-Breuer reflex (stretch receptors in lungs)

(e) Proprioceptors in muscles and joints

(f) Receptors for touch, temperature, and pain stimuli

Input to respiratory centers in the medulla oblongata and pons modifies respiration

Figure 15.16 Major Regulatory Mechanisms of Ventilation

The major regulatory mechanisms that affect the rate and depth of ventilation are shown. A plus sign indicates an increase in ventilation, and a minus sign indicates a decrease in ventilation.

reflex functions to support rhythmic respiratory movements by limiting the extent of inspiration (see figure 15.16). As muscles of inspiration contract, the lungs fill with air. Sensory receptors that respond to stretch are located in the lungs, and, as the lungs fill with air, the stretch receptors are stimulated. Action potentials from the lung stretch receptors are then sent to the medulla oblongata, where they inhibit the respiratory center neurons and cause expiration. In infants, the Hering-Breuer reflex plays an important role in regulating the basic rhythm of breathing and in preventing overinflation of the lungs. In adults, however, the reflex is important only when the tidal volume is large, such as during heavy exercise.

Touch, thermal, and pain receptors in the skin also stimulate the respiratory center, which explains the gasp in response to being splashed with cold water or being pinched (see figure 15.16).

Chemical Control of Ventilation

During aerobic respiration, oxygen is consumed and carbon dioxide is produced (see chapter 17). The respiratory system adds oxygen and removes carbon dioxide from the blood. For the respiratory system to maintain these gases at homeostatic levels, there must be some way to monitor gas levels and respond appropriately.

Carbon dioxide levels in the blood are the major driving force for regulating respiration. Under most circumstances, carbon dioxide levels are much more important than oxygen. A small increase in carbon dioxide levels can cause an increase in ventilation. A large increase, as can occur while holding one's breath, can result in a powerful urge to take a breath. A greater-than-normal amount of carbon dioxide in the blood is called **hypercapnia** (hī-per-kap′nē-ă, *hyper,* above + *kapnos,* smoke).

Oxygen is not normally as important as carbon dioxide in regulating ventilation, because hemoglobin is very effective at

Bronchi and Lungs

Bronchitis (brong-kī'tis) is an inflammation of the bronchi caused by irritants, such as cigarette smoke, air pollution, or infections. The inflammation results in swelling of the mucous membrane lining the bronchi, increased mucus production, and decreased movement of mucus by cilia. Consequently, the diameter of the bronchi is decreased, and ventilation is impaired. Bronchitis can progress to emphysema.

Emphysema (em-fi-sē'mă, *en,* in + *physēma,* a blowing) is the destruction of the alveolar walls. Many individuals have both bronchitis and emphysema, which are often referred to as **chronic obstructive pulmonary disease (COPD).** Chronic inflammation of the bronchioles, usually caused by cigarette smoke or air pollution, probably initiates emphysema. Narrowing of the bronchioles restricts air movement, and air tends to be retained in the lungs. Coughing to remove accumulated mucus increases pressure in the alveoli, resulting in rupture and destruction of alveolar walls. Loss of alveolar walls has two important consequences: (1) the respiratory membrane has a decreased surface area, which decreases gas exchange, and (2) loss of elastic fibers decreases the ability of the lungs to recoil and expel air. Symptoms of emphysema include shortness of breath and enlargement of the thoracic cavity. Treatment involves removing sources of irritants (e.g., stopping smoking), promoting the removal of bronchial secretions, using bronchiodilators, retraining people to breathe so that expiration of air is maximized, and using antibiotics to prevent infections. The progress of emphysema can be slowed, but there is no cure.

Adult respiratory distress syndrome (ARDS) is caused by damage to the respiratory membrane. The damage stimulates an inflammatory response that further damages the respiratory membrane. Water, ions, and proteins leave the blood and enter alveoli. Surfactant in the alveoli is reduced as surfactant-producing cells are damaged and surfactant present in the alveoli is diluted. The fluid-filled alveoli reduce gas exchange and make it more difficult for the lungs to expand. ARDS usually develops rapidly following an injurious event, such as an infection, inhalation of smoke from a fire, inhalation of toxic fumes, trauma, aspiration of gastric contents associated with gastric reflux, or circulatory shock. Even with oxygen inhalation therapy, there is a high mortality rate.

Cystic fibrosis (fī-brō'sis) is an inherited disease that affects the secretory cells lining the lungs, pancreas, sweat glands, and salivary glands. An abnormal transport protein is produced that does not reach the cell surface or does not function normally if it does reach the cell surface. The result is decreased Cl^- secretion out of cells. In the lungs, water normally forms a thin fluid layer over which mucus is moved by ciliated cells. In cystic fibrosis, the decreased Cl^- diffusion results in dehydrated respiratory secretions. The mucus is more viscous, resisting movement by cilia, and it accumulates in the lungs, thereby increasing the likelihood of infections. Chronic airflow obstruction causes difficulty in breathing, and severe coughing in an attempt to remove the mucus can result in pneumothorax and bleeding within the lungs. Many victims of cystic fibrosis are now surviving into young adulthood because of improved treatments. Future treatments could include the development of drugs that correct or assist the normal ion transport mechanism. Alternatively, cystic fibrosis may some day be cured through genetic engineering by inserting functional copies of the defective gene into the cells of a person with the disease. Research on this exciting possibility is currently underway.

Pulmonary fibrosis is the replacement of lung tissue with fibrous connective tissue, making the lungs less elastic and breathing more difficult. Exposure to asbestos, silica, or coal dust is the most common cause.

Lung cancer arises from the epithelium of the respiratory tract. Lung cancer is the most common cause of cancer death in males and females in the United States, and almost all cases occur in smokers. Because of the rich lymph and blood supply in the lungs, cancer in the lung can readily spread to other parts of the lung or body. In addition, the disease is often advanced before symptoms become severe enough for the victim to seek medical aid. Typical symptoms include coughing, sputum production, and blockage of the airways. Treatments include removal of part or all of the lung, chemotherapy, and radiation.

Circulatory System

Disorders of the circulatory system can affect respiratory function. Even when ventilation is adequate, blood flow through the pulmonary capillaries may be inadequate. Disorders that reduce blood flow through lung tissue include **thrombosis of the pulmonary arteries** and reduced cardiac output resulting from **heart attack** or **shock. Anemia,** which results in a reduction of the total amount of hemoglobin, reduces the capacity of blood to transport oxygen. **Carbon monoxide** binds more strongly to the hemoglobin molecule than oxygen does. It occupies binding sites, making them unavailable for oxygen transport. Thus carbon monoxide poisoning decreases the ability of hemoglobin to transport oxygen, even though it does not affect the total hemoglobin concentration in the blood.

Nervous System

Sudden infant death syndrome (SIDS), or crib death, is the most frequent cause of death of infants between 2 weeks and 1 year of age. Death results when the infant stops breathing during sleep. Although the cause of SIDS remains controversial, there is evidence that damage to the respiratory center during development is a factor. There is no treatment, but at-risk babies can be placed on monitors that sound an alarm if the baby stops breathing.

Paralysis of the respiratory muscles can result from damage of the spinal cord in the cervical or thoracic regions. The damage interrupts nerve tracts that transmit action potentials to the muscles of respiration. Transection of the spinal cord can result from trauma such as automobile accidents or diving into water that is too shallow. Another cause of paralysis is poliomyelitis (pō'lē-ō-mī'ĕ-lī'tis), a viral infection that damages neurons of the respiratory center or motor neurons that stimulate the muscles of respiration. Finally, anesthetics or central nervous system depressants can depress the function of the respiratory center if they are taken or administered in large enough doses.

Thoracic Wall

Decreased elasticity of the thoracic wall can be caused by severe arthritis or by conditions resulting in severe curvature of the spine, such as **scoliosis** (skō-lē-ō'sis, a crookedness) and **kyphosis** (kī-fō'sis, hump-back). These conditions reduce the ability of the thoracic cavity to increase its volume when the muscles of inspiration contract, thereby increasing the muscular effort required for inspiration.

picking up oxygen in the lungs. Normally there is little variation in the oxygen content of blood leaving the lungs, despite changes in oxygen demand by the body. As long as blood carbon dioxide levels are normal, blood oxygen levels are usually normal as well.

Changes in blood carbon dioxide levels are not directly detected, however. Instead, changes in blood pH are monitored. This can occur because changes in carbon dioxide cause changes in pH (see the section, Carbon Dioxide Transport and Blood pH, on p. 434). Thus, two things are accomplished through the chemical regulation of ventilation: (1) homeostatic levels of carbon dioxide and oxygen are maintained, and (2) pH homeostasis is maintained.

PREDICT 9
Explain why a person who breathes rapidly and deeply (hyperventilates) for several seconds experiences a short period of time in which respiration does not occur (apnea) before normal breathing resumes.

Chemoreceptors in the medulla oblongata are sensitive to small changes in blood pH (see figure 15.16). An increase in blood pH, typically caused by a decrease in blood carbon dioxide, can be detected by the chemoreceptors. As a result, the respiratory center decreases ventilation, which decreases the removal of carbon dioxide from the blood. Because carbon dioxide is continually produced, carbon dioxide levels now increase, causing pH to decrease, and homeostasis is maintained (figure 15.17).

The medullary chemoreceptors also respond to a decrease in blood pH, typically caused by an increase in blood carbon dioxide levels. As a result, the respiratory center increases ventilation, which increases the removal of carbon dioxide from the blood. As blood carbon dioxide levels decrease, blood pH increases, and homeostasis is maintained (see figure 15.17).

Chemoreceptors in the carotid and aortic bodies also provide input to the respiratory center (see figure 15.16). These

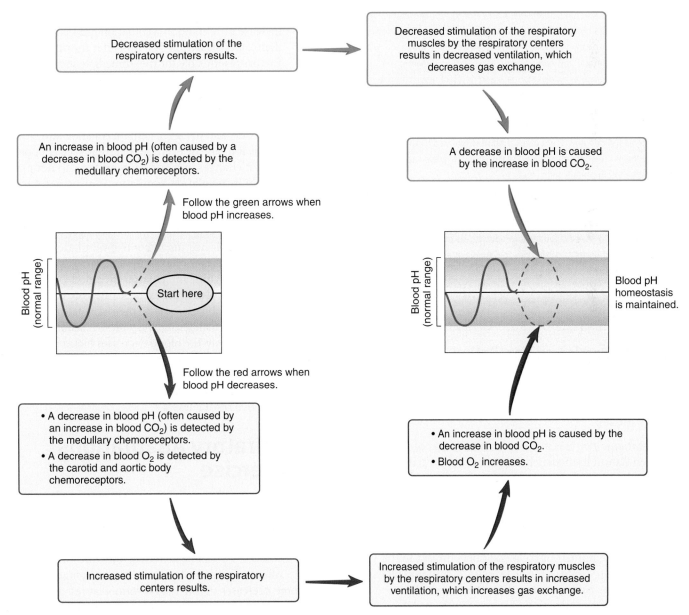

Homeostasis Figure 15.17   Regulation of Blood pH and Gases

chemoreceptors primarily respond to changes in blood oxygen. Changes in blood oxygen levels can become important when they decline to low levels, a condition called **hypoxia** (hī-pok′sē-ă, *hypo,* under + oxygen). Examples include exposure to high altitudes, emphysema, shock, and asphyxiation caused by exposure to reduced oxygen levels but normal carbon dioxide levels. In these situations, the chemoreceptors of the carotid and aortic bodies are strongly stimulated. They send action potentials to the respiratory center and produce an increase in the rate and depth of respiration (see figure 15.17). The increased ventilation increases oxygen diffusion from the alveoli into the blood, resulting in increased blood oxygen.

Effect of High Altitude and Emphysema on Ventilation

Air is composed of 21% oxygen at low and high altitudes. At sea level the atmospheric pressure is 760 mm Hg and P_{O_2} is about 160 mm Hg (760 mm Hg $\times$ 0.21 = 160 mm Hg). At higher altitudes, the atmospheric pressure is lower, and P_{O_2} is decreased. For example, at 10,000 ft above sea level, the atmospheric pressure is 523 mm Hg. Consequently, P_{O_2} is 110 mm Hg (523 mm Hg $\times$ 0.21 = 110 mm Hg). Because P_{O_2} is lower at high altitudes, the blood levels of oxygen can decline enough to stimulate the carotid and aortic bodies. Oxygen then becomes an important stimulus for an elevated rate and depth of respiration. At high altitudes, the ability of the respiratory system to eliminate carbon dioxide is not adversely affected by the low atmospheric pressure. Thus the blood carbon dioxide levels become lower than normal because of the increased rate and depth of respiration stimulated by the low oxygen blood levels. The decreased blood carbon dioxide levels cause the blood pH to rise to abnormally high levels.

A similar situation can exist in people who have emphysema. Carbon dioxide diffuses across the respiratory membrane more readily than oxygen. Thus the decrease in surface area of the respiratory membrane caused by the disease results in low blood levels of oxygen without elevated blood levels of carbon dioxide. An elevated rate and depth of respiration is the result of the stimulatory effect of low blood levels of oxygen on the carotid and aortic bodies. More severe emphysema, in which the surface area of the respiratory membrane is reduced to a minimum, can also result in elevated blood levels of carbon dioxide.

Effect of Exercise on Ventilation

The mechanisms by which ventilation is regulated during exercise is controversial, and no one factor can account for all the observed responses. Ventilation during exercise can be divided into two phases:

1. *Ventilation increases abruptly.* At the onset of exercise, ventilation immediately increases. This initial increase can be as much as 50% of the total increase that occurs during exercise. The immediate increase in ventilation occurs too quickly to be explained by changes in metabolism or blood gases. As axons pass from the motor cortex of the cerebrum through the motor pathways, numerous collateral fibers project to the respiratory center. During exercise, action potentials in the motor pathways stimulate skeletal muscle contractions and action potentials in the collateral fibers stimulate the respiratory center (see figure 15.16).

Furthermore, during exercise, body movements stimulate proprioceptors in the joints of the limbs. Nerve fibers from these proprioceptors extend to the spinal cord to connect with sensory nerve tracts ascending to the brain. Collateral fibers from these nerve tracts connect to the respiratory center, and movement of the limbs has a strong stimulatory influence on the respiratory center (see figure 15.16).

There may also be a learned component to the ventilation response during exercise. After a period of training the brain "learns" to match ventilation with the intensity of the exercise. Well-trained athletes match their respiratory movements more efficiently with their level of physical activity than do untrained individuals. Thus centers of the brain involved in learning have an indirect influence on the respiratory center, but the exact mechanism for this kind of regulation is unclear.

2. *Ventilation increases gradually.* After the immediate increase in ventilation, ventilation gradually increases and then levels off within 4–6 min after the onset of exercise. Factors responsible for the immediate increase in ventilation may play a role in the gradual increase as well.

Despite large changes in oxygen consumption and carbon dioxide production during exercise, the *average* arterial oxygen, carbon dioxide, and pH levels remain constant and close to resting levels as long as the exercise is aerobic (see chapter 7). This suggests that changes in blood gases and pH do not play an important role in regulating ventilation during aerobic exercise. During exercise, however, the values of arterial oxygen, carbon dioxide, and pH levels rise and fall more than at rest. Thus, even though their average values do not change, their oscillations may be a signal for helping to control ventilation.

The highest level of exercise that can be performed without causing a significant change in blood pH is called the **anaerobic threshold.** If the exercise intensity is high enough to exceed the anaerobic threshold, then skeletal muscles produce lactic acid through anaerobic respiration (see figure 17.5). Lactic acid released into the blood decreases blood pH, which stimulates the carotid bodies, resulting in increased ventilation. In fact, ventilation can increase so much that arterial carbon dioxide levels decrease below resting levels and arterial oxygen levels increase above resting levels.

Respiratory Adaptations to Exercise

In response to training, athletic performance increases because the cardiovascular and respiratory systems become more efficient at delivering oxygen and picking up carbon dioxide. Ventilation in most individuals does not limit performance because ventilation can increase to a greater extent than does cardiovascular function.

After training, vital capacity increases slightly and residual volume decreases slightly. Tidal volume at rest and during standardized submaximal exercise does not change. At maximal exercise, however, the tidal volume increases. After training, the

Respiratory system infections are the most common types of infections. Most are relatively mild, but some are among the most damaging types of infection. The major respiratory diseases are bacterial and viral, although some are fungal or protozoan infections.

Many respiratory infections are spread by the release of microorganisms from the respiratory tract. For example, a person who sneezes or coughs releases droplets containing microorganisms that can be inhaled by another person. Or, microorganisms released from the respiratory tract of a person can contaminate surfaces such as tabletops. If a person touches the contaminated surface, the microorganisms can then be transferred by the fingers from the contaminated surface to the mouth or nasal passages.

Infections of the Upper Respiratory Tract

Strep throat is caused by streptococcal bacteria (*Streptococcus pyogenes*) and is characterized by inflammation of the pharynx and by fever. Frequently, inflammation of the tonsils and middle ear are involved. Without a throat analysis, the infection cannot be distinguished from viral causes of pharyngeal inflammation. Current techniques allow rapid diagnosis within minutes to hours, and antibiotics are effective in treating strep throat. **Scarlet fever** occurs in response to certain strains of *S. pyogenes*. This infection is characterized by fever and a pinkish red skin rash produced by a circulating toxin released by the bacteria.

Diphtheria (dif-thēr′ē-ă, leather) was once a major cause of death among children. It is caused by a bacterium (*Corynebacterium diphteriae*). A grayish membrane forms in the throat and can block the respiratory passages totally. A vaccine against diphtheria is part of the normal immunization (DPT) program for children in the United States.

The **common cold** is the result of a viral infection. Symptoms include sneezing, excessive nasal secretions, and congestion. The infection easily can spread to sinus cavities, lower respiratory passages, and the middle ear. The common cold usually runs its course to recovery in about 1 week.

Infections of the Lower Respiratory Tract

Many of the same infections that mainly affect the upper respiratory tract also can cause **laryngitis** (inflammation of the larynx) and **bronchitis** (inflammation of the bronchi).

Whooping cough, or **pertussis** (per-tŭs′is, *per,* very (intensive) + *tussis,* cough), is a bacterial infection (caused by *Bordetella pertussis*) that causes a loss of cilia of the respiratory epithelium. Mucus accumulates, and the infected person attempts to cough up the mucous accumulations. The coughing can be severe. A vaccine for whooping cough is part of the normal vaccination procedure (DPT) for children in the United States.

Tuberculosis (tū-ber′kyū-lō′sis, *tubercle,* a knob + *-osis,* condition) is caused by the bacterium *Clostridium tuberculosis.* In the lungs, the tuberculosis bacteria form small lumplike lesions called tubercles. The lesions contain degenerating macrophages and tuberculosis bacteria. An immune reaction is directed against the bacteria, which causes the formation of larger lesions and inflammation. The tubercles can rupture, releasing additional bacteria, which infect other parts of the lung or body. A strain of tuberculosis that is resistant to treatment with antibiotics is increasing in frequency worldwide.

Pneumonia (noo-mō′nē-ă, *pneumōn,* lung + *-ia,* condition) refers to many infections of the lungs. Symptoms include fever, difficulty in breathing, and chest pain. Inflammation of the lungs results in pulmonary edema and poor inflation of the lungs with air. Most pneumonias are bacterial, but some are viral. Compared with viral pneumonia, bacterial pneumonia is more severe, produces more edema in the lungs, and is more likely to cause death. A protozoan infection (caused by *Pneumocystis carinii*) that results in pneumocystis (noo-mō-sis′tis) pneumonia is rare, except in persons who have a compromised immune system. This type of pneumonia has become one of the infections commonly suffered by persons who have AIDS.

Flu, or **influenza** (in-flū-en′ză), is a viral infection of the respiratory system and does not affect the digestive system as is commonly assumed. Flu is characterized by chills, fever, headache, and muscular aches in addition to respiratory symptoms. There are several strains of flu viruses. Flu vaccines are of limited use because it usually takes too long to develop an effective vaccine to be effective during an epidemic. The mortality rate during a flu epidemic is about 1%, and most of those deaths are among the very old and very young. During a flu epidemic, the infection rate is so rapid and the disease is so widespread that the total number of deaths is substantial, even though the percentage of deaths is relatively low.

A number of **fungal diseases** also affect the respiratory system. The fungal spores usually enter the respiratory system attached to dust particles. Spores in soil and feces of certain animals make the rate of infection high in farm workers and gardeners in certain areas of the country. The infections usually result in minor respiratory infections, but in some cases they can spread to other parts of the body.

---

respiratory rate at rest or during standardized submaximal exercise is slightly lower, but at maximal exercise, the respiratory rate is generally increased.

Minute ventilation is affected by the changes in tidal volume and respiratory rate. After training, minute ventilation is essentially unchanged or slightly reduced at rest, is slightly reduced during standardized submaximal exercise, and is greatly increased at maximal exercise. For example, an untrained person with a minute ventilation of 120 liters per minute (L/min) can increase his minute ventilation to 150 L/min after training. Increases to 180 L/min are typical of highly trained athletes.

Systems Pathology
Asthma

Mr. W. was an 18-year-old track athlete in seemingly good health. One day he came down with a common cold, resulting in the typical symptoms of nasal congestion and discomfort. After several days he began to cough and wheeze and he thought that his cold had progressed to his lungs. Determined not to get "out of shape" because of his cold, Mr. W. took a few aspirins to relieve his discomfort and went to the track to do some jogging. After a few minutes of exercise he began to wheeze very forcefully and rapidly, and he felt that he could hardly get enough air. Even though he stopped jogging his condition did not improve (figure A). Fortunately, a concerned friend who was also at the track, took him to the emergency room.

Although Mr. W. had no previous history of asthma, careful evaluation convinced the emergency room doctor that he was having an asthma attack. Mr. W. inhaled a bronchiodilator drug, which resulted in rapid improvement of his condition. He was released from the emergency room and referred to his personal physician for further treatment and education about asthma.

Background Information

Asthma (az′mă, difficult breathing) is a disease characterized by increased constriction of the trachea and bronchi in response to various stimuli, resulting in a narrowing of the air passageways and decreased ventilation efficiency. Symptoms include wheezing, coughing, and shortness of breath. In contrast to many other respiratory disorders, however, the symptoms of asthma typically reverse either spontaneously or with therapy.

It is estimated that the prevalence of asthma in the United States is from 3% to 6% of the general population. Approximately half the cases first appear before age 10, and twice as many boys as girls develop asthma. Anywhere from 25% to 50% of childhood asthmatics are symptom-free from adolescence onward.

The exact cause or causes of asthma is unknown, but asthma and allergies run strongly in some families. There is no definitive pathological feature or diagnostic test for asthma, but three important features of the disease are chronic airway inflammation, airway hyperreactivity, and airflow obstruction. The inflammatory response results in tissue damage, edema, and mucus buildup, which can block airflow through the

Figure A A Jogger with Asthma

bronchi. Airway hyperreactivity is greatly increased contraction of the smooth muscle in the trachea and bronchi in response to a stimulus. As a result of airway hyperreactivity, the diameter of the airway decreases, and resistance to airflow increases. The effects of inflammation and airway hyperreactivity combine to cause airflow obstruction.

Many cases of asthma appear to be associated with a chronic inflammatory response by the immune system. The number of immune cells in the bronchi increases, including mast cells, eosinophils, neutrophils, macrophages, and lymphocytes. These cells release chemical mediators, such as interleukins, leukotrienes, prostaglandins, platelet-activating factor, thromboxanes, and chemotactic factors. These chemical mediators promote inflammation, increase mucus secretion, and attract additional immune cells to the bronchi, resulting in chronic airway inflammation. Airway hyperreactivity and inflammation appear to be linked by some of the chemical mediators, which increase the sensitivity of the airway to stimulation and cause smooth muscle contraction.

Effects of Aging on the Respiratory System

Almost all aspects of the respiratory systems are affected by aging. Even though vital capacity, maximum ventilation rates, and gas exchange decrease with age, the elderly can engage in light to moderate exercise because the respiratory system has a large reserve capacity.

With age, mucus accumulates within the respiratory passageways. The mucus–cilia escalator is less able to move the mucus because it becomes more viscous and because the number of cilia and their rate of movement decrease. As a consequence, the elderly are more susceptible to respiratory infections and bronchitis.

Vital capacity decreases with age because of a decreased ability to fill the lungs (decreased inspiratory reserve volume)

System Interactions — Effect of Asthma on Other Systems

System	Interactions
Integumentary	Cyanosis, a bluish skin color, results from a decreased blood oxygen content.
Muscular	Skeletal muscles are necessary for respiratory movements and the cough reflex. Increased muscular work during a severe asthma attack can cause metabolic acidosis because of anaerobic respiration and excessive lactic acid production.
Skeletal	Red bone marrow is the site of production of many of the immune cells responsible for the inflammatory response of asthma. The thoracic cage is necessary for respiration.
Nervous	Emotional upset or stress can evoke an asthma attack. Peripheral and central chemoreceptor reflexes affect ventilation. The cough reflex helps to remove mucus from respiratory passages. Pain, anxiety, and death from asphyxiation can result from the altered gas exchange caused by asthma. One theory of the cause of asthma is an imbalance of the autonomic nervous system (ANS) control of bronchiolar smooth muscle, and drugs that enhance sympathetic effects or block parasympathetic effects are used in asthma treatment.
Endocrine	Steroids from the adrenal gland play a role in regulating inflammation and they are used in asthma therapy.
Cardiovascular	Increased vascular permeability of lung blood vessels results in edema. Blood carries ingested substances that provoke an asthma attack to the lungs. Blood carries immune cells from red bone marrow to the lungs. Tachycardia commonly occurs and the normal effects of respiration on venous return are exaggerated, resulting in large fluctuations of blood pressure.
Lymphatic and Immune	Immune cells release chemical mediators that promote inflammation, increase mucous production, and cause bronchiolar constriction; believed to be a major factor in asthma. Ingested allergens, such as aspirin or sulfites in food, can evoke an asthma attack.
Digestive	Ingested substances, such as aspirin, sulfiting agents (preservatives), tartrazine (tar′tră-zēn, a yellow dye), and certain foods, and reflux of stomach acid into the esophagus can evoke an asthma attack.
Urinary	Modifying hydrogen ion secretion into the urine helps to compensate for acid–base imbalances caused by asthma.

The stimuli that prompt airflow obstruction varies from one individual to another. Some asthmatics have reactions to particular allergens, which are foreign substances that evoke an inappropriate immune system response (see chapter 14). Examples include inhaled pollen, animal dander, and dust mites. Many cases of asthma may be caused by an allergic reaction to substances in the droppings and carcasses of cockroaches, which may explain the higher rate of asthma in poor, urban areas.

On the other hand, inhaled substances, such as chemicals in the workplace or cigarette smoke, can provoke an asthma attack without stimulating an allergic reaction. Over 200 substances have been associated with occupational asthma. An asthma attack can also be stimulated by ingested substances such as aspirin, nonsteroidal antiinflammatory compounds such as ibuprofen (ī-bū′prō-fen), sulfites in food preservative, and tartrazine (tar′tră-zēn) in food colorings. Asthmatics can substitute acetaminophen (as-et-ă-mē′nō-fen, a-set-ă-min′ō-fen; Tylenol) for aspirin.

Other stimuli, such as strenuous exercise, especially in cold weather, can precipitate an asthma attack. Such episodes can often be avoided by using a bronchiodilator drug prior to exercise. Viral infections, emotional upset, stress, air pollution, and even reflux of stomach acid into the esophagus are known to elicit an asthma attack.

Treatment of asthma involves avoiding the causative stimulus and drug therapy. Steroids and mast cell-stabilizing agents, which prevent the release of chemical mediators from mast cells, are used to reduce airway inflammation. Theophylline (thē-of′i-lēn) and other drugs are commonly used to cause bronchiolar dilation. Although treatment is generally effective in controlling asthma, death by asphyxiation rarely can occur. Most of these deaths probably could have been prevented by earlier and more intensive therapy.

P R E D I C T 10
It is not usually necessary to assess arterial blood gases in the diagnosis and treatment of asthma. This information, however, can sometimes be useful in cases of severe asthma attacks. Suppose that Mr. W. had a P_{O_2} of 60 mm Hg and a P_{CO_2} of 30 mm Hg when he first came to the emergency room. Explain how that could happen.

and a decreased ability to empty the lungs (decreased expiratory reserve volume). As a result, maximum minute ventilation rates decrease, which in turn decreases the ability to perform intense exercise. These changes are related to weakening of respiratory muscles and to stiffening of cartilage and ribs.

Residual volume increases with age as the alveolar ducts and many of the larger bronchioles increase in diameter. This increases the dead space, which decreases the amount of air available for gas exchange. In addition, gas exchange across the respiratory membrane is reduced because parts of the alveolar walls are lost, which decreases the surface area available for gas exchange, and the remaining walls thicken, which decreases diffusion of gases. A gradual increase in resting tidal volume with age compensates for these changes.

S U M M A R Y

Respiration includes the movement of air into and out of the lungs, the exchange of gases between the air and the blood, the transport of gases in the blood, and the exchange of gases between the blood and tissues.

Functions of the Respiratory System   (p. 417)

The respiratory system exchanges oxygen and carbon dioxide between the air and blood, regulates blood pH, produces sounds, moves air over the sensory receptors that detect smell, and protects against some microorganisms.

Anatomy of the Respiratory System   (p. 417)

Nose

1. The nose consists of the external nose and the nasal cavity.
2. The bridge of the nose is bone, and most of the external nose is cartilage.
3. The nasal cavity warms, humidifies, and cleans the air. The nares open to the outside, and the choane lead to the pharynx. The nasal cavity is divided by the nasal septum into right and left parts. The paranasal sinuses and the nasolacrimal duct open into the nasal cavity. Hairs just inside the nares trap debris. The nasal cavity is lined with pseudostratified epithelium with cilia that traps debris and moves it to the pharynx.

Pharynx

1. The nasopharynx joins the nasal cavity through the choane and contains the opening to the auditory tube and the pharyngeal tonsils.
2. The oropharynx joins the oral cavity and contains the palatine and lingual tonsils.
3. The laryngopharynx opens into the larynx and the esophagus.

Larynx

1. The larynx consists of three unpaired cartilages and six paired ones. The thyroid cartilage and cricoid cartilage form most of the larynx. The epiglottis covers the opening of the larynx during swallowing.
2. The vestibular folds can prevent air, food, and liquids from passing into the larynx.
3. The vocal folds (cords) vibrate and produce sounds when air passes through the larynx. The force of air movement controls loudness, and changes in the length and tension of the vocal folds determines pitch.

Trachea

The trachea connects the larynx to the main bronchi.

Bronchi

The main bronchi extend from the trachea to each lung.

Lungs

1. There are two lungs.
2. The airway passages of the lungs branch and decrease in size. The main bronchi form the lobar bronchi, which go to each lobe of the lungs. The lobar bronchi form the segmental bronchi, which go to each bronchopulmonary segment of the lungs. The segmental bronchi branch many times to form the bronchioles. The bronchioles branch to form the terminal bronchioles, which give rise to the respiratory bronchioles, from which alveolar ducts branch. Alveoli are air sacs connected to the alveolar ducts and respiratory bronchioles.
3. The epithelium from the trachea to the terminal bronchioles is ciliated to facilitate removal of debris. Cartilage helps to hold the tube system open (from the trachea to the bronchioles). Smooth muscle controls the diameter of the tubes (especially the bronchioles). The alveoli are formed by simple squamous epithelium, and they facilitate diffusion of gases.
4. The components of the respiratory membrane include a film of water, the walls of the alveolus and the capillary, and an interstitial space. The respiratory membranes are thin and have a large surface area that facilitates gas exchange.

Pleural Cavities

The pleural membranes surround the lungs and provide protection against friction.

Lymphatic Supply

The lungs have superficial and deep lymphatic vessels.

Ventilation and Lung Volumes   (p. 426)

Changing Thoracic Volume

1. Inspiration occurs when the diaphragm contracts and the external intercostal muscles lift the rib cage, thus increasing the volume of the thoracic cavity. During labored breathing additional muscles of inspiration increase rib movement.
2. Expiration can be passive or active. Passive expiration during quiet breathing occurs when the muscles of inspiration relax. Active expiration during labored breathing occurs when the diaphragm relaxes and the internal intercostal and abdominal muscles depress the rib cage to forcefully decrease the volume of the thoracic cavity.

Pressure Changes and Airflow

1. Respiratory muscles cause changes in thoracic volume, which cause changes in alveolar volume and pressure.
2. During inspiration, air flows into the alveoli because atmospheric pressure is greater than alveolar pressure.
3. During expiration, air flows out of the alveoli because alveolar pressure is greater than atmospheric pressure.

Lung Recoil

1. The lungs tend to collapse because of the elastic recoil of the connective tissue and surface tension of the fluid lining the alveoli.
2. The lungs normally do not collapse because surfactant reduces the surface tension of the fluid lining the alveoli and pleural pressure is less than alveolar pressure.

Changing Alveolar Volume

1. Increasing thoracic volume results in decreased pleural pressure, increased alveolar volume, decreased alveolar pressure, and air movement into the lungs.
2. Decreasing thoracic volume results in increased pleural pressure, decreased alveolar volume, increased alveolar pressure, and air movement out of the lungs.

Pulmonary Volumes and Capacities

1. There are four pulmonary volumes: tidal volume, inspiratory reserve, expiratory reserve, and residual volume.
2. Pulmonary capacities are the sum of two or more pulmonary volumes and include vital capacity and total lung capacity.
3. The forced expiratory vital capacity measures the rate at which air can be expelled from the lungs.

Gas Exchange   (p. 432)

1. The respiratory membrane is all of the areas in which gas exchange between air and blood occurs.
2. The dead space is the parts of the respiratory passageways in which gas exchange between air and blood does not occur.

Respiratory Membrane Thickness

Increases in the thickness of the respiratory membrane results in decreased gas exchange.

Surface Area

Small decreases in surface area adversely affect gas exchange during strenuous exercise; and, when the surface area is decreased to one-third to one-fourth of normal, gas exchange is inadequate under resting conditions.

Partial Pressure

1. The pressure exerted by a specific gas in a mixture of gases is reported as the partial pressure of that gas.
2. Oxygen diffuses from a higher partial pressure in the alveoli to a lower partial pressure in the pulmonary capillaries. Oxygen diffuses from a higher partial pressure in the tissue capillaries to a lower partial pressure in the tissue spaces.
3. Carbon dioxide diffuses from a higher partial pressure in the tissues to a lower partial pressure in the tissue capillaries. Carbon dioxide diffuses from a higher partial pressure in the pulmonary capillaries to a lower partial pressure in the alveoli.

Gas Transport in the Blood   (p. 434)
Oxygen Transport

1. Most (98.5%) oxygen is transported bound to hemoglobin. Some (1.5%) oxygen is transported dissolved in plasma.
2. Oxygen is released from hemoglobin in tissues when the partial pressure for oxygen is low, the partial pressure for carbon dioxide is high, pH is low, and temperature is high.

Carbon Dioxide Transport and Blood pH

1. Carbon dioxide is transported as HCO_3^- (70%), in combination with blood proteins (23%), and in solution in plasma (7%).
2. In tissue capillaries, carbon dioxide combines with water inside the red blood cells to form carbonic acid that dissociates to form HCO_3^- and H^+. This reaction promotes the transport of carbon dioxide.
3. In lung capillaries, HCO_3^- combine with H^+ to form carbonic acid. The carbonic acid dissociates to form carbon dioxide that diffuses out of the red blood cells.
4. As blood carbon dioxide levels increase, blood pH decreases; as blood carbon dioxide levels decrease, blood pH increases. Changes in ventilation change blood carbon dioxide levels and pH.

Rhythmic Ventilation   (p. 435)
Respiratory Areas in the Brainstem

1. The medullary respiratory center (dorsal respiratory and ventral respiratory groups) establishes rhythmic ventilation.

2. The pontine respiratory group is involved with the switch between inspiration and expiration.

Generation of Rhythmic Ventilation

1. Inspiration begins when stimuli from many sources, such as receptors that monitor blood gases, reach a threshold.
2. Expiration begins when the neurons causing inspiration are inhibited.

Modification of Ventilation   (p. 436)
Nervous Control of Ventilation

1. Higher brain centers allow voluntary control of ventilation. Emotions and speech production affect ventilation.
2. The Hering-Breuer reflex inhibits the respiratory center when the lungs are stretched during inspiration.
3. Touch, thermal, and pain receptors can stimulate ventilation.

Chemical Control of Ventilation

1. Chemoreceptors in the medulla oblongata respond to changes in blood pH. Usually changes in blood pH are produced by changes in blood carbon dioxide.
2. Carbon dioxide is the major chemical regulator of respiration. An increase in blood carbon dioxide causes a decrease in blood pH, resulting in increased ventilation.
3. Low blood levels of oxygen can stimulate chemoreceptors in the carotid and aortic bodies, resulting in increased ventilation.

Effect of Exercise on Ventilation

Input from higher brain centers and from proprioceptors stimulates the respiratory center during exercise.

Respiratory Adaptations to Exercise   (p. 440)

Training results in increased minute volume at maximal exercise because of increased tidal volume and respiratory rate.

Effects of Aging on the Respiratory System   (p. 442)

1. Vital capacity and maximum minute ventilation decrease with age because of weakening of respiratory muscles and stiffening of the thoracic cage.
2. Residual volume and dead space increase because of increased diameter of respiratory passageways.
3. An increase in resting tidal volume compensates for increased dead space, loss of alveolar walls (surface area), and thickening of alveolar walls.
4. The ability to remove mucus from the respiratory passageways decreases with age.

REVIEW AND COMPREHENSION

1. Define respiration.
2. What are the functions of the respiratory system?
3. Describe the structures of the nasal cavity and their functions.
4. Name the three parts of the pharynx. With what structures does each part communicate?
5. Name and give the functions of the three unpaired cartilages of the larynx.
6. What are the functions of the vestibular and vocal folds? How are sounds of different loudness and pitch produced?
7. Starting at the larynx, name in order all the tubes air passes through to reach an alveolus.
8. What is the function of the C-shaped cartilages in the trachea? What happens to the amount of cartilage in the tube system of the respiratory system as the tubes become smaller? Explain why breathing becomes more difficult during an asthma attack.
9. What is the function of the ciliated epithelium in the tracheobronchial tree?
10. Distinguish between the lungs, a lobe of the lung, and a bronchopulmonary segment.
11. List the components of the respiratory membrane.
12. Describe the pleurae of the lungs. What is their function?
13. Describe the lymphatic supply of the lungs. What is its function?

14. Explain how the muscles of respiration change thoracic volume.

15. Describe the pressure changes that cause air to move into and out of the lungs. What causes these pressure changes?

16. Give two reasons why the lungs tend to recoil or collapse. What two factors keep the lungs from collapsing?

17. Explain how changes in thoracic volume result in changes in pleural pressure, alveolar volume, alveolar pressure, and airflow during inspiration and expiration.

18. Define tidal volume, inspiratory reserve, expiratory reserve, and residual volume. Define vital capacity, total lung capacity, and forced expiratory vital capacity.

19. Describe the factors that affect the diffusion of gases across the respiratory membrane. Give some examples of diseases that decrease diffusion by altering these factors.

20. What is the partial pressure of a gas? Describe the diffusion of oxygen and carbon dioxide between the alveoli and pulmonary capillaries and between the tissue capillaries and tissues in terms of partial pressures.

21. List the ways in which oxygen is transported in the blood. What factors promote the release of oxygen in tissues?

22. List the ways in which carbon dioxide is transported in the blood.

23. How does carbon dioxide affect blood pH? How can changes in ventilation affect blood pH?

24. Name the respiratory areas of the brainstem, and explain how rhythmic ventilation is generated.

25. Describe how higher brain centers and the Hering-Breuer reflex can modify ventilation.

26. Explain the role of blood pH, carbon dioxide, and oxygen in modifying ventilation.

27. How is ventilation regulated during exercise?

28. What effect does exercise training have on the respiratory system?

29. Why do vital capacity, alveolar ventilation, and diffusion of gases across the respiratory membrane decrease with age? Why are the elderly more likely to develop respiratory infections and bronchitis?

CRITICAL THINKING

1. Cardiopulmonary resuscitation (CPR) has replaced older, less efficient methods of sustaining respiration. The back pressure/arm lift method is one such technique that is no longer used. This procedure is performed with the victim lying face down. The rescuer presses firmly on the base of the scapulae for several seconds, then grasps the arms and lifts them. The sequence is then repeated. Explain why this procedure results in ventilation of the lungs.

2. Another technique for artificial respiration is mouth-to-mouth resuscitation. The rescuer takes a deep breath, blows air into the victim's mouth, and then lets air flow out of the victim. The process is repeated. Explain the following:
   a. Why do the victim's lungs expand?
   b. Why does air move out of the victim's lungs?

3. A person's vital capacity was measured while she was standing and while she was lying down. What difference, if any, in the measurement would you predict and why?

4. If water vapor forms 10% of the gases in air at sea level, what is the partial pressure of water?

5. A patient has pneumonia, and fluids accumulate within the alveoli. Explain why this results in an increased rate of respiration that can be returned to normal with oxygen therapy.

6. A patient has severe emphysema that has extensively damaged the alveoli and reduced the surface area of the respiratory membrane. Although the patient is receiving oxygen therapy, he still has a tremendous urge to take a breath (i.e., he does not feel as if he is getting enough air). Why does this occur?

7. Patients with diabetes mellitus who are not being treated with insulin therapy rapidly metabolize lipids, and acidic by-products of lipid metabolism accumulate in the circulatory system. What effect does this have on ventilation? Why is the change in ventilation beneficial?

8. Ima Anxious was hysterical and was hyperventilating. The doctor made her breathe into a paper bag. Because you are an especially astute student, you say to the doctor, "When Ima was hyperventilating, she was reducing blood carbon dioxide levels; and, when she breathed into the paper bag, carbon dioxide was trapped in the bag, and she was rebreathing it, thus causing blood carbon dioxide levels to increase. As blood carbon dioxide levels increase, the urge to breathe should have increased. Instead, she began to breathe more slowly. Please explain." How do you think the doctor would respond? (*Hint:* Recall that the effect of decreased blood carbon dioxide on the vasomotor center results in vasodilation and a sudden decrease in blood pressure.)

9. Hyperventilating before swimming underwater can increase the time spent underwater. Explain how that could happen. Sometimes, a person who has hyperventilated before swimming underwater, passes out while still under water and drowns. Explain.

10. The blood pH of a runner was monitored during a race. It was noticed that, shortly after the beginning of the race, her blood pH increased for a short time. Propose an explanation that would account for the increased pH values following the start of the race.

Answers in Appendix D

Visit this textbook's website at www.mhhe.com/seeleyess6 for practice quizzes, animations, interactive learning exercises, and other study tools.

Digestive System

Chapter Outline and Objectives

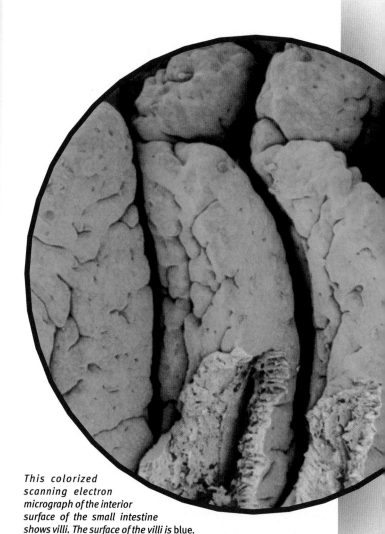

This colorized scanning electron micrograph of the interior surface of the small intestine shows villi. The surface of the villi is blue. The orange *color shows two places where the surface has been broken open to reveal the interior of the villi. The small intestine is the portion of the digestive system where nutrients are absorbed from the food we eat. This absorption is facilitated by the fingerlike villi, which greatly enlarge the internal surface area of the small intestine.*

very cell of the body needs nourishment, yet most cells cannot leave their position in the body and travel to a food source, so the food must be delivered. The **digestive system** (figure 16.1), with the help of the circulatory system, is like a gigantic "meals on wheels" system, serving over 100 trillion customers the nutrients they need. It also has its own quality control and waste disposal system. Food is taken into the digestive system, where it is broken down into smaller and smaller particles. Enzymes in the digestive system break the particles down into very small molecules, which are absorbed into the circulation and transported all over the body. Those molecules are broken down by other enzymes to release energy or are assembled into new molecules to build tissues and organs. In this chapter, the structure and function of the digestive organs and their accessory glands are described.

Functions of the Digestive System

The functions of the digestive system are to

1. *Take in food.* Food and water are taken into the body through the mouth.
2. *Break down the food.* The food that is taken into the body is broken down during the process of digestion from complex molecules to smaller molecules that can be absorbed.
3. *Absorb digested molecules.* The small molecules that result from digestion are absorbed through the walls of the intestine for use in the body.
4. *Provide nutrients.* The process of digestion and absorption provides the body with water, electrolytes, and other nutrients such as vitamins and minerals.
5. *Eliminate wastes.* Undigested material, such as fiber from food, plus waste products excreted into the digestive tract are eliminated in the feces.

Anatomy and Histology of the Digestive System

The digestive system consists of the **digestive tract,** a tube extending from the mouth to the anus, plus the associated organs, which secrete fluids into the digestive tract. The term **gastrointestinal** (**GI;** gas′trō-in-tes′tin-ăl) tract technically only refers to the stomach and intestines but is often used as a synonym for the digestive tract. The inside of the digestive tract is continuous with the outside environment, where it opens at the mouth and anus. Nutrients cross the wall of the digestive tract to enter the circulation.

The digestive tract consists of the oral cavity, pharynx, esophagus, stomach, small intestine, large intestine, and anus. Accessory glands are associated with the digestive tract (see figure 16.1). The salivary glands empty into the oral cavity, and the liver and pancreas are connected to the small intestine.

Various parts of the digestive tract are specialized for different functions, but nearly all parts consist of four layers, or tunics: the mucosa, submucosa, muscularis, and serosa or adventitia (figure 16.2). These will be described in order from the inside of the tube.

1. The innermost tunic, the **mucosa** (mū-kō′să, a tissue producing mucus), consists of **mucous epithelium,** a loose connective tissue called the **lamina propria,** and a thin smooth muscle layer, the **muscularis mucosa.** The epithelium in the mouth, esophagus, and anus resists abrasion, and epithelium in the stomach and intestine absorbs and secretes.
2. The **submucosa** lies just outside the mucosa. It is a thick layer of loose connective tissue containing nerves, blood vessels, and small glands. An extensive network of nerve cell processes forms a **plexus** (network). The plexus is innervated by autonomic nerves.
3. The next tunic is the **muscularis,** which in most parts of the digestive tube consists of an inner layer of **circular smooth muscle** and an outer layer of **longitudinal smooth muscle.** Another nerve plexus, also innervated by autonomic nerves, lies between the two muscle layers. Together the nerve plexuses of the submucosa and muscularis compose the **enteric** (en-tĕr′ik, relating to the intestine) **plexus.** This plexus is extremely important in the control of movement and secretion within the tract.

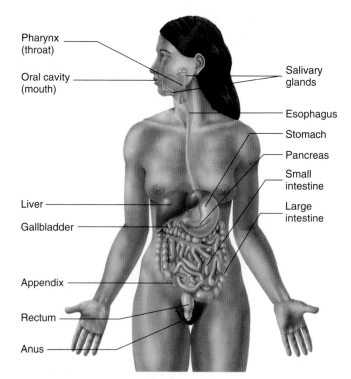

Pharynx (throat)

Oral cavity (mouth)

Liver

Gallbladder

Appendix

Rectum

Anus

Salivary glands

Esophagus

Stomach

Pancreas

Small intestine

Large intestine

Figure 16.1 The Digestive System

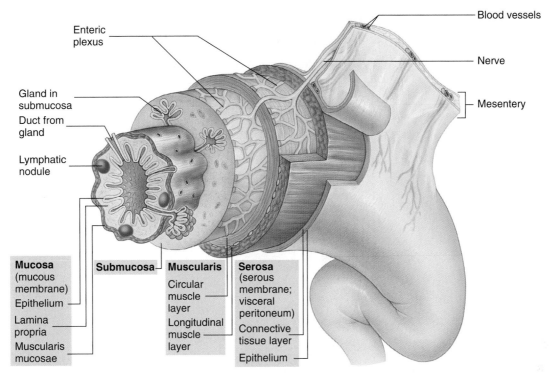

Figure 16.2   Digestive Tract Histology

The four tunics are the mucosa, submucosa, muscularis, and serosa or adventitia. Glands may exist along the digestive tract as part of the epithelium, within the submucosa, or as large glands that are outside the digestive tract.

4. The fourth, or outermost, layer of the digestive tract is either a serosa or an adventitia. Some regions of the digestive tract are covered by peritoneum, and other regions are not. The peritoneum, which is a smooth epithelial layer, and its underlying connective tissue are referred to histologically as the **serosa.** In regions of the digestive tract not covered by peritoneum, the digestive tract is covered by a connective tissue layer called the **adventitia** (ad′ven-tish′ă, foreign, coming from outside), which is continuous with the surrounding connective tissue.

Peritoneum

The body wall of the abdominal cavity and the abdominal organs are covered with **serous membranes** (figure 16.3). The serous membrane that covers the organs is the **visceral peritoneum** (per′i-tō-nē′ŭm, to stretch over), or **serosa.** The serous membrane that lines the wall of the abdominal cavity is the **parietal peritoneum.**

Peritonitis

Peritonitis (per′i-tō-nī′tis) is the inflammation of the peritoneal membranes. The inflammation may result from chemical irritation by substances such as bile that have escaped from the digestive tract; or it may result from infection originating in the digestive tract, such as may occur when an infected appendix ruptures. Peritonitis can be life-threatening.

Many of the organs of the abdominal cavity are held in place by connective tissue sheets called **mesenteries** (mes′en-ter-ēz,

middle intestine). The mesenteries consist of two layers of serous membranes with a thin layer of loose connective tissue between them. Specific mesenteries are given names. The mesentery connecting the lesser curvature of the stomach to the liver and diaphragm is the **lesser omentum** (ō-men′tŭm, membrane of the bowels), and the mesentery connecting the greater curvature of the stomach to the transverse colon and posterior body wall is the **greater omentum.** The greater omentum is unusual in that it is a long, double fold of mesentery that extends inferiorly from the stomach before looping back to the transverse colon to create a cavity, or pocket, called the **omental bursa** (ber′să, pocket). Fat accumulates in the greater omentum, giving it the appearance of a fat-filled apron that covers the anterior surface of the abdominal viscera. Mesentery is a general term referring to the serous membranes attached to the abdominal organs. The term is also used specifically to refer to the mesentery that attaches the small intestine to the posterior abdominal wall. This mesentery is also called the **mesentery proper.**

P R E D I C T
If you placed a pin completely through both folds of the greater omentum, through how many layers of simple squamous epithelium would the pin pass?

Other abdominal organs lie against the abdominal wall, have no mesenteries, and are described as **retroperitoneal** (re′trō-per′i-tō-nē′ăl, behind the peritoneum). The retroperitoneal organs include the duodenum, pancreas, ascending colon, descending colon, rectum, kidneys, adrenal glands, and urinary bladder.

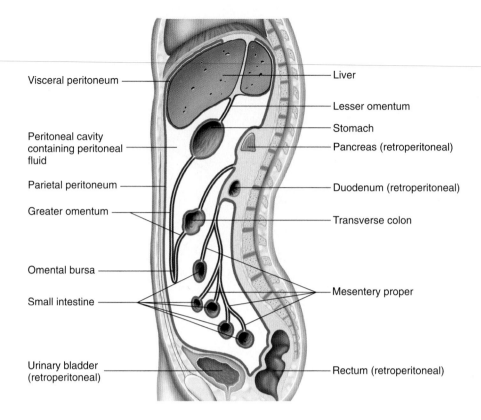

Medial view

Figure 16.3 Peritoneum and Mesenteries

The parietal peritoneum lines the abdominal cavity (*blue*), and the visceral peritoneum covers abdominal organs (*red*). Retroperitoneal organs are behind the parietal peritoneum. The mesenteries are membranes that connect abdominal organs to each other and to the body wall.

Oral Cavity, Pharynx, and Esophagus

Anatomy of the Oral Cavity

The **oral cavity** (figure 16.4), or mouth, is the first part of the digestive tract. It is bounded by the lips and cheeks and contains the teeth and tongue. The **lips** are muscular structures, formed mostly by the **orbicularis oris** (ōr-bik′ū-lā′ris ōr′is) **muscle** (see figure 7.15). The outer surfaces of the lips are covered by skin. The keratinized stratified epithelium of the skin becomes thin at the margin of the lips. The color from the underlying blood vessels can be seen through the thin, transparent epithelium, giving the lips a reddish-pink appearance. At the internal margin of the lips, the epithelium is continuous with the moist stratified squamous epithelium of the mucosa in the oral cavity. The cheeks form the lateral walls of the oral cavity. The **buccinator** (bŭk′si-nā-tōr) **muscles** (see figure 7.15) are located within the cheeks and flatten the cheeks against the teeth. The lips and cheeks are important in the process of **mastication** (mas-ti-kā′shŭn), or chewing. They help manipulate the food within the mouth and hold the food in place while the teeth crush or tear it. Mastication begins the process of mechanical digestion, in which large food particles are broken down into smaller ones. The cheeks also help form words during the speech process.

The **tongue** is a large, muscular organ that occupies most of the oral cavity. The major attachment of the tongue is in the posterior part of the oral cavity. The anterior part of the tongue is relatively free. There is an anterior attachment to the floor of the mouth by a thin fold of tissue called the **frenulum** (fren′ū-lŭm, *frenum,* bridle) (see figure 16.4). The muscles associated with the tongue are described in chapter 7.

The tongue moves food in the mouth and, in cooperation with the lips and cheeks, holds the food in place during mastication. It also plays a major role in the process of swallowing. The tongue is a major sensory organ for taste, as well as being one of the major organs of speech.

Teeth

There are 32 **teeth** in the normal adult mouth, located in the mandible and maxillae. The teeth can be divided into quadrants: right upper, left upper, right lower, and left lower. In adults, each quadrant contains one central and one lateral **incisor** (in-sī′zŏr, to cut); one **canine** (kā′nīn, dog); first and second **premolars** (prē-mō′lărz, *molaris,* a millstone); and first, second, and third **molars** (mō′lărz). The third molars are called **wisdom teeth** because they usually appear in a person's late teens or early twenties, when the person is old enough to have acquired some degree of wisdom.

The teeth of adults are **permanent,** or **secondary, teeth** (figure 16.5a). Most of them are replacements of the 20 **primary,**

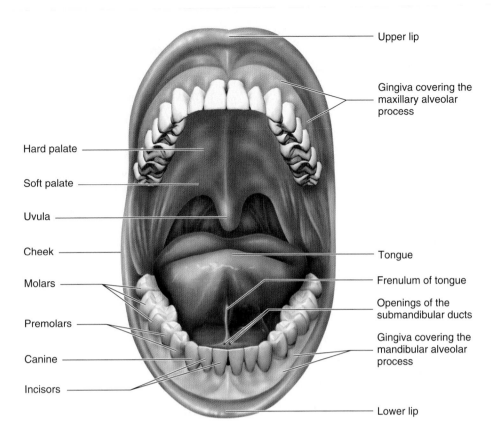

Upper lip

Gingiva covering the
maxillary alveolar
process

Hard palate

Soft palate

Uvula

Cheek

Tongue

Molars

Frenulum of tongue

Openings of the
submandibular ducts

Premolars

Gingiva covering the
mandibular alveolar
process

Canine

Incisors

Lower lip

Figure 16.4

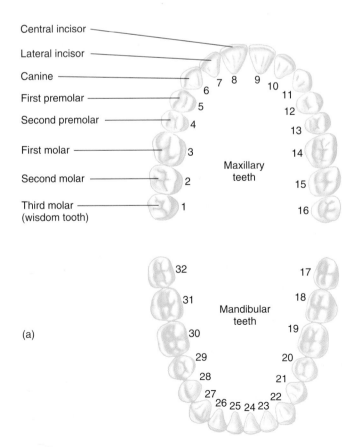

Central incisor

Lateral incisor

Canine

First premolar

Second premolar

First molar

Second molar

Third molar
(wisdom tooth)

Maxillary
teeth

Mandibular
teeth

(a)

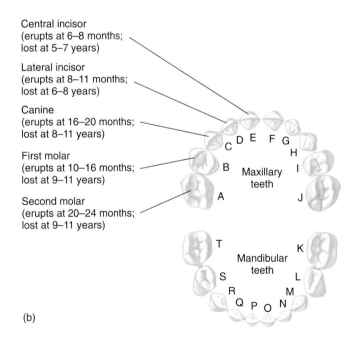

Central incisor
(erupts at 6–8 months;
lost at 5–7 years)

Lateral incisor
(erupts at 8–11 months;
lost at 6–8 years)

Canine
(erupts at 16–20 months;
lost at 8–11 years)

First molar
(erupts at 10–16 months;
lost at 9–11 years)

Second molar
(erupts at 20–24 months;
lost at 9–11 years)

Maxillary
teeth

Mandibular
teeth

(b)

Figure 16.5

(*a*) Permanent teeth. (*b*) Deciduous teeth. A "universal" numbering and
lettering system has been developed by dental professionals for convenience
in identifying individual teeth.

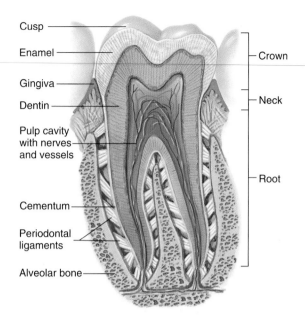

Figure 16.6 Molar Tooth in Place in the Alveolar Bone

The tooth consists of a crown and root. The root is covered with cementum, and the tooth is held in the socket by periodontal ligaments. Nerves and vessels enter and exit the tooth through a foramen in the part of the root deepest in the alveolus.

or **deciduous, teeth** (dē-sid′ū-ŭs, those that fall out; also called milk teeth) which are lost during childhood (figure 16.5b).

Each tooth (figure 16.6) consists of a **crown** with one or more **cusps** (points), a **neck,** and a **root.** The center of the tooth is a **pulp cavity,** which is filled with blood vessels, nerves, and connective tissue, called **pulp.** The pulp cavity is surrounded by a living, cellular, bonelike tissue called **dentin** (den′tin, *dens,* tooth). The dentin of the tooth crown is covered by an extremely hard, acellular substance called **enamel** (ē-nam′ĕl, to apply a glossy or enamel surface), which protects the tooth against abrasion and acids produced by bacteria in the mouth. The surface of the dentin in the root is covered with **cementum** (se-men′tŭm), which helps anchor the tooth in the jaw.

The teeth are rooted within **alveoli** (al-vē′ō-lī, sockets) along the alveolar processes of the mandible and maxillae. The alveolar processes are covered by dense fibrous connective tissue and moist stratified squamous epithelium, referred to as the **gingiva** (jin′ji-vă), or gums. The teeth are held in place by **periodontal** (per′ē-ō-don′tăl, around the teeth) **ligaments,** which are connective tissue fibers that extend from the alveolar walls and are embedded into the cementum.

Tooth Diseases

Formation of **dental caries** (kār′ēz), or tooth decay, is the result of the breakdown of enamel by acids produced by bacteria on the tooth surface. Enamel is nonliving and cannot repair itself. Consequently, a dental filling is necessary to prevent further damage. **Periodontal disease** is inflammation and degeneration of the periodontal ligaments, gingiva, and alveolar bone. This disease is the most common cause of tooth loss in adults.

Tu Thake went to his dentist for a check-up because he had experienced pain in his right mandible for several days. During the routine exam, the dentist felt along the anterior and posterior edges of his sternocleidomastiod muscle on each side, and along the inferior edges of the mandible on each side. Just inferior to the angle of the right mandible, the dentist noted a lump in the area, suggesting an enlargement of the superior cervical lymph nodes. Enlargement of the superior cervical lymph nodes indicates some problem in the face, as all lymphatic drainage from the face goes through these nodes. Additional swellings along the inferior edge of the anterior or central part of the body of the mandible suggests some problem in the mandible. The "problem" could be an infection, a cancerous growth, or it could be idiopathic (of unknown origin). An infection may occur in a tooth, in the bone, or in the soft tissues of the area. In Tu's case, further examination revealed a small abscess near his right, first mandibular molar. The dentist opened the abscess and treated it with topical and systemic antibiotics. The infection disappeared as did the swelling in the superior cervical lymph nodes.

Palate and Tonsils

The **palate** (pal′ăt), or roof of the oral cavity, consists of two parts. The anterior part contains bone and is called the **hard palate,** whereas the posterior portion consists of skeletal muscle and connective tissue and is called the **soft palate** (see figure 16.4). The **uvula** (ū′vū-lă, a grape) is a posterior extension of the soft palate. The palate separates the oral cavity from the nasal cavity and prevents food from passing into the nasal cavity during chewing and swallowing.

The **tonsils** (ton′silz) are located in the lateral posterior walls of the oral cavity, in the nasopharynx, and in the posterior surface of the tongue. The tonsils are described in chapter 14.

Salivary Glands

There are three pairs of **salivary** (sal′i-vār-ē) **glands:** the parotid, submandibular, and sublingual glands (figure 16.7). They produce **saliva** (să-lī′vă), which is a mixture of **serous** (watery) and **mucous** fluids. Saliva helps keep the oral cavity moist and contains enzymes that begin the process of chemical digestion. The salivary glands are compound alveolar glands. They have branching ducts with clusters of alveoli, resembling grapes, at the ends of the ducts (see chapter 4).

The largest of the salivary glands, the **parotid** (pă-rot′id, beside the ear) **glands,** are serous glands located just anterior to each ear. Parotid ducts enter the oral cavity adjacent to the second upper molars.

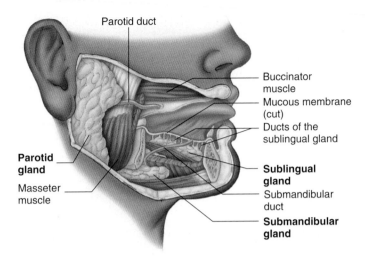

Parotid duct

Buccinator muscle

Mucous membrane (cut)

Ducts of the sublingual gland

Parotid gland

Masseter muscle

Sublingual gland

Submandibular duct

Submandibular gland

Figure 16.7 Salivary Glands

The large salivary glands are the parotid glands, the submandibular glands, and the sublingual glands.

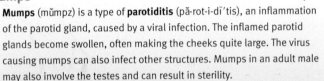

Mumps

Mumps (mŭmpz) is a type of **parotiditis** (pă-rot-i-dī′tis), an inflammation of the parotid gland, caused by a viral infection. The inflamed parotid glands become swollen, often making the cheeks quite large. The virus causing mumps can also infect other structures. Mumps in an adult male may also involve the testes and can result in sterility.

The **submandibular** (sŭb-man-dib′ū-lăr, below the mandible) **glands** produce more serous than mucous secretions. Each gland can be felt as a soft lump along the inferior border of the mandible. The submandibular ducts open into the oral cavity on each side of the frenulum of the tongue (see figure 16.4). In certain people, if the mouth is opened and the tip of the tongue is elevated, saliva can squirt out of the mouth from the ducts of these glands.

The **sublingual** (sŭb-ling′gwăl, below the tongue) **glands,** the smallest of the three paired salivary glands, produce primarily mucous secretions. They lie immediately below the mucous membrane in the floor of the oral cavity. Each sublingual gland has 10–12 small ducts opening onto the floor of the oral cavity.

Secretions of the Oral Cavity

Saliva is secreted at the rate of approximately 1 liter (L) per day. The serous part of saliva, produced mainly by the parotid and submandibular glands, contains a digestive enzyme called **salivary amylase** (am′il-ās, starch-splitting enzyme) (table 16.1), which breaks the covalent bonds between glucose molecules in starch and other polysaccharides to produce the disaccharides maltose and isomaltose. Maltose and isomaltose have a sweet taste; thus the digestion of polysaccharides by salivary amylase enhances the sweet taste of food.

Food spends very little time in the mouth. Consequently, only about 5% of the total carbohydrates humans absorb are digested in the mouth. Also, most starches are contained in plant cells, which are surrounded by cell walls composed primarily of the polysaccharide **cellulose** (sel′ū-lōs). Humans lack the necessary enzymes to digest cellulose. Cooking and thorough

chewing of food disrupt the cellulose covering and increase the efficiency of the digestive process.

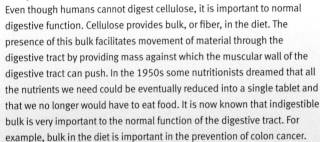

Dietary Fiber

Even though humans cannot digest cellulose, it is important to normal digestive function. Cellulose provides bulk, or fiber, in the diet. The presence of this bulk facilitates movement of material through the digestive tract by providing mass against which the muscular wall of the digestive tract can push. In the 1950s some nutritionists dreamed that all the nutrients we need could be eventually reduced into a single tablet and that we no longer would have to eat food. It is now known that indigestible bulk is very important to the normal function of the digestive tract. For example, bulk in the diet is important in the prevention of colon cancer.

Saliva prevents bacterial infection in the mouth by washing the oral cavity, and it contains **lysozyme** (lī′sō-zīm, loosening enzyme), which has a weak antibacterial action. A lack of salivary gland secretion (which can result from radiation therapy) increases the chance of ulceration and infection of the oral mucosa and caries formation in the teeth.

The serous part of saliva dissolves molecules, which must be in solution to stimulate taste receptors. The mucous secretions of the submandibular and sublingual glands contain a large amount of **mucin** (mū′sin), a proteoglycan that gives a lubricating quality to the secretions of the salivary glands.

Salivary gland secretion is regulated primarily by the autonomic nervous system, with parasympathetic stimulation being the most important. Salivary secretions increase in response to a variety of stimuli, such as tactile stimulation in the oral cavity and certain tastes, especially sour. Higher brain centers can stimulate parasympathetic activity and thus increase the activity of the salivary glands in response to the thought of food, to odors, or to the sensation of hunger. Sympathetic stimulation increases the mucus content of saliva. When a person becomes frightened and the sympathetic division of the autonomic nervous system is stimulated, the person may have a dry mouth with thick mucus.

Mastication

Food taken into the mouth is chewed, or masticated, by the teeth. The incisors and canines primarily cut and tear food, whereas the premolars and molars primarily crush and grind it. Mastication breaks large food particles into many small ones, which have a much larger total surface area than a few large particles would have. Because digestive enzymes act on molecules only at the surface of the food particles, mastication increases the efficiency of digestion.

Pharynx

The **pharynx** (far′ingks), or throat, which connects the mouth with the esophagus, consists of three parts: the nasopharynx, oropharynx, and laryngopharynx (see chapter 15). Normally, only the oropharynx and laryngopharynx transmit food. The posterior walls of the oropharynx and laryngopharynx are formed by the superior, middle, and inferior **pharyngeal constrictor muscles.**

Esophagus

The **esophagus** (ē-sof′ă-gŭs, gullet) is a muscular tube, lined with moist stratified squamous epithelium, that extends from the

Table 16.1 Functions of Digestive Secretions

Fluid or Enzyme	Source	Function
Mouth		
Saliva	Salivary glands	Moistens and lubricates food
Salivary amylase	Salivary glands	Digests starch
Lipase	Salivary glands	Begins lipid digestion
Lysozyme	Salivary glands	Weak antibacterial action
Stomach		
Hydrochloric acid	Gastric glands	Kills bacteria, activates pepsin
Pepsinogen	Gastric glands	Active form, pepsin, digests protein
Mucus	Mucous cells	Protects stomach lining
Intrinsic factor	Gastric glands	Binds to vitamin B_{12}, aiding in its absorption
Gastrin	Gastric glands	Increases stomach secretions
Small Intestine and Associated Glands		
Bile salts	Liver	Emulsify fats
Bicarbonate ions	Pancreas	Neutralize stomach acid
Trypsin, chymotrypsin	Pancreas	Digest protein
Carboxypeptidase	Pancreas	Digests protein
Pancreatic amylase	Pancreas	Digests starch
Pancreatic lipase	Pancreas	Digests lipid
Nucleases	Pancreas	Digest nucleic acid
Mucus	Duodenal glands and goblet cells	Protects duodenum from stomach acid and digestive enzymes
Secretin	Duodenum	Inhibits gastric secretions
		Stimulates sodium bicarbonate secretion from the pancreas and bile secretion from the liver
Cholecystokinin	Duodenum	Inhibits gastric secretion, stimulates gallbladder contraction and pancreas secretion (enzymes)
Gastric inhibitory polypeptide	Duodenum	Inhibits gastric motility and secretion, stimulates gallbladder contraction
Peptidases	Small intestine	Digest polypeptide
Amylase	Small intestine	Digests starch
Lipase	Small intestine	Digests lipid
Sucrase	Small intestine	Digests sucrose
Lactase	Small intestine	Digests lactose
Maltase	Small intestine	Digests maltose

pharynx to the stomach. It is about 25 centimeters (cm) long and lies anterior to the vertebrae and posterior to the trachea within the mediastinum. It passes through the diaphragm and ends at the stomach. The esophagus transports food from the pharynx to the stomach. Upper and lower **esophageal sphincters,** located at the upper and lower ends of the esophagus, respectively, regulate the movement of food into and out of the esophagus. The lower esophageal sphincter is sometimes called the **cardiac sphincter.** Numerous mucous glands produce a thick, lubricating mucus that coats the inner surface of the esophagus.

Hiatal Hernia

A **hiatal** (hī-ā′tŭl, to yawn) **hernia** is a widening of the esophageal hiatus, the opening in the diaphragm through which the esophagus passes. Widening of the hiatus allows part of the stomach to extend through the opening into the thorax. The hernia can decrease some of the pressure in the lower esophageal sphincter, allowing gastroesophageal reflux (movement of stomach contents back into the esophagus) and subsequent esophagitis (inflammation of the esophagus) to occur. Hiatal herniation can also compress blood vessels in the stomach mucosa when the stomach extends through the hernia, which can lead to gastritis (inflammation of the stomach) or ulcer formation. Esophagitis, gastritis, and ulcers are very painful.

Deglutition

Deglutition (dē-gloo-tish′ŭn), or swallowing, can be divided into three separate phases: the voluntary phase, the pharyngeal phase, and the esophageal phase (figure 16.8). During the **voluntary phase,** a bolus, or mass of food, is formed in the mouth. The bolus is pushed by the tongue against the hard

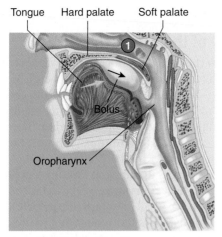

Tongue   Hard palate   Soft palate

Bolus

Oropharynx

1. During the voluntary phase, a bolus of food (*yellow*) is pushed by the tongue against the hard and soft palates and posteriorly toward the oropharynx (*blue arrow* indicates tongue movement; *black arrow* indicates movement of the bolus). *Tan*: bone, *purple*: cartilage, *red*: muscle.

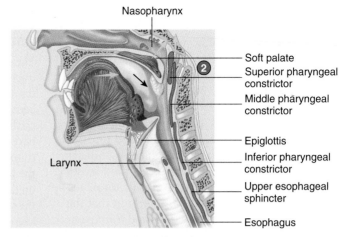

Nasopharynx

Soft palate
Superior pharyngeal constrictor
Middle pharyngeal constrictor
Epiglottis
Inferior pharyngeal constrictor
Upper esophageal sphincter
Esophagus

Larynx

2. During the pharyngeal phase, the soft palate is elevated, closing off the nasopharynx. The pharynx and larynx are elevated (*blue arrows* indicate muscle movement).

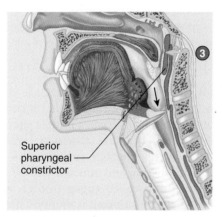

Superior pharyngeal constrictor

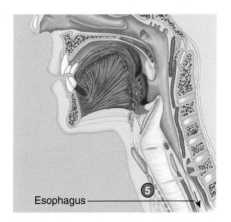

Middle pharyngeal constrictor

Epiglottis
Opening of larynx

3. Successive constriction of the pharyngeal constrictors from superior to inferior (*blue arrows*) forces the bolus through the pharynx and into the esophagus. As this occurs, the epiglottis is bent down over the opening of the larynx largely by the force of the bolus pressing against it.

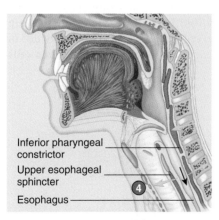

Inferior pharyngeal constrictor
Upper esophageal sphincter
Esophagus

4. As the inferior pharyngeal constrictor contracts, the upper esophageal sphincter relaxes (outwardly directed *blue arrows*), allowing the bolus to enter the esophagus.

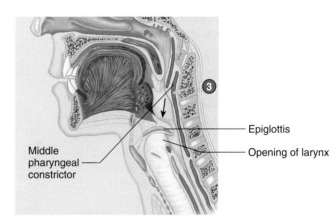

Esophagus

5. During the esophageal phase, the bolus is moved by peristaltic contractions of the esophagus toward the stomach (inwardly directed *blue arrows*).

Process Figure 16.8

palate, forcing the bolus toward the posterior part of the mouth and into the oropharynx.

The **pharyngeal phase** of swallowing is a reflex that is initiated when a bolus of food stimulates receptors in the oropharynx. This phase of swallowing begins with the elevation of the soft palate, which closes the passage between the nasopharynx and oropharynx. The pharynx elevates to receive the bolus of food from the mouth. The three **pharyngeal constrictor muscles** then contract in succession, forcing the food through the pharynx. At the same time, the upper esophageal sphincter relaxes, and food is pushed into the esophagus. As food passes through the pharynx, the **epiglottis** (ep-i-glot′is, upon the glottis, opening of the larynx) is tipped posteriorly so that the opening into the larynx is covered, preventing food from passing into the larynx.

P R E D I C T 2

What would happen if a person had a cleft of the soft palate, so that the soft palate did not completely close the passage between the nasopharynx and oropharynx during swallowing? What may happen if a person has an explosive burst of laughter while trying to swallow a liquid? What happens if a person tries to swallow and speak at the same time?

The **esophageal phase** of swallowing is responsible for moving food from the pharynx to the stomach. Muscular contractions of the esophagus occur in **peristaltic** (per-i-stal′tik, *peri*, around + *stalsis*, constriction) **waves** (figure 16.9). A wave of relaxation of the circular esophageal muscles precedes the bolus of food down the esophagus, and a wave of strong contraction of the circular muscles follows and propels the bolus through the esophagus. The peristaltic contractions associated with swallowing cause relaxation of the lower esophageal sphincter in the esophagus as the peristaltic waves approach the stomach.

Peristaltic Contractions of the Esophagus

Gravity assists the movement of material through the esophagus, especially when liquids are swallowed. The peristaltic contractions that move material through the esophagus are sufficiently forceful, however, to allow a person to swallow even while doing a headstand or floating in the zero-gravity environment of space.

Stomach

Anatomy of the Stomach

The **stomach** (figure 16.10) is an enlarged segment of the digestive tract in the left superior part of the abdomen. The opening from the esophagus into the stomach is called the **cardiac opening** because it is near the heart. The region of the stomach around the cardiac opening is called the **cardiac region.** The most superior part of the stomach is the **fundus** (fŭn′dŭs, the bottom of a round-bottomed leather bottle). The largest part of the stomach is the **body,** which turns to the right, forming a **greater curvature** on the left, and a **lesser curvature** on the right. The opening from the stomach into the small intestine is the **pyloric** (pī-lōr′ik, gatekeeper) **opening,** which is surrounded by a relatively thick ring of smooth muscle called the **pyloric sphincter.** The region of the stomach near the pyloric opening is the **pyloric region.**

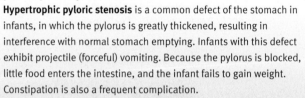

Hypertrophic Pyloric Stenosis

Hypertrophic pyloric stenosis is a common defect of the stomach in infants, in which the pylorus is greatly thickened, resulting in interference with normal stomach emptying. Infants with this defect exhibit projectile (forceful) vomiting. Because the pylorus is blocked, little food enters the intestine, and the infant fails to gain weight. Constipation is also a frequent complication.

The muscular layer of the stomach is different from other regions of the digestive tract in that it consists of three layers: an outer longitudinal layer, a middle circular layer, and an inner oblique layer. These muscular layers produce a churning action in the stomach, important in the digestive process. The submucosa and mucosa of the stomach are thrown into large folds called **rugae** (roo′gē, wrinkles) (see figure 16.10*a*) when the stomach is empty. These folds allow the mucosa and submucosa to stretch, and the folds disappear as the stomach is filled.

The stomach is lined with simple columnar epithelium. The mucosal surface forms numerous, tubelike **gastric pits** (see figure 16.10*b*), which are the openings for the **gastric glands.** The epithelial cells of the stomach can be divided into five groups. The first group consists of **surface mucous cells** on the inner

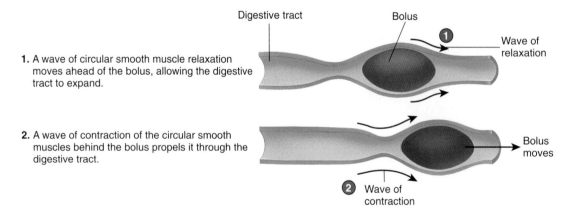

1. A wave of circular smooth muscle relaxation moves ahead of the bolus, allowing the digestive tract to expand.

2. A wave of contraction of the circular smooth muscles behind the bolus propels it through the digestive tract.

Process Figure 16.9   Peristalsis

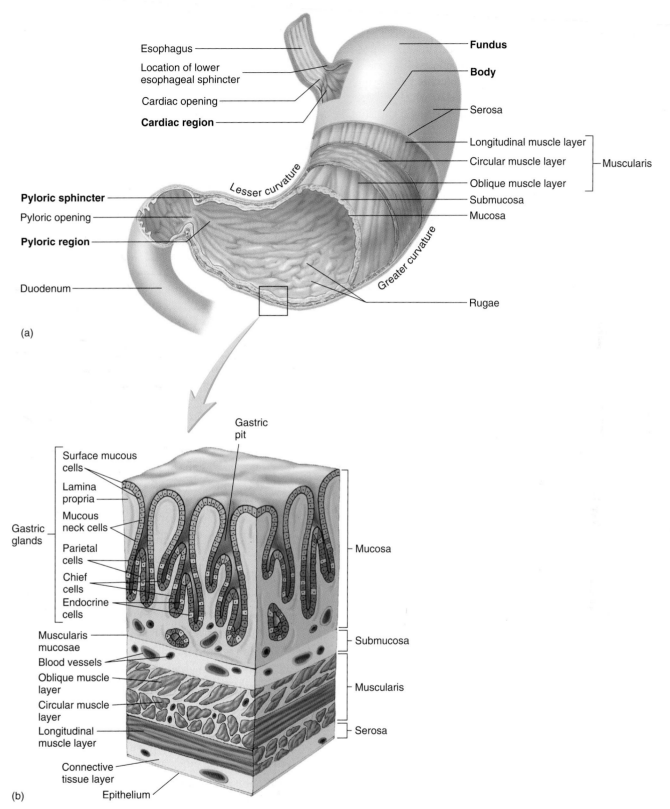

Esophagus
Location of lower esophageal sphincter
Cardiac opening
Cardiac region
Lesser curvature
Pyloric sphincter
Pyloric opening
Pyloric region
Duodenum

Fundus
Body
Serosa
Longitudinal muscle layer
Circular muscle layer ⎤ Muscularis
Oblique muscle layer
Submucosa
Mucosa
Greater curvature
Rugae

(a)

Gastric pit

Surface mucous cells
Lamina propria
Mucous neck cells
Gastric glands
Parietal cells
Chief cells
Endocrine cells
Mucosa

Muscularis mucosae
Blood vessels
Oblique muscle layer
Circular muscle layer
Longitudinal muscle layer
Connective tissue layer
Epithelium

Submucosa
Muscularis
Serosa

(b)

Figure 16.10 Anatomy and Histology of the Stomach

(a) Cutaway section reveals muscular layers and internal anatomy. (b) A section of the stomach wall that illustrates its histology, including several gastric pits and glands.

surface of the stomach and lining the gastric pits. Those cells produce mucus, which coats and protects the stomach lining. The remaining four cell types are in the gastric glands. They are **mucous neck cells,** which produce mucus; **parietal cells,** which produce hydrochloric acid and intrinsic factor; **endocrine cells,** which produce regulatory hormones; and **chief cells,** which produce **pepsinogen** (pep-sin′ō-jen), a precursor of the protein-digesting enzyme **pepsin** (pep′sin, *pepsis,* digestion).

Secretions of the Stomach

The stomach functions primarily as a storage and mixing chamber for ingested food. As food enters the stomach, it is mixed with stomach secretions to become a semifluid mixture called **chyme** (kīm, juice). Although some digestion and a small amount of absorption occur in the stomach, they are not its principal functions.

Stomach secretions from the gastric glands include mucus, hydrochloric acid, pepsinogen, intrinsic factor, and gastrin (see table 16.1). A thick layer of **mucus** lubricates and protects the epithelial cells of the stomach wall from the damaging effect of the acidic chyme and pepsin. Irritation of the stomach mucosa stimulates the secretion of a greater volume of mucus. **Hydrochloric acid** produces a pH of about 2.0 in the stomach. **Pepsinogen** is converted by hydrochloric acid to the active enzyme pepsin. **Pepsin** breaks covalent bonds of proteins to form smaller peptide chains. Pepsin exhibits optimum enzymatic activity at a pH of about 2.0. The low pH also kills microorganisms. **Intrinsic** (in-trin′sik) **factor** binds with vitamin B_{12} and makes it more readily absorbed in the small intestine. Vitamin B_{12} is important in deoxyribonucleic acid (DNA) synthesis and is important to red blood cell production. **Gastrin** (gas′trin, *gaster,* belly or stomach) is a hormone that helps regulate stomach secretions.

Regulation of Stomach Secretions

Approximately 2 L of gastric secretions (gastric juice) is produced each day. Both nervous and hormonal mechanisms regulate gastric secretions. The neural mechanisms involve central nervous system (CNS) reflexes integrated within the medulla oblongata. Higher brain centers can influence these reflexes. Local reflexes are integrated within the enteric plexus in the wall of the digestive tract and do not involve the CNS. Hormones produced by the stomach and intestine help regulate stomach secretions. Regulation of stomach secretions can be divided into three phases: the cephalic, gastric, and intestinal phases.

The **cephalic** (se-fal′ik, *kephale,* head) **phase** of stomach secretion (figure 16.11*a*) is anticipatory and prepares the stomach to receive food. In the cephalic phase, sensations of taste, the smell of food, stimulation of tactile receptors during the process of chewing and swallowing, and pleasant thoughts of food stimulate centers within the medulla oblongata that influence gastric secretions. Action potentials are sent from the medulla oblongata along parasympathetic axons within the vagus nerves to the stomach. Within the stomach wall, the preganglionic neurons stimulate postganglionic neurons in the enteric plexus. The postganglionic neurons stimulate secretory activity in the cells of the stomach mucosa, causing the release of mucus, hydrochloric acid, pepsinogen, intrinsic factor, and gastrin. The gastrin enters the circulation and is carried back to the stomach, where it stimulates additional secretory activity.

The **gastric phase** is the period of greatest gastric secretion (figure 16.11*b*). The gastric phase is responsible for the greatest volume of gastric secretions, and it is activated by the presence of food in the stomach. During the gastric phase, food is present in the stomach and is being mixed with gastric secretions. Distention of the stomach results in the stimulation of stretch receptors. Action potentials generated by these receptors activate CNS reflexes (in the medulla oblongata, by way of the vagus nerve) and local reflexes, resulting in secretion of hydrochloric acid and pepsinogen by the gastric glands. Peptides, produced by the action of pepsin on proteins, stimulate the secretion of gastrin, which in turn stimulates additional hydrochloric acid secretion.

The **intestinal phase** of gastric secretion primarily inhibits gastric secretions (figure 16.11*c*). It is controlled by the entrance of acidic chyme into the duodenum. The presence of chyme in the duodenum initiates both neural and hormonal mechanisms. When the pH of the chyme entering the duodenum drops to 2.0 or below, the inhibitory influence of the intestinal phase is greatest (see figure 16.11*c*). The hormone **secretin** (se-krē′tin), which inhibits gastric secretions, is released from the duodenum. Fatty acids and certain other lipids in the duodenum initiate the release of two hormones: **cholecystokinin** (kō′lē-sis-tō-kī′nin, *chole,* bile + *kystis,* bladder + *kineo,* to move) and **gastric inhibitory polypeptide,** which also inhibit gastric secretions. Acidic chyme (pH < 2.0) in the duodenum also inhibits CNS stimulation and initiates local reflexes that inhibit gastric secretion.

A CASE IN POINT | Heartburn

Hart Burne had a large meat-lover's pizza delivered to his apartment. He had been eating chips and drinking beer before the pizza came. Hart consumed the whole pizza and two more bottles of beer as he watched the last half of a game on TV. By the end of the game he was feeling uncomfortably full, so he laid down on the couch. Within half an hour Hart had a severe pain in his chest and headed for the bathroom to find an antacid.

Heartburn, or **gastritis,** is a painful or burning sensation in the chest usually associated with an increase in gastric acid secretion and/or backflush of acidic chyme into the esophagus. Overeating, eating fatty foods, lying down immediately after a meal, consuming too much alcohol or caffeine, smoking, or wearing extremely tight clothing can all cause heartburn. Cimetidine (si-met′i-dēn, Tagamet), ranitidine (ră-nī′ti-dēn; Zantac), and esomeprazole (eh-sō-meh′pra-zōl, Nexium) are inhibitors of gastric acid secretion by the parietal cells of the stomach. Antacids neutralize acids already secreted into the stomach.

Cephalic Phase

1. The taste, smell, or thought of food or tactile sensations of food in the mouth stimulate the medulla oblongata (*green arrows*).

2. Parasympathetic action potentials are carried by the vagus nerves to the stomach (*pink arrow*).

3. Preganglionic parasympathetic vagus nerve fibers stimulate postganglionic neurons in the enteric plexus of the stomach.

4. Postganglionic neurons stimulate secretion by parietal and chief cells and stimulate gastrin secretion by endocrine cells.

5. Gastrin is carried through the circulation back to the stomach (*purple arrow*), where it stimulates secretion by parietal and chief cells.

(a)

Gastric Phase

1. Distention of the stomach stimulates mechanoreceptors (stretch receptors) and activates a parasympathetic reflex. Action potentials generated by the mechanoreceptors are carried by the vagus nerves to the medulla oblongata (*green arrow*).

2. The medulla oblongata increases action potentials in the vagus nerves that stimulate stomach secretions (*pink arrow*).

3. Distention of the stomach also activates local reflexes that increase stomach secretions (*purple arrow*).

(b)

Gastrointestinal Phase

1. Chyme in the duodenum with a pH less than 2 or containing fat digestion products (lipids) inhibits gastric secretions by three mechanisms (2–4).

2. Chemoreceptors in the duodenum are stimulated by H^+ (low pH) or lipids. Action potentials generated by the chemoreceptors are carried by the vagus nerves to the medulla oblongata (*green arrow*), where they inhibit parasympathetic action potentials (*pink arrow*).

3. Local reflexes activated by H^+ or lipids also inhibit gastric secretion (*purple arrows*).

4. Secretin, gastric inhibitory polypeptide, and cholecystokinin produced by the duodenum (*brown arrows*) inhibit gastric secretions in the stomach.

(c)

Process Figure 16.11

(*a*) Cephalic phase. (*b*) Gastric phase. (*c*) Gastrointestinal phase.

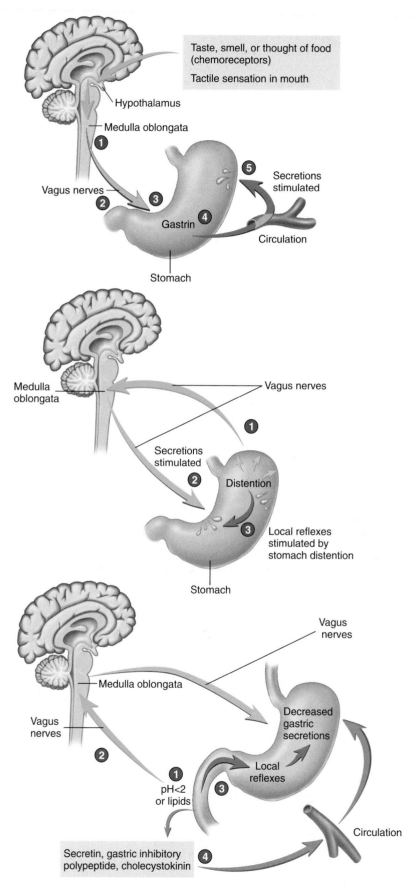

Movement in the Stomach

Two types of stomach movement occur: mixing waves and peristaltic waves (figure 16.12). Both types of movements result from smooth muscle contractions in the stomach wall. The contractions occur about every 20 seconds and proceed from the body of the stomach toward the pyloric sphincter. Relatively weak contractions result in **mixing waves,** which thoroughly mix ingested food with stomach secretions to form chyme. The more fluid part of the chyme is pushed toward the pyloric sphincter, whereas the more solid center moves back toward the body of the stomach. Stronger contractions result in **peristaltic waves,** which force the chyme toward and through the pyloric sphincter. The pyloric sphincter usually remains closed because of mild tonic contraction. Each peristaltic contraction is sufficiently strong to cause partial relaxation of the pyloric sphincter and to pump a few milliliters of chyme through the pyloric opening and into the duodenum.

If the stomach empties too fast, the efficiency of digestion and absorption in the small intestine is reduced. If the rate of emptying is too slow, however, the highly acidic contents of the stomach may damage the stomach wall. Stomach emptying is regulated to prevent these two extremes. The hormonal and neural mechanisms that increase stomach secretions also increase stomach motility so that the increased secretions are effectively mixed with the stomach contents.

1. A mixing wave initiated in the body of the stomach progresses toward the pyloric sphincter (*pink arrows directed inward*).

2. The more fluid part of the chyme is pushed toward the pyloric sphincter (*blue arrows*), whereas the more solid center of the chyme squeezes past the peristaltic constriction back toward the body of the stomach (*orange arrow*).

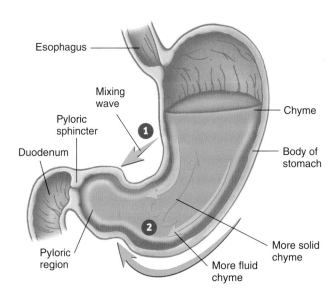

3. Peristaltic waves (*purple arrows*) move in the same direction and in the same way as the mixing waves but are stronger.

4. Again, the more fluid part of the chyme is pushed toward the pyloric region (*blue arrows*), whereas the more solid center of the chyme squeezes past the peristaltic constriction back toward the body of the stomach (*orange arrow*).

5. Peristaltic contractions force a few milliliters of the most fluid chyme through the pyloric opening into the duodenum (*small red arrows*). Most of the chyme, including the more solid portion, is forced back toward the body of the stomach for further mixing (*yellow arrow*).

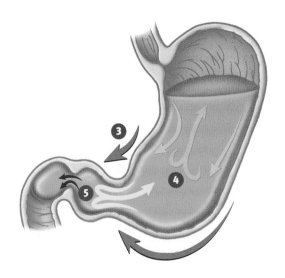

Process Figure 16.12

Small Intestine

Anatomy of the Small Intestine

The **small intestine** is about 6 meters (m) long and consists of three parts: the duodenum, jejunum, and ileum (figure 16.13). The **duodenum** (doo-od′ĕ-nŭm, or doo-ō-dē′nŭm) is about 25 cm long (the term duodenum means 12, suggesting that it is 12 in. long). The **jejunum** (jĕ-joo′nŭm, empty) is about 2.5 m long and makes up two-fifths of the total length of the small intestine. The **ileum** (il′ē-ŭm, twist) is about 3.5 m long and makes up three-fifths of the small intestine.

The duodenum nearly completes a 180-degree arc as it curves within the abdominal cavity. Part of the pancreas lies within this arc. The **common bile duct** from the liver and the **pancreatic duct** from the pancreas join each other and empty into the duodenum (see figure 16.17).

The small intestine is the major site of digestion and absorption of food, which are accomplished by the presence of a large surface area. The surface of the small intestine has three modifications that increase surface area about 600-fold:

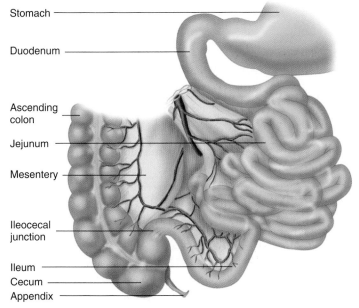

Stomach
Duodenum
Ascending colon
Jejunum
Mesentery
Ileocecal junction
Ileum
Cecum
Appendix

Figure 16.13   The Small Intestine

circular folds, villi, and microvilli. The mucosa and submucosa form a series of **circular folds** that run perpendicular to the long axis of the digestive tract (figure 16.14a). Tiny fingerlike projections of the mucosa form numerous **villi** (vil′i, sing. villus, shaggy hair), which are 0.5–1.5 mm long (figure 16.14b). Most of the cells composing the surface of the villi have numerous cytoplasmic extensions, called **microvilli** (mī′krō-vil′ī) (figure 16.14c and d). Each villus is covered by simple columnar epithelium. Within the loose connective tissue core of each villus is a blood capillary network and a lymphatic capillary called a **lacteal** (lak′tē-ăl, resembling milk) (see figure 16.14c). The blood capillary network and the lacteal are very important in transporting absorbed nutrients.

The mucosa of the small intestine is simple columnar epithelium with four major cell types: (1) **absorptive cells,** which have microvilli, produce digestive enzymes, and absorb digested food; (2) **goblet cells,** which produce a protective mucus; (3) **granular cells** (Paneth's cells), which may help protect the intestinal epithelium from bacteria; and (4) **endocrine cells,** which produce regulatory hormones.

The epithelial cells are produced within tubular glands of the mucosa, called **intestinal glands,** at the base of the villi. Granular and endocrine cells are located in the bottom of the glands. The submucosa of the duodenum contains mucous glands, called **duodenal glands,** which open into the base of the intestinal glands.

The duodenum, jejunum, and ileum are similar in structure except that there is a gradual decrease in the diameter of the small intestine, in the thickness of the intestinal wall, in the number of circular folds, and in the number of villi as one progresses through the small intestine. Lymph nodules are common along the entire length of the digestive tract. Clusters of lymph nodules, called **Peyer's patches,** are numerous in the ileum. These lymphatic tissues in the intestine help protect the intestinal tract from harmful microorganisms.

The junction between the ileum and the large intestine is the **ileocecal** (il′ē-ō-sē′kăl) **junction.** It has a ring of smooth muscle, the **ileocecal sphincter,** and an **ileocecal valve** (this can be seen in figure 16.21a), which allows material contained in the intestine to move from the ileum to the large intestine, but not in the opposite direction.

Secretions of the Small Intestine

Secretions from the mucosa of the small intestine mainly contain mucus, ions, and water. Intestinal secretions lubricate and protect the intestinal wall from the acidic chyme and the action of digestive enzymes. They also keep the chyme in the small intestine in a liquid form to facilitate the digestive process. Most of the secretions entering the small intestine are produced by the intestinal mucosa, but the secretions of the liver and the pancreas also enter the small intestine and play important roles in the process of digestion.

The epithelial cells in the walls of the small intestine have enzymes bound to their free surfaces that play a significant role

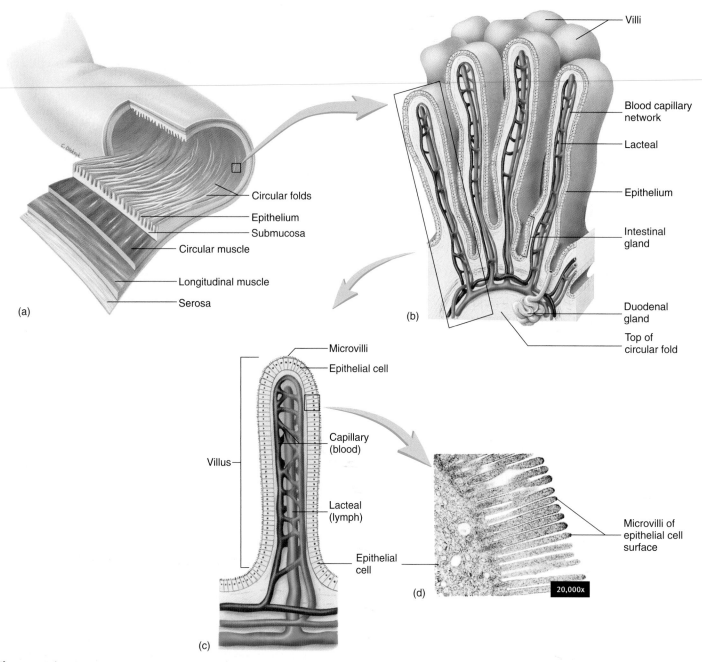

Figure 16.14 Anatomy and Histology of the Duodenum

(*a*) Wall of the duodenum, showing the circular folds. (*b*) The villi on a circular fold. (*c*) A single villus, showing the lacteal and capillary network. (*d*) Transmission electron micrograph of microvilli on the surface of a villus.

in the final steps of digestion. **Peptidases** (pep′ti-dās-ez) break the peptide bonds in proteins to form amino acids. **Disaccharidases** (dī-sak′ă-rid-ās-ez) break down disaccharides, such as maltose and isomaltose, into monosaccharides. The amino acids and monosaccharides can be absorbed by the intestinal epithelium (see table 16.1).

Mucus is produced by duodenal glands and by goblet cells, which are dispersed throughout the epithelial lining of the entire small intestine and within intestinal glands. Hormones released from the intestinal mucosa stimulate liver and pancreatic secretions. Secretion by duodenal glands is stimulated

by the vagus nerve, secretin release, and chemical or tactile irritation of the duodenal mucosa.

Lactose Intolerance

Lactase (lak′tās) is a digestive enzyme bound to the epithelium of the small intestine that digests lactose (milk sugar). A deficiency in this enzyme leads to **lactose, or milk, intolerance.** This primarily hereditary disorder affects 5–15% of the European-American population in the United States and 80–90% of the African-American and Asian-American populations. Symptoms include cramps, bloating, and diarrhea after the ingestion of milk or milk products.

1. A secretion introduced into the digestive tract or food within the tract begins in one location.

2. Segments of the digestive tract alternate between contraction and relaxation.

3. Material (*brown*) in the intestine is spread out in both directions from the site of introduction.

4. The secretion or food is spread out in the digestive tract and becomes more diffuse (*lighter color*) through time.

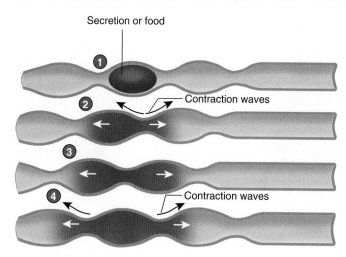

Secretion or food

Contraction waves

Contraction waves

Process Figure 16.15 Segmental Contractions in the Small Intestine

Movement in the Small Intestine

Mixing and propulsion of chyme are the primary mechanical events that occur in the small intestine. **Peristaltic contractions** proceed along the length of the intestine for variable distances and cause the chyme to move along the small intestine (see figure 16.9). **Segmental contractions** are propagated for only short distances and function to mix intestinal contents (figure 16.15).

The ileocecal sphincter at the juncture of the ileum and the large intestine remains mildly contracted most of the time, but peristaltic contractions reaching the ileocecal sphincter from the small intestine cause the sphincter to relax and allow movement of chyme from the small intestine into the cecum. The ileocecal valve allows chyme to move from the ileum into the large intestine, but tends to prevent movement from the large intestine back into the ileum.

Absorption in the Small Intestine

A major function of the small intestine is the absorption of nutrients. Most **absorption** occurs in the duodenum and jejunum, although some absorption also occurs in the ileum (this topic is discussed more fully on p. 470).

Liver and Pancreas

Two large accessory glands, the liver and the pancreas, produce secretions that empty into the duodenum.

Anatomy of the Liver

The **liver** (figure 16.16*a* and *b*; see figure 16.1) weighs about 1.36 kilograms (kg) (3 lb) and is located in the right upper quadrant of the abdomen, tucked against the inferior surface of the diaphragm. The posterior surface of the liver is in contact with the right ribs 5–12. It is divided into two major lobes, the **right** and **left lobes,** separated by a connective tissue septum, the **falciform** (fal′si-fōrm, sickle-shaped) **ligament.** Two smaller

lobes, the **caudate** (kaw′dāt, having a tail) and **quadrate** (kwah′drāt, square), can be seen from an inferior view. Also seen from the inferior view is the **porta** (gate), which is the "gate" through which blood vessels, ducts, and nerves enter or exit the liver.

The liver receives blood from two sources (see chapter 13). The **hepatic** (he-pa′tik, associated with the liver) **artery** brings oxygen-rich blood to the liver, which supplies liver cells with oxygen. The **hepatic portal vein** carries blood that is oxygen-poor but rich in absorbed nutrients and other substances from the digestive tract to the liver. Liver cells process nutrients and detoxify harmful substances from the blood. Blood exits the liver through **hepatic veins,** which empty into the inferior vena cava.

Many delicate connective tissue septa divide the liver into **lobules** with portal triads at the corners of the lobules. The **portal triads** (three) contain three structures: the hepatic artery, hepatic portal vein, and hepatic duct (figure 16.16*c*). **Hepatic** (he-pa′tik) **cords,** formed by platelike groups of cells called **hepatocytes** (hep′ă-tō-sīts), are located between the center and the margins of each lobule. The hepatic cords are separated from one another by blood channels called **hepatic sinusoids** (si′nŭ-soydz or sī′nū-soydz, resembling cavities). The sinusoid epithelium contains phagocytic cells that help remove foreign particles from the blood. Blood from the hepatic portal vein and the hepatic artery flows into the sinusoids and becomes mixed. The mixed blood flows toward the center of each lobule into a **central vein.** The central veins from all the lobes unite to form the hepatic veins, which carry blood out of the liver to the inferior vena cava.

A cleftlike lumen, the **bile canaliculus** (kan′ă-lik′ū-lŭs, little canal), is between the cells of each hepatic cord. **Bile,** produced by the hepatocytes, flows through the bile canaliculi to the hepatic ducts in the portal triads. The hepatic ducts converge and empty into the right and left **hepatic ducts,** which transport bile out of the liver. The right and left hepatic ducts unite to form a single **common hepatic duct.** The common

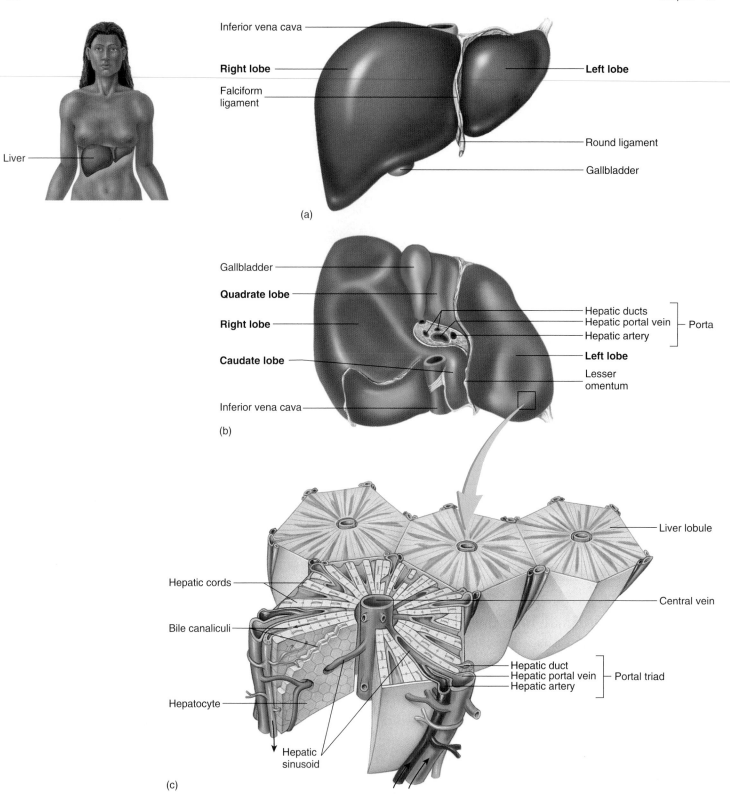

Figure 16.16   Liver

(*a*) Anterior view. (*b*) Inferior view. (*c*) Histology.

1. The hepatic ducts from the liver lobes combine to form the common hepatic duct.

2. The common hepatic duct combines with the cystic duct from the gallbladder to form the common bile duct.

3. The common bile duct joins the pancreatic duct.

4. The combined duct empties into the duodenum at the duodenal papilla.

5. Pancreatic secretions may also enter the duodenum through an accessory pancreatic duct, which also empties into the duodenum.

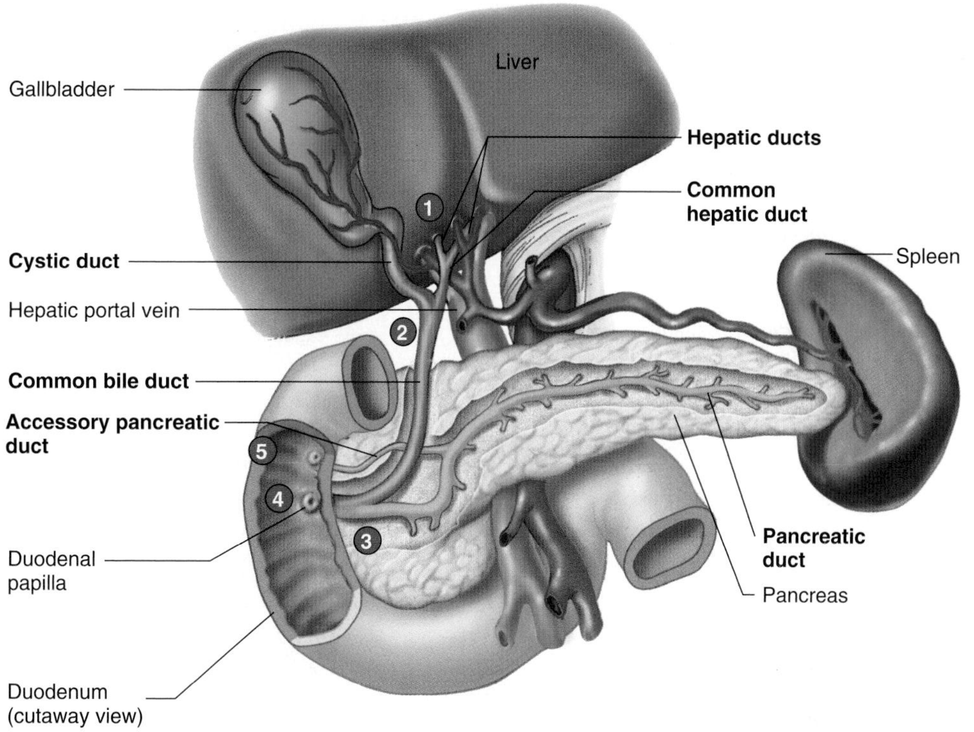

Figure 16.17   The Liver, Gallbladder, Pancreas, and Duct System

hepatic duct is joined by the **cystic** (sis′tik, *kystis,* bladder) **duct** from the gallbladder to form the **common bile duct.** The **gallbladder** is a small sac on the inferior surface of the liver that stores and concentrates bile (see figure 16.16*a* and *b*). The common bile duct joins the pancreatic duct and opens into the duodenum at the **duodenal papilla** (pă-pil′ă) (figure 16.17). The opening into the duodenum is regulated by a sphincter.

Functions of the Liver

The liver performs important digestive and excretory functions, stores and processes nutrients, synthesizes new molecules, and detoxifies harmful chemicals (table 16.2).

The liver secretes about 700 mL of **bile** each day. Bile contains no digestive enzymes, but it plays an important role in digestion by diluting and neutralizing stomach acid and by dramatically increasing the efficiency of fat digestion and absorption. Digestive enzymes cannot act efficiently on large fat globules. **Bile** (bīl) **salts** emulsify fats, breaking the fat globules into smaller droplets, much like the action of detergents in dishwater (see tables 16.1 and 16.2). The small droplets are more easily digested by digestive enzymes. Bile also contains excretory products such as bile pigments, cholesterol, and fats. **Bilirubin** (bil-i-roo′bin, *rubidus,* reddish) is a bile pigment that results from the breakdown of hemoglobin.

Table 16.2	Functions of the Liver
Function	**Explanation**
Digestion	Bile neutralizes stomach acid and emulsifies fats, which facilitates fat digestion
Excretion	Bile contains excretory products such as cholesterol, fats, and bile pigments, such as bilirubin, that result from hemoglobin breakdown
Nutrient storage	Liver cells remove sugar from the blood and store it in the form of glycogen; also store fat, vitamins (A, B_{12}, D, E, and K), copper, and iron
Nutrient conversion	Liver cells convert some nutrients into others; for example, amino acids can be converted to lipids or glucose; fats can be converted to phospholipids; vitamin D is converted to its active form
Detoxification of harmful chemicals	Liver cells remove ammonia from the circulation and convert it to urea, which is eliminated in the urine; other substances are detoxified and secreted in the bile or excreted in the urine
Synthesis of new molecules	Synthesizes blood proteins such as albumin, fibrinogen, globulins, and clotting factors

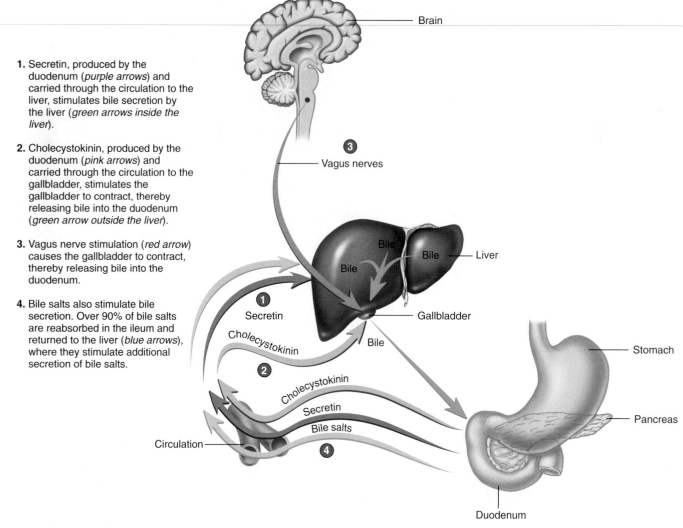

1. Secretin, produced by the duodenum (*purple arrows*) and carried through the circulation to the liver, stimulates bile secretion by the liver (*green arrows inside the liver*).

2. Cholecystokinin, produced by the duodenum (*pink arrows*) and carried through the circulation to the gallbladder, stimulates the gallbladder to contract, thereby releasing bile into the duodenum (*green arrow outside the liver*).

3. Vagus nerve stimulation (*red arrow*) causes the gallbladder to contract, thereby releasing bile into the duodenum.

4. Bile salts also stimulate bile secretion. Over 90% of bile salts are reabsorbed in the ileum and returned to the liver (*blue arrows*), where they stimulate additional secretion of bile salts.

Process Figure 16.18   Control of Bile Secretion and Release

Bile secretion by the liver is stimulated by **secretin,** which is released from the duodenum (figure 16.18). **Cholecystokinin** (kō′lē-sis-tō-kī′nin) stimulates the gallbladder to contract and release bile into the duodenum. Parasympathetic stimulation through the vagus nerve also stimulates bile secretion and release.

Most (90%) bile salts are reabsorbed in the ileum, and the blood carries them back to the liver, where they stimulate additional bile salt secretion and are once again secreted into the bile. The loss of bile salts in the feces is reduced by this recycling process.

The liver can remove sugar from the blood and store it in the form of glycogen (see table 16.2). It can also store fat, vitamins, copper, and iron. This storage function is usually short term.

Foods are not always ingested in the proportion needed by the tissues. If this is the case, the liver can convert some nutrients into others (see table 16.2). For example, if a person eats a meal that is very high in protein, a large amount of amino acids and only a small amount of lipids and carbohydrates are delivered to the liver. The liver can break down the amino acids and cycle many of them through metabolic pathways to produce ATP and to synthesize lipids and glucose (see chapter 17).

The liver also transforms some nutrients into more readily usable substances. Ingested fats, for example, can be combined with choline and phosphorus in the liver to produce phospholipids, which are essential components of cell membranes.

Many ingested substances are harmful to the cells of the body. In addition, the body itself produces many by-products of metabolism that, if accumulated, are toxic. The liver is an important line of defense against many of those harmful substances. It detoxifies them by altering their structure, making their excretion easier (see table 16.2). For example, the liver removes ammonia, which is a toxic by-product of amino acid metabolism, from the circulation and converts it to urea, which is then secreted into the circulation and eliminated by the kidneys in the urine. Other substances are removed from the circulation and excreted by the liver into the bile.

The liver can also produce its own unique new compounds (see table 16.2). Many of the blood proteins, such as albumins, fibrinogen, globulins, and clotting factors, are synthesized in the liver and released into the circulation.

Anatomy of the Pancreas

The **pancreas** is located retroperitoneal, posterior to the stomach in the inferior part of the left upper quadrant (see figure 16.1). It has a **head** near the midline of the body and a **tail** that extends to the left where it touches the spleen (figure 16.19, and see figure 16.17). It is a complex organ composed of both endocrine and exocrine tissues that perform several functions. The endocrine part of the pancreas consists of **pancreatic islets** (islets of Langerhans). The islet cells produce the hormones insulin and glucagon, which enter the blood. These hormones are very important in controlling blood levels of nutrients such as glucose and amino acids (see chapter 10).

✳The exocrine part of the pancreas is a compound acinar gland (see discussion of glands in chapter 4). The **acini** (as′i-nī, grapes) produce digestive enzymes. Clusters of acini are connected by small ducts, which join to form larger ducts, and the larger ducts join to form the **pancreatic duct.** The pancreatic duct joins the common bile duct and empties into the duodenum.

Functions of the Pancreas

The exocrine secretions of the pancreas include HCO_3^-, which neutralize the acidic chyme that enters the small intestine from the stomach. The increased pH resulting from the secretion of HCO_3^- stops pepsin digestion but provides the proper environment for the function of pancreatic enzymes. Pancreatic enzymes are also present in the exocrine secretions and are important for the digestion of all major classes of food (see table 16.1). Without the enzymes produced by the pancreas, lipids, proteins, and carbohydrates are not adequately digested.

The major proteolytic (protein-digesting) enzymes are **trypsin** (trip′sin), **chymotrypsin** (kī-mō-trip′sin), and **carboxypeptidase** (kar-box′ē-pep′ti-dās). These enzymes continue the protein digestion that started in the stomach, and **pancreatic amylase** (am′il-ās′) continues the polysaccharide digestion that began in the oral cavity. The pancreatic enzymes also include a group of lipid-digesting enzymes called pancreatic **lipases** (lip′ās-ez). **Nucleases** (noo′klē-ās-ez) are pancreatic enzymes that reduce DNA and ribonucleic acid to their component nucleotides.

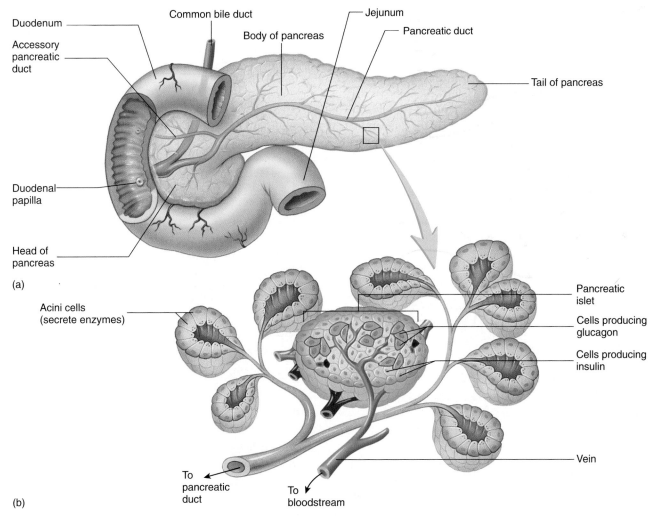

Figure 16.19 Anatomy and Histology of the Duodenum and Pancreas

(*a*) The head of the pancreas lies within the duodenal curvature, with the pancreatic duct emptying into the duodenum. (*b*) Histology of the pancreas showing both the acini and the pancreatic duct system.

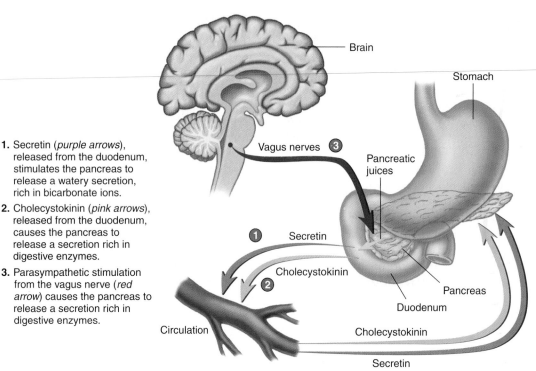

1. Secretin (*purple arrows*), released from the duodenum, stimulates the pancreas to release a watery secretion, rich in bicarbonate ions.
2. Cholecystokinin (*pink arrows*), released from the duodenum, causes the pancreas to release a secretion rich in digestive enzymes.
3. Parasympathetic stimulation from the vagus nerve (*red arrow*) causes the pancreas to release a secretion rich in digestive enzymes.

Process Figure 16.20   Control of Pancreatic Secretion

The exocrine secretory activity of the pancreas is controlled by both hormonal and neural mechanisms (figure 16.20). **Secretin** initiates the release of a watery pancreatic solution that contains a large amount of HCO_3^-. The primary stimulus for secretin release is the presence of acidic chyme in the duodenum. **Cholecystokinin** stimulates the pancreas to release an enzyme-rich solution. The primary stimulus for cholecystokinin release is the presence of fatty acids and amino acids in the duodenum, and the enzymes secreted by the pancreas digest fatty acids and amino acids. Parasympathetic stimulation through the vagus nerves also stimulates the secretion of pancreatic juices rich in pancreatic enzymes. Sympathetic action potentials inhibit pancreatic secretion.

PREDICT ③
Explain how secretin production in response to acidic chyme and bicarbonate ion secretion in response to secretin constitute a negative-feedback mechanism.

Large Intestine
Anatomy of the Large Intestine

The **large intestine** (figure 16.21; see figure 16.1) consists of the cecum, colon, rectum, and anal canal.

Cecum

The **cecum** (sē′kŭm, blind) (see figure 16.21) is the proximal end of the large intestine and is where the large and small intestines meet at the ileocecal junction. The cecum is located in the right lower quadrant of the abdomen near the iliac fossa (see figure 16.1). The cecum is a sac that extends inferiorly about 6 cm past the ileocecal junction. Attached to the cecum is a tube about 9 cm long called the **appendix.**

A CASE IN POINT | Appendicitis

Lowe Payne had been feeling nauseous for a couple of days and had lost his appetite. Suddenly he felt a sharp pain in his lower right abdomen. The pain was so intense that his mother, Lotta, took Lowe to the hospital. There, he was diagnosed as having **appendicitis.** Part of that diagnosis includes applying slight pressure, such as pushing with the fingertips, to a specific point in the right lower quadrant of the abdomen. That point, called the McBurney point, is midway between the umbilicus and the right anterior superior iliac spine of the coxa.

Appendicitis is an inflammation of the appendix and usually occurs because of obstruction. Secretions from the appendix cannot pass the obstruction and accumulate, causing enlargement and pain. Bacteria in the area can cause infection. Symptoms include sudden abdominal pain, particularly in the right lower quadrant of the abdomen, along with a slight fever, loss of appetite, constipation or diarrhea, nausea, and vomiting. If the appendix bursts, the infection can spread throughout the peritoneal cavity, causing **peritonitis,** with life-threatening results. Each year, 500,000 people in the United States suffer from appendicitis. An **appendectomy** is removal of the appendix.

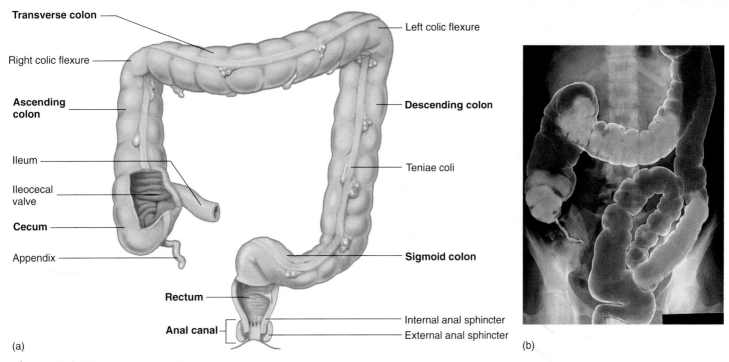

Figure 16.21 Large Intestine

(*a*) The large intestine consists of the cecum, colon, rectum, and anal canal. The teniae coli are bands of smooth muscle along the length of the colon.
(*b*) A radiograph of the large intestine following a barium enema.

Colon

The **colon** (kō′lon) (see figure 16.21) is about 1.5–1.8 m long and consists of four parts: the ascending colon, the transverse colon, the descending colon, and the sigmoid colon. The **ascending colon** extends superiorly from the cecum to the right colic flexure, near the liver, where it turns to the left. The **transverse colon** extends from the right colic flexure to the left colic flexure near the spleen, where the colon turns inferiorly; and the **descending colon** extends from the left colic flexure to the pelvis, where it becomes the **sigmoid colon.** The sigmoid colon forms an S-shaped tube that extends medially and then inferiorly into the pelvic cavity and ends at the rectum.

The mucosal lining of the colon contains numerous straight tubular glands called **crypts,** which contain many mucus-producing goblet cells. The longitudinal smooth muscle layer of the colon does not completely envelope the intestinal wall but forms three bands called **teniae coli** (tē′nē-ē kō′lī, tape on the colon).

Rectum

The **rectum** is a straight, muscular tube that begins at the termination of the sigmoid colon and ends at the anal canal (see figure 16.20). The muscular tunic is smooth muscle and it is relatively thick in the rectum compared with the rest of the digestive tract.

Anal Canal

The last 2–3 cm of the digestive tract is the **anal canal.** It begins at the inferior end of the rectum and ends at the **anus**

(external GI tract opening). The smooth muscle layer of the anal canal is even thicker than that of the rectum and forms the **internal anal sphincter** at the superior end of the anal canal. The **external anal sphincter** at the inferior end of the anal canal is formed by skeletal muscle.

Hemorrhoids

Hemorrhoids are the enlargement or inflammation of the rectal, or hemorrhoidal, veins, which supply the anal canal. Hemorrhoids may cause pain, itching, and/or bleeding around the anus. Treatments include increasing bulk in the diet, taking sitz baths, and using hydrocortisone suppositories. Surgery may be necessary if the condition is extreme and does not respond to other treatments.

Functions of the Large Intestine

Normally 18–24 hours is required for material to pass through the large intestine in contrast to the 3–5 hours required for movement of chyme through the small intestine. While in the colon, chyme is converted to **feces** (fē′sēz). Absorption of water and salts, the secretion of mucus, and extensive action of microorganisms are involved in the formation of feces. The colon stores the feces until they are eliminated by the process of **defecation** (def-ĕ-kā′shŭn, to remove feces).

Numerous microorganisms inhabit the colon. They reproduce rapidly and ultimately constitute about 30% of the dry weight of the feces. Some bacteria in the intestine synthesize vitamin K and other vitamins, which is passively absorbed in the colon.

Every 8–12 hours, large parts of the colon undergo several strong contractions called **mass movements,** which propel the colon contents a considerable distance toward the anus. Each mass movement contraction extends over 20 or more centimeters of the large intestine, which is a much longer part of the digestive tract than that covered by a peristaltic contraction. These mass movements are very common following some meals, especially breakfast.

Distention of the rectal wall by feces acts as a stimulus that initiates the **defecation reflex,** which involves local and parasympathetic reflexes. Local reflexes cause weak contractions, whereas parasympathetic reflexes cause strong contractions and are normally responsible for most of the defecation reflex. Action potentials produced in response to the distention travel along sensory nerve fibers to the sacral region of the spinal cord, where motor action potentials are initiated that reinforce peristaltic contractions in the lower colon and rectum. Action potentials from the spinal cord also cause the internal anal sphincter to relax. The external anal sphincter, which is composed of skeletal muscle and is under conscious cerebral control, prevents the movement of feces out of the rectum and through the anal opening. If this sphincter is relaxed voluntarily, feces are expelled. The defecation reflex persists for only a few minutes and quickly subsides. Generally the reflex is reinitiated after a period that may be as long as several hours. Mass movements in the colon are usually the reason for the reinitiation of the defecation reflex.

Defecation can be initiated by voluntary actions that stimulate a defecation reflex. These actions include a large inspiration of air, followed by closure of the larynx and forceful contraction of the abdominal muscles. As a consequence, the pressure in the abdominal cavity increases and forces feces into the rectum. Stretch of the rectum initiates a defecation reflex. The increased abdominal pressure also helps to push feces through the rectum.

P R E D I C T   4
Explain how an enema stimulates defecation.

Digestion, Absorption, and Transport

Digestion is the breakdown of food to molecules that are small enough to be absorbed into the circulation. **Mechanical digestion** breaks large food particles down into smaller ones. **Chemical digestion** involves the breaking of covalent chemical bonds in organic molecules by digestive enzymes (figure 16.22). Carbohydrates are broken down into monosaccharides, proteins are broken down into amino acids, and fats are broken down into fatty acids and glycerol.

Absorption begins in the stomach, where some small, lipid-soluble molecules, such as alcohol and aspirin, can diffuse through the stomach epithelium into the circulation. Most absorption occurs in the duodenum and jejunum, although some occurs in the ileum. Some molecules can diffuse through the intestinal wall. Others must be transported across the intestinal wall. **Transport** requires carrier molecules and includes facilitated diffusion, cotransport, and active transport. Cotransport and active transport require energy to move the transported molecules across the intestinal wall.

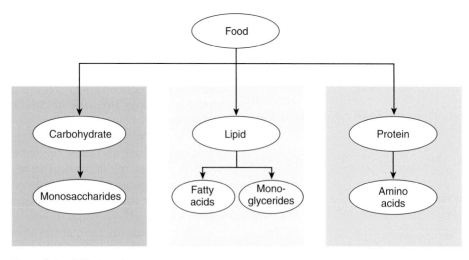

Figure 16.22   Digestion of Food Molecules
Food consists primarily of carbohydrate, lipid, and protein. Carbohydrate is broken down into monosaccharides, lipid into fatty acids and monoglycerides, and protein into amino acids.

Carbohydrates

Ingested **carbohydrates** (kar-bō-hī′drātz) consist primarily of starches, cellulose, sucrose (table sugar), and small amounts of fructose (fruit sugar) and lactose (milk sugar). Starches, cellulose, sucrose, and fructose are derived from plants, and lactose is derived from animals. **Polysaccharides** (pol-ē-sak′ă-rīdz) are large carbohydrates, such as starches, cellulose, and glycogen, which consist of many sugars linked by chemical bonds. Starch is an energy storage molecule in plants. Cellulose forms the walls of plant cells. Glycogen is an energy-storage molecule in animals and is contained in muscle and in the liver. When uncooked meats are processed or stored, the glycogen is broken

down to glucose, which in turn is further broken down, so that little, if any, glycogen remains. Therefore, almost all dietary carbohydrates come from plants. Starch is broken down by enzymes. Cellulose is a polysaccharide that is not digested but is important for providing fiber in the diet.

Salivary amylase begins the digestion of carbohydrates in the mouth (figure 16.23). The carbohydrates then pass to the stomach, where almost no carbohydrate digestion occurs. In the duodenum, **pancreatic amylase** continues the digestion of carbohydrates, and absorption begins. The amylases break down polysaccharides to **disaccharides** (dī-sak′ă-rīdz, two sugars chemically linked; see chapter 2). A group of enzymes called

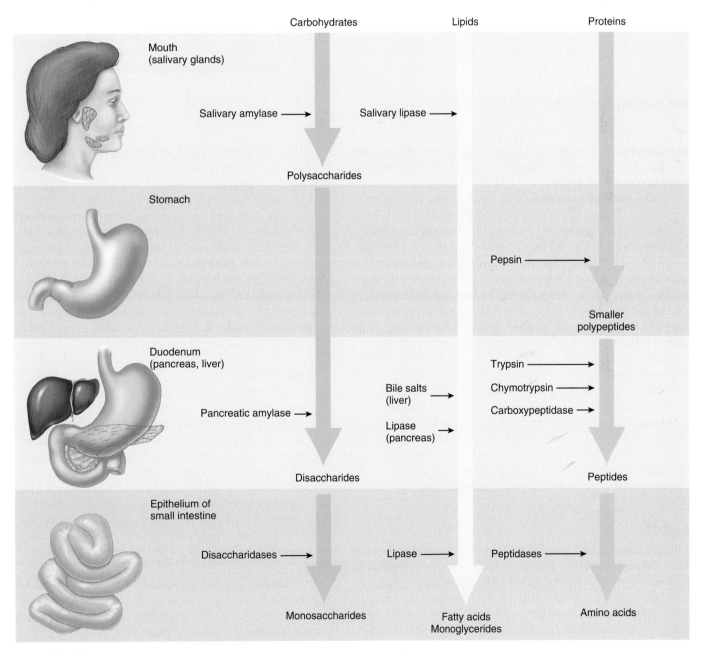

Figure 16.23   Digestion of Carbohydrates, Lipids, and Proteins

The enzymes involved in digesting carbohydrates, lipids, and proteins are depicted as well as the region of the digestive tract where each functions.

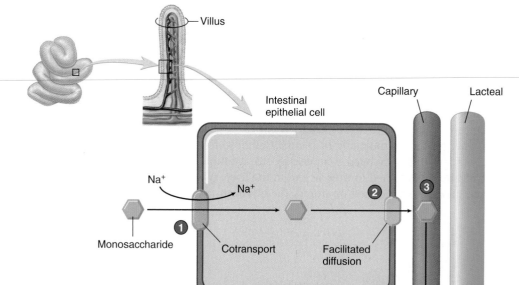

Process Figure 16.24

disaccharidases (dī-sak′ă-rid-ăs-ez) that are bound to the microvilli of the intestinal epithelium break down the disaccharides to monosaccharides.

The **monosaccharides** (mon-ō-sak′ă-rīdz, single sugars) are taken up by cotransport through the intestinal epithelial cells (figure 16.24, and see chapter 3), enter capillaries of the intestinal villi, and are carried by the hepatic portal system to the liver. Different types of monosaccharides are converted to **glucose** by liver cells. Glucose is carried from the liver by the circulation throughout the body. Glucose enters the cells by facilitated diffusion. The rate of glucose transport into most types of cells is greatly influenced by **insulin** and can increase 10-fold in the presence of insulin. Without insulin, glucose enters most cells very slowly.

Diabetes Mellitus

Insulin normally binds to receptors on target cells and increases the ability of the target cell to take up glucose. In people with **diabetes mellitus,** insulin either is lacking or does not have the normal effect on target cells. As a result, not enough glucose is transported into many cells of the body. Thus, in untreated diabetes, the cells do not have enough energy for normal function, blood glucose levels become elevated, and large amounts of glucose pass into the urine.

Lipids

Lipid molecules are insoluble or only slightly soluble in water (see chapter 2). They include triglycerides, phospholipids, steroids, and fat-soluble vitamins. **Triglycerides** (trī-glis′er-īdz), also called triacylglycerol, are the most common type of lipid. They consist of three fatty acids bound to glycerol. Triglycerides are often referred to as fats. Fats are **saturated** if their fatty acids have only single bonds between carbons and **unsaturated** if they have one (monounsaturated) or more (polyunsaturated) double bonds between carbons (see chapter 2). Saturated fats are solid at room temperature, whereas polyunsaturated fats are liquid at room temperature. Saturated fats are found in meat, dairy products, eggs, nuts, coconut oil, and palm oil. Unsaturated fats are found in fish and most plant oils.

The first step in lipid digestion is **emulsification** (ē-mŭl′si-fi-kā′shŭn), which is the transformation of large lipid droplets into much smaller droplets. Emulsification is accomplished by **bile salts** secreted by the liver. The enzymes that digest lipids are soluble in water and can digest the lipids only by acting at the surface of the droplets. The emulsification process increases the surface area of the lipid droplets exposed to the digestive enzymes by increasing the number of lipid droplets and by decreasing the size of each droplet.

Cystic Fibrosis

Cystic Fibrosis is a hereditary disorder that affects both the respiratory and digestive systems. It results from defective chloride channels, which cause cells to produce thick, viscous mucous secretions. Blockage of the pancreatic ducts often occurs so that the pancreatic digestive enzymes are prevented from reaching the duodenum. As a result, fat digestion, which depends on pancreatic enzymes, is slowed or even stopped. Consequently, fats and fat-soluble vitamins are not absorbed, and the patient suffers from vitamin A, D, E, and K deficiencies. These deficiencies result in conditions such as night blindness, skin disorders, rickets, and excessive bleeding. Therapy consists of administering the missing vitamins to the patient and reducing dietary fat intake.

Lipases (lip′ās), secreted by the pancreas and intestinal absorptive cells, digest lipid molecules (see figure 16.23). The primary products of this digestive process are fatty acids and monoglycerides (monoacylglycerol).

In the intestine, bile salts aggregate around small droplets of digested lipids to form **micelles** (mi-selz′, mī-selz′, a small morsel) (figure 16.25). The hydrophobic (water-fearing) ends

Lipid transport

1. Bile salts surround fatty acids and monoglycerides to form micelles.

2. Micelles attach to the cell membranes of intestinal epithelial cells, and the fatty acids and monoglycerides pass by simple diffusion into the intestinal epithelial cells.

3. Within the intestinal epithelial cell, the fatty acids and monoglycerides are converted to triglycerides; proteins coat the triglycerides to form chylomicrons, which move out of the intestinal epithelial cells by exocytosis.

4. The chylomicrons enter the lacteals of the intestinal villi and are carried through the lymphatic system to the general circulation.

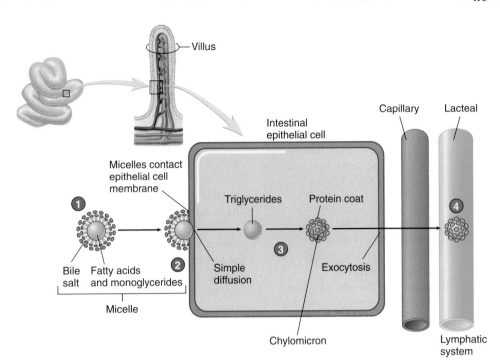

Process Figure 16.25

of the bile salts are directed toward the lipid particles, and the hydrophilic (water-loving) ends are directed outward toward the water environment. When a micelle comes into contact with the epithelial cells of the small intestine, the lipids, fatty acids, and monoglyceride molecules pass, by simple diffusion, from the micelles through cell membranes of the epithelial cells.

Once inside the intestinal epithelial cells, the fatty acids and monoglycerides are recombined to form triglycerides. These, and other lipids, are packaged inside a protein coat within the epithelial cells of the intestinal villi. The packaged lipids, called **chylomicrons** (kī-lō-mi′kron, *chylo,* juice + *micros,* small), leave the epithelial cells and enter the lacteals. Lacteals are lymphatic capillaries located within the intestinal villi. Lymph containing large amounts of absorbed lipid is called **chyle** (kīl, milky lymph). The lymphatic system carries the chyle to the bloodstream. Chylomicrons are transported to the liver, where the lipids are stored, converted into other molecules, or used as energy. They are also transported to adipose tissue, where they are stored until an energy source is needed elsewhere in the body.

High- and Low-Density Lipids

Cholesterol levels in the blood are of great concern to many adults. Cholesterol levels of less than 180 milligrams (mg)/dL are considered low, which is usually good, although extremely low cholesterol levels can be harmful. Cholesterol levels above 200 mg/dL are considered to be too high. People with high blood cholesterol levels run a much greater risk of heart disease and stroke than people with low cholesterol levels. People with high levels should seek advice from their physician, reduce their intake of foods rich in cholesterol and other fats, and increase their level of exercise. Some people with very high cholesterol levels may have to take medication to reduce their cholesterol levels.

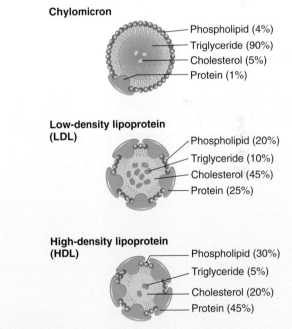

Figure 16.26

Fats are not soluble in water; thus they are transported in the blood as lipid–protein complexes, or **lipoproteins** (figure 16.26). **Low-density lipoproteins (LDLs)** carry cholesterol to the tissues for use by the cells. When LDLs are in excess, cholesterol is deposited on arterial walls. **High-density lipoproteins (HDLs),** on the other hand, transport cholesterol from the tissues to the liver, where cholesterol is removed from the bloodstream and broken down or excreted in bile. A high HDL/LDL ratio in the bloodstream is related to a lower risk of heart disease. Aerobic exercise is one way to elevate blood levels of HDL.

Clinical Focus Disorders of the Digestive Tract

Stomach

Vomiting results from activation of the vomiting reflex primarily by irritation of the stomach and small intestine. Action potentials travel through visceral sensory neurons to the vomiting center in the medulla oblongata. After the vomiting center is stimulated the following events occur:

1. A deep breath is taken.
2. The hyoid bone and larynx are elevated, opening the upper esophageal sphincter.
3. The opening of the larynx is closed.
4. The soft palate is elevated, closing the posterior opening into the nasal cavity.
5. The diaphragm and abdominal muscles are forcefully contracted, strongly compressing the stomach and increasing pressure in the stomach.
6. The lower esophageal sphincter is relaxed, and the gastric contents are forcefully expelled through the esophagus.

Ulcers

Approximately 5–12% of the population is affected by **peptic ulcers,** which are lesions in the lining of the stomach or duodenum. Most cases of peptic ulcer apparently result from the infection of a specific bacterium—*Helicobacter pylori.* It is also thought that the bacterium is involved in many cases of gastritis and gastric cancer. Conventional wisdom has focused for years on the notion that stress, diet, smoking, or alcohol cause excess acid secretion in the stomach, resulting in ulcers. Even today, antacids are used to treat 90% of all ulcers, with several billion dollars spent on antacids in the United States each year. Antacid therapy does cause the ulcer to heal in most cases. With antacid treatment, however, there is a 50% incidence of relapse within 6 months and a 95% incidence of relapse after 2 years. On the other hand, studies using antibiotic therapy together with bismuth, which increases mucus and HCO_3^- and decreases pepsin activity, and ranitidine (ră-nĭ′ti-dēn, Tritec) have demonstrated an eradication of 95% of gastric ulcers and 74% of duodenal ulcers within 2 months. Dramatically reduced relapse rates have also been obtained. One reason that bacterial involvement in ulcers was dismissed for such a long time is that it was mistakingly assumed that the extreme acid environment in the stomach would not allow bacteria to survive.

The infection rate from *H. pylori* in the United States population is about 1% per year of age: 30% of people that are 30 years old have the bacterium, and 80% of those age 80 are infected. Very little is known concerning how people become infected. Also, with such high rates of infection it is not known why only a small fraction of those infected actually develop ulcers. It may be that the several factors that increase gastric acid secretion predispose a person who is infected by the bacterium to actually develop an ulcer.

Peptic ulcer is classically viewed as a condition in which the stomach acids digest the mucosal lining of the GI tract itself. The most common site of a peptic ulcer is near the pylorus, usually on the duodenal side. This is sometimes called a **duodenal ulcer.** Approximately 80% of all peptic ulcers are actually duodenal. Ulcers that occur in the stomach are often called **gastric ulcers.** Gastric ulcers usually occur along the lesser curvature of the stomach or at the point at which the esophagus enters the stomach. The most common contributing factor to developing peptic ulcers is the oversecretion of gastric juices relative to the degree of mucous and alkaline protection of the small intestine.

People who experience severe anxiety over a long period are the most prone to develop duodenal ulcers. They often have a rate of gastric secretion between meals that is as much as 15 times the normal amount. This secretion results in highly acidic chyme entering the duodenum. The duodenum is usually protected by sodium bicarbonate, secreted mainly by the pancreas, which neutralizes the chyme. When large amounts of acid enter the duodenum, however, the sodium bicarbonate is not adequate to neutralize it. The acid tends to reduce the mucous protection of the duodenum, perhaps leaving that part of the digestive tract open to action of *H. pylori,* which may further destroy the mucous lining.

In some patients with gastric ulcers, normal or even low levels of gastric hydrochloric acid secretion often occur. The stomachs of these patients, however, have reduced resistance to their own acid. Such inhibited resistance can result from excessive ingestion of alcohol or aspirin.

Reflux of duodenal contents into the pylorus can also cause **gastric ulcers.** In this case, bile, which is present in the reflux, has a detergent effect that reduces gastric mucosal resistance to acid, and to bacteria.

Liver

Cirrhosis (sir-ō′sis) is a major disease of the liver. It is characterized by damage and death of hepatic cells and replacement by connective tissue. Cirrhosis results in loss of normal liver function and interference with blood flow through the liver. Cirrhosis is a common consequence of alcoholism.

Hepatitis (hep-ă-tī′tis) is an inflammation of the liver that can result from alcohol consumption or viral infection. If not corrected, liver cells can die and be replaced by scar tissue, resulting in loss of liver function. Death can result from liver failure.

Viral hepatitis is the second most frequently reported infectious disease in the United States. It is caused by any of seven immunologically distinct viruses. The most common are hepatitis types A, B, and C. **Hepatitis A** (infectious hepatitis) is usually transmitted by poor sanitation practices or from mollusks, such as oysters, living in contaminated waters. **Hepatitis B** (serum hepatitis) is usually transmitted through blood or other body fluids, such as through sexual contact or when blood is transferred through a contaminated hypodermic needle.

Symptoms of hepatitis include nausea, diarrhea, loss of appetite, abdominal pain, fever, chills, and malaise. **Jaundice** is seen in about two-thirds of the cases, with yellowing of the skin and sclera of the eyes resulting from the accumulation of bile pigments in those tissues (see chapter 11). **Hepatitis C** is often a chronic disease leading to cirrhosis and possibly cancer of the liver.

Cholesterol, secreted by the liver into the bile, can precipitate in the gallbladder to produce **gallstones.** Occasionally a gallstone may pass out of the gallbladder and enter the cystic duct, blocking release of the bile. Such a condition interferes with normal digestion, and often the gallbladder must be removed surgically.

Intestine

Inflammatory bowel disease (IBD) is the general name given to either Crohn's disease or ulcerative colitis. IBD occurs at a rate in Europe and North America of approximately 4–8 new cases per 100,000 people per year, which is much higher than in Asia and Africa. Males and females are affected about equally. IBD is of unknown cause, but infectious, autoimmune, and hereditary factors have been implicated. **Crohn's disease** involves localized inflammatory degeneration that may occur anywhere along the digestive tract but most commonly involves the distal ileum and proximal colon. The degeneration involves the entire thickness of the digestive tract wall. The intestinal wall often becomes thickened, constricting the lumen, with ulcerations and fissures in the damaged areas. The disease causes diarrhea, abdominal pain, fever, and weight loss. Treatment centers around antiinflammatory drugs, but other treatments, including avoiding foods that increase symptoms and even surgery are employed. **Ulcerative colitis** is limited to the mucosa of the large intestine. The involved mucosa exhibits inflammation, including edema, vascular congestion, hemorrhage, and the accumulation of plasma cells, lymphocytes, neutrophils, and eosinophils. Patients may experience abdominal pain, fever, malaise, fatigue, and weight loss, as well as diarrhea and hemorrhage. In rare cases, severe diarrhea and hemorrhage may require transfusions. Treatment includes the use of antiinflammatory drugs and, in some cases, avoiding foods that increase symptoms.

Irritable bowel syndrome (IBS) is a disorder of unknown cause in which intestinal mobility is abnormal. The disorder accounts for over half of all referrals to gastroenterologists. Male and female children are affected equally, but adult females are affected twice as often as males. IBS patients experience abdominal pain mainly in the left lower quadrant, especially after eating. They also have alternating bouts of constipation and diarrhea. There is no specific histopathology in the digestive tracts of IBS patients. There are no anatomic abnormalities, no indication of infection, and no sign of metabolic causes. Patients with IBS appear to exhibit greater than normal levels of psychological stress or depression, and show increased contractions of the esophagus and small intestine during times of stress. There is a high familial incidence. Some patients might present with a history of traumatic events such as physical or sexual abuse. Treatments include psychiatric counseling and stress management, diets with increased fiber and limited gas-producing foods, loose clothing, and in some patients, drugs that reduce parasympathetic stimulation of the digestive system may be useful.

Malabsorption syndrome (sprue) is a spectrum of disorders of the small intestine that results in abnormal nutrient absorption. In some people, one type of malabsorption results from the effects of gluten, an insoluble protein present in certain types of grains. The reaction to gluten can destroy newly formed epithelial cells, causing the villi to become blunted and the intestinal surface area to decrease. As a result, the intestinal epithelium is less capable of absorbing nutrients. Tropical malabsorption is apparently caused by bacteria, although no specific bacterium has been identified.

Infections of the Digestive Tract

Staphylococcal (staf'i-lō-kok'ăl) **food poisoning** occurs when toxin from the bacteria *Staphylococcus aureus* is ingested. The bacteria usually come from the hands of a person preparing the food. If food is cooked in large volumes at low temperatures (below 60°C) or is allowed to sit for an extended time, the bacteria can reproduce and form toxins. Reheating can eliminate the bacteria but not the toxins. Staphylococcal food poisoning is characterized by nausea, vomiting, and diarrhea from 1 to 6 hours after the contaminated food is ingested.

Salmonellosis (sal'mō-nel-ō'sis) is a disease caused by *Salmonella* bacteria. They are ingested with contaminated food (usually meat, poultry, or milk) and grow in the digestive tract. The disease symptoms are usually seen 18–36 hours after the contaminated food has been consumed. Symptoms include nausea, fever, abdominal pain, and diarrhea. The bacteria are generally destroyed by cooking food to temperatures greater than 68°C.

Typhoid (tī'foyd) **fever** is caused by a particularly virulent strain of salmonella bacteria, *Salmonella typhi*. The bacteria can cross the intestinal wall and invade other tissues. The incubation period is normally about 2 weeks. Symptoms include severe fever and headaches, as well as diarrhea. Poor sanitation practices are the main source of contamination, and typhoid fever is still a leading cause of death in many underdeveloped countries.

Cholera (kol'er-ă) is caused by a bacterium (*Vibrio cholerae*) obtained from contaminated water that infects the small intestine. The bacteria produce a toxin that stimulates the secretion of chlorides, bicarbonates, and water into the intestinal tract, resulting in severe diarrhea. The loss of as much as 12–20 L of fluid and ions per day causes shock, circulatory collapse, and even death. Cholera was common in the United States and Europe in the 1800s but is not very common in western countries today. Cholera is still a major problem in Asia, particularly in India.

Giardiasis (jē-ar-dī'ă-sis) is a disease caused by a protozoan (*Giardia lamblia*)

continued

continued

that invades the intestine. Symptoms include nausea, abdominal cramps, weakness, weight loss, and malaise and can last for several weeks. The disease is transmitted in the form of spores in the feces of humans and other animals. People who drink unfiltered water from wilderness streams containing spores are often infected.

Intestinal parasites are not uncommon in humans, especially under conditions of poor sanitation. **Tapeworms** (several genera) can infect the digestive tract by way of undercooked beef, pork, or fish. The tapeworms attach to the intestinal wall and may live in the intestine for 25 years, reaching lengths of 6 m. There are few symptoms beyond a vague abdominal discomfort.

Pinworms (*Enterobius vermicularis*) are common in humans. The tiny worm lives in the digestive tract but migrates out of the anus to lay its eggs. This causes a local itching, and the eggs can be spread by contaminated fingers to numerous surfaces. Eggs resist dehydration and can be picked up from contaminated surfaces by other people. It is common for entire households to

be contaminated if one child contracts the disease.

Hookworms (*Ancylostoma*) attach to the intestinal wall and feed on the blood and tissue of the host, rather than on partially digested food as other parasites do. Infection can cause anemia and lethargy. Because hookworms are spread through fecal contamination of the soil and bare skin contact with contaminated soil, improved sanitation and the practice of wearing shoes has greatly decreased the incidence of hookworm infection.

Ascariasis (as′kă-rī′ă-sis) is caused by a roundworm (*Ascaris lumbricoides*) and is fairly common in the United States. Ingested eggs hatch in the upper intestine into wormlike larvae that pass into the bloodstream and then into the lungs, where they can cause pulmonary symptoms. Extremely large numbers can cause pneumonia. The larvae enter the throat and are swallowed, thereby returning to the intestinal tract. Adults in the intestinal tract cause few symptoms. The adult worms, however, measuring up to 30 cm, migrate. In some cases, they may emerge from the anus, or they may cut their way through the intestinal wall and infect the abdominal cavity.

Constipation (kon-sti-pā′shŭn) is the slow movement of feces through the large intestine. The feces often become dry and hard because of the increased fluid absorption during the extended time they are retained in the large intestine. Constipation often results from irregular defecation patterns that develop after a prolonged time of inhibiting normal defecation reflexes. Spasms of the sigmoid colon resulting from irritation also can result in slow feces movement and constipation. A diet high in fiber can help prevent constipation.

Diarrhea (dī-ă-rē′ă) is a condition in which the intestinal mucosa secretes large amounts of water and ions in addition to mucus. This condition occurs when the large intestine is irritated and inflamed, such as in patients with **enteritis** (en-ter-ī′tis) bacterial infection of the bowel. Although diarrhea increases fluid and ion loss, it also moves the infected feces out of the intestine more rapidly and speeds recovery from the disease.

Dysentery (dis′en-tār-ē) is a severe form of diarrhea in which blood or mucus is present in the feces. Dysentery can be caused by bacteria, protozoa, or amoebae.

Proteins

Proteins are chains of amino acids. They are found in most of the plant and animal products we eat. **Pepsin** is an enzyme secreted by the stomach, which breaks down **proteins,** producing shorter amino acid chains called **polypeptides** (see figure 16.23). Only about 10–20% of the total ingested protein is digested by pepsin. After the remaining proteins and polypeptide chains leave the stomach and enter the small intestine, the enzymes **trypsin, chymotrypsin,** and **carboxypeptidase** produced by the pancreas in their inactive forms and activated in the intestine, continue the digestive process. These enzymes produce polypeptides, which are broken down further into tripeptides (three amino acids), dipeptides (two amino acids) or single **amino acids** by **peptidases.** Peptidases are digestive enzymes bound to the microvilli of the small intestine. They act on smaller peptides to release amino acids.

Absorption of tripeptides, dipeptides, or individual amino acids occurs through the intestinal epithelial cells by cotransport (figure 16.27). Within the intestinal epithelial cells, tripeptides and dipeptides are broken down into amino acids. The amino acids then enter blood capillaries in the villi

and are carried by the hepatic portal system to the liver. The amino acids may be modified in the liver, or they may be released into the bloodstream and distributed throughout the body.

Amino acids are actively transported into the various cells of the body. This transport is stimulated by growth hormone and insulin. Most amino acids are used as building blocks to form new proteins, but some may be metabolized, with some of the released energy used to produce ATP. The body cannot store excess amino acids. Instead, they are partially broken down and used to synthesize glycogen or fat, which can be stored. The body can store only small amounts of glycogen, so most of the excess amino acids are converted to fat.

Water and Minerals

Approximately 9 L of water enters the digestive tract each day (figure 16.28). We ingest about 2 L in food and drink, and the remaining 7 L is from digestive secretions. Approximately 92% of that water is absorbed in the small intestine, about 7% is absorbed in the large intestine, and about 1% leaves the body in the feces. Water can move in either direction by osmosis across the wall of the gastrointestinal tract. The direction of its

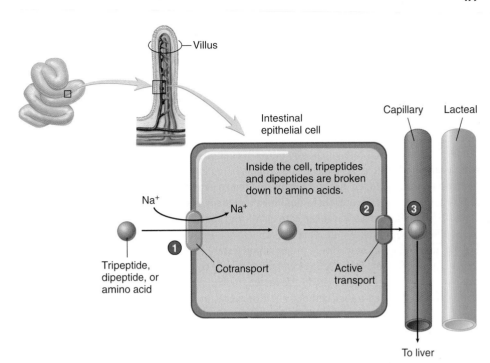

Amino acid transport

1. Tripeptides, dipeptides, and amino acids are absorbed by cotransport into intestinal epithelial cells.

2. Amino acids move out of intestinal epithelial cells by active transport.

3. They enter the capillaries of the intestinal villi and are carried through the hepatic portal vein to the liver.

Process Figure 16.27 Transport of Proteins Across the Intestinal Epithelium

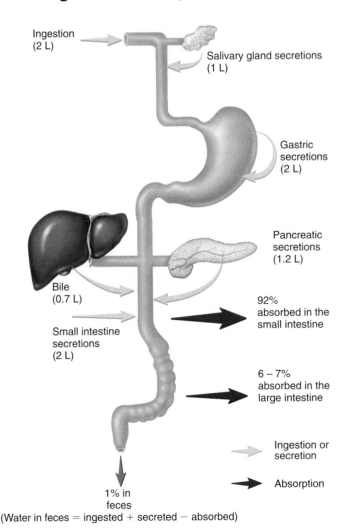

(Water in feces = ingested + secreted − absorbed)

Figure 16.28 Fluid Volumes in the Digestive Tract

movement is determined by osmotic gradients across the epithelium. When the chyme is diluted, water moves out of the intestine into the blood. If the chyme is concentrated and contains little water, water moves into the lumen of the small intestine.

Sodium, potassium, calcium, magnesium, and phosphate ions are actively transported from the small intestine. Vitamin D is required for the transport of Ca^{2+}. Negatively charged Cl^- move passively through the wall of the duodenum and jejunum with the positively charged Na^+, but Cl^- are actively transported from the ileum.

Effects of Aging on the Digestive System

As a person ages, the connective tissue layers of the digestive tract, the submucosa and serosa, tend to thin. The blood supply to the digestive tract decreases. There is also a decrease in the number of smooth muscle cells in the muscularis, resulting in decreased motility in the digestive tract. In addition, goblet cells within the mucosa secrete less mucus. Glands along the digestive tract, such as the gastric glands, the liver, and the pancreas, also tend to secrete less with age. These changes by themselves don't appreciably decrease the function of the digestive system.

Through the years the digestive tract, like the skin and lungs, is directly exposed to materials from the outside environment. Some of those substances can cause mechanical damage to the digestive tract and others may be toxic to the tissues. Because the connective tissue of the digestive tract becomes thin with age and because the protective mucus covering is reduced, the digestive tract of elderly people becomes less and less protected from

Systems Pathology
Diarrhea

While on vacation in a foreign country, Mr. T. was shopping with his wife when he started to experience sharp pains in his abdominal region. He also began to feel hot and sweaty and felt an extreme urge to defecate (figure A). Mr. T. anxiously inquired about the nearest facility. Once his immediate crisis was taken care of, Mr. and Mrs. T. went back to their hotel room, where they remained while Mr. T. recovered. During the next 2 days his stools were frequent and watery. He also vomited a couple of times. Because they were in a foreign country, Mr. T. did not consult a physician. Instead, he rested, took plenty of fluids, and was feeling much better, although a little weak, in a couple of days.

Background

Diarrhea is one of the most common complaints in clinical medicine, and diarrhea affects more than half of the tourists in developing countries. Diarrhea is defined as any change in bowel habits in which stool frequency or volume is increased or in which stool fluidity is increased. Diarrhea is not itself a disease, but is a symptom of a wide variety of disorders. In foreign countries, diarrhea may be caused by microorganisms such as bacteria or amebae, or may result from the ingestion of food to which the digestive tract is not accustomed. Normally, about 600 mL of fluid enters the colon each day and all but 150 mL is reabsorbed. The loss of more than 200 mL of stool per day is considered abnormal.

Mucus secretion by the colon increases dramatically in response to diarrhea. This mucus contains large quantities of bicarbonate ions, which come from the dissociation of carbonic acid into bicarbonate ions and hydrogen ions within the blood supply to the colon. The bicarbonate ions enter the mucus secreted by the colon, whereas the hydrogen ions remain in the circulation and, as a result, the blood pH decreases. Thus, a condition called metabolic acidosis can develop (see chapter 18).

Diarrhea in tourists usually results from the ingestion of food or water contaminated with bacteria or bacterial toxins. Acute diarrhea is

Figure A   Many Tourists Develop Diarrhea.

defined as lasting less than 2–3 weeks, and diarrhea lasting longer than that is considered chronic. Acute diarrhea is usually self-limiting, but some forms of diarrhea can be fatal if not treated. Diarrhea results from either a decrease in fluid absorption in the gut or an increase in fluid secretion. Some bacterial toxins and other chemicals can also cause an increase in bowel motor activity. As a result, chyme is moved more rapidly through the digestive tract, less nutrients and water are absorbed out of the small intestine, and more water enters the colon. Symptoms can occur in as little as 1–2 hours after bacterial

these outside influences. In addition, the mucosa of elderly people tends to heal more slowly following injury. The liver's ability to detoxify certain chemicals tends to decline, the ability of the hepatic phagocytic cells to remove particulate contaminants decreases, and the liver's ability to store glycogen decreases.

This overall decline in the defenses of the digestive tract with advancing age leaves elderly people more susceptible to in-

fections and to the effects of toxic agents. Elderly people are more likely to develop ulcerations and cancers of the digestive tract. Colorectal cancers, for example, are the second leading cause of cancer deaths in the United States, with an estimated 135,000 new cases and 57,000 deaths each year.

Gastroesophageal reflux disorder (GERD) increases with advancing age. It is probably the main reason that elderly people

System	Interactions with Digestive System
Integumentary	Pallor due to vasoconstriction of blood vessels in the skin, resulting from a decrease in blood volume. Pallor and sweating increase in response to abdominal pain and anxiety.
Muscular	Muscular weakness may result due to ion loss, metabolic acidosis, fever, and general malaise. The stimulus to defecate may become so strong that it overcomes the voluntary control mechanisms.
Nervous	Local reflexes in the colon respond to increased colon fluid volume by stimulating mass movements and the defecation reflex. Abdominal pain, much of which is felt as referred pain, can occur as the result of inflammation and distention of the colon. Decreased function due to ion loss. Reduced blood volume stimulates a sensation of thirst in the CNS.
Endocrine	A decrease in extracellular fluid volume, due to the loss of fluid in the feces, stimulates the release of hormones (antidiuretic hormone from the posterior pituitary and aldosterone from the adrenal cortex) that increase water retention and sodium reabsorption in the kidney. In addition, decreased extracellular fluid volume and anxiety result in increased release of epinephrine and norepinephrine from the adrenal medulla.
Cardiovascular	Movement of extracellular fluid into the colon results in a decreased blood volume. The reduced blood volume activates the baroreceptor reflex, which functions to maintain blood volume and blood pressure.
Lymphatic and Immune	White blood cells migrate to the colon in response to infection and inflammation. In the case of bacterial diarrhea, the immune response is initiated to begin production of antibodies against bacteria and bacterial toxins.
Respiratory	Increased HCO_3^- secretion and H^+ absorption reduces blood pH. As the result of reduced blood pH, the rate of respiration increases to eliminate carbon dioxide, which helps eliminate excess H^+.
Urinary	A decrease in urine volume and an increase in urine concentration results from activation of the baroreceptor reflex, which decreases blood flow to the kidney; antidiuretic hormone secretion, which increases water reabsorption in the kidney; and aldosterone secretion, which increases Na^+ and water reabsorption in the kidney. After a period of approximately 24 hours, the kidney is activated to compensate for metabolic acidosis by increasing hydrogen ion secretion and bicarbonate ion reabsorption.

toxins are ingested to as long as 24 hours or more for some strains of bacteria.

In cases of short-term acute diarrhea, the infectious agent is seldom identified. Nearly any bacterial species is capable of causing diarrhea. Some types of bacterial diarrhea include severe vomiting, whereas others do not. Some bacterial toxins also induce fever. Some viruses and amebic parasites can also cause diarrhea. In most cases, laboratory analysis of food or stool is necessary to identify the causal organism. In cases of mild diarrhea away from home, laboratory evaluation is not practical and therapy consists of treating the symptoms. Fluids and ions must be replaced, and consumption of fluids with ions is important. The diet should be limited to clear fluids during at least the first day or so. Bismuth subsalicylate (sŭb-să-lis′i-lāt), which increases mucus and HCO_3^- secretion and decreases pepsin activity, or loperamide (lō-per′ă-mīd), which slows intestinal motility, (except in cases of fever) may also be used to help combat diarrhea. Milk and milk products should be avoided. Breads, toast, rice, and baked fish or chicken can be added to the diet as the person improves. A normal diet can be resumed after 2–3 days.

P R E D I C T   5
Predict the effects of prolonged diarrhea on the circulatory system.

take antacids and inhibitors of hydrocloric acid secretion. Disorders that are not necessarily age-induced, such as hiatal hernia and irregular or inadequate esophageal motility, may be worsened by the effects of aging, because of a general decreased motility in the digestive tract.

The enamel on the surface of elderly people's teeth becomes thinner with age and may expose the underlying dentin.

In addition, the gingiva covering the tooth root recedes, exposing additional dentin. Exposed dentin may become painful and change the person's eating habits. Many elderly people also lose teeth, which can have a marked effect on eating habits unless artificial teeth are provided. The muscles of mastication tend to become weaker and, as a result, older people tend to chew their food less before swallowing.

Functions of the Digestive System   (p. 448)

The functions of the digestive system are to take in food, break down the food, absorb the digested molecules, and, thus, provide nutrients to the body.

Anatomy and Histology of the Digestive System   (p. 448)

The GI tract is composed of four tunics: mucosa, submucosa, muscularis, and serosa or adventitia.

Peritoneum   (p. 449)

1. The peritoneum is a serous membrane that lines the abdominal cavity and covers the organs.
2. Mesenteries are double layers of peritoneum that extend from the body wall to many of the abdominal organs.
3. Retroperitoneal organs are located behind the parietal peritoneum.

Oral Cavity, Pharynx, and Esophagus   (p. 450)
Anatomy of the Oral Cavity

1. The lips and cheeks are involved in mastication and speech.
2. The tongue is involved in speech, taste, mastication, and swallowing.
3. There are 32 permanent teeth, including incisors, canines, premolars, and molars. Each tooth consists of a crown, neck, and root.
4. The roof of the oral cavity is divided into the hard and soft palates.
5. Salivary glands produce serous and mucous secretions. The three pairs of large salivary glands are the parotid, submandibular, and sublingual glands.

Secretions of the Oral Cavity

Amylase in saliva starts starch digestion. Mucin provides lubrication.

Mastication

Mastication is accomplished by the teeth, which cut, tear, and crush the food.

Pharynx

The pharynx consists of the nasopharynx, oropharynx, and laryngopharynx.

Esophagus

The esophagus connects the pharynx to the stomach. The upper and lower esophageal sphincters regulate movement.

Deglutition

1. During the voluntary phase of deglutition, a bolus of food is moved by the tongue from the oral cavity to the pharynx.
2. During the pharyngeal phase of deglutition, the soft palate closes the nasopharynx, and the epiglottis closes the opening into the larynx. Pharyngeal muscles elevate the pharynx and larynx and then move the bolus to the esophagus.
3. During the esophageal phase of deglutition, a wave of constriction (peristalsis) moves the food down the esophagus to the stomach.

Stomach   (p. 456)
Anatomy of the Stomach

1. The stomach has a cardiac opening from the esophagus and a pyloric opening into the duodenum.
2. The wall of the stomach consists of three muscle layers: longitudinal, circular, and oblique.
3. Gastric glands produce mucus, hydrochloric acid, pepsin, gastrin, and intrinsic factor.

Secretions of the Stomach

1. Mucus protects the stomach lining.
2. Hydrochloric acid kills microorganisms and activates pepsin.
3. Pepsin starts protein digestion.
4. Intrinsic factor aids in vitamin B_{12} absorption.
5. Gastrin helps regulate stomach secretions and movements.

Regulation of Stomach Secretions

1. During the cephalic phase, the stomach secretions are initiated by the sight, smell, taste, or thought of food.
2. During the gastric phase, partially digested proteins or distention of the stomach also promotes secretion.
3. During the intestinal phase, acidic chyme in the duodenum stimulates neuronal reflexes and the secretion of hormones that induce and then inhibit gastric secretions. Secretin, gastric inhibitory polypeptide, and cholecystokinin inhibit gastric secretion.

Movement in the Stomach

1. Mixing waves mix the stomach contents with the stomach secretions to form chyme.
2. Peristaltic waves move the chyme into the duodenum.

Small Intestine   (p. 461)
Anatomy of the Small Intestine

1. The small intestine is divided into the duodenum, jejunum, and ileum.
2. Circular folds, villi, and microvilli greatly increase the surface area of the intestinal lining.
3. Goblet cells and duodenal glands produce mucus.

Secretions of the Small Intestine

1. Mucus protects against digestive enzymes and stomach acids.
2. Chemical or tactile irritation, vagal stimulation, and secretion stimulate intestinal secretion.

Movement in the Small Intestine

1. Segmental contractions occur over short distances and mix the intestinal contents.
2. Peristaltic contractions occur the length of the intestine and propel chyme through the intestine.

Absorption in the Small Intestine

Most absorption occurs in the duodenum and jejunum.

Liver and Pancreas   (p. 463)
Anatomy of the Liver

1. The liver consists of four lobes. It receives blood from the hepatic artery and the hepatic portal vein.
2. Branches of the hepatic artery and hepatic portal vein empty into hepatic sinusoids, which empty into a central vein in the center of each lobe. The central veins empty into hepatic veins, which exit the liver.
3. The liver is divided into lobules with portal triads at the corners. Portal triads contain branches of the hepatic portal vein, hepatic artery, and hepatic duct.
4. Hepatic cords, formed by hepatocytes, form the substance of each lobule. A bile canaliculus, between the cells of each cord, joins the hepatic duct system.
5. Bile leaves the liver through the hepatic duct system. The right and left hepatic ducts join to form the common hepatic duct. The gallbladder stores bile. The cystic duct joins the common hepatic duct to form the common bile duct. The common bile duct joins the pancreatic duct and empties into the duodenum.

Functions of the Liver

1. The liver produces bile, which contains bile salts that emulsify fats.
2. The liver stores and processes nutrients, produces new molecules, and detoxifies molecules.
3. The liver produces blood proteins.

Anatomy of the Pancreas

The pancreas is an endocrine and an exocrine gland. Its endocrine function is to control blood nutrient levels. Its exocrine function is to produce bicarbonate ions and digestive enzymes.

Functions of the Pancreas

1. The pancreas produces HCO_3^- and digestive enzymes.
2. Acidic chyme stimulates the release of a watery bicarbonate solution that neutralizes acidic chyme. Fatty acids and amino acids in the duodenum stimulate the release of pancreatic enzymes.

Large Intestine (p. 468)

Anatomy of the Large Intestine

1. The cecum forms a blind sac at the junction of the small and large intestines. The appendix is a blind sac off the cecum.
2. The colon consists of ascending, transverse, descending, and sigmoid portions.
3. The large intestine contains mucus-producing crypts.
4. The rectum is a straight tube that ends at the anal canal.
5. The anal canal is surrounded by an internal anal sphincter (smooth muscle) and an external anal sphincter (skeletal muscle).

Functions of the Large Intestine

1. The function of the large intestine is feces production and water absorption.
2. It takes much longer for material to move through the large intestine than the small intestine.
3. In the colon, chyme is converted to feces.
4. Mass movements occur three to four times a day.
5. Defecation is the elimination of feces. Reflex activity moves feces through the internal anal sphincter. Voluntary activity regulates movement through the external anal sphincter.

Digestion, Absorption, and Transport (p. 470)

Digestion is the chemical breakdown of organic molecules into their component parts. After the molecules are digested, some diffuse through the intestinal wall; others must be transported across the intestinal wall.

Carbohydrates

1. Polysaccharides are split into disaccharides by salivary and pancreatic amylases.
2. Disaccharides are broken down to monosaccharides by disaccharidases on the surface of the intestinal epithelium.
3. Monosaccharides are absorbed by active transport into the blood and carried by the hepatic portal vein to the liver.
4. Glucose is carried in the blood and enters most cells by facilitated diffusion. Insulin increases the rate of glucose transport into most cells.

Lipids

1. Bile salts emulsify lipids.
2. Pancreatic lipase breaks down lipids. The breakdown products aggregate with bile salts to form micelles.
3. Micelles come into contact with the intestinal epithelium, and their contents diffuse into the cells, where they are packaged and released into the lacteals.
4. Lipids are stored in adipose tissue and in the liver, which release the lipids into the blood when energy sources are needed elsewhere in the body.

Proteins

1. Proteins are split into small polypeptides by enzymes secreted by the stomach and pancreas, and on the surface of intestinal cells.
2. Peptidases on the surface of intestinal epithelial cells complete the digestive process.
3. Amino acids are absorbed into intestinal epithelial cells.
4. Amino acids are actively transported into cells under the influence of growth hormone and insulin.
5. Amino acids are used to build new proteins or as a source of energy.

Water and Minerals

Water can move either direction across the intestinal wall, depending on osmotic conditions. Approximately 99% of the water entering the intestine is absorbed. Most minerals are actively transported across the intestinal wall.

Effects of Aging on the Digestive System (p. 477)

1. With advancing age, the layers of the GI tract thin, and the blood supply decreases.
2. There is also decreased mucus secretion and decreased motility in the GI tract.
3. There is also a gradual decline in the defenses of the digestive tract, leaving it more sensitive to infection and the effects of toxic agents.
4. Enamel and gingiva are reduced with age, exposing dentin, which may become painful and affect eating habits.

R E V I E W   A N D   C O M P R E H E N S I O N

1. What are the functions of the digestive system?
2. What are the major layers, or tunics, of the digestive tract?
3. What are the peritoneum, mesenteries, and retroperitoneal organs?
4. List the functions of the lips, cheeks, and tongue.
5. What are the deciduous and permanent teeth? Name the different kinds of teeth.
6. Describe the parts of a tooth. What are dentin, enamel, and pulp?
7. What are the hard and soft palates? What is the function of the palate?
8. Name and give the location of the three pairs of salivary glands.
9. What are the functions of saliva?
10. Where is the esophagus located?
11. Describe the three phases of swallowing.
12. Describe the parts of the stomach. How are the stomach muscles different from those in the esophagus?
13. What are gastric pits and gastric glands? Name the secretions they produce.
14. List the stomach secretions, and give their functions.
15. Describe the three phases of stomach secretion.
16. What are the two kinds of stomach movements? What do they accomplish?
17. Name and describe the three parts of the small intestine.
18. What are circular folds, villi, and microvilli in the small intestine? What are their functions?

19. List the secretions of the small intestine, and give their functions.

20. Describe the kinds of movements in the small intestine, and explain what they accomplish.

21. Describe the anatomy and location of the liver and pancreas. Describe their duct systems.

22. Describe the functions of the liver.

23. Name the exocrine secretions of the pancreas. What are their functions?

24. Describe the parts of the large intestine.

25. How is chyme converted to feces?

26. Describe the defecation reflex.

27. Describe carbohydrate digestion, absorption, and transport.

28. Describe the role of bile salts in lipid digestion and absorption.

29. Describe protein digestion and amino acid absorption. What enzymes are responsible for the digestion?

30. Describe the movement of water into and out of the digestive tract.

31. Describe the effects of aging on the digestive system.

C R I T I C A L   T H I N K I N G

1. While anesthetized, patients sometimes vomit. Given that the anesthetic eliminates the swallowing reflex, explain why it is dangerous for an anesthetized patient to vomit.

2. Achlorhydria (ā-klōr-hī′drē-ă) is a condition in which the stomach stops producing hydrochloric acid and other secretions. What effect does achlorhydria have on the digestive process?

3. Victor Worrystudent developed a duodenal ulcer during final examination week. Describe the possible reasons. Explain what

habits could have contributed to the ulcer, and recommend a reasonable remedy.

4. Gallstones sometimes obstruct the common bile duct. What would be the consequences of such a blockage?

5. Many people have a bowel movement shortly after a meal, especially breakfast. Why does this occur?

Answers in Appendix D

Visit this textbook's website at www.mhhe.com/seeleyess6 for practice quizzes, animations, interactive learning exercises, and other study tools.

McGraw-Hill offers a study CD that features interactive cadaver dissection. *Anatomy & Physiology Revealed* includes cadaver photos that allow you to peel away layers of the human body to reveal structures beneath the surface. This program also includes animations, radiologic imaging, audio pronunciations, and practice quizzing.

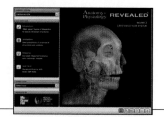

Chapter Outline and Objectives

Nutrition   (p. 484)

1. Define nutrition, essential nutrient, and kilocalorie.

2. For carbohydrates, lipids, and proteins, describe their dietary sources, their uses in the body, and the daily recommended amounts of each in the diet.

3. List the common vitamins and minerals and give a function for each.

Metabolism   (p. 492)

4. Define metabolism, anabolism, and catabolism.

5. List three ways in which enzyme activity is controlled.

6. Describe glycolysis and name its products.

7. Describe the citric acid cycle and its products.

8. Describe the electron-transport chain and how ATP is produced in the process.

9. Explain how the breakdown of glucose yields two ATP molecules in anaerobic respiration and 38 ATP molecules in aerobic respiration.

10. Describe the basic steps involved in using lipids and amino acids as an energy source.

11. Differentiate between the absorptive and postabsorptive metabolic states.

12. Define metabolic rate.

Body Temperature Regulation   (p. 502)

13. Describe heat production and regulation in the body.

Nutrition, Metabolism, and Body Temperature Regulation

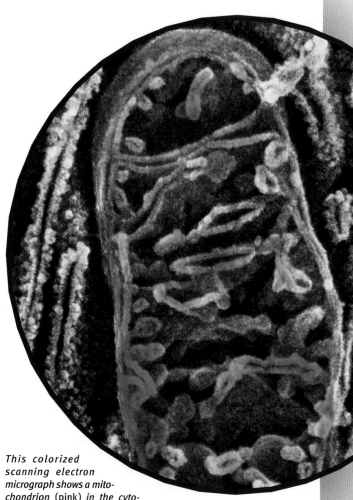

This colorized scanning electron micrograph shows a mito-chondrion (pink) *in the cyto-plasm of an intestinal epithelial cell. The mitochondrion has an outer and inner membrane. The inner membrane has numerous folds that project into the interior of the mito-chondrion. Enzymes, necessary for producing ATP, are located in these folds.*

We are often more concerned with the taste of food than with its nutritional value when choosing from a menu or when selecting food to prepare. Knowing about nutrition is important, however, because the food we eat provides us with the energy and the building blocks necessary to synthesize new molecules. What happens if we don't obtain enough vitamins, or if we eat too much sugar and fats? Health claims about foods and food supplements bombard us every day. Which ones are ridiculous, and which ones have merit? A basic understanding of nutrition can help us to answer these and other questions so that we can develop a healthy diet, and it allows us to know which questions currently do not have good answers.

Nutrition

Nutrition (noo-trish′ŭn, to nourish) is the process by which food is taken into and used by the body, and it includes digestion, absorption, transport, and metabolism. Nutrition is also the study of food and drink requirements for normal body function.

Nutrients

Nutrients are the chemicals taken into the body that provide energy and building blocks for new molecules. Some substances in food are not nutrients but provide bulk (fiber) in the diet. Nutrients can be divided into six major classes: carbohydrates, lipids, proteins, vitamins, minerals, and water. Carbohydrates, proteins, and lipids are the major organic nutrients and are broken down by enzymes into their individual subunits during digestion. Subsequently, many of these subunits are broken down further to supply energy, whereas other subunits are used as building blocks for making new carbohydrates, proteins, and lipids. Vitamins, minerals, and water are taken into the body without being broken down. They are essential participants in the chemical reactions necessary to maintain life. Some nutrients are required in fairly substantial quantities, and others, called **trace elements,** are required in only minute amounts.

Essential nutrients are nutrients that must be ingested because the body cannot manufacture them or is unable to manufacture adequate amounts of them. The essential nutrients include certain amino acids, certain fatty acids, most vitamins, minerals, water, and some carbohydrates. The term essential does not mean that only the essential nutrients are required by the body. Other nutrients are necessary, but if they are not ingested, they can be synthesized from the essential nutrients. Most of this synthesis takes place in the liver, which has a remarkable ability to transform and manufacture molecules. A balanced diet consists of enough nutrients in the correct proportions to support normal body functions.

Every 5 years the Dietary Guidelines Advisory Committee makes its recommendations on what we should eat to be healthy. The latest recommendations, "The Dietary Guidelines for Americans 2005," were published in January 2005. Unlike the previous single "food guide pyramid," there are now 12 pyramids that take into account a person's age, sex, and activity level. Thus, you can pick the pyramid, called MyPyramid, that best describes you (see www.mypyramid.gov). All of the new pyramids have the same form (figure 17.1). Six colored bands represent the approximate, recommended proportions of

Figure 17.1   **MyPyramid**

The pyramid suggests the approaches to a healthy diet: eat different foods (different colored bands), eat different amounts of each food type (band width), eat in moderation (bands narrow from bottom to top), use fats and sugars sparingly (the wide base stands for foods with little or no solid fats or added sugars), and exercise (stick figure climbing stairs).

Source: U.S. Department of Agriculture

grains (orange), vegetables (green), fruits (red), fats and oils (yellow), milk and milk products (blue), and meat and beans (purple). A balanced diet includes a variety of foods from each of the major food groups. Variety is necessary because no one food contains all of the nutrients necessary for health. Moderation is indicated by the narrowing of each food group from bottom to top. The wider base represents foods with little or no solid fats, added sugars, or caloric sweeteners. The climbing stick figure stresses the importance of daily exercise.

Benefits of a Healthy Diet

Two studies completed in 2000 compared the eating habits of 51,529 men and 67,272 women to the government's Healthy Eating index, a measure of how well a diet conformed to recommended dietary guidelines. Those who ate the best, according to the index, were compared with those who ate the worst. Men who ate best had a 28% reduction in heart disease and an 11% decrease in chronic diseases compared with men who ate worst. Women who ate best had a 14% reduction in heart disease but no significant decrease in chronic diseases compared with women who ate worst. There was no significant difference in cancer rates between the men and women who ate best compared with those who ate worst.

Kilocalories

The energy stored within the chemical bonds of certain nutrients can be used by the body. A **calorie** (kal′ō-rē, heat) **(cal)** is the amount of energy (heat) necessary to raise the temperature of 1 gram (g) of water 1°C. A **kilocalorie** (kil′ō-kal-ō-rē) **(kcal)** is 1000 cal and is used to express the larger amounts of energy supplied by foods and released through metabolism. For example, one slice of white bread contains about 75 kcal, one cup of whole milk contains 150 kcal, a banana contains 100 kcal, a hot dog contains 170 kcal (not counting the bun and dressings), a McDonald's Big Mac has 563 kcal, and a soft drink adds another 145 kcal. For each gram of carbohydrate or protein metabolized by the body, about 4 kcal of energy is released. Fats contain more energy per unit of weight than carbohydrates and proteins, and yield about 9 kcal/g. Table 17.1 lists the kilocalories supplied by some typical foods. A typical diet in the United States consists of 50% to 60% carbohydrates, 35% to 45% fats, and 10% to 15% protein. Table 17.1 also lists the carbohydrate, fat, and protein composition of some foods.

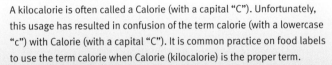

What Is a Calorie?

A kilocalorie is often called a Calorie (with a capital "C"). Unfortunately, this usage has resulted in confusion of the term calorie (with a lowercase "c") with Calorie (with a capital "C"). It is common practice on food labels to use the term calorie when Calorie (kilocalorie) is the proper term.

Carbohydrates

Sources in the Diet

Carbohydrates (kar-bō-hī′drātz) include monosaccharides, disaccharides, and polysaccharides (see chapter 2). Although most of the carbohydrates we ingest are derived from plants, lactose is derived from animals. The most common monosaccharides in the diet are glucose and fructose. Plants capture energy from sunlight and use the energy to produce glucose, which can be found in vegetables, fruits, molasses, honey, and syrup. Fructose is most often derived from fruits and berries.

The disaccharide sucrose (table sugar) is what most people think of when they use the term sugar. Sucrose consists of one glucose and one fructose molecule joined together, and its principal sources are sugarcane and sugar beets. Maltose (malt sugar), derived from germinating cereals, is a combination of two glucose molecules, and lactose (milk sugar) consists of one glucose molecule and one galactose molecule.

Complex carbohydrates are large polysaccharides, which are composed of long chains of glucose (see figure 2.11). Examples are starch, glycogen, and cellulose, which differ from one another in the arrangement of the glucose molecules and the structure of the chemical bonds holding them together.

Starch is an energy-storage molecule in plants and is found primarily in vegetables, fruits, and grains. Glycogen is an energy-storage molecule in animals and is located primarily in muscle and in the liver. By the time meats are processed, they contain little, if any, glycogen because it is used up by the dying muscle cells (see Anaerobic Respiration on p. 494). Cellulose forms the cell walls surrounding plant cells.

Uses in the Body

During digestion, polysaccharides and disaccharides are split into monosaccharides, which are absorbed into the blood (see chapter 16). Humans have enzymes that can break the bonds between the glucose molecules of starch and glycogen but do not have enzymes necessary to digest cellulose. It is important to thoroughly cook or chew plant matter. Cooking and chewing break down the plant cell walls and expose the starches that are contained inside the cells to digestive enzymes. The undigested cellulose provides fiber, or "roughage," which increases the bulk of feces, making it easier to defecate.

Fructose and other monosaccharides absorbed into the blood are converted into glucose by the liver. **Glucose,** whether absorbed directly from the digestive tract or synthesized by the liver, is an energy source used to produce ATP. Because the brain relies almost entirely on glucose for its energy, blood glucose levels are carefully regulated.

If an excess amount of glucose is present in the diet, the glucose is used to make glycogen, which is stored in muscle and in the liver. Glycogen can be rapidly converted back to glucose when energy is needed. Because cells can store only a limited amount of glycogen, additional glucose that is ingested is converted into fat for long-term storage in adipose tissue.

In addition to being used as an energy source, sugars form part of deoxyribonucleic acid (DNA), ribonucleic acid (RNA), and ATP molecules. Sugars also combine with proteins to form glycoproteins, some of which function as receptor molecules on the outer surface of the plasma membrane.

Recommended Consumption

According to the Dietary Guidelines Advisory Committee, the **Acceptable Macronutrient Distribution Range (AMDR)** for carbohydrates is 45% to 65% of total kilocalories. Although a minimum level of carbohydrates is not known, it is assumed that amounts of 100 g or less per day result in overuse of the body's proteins and fats for energy sources. Because muscles are primarily protein, the use of proteins for energy can result in the breakdown of muscle tissue. The extensive use of fats as an energy source can result in acidosis (see chapter 18).

Complex carbohydrates are recommended in the diet because starchy foods often contain other valuable nutrients, such as vitamins and minerals, and because the slower rate of digestion and absorption of complex carbohydrates does not result in large increases and decreases in blood glucose levels as the consumption of large amounts of simple sugars does. Foods containing large amounts of simple sugars, such as soft drinks and candy, are rich in carbohydrates, but they have few other nutrients. For example, a typical soft drink is mostly sucrose, containing 9 teaspoons of sugar per 12-oz container. In excess, the consumption of these kinds of foods usually results in obesity and tooth decay.

Lipids

Sources in the Diet

Lipids (lip′idz) include triacylglycerides, steroids, phospholipids, and fat-soluble vitamins. **Triglycerides** (trī-glis′er-īdz), also called **triacylglycerols** (trī-as′il-glis′er-olz), are the most

Table 17.1   Food Composition

Food	Quantity	Food Energy (kcal)	Carbohydrate (g)	Fat (g)	Protein (g)
Dairy Products					
Whole milk (3.3% fat)	1 cup	150	11	8	8
Low fat milk (2% fat)	1 cup	120	12	5	8
Butter	1 T	100	—	12	—
Grain					
Bread, white enriched	1 slice	75	24	1	2
Bread, whole wheat	1 slice	65	14	1	3
Fruit					
Apple	1	80	20	1	—
Banana	1	100	25	—	1
Orange	1	65	16	—	1
Vegetables					
Corn, canned	1 cup	140	33	1	4
Peas, canned	1 cup	150	29	1	8
Lettuce	1 cup	5	2	—	—
Celery	1 cup	20	5	—	1
Potato, baked	1 large	145	33	—	4
Meat, Fish, and Poultry					
Lean ground beef (10% fat)	3 oz	185	—	10	23
Shrimp, french fried	3 oz	190	9	9	17
Tuna, canned	3 oz	170	—	7	24
Chicken breast, fried	3 oz	160	1	5	26
Bacon	2 slices	85	—	8	4
Hot dog	1	170	1	15	7
Fast Foods					
McDonald's Egg McMuffin	1	327	31	15	19
McDonald's Big Mac	1	563	41	33	26
Taco Bell's beef burrito	1	466	37	21	30
Arby's roast beef	1	350	32	15	22
Pizza Hut Super Supreme	1 slice	260	23	13	15
Long John Silver's fish	2 pieces	366	21	22	22
Dairy Queen malt, large	1	840	125	28	22
Desserts					
Chocolate chip cookie	1	50	7	2	1
Apple pie	1 piece	135	49	14	3
Soft ice cream	1 cup	377	38	23	7
Beverage					
Cola soft drink	12 oz	145	37	—	—
Beer	12 oz	144	13	—	—
Wine	3 $\frac{1}{2}$ oz	73	2	—	—
Hard liquor (86 proof)	1 $\frac{1}{2}$ oz	105	—	—	—
Miscellaneous					
Egg	1	80	1	6	6
Mayonnaise	1 T	100	—	11	—
Sugar	1 T	45	12	—	—

common type of lipid in the diet, accounting for about 95% of the total lipid intake. Triglyceride molecules consist of three fatty acids bound to one glycerol molecule (see figure 2.12). Triglycerides are often referred to as fats. If the fat is a liquid at room temperature, it is referred to as an oil. Fats are **saturated** if their fatty acids have only single covalent bonds between carbon atoms and **unsaturated** if they have one or more double bonds (see figure 2.13). **Monounsaturated fats** have one double bond and **polyunsaturated fats** have two or more double bonds. Saturated fats are found in meat, dairy products, eggs, nuts, coconut oil, and palm oil (see table 17.1). Monounsaturated fats include olive and peanut oils; and polyunsaturated fats are found in fish, safflower, sunflower, and corn oils.

Saturating Fats

Solid fats, such as shortening and margarine, work better than liquid oils in preparing some foods such as pastries. Polyunsaturated vegetable oils can be changed from a liquid to a solid by making them more saturated, that is, by decreasing the number of double covalent bonds in their polyunsaturated fatty acids. To saturate an unsaturated oil, hydrogen gas is bubbled through it. As hydrogen binds to the fatty acids, double covalent bonds are converted to single covalent bonds, producing a change in molecular shape that solidifies the oil. The more saturated the product, the harder it becomes at room temperature.

Unprocessed polyunsaturated fats mostly occur in the *cis* form, which means the hydrogen atoms are on the same side of the carbon-carbon double bond in their fatty acids (see figure 2.13). During hydrogenation, some of the hydrogen atoms are transferred to the opposite side of the double bond to make the *trans* form, in which one hydrogen atom is on one side of the double bond and another is on the opposite side. Processed foods and oils account for most of the *trans* fats in the American diet, although some *trans* fats occur naturally in food from animal sources. *Trans* fatty acids raise the concentration of low-density lipoproteins and lower the concentration of high-density lipoproteins in the blood (see chapter 16). These changes are associated with a greater risk for cardiovascular disease.

The remaining 5% of ingested lipids include steroids and phospholipids. **Cholesterol** (kō-les′ter-ol) is a steroid (see chapter 2) found in high concentrations in the brain, the liver, and egg yolks; but it is also present in whole milk, cheese, butter, and meats. Cholesterol is not found in plants. Phospholipids, such as **lecithin** (les′i-thin, *lekithos*, egg yolk), are major components of cell membranes, and they are found in a variety of foods. A good source of lecithin is egg yolks.

Uses in the Body

Triglycerides are an important source of energy that can be used to produce ATP. A gram of triglyceride delivers over twice as many calories as does a gram of carbohydrate or protein. Some cells, such as skeletal muscle cells, derive most of their energy from triglycerides.

Ingested triglyceride molecules not immediately used are stored in adipose tissue or in the liver. When energy is required, the stored triglycerides are broken down, and the fatty acids are released into the blood. The fatty acids can be taken up and used by various tissues. In addition to storing energy, adipose tissue surrounds, pads, and protects organs. Adipose tissue located under the skin is an insulator, which helps to reduce heat loss.

Cholesterol is an important molecule with many functions in the body. It is obtained in food or it can be manufactured by the liver and most other tissues. Cholesterol is a component of the cell membrane, and it can be modified to form other useful molecules such as bile salts and steroid hormones. Bile salts emulsify fats, which is important for fat digestion and absorption (see chapter 16). Steroid hormones include the sex hormones estrogen, progesterone, and testosterone, which regulate the reproductive system. Eicosanoids, which are derived from fatty acids, are involved in inflammation, tissue repair, smooth muscle contraction, and other functions.

Phospholipids (see chapter 2) are part of the cell membrane and are used to construct myelin sheaths around the axons of nerve cells. Lecithin is found in bile and helps emulsify fats.

Recommended Consumption

The AMDR for fats is 20% to 35% for adults, 25% to 35% for children and adolescents 4 to 18 years of age, and 30% to 35% for children 2 to 3 years of age. Saturated fats should be 10% of total kilocalories or as low as possible. Most dietary fat should come from sources of polyunsaturated and monounsaturated fats. Cholesterol should be limited to 300 mg (the amount in one egg yolk) or less per day and *trans* fat consumption should be as low as possible. These guidelines reflect the belief that excess amounts of fats, especially saturated fats, *trans* fats, and cholesterol, contribute to cardiovascular disease. The typical American diet derives 35% to 45% of its kilocalories from fats, indicating that most Americans need to reduce fat consumption. See table 17.1 for a sampling of fat composition in foods.

If insufficient amounts of fats are consumed, the body can synthesize fats from carbohydrates and proteins. **Linoleic** (lin-ō-lē′ik, *linum*, flax + *oleum*, oil) **acid** and **alpha linolenic** (lin-ō-len′ik) **acid** are **essential fatty acids** because the body cannot synthesize them and they must be ingested. They are found in plant oils, such as canola or soybean oils.

Fatty Acids and Blood Clotting

The essential fatty acids are used to synthesize molecules that affect blood clotting. Linoleic acid can be converted to **arachidonic** (ă-rak-i-don′ik, *arakis*, leguminous weed) **acid,** which is used to produce prostaglandins that *increase* blood clotting. Alpha-linolenic acid can be converted to **eicosapentaenoic** (ī′kō-să-pen-tă-nō′ik) **acid (EPA),** which is used to produce prostaglandins that *decrease* blood clotting. Normally, most prostaglandins are synthesized from linoleic acid because it is more plentiful in the diet. Individuals, however, who consume foods rich in EPA, such as herring, salmon, tuna, and sardines, increase the synthesis of prostaglandins from EPA. Individuals who eat these fish two or more times per week have a lower risk of heart attack than those who don't, probably because of reduced blood clotting. Although EPA can be obtained using fish oil supplements, this is not currently recommended because fish oil supplements contain high amounts of cholesterol, vitamins A and D, and uncommon fatty acids, all of which can cause health problems when taken in large amounts.

Proteins

Sources in the Diet

Proteins (prō′tēnz) are chains of amino acids (see figure 2.16). They are found in most of the plant and animal products we eat (see table 17.1). Proteins in the body are constructed of 20 different kinds of amino acids, which are divided into two groups: essential and nonessential amino acids. The body cannot synthesize **essential amino acids,** which must be obtained in the diet. The nine essential amino acids are histidine, isoleucine, leucine, lysine, methionine, phenylalanine, threonine, tryptophane, and valine. Although the **nonessential amino acids** are necessary to construct our proteins, they are nonessential in the sense that it is not necessary to ingest them because they can be synthesized from the essential amino acids. A **complete protein** food contains all nine essential amino acids in the needed proportions, whereas an incomplete protein food does not. Animal proteins tend to be complete proteins, whereas plant proteins tend to be incomplete. Examples of complete proteins are red meat, fish, poultry, milk, cheese, and eggs. Examples of incomplete proteins are leafy green vegetables, grains, and legumes (peas and beans). If two incomplete proteins such as rice and beans are ingested, each can provide amino acids lacking in the other. Thus a vegetarian diet, if balanced correctly, provides all of the essential amino acids.

Uses in the Body

Proteins perform numerous functions in the human body, as the following examples illustrate. Collagen provides structural strength in connective tissue, as does keratin in the skin. The combination of actin and myosin makes muscle contraction possible. Enzymes are responsible for regulating the rate of chemical reactions in the body, and protein hormones regulate many physiological processes (see chapter 10). Proteins in the blood act as clotting factors, transport molecules, and buffers (which prevent changes in pH). Proteins also function as ion channels, carrier molecules, and receptor molecules in the cell membrane. Antibodies, lymphokines, and complement are all proteins that function in the immune system.

Proteins can also be used as a source of energy, yielding approximately the same amount of energy as that derived from carbohydrates. If excess proteins are ingested, the energy from the proteins can be used for the production of glycogen and fat molecules, which can be stored. When protein intake is adequate in a healthy adult, the synthesis and breakdown of proteins occur at the same rate.

Recommended Consumption

The AMDR for protein is 10% to 35% of total kilocalories. See table 17.1 for a sampling of protein composition in foods.

Vitamins

Vitamins (vīt′ă-minz, life-giving chemicals) are organic molecules that exist in minute quantities in food and are essential to normal metabolism (table 17.2). **Essential vitamins** cannot be produced by the body and must be obtained through the diet. Because no single food item or nutrient class provides all the essential vitamins, it is necessary to maintain a balanced diet by eating a variety of foods. The absence of an essential vitamin in the diet can result in a specific deficiency disease. A few vitamins, such as vitamin K, are produced by intestinal bacteria, and a few can be formed by the body from substances called provitamins. A **provitamin** is a part of a vitamin that can be assembled or modified by the body into a functional vitamin. Beta carotene is an example of a provitamin that can be modified by the body to form vitamin A. The other provitamins are **7-dehydrocholesterol** (dē-hī′drō-kō-les′ter-ol), which can be converted to vitamin D, and **tryptophan** (trip′tō-fan), which can be converted to niacin.

Vitamins are not broken down by catabolism but are used by the body in their original or slightly modified forms. After the chemical structure of a vitamin is destroyed, its function is lost. The chemical structure of many vitamins is destroyed by heat, such as when food is overcooked.

Most vitamins function as **coenzymes,** which combine with enzymes to make the enzymes functional (see chapter 2). Vitamins B_2 and B_3, biotin (bī′ō-tin), and pantothenic (pan-tō-then′ik) acid are critical for some of the chemical reactions involved in the production of ATP. Folate (fō′lāt) and vitamin B_{12} are involved in nucleic acid synthesis. Vitamins A, B_1, B_6, B_{12}, C, and D are necessary for growth. Vitamin K is necessary for the synthesis of proteins involved in blood clotting (see table 17.2).

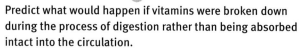

P R E D I C T ①
Predict what would happen if vitamins were broken down during the process of digestion rather than being absorbed intact into the circulation.

Vitamins are either fat-soluble or water-soluble. **Fat-soluble vitamins,** such as vitamins A, D, E, and K, are absorbed from the intestine along with lipids. Some of them can be stored in the body for a long time. Because they can be stored, it is possible to accumulate these vitamins in the body to the point of toxicity. **Water-soluble vitamins,** such as the B vitamins and vitamin C, are absorbed with water from the intestinal tract and typically remain in the body only a short time before being excreted in the urine.

Vitamins were first identified at the beginning of the twentieth century. They were found to be associated with certain foods that were known to protect people from diseases like rickets and beriberi. In 1941, the first Food and Nutrition Board established the **recommended dietary allowances (RDAs),** which are the nutrient intakes that are sufficient to meet the needs of nearly all people in certain age and gender groups. RDAs have been established for different-aged males and females, starting with infants and continuing on to adults. RDAs are also set for pregnant and lactating women. The RDAs have been reevaluated every 4–5 years and updated, when necessary, on the basis of new information.

The RDAs establish a minimum intake of vitamins and minerals that should protect almost everyone (97%) in a given group from diseases caused by vitamin or mineral deficiencies. Although personal requirements can vary, the RDAs are a good benchmark. The further dietary intake is below the RDAs, the more likely a nutritional deficiency can occur. On the other hand, the consumption of too large a quantity of some nutrients can have harmful effects. For example, the long-term ingestion

Table 17.2	The Principal Vitamins				
Vitamin	**Fat (F)- or Water (W)-Soluble**	**Source**	**Function**	**Symptoms of Deficiency**	**Reference Daily Intake (RDI)s[a]**
A (retinol)	F	From provitamin carotene found in yellow and green vegetables: preformed in liver, egg yolk, butter, and milk	Necessary for rhodopsin synthesis, normal health of epithelial cells, and bone and tooth growth	Rhodopsin deficiency, night blindness, retarded growth, skin disorders and increase in infection risk	900 RE[b]
B_1 (thiamine)	W	Yeast, grains, and milk	Involved in carbohydrate and amino acid metabolism, necessary for growth	Beriberi—muscle weakness (including cardiac muscle), neuritis, and paralysis	1.2 mg
B_2 (riboflavin)	W	Green vegetables, liver, wheat germ, milk, and eggs	Component of flavin adenine dinucleotide; involved in citric acid cycle	Eye disorders and skin cracking, especially at corners of the mouth	1.3 mg
B_3 (niacin)	W	Fish, liver, red meat, yeast, grains, peas, beans, and nuts	Component of nicotinamide adenine dinucleotide; involved in glycolysis and citric acid cycle	Pellagra—diarrhea, dermatitis, and nervous system disorder	16 mg
Pantothenic acid	W	Liver, yeast, green vegetables, grains, and intestinal bacteria	Constituent of coenzyme-A, glucose production from lipids and amino acids, and steroid hormone synthesis	Neuromuscular dysfunction and fatigue	5 mg
Biotin	W	Liver, yeast, eggs, and intestinal bacteria	Fatty acid and nucleic acid synthesis; movement of pyruvic acid into citric acid cycle	Mental and muscle dysfunction, fatigue, and nausea	30 μg
B_6 (pyridoxine)	W	Fish, liver, yeast, tomatoes, and intestinal bacteria	Involved in amino acid metabolism	Dermatitis, retarded growth, and nausea	1.7 mg
Folate	W	Liver, green leafy vegetables, and intestinal bacteria	Nucleic acid synthesis, hematopoiesis, prevents birth defects	Macrocytic anemia (enlarged red blood cells) and spina bifida	0.4 mg
B_{12} (cobalamins)	W	Liver, red meat, milk, and eggs	Necessary for red blood cell production, some nucleic acid and amino acid metabolism	Pernicious anemia and nervous system disorders	2.4 μg
C (ascorbic acid)	W	Citrus fruit, tomatoes, and green vegetables	Collagen synthesis; general protein metabolism	Scurvy—defective bone formation and poor wound healing	90 mg
D (cholecalciferol, ergosterol)	F	Fish liver oil, enriched milk, and eggs; provitamin D converted by sunlight to cholecalciferol in the skin	Promotes calcium and phosphorus use; normal growth and bone and teeth formation	Rickets—poorly developed, weak bones, osteomalacia; bone reabsorption	10 μg[c]
E (tocopherol, tocotrienols)	F	Wheat germ, cotton seed, palm, and rice oils; grain, liver, and lettuce	Prevents the oxidation of cell membranes and DNA	Hemolysis of red blood cells	15 mg
K (phylloquinone)	F	Alfalfa, liver, spinach, vegetable oils, cabbage, and intestinal bacteria	Required for synthesis of a number of clotting factors	Excessive bleeding due to retarded blood clotting	120 μg

[a] RDIs for people over 4 years of age; IU = international units.

[b] Retinol equivalents (RE). 1 retinol equivalent = 1 μg retinol or 6 μg β-carotene.

[c] As cholecalciferol. 1 μg cholecalciferol = 40 IU (international units) vitamin D.

of 3–10 times the RDA for vitamin A can cause bone and muscle pain, skin disorders, hair loss, and increased liver size. The long-term consumption of 5–10 times the RDA of vitamin D can result in the deposition of calcium in the kidneys, heart, and blood vessels, and the regular consumption of more than 2 g of vitamin C daily can cause stomach inflammation and diarrhea.

Free Radicals and Antioxidants

Damage from free radicals may contribute to aging and certain diseases, such as atherosclerosis and cancer. **Free radicals** are molecules, produced as part of normal metabolism, that are missing an electron. Free radicals can replace the missing electron by taking an electron from cell molecules, such as fats, proteins, or DNA, resulting in damage to the cell. The loss of an electron from a molecule is called oxidation. **Antioxidants** are substances that prevent oxidation of cell components by donating an electron to free radicals. Examples of antioxidants include beta carotene (provitamin A), vitamin C, and vitamin E.

Many studies have been done to determine whether or not taking large doses of antioxidants is beneficial. Although future research may suggest otherwise, the consensus among scientists establishing the RDAs is that the best evidence presently available does not support the claims that taking large doses of antioxidants prevents chronic disease or otherwise improves health. On the other hand, the amount of antioxidants normally found in a balanced diet that includes fruits and vegetables rich in antioxidants, combined with the complex mix of other chemicals found in food, can be beneficial.

Minerals

Minerals (min′er-ălz) are inorganic nutrients that are essential for normal metabolic functions. Minerals are taken into the body by themselves or in combination with organic molecules. They constitute about 4–5% of the total body weight and are involved in a number of important functions, such as establishing resting membrane potentials and generating action potentials; adding mechanical strength to bones and teeth; combining with organic molecules; or acting as coenzymes, buffers, or regulators of osmotic pressure. A balanced diet can provide all the necessary minerals, with a few possible exceptions. For example, anyone suffering from chronic bleeding or women who have excessive menstrual bleeding may need an iron supplement. Table 17.3 lists some minerals and their functions.

Daily Values

Daily Values appear on food labels to help consumers plan a healthy diet and to minimize confusion. Not all possible Daily Values are required to be listed on food labels. Daily Values are based on two other sets of reference values: Reference Daily Intakes and Daily Reference Values.

The **Reference Daily Intakes (RDIs)** are based on the 1968 RDAs for certain vitamins and minerals. RDIs have been set for four categories of people: infants, toddlers, people over 4 years of age, and pregnant or lactating women. Generally, the RDIs are set to the highest 1968 RDA value of an age category. For example, the highest RDA for iron in males over 4 years of age is 10 mg/day and for females over 4 years of age is 18 mg/day. Thus, the RDI for iron is set at 18 mg/day.

The **Daily Reference Values (DRVs)** are set for total fat, saturated fat, cholesterol, total carbohydrate, dietary fiber, sodium, potassium, and protein.

The Daily Values appearing on food labels are based on a 2000 kcal reference diet, which approximates the weight maintenance requirements of postmenopausal women, women who exercise moderately, teenage girls, and sedentary men (figure 17.2). On large food labels, additional information is listed based on a daily intake of 2500 kcal, which is adequate for young men.

The Daily Values for energy-producing nutrients are determined as a percentage of daily kilocaloric intake: 60% for carbohydrates, 30% for total fats, 10% for saturated fats, and 10% for proteins. The Daily Value for fiber is 14 g for each 1000 kcal of intake. The Daily Values for a nutrient in a 2000 kcal/day diet can be calculated on the basis of the recommended daily percentage of the nutrient and the kilocalories in a gram of the nutrient. For example, carbohydrates should be 60% of a 2000 kcal/day diet, or 1200 kcal/day (0.60 × 2000). Because there are 4 kilocalories in a gram of carbohydrate, the Daily Value for carbohydrate is 300 g/day (1200/4).

The Daily Values for some nutrients are the uppermost limit considered desirable because of the link between these nutrients and certain diseases. Thus, the Daily Values for total fats are less than 65 g; saturated fats, less than 20 g; and cholesterol, less than 300 mg because of their association with increased risk of heart disease. The Daily Value for sodium is less than 2400 mg because of its association with high blood pressure in some people.

For a particular food, the Daily Values are used to calculate the **Percent Daily Value (% Daily Value)** for some of the nutrients in one serving of the food (see figure 17.2). For

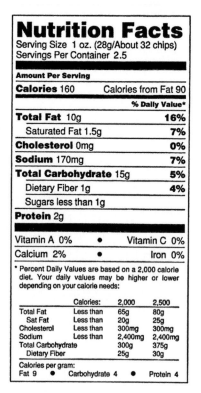

Figure 17.2   Food Label

Source: U.S. Food and Drug Administration

Table 17.3 Important Minerals

Mineral	Function	Symptoms of Deficiency	Reference Daily Intake (RDIs)[a]
Calcium	Bone and teeth formation, blood clotting, muscle activity, and nerve function	Spontaneous action potential generation in neurons and tetany	1300 mg
Chlorine	Blood acid–base balance; hydrochloric acid production in stomach	Acid–base imbalance	2.3 g[b]
Chromium	Associated with enzymes in glucose metabolism	Unknown	35 μg
Cobalt	Component of vitamin B_{12}; red blood cell production	Anemia	Unknown
Copper	Hemoglobin and melanin production, electron-transport system	Anemia and loss of energy	0.9 mg
Fluorine	Provides extra strength in teeth; prevents dental caries	No real pathology	4 mg
Iodine	Thyroid hormone production, maintenance of normal metabolic rate	Goiter and decrease in normal metabolism	150 μg
Iron	Component of hemoglobin; ATP production in electron-transport system	Anemia, decreased oxygen transport, and energy loss	18 mg
Magnesium	Coenzyme constituent; bone formation; muscle and nerve function	Increased nervous system irritability, vasodilation, and arrhythmias	420 mg
Manganese	Hemoglobin synthesis; growth; activation of several enzymes	Tremors and convulsions	2.3 mg
Molybdenum	Enzyme component	Unknown	45 μg
Phosphorus	Bone and teeth formation; important in energy transfer (ATP); component of nucleic acids	Loss of energy and cellular function	1250 mg
Potassium	Muscle and nerve function	Muscle weakness, abnormal electrocardiogram, and alkaline urine	4.7 g
Selenium	Component of many enzymes	Unknown	55 μg
Sodium	Osmotic pressure regulation; nerve and muscle function	Nausea, vomiting, exhaustion, and dizziness	1.5 g[b]
Sulfur	Component of hormones; several vitamins, and proteins	Unknown	Unknown
Zinc	Component of several enzymes; carbon dioxide transport and metabolism; necessary for protein metabolism	Deficient carbon dioxide transport and deficient protein metabolism	11 mg

[a] RDIs for people over 4 years of age, except for sodium.

[b] 3.8 g sodium chloride (table salt)

example, if a serving of food has 3 g of fat and the Daily Value for total fat is 65 g, then the % Daily Value is 5% (3/65 = 0.05, or 5%). The Food and Drug Administration (FDA) requires % Daily Values to be on food labels so that the public has useful and accurate dietary information.

P R E D I C T 2
One serving of a food has 30 g of carbohydrate. What % Daily Value for carbohydrate is on the food label for this food?

The % Daily Values for nutrients related to energy consumption are based on a 2000 kcal/day diet. For people who maintain their weight on a 2000 kcal/day diet, the total of the % Daily Values for each of these nutrients should add up to no more than 100%. For individuals consuming more or fewer kilocalories per day than 2000 kcal, however, the total of the % Daily Values can be more than 100%. For example, for a

person consuming 2200 kcal/day, the total of the % Daily Values for each of these nutrients should add up to no more than 110% because 2200/2000 = 1.10 or 110%.

P R E D I C T 3
Suppose a person consumes 1800 kcal/day. What total % Daily Values for energy-producing nutrients is recommended?

When using the % Daily Values of a food to determine how the amounts of certain nutrients in the food fit into the overall diet, the number of servings in a container or package needs to be considered. For example, suppose a small (2.25 oz) bag of corn chips has a % Daily Value of 16% for total fat. One might suppose that eating the bag of chips accounts for 16% of total fat for the day. The bag, however, contains 2.5 servings. All of the chips in the bag account for 40% (16% × 2.5) of the maximum recommended total fat.

Hazel Nutt decides to switch from a typical American diet to a vegetarian diet because she believes a vegetarian diet is better for her health and she no longer wants to eat animals. A strict vegetarian or vegan diet includes only plant foods.

Plants alone can provide all of the protein required for good health. In order to get adequate amounts of the essential amino acids, a variety of protein sources, such as grains and legumes, should be consumed.

The Dietary Guidelines for Americans recommends that vegan diets be supplemented with vitamin B_{12}, vitamin D, calcium, iron, and zinc. This is especially important for children and pregnant and lactating women. Plant sources do not supply vitamin B_{12} or sufficient amounts of vitamin D, although the body can produce vitamin D with adequate exposure to sunlight (see chapter 5). Calcium is found in green leafy vegetables and nuts. Iron and zinc are in whole grains, nuts, and legumes. However, these minerals are either in low amounts or they are not easily absorbed.

Metabolism

Metabolism (mĕ-tab′ō-lizm, change) is the total of all the chemical reactions that occur in the body. It consists of **anabolism** (ă-nab′ō-lizm), the energy-requiring process by which small molecules are joined to form larger ones, and **catabolism** (kă-tab′ō-lizm), the energy-releasing process by which large molecules are broken down into smaller ones. Anabolism occurs in all cells of the body as they divide to form new cells, maintain their own intracellular structure, and produce molecules such as hormones, neurotransmitters, or extracellular matrix molecules for export. Catabolism begins during the process of digestion and is concluded within individual cells. The energy derived from catabolism is used to drive anabolic reactions.

Metabolism can be divided into the chemical reactions that occur during digestion and the chemical reactions that occur after the products of digestion are taken up by cells. The chemical reactions that occur within cells are often referred to as **cellular metabolism.** The digestive products of carbohydrates, proteins, and lipids can be further broken down inside cells. The energy released during this breakdown can be used to combine **ADP** and an inorganic phosphate group (P_i) to form **ATP** (figure 17.3).

ATP is often called the energy currency of the cell. When ATP is broken down to ADP, the released energy can be used by cells for active transport, muscle contraction, and the synthesis of molecules. Because the body has high energy demands, it uses ATP rapidly.

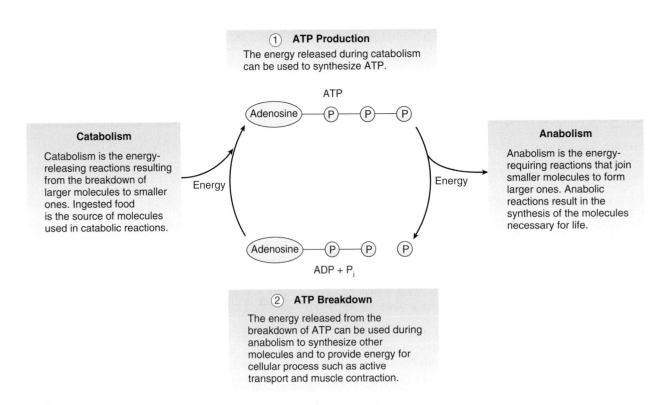

① ATP Production
The energy released during catabolism can be used to synthesize ATP.

Catabolism

Catabolism is the energy-releasing reactions resulting from the breakdown of larger molecules to smaller ones. Ingested food is the source of molecules used in catabolic reactions.

Anabolism

Anabolism is the energy-requiring reactions that join smaller molecules to form larger ones. Anabolic reactions result in the synthesis of the molecules necessary for life.

② ATP Breakdown
The energy released from the breakdown of ATP can be used during anabolism to synthesize other molecules and to provide energy for cellular process such as active transport and muscle contraction.

Process Figure 17.3   **ATP Coupling of Catabolic and Anabolic Reactions**

Energy released by catabolic reactions is used to form ATP, which releases the energy for use in anabolic reactions.

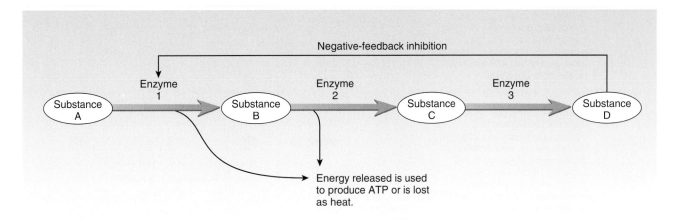

Figure 17.4   Biochemical Pathway

Each step in the pathway is regulated by a specific enzyme. Substance D can inhibit enzyme 1, thus regulating its own production. Some of the energy released by reactions in the pathway can be used to synthesize ATP.

Regulation of Metabolism

The products of digestion, such as glucose, fatty acids, and amino acids, are molecules containing energy within their chemical bonds. A series of chemical reactions, called a **biochemical pathway,** controls the energy release from these molecules. At some of the steps, small amounts of energy are released, part of which is used to synthesize ATP (figure 17.4). About 40% of the energy in foods is incorporated into ATP; the rest is lost as heat.

There are several different biochemical pathways inside cells. Which pathways function and how much each pathway is used is determined by enzymes, because each step in the pathway requires a specific enzyme (see chapter 2). In turn, enzymes are regulated in several ways:

1. *Enzyme synthesis.* Enzymes are proteins and their synthesis depends on DNA (see chapter 3). Thus, the type and amount of enzymes present in cells is under genetic control.
2. *Receptor-mediated enzyme activity.* The combination of a chemical signal, such as a neurotransmitter or hormone, with a membrane-bound or intracellular receptor can activate or inhibit enzyme activity (see chapter 10).
3. *Product control of enzyme activity.* The end product of a biochemical pathway can inhibit the enzyme responsible for the first reaction in the pathway. This is negative-feedback regulation that prevents accumulation of the intermediate products and the end product of the pathway (see figure 17.4).

Enzymes and Disease

Many metabolic disorders result from missing or dysfunctional enzymes. For example, in **Tay-Sachs disease,** the breakdown of lipids within lysosomes is impaired. The abnormal accumulation of the intermediate products of lipid metabolism results in the destruction of neurons and death by age 3–4 years. **Phenylketonuria** (fen′il·kē′tō-noo′rē-a) **(PKU)** results from the inability to convert the amino acid phenylalanine to tyrosine. Accumulation of phenylalanine in nerve cells causes brain damage. Fortunately, restricting the intake of phenylalanine in the diet is an effective treatment. In **albinism** (al′bi-nizm), the enzyme necessary to convert tyrosine to melanin is missing, resulting in lack of skin pigmentation (see chapter 5).

Carbohydrate Metabolism

Monosaccharides are the breakdown products of carbohydrate digestion. Of these, glucose is the most important as far as cellular metabolism is concerned. Glucose is transported in the circulation to all tissues of the body, where it is used as a source of energy. Any excess glucose in the blood following a meal can be used to form **glycogen** (glī′kō-jen, *glyks,* sweet), or it can be partially broken down, and the components used to form fat. Glycogen is a short-term energy storage molecule, which can only be stored by the body in limited amounts, whereas fat is a long-term energy storage molecule, which can be stored in the body in large amounts. Most of the body's glycogen is in skeletal muscle and in the liver.

Glycolysis

Glycolysis (glī-kol′i-sis) is a series of chemical reactions that occurs in the fluid part of cytoplasm surrounding the organelles. It results in the breakdown of glucose to two **pyruvic** (pī-roo′vik) **acid** molecules (figure 17.5). When glucose is converted to pyruvic acid, two ATP molecules are used and four ATP molecules are produced, for a net gain of two ATP molecules.

Glucose consists of many hydrogen atoms covalently bonded to the carbon atoms of the molecule. During the breakdown of glucose, a hydrogen ion (H^+) and two electrons (e^-) are released and can attach to a **carrier molecule,** which functions to move the H^+ and electrons to other parts of the cell. A very common carrier molecule in cells is **nicotinamide adenine dinucleotide** (nik-ō-tin′a-mīd ad′ĕ-nēn dī-noo′klē-ō-tīd) **(NADH).**

$$NAD^+ + 2\,e^- + H^+ \rightarrow NADH$$

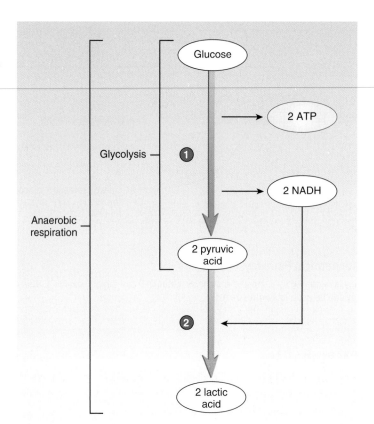

1. Glycolysis converts glucose to two pyruvic acid molecules. The many reactions within the pathway are not shown. There is a net gain of two ATP and two NADH from glycolysis.

2. Anaerobic respiration, which does not require oxygen, includes glycolysis and converts the two pyruvic acid molecules produced by glycolysis to two lactic acid molecules. This conversion requires energy, which is derived from the NADH generated in glycolysis.

Process Figure 17.5   Glycolysis and Anaerobic Respiration

The H^+ and high-energy electrons in the NADH molecules can be used in other chemical reactions or in the production of ATP molecules in the electron-transport chain (described on p. 496).

The pyruvic acid and NADH produced in glycolysis can be used in two different biochemical pathways, depending on the availability of oxygen. If the cell has adequate amounts of oxygen, the pyruvic acid and NADH produced in glycolysis are used in aerobic respiration to produce many more ATP. When the amounts of oxygen are limited, anaerobic respiration can take place. Anaerobic respiration does not require oxygen and functions to quickly produce a few ATP molecules for a short time.

Anaerobic Respiration

Anaerobic (an-ār-ō′bik) **respiration** is the breakdown of glucose in the absence of oxygen to produce two molecules of **lactic** (lak′tik) **acid** and two molecules of ATP (see figure 17.5). The ATP thus produced is a source of energy during activities such as intense exercise when insufficient oxygen is delivered to tissues. Anaerobic respiration can be divided into two phases:

1. *Glycolysis.* The first phase of anaerobic respiration is glycolysis, in which glucose undergoes several reactions to produce two pyruvic acid molecules, two ATP, and two NADH.

2. *Lactic acid formation.* The second phase is the conversion of pyruvic acid to lactic acid, a reaction that requires the input of energy from the NADH produced in phase 1 of anaerobic respiration.

Lactic acid is released from the cells that produce it and is transported by the blood to the liver. When oxygen becomes available, the lactic acid in the liver can be converted through a series of chemical reactions into glucose. The glucose then can be released from the liver and transported in the blood to cells that use glucose as an energy source. Some of the reactions that convert lactic acid into glucose require the input of energy derived from ATP that is produced by aerobic respiration. The oxygen necessary to make enough ATP for the synthesis of glucose from lactic acid contributes to the **oxygen debt** (see chapter 7).

Aerobic Respiration

Aerobic (ār-ō′bik) **respiration** (figure 17.6) is the breakdown of glucose in the presence of oxygen to produce carbon dioxide, water, and 38 molecules of ATP. Aerobic respiration can be divided into four phases:

1. *Glycolysis.* The first phase of aerobic respiration, as in anaerobic respiration, is glycolysis. The six-carbon glucose molecule is broken down to form two molecules of pyruvic acid, each of which consists of three carbon

1. Glycolysis in the cytoplasm converts glucose to two pyruvic acid molecules and produces two ATP and two NADH. The NADH can go to the electron-transport chain in the inner mitochondrial membrane.

2. The two pyruvic acid molecules produced in glycolysis are converted to two acetyl-CoA molecules, producing two CO_2 and two NADH. The NADH can go to the electron-transport chain.

3. The two acetyl-CoA molecules enter the citric acid cycle, which produces four CO_2, six NADH, two $FADH_2$, and two ATP. The NADH and $FADH_2$ can go to the electron-transport chain.

4. The electron-transport chain uses NADH and $FADH_2$ to produce 34 ATP. This process requires O_2, which combines with H^+ to form H_2O.

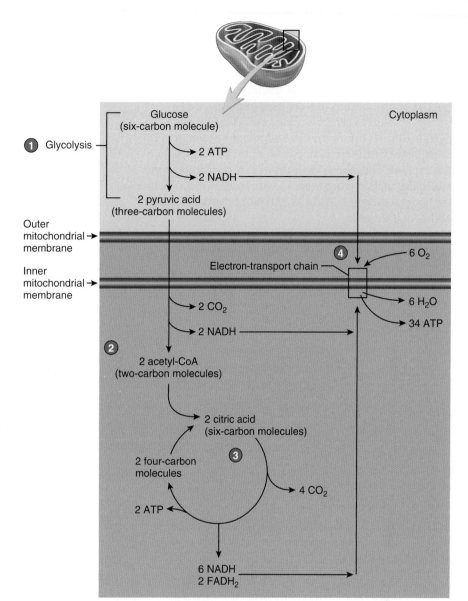

Process Figure 17.6 Aerobic Respiration

Aerobic respiration involves four steps: (1) glycolysis, (2) acetyl-CoA formation, (3) citric acid cycle, and (4) electron-transport chain. The number of carbon atoms in a molecule is indicated after the molecule's name. As glucose is broken down, the carbon atoms from glucose are incorporated into carbon dioxide.

atoms. Two ATP and two NADH molecules are also produced.

2. *Acetyl-coenzyme A formation.* In the second phase, each pyruvic acid moves from the cytoplasm into a mitochondrion, where enzymes remove a carbon atom from the three-carbon pyruvic acid molecule to form carbon dioxide and a two-carbon acetyl (as′e-til, a-set′il) group. Hydrogen ions and electrons are released in the reaction, which can be used to produce NADH. Each acetyl group combines with coenzyme A (CoA), derived from vitamin B_2, to form **acetyl-CoA.** Because two pyruvic molecules are produced in phase 1, phase 2 results in two acetyl-CoA's, two carbon dioxide, and two NADH molecules for each glucose molecule.

3. *Citric acid cycle.* In the third phase, each acetyl-CoA combines with a four-carbon molecule to form a six-carbon citric acid molecule, which enters the citric acid cycle. The **citric** (sit′rik) **acid cycle** is also called the **tricarboxylic** (trī-kar-bok′sil-ik) **acid (TCA) cycle** (citric acid has three carboxylic acid groups) or the **Krebs' cycle,** after its discoverer, the British biochemist Sir Hans Krebs (1900–1981). The citric acid cycle is a series of reactions wherein the six-carbon citric acid molecule is converted, in a number of steps, into a four-carbon molecule (see figure 17.6). The four-carbon molecule can then combine with another acetyl-CoA molecule to form another citric acid molecule, and reinitiate the cycle. During the cycle, two carbon atoms

are used to form carbon dioxide; and energy, H^+, and electrons are released. Some of the energy can be used to produce ATP. Most of the energy, H^+, and electrons are used to form NADH molecules and another carrier molecule called **flavin** (flā'vin) **adenine dinucleotide (FADH₂).** These molecules are used in the electron-transport chain to generate additional ATP. Carbon dioxide diffuses out of the cell and into the blood. It is transported by the circulatory system to the lungs, where it is expired. Thus the carbon atoms that constitute food molecules such as glucose are eventually eliminated from the body as carbon dioxide. We literally breathe out part of the food we eat!

4. *Electron-transport chain.* The **electron-transport chain** is a series of electron-transport molecules attached to the inner mitochondrial membrane (figure 17.7). This membrane divides the interior of the

mitochondrion into an inner and outer compartment. Electrons are transferred from NADH and $FADH_2$ to the electron-transport carriers, and H^+ are released into the inner mitochondrial compartment. Some of the electron-transport carriers are also H^+ pumps, which use some of the energy from the transported electrons to pump H^+ from the inner to the outer mitochondrial compartment. Because of an increased H^+ concentration in the outer compartment, the H^+ pass by diffusion back into the inner compartment. The H^+ pass through special channels in the inner mitochondrial membrane that couple the movement of the H^+ to ATP production. In the last step of the electron-transport chain, two H^+ and two electrons combine with an O_2 atom to form H_2O.

$$2\,H^+ + 2\,e^- + \tfrac{1}{2}O_2 \rightarrow H_2O$$

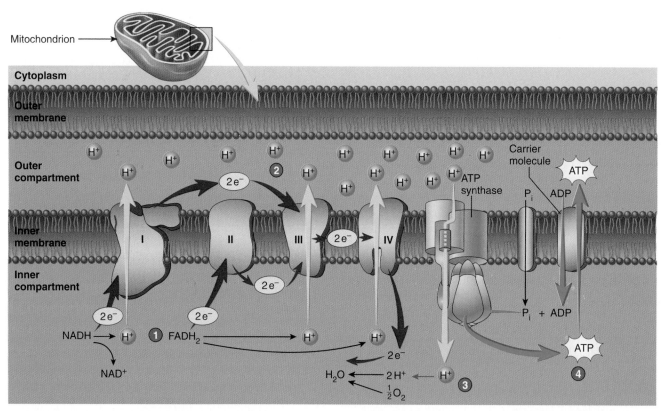

1. NADH or FADH₂ transfer their electrons to the electron-transport chain.

2. As the electrons move through the electron-transport chain, some of their energy is used to pump H⁺ into the outer compartment, resulting in a higher concentration of H⁺ in the outer than in the inner compartment.

3. The H⁺ diffuses back into the inner compartment through special channels (ATP synthase) that couple the H⁺ movement with the production of ATP. The electrons, H⁺, and O₂ combine to form H₂O.

4. ATP is transported out of the inner compartment by a carrier molecule that exchanges ATP for ADP. A different carrier molecule moves phosphate into the inner compartment.

Process Figure 17.7   The Electron-Transport Chain

The electron-transport chain in the inner membrane consists of four protein complexes (*purple;* numbered I to IV) with carrier molecules. As electrons are transferred from one carrier molecule to another, they lose energy that is used to move H^+ out of the inner compartment. Hydrogen ions move back into the inner compartment through special channels (ATP synthase; *green*), which produce ATP. Carrier molecules (*brown*) move ATP out of and ADP and P_i into the inner compartment.

Without O_2 to accept the H^+ and electrons, the citric acid cycle and the electron-transport chain cannot function. Note that the O_2 we breathe in is eventually bound to two hydrogen atoms to become water, which has many uses in the body (see chapter 2).

P R E D I C T
Many poisons function by blocking certain steps in the metabolic pathways. For example, cyanide blocks the last step in the electron-transport chain. Explain why this blockage causes death.

Summary of Anaerobic and Aerobic Respiration

In anaerobic respiration, each glucose molecule ($C_6H_{12}O_6$) yields two ATP and two lactic acid molecules ($C_3H_6O_3$) through glycolysis:

$$C_6H_{12}O_6 + 2\,ADP + P_i \rightarrow 2\,C_3H_6O_3 + 2\,ATP$$

In contrast, in aerobic respiration, each glucose molecule yields 38 ATP:

$$C_6H_{12}O_6 + 6\,O_2 + 38\,ADP + 38\,P_i \rightarrow 6\,CO_2 + 6\,H_2O + 38\,ATP$$

Of the 38 ATP molecules, 2 are produced in glycolysis, 2 are produced in the citric acid cycle, and 34 are formed through the electron-transport chain. Thus, aerobic respiration is much more efficient at producing ATP than is anaerobic respiration. In addition, many of the chemical reactions of aerobic respiration can be used to produce energy from other food molecules, such as lipids and proteins (see next sections on Lipid and Protein Metabolism).

The number of ATP molecules produced during aerobic respiration can also be reported as 36 ATP molecules. The two NADH molecules produced by glycolysis cannot cross the inner mitochondrial membrane; thus their electrons are donated to a shuttle molecule that carries the electrons to the electron-transport chain. Depending on the shuttle molecule, each glycolytic NADH molecule can produce two or three ATP molecules. In skeletal muscle and in the brain, only 2 ATP molecules are produced for each NADH molecule formed during glycolysis, resulting in a total number of 36 ATP molecules; but in the liver, kidneys, and heart, 3 ATP molecules are produced for each NADH molecule, and the total number of ATP molecules formed is 38.

The Quantity of ATP Produced from Glucose

The number of ATP molecules produced per glucose molecule is a theoretical number that assumes two H^+ are necessary for the formation of each ATP. If the number required is more than two, the efficiency of aerobic respiration decreases. In addition, it is now understood that it costs energy to get ADP and phosphates into the mitochondria and to get ATP out. Considering all these factors, it is currently estimated that each glucose molecule yields about 25 ATP molecules instead of the theoretical 38 ATP molecules.

Lipid Metabolism

Triglycerides, or fat, is the body's main energy storage molecule. In a normal person, fat is responsible for about 99%

of the body's energy storage, and glycogen accounts for the remaining 1%.

Between meals, triglycerides in adipose tissue is broken down into fatty acids and glycerol. Some of the fatty acids produced are released into the blood. Other tissues, especially skeletal muscle and the liver, use the fatty acids as a source of energy.

The metabolism of fatty acids takes place in the mitochondria. It occurs by a series of reactions wherein two carbon atoms are removed from the end of a fatty acid chain to form acetyl-CoA (figure 17.8). As the process continues, carbon atoms are removed two at a time until the entire fatty acid chain is converted into acetyl-CoA. Acetyl-CoA can enter the citric acid cycle and be used to generate ATP. In the liver, two acetyl-CoA molecules can also combine to form **ketones** (kē′tōnz). The ketones are released into the blood and travel to other tissues, especially skeletal muscle. In these tissues, the ketones are converted back to acetyl-CoA, which can enter the citric acid cycle to produce ATP.

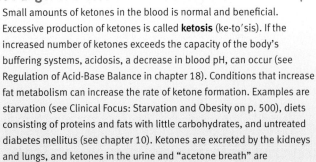

The Danger of Excessive Amounts of Ketones

Small amounts of ketones in the blood is normal and beneficial. Excessive production of ketones is called **ketosis** (ke-to′sis). If the increased number of ketones exceeds the capacity of the body's buffering systems, acidosis, a decrease in blood pH, can occur (see Regulation of Acid-Base Balance in chapter 18). Conditions that increase fat metabolism can increase the rate of ketone formation. Examples are starvation (see Clinical Focus: Starvation and Obesity on p. 500), diets consisting of proteins and fats with little carbohydrates, and untreated diabetes mellitus (see chapter 10). Ketones are excreted by the kidneys and lungs, and ketones in the urine and "acetone breath" are characteristic of untreated diabetes mellitus.

Protein Metabolism

Amino acids are the products of protein digestion. Once amino acids are absorbed into the body, they are quickly taken up by cells, especially in the liver. Amino acids are used primarily to synthesize needed proteins and only secondarily as a source of energy. If amino acids are used as a source of energy, they can be used in two ways (see figure 17.8): (1) The amino acids can be converted into the molecules of carbohydrate metabolism, such as pyruvic acid and acetyl-CoA. These molecules can be metabolized to yield ATP. (2) The amine group ($-NH_2$) can be removed from the amino acid, leaving ammonia and an α-keto acid. In this process, NADH is produced, which can enter the electron-transport chain to produce ATP. Ammonia is toxic to cells, so it is converted by the liver into urea, which is carried by the blood to the kidneys, where it is eliminated. The α-keto acid can enter the citric acid cycle or can be converted into pyruvic acid, acetyl-CoA, or glucose. Although proteins can be used as an energy source, they are not considered to be major storage molecules.

Metabolic States

There are two major metabolic states in the body. The first is the **absorptive state,** the period immediately after a meal when nutrients are being absorbed through the intestinal wall into the

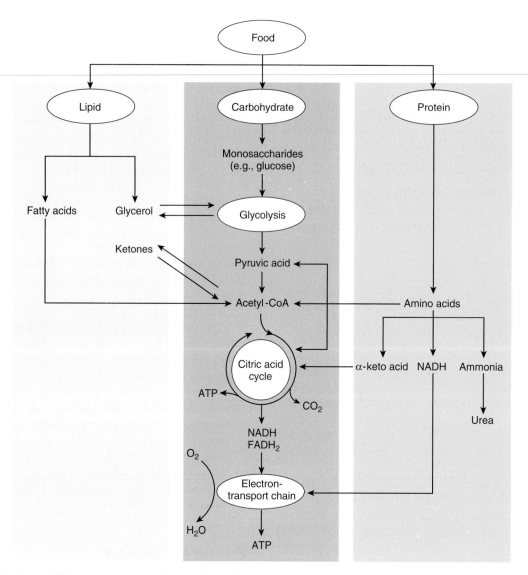

Figure 17.8 The Overall Pathways for the Metabolism of Food

Carbohydrates, lipids, and proteins enter biochemical pathways to produce energy in the cell.

circulatory and lymphatic systems (figure 17.9). The absorptive state usually lasts about 4 hours after each meal, and most of the glucose that enters the circulation is used by cells to provide the energy they require. The remainder of the glucose is converted into glycogen or fats. Most of the absorbed fats are deposited in adipose tissue. Many of the absorbed amino acids are used by cells in protein synthesis, some are used for energy, and others enter the liver and are converted to fats or carbohydrates.

The second state, the **postabsorptive state,** occurs late in the morning, late in the afternoon, or during the night after each absorptive state is concluded (figure 17.10). Normal blood glucose levels range between 70 and 110 mg/dL, and it is vital to the body's homeostasis that this range be maintained. During the postabsorptive state, blood glucose levels are maintained by the conversion of other molecules to glucose. The first source of blood glucose during the postabsorptive state is the glycogen stored in the liver. This glycogen supply can provide glucose for only about 4 hours, however. The glycogen stored in skeletal muscles can also be used during times of vigorous exercise. As the glycogen stores are depleted, fats are used as an energy source. The glycerol from triglycerides can be converted to glucose. The fatty acids from fat can be converted to acetyl-CoA, moved into the citric acid cycle, and used as a source of energy to produce ATP. In the liver, acetyl-CoA can be used to produce ketone bodies that other tissues can use for energy. The use of fatty acids as an energy source can partly eliminate the need to use glucose for energy, resulting in reduced glucose removal from the blood and maintaining blood glucose levels at homeostatic levels. The amino acids of proteins can be converted into glucose or can be used for energy production, again sparing blood glucose.

Metabolic Rate

The **metabolic rate** is the total amount of energy produced and used by the body per unit of time. Metabolic rate is usually estimated by measuring the amount of oxygen used per minute.

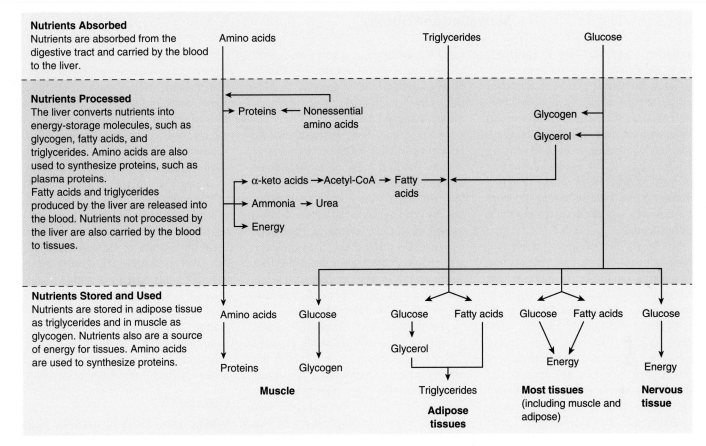

Figure 17.9 Events of the Absorptive State

Absorbed molecules, especially glucose, are used as sources of energy. Molecules not immediately needed for energy are stored: glucose is converted to glycogen or triglycerides, triglycerides are deposited in adipose tissue, and amino acids are converted to triglycerides or carbohydrates.

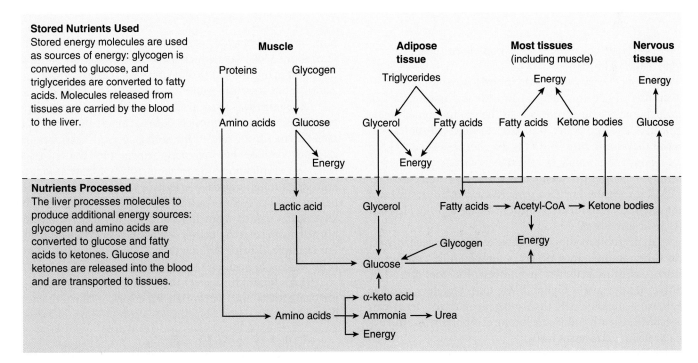

Figure 17.10 Events of the Postabsorptive State

Stored energy molecules are used as sources of energy: glycogen is converted to glucose; triglycerides are broken down to fatty acids, some of which are converted to ketones; and proteins are converted to glucose.

Clinical Focus   Starvation and Obesity

Starvation is the inadequate intake of nutrients or the inability to metabolize or absorb nutrients. Starvation can be caused by a number of factors, such as prolonged fasting, anorexia, deprivation, or disease. No matter what the cause, starvation takes approximately the same course and consists of three phases. The events of the first two phases occur even during relatively short periods of fasting or dieting. The third phase occurs in prolonged starvation and ends in death.

During the first phase of starvation, blood glucose levels are maintained through the production of glucose from glycogen, proteins, and fats. At first, glycogen is broken down into glucose. Enough glycogen is stored in the liver to last only a few hours, however. Thereafter, blood glucose levels are maintained by the breakdown of proteins and fats. Fats are decomposed into fatty acids and glycerol. Fatty acids can be used as a source of energy, especially by skeletal muscle, thus decreasing the use of glucose by tissues other than the brain. The brain cannot use fatty acids as an energy source, so the conservation of glucose is critical to normal brain function. Glycerol

can be used to make a small amount of glucose, but most of the glucose is formed from the amino acids of proteins. In addition, some amino acids can be used directly for energy.

In the second stage, which can last for several weeks, fats are the primary energy source. The liver metabolizes fatty acids into ketone bodies that can be used as a source of energy. After about a week of fasting, the brain begins to use ketone bodies, as well as glucose, for energy. This usage decreases the demand for glucose, and the rate of protein breakdown diminishes but does not stop. In addition, there is a selective use of proteins; proteins not essential for survival are used first.

The third stage of starvation begins when the fat reserves are nearly depleted and the body switches to proteins as the major energy source. Muscles, the largest source of protein in the body, are rapidly depleted. At the end of this stage, proteins essential for cellular functions are broken down, and cell function degenerates. Death can occur very rapidly.

Symptoms of starvation, in addition to weight loss, include apathy, listlessness,

withdrawal, and increased susceptibility to infectious diseases. Few people die directly from starvation because they usually die of some infectious disease first. Other signs of starvation may include changes in hair color, flaky skin, and massive edema in the abdomen and lower limbs, causing the abdomen to appear bloated.

During the process of starvation, the body's ability to consume normal volumes of food also decreases. Foods high in bulk but low in protein content often cannot reverse the process of starvation. Intervention involves feeding the starving person low-bulk foods containing ample protein, calories, vitamins, and minerals. The process of starvation also results in dehydration, and rehydration is an important part of intervention. Even with intervention, a victim may be so affected by disease or weakness that he or she cannot recover.

Obesity

Obesity is the storage of excess fat, and it results from the ingestion of more food than is necessary for the body's energy needs. Obesity can be defined on the basis of body weight, body mass index, or body

---

One liter of oxygen consumed by the body is estimated to produce 4.825 kcal of energy.

Metabolic energy can be used in three ways: for basal metabolism, for muscle contraction, and for the assimilation of food, which involves such processes as the production of digestive enzymes and the active transport of digested molecules. The **basal metabolic rate (BMR)** is the energy needed to keep the resting body functional. It is the metabolic rate calculated in expended kilocalories per square meter of body surface area per hour. BMR is measured when a person is awake but restful and has not eaten for 12 hours. A typical BMR for a 70-kg (154-lb) male is 38 kcal/m^2/hour.

Basal metabolism supports active transport mechanisms, muscle tone, maintenance of body temperature, beating of the heart, and other activities. A number of factors can affect the BMR. Males have a higher BMR than females, younger people have a higher BMR than older people, and fever can increase BMR. Greatly reduced kilocaloric input, such as during dieting or fasting, depresses BMR.

The daily input of energy should equal the energy demand of metabolism; otherwise, a person will gain or lose weight. For a 23-year-old, 70-kg (154-lb) male to maintain his weight, the input should be 2700 kcal/day; for a 58-kg (128-lb) female of the

same age, 2000 kcal/day is necessary. A pound of body fat contains about 3500 kcal. Reducing kilocaloric intake by 500 kcal/day can result in the loss of 1 lb of fat per week. Clearly, adjusting kilocaloric input is an important way to control body weight.

The other way to control weight is through energy expenditure. Physical activity through skeletal muscle movement greatly increases metabolic rate. In the average person, basal metabolism accounts for about 60% of energy expenditure, muscular activity 30%, and assimilation of food about 10%. Of these amounts, energy loss through muscular activity is the only component that a person reasonably can control. A comparison of the number of kilocalories gained from food and the number of kilocalories burned during exercise reveals why losing weight is a difficult task. For example, if brisk walking uses 225 kcal/h, it takes 20 min of brisk walking to burn off the 75 kcal in one slice of bread (75/225 = 0.33 h). Research suggests that a combination of appropriate physical activity and appropriate kilocaloric intake is the best approach to maintaining a healthy body composition and weight.

P R E D I C T   5
If watching TV uses 95 kcal/h, how long does it take to burn off the kilocalories in one cola or beer (see table 17.1)? If jogging at a pace of 6 mph uses 580 kcal/h, how long does it take to use the kilocalories in one cola or beer?

fat. "Desirable body weight" is listed in a table produced by the Metropolitan Life Insurance Company and indicates, for any height, the weight associated with a maximum life span. Overweight is defined as weighing 10% more than the "desirable weight," and obesity is weighing 20% more than the "desirable weight." **Body mass index (BMI)** can be calculated by dividing a person's weight (Wt) in kilograms by the square of his or her height (Ht) in meters: $BMI = Wt/Ht^2$. A BMI greater than 25–27 is overweight, and a value greater than 30 is defined as obese. About 10% of Americans have a BMI of 30 or greater. In terms of the percent of the total body weight contributed by fat, 15% body fat in men and 25% body fat in women is associated with reduced health risks. Obesity is defined to be more than 25% body fat in men and 30% to 35% in women.

The distribution of fat in obese individuals can vary. Fat can be found mainly in the upper body, such as in the abdominal region, or it can be associated with the hips and buttocks. These distribution differences can be clinically significant because upper body obesity is associated with an increased likelihood of diabetes mellitus, cardiovascular disease, stroke, and death.

In some cases, a specific cause of obesity can be identified. For example, a tumor in the hypothalamus can stimulate overeating. In most cases, however, no specific cause can be recognized. In fact, obesity can occur for many reasons, and obesity in an individual can have more than one cause. There seems to be a genetic component for obesity, and, if one or both parents are obese, their children are more likely to also be obese. Environmental factors such as eating habits, however, can also play an important role. For example, adopted children can exhibit similarities in obesity to their adoptive parents. In addition, psychological factors such as overeating as a means for dealing with stress can contribute to obesity.

In **hypertrophic** (hī-per-trof′ik, *hyper-*, above normal + *trophē*, nourishment) **obesity,** the number of adipose cells is usually normal, but the amount of fat contained in each adipose cell is increased. This type of obesity is characteristic of adult-onset obesity. People who were thin or of average weight and quite active when young become less active as they age. Although they no longer use as many kilocalories, they still consume the same amount of food as when they were younger. The excess kilocalories (see discussion, Metabolic Rate, on p. 500) are used to synthesize fat. In this type of obesity, the amount of fat in each adipose cell increases, and, if the amount of stored fat continues to increase, the total number of adipose cells may also increase. It is estimated that the average U.S. resident gains 1.25–1.5 lb of fat per year after age 25 and, at the same time, loses 0.25–0.5 lb of lean body weight (muscle mass) per year.

In **hyperplastic** (hī-per-plas′tik, *hyper-* + *plasis,* a molding) **obesity,** which is characteristic of juvenile-onset obesity, the number of adipose cells is increased. This condition may also be accompanied by an increase in cell size (hypertrophic obesity). Hyperplastic obesity has a very strong hereditary component, but family eating habits can also have a great influence. People with hyperplastic obesity are obese as children and become more obese with age. This type of obesity is a major health problem in school-aged children.

A CASE IN POINT | Gastric Bypass Surgery

Les Moore is a 35-year-old male who weighs 420 pounds. Although he has weighed less in the past as a result of weight loss from dieting, he not only regained the lost weight, but also additional weight, so that he ended up weighing more than he did before the dieting. Les has type II diabetes, high blood cholesterol, and high blood pressure. He elects to have a Roux-en-Y gastric bypass. In this procedure, the digestive tract is surgically rearranged to form a Y-shaped structure. One arm of the Y is formed by separating the superior part of the stomach to form a small pouch that is connected to the jejunum. The small size of the stomach pouch dramatically reduces the amount of food a person can eat. The other arm of the Y consists of the remainder of the stomach, the duodenum, and part of the jejunum. The body of the Y is the distal jejunum. Food from the esophagus enters the small stomach pouch and bypasses the rest of the stomach and the duodenum by entering the jejunum. Bile and pancreatic juices enter the duodenum and pass into the jejunum.

Roux-en-Y is the most commonly performed gastric bypass surgery in the United States. It results in permanent weight loss of over 50% of pre-surgery body weight for over 90% of patients. It also cures type II diabetes, lowers blood pressure, decreases cholesterol levels, reduces the risk of premature death, and improves the quality of life in most patients. Despite the benefits, gastric bypass has some negatives. It is major surgery—complications and even death are possible. Also, the normal digestive functions of the stomach and duodenum are bypassed as food moves quickly through the stomach pouch into the jejunum. As a result, vitamin and mineral deficiencies can develop. Of biggest concern are vitamin B_{12}, calcium, and iron deficiencies.

Body Temperature Regulation

Humans can maintain a relatively constant internal body temperature despite changes in the temperature of the surrounding environment. Maintenance of a constant body temperature is very important to homeostasis. Most enzymes are very temperature-sensitive and function only within narrow temperature ranges. Environmental temperatures are too low for normal enzyme function. The heat produced by metabolism and muscle contraction helps maintain the body temperature at a steady, elevated level that is high enough for normal enzyme function. Excessively high temperatures can alter enzyme structure, resulting in the loss of the enzyme's function.

Free energy is the total amount of energy that can be liberated by the complete catabolism of food. About 40% of the total energy released by catabolism is used to accomplish biological work such as anabolism, muscular contraction, and other cellular activities. The remaining energy is lost as **heat.**

P R E D I C T   6
Explain why we become warm during exercise, and explain the usefulness of shivering when it is cold.

Normal body temperature is regulated like other homeostatic conditions in the body. The average normal temperature is usually considered to be 37°C (98.6°F) when it is measured orally, and 37.6°C (99.7°F) when it is measured rectally. Rectal temperature comes closer to the true core body temperature, but an oral temperature is more easily obtained in older children and adults and therefore is the preferred measure. The normal oral temperature may vary from person to person, with a range of approximately 36.1–37.2°C (97–99°F).

Body temperature is maintained by balancing heat input with heat loss. Heat exchanged between the body and the environment occurs in a number of ways (figure 17.11). **Radiation** is the gain or loss of heat as infrared energy between two objects that are not in physical contact with each other. For example, heat can be gained by radiation from the sun, a hot coal, or the hot sand of a beach. On the other hand, heat can be lost as radiation to cool vegetation, water in the ocean, and snow on the ground. **Conduction** is the exchange of heat between objects that are in direct contact with each other, such as the bottom of the feet and the ground. **Convection** is a transfer of heat between the body and the air or water. A cool breeze results in movement of air over the body and loss of heat from the body. **Evaporation** is the conversion of water from a liquid to a gaseous form. As water evaporates from body surfaces, heat is lost.

The amount of heat exchanged between the environment and the body is determined by the difference in temperature between the body and the environment. The greater the temperature difference, the greater the rate of heat exchange. Control of the temperature difference can be used to regulate body

Figure 17.11   Heat Exchange

Heat exchange between a person and the environment occurs by convection, radiation, evaporation, and conduction. *Arrows* show the direction of net heat gain or loss in this environment.

temperature. If environmental temperature is very cold, such as on a winter day, there is a large temperature difference between the body and the environment, and there is a large loss of heat. The loss of heat can be decreased by behaviorally selecting a warmer environment, such as going inside a heated house, or by insulating the exchange surface by putting on extra clothes. Physiologically, temperature difference can be controlled through dilation and constriction of blood vessels in the skin. When these blood vessels dilate, they bring warm blood to the surface of the body, raising skin temperature, whereas blood vessel constriction decreases blood flow and lowers skin temperature (see figure 5.8).

P R E D I C T   7
Explain why constriction of skin blood vessels on a cold winter day is beneficial.

When environmental temperature is greater than body temperature, dilation of blood vessels to the skin brings blood

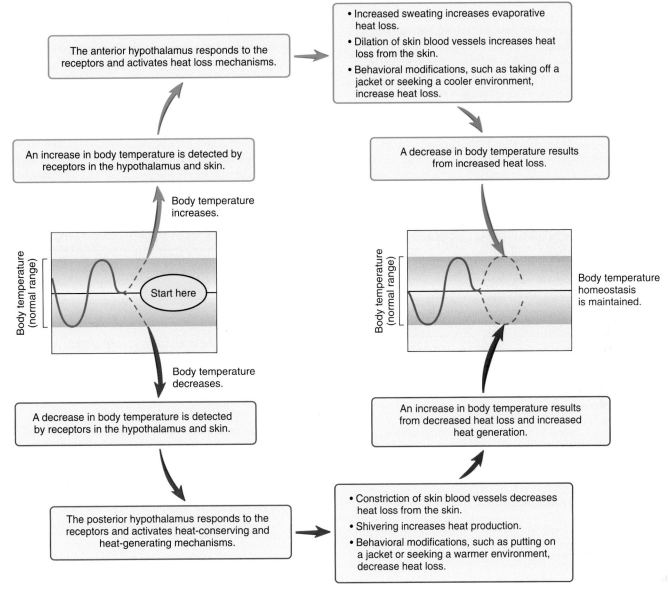

Homeostasis Figure 17.12 Temperature Regulation

to the skin, causing an increase in skin temperature that decreases the gain of heat from the environment. At the same time, evaporation carries away excess heat to prevent heat gain and overheating.

Body temperature regulation is an example of a negative-feedback system (figure 17.12). Maintenance of a specific body temperature is accomplished by neurons in the hypothalamus, which regulate body temperature around a set point. A small area in the anterior part of the hypothalamus can detect slight increases in body temperature through changes in blood temperature. As a result, mechanisms that cause heat loss, such as

dilation of blood vessels to the skin and sweating, are activated, and body temperature decreases. A small area in the posterior hypothalamus can detect slight decreases in body temperature and can initiate heat gain by increasing muscular activity (shivering) and by initiating constriction of blood vessels in the skin.

Under some conditions, the hypothalamus set point is actually changed. For example, during a fever, the set point is raised. Heat-conserving and heat-producing mechanisms are stimulated, and body temperature increases. In recovery from a fever, the set point is reduced to normal. Heat loss mechanisms are initiated, and body temperature decreases.

Too Hot or Too Cold

Hyperthermia is a condition in which heat gain exceeds heat loss in the body. Hyperthermia can result from exposure to a hot environment, exercise, or fever. Prolonged exposure to a hot environment can lead to **heat exhaustion,** a condition in which normal temperature reduction mechanisms are working, but they cannot keep pace with the excessive environmental heat, and the body temperature rises. Heat exhaustion is characterized by cool, wet skin caused by heavy sweating. Weakness, dizziness, and nausea usually occur as well. The heavy sweating can lead to dehydration, decreased blood volume, decreased blood pressure, and increased heart rate. Treatment involves increasing heat loss by moving the person to a cooler environment, decreasing heat production by decreasing muscular activity, and replacing lost body fluids. **Heat stroke** results from an increase in the level of the hypothalamic set point and is characterized by a dry, flushed skin

because sweating is inhibited. The person becomes confused and irritable, and can even become comatose. Treatment is the same as for heat exhaustion, but also involves increasing evaporation from the skin by applying water to the skin or by placing the person into cool water.

Hypothermia is a condition in which heat loss exceeds heat gain. The normal temperature increase mechanisms are working in the body, but they cannot keep pace with heat loss, and the body temperature decreases. Hypothermia usually results from a prolonged exposure to a cold environment or even to a cool, damp environment, because the moisture draws heat away from the body. Treatment for hypothermia is to rewarm the body at a rate of a few degrees per hour. **Frostbite** is local damage to the skin or deeper tissues resulting from prolonged exposure to a cold environment. The best treatment for frostbite is immersion in a warm water bath. Rubbing the affected area and local, dry heat should be avoided.

S U M M A R Y

Nutrition (p. 484)

1. Nutrition is the taking in and the use of food.

Nutrients

1. Nutrients are the chemicals used by the body and consist of carbohydrates, lipids, proteins, vitamins, minerals, and water.
2. Essential nutrients cannot be produced by the body or cannot be produced in adequate amounts.
3. The MyPyramid guide recommends the proper amounts of different food types and fiber necessary for good health, based on a person's age, sex, and physical activity.

Kilocalories

A kilocalorie is the energy required to raise the temperature of 1000 g of water from 14°C to 15°C. A kilocalorie (Calorie) is the unit of measurement used to express the energy content of food.

Carbohydrates

1. Carbohydrates include monosaccharides, disaccharides, and polysaccharides.
2. Most carbohydrates we ingest are from plants.
3. Carbohydrates are used as an energy source and for making DNA, RNA, and ATP.
4. The Acceptable Macronutrient Distribution Range (AMDR) for carbohydrates is 45% to 65% of total kilocalories.

Lipids

1. Lipids include triglycerides, phospholipids, steroids, and fat-soluble vitamins. Triglyceride is a major source of energy. Eicosanoids are involved in inflammation, tissue repair, and smooth muscle contraction. Cholesterol and phospholipids are part of the cell membrane. Some steroid hormones regulate the reproductive system.
2. The AMDR for lipids is 20% to 35%.

Proteins

1. Proteins are chains of amino acids.
2. Animal proteins tend to be complete proteins, whereas plant proteins tend to be incomplete.
3. Proteins are involved in structural strength, muscle contraction, regulation, buffering, clotting, transport, ion channels, receptors, and the immune system.
4. The AMDR for protein is 10% to 35% of total kilocalories.

Vitamins

1. Most vitamins are not produced by the body and must be obtained in the diet. Some vitamins can be formed from provitamins.
2. Vitamins are important in energy production, nucleic acid synthesis, growth, and blood clotting.
3. Vitamins are classified as either fat-soluble or water-soluble.
4. Recommended dietary allowances (RDAs) are a guide for estimating the nutritional needs of groups of people on the basis of their age, sex, and other factors.

Minerals

Minerals are essential for normal metabolic functions. They are involved with establishing the resting membrane potential; generating action potentials; adding mechanical strength to bones and teeth; combining with organic molecules; or acting as coenzymes, buffers, or regulators of osmotic pressure.

Daily Values

1. Daily Values are dietary references that can be used to help plan a healthy diet.
2. Daily Values for vitamins and minerals are based on Reference Daily Intakes (RDIs), which are generally the highest 1968 RDA value of an age category.
3. Daily Values are based on Daily Reference Values. The Daily Reference Values for energy-producing nutrients (carbohydrates, total fat, saturated fat, and proteins) and dietary fiber are recommended percentages of the total kilocalories ingested daily for each nutrient. The Daily Reference Values for total fats, saturated fats, cholesterol, and sodium are the uppermost limit considered desirable because of their link to diseases.
4. The % Daily Value is the percent of the recommended Daily Value of a nutrient found in one serving of a particular food.

Metabolism (p. 492)

1. Metabolism consists of anabolism and catabolism. Anabolism is the synthesis of molecules and requires energy. Catabolism is the breaking down of molecules and gives off energy.
2. The energy in carbohydrates, lipids, and proteins is used to produce ATP.
3. The energy from ATP can be used for active transport, muscle contraction, and the synthesis of molecules.

Regulation of Metabolism

1. Biochemical pathways are a series of chemical reactions, some of which release energy that can be used to synthesize ATP.
2. Each step in a biochemical pathway requires enzymes.
3. Enzyme synthesis is determined by DNA. Enzyme activity is modified by receptor-mediated and end-product processes.

Carbohydrate Metabolism

1. Glycolysis is the breakdown of glucose to two pyruvic acid molecules. Two ATP molecules are also produced.
2. Anaerobic respiration is the breakdown of glucose in the absence of oxygen to two lactic acid molecules and two ATP molecules.
3. Lactic acid can be converted to glucose using aerobically produced ATP; the necessary oxygen is the oxygen debt.
4. Aerobic respiration is the breakdown of glucose in the presence of oxygen to produce carbon dioxide, water, and 38 molecules of ATP. The first phase of aerobic metabolism is glycolysis. The second phase of aerobic metabolism is the conversion of pyruvic acid to acetyl-CoA. The third phase of aerobic metabolism is the citric acid cycle. The fourth phase is the electron-transport chain, which uses carrier molecules such as NADH to synthesize ATP.

Lipid Metabolism

1. Lipids are broken down in adipose tissue, and fatty acids are released into the blood.

2. Fatty acids are taken up by cells and broken down into acetyl-CoA, which can enter the citric acid cycle. Acetyl-CoA can also be converted into ketones by the liver. Ketones released from the liver into the blood are used as energy sources by other cells.

Protein Metabolism

1. Amino acids are used to synthesize proteins.
2. Amino acids can be used for energy, and ammonia is produced as a by-product. Ammonia is converted to urea and excreted by the kidneys.

Metabolic States

1. In the absorptive state, nutrients are used as energy or stored.
2. In the postabsorptive state, stored nutrients are used for energy.

Metabolic Rate

1. Metabolic rate is the total energy expenditure per unit of time.
2. Metabolic energy is used for basal metabolism, muscular activity, and the assimilation of food.

Body Temperature Regulation (p. 502)

1. Body temperature is a balance between heat gain and heat loss.
2. Heat is produced through metabolism.
3. Heat is exchanged through radiation, conduction, convection, and evaporation.
4. The greater the temperature difference, the greater the rate of heat exchange.
5. Body temperature is regulated at a "set point" by neural circuits in the hypothalamus.
6. Dilation of blood vessels in the skin and sweating can increase heat loss from the body.
7. Constriction of blood vessels in the skin and shivering promote heat gain by the body.

R E V I E W   A N D   C O M P R E H E N S I O N

1. Define a nutrient, and list the six major classes of nutrients. What is an essential nutrient?
2. What is a kilocalorie (Calorie)? Distinguish between a Calorie and a calorie.
3. List some sources of carbohydrates, fats, and proteins in the diet.
4. List the recommended consumption amounts of carbohydrates, fats, and proteins.
5. What are vitamins and provitamins? Name the water-soluble vitamins and the fat-soluble vitamins. List some of the functions of vitamins.
6. What are the Recommended Dietary Allowances (RDAs)?
7. List some of the minerals, and give their functions.
8. What are the Daily Values? How are the Daily Values related to total daily kilocaloric intake? Why are some Daily Values considered to be the uppermost amount that should be consumed?
9. Define a % Daily Value.
10. Define a biochemical pathway. How are the steps in a biochemical pathway controlled? What are three ways in which enzymes are regulated?
11. Describe glycolysis. What molecule is the end product of glycolysis? How many ATP and NADH molecules are produced?

12. What determines whether the pyruvic acid produced in glycolysis becomes lactic acid or acetyl-CoA?
13. Describe the two phases of anaerobic respiration. How many ATP molecules are produced? What happens to the lactic acid produced when oxygen becomes available?
14. Define aerobic respiration, and state how many ATP molecules are produced.
15. Describe the citric acid cycle.
16. What is the function of the electron-transport chain?
17. What happens to the carbon atoms in ingested food during metabolism? What happens to the oxygen we breathe in during metabolism?
18. Describe the events occurring during the absorptive and postabsorptive metabolic states.
19. What is meant by metabolic rate? Name three ways that the body uses metabolic energy.
20. Describe how heat is produced by and lost from the body. How is body temperature regulated?

CRITICAL   THINKING

1. One serving of a food has 2 g of saturated fat. What % Daily Value for saturated fat would appear on a food label for this food? (See bottom of figure 17.2 for information needed to answer this question.)

2. An active teenage boy has a daily intake of 3000 kcal/day. What is the maximum amount (weight) of total fats he should consume according to the Daily Values?

3. If the teenager in question 2 eats a food that has a total fat content of 10 g/serving, what is his total fat % Daily Value?

4. Suppose the food in question 3 is in a package that lists a serving size of $\frac{1}{2}$ cup with four servings in the package. If the teenager eats half the contents of the package (1 cup), how much of his % Daily Value does he consume?

5. Why can some people lose weight on a 1200 cal/k day diet and other people cannot?

6. Lotta Bulk, a bodybuilder, wanted to increase her muscle mass. Knowing that proteins are the main components of muscle, she consumed large amounts of protein daily (high-protein diet), along with small amounts of lipid and carbohydrate. Explain why this strategy will or will not work.

7. After consuming a high-protein diet for several days, does Lotta Bulk's urine contain less, the same, or more urea than before she consumed the proteins? Explain.

8. Thyroid hormone is known to increase the activity of the sodium–potassium exchange pump, which is an active-transport mechanism, therefore increasing the breakdown of ATP. If a person produced excess amounts of thyroid hormone, what effect would this have on basal metabolic rate, body weight, and body temperature? How might the body attempt to compensate for the changes in body weight and temperature?

9. On learning that sweat evaporation results in the loss of calories, an anatomy and physiology student decides that sweating is an easier way to lose weight than dieting. He knows that a liter (about a quart) of water weighs 1000 g, which is equivalent to 580,000 cal (or 580 kcal) of heat when lost as sweat. Instead of reducing his caloric intake by 580 kcal/day, if he loses a liter of sweat every day in the sauna he believes he will lose about a pound of fat a week. Will this approach work? Explain.

10. It is recommended that a person on a diet drink six to eight glasses of cool water per day. How could this practice help a person to lose weight?

11. In some diseases an infection results in a high fever. The patient is on the way to recovery when the crisis is over and body temperature begins to return to normal. If you were looking for symptoms in a patient who had just passed through the crisis state, would you look for a dry, pale skin or for a wet, flushed skin? Explain.

Answers in Appendix D

Visit this textbook's website at www.mhhe.com/seeleyess6 for practice quizzes, animations, interactive learning exercises, and other study tools.

Chapter Outline and Objectives

Functions of the Urinary System (p. 508)

Urinary System (p. 508)

Urine Production (p. 515)

Regulation of Urine Concentration and Volume (p. 522)

Urine Movement (p. 526)

Body Fluid Compartments (p. 526)

Regulation of Extracellular Fluid Composition (p. 528)

Regulation of Acid–Base Balance (p. 532)

Urinary System and Fluid Balance

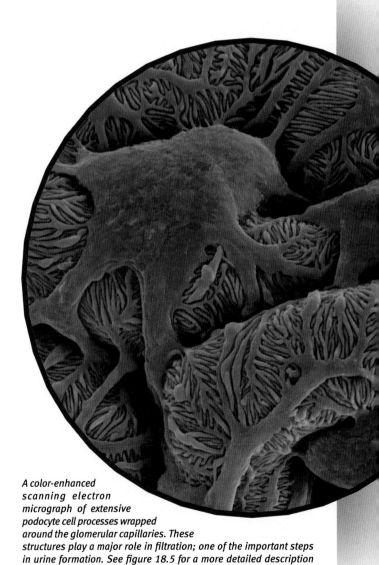

A color-enhanced scanning electron micrograph of extensive podocyte cell processes wrapped around the glomerular capillaries. These structures play a major role in filtration; one of the important steps in urine formation. See figure 18.5 for a more detailed description of the filtration membranes in the kidneys.

most people know that each person has two kidneys, their general location, and that they are essential to the maintenance of life, but fewer people are aware of the many functions the kidneys perform. Most people have a much better understanding of the function of the urinary bladder, and a great appreciation for the attention that is required when it is filled with the urine produced by the kidneys.

The **urinary** (ūr′i-nār-ē) **system** consists of two kidneys, two ureters, the urinary bladder, and the urethra (figure 18.1). A large volume of blood flows through the kidneys, which removes substances from the blood to form urine. The urine contains excess water and ions, metabolic wastes such as urea, and toxic substances consumed with food. The urine produced by the kidneys flows through the ureters to the urinary bladder, where it is stored until it is eliminated through the urethra.

The kidneys can suffer extensive damage and still maintain their extremely important role in the maintenance of homeostasis. As long as about one-third of one kidney remains functional, survival is possible. If the functional ability of the kidneys fails completely, however, death will result without special medical treatment.

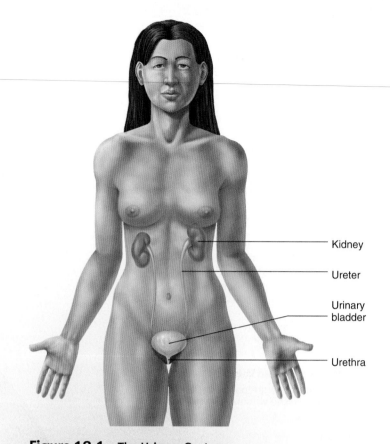

Figure 18.1   The Urinary System

The urinary system consists of two kidneys, two ureters, a urinary bladder, and a urethra.

Functions of the Urinary System

The major functions of the urinary system are performed by the kidneys, and the kidneys play the following essential roles in controlling the composition and volume of body fluids:

1. *Excretion.* The kidneys are the major excretory organs of the body. They remove waste products, many of which are toxic, from the blood. Most waste products are metabolic by-products of cells and substances absorbed from the intestine. The skin, liver, lungs, and intestines eliminate some of these waste products, but they cannot compensate if the kidneys fail to function.

2. *Regulation of blood volume and pressure.* The kidneys play a major role in controlling the extracellular fluid volume in the body by producing either a large volume of dilute urine or a small volume of concentrated urine. Consequently, blood volume and blood pressure are regulated by the kidneys.

3. *Regulation of the concentration of solutes in the blood.* The kidneys help regulate the concentration of the major molecules and ions such as glucose, Na^+, Cl^-, K^+, Ca^{2+}, HCO_3^-, and HPO_4^{2-}.

4. *Regulation of extracellular fluid pH.* The kidneys secrete variable amounts of H^+ to help regulate extracellular fluid pH.

5. *Regulation of red blood cell synthesis.* The kidneys secrete a hormone, erythropoietin, which regulates the synthesis of red blood cells in bone marrow (see chapter 11).

6. *Vitamin D synthesis.* The kidneys play an important role in controlling blood levels of Ca^{2+} by regulating the synthesis of vitamin D (see chapter 6).

Urinary System
Kidneys

The **kidneys** are bean-shaped organs, each about the size of a tightly clenched fist. They lie on the posterior abdominal wall, behind the peritoneum, with one kidney on either side of the vertebral column (figure 18.2). Structures that are behind the peritoneum are said to be **retroperitoneal** (re′trō-per′i-tō-nē′ăl). A connective tissue **renal** (Latin for kidney) **capsule** surrounds each kidney. Around the renal capsule is a thick layer of

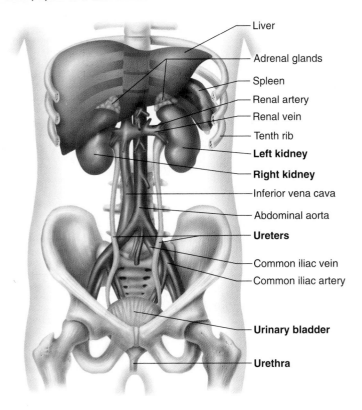

Anterior view

(a) The kidneys are located in the abdominal cavity, with the right kidney just below the liver and the left kidney below the spleen. The ureters extend from the kidneys to the urinary bladder within the pelvic cavity. An adrenal gland is located at the superior pole of each kidney.

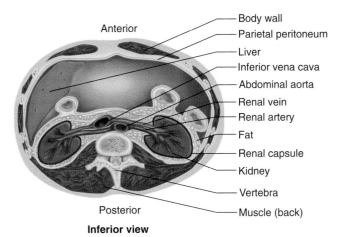

Inferior view

(b) The kidneys are located behind the parietal peritoneum. Fat surrounds each kidney. The renal arteries extend from the abdominal aorta to each kidney, and the renal veins extend from the kidneys to the inferior vena cava.

Figure 18.2 Anatomy of the Urinary System

fat, which protects the kidney from mechanical shock. On the medial side of each kidney is the **hilum** (hī′lŭm, a small amount), where the renal artery and nerves enter and where the renal vein and ureter exit the kidney (figure 18.3). The hilum opens into a cavity called the **renal sinus,** which contains blood

vessels, part of the system for collecting urine (see following discussion of calyces and renal pelvis), and fat.

The kidney is divided into an outer **cortex** and an inner **medulla,** which surround the renal sinus. The bases of several cone-shaped **renal pyramids** are located at the boundary between the cortex and the medulla, and the tips of the renal pyramids project toward the center of the kidney. A funnel-shaped structure called a **calyx** (kā′liks, cup of a flower) surrounds the tip of each renal pyramid. The calyces from all the renal pyramids join to form a larger funnel called the **renal pelvis.** The renal pelvis then narrows to form a small tube, the **ureter** (ū-rē′ter or ū′re-ter, urinary canal), which exits the kidney and connects to the urinary bladder. Urine passes from the tips of the renal pyramids into the calyces. From the calyces, urine collects in the renal pelvis and exits the kidney through the ureter (see figure 18.3).

The functional unit of the kidney is the **nephron** (nef′ron, Greek for kidney), and there are approximately 1.3 million of them in each kidney. Each nephron consists of a **renal corpuscle,** a **proximal tubule,** a **loop of Henle,** or nephronic loop, and a **distal tubule** (figure 18.4). Fluid enters the renal corpuscle and then flows into the proximal tubule. From there, it flows into the loop of Henle. Each loop of Henle has a descending limb, which extends toward the renal sinus, and an ascending limb, which extends back toward the cortex. The fluid flows through the ascending limb of the loop of Henle to the distal tubule. Many distal tubules empty into a **collecting duct,** which carries the fluid from the cortex, through the medulla. Many collecting ducts empty into a **papillary duct,** and the papillary ducts empty their contents into a calyx.

The renal corpuscle and both convoluted tubules are in the renal cortex (see figure 18.4). The collecting duct and loop of Henle enter the medulla. Approximately 15% of the nephrons, called **juxtamedullary** (next to the medulla) **nephrons,** have loops of Henle that extend deep into the medulla of the kidney. The other nephrons (85%), called **cortical nephrons,** have loops of Henle that do not extend deep into the medulla.

The renal corpuscle of the nephron consists of Bowman's capsule and the glomerulus (see figure 18.4; figure 18.5). **Bowman's capsule** consists of the enlarged end of the nephron, which is indented to form a double-walled chamber. The indentation is occupied by a tuft of capillaries called the **glomerulus** (glō-mār′ū-lŭs, *glomus,* a ball of yarn), which resembles a ball of yarn. The cavity of Bowman's capsule opens into the proximal tubule, which carries fluid away from the capsule. The inner layer of Bowman's capsule consists of specialized cells called **podocytes** (pod′ō-sīts, *pod,* foot + *kytos,* a hollow cell), which wrap around the glomerular capillaries. The outer layer of Bowman's capsule consists of simple squamous epithelial cells.

The glomerular capillaries have pores in their walls, and the podocytes have numerous cell processes with gaps between them. The endothelium of the glomerular capillaries, the podocytes, and the basement membrane between them form a **filtration membrane** (see figure 18.5*d*). In the first step of urine formation, fluid, consisting of water and solutes smaller than proteins, pass from the blood in the glomerular capillaries through the **filtration membrane** into Bowman's capsule (see figure 18.5*d*). The fluid that passes across the filtration membrane is called **filtrate.**

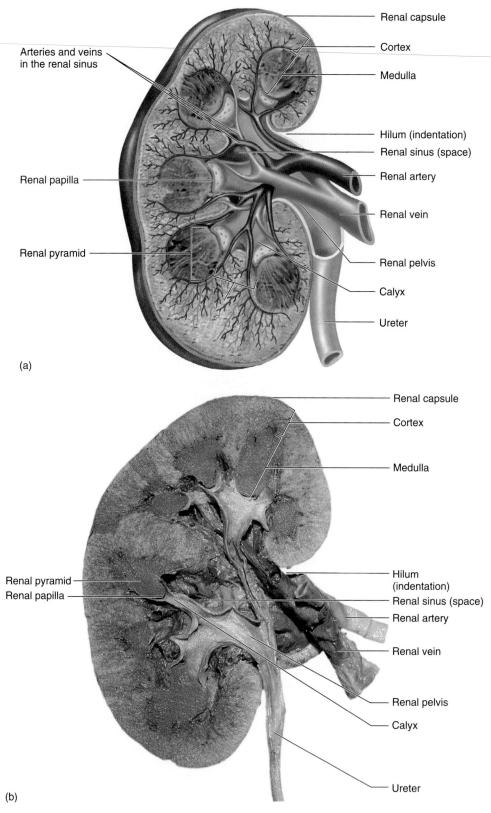

Figure 18.3 **Longitudinal Section of the Kidney**

(a) The cortex and the medulla of the kidney surround the renal sinus. The renal sinus is a space containing the renal pelvis, calyces, blood vessels, fat, and other connective tissues. The renal pyramids extend from the cortex of the kidney to the renal sinus. The tip of each renal pyramid is surrounded by a calyx. The calyces connect to the renal pelvis. Urine flows from the tip of the renal pyramid through the calyx and renal pelvis into the ureter. (b) Photograph of a longitudinal section through the kidney.

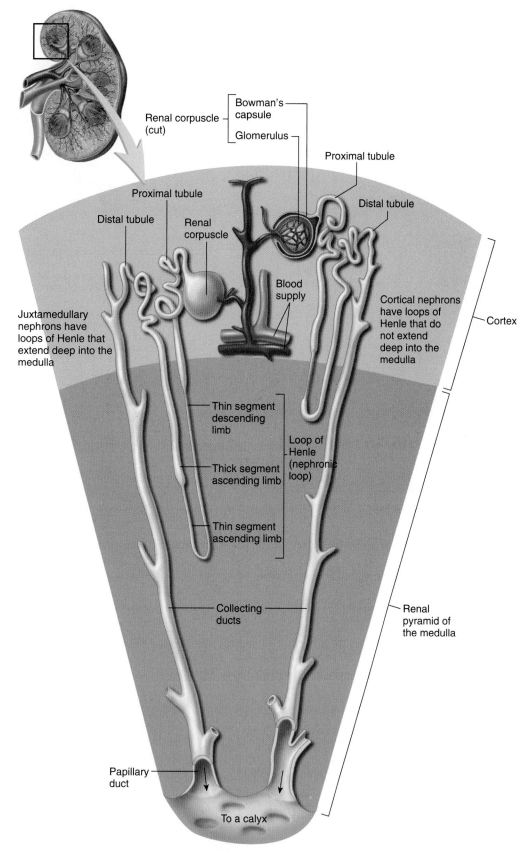

Renal corpuscle (cut)

Bowman's capsule

Glomerulus

Proximal tubule

Proximal tubule

Distal tubule

Distal tubule

Renal corpuscle

Blood supply

Juxtamedullary nephrons have loops of Henle that extend deep into the medulla

Cortical nephrons have loops of Henle that do not extend deep into the medulla

Cortex

Thin segment descending limb

Thick segment ascending limb

Loop of Henle (nephronic loop)

Thin segment ascending limb

Collecting ducts

Renal pyramid of the medulla

Papillary duct

To a calyx

Figure 18.4   Functional Unit of the Kidney—The Nephron

A nephron consists of a renal corpuscle, a proximal tubule, loop of Henle, and distal tubule. The distal tubule empties into a collecting duct. Juxtamedullary nephrons (those near the medulla of the kidney) have loops of Henle that extend deep into the medulla of the kidney, whereas cortical nephrons do not. Collecting ducts merge into larger papillary ducts, which empty into a calyx.

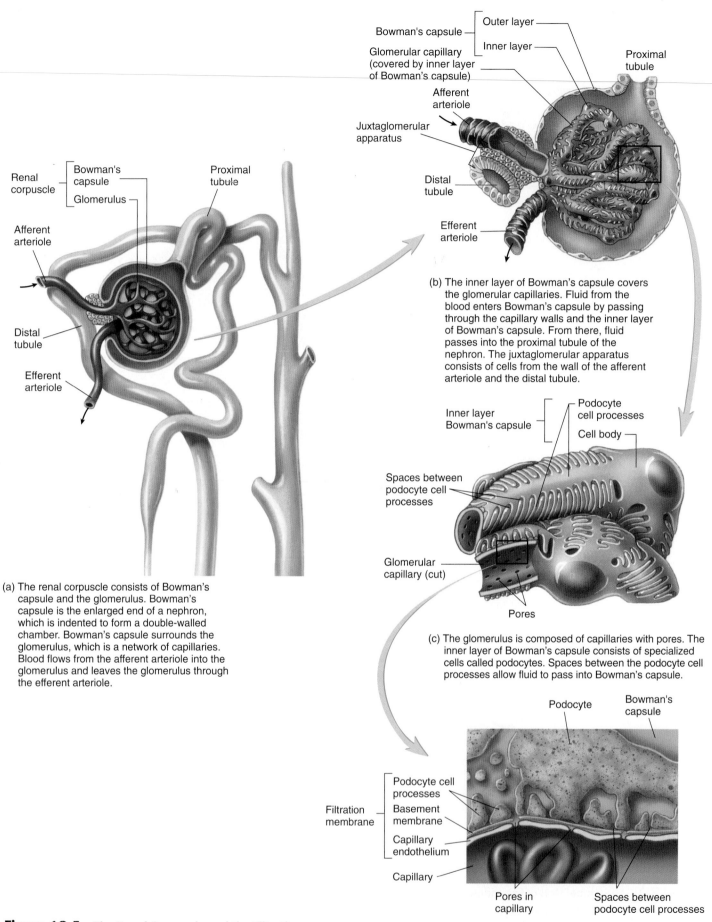

Bowman's capsule ⎯ Outer layer

Glomerular capillary (covered by inner layer of Bowman's capsule) ⎯ Inner layer

Proximal tubule

Afferent arteriole

Juxtaglomerular apparatus

Distal tubule

Efferent arteriole

(b) The inner layer of Bowman's capsule covers the glomerular capillaries. Fluid from the blood enters Bowman's capsule by passing through the capillary walls and the inner layer of Bowman's capsule. From there, fluid passes into the proximal tubule of the nephron. The juxtaglomerular apparatus consists of cells from the wall of the afferent arteriole and the distal tubule.

Renal corpuscle ⎯ Bowman's capsule / Glomerulus

Proximal tubule

Afferent arteriole

Distal tubule

Efferent arteriole

(a) The renal corpuscle consists of Bowman's capsule and the glomerulus. Bowman's capsule is the enlarged end of a nephron, which is indented to form a double-walled chamber. Bowman's capsule surrounds the glomerulus, which is a network of capillaries. Blood flows from the afferent arteriole into the glomerulus and leaves the glomerulus through the efferent arteriole.

Inner layer Bowman's capsule

Podocyte cell processes

Cell body

Spaces between podocyte cell processes

Glomerular capillary (cut)

Pores

(c) The glomerulus is composed of capillaries with pores. The inner layer of Bowman's capsule consists of specialized cells called podocytes. Spaces between the podocyte cell processes allow fluid to pass into Bowman's capsule.

Podocyte

Bowman's capsule

Filtration membrane

Podocyte cell processes

Basement membrane

Capillary endothelium

Capillary

Pores in capillary

Spaces between podocyte cell processes

(d) The filtration membrane consists of the capillary endothelial cells, a basement membrane, and the podocytes.

Figure 18.5 The Renal Corpuscle and the Filtration Membrane

The proximal tubules, the thick segment of the Henle's loops, the distal tubules, and the collecting ducts consist of simple cuboidal epithelium. The cuboidal epithelial cells have microvilli and many mitochondria. These portions of the nephron actively transport molecules and ions across the wall of the nephron. The thin segments of the descending and ascending limbs of Henle's loops have very thin walls made up of simple squamous epithelium. Water and solutes pass through the walls of these portions of the nephron by diffusion. The thin segment of the descending limb of Henle's loop is permeable to water and solutes, and the thin segment of the ascending limb is permeable to solutes, but not to water.

Arteries and Veins

The **renal arteries** branch off the abdominal aorta and enter the kidneys (figure 18.6a). They give rise to several branches. The **interlobar** (in-ter-lō′bar, between the lobes) **arteries** pass between the renal pyramids and give rise to the **arcuate** (ar′kū-āt, bowed or arched) **arteries,** which arch between the cortex and medulla. **Interlobular arteries** branch off the arcuate arteries to project into the cortex. The **afferent arterioles** arise from branches of the interlobular arteries and extend to the glomerular capillaries (figure 18.6b). **Efferent arterioles** extend from the glomerular capillaries to the **peritubular** (around the tubes) **capillaries,** which surround the proximal and distal tubules and the loops of Henle. The **vasa recta** (vā′să rek′tă, straight vessels) are specialized portions of the peritubular capillaries that extend deep into the medulla of the kidney and surround the loops of Henle and collecting ducts. Blood from the peritubular capillaries, including the vasa recta, enters the interlobular veins. The veins of the kidney run parallel to the arteries and have similar names (see figure 18.6).

A structure called the **juxtaglomerular** (jŭks′-tă-glŏ-mer′ū-lăr, close to the medulla) **apparatus** is formed where the distal tubule comes into contact with the afferent arteriole

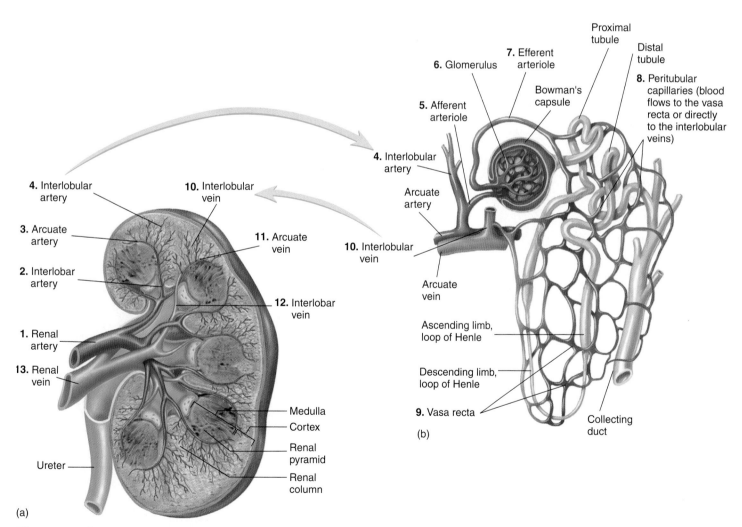

Figure 18.6 Blood Flow Through the Kidney

(a) Blood flow through the larger arteries and veins of the kidney. (b) Blood flow through the arteries, capillaries, and veins that provide circulation to the nephrons.

next to Bowman's capsule (see figure 18.5*b*). The juxtaglomerular apparatus consists of specialized cells of the walls of the distal tubules and collecting ducts. These cells secrete renin and play an important role in blood pressure regulation (see p. 523).

Ureters, Urinary Bladder, and Urethra

The **ureters** are small tubes that carry urine from the renal pelvis of the kidney to the posterior inferior portion of the urinary bladder (figure 18.7). The **urinary bladder** is a hollow muscular container that lies in the pelvic cavity just posterior to the pubic symphysis. It functions to store urine, and its size depends on the quantity of urine present. The urinary bladder can hold from a few milliliters (mL) to a maximum of about 1000 mL of urine. When the urinary bladder reaches a volume of a few hundred mL, the wall of the urinary bladder is stretched enough to activate a reflex that causes the smooth muscle of the urinary bladder to contract, and most of the urine flows out of the urinary bladder through the urethra.

The **urethra** is a tube that exits the urinary bladder inferiorly and anteriorly. The triangle-shaped portion of the urinary bladder located between the opening of the ureters and the opening of the urethra is called the **trigone** (trī′gōn, triangle). The urethra carries urine from the urinary bladder to the outside of the body.

The ureters and the urinary bladder are lined with transitional epithelium, which is specialized to stretch (see chapter 4). As the volume of the urinary bladder increases, the epithelial cells change from columnar to flat epithelial cells, and the number of epithelial cell layers decreases. As the volume of the urinary bladder decreases, transitional epithelial cells assume their columnar shape and form a greater number of cell layers.

The walls of the ureter and urinary bladder are composed of layers of smooth muscle and connective tissue. Regular waves of smooth muscle contractions in the ureters produce the force that causes urine to flow from the kidneys to the urinary bladder. Contractions of smooth muscle in the urinary bladder force urine to flow from the bladder through the urethra.

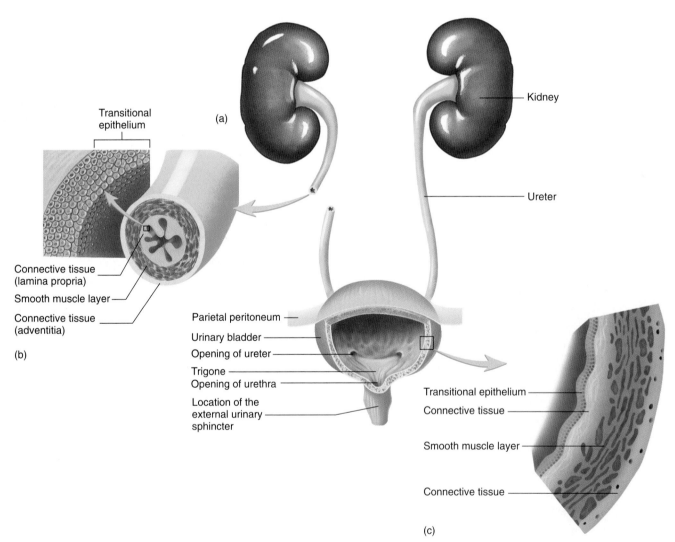

Figure 18.7 **Ureters and Urinary Bladder**

(*a*) Ureters extend from the renal pelvis to the urinary bladder. (*b*) The walls of the ureters and the urinary bladder are lined with transitional epithelium, which is surrounded by a connective tissue layer (lamina propria), smooth muscle layers, and a fibrous adventitia. (*c*) Section through the wall of the urinary bladder.

At the junction of the urinary bladder and urethra, the smooth muscle of the bladder wall forms the **internal urinary sphincter** in males. No well-defined internal urinary sphincter is found in females. Elastic fibers at the junction of the urinary bladder and urethra keep urine from passing through the urethra until the urinary bladder pressure increases. The internal urinary sphincter of males is under involuntary control. Contraction of the internal urinary sphincter during ejaculation prevents semen from entering the urinary bladder and keeps urine from flowing through the urethra. The **external urinary sphincter** is formed of skeletal muscle that surrounds the urethra as the urethra extends through the pelvic floor. The external urinary sphincter is under voluntary control. It controls the flow of urine through the urethra.

In the male, the urethra extends to the end of the penis, where it opens to the outside. The female urethra is much shorter (approximately 4 cm) than the male urethra (approximately 20 cm) and opens into the vestibule anterior to the vaginal opening.

P R E D I C T ①

Cystitis (sis-tī′tis), which is an inflammation of the urinary bladder, is often caused by bacterial infections. Typically bacteria from outside the body enter the bladder. Are males or females more prone to cystitis caused by urinary bladder infections? Explain.

A CASE IN POINT | Cystitis

Ima Burning was attending a 3-day business meeting out of town. On the last day of the meeting, she noticed a frequent urge to urinate, but her urine volume was small and a burning sensation accompanied urination. By the time she returned home, she had lower abdominal pain. Her urine appeared cloudy and it had an unpleasant odor. Ima made an appointment with her physician, Dr. Blatter, who had Ima collect a urine sample. Urine is normally sterile, but Ima's urine contained abundant bacteria. She has **cystitis** (sis-tī′tis) *cyst,* bladder + *itis,* inflammation), which is an inflammation of the urinary bladder usually resulting from a bacterial infection. Infection by the bacterium *E. coli* is the most common cause of cystitis. Dr. Blatter was unable to identify a specific cause of Ima's infection, and she explained to Ima that 30% of women experience cystitis during their lifetime. She prescribed an antibiotic. Within 3 days Ima was feeling normal again. A urine sample, taken several days later, showed no sign of infection. It is important to recognize cystitis early and treat it because the infection can ascend along the ureters and result in kidney infection.

Urine Production

Urine is mostly water and contains organic waste products such as urea, uric acid, and creatinine (krē-at′i-nēn), as well as excess ions, such as sodium (Na^+), potassium (K^+), chloride (Cl^-), bicarbonate (HCO_3^-), and hydrogen (H^+) (table 18.1). The three processes critical to the formation of urine are filtration, reabsorption, and secretion (figure 18.8).

Filtration is the movement of water, ions, and small molecules through the filtration membrane into Bowman's capsule. The portion of the plasma entering the nephron is called the **filtrate. Tubular reabsorption** is the movement of substances from the filtrate across the wall of the nephron back into the blood of the peritubular capillaries. Certain solute molecules and ions are reabsorbed by processes such as active transport or cotransport into the cells of the nephron wall and then from the cells of the nephron wall into the interstitial fluid. Water reabsorption occurs by osmosis across the wall of the nephron. The molecules and ions that enter interstitial fluid surrounding the nephron pass into the peritubular capillaries. In general, the useful substances that enter the filtrate are reabsorbed, and metabolic waste products remain in the filtrate and are eliminated. For example, when proteins are metabolized, ammonia is a by-product. Ammonia, which is toxic to humans, is converted into urea by the liver. Urea forms part of the filtrate and, although some of it is reabsorbed, much of it is eliminated in the urine. **Tubular secretion** is the transport of substances, usually waste products, from the interstitial fluid across the wall of the nephron into the filtrate. Urine produced by the nephrons therefore consists

Table 18.1	Concentrations of Major Substances in Urine (Average Values)	
Substance	**Plasma**	**Urine**
Water (L/day)		1.4
Organic molecules (mg/dL)		
Protein	3900–5000	0*
Glucose	100	0
Urea	26	1820
Uric acid	3	42
Creatinine	1	196
Ions (mEq/L)		
Na^+	142	128
K^+	5	60
Cl^-	103	134
HCO_3^-	28	14
Specific gravity (g/ml)†		1.005–1.030
pH		4.5–8.0

*Trace amounts of protein can be found in the urine.

†The specific gravity increases as the concentration of solutes in urine increase.

Urine formation results from the following three processes:

1. **Filtration**   Filtration (*blue arrow*) is the movement of materials across the filtration membrane into Bowman's capsule to form filtrate.

2. **Reabsorption**   Solutes are reabsorbed (*purple arrow*) across the wall of the nephron into the interstitial fluid by transport processes, such as active transport and cotransport.
       Water is reabsorbed (*green arrow*) across the wall of the nephron by osmosis.
       Water and solutes pass from the interstitial fluid into the peritubular capillaries.

3. **Secretion**   Solutes are secreted (*orange arrow*) across the wall of the nephron into the filtrate.

Process Figure 18.8   Urine Formation

of the substances that are filtered and secreted from the peritubular capillaries into the nephron, minus those substances that are reabsorbed.

Filtration

An average of 21% of the blood pumped by the heart each minute flows through the kidneys. Of the total volume of blood plasma that flows through the glomerular capillaries, about 19% passes through the filtration membrane into Bowman's capsule to become filtrate. In all of the nephrons of both kidneys, about 180 liters (L) of filtrate are produced each day, but only about 1% or less of the filtrate becomes urine because most of the filtrate is reabsorbed.

The filtration membrane allows some substances, but not others, to pass from the blood into Bowman's capsule. Water and solutes of small size readily pass through the openings of the filtration membrane, but blood cells and proteins, which are too large to pass through the filtration membrane, do not enter Bowman's capsule. Albumin, a small blood protein with a diameter slightly less than the openings in the filtration membrane, enters the filtrate in very small amounts. Negative charges on the albumin proteins are repelled by the negative charges of the filtration membrane. Consequently, the filtrate contains no cells and only a small amount of protein.

The formation of filtrate depends on a pressure gradient, called the **filtration pressure,** which forces fluid from the glomerular capillary across the filtration membrane into Bowman's capsule (figure 18.9). The filtration pressure results from forces that move fluid out of the glomerular capillary into Bowman's capsule minus the forces that move fluid out of Bowman's capsule into the glomerular capillary. The

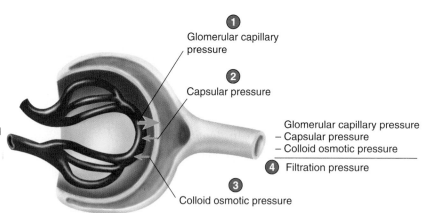

1. Glomerular capillary pressure, the blood pressure within the glomerulus, moves fluid from the blood into Bowman's capsule.

2. Capsular pressure, the pressure inside Bowman's capsule, moves fluid from the capsule into the blood.

3. Colloid osmotic pressure, produced by the concentration of blood proteins, moves fluid from Bowman's capsule into the blood by osmosis.

4. Filtration pressure is equal to the glomerular capillary pressure minus the capsular and colloid osmotic pressures.

Process Figure 18.9   Filtration Pressure

Filtration pressure across the filtration membrane is equal to the glomerular capillary pressure minus the capsular pressure and minus the colloid osmotic pressure.

glomerular capillary pressure is the blood pressure in the glomerular capillary. It is the major force causing fluid to move from the glomerular capillary across the filtration membrane into Bowman's capsule. Opposing the movement of fluid into the lumen of Bowman's capsule is the **capsular pressure** caused by the pressure of filtrate already inside Bowman's capsule, and the **colloid osmotic pressure** within the glomerular capillary. The colloid osmotic pressure exists because the plasma proteins do not pass through the filtration membrane. Instead, they remain within the glomerular capillary and produce an osmotic pressure that favors fluid movement to the glomerular capillary from Bowman's capsule. The filtration pressure is, therefore:

Glomerular capillary pressure
– Capsular pressure
– Colloid osmotic pressure
_____

Filtration pressure

The filtration pressure forces fluid from the glomerulus into Bowman's capsule because the glomerular capillary pressure is greater than both the capsular and colloid osmotic pressures. Under most conditions, the filtration pressure remains within a narrow range of values. However, when the filtration pressure increases, both the filtrate and urine volumes increase, and when the filtration pressure decreases, both the filtrate and urine volumes decrease.

The filtration pressure is influenced by the blood pressure in the glomerular capillaries, the blood protein concentration, and the pressure in Bowman's capsules. The blood pressure is normally higher in the glomerular capillaries than it is in most capillaries. The filtration pressure increases if the blood pressure in the glomerular capillaries increases further. The filtration pressure decreases if the blood pressure in the glomerular capillary decreases.

The concentration of proteins in the blood opposes the effect of blood pressure on the filtration pressure because of osmosis (see chapter 3). An increase in blood protein concentration increases the movement of water by osmosis into the blood and, therefore, reduces the filtration pressure. A decrease in blood protein concentration decreases the movement of water by osmosis into the blood, which increases the filtration pressure.

The blood pressure within the glomerular capillaries is fairly constant because the afferent and efferent arterioles either dilate or constrict to regulate the blood pressure in the glomerular capillaries even though the systemic blood pressure may fluctuate substantially. Also, the concentration of blood proteins and the pressure inside Bowman's capsule are fairly constant. As a consequence, the filtration pressure and the rate of filtrate formation are maintained within a narrow range of values most of the time.

The filtration pressure does change dramatically under some conditions. Strong sympathetic stimulation in response to periods of excitement, rigorous physical activity, or emergency conditions cause renal blood vessels to undergo vasoconstriction. The blood pressure in the glomerular

capillaries decreases, causing the filtration pressure to decrease. The rate of filtrate and urine formation can be reduced to nearly zero.

Decreases in the concentration of plasma proteins, caused by conditions such as inflammation of the liver where most of the blood proteins are produced, increase the filtration pressure. The increased filtration pressure causes the filtrate and urine volume to increase.

Cardiovascular Shock

During **cardiovascular shock,** renal blood vessels constrict, and blood flow to the kidneys is decreased to a very low rate. The filtration pressure and filtrate formation fall to very low levels. One of the dangers of cardiovascular shock is that the renal blood flow can be so low that the kidneys suffer from a lack of oxygen. If the oxygen level remains too low for a long enough time, permanent kidney damage or complete failure of the kidneys results. One important reason for treating cardiovascular shock quickly is to avoid damage to the kidneys.

Reabsorption

As the filtrate flows from Bowman's capsule through the proximal tubule, loop of Henle, distal tubule, and collecting duct, many of the solutes in the filtrate are reabsorbed. About 99% of the original filtrate volume is reabsorbed and enters the peritubular capillaries. The reabsorbed filtrate flows through the renal veins to enter the general circulation. Only 1% of the original filtrate volume becomes urine (figure 18.10). Because excess ions and metabolic waste products are not readily reabsorbed, the small volume of urine produced contains a high concentration of ions and metabolic waste products.

The proximal tubule is the primary site for the reabsorption of solutes and water. The cuboidal cells of the proximal tubule have numerous microvilli and mitochrondria, and they are well adapted to transport molecules and ions across the wall of the nephron by active transport and cotransport. Substances transported from the proximal tubule include proteins; amino acids; glucose; and fructose molecules; as well as Na^+, K^+, Ca^{2+}, HCO_3^-, and Cl^-. The proximal tubule is permeable to water. As solute molecules are transported out of the proximal tubule into the interstial fluid, water moves by osmosis in the same direction. The solutes and water then enter the peritubular capillaries. Consequently, 65% of the filtrate volume is reabsorbed from the proximal tubule (see figures 18.10, step 1, and 18.11).

The descending limb of the loop of Henle functions to further concentrate the filtrate. The renal medulla contains very concentrated interstitial fluid that has large amounts of Na^+, Cl^-, and urea. The wall of the thin segment of the descending limb is permeable to water and moderately permeable to solutes. As the filtrate passes through the descending limb of the loop of Henle into the medulla of the kidney, water moves out of the nephron by osmosis, and some solutes move into the nephron by diffusion. By the time the filtrate

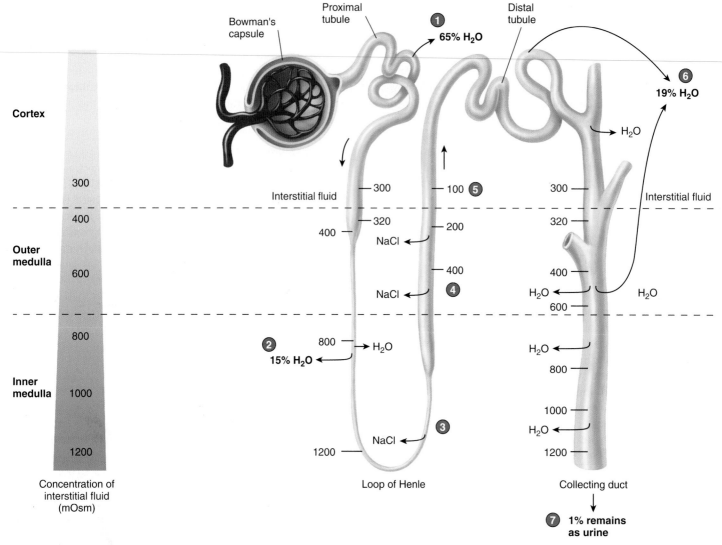

Cortex

Outer medulla

Inner medulla

Concentration of
interstitial fluid
(mOsm)

Loop of Henle

Collecting duct

7 **1% remains
as urine**

Process Figure 18.10   Summary of Filtrate Volume Reabsorption from the Filtrate

The concentration gradient from the cortex to the inner medulla is shown on the left. Interstitial fluid increases in concentration from 300 mOsm/L in the cortex to 1200 mOsm/L in the medulla. The concentration of the filtrate in different parts of the nephron are also shown.

1. Approximately 180 L of filtrate enters the nephrons each day; of that volume 65% is reabsorbed in the proximal tubule. In the proximal tubule, solute molecules move by active transport and cotransport from the lumen of the tubule into the interstitial fluid. Water moves by osmosis because the cells of the tubule wall are permeable to water (see figure 18.11).

2. Approximatley 15% of the filtrate volume is reabsorbed in this segment of the descending limb of the loop of Henle. The descending limb passes through the concentrated interstitial fluid of the medulla. Because the wall of the descending limb is permeable to water, water moves by osmosis from the tubule into the more concentrated interstitial fluid (see figure 18.12a). By the time the filtrate reaches the tip of the renal pyramid, the concentration of the filtrate is equal to the concentration of the interstitial fluid.

3. The ascending limb of the loop of Henle is not permeable to water. Solutes diffuse out of the thin segment (see figure 18.12b).

4. Na$^+$ are actively transported, and K$^+$ and Cl$^-$ are cotransported, from the filtrate of the thick segment into the interstitial fluid (see figure 18.13).

5. The volume of the filtrate doesn't change as it passes through the ascending limb, but the concentration is greatly reduced (see figure 18.13). By the time the filtrate reaches the cortex of the kidney, the concentration is approximately 100 mOsm/L, which is less concentrated than the interstitial fluid of the cortex (300 mOsm/L).

6. The distal tubule and collecting duct are permeable to water if ADH is present. If ADH is present, water moves by osmosis from the less concentrated filtrate into the more concentrated interstitial fluid (see figure 18.14). By the time the filtrate reaches the tip of the renal pyramid an additional 19% of the filtrate is reabsorbed.

7. One percent or less remains as urine.

- Solute molecules such as amino acids, glucose, and fructose, as well as Na$^+$, K$^+$, Ca^{2+}, HCO$_3^-$, and Cl$^-$ are reabsorbed by active transport and cotransport out of the proximal tubule by the epithelial cells.

- Water moves by osmosis out of the proximal tubule.

- Solutes plus 65% of the filtrate volume are reabsorbed from the proximal tubule and enter the peritubular capillary.

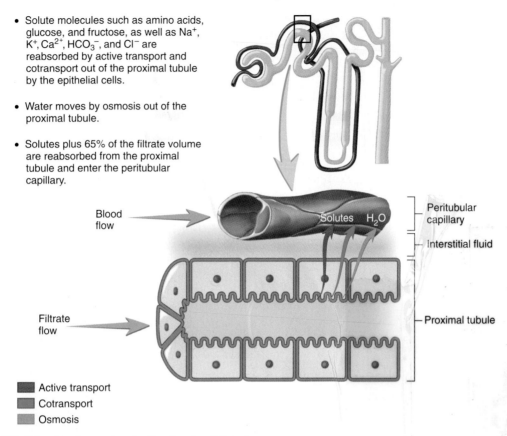

Blood flow

Solutes   H$_2$O

Peritubular capillary

Interstitial fluid

Filtrate flow

Proximal tubule

Active transport
Cotransport
Osmosis

Process Figure 18.11   Reabsorption in the Proximal Tubule

has passed through the descending limb, another 15% of the filtrate volume has been reabsorbed, and the filtrate is as concentrated as the interstitial fluid of the medulla. The reabsorbed filtrate enters the vasa recta (see figures 18.10, step 2, and 18.12).

The ascending limb of the loop of Henle functions to dilute the filtrate by removing solutes. The thin segment of the ascending limb is not permeable to water, but it is permeable to solutes. Consequently solutes diffuse out of the nephron (see figures 18.10, step 3, and 18.12).

The cuboidal epithelial cells of the thick segment of the ascending limb actively transport Na$^+$ out of the nephron, and K$^+$ and Cl$^-$ are cotransported with Na$^+$. The thick segment of the ascending limb is not permeable to water. As a result, Na$^+$, K$^+$, and Cl$^-$, but little water, are removed from the filtrate (see figure 18.10, step 4). Because of the efficient removal of these solutes, the highly concentrated filtrate that enters the ascending limb of Henle's loop is converted to a dilute solution by the time it reaches the distal tubule (see figures 18.10, step 5, and 18.13). As the filtrate enters the distal tubule,

it is more dilute than the interstitial fluid of the renal cortex. Also, because of the volume of filtrate reabsorbed in the proximal tubule and the descending limb of Henle's loop, only about 20% of the original filtrate volume remains. The solutes transported from the ascending limb of the loop of Henle enter the interstitial fluid of the medulla and help keep the concentration of solutes in the medulla high. Excess solutes enter the vasa recta.

The cuboidal cells of the distal tubule and collecting duct function to remove water and additional solutes. Na$^+$ and Cl$^-$ are reabsorbed. Sodium ions are actively transported and Cl$^-$ are cotransported. Also, 19% of the original filtrate volume is reabsorbed by osmosis, leaving about 1% of the original filtrate as urine (see figures 18.10, steps 6 and 7, and 18.14). The reabsorbed water and solutes from the distal tubule enter the peritubular capillaries and enter the vasa recta from the collecting ducts.

The reabsorption of water and solutes from the distal tubule and collecting duct is controlled by hormones, which have a great influence on urine concentration and volume (see

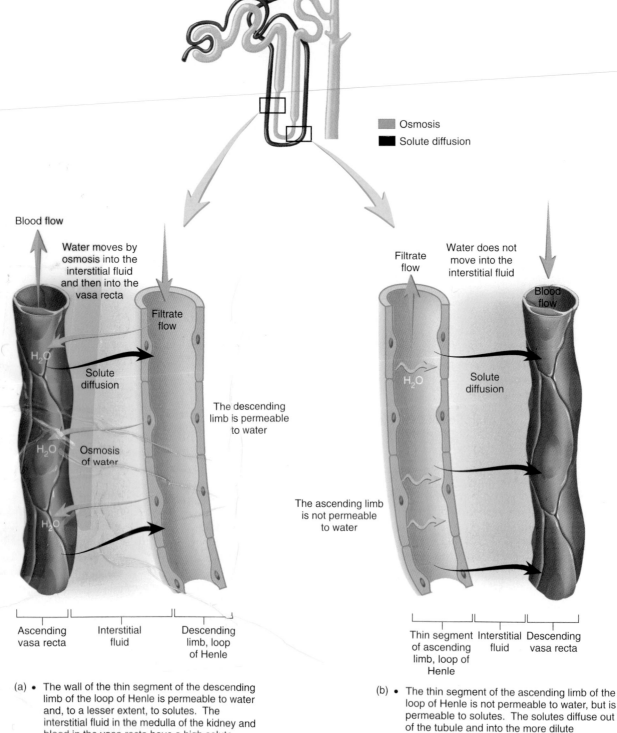

Osmosis
Solute diffusion

Blood flow

Water moves by osmosis into the interstitial fluid and then into the vasa recta

H₂O

Solute diffusion

H₂O   Osmosis of water

H₂O

Filtrate flow

The descending limb is permeable to water

| Ascending vasa recta | Interstitial fluid | Descending limb, loop of Henle |

(a) • The wall of the thin segment of the descending limb of the loop of Henle is permeable to water and, to a lesser extent, to solutes. The interstitial fluid in the medulla of the kidney and blood in the vasa recta have a high solute concentration (high osmolality). Water therefore moves by osmosis from the tubule into the interstitial fluid and into the vasa recta. An additional 15% of the filtrate volume is reabsorbed.

 • To a lesser extent, solutes diffuse from the vasa recta and interstitial fluid into the tubule.

Filtrate flow

Water does not move into the interstitial fluid

Blood flow

H₂O

Solute diffusion

The ascending limb is not permeable to water

| Thin segment of ascending limb, loop of Henle | Interstitial fluid | Descending vasa recta |

(b) • The thin segment of the ascending limb of the loop of Henle is not permeable to water, but is permeable to solutes. The solutes diffuse out of the tubule and into the more dilute interstitial fluid as the ascending limb projects toward the cortex. The solutes diffuse into the descending vasa recta.

Process Figure 18.12 Reabsorption in the Loop of Henle: The Descending Limb and the Thin Segment of the Ascending Limb

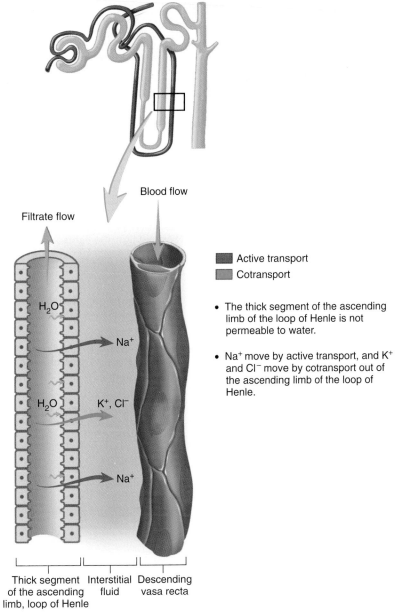

Blood flow

Filtrate flow

■ Active transport
■ Cotransport

- The thick segment of the ascending limb of the loop of Henle is not permeable to water.

- Na⁺ move by active transport, and K⁺ and Cl⁻ move by cotransport out of the ascending limb of the loop of Henle.

H_2O

Na^+

H_2O

K^+, Cl^-

Na^+

| Thick segment of the ascending limb, loop of Henle | Interstitial fluid | Descending vasa recta |

Process Figure 18.13   Reabsorption in the Loop of Henle: The Thick Segment of the Ascending Limb

under Regulation of Urine Concentration and Volume, which follows).

In summary, most of the useful solutes that pass through the filtration membrane into Bowman's capsule are reabsorbed in the proximal tubule. Filtrate volume is reduced by 65% in the proximal tubule and by 15% in the descending limb of the loop of Henle. In the ascending limb of the loop of Henle, Na^+, K^+, and Cl^-, but little water, is removed from the filtrate. Consequently, the filtrate becomes dilute. In the distal tubule and the collecting duct, additional Na^+ and Cl^- are removed, water moves out by osmosis, and the filtrate volume is reduced by another 19%, leaving 1% of the original filtrate volume as urine.

P R E D I C T
People who suffer from untreated diabetes mellitus can experience very high levels of glucose in the blood. The glucose can easily cross the filtration membrane into Bowman's capsule. Normally all the glucose is reabsorbed from the nephron. If the concentration of glucose in the nephron becomes too high, however, not all the glucose can be reabsorbed because the number of transport molecules in the cells of the proximal tubule is limited. How does the volume of urine produced by a person with untreated diabetes mellitus differ from that of a normal person, and how does the concentration of the urine differ from that of a normal person?

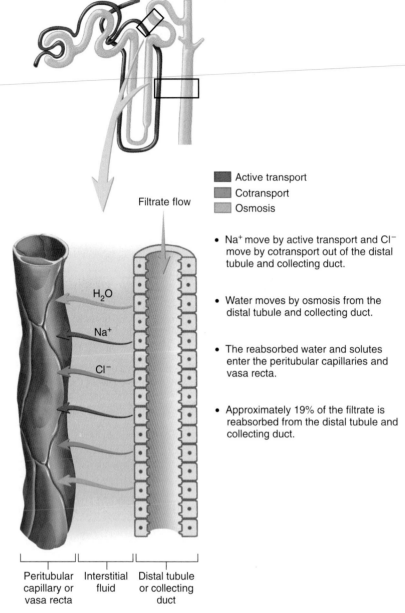

Active transport
Cotransport
Osmosis

- Na^+ move by active transport and Cl^- move by cotransport out of the distal tubule and collecting duct.

- Water moves by osmosis from the distal tubule and collecting duct.

- The reabsorbed water and solutes enter the peritubular capillaries and vasa recta.

- Approximately 19% of the filtrate is reabsorbed from the distal tubule and collecting duct.

Filtrate flow

H_2O

Na^+

Cl^-

Peritubular capillary or vasa recta | Interstitial fluid | Distal tubule or collecting duct

Process Figure 18.14   Reabsorption in the Distal Tubule and Collecting Duct

Secretion

Some substances, including by-products of metabolism that become toxic in high concentrations and drugs or molecules not normally produced by the body, are secreted into the nephron from the peritubular capillaries. As with tubular reabsorption, tubular secretion can be either active or passive. For example, ammonia diffuses into the lumen of the nephron, whereas H^+, K^+, creatinine, histamine, and penicillin are actively transported into the nephron.

Hydrogen ions are actively transported into the proximal tubule. The epithelial cells actively transport large quantities of H^+ across the wall of the nephron into the filtrate. The secretion of H^+ plays an important role in the regulation of the body fluid pH.

In the proximal tubule, K^+ are reabsorbed. In the distal tubule and collecting duct, K^+ are secreted. Overall, there is a net loss of K^+ in the urine.

Regulation of Urine Concentration and Volume

Given a solution in a container, such as a pan on a stove, it is possible to decrease the concentration of the solution by adding water to it. It is also possible to increase the concentration of the solution by boiling the water in the pan, thus removing water from the solution by evaporation. Similarly, the kidneys function to maintain the concentration of the body fluids by increasing water reabsorption from the filtrate when the body fluid concentration increases and by reducing water reabsorption from the filtrate when the body fluid concentration decreases. The volume and composition of urine therefore changes, depending on conditions in the body. If body fluid concentration increases above normal levels, the kidneys produce a smaller than normal amount of concentrated urine. This eliminates solutes and conserves water, both of which help to lower the body fluid concentration back to normal. On the other hand, if the body fluid concentration decreases, the kidneys produce a large volume of dilute urine. As a result, water is lost, solutes are conserved, and the body fluid concentration increases.

Urine production also maintains blood volume and therefore blood pressure. An increase in blood volume can increase blood pressure, and a decrease in blood volume can decrease blood pressure. When blood volume increases above normal, the kidneys produce a large volume of urine. The loss of water in the urine lowers blood volume.

Conversely, if blood volume decreases below normal, the kidneys produce a small volume of urine to conserve water and maintain blood volume.

Hormonal Mechanisms

Antidiuretic Hormone

Antidiuretic (an′tē-dī-ū-ret′ik) **hormone (ADH),** secreted by the posterior pituitary gland, passes through the circulatory system to the kidneys. ADH regulates the amount of water reabsorbed by the distal tubules and collecting ducts. When ADH levels increase, the permeability of the distal tubules and collecting ducts to water increases, and more water is reabsorbed from the filtrate. Consequently, an increase in ADH results in the production of a small volume of concentrated urine. On the other hand, when ADH levels decrease, the distal tubules and collecting ducts become less permeable to water. As a result, less water is reabsorbed, and a large volume of dilute urine is produced (figure 18.15 and see figure 18.18).

The release of ADH from the posterior pituitary is regulated by the hypothalamus. Certain cells of the hypothalamus are sensitive to changes in the solute concentration of the interstitial fluid within the hypothalamus. An increased solute concentration (increased osmolality) of the blood and interstitial fluid results in action potentials being sent along the axons of the ADH-secreting neurons of the hypothalamus to the posterior pituitary, causing ADH to be released from the ends of the axons (see figure 18.15 and chapter 10). A reduced solute concentration (reduced osmolality) in the blood and interstitial fluid within the hypothalamus causes inhibition of ADH release.

Baroreceptors that monitor blood pressure also influence ADH secretion. A large decrease in blood pressure causes an increase in ADH secretion (see figure 18.15 and figure 18.18), and a large increase in blood pressure decreases ADH secretion.

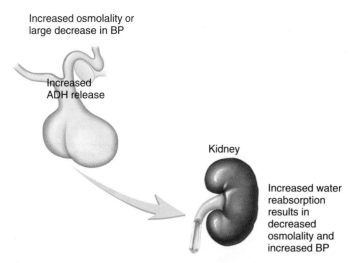

Increased osmolality or large decrease in BP

Increased ADH release

Kidney

Increased water reabsorption results in decreased osmolality and increased BP

Figure 18.15 **ADH and the Regulation of Extracellular Fluid Osmolality**

Increased blood osmolality affects hypothalamic neurons, and decreased blood pressure affects baroreceptors. As a result of these stimuli, an increased rate of antidiuretic hormone (ADH) secretion from the posterior pituitary results, which increases water reabsorption by the kidney.

Diabetes Insipidus

Diabetes insipidus is a pathological condition in which the posterior pituitary fails to secrete ADH or the kidney tubules have abnormal receptors for ADH and do not respond to the presence of ADH. In people suffering from diabetes insipidus, much of the filtrate entering the proximal and distal tubules becomes urine. People with this condition can produce as much as 20–30 L of urine each day. Because they lose so much water, they are continually in danger of severe dehydration. Even though they produce dilute urine, producing such a large volume of urine also results in the loss of Na$^+$, Ca^{2+}, and other ions. The resulting ionic imbalances cause the nervous system and cardiac muscle to function abnormally. People suffering from diabetes insipidus due to reduced ADH secretion can be successfully treated by taking ADH in the form of either a nasal spray or an injection.

Renin–Angiotensin–Aldosterone

Renin (rē′nin or ren′in) and **angiotensin** (an′jē-ō-ten′sin) help regulate **aldosterone** (al-dos′ter-ōn) secretion. Renin is secreted by cells of the juxtaglomerular apparati in the kidneys (see figure 18.5b). Renin is an enzyme that acts on a protein produced by the liver called **angiotensinogen** (an′jē-ō-ten-sin′ō-jen). Amino acids are removed from angiotensinogen, leaving **angiotensin I.** Angiotensin I is rapidly converted to a smaller peptide called **angiotensin II** by **angiotensin-converting enzyme (ACE).** Angiotensin II acts on the adrenal cortex, causing it to secrete aldosterone (see chapter 13).

Aldosterone increases the rate of active transport of Na$^+$ in the distal tubules and collecting ducts. In the absence of aldosterone, large amounts of Na$^+$ remain in the nephron and become part of the urine. A high Na$^+$ concentration in

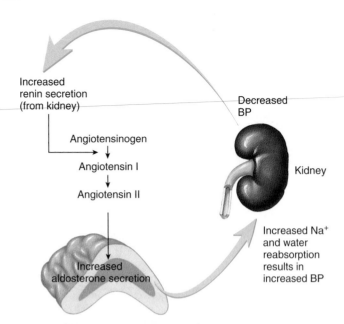

Figure 18.16 Aldosterone and the Regulation of Na⁺ and Water in Extracellular Fluid

Low blood pressure (BP) stimulates renin secretion from the kidney. Renin stimulates the production of angiotensin I, which is converted to angiotensin II, which in turn stimulates aldosterone secretion from the adrenal cortex. Aldosterone increases Na⁺ and water reabsorption in the kidney.

the filtrate causes water to remain in the nephrons and increases urine volume. When the rate of active transport of Na^+ is slow, urine volume therefore, increases, and the urine contains a high concentration of Na^+. Because Cl^- are attracted by the positive charge on Na^+, Cl^- are cotransported with Na^+.

When blood pressure suddenly decreases (figures 18.16 and 18.18) or when the concentration of Na^+ in the blood becomes too low, renin is released from the kidney. The resultant increase in aldosterone causes an increase in Na^+ and Cl^- reabsorption from the nephrons. Water follows the Na^+ and Cl^-. Thus the volume of water lost in the form of urine declines. This method of conserving water helps prevent a further decline in blood pressure (see figures 18.16 and 18.18).

P R E D I C T   ③
Drugs that increase the urine volume are called **diuretics** (dī-ū-ret′iks). Some diuretics inhibit the active transport of Na⁺ in the nephron. Explain how these diuretic drugs could cause an increase in urine volume.

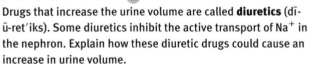

Alcohol and Caffeine

Alcohol and caffeine are examples of diuretics. Alcohol inhibits the secretion of ADH from the posterior pituitary. Consequently, the consumption of alcoholic beverages results in the production of a large volume of dilute urine. The volume of urine lost easily can exceed the

volume of water consumed with the alcohol. Consequently, several hours after drinking alcoholic beverages, such as the morning after an intense celebration, dehydration and intense thirst can occur. Caffeine also is a diuretic. Caffeine and related substances act on the kidneys by increasing blood flow to the kidney and by increasing the loss of Na⁺ and Cl⁻ in the urine. Both the increased blood flow to the kidney and the increased loss of Na⁺ and Cl⁻ in the urine increase urine volume.

Atrial Natriuretic Hormone

Atrial natriuretic (nā′trē-ū-ret′ik) **hormone (ANH)** is secreted from cardiac muscle cells in the right atrium of the heart when blood pressure in the right atrium increases above normal values (see figures 18.17 and 18.18). Atrial natriuretic hormone acts on the kidney to decrease Na^+ reabsorption. Sodium ions and water, therefore, remain in the nephron to become urine. The increased loss of Na^+ and water as urine reduces the blood volume and the blood pressure.

P R E D I C T   ④
Ivy Saline worked as a nurse in a hospital. Because she was very observant, she recognized that one of her patients received a much larger volume of an intravenous (IV) saline solution than she was supposed to receive. A saline solution consists of NaCl but it also contains other solutes such as small amounts of KCl. Saline solutions have the same concentration as body fluids. Predict the effect of the large volume of IV saline solution on the rate of urine production, and describe the role of ADH, ANH, and aldosterone in the control of the change in urine production.

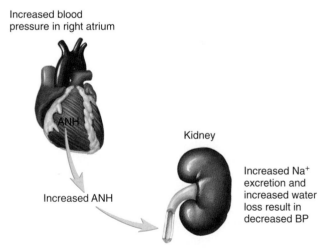

Figure 18.17 ANH and the Regulation of Na⁺ and Water in Extracellular Fluid

Increased blood pressure in the right atrium of the heart causes increased secretion of atrial natriuretic hormone (ANH), which increases Na⁺ excretion and water loss in the form of urine.

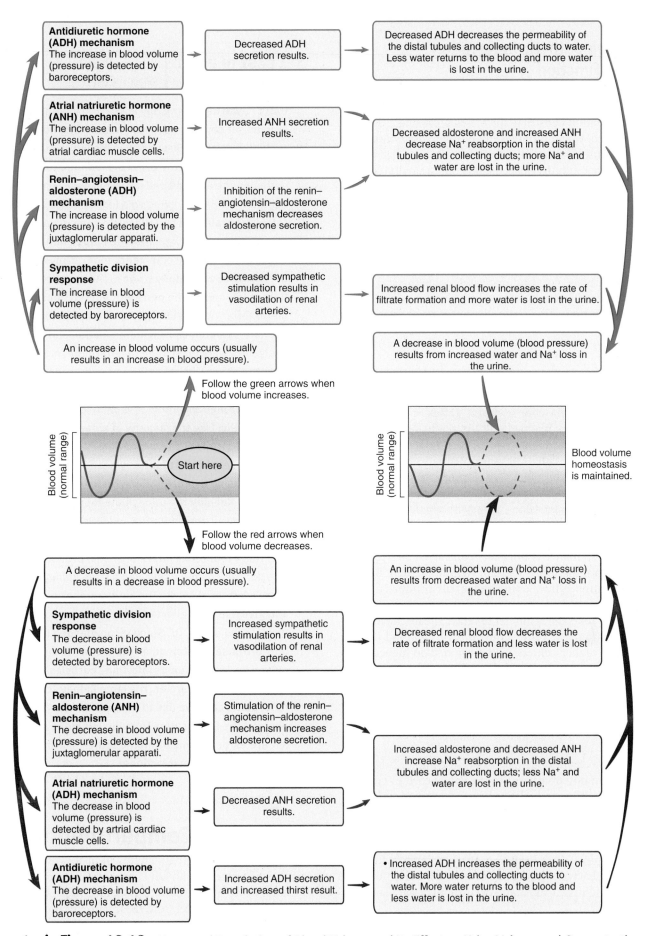

Antidiuretic hormone (ADH) mechanism
The increase in blood volume (pressure) is detected by baroreceptors.

Decreased ADH secretion results.

Decreased ADH decreases the permeability of the distal tubules and collecting ducts to water. Less water returns to the blood and more water is lost in the urine.

Atrial natriuretic hormone (ANH) mechanism
The increase in blood volume (pressure) is detected by atrial cardiac muscle cells.

Increased ANH secretion results.

Decreased aldosterone and increased ANH decrease Na⁺ reabsorption in the distal tubules and collecting ducts; more Na⁺ and water are lost in the urine.

Renin–angiotensin–aldosterone (ADH) mechanism
The increase in blood volume (pressure) is detected by the juxtaglomerular apparati.

Inhibition of the renin–angiotensin–aldosterone mechanism decreases aldosterone secretion.

Sympathetic division response
The increase in blood volume (pressure) is detected by baroreceptors.

Decreased sympathetic stimulation results in vasodilation of renal arteries.

Increased renal blood flow increases the rate of filtrate formation and more water is lost in the urine.

An increase in blood volume occurs (usually results in an increase in blood pressure).

A decrease in blood volume (blood pressure) results from increased water and Na⁺ loss in the urine.

Follow the green arrows when blood volume increases.

Blood volume (normal range)

Start here

Blood volume (normal range)

Blood volume homeostasis is maintained.

Follow the red arrows when blood volume decreases.

A decrease in blood volume occurs (usually results in a decrease in blood pressure).

An increase in blood volume (blood pressure) results from decreased water and Na⁺ loss in the urine.

Sympathetic division response
The decrease in blood volume (pressure) is detected by baroreceptors.

Increased sympathetic stimulation results in vasodilation of renal arteries.

Decreased renal blood flow decreases the rate of filtrate formation and less water is lost in the urine.

Renin–angiotensin–aldosterone (ANH) mechanism
The decrease in blood volume (pressure) is detected by the juxtaglomerular apparati.

Stimulation of the renin–angiotensin–aldosterone mechanism increases aldosterone secretion.

Increased aldosterone and decreased ANH increase Na⁺ reabsorption in the distal tubules and collecting ducts; less Na⁺ and water are lost in the urine.

Atrial natriuretic hormone (ADH) mechanism
The decrease in blood volume (pressure) is detected by atrial cardiac muscle cells.

Decreased ANH secretion results.

Antidiuretic hormone (ADH) mechanism
The decrease in blood volume (pressure) is detected by baroreceptors.

Increased ADH secretion and increased thirst result.

• Increased ADH increases the permeability of the distal tubules and collecting ducts to water. More water returns to the blood and less water is lost in the urine.

Homeostasis Figure 18.18   Hormonal Regulation of Blood Volume and Its Effect on Urine Volume and Concentration

Effect of Sympathetic Innervation on Kidney Function

Sympathetic neurons with norepinephrine as their neurotransmitter substance innervate the blood vessels of the kidney. Sympathetic stimulation constricts the arteries, causing a decrease in renal blood flow and filtrate formation. Intense sympathetic stimulation causes the rate of filtrate formation to decrease to only a few mL per minute. Consequently only a small volume of urine is produced (see figure 18.18). Decreases in blood pressure, such as during shock, are detected by baroreceptors and the result is to increase sympathetic stimulation of renal arteries. Other conditions such as intense physical activity or trauma increase sympathetic stimulation of renal arteries and decrease urine production to very low levels. Increased blood pressure is detected by baroreceptors and decreases sympathetic stimulation of renal blood arteries. Urine volume increases in response to a decrease in sympathetic stimulation of renal arteries (see figure 18.18).

Urine Movement

The **micturition** (mik-choo-rish′ŭn) **reflex** is activated by stretch of the urinary bladder wall. As the bladder fills with urine, pressure increases, and stretch receptors in the wall of the bladder are stimulated. Action potentials are conducted from the bladder to the spinal cord through the pelvic nerves. Integration of the reflex occurs in the spinal cord, and action potentials are conducted along parasympathetic nerve fibers to the urinary bladder. Parasympathetic action potentials cause the urinary bladder to contract (figure 18.19). The external urinary sphincter is normally contracted as a result of stimulation from the somatic motor nervous system. Because of the micturition reflex, action potentials conducted along somatic motor nerves to the external urinary sphincter decrease, which causes the sphincter to relax. The micturition reflex is an automatic reflex, but it can be inhibited or stimulated by higher centers in the brain. The higher brain centers prevent micturition by sending action potentials through the spinal cord to decrease the intensity of the autonomic reflex that stimulates urinary bladder contractions and to stimulate nerve fibers that keep the external urinary sphincter contracted. The ability to voluntarily inhibit micturition develops at the age of 2–3 years.

When the desire to urinate exists, the higher brain centers alter action potentials sent to the spinal cord to facilitate the micturition reflex and relax the external urinary sphincter. Awareness of the need to urinate occurs because stretch of the urinary bladder stimulates sensory nerve fibers that increase action potentials carried to the brain by ascending tracts in the spinal cord. Irritation of the urinary bladder or the urethra by bacterial infections or by other conditions can also initiate the urge to urinate, even though the bladder is nearly empty.

A CASE IN POINT | Kidney Stones (Renal Calculi)

Harry Payne had a long day at work. Just before going to bed, he sat down to watch the news. Harry noticed a rapidly developing discomfort in his left lateral abdominal region. Before long, the pain also radiated to his inguinal region on the left side and it had become excruciating and debilitating. Harry suspected a kidney stone or renal calculus because he had the misfortune to pass a kidney stone about 2 years ago. His wife helped him to the car and drove him to the emergency room. X-rays confirmed the presence of a kidney stone in the upper portion of Harry's left ureter. The peristaltic contractions of the ureter force the irregularly shaped kidney stone through the narrow ureter, causing inflammation and pain. Strong analgesics were prescribed to reduce the intense pain. Over the next 24 hours, urine was collected and finally a kidney stone was recovered. Once the kidney stone passed through the ureter, the pain decreased dramatically. Like the majority of kidney stones, this one consisted mainly of calcium oxalate. Dr. Stone, Harry's physician, explained that Harry needed to increase his fluid intake so that he produced at least 2 liters of urine each day. Keeping the urine dilute reduces the likelihood of calcium salts precipitating and forming additional kidney stones. Also, as a precaution, she recommended that Harry reduce his intake of animal protein. A CT scan was performed, and it revealed a large kidney stone in Harry's left renal pelvis. The kidney stone could fragment and give rise to smaller kidney stones that could pass through the ureter. Dr. Stone recommended lithotripsy (lith′ō-trip-sē) treatment, an ultrasound technique that pulverizes kidney stones into small particles that pass easily through the ureter. The large kidney stone in the renal pelvis can give rise to additional kidney stones, obstruct the flow of urine from the left kidney through the left ureter, or become a chronic source of irritation, which could lead to kidney infections.

Body Fluid Compartments

For an adult male, approximately 60% of the total body weight consists of water. For an adult female, approximately 50% of the total body weight is water. Because the water content of adipose tissue is relatively low, the fraction of the body's weight

Control of the micturition reflex by higher brain centers

A. Ascending tracts carry an increased frequency of action potentials up the spinal cord to the pons and cerebrum when the urinary bladder becomes stretched. This increases the conscious desire to urinate.

B. Descending tracts carry action potentials to the sacral region of the spinal cord to tonically inhibit the micturition reflex, preventing automatic urination when the bladder is full. Descending tracts carry action potentials from the cerebrum to the sacral region of the spinal cord to stimulate the reflex when stretch of the urinary bladder produces the conscious urge to urinate and when one voluntarily chooses to urinate. This reinforces the micturition reflex.

Micturition reflex

1. Urine in the urinary bladder stretches the bladder wall.

2. Action potentials produced by stretch receptors are carried along pelvic nerves (*green line*) to the sacral region of the spinal cord.

3. Action potentials are carried by parasympathetic nerves (*red line*) to contract the smooth muscles of the urinary bladder. Decreased action potentials carried by somatic motor nerves (*purple line*) cause the external urinary sphincter to relax.

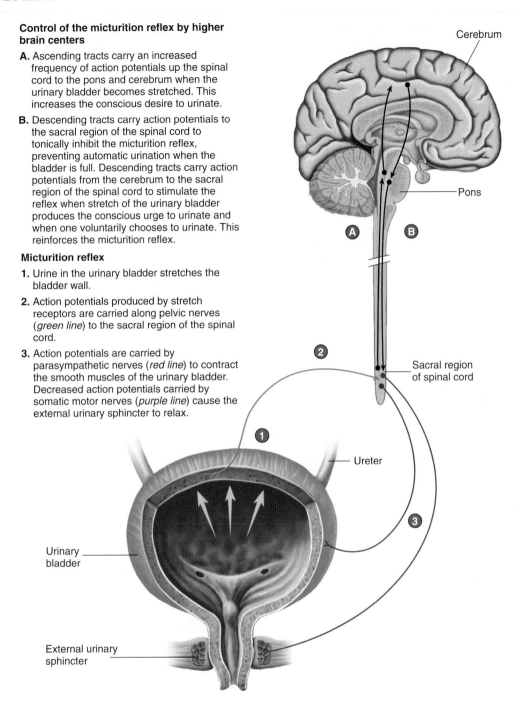

Process Figure 18.19 Micturition Reflex

composed of water decreases as the amount of adipose tissue increases. A smaller percentage of the body weight of the adult female consists of water because females generally have a greater percentage of body fat than males. Water and ions dissolved in it are distributed in two major compartments (table 18.2). Water and ions move between these compartments, but their movement is regulated.

The **intracellular fluid compartment** includes the fluid inside all the cells of the body. The cell membranes of the individual cells enclose the intracellular compartment, which actually consists of trillions of small compartments. The composition of the fluid in these compartments and the regulation of fluid movement across cell membranes are similar. Approximately two-thirds of all the water in the body is in the intracellular fluid compartment.

The **extracellular fluid compartment** includes all the fluid outside the cells. It constitutes approximately one-third of the total body water. The extracellular fluid compartment

Table 18.2 Approximate Volumes of Body Fluid Compartments*

Age of Person	Total Body Water	Intracellular Fluid	Extracellular Fluid		
			Plasma	Interstitial	Total
Infants	75	45	4	26	30
Adult males	60	40	5	15	20
Adult females	50	35	5	10	15

*Expressed as percentage of body weight.

includes the interstitial fluid, the plasma within blood vessels, and fluid in the lymphatic vessels. A small portion of the extracellular fluid volume is separated by membranes into subcompartments. These special subcompartments contain fluid with a composition different from that of the remainder of the extracellular fluid. Included among the subcompartments are the aqueous and vitreous humor of the eye, cerebrospinal fluid, synovial fluid in joint cavities, serous fluid in the body cavities, and fluid secreted by glands.

Composition of the Fluid in the Body Fluid Compartments

Intracellular fluid has a similar composition from cell to cell. The intracellular fluid contains a relatively high concentration of ions, such as K^+, magnesium (Mg^{2+}), phosphate (PO_4^{3-}), and sulfate ions (SO_4^{2-}), compared with the extracellular fluid. It has a lower concentration of Na^+, Ca^{2+}, Cl^-, and HCO_3^- than that of the extracellular fluid. The concentration of protein in the intracellular fluid is also greater than that in the extracellular fluid. Like intracellular fluid, the extracellular fluid also has a fairly consistent composition from one area of the body to another.

Exchange Between Body Fluid Compartments

The cell membranes that separate the body fluid compartments are selectively permeable. Water continually passes through them, but they are much less permeable to ions dissolved in the water. Water movement is regulated mainly by hydrostatic pressure differences and osmotic differences between the compartments. For example, water moves across the wall of the capillary at the arteriolar end of the capillary because the blood pressure is great enough to force fluid through the wall of the capillary into the interstitial space. At the venous end of the capillary, the blood pressure is much lower, and fluid returns to the capillary because the osmotic pressure is higher inside the capillary than outside it (see chapter 13).

The major influence controlling the movement of water between the intracellular and extracellular spaces is osmosis. For example, if the extracellular concentration of ions increases, water moves by osmosis from cells into the extracellular fluid.

The intracellular fluid can help maintain the extracellular fluid volume if it is depleted. When a person becomes dehydrated, the concentration of ions in the extracellular fluid increases. As a consequence, water moves from the intracellular fluid to the extracellular fluid, thus maintaining the extracellular fluid volume. Because blood is an important component of the extracellular fluid volume, this process helps to maintain blood volume. Movement of water from the intracellular fluid compartment to the extracellular fluid compartment can help prolong the time a person can survive conditions such as dehydration or cardiovascular shock.

If the concentration of ions in the extracellular fluid decreases, water moves, by osmosis, from the extracellular fluid into the cells. The water movement can cause the cells to swell. Under most conditions, the movement of water between the intracellular and extracellular fluid compartments is maintained within limits that are consistent with survival of the individual.

Regulation of Extracellular Fluid Composition

Homeostasis requires that the intake of substances such as water and ions equals their elimination. Ingestion of water and ions adds these substances to the body; they are excreted from the body by organs such as the kidneys and, to a lesser degree, by the skin, liver, and gastrointestinal tract. Greater quantities of water and ions are lost from the body in the form of perspiration on warm days than on cool days, and varying amounts of water and ions can be lost in the form of feces. Over a long period, the total amount of water and ions in the body does not change unless the individual is growing, gaining weight, or losing weight. The regulation of water and ions involves the coordinated participation of several organ systems, but the most important organ in regulating the loss of water and ions from the body is the kidney.

Thirst

Water intake is controlled by neurons in the hypothalamus, collectively called the **thirst center.** When the concentration of blood increases (increased osmolality), the thirst center responds by initiating the sensation of thirst (figure 18.20). When water or some other dilute solution is consumed, the

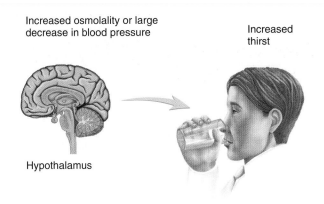

Increased osmolality or large
decrease in blood pressure

Increased
thirst

Hypothalamus

Figure 18.20   **Thirst and the Regulation of Extracellular
Fluid Concentration**

Increased blood osmolality affects hypothalamic neurons, and large
decreases in blood pressure affect baroreceptors in the aortic arch, carotid
sinuses, and atrium. As a result of these stimuli, an increase in thirst results,
which increases water intake. Increased water intake reduces blood
osmolality and increases blood volume.

concentration of the blood decreases, and the sensation of
thirst also decreases. When blood pressure decreases, such as
during shock, the thirst center is activated, and the sensation
of thirst is triggered. Consumption of water increases the
blood volume and allows the blood pressure to increase toward its normal value. Other stimuli in addition to changes in
solute concentration and blood pressure can trigger the sensation of thirst. For example, if the mucosa of the mouth becomes dry, the thirst center is activated. Thirst is one of the
important means of regulating extracellular fluid volume and
concentration.

Ions

The kidneys and other organ systems function to regulate the
composition of the extracellular fluid. If the water content or
concentration of ions in the extracellular fluid deviates from
its normal range, cells cannot control the movement of substances across their cell membranes or the composition of
their intracellular fluid. The consequence is abnormal cell
function or even cell death. Keeping the extracellular fluid
composition within a normal range of values is therefore required to sustain life.

Regulating the concentrations of positively charged ions
such as Na^+, K^+, and Ca^{2+} in the body fluids is particularly
important. Action potentials, contraction of muscles, and
maintenance of normal cell membrane permeability depend
on the maintenance of a narrow range of concentrations for
these ions. Important mechanisms control the concentrations
of these ions in the body. Negatively charged ions such as Cl^-
are secondarily regulated by the mechanisms that control
the positively charged ions. The negatively charged ions are attracted to positively charged ions; when the positively charged
ions are transported, the negatively charged ions move with
them.

Sodium Ions

Sodium ions are major ions in the extracellular fluid. About
90–95% of the osmotic pressure of the extracellular fluid results from Na^+ and from the negative ions associated with
them.

The recommended intake of Na^+ is 2.4 grams per day
(g/day), because of its association with high blood pressure in
some people. Most people in the United States consume two
to three times the recommended amount of Na^+. The kidneys provide the major route by which the excess Na^+ are
excreted.

Stimuli that control aldosterone secretion influence the reabsorption of Na^+ from nephrons of the kidneys and the total
amount of Na^+ in the body fluids. Reabsorption of Na^+ from the
distal tubules and collecting duct is very efficient, and little Na^+ is
lost in the urine when aldosterone is present. When aldosterone is
absent, reabsorption of Na^+ in the nephron is greatly reduced,
and the amount of Na^+ lost in the urine increases. Aldosterone
also plays an essential role in regulating the extracellular K^+ concentration (figure 18.21).

Sodium ions are also excreted from the body in **sweat.**
Normally only a small quantity of Na^+ is lost each day in the
form of sweat, but the amount increases during conditions of
heavy exercise in a warm environment.

Because Na^+ have such a large effect on the osmotic
pressure of the extracellular fluid, mechanisms that influence
Na^+ concentrations in the extracellular fluid also influence
the extracellular fluid volume. The mechanisms that play important roles in controlling the Na^+ concentration in the extracellular fluid and the extracellular fluid volume are the
renin–angiotensin–aldosterone mechanism, the atrial natriuretic mechanism, and antidiuretic hormone. These mechanisms are illustrated in figures 18.15, 18.16, 18.17, and 18.18.
For example, low blood pressure increases renin and ADH secretion. The result is an increase in Na^+ and water reabsorption in the kidney to bring blood pressure and the Na^+
concentration back to their normal ranges. Increased blood
pressure inhibits renin and ADH secretion, and it stimulates
ANH secretion. The result is a decrease in Na^+ reabsorption
and an increase in urine production to bring blood pressure
and the Na^+ concentration in the blood to their normal
ranges.

Potassium Ions

Electrically excitable tissues such as muscle and nerve are
highly sensitive to slight changes in the extracellular K^+ concentration. The extracellular concentration of K^+ must be
maintained within a narrow range for these tissues to function
normally.

Aldosterone plays a major role in regulating the concentration of K^+ in the extracellular fluid. Dehydration, circulatory
system shock resulting from plasma loss, and tissue damage due
to injuries such as severe burns, all cause extracellular K^+ to become more concentrated than normal. In response, aldosterone
secretion from the adrenal cortex increases and causes K^+ secretion to increase (see figure 18.21).

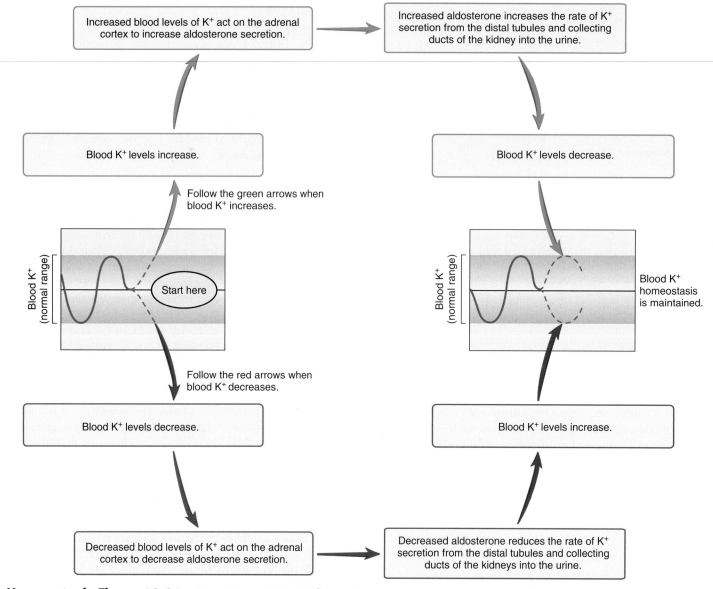

Homeostasis Figure 18.21   Regulation of Blood K⁺ Levels

If the K^+ concentration in the extracellular fluid becomes reduced, aldosterone secretion from the adrenal cortex decreases. In response, the rate of K^+ secretion by the kidney is reduced (see figure 18.21).

Calcium Ions

The extracellular concentration of Ca^{2+}, like that of other ions, is maintained within a narrow range. Increases and decreases in the extracellular concentration of Ca^{2+} have dramatic effects on the electrical properties of excitable tissues. Decreased extracellular Ca^{2+} concentrations make cell membranes more permeable to Na^+, thus making them more electrically excitable. Decreased extracellular concentrations of Ca^{2+} cause sponta-

neous action potentials in nerve and muscle cells, resulting in hyperexcitability and tetany of muscles. Increased extracellular Ca^{2+} concentrations make cell membranes less permeable to Na^+, thus making them less electrically excitable. Increased extracellular concentrations of Ca^{2+} inhibit action potentials in nerve and muscle cells, resulting in reduced excitability and either muscle weakness or paralysis.

Parathyroid hormone (PTH), secreted by the parathyroid glands, increases extracellular Ca^{2+} concentration. The rate of PTH secretion is regulated by the extracellular Ca^{2+} concentration (figure 18.22). An elevated Ca^{2+} concentration inhibits, and a reduced Ca^{2+} concentration stimulates, the secretion of PTH. PTH causes osteoclasts to degrade bone and

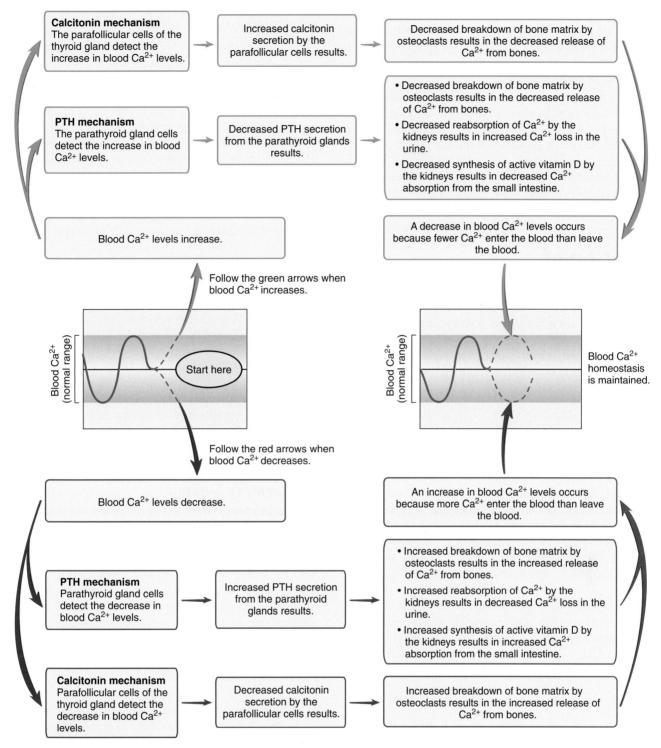

Homeostasis Figure 18.22   Regulation of Blood Ca^{2+} Levels

release Ca^{2+} into the body fluids. PTH also increases the rate of Ca^{2+} reabsorption from kidney nephrons.

Vitamin D increases Ca^{2+} concentration in the blood by increasing the rate of Ca^{2+} absorption by the intestine. Some vitamin D is consumed in food and the rest is produced by the body (see chapter 5). PTH affects the intestinal uptake

of Ca^{2+} because PTH increases the rate of vitamin D production in the body.

Calcitonin (kal-si-tō′nin) is secreted by the thyroid gland. Calcitonin reduces the blood Ca^{2+} concentration when it is too high. An elevated blood Ca^{2+} concentration causes the thyroid gland to secrete calcitonin, and a low blood Ca^{2+}

concentration inhibits calcitonin secretion. Calcitonin reduces the rate at which bone is broken down and decreases the release of Ca^{2+} from bone (see figure 18.22).

Phosphate and Sulfate Ions

Some ions, such as **phosphate ions** (PO_4^{3-}) and **sulfate ions** (SO_4^{2-}), are reabsorbed by active transport in the kidneys. The rate of reabsorption is slow so that, if the concentration of these ions in the filtrate exceeds the ability of the nephron to reabsorb them, the excess is excreted into the urine. As long as the concentration of these ions is low, nearly all of them are reabsorbed by active transport. This mechanism plays a major role in regulating the concentration of PO_4^{3-} and SO_4^{2-} in the body fluid.

Regulation of Acid–Base Balance

The concentration of H^+ in the body fluids is reported as the pH of body fluids. The body fluid pH is maintained between 7.35 and 7.45, and any deviation from that range is life-threatening. Consequently, the mechanisms that regulate body pH are critical for survival. The pH of body fluids is controlled by buffers, by the respiratory system, and by the kidneys.

Buffers

Buffers are chemicals that resist a change in the pH of a solution when either acids or bases are added to the solution. The buffers found in the body fluids contain salts of either weak acids or weak bases that combine with H^+ when H^+ increase in those fluids or release H^+ when H^+ decrease in those fluids. Buffers tend to keep the H^+ concentration, and thus the pH, within a narrow range of values (figure 18.23) because of these characteristics. The three major buffers in the body fluids are the proteins, the PO_4^{3-} buffer system, and the HCO_3^- buffer system.

Proteins and PO_4^{3-} ions in the body fluids combine with a large number of H^+. When the H^+ concentration increases, proteins and PO_4^{3-} prevent a decrease in pH by combining with the H^+. Conversely, when the H^+ concentration decreases, proteins and PO_4^{3-} release H^+, preventing an increase in pH.

The following reaction illustrates how PO_4^{3-} buffers work:

$$HPO_4^{2-} \quad + \quad H^+ \quad \rightleftarrows \quad H_2PO_4^-$$

monohydrogen phosphate ion　　hydrogen ion　　dihydrogen phosphate ion

Monohydrogen phosphate (HPO_4^{2-}) combines with H^+ to form dihydrogen phosphate ($H_2PO_4^-$) when excess H^+ are present. When H^+ concentration declines, some of the H^+ separate from the $H_2PO_4^-$.

Proteins are able to function as buffers because amino acids in the proteins have side chains that function as weak acids and weak bases. Many side chains contain carboxyl groups (–COOH) or amine groups (–NH$_2$). Both of these groups are able to function as buffers because of the following reactions:

$$-COO^- \quad + \quad H^+ \quad \rightleftarrows \quad -COOH$$

carboxyl group (ionized)　　hydrogen ion　　carboxyl group

$$-NH_2 \quad + \quad H^+ \quad \rightleftarrows \quad -NH_3$$

amine group　　hydrogen ion　　ammonium group

The bicarbonate (HCO_3^-) buffer system is unable to combine with as many H^+ as can proteins and PO_4^{3-} buffers, but the HCO_3^- buffer system is critical because it can be regulated by the respiratory and urinary systems. Carbon dioxide (CO_2) combines with water (H_2O) to form carbonic acid (H_2CO_3), which in turn forms H^+ and HCO_3^- as follows:

$$H_2O \quad + \quad CO_2 \quad \rightleftarrows \quad H_2CO_3 \quad \rightleftarrows \quad H^+ \quad + \quad HCO_3^-$$

water　　carbon dioxide　　carbonic acid　　hydrogen ion　　bicarbonate ion

The reaction between CO_2 and H_2O is catalyzed by an enzyme, called carbonic anhydrase, which is found in red blood cells and on the surface of capillary epithelial cells (see chapter 15). The enzyme accelerates the rate at which the reaction proceeds in either direction.

The higher the concentration of CO_2, the greater the amount of H_2CO_3 formed, and the greater the number of H^+ and HCO_3^- formed. This results in a decreased pH. The reaction is reversible, however. If CO_2 levels decline, the equilibrium shifts in the opposite direction. That is, H^+ and HCO_3^- combine to form H_2CO_3, which then forms CO_2 and H_2O, and the pH increases.

Respiratory System

The **respiratory system** responds rapidly to a change in pH and functions to bring the pH of body fluids back toward its normal range. Increasing carbon dioxide levels and decreasing body fluid pH stimulate neurons in the respiratory center of the brain and cause the rate and depth of ventilation to increase. As a result of the increased rate and depth of ventilation, carbon dioxide is eliminated from the body through the lungs at a greater rate, and the concentration of carbon dioxide in the body fluids decreases. As carbon dioxide levels decline, the concentration of H^+ also declines. The pH therefore increases back to its normal range (see figure 18.23).

P R E D I C T 　5

Under stressful conditions some people hyperventilate. Predict the effect of the rapid rate of ventilation on the pH of body fluids. In addition, explain why a person who is hyperventilating may benefit from breathing into a paper bag.

If carbon dioxide levels become too low or the pH of the body fluids is elevated, the rate and depth of respiration decline. As a consequence, the rate at which carbon dioxide is eliminated from the body is reduced. Carbon dioxide then accumulates in the body fluids because it is continually produced as a by-product of metabolism. As carbon dioxide accumulates in the body fluids, so do H^+, resulting in a decreased pH.

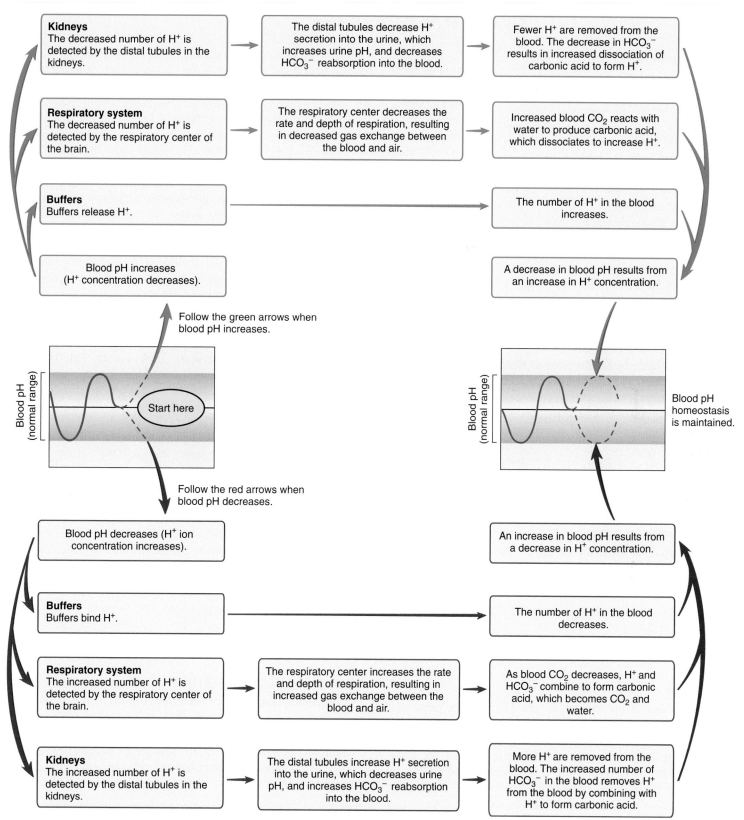

Kidneys
The decreased number of H⁺ is detected by the distal tubules in the kidneys.

The distal tubules decrease H⁺ secretion into the urine, which increases urine pH, and decreases HCO_3^- reabsorption into the blood.

Fewer H⁺ are removed from the blood. The decrease in HCO_3^- results in increased dissociation of carbonic acid to form H⁺.

Respiratory system
The decreased number of H⁺ is detected by the respiratory center of the brain.

The respiratory center decreases the rate and depth of respiration, resulting in decreased gas exchange between the blood and air.

Increased blood CO_2 reacts with water to produce carbonic acid, which dissociates to increase H⁺.

Buffers
Buffers release H⁺.

The number of H⁺ in the blood increases.

Blood pH increases (H⁺ concentration decreases).

A decrease in blood pH results from an increase in H⁺ concentration.

Follow the green arrows when blood pH increases.

Blood pH (normal range)

Start here

Follow the red arrows when blood pH decreases.

Blood pH (normal range)

Blood pH homeostasis is maintained.

Blood pH decreases (H⁺ ion concentration increases).

An increase in blood pH results from a decrease in H⁺ concentration.

Buffers
Buffers bind H⁺.

The number of H⁺ in the blood decreases.

Respiratory system
The increased number of H⁺ is detected by the respiratory center of the brain.

The respiratory center increases the rate and depth of respiration, resulting in increased gas exchange between the blood and air.

As blood CO_2 decreases, H⁺ and HCO_3^- combine to form carbonic acid, which becomes CO_2 and water.

Kidneys
The increased number of H⁺ is detected by the distal tubules in the kidneys.

The distal tubules increase H⁺ secretion into the urine, which decreases urine pH, and increases HCO_3^- reabsorption into the blood.

More H⁺ are removed from the blood. The increased number of HCO_3^- in the blood removes H⁺ from the blood by combining with H⁺ to form carbonic acid.

Homeostasis Figure 18.23 Regulation of Acid–Base Balance

Clinical Focus Disorders of the Urinary System

Inflammation of the Kidneys

Glomerulonephritis (glō-mār′ū-lō-nef′rī-tis) is an inflammation of the filtration membrane within the renal corpuscle. The permeability of the filtration membrane increases, and plasma proteins and white blood cells enter the filtrate. The plasma proteins cause the urine volume to increase because they increase the osmotic concentration of the filtrate.

Acute glomerulonephritis often occurs 1–3 weeks after a severe bacterial infection such as streptococcal sore throat or scarlet fever. Antigen–antibody complexes associated with the infection become deposited in the filtration membrane and cause inflammation. Acute glomerulonephritis normally subsides after several days.

Chronic glomerulonephritis is long term and usually progressive. The filtration membrane thickens and eventually is replaced by connective tissue. In its early stages, chronic glomerulonephritis resembles the acute form. In the advanced stages, many of the renal corpuscles are replaced by connective tissue, and the kidneys eventually become nonfunctional.

Renal Failure

Renal failure can result from any condition that interferes with kidney function. **Acute renal failure** occurs when damage to the kidney is rapid and extensive. It leads to the accumulation of urea and other metabolites in the blood and to acidosis. If renal failure is complete, death can occur in 1–2 weeks. Acute renal failure can result from acute glomerulonephritis, or it can be caused by damage to or blockage of the renal tubules. Lack of blood supply or exposure to certain toxic substances can cause damage to the epithelial cells of the nephron and lead to acute renal failure.

Chronic renal failure is the result of permanent damage to so many nephrons that the remaining nephrons are inadequate for normal kidney function. Chronic renal failure can result from chronic glomerulonephritis, trauma to the kidneys, tumors, urinary tract obstruction by kidney stones, or severe lack of blood supply resulting from arteriosclerosis. Chronic renal failure leads to the inability to eliminate toxic metabolic by-products. Water retention and edema result from the accumulation of solutes in the body fluids. Potassium levels become elevated, and acidosis develops. The toxic effects of accumulated metabolic waste products are mental confusion, coma, and finally death, when chronic renal failure is severe.

Treatments for Renal Failure

Hemodialysis (hē′mō-dī-al′i-sis) is used when a person is suffering from severe acute or chronic kidney failure. The procedure substitutes for the excretory functions of the kidney. Hemodialysis is based on blood flow through tubes composed of a selectively permeable membrane. Blood is usually taken from an artery, passed through the tubes of the dialysis machine, and then returned to a vein (figure A). On the outside of the dialysis tubes is a fluid, called dialysis fluid, which contains the same concentration of solutes as normal plasma except for the metabolic waste products. As a consequence, the metabolic wastes diffuse from the blood to the dialysis fluid. The dialysis membrane has pores that are too small to allow plasma proteins to pass through them, and, because the dialysis fluid contains the same beneficial solutes as the plasma, the net movement of these substances is zero.

Peritoneal (per′i-tō-nē′ăl) **dialysis** is sometimes used to treat people suffering from kidney failure. The principles by which peritoneal dialysis works are the same as those for hemodialysis. The dialysis fluid flows through a tube inserted into the peritoneal cavity, however. The visceral and parietal peritoneum act as the dialysis membrane. Waste products diffuse from the blood vessels beneath the peritoneum, across the peritoneum, and into the dialysis fluid.

Kidney transplants are sometimes performed on people who suffer from severe

Kidneys

The nephrons of the kidneys secrete H^+ into the urine and therefore can directly regulate pH of the body fluids. The kidney is a powerful regulator of pH, but it responds more slowly than does the respiratory system. Cells in the walls of the distal tubule are primarily responsible for the secretion of H^+. As the pH of the body fluids decreases below normal, the rate at which the distal tubules secrete H^+ increases (see figure 18.23). At the same time, reabsorption of HCO_3^- increases. The increased rate of H^+ secretion and the increased rate of HCO_3^- reabsorption both cause the blood pH to increase toward its normal value. On the other hand, as the body fluid pH increases above normal, the rate of H^+ secretion by the distal tubules declines, and the amount of HCO_3^- lost in the urine increases. Consequently, the blood pH decreases toward its normal value.

Acidosis and Alkalosis

Failure of the buffer systems, the respiratory system, or the urinary system to maintain normal pH levels can result in acidosis or alkalosis.

Acidosis

Acidosis (as-i-dō′sis) occurs when the blood pH falls below 7.35. The central nervous system malfunctions, and the individual becomes disoriented and, as the condition worsens, can become comatose. Acidosis is separated into two categories.

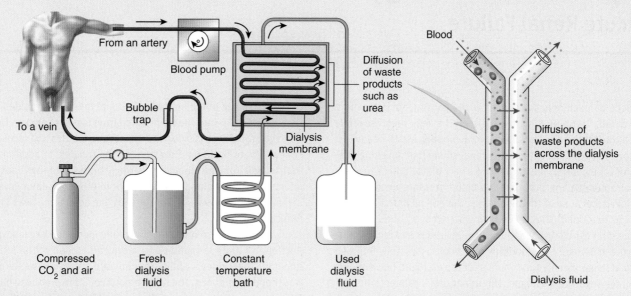

Figure A Hemodialysis

During hemodialysis, blood flows through a system of tubes composed of a selectively permeable membrane. Dialysis fluid, the composition of which is similar to that of normal blood, except that the concentration of waste products is very low, flows in the opposite direction on the outside of the dialysis tubes. Waste products, such as urea, diffuse from the blood into the dialysis fluid. Other substances, such as Na^+, K^+, and glucose, can diffuse from the blood into the dialysis fluid if they are present in higher than normal concentrations, because these substances are present in the dialysis fluid at the same concentrations found in normal blood.

renal failure. Usually the donor has suffered an accidental death and had granted permission to have his or her kidneys used for transplantation. An attempt is made to match the immune characteristics of the donor and recipient to reduce the tendency for the recipient's immune system to reject the transplanted kidney. Even with careful matching, recipients have to take medication for the rest of their lives to suppress their immune systems so that rejection is less likely. The major cause of kidney transplant failure is rejection by the recipient's immune system. In most cases, the transplanted kidney functions well, and the tendency for the recipient's immune system to reject the transplanted kidney can be controlled.

Respiratory acidosis results when the respiratory system is unable to eliminate adequate amounts of carbon dioxide. Carbon dioxide accumulates in the circulatory system, causing the pH of the body fluids to decline. **Metabolic acidosis** results from excess production of acidic substances, such as lactic acid and ketone bodies, because of increased metabolism or a decreased ability of the kidneys to eliminate H^+ in the urine.

P R E D I C T

Mr. Puffer suffers from severe emphysema. Gas exchange in his lungs is not adequate, and he must have a supply of oxygen. His blood carbon dioxide levels are elevated. Nevertheless his blood pH is close to normal. Explain.

Alkalosis

Alkalosis (al-kă-lō′sis) occurs when the blood pH increases above 7.45. A major effect of alkalosis is hyperexcitability of the nervous system. Peripheral nerves are affected first, resulting in spontaneous nervous stimulation of muscles. Spasms and tetanic contractions result, as can extreme nervousness or convulsions. Tetany of respiratory muscles can cause death. **Respiratory alkalosis** results from hyperventilation, such as can occur in response to stress. **Metabolic alkalosis** usually results from the rapid elimination of H^+ from the body, such as during severe vomiting, or when excess aldosterone is secreted by the adrenal cortex.

Systems Pathology
Acute Renal Failure

A large piece of machinery overturned at the construction site where Mr. H. worked, trapping him beneath it. His legs were severely crushed, although they healed after several months. Mr. H. nearly lost his life, however, because of the acute renal failure that developed because of his injury. Mr. H. was trapped in a very difficult place to reach for several hours. During that time his blood pressure decreased to very low levels because of the blood loss, the edema in the inflamed tissues, and emotional shock. After he was removed, fluid replacement in the form of both intravenous saline solutions and blood transfusions were given and his blood pressure was successfully returned to its normal range. Twenty-four hours after the accident, however, his urine volume began to decrease. His urinary Na^+ concentration increased, but his urine osmolality decreased. Casts and cellular debris were evident in his urine.

For approximately 7 days Mr. H. exhibited reduced urine production. During this period, renal dialysis was required to maintain his blood volume and ion concentrations within normal ranges. After approximately 7 days, his kidneys gradually began to produce large quantities of urine. Careful observation was required to keep his blood pressure and ion concentrations within normal ranges. Substantial quantities of water, Na^+, and K^+ had to be administered to him. After about 3 weeks, the functions of his kidneys slowly began to improve, although many months passed before his kidney functions had returned to normal.

Background Information

The events after 24 hours are consistent with acute renal failure caused by prolonged low blood pressure and lack of blood flow to the kidneys. The reduced blood flow to the kidneys was severe enough to result in damage to the epithelial lining of the kidney tubules. The period of reduced urine volume resulted from tubular damage. Dead and damaged tubular cells sloughed off into the tubules and blocked them so that filtrate could not flow through the tubules. In addition, the filtrate leaked from the blocked or partially blocked tubules back into the interstitial spaces and, therefore, back into the circulatory system. As a result, the amount of filtrate that became urine was markedly reduced.

Blood levels of urea and of creatine increased because of the reduction in filtrate formation and reduced function of the tubular epithelium. The kidney's ability to eliminate metabolic waste products was therefore reduced. The small amount of urine produced had a high Na^+ concentration, although the osmolality was close to the concentration of the body fluids because the kidney was not able to reabsorb Na^+ and because the urine-concentrating ability of the kidney was severely damaged.

After 7 days, the nephrons were partially healed and could produce urine, but the ability of the nephrons to concentrate urine was not yet normal. Large volumes of urine that contained significant amounts of Na^+ and K^+ were therefore produced. The kidneys were able to produce urine that was more concentrated than the body fluids, but the concentrating ability of the kidneys was still below normal. As time passed, the concentrating ability of the kidneys improved and eventually became normal once again.

P R E D I C T 7

Nine days after the accident, Mr. H. began to appear pale, he became dizzy on standing and lethargic. His hematocrit was elevated and his heart was arrythmic. He was very weak. Explain these manifestations.

S U M M A R Y

The urinary system consists of two kidneys, two ureters, the urinary bladder, and the urethra.

Functions of the Urinary System   (p. 508)

1. The kidneys excrete waste products.
2. The kidneys control blood volume by regulating the volume of urine produced.
3. The kidneys help regulate the concentration of major ions in the body fluids.
4. The kidneys help regulate pH of the body fluids.
5. The kidneys regulate the concentration of red blood cells in the blood.
6. The kidneys participate, with the skin and liver, in vitamin D synthesis.

Urinary System   (p. 508)
Kidneys

1. Each kidney is behind the peritoneum and surrounded by a renal capsule and adipose tissue.
2. The kidney is divided into an outer cortex and an inner medulla.
3. Each renal pyramid has a base located at the boundary between the cortex and medulla, and the tip extends toward the center of the kidney and is surrounded by a calyx.
4. Calyces are extensions of the renal pelvis, which is the expanded end of the ureter within the renal sinus.
5. The functional unit of the kidney is the nephron. The parts of a nephron are the renal corpuscle, the proximal tubule, the loop of Henle, and the distal tubule.

System Interactions: The Effect of Renal Failure on Other Systems

System	Interactions
Integumentary	Pallor results from anemia, and bruising results from reduced clotting proteins in the blood because they are lost in the urine. A waxy yellow coloring develops in the skin of light-skinned people, an ashen gray caste in black-skinned people, or a yellowish brown caste in brown-skinned people due to accumulation of urinary pigments. When the urea concentration in the blood is very high it can give a yellow caste to light-skinned people, and white crystals of urea, called uremic frost, may appear on areas of the skin where there is heavy perspiration.
Skeletal	Changes in the skeletal system is not significant unless kidney damage results in chronic kidney failure. Bone resorption can result because of excessive and chronic loss of Ca^{2+} in the urine when the kidneys produce large volumes of urine. Also, vitamin D levels may be reduced.
Muscular	Neuromuscular irritability results from the toxic effect of metabolic wastes on the central nervous system and ionic imbalances such as elevated blood K^+ levels. Involuntary jerking and twitching can occur as neuromuscular irritability develops. Tremor of the hands is an indication of the toxic effects of metabolic wastes on the cerebrum.
Nervous	Elevated blood K^+ levels and the toxic effects of metabolic wastes result in depolarization of neurons. Slowing of action potential conduction, burning sensations, pain, numbness, or tingling result. Also, decreased mental acuity, reduced ability to concentrate, apathy, and lethargy result. Periods of lethargy can alternate with restlessness and insomnia. In severe cases, the patient can become confused and comatose.
Endocrine	Major predictable hormone deficiencies include vitamin D deficiency. In addition, the secretion of reproductive hormones decreases due to the effects of metabolic wastes and ionic imbalances on the hypothalamus.
Cardiovascular	Water and Na^+ retention can result in edema in peripheral tissues and in the lung. Also, increased blood pressure and congestive heart failure may result. Elevated blood K^+ levels result in dysrhythmias and can cause cardiac arrest. Anemia due to decreased erythropoietin production by the damaged kidney and decreased half-life of red blood cells can result. Nosebleeds and bruising occur due to reduced concentration of clotting factors because they are lost in the urine.
Lymphatic	There are no major direct effects on the lymphatic system with the exception that increased lymph flow occurs as a result of edema.
Respiratory	Early during acute renal failure the depth of breathing increases and becomes labored as acidosis develops because the kidneys are not able to secrete H^+. Pulmonary edema often develops because of water and Na^+ retention as a result of reduced urine production. The likelihood of infection increases, as a result of pulmonary edema.
Digestive	Decreased appetite, nausea, and vomiting result from altered digestive tract functions due to the effects of ionic imbalances on the nervous system. The breath can have the odor of ammonia, and there can be a metallic taste in the mouth. These effects are the result of the accumulation of metabolic waste products in the digestive tract and the action of the normal digestive tract organisms on the waste products, which convert urea to ammonia. The ammonia and other metabolic waste products predispose the mouth to inflammation and infection.

6. The filtration membrane is formed by the glomerular capillaries, the basement membrane, and the podocytes of Bowman's capsule.

Arteries and Veins

1. Renal arteries give rise to branches that lead to afferent arterioles.
2. Afferent arterioles supply the glomeruli.
3. Efferent arterioles carry blood from the glomeruli to the peritubular capillaries.
4. Blood from the peritubular capillaries flows to the renal veins.

Ureters, Urinary Bladder, and Urethra

1. Each ureter carries urine from a renal pelvis to the urinary bladder.
2. The urethra carries urine from the urinary bladder to the outside of the body.
3. The ureters and urinary bladder are lined with transitional epithelium and have smooth muscle in their walls.
4. The external urinary sphincter regulates the flow of urine through the urethra.

Urine Production   (p. 515)

Urine is produced by the processes of filtration, reabsorption, and secretion.

Filtration

1. The renal filtrate passes from the glomerulus into Bowman's capsule and contains no blood cells and few blood proteins.
2. Filtration pressure is responsible for filtrate formation.

Reabsorption

1. About 99% of the filtrate volume is reabsorbed; 1% becomes urine.
2. Proteins; amino acids; glucose; fructose; and Na^+, K^+, Ca^{2+}, HCO_3^-, and Cl^- ions are among the substances reabsorbed.
3. About 65% of the filtrate volume is reabsorbed in the proximal tubule, 15% is reabsorbed in the descending limb of the loop of Henle, and another 19% is reabsorbed in the distal tubule and collecting duct.

Secretion

Hydrogen ions, some by-products of metabolism, and some drugs are actively secreted into the nephron.

Regulation of Urine Concentration and Volume   (p. 522)

Hormonal Mechanisms

1. ADH is secreted from the posterior pituitary when the concentration of blood increases or when blood pressure decreases. ADH increases the permeability to water of the distal convoluted tubule and collecting duct. It increases water reabsorption by the kidney.
2. Renin is secreted from the kidney when the blood pressure decreases. Renin converts angiotensinogen to angiotensin I which is then converted to angiotensin II by angiotensin-converting enzyme. Angiotensin II stimulates aldosterone secretion, and aldosterone increases the rate of Na^+ and Cl^- reabsorption from the nephron.
3. Atrial natriuretic hormone, secreted from the right atrium in response to increases in blood pressure, acts on the kidney to increase Na^+ and water loss in the urine.

Effect of Sympathetic Innervation on Kidney Function

Increased sympathetic activity decreases blood flow to the kidney, decreases filtrate formation, and decreases urine formation.

Urine Movement   (p. 526)

1. Increased volume in the urinary bladder stretches its wall and activates the micturition reflex.
2. Parasympathetic action potentials cause contraction of the urinary bladder. Reduced somatic motor action potentials cause relaxation of the external urinary sphincter.
3. Higher brain centers control the micturition reflex. Stretch of the urinary bladder stimulates sensory neurons that carry impulses to the brain and inform the brain of the need to urinate.

Body Fluid Compartments   (p. 526)

1. Water and ions dissolved in it are distributed in the intracellular and extracellular fluid compartments.
2. Approximately 60% of the total body water is found within cells.
3. Approximately 40% of the total body water is found outside cells, mainly in interstitial fluid, plasma of blood, and lymph.

Composition of the Fluid in the Body Fluid Compartments

1. Intracellular fluid contains more K^+, Mg^{2+}, PO_4^{3-}, SO_4^{2-}, and protein than extracellular fluid.
2. Extracellular fluid contains more Na^+, Cl^-, and HCO_3^-, and Ca^{2+} than intracellular fluid.

Exchange Between Body Fluid Compartments

Water moves between compartments continually in response to hydrostatic pressure differences and osmotic differences between the compartments.

Regulation of Extracellular Fluid Composition   (p. 528)

The total amount of water and electrolytes in the body does not change unless the person is growing, gaining weight, or losing weight.

Thirst

The sensation of thirst increases if extracellular fluid becomes more concentrated or if blood pressure decreases.

Ions

1. Sodium ions are dominant extracellular ions. Aldosterone increases Na^+ reabsorption from the filtrate. ADH increases water reabsorption from the nephron, and atrial natriuretic hormone increases Na^+ loss in the urine.
2. Aldosterone increases K^+ secretion in the urine. Increased blood levels of K^+ stimulate, and decreased blood levels of K^+ inhibit, aldosterone secretion.
3. Parathyroid hormone secreted from the parathyroid glands increases extracellular Ca^{2+} levels by causing bone resorption and increased Ca^{2+} uptake in the kidney. Parathyroid hormone increases vitamin D synthesis. Calcitonin, secreted by the thyroid gland, inhibits bone resorption and lowers blood Ca^{2+} levels when they are too high.
4. When PO_4^{3-} and SO_4^{2-} levels in the filtrate are low, nearly all PO_4^{3-} and SO_4^{2-} are reabsorbed. When levels are high, excess is lost in the urine.

Regulation of Acid–Base Balance   (p. 532)

Buffers

Three principal classes of buffers in the circulatory system resist changes in the pH: protein, phosphate, and bicarbonate buffers.

Respiratory System

The respiratory system rapidly regulates pH. An increased respiratory rate raises the pH because the rate of carbon dioxide elimination is increased, and a reduced respiratory rate reduces the pH because the rate of carbon dioxide elimination is reduced.

Kidneys

The kidneys excrete H^+ in response to a decreasing blood pH, and they reabsorb H^+ in response to an increasing blood pH.

Acidosis and Alkalosis

1. Acidosis occurs when the pH of the blood falls below 7.35. The two major types are respiratory acidosis and metabolic acidosis.
2. Alkalosis occurs when the pH of the blood increases above 7.45. The two major types are respiratory alkalosis and metabolic alkalosis.

R E V I E W   A N D   C O M P R E H E N S I O N

1. Name the structures that make up the urinary system. List the functions of the urinary system.
2. What structures surround the kidney?
3. Describe the relationships of the renal pyramids, calyces, renal pelvis, and ureter.
4. What is the functional unit of the kidney? Name its parts.
5. Describe the blood supply of the kidney.
6. What are the functions of the ureters, urinary bladder, and urethra? Describe their structure.
7. Name the three general processes that are involved in the production of urine.
8. Describe the filtration membrane. What substances do not pass through it?
9. How do changes in blood pressure in the glomerulus affect the volume of filtrate produced?
10. What substances are reabsorbed in the nephron? What happens to most of the filtrate volume that enters the nephron?

11. In what parts of the nephron are large volumes of filtrate reabsorbed? In what part of the nephron is no filtrate reabsorbed?

12. In general, what substances are secreted into the nephron?

13. What effect does ADH have on urine volume? Name the factors that cause an increase in ADH secretion.

14. Where is renin produced, and what stimulates its secretion?

15. Explain how renin controls the synthesis of angiotensin I. What enzyme regulates the conversion of antiotensin I to angiotensin II?

16. Describe the effect of angiotensin II on aldosterone secretion.

17. Where is aldosterone produced, and what effect does it have on urine volume? What factors stimulate aldosterone secretion?

18. Where is atrial natriuretic hormone produced, and what effect does it have on urine production?

19. What effect does sympathetic stimulation have on the kidneys?

20. Describe the micturition reflex. How is voluntary control of micturition accomplished?

21. What stimuli result in an increased sensation of thirst?

22. Describe how Na^+ levels are regulated in the body fluids.

23. Describe how K^+ levels are regulated in the body fluids.

24. Describe how Ca^{2+} levels are regulated in the body fluids.

25. Explain how buffers respond to changes in the pH of body fluids.

26. Explain how the respiratory system and the kidneys respond to changes in the pH of body fluids.

27. Define respiratory acidosis, metabolic acidosis, respiratory alkalosis, and metabolic alkalosis.

C R I T I C A L   T H I N K I N G   Q U E S T I O N S

1. Mucho McPhee decided to do an experiment after reading the urinary system chapter in his favorite anatomy and physiology textbook. He drank 2 L of water in 15 min and then monitored the rate of urine production and urine concentration over the next 2 h. What did he observe? Explain the major mechanism involved.

2. A student ate a full bag of salty (NaCl) potato chips but drank no liquids. What effect did this have on urine concentration and the rate of urine production? Explain the mechanisms involved.

3. During severe exertion in a hot environment, a person can lose up to 4 L of sweat per hour (sweat is less concentrated than extracellular fluid in the body). What effect would this loss have on urine concentration and rate of production? Explain the mechanisms involved.

4. Which of the following symptoms are consistent with reduced secretion of aldosterone: excessive urine production, low blood pressure, high plasma potassium levels, and high plasma sodium levels? Explain.

5. Propose as many ways as you can to decrease the rate filtrate enters Bowman's capsule.

6. Swifty Trotts has an enteropathogenic *Escherichia coli* infection that produces severe diarrhea. Diarrhea causes the production of a large volume of mucus that contains high concentrations of HCO_3^-. What would this diarrhea do to his blood pH, urine pH, and respiration rate?

7. Spanky and his mother went to a grocery store where Spanky eyed some candy he wanted. His mother refused to buy it, so Spanky became angry. He held his breath for 2 min. What effect did this have on his body fluid pH? After the 2 min, what mechanisms were most important in reestablishing the normal body fluid pH?

8. Martha suffered from severe nausea for 2 days. She vomited frequently, but she was so nauseous she could not stand to consume anything. Explain how each of the following changed in her by the second day: blood pH, blood ADH levels, blood aldosterone levels, and urine pH.

Answers in Appendix D

Visit this textbook's website at www.mhhe.com/seeleyess6 for practice quizzes, animations, interactive learning exercises, and other study tools.

19

Reproductive System

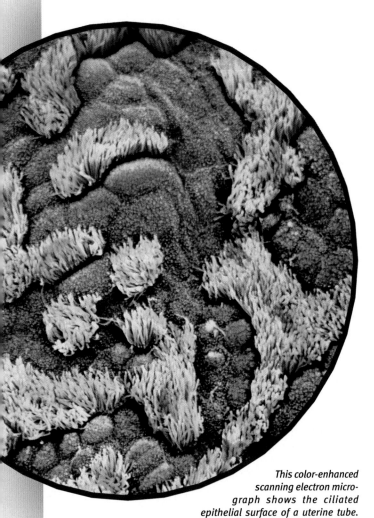

This color-enhanced scanning electron micrograph shows the ciliated epithelial surface of a uterine tube. The epithelial cell secretions and movements of the cilia help move the ovulated oocyte from the ovary to the uterus.

Chapter Outline and Objectives

the human species could not survive without functional reproductive systems. The reproductive systems play essential roles in the development of the structural and functional differences between males and females, influence human behavior, and produce offspring. However, reproductive systems, unlike other organs systems, are not necessary for the survival of individual humans. Most organ systems of the body show little difference between males and females, but the male and female reproductive systems exhibit striking differences. They also share a number of similarities. Many reproductive organs of males and females are derived from the same embryological structures (see chapter 20), and some hormones are the same in males and females, even though they produce different responses.

Functions of the Reproductive System

The male reproductive system performs the following functions:

1. *Production of sperm cells.* The reproductive system produces male sex cells, or sperm cells, in the testes.
2. *Sustaining and transfer of the sperm cells to the female.* The duct system provides nutrients for the sperm cells produced in the testes, provides an environment in which the sperm cells mature, provides secretions that form most of the volume of the semen transferred to the female, and transports the sperm cells from the testes through the penis, which is a specialized organ that functions to deposit the sperm cells in the female reproductive system.
3. *Production of male sex hormones.* Hormones produced by the male reproductive system control the development of the reproductive system itself and of the male body form. These hormones are also essential for the normal function of the reproductive system and reproductive behavior.

The female reproductive system performs the following functions:

1. *Production of female sex cells.* The reproductive system produces female sex cells, or oocytes, in the ovaries.
2. *Reception of sperm cells from the male.* The female reproductive system includes structures that receive sperm cells from the male and transport the sperm cells to the site of fertilization.
3. *Nurturing the development of and providing nourishment for the new individual.* The female reproductive system nurtures the development of a new individual in the uterus until birth and provides nourishment in the form of milk after birth.
4. *Production of female sex hormones.* Hormones produced by the female reproductive system control the development of the reproductive system itself and of the female body form. These hormones are also essential for the normal function of the reproductive system and reproductive behavior.

Formation of Sex Cells

The testes in males and ovaries in females (figure 19.1) produce **sex cells,** or **gametes** (gam′ētz, *gametēs,* husband; *gametē,* wife). The formation of sex cells in males and females occurs by a special type of cell division called **meiosis** (mī-ō′sis, a lessening) (see Clinical Focus: Meiosis on p. 544). For both males and females, meiosis begins in cells that contain 23 pairs of chromosomes (46 chromosomes) and ends with gametes containing 23 chromosomes.

Male Reproductive System

The male reproductive system consists of the testes (sing., testis), a series of ducts, accessory glands, and supporting structures. The ducts include the epididymides (sing., epididymis), ductus deferentia (sing., deferens; also vas deferens), and urethra. Accessory glands include the seminal vesicles, prostate gland, and bulbourethral glands. Supporting structures include the scrotum and penis (figure 19.2). The sperm cells are very heat-sensitive and must develop at a temperature slightly less than normal body temperature. The testes, in which the sperm cells develop, are located outside the body cavity in the scrotum, where the temperature is lower. Sperm cells are transported from each testis to an epididymis, which lies on the posterior surface of the testis, and then through a ductus deferens into the pelvic cavity. Just before the ductus deferens enters the prostate gland, the ductus deferens increases in diameter to become the ampulla of the ductus deferens. A short duct of a seminal vesicle joins the ampulla of the ductus deferens to form the ejaculatory duct at the prostate, which then enters the prostate gland and empties into the urethra within the prostate gland. The urethra exits from the pelvis, passes through the penis, and opens to the outside of the body.

Scrotum

The **scrotum** (skrō′tum) is a saclike structure containing the testes. It is divided into right and left internal compartments by an incomplete connective tissue septum. Externally the scrotum consists of skin. Beneath the skin is a layer of loose connective tissue and a layer of smooth muscle, called the **dartos** (dar′tōs, to skin) **muscle.**

In cold temperatures, the dartos muscle contracts, causing the skin of the scrotum to become firm and wrinkled and reducing the overall size of the scrotum. At the same time, extensions of abdominal muscles into the scrotum, called **cremaster** (krē-mas′ter, to hang) **muscles,** contract (see figure 19.5). Consequently, the testes are pulled nearer to the body, and their temperature is raised. During warm weather or exercise, the dartos and cremaster muscles relax, the skin of the

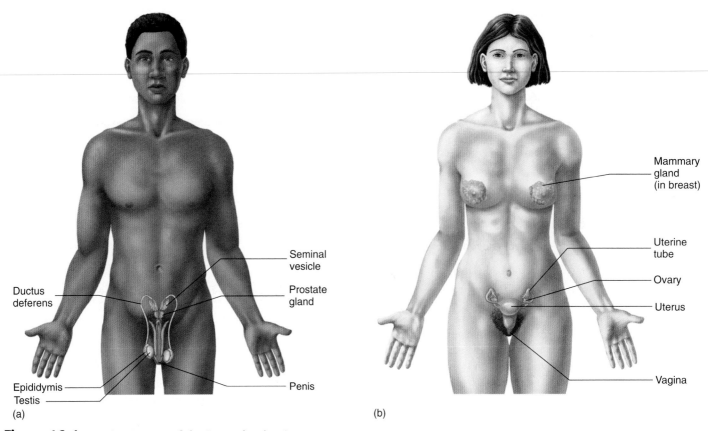

Ductus
deferens

Seminal
vesicle

Prostate
gland

Epididymis
Testis
(a)

Penis

Mammary
gland
(in breast)

Uterine
tube

Ovary

Uterus

Vagina

(b)

Figure 19.1 Major Organs of the Reproductive System

(a) The male reproductive system: testes, epididymis, ductus deferens, seminal vesicles, prostate gland, and penis. (b) The female reproductive system: ovaries, uterine tubes, uterus, vagina, and mammary glands.

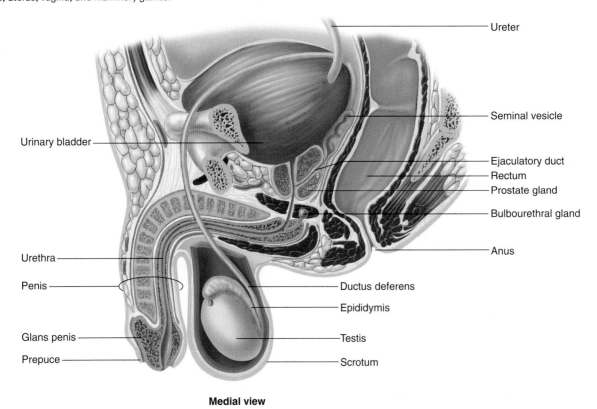

Ureter

Urinary bladder

Seminal vesicle

Ejaculatory duct
Rectum
Prostate gland

Bulbourethral gland

Anus

Urethra

Penis

Glans penis

Prepuce

Ductus deferens

Epididymis

Testis

Scrotum

Medial view

Figure 19.2 Male Reproductive Structures

Medial view of the male pelvis showing the male reproductive structures.

scrotum becomes loose and thin, and the testes descend away from the body, which lowers their temperature. The response of the dartos and cremaster muscles is important in the regulation of temperature in the testes. If the testes become too warm or too cold, normal sperm cell development does not occur.

Testes

The **testes** (tes′tēz), or male gonads, (gō′nădz, *gonē*, seed) are oval organs, each about 4–5 cm long, within the scrotum (see figure 19.2). The outer part of each testis consists of a thick, white connective tissue capsule. Extensions of the capsule project into the interior of the testis and divide each testis into about 250 cone-shaped lobules (figure 19.3a). The lobules contain **seminiferous** (sem′ĭ-nif′er-ŭs, seed carriers) **tubules,** in which sperm cells develop. Delicate connective tissue surrounding the seminiferous tubules contains clusters of endocrine cells called **interstitial** (in-ter-stish′ăl) **cells,** or **cells of Leydig** (lī′dig, named after Franz von Leydig [1821–1908], a German anatomist), which secrete testosterone.

Descent of the Testes

The testes develop in the abdominopelvic cavity. They move from the abdominopelvic cavity through the **inguinal** (ing′gwi-năl, *inguen*, groin) **canal** to the scrotum. The inguinal canals and the internal layers of the scrotum originate as outpocketings of the abdominal cavity along the lateral, superior margin of the pubis. The descent of the testes occurs during the seventh or eighth month of fetal development or, in some cases, shortly after birth. Failure of the testes to descend into the scrotal sac is called **cryptorchidism** (krip-tōr′ki-dizm, *crypto*, concealed + *orchis*, testis). It results in sterility because of the inhibiting effect of normal body temperature on sperm cell development.

After the testes descend, the inguinal canals narrow permanently, but they remain as weak spots in the abdominal wall. If an inguinal canal enlarges or ruptures, this can result in an **inguinal hernia** (her′nē-ă, rupture) through which a loop of intestine can protrude. This herniation can be quite painful and even very dangerous, especially if the inguinal canal compresses the intestine and cuts off its blood supply. Fortunately, inguinal hernias can be repaired surgically.

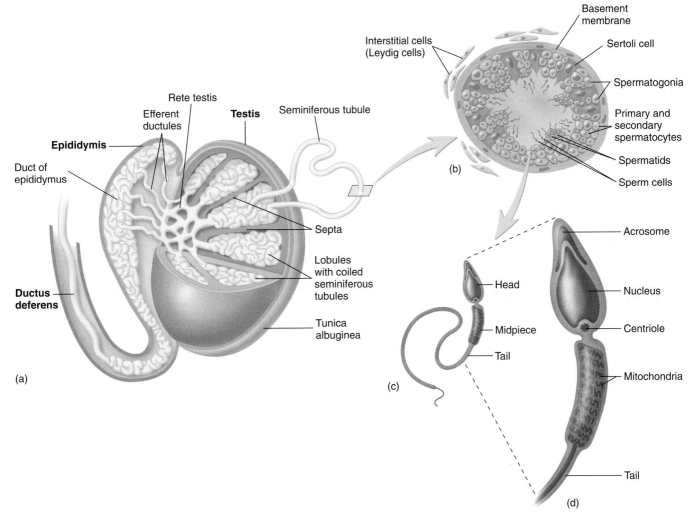

Figure 19.3 Structure of the Testis and Sperm Cell

(a) Gross anatomy of the testis with a section cut away to reveal internal structures. The epididymis is also shown. (b) Cross-section of a seminiferous tubule. Spermatogonia are near the periphery, and mature sperm cells are near the lumen of the seminiferous tubule. (c) The head, midpiece, and tail of a sperm cell. (d) Enlargement of the head and midpiece of a sperm cell to show the nucleus and acrosome in the head and the mitochondria in the midpiece.

Clinical Focus Meiosis

First Meiotic Division (Meiosis I)

Second Meiotic Division (Meiosis II)
(continued from the bottom of previous column)

1. Early prophase I
The duplicated chromosomes become visible chromatids (shown separated for emphasis; they actually are so close together that they appear as a single strand).

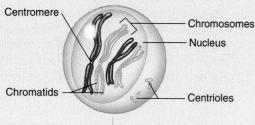

Centromere

Chromosomes

Nucleus

Chromatids

Centrioles

6. Prophase II
Each chromosome consists of two chromatids.

2. Middle prophase I
Pairs of chromosomes synapse. Crossing over may occur at this stage.

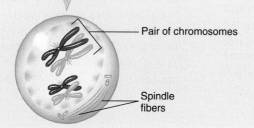

Pair of chromosomes

Spindle fibers

3. Metaphase I
Pairs of chromosomes align along the center of the cell. Random assortment of chromosomes occurs.

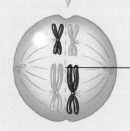

Equatorial plane

7. Metaphase II
Chromosomes align along the center of the cell.

4. Anaphase I
Chromosomes move apart to opposite sides of the cell.

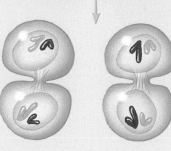

8. Anaphase II
Chromatids separate and each is now called a chromosome.

5. Telophase I
New nuclei form, and the cell divides.

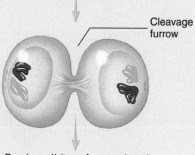

Cleavage furrow

9. Telophase II
New nuclei form around the chromosomes. The cells divide to form four daughter cells with half as many chromosomes as the parent cell.

Prophase II (top of next column)

Process Figure A Meiosis

The formation of sperm cells and female sex cells involves meiosis (see chapter 3). This kind of cell division occurs only in the testis and ovary. It consists of two consecutive cell divisions, and the production of four daughter cells, each having half as many chromosomes as the parent cell.

The two divisions of meiosis are called meiosis I and meiosis II. Like mitosis, each division of meiosis has prophase, metaphase, anaphase, and telophase. Distinct differences exist between meiosis and mitosis, however.

Before meiosis begins, all the chromosomes are duplicated. At the beginning of meiosis, each of the 46 chromosomes consists of two chromatids connected by a centromere (figure A, step 1). The chromosomes align as pairs in a process called **synapsis** (si-nap′sis, a connection) (figure A, step 2). Because each chromosome consists of two chromatids, the pairing of the chromosomes brings two chromatids of each chromosome close together. Occasionally, part of a chromatid of one chromosome will break off and be exchanged with part of another chromatid from the other chromosome. This exchange is called **crossing over.** Crossing over allows the exchange of genetic material between chromosomes.

The chromosomes align along the center of the cell (figure A, step 3) and then the pairs of chromosomes are separated to each side of the cell (figure A, step 4). As a consequence, when meiosis I is complete, each daughter cell has one chromosome from each of the pairs (figure A, step 5) or 23 chromosomes. Each of the 23 chromosomes in each daughter cell consists of two chromatids joined by a centromere.

It is during the first meiotic division that the chromosome number is reduced from 46 (23 pairs) to 23 chromosomes. The first meiotic division is therefore called a reduction division.

The second meiotic division is similar to mitosis. The chromosomes, each consisting of two chromatids (figure A, steps 6 and 7) align along the center of the cell. Then the chromatids separate at the centromere, and each daughter cell receives one of the chromatids from each chromosome (figure A, steps 8 and 9). When the centromere separates, each of the chromatids is called a chromosome. Consequently, each of the four daughter cells produced by meiosis contains 23 chromosomes.

During fertilization, the zygote receives one chromosome of each pair of chromosomes from each parent. Although half of the genetic material of a zygote comes from each parent, the genetic makeup of the zygote is unique.

Spermatogenesis

Spermatogenesis (sper′mă-tō-jen′ĕ-sis) is the formation of sperm cells. Before puberty, the testes remain relatively simple and unchanged from the time of their initial development. The interstitial cells are not prominent, and the seminiferous tubules are small and not yet functional. At the time of puberty, the interstitial cells increase in number and size, the seminiferous tubules enlarge, and spermatogenesis begins.

The seminiferous tubules contain **germ cells** and **Sertoli** (ser-tō′lē, named for the Italian histologist, Enrico Sertoli [1842–1910]) **cells** (figure 19.3b). Sertoli cells are large and extend from the periphery to the lumen of the seminiferous tubule. They nourish the germ cells and produce a number of hormones.

Germ cells are scattered between the Sertoli cells. The most peripheral germ cells are **spermatogonia** (sper′mă-tō-gō′nē-ă, primitive sperm cell), which divide through mitosis (figure 19.4). Some daughter cells produced from these mitotic divisions remain as spermatogonia and continue to divide by mitosis. Other daughter cells form **primary spermatocytes** (sper′mă-tō-sītz, sperm cell), which divide by meiosis.

A primary spermatocyte contains 46 chromosomes, each consisting of two chromatids. Each primary spermatocyte passes through the first meiotic division to produce two **secondary spermatocytes.** Each secondary spermatocyte undergoes a second meiotic division to produce two smaller cells called **spermatids** (sper′mă-tidz), each having 23 chromosomes. After the second meiotic division, the spermatids undergo major structural changes to form sperm cells (see figures 19.3b,c, and 19.4). Much of the cytoplasm of the spermatids is eliminated, and each spermatid develops a head, midpiece, and flagellum (tail) to become a **sperm cell,** or **spermatozoon** (see figures 19.3b to d, and figure 19.4). The nucleus of the sperm cell is located in the head of the sperm cell. Just anterior to the nucleus is a vesicle called the **acrosome** (ak′rō-sōm, acro, tip + soma, body), which contains enzymes that are released during the process of fertilization and are necessary for the sperm cell to penetrate the oocyte (ōon, egg + kytos, cell).

At the end of spermatogenesis, the developing sperm cells are located around the lumen of the seminiferous tubules with their heads directed toward the surrounding Sertoli cells and their tails directed toward the center of the lumen (see figures 19.3b and 19.4). Finally, sperm cells are released into the lumen of the seminiferous tubules.

Ducts

After their production, the sperm cells are transported through the seminiferous tubules and a series of ducts to reach the exterior of the body.

Epididymis

The seminiferous tubules of each testis empty into a tubular network called the **rete** (rē′tē, net) **testis** (see figure 19.3a). The rete testis empties into 15–20 tubules called the **efferent**

1. Spermatogonia are the cells from which sperm cells arise. The spermatogonia divide by mitosis. One daughter cell remains a spermatogonium that can divide again by mitosis. One daughter cell becomes a primary spermatocyte.

2. The primary spermatocyte divides by meiosis to form secondary spermatocytes.

3. The secondary spermatocytes divide by meiosis to form spermatids.

4. The spermatids differentiate to form sperm cells.

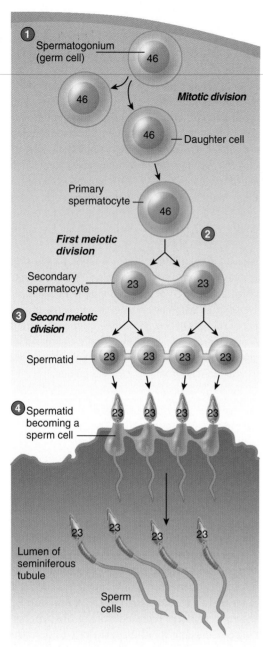

Process Figure 19.4 Spermatogenesis

A section of the seminiferous tubule illustrating the process of meiosis and sperm cell formation.

ductules (ef′er-ent dŭk′toolz). The efferent ductules carry sperm cells from the testis to a tightly coiled series of threadlike tubules that form a comma-shaped structure on the posterior side of the testis called the **epididymis** (ep-i-did′i-mis, upon twins) (see figures 19.2, and 19.3a, and figure 19.5a). The sperm cells continue to mature within the epididymis. Within the epididymis, sperm cells develop the capacity to swim and the ability to bind to the secondary oocyte. Sperm cells taken directly

from the testes are not capable of fertilizing secondary oocytes, but after maturing for several days in the epididymis, the sperm cells develop the capacity to function as sex cells. Final changes in sperm cells, called **capacitation** (kă-pas′i-tă′shun, capable of) occur after ejaculation of semen into the vagina and prior to fertilization.

Ductus Deferens

The **ductus deferens** (dŭk′tŭs def′er-enz, *defero,* to carry away), or **vas deferens,** emerges from the epididymis and ascends along the posterior side of the testis to become associated with the blood vessels and nerves that supply the testis. These structures form the **spermatic cord** (see figure 19.5a). Each spermatic cord consists of the ductus deferens, testicular artery and veins, lymphatic vessels, and testicular nerve. It is surrounded by the cremaster muscle, and two connective tissue sheaths.

Each ductus deferens extends, in the spermatic cord, through the abdominal wall by way of the inguinal canal. Each ductus deferens then crosses the lateral wall of the pelvic cavity and loops behind the posterior surface of the urinary bladder to approach the prostate gland (see figures 19.2 and 19.5a). The total length of the ductus deferens is about 45 cm. Just before reaching the prostate gland, the ductus deferens increases in diameter to become the **ampulla of the ductus deferens** (see figure 19.5a). The wall of the ductus deferens contains smooth muscle. Peristaltic contractions of the smooth muscle propel the sperm cells from the epididymis through the ductus deferens.

Seminal Vesicle and Ejaculatory Duct

Near the ampulla of each ductus deferens is a sac-shaped gland called the **seminal vesicle** (sem′i-năl ves′i-kl, relating to semen + *vesica,* bladder). A short duct extends from the seminal vesicle to the ampulla of the ductus deferens. The duct from the seminal vesicle and the ampulla of the ductus deferens join at the prostate gland to form the **ejaculatory** (ē-jak′ū-lă-tōr-ē, to shoot out) **duct.** Each ejaculatory duct extends into the prostate gland and ends by joining the urethra within the prostate gland (see figure 19.5a).

Urethra

The male **urethra** (ū-rē′thră) extends from the urinary bladder to the distal end of the penis (see figures 19.2 and 19.5a). The urethra can be divided into three parts: the **prostatic urethra,** which passes through the prostate gland; the **membranous urethra,** which passes through the floor of the pelvis and is surrounded by the external urinary sphincter; and the **spongy urethra,** which extends the length of the penis and opens at its end. The urethra is a passageway for both urine and male reproductive fluids. Urine and the reproductive fluids, however, do not exit the urethra at the same time. While male reproductive fluids are passing through the urethra, a sympathetic reflex causes the internal urinary sphincter to contract,

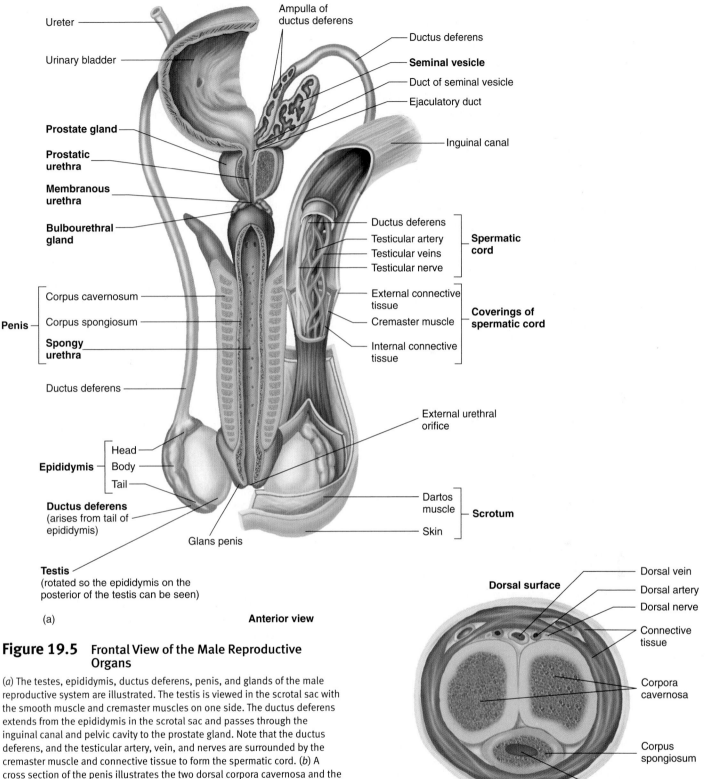

Ureter

Urinary bladder

Ampulla of ductus deferens

Ductus deferens

Seminal vesicle

Duct of seminal vesicle

Ejaculatory duct

Inguinal canal

Prostate gland

Prostatic urethra

Membranous urethra

Bulbourethral gland

Ductus deferens
Testicular artery
Testicular veins
Testicular nerve

Spermatic cord

External connective tissue

Cremaster muscle

Internal connective tissue

Coverings of spermatic cord

Corpus cavernosum

Corpus spongiosum

Penis

Spongy urethra

Ductus deferens

External urethral orifice

Epididymis
Head
Body
Tail

Dartos muscle

Skin

Scrotum

Ductus deferens (arises from tail of epididymis)

Glans penis

Testis (rotated so the epididymis on the posterior of the testis can be seen)

(a)

Anterior view

Dorsal vein
Dorsal artery
Dorsal nerve
Connective tissue

Dorsal surface

Corpora cavernosa

Corpus spongiosum

Spongy urethra

(b)

Ventral surface

Figure 19.5 Frontal View of the Male Reproductive Organs

(a) The testes, epididymis, ductus deferens, penis, and glands of the male reproductive system are illustrated. The testis is viewed in the scrotal sac with the smooth muscle and cremaster muscles on one side. The ductus deferens extends from the epididymis in the scrotal sac and passes through the inguinal canal and pelvic cavity to the prostate gland. Note that the ductus deferens, and the testicular artery, vein, and nerves are surrounded by the cremaster muscle and connective tissue to form the spermatic cord. (b) A cross section of the penis illustrates the two dorsal corpora cavernosa and the ventral corpus spongiosum. Connective tissue sheaths and skin cover the three erectile bodies. Blood vessels, including the dorsal artery and vein and the dorsal nerve of the penis, are visible.

which keeps semen from passing into the urinary bladder and prevents urine from entering the urethra. Many minute, mucus-secreting glands are located in the epithelial lining of the urethra.

Penis

The **penis** (pē′nis, tail) contains three columns of erectile tissue (figure 19.5b and see figure 19.5a). Engorgement of this erectile tissue with blood causes the penis to enlarge and become firm, a process called **erection** (ē-rek′shŭn, *erectio*, to set up). The penis is the male organ of copulation and functions in the transfer of sperm cells from the male to the female. Two columns of erectile tissue form the dorsal portion and the sides of the penis and are called the **corpora cavernosa** (kōr′pōr-ă kav-er-nōs′ă, cavernous body). The third and smaller erectile column occupies the ventral portion of the penis and is called the **corpus spongiosum** (kōr′pŭs spŭng-gē-ō′sŭm, spongy body). It expands over the distal end of the penis to form a cap, the **glans** (glanz, acorn) **penis.** The spongy urethra passes through the corpus spongiosum, including the glans penis, and opens to the exterior as the **external urethral orifice.**

The shaft of the penis is covered by skin that is loosely attached to the connective tissue surrounding the penis. The skin is firmly attached at the base of the glans penis, and a thinner layer of skin tightly covers the glans penis. The skin of the penis, especially the glans penis, is well supplied with sensory receptors. A loose fold of skin, called the **prepuce** (prē′pūs), or **foreskin,** covers the glans penis (see figure 19.2).

Circumcision

Circumcision (ser-kŭm-sizh′ŭn) is the surgical removal of the prepuce, usually near the time of birth. There are few compelling medical reasons for circumcision. Uncircumcised males have a higher incidence of penile cancer, but the underlying cause appears to be related to chronic infections and poor hygiene. In those few cases in which the prepuce is "too tight" to be moved over the glans penis, circumcision can be necessary to avoid chronic infections and maintain normal circulation.

Glands

The **seminal vesicles** are glands consisting of many saclike structures located next to the ampulla of the ductus deferens (see figures 19.2 and 19.5a). There are two seminal vesicles. Each is about 5 cm long and tapers into a short duct that joins the ampulla of the ductus deferens to form the ejaculatory duct.

The **prostate** (pros′tāt, to stand before, protect) **gland** consists of both glandular and muscular tissue and is about the size and shape of a walnut (see figures 19.2 and 19.5a). The prostate gland surrounds the urethra and the two ejaculatory ducts. It consists of a capsule and numerous partitions. The cells lining the partitions secrete prostatic fluid. There are 10–20 short ducts (not seen in figure 19.5a) that carry secretions of the prostate gland to the prostatic part of the urethra.

A CASE IN POINT | Prostate Cancer

Willy Dye, who is 65 years old, has a physical examination every year. Ten years ago, a prostate-specific antigen (PSA) test indicated that Willy's PSA levels were higher than the results from his previous tests. His physician reported moderate enlargement of the prostate gland, but he detected no obvious tumor-like structures in a digital examination. Because of the increasing PSA levels, Willy's physician recommended a needle biopsy of the prostate gland through the rectum. The pathology report described suspicious cells consistent with prostate cancer in one of the tissue samples. Willy's physician had the biopsy samples examined by another pathology laboratory. The second pathology report did not confirm the first pathology report. As a consequence, Willy's physician explained that one option was to do nothing and continue having regular checkups because prostate cancer typically develops slowly. Eight years later, a PSA test showed another substantial increase in Willy's PSA levels, although no tumor could be detected by a digital exam, and Willy had no complaints, such as difficulty in urinating. This time a needle biopsy of the prostate gland indicated cancer cells were present in two of the six biopsy samples. Willy's physician explained that his chance of surviving were high because it appears that the cancer was discovered before it metastasized to areas outside of the prostate gland. Willy could choose to do nothing, have his prostate gland surgically removed, or treat the cancer with radiation therapy, hormonal therapy, or chemotherapy. Willy elected to have radiation therapy, which focuses radiation on the prostate gland to kill the cancer cells. Statistics indicate that surgery and radiation therapy have similar success rates for small, localized tumors like Willy's. The trauma of surgery and the higher probability of erectile dysfunction following surgery convinced Willy that radiation therapy was preferable. Willy's physician indicated that doing nothing is a reasonable option for men who are significantly older than Willy because older men diagnosed with prostate cancer often die of other conditions before they succumb to prostate cancer. Willy's physician explained that, for patients like him, approximately 85% are cancer free after 5 years. Willy was grateful that he had annual physical examinations. Prostate cancer represents 29% of cancers in males in the United States and 14% of the deaths due to cancer. Only lung cancer results in more cancer deaths in men.

Changes in the size and texture of the prostate gland can be an indication of developing prostate cancer. Suggest a way that the size and texture of the prostate gland can be examined by palpation without surgical techniques (see figure 19.2).

The **bulbourethral** (bul′bō-ū-rē′thrăl, refering to bulbous penis and urethra) **glands,** also called **Cowper's glands** (William Cowper, English anatomist, 1666–1709), are a pair of small mucus-secreting glands located near the base of the penis (see figures 19.2 and 19.5a). In young adults, each is about the size of a pea, but they decrease in size with age. A single duct from each gland enters the urethra.

Secretions

Semen (sē′men, *semen*, seed) is a mixture of sperm cells and secretions from the male reproductive glands. The seminal vesicles produce about 60% of the fluid, the prostate gland contributes approximately 30%, the testes contribute 5%, and the bulbourethral glands contribute 5%.

The bulbourethral glands and the mucous glands of the urethra produce a mucous secretion, which lubricates the urethra, helps neutralize the contents of the normally acidic urethra, provides a small amount of lubrication during intercourse, and helps reduce the acidity in the vagina.

Testicular secretions include sperm cells and a small amount of fluid. The thick, mucuslike secretion of the seminal vesicles contains the sugar fructose and other nutrients that provide nourishment to sperm cells. The seminal vesicle secretions also contain proteins that weakly coagulate after ejaculation and enzymes that are thought to help destroy abnormal sperm cells. Prostaglandins, which stimulate smooth muscle contractions, are present in high concentrations in the secretions of the seminal vesicles and can cause contractions of the female reproductive tract, which help transport sperm cells through the female reproductive tract.

The thin, milky secretions of the prostate have an alkaline pH and help neutralize the acidic urethra, as well as the acidic secretions of the testes, the seminal vesicles, and the vagina. The increased pH is important for normal sperm cell function. The movement of sperm cells is not optimal until the pH is increased to between 6.0 and 6.5. In contrast, the secretions of the vagina have a pH between 3.5 and 4.0. Prostatic secretions also contain proteolytic enzymes that break down the coagulated proteins of the seminal vesicles and make the semen more liquid. The normal volume of semen is 2–5 milliliters (mL). The normal sperm cell count is about 100 million sperm cells per mL of semen.

Physiology of Male Reproduction

The male reproductive system depends on both hormonal and neural mechanisms to function normally. Hormones control the development of reproductive structures, the development of secondary sexual characteristics, spermatogenesis, and, in part, sexual behavior. The mature neural mechanisms are primarily involved in controlling the sexual act and in the expression of sexual behavior.

Regulation of Sex Hormone Secretion

The hypothalamus of the brain, the anterior pituitary gland, and the testes (figure 19.6) produce hormones that influence the male reproductive system. **Gonadotropin-releasing** (gō′nad-ō-trō′pin, *gonad* + *trophe,* nourishment) **hormone (GnRH)** is released from neurons in the hypothalamus and passes to the anterior pituitary gland (table 19.1). GnRH causes cells in the anterior pituitary gland to secrete two hormones, **luteinizing** (loo′tē-ĭ-nīz-ing) **hormone (LH)** and **follicle-stimulating hormone (FSH),** into the blood (see figure 19.6 and table 19.1). LH and FSH are named for their functions in females, but they are also essential reproductive hormones in males.

LH binds to the interstitial cells in the testes and causes them to secrete testosterone. LH was once referred to as interstitial cell-stimulating hormone (ICSH) because it stimulates interstitial cells of the testes to secrete testosterone, but it was later discovered to be identical to LH found in the female. Consequently, it is now simply called LH. FSH binds primarily to Sertoli cells in the seminiferous tubules and promotes sperm cell development. It also increases the secretion of a hormone called **inhibin** (in-hib′in, to inhibit).

Testosterone has a negative-feedback effect on the secretion of GnRH from the hypothalamus, and on LH and FSH from the anterior pituitary gland. Inhibin has a negative-feedback effect on the secretion of FSH from the anterior pituitary gland.

GnRH Control of LH and FSH Release

For GnRH to stimulate LH and FSH release, the pituitary gland must be exposed to a series of brief increases and decreases in GnRH. If GnRH is maintained at a high level in the circulatory system for days or weeks, the anterior pituitary cells become insensitive to it. GnRH can be produced synthetically and is useful in treating some people who are infertile. Synthetic GnRH must be administered in small amounts in frequent pulses or surges. GnRH can also inhibit reproduction, because long-term administration of GnRH can sufficiently reduce LH and FSH levels to prevent sperm cell production in males or ovulation in females.

Puberty

Puberty (pū′ber-tē) is the sequence of events by which a child is transformed into a young adult. The reproductive system matures and assumes its adult functions, and the structural differences between adult males and females become more apparent. In boys, puberty commonly begins at ages 12–14 and is largely completed by age 18. Before puberty, small amounts of testosterone, secreted by the testes and the adrenal cortex, inhibit GnRH, LH, and FSH secretion. Beginning just before puberty and continuing throughout

1. Gonadotropin-releasing hormone(GnRH) from the hypothalamus stimulates the secretion of luteinizing hormone (LH) and follicle-stimulating hormone(FSH) from the anterior pituitary.

2. LH stimulates testosterone secretion from the interstitial cells.

3. FSH stimulates sertoli cells of the seminiferous tubules to increase spermatogenesis and to secrete inhibin.

4. Testosterone has a Stimulatory effect on the Sertoli cells of the seminiferous tubules, and it has a stimulatory effect on the development of sex organs and secondary sex characteristics.

5. Testosterone has a negative-feedback effect on the hypothalamus and pituitary to reduce GnRH, LH, and FSH secretion.

6. Inhibin has a negative-feedback effect on the anterior pituitary to reduce FSH secretion.

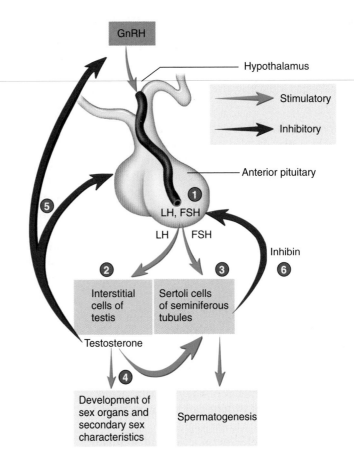

Process Figure 19.6   Regulation of Reproductive Hormone Secretion in Males

puberty, developmental changes in the hypothalamus cause the hypothalamus and the anterior pituitary gland to become much less sensitive to the inhibitory effect of testosterone, and the rate of GnRH, LH, and FSH secretion increases. Consequently, elevated FSH levels promote spermatogenesis, and elevated LH levels cause the interstitial cells to secrete larger amounts of testosterone. Testosterone still has a negative-feedback effect on the hypothalamus and anterior pituitary gland, but GnRH, LH, and FSH secretion occur at substantially higher levels.

Anabolic Steroids

Some athletes, especially those who depend on muscle strength, may either ingest or inject synthetic **androgens** (an'drō-jenz, *andros,* male), which are hormones that have testosteronelike effects, such as stimulating the development of male sexual characteristics. The synthetic androgens are commonly called **anabolic steroids,** or simply **steroids,** and they are used in an attempt to increase muscle mass. Many of the synthetic androgens are structurally different from testosterone. Their effect on muscle is greater than their effect on the reproductive organs. They are often taken in large amounts, however, and they can influence the reproductive system. Large doses of synthetic androgens have a negative-feedback effect on the hypothalamus and pituitary, causing a reduction in GnRH, LH, and FSH levels. As a result, the testes can atrophy, and sterility can develop. Other side effects of large doses of synthetic androgens include kidney and liver damage, heart attack, and stroke. Taking synthetic androgens is highly discouraged by the medical profession, is a violation of the rules of most athletic organizations, and is illegal. For people who take anabolic steroids by injection, the risk of contracting hepatitis B or HIV (the AIDS virus) is increased if they share needles with other people.

Effects of Testosterone

Testosterone (tes'tos'tě-rōn) is the major male hormone secreted by the testes. Testosterone influences reproductive organs and nonreproductive structures. During puberty, testosterone causes the enlargement and differentiation of the male genitals and reproductive duct system. It is necessary for spermatogenesis and for the development of male secondary sexual characteristics. The **secondary sexual characteristics** are those structural and behavioral changes, other than in the reproductive organs, that develop at puberty and distinguish males from

Table 19.1 Major Reproductive Hormones in Males and Females

Hormone	Source	Target Tissue	Response
Male Reproductive System			
Gonadotropin-releasing hormone (GnRH)	Hypothalamus	Anterior pituitary	Stimulates secretion of LH and FSH
Luteinizing hormone (LH)	Anterior pituitary	Interstitial cells of the testes	Stimulates synthesis and secretion of testosterone
Follicle-stimulating hormone (FSH)	Anterior pituitary	Seminiferous tubules (Sertoli cells)	Supports spermatogenesis and inhibin secretion
Testosterone	Interstitial cells of the testes	Testes; body tissues	Development and maintenance of reproductive organs; supports spermatogenesis and causes the development and maintenance of secondary sexual characteristics
		Anterior pituitary and hypothalamus	Inhibits GnRH, LH, and FSH secretion through negative feedback
Inhibin	Sertoli cells	Anterior pituitary	Inhibits FSH secretion through negative feedback
Female Reproductive System			
Gonadotropin-releasing hormone (GnRH)	Hypothalamus	Anterior pituitary	Stimulates secretion of LH and FSH
Luteinizing hormone (LH)	Anterior pituitary	Ovaries	Causes follicles to complete maturation and undergo ovulation; causes ovulation, causes the ovulated follicle to become the corpus luteum
Follicle-stimulating hormone (FSH)	Anterior pituitary	Ovaries	Causes follicles to begin development
Estrogen	Follicles of ovaries and corpus luteum	Uterus	Proliferation of endometrial cells
		Breasts	Development of the mammary glands (especially duct systems)
		Anterior pituitary and hypothalamus	Positive feedback before ovulation, resulting in increased LH and FSH secretion; negative feedback with progesterone on the hypothalamus and anterior pituitary after ovulation, resulting in decreased LH and FSH secretion
		Other tissues	Development and maintenance of secondary sexual characteristics
Progesterone	Corpus luteum of ovaries	Uterus	Enlargement of endometrial cells and secretion of fluid from uterine glands
		Breasts	Development of the mammary glands (especially alveoli)
		Anterior pituitary	Negative feedback, with estrogen, on the hypothalamus and anterior pituitary after ovulation, resulting in decreased LH and FSH secretion
		Other tissues	Secondary sexual characteristics
Oxytocin	Posterior pituitary	Uterus and mammary glands	Contraction of uterine smooth muscle and contraction of cells in the breast, resulting in milk letdown in lactating women
Human chorionic gonadotropin	Placenta	Corpus luteum of ovaries	Maintains the corpus luteum and increases its rate of progesterone secretion during the first one-third (first trimester) of pregnancy. Increases testosterone production in testes of male fetuses.

Table 19.2   Effects of Testosterone on Target Tissues

Target Tissue	Response
Penis and scrotum	Enlargement and differentiation
Hair follicles	Hair growth and coarser hair in pubic area, legs, chest, axillary region, the face, and occasionally the back; male pattern baldness on the head if the person has the appropriate genetic makeup
Skin	Coarser texture of skin; increased rate of secretion of sebaceous glands, frequently resulting in acne at the time of puberty; increased secretion of sweat glands in axillary regions
Larynx	Enlargement of larynx and deeper masculine voice
Most tissues	Increased rate of metabolism
Red blood cells	Increased rate of red blood cell production; red blood cell count increased by about 20% as a result of increased erythropoietin secretion
Kidney	Retention of sodium and water to a small degree, resulting in increased extracellular fluid volume
Skeletal muscle	Skeletal muscle mass increases at puberty; the average increase is greater in males than in females
Bone	Rapid bone growth resulting in increased rate of growth and in early cessation of bone growth; males who mature sexually at a later age do not exhibit a rapid period of growth, but they grow for a longer time and can become taller than men who mature earlier

females (table 19.2). Secondary sexual characteristics in males include hair distribution and growth, skin texture, fat distribution, skeletal muscle growth, and changes in the larynx. After puberty, testosterone maintains the adult structure of the male genitals, reproductive ducts, and secondary sex characteristics.

Male Pattern Baldness

Some men have a genetic tendency called **male pattern baldness,** which develops in response to testosterone and other androgens. When testosterone levels increase at puberty, the density of hair on top of the head begins to decrease. Baldness usually reaches its maximum rate of development when the individual is in the third or fourth decade of life. Minoxidil (mi-noks′si-dil, Rogaine) is a drug that effectively prevents a decrease in hair growth in many men who exhibit male pattern baldness. Its effectiveness is increased in those who are young and starting to show evidence of baldness. Minoxidil causes blood vessels to dilate, including those close to hair follicles, and this may explain how it works, but the mechanism by which the drug works has not been confirmed.

Male Sexual Behavior and the Male Sex Act

Testosterone is required for normal sexual behavior. Testosterone enters certain cells within the brain, especially within the hypothalamus, and influences their functions. The blood levels of testosterone remain relatively constant throughout the lifetime of a male, from puberty until about 40 years of age. Thereafter the levels slowly decline to approximately 20% of this value by 80 years of age, causing a slow decrease in sex drive and fertility.

P R E D I C T   2
Predict the effect on secondary sexual characteristics, external genitalia, and sexual behavior if the testes fail to produce normal amounts of testosterone at puberty.

The male sexual act is a complex series of reflexes that result in erection of the penis, secretion of mucus into the urethra, emission, and ejaculation. **Emission** (ē-mish′ŭn, *emitto,* to send out) is the movement of sperm cells, mucus, prostatic secretions, and seminal vesicle secretions into the prostatic, membranous, and spongy urethra. **Ejaculation** (ē-jak′ū-lā′shŭn) is the forceful expulsion of the secretions that have accumulated in the urethra to the exterior. Sensations, normally interpreted as pleasurable, occur during the male sexual act and result in an intense sensation called an **orgasm** (ōr′gazm, *orgaō,* to swell or be excited), or **climax.** In males, orgasm is closely associated with ejaculation, although they are separate functions and do not always occur simultaneously. A phase called **resolution** occurs after ejaculation. During resolution the penis becomes flaccid, an overall feeling of satisfaction exists, and the male is unable to achieve erection and a second ejaculation.

Sensory Impulses and Integration

Sensory action potentials from the genitals are carried to the sacral region of the spinal cord, in which reflexes that result in the male sexual act are integrated. Action potentials also travel from the spinal cord to the cerebrum to produce conscious sexual sensations.

Rhythmic massage of the penis, especially the glans, and surrounding tissues, such as the scrotal, anal, and pubic regions, provide important sources of sensory action potentials. Engorgement of the prostate gland and seminal vesicles with secretions or irritation of the urethra, urinary bladder, ductus deferens, and testes can also cause sexual sensations.

Psychic stimuli such as sight, sound, odor, or thoughts have a major effect on male sexual reflexes. Ejaculation while sleeping (nocturnal emission) is a relatively common event in young males and is thought to be triggered by psychic stimuli associated with dreaming. The inability to concentrate on sexual sensations can result in **impotence** (im′pŏ-tens, lack of power), or the inability to achieve an erection of the penis. Impotence can also be caused by nervous system lesions or physical factors such as inability of the erectile tissue to fill with blood.

Erection, Emission, and Ejaculation

Erection is the first major component of the male sexual act. Parasympathetic action potentials from the sacral region of the spinal cord cause the arteries that supply blood to the erectile tissues to dilate. Blood then fills small venous sinuses called **sinusoids** in the erectile tissue and compresses the veins, which reduces blood flow from the penis. The increased blood pressure in the sinusoids causes the erectile tissue to become inflated and rigid. Parasympathetic action potentials also cause the mucous glands within the urethra and the bulbourethral glands to secrete mucus.

Treatment of Erectile Dysfunction

Failure to achieve erections, or erectile dysfunction (ED), can be a major source of frustration and can contribute to disharmony in relationships. The inability to achieve erections can be due to reduced testosterone secretion resulting from hypothalamic, pituitary, or testicular complications. In other cases the inability to achieve erections can be due to defective stimulation of the erectile tissue by nerve fibers or reduced response of the blood vessels to neural stimulation. Erection can be achieved in some people by oral medications, such as sildenafil (Viagra), tadalafil (Cialis), or verdenafil (Livitra), or by the injection of specific drugs into the base of the penis. These drugs function to increase blood flow into the erectile tissue of the penis, resulting in erection for many minutes. Sildenafil blocks the activity of an enzyme that converts cGMP to GMP. This allows cGMP to accumulate in smooth muscle cells in the arteries of erectile tissues and causes them to relax. This response is effective in enhancing erection of the penis in males. Sildenafil's action is not specific to erectile tissue of the penis. It also causes vasodilation in other tissues and can increase the workload of the heart.

Before ejaculation, the ductus deferens begins to contract rhythmically, propelling sperm cells and testicular fluid from the epididymis through the ductus deferens. Contractions of the ductus deferens, seminal vesicles, and ejaculatory ducts cause the sperm cells, testicular secretions, and seminal fluid to move into the urethra, in which they mix with prostatic secretions released as a result of contraction of the prostate.

Emission is stimulated by sympathetic action potentials that originate in the lumbar region of the spinal cord. Action potentials cause contractions of the reproductive ducts and stimulate the seminal vesicles and prostate gland to release secretions. Consequently, semen accumulates in the urethra.

Ejaculation results from the contraction of smooth muscle in the wall of the urethra and skeletal muscles surrounding the base of the penis. Just before ejaculation, action potentials are sent to the skeletal muscles that surround the base of the penis. Rhythmic contractions are produced that force the semen out of the urethra, resulting in ejaculation. In addition, muscle tension increases throughout the body.

Infertility in Males

Infertility (in-fer-til′i-tē) is reduced or diminished fertility. The most common cause of infertility in males is a low sperm cell count. If the sperm cell count drops to below 20 million sperm cells per mL, the male is usually sterile.

Decreased sperm cell count can occur because of damage to the testes as a result of trauma, radiation, cryptorchidism, or infections such as mumps, which block the ducts in the epididymis. Reduced sperm cell counts can result from inadequate secretion of luteinizing hormone and follicle-stimulating hormone, which can be caused by hypothyroidism, trauma to the hypothalamus, infarctions of the hypothalamus or anterior pituitary gland, or tumors. Decreased testosterone secretion also reduces the sperm cell count.

Fertility is reduced if the sperm cell count is normal but sperm cell structure is abnormal, such as chromosomal abnormalities caused by genetic factors. Reduced sperm cell motility also results in infertility. A major cause of reduced sperm cell motility is antisperm antibodies produced by the immune system, which bind to sperm cells.

Fertility can sometimes be achieved by collecting several ejaculations, concentrating the sperm cells, and inserting them into the female's reproductive tract, a process called **artificial insemination** (in-sem-i-nā′shŭn).

Female Reproductive System

The female reproductive organs consist of the ovaries, uterine tubes (or fallopian tubes), uterus, vagina, external genitalia, and mammary glands (see figure 19.1*b*). The internal reproductive organs of the female are located within the pelvis, between the urinary bladder and the rectum (figure 19.7). The uterus and the vagina are in the midline, with an ovary to each side of the uterus (figure 19.8). The internal reproductive organs are held in place within the pelvis by a group of ligaments. The most conspicuous is the **broad ligament,** which spreads out on both sides of the uterus and to which the ovaries and uterine tubes attach.

Ovaries

The two **ovaries** (ō′vă-rēz, *ovum,* egg) are small organs suspended in the pelvic cavity by ligaments. The **suspensory ligament** extends from each ovary to the lateral body wall, and the **ovarian ligament** attaches the ovary to the superior margin of the uterus (see figure 19.8). In addition, the ovaries are attached to the posterior surface of the broad ligament by folds of peritoneum called the **mesovarium** (mez′ō-vā′rē-ŭm, mesentery of the ovary). The ovarian arteries, veins, and nerves traverse the suspensory ligament and enter the ovary through the mesovarium.

A layer of visceral peritoneum covers the surface of the ovary. The outer part of the ovary is made up of dense connective tissue and contains **ovarian follicles** (figure 19.9). Each of the ovarian follicles contains an **oocyte** (ō′ō-sīt, *ōon,* egg + *kytos,* cell), the female sex cell. Loose connective tissue makes up the inner part of the ovary, where blood vessels, lymphatic vessels, and nerves are located.

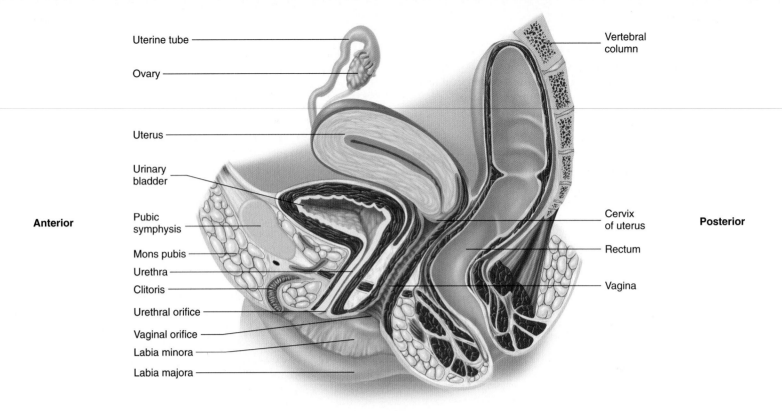

Medial view

Figure 19.7 **Medial View of the Female Pelvis**

The female reproductive tract, including the uterus, vagina, and surrounding pelvic structures are shown in the medial view of the female pelvis. Note that the female reproductive and female urinary tracts open separately to the exterior in the vestibule.

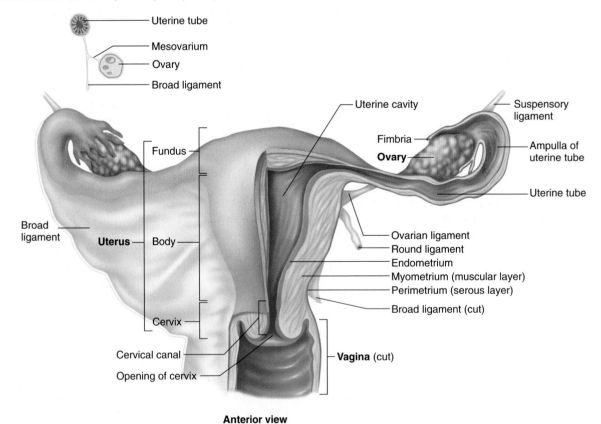

Anterior view

Figure 19.8 **Uterus, Vagina, Uterine Tubes, Ovaries, and Supporting Ligaments**

Anterior view of the uterus, uterine tubes, and associated ligaments. The uterus and uterine tubes are cut in section (on the left side), and the vagina is cut to show the internal anatomy. The inset shows the relationships between the ovary, uterine tube, and the ligaments that suspend them in the pelvic cavity.

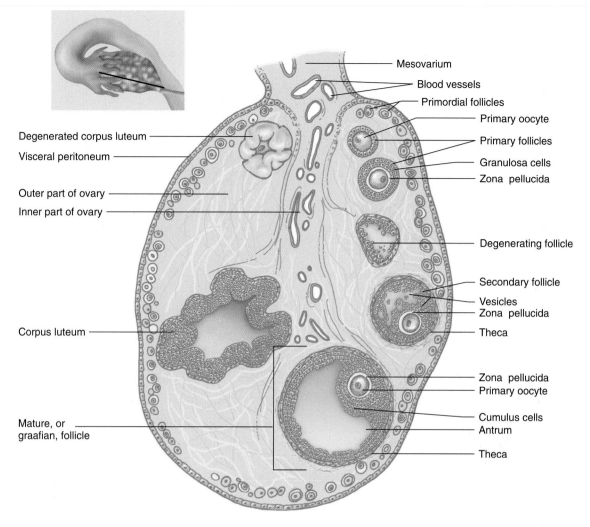

Figure 19.9 **Structure of the Ovary**

The ovary is sectioned to illustrate its internal structure (inset shows plane of section). Ovarian follicles from each major stage of development are presented, and a corpus luteum is also shown.

Oocyte Development and Fertilization

The formation of female sex cells begins in the fetus. By the fourth month of development, the ovaries contain 5 million **oogonia** (ō-ō-gō′nē-ă, *oon*, egg + *gone*, generation), the cells from which oocytes develop (figure 19.10). By the time of birth, many of the oogonia have degenerated, and the remaining have begun meiosis. Also, some data indicate that oogonia can form after birth from stem cells, but the extent to which this occurs, and how long it occurs, is not clear. As in meiosis in males, the genetic material is duplicated and two cell divisions occur (see Clinical Focus: Meiosis on p. 544). Meiosis stops, however, during the first meiotic division at a stage called prophase I. The cell at this stage is called a **primary oocyte,** and at birth there are about 2 million of them. From birth to puberty, many primary oocytes degenerate. The number of primary oocytes decreases to around 300,000 to 400,000; of these only about 400 will complete development and be released from the ovaries. Nearly all others degenerate after partial development.

Ovulation is the release of an oocyte from an ovary (see figure 19.10). Just before ovulation, the primary oocyte completes the first meiotic division to produce a **secondary oocyte** and a **polar body.** Unlike meiosis in males, cytoplasm is not split evenly between the two cells. Most of the cytoplasm of the primary oocyte remains with the secondary oocyte. The polar body either degenerates or divides to form two polar bodies. The secondary oocyte begins the second meiotic division, but stops in metaphase II.

After ovulation, the secondary oocyte may be fertilized by a sperm cell (see figure 19.10). **Fertilization** (fer′til-i-zā-shŭn) begins when a sperm cell penetrates into the cytoplasm of a secondary oocyte. Subsequently, the secondary oocyte completes the second meiotic division to form two cells, each containing 23 chromosomes. One of these cells has very little cytoplasm and is another polar body that degenerates. In the other, larger cell, the 23 chromosomes from the male join with the 23 from the female to form a **zygote** (zī′gōt, *zygotos*, yolked) and complete fertilization. The zygote has 23 pairs or 46 chromosomes. All cells of the human body contain 23 pairs of chromosomes, except for the male and female sex cells. The zygote divides by mitosis to form two cells, which divide to form four cells, and so on. The mass of cells formed may eventually implant, or attach, to the wall of the uterus and develop into a new individual (see chapter 20).

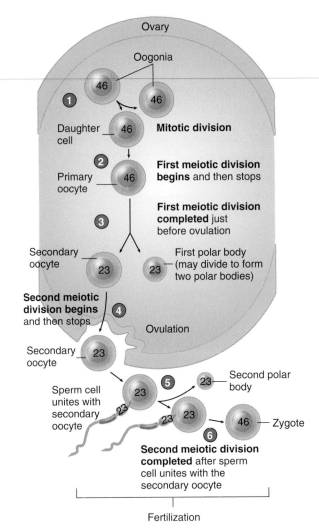

1. Oogonia are the cells from which oocytes arise. The oogonia divide by mitosis to produce other oogonia and primary oocytes.

2. Five million oocytes may be produced by the fourth month of prenatal life. Primary oocytes begin the first meiotic division but stop at prophase I. All of the primary oocytes remain in this state until puberty.

3. The first meiotic division is completed in a single mature follicle just before ovulation during each menstrual cycle. A secondary oocyte and the first polar body result from the unequal division of the cytoplasm.

4. The secondary oocyte begins the second meiotic division but stops at metaphase II.

5. The second meiotic division is completed after ovulation and after a sperm cell unites with the secondary oocyte. A secondary oocyte and a second polar body are formed.

6. Fertilization is completed after the nuclei of the secondary oocyte and the sperm cell unite. The resulting cell is called a zygote.

Process Figure 19.10 Maturation and Fertilization of the Oocyte

The primary oocyte undergoes meiosis and gives off the first polar body to become a secondary oocyte just before ovulation. Sperm cell penetration initiates the completion of the second meiotic division and the expulsion of a second polar body. The nuclei of the oocyte and the sperm cell unite. Fertilization results in the formation of a zygote.

Follicle Development

A **primordial follicle** is a primary oocyte surrounded by a single layer of flat cells, called **granulosa cells** (figure 19.11). Beginning with puberty, some of the primordial follicles are converted to **primary follicles** when the oocyte enlarges and the single layer of granulosa cells becomes enlarged and cuboidal. Subsequently, several layers of granulosa cells form and a layer of clear material is deposited around the primary oocyte called the **zona pellucida** (zō′nă pellū′sid-dă, *zone*, girdle + *pellucidus*, passage of light).

Approximately every 28 days, hormonal changes stimulate some of the primary follicles to continue to develop (see figure 19.11). The primary follicle becomes a **secondary follicle** as fluid-filled spaces called **vesicles** form among the granulosa cells, and a capsule called the **theca** (thē′kă, a box) forms around the follicle.

The secondary follicle continues to enlarge, and, when the fluid-filled vesicles fuse to form a single, fluid-filled chamber called the **antrum** (an′trŭm, *antron*, a cave), the follicle is called the **mature** or **graafian** (graf′ē-ăn, named for the Dutch histologist, Reijnier de Graaf, 1641), **follicle.** The primary oocyte is pushed off to one side and lies in a mass of granulosa cells called the **cumulus cells.**

The mature follicle forms a lump on the surface of the ovary. During ovulation, the mature follicle ruptures, forcing a small amount of blood, follicular fluid, and the primary oocyte, surrounded by the cumulus cells, into the peritoneal cavity. In most cases, only one of the follicles that begin to develop forms a mature follicle and undergoes ovulation. The other follicles degenerate. After ovulation, the remaining cells of the ruptured follicle are transformed into a glandular structure called the **corpus luteum** (kōr′pŭs, body; loo′tē-ŭm, yellow). If pregnancy occurs, the corpus luteum enlarges in response to a hormone secreted by the placenta called **human chorionic gonadotropin hormone (HCG)** (kō-rē-on′ik gō′nad-o-trō′pin) (see table 19.1). If pregnancy does not occur, the corpus luteum lasts for 10–12 days and then begins to degenerate.

Uterine Tubes

A **uterine tube, fallopian** (fa-lō′pē-an, named for the Italian anatomist, Gabriele Fallopio, 1523–1562) **tube,** or **oviduct**

1. The primordial follicle consists of an oocyte surrounded by a single layer of flat granulosa cells.

2. A primordial follicle becomes a primary follicle as the granulosa cells become enlarged and cuboidal.

3. The primary follicle enlarges. Granulosa cells form more than one layer of cells. The zona pellucida forms around the oocyte.

4. A secondary follicle forms when fluid-filled vesicles (spaces) develop among the granulosa cells and a well-developed theca becomes apparent around the granulosa cells.

5. A mature follicle forms when the fluid-filled vesicles form a single antrum. When a follicle becomes fully mature, it is enlarged to its maximum size, a large antrum is present, and the oocyte is located in the cumulus cells.

6. During ovulation the oocyte is released from the follicle, along with some surrounding granulosa cells called cumulus cells.

7. Following ovulation, the granulosa cells divide rapidly and enlarge to form the corpus luteum.

8. The corpus luteum degenerates to form a scar or corpus albicans.

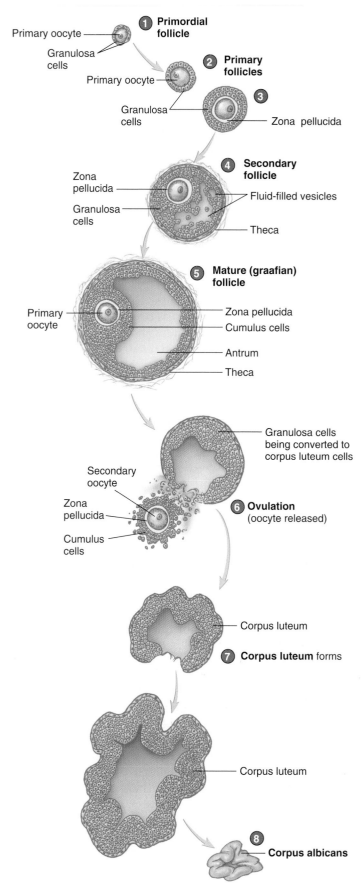

Figure 19.11 Maturation of the Follicle and Corpus Luteum

(ō'vi-dŭct), is associated with each ovary. The uterine tubes extend from the area of the ovaries to the uterus. They open directly into the peritoneal cavity near each ovary and receive the oocyte. The opening of each uterine tube is surrounded by long, thin processes called **fimbriae** (fim'brē-ă, fringe) (see figure 19.8).

The fimbriae nearly surround the surface of the ovary. As a result, as soon as the oocyte is ovulated, it comes into contact with the surface of the fimbriae. Cilia on the fimbriae surface sweep the oocyte into the uterine tube. Fertilization usually occurs in the part of the uterine tube near the ovary, called the **ampulla** (am-pul'lă).

Uterus

The **uterus** (ū'ter-ŭs, womb) is as big as a medium-sized pear (see figures 19.7 and 19.8). It is oriented in the pelvic cavity with the larger, rounded part directed superiorly. The part of the uterus superior to the entrance of the uterine tubes is called the **fundus** (fŭn'dŭs). The main part of the uterus is called the **body,** and the narrower part, the **cervix** (ser'viks, neck), is directed inferiorly. Internally, the **uterine cavity** in the fundus and uterine body continues through the cervix as the **cervical canal,** which opens into the vagina. The cervical canal is lined by mucous glands.

Cancer of the Cervix

Cancer of the cervix is relatively common in females and fortunately can be detected and treated. A **Papanicolaou (Pap)**, named for the Greek-American physician, George Papanicolaou, (1883–1962) **smear** is a diagnostic test used to determine if a woman is suffering from cancer of the uterine cervix. A smear of epithelial cells is taken from the area of the cervix by inserting a swab through the vagina. Cells from the wall of the vagina are also included in the smear. The smear is placed on a glass slide and stained. The cells are then examined microscopically to determine whether some of them show signs of being cancerous. Early in the development of cervical cancer, the cells of the cervix change in a characteristic way. Cells that are cancerous appear to be less mature than the characteristic epithelial cells of the cervix or vaginal wall.

The uterine wall is composed of three layers: a serous layer, a muscular layer, and a layer of endometrium (see figure 19.8). The outer layer, called the **serous layer,** or **perimetrium** (per-i-mē'trē-ŭm), of the uterus, is formed from visceral peritoneum. The middle layer, called the **muscular layer,** or **myometrium** (mī'ō-mē'trē-ŭm), consists of smooth muscle, is quite thick, and accounts for the bulk of the uterine wall. The innermost layer of the uterus is the **endometrium** (en'dō-mē'trē-ŭm). The endometrium consists of simple columnar epithelial cells with an underlying connective tissue layer. Simple tubular glands, called endometrial glands, are formed by folds of the endometrium. The superficial part of the endometrium is sloughed off during menstruation.

The uterus is supported by the broad ligament and the **round ligament.** In addition to these ligaments that support the uterus, much support is provided inferiorly to the uterus by skeletal muscles of the pelvic floor. If ligaments that support the uterus or muscles of the pelvic floor are weakened such as in childbirth, the uterus can extend inferiorly into the vagina, a condition called a **prolapsed uterus.** Severe cases require surgical correction.

Ectopic Pregnancy

An **ectopic pregnancy** results if implantation occurs anywhere other than in the uterine cavity. The most common site of ectopic pregnancy is the uterine tube. Implantation in the uterine tube eventually is fatal to the fetus and can cause the tube to rupture. In some cases, implantation can occur in the mesenteries of the abdominal cavity, and the fetus can develop normally but must be delivered by cesarean section.

Vagina

The **vagina** (vă-jī'nă) is the female organ of copulation and functions to receive the penis during intercourse. It also allows menstrual flow and childbirth. The vagina extends from the uterus to the outside of the body (see figures 19.7 and 19.8). The superior portion of the vagina is attached to the sides of the cervix so that a part of the cervix extends into the vagina.

The wall of the vagina consists of an outer muscular layer and an inner mucous membrane. The muscular layer is smooth muscle and contains many elastic fibers. Thus the vagina can increase in size to accommodate the penis during intercourse, and it can stretch greatly during childbirth. The mucous membrane is moist stratified squamous epithelium that forms a protective surface layer. Lubricating fluid passes through the vaginal epithelium into the vagina.

In young females, the vaginal opening is covered by a thin mucous membrane called the **hymen** (hī'men, membrane). The hymen can completely close the vaginal orifice, in which case it must be removed to allow menstrual flow. More commonly, the hymen is perforated by one or several holes. The openings in the hymen are usually greatly enlarged during the first sexual intercourse. The hymen can also be perforated or torn at some earlier time in a young female's life during a variety of activities including strenuous exercise. The condition of the hymen is therefore not a reliable indicator of virginity.

External Genitalia

The external female genitalia, also called the **vulva** (vŭl'vă, a wrapper or covering), or **pudendum** (pū-den'dŭm), consist of the vestibule and its surrounding structures (figure 19.12). The **vestibule** (ves'ti-bool, entrance court) is the space into which the vagina and urethra open. The urethra opens just anterior to the vagina. The vestibule is bordered by a pair of thin, longitudinal skin folds called the **labia minora** (lā'bē-ă, lips; mī-nō'ră, small). A small erectile structure called the **clitoris** (klit'ŏ-ris, klī'tŏ-ris) is located in the anterior margin of the vestibule. The two labia minora unite over the clitoris to form a fold of skin called the **prepuce.**

The clitoris (see figure 19.7) consists of a shaft and a distal glans. Like the glans penis, the clitoris is well supplied with sensory receptors, and it is made up of erectile tissue. Additional erectile tissue is located on either side of the vaginal opening.

On each side of the vestibule, between the vaginal opening and the labia minora, are openings of the **greater vestibular glands.** These glands produce a lubricating fluid that helps maintain the moistness of the vestibule.

Lateral to the labia minora are two prominent, rounded folds of skin called the **labia majora** (mă-jō'ră, large). The two labia majora unite anteriorly at an elevation of tissue over the

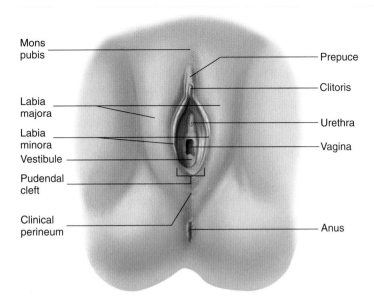

Figure 19.12 Female External Genitalia

pubic symphysis called the **mons pubis** (monz pū′bis, an elevation + symphysis pubis) (see figure 19.12). The lateral surfaces of the labia majora and the surface of the mons pubis are covered with coarse hair. The medial surfaces of the labia majora are covered with numerous sebaceous and sweat glands. The space between the labia majora is called the **pudendal** (pū-den′dal *pudeo*, to feel ashamed) **cleft.** Most of the time, the labia majora are in contact with each other across the midline, closing the pudendal cleft and covering the deeper structures within the vestibule.

The region between the vagina and the anus is the **clinical perineum** (per′i-nē′um, area between the thighs). The skin and muscle of this region can tear during childbirth. To prevent such tearing, an incision called an **episiotomy** (e-piz-ē-ot′ō-mē, pubic region + *tōme*, incision) is sometimes made in the clinical perineum. Traditionally, this clean, straight incision is thought to result in less injury, less trouble in healing, and less pain. However, many studies indicate that there is less injury and pain when no episiotomy is performed.

Mammary Glands

The **mammary** (mam′ă-rē, relating to breasts) **glands** are the organs of milk production and are located in the **breasts,** or **mammae** (mam′ē) (figure 19.13). The mammary glands are

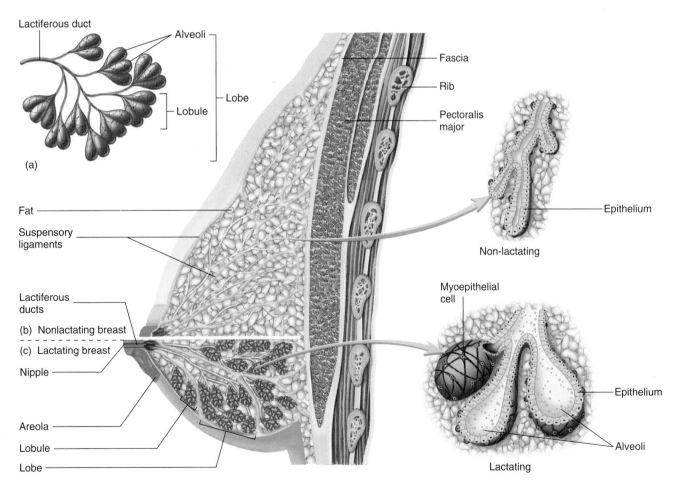

Figure 19.13 Anatomy of the Breast

(*a*) Each lactiferous duct of the mammary gland branches. At the end of each branch is one or more alveoli. (*b*) The nonlactating mammary gland has a duct system that is not developed extensively. The branches of the lactiferous ducts end as small, tube-like structures. (*c*) The lactating mammary gland has a well-developed duct system with many branches. The branches of the lactiferous duct end with well-developed alveoli. Adipose tissue is abundant in the nonlactating and lactating mammary gland.

Clinical Focus **Reproductive Disorders**

Infectious Diseases

Sexually Transmitted Diseases

Sexually transmitted diseases (STDs) are a class of infectious diseases spread by intimate sexual contact between individuals. These diseases include the major venereal diseases such as nongonococcal urethritis, trichomoniasis, gonorrhea, genital herpes, genital warts, syphilis, and acquired immunodeficiency syndrome (AIDS).

Nongonococcal urethritis (non-gon'ō-kok'ăl ū-rē-thrī'tis) refers to any inflammation of the urethra that is not caused by gonorrhea. Factors such as trauma or passage of a nonsterile catheter through the urethra can cause this condition, but many cases are acquired through sexual contact. In most cases, a bacterium such as *Chlamydia trachomatis* (kla-mid'ē-ă tra-kō'mă-tis) is responsible, but other bacteria can be involved. *C. trachomatis* infection is one of the most common sexually transmitted diseases. It is often unrecognized in people who have it, and it is responsible for many cases of pelvic inflammatory disease. It can also result in sterility. Antibiotics are usually effective in treating the condition.

Trichomonas (trik'ō-mō'nas) is a protozoan commonly found in the vagina of females and the urethra of males. If the normal acidity of the vagina is disturbed, *Trichomonas* can grow rapidly. **Trichomoniasis** (trik-ō-mō-nīă-sis) is more common in females than in males. The rapid growth of these organisms results in inflammation and a greenish yellow discharge characterized by a foul odor.

Gonorrhea (gon-ō-rē'ă, *gonc*, seed + *rhoia*, a flow) is caused by *Neisseria gonorrhoeae* (nī-sē'rē-ă gon-ō-rē'ă). The organisms attach to the epithelial cells of the vagina or to the male urethra. The invasion of bacteria establishes an inflammatory response in which pus is formed. Males become aware of a gonorrheal infection because of painful urination and the discharge of pus-containing material from the urethra. Symptoms appear within a few days to a week. Recovery can eventually occur without complication, but, when complications do occur, they can be serious. The urethra can become partially blocked, or sterility can result from scar tissue blocking the reproductive ducts. In some cases, other organ systems, such as the heart, meninges of the brain, or joints, can become infected. In females, the early stages of infection may not be noticeable, but the infection can lead to pelvic inflammatory disease. Gonorrheal eye infections can occur in newborn children of women with gonorrheal infections. Antibiotics are usually effective in treating gonorrheal infections, and the immune system often successfully combats gonorrheal infections in untreated individuals.

Genital herpes (her'pēz, *herpō*, to creep) is an infection caused by herpes simplex type 2 virus. Lesions appear after an incubation period of about 1 week and cause a burning sensation. After this, blisterlike areas of inflammation appear. In males and females, urination can be painful, and walking or sitting can be unpleasant, depending on the location of the lesions. The blisterlike areas heal in about 2 weeks. The lesions can recur. The viruses exist in a latent condition in infected sensory neurons and can initiate periods of inflammation in response to factors such as menstruation, emotional stress, or illness. If active lesions are present in the mother's vagina or external genitalia, a cesarean delivery should be performed to prevent newborns from becoming infected with the herpes virus. Antibiotics are not effective against

modified sweat glands. Externally, each of the breasts of both males and females have a raised **nipple** surrounded by a circular, pigmented area called the **areola** (ă-rē'ō-lă, area).

In prepubescent children, the general structure of the male and female breasts is similar, and both males and females possess a rudimentary duct system. The female breasts begin to enlarge during puberty, under the influence of estrogen and progesterone. Some males also experience a minor and temporary enlargement of the breasts at puberty. The breasts of a male can become permanently enlarged, however, a condition called **gynecomastia** (gī'nĕ-kō-mas'tē-ă). Causes of gynecomastia include hormonal imbalances and the abuse of anabolic steroids.

Each adult female breast contains mammary glands consisting of usually 15–20 glandular **lobes** covered by a considerable amount of fat tissue (see figure 19.13*a* and *b*). It is primarily this superficial fat that gives the breast its form. Each lobe possesses a single **lactiferous duct** that opens independently to the surface of the nipple. The duct of each lobe is formed as several smaller ducts that originate from **lobules** converge. Within a lobule, the ducts branch and become even smaller. In the milk-producing, or lactating, mammary gland, the ends of these small ducts expand to form secretory sacs called **alveoli. Myoepithelial cells** surround the alveoli and contract to expel milk from the alveoli (see figure 19.13*c*).

The breasts are supported by the suspensory ligaments, which extend from the fascia over the pectoralis major muscles to the skin over the breasts and prevent them from excessive sagging (see figure 19.13*b*). In older adults, the suspensory ligaments can weaken and elongate, increasing the tendency for the breasts to sag.

The nipples are very sensitive to tactile stimulation and contain smooth muscle. When the smooth muscle contracts, the nipple becomes erect. The smooth muscle cells contract in response to stimuli such as touch, cold, and sexual arousal.

genital herpes but it can be treated with antiviral drugs.

Genital warts (wōrtz) result from a viral infection and are quite contagious. Genital warts are common, and their frequency is increasing. Genital warts can also be transmitted from infected mothers to their infants. Genital warts vary from separate small warty growths to large cauliflowerlike clusters. The lesions are usually not painful, but they can cause painful intercourse and they bleed easily. Women who have genital warts have an increased risk of developing cervical cancer. Treatments for genital warts include topical agents, cryosurgery, or other surgical methods.

Syphilis (sif'i-lis) is caused by the bacterium *Treponema pallidum* (trep-ō-nē'mă pal'i-dŭm), which can be spread by sexual contact of all kinds. Syphilis exhibits an incubation period from 2 weeks to several months. The disease progresses through several recognized stages. In the primary stage, the initial symptom is a small, **chancre** (shang'ker) or hard and insensitive sore, which usually appears at the site of infection. Several weeks after the primary stage, the disease enters the secondary stage, characterized mainly by skin rashes and mild fever. The symptoms of secondary syphilis usually subside after a few weeks, and the disease enters a latent period in which no symptoms are present. In less than half the cases, a tertiary stage develops after many years. In the tertiary stage, many neural lesions develop that can cause extensive tissue damage and can lead to paralysis, insanity, and even death. Syphilis can be passed on to newborns if the mother is infected. Damage to mental development and other neurological symptoms are among the more serious consequences. Females who have syphilis in the latent phase are most likely to have babies who are infected. Antibiotics are used to treat syphilis, although some strains are very resistant to certain antibiotics.

Acquired immunodeficiency syndrome (AIDS) is caused by infection with the human immunodeficiency virus (HIV), which appears to ultimately result in destruction of the immune system (see chapter 14). The most common mechanisms of transmission of the virus are through sexual contact with a person infected with HIV or through sharing needles with an infected person during the administration of illicit drugs. Screening techniques now implemented make the transmission of HIV through blood transfusions very rare. Some cases of transmission of HIV through accidental needlesticks in hospitals and other health care facilities have been documented, but the frequency is rare. There is no evidence that casual contact with a person who has AIDS or who is infected with HIV results in transmission of the disease. Transmission appears to require exposure to body fluids of an infected person in a way that allows HIV into the interior of another person. Normal casual contact, including touching an HIV-infected person, does not increase the risk of infection.

Other Infectious Diseases

Pelvic inflammatory disease (PID) is a bacterial infection of the female pelvic organs. It usually involves the uterus, uterine tubes, or ovaries. A vaginal or uterine infection can spread throughout the pelvis. PID is commonly caused by the gonorrhea or chlamydia bacteria, but other bacteria can be involved. Early symptoms of PID include increased vaginal discharge and pelvic pain. Early treatment with antibiotics can stop the spread of PID, but lack of treatment results in a life-threatening infection. PID can also lead to sterility.

Cancer of the Breast

Cancer of the breast is a serious, often fatal disease most often occurring in women. **Mammography** (ma-mog'ră-fē) uses low-intensity x-rays to detect tumors in the soft tissue of the breast. The use of mammography and regular self-examination of the breast can lead to early detection of breast cancer and effective treatment. With mammography, however, tumors can often be identified before they can be detected by palpation. Once a tumor is identified, a biopsy is normally performed to determine whether the tumor is benign or malignant. Most tumors of the mammary glands are benign. Those that are malignant have the potential to spread to other areas of the body and ultimately lead to death.

Physiology of Female Reproduction

As in the male, female reproduction is controlled by hormonal and nervous mechanisms.

Puberty

The first signs of puberty typically appear between 11 and 13 years of age in girls, and the process is largely completed by age 16. Puberty in females is marked by the first episode of menstrual bleeding, which is called **menarche** (me-nar'kē, *mēn*, month + *archē*, beginning). During puberty, the vagina, uterus, uterine tubes, and external genitalia begin to enlarge. Fat is deposited in the breasts and around the hips, causing them to enlarge and assume an adult form. In addition, pubic and axillary hair grows. Development of sexual drive is also associated with puberty.

The changes associated with puberty are primarily the result of the increasing rate of estrogen and progesterone secretion by the ovaries. Before puberty, estrogen and progesterone are secreted in very small amounts. At puberty, the cyclical adult pattern of hormone secretion is gradually established.

Before puberty, the rate of GnRH secretion from the hypothalamus and the rate of LH and FSH secretion from the anterior pituitary are very low. Estrogen and progesterone from the ovaries have a strong negative-feedback effect on the hypothalamus and

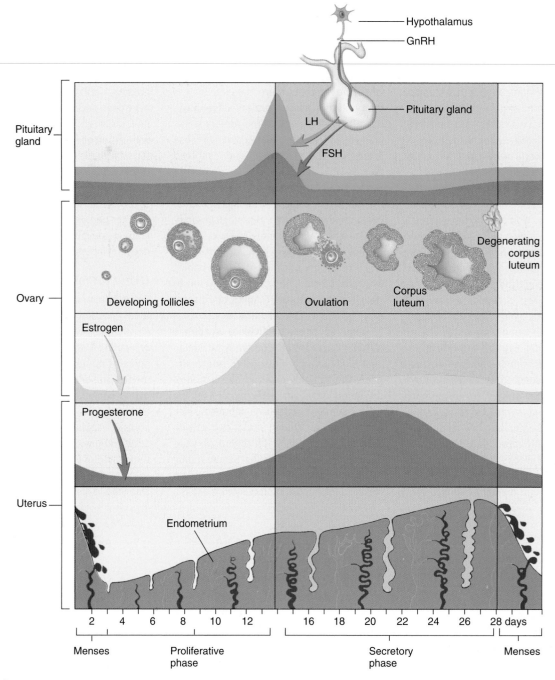

Figure 19.14   The Menstrual Cycle

The figure illustrates changes in follicle-stimulating hormone (FSH) and luteinizing hormone (LH) secretion from the anterior pituitary gland and changes in estrogen and progesterone secretion from the ovary. In addition, changes in the ovary and changes in the endometrium of the uterus are correlated with the changes in hormone secretion throughout the menstrual cycle.

pituitary. After the onset of puberty, the hypothalamus and anterior pituitary secrete larger amounts of GnRH, LH, and FSH. Estrogen and progesterone have less of a negative-feedback effect on the hypothalamus and pituitary, and a sustained increase in estrogen concentration has a positive-feedback effect. The normal cyclical pattern of reproductive hormone secretion that occurs during the menstrual cycle becomes established. The initial change that results in puberty appears to be maturation of the hypothalamus.

Menstrual Cycle

The term **menstrual** (men′stroo-ăl) **cycle** refers to the series of changes that occur in sexually mature, nonpregnant females and that culminate in menses. **Menses** (men′sēz, month) is a period of mild hemorrhage during which part of the endometrium is sloughed and expelled from the uterus. Typically the menstrual cycle is about 28 days long, although it can be as short as 18 days or as long as 40 days (figure 19.14

Table 19.3 Events During the Menstrual Cycle

Menses (Day 1 to Day 4 or 5 of the Menstrual Cycle)

Pituitary gland	The rate of FSH and LH secretion is low, but the rate of FSH secretion increases as progesterone levels decline.
Ovary	The rate of estrogen and progesterone secretion is low after degeneration of the corpus luteum produced during the previous menstrual cycle.
Uterus	In response to declining progesterone levels, the endometrial lining of the uterus sloughs off, resulting in menses followed by repair of the endometrium.

Proliferative Phase (from Day 4 or 5 Until Ovulation on About Day 14)

Pituitary gland	The rate of FSH and LH secretion is only slightly elevated during most of the proliferative phase; FSH and LH secretion increase near the end of the proliferative phase in response to increasing estrogen secretion from the ovaries.
Ovary	Developing follicles secrete increasing amounts of estrogen, especially near the end of the proliferative phase; increasing FSH and LH cause additional estrogen secretion from the ovaries near the end of the proliferative phase.
Uterus	Estrogen causes endometrial cells of the uterus to divide. The endometrium of the uterus thickens and tubelike glands form. Estrogen causes the cells of the uterus to be more sensitive to progesterone by increasing the number of progesterone receptors in uterine tissues.

Ovulation (About Day 14)

Pituitary gland	The rate of FSH and LH secretion increases rapidly just before ovulation in response to increasing estrogen levels. Increasing FSH and LH levels stimulate estrogen secretion, resulting in a positive-feedback cycle.
Ovary	LH causes final maturation of a mature follicle and initiates the process of ovulation. FSH acts on immature follicles and causes several of them to begin to enlarge.
Uterus	The endometrium continues to divide in response to estrogen.

Secretory Phase (from About Day 14 to Day 28)

Pituitary gland	Estrogen and progesterone reach levels high enough to inhibit FSH and LH secretion from the pituitary gland.
Ovary	After ovulation, the follicle is converted to the corpus luteum; the corpus luteum secretes large amounts of progesterone and smaller amounts of estrogen from shortly after ovulation until about day 24 or 25. If fertilization does not occur, the corpus luteum degenerates after about day 25, and the rate of progesterone secretion rapidly declines to low levels.
Uterus	In response to progesterone, the endometrial cells enlarge, the endometrial layer thickens, and the glands of the endometrium reach their greatest degree of development; the endometrial cells secrete a small amount of fluid. After progesterone levels decline, the endometrium begins to degenerate.

Menses (Day 1 to Day 4 or 5 of the Next Menstrual Cycle)

Pituitary gland	The rate of LH remains low and the rate of FSH secretion increases as progesterone levels decline.
Ovary	The rate of estrogen and progesterone secretion is low.
Uterus	In response to declining progesterone levels, the endometrial lining of the uterus sloughs off, resulting in menses followed by repair of the endometrium.

and table 19.3). The menstrual cycle results from the cyclical changes that occur in the endometrium of the uterus. These changes result from the cyclical changes that occur in the ovary and are controlled by the secretions of FSH and LH from the anterior pituitary gland.

The first day of menstrual bleeding (menses) is considered to be day 1 of the menstrual cycle. Menses typically lasts 4 or 5 days. Ovulation occurs on about day 14 of the menstrual cycle, although the timing of ovulation varies from individual to individual and can vary within an individual from one menstrual cycle to the next. A small increase in FSH secretion occurs just prior to menses as a result of the end of the previous menstrual cycle (see following discussion). This increase in FSH stimulates follicle development (see figure 19.11).

The time between the ending of menses and ovulation is called the **proliferative phase,** which refers to proliferation of the endometrium. During the proliferative phase, the maturing follicles in the ovary mature, and, as they do so, they secrete increasing amounts of estrogen. Estrogen acts on the uterus and causes the epithelial cells of the endometrium to divide rapidly. The endometrium thickens, and endometrial glands form.

The sustained increase of estrogen secreted by the developing follicles stimulates GnRH secretion from the hypothalamus. GnRH, in turn, triggers FSH and LH secretion from the anterior pituitary gland. FSH stimulates estrogen secretion at an increasing rate from the developing follicles. This positive-feedback loop produces a series of larger and larger surges of FSH and LH secretion. Ovulation occurs in response to the large increases in LH levels that normally occur on about day 14 of the menstrual cycle. This large increase in LH is also responsible for the development of the corpus luteum.

Following ovulation, the corpus luteum begins to secrete progesterone and smaller amounts of estrogen. Progesterone acts on the uterus, causing the cells of the endometrium to become larger and to secrete a small amount of fluid. Together, progesterone and estrogen act on the hypothalamus and anterior pituitary gland to inhibit GnRH, LH, and FSH secretion. Thus LH and FSH levels decline to low levels after ovulation.

The time between ovulation and the next menses is called the **secretory phase** of the menstrual cycle because of the small amount of fluid secreted by the cells of the endometrium. During the secretory phase, the lining of the uterus reaches its greatest degree of development.

If fertilization occurs, the zygote undergoes several cell divisions to produce a collection of cells called the **blastocyst** (blas'tō-sist). The blastocyst passes through the uterine tube and arrives in the uterus by 7 or 8 days after ovulation. The endometrium is prepared to receive the blastocyst, which becomes implanted in the endometrium, where development continues. If the secondary oocyte is not fertilized, the endometrium sloughs away as a result of declining blood progesterone levels. Unless the secondary oocyte is fertilized, the corpus luteum begins to produce less progesterone by day 24 or 25 of the menstrual cycle. By day 28, the declining progesterone causes the endometrium to slough away to begin menses and the next menstrual cycle. The declining progesterone secretion results in a small increase in FSH secretion at the beginning of the next menses.

P R E D I C T 3

Predict the effect of administering a relatively large amount of progesterone and estrogen just before the increase in LH that precedes ovulation.

A topic of continuing research concerns a phenomenon called the **premenstrual syndrome (PMS).** Some women suffer from severe changes in mood that can result in aggression and other socially unacceptable behaviors before menses. It has been hypothesized that hormonal changes associated with the menstrual cycle trigger these mood changes. Some women with severe cases of PMS can be treated with drugs used to treat depression, such as fluoxetine (Prozac). Although some women with PMS appear to have been successfully treated with progesteronelike hormones, and a variety of other drugs, these treatments do not appear to be effective for all people. Similarly, reduced caffeine, alcohol, sugar, or animal fat consumption helps some people. It is unclear how many women are affected by PMS. The definition of PMS is not well established, the symptoms of the condition are not easily monitored, and its precise cause and physiological mechanisms are unknown. In addition, it is not clear that all women diagnosed as having PMS are suffering from the same condition. Additional research is needed to resolve these uncertainties.

Menstruation

Menstrual cramps are the result of strong myometrial contractions that occur before and during menstruation. The cramps can result from excessive secretion of prostaglandins. Sloughing of the endometrium of the uterus results in an inflammation in the endometrial layer of the uterus, and prostaglandins are produced as part of the inflammation.

Sloughing of the endometrium is inhibited by progesterone but stimulated by estrogen. In some women, menstrual cramps are extremely uncomfortable. Many women can alleviate painful menstruation by taking drugs, such as aspirinlike drugs, that inhibit prostaglandin biosynthesis just before the onset of menstruation. These treatments, however, are not effective in treating all painful menstruation, especially when the causes of pain, such as those in some women who have tumors of the myometrium, differ from the ones caused by the inflammatory response.

The absence of a menstrual cycle is called **amenorrhea** (ă-men-ō-rē'ă, without menses). If the pituitary gland does not function properly because of abnormal development, the woman will not begin to menstruate at puberty. This condition is called **primary amenorrhea.** In contrast, if a woman has had normal menstrual cycles and later stops menstruating, the condition is called **secondary amenorrhea.** One cause of secondary amenorrhea is anorexia, a condition in which the lack of food causes the hypothalamus of the brain to decrease GnRH secretion to levels so low that the menstrual cycle cannot occur. Female athletes or ballet dancers who have rigorous training schedules often have secondary amenorrhea. The physical stress that can be coupled with an inadequate food intake also results in very low GnRH secretion. Increased food intake for anorexic women or reduced training for dancers and athletes generally restores normal hormone secretion and normal menstrual cycles.

Secondary amenorrhea can also be the result of pituitary tumors, which decrease FSH and LH secretion, or from a lack of GnRH secretion from the hypothalamus. Head trauma and tumors that affect the hypothalamus can result in lack of GnRH secretion.

Secondary amenorrhea can result from a lack of normal hormone secretion from the ovaries, which can result from autoimmune diseases that attack the ovary, or from **polycystic ovarian disease,** in which cysts in the ovary produce large amounts of androgen that are converted to estrogen by other tissues in the body. The increased estrogen prevents the normal cycle of FSH and LH secretion required for ovulation to occur. Other hormone-secreting tumors of the ovary can also disrupt the normal menstrual cycle and result in amenorrhea.

Menopause

When a woman is 40–50 years old, the menstrual cycles become less regular, and ovulation does not consistently occur during each cycle. Eventually the cycles stop completely. The cessation of menstrual cycles is called **menopause** (men'ō-pawz, *mēn*, month + *pausis*, cessation), and the whole time period from the onset of irregular cycles to their complete cessation is called the female **climacteric** (klī-mak'ter-ik).

The major cause of menopause is age-related changes in the ovaries. The number of follicles remaining in the ovaries of menopausal women is small. In addition, the follicles that remain become less sensitive to stimulation by FSH and LH. As the ovaries become less responsive to stimulation by FSH and LH, fewer mature follicles and corpora lutea are produced. Gradual changes occur in women in response to the reduced amount of estrogen and progesterone produced by the ovaries (table 19.4).

Table 19.4	Possible Changes in Postmenopausal Women Caused by Decreased Ovarian Hormone Secretion
	Changes
Menstrual cycle	Five to seven years before menopause the cycle becomes irregular; the number of cycles in which ovulation does not occur and in which corpora lutea do not develop increases.
Uterus	Gradual increase in irregular menstruations is followed by no menstruation; the endometrium finally atrophies, and the uterus becomes smaller.
Vagina and external genitalia	The epithelial lining becomes thinner; the external genitalia become thinner and less elastic; the labia majora becomes smaller; the pubic hair decreases; reduced secretion leads to dryness; the vagina is more easily inflamed and infected.
Skin	The epidermis becomes thinner.
Cardiovascular system	Hypertension and atherosclerosis occur more frequently.
Vasomotor instability	Hot flashes and increased sweating are correlated with vasodilation of cutaneous blood vessels; the hot flashes are related to decreased estrogen levels.
Libido	Temporary changes, usually a decrease in libido are associated with the onset of menopause.
Fertility	Fertility begins to decline about 10 years before the onset of menopause; by age 50 almost all the oocytes and follicles have degenerated.
Pituitary function	Low levels of estrogen and progesterone produced by the ovaries cause the pituitary gland to secrete larger than normal amounts of LH and FSH; increased levels of these hormones have little effect on the postmenopausal ovaries.

During the climacteric, some women experience "hot flashes," irritability, fatigue, anxiety, temporary decrease in libido, and occasionally severe emotional disturbances. Many of these symptoms can be treated successfully with hormone replacement therapy (HRT), which usually consists of small amounts of estrogen and progesterone. A potential side effect of HRT is a slightly increased possibility of the development of breast cancer, uterine cancer, heart attacks, strokes, and blood clots. HRT does slow the decrease in bone density that can become severe in some women after menopause, and decreases the risk of developing colorectal cancer.

Female Sexual Behavior and the Female Sex Act

Sexual drive in females, like sexual drive in males, is dependent on hormones. Testosteronelike hormones, and possibly estrogen, affect brain cells (especially in the area of the hypothalamus) and influence sexual behavior. Testosteronelike hormones are produced primarily in the adrenal cortex. Psychological factors also play a role in sexual behavior. The sensory and motor neural pathways involved in controlling female sexual responses are similar to those found in the male.

Erectile tissue within the clitoris and around the vaginal opening becomes engorged with blood during sexual excitement. The mucous glands within the vestibule, especially the greater vestibular glands, secrete small amounts of mucus. Larger amounts of mucuslike fluid are also extruded into the vagina through its wall. These secretions provide lubrication to allow easy entry and movement of the penis in the vagina during intercourse. Tactile stimulation of the female's genitals, during sexual intercourse and psychological stimuli, normally trigger an **orgasm,** or **climax.** The vaginal and uterine smooth muscle, as well as the surrounding skeletal muscles, contract rhythmically; and muscle tension increases throughout much of the body. After the sexual act, there is a period of **resolution,** which is characterized by an overall sense of satisfaction and relaxation. Females are sometimes receptive to further immediate stimulation, however, and can experience successive orgasms. Orgasm is not necessary for fertilization to occur. Ovulation occurs as a result of hormonal stimuli and is not dependent on the female sexual act.

Infertility

Causes of infertility in females include malfunctions of the uterine tubes, reduced hormone secretion from the pituitary or ovary, and interruption of implantation. Adhesions from pelvic inflammatory conditions, caused by a variety of infections, can cause blockage of one or more uterine tubes and is a relatively common cause of infertility in women. Reduced ovulation can result from inadequate secretion of LH and FSH, which can be caused by hypothyroidism, trauma to the hypothalamus, infarctions of the hypothalamus or anterior pituitary gland, and tumors.

Interruption of implantation can result from uterine tumors or conditions causing abnormal ovarian hormone secretion. For example, premature degeneration of the corpus luteum causes progesterone levels to decline and menses to occur. If the corpus luteum degenerates before the placenta begins to secrete progesterone, the endometrium and the developing embryonic mass will degenerate and be eliminated from the uterus. The conditions that result in secondary amenorrhea also reduce fertility.

Many methods are used to prevent or terminate pregnancy (figure B), including methods that prevent fertilization (contraception), prevent implantation of the developing embryo (IUDs), or remove the implanted embryo or fetus (abortion). Many of these techniques are quite effective when done properly (table A) and used consistently. Some disadvantages are associated with each of them, and the use of some of them is controversial.

Behavioral Methods

Abstinence, or refraining from sexual intercourse, is a sure way to prevent pregnancy when it is practiced consistently. It is not an effective method when used only occasionally.

Coitus interruptus (kō′i-tŭs int-ĕ-rŭp′tŭs), or **withdrawal,** is removal of the penis from the vagina just before ejaculation. This is a very unreliable method of preventing pregnancy because it requires perfect awareness and willingness to withdraw the penis at the correct time. It also ignores the fact that some sperm cells are found in preejaculatory emissions.

The **rhythm method** requires abstaining from sexual intercourse near the time of ovulation. A major factor in the success of this method is the ability to predict accurately the time of ovulation. Although the rhythm method provides some protection against becoming pregnant, it has a relatively high rate of failure because of both the inability to predict the time of ovulation and the failure to abstain from intercourse around the time of ovulation.

Barrier Methods

A male **condom** (kon′dom) is a sheath of animal membrane, rubber, or plastic. A condom, placed over the erect penis, is a barrier device because the semen is collected within the condom instead of being released into the vagina. Condoms also provide some protection against sexually transmitted diseases.

A **vaginal** (or female) **condom** also acts as a barrier. The vaginal condom can be placed into the vagina by the female before sexual intercourse.

Methods to prevent sperm cells from reaching the oocyte once they are in the vagina include use of a diaphragm, spermicidal agents, and a vaginal sponge. The **di-**

Table A	Effectiveness of Various Methods for Preventing Pregnancy	
Technique		**Effectiveness When Used Properly (%)**
Abortion		100
Sterilization		100
Combination (estrogen and progesterone) pill		99.9
Pill (low dose of estrogen and progesterone)		99
Mini-pill		95
Depo-Provera		99.7
Intrauterine device		98
Intrauterine device (copper)		99
Intrauterine device (progesterone)		98
Male condom		97
Male condom plus spermicide		99
Female condom		79
Diaphragm plus spermicide		97
Cervical cap		60–90
Foam		97
Rhythm		97

aphragm and the **cervical cap** are flexible latex domes that are placed over the cervix within the vagina, where they prevent passage of sperm cells from the vagina through the cervical canal of the uterus. The diaphragm is a larger, shallow latex cup and the cervical cap is a smaller, thimble-shaped latex cup. The most commonly used **spermicidal agents** are foams or creams that kill sperm cells. They are inserted into the vagina before sexual intercourse. **Spermicidal douches** (dūsh′ez), which remove and kill sperm cells, are sometimes used. Spermicidal douches used alone are not very effective.

Lactation (lak-tā′shŭn) often prevents the menstrual cycle for a few months after childbirth. Action potentials sent to the hypothalamus in response to suckling inhibit GnRH release from the hypothalamus. Reduced GnRH reduces LH, which prevents ovulation. Despite continual lactation, the menstrual cycle eventually resumes. Because ovulation normally precedes menstruation, relying on lactation to prevent pregnancy is not consistently effective.

Chemical Methods

Synthetic estrogen and progesterone in **oral contraceptives** (birth control pills) effectively suppress fertility in females. These substances can have more than one action, but they reduce LH and FSH release from the anterior pituitary. Estrogen and progesterone are present in high enough concentrations to have a negative-feedback effect on the pituitary, which prevents the large increase in LH and FSH secretion that triggers ovulation. Over the years, the dose of estrogen and progesterone in birth control pills has been reduced. The current lower dose birth control pills have fewer side effects than earlier dosages. There is an increased risk of heart attack or stroke in females using oral contraceptives who smoke or who have a history of hypertension or coagulation disorders. For most females, the pill is effective and has a minimum frequency of complications, until at least age 35. The **mini-pill** is an oral contraceptive that contains only synthetic progesterone. It reduces and thickens mucus of the cervix, which prevents sperm cells from reaching the egg. It also prevents blastocytes from implanting in the uterus.

Progesteronelike chemicals, such as medroxyprogesterone (med-rok′-sē-prō-jes′-ter-ōn) (Depo-Provera), which are injected intramuscularly and slowly released into the circulatory system, can act as effective

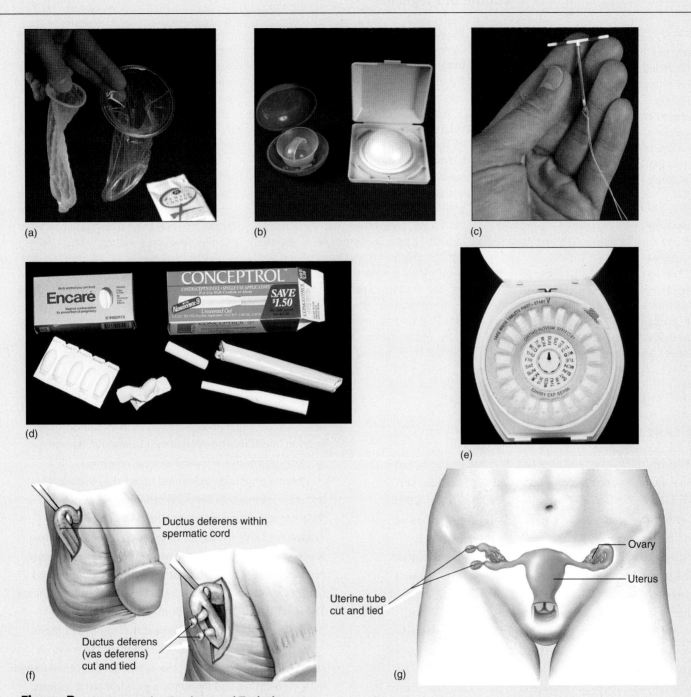

Figure B Contraceptive Devices and Techniques

(*a*) Condom. (*b*) Cervical cap and diaphragm used with spermicidal jelly. (*c*) Intrauterine device (IUD). (*d*) Spermicidal foam. (*e*) Oral contraceptives. (*f*) Vasectomy. (*g*) Tubal ligation.

contraceptives. Injected progesteronelike chemicals can provide protection from pregnancy for approximately 1 month, depending on the amount injected. The **patch** is an adhesive skin patch containing synthetic estrogen and progesterone. It is worn on the lower abdomen, buttocks, or upper body. The **vaginal contraceptive ring** is inserted into the vagina and releases synthetic estrogen and progesterone.

A drug called **RU486** or **mifepristone** (mif′pris-tōn), blocks the action of progesterone, causing the endometrium of the uterus to slough off as it does at the time of menstruation. It can, therefore, be used to induce menstruation and reduce the possibil-ity of implantation when sexual intercourse has occurred near the time of ovulation. It can also be used to terminate pregnancies.

Morning-after pills, similar in composition to birth control pills, are available. Doubling the number of birth control pills after sexual intercourse within 3 days and

continued

continued

again after 12 more hours is sometimes recommended. These techniques can be used after intercourse, but they are only about 75% effective. The elevated blood levels of estrogen and progesterone may inhibit the increase in LH that causes ovulation, may alter the rate at which the feritlized oocyte is transported through the uterine tube to the uterus, or may inhibit implantation.

Surgical Methods

Vasectomy (va-sek′tō-mē) is a common method used to render males permanently incapable of fertilization without affecting the performance of the sexual act. Vasectomy is a surgical procedure in which the ductus deferens from each testis is cut and tied off within the scrotal sac. This procedure prevents sperm cells from passing through the ductus deferens and becoming part of the ejaculate.

Because such a small volume of ejaculate comes from the testis and epididymis, vasectomy has little effect on the volume of the ejaculated semen. The sequestered sperm cells are reabsorbed in the epididymis.

A common method of permanent birth control in females is **tubal ligation** (lī-gā′shŭn), a procedure in which the uterine tubes are tied and cut or clamped by means of an incision made through the wall of the abdomen. This procedure closes off the path between the sperm cells and the oocyte. **Laparoscopy** (lap-ă-ros′kŏ-pē), in which a special instrument is inserted into the abdomen through a small incision, is commonly used so that only small openings are required to perform the operation.

In some cases, pregnancies are terminated by surgical procedures called **abortions** (ă-bōr′shŭnz). The most common method for performing abortions is the in-sertion of an instrument through the cervix into the uterus. The instrument scrapes the endometrial surface, and at the same time a strong suction is applied. The endometrium and the embedded embryo are disrupted and sucked out of the uterus. This technique is normally used only in pregnancies that have progressed less than 3 months.

Prevention of Implantation

Intrauterine (in′tră-yū′ter-in) **devices (IUDs)** are inserted into the uterus through the cervix, and they prevent normal implantation of the developing embryo within the endometrium. Some early IUD designs produced serious side effects such as perforation of the uterus, and, as a result, many IUDs have been removed from the market. Data indicate, however, that IUDs are effective in preventing pregnancy. Effective IUD's include those containing copper and progesterone.

A CASE IN POINT | Endometriosis

Helen Hurtz is in her mid twenties. She has been married for 4 years. She has become very frustrated because she experiences pain during and after sexual intercourse that has become worse over the past 2 years. She also has an increasing, persistent pain in her pelvic region that is especially uncomfortable before and during menstruation. The pain is also more intense during urination and bowel movements. She recently has developed periodic bouts of diarrhea. She reported her symptoms to her physician.

Helen's physician suspects **endometriosis** (en′dō-mē-trē-ō′sis), a condition in which endometrial tissue migrates from the lining of the uterus into the peritoneal cavity where it attaches to the surface of organs. Common sites of attachment are on the ovaries and pelvic peritoneum. Other sites of attachment include the intestines, uterus, urinary bladder, and vagina. If the attached endometrial tissue has an adequate blood supply, it proliferates, breaks down, and bleeds in response to the hormones produced during the menstrual cycle. Unlike the normal endometrium, which is shed each month during menstruation, endometrial tissue attached outside of the uterus causes lesions or tumors to develop, resulting in internal bleeding, scar tissue formation, inflammation,

and pain. Other major complications of endometriosis include infertility and ovarian cyst formation. Between 40–50% of infertile women have endometriosis.

Helen's physician explained to her that the most accurate method of confirming the diagnosis is by laparoscopy. This procedure allows the physician to visually observe the abdominopelvic cavity. After a few weeks of thinking about the procedure, Helen agreed to the laparoscopic examination. The physician observed several lesions, which are characteristic of endometriosis. The lesions were removed during the laparoscopic procedure with a laser instrument that vaporizes them.

Helen's physician explained to her that there is no cure for endometriosis, but the condition can be managed with medication and by removing endometrial lesions.

Effects of Aging on the Reproductive System

Benign prostatic enlargement is common in men over 50 years of age. A major consequence of prostatic enlargement is blockage of the prostatic urethra. The frequency of prostate cancer also increases as men age and is a significant cause of death in men. There is an increased tendency for erectile dysfunction to occur as men age.

Mrs. M. had four children and was 43 years of age. She noticed that menstruation was becoming gradually more severe and lasting up to several days longer each time menstruation started. After Mrs. M. menstruated almost continuously for 2 months she made an appointment with her physician. The physician performed a pelvic examination on

Mrs. M., including tests to check for conditions such as cervical and uterine cancer. Palpation of the uterus indicated the presence of enlarged masses in Mrs. M.'s uterus. Dilation and curettage (D&C), which is dilation of the cervix and scraping (curettage) of the endometrium to remove growths or other abnormal tissues, was performed. The results of the D&C indicated that Mrs. M. suffered from leiomyomas.

Background Information

Leiomyomas (lī′ō-mī-ō′măz), also called uterine fibroids, are fibroid tumors of the uterus (figure C). They are one of the most common disorders of the uterus, and the most frequent tumor in women, affecting one of every four. Three-fourths of the women with this condition, however, experience no symptoms. The enlarged masses that originate from smooth muscle tissue compresses the uterine lining (endometrium), resulting in ischemia and inflammation. The increased inflammation, which shares some characteristics with menstruation, results in frequent and severe menstruations. Abdominal cramping due to strong uterine contractions, in response to chemical mediators of inflammation, can be present. Constant menstruation is a frequent manifestation of these tumors, and it is one of the most common reasons why women elect to have the uterus removed, a procedure called a **hysterectomy** (his-ter-ek′tō-mē).

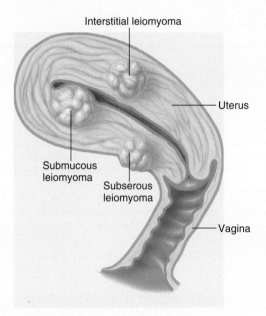

Figure C Leiomyomas or Fibroid Tumors

Leiomyomas, or fibroid tumors, are enlarged masses of smooth muscle. They are located near the mucosa (submucous), within the myometrium (interstitial), or near the serosa (subserous).

P R E D I C T   4

When discussing her condition with her mother, Mrs. M. discovered that her mother recalled frequent menstruations that were irregular and prolonged when she was in her late forties. Mrs. M.'s mother did not have a hysterectomy and in a few years, the frequency of menstruation began to gradually subside. Explain.

System Interactions — Effects of Benign Uterine Tumors on other Systems

System	Interactions
Integumentary	If anemia does not develop, skin appearance is normal. But if anemia does develop, the skin can appear pale because of the reduced hemoglobin in red blood cells.
Muscular	If anemia develops and is severe, muscle weakness results because of the reduced ability of the cardiovascular system to deliver adequate oxygen to muscles.
Skeletal	The rate of red blood cell synthesis in the red bone marrow increases.
Cardiovascular	A chronic loss of blood, as in prolonged menstruation over many months to years, frequently results in iron-deficiency anemia. Manifestations of anemia include pale skin, reduced hematocrit, reduced hemoglobin concentration, smaller than normal red blood cells (microcytic anemia), and increased heart rate.
Respiratory	Because of anemia, the oxygen-carrying capacity of the blood is reduced. Increased respiration during physical exertion and rapid fatigue are likely to occur if anemia develops.
Urinary	The kidney increases erythropoietin secretion in response to the reduced oxygen-carrying capacity of blood. The erythropoietin increases red blood cell synthesis in red bone marrow. An enlarged tumor can put pressure on the urinary bladder, resulting in increased frequency of and painful urination.

Although sexual activity is often maintained in men and women as they age, there is usually a gradual decrease in the frequency of sexual intercourse.

The most significant age-related change in females is menopause. By age 50, few viable follicles remain in the ovaries. As a result, there is a decrease in the estrogen and progesterone produced by the ovaries. The uterus decreases in size, and the endometrium decreases in thickness. The times between menstruations become irregular and longer. Finally menstruation stops. The vaginal wall becomes thinner and less elastic. There is less lubrication of the vagina and the epithelial lining is more fragile. There is an increased tendency for vaginal infections.

Approximately 10% of all women will have breast cancer. The incidence of breast cancer is greatest between 45 and 65 years of age, and the incidence is greater for those who have a history of breast cancer in their families than for those who do not. Cancer of the uterus and cancer of the uterine cervix increase between 50 and 65 years of age. Ovarian cancer also increases in frequency in older women. Occasionally, a prolapsed uterus develops when the ligaments of the uterus allow it to descend and protrude into the vagina.

S U M M A R Y

Functions of the Reproductive System   (p. 541)

1. The male reproductive system produces sperm cells, provides nutrients for the sperm cells and secretions, transfers the sperm cells to the female, and makes male sex hormones.
2. The female reproductive system produces oocytes, receives sex cells from the male, provides nourishment for the developing individual before and after birth, and produces female sex hormones.

Formation of Sex Cells   (p. 541)

The reproductive organs in males and females produce sex cells by meiosis.

Male Reproductive System   (p. 541)

Scrotum

1. The scrotum is a sac containing the testes.
2. The dartos and cremaster muscles help to regulate testes temperature.

Testes

1. The testes are divided into lobules containing the seminiferous tubules and interstitial cells.
2. During development the testes pass from the abdominal cavity through the inguinal canal to the scrotum.

Spermatogenesis

1. Spermatogenesis begins in the seminiferous tubules at the time of puberty.
2. Sperm cells are produced in the seminiferous tubules.
3. Sertoli cells nourish the sperm cells and produce small amounts of hormones.
4. Spermatogonia divide (mitosis) to form primary spermatocytes.
5. Primary spermatocytes divide by meiosis to first produce secondary spermatocytes and then spermatids. The spermatids then mature to form sperm cells.
6. Spermatids develop a head, midpiece, and flagellum to become a sperm cell. The head contains the acrosome and the nucleus.

Ducts

1. The epididymis is a coiled tube system, located on the testis, that is the site of sperm maturation. Final changes, called capacitation of sperm cells, occur after ejaculation.
2. The seminiferous tubules lead to the rete testis.
3. The rete testis opens into the efferent ductules that extend to the epididymis.
4. The ductus deferens passes from the epididymis into the abdominal cavity.
5. The ejaculatory duct is formed by the joining of the ductus deferens and the duct from the seminal vesicle.

6. The ejaculatory ducts join the prostatic urethra in the prostate gland.
7. The urethra extends from the urinary bladder through the penis to the outside of the body.

Penis

1. The penis consists of erectile tissue.
2. The two corpora cavernosa form the dorsum and the sides.
3. The corpus spongiosum forms the ventral portion and the glans penis, and it encloses the spongy urethra. The prepuce covers the glans penis.

Glands

1. The seminal vesicles empty into the ejaculatory duct.
2. The prostate gland consists of glandular and muscular tissue and empties into the urethra.
3. The bulbourethral glands empty into the urethra.

Secretions

1. Semen is a mixture of gland secretions and sperm cells.
2. The bulbourethral glands and the urethral mucous glands produce mucus that neutralizes the acidic pH of the urethra.
3. The testicular secretions contain sperm cells.
4. The seminal vesicle fluid contains nutrients, prostaglandins, and proteins that coagulate.
5. The prostate fluid contains nutrients and proteolytic enzymes, and it neutralizes the pH of the vagina.

Physiology of Male Reproduction   (p. 549)

Regulation of Sex Hormone Secretion

1. GnRH is produced in the hypothalamus and is released in surges.
2. GnRH stimulates release of LH and FSH from the anterior pituitary.
3. LH stimulates the interstitial cells to produce testosterone.
4. FSH binds to Sertoli cells and stimulates spermatogenesis and secretion of inhibin.
5. Testosterone has a negative-feedback effect on GnRH, LH, and FSH secretion.
6. Inhibin has a negative-feedback effect on FSH secretion.

Puberty

1. Before puberty small amounts of testosterone inhibit GnRH release.
2. During puberty testosterone does not completely suppress GnRH release, resulting in increased production of FSH, LH, and testosterone.

Effects of Testosterone

1. Testosterone causes enlargement of the genitals and is necessary for spermatogenesis.
2. Testosterone is responsible for the development of secondary sex characteristics.

Male Sexual Behavior and the Male Sex Act

1. Testosterone is required for normal sex drive.
2. Stimulation of the sexual act can be tactile or psychological.
3. Sensory impulses pass to the sacral region of the spinal cord.
4. Motor stimulation causes erection, mucus production, emission, and ejaculation.
5. The most common cause of infertility is a low sperm cell count.

Female Reproductive System (p. 553)
Ovaries

1. By the fourth month of development the ovaries contain 5 million oogonia.
2. By birth many oogonia have degenerated, and for the remaining oogonia meiosis has stopped in prophase I, causing them to become primary oocytes.
3. By puberty 300,000 to 400,000 primary oocytes remain and about 400 will be released from the ovaries.
4. Ovulation is the release of an oocyte from an ovary. The first meiotic division is completed and a secondary oocyte is released.
5. A sperm cell penetrates the secondary oocyte, the second meiotic division is completed, and the nucleus of the oocyte and sperm cell are united to complete fertilization.
6. A primordial follicle is a primary oocyte surrounded by a single layer of flat granulosa cells.
7. In primary follicles, the oocyte enlarges, granulosa cells are cuboidal, and they make up more than one layer. A zona pellucida is present.
8. In a secondary follicle, fluid-filled vesicles appear and a theca forms around the follicle.
9. In a mature follicle, vesicles fuse to form an antrum and the primary oocyte is surrounded by cumulus cells.
10. During ovulation, the mature follicle ruptures releasing the secondary oocyte, surrounded by cumulus cells, into the peritoneal cavity.
11. The remaining granulosa cells in the follicle develop into the corpus luteum.
12. If fertilization occurs, the corpus luteum persists. If there is no fertilization, it degenerates.

Uterine Tubes

1. The ovarian end of the uterine tube is surrounded by fimbriae.
2. Cilia on the fimbriae move the oocyte into the uterine tube.
3. Fertilization usually occurs in the ampulla of the uterine tube, which is near the ovary.

Uterus

1. The uterus is a pear-shaped organ. The uterine cavity and the cervical canal are the spaces formed by the uterus.
2. The wall of the uterus consists of the perimetrium or serous layer, the myometrium (smooth muscle), and the endometrium.

Vagina

1. The vagina connects the uterus (cervix) to the vestibule.
2. The vagina consists of a layer of smooth muscle and an inner lining of moist stratified squamous epithelium.
3. Lubricating fluid is produced by the wall of the vagina.
4. The hymen covers the vestibular opening of the vagina.

External Genitalia

1. The vestibule is a space into which the vagina and the urethra open.
2. The clitoris is composed of erectile tissue and contains many sensory receptors important in detecting sexual stimuli.
3. The labia minora are folds that cover the vestibule and form the prepuce.
4. The greater vestibular glands produce a mucous fluid.
5. The labia majora cover the labia minora, and the pudendal cleft is a space between the labia majora.
6. The mons pubis is an elevated area superior to the labia majora.

Mammary Glands

1. Mammary glands are the organs of milk production.
2. The mammary glands are modified sweat glands that consist of glandular lobes and adipose tissue.
3. The lobes connect to the nipple through ducts. The nipple is surrounded by the areola.
4. The female breast enlarges during puberty under the influence of estrogen and progesterone.

Physiology of Female Reproduction (p. 561)
Puberty

1. Puberty begins with the first menstrual bleeding (menarche).
2. Puberty begins when GnRH, LH, and FSH levels increase.

Menstrual Cycle

1. The cyclical changes in the uterus are controlled by estrogen and progesterone produced by the ovary.
2. Menses (day 1 to days 4 or 5): menstrual fluid is produced by degeneration of the endometrium.
3. Proliferative phase (day 5 to day of ovulation): epithelial cells multiply and form glands.
4. Secretory phase (from day of ovulation to day 28): the endometrium becomes thicker, and endometrial glands secrete. The uterus is prepared for implantation of the developing blastocyst by day 21.
5. Estrogen stimulates proliferation of the endometrium, and progesterone causes thickening of the endometrium. Decreased progesterone causes menses.
6. FSH initiates the development of the follicles.
7. Estrogen produced by the follicles stimulates GnRH, FSH, and LH secretion, and FSH and LH stimulate more estrogen secretion. This positive-feedback mechanism causes FSH and LH levels to increase near the time of ovulation.
8. LH stimulates ovulation and formation of the corpus luteum.
9. Estrogen and progesterone inhibit LH and FSH secretion following ovulation.
10. If fertilization does not occur, progesterone secretion by the corpus luteum decreases and menses begins.
11. If fertilization does occur, the corpus luteum continues to secrete progesterone and menses does not occur.

Menopause

The cessation of the menstrual cycle is called menopause.

Female Sexual Behavior and the Female Sex Act

1. Female sex drive is partially influenced by testosteronelike hormones (produced by the adrenal cortex) and estrogen produced by the ovary.
2. Autonomic nerves cause erectile tissue to become engorged with blood, the vestibular glands to secrete mucus, and the vagina to produce a lubricating fluid.

Infertility

Causes of infertility in females include malfunctions of the uterine tubes, reduced hormone secretion from the pituitary or ovary, and interruption of implantation.

Effects of Aging on the Reproductive System (p. 568)

1. Benign prostatic enlargement affects men as they age and it blocks urine flow through the prostatic urethra.
2. Prostatic cancer is more common in elderly men.
3. Menopause is the most common, age-related change in females.
4. Cancers of the breast, the uterine cervix, and ovaries increase in elderly women.
5. Occasionally prolapsed uterus develops in elderly women.

CONTENT REVIEW

1. List the functions of the male and female reproductive systems.

2. What is the scrotum? Explain the function of the dartos and cremaster muscles.

3. Where, specifically, are sperm cells produced in the testes? Describe the process of spermatogenesis.

4. Name the ducts the sperm cells traverse to go from their site of production to the outside of the body.

5. Where do sperm cells develop their ability to fertilize?

6. Name the erectile tissues of the penis, and describe how erectile tissue becomes erect.

7. State where the seminal vesicles, prostate gland, and bulbourethral glands empty into the male reproductive duct system.

8. Define emission and ejaculation.

9. Define semen. What structures give rise to secretions that make up the semen? Describe the composition of the secretions of each gland.

10. Describe where GnRH, FSH, LH, and testosterone are produced and describe the regulation of their secretion.

11. Describe the effects of testosterone during puberty and on the adult male.

12. Describe the regulation of the male sexual act.

13. Describe the process of follicle development and ovulation.

14. What is the corpus luteum? What happens to the corpus luteum if fertilization occurs? If fertilization does not occur?

15. Describe the normal pathway followed by the oocyte after ovulation. Where does fertilization usually take place?

16. Describe the relationship between the uterus, vagina, vestibule, and external genitalia.

17. Describe the labia minora, the prepuce, the labia majora, the pudendal cleft, and the mons pubis.

18. What are the effects of estrogen and progesterone on the uterus?

19. Describe the hormonal changes that result in ovulation. Explain the sequence of events during each phase of the menstrual cycle.

20. Define menopause and female climacteric. What causes these changes?

21. List the major age-related changes that occur in the female reproductive system.

22. List the major age-related diseases that occur in the female reproductive system.

23. List the major age-related diseases that occur in the male reproductive system.

CRITICAL THINKING QUESTIONS

1. If an adult male was castrated by having his testes removed, what would happen to the levels of GnRH, FSH, LH, and testosterone in his blood?

2. Birth control pills for women contain estrogen and progesterone compounds. Explain how these hormones can prevent pregnancy.

3. During the secretory phase of the menstrual cycle, you would normally expect:
   a. The highest levels of progesterone that occur during the menstrual cycle
   b. A follicle present in the ovary that is ready to undergo ovulation
   c. That the endometrium reaches its greatest degree of development
   d. a and b
   e. a and c

4. Between approximately days 12–14 of the menstrual cycle, you would normally expect:
   a. Increasing blood levels of estrogen
   b. Increasing blood levels of LH

   c. Blood levels of progesterone are near their maximum
   d. a and b
   e. a, b, and c

5. On day 15 of the menstrual cycle, you would normally expect:
   a. Decreasing blood levels of LH
   b. Decreasing blood levels of estrogen
   c. Increasing blood levels of progesterone
   d. a and b
   e. a, b, and c

6. Predict the consequences if a drug that blocks the effect of progesterone is taken by a woman 2 or 3 days following ovulation or by a woman who is pregnant.

7. During menopause, which reproductive hormones are reduced in the blood and which are increased?

Answers in Appendix D

Visit this textbook's website at www.mhhe.com/seeleyess6 for practice quizzes, animations, interactive learning exercises, and other study tools.

Chapter Outline and Objectives

Prenatal Development (p. 574)

1. List the prenatal periods and state the major developmental events associated with each.
2. Describe the process of fertilization.
3. Describe the blastocyst, the process of implantation, and placental formation.
4. Describe the maternal changes that occur during pregnancy.
5. List the three germ layers, describe their formation, and list the adult derivatives of each layer.
6. Describe the formation of the neural tube and neural crest cells.
7. Describe the formation of the gastrointestinal tract, the limbs, and the face.
8. Explain how the single heart tube is divided into four chambers.

Parturition (p. 586)

9. Explain the events that occur during parturition.

The Newborn (p. 590)

10. Discuss the respiratory, circulatory, and digestive changes that occur in the newborn at the time of birth.

Lactation (p. 592)

11. Describe the events of lactation.

First Year Following Birth (p. 593)

12. Describe the changes that occur during the first year after birth.

Life Stages (p. 593)

13. List the stages of life and describe the major events that are associated with each stage. Describe the changes that occur during the aging process.

Aging (p. 593)

14. Describe the process of aging.

Death (p. 595)

15. Describe the events that occur at the time of death.

Genetics (p. 595)

16. Define genetics and explain how chromosomes are related to genetics.
17. Describe the major types of inheritance.

Development, Heredity, and Aging

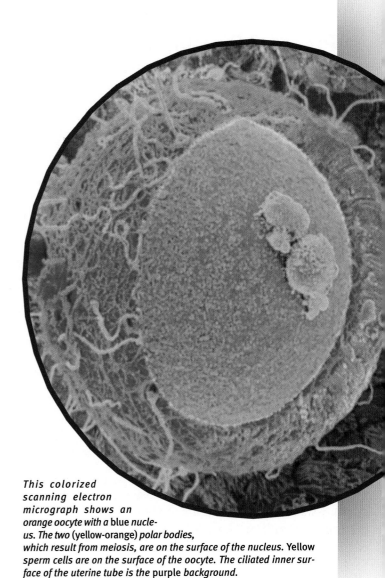

This colorized scanning electron micrograph shows an orange oocyte with a blue *nucleus. The two* (yellow-orange) *polar bodies, which result from meiosis, are on the surface of the nucleus. Yellow sperm cells are on the surface of the oocyte. The ciliated inner surface of the uterine tube is the* purple *background.*

the stages of life and associated activities are issues of great interest in today's society. Life stages are viewed very differently today than a few years ago. For example, in 1960, 20% of males and 12% of females graduating from high school attended college. Today, over half of all people 25 and older have attended college for some period of time. In addition, there are many more nontraditional college students than there were just a few years ago.

In 1900, only 5% of the U.S. population was over age 65. Today, about 16% of the population is over age 65, and by 2030 more than 20% will be older than 65. The average life expectancy in 1900 was about 47 years, in 1940 it was about 63 years, and today it is about 78 years. In 1900, nearly 70% of all males over age 65 were still working; today only about 20% are still working past age 65. Older people are healthier and more active than they have ever been.

The life span is usually considered the period between birth and death; however, the 9 months before birth are a critical part of a person's existence. What happens in these 9 months profoundly affects the rest of a person's life. Although most people develop normally and are born without defects, approximately three out of every 100 people are born with a birth defect so severe that it requires medical attention during the first year of life. Later in life, many more people discover previously unknown problems, such as the tendency to develop asthma, certain brain disorders, or cancer.

Prenatal Development

The **prenatal** (prē-nā′tăl, before birth) **period,** the period from conception to birth, can be divided into three parts: (1) during approximately the first 2 weeks of development the primitive germ layers are formed; (2) from about the second to the eighth week of development the major organ systems come into existence; and (3) during the last 7 months of the prenatal period the organ systems grow and become more mature. The developing human between the time of fertilization and 8 weeks of development is called an **embryo** (em′brē-ō, *en,* in + *bryō,* to swell). From 8 weeks to birth, the developing human is called a **fetus** (fē′tus, offspring).

The medical community in general uses the **last menstrual** (men′stroo-ăl, *mensis,* month) **period (LMP)** to calculate the **clinical age** of the unborn child. An embryo or fetus is therefore considered to be a certain number of days post-LMP. Most embryologists, on the other hand, use **developmental age,** which begins with fertilization, to describe the timing of developmental events. Because fertilization is assumed to occur approximately 14 days after LMP, it is assumed that developmental age is 14 days less than clinical age. The times presented in this chapter are based on developmental age.

Fertilization

After **sperm cells** are ejaculated into the vagina, they are transported through the cervix, the body of the uterus, and the uterine tubes, where fertilization occurs. The swimming ability of the sperm cells and muscular contractions of the uterus and uterine tubes are responsible for the movement of sperm cells through the female reproductive tract. Oxytocin released by the female posterior pituitary and prostaglandins within the semen both stimulate contractions in the uterus and uterine tubes.

While passing through the uterus and the uterine tubes, the sperm cells undergo capacitation. **Capacitation** (kă-pas′i-tā′shŭn) makes the sperm cells capable of releasing enzymes contained in the acrosome, a region of concentrated enzymes located between the cell membrane and nuclear membrane within the leading edge of the sperm cell head. The enzymes digest a pathway through the cumulus cells and the zona pellucida of the **secondary oocyte** (ō′ō-sīt, egg cell). One sperm cell attaches to the oocyte cell membrane and enters the oocyte (figure 20.1).

The secondary oocyte is capable of being fertilized for perhaps up to about 1 day after ovulation, and some sperm cells remain viable in the female reproductive tract for up to 6 days, although most of them degenerate before that time.

> **PREDICT** 1
> During what days of the menstrual cycle is sexual intercourse most likely to result in pregnancy?

Hundreds of sperm cells reach the secondary oocyte, but normally a change occurs in the oocyte cell membrane that prevents more than one sperm cell from entering the secondary oocyte. The secondary oocyte undergoes the second meiotic division only after a sperm cell enters it. After the second meiotic division, the oocyte nucleus moves to the center of the cell, where it meets the nucleus of the sperm cell. Each of these nuclei has 23 chromosomes, each having one chromosome from each chromosome pair. Their fusion, which completes the process of fertilization, restores the number of chromosomes to 46 (see chapter 19).

Fertilization (fer′til-i-zā′shŭn, *fero,* to bear + process; see figure 20.1) is defined as the union of the sperm cell and secondary oocyte, with the union of their genetic material, the chromosomes. The product of fertilization is a single cell called the **zygote** (zī′gōt, having a yoke).

Early Cell Division

About 18–36 hours after fertilization, the zygote divides to form two cells (figure 20.2). Those two cells divide to form four cells, which divide to form eight, and so on. Even though the number of cells increases, the size of each cell decreases, so that the total mass of cells remains about the same size as the zygote. These cells have the ability to develop into a wide range of tissues. As a result, the total number of cells can be decreased, increased, or reorganized during this period without affecting normal development.

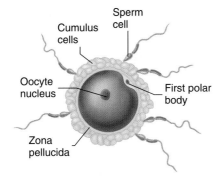

1. Many sperm cells attach to the cumulus cells of a secondary oocyte.

Cumulus cells • Sperm cell • Oocyte nucleus • First polar body • Zona pellucida

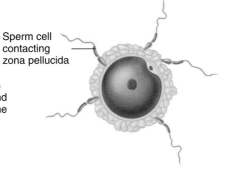

2. One sperm cell attaches to the zona pellucida, and enzymes in the acrosome digest through the zona pellucida.

Sperm cell contacting zona pellucida

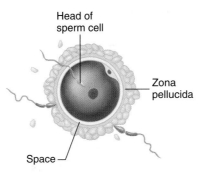

3. The head of one sperm cell penetrates the zona pellucida and oocyte cell membrane to enter the oocyte cytoplasm. Changes in the zona pellucida (moving away from the oocyte) form a space that prevents additional sperm cells from entering the oocyte.

Head of sperm cell • Zona pellucida • Space

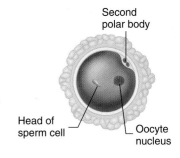

4. In response to the entry of the sperm head into the oocyte, the oocyte nucleus moves to one side of the oocyte where it completes the second meiotic division and gives off a second polar body.

Second polar body • Head of sperm cell • Oocyte nucleus

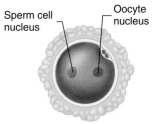

5. The sperm head enlarges and gives rise to the sperm cell nucleus.

Sperm cell nucleus • Oocyte nucleus

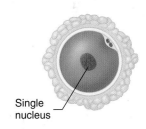

6. The two nuclei fuse to form a single nucleus. Fertilization is complete and a zygote results.

Single nucleus

Process Figure 20.1 Fertilization

A CASE IN POINT | Twins

The parents of twin boys named the babies Juan and Hamall because, at birth, they looked so much alike that if you've seen Juan, you've seen Hamall. As the twins grew older, they continued to look very much alike but not exactly the same. Their parents want to know if they are identical or fraternal twins. The parents have an expensive genetic analysis performed on their twin boys and learn that they are identical, in spite of their slight differences in appearance.

In rare cases, following early cell divisions, the cells may separate and develop to form two individuals, called **"identical,"** or **monozygotic** (mon-ō-zī-got′ik), **twins.** Identical twins therefore have identical genetic information in their cells. Identical twins can also occur by other mechanisms, which occur a little later in development.

Occasionally a woman can ovulate two or more secondary oocytes at the same time. Fertilization of multiple oocytes by different sperm cells results in **"fraternal"** (fră-ter′năl), or **dizygotic** (dī′zī-got′ik), **twins.**

Even though who we are and what we look like are determined in large part by genes, identical twins—although they have the same genes—may not look exactly alike. In fact, identical twins often look more like mirror images of each other. Genes interact with minute environmental cues in the embryo and in the child to determine the final form of the individual.

Blastocyst

About 2 or 3 days after fertilization, multiple cell divisions have produced an embryonic mass of about 12–16 cells, called a **morula** (mōr′ū-lă, mulberry). Unlike the early dividing mass of cells, the embryo after day 14 is developing a complex form, and the cells within the embryo are developing into tissues and organs. Indeed, most of the cells of the morula will not form the embryo proper but will form support structures such as the placenta.

When a cavity begins to appear within the mass of cells, the whole structure is then called a **blastocyst** (blas′tō-sist, *blastos,* germ + *kystis,* bladder) (see figure 20.2). The fluid-filled cavity is called the **blastocele** (blas′tō-sēl, *koilos,* hollow). Most of the blastocele is surrounded by a single layer of cells, but at one end of the blastocyst the cells are several layers thick. The thickened area is the **inner cell mass.** Not all cells of the blastocyst give rise to the new individual. The embryo proper, which will become the new individual, will develop from only a few cells of the inner cell mass. These cells are commonly referred to as **stem cells** because they give rise to all cell types within the body. The remaining cells of the blastocyst are called the **trophoblast** (trof′ō-blast, trō′fō-blast, feeding layer), which forms the embryonic part of the placenta and the membranes (chorion and amnion) surrounding the embryo.

Implantation of the Blastocyst and Development of the Placenta

All these early events, from the first cell division to formation of the blastocele, occur as the embryonic mass moves from the site of fertilization in the uterine tube to the site of implantation in the uterus. By 7 or 8 days after ovulation (day 21 or 22 post-LMP), the endometrium of the uterus is prepared for implantation. About 7 days after fertilization, the blastocyst attaches itself to the uterine wall and begins the process of **implantation** (im-plan-tā′shŭn). The trophoblast cells of the blastocyst digest the uterine tissues as the blastocyst burrows into the uterine wall. Before implantation, and for a short time after implantation, the embryo is insensitive to environmental toxins. During the first few days of development, each cell has enough yolk to supply its own energy needs and few external nutrients are needed. Furthermore, during the first couple of weeks of development, large numbers of cells can be killed and the embryo can fully recover.

As the blastocyst burrows into the uterine wall, trophoblast cells, now called the **chorion** (kō′rē-on, membrane of the fetus), form the embryonic portion of the **placenta** (plă-sen′tă, a cake). Fingerlike projections, called **chorionic villi,** protrude into cavities formed within the maternal endometrium. Those cavities, called **lacunae** (lă-koo′nē, a lake), are filled with maternal blood (figure 20.3). In the mature placenta, the embryonic blood supply is separated from the maternal blood supply by the

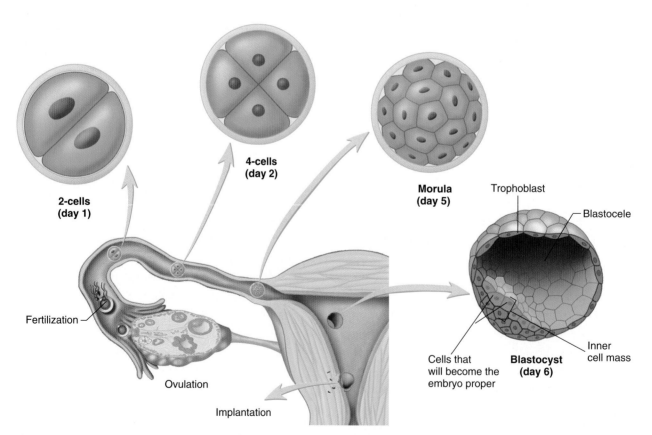

Figure 20.2   **Development of the Blastocyst and Implantation**

In the blastocyst, *green* cells are trophoblastic, and *orange* cells are embryonic.

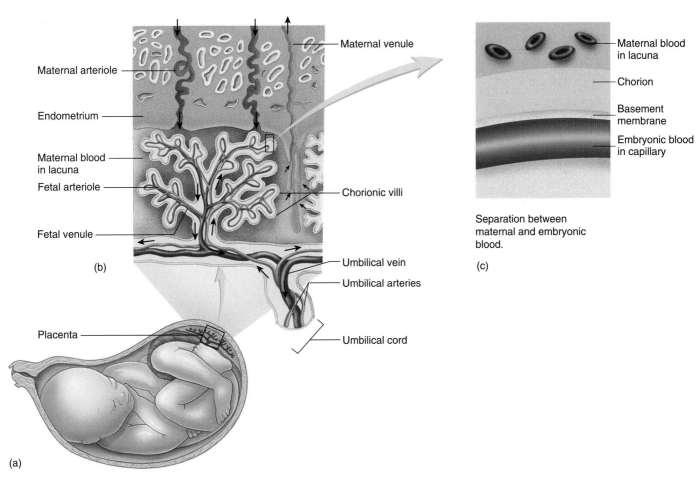

Figure 20.3   The Interface Between Maternal and Fetal Circulation

(*a*) Location of the placenta and umbilical cord. (*b*) As maternal blood vessels are encountered by the chorion (chorionic villi), lacunae (cavities) are formed and filled with maternal blood. In the mature placenta, the embryonic blood vessels and other tissue form chorionic villi, and nutrients are exchanged between embryonic and maternal blood. (*c*) Under normal conditions, the maternal and fetal blood are separated by the chorion and a basement membrane and do not mix.

embryonic capillary wall, a basement membrane, and a thin layer of chorion. As a result, the embryonic and maternal blood do not mix. Nutrients and waste products must cross this semipermeable barrier between the two circulations.

Initially the embryo is attached to the placenta by a connecting stalk. As the embryo matures, the connecting stalk elongates and becomes known as the **umbilical** (ŭm-bil′i-kăl, naval) **cord** (see figure 20.3). Within the umbilical cord, blood vessels carry blood from the embryo to the placenta and from the placenta to the embryo.

Maternal Changes

The chorion secretes **human chorionic gonadotropin** (gō′nad-ō-trō′pin) **(HCG),** which is transported in the blood to the maternal ovary and causes the corpus luteum to remain

functional. The secretion of HCG begins shortly after implantation, increases rapidly, and reaches a peak about 8 or 9 weeks after fertilization. Subsequently, HCG levels decline to a lower level and are maintained at a low level throughout the remainder of the pregnancy (figure 20.4). Most pregnancy tests are designed to detect HCG in either urine or blood.

The estrogen and progesterone secreted by the corpus luteum (see chapter 19) are essential for the maintenance of the endometrium for the first 3 months of pregnancy. After the placenta forms, it also begins to secrete estrogens and progesterone. By the third month of pregnancy, the placenta has become an endocrine gland that secretes sufficient quantities of estrogen and progesterone to maintain pregnancy and the corpus luteum is no longer needed. Estrogen and progesterone levels increase in the mother's blood throughout pregnancy.

1. Human chorionic gonadotropin (HCG) increases until it reaches a maximum concentration near the end of the first 3 months of pregnancy and then decreases to a low level thereafter.

2. Progesterone continues to increase until it levels off near the end of pregnancy. Early in pregnancy, progesterone is produced by the corpus luteum in the ovary; by the second trimester production shifts to the placenta.

3. Estrogen levels increase slowly throughout pregnancy, but they increase more rapidly as the end of pregnancy approaches. Early in pregnancy, estrogen is produced only in the ovary; by the second trimester production shifts to the placenta.

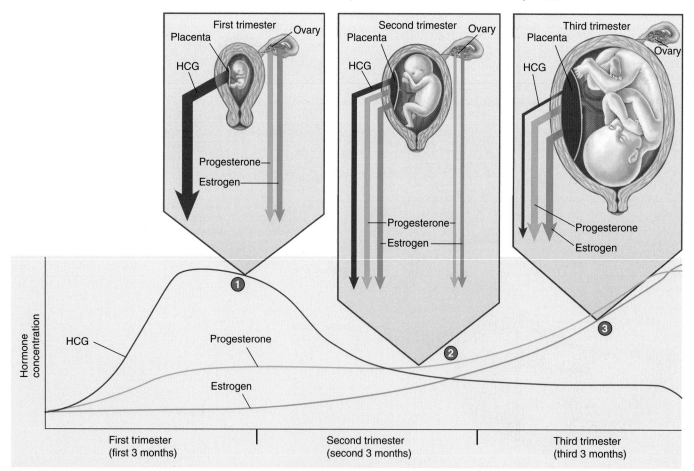

Process Figure 20.4   Changes in Hormone Concentration During Pregnancy

HCG, progesterone, and estrogen are secreted from the placenta during pregnancy. Early in pregnancy estrogen and progesterone are secreted by the ovary. During midpregnancy there is a shift toward estrogen and progesterone secretion by the placenta. Late in pregnancy these two hormones are secreted by the placenta.

Formation of the Germ Layers

After implantation, a new cavity, called the **amniotic** (am-nē-ot'ik, *amnios*, lamb) **cavity,** forms inside the inner cell mass and causes the part of the inner cell mass nearest the blastocele to separate as a flat disk of tissue called the **embryonic disk** (figure 20.5). The amniotic cavity is bounded by a membrane called the **amniotic sac** and is filled with **amniotic fluid.** The embryo will grow into the amniotic cavity, where the amniotic fluid forms a protective cushion. The embryonic disk at first is composed of two layers of cells: an **ectoderm** (ek'tō-derm, outside layer) adjacent to the amniotic cavity and an **endoderm** (en'dō-derm, inside layer) on the side of the disk opposite the amniotic cavity. A third cavity, the **yolk sac,** forms inside the blastocele from the endoderm.

At about 14 days after fertilization, the embryonic disk has become a slightly elongated, oval structure. Some of the ectoderm cells migrate toward the center of the disk, forming a thickened line called the **primitive streak.** The formation of the primitive streak establishes the central axis of the embryo. Some of the ectoderm cells migrate through the primitive streak and emerge between the ectoderm and endoderm as a new germ layer, called the **mesoderm** (mez'ō-derm, middle layer) (figure 20.6). The embryo is now three-layered, having ectoderm, mesoderm, and endoderm. All tissues of the adult can be traced to these three germ layers (table 20.1).

From about day 14 until about day 35, the embryo is at maximum risk from environmental toxins and drugs that can cause birth defects. The causes of birth defects are a major unsolved issue in biology at present. However, it appears that oxidative damage to key molecules in certain developing cells and/or changes to the DNA function within those cells may be involved.

A specialized group of cells at the cephalic end of the primitive streak moves from one end of the primitive streak to the other and, in some as yet unknown way, organizes the embryo. A cordlike structure called the **notochord** (nō'tō-kōrd, *notos*, back + *chorde*, cord or string) is formed by these cells as they move down the primitive streak. The notochord marks the central axis of the developing embryo (see figure 20.6).

P R E D I C T ②
Predict the result if two primitive streaks form in one embryonic disk. What if the two primitive streaks are touching each other?

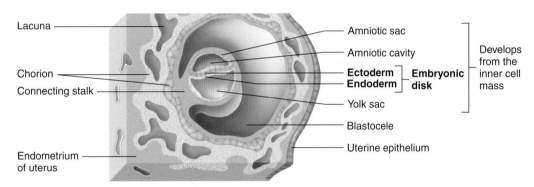

Lacuna — **Amniotic sac** / **Amniotic cavity** — Develops from the inner cell mass / **Ectoderm** / **Endoderm** **Embryonic disk**

Chorion — **Connecting stalk** — **Yolk sac** — **Blastocele** — **Uterine epithelium**

Endometrium of uterus

Figure 20.5 **Early Embryo and Surrounding Structures in the Placenta**

The embryonic disk consisting of ectoderm and endoderm, with amniotic cavity and yolk sac. The connecting stalk, which attaches the embryo to the uterus, will become part of the umbilical cord.

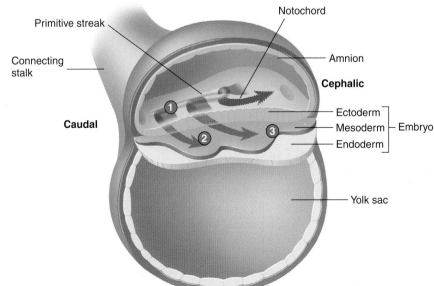

1. Cells in the surface ectoderm move toward the primitive streak and migrate through the streak (*blue arrow tails*).

2. Cells of the ectoderm that migrate through the primitive streak become mesodermal cells (*red arrows*).

3. The mesoderm (*red*) lies between the ectoderm (*blue*) and endoderm (*yellow*).

Primitive streak — Notochord — Amnion — **Cephalic** — Ectoderm / Mesoderm / Endoderm — Embryo

Connecting stalk — **Caudal** — Yolk sac

Process Figure 20.6 **Primitive Streak**

The head of the embryo will develop over the notochord.

Table 20.1   Tissues Derived from Each Germ Layer

Endoderm	Ectoderm	Mesoderm
Lining of gastrointestinal tract	Epidermis of skin	Dermis of skin
Lining of lungs	Tooth enamel	Circulatory system
Lining of hepatic, pancreatic, and other exocrine ducts	Lens and cornea of eye	Parenchyma (substance) of glands
	Outer ear	Kidneys
Kidney ducts and bladder	Nasal cavity	Gonads
Anterior pituitary	Neuroectoderm	Muscle
Thymus gland	Brain and spinal cord	Bones (except facial)
Thyroid gland	Somatic motor neurons	
Parathyroid gland	Preganglionic autonomic neurons	
Tonsils	Neural crest cells	
	Melanocytes	
	Sensory neurons	
	Postganglionic autonomic neurons	
	Adrenal medulla	
	Facial bones	
	Teeth: dentin and pulp	
	Skeletal muscles in head	

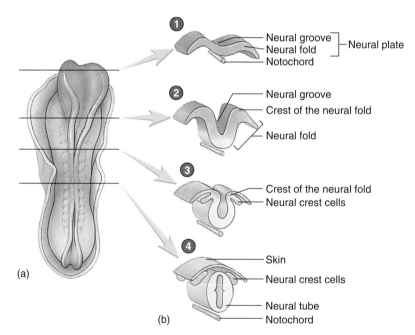

1. The neural plate is formed from ectoderm.

2. Neural folds form as parallel ridges along the embryo.

3. Neural crest cells begin to form from the crest of the neural folds.

4. The neural folds meet at the midline to form the neural tube and neural crest cells separate from the neural folds.

Process Figure 20.7   Formation of the Neural Tube

The neural folds, which consist of neuroectoderm, come together in the midline and fuse to form a neural tube. This fusion begins in the center and moves both cranially and caudally. (a) The embryo shown is about 21 days after fertilization. (b) The insets to the right show progressive closure of the neural tube.

Neural Tube and Neural Crest Formation

At about 18 days after fertilization, the ectoderm overlying the notochord thickens to form the **neural plate** (figure 20.7). The lateral edges of the plate begin to rise like two ocean waves coming together. These edges are called the **neural folds,** and a **neural groove** lies between them. The neural folds begin to meet in the midline and fuse into a neural tube, which is completely closed by day 26. The cells of the **neural tube** are called **neuroectoderm** (noor-ō-ek′tō-derm) (see table 20.1). Neuroectoderm

becomes the brain, the spinal cord, and parts of the peripheral nervous system. If the neural tube fails to close, major defects of the central nervous system can result.

As the neural folds come together and fuse, a population of cells breaks away from the neuroectoderm all along the crests of the folds. Most of these **neural crest cells** become part of the peripheral nervous system or become melanocytes of the skin. In the head, neural crest cells also contribute to the skull, the dentin of teeth, blood vessels, and general connective tissue.

Neural Tube Defects

Anencephaly (an'en-sef'ă-lē, no brain) is a birth defect wherein much of the brain fails to form because the neural tube fails to close in the region of the head. A baby born with anencephaly cannot survive. **Spina bifida** (spī'nă bif'i-dă, split spine) is a general term describing defects of the spinal cord or vertebral column. Spina bifida can range from a simple defect with no clinical manifestation and with one or more vertebral spinous processes split or missing, to a more severe defect that can result in paralysis of the limbs or the bowels and bladder, depending on where the defect occurs. More severe forms of spina bifida result from failure of the neural tube in the area of the spinal cord to close. It has been demonstrated that adequate amounts of folic acid, the B vitamin folate, in the diet during pregnancy can reduce the risk of such defects.

Formation of the General Body Structure

Arms and legs first appear at about 28 days after fertilization as **limb buds** (figure 20.8) and quickly begin to elongate. At about 35 days, expansions called hand and foot plates form at the ends of the limb buds. Zones of cell death between the future fingers and toes of the hand and foot plates help sculpture the fingers and toes.

The face develops by fusion of five growing masses of tissue, called **processes.** One, the **frontonasal process,** forms the forehead, nose, and center of the upper jaw and lip. Two **maxillary processes** form the maxillae (upper lip and jaw) and two **mondibular processes** form the mandible (lower lip and jaw). The nose begins as two structures, one on each side of the forehead mass (figure 20.9).

As the brain enlarges and the face matures, the two parts of the nose approach each other in the midline and fuse (see figure 20.9). The two masses forming the upper jaw expand toward the midline and fuse with part of the nose to form the

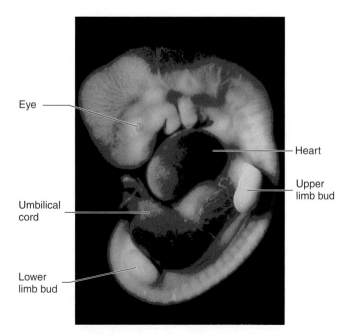

Figure 20.8 Human Embryo 35 Days After Fertilization

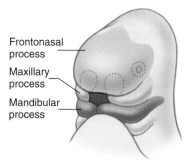

1. **28 days after fertilization**
   The face develops from five processes: frontonasal (*blue*), two maxillary (*yellow*), and two mandibular (*orange*; already fused).

   Frontonasal process
   Maxillary process
   Mandibular process

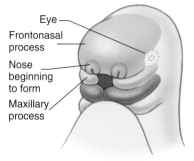

2. **33 days after fertilization**
   Nasal placodes appear in the frontonasal process.

   Eye
   Frontonasal process
   Nose beginning to form
   Maxillary process

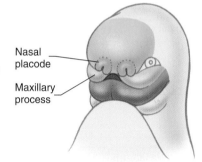

3. **40 days after fertilization**
   Maxillary processes extend toward the midline. The nasal placodes also move toward the midline and fuse with the maxillary processes to form the jaw and lip.

   Nasal placode
   Maxillary process

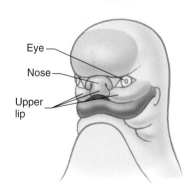

4. **48 days after fertilization**
   Continued growth brings structures more toward the midline.

   Eye
   Nose
   Upper lip

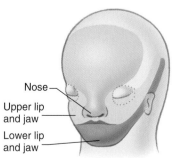

5. **14 weeks after fertilization**
   Colors show the contributions of each process to the adult face.

   Nose
   Upper lip and jaw
   Lower lip and jaw

Process Figure 20.9   Development of the Face

Ages indicate developmental days; colors show the contributions of each process to the adult face.

Table 20.2 Development of the Organ Systems

	Age (Days Since Fertilization)					
	1–5	**6–10**	**11–15**	**16–20**	**21–25**	**26–30**
General Features	Fertilization Blastocyst	Blastocyst implants	Primitive streak Three germ layers	Neural plate	Neural tube closed	Limb buds and other "buds" appear
Integumentary System			Ectoderm Mesoderm		Melanocytes from neural crest	
Skeletal System			Mesoderm		Neural crest cells	Limb buds
Muscular System			Mesoderm	Somites (body segments) begin to form		Somites all formed
Nervous System			Ectoderm	Neural plate	Neural tube complete Neural crest Eyes and ears begin to form	Lens appears
Endocrine System			Ectoderm Mesoderm Endoderm	Thyroid gland begins to develop		Parathyroid glands and pancreas appear
Cardiovascular System			Mesoderm	Blood islands form Two-tubed heart	Single-tubed heart begins to beat	Interatrial septum forms
Lymphatic System			Mesoderm			Thymus appears
Respiratory System			Mesoderm Endoderm		Diaphragm begins to form	Trachea Lung buds
Digestive System			Endoderm		Foregut and hindgut form	Liver and pancreas appear as buds
Urinary System			Mesoderm Endoderm		Embryonic kidneys appear	Embryonic kidneys elongate
Reproductive System			Mesoderm Endoderm	Primordial germ cells on yolk sac	Male reproductive ducts appear External genital structures begin to form	

upper jaw and lip. A **cleft lip** results from failure of these structures to fuse. Cleft lips usually do not occur in the midline, but to one side (or both sides). The cleft can vary in severity from a slight indentation in the lip to a fissure that extends from the mouth to the naris (nostril).

The roof of the mouth, or palate, begins to form as vertical shelves of tissue that grow on the inside of the maxillary masses. These shelves swing to a horizontal position and begin to fuse with each other at about 56 days of development. If the palate does not fuse, a midline cleft in the roof of the mouth called a **cleft palate** results. A cleft palate can range in severity from a slight cleft of the uvula to a fissure extending the entire length of the palate. A cleft lip and cleft palate can occur together, forming a continuous fissure.

Age (Days Since Fertilization)					
31–35	36–40	41–45	46–50	51–55	56–60
Hand and foot plates on limbs	Fingers and toes appear Lips formed Embryo 15 mm	External ear forming Embryo 20 mm	Embryo 25 mm	Limbs elongate to adult proportions Embryo 35 mm	Face is distinctly human in appearance
Sensory receptors appear in skin		Collagen fibers clearly present in skin		Extensive sensory nerve endings in skin	
Mesoderm condensation in areas of future bone	Cartilage in site of future humerus	Cartilage in site of future ulna and radius	Cartilage in site of future hand and fingers		Ossification begins in clavicle and then in other bones
Muscle precursor cells enter limb buds			Functional muscle		Nearly all muscles appear in adult form
Nerve processes enter limb buds		External ear forming Olfactory nerve begins to form		Semicircular canals in inner ear complete	Eyelids form Cochlea in inner ear complete
Pituitary appears as evaginations from brain and mouth	Gonads begin to form Adrenal glands form		Pineal body appears	Thyroid gland in adult position	Anterior pituitary loses its connection to mouth
Interventricular septum begins to form		Interventricular septum complete	Interatrial septum complete but foramen ovale remains until birth		
Large lymphatic vessels form in neck	Spleen appears			Adult lymph pattern formed	
Secondary bronchi to lobes form	Tertiary bronchi to lobules form		Tracheal cartilage begins to form		
Mouth opens to outside		Palate begins to form Tooth buds begin to form			Palate begins to fuse (fusion complete by 90 days); anus opens
Adult kidneys begin to develop				Embryonic kidneys degenerate	
	Gonads begin to form	Primordial germ cells enter gonads	Female reproductive ducts appear		Uterus forming External genitalia begins to differentiate in male and female

Development of the Organ Systems

The major organ systems appear and begin to develop during the embryonic period (second to eighth week of development). This period is therefore also called the period of **organogenesis** (ōr′gă-nō-jen′ĕ-sis). The individual organ systems are not described in the text but are listed in table 20.2.

Only general comments about a few select systems are presented in the text.

At the same time the neural tube is forming (18–26 days), the remainder of the embryo is folding to form a tube along the upper part of the yolk sac (figure 20.10). Another tube, which will form the gastrointestinal (GI) tract, develops inside the embryo from the upper part of the yolk sac.

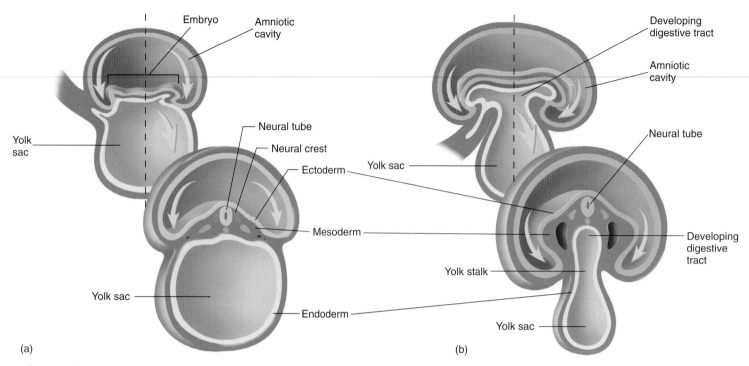

Figure 20.10   **Development of the Digestive Tract**

The digestive tract develops along the dorsal side of the yolk sac (*yellow*) as the body folds into a tube (*blue arrows*). The figures in back are shown in sagittal section. The figures in front are shown in cross section. The *dashed line* on the figures in back shows the plane of section in the figures in front. (*a*) An early embryo (about 24 days). (*b*) A slightly older embryo (about 28 days).

A considerable number of outpocketings appear at about 28 days after fertilization along the entire length of the GI tract (figure 20.11). A surprisingly large number of important internal organs develops from those outpocketings, including the auditory tubes, tonsils, thymus gland, anterior pituitary gland, thyroid gland, parathyroid glands, lungs, liver, pancreas, and urinary bladder.

The heart develops from two blood vessels, lying side by side in the early embryo, which fuse about 21 days after fertilization into a single, midline heart. At about this time, the primitive heart begins to beat. Blood vessels form from "blood islands" on the surface of the yolk sac and inside the embryo. These islands expand and fuse to form the circulatory system.

The major chambers of the heart, the atrium and ventricle, expand rapidly. The single ventricle is subdivided into two chambers by the development of an **interventricular** (in-ter-ven-trik′ŭ-lăr) **septum** (figure 20.12). If the interventricular septum does not grow enough to completely separate the ventricles, a ventricular septal defect results.

An **interatrial** (in-ter-ā′trē-ăl) **septum** forms to separate the two atria. An opening in the interatrial septum called the **foramen ovale** (ō-val′ē, oval, egg-shaped) connects the two atria and allows blood to flow from the right to the left atrium in the fetus. Because of the presence of the foramen ovale, some blood in the fetus passes from the right atrium to the left atrium and bypasses the right ventricle and the lungs. The foramen ovale normally closes off at the time of birth, and blood then circulates through the right ventricle and the lungs. If this does not occur, an interatrial septal defect occurs. An interatrial

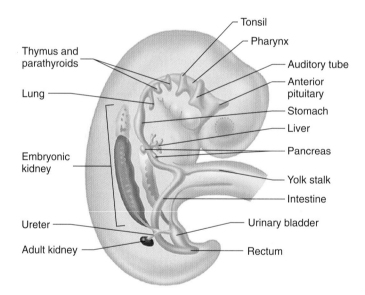

Figure 20.11   **The Embryonic Digestive and Urinary Systems**

Outpocketings of the digestive tract (*yellow*), which will form many adult structures such as the lungs and glands, are depicted. The embryonic and adult kidneys are also shown (*purple*).

septal defect or a ventricular septal defect usually results in a heart murmur.

The kidneys develop from mesoderm located along the lateral wall of the body cavity (see figure 20.11). The embryonic kidneys are much more extensive than the adult kidneys,

1. **20 days after fertilization**
   At this age, the heart consists of two parallel tubes.

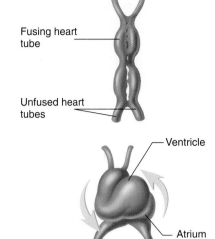

Fusing heart tube

Unfused heart tubes

2. **22 days after fertilization**
   The two parallel tubes have fused to form one tube. This tube bends as it elongates (*blue arrows* suggest the direction of bending) within the confined space of the pericardium.

Ventricle

Atrium

3. **31 days after fertilization**
   The interatrial septum (*green*) and the interventricular septum grow toward the center of the heart.

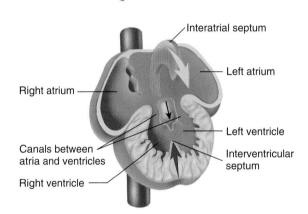

Interatrial septum

Left atrium

Right atrium

Left ventricle

Canals between atria and ventricles

Interventricular septum

Right ventricle

4. **35 days after fertilization**
   The interventricular septum is nearly complete. A foramen, which will become the left side of the foramen ovale, opens in the left side of the interatrial septum (*green*) as the right side of the interatrial septum begins to form (*blue*).

Interatrial septum

Foramen

Interventricular septum

5. **The final embryonic condition of the interatrial septum**
   A foramen remains in the right side of the interatrial septum (*blue*), which forms the right part of the foramen ovale. Blood from the right atrium can flow through the foramen ovale into the left atrium. After birth, as blood begins to flow in the other direction, the left side of the interatrial septum is forced against the right side, closing the foramen ovale.

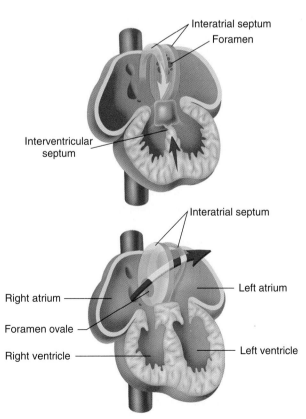

Interatrial septum

Right atrium

Left atrium

Foramen ovale

Right ventricle

Left ventricle

Process Figure 20.12 Formation of the Heart

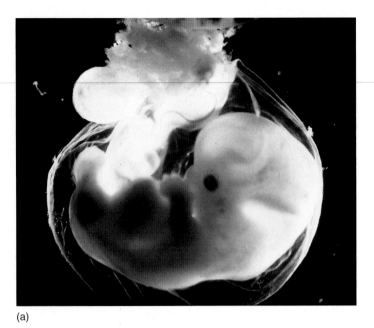

(a)

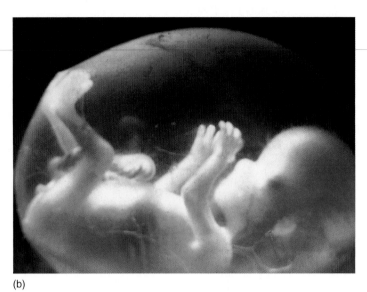

(b)

Figure 20.13   Late Embryo and Fetus

(*a*) Embryo at 50 days of development. (*b*) Fetus at 3 months of development.

extending the entire length of the body cavity. They are closely associated with internal reproductive organs such as the ovaries or testes and reproductive ducts such as the uterine tubes or ductus deferens. Most of the embryonic kidneys degenerate, with only a very small part forming the adult kidney.

Growth of the Fetus

The embryo becomes a fetus about 8 weeks after fertilization (figure 20.13). The beginning of the fetal period is marked by the beginning of bone ossification. In the embryo, most of the organ systems are developing, whereas in the fetus the organs are present. During the fetal period, the organ systems enlarge and mature. The fetus grows on average from about 3 cm and 2.5 g (0.09 oz) at 8 weeks to 50 cm and 3300 g (7 lb, 4 oz) at the end of pregnancy. The growth during the fetal period represents more than a 15-fold increase in length and a 1400-fold increase in weight.

Fine, soft hair called **lanugo** (lă-noo′gō, *lana*, wool) covers the fetus, and a waxy coat of loose epithelial cells called **vernix caseosa** (ver′niks kā′sē-ō′să, varnish, cheesy) form a protective layer between the fetus and the amniotic fluid. The amniotic fluid contains toxic waste products from the digestive tract and kidneys of the fetus.

Subcutaneous fat that accumulates in the fetus provides a nutrient reserve, helps insulate, and aids the newborn in sucking by strengthening and supporting the cheeks so that a small vacuum can be developed in the oral cavity.

Peak body growth occurs late in gestation, but as placental size reaches a maximum, the oxygen and nutrient supply to the fetus becomes limited. Growth of the placenta essentially stops at about 35 weeks, limiting fetal growth.

At approximately 38 weeks of development, the fetus has progressed to the point at which it is ready to be delivered. The average weight at this point is 3250 g (7 lb, 2 oz) for a female fetus and 3300 g (7 lb, 4 oz) for a male fetus.

Parturition

Physicians usually calculate the **gestation** (jes-tā′shŭn, to bear) **period** (length of pregnancy) as 280 days (40 weeks or 10 lunar months) from the LMP to the date of delivery of the fetus. **Parturition** (par-toor-ish′ŭn, to be in labor) refers to the process by which the baby is born (figure 20.14). Near the end of pregnancy, the uterus becomes progressively more irritable and usually exhibits occasional contractions that become stronger and more frequent until parturition is initiated. The cervix gradually dilates, and strong uterine contractions help expel the fetus from the uterus through the vagina. Just before expulsion of the fetus from the uterus, the amniotic sac surrounding the fetus ruptures, and amniotic fluid flows through the vagina to the exterior. This event is commonly referred to as the woman's "water breaking."

Labor is the period during which uterine contractions occur that result in expulsion of the fetus. Although labor may differ greatly from woman to woman and from one pregnancy to another for the same woman, it can usually be divided into three stages.

P R E D I C T   3
Compare and contrast clinical age and developmental age for fertilization, implantation, beginning of the fetal period, and parturition.

1. The **first stage** begins with the onset of regular uterine contractions and extends until the cervix dilates to a diameter about the size of the fetus' head (10 cm) (see figure 20.14). This stage takes approximately 24 hours, but it may be as short as a few minutes in some women who have had more than one child.
2. The **second stage** of labor lasts from the time of maximum cervical dilation until the time that the baby exits the vagina. This stage may last from 1 minute to up

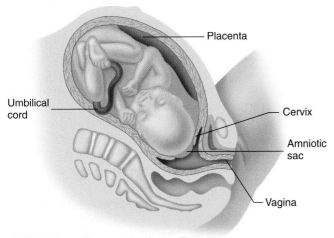

1. First stage. The cervix begins to dilate.

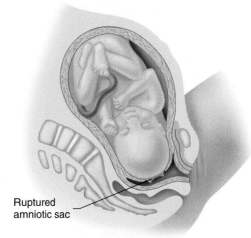

2. First stage. Further dilation of the cervix and rupture of the amniotic sac occur.

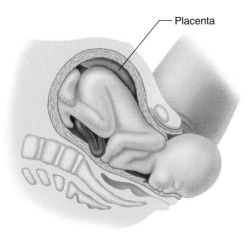

3. Second stage. The fetus is expelled from the uterus.

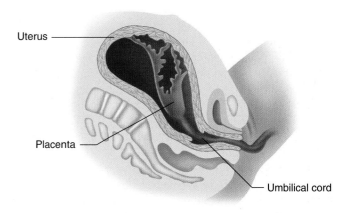

4. Third stage. The placenta is then expelled.

Process Figure 20.14 Parturition

to 1 hour. During this stage, contraction of the woman's abdominal muscles assists the uterine contractions.

3. The **third stage** of labor involves the expulsion of the placenta from the uterus. Contractions of the uterus cause the placenta to tear away from the wall of the uterus. Some bleeding from the wall of the uterus occurs because of the intimate contact between the placenta and the uterus. Bleeding is normally limited, however, because uterine smooth muscle contractions compress the blood vessels.

During the 4 or 5 weeks following parturition, the uterus becomes much smaller, but it remains somewhat larger than it was before pregnancy. A vaginal discharge composed of small amounts of blood and degenerating endometrium can persist for several days after parturition.

The precise signal that triggers parturition is unknown, but many factors that support it have been identified (figure 20.15). Before parturition, the progesterone concentration in the mother's blood has reached its highest level.

Progesterone has an inhibitory effect on uterine smooth muscle cells. Estrogen levels continually increase in the maternal circulation, however, exciting uterine smooth muscle. The inhibitory influence of progesterone on smooth muscle can be overcome by the stimulatory effect of estrogen near the end of pregnancy.

The fetus also plays a role in stimulating parturition. Stress on the fetus triggers the secretion of a releasing hormone from the fetal hypothalamus, which, in turn, causes adrenocorticotropic hormone (ACTH) release from the fetal anterior pituitary. ACTH stimulates the fetal adrenal gland to secrete hormones from the adrenal cortex that reduce progesterone secretion, increase estrogen secretion, and increase prostaglandin production by the placenta. Prostaglandins strongly stimulate uterine contractions.

During parturition, oxytocin is released from the mother's posterior pituitary. Stretching of the cervix produces action potentials that are sent to the hypothalamus and cause the release of oxytocin. Oxytocin also stimulates uterine contractions.

Ectopic Pregnancy

The term **ectopic** (ek-top′ik) means out of place, and an ectopic pregnancy is one that occurs outside the uterus. The most common site of an ectopic pregnancy, the uterine tube, produces a tubal pregnancy. The blastocyst may not have reached the uterine cavity by the time it is ready to implant and may implant into the wall of the uterine tube. The uterine tube cannot expand enough to accommodate the growing fetus, and, if the fetus is not removed, the tube eventually ruptures. The ruptured uterine tube causes life-threatening internal bleeding.

Miscarriage

It is estimated that as many as 50% of all zygotes are lost before delivery. Most are lost before implantation. Approximately 15% of all pregnancies end in **miscarriage** (mis-kar′ij), or **spontaneous abortion** (ă-bōr′shŭn), which results from the death or early delivery of the fetus. Before about 24 weeks post-LMP, the fetus is not viable outside the uterus. After 24 weeks, but before 37 weeks post-LMP, the infant is referred to as **premature.**

Although there is a higher incidence of birth defects among aborted fetuses, the vast majority of miscarried fetuses appear to be normal. Many factors can cause a miscarriage, many of which do not directly involve the fetus, and many of which are unknown. One common cause of miscarriage is improper implantation of the blastocyst in the uterus. In most cases, the blastocyst implants in the upper part of the uterus, but occasionally a blastocyst can implant near the opening into the cervical canal, a condition called **placenta previa** (prē′vē-ă). As the fetus grows and the uterus stretches, the previa placenta may tear away from the uterine wall, a condition called **placental abruption** (ab-rŭp′shŭn). When this occurs, the fetus often dies. The associated hemorrhaging can be life-threatening to the mother as well.

Pregnancy-Induced Hypertension

One reason the mother's weight is carefully monitored during pregnancy is that a sudden weight gain associated with edema and high blood pressure can be a sign of **pregnancy-induced hypertension,** or **toxemia** (tok-sē′mē-ă) of **pregnancy.** The cause of the disorder is unknown, but it can result in convulsions, kidney failure, and death of both the mother and the fetus.

Teratogens

Teratogens (ter′ă-tō-jenz) are drugs or other chemicals that can cross the placenta and cause birth defects in the developing embryo. The most famous teratogen is thalidomide (thă-lid′ō-mīd), an over-the-counter drug that was given to thousands of pregnant women in the early 1960s. The drug inhibited normal limb development and resulted in several thousand children being born, mostly in Germany and England, with severely reduced or even absent arms or legs. Thalidomide has been recently discovered effective in treating a variety of diseases, including leprosy, rheumatoid arthritis, tuberculosis, some cancers and some side effects of AIDS. So although thalidomide was withdrawn from the market in the early 1960s, it was approved by the FDA in 1998 for use in the United States. Its use is tightly controlled in an attempt to prevent embryonic exposure.

Fetal alcohol syndrome (FAS) is a major concern today. This syndrome, which consists of brain dysfunction, growth retardation, and facial peculiarities, is seen in children of women who consumed substantial amounts of alcohol while they were pregnant. It has been estimated that FAS may occur as often as 1 in 350 births and may account for as much as 33% of all mental retardation. **Fetal alcohol effect** includes brain dysfunction without the facial characteristics and may be three times as common as FAS.

Cocaine addiction can occur in babies whose mothers were cocaine users during pregnancy. A fetus can also suffer stroke-like symptoms if the mother ingests cocaine during the latter part of pregnancy.

Infections

Infections can occur in the mother or infant, or both. Maternal death associated with childbirth can result from infections. Cleanliness associated with childbirth procedures can reduce the rate of infection, and antibiotics have greatly reduced the number of fatal infections.

If a pregnant woman contracts **German measles** (mē′zlz), or **rubella** (rū-bel′ă), during pregnancy, the fetus may be severely affected. These effects can occur even if the mother suffers only a mild case of measles. Defects in the newborn can include visual and hearing defects, as well as mental retardation.

Neonatal gonorrheal ophthalmia (of-thal′mē-ă) is a severe form of conjunctivitis that is contracted by an infant as it passes through the birth canal of a mother with gonorrhea. This infection carries a high risk of blindness. The treatment of newborn eyes with silver nitrate or antibiotics is effective in preventing the disease.

Chlamydial conjunctivitis (kon-jŭnk-ti-vī′tis) is also contracted as an infant passes through the birth canal if the mother has a chlamydial infection. This infection is not affected by silver nitrate eyedrops; thus in many places newborns are treated with antibiotics against both chlamydia and gonorrhea.

If a woman has **genital herpes** (her′pēz) and has open lesions in the birth canal near the time of parturition, the baby can be removed by cesarean section to prevent the infection spreading to the baby.

Human immunodeficiency virus (HIV), the virus that causes **acquired immune deficiency syndrome (AIDS),** can cross the placenta and infect the fetus in utero, can infect the infant during parturition, or can infect the infant during breast-feeding. Approximately 30–50% of the infants born to HIV-infected mothers are infected. About 20% of those infected die of AIDS within the first 18 postnatal months. Azidothymidine (az′i-dō-thi′mi-dēn) (AZT) is a drug that inhibits HIV replication. The number of infants who contract AIDS from their mothers can be dramatically reduced by giving AZT to HIV-infected women and to their newborn infants.

Fetal Monitoring

Amniocentesis (am′nē-ō-sen-tē′sis) is the removal of amniotic fluid from the amniotic cavity (figure A). As the fetus develops, it expels molecules of various types, as well as living cells, into the amniotic fluid. These molecules and cells can be collected and analyzed. A number of normal conditions can be evaluated, and a number of metabolic disorders can be detected by analysis of the types of molecules that the fetus expels.

One of the molecules normally produced by the fetus and released into the amniotic fluid is **α-fetoprotein.** If the fetus has tissues exposed to the amniotic fluid that are normally covered by skin, an excessive amount of α-fetoprotein is lost into the amniotic fluid. For example, failure of the neural tube to close results in exposure of neural tissue to amniotic fluid, and failure of the abdominal wall to fully form exposes abdominal organs to amniotic fluid.

Some of the metabolic by-products from the fetus, such as α-fetoprotein can enter the maternal blood. In some cases, the by-products are processed and passed to the maternal urine. The levels of these fetal products can then be measured in the mother's blood or urine.

The fetus can be seen within the uterus by **ultrasound,** which uses sound waves

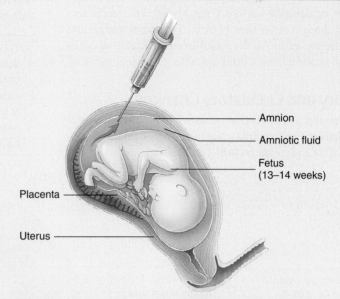

Figure A Removal of Amniotic Fluid for Amniocentesis

that are bounced off the fetus like sonar and then analyzed and enhanced by computer.

Fetal heart rate can be detected with an ultrasound stethoscope by the tenth week after fertilization and with a conventional stethoscope by 20 weeks. The normal fetal heart rate is 140 bpm (normal range is 110–160).

1. The fetal hypothalamus secretes a releasing hormone that stimulates adrenocorticotropic hormone (ACTH) secretion from the pituitary. The fetal pituitary secretes ACTH in greater amounts near parturition.

2. ACTH causes the fetal adrenal gland to secrete greater quantities of adrenal cortical steroids.

3. Adrenal cortical steroids travel in the umbilical blood to the placenta.

4. In the placenta the adrenal cortical steroids cause progesterone synthesis to level off and estrogen and prostaglandin synthesis to increase, making the uterus more irritable.

5. The stretching of the uterus produces action potentials that are transmitted to the brain through ascending pathways.

6. Action potentials stimulate the secretion of oxytocin from the mother's posterior pituitary.

7. Oxytocin causes the uterine smooth muscle to contract.

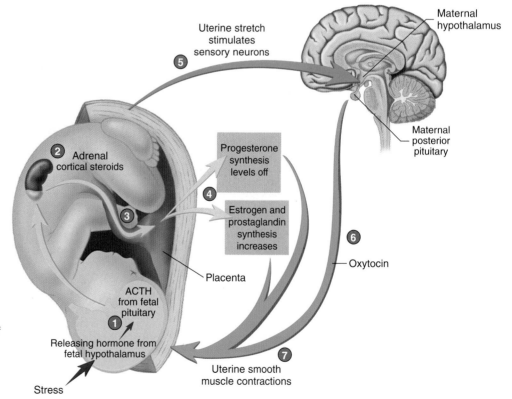

Process Figure 20.15   Factors That Influence the Process of Parturition

Although the precise control of parturition in humans is unknown, these changes appear to play a role.

The Newborn

The newborn, or **neonate** (nē′ō-nāt, newborn), experiences several dramatic changes at the time of birth. The major and earliest changes are the separation of the infant from the maternal circulation and the transfer from a fluid to a gaseous environment.

Respiratory and Circulatory Changes

The large, forced gasps of air that occur when the infant cries at the time of delivery help to inflate the lungs. The fetal lungs produce a substance called **surfactant** (ser-fak′tănt, surface active agent), which coats the inner surface of the alveoli, reduces surface tension in the lungs, and allows the newborn lungs to inflate (see chapter 15).

Surfactant and Respiratory Distress

Surfactant is not manufactured in the fetal lungs before about 6 months after fertilization. If a fetus is born before the lungs can produce surfactant, the surface tension inside the lungs is too great for the lungs to inflate. Under these conditions, the newborn may die of respiratory distress. Therefore, premature newborns are treated with bovine or synthetic surfactant.

1. Blood bypasses the lungs by flowing from the pulmonary trunk through the ductus arteriosus to the aorta.

2. Blood also bypasses the lungs by flowing from the right to the left atrium through the foramen ovale.

3. Blood bypasses the liver sinusoids by flowing through the ductus venosus.

4. Oxygen-rich blood is returned to the fetus from the placenta by the umbilical vein.

5. Oxygen-poor blood is carried from the fetus to the placenta through the umbilical arteries.

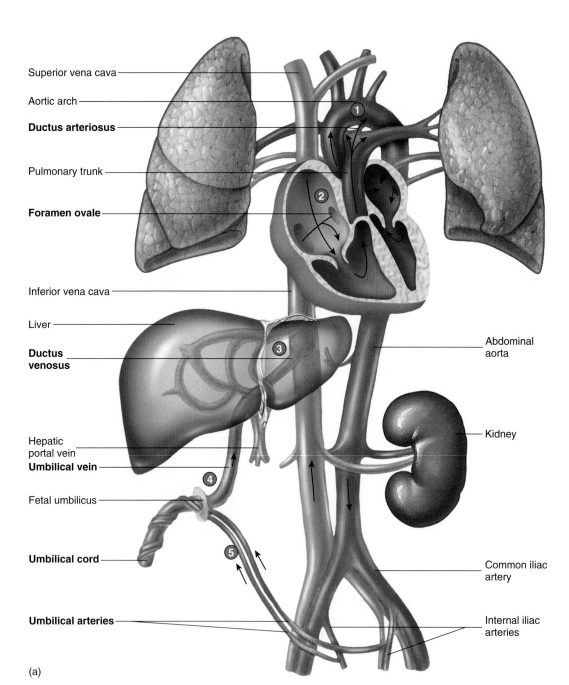

(a)

Process Figure 20.16   Circulation in the Fetus and Newborn

(*a*) Circulatory conditions in the fetus.

The initial inflation of the lungs causes important changes in the circulatory system (figure 20.16). Expansion of the lungs reduces the resistance to blood flow through the lungs, resulting in increased blood flow from the right ventricle of the heart through the pulmonary arteries. Consequently, an increased amount of blood flows from the right atrium to the right ventricle and into the pulmonary arteries, and less blood flows from the right atrium through the foramen ovale to the left atrium. The reduced resistance to blood flow through the lungs and the increasing volume of blood returning from the lungs through the pulmonary veins to the left atrium makes the pressure in the left atrium greater than that in the right atrium. This pressure difference forces blood against the interatrial septum, closing a flap of tissue that develops in that region over the foramen ovale. This action completes the separation of the heart into two pumps: the right side and the left side of the heart.

A short artery, the **ductus arteriosus** (ar-tēr′ē-ō-sŭs), connects the pulmonary trunk to the aorta. Before birth, the ductus arteriosus carries blood from the pulmonary trunk to the aorta, bypassing the fetal lungs. This artery closes off shortly after birth, forcing blood to flow through the lungs.

The fetal blood supply passes to the placenta through **umbilical** (ŭm-bil′i-kăl) **arteries,** which originate in the internal iliac arteries, and returns through an **umbilical vein.**

1. When air enters the lungs, blood is forced through the pulmonary arteries to the lungs. The ductus arteriosus closes (*gray*).

2. The foramen ovale closes and becomes the fossa ovalis. Blood can no longer flow from the right to the left atrium.

3. The ductus venosus degenerates and becomes the ligamentum venosum (*gray*).

4. The umbilical arteries and vein are cut. The umbilical vein becomes the round ligament of the liver (*gray*).

5. The umbilical arteries also degenerate (*gray*).

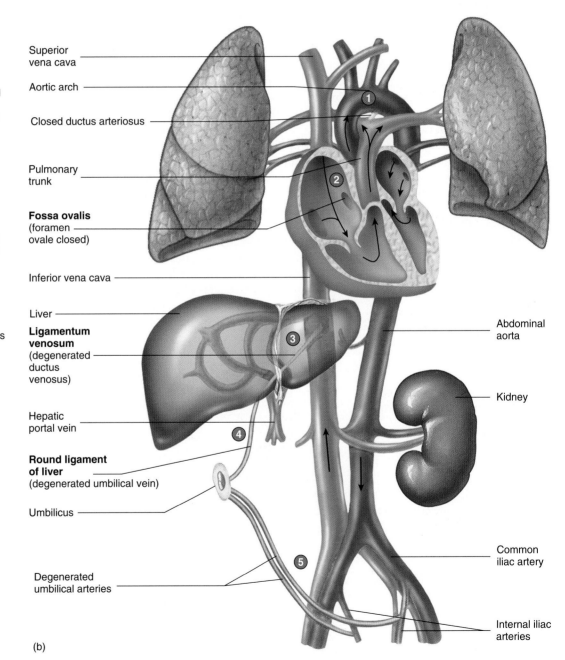

Process Figure 20.16   *continued*

(*b*) Circulatory changes that occur after birth.

The umbilical vein passes through the liver but bypasses the sinusoids of the liver by way of the **ductus venosus** (vē-nō′sŭs) and joins the inferior vena cava. When the umbilical cord is tied and cut, no more blood flows through the umbilical vein and arteries, and they degenerate. The remnant of the umbilical vein becomes the round ligament of the liver.

Digestive Changes

The fetus swallows amniotic fluid from time to time late in gestation. Shortly after birth, this swallowed fluid plus cells sloughed from the mucosal lining of the GI tract, mucus produced by intestinal mucous glands, and bile from the liver are eliminated as a greenish anal discharge called **meconium** (mē-kō′nē-ŭm, *mekon,* poppy).

After birth, the neonate is suddenly separated from its source of nutrients provided by the maternal circulation. Because of this separation and the shock of birth, the neonate usually loses 5–10% of its total body weight during the first few days of life. Although the digestive system of the fetus becomes somewhat functional late in development, it is still very immature in comparison with that of the adult, and only a limited number of food types can be digested.

The newborn digestive system is capable of digesting lactose (milk sugar) from the time of birth. The pancreatic secretions are sufficiently mature for a milk diet, but the digestive system only gradually develops the ability to digest more solid foods over the first year or two. New foods should therefore be introduced gradually during the first 2 years. It is also advised that only one new food at a time be introduced into the infant's diet so that, if an allergic reaction occurs, the cause is more easily determined.

Amylase secretion by the salivary glands and the pancreas remains low until after the first year. Lactase activity in the small intestine is high at birth but declines during infancy, although the levels still exceed those in adults. In many adults, lactase activity is lost, and an intolerance to milk can develop.

Lactation

Lactation (lak-tā′shŭn, *lactatio,* suckle) is the production of milk by the mammary glands (figure 20.17). It normally occurs in women following parturition and may continue for up to 2 or 3 years.

During pregnancy, the high concentration and continuous presence of estrogen and progesterone cause development of the duct system and the secretory units within the breast. Other hormones, including a prolactinlike hormone produced by the placenta, help support the development of the breasts. Also, additional adipose tissue is deposited; thus the size of the breasts increases substantially throughout pregnancy. Estrogen and progesterone prevent the secretory part of the breast from producing milk during pregnancy.

Blood levels of estrogen and progesterone fall dramatically after parturition. Once the placenta has been dislodged from the uterus, the source of these hormones is gone. After parturition, in the absence of estrogen and progesterone, prolactin produced by the anterior pituitary stimulates milk

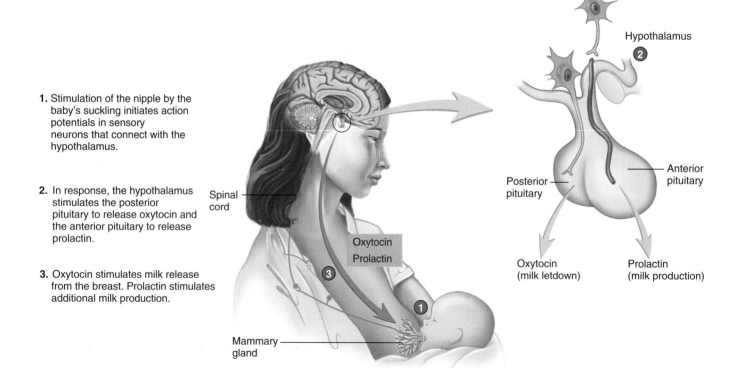

1. Stimulation of the nipple by the baby's suckling initiates action potentials in sensory neurons that connect with the hypothalamus.

2. In response, the hypothalamus stimulates the posterior pituitary to release oxytocin and the anterior pituitary to release prolactin.

3. Oxytocin stimulates milk release from the breast. Prolactin stimulates additional milk production.

Process Figure 20.17   Hormonal Control of Lactation

production. During suckling, sensory action potentials are sent from the nipple to the brain and result in the release of prolactin from the anterior pituitary (see figure 20.17). For the first few days following parturition, the mammary glands secrete **colostrum** (kō-los′trŭm), which contains little fat and less lactose than milk. Eventually, more nutritious milk is produced. Colostrum and milk provide nutrition and antibodies that help protect the nursing baby from infections.

Repeated stimulation of prolactin release makes nursing possible for several years. If nursing is stopped, prolactin release stops and within a few days the ability of the breast to respond to prolactin is lost, and milk production ceases.

At the time of nursing, milk contained in the alveoli and ducts of the breast is forced out of the breast by contractions of cells surrounding the alveoli. Suckling produces action potentials that are carried to the hypothalamus, where they cause the release of oxytocin from the posterior pituitary (see figure 20.17). Oxytocin stimulates cells surrounding the alveoli to contract. As a result, milk flows from the breasts, a process that is called **milk "letdown."** Higher brain centers can cause the release of oxytocin as a result of a conditioned reflex in response to such things as hearing an infant cry or thinking about breastfeeding.

P R E D I C T   4
While nursing her baby, a woman noticed that she developed cramps in her abdomen. Explain what was happening.

First Year Following Birth

A great number of changes occur in the infant from the time of birth until 1 year of age. The time when these changes occur may vary considerably from child to child, and the dates given are only rough estimates. The brain is still developing at this time, and much of what the infant can accomplish depends on the amount of brain development achieved. It is estimated that nearly all the adult number of neurons is present in the central nervous system at birth, but subsequent growth and maturation of the brain involves the addition of new neuroglial cells, myelin sheaths, and new connections between neurons, which may continue throughout life.

By 6 weeks, the baby is usually able to hold up his or her head when placed in a prone position and begins to smile in response to people or objects. At 3 months of age, the infant's limbs are exercised aimlessly. The arms and hands are controlled enough, however, that voluntary thumb-sucking can occur. The infant can follow a moving person with his eyes. At 4 months the infant begins to do push-ups; that is, raises himself by the arms. The infant can begin to grasp things placed in his hands, coo and gurgle, roll from back to side, listen quietly when hearing a person's voice or music, hold the head erect, and play with his hands. At 5 months the infant can usually laugh out loud, reach for objects, turn the head to follow an object, lift the head and shoulders, sit with support, and roll over. At 8 months the infant can recognize familiar people, sit up without support, and reach for specific objects. At 12 months the infant may pull herself to a standing position and may be able to walk without support. The child can pick up objects in her hands and examine them carefully. A 12-month-old child can understand much of what is said and may say several words.

Life Stages

The stages of life from fertilization to death are divided into three prenatal stages and five postnatal stages as follows: (1) the **germinal** (jer′mi-năl) **period**—fertilization to 14 days; (2) the **embryo**—14–56 days after fertilization; (3) the **fetus**—56 days after fertilization to birth; (4) **neonate**—birth to 1 month after birth; (5) **infant** (in′fănt)—1 month to 1 or 2 years after birth (the end of infancy is sometimes set at the time that the child begins to walk); (6) **child**—1 or 2 years old to puberty (about 11–14 years); (7) **adolescent** (ad-ō-les′ent)—teenage years, from puberty to 20 years old; (8) **adult**—20 years old to death. Adulthood is sometimes divided into three periods: **young adult,** 20–40 years old; **middle age,** 40–65 years old; and **older adult,** or **senior citizen,** 65 years old to death. Much of this designation is associated more with social norms than with physiology.

During childhood the individual grows in size and develops considerably. Many of the emotional characteristics that a person possesses throughout life are formed during early childhood.

Major physical and physiological changes occur during adolescence, and many of these changes also affect the emotions and behavior of the individual. Other emotional changes occur as the adolescent attempts to fit into an adult world. **Puberty** (pū′ber-tē) is the time when maturation of reproductive cells begins and when gonadal hormones are first secreted in substantial amounts. These hormones stimulate the development and maturation of secondary sex characteristics, such as enlargement of the female breasts and growth of body hair in both sexes. Puberty usually occurs in females who are about 11–13 years old and usually begins in males who are about 12–14 years old. The onset of puberty is usually accompanied by a growth spurt, followed by a period of slower growth. Full adult stature is usually achieved before age 17 or 18 in females and before age 19 or 20 in males.

Aging

Development of a new human being begins at fertilization, as does the process of aging. Cell division occurs at an extremely rapid rate during early development and then begins to slow as various cells become committed to specific functions within the body.

Many cells of the body continue to divide throughout life, replacing dead or damaged tissue; but other cells, such as the neurons in the brain, cease to divide once they have reached a certain number. Dead neurons tend not to be replaced. After the number of neurons reaches a peak (at approximately the time of birth), the number begins to decline. Neuronal loss is most rapid early in life and decreases to a slower, steady rate.

Young embryonic tissue is very flexible and elastic. It has relatively small amounts of collagen, and the collagen that is present is not highly cross-linked. Many of the collagen fibers produced during development, however, are permanent components of the individual. As the individual ages, more and more cross-links form between the collagen molecules, rendering the tissues more rigid and less flexible.

The tissues with the highest collagen content and the greatest dependency on collagen for their function are the most severely affected by the collagen cross-linking and tissue rigidity associated with aging. The lens of the eye is one of the first structures to exhibit age-related changes as a result of this increased rigidity. Vision of close objects becomes more difficult with advancing age until most middle-aged people require reading glasses (see chapter 9). Loss of elasticity also affects other tissues, including the joints, kidneys, lungs, and heart, and greatly reduces the functional ability of these organs.

As with nervous tissue, the number of skeletal muscle fibers declines with age. The strength of skeletal muscle reaches a peak between 20 to 35 years of life and usually declines steadily thereafter. Skeletal muscle strength depends primarily on the size of the muscle fibers, but the total number of fibers is probably also important to muscle strength. As most people age, both the number of fibers and the size of each tend to decline. The decline in muscle fiber size may be more related to a general decrease in activity, rather than to any specific age-related changes. Like the collagen of connective tissue, however, the macromolecules of skeletal muscle cells undergo biochemical changes during aging, rendering the muscle tissue less functional. A good exercise program can slow or even partly reverse the process of muscular aging.

Cardiac muscle cells also do not normally divide after birth. Age-related changes in cardiac muscle cell function probably contribute to a decline in cardiac function with advancing age. The heart loses elastic recoil ability and muscular contractility. As a result, total cardiac output declines, and less oxygen and nutrients reach the cells of the body supplied by the cardiovascular system. This decline in nutrition can be particularly harmful to cells that require high oxygen levels, such as neurons of the brain, and cells that are already compromised, such as cartilage cells of the joints, contributing to the general decline in these tissues.

Reduced cardiac function also can result in decreased blood flow to the kidneys, contributing to decreases in the filtration ability of the kidney. Degeneration of the connective tissues as a result of collagen cross-linking and other factors also decreases the filtration efficiency of the glomerular basement membrane.

Arteriosclerosis (ar-tēr′ē-ō-skler-ō′sis, hardening of the arteries) is a general hardening of the arteries affecting mainly arterioles. **Atherosclerosis** (ath′er-ō-skler-ō′sis, gruel + hardness) is the gradual formation of lipid-containing plaques in the arterial wall of large and medium-sized arteries (see chapter 13). These plaques then may become fibrotic and calcified, resulting in **arteriosclerosis.** Atherosclerosis interferes with normal blood flow and may result in a **thrombosis** (throm-bō′sis, clotting), which is a clot or plaque formed inside a vessel. An **embolus** (em′bō-lŭs, a patch) is a piece of the clot that has broken loose and floats through the circulation. An embolus can lodge in smaller arteries to cause heart attacks or strokes. Although atherosclerosis occurs to some extent in all middle-aged and elderly people and can occur even in certain young people, some people appear more at risk because of high blood cholesterol levels. This condition seems to have a hereditary

component, and blood tests are available to screen people for high blood cholesterol levels.

Many other organs, such as the liver, pancreas, stomach, and colon, undergo degenerative changes with age. The ingestion of harmful agents can accelerate such changes. Examples include the degenerative changes induced in the lungs (aside from lung cancer) by cigarette smoke and sclerotic changes in the liver as a result of excessive alcohol consumption.

In addition to the previously described changes associated with aging, cellular wear and tear, or cellular aging, is another factor that contributes to aging. Progressive damage from many sources such as radiation and toxic substances can result in irreversible cellular insults and may be one of the major factors leading to aging. Although the data are mixed, and their interpretation controversial, it has been suggested that ingestion of moderate amounts of vitamins E and C in combination may help slow aging due to cellular insult by stimulating cell repair. Vitamin C also stimulates collagen production and may slow the loss of tissue elasticity associated with aging collagen.

According to the **free radical theory of aging,** free radicals, which are atoms or molecules with an unpaired electron, can react with and alter the structure of molecules that are critical for normal cell function. Alteration of these molecules can result in cell dysfunction, cancer, or other types of cellular damage. Free radicals are produced as a normal part of metabolism and are introduced into the body from the environment through the air we breathe and the food we eat. The damage caused by free radicals may accumulate with age. Antioxidants, such as beta carotene (provitamin A), vitamin C, and vitamin E, can donate electrons to free radicals, without themselves becoming harmful. Thus, antioxidants may prevent the damage caused by the free radicals and may ward off age-related disorders, ranging from wrinkles to cancer. Again, the data are mixed and their interpretation controversial.

One characteristic of aging is an overall decrease in ATP production. This decline in ATP production is associated with a decrease in oxidative phosphorylation. This decline in oxidation has been shown in many cases to be associated with **mitochondrial DNA mutations.** Such mutations are often associated with Alzheimer's disease. There are also genes in the nuclear DNA associated with Alzheimer's.

Immune system changes may also be a major contributing factor to aging. The aging immune system loses its ability to respond to outside antigens and begins to be more sensitive to the body's own antigens. Immune responses to one's own tissues can result in the degeneration of the tissues and may be responsible for such things as arthritic joint disorders, chronic glomerular nephritis, and hyperthyroidism. In addition, T lymphocytes tend to lose their functional capacity with aging and cannot destroy abnormal cells as efficiently. This change may be one reason that certain types of cancer occur more frequently in older people.

One of the greatest disadvantages of aging is the increasing lack of ability to adjust to stress. Homeostasis is far more precarious in elderly people, and eventually some stress is encountered that is so great that the body's ability to recover is surpassed and death results.

Death

Death is usually not attributed to old age. Some other problem such as heart failure, renal failure, or stroke is usually listed as the cause of death.

Death was once defined as the loss of heartbeat and respiration. In recent years, however, more precise definitions of death have been developed, because both the heart and the lungs can be kept working artificially, such as during cardiopulmonary resuscitation, and the heart can even temporarily be replaced by an artificial device. Modern definitions of death are based on the permanent cessation of life functions and the cessation of integrated tissue and organ function. The most widely accepted indication of death in humans is whole brain death, which is manifested clinically by the absence of response to stimulation, the absence of natural respiration and heart function, absence of brainstem reflexes, and an isoelectric ("flat") electroencephalogram for at least 30 minutes, in the absence of known central nervous system poisoning or hypothermia. Some central nervous system poisons can cause a flat electroencephalogram, but the patient can be revived if the effects of the poison are eliminated. Hypothermia slows down all chemical reactions, including those involved in degenerative changes that begin at the time of death. As a result, a person suffering from hypothermia can exhibit no response to stimulation, exhibit no respiration or heart beat, and have a flat electroencephalogram for more than 30 minutes and still be revived.

Death

Neocortical (nē-ō-kōr'ti-kăl) **death** is a condition in which major portions of the cerebrum are no longer functioning. Under these conditions, the patient is comatose and incapable of responding to stimuli. Heartbeat and respiration still continue, however, because of some relatively unimpaired brainstem functions. Also, because some brainstem function still occurs, the electroencephalogram is not flat but exhibits some level of activity. Some state laws require that under these conditions the patient be kept alive by intravenous feeding and by other support equipment. The patient may have previously stated in a "living will" that, if neocortical death occurs and the patient cannot be returned to a reasonably normal level of function, no artificial support should be applied in an attempt to keep the patient's body alive.

Genetics

Genetics is the study of heredity; that is, those characteristics inherited by children from their parents. Many of a person's abilities, susceptibility to disease, and even life span are influenced by heredity.

Chromosomes

Deoxyribonucleic (dē-oks'ē-rī'bō-noo-klē'ic) **acid (DNA)** molecules and their associated proteins become visible as densely stained bodies, called **chromosomes** (krō'mō-sōmz, colored bodies), during cell division (see chapter 3). **Somatic** (sō-mat'ik) **cells,** all the cells of the body except the sex cells, contain 23 pairs

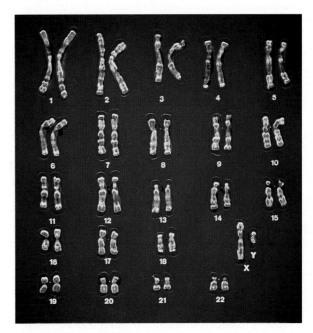

Figure 20.18   Human Karyotype

The 23 pairs of chromosomes in humans consist of 22 pairs of autosomal chromosomes (numbered 1–22) and 1 pair of sex chromosomes. The autosomal chromosome pairs are numbered in order from the largest to smallest. This karyotype is of a male and has an X and a Y sex chromosome. A female karyotype would have two X chromosomes.

of chromosomes, or 46 total chromosomes. The sex cells, or the **gametes** (gam'ētz) contain 23 unpaired chromosomes.

A **karyotype** (kar'ē-ō-tīp, *karyon,* nucleus + *typos,* model) is a display of the chromosomes in a somatic cell (figure 20.18; see chapter 3). There are 22 pairs of **autosomal** (aw-tō-sō'măl) **chromosomes,** which are all the chromosomes but the sex chromosomes, and there is one pair of **sex chromosomes.** A normal female has two **X chromosomes** (XX) in each somatic cell, whereas a normal male has one X and one **Y chromosome** (XY).

Sex Chromosome Abnormalities

There is a wide range of sex chromosome abnormalities. The presence of a Y chromosome makes a person male, and the absence of a Y chromosome makes a person female, regardless of the number of X chromosomes. The following combinations, therefore, are female: XO (Turner's syndrome), XX, XXX, or XXXX. Any combinations that include a Y are male: XY, XXY, XXXY, or XYY. A YO condition is lethal, because the genes on the X chromosome are necessary for survival. Secondary sexual characteristics are sometimes underdeveloped in both the XXX female and the XXY male (called Klinefelter's syndrome), and additional X chromosomes (XXXX or XXXY) are often associated with some degree of mental retardation.

Gametes are produced by **meiosis** (mī-ō'sis, a lessening) (see chapter 19). Meiosis is called a reduction division because the number of chromosomes in the gametes is half the number in the somatic cells. When a sperm cell and an oocyte fuse during fertilization, each contributes one-half of the chromosomes necessary to produce new somatic cells. Half of an individual's genetic makeup therefore comes from the father, and half comes from the mother.

During meiosis, the chromosomes are distributed in such a way that each gamete receives only one chromosome from each **homologous** (hŏ-mol′ō-gŭs) pair of chromosomes (see chapter 19). Homologous chromosomes contain the same compliment of genetic information. The inheritance of sex illustrates, in part, how chromosomes are distributed during gamete formation and fertilization. During meiosis and gamete formation, the pair of sex chromosomes separates so that each oocyte receives one of a homologous pair of X chromosomes, whereas each sperm cell receives either an X chromosome or a Y chromosome (figure 20.19). When a sperm cell fertilizes an oocyte to form a single cell, the sex of the individual is determined randomly. If the oocyte is fertilized by a sperm cell with a Y chromosome, a male results, but if the oocyte is fertilized by a sperm cell with an X chromosome, a female results. Estimating the probability of any given zygote being male or female is much like flipping a coin. When all the possible combinations of sperm cells with oocytes are considered, half the individuals should be female and the other half should be male.

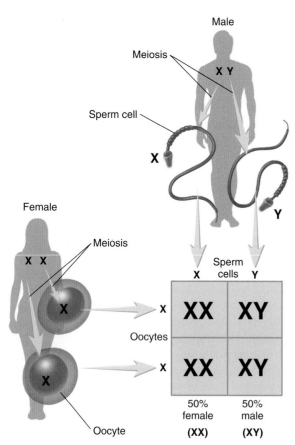

Figure 20.19   Inheritance of Sex

The female produces oocytes containing one X chromosome, whereas the male produces sperm cells with either an X or a Y chromosome. There are four possible combinations of an oocyte with a sperm cell, half of which produce females and half of which produce males.

Genes

The functional unit of heredity is the gene. Each **gene** consists of a certain portion of a DNA molecule but not necessarily a continuous stretch of DNA. Each chromosome contains thousands of genes. Both chromosomes of a given pair contain similar but not necessarily identical genes. Similar genes on homologous chromosomes are called **alleles** (ă-lēlz′, *allelon*, reciprocally). If the two allelic genes are identical, the person is **homozygous** (hō-mō-zī′gŭs) for the trait specified by that gene. If the two alleles are slightly different, the person is **heterozygous** (het′er-ō-zī′gŭs) for the trait. All the genes in one homologous set of 23 chromosomes in one individual, taken together is called the **genome.**

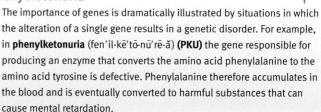

Phenylketonuria

The importance of genes is dramatically illustrated by situations in which the alteration of a single gene results in a genetic disorder. For example, in **phenylketonuria** (fen′il-kē′tō-nū′rē-ă) **(PKU)** the gene responsible for producing an enzyme that converts the amino acid phenylalanine to the amino acid tyrosine is defective. Phenylalanine therefore accumulates in the blood and is eventually converted to harmful substances that can cause mental retardation.

Through the processes of meiosis, gamete formation, and fertilization the distribution of genes received from each parent is essentially random. This random distribution is influenced by several factors, however. For example, all of the genes on a given chromosome are **linked;** that is, they tend to be inherited as a set rather than as individual genes because chromosomes, not individual genes, segregate during meiosis. Also during meiosis, however, homologous chromosomes may exchange genetic information by **crossing over.**

Furthermore, segregation errors can occur during meiosis. As the chromosomes separate during meiosis, the two members of a homologous pair may not segregate. As a result, one of the daughter cells receives both chromosome pairs and the other daughter cell receives none. When the gametes are fertilized, the resulting zygote has either 47 chromosomes or 45 chromosomes rather than the normal 46. When this condition results in an abnormal autosomal chromosome number, it is usually, but not always, lethal and is one reason for a high rate of early embryo loss. **Down's syndrome,** or **trisomy 21,** in which there are three #21 chromosomes is one of the few autosomal trisomies that is not lethal. In contrast, sex chromosome abnormalities (described in the section dealing with chromosomes) are not usually lethal.

Dominant and Recessive Genes

Most human genetic traits are recognized because defective alleles for those traits exist in the population. For example, on chromosome 11 is a gene that produces an enzyme necessary for the synthesis of melanin, the pigment responsible for skin, hair, and eye color (see chapter 5). An abnormal allele, however, produces a defective enzyme not capable of catalyzing one of the steps in melanin synthesis. If a given person inherits two defective alleles, a homozygous condition, the person is unable to produce melanin, and therefore lacks normal pigment. This condition is referred to as **albinism** (al′bi-nizm, *albo,* white).

For many genetic traits, the effects of one allele for that trait can mask the effect of another allele for that same trait. For example, a person who is heterozygous for the melanin-producing enzyme gene has one normal gene for melanin production and one defective gene for melanin production. One copy of the gene and its resulting enzymes are enough to make normal melanin. As a result, the person who is heterozygous produces melanin and appears normal. In this case, the allele that produces the normal enzyme is said to be **dominant,** whereas the allele producing the abnormal enzyme is **recessive.** By convention, dominant traits are indicated by uppercase letters and recessive traits are indicated by lowercase letters. In this example, the letter "*A*" designates the dominant normal, pigmented condition, and the letter "*a*" the recessive albino condition. It is important to note that not all dominant traits are the normal condition and that not all recessive traits are abnormal. Many examples exist of abnormal dominant traits.

The possible combinations of dominant and recessive alleles for normal melanin production versus albanism are *AA* (homozygous dominant, normal), *Aa* (heterozygous, normal), and *aa* (homozygous recessive, albino). The alleles a person has for a given trait are called the **genotype** (jen′ō-tīp). The person's appearance is called the **phenotype** (fē′nō-tīp). A person with the genotype *AA* or *Aa* would have the phenotype of normal pigmentation, whereas a person with the genotype *aa* would have the phenotype of albinism. Note that the recessive trait is expressed only when no allele for the dominant trait is present.

PREDICT 5

Polydactyly (pol-ē-dak′ti-lē) is a condition in which a person has extra fingers or toes. Given that polydactyly is a dominant trait, list all the possible genotypes and phenotypes for polydactyly. Use the letters "D" and "d" for the genotypes.

The inheritance of dominant and recessive traits can be determined if the genotypes of the parents are known. For example, if an albino person (*aa*) mates with a heterozygous normal person (*Aa*), the probability is that half of the children will be albino (*aa*), and half will be normal heterozygous carriers (*Aa*). If two carriers (*Aa*) mate, the probability is that 1/4 will be homozygous dominant (*AA*), 1/4 will be homozygous recessive (*aa*), and 1/2 will be heterozygous (*Aa*). Such a probability can be easily determined by the use of a table called a **Punnett square** (figure 20.20). A **carrier** is a heterozygous person with an abnormal recessive gene, but with a normal phenotype because they also have a normal dominant allele for that gene.

Sex-Linked Traits

Traits affected by genes on the sex chromosomes are called **sex-linked traits.** Most sex-linked traits are **X-linked;** that is, they are on the X chromosome, whereas, only a few **Y-linked** traits exist, largely because the Y chromosome is very small. An example of an X-linked trait is **hemophilia A** (classic hemophilia) in which the person is unable to produce one of the clotting factors (see chapter 11). Consequently, clotting is impaired and persistent bleeding can occur either spontaneously or as a result of an injury. Hemophilia A is a recessive trait and the allele for the trait is located on the X chromosome. The possible

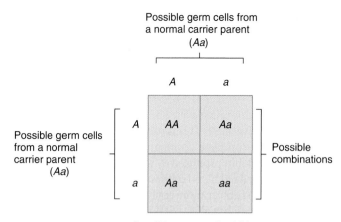

Possible outcome in children:
$\frac{1}{4}$ *AA* (normal) : $\frac{1}{2}$ *Aa* (normal carrier) : $\frac{1}{4}$ *aa* (albino)

Figure 20.20   **Inheritance of a Recessive Trait: Albinism**

A represents the normal pigmented condition, and *a* represents the recessive unpigmented condition. The figure shows a Punnett square of a mating between two normal carriers.

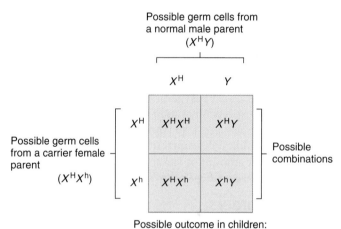

Possible outcome in children:
$\frac{1}{4}$ X^{H}X^H (normal female) : $\frac{1}{4}$ X^{H}X^h (carrier female) :
$\frac{1}{4}$ X^{H}Y (normal male) : $\frac{1}{4}$ X^{h}Y (male with hemophilia)

Figure 20.21   **Inheritance of an X-Linked Trait: Hemophilia**

X^H represents the normal X chromosome condition with all clotting factors, and X^h represents the X chromosome lacking an allele for one clotting factor. The figure shows a Punnett square of a mating between a normal male and a normal carrier female.

genotypes and phenotypes (in parentheses) are therefore X^{H}X^H (normal homozygous female), X^{H}X^h (normal heterozygous female), X^{h}X^h (hemophiliac homozygous female), X^{H}Y (normal male), and X^{h}Y (hemophiliac male). Note that a female must have both recessive alleles to exhibit hemophilia, whereas a male, because he has only one X chromosome, has hemophilia if he has one recessive allele. An example of the inheritance of hemophilia is illustrated in figure 20.21. If a woman who is a carrier for hemophilia mates with a man who does not have hemophilia, none of their daughters but half of their sons will have hemophilia. Half of their daughters will be carriers.

P R E D I C T **6**
Predict the probability of a girl with Turner's syndrome (refer to the box on sex chromosome abnormalities on p. 595) having hemophilia if her mother is a carrier for hemophilia.

Other Types of Gene Expression

In some cases the dominant allele does not completely mask the effects of the recessive allele. This is called **incomplete dominance.** An example of incomplete dominance is **sickle-cell anemia,** in which the hemoglobin produced by the gene is abnormal. The result is sickle-shaped red blood cells, which are likely to stick in capillaries and tend to rupture more easily than normal red blood cells. The normal hemoglobin allele (S) is dominant over the sickle-cell allele (s). A normal person (SS) has normal hemoglobin, and person with sickle cell anemia (ss) has abnormal hemoglobin. A person who is heterozygous (Ss) has half normal hemoglobin and half abnormal hemoglobin, and usually has only a few sickle-shaped red blood cells. This condition is called **sickle-cell trait.**

In another type of gene expression, called **codominance** (kō-dom′i-năns), two alleles can combine to produce an effect without either of them being dominant or recessive. For example, a person with type AB blood has A antigens and B antigens on the surface of his red blood cells (see chapter 11). The antigens result from a gene that causes the production of the A antigen and a different gene that causes the production of the B antigen, and neither gene is dominant or recessive in relation to each other.

Many traits, called **polygenic** (pol-ē-jen′ik) **traits,** are determined by the expression of multiple genes on different chromosomes. Examples are a person's height, intelligence, eye color, and skin color. Polygenic traits typically are characterized by having a great amount of variability. For example, there are many different shades of eye color and skin color (figure 20.22).

Genetic Disorders

Genetic disorders are caused by abnormalities in a person's genetic makeup, that is, in his or her DNA. They may involve a single gene or an entire chromosome. Some genetic disorders result from a **mutation** (mū-tā′shŭn, to change), a change in a gene that usually involves a change in the nucleotides composing the DNA (see chapter 2). Mutations occur by chance or can be caused by chemicals, radiation, or viruses. Once a mutation has occurred, the abnormal trait can be passed from one generation to the next.

Cancer is a tumor resulting from uncontrolled cell divisions. **Oncogenes** (ong′kō-jēnz) are genes associated with cancer. Many oncogenes are actually control genes involved in regulating cell division and differentiation in the embryo and fetus. A change in an oncogene or in the regulation of an oncogene can result in uncontrolled cell division and the development of cancerous tumors. The normal control of oncogenes involves other genes, called **tumor suppressor genes.** Cancer can occur when a mutation activates an oncogene or inactivates a tumor suppressor gene. An accumulation of several mutations is necessary for cancer to occur. It is believed that exposure to ionizing radiation or to certain chemicals called **carcinogens** (kar-sin′ō-jenz) can induce such mutations and thereby initiate the development of cancer. For example, chemicals in cigarette smoke are known to cause lung cancer.

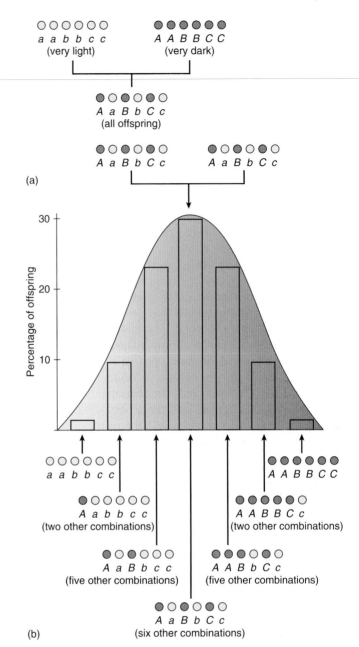

Figure 20.22 Inheritance of a Polygenic Trait: Skin Color

In this example, three genes for skin color are shown. The dominant alleles (A, B, C,) each of which contributes one "unit of dark color" to the offspring (indicated by a *dark dot*), are incompletely dominant over the recessive alleles (a, b, c), each of which contributes one "unit of light color" to the offspring (indicated by a *light dot*). (a) A mating between a very light-skinned person (aabbcc) and a very dark-skinned person (AABBCC) is shown. All the offspring are of intermediate color. (b) A mating between two people of intermediate skin color (AaBbCc). The possible offspring skin color falls within a normal distribution in which a very low percentage (less than 2%) are either very light or very dark, and most of the offspring are of intermediate color.

A change in cells that results in cancer is not usually inherited. Nonetheless, there may be a genetic basis that allows cancer development, especially under the right environmental conditions. In this sense, the inheritance of cancer and other abnormalities has been described as **genetic susceptibility,** or **genetic predisposition.** For example, if a woman's close relatives, such as her mother

The **human genome** (jē'nōm) is all of the genes found in one homologous set of human chromosomes. It is estimated that humans have 26,500–30,000 genes. A **genomic** (jĕ-nom'ik or je-nōm'ik) **map** is a description of the DNA nucleotide sequences of the genes and their locations on the chromosomes (figure B). The **Human Genome Project** was completed in February 2003.

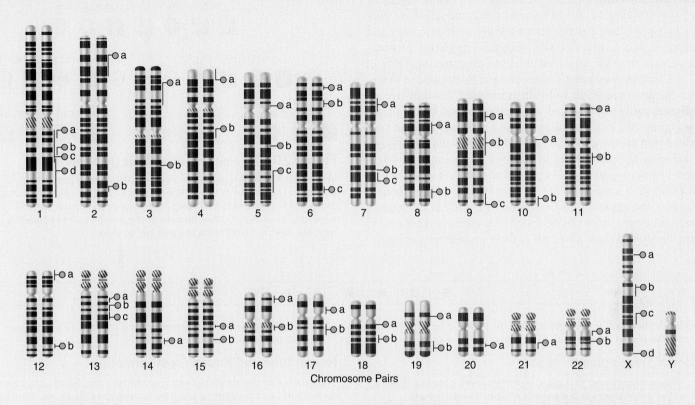

Chromosome Pairs

1. a. Gaucher disease
   b. Prostate cancer
   c. Glaucoma
   d. Alzheimer disease*
2. a. Familial colon cancer*
   b. Waardenburg syndrome
3. a. Lung cancer
   b. Retinitis pigmentosa*
4. a. Huntington chorea
   b. Parkinson disease
5. a. Cockayne syndrome
   b. Familial polyposis of the colon
   c. Asthma
6. a. Spinocerebellar ataxia
   b. Diabetes*
   c. Epilepsy*

7. a. Diabetes*
   b. Cystic fibrosis
   c. Obesity*
8. a. Werner syndrome
   b. Burkitt lymphoma
9. a. Malignant melanoma
   b. Friedreich ataxia
   c. Tuberous sclerosis
10. a. Multiple endocrine neoplasia, type 2
    b. Gyrate atrophy
11. a. Sickle-cell anemia
    b. Multiple endocrine neoplasia
12. a. Zellweger syndrome
    b. Phenylketonuria (PKU)

13. a. Breast cancer*
    b. Retinoblastoma
    c. Wilson disease
14. a. Alzheimer disease*
15. a. Marfan syndrome
    b. Tay-Sachs disease
16. a. Polycystic kidney disease
    b. Crohn disease*
17. a. Peroneal muscular atrophy
    b. Breast cancer*
18. a. Amyloidosis
    b. Pancreatic cancer*
19. a. Familial hypercholesterolemia
    b. Myotonic dystrophy

20. a. Severe combined immunodeficiency
21. a. Amyotrophic lateral sclerosis*
22. a. DiGeorge syndrome
    b. Neurofibromatosis, type 2
X a. Duchenne dystrophy
    b. Menkes syndrome
    c. X-linked severe combined immunodeficiency
    d. Factor VIII deficiency (hemophilia A)

*Gene responsible for only some cases.

Figure B The Human Genomic Map

Representative genetic defects mapped to date. The *green circles* indicate the location of the genes listed for each chromosome.

or sister, have breast cancer, she has a greater than average risk of developing it herself. Similar genetic susceptibilities have been found for diabetes mellitus, schizophrenia, and other disorders.

Genetic Counseling

Genetic counseling includes predicting the possible results of matings involving carriers of harmful genes and talking to parents or prospective parents about the possible outcomes and treatments of a genetic disorder. With this knowledge, prospective parents can make informed decisions about having children.

A first step in genetic counseling is to attempt to determine the genotype of the individuals involved. A family tree, or **pedigree,** provides historical information about family members (figure 20.23), for example in the case of a dominant trait such as Huntington chorea, a neurological disorder. Sometimes by knowing the phenotypes of relatives it is possible to determine a person's genotype. As part of the process of collecting information, a karyotype can be prepared. For some genetic disorders, the amount of a given substance, such as an enzyme, produced by a carrier can be tested. For example, carriers for cystic fibrosis produce more salt in their sweat than is normal.

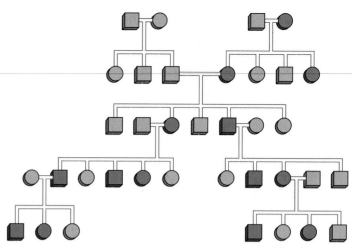

Figure 20.23 Pedigree of a Simple Dominant Trait
Males are indicated by squares, females by circles. Affected people are indicated in *purple*. The horizontal line between symbols represents a mating. The symbols connected to the mating line by vertical and horizontal lines represent the children resulting from the mating in order of birth from left to right. Matings not related to the pedigree are not shown.

S U M M A R Y

Prenatal development is an important part of an individual's life. About 7 of every 100 people are born with some type of birth defect.

Prenatal Development (p. 574)

1. The developing human is called an embryo during the first 8 weeks of the prenatal period and an embryo from 8 weeks to birth.
2. Developmental age is 14 days less than clinical age.

Fertilization

Fertilization, the union of the oocyte and sperm cell, results in a zygote.

Early Cell Division

The zygote undergoes divisions until it becomes a mass of cells.

Blastocyst

1. The embryonic mass develops a cavity and is known as the blastocyst.
2. The blastocyst consists of a trophoblast and an inner cell mass, where the embryo forms.

Implantation of the Blastocyst and Development of the Placenta

1. The blastocyst implants into the uterus about 7 days after fertilization.
2. The embryonic portion of the placenta is derived from the trophoblast of the blastocyst.

Maternal Changes

1. Human chorionic gonadotropin levels are high in early pregnancy then decline to low levels.
2. Progesterone and estrogen levels are low in early pregnancy but increase to high levels late in pregnancy.

Formation of the Germ Layers

1. The embryo forms around the primitive streak, which forms about 14 days after fertilization.

2. All tissues of the body are derived from three primary germ layers: ectoderm, mesoderm, and endoderm.

Neural Tube and Neural Crest Formation

The nervous system develops from a neural tube that forms in the ectodermal surface of the embryo and from neural crest cells derived from the developing neural tube.

Formation of the General Body Structure

1. The limbs develop as outgrowths called limb buds.
2. The face develops by the fusion of five tissue masses.

Development of the Organ Systems

1. The GI tract develops as the developing embryo closes off part of the yolk sac.
2. The heart develops as two tubes fuse into a single tube that develops septa to form four chambers.
3. The kidneys and reproductive system are closely related in their development.

Growth of the Fetus

The fetus increases 15-fold in length and 1400-fold in weight.

Parturition (p. 586)

1. Uterine contractions force the baby out of the uterus during labor.
2. Increased estrogen, decreased progesterone levels, and secretions from the fetal adrenal cortex initiate parturition.
3. Stretching of the uterus stimulates oxytocin secretion, which stimulates uterine contractions.

The Newborn (p. 590)

Respiratory and Circulatory Changes

1. Inflation of the lungs at birth results in closure of the foramen ovale and the ductus arteriosus.
2. When the umbilical cord is cut, blood no longer flows through the umbilical vessels.

Digestive Changes

The newborn digestive system only gradually develops the ability to digest a variety of foods.

Lactation (p. 592)

1. Estrogen and progesterone help stimulate the growth of the breasts during pregnancy.
2. Suckling stimulates prolactin and oxytocin synthesis. Prolactin stimulates milk production, and oxytocin stimulates milk "letdown."

First Year Following Birth (p. 593)

Many important changes occur during the first year after birth. Many of these changes are linked to continued development of the brain.

Life Stages (p. 593)

The eight stages of life are: germinal period (fertilization to 14 days), embryo (14–56 days after fertilization), fetus (56 days after fertilization to birth), neonate (birth–1 month), infant (1 month–1 or 2 years), child (1 or 2 years–puberty), adolescent (puberty–20 years), adult (20 years–death).

Aging (p. 593)

1. Aging occurs as irreplaceable cells wear out and the tissue becomes more brittle and less able to repair damage.
2. Atherosclerosis is the deposit of lipids in the arteries. Arteriosclerosis is hardening of the arteries.

Death (p. 595)

Death is defined as the absence of brain response to stimulation, the absence of natural respiration and heart function, and a flat electroencephalogram for 30 minutes.

Genetics (p. 595)

Chromosomes

1. Humans have 46 chromosomes in 23 pairs.
2. Males have the sex chromosomes XY and females have XX.
3. During gamete formation, the chromosomes of each pair of chromosomes separate; therefore half of a person's genetic makeup comes from the father and half from the mother.

Genes

1. A gene is a portion of a DNA molecule. Genes determine the proteins in a cell.
2. Genes are paired (located on the paired chromosomes).
3. Dominant genes mask the effects of recessive genes.
4. Sex-linked traits result from genes on the sex chromosomes.
5. In incomplete dominance, the heterozygote expresses a trait that is intermediate between the two homozygous traits.
6. In codominance, neither gene is dominant or recessive, but both are fully expressed.
7. Polygenic traits result from the expression of multiple genes.

Genetic Disorders

1. A mutation is a change in the DNA.
2. Some genetic disorders result from an abnormal distribution of chromosomes during gamete formation.
3. Oncogenes are genes associated with cancer.
4. Genetic predisposition makes it more likely a person will develop a disorder.

Genetic Counseling

1. A pedigree (family history) can be used to determine the risk of having children with a genetic disorder.
2. Specific chemical tests or examination of a person's karyotype can be used to determine a person's genotype.

R E V I E W   A N D   C O M P R E H E N S I O N

1. Define clinical age and developmental age, and state the difference between the two in number of days. Define embryo and fetus.
2. What are the events during the first week after fertilization? Define zygote, morula, and blastocyst.
3. How does the placenta develop?
4. Describe the formation of the germ layers and the role of the primitive streak.
5. How are the neural tube and neural crest cells formed? What do they become?
6. Describe the formation of the limbs and face.
7. Describe the formation of the digestive tract.
8. How does the single heart tube become four-chambered?
9. What major events distinguish embryonic and fetal development?
10. Describe the hormonal changes that take place before and during parturition.
11. What changes take place in the newborn's circulatory system and digestive system shortly after birth?
12. What hormones are involved in preparing the breasts for lactation? What hormones are involved in milk production and milk "letdown"?
13. Describe the changes in motor and language skills that take place during the first year of life.
14. List the different life stages.
15. How does the loss of cells that are not replaced affect the aging process? Give examples.

16. How does the loss of tissue elasticity affect the aging process? Give examples.
17. How does aging affect the immune system?
18. Define death.
19. Give the number and type of chromosomes in the karyotype of a human somatic cell. How do the chromosomes of a male and female differ from each other?
20. How do the chromosomes in somatic cells and gametes differ from each other?
21. What is a gene, and how are genes responsible for the structure and function of cells?
22. Define homozygous dominant, heterozygous, and homozygous recessive.
23. What is a sex-linked trait? Give an example.
24. How does sickle-cell anemia, type AB blood, and a person's height result from the expression of genes?
25. What is a mutation?
26. What is the cause of the genetic disorder called Down syndrome?
27. What are oncogenes?
28. What is genetic susceptibility?
29. How are pedigrees, karyotypes, and chemical tests used in genetic counseling?

CRITICAL THINKING

1. A woman is told by her physician that her pregnancy has progressed 44 days since her last menstrual period (LMP). How many days has the embryo been developing, and what developmental events are occurring?

2. A high fever can prevent neural tube closure. If a woman has a high fever about 35 to 45 days post-LMP, what kinds of birth defects may be seen in the developing embryo?

3. A drug that would stop the production of milk in the breast after a few days probably has which effect?
   a. Inhibits prolactin secretion
   b. Inhibits oxytocin secretion
   c. Increases estrogen secretion
   d. Increases progesterone secretion
   e. Increases prolactin secretion

4. Dimpled cheeks are inherited as a dominant trait. Is it possible for two parents, each of whom has dimpled cheeks, to have a child without dimpled cheeks? Explain.

5. The ability to roll the tongue to form a "tube" results from a dominant gene. Suppose that a woman and her son can both roll their tongues, but her husband cannot. Is it possible to determine if the husband is the father of her son?

6. A woman who does not have hemophilia marries a man who has the disorder. Determine the genotype of both parents if half of their children have hemophilia.

Answers in Appendix D

Visit this textbook's website at www.mhhe.com/seeleyess6 for practice quizzes, animations, interactive learning exercises, and other study tools.

McGraw-Hill offers a study CD that features interactive cadaver dissection. *Anatomy & Physiology Revealed* includes cadaver photos that allow you to peel away layers of the human body to reveal structures beneath the surface. This program also includes animations, radiologic imaging, audio pronunciations, and practice quizzing.

Appendix A

Table of Measurements

Table A.1 — Table of Measurements

Unit	Metric Equivalent	Symbol	U.S. Equivalent
Measures of Length			
1 kilometer	= 1000 meters	km	0.62137 mile
1 meter	= 10 decimeters or 100 centimeters	m	39.37 inches
1 decimeter	= 10 centimeters	dm	3.937 inches
1 centimeter	= 10 millimeters	cm	0.3937 inch
1 millimeter	= 1000 micrometers	mm	
1 micrometer*	= 1/1000 millimeter or 1000 nanometers	µm	
1 nanometer*	= 10 angstroms or 1000 picometers	nm	No U.S. equivalent
1 angstrom	= 1/10,000,000 millimeter	Å	
1 picometer	= 1/1,000,000,000 millimeter	pm	
Measures of Volume			
1 cubic meter	= 1000 cubic decimeters	m^3	1.308 cubic yards
1 cubic decimeter	= 1000 cubic centimeters	dm^3	0.03531 cubic foot
1 cubic centimeter	= 1000 cubic millimeters or 1 milliliter	cm^3 (cc)	0.06102 cubic inch
Measures of Capacity			
1 kiloliter	= 1000 liters	kL	264.18 gallons
1 liter	= 10 deciliters	L	1.0567 quarts
1 deciliter	= 100 milliliters	dL	0.4227 cup
1 milliliter	= volume of 1 gram of water at standard temperature and pressure	mL	0.3381 ounces
1 microliter	= $1mm^3$ (cubic millimeter) or 10^{-6} L	µL	
Measures of Mass			
1 kilogram	= 1000 grams	kg	2.2046 pounds
1 gram	= 100 centigrams or 1000 milligrams	g	0.0353 ounces
1 centigram	= 10 milligrams	cg	0.1543 grain
1 milligram	= 1/1000 gram	mg	

*Note that a micrometer was formerly called a micron (µm), and a nanometer was formerly called a millimicron (mµ).

Table A.2 — Comparative Temperature Scales

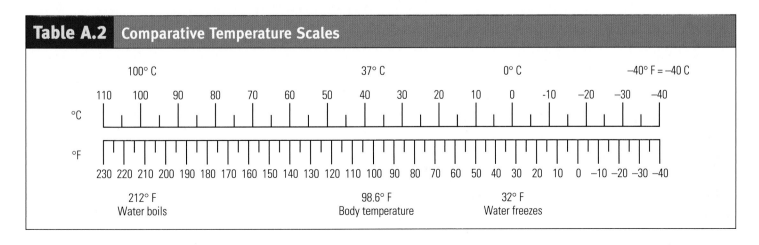

Appendix B

Some Reference Laboratory Values

Table B.1	Blood, Plasma, or Serum Values	
Test	**Normal Values**	**Clinical Significance**
Acetoacetate plus acetone	0.32–2 mg/100 mL	Values increase in diabetic acidosis, fasting, high-fat diet, and toxemia of pregnancy
Ammonia	9–33 µmol/L (micromol/L)	Values decrease with proteinuria and as a result of severe burns and increase in multiple myeloma
Amylase	4–25 U/mL*	Values increase in acute pancreatitis, intestinal obstruction, and mumps; values decrease in cirrhosis of the liver, toxemia of pregnancy, and chronic pancreatitis
Barbiturate	0	Coma level: phenobarbital, approximately 10 mg/100 mL; most other drugs, 1–3 mg/100 mL
Bilirubin	0.4 mg/100 mL	Values increase in conditions causing red blood cell destruction or biliary obstruction or liver inflammation
Blood volume	8.5–9% of body weight in kilograms	
Calcium	8.5–10.5 mg/100 mL	Values increase in hyperparathyroidism, vitamin D hypervitaminosis; values decrease in hypoparathyroidism, malnutrition, and severe diarrhea
Carbon dioxide content	24–30 mEq/L 20–26 mEq/L in infants (as HCO_3^-)	Values increase in respiratory diseases, vomiting, and intestinal obstruction; they decrease in acidosis, nephritis, and diarrhea
Carbon monoxide	0	Symptoms with over 20% saturation
Chloride	100–106 mEq/L	Values increase in Cushing's syndrome, nephritis, and hyperventilation; they decrease in diabetic acidosis, Addison's disease, and diarrhea and after severe burns
Creatine phosphokinase (CPK)	Female 5–35 mU/mL Male 5–55 mU/mL	Values increase in myocardial infarction and skeletal muscle diseases such as muscular dystrophy
Creatinine	0.6–1.5 mg/100 mL	Values increase in certain kidney diseases
Ethanol	0	0.3–0.4%, marked intoxication 0.4–0.5%, alcoholic stupor 0.5% or over, alcoholic coma
Glucose	Fasting 70–110 mg/100 mL	Values increase in diabetes mellitus, liver diseases, nephritis, hyperthyroidism, and pregnancy; they decrease in hyperinsulinsim, hypothyroidism, and Addison's disease
Iron	50–150 µg/100 mL	Values increase in various anemias and liver disease; they decrease in iron-deficiency anemia
Lactic acid	0.6–1.8 mEq/L	Values increase with muscular activity and in congestive heart failure, severe hemorrhage, shock, and anaerobic exercise
Lactic dehydrogenase	60–120 U/mL	Values increase in pernicious anemia, myocardial infarction, liver diseases, acute leukemia, and widespread carcinoma

*A unit (U) is the quantity of a substance that has a physiological effect.

Continued on next page

Table B.1 — Blood, Plasma, or Serum Values (continued)

Test	Normal Values	Clinical Significance
Lipids	Cholesterol 120–220 mg/100 mL; Cholesterol esters 60–75% of cholesterol; Phospholipids 9–16 mg/100 mL as lipid phosphorus; Total fatty acids 190–420 mg/100 mL; Total lipids 450–1000 mg/100 mL; Triglycerides 40–150 mg/100 mL	Increased values for cholesterol and triglycerides are connected with increased risk of cardiovascular disease, such as heart attack and stroke
Lithium	Toxic levels 2 mEq/L	
Osmolality	285–295 mOsm/kg water	
Oxygen saturation (arterial) (see Po$_2$)	96–100%	
Pco$_2$	35–43 mm Hg	Values decrease in acidosis, nephritis, and diarrhea; they increase in respiratory diseases, intestinal obstruction, and vomiting
pH	7.35–7.45	Values decrease as a result of hypoventilation, severe diarrhea, Addison's disease, and diabetic acidosis; values increase due to hyperventilation, Cushing's syndrome, and vomiting
Po$_2$	75–100 mm Hg (breathing room air)	Values increase in polycythemia and decrease in anemia and obstructive pulmonary diseases
Phosphatase (acid)	Male: total 0.13–0.63 U/mL Female: total 0.01–0.56 U/mL	Values increase in cancer of the prostate gland, hyperparathyroidism, some liver diseases, myocardial infarction, and pulmonary embolism
Phosphatase (alkaline)	13–39 IU/L* (infants and adolescents up to 104 IU/L)	Values increase in hyperparathyroidism, some liver diseases, and pregnancy
Phosphorus (inorganic)	3–4.5 mg/100 ml (infants in first year up to 6 mg/100 mL)	Values increase in hypoparathyroidism, acromegaly, vitamin D hypervitaminosis, and kidney diseases; they decrease in hyperparathyroidism
Potassium	3.5–5 mEq/100 mL	
Protein	Total 6–8.4 g/100 mL Albumin 3.5–5 g/100 mL Globulin 2.3–3.5 g/100 mL	Total protein values increase in severe dehydration and shock; they decrease in severe malnutrition and hemorrhage
Salicylate	0	
Therapeutic		20–25 mg/100 mL
Toxic		Over 30 mg/100 mL Over 20 mg/100 mL after age 60
Sodium	135–145 mEq/L	Values increase in nephritis and severe dehydration; they decrease in Addison's disease, myxedema, kidney disease, and diarrhea
Sulfonamide	0	
Therapeutic		5–15 mg/100 mL
Urea nitrogen	8–25 mg/100 mL	Values increase in response to increased dietary protein intake; values decrease in impaired renal function
Uric acid	3–7 mg/100 mL	Values increase in gout and toxemia of pregnancy and as a result of tissue damage

*A unit (U) is the quantity of a substance that has a physiological effect.

Table B.2 Blood Count Values

Test	Normal Values	Clinical Significance
Clotting (coagulation) time	5–10 min	Values increase in afibrinogenemia and hyperheparinemia, severe liver damage
Fetal hemoglobin	Newborns: 60–90% Before age 2: 0–4% Adults: 0–2%	Values increase in thalassemia, sickle-cell anemia, and leakage of fetal blood into maternal bloodstream during pregnancy
Hemoglobin	Male: 14–16.5 g/100 mL Female: 12–15 g/100 mL Newborn: 14–20 g/100 mL	Values decrease in anemia, hyperthyroidism, cirrhosis of the liver, and severe hemorrhage; values increase in polycythemia, congestive heart failure, obstructive pulmonary disease, high altitudes
Hematocrit	Male: 40–54% Female: 38–47%	Values increase in polycythemia, severe dehydration, and shock; values decrease in anemia, leukemia, cirrhosis, and hyperthyroidism
Ketone bodies	0.3–2 mg/100 mL Toxic level: 20 mg/100 mL	Values increase in ketoacidosis, fever, anorexia, fasting, starvation, high-fat diet
Platelet count	250,000–400,000/μL	Values decrease in anemias and allergic conditions and during cancer chemotherapy; values increase in cancer, trauma, heart disease and cirrhosis
Prothrombin time	11–15 s	Values increase in prothrombin and vitamin deficiency, liver disease, and hypervitaminosis A
Red blood cell count	Males: 4.6–6.2 million/μL Females: 4.2–5.4 million/μL	Values decrease in systemic lupus erythematosus, anemias, and Addison's disease; values increase in polycythemia and dehydration and following hemorrhage
Reticulocyte count	1–3%	Values decrease in iron-deficiency and pernicious anemia and radiation therapy; values increase in hemolytic anemia, leukemia, and metastatic carcinoma
White blood cell count, differential	Neutrophils 60–70% Eosinophils 2–4% Basophils 0.5–1% Lymphocytes 20–25% Monocytes 3–8%	Neutrophils increase in acute infections; eosinophils and basophils increase in allergic reactions; monocytes increase in chronic infections; lymphocytes increase during antigen–antibody reactions
White blood cell count, total	5000–9000/μL	Values decrease in diabetes mellitus, anemias, and following cancer chemotherapy; values increase in acute infections, trauma, some malignant diseases, and some cardiovascular diseases

Table B.3 Urine Values

Test	Normal Values	Clinical Significance
Acetone and acetoacetate	0	Values increase in diabetic acidosis and during fasting
Albumin	0 to trace	Values increase in glomerular nephritis and hypertension
Ammonia	20–70 mEq/L	Values increase in diabetes mellitus and liver disease
Bacterial count	Under 10,000/mL	Values increase in urinary tract infection
Bile and bilirubin	0	Values increase in biliary tract obstruction
Calcium	Under 250 mg/24 h	Values increase in hyperparathyroidism and decrease in hypoparathyroidism
Chloride	110–254 mEq/24 h	Values decrease in pyloric obstruction, diarrhea; values increase in Addison's disease and dehydration
Potassium	25–100 mEq/L	Values decrease in diarrhea, malabsorption syndrome, and adrenal cortical insufficiency; values increase in chronic renal failure, dehydration, and Cushing's syndrome
Sodium	75–200 mg/24 h	Values decrease in diarrhea, acute renal failure, and Cushing's syndrome; values increase in dehydration, starvation, and diabetic acidosis
Creatinine clearance	100–140 mL/minute	Values increase in renal diseases
Creatinine	1–2 g/24 h	Values increase in infections and decrease in muscular atrophy, anemia, and certain kidney diseases
Glucose	0	Values increase in diabetes mellitus and certain pituitary gland disorders
Urea clearance	Over 40 mL of blood cleared of urea per minute	Values increase in certain kidney diseases
Urea	25–35 g/24 h	Values decrease in complete biliary obstruction and severe diarrhea; values increase in liver diseases and hemolytic anemia
Uric acid	0.6–1 g/24 h	Values increase in gout and decrease in certain kidney diseases
Casts		
Epithelial	Occasional	Increase in nephrosis and heavy metal poisoning
Granular	Occasional	Increase in nephritis and pyelonephritis
Hyaline	Occasional	Increase in glomerular membrane damage and fever
Red blood cell	Occasional	Values increase in pyelonephritis; blood cells appear in urine in response to kidney stones and cystitis
White blood cell	Occasional	Values increase in kidney infections
Color	Amber, straw, transparent yellow	Varies with hydration, diet, and disease states
Odor	Aromatic	Becomes acetonelike in diabetic ketosis
Osmolality	500–800 mOsm/kg water	Values decrease in aldosteronism and diabetes insipidus; values increase in high-protein diets, heart failure, and dehydration
pH	4.6–8	Values decrease in acidosis, emphysema, starvation, and dehydration; values increase in urinary tract infections and severe alkalosis

Table B.4 Hormone Levels

Test	Normal Values
Steroid hormones	
Aldosterone	Excretion: 5–19 µg/24 h*
Fasting at rest, 210 mEq sodium diet	Supine: 48 ± 29 pg/mL†
	Upright: 65 ± 23 pg/mL
Fasting at rest, 10 mEq sodium diet	Supine: 175 ± 75 pg/mL
	Upright: 532 ± 228 pg/mL
Cortisol	
Fasting	8 AM: 5–25 µg/100 mL
At rest	8 PM: Below 10 µg/100 mL
Testosterone	Adult male: 300–1100 ng/100 mL‡
	Adolescent male: over 100 ng/100 mL
	Female: 25–90 ng/100 mL
Peptide hormones	
Adrenocorticotropin (ACTH)	15–170 pg/mL
Calcitonin	Undetectable in normals
Growth hormone (GH)	
Fasting, at rest	Below 5 ng/mL
After exercise	Children: over 10 ng/mL
	Male: below 5 ng/mL
	Female: up to 30 ng/mL
Insulin	
Fasting	6–26 µU/mL
During hypoglycemia	Below 20 µU/mL
After glucose	Up to 150 µU/mL
Luteinizing hormone (LH)	Male: 6–18 mU/mL
	Preovulatory or postovulatory female: 5–22 mU/mL
	Midcycle peak 30–250 mU/mL
Parathyroid hormone	Less than 10 µEq/L§
Prolactin	2–15 ng/mL
Renin activity	
Normal diet	
Supine	1.1 ± 0.8 ng/mL/h
Upright	1.9 ± 1.7 ng/mL/h
Low-sodium diet	
Supine	2.7 ± 1.8 ng/mL/h
Upright	6.6 ± 2.5 ng/mL/h
Thyroid-stimulating hormone (TSH)	0.5–3.5 µU/mL
Thyroxine-binding globulin	15.25 µg T_4/100 mL
Total thyroxine	4–12 µg/100 mL

*1 microgram (1 µg) is equal to 10^{-6} g.

†1 picogram (1 pg) is equal to 10^{-12} g.

‡1 nanogram (1 ng) is equal to 10^{-9} g.

§µEq = microequivalent; 1/1000 m Eq, which is defined in Appendix C.

Appendix C

Solution Concentrations

Physiologists often express solution concentration in terms of percent, molarity, molality, and equivalents.

Percent

The weight-volume method of expressing percent concentrations states the weight of a solute in a given volume of solvent. For example, to prepare a 10% solution of sodium chloride, 10 g of sodium chloride is dissolved in a small amount of water (solvent) to form a salt solution. Then additional water is added to the salt solution to form 100 mL of salt solution. Note that the sodium chloride was dissolved in water and then diluted to the required volume. The sodium chloride was not dissolved directly in 100 mL of water.

Molarity

Molarity determines the number of moles of solute dissolved in a given volume of solvent. A **mole** (mol) of a substance contains 6.023×10^{23} number (Avogadro's number) of particles, such as atoms, ions, compounds, or molecules. A 1-molar (1 M) solution is made by dissolving 1 mol of a substance in enough water to make 1 L of solution. For example, 1 mol of sodium chloride solution is made by dissolving 58.44 g of sodium chloride in enough water to make 1 L of solution. One mole of glucose solution is made by dissolving 180.2 g of glucose in enough water to make 1 L of solution. Both solutions have the same number (Avogadro's number) of sodium chloride or glucose compounds in solution.

Molality

Although 1-M solutions have the same number of solute particles, they do not have the same number of solvent (water) molecules. Because 58.5 g of sodium chloride occupies less volume than 180 g of glucose, the sodium chloride solution has more water molecules. **Molality** is a method of calculating concentrations that takes into account the number of solute and solvent molecules. A 1-molal solution (1 m) is 1 mol of a substance dissolved in 1 kg of water. Thus, a 1-m solution of sodium chloride and a 1-m solution of glucose contain the same number of sodium chloride and glucose compounds dissolved in the same amount of water.

When sodium chloride (NaCl) is dissolved in water it dissociates, or separates, to form two ions: an Na^+ and a Cl^-. Glucose, however, does not dissociate when dissolved in water. Although 1-m solutions of sodium chloride and of glucose have the same number of sodium chloride and glucose compounds, because of dissociation the sodium chloride solution contains twice as many particles as the glucose solution (one Na^+ and one Cl^- for each glucose molecule). To report the concentration of these substances in a way that reflects the number of particles in a given mass of solvent, the concept of **osmolality** is used. The osmolality of a solution is the molality of the solution times the number of particles into which the solute dissociates in 1 kg of solvent. Thus 1 mol of sodium chloride in 1 kg of water is a 2-osmolal (2-osm) solution because sodium chloride dissociates to form two ions.

The osmolality of a solution is a reflection of the number, not the type, of particles in a solution. Thus, a 1-osm solution contains 1 osm of particles per kilogram of solvent, but the particles may be all one type or a complex mixture of different types.

The concentration of particles in body fluids is so low that the measurement milliosmole (mOsm), 1/1000 of an osmole, is used. Most body fluids have an osmotic concentration of approximately 300 mOsm and consist of many different ions and molecules. The osmotic concentration of body fluids is important because it influences the movement of water into or out of cells (see chapter 3).

Equivalents

Equivalents are a measure of the concentrations of ionized substances. One equivalent (Eq) is 1 mol of an ionized substance multiplied by the absolute value of its charge. For example, 1 mol of NaCl dissociates into 1 mol of Na^+ and 1 mol of Cl^-. Thus, there is 1 Eq of Na^+ (1 mol × 1) and 1 Eq of Cl^- (1 mol × 1). One mole of $CaCl_2$ dissociates into 1 mol of Ca^{2+} and 2 mol of Cl^-. Thus, there are 2 Eq of Ca^{2+} (1 mol × 2) and 2 Eq of Cl^- (2 mol × 1). In an electrically neutral solution, the equivalent concentration of positively charged ions is equal to the equivalent concentration of the negatively charged ions. One milliequivalent (mEq) is 1/1000 of an equivalent.

Appendix D

Answers to Critical Thinking Questions

Chapter One

1. D is correct. Positive-feedback mechanisms result in movement away from homeostasis and are usually harmful. The continually decreasing blood pressure is an example. Negative-feedback mechanisms result in a return to homeostasis. The elevated heart rate is a negative-feedback mechanism that attempts to return blood pressure to a normal value. In this case, the negative-feedback mechanism was inadequate to restore homeostasis, and blood pressure continued to decrease. Medical intervention (a transfusion) increased blood volume and blood pressure. With the increase in blood pressure, the positive-feedback mechanism is interrupted and the negative-feedback mechanism is able to maintain blood pressure.

2. Student B is correct. As the muscles become more active, they use more oxygen. The increased rate of respiration maintains blood and muscle tissue oxygen levels (homeostasis) and is a negative-feedback mechanism.

3. a. Inferior
   b. Posterior (dorsal) or deep
   c. Distal or inferior
   d. Lateral

4. The heel is distal, inferior, and posterior to the kneecap. The kneecap is proximal, superior, and anterior to the heel.

5. The wedding band should be worn proximal to the engagement ring.

6. The pancreas is located in the right-upper and left-upper quadrants; it is located in the epigastric and left hypochondriac regions. The top of the urinary bladder is located in the right-lower quadrant and the left-lower quadrant; it is located in the hypogastric region. The rest of the urinary bladder is located in the pelvic cavity.

7. The pelvic cavity contains the uterus, which increases greatly in size during pregnancy as the fetus within the uterus grows. However, the pelvic cavity is surrounded by bones, which are not expandable. Therefore the uterus moves superiorly into the abdominal cavity, crowding abdominal organs and dramatically increasing the size of the abdominal cavity.

8. After passing through the left thoracic wall, the first membrane to be encountered is the parietal pleura. Continuing through the pleural cavity, the visceral peritoneum and the left lung are pierced. Leaving the lung the bullet penetrates the visceral pleura, the pleural cavity, and the parietal pleura (remember that the lung is surrounded by a double membrane sac). Next the parietal pericardium, the pericardial cavity, the visceral pericardium, and the heart are encountered.

9. The kidneys are located in the abdominal cavity, but are retroperitoneal. When a person is lying prone, it is possible to cut through the posterior abdominal wall and remove a kidney without cutting through parietal peritoneum.

Chapter Two

1. Because atoms are electrically neutral, the iodine atom has the same number of protons and electrons. The gain of an electron means the iodine ion has one more electron than protons and therefore a charge of minus one. The correct symbol is I^{-1}.

2. a. Dissociation
   b. Synthesis
   c. Decomposition

3. Muscle contains proteins and to increase muscle mass proteins must be synthesized. The synthesis of molecules in living organisms results from chemical reactions that require the input of energy. The energy needed to drive these synthesis reactions comes from chemical reactions that release energy. The energy-releasing reactions occur during the decomposition of food molecules.

4. The sodium bicarbonate dissociates to form sodium ions (Na^+) and bicarbonate ions (HCO_3^-). Because this is a reversible reaction, The HCO_3^- added to the solution bind with H^+ to form carbon dioxide and water. The decrease in H^+ causes the pH to increase (become more basic).

5. The slight amount of heat functioned as activation energy and started a chemical reaction. The reaction released energy, especially heat, which caused the solution to become very hot.

6. There was most likely at least one enzyme present that was required for one of the initial reactions. Boiling denatured any enzymes present in the solution. Without enzymes, the reaction(s) will not occur.

Chapter Three

1. B is the most logical conclusion. Swollen lung tissue suggests that the tissues had been submerged in a hypotonic solution such as fresh water. Because the bay contains salt water, which is slightly hypertonic to blood, it is unlikely he drowned in the bay. It is more likely he drowned in fresh water and was later placed in the bay. Although this is the most logical conclusion of those given, there are possibilities other than murder. For example, he may have accidentally drowned in a fresh water stream and then been washed into the bay.

2. B is correct. A solution that is isotonic causes cells to neither shrink nor swell. Therefore there is no net movement of water between the blood and the dialysis fluid. Also, if the solution has the same composition as blood except for having no urea, only urea diffuses from a higher concentration in the blood to the lower concentration in the dialysis fluid. A solution that is isotonic and contains only protein will cause many different small molecules (not just urea) that are found in blood to diffuse into the dialysis fluid. Distilled water is hypotonic and causes blood cells to swell and undergo lysis. Also, molecules that are able to pass through the dialysis membrane diffuse from the blood into the distilled water. Blood is not a good solution to use because it would have to be blood that has already had urea removed in order to be useful as dialysis fluid.

3. Because the plasma membrane stays the same size, even though small pieces of membrane from secretory vesicles are continually added to the plasma membrane, one must conclude that some process is removing small pieces of

membrane at the same rate that they are added. The cell membrane is constantly being recycled.

4. The main function of this cell is to secrete proteins. Well-developed rough endoplasmic reticulum for protein production, well developed Golgi apparatuses for the packaging of proteins for secretion, and many vesicles for exocytosis are typical of many cells that secrete proteins. Many mitochondria produce large amounts of ATP. The ATP serves as an energy source for forming the bonds between the amino acids, for the active transport of amino acids into the cell, and for exocytosis of the completed proteins. Many microvilli, which greatly increase surface area for transport, can also be found in cells that produce and secrete proteins. However, large numbers of mitochondria or large amounts of microvilli can be found in other types of cells as well.

5. The sickle-cell gene for producing the protein has a different nucleotide than the normal gene for producing the protein. Because the nucleotides are the set of instructions for making the protein, the change means the instructions contain an error. When mRNA copies the instructions from the faulty DNA, the error is also copied. Consequently, when the amino acids are joined together to form the protein at the ribosome, the instructions are wrong and the protein is incorrectly assembled. A substitution of a single nucleotide results in a different amino acid in the chain of amino acids that make up the protein. This change alters the shape of the protein. Just as the function of enzymes depends on their shape, the functions of other proteins depends on their shape. The incorrectly manufactured proteins in the red blood cells have the wrong shape and do not stack correctly. As a result of the abnormal stacking, the red blood cells have an abnormal, sickle shape.

Chapter Four

1. Pseudostratified columnar ciliated epithelium is found in the trachea. It produces mucus that traps dust and debris in air. The cilia move the mucus with entrapped dust and debris to the throat where it is swallowed. In heavy smokers the pseudostratified columnar epithelium is replaced by stratified squamous epithelium that serves a protective function against the irritating materials in the smoke. Unfortunately this type of epithelium is not ciliated, so removal of foreign materials from the trachea is more difficult. After 2 years without smoking, the original pseudostratified columnar ciliated epithelium replaces the stratified squamous epithelium and debris removal resumes.

2. Tight junctions prevent the passage of materials between the epithelial cells.

3. A secretory epithelium is generally a simple epithelium. To manufacture large amounts of enzymes, a cuboidal or columnar cell with the appropriate organelles such as rough endoplasmic reticulum and Golgi aparatuses would be expected. The pancreas is formed of simple columnar epithelium. The epithelium has microvilli on its free surface, which increase the surface area and facilitate secretion. The tight junctions that connect the epithelial cells of the pancreatic glands and duct system to each other prevent damage to the underlying tissues (by the action of pancreatic digestive enzymes).

4. a. The stratified squamous epithelium that lines the mouth provides protection. Replacement of it with simple columnar epithelium makes the lining of the mouth much more susceptible to damage because the single layer of epithelial cells is easier to damage.
   b. Tendons attach muscles to bone. When muscles contract, muscles pull on the tendons and, because the tendons are attached to bone, the bones move. If the tendons contained elastic tissue, the tendons would be more like elastic bands. The muscles would contract and stretch the tendons. Not all of the force of muscle contraction would be transferred to the bones to cause them to move.
   c. If bones were made of elastic cartilage they would be much more flexible and could bend and then return to their original shape. They would not be rigid structures, like bone, that support our weight and result in efficient movement.

5. The fibers are organized along the direction of pull on the ligaments; that is, parallel to the length of the ligament. When the back is bent the ligaments are stretched. The elastic recoil of the stretched ligaments helps to straighten the back.

6. When blood is ejected into the aorta, the aortic wall expands and the diameter of the aorta increases. The elastic fibers are arranged in a circular fashion, so that when the aortic wall expands, the elastic fibers are stretched. Recoil of the fibers causes the aorta to resume its original shape, which helps to force blood to flow through the aorta.

7. Chemical mediators of inflammation normally produce beneficial responses, such as dilation of small diameter blood vessels and increased blood vessel permeability. Blocking these effects could reduce the ability of the body to deal with harmful agents such as bacteria. On the other hand, antihistamines also reduce many of the unpleasant symptoms of inflammation, making the patient more comfortable and this can be considered beneficial. Antihistamines are commonly taken to prevent allergy symptoms that are often an overreaction of the inflammatory response to foreign substances such as pollen.

8. Collagen synthesis is required for granulation tissue and scar formation. If collagen synthesis does not occur because of a lack of vitamin C, or if collagen synthesis is slowed, wound healing does not occur or is slower than normal. One might expect that the density of collagen fibers in a scar is reduced and the scar is not as durable as a normal scar.

Chapter Five

1. Yes, the skin (dermis) can be over stretched as a result of obesity.

2. Alcohol is a solvent that dissolves lipids (see chapter 2). It removes the lipids from the skin, especially in the stratum corneum. The rate of water loss increases after soaking the hand in alcohol because of the removal of the lipids that normally prevent water loss.

3. Carotene, a yellow pigment from ingested plants, accumulates in lipids. The stratum corneum of a callus has more layers of cells than other noncallused parts of the skin and the cells in each layer are surrounded by lipids. The carotene in the lipids make the callus appear yellow.

4. The vermillion border is covered by keratinized epithelium that is a transition between the nonkeratinized stratified epithelium of the mucous membrane and the keratinized stratified epithelium of the facial skin. The mucous membrane has mucous glands, which secrete mucus (see chapter 4). The mucus helps to keep the inner surface of the lips moist. In addition, the inner surface of the lips is "sealed off" most of the time and is moistened by saliva. In contrast to facial skin, the skin of the vermillion border does not have sebaceous or sweat glands. Without sebaceous glands, the surface of the vermillion border is not protected against drying by sebum. Also, the vermillion border is not as heavily keratinized as

facial skin; that is, the number of cell layers with surrounding lipids is less. Consequently, the vermillion border dries out more easily.

5. The hair follicle, but not the hair, is surrounded with nerve endings that can detect movement or pulling of the hair. The hair is dead, keratinized epithelium, so cutting the hair is not painful.

6. Several methods have some degree of success in treating acne: (1) Kill the bacteria. One effective agent is benzoyl peroxide in some acne medications. Antibiotics have also been used to treat severe cases of acne. (2) Prevent sebum production. Isotretivioin, or Accutane (a derivative of vitamin A), has proven effective in preventing sebum production. However, Accutane produces birth defects and should be used with caution. Estrogen, the female sex hormone, has also been used to treat severe cases of acne. (3) Unplug the follicle. Some sulfur compounds speed up peeling of the skin and unplug the follicle.

7. Rickets is a disease of children resulting from inadequate vitamin D. With inadequate vitamin D there is insufficient absorption of calcium from the intestine, resulting in soft bones. If adequate vitamin D is ingested, rickets is prevented, whether one is dark or fair skinned. However, if dietary vitamin D is inadequate, when the skin is exposed to ultraviolet light, it can produce a precursor molecule that can be converted to vitamin D. Dark-skinned children are more susceptible to rickets because the additional melanin in their skin screens out the ultraviolet light and they produce less vitamin D.

8. When first exposed to the cold temperature just before starting the run, the blood vessels in the skin constrict to conserve heat. This produces a pale skin color. Dilation of skin blood vessels does not occur at this time because the skin has not been exposed to the cold long enough to cause skin temperature to fall below 15°C (27°F). After running awhile, as a result of the excess heat generated by the exercise, the blood vessels in the skin dilate. This results in heat loss and helps to prevent overheating. Increased blood flow through the skin causes it to turn red. After the run, the body still has excess heat to eliminate, so for awhile the skin remains red.

Chapter Six

1. The injury separated the head (epiphysis) from the shaft (diaphysis) at the epiphyseal plate, which is cartilage. Because the epiphyseal plate is the site of bone elongation, damage to the epiphyseal plate can interfere with bone elongation, resulting in a shortened limb.

2. If Justin landed on his heels, it is likely that he could have broken one or both of his heel bones, the calcaneus. When Hefty Stomper stepped on Justin's foot, the most likely bones broken would be the metatarsals, which make up the main structure of the foot.

3. The humerus articulating with the ulna and radius forms the elbow joint. The ulna fits tightly over the end of the humerus. The radius attaches to both the humerus and the ulna, but not as tightly to the humerus as does the ulna. Pulling forcefully on the forearm, would, therefore, be more likely to pull the radius away from the humerus than the ulna.

4. Because the female pelvis is relatively wider that the male pelvis, the heads of the femora (pl. of femur) are placed wider apart. For the lower limbs to be placed directly below the center of gravity in the body, the femora must be angled in more sharply in females than in males. This angle may bring the knees closer together, resulting in the condition called knock-kneed.

5. All skeletons, including those of humans, have features unique to each species. A human skeleton can be identified based on its unique anatomic features. In addition, female skeletons have several anatomic features that distinguish them from male skeletons. For example, the female pelvis differs in several features from the male pelvis (see figure 6.32).

Chapter Seven

1. Botulism toxin decreases acetylcholine release in the neuromuscular junction. This prevents action potentials in skeletal muscle cells. Thus the respiratory muscles (e.g., the diaphragm) relax and do not contract. Other explanations that you could have proposed because they would also lead to respiratory failure, would be that the toxin prevents acetylcholine from binding to its receptor on skeletal muscle cells, or that the toxin inhibits the production of acetylcholine.

2. Harvey's hands became fatigued as a result of ATP depletion. Without ATP, the muscles moving the fingers could not contract and so the fingers could not flex or extend. In addition, his inability to release his grip is an indication of physiological contracture, where too little ATP is present to allow cross-bridge release.

3. Both the 100-m dash and weight lifting (to a lesser extent) involve short bursts of anaerobic activity, so, in both cases the researcher should expect to find more fast-twitch muscle in the gastrocnemius of these athletes. The 10,000-m run involves aerobic metabolism, and the outstanding athlete should have more slow-twitch fibers.

4. The exercise that best builds the anterior arm muscles is flexion of the elbow against force, such as in pull-ups. A pull-up with the hand supinated builds both the biceps and brachialis, but a curl with the hand pronated builds the brachialis to a greater extent. The posterior arm is best built by extension of the elbow, as in push-ups or dips. The anterior forearm muscles are best built up by forcefully flexing the wrist and fingers. The anterior thigh is best built by extending the knee, as in partial knee bends. The posterior leg muscles are best built by forcefully plantar flexing the ankle and toes, as in standing on ones toes. The abdominal muscles are best built by flexing the vertebral column against resistance, as in leg lifts or partial sit-ups. Of course, each exercise has to be repeated many times to increase the size of the muscles.

5. This is a typical hamstring injury (pulled hamstring), in which one or more of the hamstring muscles are either pulled away from their attachment to bone or there is a tear in the muscle. The slight flexion of the knee and the bulge in the muscle occur because muscle fibers contract in the hamstrings, but the tear does not allow antagonist muscles in the anterior thigh to pull on the muscle fibers to lengthen them. The pain results from the damaged tissue in the muscle (flexion and pain are also due to spasmodic contractions of the muscle in response to the injury). Because of the tissue damage and cramps, the muscles could not respond to voluntary nervous control and could not contract.

Chapter Eight

1. If one series of neurons had more neurons, it would have more synapses, which should slow down the rate of action potential propagation. Also, if one series were unmyelinated and the other myelinated, the unmyelinated series would be slower.

2. The phrenic nerves are the nerves to the diaphragm. The left phrenic nerve would be cut to paralyze the left side of the diaphragm.

3. a. Radial nerve. This nerve supplies posterior (extensor) muscles. When these muscles are paralyzed, the flexor muscles are unopposed and result in the flexed limb described.
   b. Femoral nerve

4. The cerebellum controls motor functions such as balance, muscle tone, and fine motor movements. Cerebellar dysfunction is characterized as a loss of balance, loss of motor tone, and the inability to control fine motor movements, such as touching one's nose with the eyes closed. Such patients appear to be drunk (alcohol apparently most directly affects the cerebellum).

5. The blow to the head erased Louis' short-term memory, which would have extended to about 10 minutes before the blow, and that information was never transferred to long-term memory.

6. a. Hypoglossal nerve (cranial nerve XII)
   b. Optic nerve (cranial nerve II)
   c. Trigeminal nerve (cranial nerve V)
   d. Facial nerve (cranial nerve VII)
   e. Oculomotor nerve (cranial nerve III)

7. Nervous system stimulation of the digestive system occurs primarily through the vagus nerve (cranial nerve X), which arises from the brain. Therefore, an injury to the spinal cord at the level of C6 does not affect the functioning of this nerve.

Chapter Nine

1. The lenses of the eye are biconvex structures, and are most important in close vision as they are thickened (made more convex) by relaxation of the suspensory ligaments (accommodation). Therefore, if they are removed, close vision would be most affected, and convex glasses would do the best job of replacing the lenses for near vision, such as when reading.

2. The fovea centralis, where a person focuses on an object, contains almost entirely cones. Cones do not detect faint light, so that dim stars, which can be detected by the rods, are lost from vision when the person looks directly at them. When the person looks slightly to the side, the more peripherally located rods can again detect the faint light from the stars.

3. Under normal conditions, pressure changes in the middle ear occur through the auditory tube. The auditory tube connects the middle ear to the pharynx, which opens to the outside of the body through the oral and nasal cavities. This allows air pressure to equalize on both sides of the eardrum. When a person is under water, the increased water pressure on the outside of the eardrum forces the eardrum toward the middle ear because the external water pressure exceeds the air pressure in the middle ear. As a result, the eardrum cannot vibrate as freely as when the pressure is equalized, and sound transmission is dampened. If the pressure

difference becomes too great, it can rupture the eardrum.

4. Sound is normally transmitted through the middle ear by vibration of the auditory ossicles. A vibrating tuning fork touching the mastoid process sets up vibrations in the temporal bone, causing the perilymph and endolymph of the inner ear to vibrate. These vibrations cause the basilar membrane to vibrate and are therefore perceived as sound.

5. When a contaminated hand rubs the eyes, the virus can be introduced into tears on the conjunctiva. From there the virus can spread into the lacrimal canaliculi, and pass through the nasolacrimal duct into the nasal cavity.

Chapter Ten

1. The hormone binds to a membrane-bound receptor and activates a G protein mechanism. Because a drug inhibits the binding of GTP to a protein suggests that GTP cannot bind to the α subunit of the G protein complex. Also, because a drug that inhibits the breakdown of cyclic-AMP causes an increased response, suggests that the α subunit of the G protein complex, with GTP bound to it, activates the enzymes that are responsible for the synthesis of cyclic-AMP.

2. The effect of ADH on cells is mediated through membrane-bound receptors, whereas the effects of aldosterone on cells is mediated through intracellular receptors. In addition, ADH levels increase more rapidly than aldosterone levels in response to a decrease in blood volume. ADH is secreted within minutes by the posterior pituitary gland in response to dehydration of the body. The hypothalamus detects a decrease in blood volume and an increase in the concentration of the blood, which stimulates neurons in the hypothalamus to secrete ADH. ADH binds to membrane-bound receptors in the kidney. The cells of the kidney respond within minutes. As a result the kidney produces a small volume of very concentrated urine. Aldosterone is secreted in response to decreased blood pressure (caused by decreased blood volume), but the process involves other compounds. The kidneys respond to a decrease in blood pressure by secreting renin. Renin acts as an enzyme to break down angiotensinogen into angiotensin I. Angiotensin-converting enzyme converts angiotensin I to angiotensin II. Finally, angiotensin II stimulates the adrenal cortex to increase aldosterone secretion. Aldosterone binds to intracellular

receptors in cells of the kidney. The response of these cells is to increase specific mRNA and protein synthesis. The new mRNA and protein synthesis requires a significant amount of time. The response to the synthesis of new proteins is to increase the rate of solute reabsorption in the kidney, including Na^+. This enhances water reabsorption by the kidney.

3. Vitamin D increases the rate of transport of Ca^{2+} across the wall of the intestine. Therefore, in response to high concentrations of vitamin D, blood Ca^{2+} levels can become abnormally high. Because the blood Ca^{2+} levels are elevated, the rate of PTH secretion decreases to low levels. Also, the elevated blood Ca^{2+} levels stimulate calcitonin secretion.

4. If the adrenal cortex degenerated, glucocorticoids, mineralocorticoids, and the androgens normally secreted by the adrenal cortex would no longer be secreted. A lack of glucocorticoid hormone secretion results in a reduced ability to maintain blood nutrient levels, such as glucose, between meals or during periods when no food is available. The ability to inhibit inflammation is also lost. Reduced levels of mineralocorticoids result in the loss of Na^+ in the urine and the loss of water in the urine. Thus, the ability to regulate blood pressure is reduced. Reduced secretion of androgens also occurs. Because the testes secrete androgens, the reduced androgen secretion in males is not significant. The effect of reduced androgens in females is unclear although adrenal androgens may influence sexual behavior, to some degree, in females.

5. Aldosterone increases Na^+ reabsorption from the distal nephron and it increases K^+ secretion into the nephron. Elevated aldosterone secretion results in the increased retention of Na^+ and H_2O in the kidney and an increased rate of K^+ elimination in the urine. As a result, blood pressure increases and blood K^+ levels can get very low.

6. After 24 hours without food, blood glucose levels will be decreasing. This will activate mechanisms that cause blood glucose levels to increase. The decreasing blood glucose levels will result in an increased rate of secretion of cortisol, a glucocorticoid, from the adrenal cortex, epinephrine from the adrenal medulla, and glucagon from pancreatic islets. The decreasing blood glucose levels result in increased ACTH secretion from the anterior pituitary gland, increased sympathetic nervous system stimulation of the adrenal medulla resulting in increased

epinephrine secretion, and increased secretion of glucagon from the pancreatic islets. Cortisol increases fat and protein breakdown so that these substances can be used as energy sources. It also causes increased glucose synthesis, primarily from, amino acids. Epinephrine and glucagon both bind to receptors on the liver and increase the release of glucose from the liver. The decreasing blood glucose levels inhibit insulin secretion. The decreased insulin secretion slows the uptake of glucose by tissues. Thus, blood glucose levels are maintained within normal levels.

7. Stressful conditions cause increased ACTH secretion. Increased ACTH stimulates the secretion of glucocorticoids, such as cortisol. One of the effects of increased cortisol secretion is inhibition of the immune response. This could lead to an increased likelihood of a variety of infections, including colds and other diseases. Stomach pains may be due to inflammation of the stomach. Many stomach ulcers result from inflammation caused by bacteria.

Chapter Eleven

1. Blood doping increases the number of red blood cells in the blood, increasing its oxygen-carrying capacity. However, the increased number of red blood cells also makes it more difficult for the blood to flow through the blood vessels, increasing the workload on the heart. This same effect is seen in erythrocytosis (polycythemia). In addition, if the process is not done properly there is a risk of infection.

2. Symptoms resulting from decreased red blood cells are associated with a decreased ability of the blood to carry oxygen: shortness of breath, weakness, fatigue, and pallor. Symptoms resulting from decreased platelets are associated with a decreased ability to form platelet plugs and clots: small areas of hemorrhage in the skin, bruises, and decreased ability to stop bleeding. Symptoms resulting from decreased lymphocytes include an increased susceptibility for infections and diseases.

3. The hypoventilation results in decreased blood oxygen, which stimulates the release of erythropoietin from the kidneys. The erythropoietin stimulates red blood cell production in the red bone marrow. Consequently, red blood cell numbers increase.

4. Removal of the stomach results in removal of intrinsic factor, which is necessary for adequate vitamin B_{12} absorption in the small intestine. Therefore the patient would develop pernicious anemia.

Vitamin B_{12} injections can be used to prevent pernicious anemia.

5. Vitamin B_{12} and folate are necessary for blood cell division. Lack of these vitamins results in anemia. Iron is necessary for the production of hemoglobin. Lack of iron results in iron-deficiency anemia. Vitamin K is necessary for the production of many blood clotting factors. Lack of vitamin K can greatly increase blood-clotting time.

6. The anemia results from too little hemoglobin. Because there is less hemoglobin, less hemoglobin is broken down into bilirubin. Consequently, less bilirubin is excreted as part of the bile into the small intestine. With decreased bilirubin in the small intestine, bacteria produce fewer of the pigments that normally color the feces.

7. Reddie Popper has hemolytic anemia. The RBC is lower than normal because the red blood cells are being destroyed faster than they are replaced. With fewer red blood cells, hemoglobin and hematocrit are lower than normal. Bilirubin levels are above normal because of the breakdown of the hemoglobin released from the ruptured red blood cells.

Chapter Twelve

1. A heart murmur is an abnormal heart sound. It is unlikely that an ECG will reveal that he has a heart murmur because an ECG monitors electrical activity of the heart and not heart sounds. A stethoscope is a more likely instrument for detecting a heart murmur. It is possible that a condition that causes a heart murmur could also cause an obvious abnormality in the ECG, but that is not always the case.

2. Starling's law of the heart is an intrinsic regulatory mechanism, whereas parasympathetic innervation of the heart is a component of the extrinsic regulation of the heart. Cutting the vagus nerve does not significantly affect the ability of the Starling's law of the heart to operate.

3. Cutting sensory nerve fibers from baroreceptors would reduce the frequency of action potentials delivered to the medulla oblongata from the baroreceptors. This results because the normal blood pressure stimulates the baroreceptors. An increase in blood pressure increases the action potential frequency and a decrease in blood pressure decreases the action potential frequency. Because cutting the sensory nerve fibers decreases the action potential frequency, this acts as a signal to the medulla oblongata that a decrease in blood

pressure has occurred, even though the blood pressure has not decreased. The medulla oblongata responds by increasing sympathetic action potentials and reducing parasympathetic action potentials delivered to the heart. Consequently, the heart rate increases.

4. When the internal carotid arteries are clamped, blood flow and blood pressure in the clamped internal carotid arteries decrease dramatically. Thus, blood pressure in the area of the baroreceptors in the internal carotid arteries is low. Sensory neurons carry a lower frequency of action potentials from the baroreceptors of the internal carotid arteries to the medulla oblongata. The cardioregulatory center within the medulla oblongata responds as if the blood pressure has decreased by causing an increase in sympathetic stimulation and a decrease in parasympathetic stimulation of the heart. In addition, epinephrine and norepinephrine are released from the adrenal medulla. Consequently, there is an increased heart rate and stroke volume. Therefore, blood pressure in the aorta increases.

5. A drug that blocks Ca^{2+} channels decreases the heart rate and the force of contraction of the heart. This occurs because Ca^{2+} are involved in the depolarization of the cardiac muscle cells. If fewer Ca^{2+} flow into the cardiac muscle cells, the rate and degree of depolarization decrease. The result is that action potentials develop more slowly and depolarization does not occur to the same degree as normal. Slower development of action potentials slows the heart rate. The degree of depolarization causes less Ca^{2+} to enter the cell and, consequently, the force of contraction decreases.

6. Consuming a large amount of fluid increases the total volume of the blood, at least until the mechanisms that regulate blood volume decrease the blood volume to normal values. If the blood volume is increased, it causes an increase in venous return to the heart. Because of Starling's law of the heart, the increased venous return results in an increased stroke volume. A slight increase in the heart rate also occurs. Mechanisms that regulate blood pressure, such as the baroreceptor reflex, would prevent a large increase in blood pressure. Therefore, there may be an increased stroke volume but the blood pressure would not increase dramatically.

7. Cardiac output is influenced by the heart rate and stroke volume (CO = SV × HR). An athlete's cardiac output can be equal to

a nonathlete's cardiac output while they are both at rest, even though the athlete's heart rate is lower than the nonathlete's heart rate because the stroke volume of the athlete's heart is greater than the stroke volume of the nonathlete's heart. Athletic training causes a gradual hypertrophy (enlargement) of the heart. Therefore, athletes can maintain a cardiac output that is equal to a nonathlete's because they have an increased stroke volume, but a decreased heart rate. During exercise the athlete's heart rate and stroke volume increase. The stroke volume for the athlete's heart is much greater than the stroke volume of the nonathlete's. Therefore, the cardiac output of the athletes's heart is greater than the cardiac output of the nonathlete's during exercise.

8. The walls of the ventricles are thicker than the walls of the atria because the force that the ventricles must produce is greater than the force the atria must produce. The pressure in the ventricles during ventricular systole is substantially higher than the pressure in the atria during atrial systole. In addition, most of the ventricular filling (approximately 70%) occurs before contraction of the atria. Contraction of the atria is responsible for only about 30% of ventricular filling. In contrast, all of the blood ejected from the ventricles during ventricular systole is ejected because of contraction of the ventricles.

9. An incompetent aortic semilunar valve is leaky. Thus, when it is closed, blood is able to leak through it back into the left ventricle from the aorta. The aortic semilunar valve is normally closed from the beginning of ventricular diastole until just after the beginning of ventricular systole. During ventricular diastole, blood flows from the left atrium into the left ventricle. If the aortic semilunar valve is leaky, blood will also flow through the aortic semilunar valve into the left ventricle. Consequently, the volume of the left ventricle and the pressure in the left ventricle become greater than normal during ventricular diastole. Because blood is leaking from the aorta, the aortic pressure, during ventricular diastole, is lower than normal in the aorta. Because of Starling's law of the heart during ventricular systole, the ventricular muscle contracts with a greater force and forces a greater volume of blood into the aorta. Consequently, the pressure in the left ventricle is greater than normal during ventricular systole. Also, pressure in the aorta is greater than normal during ventricular systole.

Chapter Thirteen

1. a. For the right side of the brain: the aorta, the brachiocephalic artery, the right common carotid artery, the right internal carotid artery, the cerebral arterial circle, and then the brain tissue. For the left side of the brain: the aorta, the left common carotid artery, the left internal carotid artery, the cerebral arterial circle, and then the brain tissue

   b. For the left external portion of the skull: the aorta, the brachiocephalic artery, the right common carotid artery, the right external carotid artery, and then the external portion of the skull. For the left side of the external portion of the skull: the aorta, the left common carotid artery, the left external carotid arery, and then the external portion of the skull

   c. The aorta, the left subclavian artery, the left axillary artery, the left brachial artery, either the left radial or ulnar artery, and the left hand

   d. The aorta, the right common iliac artery, the external iliac artery, femoral artery, popliteal artery, the anterior tibial artery, and the anterior portion of the right leg

2. a. The left internal jugular vein, the left brachiocephalic vein, superior vena cava, right atrium of the heart

   b. The right external jugular vein, the right subclavian vein, the right brachiocephalic vein, superior vena cava, right atrium of the heart

   c. The superficial veins of the left hand and forearm; either the left cephalic or left basilic vein; the left cephalic vein and the left basilic vein empty into the axillary vein; the left axillary vein, left subclavian vein, left brachiocephalic vein, the superior vena cava, and the right atrium of the heart

   d. Right great saphenous vein, right femoral vein, right external iliac vein, right common iliac vein, inferior vena cava, and the right atrium of the heart

   e. Renal vein, inferior vena cava, and right atrium of the heart

   f. Superior mesenteric vein, hepatic portal vein, liver, hepatic veins, inferior vena cava, and right atrium of the heart

3. The femoral artery and vein are close to the surface in the femoral triangle, which is in the superior and medial part of the thigh. The femoral artery is the vessel into which the catheter will be placed. The femoral vein cannot be used because extending a catheter superiorly in the femoral vein will deliver the catheter to the right side of the heart. The anterior interventricular artery is a branch of the left coronary artery, which originates from the ascending aorta, just superior to the aortic semilunar valve. The catheter is inserted into the femoral artery. From there it passes through the external iliac, through the common iliac artery, and through the aorta. It passes through the abdominal aorta, the thoracic aorta, and the aortic arch to the beginning of the left coronary artery and then to the anterior interventricular artery.

4. Cells from the tumor can spread from the colon to the liver through the hepatic portal vein. Cells from the tumor can enter the superior mesenteric or the inferior mesenteric vein. The cells can pass from the superior mesenteric vein to the hepatic portal vein. The cells can also pass from the inferior mesenteric vein to the splenic vein and then the heptic portal vein. The cells can pass through the heptic portal vein to the liver.

5. Reduced blood flow to the kidney stimulates renin secretion. Renin acts on angiotensinogen to produce angiotensin I. Angiotensin I is converted to angiotensin II by the action of angiotensin-converting enzyme. Angiotensin II causes vasoconstriction, which increases blood pressure. In addition, angiotensin II stimulates the secretion of aldosterone from the adrenal cortex. Aldosterone acts on the kidney causing the reabsorption of Na^+ and water, thus increasing the blood volume. The increased blood volume results in increased blood pressure.

6. During exercise, vasoconstriction occurs in the viscera, but vasodilation occurs in the exercising muscles. Even though the cardiac output is increased because of the increase in stoke volume and heart rate, the blood pressure does not go up as much as it would if vasoconstriction occurred in the viscera without vasodilation of the blood vessels in the exercising skeletal muscles.

7. Dilation of arteries and veins allows blood to accumulate in the dilated blood vessels. Consequently, less blood is returned to the heart (decreased venous return). Because the venous return is reduced, the stroke volume of the heart decreases (see Starling's law of the heart—chapter 12). Consequently, the heart does less work and less oxygen is required to support the contraction of the heart. Therefore, angina pains, which are caused by inadequate oxygen delivery to the heart muscle, are reduced.

Chapter Fourteen

1. Massage moves lymph through the lymphatic vessels in the same fashion as does contraction of skeletal muscle. The application of pressure periodically to lymphatic vessels forces lymph to flow toward the trunk of the body, but valves prevent the flow of lymph in the reverse direction. The removal of lymph from the tissue helps to relieve edema. Elevation of the limb allows gravity to assist the movement of lymph toward the heart.

2. Normally T cells are processed in the thymus and then migrate to other lymphatic tissues. Without the thymus, this processing is prevented. Because there are normally five T cells for every B cell, the number of lymphocytes is greatly reduced. The loss of T cells results in an increased susceptibility to infection and inability to reject grafts because of the loss of cell-mediated immunity. In addition, because T cells are involved with the activation of B cells, antibody-mediated immunity is also depressed.

3. Injection B results in the greatest amount of antibody production. At first, the antigen causes a primary response. A few weeks later, the slowly released antigen causes a secondary response, resulting in a greatly increased production of antibodies. Injection A does not cause a secondary response because all of the antigen is eliminated by the primary response.

4. Active immunity varies from a few weeks (e.g., the common cold) to lifelong (e.g., measles). Immunity can be long lasting if enough memory cells (B or T) are produced and persist to respond to later antigen exposure. Passive immunity is not long lasting because the individual does not produce his or her own memory cells. Because active immunity can last longer than passive immunity, it is the preferred method in most cases. However, passive immunity is preferred in situations in which immediate protection is needed, because it takes time for active immunity to develop.

5. If the patient has already been vaccinated, the booster shot stimulates a memory (secondary) response and rapid production of antibodies against the toxin. If the patient has never been vaccinated, vaccinating him is not effective because there is not enough time for the patient to develop his own primary response. Therefore, an antiserum is given to provide immediate, but temporary, protection. Sometimes both are given: the antiserum provides short-term protection and the tetanus vaccine stimulates the patient's immune system to provide long-term protection. If the shots are given at the same location in the body, the antiserum (antibodies against the tetanus toxin) could cancel the effects of the tetanus vaccine (tetanus toxin altered to be nonharmful).

6. Resistance to extracellular bacterial infections is accomplished by antibody-mediated immunity, which is not functioning properly here. Maternal antibodies (IGg) that crossed the placenta provided protection following birth, but eventually were degraded. Resistance to intracellular viral infections is accomplished by cell-mediated immunity, which appears to be working normally.

7. At the first location, an antibody-mediated response results in an inflammatory response (immediate hypersensitivity). The combination of antibodies with the allergen triggers the release and activation of inflammatory chemicals. At the second location, a cell-mediated response also results in an inflammatory response (delayed hypersensitivity). This probably involves the release of lymphokines and the lysis of cells. At the other locations, there is neither an antibody-mediated nor a cell-mediated response.

8. The ointment was a good idea for the poison ivy, which caused a delayed hypersensitivity reaction—that is, too much inflammation. For the scrape, it is a bad idea because a normal amount of inflammation is beneficial and helps to fight infection in the scrape.

9. Because antibodies and cytokines both produce inflammation, the fact that the metal in the jewelry resulted in inflammation is not enough information to answer the question. However, the fact that it took most of the day (many hours) to develop the reaction indicates a delayed hypersensitivity reaction and, therefore, cytokines.

Chapter Fifteen

1. Pressing firmly on the base of the scapulae decreases the volume of the thoracic cavity, pleural pressure increases, and alveolar pressure increases to a value greater than atmospheric pressure. Thus, air flows from the lungs. When the arms are lifted, gravity causes the thoracic cavity to sag downward and expand. The volume of the thoracic cavity increases, pleural pressure decreases, and alveolar pressure decreases to less than atmospheric pressure, causing air to flow into the lungs.

2. a. The victim's lungs expand because the rescuer is blowing air into the lungs at a pressure higher than alveolar pressure. When alveolar pressure exceeds pleural pressure, the lungs tend to expand.

   b. Air flows out of the lungs because of the natural recoil of the lungs. Elastic fibers in the lungs and surface tension of water in the alveoli cause the recoil of the lungs.

3. The best prediction would be that her vital capacity would be greatest when she is standing up. In the upright position, gravity tends to pull the abdominal organs downward. As a result, movement of the diaphragm is not as restricted and thoracic volume is increased. Lying down allows abdominal organs to exert pressure on the diaphragm, decreasing the thoracic volume.

4. At sea level, all the gases in the atmosphere exert a pressure of 760 mm Hg. If water vapor is 10% of that gas mixture, then water vapor must have a partial pressure of 760×0.10, or 76 mm Hg.

5. As fluid accumulates in the alveoli, the layer through which oxygen and carbon dioxide must diffuse in the alveoli becomes thicker. As the layer becomes thicker, the rate at which gases diffuse slows. Consequently, the amount of oxygen that diffuses into the pulmonary capillaries is reduced and the amount of carbon dioxide that diffuses out of the pulmonary capillaries decreases. The decreased blood oxygen levels and the increased carbon dioxide levels stimulate the respiratory center and cause the rate and depth of respiration to increase. Oxygen therapy increases the P_{O_2} in the alveoli and reduces the P_{CO_2} in the alveoli. Therefore, oxygen diffuses more rapidly from the alveoli into the pulmonary capillaries and carbon dioxide diffuses more rapidly from the pulmonary capillaries into the alveoli. This establishes more normal blood levels of oxygen and carbon dioxide. Because of increased levels of oxygen, and decreased levels of carbon dioxide in the blood, the rate of respiratory movements is decreased.

6. In severe emphysema, the surface area for gas exchange is reduced so that not enough oxygen can diffuse from the alveoli into the pulmonary capillaries and not enough carbon dioxide can diffuse from the pulmonary capillaries into the alveoli even when oxygen therapy is given. The elevated carbon dioxide and the reduced oxygen levels in the blood both stimulate the respiratory center to produce an urge to take a breath.

7. The pH of the body fluids in patients who suffer from untreated diabetes mellitus decreases (becomes more acidic). A decreasing pH acts as a strong stimulus to the respiratory center. Consequently, the rate and depth of respiration increases. The increased rate and depth of respiration causes CO_2 to be released from the circulation at a greater rate. The lower blood level of CO_2 opposes the reduced pH because of the following reaction:

$$CO_2 + H_2O \rightleftharpoons H_2CO_3 \rightleftharpoons H^+ + HCO_3^-$$

As more CO_2 is removed from the blood, more H^+ and HCO_3^- combine to form H_2CO_3. Removal of the H^+ from the circulation resists a further reduction in the body fluid pH.

8. When Ima is hyperventilating, the stimulus for the hyperventilation is anxiousness, and the anxiousness is more important than the carbon dioxide in controlling respiratory movements. As the blood levels of carbon dioxide decrease during hyperventilation, vasodilation occurs in the periphery. As a result, the systemic blood pressure decreases. The systemic blood pressure can decrease enough so that the blood flow to the brain decreases. Decreased blood flow to the brain results in a reduced oxygen level in the brain tissue, thus causing dizziness. Breathing into a paper bag increases Ima's blood levels of carbon dioxide toward normal. Because the carbon dioxide does not increase above normal, it does not increase the urge to breathe. The more normal level of carbon dioxide prevents the peripheral vasodilation. As Ima breathes into the paper bag, the anxiousness is likely to subside. As the anxiousness subsides, the normal regulation of respiration resumes.

9. Hyperventilating before swimming decreases the amount of carbon dioxide in the blood. The decreased amount of carbon dioxide allows a longer than normal period of time before the swimmer has a strong urge to breathe. This can be dangerous, however, because the reduced level of carbon dioxide causes vasodilation in the periphery and reduced blood flow to the brain, and could cause a person to pass out. Passing out underwater could lead to drowning.

10. Immediately after the beginning of a race, runners increase their rate and depth of respiration before blood carbon dioxide levels have a chance to increase. The rate and depth of respiration increase in anticipation of increased muscular activity. As a result, for a short time the increased rate and depth of respiration can actually cause blood levels of carbon dioxide to decrease. The lower levels of blood carbon dioxide result in a slightly increased blood pH. After the race has progressed, the increased metabolic activity of the muscles can produce enough extra carbon dioxide to cause blood carbon dioxide levels to increase above resting levels, even with an increased rate and depth of respiration.

Chapter Sixteen

1. With the loss of the swallowing reflex, vomit can enter the larynx and trachea, block the respiratory tract, and keep the patient from breathing. Acidic stomach secretions cause severe inflammation and swelling of the respiratory passages.

2. Without adequate amounts of hydrochloric acid, the pH in the stomach is not low enough for the activation of pepsin. This loss of pepsin function results in inadequate protein digestion in the stomach. However, if the food is well chewed, proteolytic enzymes in the small intestine (e.g., trypsin, chymotrypsin) can still digest the protein. If the stomach secretion of intrinsic factor decreases, the absorption of vitamin B_{12} is hindered.

3. Even though bacteria apparently cause most ulcers, overproduction of hydrochloric acid because of stress is a possible contributing factor. Antibiotic therapy is an effective treatment, but reducing hydrochloric acid production is recommended. Possible solutions also include drugs that reduce stomach acid secretion, antacids to neutralize the stomach acid, smaller meals (distention of the stomach stimulates acid production), and proper diet. The person should avoid alcohol, caffeine, and large amounts of protein because they stimulate acid production. He should ingest fatty acids because they inhibit acid production, causing the release of gastric inhibitory polypeptide and cholecystokinin. Stress also stimulates the sympathetic nervous system, which inhibits duodenal gland secretion. As a result, the duodenum has less of a mucous coating and is more susceptible to gastric acid and enzymes. Relaxing after a meal helps decrease sympathetic activities and increase parasympathetic activities.

4. Obstruction of the common bile duct blocks the flow of bile from the gallbladder to the small intestine. As a result, stomach acids are not diluted and neutralized to as great an extent as if bile were present. Fats are not emulsified by the bile, resulting in decreased fat digestion and absorption. Excretory products such as bile pigments, cholesterol, and fats are not as readily removed from the body.

5. Introduction of food into the stomach increases the frequency of mass movements. The mass movements move feces into the rectum. Also, during the night, much of the material in the intestine has moved to the lower part of the large intestine. Mass movements following breakfast, therefore, are more likely to move feces into the rectum. Stretch of the rectum triggers the defecation reflex.

Chapter Seventeen

1. The recommended daily consumption of saturated fats is less than 20 g. Therefore, the % Daily Value for 2 g of fat is 10% (2/20 = 0.10 or 10%).

2. It is recommended that no more than 30% of the daily total kilocaloric intake be fats. For a 3000 kcal/day diet, that is 900 kcal (3000 × 0.30). There is 9 kcal/g of fat. Therefore, the maximum amount (weight) of total fat consumed per day is 100 g (900/9).

3. From question 2, the teenager's daily maximum amount (weight) of fat consumed per day is 100 g. If he consumes a 10 g serving of fat, his % Daily Value is 10% (10/100 = 0.10 or 10%).

4. From question 3, one serving equals 10% Daily Value. If he eats half of the contents of the package, he consumes two servings or 20% Daily Value.

5. An active person uses many more kilocalories in a day than a person who is not active, and so can lose weight on a higher kilocalorie diet. In addition, some people have a higher basal metabolic rate than others and tend to burn up more kilocalories.

6. Ingested proteins are digested to amino acids in the stomach and intestine and are then absorbed into the circulation. The amino acids can be used by cells as building blocks for proteins, which is very important if a person is attempting to build up muscle mass. However, beyond a certain point, the excess amino acids are broken down in the liver to make glucose. These amino acids are, therefore, no more helpful than any other source of energy. If the amount of amino acids is excessive, ammonia or keto acids, the breakdown products of amino acid metabolism, can accumulate to toxic levels.

7. Excess protein is metabolized and used as a source of energy. One of the breakdown products of protein is ammonia. Ammonia is toxic, so it is converted in the liver to urea, which is carried by the blood to the kidneys and eliminated.

8. As ATP breakdown increases, more ATP are produced to replace the ones used. Over an extended time, the ATP must be produced through aerobic respiration. Therefore, oxygen consumption and basal metabolic rate increase. The production of ATP requires the metabolism of carbohydrates, lipids, or proteins. As these molecules are used at a faster than normal rate, body weight decreases. Increased appetite and increased food consumption resist the loss in body weight. As ATP are produced and used, heat is released as a by-product. The heat raises body temperature, which is resisted by the dilation of blood vessels in the skin and by sweating.

9. No, this approach does not work because he is not losing stored energy from adipose tissue. In the sauna, he gains heat, primarily by convection from the hot air and by radiation from the hot walls. The evaporating sweat is removing heat gained from the sauna. The loss of water will make him thirsty, and he will regain the lost weight from drinking fluids and eating foods containing water.

10. Drinking cool water could help in two ways. Because the water is cool, raising the water to body temperature requires the expenditure of energy. Also, stretch of the stomach decreases appetite (see chapter 16).

11. During fever production, the body produces heat by shivering. The body also conserves heat by constriction of blood vessels in the skin (producing pale skin) and by reduction in sweat loss (producing dry skin). When the fever breaks; that is, "the crisis is over," heat is lost from the body to lower body temperature to normal. This is accomplished by dilation of blood vessels in the skin (producing flushed skin) and increased sweat loss (producing wet skin).

Chapter Eighteen

1. The rate of urine production increased over the next 2 hours, and the urine was dilute. The water he consumed increased his blood volume and decreased the concentration (osmolality) of his blood. Both of these changes inhibit ADH secretion from the posterior pituitary gland. The reduced ADH levels in the blood cause the kidney to produce a large volume of dilute urine.

2. The sodium chloride increased the concentration (osmolality) of his blood without affecting the volume of blood. The increased concentration (osmolality) of his blood stimulated ADH secretion from the posterior pituitary gland. The ADH caused the kidney to produce a small volume of concentrated urine.

3. The excessive sweating results in a reduced blood volume and the blood has a greater than normal concentration (osmolality). Both of these changes stimulate ADH secretion from the posterior pituitary gland. The ADH acts on the kidney to produce a small volume of concentrated urine.

4. Excessive urine production, low blood pressure, high plasma K^+ levels, but not high plasma Na^+ levels, result. Low aldosterone levels result in excessive loss of Na^+ in the urine. The Na^+ attract water molecules from the peritubular capillaries into the nephron, which results in a greater than normal urine volume. The loss of Na^+ in the urine reduces the plasma Na^+ concentration. The loss of water in the urine reduces the blood volume, and therefore, reduces the blood pressure. Because aldosterone causes Na^+ reabsorption from the nephron and K^+ secretion into the nephron, K^+ accumulate in the plasma.

5. The following all decrease the blood pressure in the glomerular capillaries and, therefore, decrease the glomerular filtration rate: decreased blood pressure and increased sympathetic stimulation that constricts the renal arteries and the afferent arterioles. Blocking the nephron causes the capsule pressure in the nephron to increase so it is equal to the pressure in the glomerular capillaries, which reduces the filtration rate. An increase in plasma proteins also decreases the filtration rate. The increased plasma protein concentration attracts water by osmosis and, therefore, reduces the tendency for water to pass from the glomerulus to Bowman's capsule.

6. Removal of a large amount of bicarbonate ion (HCO_3^-) from the body fluids causes a decrease in the body fluid pH because there are fewer HCO_3^- to combine with hydrogen ions (H^+) according to the following formula:

$$CO_2 + H_2O \rightleftharpoons H_2CO_3 \rightleftharpoons H^+ + HCO_3^-$$

carbon   water   carbonic   hydro-   bicarbon-
dioxide   acid   gen ion ate ion

Consequently, the H^+ accumulate in the body fluids and decrease the body fluid pH. The rate and depth of respiration increase because a decreased body fluid pH stimulates the respiratory center. The decrease in the body fluid pH causes the kidneys to secrete additional H^+ into the nephron. Therefore, the pH of the urine decreases.

7. While Spanky held his breath, carbon dioxide accumulated in his body fluids. The increased carbon dioxide combined with water to form carbonic acid that, in turn, produced H^+ and HCO_3^-. The increased concentration of H^+ in the body fluids caused the pH of the body fluids to decrease. After 2 minutes, the reduced pH and increased body fluid carbon dioxide levels strongly stimulated the respiratory center to increase the rate and depth of respiration. The increased rate of respiration reduced the body fluid carbon dioxide levels and increased the body fluid pH back toward its normal value. The kidneys did not respond to the change in pH within 2 minutes. The kidneys are more powerful regulators of body fluid pH, but they require several hours to become maximally active.

8. Martha's blood pH increased. Her frequent vomiting resulted in the loss of H^+ from her stomach and an increase in HCO_3^- in her blood. Since Martha lost fluid from her body and she consumed no fluid, the concentration of her body fluids increased and her blood volume decreased. Consequently, ADH secretion increased. Because blood volume decreases, the blood flow to the kidneys would decrease. This stimulates renin secretion from the kidney. This, in turn, converts angiotensinogen to angiotensin I. Angiotensin I is converted to angiotensin II, which stimulates aldosterone secretion from the adrenal cortex. In response to an increase in blood pH, the kidneys decrease their rate of H^+ secretion and HCO_3^- reabsorption. Consequently, this causes an increase in blood H^+ concentration and a decrease in blood HCO_3^-. The kidney's response to changes in pH takes many hours, but it has a large capacity to maintain pH homeostasis.

Chapter Nineteen

1. Removing the testes eliminates the source of testosterone. Therefore, blood levels of testosterone would decrease. Because testosterone has a negative-feedback effect on the hypothalamus and pituitary gland, GnRH, FSH, and LH secretion increase and the blood levels of these hormones increase.

2. The estrogen and progesterone in birth control pills inhibit the large increase in LH secretion from the anterior pituitary, which is responsible for ovulation. Without the large increase in LH secretion, ovulation cannot occur.

3. (E) is the correct answer. The secretory phase of the menstrual cycle occurs after ovulation. It is following ovulation that the corpus luteum forms and produces progesterone. In addition, the progesterone acts on the endometrium of the uterus to cause its maximum development. Therefore, progesterone secretion reaches its maximum levels and the endometrium reaches its greatest degree of development during the secretory phase of the menstrual cycle.

4. Between approximately 12–14 days of the menstrual cycle, you would normally expect increasing blood levels of estrogen and LH. In the average menstrual cycle, ovulation occurs on day 14. Prior to that time estrogen levels are increasing. The increasing estrogen stimulates LH secretion. The increasing LH, in turn, causes increased estrogen secretion from the developing follicle. Blood progesterone levels are low at this time. Progesterone is not secreted in large amounts until the corpus luteum is formed after ovulation.

5. On day 15 of the menstrual cycle, one would expect decreasing blood levels of LH, decreasing blood levels of estrogen, and increasing blood levels of progesterone. In the average menstrual cycle, ovulation occurs on day 14. After ovulation, the ovulated follicle develops into the corpus luteum. Estrogen levels decrease as the follicle is converted to the corpus luteum. The corpus luteum secretes progesterone in increasing quantities and small amounts of estrogen as the corpus luteum develops. The increasing blood levels of progesterone inhibit LH secretion from the pituitary gland. Consequently, blood levels of LH fall to low levels.

6. A drug that blocks the effect of progesterone on its target tissues will cause the tissue to respond as if no progesterone is present.

Progesterone is secreted by the corpus luteum for about 7 days after ovulation. The progesterone affects the endometrium of the uterus and, in response, the endometrium becomes prepared for implantation. A decline in progesterone causes menstruation to occur. If a drug is taken 3 to 4 days following ovulation that blocks the effect of progesterone, the endometrium will not become fully prepared for implantation and menstruation will begin early. If a drug that blocks progesterone is taken by a woman who is pregnant the effect will be to cause the endometrium to sluff. This is similar to the events that occur during menstruation. If this occurs it will terminate the pregnancy. Progesterone is necessary to maintain pregnancy. The corpus luteum of the ovary secretes progesterone until the end of the first three months of pregnancy, prior to the third month of pregnancy the placenta begins to secreting progesterone. The placenta becomes the primary source of progesterone after the third month of pregnancy.

7. After menopause estrogen and progesterone are produced by the ovary in only small amounts. Consequently estrogen and progesterone levels in the blood are low. These hormones have a negative feedback effect on the secretion of FSH and LH from the anterior pituitary gland. Therefore, the absence of estrogen and progesterone, FSH and LH are secreted in greater amounts and the blood levels of FSH and LH increase. However, there is no large increase in LH, like the increases that occur prior to ovulation. The average concentration of LH and FSH is greater than the levels that occur either before or after ovulation.

Chapter Twenty

1. Fertilization occurs approximately 14 days after the LMP, so that the embryo in this case is 30 days old. It has just completed the "budding" period so that limb buds and other "buds" have just formed (see table 20.2).

2. At 35–45 days post-LMP, the embryo would be 21–31 days old, which overlaps the period of neural tube closure (18–26 days). A high fever during this period could prevent neural tube closure in the embryo so that the newborn baby may have neural tube defects such as anencephaly (open neural tube in the area of the brain, resulting in absence of the upper brain) or spina bifida (open spinal cord resulting from failure of the neural tube to close in that area).

3. The correct answer is (A). Prolactin is responsible for the production of milk by the breast. Surges of prolactin are stimulated by the process of suckling and are required to maintain milk production. Drugs that inhibit prolactin release will cause the breast to cease milk production after a few days.

4. Yes. Because dimpled cheeks are dominant, a person with dimpled cheeks may have a genotype of Dd ("D" being the dominant gene for dimples and "d" being the normal gene). If two Dd, dimpled people have children, it may be expected that approximately 1/4 of their children would not have dimples (dd).

5. No, not without additional information. To be able to roll their tongue into a tube, both mother and son need to have only one dominant gene. The son could inherit this dominant gene from his mother, so it is possible that his father is a nontongue roller. It is also possible that his father is a tongue roller, but there is no proof for either hypothesis.

6. Hemophilia is an X-linked recessive gene. A female must have the genotype $X^h X^h$ to have hemophilia, whereas because males have only one X chromosome, a single hemophilia gene will cause them to have the disorder ($X^h Y$). For half their children to have hemophilia, the mother's genotype must be $X X^h$ (she does not have hemophilia) and the father's genotype must be $X^h Y$ (he has hemophilia). If the woman did not carry the hemophilia gene, none of her daughters or sons could have the disorder.

Appendix E

Answers to Predict Questions

Chapter One

1. Donating a pint of blood results in a decrease in blood pressure. Negative-feedback mechanisms, such as an increase in heart rate, return blood pressure toward a normal value. When a negative-feedback mechanism fails to return a value to its normal level, the value can continue to deviate from its normal range. Homeostasis is not maintained in this situation, and the person's health or life can be threatened.

2. The thirst sensation is associated with a decrease in body fluid levels. The thirst mechanism causes the person to drink fluids, which returns the fluid level to normal. Thirst is therefore a sensation involved in negative-feedback control of body fluids.

3. When a boy is standing on his head, his nose is superior to his mouth. Remember that directional terms refer to a person in the anatomical position and not to the body's current position.

4. The spleen is in the left-upper quadrant, the gallbladder is in the right-upper quadrant, the left kidney is in the left-upper quadrant, the right kidney is in the right-upper quadrant, the stomach is mostly in the left-upper quadrant, and the liver is mostly in the right-upper quadrant.

5. There are two ways. First, the visceral peritoneum wraps around organs. Thus the peritoneal cavity surrounds the organ, but the organ is not inside the peritoneal cavity. The peritoneal cavity contains only peritoneal fluid. Second, retroperitoneal organs are in the abdominopelvic cavity, but they are between the wall of the abdominopelvic cavity and the parietal peritoneal membrane.

Chapter Two

1. The mass (amount of matter) of the astronaut on the surface of earth and in outer space does not change. In outer space, where the force of gravity from earth is very small, the astronaut is "weightless" compared with his weight on earth's surface.

2. Fluorine has 9 protons (the atomic number), 10 neutrons (the mass number minus the atomic number), and 9 electrons (equal to the number of protons).

3. Because atoms are electrically neutral, the iron (Fe) atom has the same number of protons and electrons. The loss of three electrons results in an iron ion that has three more protons than electrons and therefore a charge of $+3$. The correct symbol is Fe^{3+}.

4. Carbon dioxide and water are in equilibrium with H^+ and HCO_3^- ions. A decrease in CO_2 causes some H^+ to react with HCO_3^- to form CO_2 and H_2O. Consequently, the H^+ concentration decreases.

5. During exercise, muscle contractions increase. This requires the release of potential energy during chemical reactions. For example, the potential energy in the phosphate bond of ATP is released when ATP is broken down into ADP and P_i. Some of the energy is used to drive muscle contractions, and some of it is released as heat. Because the rate of these reactions increases during exercise, more heat is produced than when at rest, and body temperature increases.

6. A base added to a solution combines with H^+, which decreases the H^+ concentration. By definition, a decrease in H^+ concentration is an increase in pH. Note that buffers take up H^+ when an acid is added to a buffered solution. It is reasonable to expect a buffer to release H^+ when a base is added to a buffered solution. The released H^+ can combine with the base to prevent an increase in pH. For example, H^+ can combine with OH^- to form H_2O.

Chapter Three

1. Urea is produced continually by liver cells and diffuses from the cells into the blood. If the kidneys stop eliminating urea, it begins to accumulate in the blood and in the liver cells. The urea finally reaches concentrations high enough to be toxic to cells, causing cell damage followed by cell death.

2. Glucose transported by facilitated diffusion across the cell membrane moves from a higher to a lower concentration. If glucose molecules are converted quickly to some other molecule as they enter the cell, a large concentration difference is maintained, and thus glucose transport into the cell continues proportional to the magnitude of the concentration difference.

3. a. Cells specialized to synthesize and secrete proteins have abundant rough ER, because this is an important site of protein synthesis. Well-developed Golgi apparatuses exist to package proteins in secretory vesicles, and numerous secretory vesicles are present.

   b. Cells highly specialized to actively transport substances into the cell have a large surface area, such as microvilli, exposed to the fluid from which substances are actively transported. Numerous mitochondria are present near the membrane across which active transport occurs because the process requires energy.

   c. Cells highly specialized to ingest foreign substances by endocytosis have numerous lysosomes in their cytoplasm and evidence of vesicles containing foreign substances.

4. Changing a single nucleotide within a DNA molecule also changes the nucleotide sequence of the mRNA produced from that segment of DNA. The change in mRNA results in a different codon, and a different amino acid is placed in the amino acid chain for which the mRNA codes. Because a change in the amino acid sequence of a protein can change its structure, one substitution of a nucleotide in a DNA chain can result in altered protein structure and function.

5. Chloride ions do not move in normal amounts out of the cells of people with cystic fibrosis because Cl^- channels are defective. Instead, the Cl^- tend to accumulate inside the cell. Potassium ions tend to move out of muscle and nerve cells down their concentration gradient. The positively charged K^+, however, are attracted by the negatively charged Cl^- accumulated inside the cell. This attraction reduces the movement of K^+ out of the cell and causes more K^+ to accumulate inside the cell.

6. Cancer cells generally appear to be undifferentiated. Instead of dividing and

then undergoing differentiation, they continue to divide and do not differentiate. One measure of the severity of cancer is related to the degree of differentiation the cancer cells have undergone. Those that are more differentiated divide more slowly and are less dangerous than those that differentiate little.

Chapter Four

1. a. If nonkeratinized stratified squamous epithelium lined the digestive tract, it would provide protection against abrasion but would hinder the secretion of digestive enzymes and the absorption of digested food. Nonkeratinized stratified squamous epithelium is not specialized to absorb or secrete because the many layers of cells hinder the movement of materials across the epithelium. Squamous cells, as opposed to cuboidal or columnar cells, do not contain the number of organelles necessary to support the production and transport of large numbers of complex molecules.

   b. Keratinized stratified epithelium forms a tough layer that is a barrier to the movement of water. Replacing the epithelium of skin with moist stratified squamous epithelium would increase the loss of water across the skin because water can diffuse through nonkeratinized stratified squamous epithelium, and it is more delicate and provides less protection than keratinized stratified squamous epithelium.

2. When a muscle contracts, the pull it exerts is transmitted along the length of its tendons. The tendons need to be very strong in that direction but not as strong in others. The collagen fibers, which are like microscopic ropes, are therefore all arranged in the same direction to maximize their strength. In the skin, collagen fibers are oriented in many directions because the skin can be pulled in many directions. The collagen fibers can be somewhat randomly oriented, or they can be organized into alternating layers. The fibers within a layer run in the same direction, but the fibers of different layers run in different directions.

3. Collagen synthesis is required for scar formation. If collagen synthesis does not occur because of a lack of vitamin C or if collagen synthesis is slowed, wound healing does not occur or is slower than normal. One might expect that the density of collagen fibers in a scar is reduced and the scar is not as durable as a normal scar.

4. There is more than one way to organize a table that summarizes the characteristics of the major muscle types. See the following table.

Table D.1 Major Characteristics of the Three Muscle Types

Muscle Type	Nuclei in Each Cell	Location of Nuclei	Control	Cell Shape	Striated	Branching Fibers
Skeletal	Many	Peripheral	Voluntary	Long and cylindrical	Yes	No
Cardiac	One	Central	Involuntary	Branching cylinders joined by intercalated disks	Yes	Yes
Smooth	One	Central	Involuntary	Spindle-shaped cells	No	No

5. In severely damaged tissue in which cells are killed and blood vessels are destroyed, the usual symptoms of inflammation cannot occur. Surrounding these areas of severe tissue damage, however, where blood vessels are still intact and cells are still living, the classic signs of inflammation do develop. The signs of inflammation therefore appear around the periphery of severely injured tissues.

6. Suturing large wounds brings the edges of the wounds close together. Healing is therefore more rapid, there is less danger of infections, less scar tissue is formed, and wound contracture is greatly reduced.

7. The circulatory system and the lymphatic system are routes by which malignant cells can spread to distant sites. Cells can enter the blood and pass through the circulatory system to distant sites, such as the brain and lungs. Cells can also enter lymph and spread to lymph nodes or distant sites where they initiate tumors. The lymph nodes closest to the original tumor are affected first, and those more distant are affected later.

Metastasis to the nervous system can result in tumors that affect the function of the brain. Pain, paralysis, loss of sensation, and death may result as tumors compress and destroy nervous tissue in the brain or spinal cord.

The lungs are common sites for metastasis. Malignant tumors destroy lung tissue and block air passageways, resulting in reduced gas exchange. Irritation and inflammation of lung tissue caused by the cancer as well as the possibility of secondary infections can cause coughing.

Chapter Five

1. Because the permeability barrier is composed mainly of lipids surrounding the epidermal cells, substances that are lipid-soluble can easily diffuse through the barrier. This fact is used as a basis for administering some medications through the skin. On the other hand, water-soluble substances have difficulty diffusing through the skin. The lipid barrier of the skin prevents water loss from the body.

2. a. The lips are pinker or redder than the palms of the hand. Several explanations for this are possible. There could be more blood vessels in the lips, there could be increased blood flow in the lips, or the blood vessels could be easier to see through the epidermis of the lips. The last possibility actually explains most of the difference in color between the lips and palms. The epidermis of the lips is thinner and not as heavily keratinized as that of the palms. In addition, the dermal papillae containing the blood vessels in the lips are "high" and closer to the surface.

   b. A person who does manual labor has a thicker stratum corneum (and possibly calluses) than a person who does not perform manual labor. The thicker epidermis masks the underlying blood vessels, and the palms do not appear as pink. In addition, carotene accumulating in the lipids of the stratum corneum might give the palms a yellowish cast.

   c. The posterior surface of the forearm appears darker because of the tanning effect of ultraviolet light from the sun. The posterior surface of the forearm is usually exposed to more sunlight than the anterior surface of the forearm.

3. The story is impossible. Hair color results from melanin that is added to the hair in the hair bulb as the hair grows. The hair itself is dead. To turn white, the hair must grow out without the addition of melanin. This, of course, takes considerably more time than one night.

4. On cold days, skin blood vessels of the ears and nose can dilate, bringing warm blood to the ears and nose, thus preventing tissue damage from the cold. The increased blood flow makes the ears and nose appear red.

5. Reducing water loss is one of the normal functions of the skin. Loss of or damage to the skin can greatly increase water loss, resulting in dehydration and reduced urine production. To counteract the increased loss of fluid from burns, during the first 24 hours following the injury large volumes of fluid are administered. But, how much fluid should be given? The amount of fluid given should be sufficient to match that lost plus enough to allow the kidneys to function. Urine output is therefore monitored. If it is too low, more fluid is administered, and if it is too high, less fluid is administered. An adult receiving intravenous fluids should produce 30–50 mL of urine per hour, and children should produce 1 mL/kg of body weight per hour.

Chapter Six

1. If all the mineral is removed, the bone becomes so flexible that it can be tied into a knot. This can be accomplished by soaking a bone in vinegar (a weak acid) for an extended time. The bone will not be rigid enough to support weight.

   If all the collagen is removed, the bone becomes very brittle and can be easily broken. Because of collagen loss, the bones of many older people break easily.

2. If cartilage growth fails to occur, the bone is normal in diameter (or even greater in diameter than normal), because growth in diameter does not require cartilage growth, but much shorter than normal. This is the condition seen in one type of dwarfism, in which the head and trunk are normal in size, but the long bones of the limbs are very short.

3. Tears are produced in lacrimal glands in the superior lateral corner of the orbit. The tears run across the surface of the eye and enter the duct that passes through the nasolacrimal canal to the nasal cavity.

The extra moisture in the nasal cavity causes a "runny nose."

4. Just before the swimmer begins the power stroke, the arm is flexed and medially rotated, and the forearm is extended and pronated. During the power stroke, the arm is powerfully extended, slightly abducted, and medially rotated. During the recovery stroke, the arm is circumducted, laterally rotated, and flexed in preparation for the next stroke. The forearm is flexed during the first part of the recovery stroke and extended during the last part.

5. Taking in adequate calcium and vitamin D through the digestive system during adulthood increases calcium absorption from the small intestine. The increased calcium is used to increase bone mass. The greater the bone mass before the onset of osteoporosis, the greater the tolerance for bone loss later in life. For this reason it is important for adults, especially women in their 20s and 30s, to ingest adequate amounts of calcium. Exercising the muscular system places stress on bone, which also increases bone density. The granddaughter should not smoke because this reduces estrogen levels.

Chapter Seven

1. a. If ATP levels are low in a muscle fiber before stimulation, the force of contraction is reduced because there is insufficient ATP for all motor units to contract.

   b. If action potentials occur at a frequency so great that calcium is not transported back into the sarcoplasmic reticulum between individual action potentials, the muscle continues to contract and does not relax.

2. During a 10-mile run, aerobic metabolism is the primary source of ATP production for muscle contraction. Anaerobic metabolism provides the short (15–20 s) burst of energy for the sprint at the finish. After the race, aerobic metabolism is elevated for a time to repay the oxygen debt, causing the heavy breathing after the race.

3. Raising eyebrows—occipitofrontalis; winking—orbicularis oculi; whistling—orbicularis oris and buccinator; smiling—zygomaticus; frowning—depressor anguli oris; sneering—levator labii superioris.

4. Shortening the right sternocleidomastoid muscle rotates the head to the left and also slightly elevates the chin.

5. In the sprinter's stance and the bicyclist's racing posture, the thigh is flexed at a

45-degree angle because at that angle the gluteus maximus functions at its maximum in extending the thigh, thus providing maximum force.

6. DMD affects the muscles of respiration and causes deformity of the thoracic cavity. The reduced capacity of muscle tissue to contract is one factor that reduces the ability to breathe deeply or cough effectively. In addition, the thoracic cavity can become severely deformed because of the replacement of skeletal muscle with connective tissue. The deformity can reduce the ability to breathe deeply. DMD can also affect the muscle of the heart and cause heart failure.

Chapter Eight

1. Dorsal root ganglia are larger in diameter because they contain sensory neuron cell bodies, which are larger than the axons of the spinal nerves.

2. Damage to the right phrenic nerve results in the absence of muscular contraction in the right half of the diaphragm. Because the phrenic nerves originate in the cervical region of the spinal cord, damage to the spinal cord in the thoracic region does not affect the diaphragm. Damage to the upper cervical region, however, cuts the connection between the upper and lower motor neurons. This eliminates phrenic nerve stimulation of the diaphragm and dramatically interferes with breathing. Death is likely to occur.

3. Nuclei within the medulla oblongata regulate heart rate, blood vessel diameter, breathing, swallowing, vomiting, coughing, sneezing, balance, and coordination. Even though all of these functions are important, loss of some of them may not necessarily result in death. Loss of cardiovascular regulation or loss of breathing regulation, however, could result in death. Because both blood flow and respiration are vital functions, interference with either of them may inhibit normal homeostatic functions. If not corrected, the loss of homeostasis results in death. Neuronal control of breathing is more critical than cardiovascular control. Death can occur in minutes if neuronal control of respiration is lost. Neuronal control is not absolutely necessary for the heart to continue beating. Without cardiovascular regulation, however, the blood pressure is reduced, resulting in shock, which can ultimately result in death (see chapters 12 and 13).

4. If a person holds an object in her hand, sensations from the skin of the hand are sent to the primary somatic sensory

cortex. The information is then passed to the somatic sensory association area, where the object is recognized. Action potentials then travel to the sensory speech area, where the object is given a name. From there, action potentials travel to the motor speech area, where the spoken word is initiated. Action potentials from the motor speech area travel to the premotor area and then to the primary motor cortex, where action potentials are initiated that stimulate the muscles necessary to formulate the word.

5. a. In a person who is extremely angry, the sympathetic division of the autonomic nervous system is activated, and the expected responses include increased heart rate, increased blood pressure, dilated pupils, and perspiration.
   b. In a person who has just finished eating and is now relaxing, the parasympathetic division of the autonomic nervous system is the primary autonomic division functioning. The responses include decreased heart rate, decreased blood pressure, and increased digestive activities.

6. The stroke was on the left side of the brainstem. Both the motor and sensory neurons to the right side of the body are located in the left cerebral cortex. At the level of the upper medulla oblongata, neither the motor or sensory pathways to the limbs have yet crossed over to the left side of the CNS. Loss of pain and temperature to the left side of the face indicates that the lesion occurred at a level where the nerve fibers from the face had entered the CNS but had not yet crossed (in the brainstem).

Chapter Nine

1. The pain is diffuse because deep or visceral pain, such as in the colon, is not highly localized because of the absence of tactile receptors in the deeper structures. The pain is referred to the skin over the lower central abdomen (see figure 9.2). This occurs because sensory neurons from the superficial area to which the pain is referred and the neurons from the colon, where the pain stimulation originated, converge onto the same ascending neurons in the spinal cord.

2. Much of taste is based on olfactory function. A cold may include a stuffy nose, which decreases airflow and increases the thickness of the mucus covering the olfactory epithelium and may interfere with olfaction and, thus, with taste.

3. Medications placed into the eyes can pass through the nasolacrimal duct into the nasal cavity, where their odor can be detected. Because much of our taste sensation is actually smell, the medication is perceived to have a taste.

4. Cones are responsible for color vision, but they are not nearly as sensitive to light as the rods. Rods are much more sensitive to light, but they do not detect differences in color. In dim light, when rods rather than cones are responsible for vision, objects therefore appear only in shades of gray.

5. While you are driving, your vision is focused out on the road, some distance in front of the car. As a result, the ciliary muscles are relaxed and the lens is relatively flat, allowing for distant vision. When you look down at the speedometer, the eye accommodates for close vision, the ciliary muscles contract, pulling the ciliary body toward the lens. This reduces the tension on the suspensory ligaments of the lens and allows the lens to assume a more spherical form. When you look back up at the road, the ciliary muscles again relax and the lens flattens, allowing for distant vision. Even though these changes occur very quickly, there is a brief moment, before accommodation occurs, when you first look at the speedometer that it is out of focus. When you look back at the road, it is also briefly out of focus, as the lens changes back for distant vision.

6. When you hear a faint sound, you turn your head toward it because sound waves are collected by the auricle and conducted through the external acoustic meatus toward the tympanic membrane. The auricle is shaped in such a way that sound waves coming from the side and front of the head are most efficiently conducted into the external acoustic meatus. Because a specific sound reaches each ear at slightly different times, a person can localize the origin of the sound. Turning the head toward the sound facilitates maximum accumulation of the sound waves by the ear. In addition, reflexes integrated in the superior colliculi cause the head and eyes to turn toward a sound so that you can see what is making the sound.

Chapter Ten

1. Because the drug binds to a receptor and prevents the response of a target tissue to a chemical signal and because the drug is lipid-soluble, it is likely that the drug diffuses across the cell membrane of the cell and binds to the receptor for the chemical signal, which is inside the cell, and prevents the chemical signal from

binding to its receptor site. Thus, the chemical signal functions by diffusing across the cell membrane and binding to an intracellular receptor. The response of the intracellular receptor is to produce new messenger RNA, which leads to the synthesis of new proteins (see chapter 3 for a description of the role of messenger RNA in protein synthesis). The new proteins produce the response of the cell to the chemical signal.

2. The doctor might explain that GH administration would cause his son to grow taller. He might also explain that he might develop unwanted changes in his skeletal structure consistent with acromegaly. Other side effects are also possible. Examples include abnormal joint formation and diabetes mellitus.

3. The protein that is similar to TSH causes oversecretion of the thyroid gland (hyperthyroidism). Because the production of the protein cannot be inhibited by thyroid hormones, oversecretion of the thyroid gland is prolonged and symptoms associated with hypersecretion of thyroid hormones become obvious and the thyroid gland enlarges. In addition, the increased thyroid hormones have a negative-feedback effect on the hypothalamus and pituitary gland. TSH-releasing hormone secretion from the hypothalamus and TSH secretion from the anterior pituitary gland are inhibited.

4. Insufficient vitamin D results in insufficient Ca^{2+} absorption by the intestine. As a result, blood Ca^{2+} levels begin to fall. In response to the low blood Ca^{2+} levels, PTH is secreted from the parathyroid glands. PTH acts primarily on bone, causing bone to be broken down and Ca^{2+} to be released into the blood to maintain blood Ca^{2+} levels within the normal range. In adults, so much Ca^{2+} can eventually be removed from bones that they become soft, fragile, and easily broken. In adults the condition is called osteomalacia, and in children, the condition is called ricketts. The bones of children can become bent and deformed.

5. Large doses of cortisone can damage the adrenal cortex because cortisone inhibits ACTH secretion from the anterior pituitary gland. ACTH is required to keep the adrenal cortex from undergoing atrophy. Prolonged use of large doses of cortisone can cause the adrenal gland to atrophy so much that it cannot recover if ACTH secretion does increase again.

6. Reduced aldosterone secretion causes Na^+ and water to not be retained. Consequently, blood Na^+ levels and blood volume

decrease, resulting in low blood pressure. Also, blood K^+ levels increase. High blood K^+ levels lead to altered nerve and muscle function. These changes can be life-threatening.

7. After a large meal, glucose enters the blood from the intestine. The increasing blood glucose stimulates insulin secretion and decreases glucagon secretion. Well before 12 hours without eating, blood glucose would have started to decrease. Decreasing blood glucose levels result in a decreased rate of insulin secretion and a stimulation of glucagon secretion.

8. The pineal body secretes melatonin, which inhibits the release of reproductive hormones by acting on the hypothalamus of the brain. If the pineal body secretes less melatonin, it no longer should have an inhibitory effect on the hypothalamus. As a result, reproductive hormones could be secreted in greater amounts, which would result in exaggerated development of the reproductive system in young people with this condition. The evidence for this mechanism is not as clear in humans as it is in other animals.

9. Removal of part of the thyroid gland reduces the amount of thyroid hormone secreted by the gland. Usually enough thyroid tissue can be removed to cause the amount of thyroid hormone secreted to be reduced to a normal range of values. In addition, the remaining thyroid tissue normally does not hypertrophy enough to cause the thyroid tissue to produce more than enough thyroid hormone, although there are exceptions. The removal of the thyroid tissue does not remove the influence of the abnormal antibodies on the tissues behind the eyes. Thus, in many cases the effect of the condition on the eyes is not improved.

Chapter Eleven

1. Carbon monoxide binds to the iron of hemoglobin and prevents the transport of oxygen. The decreased oxygen stimulates the release of erythropoietin from the kidneys. Erythropoietin increases red blood cell production in red bone marrow, causing the number of red blood cells in the blood to increase.

2. The white blood cells shown in figure 11.7 are (a) lymphocyte, (b) basophil, (c) monocyte, (d) neutrophil, and (e) eosinophil.

3. People with type AB blood were called universal recipients because they could receive type A, B, AB, or O blood with little likelihood of a transfusion reaction. Type AB blood does not have antibodies

against type A or B antigens. Transfusion of these antigens in type A, B, or AB blood does not therefore cause a transfusion reaction in a person with type AB blood. The term is misleading, however, for two reasons. First, other blood groups can cause a transfusion reaction. Second, antibodies in the donor's blood can cause a transfusion reaction. For example, type O blood contains anti-A and anti-B antibodies that can react against the A and B antigens in type AB blood.

4. The donor blood used in exchange transfusions for the treatment of HDN should be Rh-negative, even though the newborn is Rh-positive. Rh-negative red blood cells do not have Rh antigens. Therefore, any anti-Rh antibodies in the newborn's blood do not react with the transfused Rh-negative red blood cells. Giving Rh-negative blood to a Rh-positive newborn does not change the blood type of the newborn because blood type is determined genetically. Eventually, all of the Rh-negative red blood cells die, and only Rh-positive red blood cells are produced by the newborn.

5. An increase in the white blood cell count often indicates a bacterial infection. A white blood cell differential count with an abnormally high neutrophil percentage supports the diagnosis.

Chapter Twelve

1. The anterior interventricular artery supplies blood to the anterior wall of the heart and to much of the left ventricle. A blocked anterior interventricular artery reduces the oxygen supply to the portion of the heart that is supplied by that artery, and the cardiac muscle in that area is not able to contract effectively. Thus, the left ventricle on the anterior surface of the heart does not contract normally.

2. It is important to prevent tetanic contractions in cardiac muscle because the cycle of contraction and relaxation stops during tetanic contractions. This would cause the pumping action of the heart to stop. In skeletal muscle, the cycle of contraction and relaxation is not important as a pump, but it is important to maintain a static contracted state. This is essential to maintaining posture or to holding a limb in a specific position.

3. If the normal blood supply is reduced in a small area of the heart through which the left bundle branch passes, conduction of action potentials through that side of the heart is slowed or blocked. As a consequence, the left side of the heart contracts more slowly. The right side of

the heart contracts more normally. The reduced rate of contraction of the left ventricle reduces the pumping effectiveness of the left ventricle.

4. A person who has a damaged left bundle branch will exhibit the consequences outlined for Predict question 3, but the electrocardiogram will also be altered. The QRS complex results from depolarization of the ventricles. Action potentials pass through the right bundle branch normally but conduction of action potentials through the left bundle branch is slowed because of damage. The QRS complex has an abnormal shape and it is prolonged. If many ectopic action potentials arise in the atria, heart rate increases. Each ectopic action potential initiates a new heart beat. It is possible for some ectopic action potentials arising in the atria to occur while the ventricle is depolarized; but these action potentials do not initiate ventricular contractions. There can, therefore, be more P waves than QRS complexes in the electrocardiogram. If ectopic action potentials do not occur in a regular fashion, they can cause the heart to beat at an irregular rate, or arrhythmically.

5. A leaky aortic semilunar valve results in an increased left ventricular volume just before ventricular contraction. During ventricular relaxation, the aortic semilunar valve closes in a normal person, and blood flows out of the left ventricle into the aorta. When the aortic semilunar valve is incompetent, some blood leaks back into the left ventricle from the aorta during ventricular relaxation. When this blood is added to the blood that normally enters the left ventricle from the left atrium, there is a greater than normal volume of blood in the left ventricle just before ventricular contraction.

A severely narrowed opening through the aortic semilunar valve increases the amount of work the heart must do to pump the normal volume of blood into the aorta. A greater pressure is required in the ventricle to force the same amount of blood through the narrowed opening during ventricular contraction.

6. Most of the ventricular contraction occurs between the first and second heart sounds of the same beat. Between the first and second heart sounds, blood therefore is ejected from the ventricles into the pulmonary trunk and the aorta. Between the second heart sound of one beat and the first heart sound of the next beat, the ventricles are relaxing and the semilunar valves are closed. No blood passes from the ventricles into the aorta or pulmonary trunk during that period.

7. The shhh sound made after a heart sound is created by the backward flow of blood after closure of a leaky or incompetent valve. A swishing sound immediately after the second heart sound (lubb–duppshhh) represents a leaky aortic semilunar or pulmonary semilunar valve. The shhh sound before a heart sound is created by blood being forced through a narrowed, or stenosed, valve just before the valve closes. The lubb–shhhdupp suggests that there is a swishing sound immediately before the second heart sound; thus indicating a stenosed aortic or pulmonary semilunar valve.

8. In response to severe hemorrhage, blood pressure decreases, which is detected by baroreceptors. A reduced frequency of action potentials is sent from the baroreceptors to the medulla oblongata. This causes the cardioregulatory center to increase sympathetic stimulation of the heart and increase the heart rate. Sympathetic stimulation of the heart also increases stroke volume, as long as the volume of blood returned to the heart is adequate. Following hemorrhage, however, the blood volume in the body is reduced, and the venous return to the heart from the body is reduced. As a consequence, the volume of blood in the heart is lower than normal. Because of Starling's law, the stroke volume is reduced. The heart rate is increased, but the volume of blood returning to the heart is decreased; thus the ventricle does not fill with blood. As a consequence, the stroke volume is low, and the heart rate is high.

9. Rupture of the left ventricle can occur several days after a myocardial infarction. As the necrotic tissues are being removed by macrophages, the wall of the ventricle becomes thinner and may bulge during systole. If the wall of the ventricle becomes very thin before new connective tissue is deposited, it can rupture. If the left ventricle ruptures, blood flows from the left ventricle into the pericardial sac, resulting in cardiac tamponade. As blood fills the pericardial sac, it compresses the ventricle from the outside. As a consequence, the ventricle is not able to fill with blood and its pumping ability is rapidly eliminated. Rupture of the left ventricular wall quickly results in death.

Chapter Thirteen

1. Atherosclerosis slowly increases the resistance to blood flow. Blood flow through atherosclerotic carotid arteries to the brain therefore decreases. In advanced stages of arteriosclerosis, the resistance to blood flow increases so much that the blood flow to the brain is reduced significantly. Results include confusion, loss of memory, and a reduced ability to perform other normal brain functions.

2. If a thrombus in the posterior tibial vein gave rise to an embolus, it would travel through the following vessels before lodging in the lungs: posterior tibial vein, popliteal vein, femoral vein, external iliac vein, common iliac vein, inferior vena cava, right atrium, right ventricle, pulmonary trunk, pulmonary artery. The lung is the most likely place for the embolus to lodge because the pulmonary artery branches and gives rise to smaller and smaller arteries until the arteries deliver blood to the pulmonary capillaries. The embolus will lodge in one of the branches of the pulmonary artery in the lung and block it. All of the other vessels through which the embolus passes are large in diameter.

3. Premature beats of the heart result in contraction of the heart muscle before the heart has time to fill to its normal capacity. The volume of blood ejected by the left ventricle of the heart is therefore reduced (reduced stroke volume) (see chapter 12). The reduced stroke volume is responsible for the weak pulse. In response to cardiovascular shock that is due to hemorrhage, the stroke volume is also reduced because less blood flows into the heart between beats. The reduced stroke volume is responsible for the weak pulse. In a person who is exercising, both the heart rate and the stroke volume increase. The increased stroke volume results in a stronger than normal pulse.

4. a. Decreased plasma protein concentration reduces the osmotic pressure of the blood. Edema results primarily because the movement of fluid into the venous end of the capillary as a result of osmosis is reduced. Consequently, less fluid returns to the capillary at its venous end, and fluid accumulates in the interstitial space, resulting in edema.

   b. Increased blood pressure within the capillary forces more fluid to leave the capillary at its arteriolar end. There is no increased tendency for fluid to reenter the capillary at its venous end. The extra fluid that leaves the capillary can accumulate in the tissue space and result in edema.

5. When a blood vessel is blocked (when the legs are crossed, for instance), oxygen and nutrients are depleted, and waste products accumulate in tissue supplied by the blocked blood vessel. The reduced supply of oxygen and nutrients and the accumulated waste products all cause relaxation of the precapillary sphincters and greatly increase blood flow through the area when the block is removed.

6. Severe vasoconstriction that results from Raynaud's syndrome causes the digits to appear white because of the lack of blood flow through the capillary beds in the digits. The intensity of the vasoconstriction is more severe when the digits are exposed to cold temperatures. If the vasoconstriction is severe enough, there is not enough blood flow to the digits to supply nutrients to the tissues in the digits. A result may be the development of necrotic tissue and gangrene.

7. The rapid loss of a large volume of blood activates the mechanisms responsible for maintaining blood pressure. This includes the baroreceptor reflexes and, if blood oxygen levels decrease substantially and carbon dioxide levels increase (and pH decreases), the chemoreceptor reflexes are activated. In addition, the adrenal medullary mechanism, the renin-angiotensin-aldosterone mechanism, and the vasopressin mechanism are all activated. The baroreceptor mechanism, adrenal medullary mechanism, and chemoreceptor mechanism increase the heart rate and result in vasoconstriction of blood vessels, especially of the skin and the viscera. Angiotensin II is produced quickly and it causes vasoconstriction and stimulates aldosterone secretion. Aldosterone requires up to 24 hours to become maximally active. It increases water reabsorption from the kidneys and reduces the loss of water in the form of urine. All of these mechanisms function to increase blood pressure back to its normal value.

   If blood is lost over several hours, the decrease in blood pressure is not as dramatic as when blood loss occurs quickly. Consequently, mechanisms that respond to a rapid and large decrease in blood pressure are stimulated to a lesser degree. These include the chemoreceptor reflex, the vasopressin mechanism, and the adrenal medullary mechanism. The baroreceptor reflexes are most sensitive to sudden decreases in blood pressure, but the baroreceptor refex is sensitive to a decrease in blood pressure even if it occurs over a period of several hours. The baroreceptor reflexes trigger vasoconstriction if the blood loss is substantial. The kidneys detect even small decreases in blood volume. Consequently, the renin-angiotensin-aldosterone mechanism is activated and remains active until the blood pressure is returned to its normal range of values. Aldosterone secretion increase and, though it requires several hours to become maximally active, it continues to stimulate water reabsorption by the kidneys until the blood pressure returns to its normal range of values.

Chapter Fourteen

1. Cutting and tying off the lymphatic vessels prevents the movement of fluid from the affected tissue. The result is edema.
2. When the antigen is eliminated, it is no longer available for processing and combining with MHC molecules. Consequently, there is no signal to cause lymphocytes to proliferate and produce antibodies.
3. The first exposure to the disease-causing agent (antigen) evokes a primary immune system response. Gradually, however, the antibodies degrade, and memory cells die. If, before all the memory cells are eliminated, a second exposure to the antigen occurs, a secondary response results. The memory cells produced could provide immunity until the next exposure to the antigen.
4. The booster shot stimulates a memory (secondary) response, resulting in the formation of large amounts of antibodies and memory cells. Consequently, there is better, longer lasting immunity.
5. SLE is an autoimmune disorder in which self-antigens activate immune responses. Often, this results in the formation of immune complexes and inflammation. Sometimes antibodies bind to antigens on cell membranes, resulting in the rupture of the cell membranes. Purpura results from bleeding into the skin. One cause of purpura is thrombocytopenia, a condition in which the number of platelets is greatly reduced, resulting in decreased platelet plug formation and blood clotting (see chapter 11). In SLE, antibodies can stimulate the destruction of platelets.

Chapter Fifteen

1. When you sleep with your mouth open, less air passes through the nasal passages. This is especially true when nasal passages are plugged because you have a cold. As a consequence, air is not humidified and warmed. The dry air dries the throat and the trachea, thus irritating them. Breathing through the mouth while running in very cold weather results in air that is not humidified and warmed, thereby irritating the respiratory passageways.
2. When a large mouthful of food is swallowed, the esophagus is enlarged in the area through which the food passes. The bulge in the esophagus applies pressure on the trachea, which is immediately anterior to the esophagus. Because the C-shaped cartilages of the trachea have their open portion facing the esophagus, the posterior wall of the trachea collapses momentarily as the food passes. Thus, the passage of food through the esophagus is not hampered by the trachea.
3. During respiratory movements, the parietal and visceral pleurae slide over each other. Normally the pleural fluid in the pleural cavities lubricates the surfaces of these membranes. When the pleural membranes are inflamed, their surfaces become roughened. The rough surfaces rub against each other and create an intense pain. The pain is increased when a person takes a deep breath because the movement of the membranes is greater than during normal breaths.
4. Relaxation of the abdominal muscles allows the abdominal organs to move inferiorly. Thus, it is easier for the diaphragm to increase the volume of the thoracic cavity.
5. The tube should apply suction. In order for the lung to expand, pressure in the alveoli must be greater than the pressure in the pleural cavity. This can be accomplished by lowering the pressure in the pleural cavity through suction. Applying air under pressure would make the pressure in the pleural cavity greater than the pressure in the alveoli, which would keep the alveoli collapsed.
6. The resting person with a tidal volume of 500 mL and a respiratory rate of 12 respirations/min, has a minute ventilation of 6000 mL (500 mL × 12 respirations/min). The exercising person with a tidal volume of 4000 mL and a respiratory rate of 24 respirations/min, has a minute ventilation of 96,000 mL (4000 mL × 24 respirations/min). The difference between the two is 90,000 mL, which means that the exercising person respired 90,000 mL more air per minute than the person at rest.
7. During exercise, skeletal muscle cells increase oxygen use in order to produce the ATP molecules required for muscle contraction. The P_{O_2} inside the cells therefore decreases, which increases the diffusion gradient for oxygen, resulting in increased movement of oxygen into the cells. The aerobic production of ATP also produces carbon dioxide (see chapter 17). The P_{CO_2} inside the cell therefore increases, which increases the diffusion gradient for carbon dioxide, resulting in increased movement of carbon dioxide out of cells.
8. A rapid rate of respiration increases the blood pH because CO_2 is eliminated from the blood more rapidly during rapid respiration. As CO_2 is lost, H^+ and HCO_3^- combine to form H_2CO_3, which in turn dissociates to form CO_2 and H_2O. The decrease in H^+ causes an increase in blood pH. Holding one's breath results in a decrease in pH, because CO_2 accumulates in the blood. The CO_2 combines with H_2O to form H_2CO_3, which dissociates to form H^+ and HCO_3^-. The increase in H^+ causes a decrease in blood pH.
9. When a person breathes rapidly and deeply for several seconds, the carbon dioxide levels decrease, and blood pH increases. Carbon dioxide is an important regulator of respiratory movements. A decrease in blood carbon dioxide and an increase in blood pH result in a reduced stimulus to the respiratory center. As a consequence, respiratory movements stop until blood carbon dioxide levels build up again in the body fluid. This normally takes only a short time.
10. A P_{O_2} of 60 mm Hg and a P_{CO_2} of 30 mm Hg are both below normal. The movement of air into and out the lungs is restricted because of the asthma, and there is a mismatch between ventilation of the alveoli and blood flow to the alveoli. Consequently, because of the ineffective ventilation, blood oxygen levels decrease. Mr. W. hyperventilates, which helps to maintain blood oxygen levels but also results in lower than normal blood carbon dioxide levels. (If there was no hyperventilation, one would expect decreased blood oxygen but increased blood carbon dioxide.)

Chapter Sixteen

1. Four. Each portion of the mesentery has two layers, with a layer of connective tissue in between. The mesentery is folded back on itself to form the greater omentum.
2. It is important to close off the nasopharynx during swallowing so that food, and especially liquid, doesn't pass into the nasal cavity. If a person has a cleft of the soft palate, there is an opening between the oral and nasal cavities, and the nasopharynx is not closed off during swallowing. If a person has an explosive burst of laughter while trying to swallow a liquid, the liquid may be explosively

expelled from the mouth and even from the nose. Speaking requires that the epiglottis be elevated so that air can pass out of the larynx. If the epiglottis is elevated while one is swallowing, the food, and especially liquid, may pass into the larynx, causing a person to choke.

3. Secretin production in response to acidic chyme in the small intestine stimulates bicarbonate ion secretion from the pancreas, which neutralizes the acidic chyme. Thus secretin prevents the acid levels in the chyme from becoming too great. Because this mechanism keeps the pH of the intestinal contents within a normal range, this is an example of a negative-feedback system.

4. Introducing fluid into the rectum by way of an enema causes distention of the rectum. Distention stimulates the defecation reflex.

5. The effects of prolonged diarrhea result from a continued loss of fluid and ions. The major effect is on the cardiovascular system, and the effects are much like massive blood loss. Hypovolemia would continue to increase. Blood pressure would decline in a positive-feedback cycle, and, without intervention, could lead to heart failure.

Chapter Seventeen

1. If vitamins are broken down during the process of digestion, their structures are destroyed, and, as a result, their ability to function as vitamins is lost. Because vitamins cannot be synthesized in adequate amounts, vitamin deficiency diseases would occur.

2. The Daily Value for carbohydrate is 300 g/day. One serving of food with 30 g of carbohydrate has a % Daily Value of 10% (30/300 = 0.10 or 10%).

3. On a 1800 kcal/day diet, the total percentage of Daily Values for energy-producing nutrients should add up to no more than 90%, because 1800/2000 = 0.9 or 90%.

4. If electrons in the electron-transport chain cannot be donated to oxygen atoms in the last step of the electron-transport chain, the entire chain and the citric acid cycle stop, no ATP is produced aerobically, and death results because too little energy is available for the body to maintain vital functions. Anaerobic metabolism can provide energy for only very short periods of time and cannot sustain life very long.

5. The kilocalories in a beer or cola is about 145 kcal. It takes about 1.5 h to burn off these kilocalories while watching TV (145/95 = 1.5 h) and about 15 min while jogging (145/580 = 0.25 h). Although it

may be difficult to burn off kilocalories through exercise, it is clear that exercise can significantly increase kilocalorie use.

6. When muscles contract, they must use ATP as the energy source for the contractions. As more ATP is produced, heat is also produced. During exercise, the large amounts of heat can raise body temperature, and we feel warm. Shivering consists of small, rapid muscle contractions that produce heat in an effort to prevent a decrease in body temperature in the cold. When the body temperature declines below normal, shivering is initiated involuntarily.

7. Constriction of blood vessels to the skin reduces blood flow to the skin, which cools as a result. The benefit is that less heat is lost through the skin to the environment and the internal body temperature is maintained. As the difference in temperature between the skin and the environment decreases, less heat is lost. If the skin temperature decreases too much, however, dilation of blood vessels to the skin occurs, which functions to prevent the skin from becoming so cold that it is damaged.

Chapter Eighteen

1. The female urinary bladder is more accessible to bacteria from the exterior because the urethra of a female is much shorter than that of the male. For this reason urinary bladder infections are more common in females than in males.

2. If large amounts of glucose enter the nephron and are not reabsorbed, the glucose causes the concentration of solutes in the filtrate to increase. The glucose molecules attract water and, because the glucose molecules are trapped in the nephron, the amount of water that remains in the nephron is increased. A large volume of urine that contains glucose is a symptom of diabetes mellitus.

3. Without the normal active transport of Na^+ and Cl^-, their concentration within the nephron remains elevated. The normal movement of water out of the nephron cannot occur because of the osmotic effects of the Na^+ and Cl^- trapped in the nephron. The result is an increased urine volume.

4. Because the solution was a saline solution, it had the same concentration of solutes as the body fluids. The excess IV solution did not therefore change the concentration of the body fluids, but it did increase the volume of the body fluids. An increased volume of saline solution increases the blood volume and blood pressure.

Increased blood pressure stimulates baroreceptors, which results in inhibition of ADH secretion. The reduced ADH secretion causes the kidneys to produce a large volume of urine. At the same time, the increased blood volume stretches the walls of the atria, especially the right atrium, and causes the release of atrial natriuretic hormone. Atrial natriuretic hormone acts on the kidneys to reduce Na^+ reabsorption. Because Na^+ reabsorption is decreased, both Na^+ and water are lost in the urine. The increased pressure also results in less renin secretion from the kidney. The reduced renin causes less angiotensinogen to be converted to angiotensin. Consequently, less angiotensin II is formed, which reduces aldosterone secretion from the adrenal cortex. The decreased aldosterone slows Na^+ and water reabsorption causing more Na^+ and water to be lost in the urine. Consequently, the urine volume and the amount of NaCl in the urine increase until the excess saline solution is eliminated.

5. Hyperventilation results in a greater than normal rate of carbon dioxide loss from the circulatory system. Because carbon dioxide is lost from the circulatory system, H^+ concentration decreases, and the pH of body fluids increases. Breathing into a paper bag corrects for the effects of hyperventilation because the person rebreathes air that has a higher concentration of carbon dioxide. The result is an increase in carbon dioxide in the body. Consequently, the H^+ concentration increases, and pH decreases toward normal levels.

6. Elevated blood carbon dioxide levels cause an increase in H^+ and a decrease in blood pH due to the following reaction:

$$CO_2 + H_2O \rightleftarrows H_2CO_3 \rightleftarrows H^+ + HCO_3^-$$

However, the kidney plays an important role in the regulation of blood pH. The kidney's rate of H^+ secretion into the urine and reabsorption of HCO_3^- increase. This helps prevent high blood H^+ levels and low blood pH in Mr. Puffer.

7. After 7 days Mr. H.'s kidney's began to produce a large volume of urine with larger than normal Na^+ and K^+ concentrations. As a result, Mr. H. became dehydrated by day 9. Dehydration results in reduced blood volume and blood pressure. His hematocrit was increased because the volume of his blood was decreased, but there was no decrease in the number of red blood cells. The percentage of the blood made of red blood cells therefore

increased. The pale skin was the result of vasoconstriction, which was triggered by the reduced blood pressure. Dizziness resulted from reduced blood flow to the brain when Mr. H. tried to stand and walk. He was lethargic in part because of reduced blood volume, but also because of low blood levels of K^+ and Na^+, caused by the loss of these ions in the urine. Low blood levels of Na^+ and K^+ alter the electrical activity of nerve and muscle cells and results in muscular weakness. The arrythmia of his heart was due to low blood levels of K^+ and increased sympathetic stimulation, which was also triggered by low blood pressure.

Chapter Nineteen

1. The prostate gland is located just anterior to the rectum. It can be palpated through the wall of the rectum. A physician can insert a finger into the rectum and palpate the prostate through the wall of the rectum. The procedure does not require surgical procedures and involves relatively minor discomfort.

2. Because secondary sexual characteristics, external genitalia, and sexual behavior develop in response to testosterone, if the testes fail to produce normal amounts of testosterone at puberty, they do not develop normally. Secondary sexual characteristics and external genitalia remain juvenile, and normal adult sexual behavior does not develop.

3. Administration of a large amount of progesterone and estrogen just before the preovulatory LH surge inhibits the release of GnRH, LH, and FSH. Consequently, ovulation does not occur. A small amount of estrogen administered, without progesterone, before the preovulatory LH surge, however, could stimulate GnRH, LH, and FSH secretion.

4. Mrs. M.'s mother could have had leiomyomas also, although, without direct data of medical examinations, one cannot be certain. If that was the cause of her irregular menstruations, they may have become less frequent as Mrs. M.'s mother experienced menopause. During menopause the uterus gradually becomes smaller, and eventually the cyclical changes in the endometrial lining ceases. If the condition was relatively mild, the onset of menopause could explain the gradual disappearance of the irregular and prolonged menstruations (*Note:* If the tumors are large, constant and severe menstruations are likely even if regular menstrual cycles stop due to menopause.)

Chapter Twenty

1. Because some sperm cells remain viable in the female reproductive tract for up to 6 days, and the secondary oocyte is capable of being fertilized for up to 1 day after ovulation, fertilization could occur if sexual intercourse occurred between 5 days before ovulation and 1 day following ovulation. That will be between 10 days and 15 days post-LMP. Data indicate, however, that the most fertile period during the menstrual cycle is between 2 days just before ovulation and the day of ovulation.

2. Two primitive streaks forming in one embryonic disk can result in identical twins, because one embryo is formed in association with each primitive streak. If the two streaks are touching, the twins will be conjoined (Siamese), or attached to each other. This attachment can be fairly simple, and the twins can be separated fairly easily by surgery; or the attachment can be extensive, involving internal organs and may not be corrected easily.

3.

	Clinical Age	Developmental Age
Fertilization	14 days	0 days
Implantation	21 days	7 days
Fetal Period	70 days	56 days
Parturition	280 days	266 days

4. Suckling causes a reflex release of oxytocin from the mother's posterior pituitary. Oxytocin causes expulsion of milk from the breast, but it also causes contraction of the uterus. Contraction of the uterus is responsible for the sensation of cramps.

5. Genotype *DD* (homozygous dominant) has the polydactyly phenotype, genotype *Dd* (heterozygous) has the polydactyly phenotype, and genotype *dd* (homozygous recessive) has the normal phenotype.

6. The probability of a girl with Turner's syndrome (XO; that is, with one X chromosome and no other sex chromosome) having hemophilia if her mother is a carrier for hemophilia is the same as for a male because she has only one X chromosome. If her mother is a carrier for hemophilia ($X^H X^h$), the daughter will be either $X^H O$ (normal), or $X^h O$ (hemophiliac). The probability is 1/2, or 50%.

Glossary

Many of the words in this glossary and throughout the text are followed by a simplified phonetic spelling showing pronunciation. The pronunciation key reflects standard clinical usage, with minor modifications, as presented in *Stedman's Medical Dictionary* (27th edition), which has long been a leading reference volume in the health sciences.

ā as in day, ate, way
a as in mat, hat, act
ă as in alone, abortion, media
ah as in father
ar as in far
aw as in fall (fawl)
ē as in be, bee, meet
ĕ as in taken, genesis
er as in term, earn, learn
ī as in pie, pine, side
i as in pit, tip, fit
ĭ as in pencil
ō as in no, note, toe
o as in not, box, cot
ŏ as in occult, lemon, son
oo as in food, to, tool
ow as in cow, brow, plow, now
oy as in boy, toy, oil
u as in wood, foot, took
ŭ as in but, sun, bud, cup, up
ū as in pure, unit, union, future

A

abdomen (ab-dō′men, ab′dō-men) Belly, between the thorax and the pelvis.

abdominal cavity (ab-dom′i-năl) Space bounded by the diaphragm, the abdominal wall, and the pelvis.

abdominopelvic cavity (ab-dom′i-nō-pel′vik) The abdominal and pelvic cavities considered together.

abduction (ab-dŭk′shun) [abductio] Movement away from the midline.

absorption (ab-sōrp′shŭn) The taking in or reception of gases, liquids, light, heat, or solutes, such as the movement of digested molecules across the intestinal wall and into the bloodstream, the movement of substances through the skin, and the movement of fluid into the lymphatics from the interstitial fluid.

accommodation (ă-kom′ŏ-dā′shŭn) The act or state of adjustment or adaptation such as the increase in the thickness and convexity of the lens in order to focus an object on the retina as the object moves closer to the eyes; decreasing sensitivity of a nerve cell to a stimulus of constant strength.

acetabulum (as-ĕ-tab′ū-lŭm) [L., shallow vinegar vessel or cup] Cup-shaped depression on the lateral surface of the coxa, where the head of the femur articulates.

acetylcholine (as-e-til-kō′lēn) Neurotransmitter substance released from motor neurons that innervate skeletal muscle fibers, all autonomic preganglionic neurons, all postganglionic parasympathetic neurons, some postganglionic sympathetic neurons, and some central nervous system neurons.

acetylcholinesterase (as′e-til-kō-lin-es′ter-ās) An enzyme that breaks down acetylcholine to acetic acid and choline.

acetyl-CoA (as′e-til) Acetyl-coenzyme A; formed by the combination of the two-carbon acetyl group with coenzyme A; the molecule that combines with a four-carbon molecule to enter the citric acid cycle.

Achilles (ă-kil′ēz) **tendon** Common tendon of the calf muscles that attaches to the heel (calcaneus); named after a mythical Greek warrior who was vulnerable only in the heel.

acid (as′id) Any substance that is a proton donor; or any substance that releases hydrogen ions.

acidic solution Solution with more hydrogen ions than hydroxide ions; has a pH of less than 7.

acidosis (as-i-dō′sis) Condition characterized by a lower than normal blood pH (pH of 7.35 or lower).

acinus (as′i-nŭs), pl. **acini** (as′i-nī) [L., berry, grape] Grape-shaped secretory portion of a gland.

acromegaly (ak-rō-meg′ă-lē) [G. acro; megas, large] Disorder marked by progressive enlargement of the bones of the head, face, hands, feet, and thorax as a result of excessive secretion of growth hormone by the anterior pituitary gland.

acromion (ă-krō′mē-on) [fr. akron, tip + omos, shoulder] The lateral end of the spine of the scapula that projects as a broad flattened process overhanging the glenoid fossa; articulates with the clavicle.

acrosome (ak′rō-sōm) [acro, tip + G. soma, body]. A caplike organelle surrounding the anterior portion of a sperm cell, containing enzymes that facilitate entry of the sperm cell through the zonapellucida.

actin myofilament (ak′tin mī-ō-fil′ă-ment) One of the two major kinds of protein fibers that make up a sarcomere; thin filaments; resemble two minute strands of pearls twisted together.

action potential All-or-none change in membrane potential in an excitable tissue that is propagated as an electrical signal.

activation energy Energy that must be added to atoms or molecules to start a chemical reaction.

active transport Carrier-mediated process that requires ATP and can move substances into or out of cells from a lower to a higher concentration.

adaptive immunity Immune system response in which there is an ability to recognize, remember, and destroy a specific antigen.

adduction (ă-dŭk′shŭn) [L. adductus, to bring toward] Movement toward the midline.

adductor (a-dŭk′ter, -tōr) [L. adductus, to bring toward] A muscle causing movement toward the midline.

adenoid (ad′ĕ-noyd) Enlarged pharyngeal tonsil.

adenosine triphosphate (ă-den′ō-sēn trī-fos′fāt) **(ATP)** Adenosine, an organic base, with three phosphate groups attached to it; energy stored in adenosine triphosphate is used in nearly all the energy-requiring reactions in the body.

ADH See *antidiuretic hormone.*

adipose (ad′i-pōs) [L. adeps, fat] Fat; relating to fat tissue.

adrenal cortex (ă-drē′năl kōr′teks) The outer part of the adrenal gland, which secretes the following steroid hormones: glucocorticoids, mainly cortisol; mineralocorticoids, mainly aldosterone; and androgens.

adrenal gland (ă-drē′năl) [L. ad, to; ren, kidney, near or on the kidneys] One of two endocrine glands located on the superior pole of each kidney; secretes the hormones epinephrine, norepinephrine, aldosterone, cortisol, and androgens.

adrenal medulla (ă-drē′năl me-dool′ă) The inner part of the adrenal gland, which secretes mainly epinephrine but also small amounts of norepinephrine.

adrenalin (ă-dren′ă-lin) [from the adrenal gland] Synonym for epinephrine.

adrenocorticotropic hormone (ă-drē′nō-kōr′ti-kō-trō′pik) [L. ad, near + ren, kidney + cortico, cortex + trophre, nurture] **(ACTH)** Hormone of the anterior pituitary gland that stimulates the adrenal cortex to secrete cortisol.

adventitia (ad-ven-tish′ă) [L. adventicius, coming from abroad or outside; foreign] Outermost covering of an organ that is continuous with the surrounding connective tissue.

aerobic respiration (ār-ō′bik) Breakdown of glucose in the presence of oxygen to produce carbon dioxide, water, and approximately 38 ATP molecules; includes glycolysis, the citric acid cycle, and the electron-transport chain.

afferent (af′er-ent) [L. afferens, to bring to] Inflowing; conducting toward a center, denoting certain arteries, veins, lymphatics, and sensory nerves. Opposite of efferent.

afferent arteriole (ar-tēr′ē-ōl) A small artery in the renal cortex that supplies blood to the glomerulus.

afferent fiber Sensory nerve fiber going from the peripheral to the central nervous system; sensory or afferent fiber.

afterload The resistance against which the ventricles must pump blood; it is increased in people who have hypertension.

agglutination (ă-gloo′ti-nā′shŭn) [L. ad, to + gluten, glue] The process by which cells stick together to form clumps.

agonist (ăg′ŏn-ist) [G. agon, a contest] Denoting a muscle in a state of contraction, with reference to its opposing muscle, or antagonist.

agranulocyte (ă-gran′ū-lō-sīt) [G. a-, without + granular + kytos, cell] White blood cells with very small cytoplasmic granules that cannot be easily seen with the light microscope; lymphocytes and monocytes.

aldosterone (al-dos′ter-ōn) Steroid hormone produced by the adrenal cortex that facilitates potassium exchange for sodium in the distal tubule and collecting duct, causing sodium ion reabsorption and potassium and hydrogen ion secretion.

alkaline solution (al′kă-līn) See *basic solution.*

alkalosis (al-kă-lō′sis) Condition characterized by a higher than normal blood pH (pH of 7.45 or above).

alveolar duct (al-vē′ō-lăr) Part of the respiratory passages beyond a respiratory bronchiole; from it arise alveolar sacs and alveoli.

alveolar sac Two or more alveoli that share a common opening.

alveolus (al-vē′ō-lŭs), pl. **alveoli** (al-vē′ō-lī) [L., small cavity or hollow sac] Cavity; examples include the sockets into which the teeth fit and the ends of the respiratory system.

amino acid (ă-mē′nō) Class of organic acids containing an amine group (NH_2) that makes up the building blocks of proteins.

amniotic cavity (am-nē-ot′ik) [G. *amnios*, lamb] Fluid-filled cavity surrounding and protecting the developing embryo.

amylase (am′il-ās) One of a group of starch-splitting enzymes that cleave starch, glycogen, and related polysaccharides.

anabolism (ă-nab′ŏ-lizm) [G. *anabole*, a raising up] All the synthesis reactions that occur within the body; requires energy.

anaerobic respiration (an-ār-ō′bik) Breakdown of glucose in the absence of oxygen to produce lactic acid and two ATP molecules; consists of glycolysis and the reduction of pyruvic acid to lactic acid.

anaphase (an′ă-fāz) [G. *ana*, up + *phases*, appearance] The stage of mitosis or meiosis in which the chromosomes move from the center area of the cell, the equatorial plane, toward the poles of the cell.

anatomic position (an′ă-tom′i-k) Position in which a person is standing erect with the feet facing forward, arms hanging to the sides, and the palms of the hands facing forward.

anatomy (ă-nat′ŏ-mē) [G. *ana*, apart + *tome*, a cutting] Scientific discipline that investigates the structure of the body.

androgen (an′drō-jen) [G. *andros*, male] Hormone that stimulates the development of male sexual characteristics, includes testosterone.

anemia (ă-nē′mē-ă) [G. *an*, without + *haima*, blood] Any condition that results in less than normal hemoglobin in the blood or a lower than normal number of red blood cells.

anencephaly (an′en-sef′ă-lē) Defective development of the brain with absence of the cerebral and cerebellar hemispheres, and with only a rudimentary brainstem.

angina pectoris (an′ji-nă pek′tō-ris, an-jī′nă) Pain resulting from a reduced blood supply to cardiac muscle.

angioplasty (an′jē-ō-plas-tē) [G. *angio*, blood vessel] A technique used to dilate the coronary arteries by threading a small balloonlike device into a partially blocked coronary artery and then inflating the balloon to enlarge the diameter of the vessel.

angiotensin (an-jē-ō-ten′sin) [*angio*, blood vessel + *tensus*, to stretch] Angiotensin I is a peptide derived when renin acts on angiotensinogen; angiotensin II is formed from angiotensin I when angiotensin-converting enzyme acts on angiotensin I; angiotensin II is a potent vasoconstrictor, and it stimulates the secretion of aldosterone from the adrenal cortex.

angiotensinogen (an′jē-ō-ten-sin′ō-jen) A protein found in the blood that gives rise to angiotensin I after renin, an enzyme secreted from the kidney, acts on it.

ANH See *atrial natriuretic hormone.*

antagonist (an-tag′ŏ-nist) A muscle that works in opposition to another muscle.

anterior (an-tēr′ē-ōr) [L., to go before] That which goes first; in humans, toward the belly or front.

anterior horn The part of the spinal cord gray matter containing motor neurons; also called the ventral horn or motor horn.

anterior pituitary gland Portion of the pituitary gland derived from the oral epithelium.

antibody (an′tē-bod-ē) [G. *anti*, against + *body*, a thing] Protein found in the plasma that is responsible for antibody-mediated (humoral) immunity; binds specifically to an antigen.

antibody-mediated immunity Immunity resulting from B cells and the production of antibodies.

anticoagulant (an′tē-kō-ag′ū-lant) Chemical that prevents coagulation or blood clotting; an example is antithrombin.

antidiuretic hormone (an′tē-dī-ū-ret′ik) [G. *anti*, against + *uresis*, urine volume] **(ADH)** Hormone secreted from the posterior pituitary gland that acts on the kidney to reduce the output of urine; also called vasopressin.

antigen (an′ti-jen) [G., anti (body) + *-gen*, a thing] Any substance that induces a state of sensitivity or resistance to microorganisms or toxic substances after a latent period; substance that stimulates the adaptive immune system; self-antigens are produced by the body, and foreign antigens are introduced into the body.

antigen-receptor Molecule on the surface of lymphocytes that specifically binds antigens.

aorta (ā-ōr′tă) [G. *aorte*, from + *aeiro*, to lift up] Large elastic artery that is the main trunk of the systemic arterial system, which carries blood from the left ventricle of the heart and passes through the thorax and abdomen.

aortic semilunar valve The semilunar valve consisting of three cusps of tissue located at the base of the aorta where it arises from the left ventricle; the cusps overlap during ventricular diastole to prevent leakage of blood from the aorta into the left ventricle.

apex (ā′peks) [L., tip] Extremity of a conical or pyramidal structure; the apex of the heart is the rounded tip directed anteriorly and slightly inferiorly.

aphasia (ă-fā′zē-ă) [G. *a-*, without + *phasis*, speech or speechlessness] Impaired or absent communication by speech, writing, or signs, because of dysfunction of brain centers in the dominant cerebral hemisphere.

apocrine (ap′ō-krin) [G. *apo*, away from + *krino*, to separate] Gland whose cells contribute cytoplasm to its secretion; sweat glands that produce organic secretions traditionally are called apocrine. These sweat glands now are known, however, to be merocrine glands; see also *merocrine* and *holocrine.*

aponeurosis (ap′ō-noo-rō′sis) [G. *neuron*, sinew; end of a muscle where it becomes a tendon] A sheet of fibrous connective tissue, or an expanded tendon, serving as the origin or insertion of a flat muscle.

appendicular (ap′en-dik′ū-lăr) [L. *appendo*, to hang something on] Relating to an appendage, such as the limbs and their associated girdles.

appendix (ă-pen′diks), pl. **appendices** (ă-pen′di-sēs) [L. *appendo*, to hang something on] Smaller structure usually attached by one end to a larger structure; a small blind extension of the colon attached to the cecum.

appositional growth (ap-ō-zish′ŭn-ăl) [L. *ap* + *pono*, to place at or to] To place one layer of bone, cartilage, or other connective tissue against an existing layer; increases the width or diameter of bones.

aqueous humor (ak′wē-ŭs, ā′kwē-ŭs) Watery, clear fluid that fills the anterior compartment of the eye.

arachnoid mater (ă-rak′noyd ma′ter) [G. *arachne*, spiderlike, cobweb] Thin, cobweblike meningeal layer surrounding the brain and spinal cord; the middle of three layers.

areola (ă-rē′ō-lă, -ē), pl. **areolae** (ă-rē′ō-lē) [small areas] A pigmented area surrounding the nipple of a mammary gland.

areolar (ă-rē′ō-lăr) Relating to connective tissue with small spaces within it; loose connective tissue.

arrector pili (ă-rek′tōr pī′lī) [L., that which raises hair] Smooth muscle attached to the hair follicle and dermis that raises the hair when it contracts.

arteriosclerosis (ar-tēr′ē-ō-skler-ō′sis) [L. *arterio-* + G. *sklerosis*, hardness] Hardness of the arteries.

arteriosclerotic lesion (ar-tēr′ē-ō-skler-ot′ik) A lesion or growth in arteries that narrows the lumen, or passage, and makes the walls of the arteries less elastic.

artery (ar′ter-ē) [G. *arteria*, the windpipe] Blood vessel that carries blood away from the heart.

articulation (ar-tik-ū-lā′shŭn) [L. *articulatio*, a forming of vines] The place where two bones come together; a joint.

artificial heart A mechanical pump used to replace a diseased heart.

artificial pacemaker An electronic device implanted beneath the skin with an electrode that extends to the heart; provides periodic electrical stimuli to the heart and substitutes for a faulty SA node.

astrocyte (as′trō-sīt) [G. *astron*, star + *kytos*, a hollow-cell] Star-shaped neuroglial cell that helps regulate the composition of fluid around the neurons of the central nervous system.

atherosclerosis (ath′er-ō-skler-ō′sis) [G. *athere*, gruel or soft, pasty material + *sklerosis*, hardness] Lipid deposits (plaques) in the tunica intima of large and medium-sized arteries.

atom (at′ŏm) [G. *atomos*, indivisible, uncut] Smallest particle into which an element can be divided using chemical methods; composed of neutrons, protons, and electrons.

atomic number (ă-tom′ik) The number of protons in an element.

ATP See *adenosine triphosphate.*

atrial natriuretic (ā′trē-ăl nā′tre-yū-ret′ik) **hormone (ANH)** Hormone released from cells in the atrial wall of the heart when atrial blood pressure is increased; acts to lower blood pressure by increasing the rate of urine production.

atrioventricular (AV) bundle (ā-trē-ō-ven-trik′ū-lar) Bundle of modified cardiac muscle fibers that projects from the AV node through the interventricular septum; conducts action potentials from the AV node rapidly through the interventricular septum; also called the bundle of His.

atrioventricular (AV) node Small collection of specialized cardiac muscle fibers located in the inferior part of the right atrium; functions to delay action potential transmission to the atrioventricular bundle.

atrioventricular valve Valve located between the atrium and the ventricle of the heart, the tricuspid valve between the right atrium and right ventricle and the bicuspid (or mitral valve) between the left atrium and left ventricle.

atrium (ā′trē-ŭm), pl. **atria** (ā′trē-ă) [L., entrance chamber] One of the two chambers of the heart that collect blood during ventricular contraction and pump blood into the ventricles to complete ventricular filling at the end of ventricular relaxation; the right atrium receives blood from the inferior and superior venae cavae and from the coronary sinus, and delivers blood to the right ventricle; the left atrium receives blood from the pulmonary veins and delivers blood to the left ventricle.

auditory (aw′di-tōr-ē) Relating to hearing.

auditory ossicles (os′i-klz) Bones of the middle ear; the malleus, incus, and stapes.

auditory tube Air-filled passageway between the middle ear and pharynx.

auricle (aw′rĭ-kl) [L. *auris*, ear] The fleshy part of the external ear on the outside of the head; a small conical pouch projecting from the upper anterior part of each atrium of the heart.

auscultatory (aws-kŭl′tă-tō-rē) [L. *ausculto*, to listen] To listen to the sounds made by the various body structures, especially to Korotkoff sounds when determining blood pressure.

autocrine (aw′tō-krin) [G. *autos*, self + *krino*, to separate] Denoting self-stimulation through cellular production of a factor and a specific receptor for it.

autoimmune disease (aw-tō-i-mūn′) Disorder resulting from a specific immune system reaction against self-antigens.

autonomic nervous system (ANS) (aw-tō-nom′ik) That part of the peripheral nervous system composed of efferent fibers that reach from the central nervous system to smooth muscle, cardiac muscle, and glands.

autosome (aw′tō-sōm) [G. *auto-* self + *soma*, body] Any chromosome other than a sex chromosome; normally occurs in pairs in somatic cells and singly in gametes.

AV See *atrioventricular.*

axial (ak′sē-ăl) [L. *axle*, axis] Head, neck, and trunk as distinguished from the extremities.

axon (ak′son) [G., axis] Main process of a neuron; usually conducts action potentials away from the neuron cell body.

B

baroreceptor (bar′ō-rē-sep′ter) [G. *baro*, weight or pressure] Sensory nerve endings in the walls of the atria of the heart, aortic arch, and carotid sinuses; sensitive to stretching of the wall caused by increased blood pressure; also called a pressoreceptor.

baroreceptor reflex Process in which baroreceptors detect changes in blood pressure and produce changes in heart rate, force of heart contraction, and blood vessel diameter that return blood pressure toward normal levels.

basal nuclei Nuclei at the base of the cerebrum, diencephalon, and midbrain involved in controlling motor functions.

base Any substance that is a proton acceptor; or any substance that binds to hydrogen ions; lower part or bottom of a structure; the base of the heart is the flat portion directed posteriorly and superiorly; veins and arteries project into and out of the base, respectively.

basement membrane The structure that attaches most epithelia (exceptions include lymph vessels and the liver sinusoids) to underlying tissue; consists of carbohydrates and proteins secreted by the epithelia and the underlying connective tissue.

basic solution (bā′sik) Solution with fewer hydrogen ions than hydroxide ions; has a pH greater than 7.

basilar membrane (bas′i-lăr) One of two membranes forming the cochlear duct; supports the spiral organ.

basophil (bā′sō-fil) [G. *basis*, base + *phileo*, to love] White blood cell with granules that stain purple with basic dyes; promotes inflammation and prevents clot formation.

belly The largest part of a muscle between the origin and insertion.

benign (bē-nīn′) [L. *benignus*, kind] Mild in character or nonmalignant; does not spread to distant sites.

beta-adrenergic (bā′tă ad-rĕ-ner′jik) **blocking agent** Drug that binds to and prevents adrenergic receptors from responding to adrenergic compounds that normally bind to beta-adrenergic receptors and cause them to function; beta-adrenergic blocking agents are used to treat certain arrhythmias in the heart and to treat tachycardia (rapid heart rate).

biceps brachii (bī′seps brā′kē-ī) Muscle in the anterior arm with two heads or origins on the scapula and an insertion onto the radius; flexes and supinates the forearm.

bicuspid valve (bī-kŭs′pid) Valve closing the opening between the left atrium and left ventricle of the heart; has two cusps; also called the mitral valve.

bile (bīl) Fluid secreted from the liver, stored in the gallbladder, and released into the duodenum; consists of bile salts, bile pigments, bicarbonate ions, fats, and other materials.

bile salt Organic salt secreted by the liver that functions to emulsify lipids.

bilirubin (bil-i-roo′bin) [L., bile + *ruber*, red] A bile pigment formed from the heme in hemoglobin during the destruction of red blood cells by macrophages.

biopsy (bī′op-sē) The process of removing tissue from living patients for diagnostic examination, or a specimen obtained by biopsy.

blastocele (blas′tō-sēl) [G. *blastos*, germ + *koilos*, hollow] Cavity in the blastocyst.

blastocyst (blas′tō-sist) [G. *blastos*, germ + *kystis*, bladder] Early stage of mammalian embryo development consisting of a hollow ball of cells with an inner cell mass and an outer trophoblast layer.

blood–brain barrier Cellular and matrix barrier made up primarily of blood vessel endothelium, with some help from the surrounding astrocytes; it allows some (usually small) substances to pass from the circulation into the brain, but does not allow other (larger) substances to pass.

blood group A category of red blood cells based on the type of antigen on the surface of the red blood cell; for example, the ABO blood group is involved with transfusion reactions.

blood pressure [L. *pressus*, to press] The force blood exerts against the blood vessel walls; expressed relative to atmospheric pressure and reported in the form of millimeters of mercury (mm Hg) of pressure.

bony labyrinth (lab′i-rinth) The interconnecting tunnels and chambers within the temporal bone in which the inner ear is located.

Bowman's capsule The enlarged end of the nephron; Bowman's capsule and the glomerulus make up the renal corpuscle.

brachialis (brā′kē-ăl-is) Muscle of the anterior arm that originates on the humerus and inserts onto the ulna; flexes the forearm.

brachial plexus (brā′kē-ăl) [L. *brachium*, arm] The nerve plexus to the upper limb; originates from spinal nerves C5 to T1.

brainstem Portion of the brain consisting of the midbrain, pons, and medulla oblongata.

breathing (brēthing) Movement of air into and out of the lung, see *ventilation.*

bronchiole (brong′kē-ōl) One of the finer subdivisions of the bronchial tubes, less than 1 mm in diameter, that has no cartilage in its wall, but has relatively more smooth muscle and elastic fibers than do larger bronchial tubes.

bronchus (brong′kŭs), pl. **bronchi** (brong′kī) [G. *bronchos*, windpipe] Any one of the air ducts conducting air from the trachea to the bronchioles.

buccinator (buk′sĭ-nā′tōr) Muscle making up the lateral sides of the oral cavity; flattens the cheeks.

buffer (bŭf′er) Chemical that resists changes in pH when either an acid or a base is added to a solution containing the buffer.

bundle of His See *atrioventricular bundle.*

burn A lesion caused by heat, acid, or other agents; a partial-thickness burn of the skin damages only the epidermis (first-degree burn) or the epidermis and part of the dermis (second-degree burn); a full-thickness (third-degree) burn destroys the epidermis and the dermis and sometimes the underlying tissue as well.

bursa (ber′să) [L., purse or pocket] Closed sac or pocket containing synovial fluid, usually found in areas where friction occurs.

C

calcaneus (kal-kā′nē-ŭs) [L., the heel] The largest tarsal bone forming the heel.

calcitonin (kal-si-tō′nin) Hormone released from cells of the thyroid gland that acts on tissues, especially bone, to cause a decrease in blood levels of calcium ions.

calcium channel blocker (kal′sē-ŭm) A class of drugs that specifically block channels in cell membranes through which calcium ions pass; calcium channel blockers are used to treat some kinds of cardiac arrhythmias.

callus (kal′ŭs) [L., hard skin] Thickening of the stratum corneum of skin in response to friction. The zone of tissue repair between fragments of a broken bone.

calorie (kal′ō-rē) [L. *calor*, heat] Unit of heat or energy content; the quantity of energy required to raise the temperature of 1 gram of water 1°C. A Calorie (Cal), or kilocalorie (kcal), is the amount of heat or energy required to raise the temperature of 1000 grams of water from 14°C to 15°C.

calyx (kā′liks), pl. **calyces** (kal′i-sēz) [G., flower petal or cup of a flower] The small containers into which urine flows as it leaves the collecting ducts at the tip of the renal pyramids; the calyces come together to form the renal pelvis.

canaliculus (kan-ă-lik′ū-lŭs) Tiny canal in bone between osteocytes containing osteocyte cell processes; a cleftlike lumen between the cells of each hepatic cord, connects medial corner of eye to the lacrimal sac.

cancellous bone (kan′sĕ-lŭs) [L., grating or lattice] Bone with latticelike appearance; spongy bone.

cancer (kan′ser) [L., a crab, suggesting crablike movement] A malignant neoplasm, or tumor.

capacitation (kă-pas′i-tā′shŭn) The process whereby the sperm cells develop the ability to fertilize oocytes.

capillary (kap′i-lār-ē) [L. *capillaris*, relating to hair, resembling a fine hair] Minute blood vessel consisting only of simple squamous epithelium and a basement membrane; major site for the exchange of substances between the blood and tissues.

carbohydrate (kar-bō-hī′drāt) Organic molecule made up of one or more monosaccharides chemically bound together; sugars and starches.

carbonic anhydrase (kar-bon′ik an-hī′drās) An enzyme that increases the rate at which carbon dioxide reacts with water to form hydrogen ions and bicarbonate ions.

carcinoma (kar-si-nō′mă) [G. *karkinoma*, cancer + *oma*, tumor] A malignant tumor derived from epithelial tissue.

cardiac cycle (kar′dē-ak) One complete sequence of cardiac systole and diastole.

cardiac output Volume of blood pumped by either ventricle of the heart per minute; about 5 L/min for the heart of a healthy adult at rest.

cardioregulatory center Specialized area within the medulla oblongata of the brain that receives sensory input and functions to control parasympathetic and sympathetic stimulation of the heart.

carotene (kar′ō-tēn) A yellow pigment found in plants such as squash and carrots; accumulates in the lipids of the stratum corneum and in the fat cells of the dermis and hypodermis and is used as a source of vitamin A.

carotid bodies (ka-rot′id) Small organ near the carotid sinuses that detects changes in blood oxygen, carbon dioxide, and pH.

carotid sinus Enlargement of the internal carotid artery near the point where the internal carotid artery branches from the common carotid artery; contains baroreceptors.

carpal (kar′păl) [G. *karpos*, wrist] Associated with the wrist; bones of the wrist.

carrier molecule Protein that extends from one side of the plasma membrane to the other; binds to molecules to be transported and moves them from one side of the membrane to the other.

cartilage (kar′ti-lij) [L. *cartilage*, gristle] Firm, smooth, resilient, nonvascular connective tissue.

cascade (kas-kād′) [Fr. *cascare*, to fall] A series of sequential interactions, which once initiated continues to the final one; each interaction is activated by the preceding one, with cumulative effect.

catabolism (kă-tab′ō-lizm) [G. *katabole*, a casting down] All the decomposition reactions that occur in the body; releases energy.

catalyst (kat′ă-list) A substance that increases the rate of a chemical reaction; in the process the catalyst is not permanently changed or used up.

cecum (sē′kŭm) [L. *caecus*, blind] A blind sac forming the beginning of the large intestine.

cell (sel) [L. *cella*, chamber] Basic living unit of all plants and animals.

cell-mediated immunity Immunity resulting from the actions of T cells.

cell membrane Plasma membrane; outermost component of the cell, surrounding and binding the rest of the cell contents.

central canal A small canal containing blood vessels, nerves, and loose connective tissue and running parallel to the long axis of a bone. Also called a haversian canal.

central nervous system (CNS) The brain and spinal cord.

centriole (sen′trē-ōl) Small organelle that divides and migrates to each pole of the nucleus; spindle fibers extend from the centromeres to the centrioles during mitosis.

centromere (sen′trō-mēr) [G. *kentron*, center + *meros*, part] A specialized region where chromatids are linked together in a chromosome.

cerebellum (ser-e-bel′ŭm) [L., little brain] A part of the brain attached to the brainstem; important in maintaining muscle tone, balance, and coordination of movements.

cerebral aqueduct (ser′ĕ-brăl, sĕ-rē′brăl) A small connecting tube through the midbrain between the third and fourth ventricles.

cerebrospinal fluid (ser′ĕ-brō-spī′năl, sĕ-rē′brō-spī′năl) (CSF) Fluid filling the ventricles and surrounding the brain and spinal cord.

cerebrum (ser′ĕ-brŭm, sĕ-rē′brŭm) [L., brain] The largest part of the brain, consisting of two hemispheres and including the cortex, nerve tracts, and basal nuclei.

cerumen (sĕ-roo′men) [L. *cera*, wax] A specific type of sebum produced in the external auditory meatus; earwax.

cervical (ser′vĭ-kal) Neck.

cervical plexus The nerve plexus of the neck; originates from spinal nerves C1 to C4.

cervix (ser′viks) [L., neck] Lower part of the uterus extending to the vagina.

chemical (kem′i-kăl) Relating to chemistry, especially to the characteristics of atoms and molecules and to their interactions.

chemical bond An association between two atoms formed when the outermost electrons are transferred or shared between atoms.

chemical reaction Process by which atoms or molecules interact to form or break chemical bonds.

chemistry (kem′is-trē) [G. *chemeia*, alchemy] Science dealing with the atomic composition of substances and the reactions they undergo.

chemoreceptor reflex (kem′ō-rē-sep′tŏr) Process in which chemoreceptors detect changes in oxygen levels, carbon dioxide levels, and pH in the blood and produce changes in heart rate, force of heart contraction, and blood vessel diameter that return these values toward their normal levels.

cholecystokinin (kō′lē-sis-tō-kī′nin) [G. *chole*, bile + *kysis*, bladder + *kineo*, to move] Hormone released from the duodenum; inhibits gastric acid secretion and stimulates contraction of the gallbladder.

chondrocyte (kon′drō-sīt) [G. *chondrion*, gristle + cyte] Cartilage cell.

chordae tendineae (kōr′dē ten′di-nē-ē) [L., cord] Tendinous strands running from the papillary muscles to the free margin of the cusps that make up the tricuspid and bicuspid valves; prevent the cusps of these valves from extending up into the atria during ventricular contraction.

choroid (kō′royd) [G. *chorioeides*, membranelike or lacy] Portion of the vascular tunic associated with the sclera of the eye; functions to prevent scattering of light.

choroid plexus Specialized group of ependymal cells in the ventricles; secretes cerebrospinal fluid.

chromatid (krō′mă-tid) [G. *chroma*, color] One of a pair of duplicated chromosomes, joined by the centromere, which separates from its partner during cell division.

chromatin (krō′ma-tin) [G. *chroma*, color] The genetic material of the nucleus consisting of deoxyribonucleic acid (DNA) associated with proteins.

chromosome (krō′mō-sōm) [G. *chroma*, color + *soma*, body] One of the bodies (normally 46 in humans) in the cell nucleus that carry the cell's genetic information.

chyle (kīl) [G. *chylos*, juice] Milky colored lymph with a high fat content.

chylomicron (kī-lō-mi′kron) [chylo- + G. *micros*, small] A lipid droplet synthesized in the epithelial cells of the small intestine containing triglycerides, cholesterol, and lipoproteins.

chyme (kīm) [G. *chymos*, juice] Semifluid mass of partly digested food passed from the stomach into the duodenum.

ciliary body (sil′ē-ar-ē) [like an eyelash] Structure continuous with the choroid layer of the eye at its anterior margin that contains smooth muscle cells and is attached to the lens by suspensory ligaments; regulates the thickness of the lens and produces aqueous humor.

cilium (sil′ē-ŭm), pl. **cilia** (sil′ē-ă) [L., eyelid] A mobile extension of a cell surface; varies from one to thousands per cell, and contains specialized microtubules enclosed by the cell membrane.

citric acid cycle (sit′rik) Series of chemical reactions in which citric acid (six-carbon molecule) is converted into a four-carbon molecule, carbon dioxide is formed, and energy is released; the released energy is used to form ATP; the four-carbon molecule can combine with acetyl-CoA (two-carbon) to form citric acid and start the cycle again.

clavicle (klav′i-kl) [L., a small key] The bone between the sternum and shoulder; the collarbone.

climacteric (klī-mak′ter-ik, klī-mak-ter′ik) [G., the rung of a ladder] The period of endocrine, somatic, and transitory psychological changes occuring in the transition to menopause.

clitoris (klit′ō-ris) A small erectile structure located in the anterior margin of the vestibule.

clot (klot) To coagulate; a soft insoluble mass formed when blood coagulates.

clot retraction Condensation of the clot into a denser, more compact structure.

clotting factor One of many proteins found in the blood in an inactive state; activated in a series of chemical reactions that result in the formation of a blood clot.

coagulation (kō-ag-ū-lā′shŭn) The process of changing from a liquid to a solid, especially blood.

cochlea (kok′lē-ă) The portion of the inner ear involved in hearing; shaped like a snail shell.

codon (kō′don) Sequence of three nucleotides in mRNA that codes for a specific amino acid in a protein.

coenzyme (kō-en′zīm) A substance that enhances or is necessary for the function of an enzyme.

collagen (kol′lă-jen) [G. *koila*, glue + *gen*, producing] Ropelike protein of the extracellular matrix.

collecting duct Straight tubule that extends from the cortex of the kidney to the tip of the renal pyramid; filtrate from the distal tubules enter the collecting duct and is carried to the calyces.

colliculus (ko-lik′ū-lŭs) [L. *collis*, hill] One of four small mounds on the dorsal side of the midbrain; the superior two are involved in visual reflexes, and the inferior two are involved in hearing.

colon (kō′lon) Division of the large intestine that extends from the cecum to the rectum.

commissure (kom′i-shūr) [L., a joining together] A bundle of nerve fibers passing from one side to the other in the brain or spinal cord.

common bile duct Duct formed by the union of the common hepatic and cystic ducts; it joins the pancreatic duct and empties into the duodenum.

common hepatic duct Duct formed by union of the right and left hepatic ducts; it joins the cystic duct to form the common bile duct.

compact bone Bone that is denser and has fewer spaces than cancellous bone.

complement (kom′plĕ-ment) Group of serum proteins that stimulates phagocytosis, inflammation, and lysis of cells.

compound (kom′pound) [to place together] A substance containing two or more different kinds of atoms that are chemically combined.

concha (kon′kă) [L., shell] Structure resembling a shell in shape; the three bony ridges on the lateral wall of the nasal cavity.

condyle (kon′dīl) [G. *kondyles*, knuckle] Rounded articulating surface of a joint.

cone Photoreceptor cell in the retina of the eye with cone-shaped photoreceptive process; important in color vision and visual acuity.

conjunctiva (kon-jŭnk-tī′vă) [L. *conjungo*, to bind together] Mucous membrane covering the anterior surface of the eye and the inner lining of the eyelids.

connective tissue One of the four major tissue types; consists of cells usually surrounded by large amounts of extracellular material; functions to hold other tissues together and provides a supporting framework for the body.

constant region Part of an antibody that does not combine with an antigen and is the same in different antibodies; responsible for activation of complement and binding the antibody to cells such as macrophages, basophils, and mast cells.

corn [L. *cornu*, horn] Thickening of the stratum corneum of the skin over a bony projection in response to friction or pressure.

cornea (kōr′nē-ă) [hornlike] Transparent, anterior part of the fibrous tunic of the eye through which light enters the eye.

corneum (kōr′nē-ŭm) See *stratum corneum*.

coronal plane (kōr′ō-năl) [G. *korone,* crown] Plane separating the body into anterior and posterior portions; also called a frontal plane.

coronary artery (kōr′o-nār-ē) [circling like a crown] An artery that carries blood to the muscles of the heart; the left and right coronary arteries arise from the aorta.

coronary bypass Surgery in which a vein from some other part of the body is grafted to a coronary artery in such a way as to allow blood flow past a blockage in the coronary artery.

coronary vein Vein that carries blood from the heart muscle primarily to the right atrium.

corpus callosum (kōr′pus kă-lō′sŭm) [L., body; callous] A large, thick nerve fiber tract connecting the two cerebral hemispheres.

corpus luteum (loo′tē′ŭm) Yellow endocrine body formed in the ovary in the site of a ruptured follicle immediately after ovulation; secretes progesterone and estrogen.

cortex (kōr′teks), pl. **cortices** (kōr′ti-sēz) [L., bark] The outer part of an organ such as the brain, kidney, adrenal gland, or hair.

cortisol (kōr′ti-sol) Steroid hormone released by the adrenal cortex; increases blood glucose and inhibits inflammation; it is a glucocorticoid.

cotransport (kō-trans′pōrt) The transport of one substance across a plasma membrane, coupled with the simultaneous transport of another substance across the same membrane in the same direction.

covalent bond (kō-vāl′ent) Chemical bond that is formed when two atoms share one or more pairs of electrons.

coxa (kok′să), pl. **coxae** (kok′sē) [L., hip] The bone of the hip.

cranial nerve (krā′nē-ăl) Peripheral nerve originating in the brain.

cranial vault Eight skull bones that surround and protect the brain; braincase.

cremaster muscle (krē-mas′ter) Extension of abdominal muscles; in the male it raises the testis.

crenation (krē-nā′shŭn) [L. *crena,* a notch] Cell shrinkage that occurs when water moves by osmosis from a cell into a hypertonic solution.

cretinism (krē′tin-izm) Hypothyroidism in an infant; appears during the first years of life and results in stunting of bodily growth and of mental development; hypothyroid dwarfism.

cricoid cartilage (krī′koyd) Most inferior laryngeal cartilage.

cricothyrotomy (krī′kō-thī-rot′ō-mē) Formation of an artificial opening in a victim's air passageway through the membrane between the cricoid and thyroid cartilage.

crown That part of the tooth that is formed of and covered by enamel.

crypt (kript) [G. *kryptos,* hidden] A pitlike depression.

cryptorchidism (krip-tōr′ki-dizm) Failure of the testes to descend into the scrotal sac.

cupula (koo′poo-lă) [L. *cupa,* a tub] Gelatinous mass that overlies the hair cells of the cristae ampullares of the semicircular canals; responds to fluid movement.

cutaneous (kū-tā′nē-ŭs) [L. *cutis,* skin] Relating to the skin.

cuticle (kū′ti-kl) [L. *cutis,* skin] The outer thin layer, usually horny, for example, the outer covering of hair or the growth of the stratum corneum onto the nail.

cyanosis (sī-ă-nō′sis) [G., dark blue color] Blue coloration of the skin and mucous membranes caused by insufficient oxygenation of blood.

cystic duct (sis′tik) Duct from the gallbladder; it joins the common hepatic duct to form the common bile duct.

cytoplasm (sī′tō-plazm) [G. *cyto,* cell + *plasma,* a thing formed] Cellular material surrounding the nucleus.

cytoskeleton (sī-tō-skel′ĕ-ton) The collection of microtubules, microfilaments, and intermediate filaments that support the cytoplasm and organelles; also involved with cell movements.

D

dartos muscle (dar′tōs) [fr. *derō,* to skin] The layer of smooth muscle beneath the skin of the scrotum.

deciduous teeth (dē-sid′ū-ŭs) [L. *deciduus,* falling off] The primary teeth that fall out to be replaced by the permanent teeth.

decomposition reaction (dē′kom-pō-zish′ŭn) The breakdown of a larger molecule into smaller molecules, ions, or atoms.

deep [O.E. *deop,* deep] Away from the surface, internal.

defecation (def-ĕ-kā′shŭn) [L. *defaeco,* to purify] Discharge of feces from the rectum.

deglutition (dē-gloo-tish′ŭn) [L. *de-,* from, away + *glutio,* to swallow] The act of swallowing.

deltoid (del′toyd) [triangular] Triangular muscle over the shoulder; inserts onto the humerus; abducts the arm.

denaturation (dē-na-tū-rā′shŭn) The change in shape of a protein caused by breaking hydrogen bonds; agents that cause denaturation include heat and changes in pH.

dendrite (den′drīt) [G. *dendrite,* tree] Short, treelike cell process of a neuron; receives stimuli.

dentin (den′tin) Bonelike material forming the mass of the tooth.

deoxyribonucleic (dē-oks′ē-rī′bō-noo-klē′ic) **acid (DNA)** Type of nucleic acid containing the sugar deoxyribose; the genetic material of cells; DNA.

depolarize Decrease in the difference in potential (charge) between two points, as between the inside and outside of a cell membrane.

dermis (der′mis) [G. *derma,* skin] Dense connective tissue that forms the deep layer of the skin; responsible for the structural strength of the skin.

desmosome (dez′mō-sōm) [G. *desmos,* a band + *soma,* body] A point of adhesion between two cells.

diabetes mellitus (dī-ă-bē′tēz mel-lī′tŭs) A condition resulting from too little insulin secreted from the pancreatic islets, insufficient numbers of insulin receptors on target cells, or defective receptors that do not respond to insulin.

diaphragm (dī′ă-fram) [a partition wall] Muscular separation between the thoracic and abdominal cavities; its contraction results in inspiration.

diaphysis (dī-af′i-sis) [G., growing between] Shaft of a long bone.

diastole (dī-as′tō-lē) [G. *diastole,* dilation] Relaxation of the heart chambers during which they fill with blood; usually refers to ventricular relaxation.

diastolic pressure The minimum arterial blood pressure achieved during ventricular diastole.

diencephalon (dī-en-sef′ă-lon) [G. *dia,* through + *enkephalos,* brain] Part of the brain inferior to and nearly surrounded by the cerebrum, and connecting posteriorly and inferiorly to the brainstem.

diffusion (di-fū′zhŭn) [L. *diffundo,* to pour in different directions] Tendency for solute molecules to move from an area of higher concentration to an area of lower concentration in a solution; the product of the constant random motion of all atoms, ions, or molecules, in a solution.

digestion (di-jes′chŭn, dī-jes′chŭn) The breakdown of carbohydrates, lipids, proteins, and other large molecules to their component parts.

digestive tract (di-jes′tiv, dī-jes′tiv) The tract from the mouth to the anus, including the stomach and intestines, where food is taken in, broken down, and absorbed.

digitalis (dij′i-tal′is) [L., relating to fingerlike flowers] A steroid used in the treatment of heart diseases such as heart failure; increases the force of contraction of the heart; extracted from the foxglove plant (*Digitalis purpura*).

diploid (dip′loyd) The condition in which there are two copies of each autosome and two sex chromosomes (46 total chromosomes in humans).

disaccharidase (dī-sak′ă-rid-ās) An enzyme that breaks disaccharides down to monosaccharides; commonly found in the microvilli of the intestinal epithelium.

disaccharide (dī-sak′ă-rīd) [two sugars] Two monosaccharides chemically bound together; glucose and fructose chemically join to form sucrose.

dissociate (di-sō-sē-āt′) [L. *dis-* + *socio,* to disjoin, separate] The separation of positive and negative ions when they dissolve in water and are surrounded by water molecules.

distal (dis′tăl) [L. *di-* + *sto,* to be distant] Farther from the point of attachment to the body than another structure.

distal tubule Convoluted tubule of the nephron that extends from the ascending limb of the loop of Henle and ends in a collecting duct.

DNA See *deoxyribonucleic acid*.

dominant (dom′i-nant) [L. *dominus,* a master] In genetics, a gene that is expressed phenotypically to the exclusion of a contrasting recessive trait.

dorsal (dōr′săl) [L. *dorsum,* back] Back surface of the body; in humans, synonymous with posterior.

dorsal root Sensory root of a spinal nerve.

ductus arteriosus (dŭk′tŭs ar-tēr′ē-ō-sŭs) A short artery that extends from the pulmonary trunk to the aorta; in the fetus, blood flows through the ductus arteriosus from the pulmonary trunk into the aorta and bypasses the lungs.

ductus deferens (dŭk′tŭs def′er-enz) Duct of the testis, running from the epididymis to the ejaculatory duct; also called the vas deferens.

duodenum (doo-ō-dē′nŭm, doo-od′ĕ-nŭm) [L. *duodeni,* twelve] First division of the small intestine; connects to the stomach.

dura mater (doo′ră mā′ter) [L., tough mother] Tough, fibrous membrane forming the outermost meningeal covering of the brain and spinal cord.

E

eardrum See *tympanic membrane*.

eccrine (ek′rin) [G. *ek,* out + *krino,* to separate] Exocrine; refers to water-producing sweat glands; see *merocrine*.

ECG See *electrocardiogram*.

ectoderm (ek′tō-derm) Outermost of the three germ layers of the embryo.

ectopic beat (ek-top′ik) A heart beat that originates from an area of the heart other than the SA node.

edema (e-dē′mă) [G. *oidema,* a swelling] Excessive accumulation of fluid, usually causing swelling.

efferent (ef′er-ent) [L. *efferens,* to bring out] Conducting outward from a given organ or part, denoting certain arteries, veins, lymphatics, and motor nerves. Opposite of afferent.

efferent arteriole (ar-tēr′ē-ōl) Vessel that carries blood from the glomerulus to the peritubular capillaries.

efferent ductule (dŭk′tool) Small duct that leads from the testis to the epididymis.

efferent fiber Nerve fiber going from the central nervous system toward the peripheral nervous system; motor fiber.

ejaculation (ē-jak′ū-lā′shŭn) [to shoot out] Reflexive expulsion of semen from the penis.

ejaculatory duct (ē-jak′ū-lă-tōr-ē) Duct formed by the union of the ductus deferens and the excretory duct of the seminal vesicle, which opens into the urethra.

electrocardiogram (ē-lek-trō-kar′dē-ō-gram) **(ECG)** Graphic record of the heart's electrical currents obtained with an electronic recording instrument.

electrolyte (ē-lek′trō-līt) [G. electro, + lytos, soluble] Positive and negative ions that conduct electricity in solution.

electron (ē-lek′tron) Negatively charged particle found around the nucleus of atoms.

electron-transport chain Series of energy transfer molecules in the inner mitochondrial membrane; they receive energy and use it in the formation of ATP and water.

element (el′ĕ-ment) [L. elementum, a rudiment] The simplest type of matter with unique chemical properties.

embolus (em′bō-lŭs) [G. embolos, a plug] A detached clot or other foreign body that occludes a blood vessel.

embryo (em′brē-ō) [fr. en, in + bryō, to swell] In prenatal development, the developing human between the time of fertilization to approximately the end of the second month.

emission (ē-mish′ŭn) [L. emissio, to send out] Discharge; formation and accumulation of semen prior to ejaculation.

emulsification (ē-mŭl′si-fi-kā-shŭn) The dispersal of one liquid, or very small globules of the liquid, within another liquid.

emulsify (ē-mŭl′si-fī) To form an emulsion, which is one liquid dispersed in another liquid.

enamel (ē-nam′ĕl) Hard substance covering the exposed portion of the tooth.

endocardium (en-dō-kar′dē-ŭm) [G. endon, within + kardia, heart] Innermost layer of the heart, including endothelium and connective tissue.

endochondral (en-dō-kon′drăl) [endo + G. chondrion, gristle] Growth of cartilage, which is then replaced by bone.

endochondral ossification (en-dō-kon′drăl os′i-fi-kā′shŭn) Bone formation within cartilage.

endocrine (en′dō-krin) [endo + G. krino, to separate] Ductless gland that secretes internally, usually into the circulatory system.

endocytosis (en′dō-sī-tō′sis) [endo + G. kytos, cell + -osis, condition] Bulk uptake of material through the cell membrane by taking it into a vesicle.

endoderm (en′dō-derm) [endo + G. derma, skin] Innermost of the three germ layers of the embryo.

endolymph (en′dō-limf) [endo + G. lympha, clear fluid or springwater] The fluid inside the membranous labyrinth of the inner ear.

endometrium (en′dō-mē′trē-ŭm) [endo + G. mētra, uterus] Mucous membrane that constitutes the inner layer of the uterine wall; consists of a simple columnar epithelium and a lamina propria that contains simple tubular uterine glands.

endomysium (en′dō-mis′ē-ŭm, en′dō-miz′ē-ŭm,) [endo- + G. mys, muscle] The fine connective tissue sheath surrounding a muscle fiber.

endoplasmic reticulum (en′dō-plas′mik re-tik′ū-lŭm) [endo + G. plastos, formed a network] Membranous network inside the cytoplasm; rough endoplasmic reticulum has ribosomes attached to the surface; smooth endoplasmic reticulum does not have ribosomes attached.

endosteum (en-dos′tē-ŭm) [endo + G. osteon, bone] Membranous lining of the medullary cavity and the cavities of spongy bone.

endothelium, pl. endothelia (en-dō-thē′lē-ŭm) [G. endo + thēlē, nipple] A layer of flat cells lining especially blood and lymphatic vessels and the heart.

enzyme (en′zīm) [G. en, in + zyme, leaven] A protein molecule that increases the rate of a chemical reaction without being permanently altered; an organic catalyst.

eosinophil (ē-ō-sin′ō-fil) [eosin, an acidic dye + G. phileo, to love] White blood cell with granules that stain red with acidic dyes; inhibits inflammation.

ependymal (ep-en′di-măl) The neuroglial cell layer lining the ventricles of the brain.

epicardium (ep-i-kar′dē-ŭm) [G. epi-, upon + kardia, heart] Serous membrane covering the surface of the heart; also called the visceral pericardium.

epicondyle (ep′i-kon′dīl) [epi + G. kondyles, knuckle] Projection on (usually to the side of) a condyle.

epidermis (ep-i-derm′is) [epi + G. derma, skin] Outer portion of the skin formed of epithelial tissue that rests on the dermis; resists abrasion and forms a permeability barrier.

epididymis (ep-i-did′i-mis) [epi + G. didymos, twin] Elongated structure connected to the posterior surface of the testis; site of storage and maturation of the sperm cells.

epiglottis (ep-i-glot′is) [epi + G. glottis, the mouth of the windpipe] Plate of elastic cartilage, covered with mucous membrane, that serves as a valve over the opening of the larynx during swallowing to prevent materials from entering the larynx.

epimysium (ep-i-mis′ē-ŭm, -miz′ē-ŭm) [epi + G. mys, muscle] The fibrous connective tissue layer surrounding a skeletal muscle.

epinephrine (ep′i-nef′rin) [epi + G. nephros, kidney] Hormone similar in structure to the neurotransmitter norepinephrine; major hormone released from the adrenal medulla; increases cardiac output and blood glucose levels.

epiphyseal line (ep-i-fiz′ē-ăl) Dense plate of bone in a bone that is no longer growing, indicating the former site of the epiphyseal plate.

epiphyseal plate Site at which bone growth in length occurs; located between the epiphysis and diaphysis of a long bone; area of cartilage where cartilage growth is followed by ossification; also called the growth plate.

epiphysis (e-pif′-i-sis) [epi, on + G. physis, growth] The end of a bone; separated from the remainder of the bone by the epiphyseal plate or epiphyseal line.

epiploic appendage (ep′i-plō′ik) One of a number of little, fat-filled processes of peritoneum projecting from the serous coat of the large intestine.

episiotomy (e-piz-ē-ot′ō-mē, e-pis-e-ot′ō-mē) [pubic region + G. tōme, incision] An incision in the clinical perineum, performed sometimes during childbirth.

epithalamus (ep′i-thal′ă-mŭs) [G. epi, upon + thalamus] A small dorsomedial area of the thalamus corresponding to the habenula and its associated structures, the stria medullaris of the thalamus, pineal gland, and habenular commissure.

epithelial tissue (ep-i-thē′lē-ăl) One of the four major tissue types consisting of cells with a basement membrane (exceptions are lymph vessels and liver sinusoids), little extracellular material, and no blood vessels; covers the surfaces of the body and forms glands.

epithelium (ep-i-thē′lē-ŭm) [G. epi, on + thele, covering or lining] pl. **epithelia** (ep-i-thē′lē-ă) See epithelial tissue.

eponychium (ep-ō-nik′ē-ŭm) [G. epi + onyx, nail] The thin skin that attaches to the proximal part of the nail.

equilibrium (ē-kwi-lib′rē-ŭm) [G. aequus, equal + libra, a balance] A state created by a chemical reaction proceeding in opposite directions (e.g., from reactants to products and from products to reactants) at equal speed.

erection (ē-rek′shŭn) Engorgement of erectile tissue with blood such as in the erectile tissues of the penis causing the penis to enlarge and become firm.

erector spinae (ē-rek′tōr spī′nē) Common name of the muscle group of the back; holds the back erect.

erythroblastosis fetalis (ĕ-rith′rō-blas-tō′sis fē-tă′lis) [erythroblast + G. -osis, condition] See hemolytic disease of the newborn.

erythrocyte (ĕ-rith′rō-sīt) [G. erythro, red + kytos, cell] See red blood cell.

erythropoietin (ĕ-rith′rō-poy′ĕ-tin) [erythrocyte + G. poiesis, a making] Protein hormone that stimulates red blood cell formation in red bone marrow.

esophagus (ē-sof′ă-gŭs) [G. oisophagos, gullet] The part of the digestive tract between the pharynx and stomach.

estrogen (es′trō-jen) Steroid hormone secreted primarily by the ovaries; involved in the maintenance and development of female reproductive organs, secondary sexual characteristics, and the menstrual cycle.

eustachian tube (ū-stā′shŭn) See auditory tube.

exchange reaction A combination of a decomposition reaction, in which molecules are broken down, and a synthesis reaction, in which the products of the decomposition reaction are combined to form new molecules.

exocrine (ek′sō-krin) [G. exo-, outside + krino, to separate] Gland that secretes to a surface or outward through a duct.

exocytosis (ek′sō-sī-to′sis) Elimination of material from a cell through the formation of vesicles.

exophthalmia (ek-sof-thal′mē-ă) [G. ex, cut + ophthalmos, eye] Bulging of the eyes that frequently accompanies Graves' disease, due to accumulation of a type of connective tissue behind the eye.

expiration (eks-pi-rā′shŭn) To breathe out, to move air out of the lungs.

extension [L. extensio] To stretch out; usually to straighten out a joint.

extracellular (eks-tră-sel′ū-lăr) Refers to the outside of the cell.

extracellular matrix (mā′triks) Nonliving chemical substances located between cells; often consisting of protein fibers, ground substance, and fluid.

extrinsic muscle (eks-trin′sik) Muscle located outside of the structure on which it acts.

extrinsic regulation Regulation of the heart that involves mechanisms outside the heart, including nervous and hormonal regulation.

F

facet (fas′et) [Fr., little face] A small, smooth articular surface.

facilitated diffusion (fă-sil′i-tā-tĭd di-fū′zhŭn) Carrier-mediated process that does not require ATP and moves substances into or out of cells from a higher to a lower concentration.

fascia (fash′ē-ă) [L., band or fillet] Loose areolar connective tissue found beneath the skin (hypodermis), or dense connective tissue that encloses and separates muscles.

fasciculus (fă-sik′ū-lus) [L. *fascis*, bundle] Band or bundle of nerve or muscle fibers bound together by connective tissue.

fat Greasy, soft-solid lipid found in animal tissues and many plants; composed of glycerol and fatty acids.

fatty acid Straight chain of carbon atoms with a carboxyl group (–COOH) attached at one end; a building block of fats.

feces (fē′sēz) Matter discharged from the digestive tract during defecation, consisting of the undigested residue of food, epithelial cells, intestinal mucus, bacteria, and waste material.

fertilization (fer′til-i-zā′shŭn) Union of the sperm cell and oocyte to form a zygote.

fetus (fē′tŭs) In prenatal development, the developing human between approximately 56 days and birth.

fibrillation (fī-bri-lā′shŭn, fib-rī-lā′shŭn) Very rapid contraction of cardiac muscle fibers, but not of the muscle as a whole; results in dramatically reduced pumping action of the heart.

fibrin (fī′brin) [L. *fibra*, fiber] A threadlike protein fiber derived from fibrinogen by the action of thrombin; forms a clot, that is, a network of fibers that traps blood cells, platelets, and fluid, which stops bleeding.

fibrinogen (fī-brin′ō-jen) [L. *fibra*, fiber + *gen*, produce] A protein in plasma that gives rise to fibrin when acted on by thrombin to form a clot.

fibrinolysis (fī-bri-nol′i-sis) [L. *fibra*, fiber + G. *lysis*, dissolution] The breakdown of a clot by plasmin.

fibroblast (fī′brō-blast) Cell in connective tissue responsible for the production of collagen.

filtration (fil-trā′shŭn) Movement, resulting from a pressure difference, of a liquid through a filter, which prevents some or all of the substances in the liquid from passing.

filtration membrane Membrane formed by the glomerular capillary endothelium, the basement membrane, and the podocytes of Bowman's capsule.

fimbria (fim′brē-ă), pl. **fimbriae** (fim′brē-ē) Long thin process that surrounds the opening of the uterine tube.

first heart sound The heart sound that results from the simultaneous closure of the tricuspid and bicuspid valves.

flagellum (flă-jel′ŭm), pl. **flagella** (flă-jel′ă) [L., whip] Whiplike locomotor organelle similar to cilia except longer, and there is usually one per sperm cell.

flexion (flek′shŭn) [L. *flectus*] To bend.

focal point The point at which light rays cross after passing through a concave lens.

follicle-stimulating hormone (fol′i-kl) **(FSH)** Hormone of the anterior pituitary gland that, in the female, stimulates the follicles of the ovary, assists in maturation of the follicle, and causes secretion of estrogen from the follicle; in the male, stimulates the epithelium of the seminiferous tubules and is partially responsible for inducing spermatogenesis.

fontanel (fon′tă-nel′) [Fr., fountain] One of several membranous gaps between bones of the skull.

foramen (fō-rā′men) A hole; referring to a hole or opening in a bone.

foramen ovale (ō-val′ē) In the fetal heart, the oval opening in the interatrial septum with a valve that allows blood to flow from the right to left atrium but not in the opposite direction; becomes the fossa ovalis after birth.

formed element A cell, such as a red blood cell or white blood cell, or cell fragments, such as a platelet, in blood.

fossa (fos′ă) A depression, usually more or less longitudinal in shape, below the level of the surface of a bone.

fovea centralis (fō′vē-ă) Depression in the center of the macula of the eye; has the greatest visual acuity and where there are only cones.

free energy Total amount of energy that can be liberated by the complete catabolism of food.

frenulum (fren′ū-lŭm) [L. *frenum*, bridle] Fold extending from the floor of the mouth to the middle of the under surface of the tongue.

frontal plane Plane separating the body into anterior and posterior portions; also called a coronal plane.

FSH See *follicle-stimulating hormone*.

full-thickness burn Burn that destroys the epidermis and the dermis and sometimes the underlying tissue as well; sometimes called a third-degree burn.

fundus (fŭn′dŭs) [L., bottom] The bottom, or area farthest from the opening, of a hollow organ such as the stomach, uterus.

G

gamete (gam′ēt) [G. *gametēs*, husband; *gametē*, wife] Germ cell such as an oocyte or sperm cell.

gamma globulin (gam′ă glob′ū-lin) A family of proteins found in plasma.

ganglion (gang′glē-on), pl. **ganglia** (gang′glē-ă) [G., knot] A group of neuron cell bodies in the peripheral nervous system.

gap junction Small channels that allow materials to pass from one cell to an adjacent cell; provides a means of intercellular communication.

gastric gland (gas′trik) A gland within the stomach.

gastric inhibitory polypeptide Hormone released from the duodenum; inhibits gastric acid secretion.

gastrin (gas′trin) Hormone secreted in the mucosa of the stomach and duodenum that stimulates secretion of hydrochloric acid by the gastric glands.

gastrointestinal tract (gas′trō-in-tes′tin-ăl) Technically only the stomach and intestines. Often used as a synonym for digestive tract, which extends from the mouth to the anus.

gene A sequence of nucleotides in DNA that is a chemical set of instructions for making a specific protein.

genetics (jĕ-net′iks) The branch of science that deals with heredity.

genotype (jen′ō-tīp) Genetic makeup of an individual.

GH See *growth hormone*.

giantism (jī′an-tizm) Abnormal growth in young people because of hypersecretion of growth hormone by the pituitary gland.

gingiva (jin′ji-vă) Dense fibrous tissue, covered by mucous membrane, that covers the alveolar processes of the upper and lower jaws and surrounds the necks of the teeth.

girdle (ger′dl) A bony ring or belt that attaches a limb to the body such as the pectoral (shoulder) and pelvic girdles.

gland A single cell or a multicellular structure that secretes substances into the blood, into a cavity, or onto a surface.

glia (glī′ă) See *neuroglia*.

glomerulus (glō-măr′ū-lŭs) [L. *glomus*, ball of yarn] Mass of capillary loops at the beginning of each nephron, nearly surrounded by Bowman's capsule.

glucagon (gloo′kă-gon) [glucose + agō, to lead] Hormone secreted from the pancreatic islets of the pancreas that acts primarily on the liver to release glucose into the circulatory system.

glucocorticoid (gloo-kō-kōr′ti-koyd) Hormones from the adrenal cortex capable of increasing the rate at which lipids are broken down to fatty acids and proteins are broken down to amino acids; elevates blood glucose levels, and acts as an anti-inflammatory substance.

glycerol (glis′er-ol) A three-carbon molecule with a hydroxyl group attached to each carbon; a building block of fats.

glycogen (glī′kō-jen) Animal starch; composed of many glucose molecules bound together in chains that are highly branched; functions as a carbohydrate reserve; stores glucose molecules; in animal cells.

glycolysis (glī-kol′i-sis) [G. *glykys*, sweet + *lysis*, a loosening] Anaerobic process during which one glucose molecule is converted to two pyruvic acid molecules; a net of two ATP molecules is produced during glycolysis.

glycoprotein (glī-kō-prō′tēn) An organic molecule composed of a protein and a carbohydrate.

GnRH See *gonadotropin-releasing hormone*.

goblet cell Epithelial cell that has the end of the cell at the free surface distended with mucin.

goiter (goy′ter) [L. *guttur*, throat] An enlargement of the thyroid gland, not due to a neoplasm, usually caused by a lack of iodine in the diet.

Golgi apparatus (gol′jē) Named for Camillo Golgi, Italian histologist and Nobel laureate, 1843–1926; stacks of flattened sacks, formed by membranes, that collect, modify, package, and distribute proteins and lipids.

gonad (gō′nad) [L. *gonē*, seed] An organ that produces sex cells; a testis or an ovary.

gonadotropin (gō′nad-ō-trō′pin) [*gonē*, seed + *trope*, a turning] Hormone capable of promoting gonadal growth and function; two major gonadotropins are luteinizing hormone (LH) and follicle-stimulating hormone (FSH).

gonadotropin-releasing hormone (GnRH) Hypothalamic hormone that stimulates the secretion of LH and FSH from the anterior pituitary gland.

granulation tissue (gran′ū-lā′shŭn) Vascular connective tissue formed in wounds.

granulocyte (gran′ū-lō-sīt) [granular + G. *kytos*, cell] White blood cell named according to the appearance, in stained preparations, of large cytoplasmic granules; neutrophils, basophils, and eosinophils.

Graves' disease A type of hyperthyroidism resulting from abnormal proteins produced by the immune system that are similar in structure and function to thyroid-stimulating hormone, often accompanied by exophthalmia.

growth hormone (GH) Protein hormone of the anterior pituitary gland; it promotes body growth, increases fat mobilization, increases blood glucose levels because it inhibits glucose utilization.

gynecomastia (gī′nĕ-kō-mas′tē-ă) Enlarged breasts in males.

gyrus (jī′rŭs) [L. *gyros*, circle] Rounded elevation or fold on the surface of the brain.

H

hair A threadlike outgrowth of the skin consisting of columns of dead keratinized epithelial cells.

hair cell Cell of the inner ear containing hairlike processes (microvilli) that respond to bending of the hairs by depolarizing.

hamstring muscle One of the three major muscles of the posterior thigh.

haploid (hap′loyd) The condition in which a cell has one copy of each autosome and one sex chromosome (23 total chromosomes in humans); characteristic of gametes.

haustra (haw′strǎ) Sacs of the colon, formed by the teniae coli, which are slightly shorter than the gut, so that the gut forms pouches.

haversian canal (ha-ver′shan) Named for 17th-century English anatomist, Clopton Havers (1650–1702); see *central canal*.

haversian system See *osteon*.

HCG See *human chorionic gonadotropin*.

heart–lung machine A machine that pumps blood and carries out the process of gas exchange; it substitutes for the heart and lungs during heart surgery.

heart rate The number of complete cardiac cycles (heartbeats) per minute.

heart transplant The process of taking a healthy heart from a recently deceased donor and transplanting it into a recipient who has a diseased heart.

hematocrit (hē′mǎ-tō-krit, hem′a-tō-krit) [G. *hemato*, blood + *krino*, to separate] The percentage of total blood volume composed of red blood cells.

hematopoiesis (hē′mǎ-tō-poy-ē′sis) [G. *hemato*, blood + *poiesis*, a making] Production of blood cells.

hemidesmosome (hem-ē-des′mō-sōm) [G. *hemi*, one half] Half desmosome that occurs on the basal surface epithelial cells that rest on the basement membrane.

hemoglobin (hē-mō-glō′bin) [G. *hemato*, blood + *glob*, a ball] Red protein of red blood cells consisting of four globin proteins with an iron-containing red pigment, heme, bound to each globin protein; transports oxygen and carbon dioxide.

hemolysis (hē-mol′i-sis) [G. *hemo*, blood + *lysis*, destruction] The rupture of red blood cells.

hemolytic (hē-mō-lit′ik) **disease of the newborn** Destruction of red blood cells in the fetus or newborn caused by antibodies produced in the Rh-negative mother acting on the Rh-positive blood of the fetus or newborn.

hemorrhage (hem′ŏ-rij) [G. *haima*, blood + *rhegnymi*, to burst forth] Rupture or leaking of blood from vessels.

hepatic (he-pat′ik) [G. *hepar*, liver] Associated with the liver.

hepatic portal system [L. *porta*, gate] Blood flow through the veins that begin as capillary beds in the small intestine, spleen, pancreas, and stomach and carry blood to the liver, where they end as a capillary bed.

hepatic portal vein The vein that carries blood from the intestines, stomach, spleen, and pancreas to the liver.

Hering–Breuer reflex Named for the German physiologist, Heinrich E. Hering (1866–1948) and the Austrian internist, Josef Breuer (1842–1925). Process in which action potentials from stretch receptors in the lungs arrest inspiration, expiration then occurs.

heterozygous (het′er-ō-zī′gǔs) Having two different genes for a given trait.

hilum (hī′lǔm) [L., a small amount or trifle] Part of an organ where the nerves and vessels enter and leave.

histology (his-tol′ō-jē) [G. *histo*, web (tissue) + *logos*, study] The science that deals with the structure of cells, tissues, and organs in relation to their function.

holocrine (hol′ō-krin) [G. *holo*, whale + *krino*, to separate] Gland whose secretion consists of disintegrated cells of the gland. An example is a sebaceous gland.

homeostasis (hō′mē-ō-stā′sis) [G. *homoio*, like + *stasis*, a standing] Existence and maintenance of a relatively constant environment within the body with respect to functions and the composition of fluids and tissues.

homeotherm (hō′mē-ō-therm) [G. *homoiois*, like + *thermos*, warm] Any animal, including mammals and birds, that tends to maintain a constant body temperature; also referred to as warm-blooded.

homozygous (hō-mō-zī′gǔs) Having two identical genes for a given trait.

hormone (hōr′mōn) [G. *hormon*, to set into motion] Substance secreted by endocrine tissues into the blood that acts on a target tissue to produce a specific response.

human chorionic gonadotropin (HCG) (kō-rē-on′ik gō′nad-ō-trō′pin) A hormone similar to LH secreted from the placenta and is essential for the maintenance of pregnancy for the first three months; functions to prevent the corpus luteum from degenerating.

humerus (hū′mer-ǔs) [L., shoulder] The bone of the arm.

humoral immunity (hū′-mōr-ǎl i-mū′ni-tē) See *antibody-mediated immunity*.

hydrogen bond (hī′drō-jen) The weak attraction between the oppositely charged ends of two polar covalent molecules; the weak attraction between the end of a polar covalent molecule and an ion.

hydrophilic (hī-drō-fil′ik) [G. *hydor*, water + *philos*, fond or loving] Attracting or associating with water molecules; tending to dissolve or associate with water molecules; tending to dissolve in water; polar.

hydrophobic (hī-drō-fōb′ik) [G. *hydor*, water + *phobos*, fear] Lacking affinity for water molecules; tending to not dissolve in water; nonpolar.

hydroxyapatite (hī-drok′sē-ap-ǎ-tīt) The complex crystal structure that makes up the mineral portion of bones and teeth.

hymen (hī′men) [G. *hymen*, membrane] A thin membranous fold highly variable in appearance; partly occludes the opening of the vagina prior to its rupture; may occur for a variety of reasons and is frequently absent.

hyoid (hī′oyd) [G., shaped like the letter epsilon, ε] The U-shaped bone in the throat.

hypercapnia (hī-per-kap′nē-ǎ) [hyper + G. *kapnos*, smoke, vapor] Abnormally increased arterial carbon dioxide tension or levels.

hyperpolarize (hī′per-pō′lǎr-īz) [G. *hyper*, above + *polaris*, polar] An increase in polarization of membranes of nerve or muscle cells.

hypertension (hī′per-ten′shǔn) [G. *hyper*, above + *tensio*, tension] High blood pressure; generally blood pressure greater than 140/90 is considered to be too high.

hyperthyroidism (hī-per-thī′royd-izm) An abnormality of the thyroid gland in which thyroid hormone secretion is increased.

hypertonic (hī-per-ton′ik) [G. *hyper*, above + *tonos*, tension] Solution that causes cells to shrink.

hypodermis (hī-pō-der′mis) [G. *hypo*, under + *dermis*, skin] Loose connective tissue under the dermis that attaches the skin to muscle and bone.

hypophysis (hī-pof′i-sis) [*hypo* + G., an undergrowth or growth] Endocrine gland attached to the hypothalamus by the infundibulum; the pituitary gland.

hypothalamic–pituitary portal system (hī′pō-thal′ǎ-mĭk-pi-too′i-tār-ē) Series of blood vessels that carry blood from the area of the hypothalamus to the anterior pituitary gland; originates from capillary beds in the hypothalamus and terminates as a capillary bed in the anterior pituitary gland.

hypothalamus (hī′pō-thal′ǎ-mǔs) [*hypo* + G. *thalamos*, bedroom] Important autonomical and endocrine control center of the brain located beneath the thalamus of the brain.

hypothyroidism (hī′pō-thī′royd-izm) Reduced secretion of thyroid hormones from the thyroid gland, leading to cretinism in infants and symptoms of inadequate thyroid hormone secretion in adults.

hypotonic (hī-pō-ton′ik) [*hypo*, under + G. *tonos*, tension or tone] Solution that causes cells to swell.

hypoxia (hī-pok′sē-ǎ) [hypo + oxygen] Below normal levels of oxygen in arterial blood.

I

ICSH See *interstitial cell-stimulating hormone*.

Ig See *immunoglobulin*.

ileocecal junction (il′ē-ō-se′kǎl) The junction of the ileum of the small intestine and the cecum of the large intestine.

ileum (il′ē-ǔm) [L. *eileo*, to roll up, twist] The third portion of the small intestine, about 3.5 meters in length; extends from the jejunum to the ileocecal opening.

ilium (il′ē-ǔm) The broad, flaring portion of the hipbone, becomes fused with the ischium and pubis.

immunity (i-mū′ni-tē) The ability to resist damage from foreign substances such as microorganisms and harmful chemicals such as toxins released by microorganisms.

immunoglobulin (im′ū-nō-glob-ū-lin) **(Ig)** Refers to antibodies.

implantation (im-plan-tā′shǔn) Attachment of the blastocyst to the endometrium of the uterus; occurring 6 or 7 days after fertilization of the oocyte.

impotence (im′pŏ-tens) Inability of the male to achieve or maintain an erection and thus engage in sexual intercourse.

incompetent valve A leaky valve; usually refers to a leaky valve in the heart that allows blood to flow through it when it is closed.

incus (ing′kus) [L., anvil] The middle bone of the middle ear; the anvil.

infarct (in′farkt) Area of necrosis resulting from a sudden insufficiency of arterial blood supply.

inferior (in-fē′rē-ōr) [L., lower] Down, or lower, with reference to the anatomical position.

inferior vena cava (vē′nǎ kā′vǎ) Receives blood from the lower limbs, pelvis, and abdominal organs and empties into the right atrium of the heart.

inflammatory (in-flam′ǎ-tōr-ē) **response** Complex sequence of events involving chemicals and immune system cells that results in the isolation and destruction of foreign substances such as bacteria; symptoms include redness, heat, swelling, pain, and disturbance of function.

infundibulum (in-fǔn-dib′ū-lǔm) [L., a funnel] Funnel-shaped structure or passage, for example, the infundibulum that attaches the pituitary gland to the hypothalamus; funnellike expansion of the uterine tube near the ovary.

inguinal canal (ing′gwi-nǎl) The passageway through which a testis passes as it descends from the abdominopelvic cavity to the scrotum.

inguinal hernia (her′nē-ǎ) A rupture that allows the potential protrusion of abdominal organs such as the small intestine through the inguinal canal.

innate immunity (i′nāt, i-nāt′) Immune system response that is the same on each exposure to an antigen; there is no ability to remember a previous exposure to a specific antigen.

inner cell mass Group of cells at one end of the blastocyst from which the embryo develops.

inorganic (in-ōr-gan′ik) Molecules that do not contain carbon atoms; originally defined as molecules that came from nonliving sources; the original definition is no longer valid because carbon dioxide produced by living organisms is considered an inorganic molecule.

insertion (in-ser′shŭn) The more movable attachment point of a muscle.

inspiration (in-spi-rā′shŭn) To breathe in, to move air into the lungs, or inhale.

insulin (in′sŭ-lin) Protein hormone secreted from the pancreas that increases the uptake of glucose and amino acids by most tissues.

interatrial septum (in-ter-ā′trē-ăl) The cardiac muscle partition separating the right and left atria.

intercalated disk (in-ter′kă-lā-ted) [inserted between] Connection between cardiac muscle cells; important in coordinating the contractions of cardiac muscle cells; contains gap junctions that allow action potentials to pass from one cardiac muscle cell to adjacent cardiac muscle cells.

intercostal muscle (in-ter-kos′tăl) Muscle located between ribs.

interferon (in-ter-fēr′on) A protein released by virally infected cells that binds to other cells and stimulates them to produce antiviral proteins that inhibit viral replication.

interkinesis (in′ter-ki-nē′sis) The short time period between the formation of the daughter cells of the first meiotic division and the second meiotic division.

interstitial cell (in-ter-stish′ăl) Cell between the seminiferous tubules of the testes; secretes testosterone; also called Leydig cell.

interstitial cell-stimulating hormone (ICSH) A term sometimes used to refer to luteinizing hormone in males. Hormone of the anterior pituitary gland that stimulates the secretion of testosterone in the testes. See *luteinizing hormone.*

interventricular septum (in-ter-ven-trik′ū-lăr) The cardiac muscle partition separating the right and left ventricles.

intestinal glands (in-tes′ti-năl) Tubular glands in the mucous membrane of the small intestine.

intracellular (in-tră-sel′ū-lăr) Refers to the inside of the cell.

intramembranous ossification (in′tră-mem′bră-nŭs os′i-fi-kā′shŭn) Bone formation within connective tissue membranes.

intramural plexus (in′tră-mū′răl plek′sŭs) [L., within the wall] A nerve plexus within the walls of the gastrointestinal tract; involved in local and autonomic control of digestion.

intrinsic factor (in-trin′sik) Factor secreted by the gastric glands and required for adequate absorption of vitamin B_{12}.

intrinsic muscle Muscle located within the structure on which it acts.

ion (ī′on) Atom or group of atoms carrying an electrical charge because of a loss or gain of one or more electrons.

ionic bond (ī-on′ik) Chemical bond resulting from the attraction between ions of opposite charge.

iris (ī′ris) Specialized part of the vascular tunic of the eye; the "colored" part of the eye that can be seen through the cornea; consists of smooth muscles that regulate the amount of light entering the eye.

isometric contraction (ī-sō-met′rik) Muscle contraction in which the length of the muscle does not change but the amount of tension increases.

isotonic (ī′sō-ton′ik) [G. *iso,* equal + *tonos,* tension] Solution that causes cells to neither shrink nor swell.

isotonic contraction Muscle contraction in which the amount of tension is constant and the muscle shortens.

isotope (ī′sō-tōp) [G. *isos,* equal + *topos,* part] One of two or more elements that have the same number of protons and electrons, but a different number of neutrons.

J

jaundice (jawn′dis) [Fr. *jaune,* yellow] Yellowish staining of the skin, sclerae, and deeper tissues and excretions with bile pigments.

jejunum (jĕ-joo′nŭm) [L. *jejunus,* empty] A portion of the small intestine, about 2.5 meters in length, between the duodenum and ileum.

juxtaglomerular apparatus (jŭks′tă-glŏ-mer′ū-lăr) [L. *juxta;* close to + *glomerulus*] Specialized wall of the distal tubule and afferent arteriole that secretes renin.

K

keratin (ker′ă-tin) A protein that accumulates in cells of nails, hair, and the superficial layers of the epidermis of the skin.

keratinization (ker′ă-tin-i-zā′shŭn) Production of keratin and changes in the structure and shape of epithelial cells as they move to the skin surface.

kinetic (ki-net′ik) [G. *kinetikos,* of motion] Relating to motion or movement.

Korotkoff sound (kō-rot′kof) Named for the Russian physician Nickolai Korotkoff (1874–1920). Sound heard over an artery when blood pressure is determined by the auscultatory method.

kyphosis (kī-fō′sis) [G., hump-back] Abnormal posterior curvature, or flexion, of the spine.

L

labia majora (lā′bē-a) Two rounded folds of skin surrounding the labia minora and vestibule.

labia minora Two narrow, longitudinal folds of mucous membrane enclosed by the labia majora; they unite anteriorly to form the prepuce.

labyrinth (lab′i-rinth) A series of membranous and bony tunnels in the temporal bone; part of the inner ear involved in hearing and balance.

lacrimal (lak′ri-măl) [L., a tear] Relating to tears or tear production.

lactation (lak-tā′shŭn) [L. *lactatio,* suckle] Period following childbirth during which milk is formed in the breasts.

lacteal (lak′tē-ăl) [relating to milk] Lymphatic vessel in the wall of the small intestine that carries chyle from the intestine and absorbs fat.

lactic acid (lak′tik) Three-carbon molecule derived from pyruvic acid as a product of anaerobic respiration.

lacuna (lă-koo′nă), pl. **lacunae** (lă-koo′nē) [L., a pit] A small space, cavity, or depression; a space in cartilage in which a condrocyte is located; a space in bone matrix in which an osteocyte is located; cavity containing maternal blood in the placenta.

lamella (lă-mel′ă), pl. **lamellae** (lă-mel′ē) [L. *lamina,* plate, leaf] A thin sheet or layer of bone.

lamina (lam′i-nă), pl. **laminae** (lam′i-nē) [L. *lamina,* plate, leaf] A layer; a portion of the vertebra that extends from the transverse process to the spinous process.

lamina propria (prō′prē-ă) Layer of connective tissue underlying the epithelium of a mucous membrane.

lanugo (lă-noo′gō) [L. *lana,* wool] Fine, soft, fetal or embryonic hair.

laryngitis (lar-in-jī′tis) Inflammation of the mucous membrane of the larynx.

laryngopharynx (lă-ring′gō-far-ingks) Part of the pharynx lying below the tip of the epiglottis extending to the level of the cricoid cartilage of the larynx.

larynx (lar′ingks) Organ of voice production located between the pharynx and the trachea; it consists of a framework of cartilages and elastic membranes housing the vocal folds (true vocal cords) and the muscles that control the position and tension of these elements.

lateral [L. *latus,* side] Away from the middle or midline of the body.

lateral horn The small, lateral extension of spinal cord gray matter; located only in spinal cord regions T1–L2; containing preganglionic sympathetic neuron cell bodies.

lens The biconvex structure in the anterior part of the eye capable of being flattened or thickened to adjust the focus of light entering the eye.

leukemia (loo-kē′mē-ă) [G. *leukos,* white + *haima,* blood] A tumor of the red bone marrow that results in the production of large numbers of abnormal white blood cells; often accompanied by decreased production of red blood cells and platelets.

leukocyte (loo′kō-sīt) [G. *leukos,* white + *kytos,* cell] See *white blood cell.*

leukocytosis (loo′kō-sī-tō′sis) [leukocyte + G. *-osis,* a condition] A higher than normal number of white blood cells.

leukopenia (loo-kō-pē′nē-ă) [leukocyte + G. *penia,* poverty] A lower than normal number of white blood cells.

Leydig cell (lī′dig) Named for the German anatomist, Franz von Leydig (1821–1908). See *interstitial cell.*

LH See *luteinizing hormone.*

ligament (lig′ă-ment) A tough connective tissue band usually connecting bone to bone.

ligand (lig′and, līgand) [L. *ligo,* to bind] A molecule that binds to a macromolecule, e.g., a ligand binding to a receptor.

limbic system (lim′bik) [L. *limbus,* a border or boundary] A primitive part of the brain involved in visceral and emotional response and in the response to odor.

linea alba (lin′ē-ă al′bă) White line in the center of the abdomen where muscles of the abdominal wall insert.

lipase (lip′ās) An enzyme that breaks down lipids.

lipid (lip′id) [G. *lipos,* fat] Substance composed principally of carbon, oxygen, and hydrogen; generally soluble in nonpolar solvents; fats and cholesterol.

local inflammation Inflammation confined to a specific area of the body; symptoms include redness, heat, swelling, pain, and loss of function.

longitudinal section A cut made through the long axis of an organ.

loop of Henle U-shaped part of the nephron extending from the proximal to the distal tubule and consisting of descending and ascending limbs; many of the loops of Henle extend into the renal pyramids.

lordosis (lōr-dō′sis) [G., a bending backward; swayback] An abnormal anterior curvature of the spine, usually in the lumbar region; saddle back or swayback.

lower motor neuron A motor neuron located in the brainstem or spinal cord, as opposed to the cerebral cortex.

lumbosacral plexus (lŭm′bō-sā′krăl) [L. *lumbus,* loin + *sacrum,* sacred] The nerve plexus that innervates the lower limbs; originates from spinal nerves L1 to S4.

lunula (loo′noo-lă) [L. *luna,* moon] White, crescent-shaped portion of the nail matrix visible through the proximal end of the nail.

luteinizing hormone (LH) (loo′tē-ĭ-nīz-ing) Hormone of the anterior pituitary gland that, in the female, initiates final maturation of the follicles, their rupture to release the oocyte, the conversion of the ruptured follicle into the corpus luteum, and the secretion of proges-terone; in the male, stimulates the secretion of testosterone in the testes and is sometimes referred to as interstitial cell-stimulating hormone (ICSH).

lymph (limf) [L. *lympha,* clear spring water] Clear or yellowish fluid derived from interstitial fluid and found in lymphatic vessels.

lymph node Encapsulated mass of lymphatic tissue found along lymphatic vessels; functions to filter lymph and produce lymphocytes.

lymphocyte (lim′fō-sīt) [L. *lympho,* lymph + G. *kytos,* cell] Nongranulocytic white blood cell involved in the immune system; there are several types of lymphocytes with diverse functions, including antibody production, allergic reactions, graft rejections, tumor control, and regulation of the immune system.

lymphokine (lim′fō-kīn) A class of chemicals produced by T cells that activate macrophages and other immune cells; promote phagocytosis and inflammation.

lymphoma (lim-fō′mă) A neoplasm (tumor) of lymphatic tissue that is almost always malignant.

lysis (lī′sis) [G., dissolution or loosening] The rupturing or breaking of the cell membrane of a cell.

lysosome (lī′sō-sōm) [G. *lysis,* a loosening + *soma,* body] Membrane-bound vesicle containing intracellular digestive enzymes.

M

macrophage (mak′rō-fāj) [G. *makros,* large + *phago,* to eat] Any large mononuclear, phagocytic cell.

macula (mak′ū-lă) [L., a spot] One of the sensory structures in the vestibule, consisting of hair cells and a gelatinous mass embedded with otoliths; responds to gravity.

macula lutea (loo′tē-ă) [L., a yellow spot] Small yellow spot in the posterior retina of the eye where the cones are concentrated; has no red tint because it is devoid of blood vessels.

malignant (mă-lig′nănt) [L. *maligno,* to do anything malicious, with malice or intent to do harm] In reference to a neoplasm, the property of locally spreading and spreading to distant sites.

malleus (mal′ē-ŭs) [L., hammer] The most lateral of the middle ear bones, attached to the tympanic membrane; the hammer.

mamma (mam′ă) pl. **mammae** (mam′ē) See *mammary gland.*

mammary gland (mam′ă-rē) The organ of milk secretion, located in the breast or mamma.

mastication (mas-ti-kā-shŭn) [L. *mastico,* to chew] Process of chewing.

mastoid (mas′toyd) [*mastos,* breast + *eidos,* resem-blance] Resembling a breast; for example, the mastoid process of the temporal bone.

matrix (mā′triks) The substance between the cells of a tissue.

matter Anything that occupies space.

mean arterial blood pressure The average of the arterial blood pressure; it is slightly less than the average of the systolic and diastolic blood pressure, because diastole lasts longer than systole.

meatus (mē-ā′tŭs) [L., to go, pass] Passageway or tunnel.

meconium (mē-kō′nē-ŭm) [L. *mēkōn,* poppy] Greenish anal discharge from the fetus; consists of fluid swallowed, epithelial cells from the mucosa of the gut, mucus from the intestinal glands, and bile from the liver.

medial (mē′dē-ăl) [L. *medialis,* middle] Toward the middle or midline of the body.

mediastinum (me′dē-as-tī′nŭm) [L., middle septum or wall] The middle wall of the thorax consisting of the trachea, esophagus, thymus, heart, and other structures.

mediator of inflammation Chemical released or activated by injured tissues and adjacent blood vessels; produces vasodilation, increases vascular permeability, and attracts blood cells; includes histamine, kinins, prostaglandins, and leukotrienes.

medulla (me-dool′ă) [L. *medius,* middle, marrow] The center or core of an organ such as the adrenal gland, kidney, or hair.

medulla oblongata (ob-long-gah′tă) Inferior portion of the brainstem that connects the spinal cord with the brain; contains nuclei of cranial nerves plus autonomic control centers for heart rate, respiration, and so forth.

medullary cavity (med′ŭl-er-ē) Large, marrow-filled cavity in the diaphysis of a long bone.

medullary respiratory center (res′pi-ră-tōr-ē, rĕ-spir′ă-tōr-ē) Nerve cells in the medulla oblongata and pons of the brain that control inspiration and expiration.

megakaryocyte (meg-ă-kar′ē-ō-sīt) Large cell in red bone marrow that gives rise to platelets.

meiosis (mī-ō′sis) [G., a lessening] Process of cell division that results in gametes. Consists of two cell divisions that result in four cells, each of which contains half the number of chromosomes as the parent cell; occur in the testes and ovaries.

melanin (mel′ă-nin) [G. *melas,* black] Brown to black pigment responsible for skin and hair color.

melanocyte (mel′ă-nō-sīt) [G. *melas,* black + *kytos,* cell] Cells found mainly in the stratum basale of skin that produce the brown or black pigment melanin.

melanocyte-stimulating hormone (MSH) (mel′ă-nō-sīt) [G. *melas,* black + *kytos,* cell] Peptide hormone secreted by the anterior pituitary gland; increases melanin production by melanocytes, making the skin darker in color.

melanoma (mel′ă-nō′mă) [G. *melas,* black + *oma,* tumor] A malignant tumor derived from melanocytes.

melanosome (mel′ă-nō′sōm) [G. *melano,* black + *soma,* body] Pigment granule produced by melanocytes.

melatonin (mel-ă-tōn′in) Hormone secreted by the pineal body; may inhibit gonadotropin-releasing hormone secretion from the hypothalamus.

membranous labyrinth (mem′bră-nŭs lab′i-rinth) The membrane-bound set of tunnels and chambers of the inner ear.

memory cell Lymphocyte derived from a B cell or T cell that has been exposed to an antigen; when exposed to the same antigen a second time, the memory cell rapidly responds to provide immunity.

memory response Immune response that occurs when the immune system is exposed to an antigen against which it has already had a primary response; results in the production of large amounts of antibodies and memory cells; also called a secondary response.

menarche (me-nar′kē) [G. *mēn,* month + *archē,* beginning] The time of the first menstrual period or flow.

meninges (mĕ-nin′jēz) [G. *meninx,* membrane] A series of three connective tissue membranes; the dura mater, arachnoid mater, and pia mater; surround and protect the brain and spinal cord.

menopause (men′ō-pawz) [L. *mēn,* month + *pausis,* cessation] Permanent cessation of the menstrual cycle.

menses (men′sēz) [L. *mensis,* month] Loss of blood and tissue as the endometrium of the uterus sloughs away at the end of the menstrual cycle; occurring at about 28-day intervals in the nonpregnant female of reproductive age.

menstrual cycle (men′stroo-ăl) Series of changes that occur in sexually mature, nonpregnant females and result in menses; specifically includes the cyclical changes that occur in the uterus and ovary.

merocrine (mer′ō-krin) [G. *meros,* part + *krino,* to separate] Gland that secretes products with no loss of cellular material; an example is water-producing sweat glands; see *apocrine* and *holocrine.*

mesentery (mes′en-ter-ē) [G. *mesos,* middle + *enteron,* intestine] Double layer of peritoneum extending from the abdominal wall to the abdominopelvic organs; conveys blood vessels and nerves to abdominopelvic organs; holds and supports abdominopelvic organs.

mesoderm (mez′ō-derm) Middle of the three germ layers of the embryo.

mesovarium (mez′ō-vā′rē-ŭm) Mesentery of the ovary; mesentery that attaches the ovary to the posterior surface of the broad ligament.

metabolic rate The total amount of energy produced and used by the body per unit of time.

metabolism (mĕ-tab′ō-lizm) [G. *metabole,* change] Sum of the chemical changes that occur in tissues, consisting of the breakdown of molecules (catabolism) to produce energy and the buildup of molecules (anabolism), which requires energy.

metaphase (met′ă-fāz) [G. *meta,* after + *phasis,* an appearance] The stage of mitosis or meiosis in which the chromosomes become aligned near the center of the cell, at the equatorial plane, separating the centromeres, or chromosome pairs (see *meiosis*). The centromeres of each chromosome divide and the two daughter chro-mosomes are directed toward opposite poles of the cell.

metastasis (mĕ-tas′tă-sis) The shifting of a disease or a neoplasm from one part of the body to another remote from the original location.

micelle (mi-sel′, mī-sel′) [L. *micella,* small morsel] Droplet of digested lipid surrounded by bile salts in the small intestine.

microglia (mī-krog′lē-ă) [G. *micro,* small + *glia,* glue] Small neuroglial cells that become phagocytic and mobile in response to inflam-mation; considered to be macrophages of the central nervous system.

microtubule (mī-krō-too′būl) Hollow tube composed of tubulin; microtubules help provide support to the cytoplasm of the cell and are components of certain cell organelles such as cilia and flagella.

microvillus (mī′krō-vil′ŭs), pl. **microvilli** (mī′krō-vil′ī) One of the minute projections of the cell membrane that greatly increase the surface area of the cell membrane.

micturition reflex (mik-choo-rish′ŭn) Contraction of the urinary bladder stimulated by stretching of the urinary bladder wall; results in emptying of the urinary bladder.

midbrain The superior end of the brainstem; located between the pons and diencephalon; contains fibers crossing from the brain to the spinal cord and vice versa, as well as nuclei and visual reflex centers.

midsagittal (mid′saj′i-tăl) Plane running vertically through the body and dividing it into equal right and left parts.

mineral (min′er-ăl) Inorganic nutrient necessary for normal metabolic functions.

mineralocorticoid (min′er-al-ō-kōr′ti-koyd) A steroid hormone released from the adrenal cortex; acts on the kidney to increase the rate of sodium ion reabsorption from the nephron and potassium and hydrogen ion secretion into the nephron of the kidney; an example is aldosterone.

mitochondrion (mī-tō-kon′drē-on), pl. **mitochondria** (mī-tō-kon′drē-ă) [G. *mitos*, thread + *chandros*, granule] Small, spherical, rod-shaped or thin filamentous structure in the cytoplasm that is a major site of ATP production.

mitosis (mī-tō′sis) [G., thread] Division of the nucleus. Process of cell division that results in two daughter cells with exactly the same number and type of chromosomes as the parent cell.

mitral valve (mī′trăl) [resembling a bishop's miter, a two-pointed hat] See *bicuspid valve*.

molecule (mol′ĕ-kūl) Two or more atoms chemically combined to form a structure that behaves as an independent unit.

monocyte (mon′ō-sīt) [G. *mono*, one or single + *kytos*, a cell] A type of white blood cell that transforms to become a macrophage.

mononuclear phagocytic system (mon-ō-noo′klē-ăr fag-ō-sit′ik) Phagocytic cells with a single nucleus, derived from monocytes; the cells either enter a tissue by chemotaxis in response to infection or tissue damage, or are positioned to intercept microorganisms entering tissues.

monosaccharide (mon-ō-sak′ă-rīd) The basic building block from which more complex carbohydrates are constructed; for example, glucose and fructose.

mons pubis (monz pū′bis) [L., mountain] Prominence formed by a pad of fatty tissue over the symphysis pubis in the female.

morula (mōr′oo-lă) [L. *morus*, mulberry] The solid mass of cells resulting from the early cleavage divisions of the zygote.

motor neuron Neuron in the brain or spinal cord that innervates skeletal, smooth, or cardiac muscle cells or glands. Somatic motor neurons directly innervate skeletal muscle cells. Two autonomic motor neurons in series extend from the central nervous system to smooth or cardiac muscle cells or glands.

motor unit A single motor neuron and all the skeletal muscle fibers it innervates.

MSH See *melanocyte-stimulating hormone*.

mucin (mū′sin) Secretion containing mucopolysaccharides (proteoglycans), produced by mucous gland cells.

mucosa (mū-kō′să) [mucus-producing membrane] Mucous membrane consisting of the epithelium and connective tissue; in the digestive tract there is also a layer of smooth muscle.

mucous membrane (mū′kŭs) Thin sheet consisting of epithelium and connective tissue that lines cavities opening to the outside of the body; many contain mucous glands, which secrete mucus.

mucus (mū′kŭs) Viscous secretion produced by and covering mucous membranes; lubricates and protects the mucous membrane, and traps foreign substances.

murmur (mer′mer) An abnormal sound produced within the heart.

muscle fiber (mŭs′ĕl) Muscle cell.

muscle tissue One of the four major tissue types; consists of cells with the ability to contract; includes skeletal, cardiac, and smooth muscle.

muscle twitch Contraction of an entire muscle in response to a stimulus that causes an action potential in one or more muscle fibers.

muscularis (mŭs-kū-lā′ris) The outermost, smooth muscle coat of a hollow organ.

muscularis mucosa The inner, thin layer of smooth muscle found in most parts of the digestive tube outside the lamina propria.

myelinated (mī′ĕ-li-nāt-ed) [G. *myelos*, marrow] Nerve fibers having a myelin sheath.

myelin sheath (mī′ĕ-lin) A lipoprotein envelope made by wrappings of the cell membrane of a Schwann cell or oligodendrocyte around an axon.

myocardium (mī-ō-kar′dē-ŭm) [*myo*- + G. *kordin*, heart] Middle layer of the heart, consisting of cardiac muscle.

myofibril (mī-ō-fī′bril) A fine, longitudinal fibril within a skeletal muscle fiber; consisting of sarcomeres composed of thick (myosin) and thin (actin) myofilaments, placed end to end.

myofilament (mī-ō-fil′ă-ment) An ultramicroscopic protein thread helping to form myofibrils in skeletal muscle; thin myofilaments are composed of actin, and thick myofilaments are composed of myosin.

myometrium (mī-ō-mē′trē-ŭm) Muscular wall of the uterus, composed of smooth muscle.

myosin myofilament (mī′ō-sin) One of the two major kinds of protein fibers of a sarcomere; thick filament, resembles bundles of golf clubs.

myxedema (mik-se-dē′mă) Hypothyroidism characterized by edema beneath the skin due to a change in the structure of the subcutaneous connective tissue.

N

NADH See *nicotinamide adenine dinucleotide*.

nail (nāl) A thin, horny plate at the ends of the fingers and toes, consisting of several layers of dead epithelial cells containing a hard keratin.

naris (nā′ris), pl. **nares** (nā′res) Nostril, the opening into the nasal cavity.

nasal cavity (nā′zăl) Cavity divided by the nasal septum, and extending from the external nares anteriorly to the nasopharynx posteriorly; bounded inferiorly by the hard palate.

nasolacrimal duct (nā-zō-lak′ri-măl) [L. *nasus*, nose + *lacrima*, tear] Duct that leads from the lacrimal sac to the nasal cavity.

nasopharynx (nā′zō-far′ingks) Part of the pharynx that lies above the soft palate; anteriorly it opens into the nasal cavity.

negative feedback Mechanisms by which any deviation from an ideal normal value or set point is resisted or negated; returns a parameter to its normal range and thereby maintains homeostasis.

neonate (nē′ō-nāt) [G. *neos*, new + L. *natalis*, relating to birth] Newborn, from birth to 1 month.

neoplasm (nē′ō-plazm) [*neo* + G. *plasma*, thing formed] New growth, an abnormal tissue growth that grows by cellular proliferation; may be benign or malignant.

nephron (nef′ron) [G. *nephros*, kidney] Functional unit of the kidney, consisting of the renal corpuscle, the proximal tubule, the loop of Henle, and the distal tubule.

nerve (nerv) A collection of axons in the peripheral nervous system; functions to conduct action potentials to and from the central nervous system.

nerve cell A cell capable of receiving a stimulus and propagating an action potential; a neuron.

nerve tract Bundle of axons, their sheaths, and accompanying connective tissues located in the central nervous system.

nervous tissue (ner′vŭs) One of the four major tissue types; consists of neurons, which have the ability to conduct action potentials, and neuroglia, which are support cells.

neural crest cell (noor′ăl) Cell derived during embryonic development from the crests of the neural folds; gives rise to facial structures, pigment cells, and peripheral nerve ganglia.

neural tube Tube formed from the neuroectoderm in the embryo by closure of the neural groove; develops into the brain and spinal cord.

neuroectoderm (noor-ō-ek′tō-derm) That part of the ectoderm that forms the neural tube and neural crest.

neuroglia (noo-rog′lē-ă) [G. *neuro*, nerve + *glia*, glue] Cells of the nervous system other than neurons; play a support role in the nervous system; include astrocytes, ependymal cells, microglia, oligodendrocytes, and Schwann cells; also called glia.

neurolemmocyte (noor-ō-lem′ō′sīt) [G. *neuro*, nerve + *lemma*, husk + *kytos*, cell] See *Schwann cell*.

neuromuscular junction (noor-rō-mŭs′kū-lăr) The synaptic junction between a nerve axon and a muscle fiber.

neuron (noor′on) [G., nerve] A nerve cell.

neurotransmitter (noor′ō-trans-mit′er) [G. *neuro*, nerve + L. *transmitto*, to send across] A chemical that is released by a presynaptic cell into the synaptic cleft and that acts upon the postsynaptic cell to cause a response.

neutral solution (noo′trăl) Solution with equal numbers of hydrogen and hydroxide ions; has a pH of 7.0.

neutron (noo′tron) [L. *neuter*, neither] Electrically neutral particle found in the nucleus of atoms.

neutrophil (noo′trō-fil) [L. *neuter*, neither + G. *phileo*, to love] White blood cell with granules that stains equally with either basic or acidic dyes; phagocytic white blood cell.

nevus (nē′vŭs), pl. **nevi** (nē′vī) A benign, localized overgrowth of the melanin-forming cells of the skin present at birth or appearing early in life; a mole.

nicotinamide adenine dinucleotide (NADH) (nik-ō-tin′ă-mīd ad′ĕ-nēn dī-noo′klē-ō-tīd) A base-containing organic molecule capable of accepting hydrogen atoms and of transferring energy from glycolysis and the citric acid cycle to the electron-transport chain.

nitroglycerin (nī-trō-glis′er-in) Glyceryl trinitrate used as a vasodilator, especially in angina pectoris.

nociceptor (nō′si-sep′tŏr) [L. *oceo*, hurt, pain, injury + *capio*, to take] A peripheral sensory receptor or mechanism for the reception and transmission of painful or injurious stimuli.

node of Ranvier (ron′vē-ă) The unmyelinated area of an axon, every 0.1–1.0 mm, between adjacent oligodendrocytes of an axon in the central nervous system and between individual Schwann cells of the peripheral nervous system.

norepinephrine (nōr′ep-i-nef′rin) Neurotransmitter substance released from most of the postganglionic neurons of the sympathetic division; hormone released from the adrenal cortex that increases cardiac output and blood glucose levels.

notochord (nō′tō-kōrd) [G. *notor*, back + *chorde*, cord or string] Small rod of tissue lying ventral to the neural tube; characteristic of all vertebrates; in humans it becomes the nucleus pulposus of the intervertebral disks.

nuclear pore (noo′klē-er) Point where the inner and outer membranes of the nuclear envelope come together to form a hole.

nuclease (noo′klē-ās) An enzyme that breaks down nucleic acids.

nucleic acid (noo-klē′ik, -klā′ik) Molecule consisting of many nucleotides chemically bound together; deoxyribonucleic acid and ribonucleic acid.

nucleolus (noo-klē′ō-lŭs), pl. **nucleoli** (noo-klē′ō-lī) Rounded, dense, well-defined nuclear bodies with no surrounding membrane; subunits of ribosomes are manufactured within the nucleolus.

nucleotide (noo′klē-ō-tīd) Basic building block of nucleic acids consisting of a sugar molecule (either ribose or deoxyribose), one of several types of organic bases, and a phosphate group.

nucleus (noo′klē-ŭs), pl. **nuclei** (noo′klē-ī) [L., a little nut, stone of a fruit] Cell organelle containing most of the genetic material of the cell; center of an atom consisting of protons and neutrons; collection of neuron cell bodies in the central nervous system.

nutrient (noo′trē-ent) [L. nutriens, to nourish] Chemical taken into the body that is used to produce energy, provide building blocks for new molecules, or function in other chemical reactions.

nutrition (noo-trish′ŭn) Process by which nutrients are obtained and used in the body.

O

oblique section (ob-lēk′) A cut made at other than a right angle to the long axis of an organ.

obturator (ob′toor-ā-tōr) [L., to occlude or stop up] Any occluding structure or a foramen so occluded, as with the obturator foramen of the hip.

occipital (ok-sip′i-tăl) The back of the head.

odorant (ō′-dōr-ănt) A substance with an odor.

olecranon (ō-lek′ră-non) The point of the elbow.

olfaction (ol-fak′shŭn) [L., to smell] The sense of smell.

olfactory (ol-fak′tŏ-rē) Relating to the sense of smell.

oligodendrocyte (ol′i-gō-den′drō-sīt) Neuroglial cells with multiple cell processes that form myelin sheaths around axons in the central nervous system.

omental bursa (ō-men′tăl ber′să) The pocketlike sac inside the fold of the greater omentum.

omentum (ō-men′tŭm) [L., membrane of the bowels] A fold of peritoneum extending from the stomach to another organ.

oncology (ong-kol′ō-jē) [G. onco, a tumor + logos, to study] The study of cancer and its associated problems.

oocyte (ō′ō-sīt) [G. oon, egg + kytos, cell] Female gamete, or sex cell; a secondary oocyte and a polar body result from the first meiotic division, which occurs prior to the time of ovulation; a zygote and a polar body result from the second meiotic division, which occurs following union of the sperm cell with the secondary oocyte.

oogonium (ō-ō-gō′nē-ŭm), pl. **oogonia** (ō-ō-gō′nē-ă) [G. oon, egg + gone, generation] Cells that give rise to oocytes; have a diploid number of chromosomes.

opsin The protein portion of the rhodopsin molecule; at least three different opsins are located in cone cells.

optic (op′tik) Relating to vision.

optic disc The region in the posterior wall of the eye where the optic nerve exits the eye; the blind spot.

optic nerve Nerve that leaves the eye and exits the orbit through the optic foramen to enter the cranial vault.

oral cavity (ōr′ăl) Mouth; the first portion of the digestive tract.

orbit (ōr′bit) Seven skull bones that surround and protect the eye; eye socket.

organ (ōr′găn) [G. organon, tool] Part of the body composed of two or more tissue types and performing one or more specific functions.

organ of Corti Named for the Italian anatomist, Alfonso Corti (1822–1988). Specialized region of the cochlear duct consisting of hair cells; produces action potentials in response to sound waves.

organ system Group of organs classified as a unit because of a common function or set of functions.

organelle (or′gă-nel) [G. organon, a tool + L. -elle, small, a little organ] Specialized part of a cell performing one or more specific functions.

organic (ōr-gan′ik) Molecules that contain a carbon atom (carbon dioxide is an exception); originally defined as molecules extracted from living organisms; the original definition became obsolete when it became possible to manufacture these molecules in the laboratory.

organism (ōr′gă-nizm) Any living thing considered as a whole, whether composed of one cell or many.

organogenesis (ōr′gă-nō-jen′ĕ-sis) The formation of organs during embryonic development.

orgasm (ōr′gazm) [G. orgao, to swell, be excited] Climax of the sexual act, often associated with a pleasurable sensation.

origin (ōr′i-jin) The less movable attachment point of a muscle.

oropharynx (ōr′ō-far′ingks) Portion of the pharynx that lies posterior to the mouth; it is continuous above with the nasopharynx and below with the laryngopharynx.

osmosis (os-mō′sis) [G. osmos, thrusting or an impulsion] Diffusion of solvent (water) through a selectively permeable membrane from a region of higher water concentration to one of lower water concentration.

osmotic pressure (os-mot′ik) Force required to prevent the movement of water across a selectively permeable membrane.

ossification (os′i-fi-kā′shŭn) [L. os, bone + facio, to make] Bone formation.

osteoblast (os′tē-ō-blast) [G. osteo, bone] A cell that makes bone.

osteoclast (os′tē-ō-klast) [bone eating] A cell that digests and removes bone.

osteocyte (os′tē-ō-sīt) [G. osteon, bone + kytos, cell] Mature bone cell surrounded by bone matrix.

osteon (os′tē-on) A single central canal, with its contents, and the associated lamellae and osteocytes surrounding it. Also called a haversian system.

otolith (ō′tō-lith) [G. ous, ear + lithos, stone] Small protein and calcium carbonate weights in the maculae of the vestibule.

ovary (ō′vă-rē) One of two female reproductive glands located in the pelvic cavity; produces the oocyte, estrogen, and progesterone.

ovulation (ov′ū-lā′shŭn) Release of an oocyte from the mature follicle.

oxidative metabolism (ok-si-dā′tiv mĕ-tab′ō-lizm) Metabolism in which oxygen is required to produce ATP.

oxygen debt (ok′sē-jen) The amount of oxygen required to convert the lactic acid produced during anaerobic respiration to glucose and to replenish creatine phosphate stores.

oxytocin (ok′sī-tō′sin) [G., swift birth] Peptide hormone secreted by the posterior pituitary gland that increases uterine contraction and stimulates milk ejection from the mammary glands.

P

palate (pal′ăt) The roof of the oral cavity; consists of the anterior bony part, the hard palate, and the posterior soft palate that consists mainly of skeletal muscle and connective tissue.

pancreas (pan′krē-as) An elongated gland extending from the duodenum to the spleen; consists of a head, body, and a tail. There is an exocrine portion, which secretes digestive enzymes that are carried by the pancreatic duct to the duodenum, and pancreatic islets, which secrete insulin and glucagon.

pancreatic duct (pan-krē′at′ik) The duct of the pancreas; it joins the common bile duct to empty into the duodenum.

pancreatic islet (i′let) Cellular mass in the tissue of the pancreas; composed of different cell types that constitute the endocrine portion of the pancreas and are the source of insulin and glucagon.

papilla (pă-pil′ă), pl. **papillae** (pă-pil′ē) [L., nipple] A small, nipplelike process; projection of the dermis, containing blood vessels and nerves, into the epidermis; projections on the surface of the tongue.

papillary muscle (pap′i-l-ār-ē) A raised area of cardiac muscle in the ventricle to which the chordae tendineae attach.

paracrine (par′ă-krin) [G. para, alongside + krino, to separate] A kind of hormone function in which the effects of the hormone are restricted to the local environment.

parafollicular cell (par-ă-fo-lik′ū-lăr) A cell type scattered in a network of loose connective tissue between the thyroid follicles of the thyroid gland; secretes calcitonin.

paranasal sinus (par-ă-nā′săl) Air-filled cavity within certain skull bones that connects to the nasal cavity; the four sets of paranasal sinuses are the frontal, maxillary, sphenoidal, and ethmoidal.

parasympathetic (par-ă-sim-pa-thet′ik) [G. para, alongside + sympathetic] Subdivision of the autonomic nervous system with preganglionic neurons in the brainstem and sacral part of the spinal cord; involved in involuntary functions such as digestion, defecation, and urination.

parathyroid gland (par-ă-thī′royd) One of four glandular masses embedded in the posterior surface of the thyroid gland; secretes parathyroid hormone.

parathyroid hormone (PTH) (hōr′mōn) Hormone produced by the parathyroid gland; increases bone breakdown and blood calcium levels.

parietal (pă-rī′ĕ-tăl) [L. paries, wall] Relating to the wall of any cavity; parietal serous membranes are in contact with the walls of cavities. The parietal bones form part of the skull.

parietal peritoneum (pĕ′rĭ-tō-nē′ŭm) [L., wall] That portion of the serous membranes of the abdominal cavity lining the inner surface of the body wall.

parotid gland (pă-rot′id) The largest of the salivary glands; one of a pair of salivary glands located anterior and inferior to each ear.

partial pressure Pressure exerted by a single gas in a mixture of gases.

partial-thickness burn Burn that damages only the epidermis (first-degree burn) or the epidermis and part of the dermis (also called a second-degree burn).

parturition (par-toor-ish′ŭn) [L. parturio, to be in labor] Childbirth; the delivery of a baby at the end of pregnancy.

patella (pa-tel′ă) [L. patina, shallow disk] Kneecap.

pectoral (pek′tŏ-răl) [L. pectoralis, breastbone] Relating to the chest.

pedicle (ped′ĭ-kl) [L. pedicellus, foot] Portion of a vertebra that extends from the body to the transverse process.

pelvic cavity (pel′vik) Space completely surrounded by the pelvic bones.

pepsin (pep′sin) [G. pepsis, digestion] Principal digestive enzyme produced by the stomach; digests proteins into smaller peptide chains.

peptidase (pep′ti-dās) An enzyme capable of breaking peptide chains into smaller chains and amino acids.

peptide bond (pep′tīd) A covalent chemical bond between adjacent amino acids in a polypeptide chain.

pericardial cavity (per-i-kar′dē-ăl) [G. peri-, around + kardia, the heart] Space between the visceral and parietal pericardium, filled with pericardial fluid; a cavity that surrounds the heart.

pericardial fluid The serous fluid found within the pericardial cavity.

pericardial membrane Serous membranes associated with the heart.

pericardium (per-i-kar′dē-ŭm) [G. *pericardion,* the membrane around the heart] The membrane consisting of the epicardium and parietal pericardium (of the serous layers) and the outer fibrous pericardium; also called the pericardial sac.

perilymph (per′i-limf) [*peri* + G. *lympha,* clear fluid] Fluid contained between the bony labyrinth and the membranous labyrinth of the inner ear.

perimetrium (per-i-mē′trē-ŭm) The outer layer of the uterus, also called the serous layer.

perimysium (per′-i-mis′ē-ŭm, per′-i-miz′ē-ŭm) [*peri-* + G. *mys,* muscle] The fibrous sheath enveloping each of the skeletal muscle fascicles.

perineum (per′-i-nē′ŭm) Area inferior to the pelvic diaphragm between the thighs; extends from the coccyx to the pubis.

periodontal (per′ē-ō-don′tăl) [*peri-* + G. *odous,* tooth] Referring to structures surrounding the tooth, primarily in the alveolus.

periosteum (per-ē-os′tē-ŭm) [*peri-* + G. *osteon,* bone] Thick, double-layered connective tissue sheath covering the entire surface of a bone except the articular surface, which is covered with cartilage.

peripheral circulation (pĕ-rif′ĕ-răl) Blood flow through all blood vessels that carry blood away from the heart (arteries), the capillaries, and all vessels that carry blood back to the heart (veins); consists of the pulmonary circulation and the systemic circulation; includes all blood flow except that through the heart tissue itself.

peripheral nervous system (PNS) The part of the nervous system not surrounded by the skull or vertebral column; consisting of nerves and ganglia.

peristaltic waves (per-i-stal′tik) [*peri* + G. *stalsis,* constriction] Waves of relaxation followed by waves of contraction moving along a tube; propels food along the digestive tube.

peritoneal cavity (per′i-tō-nē′ăl) [to stretch over] Space between the visceral and parietal peritoneum, filled with peritoneal fluid; cavity that surrounds many abdominopelvic organs.

peritoneal membrane Serous membrane associated with the peritoneal cavity.

peritubular capillary (per′ĭ-too′bū-lăr) The capillary network located in the cortex of the kidney; associated with the distal and proximal convoluted tubules.

peroxisome (per-ok′si-sōm) Membrane-bound body similar to a lysosome in appearance but often smaller and irregular in shape; contains enzymes that either decompose or synthesize hydrogen peroxide.

Peyer's patch Named for the Swiss anatomist, Johann Peyer (1653–1712). Collection of lymph nodules found in the distal half of the small intestine and in the appendix.

pH scale A measure of the hydrogen ion concentration of a solution; the scale extends from 0 to 14.0. A pH of 7.0 being neutral, a pH of less than 7 acidic, and a pH of greater than 7 basic.

phagocytosis (fag′ō-sī-tō′sis) [G. *phagein,* to eat + L. *kytos,* cell + *osis,* condition] Process of ingestion and digestion by cells of substances such as other cells, bacteria, cell debris, and foreign particles.

pharynx (far′ingks) [G. *pharynx,* throat] The joint openings of the digestive tract and the windpipe. The part of the digestive and respiratory tubes

superior to the larynx and esophagus and inferior and posterior to the oral and nasal cavities.

phenotype (fē′nō-tīp) [G. *phaino,* to display + *typos,* model] Characteristic observed in the individual resulting from expression of the genotype.

pheromones (fer′ō-mōnz) [G. *pherō,* to carry + *hormaō,* to excite] Chemical signals secreted by an individual into the environment and perceived by a second individual of the same, or similar, species producing a change in sexual or social behavior of that individual.

phlebitis (fle-bī′tis) Inflammation of a vein.

phospholipid (fos-fō-lip′id) Lipid with phosphorus resulting in a molecule with a polar and a nonpolar end; main component of cell membranes.

physiology (fiz-ē-ol′o-jē) [G. *physis,* nature + *logos,* study] Scientific discipline that deals with the processes or functions of living things.

pia mater (pī′ă mā′ter, pē′ă ma′ter) [L., affectionate mother] The innermost meningeal layer; tightly attached to the brain and spinal cord.

pineal body (pin′ē-ăl) [L. *pineus,* pinecone-shaped] A small endocrine gland attached to the dorsal surface of the diencephalon; may influence the onset of puberty and may play a role in some long-term cycles.

pinocytosis (pī′nō-sī-tō′sis, pī′nō-sī-tō′sis) [G. *pineo,* to drink + *kytos,* cell; *osis,* condition] Cell drinking; uptake of liquid by a cell.

pituitary dwarf (dwŏrf) An individual of short stature, of relatively normal proportion, as the result of insufficient growth hormone secreted from the anterior pituitary gland.

pituitary gland (pi-too′i-tār-rē) [L. *pituita,* phlegm or a thick mucous secretion] Endocrine gland attached to the hypothalamus by the infundibulum; secretes hormones that influence the function of several other glands and tissues.

placenta (plă-sen′tă) Structure derived from embryonic and maternal tissues by which the embryo and fetus are attached to the uterus.

plasma (plaz′mă) Fluid portion of blood; blood minus the formed elements.

plasma membrane Cell membrane; outermost component of the cell, surrounding and binding the rest of the cell contents.

plasmin (plaz′min) An enzyme that breaks down the fibrin in blood clots; derived from plasminogen.

platelet (plāt′let) Minute fragments of cells derived from megakaryocytes; play an important role in preventing blood loss.

platelet plug Accumulation of platelets that stick to connective tissue and to one another and prevent blood loss from damaged blood vessels.

pleural (ploor′ăl) [G., a rib or cavity] **cavity** Space between the visceral and parietal pleura, filled with pleural fluid; a cavity that surrounds each lung.

pleural membrane Serous membranes associated with the lungs.

plexus (plek′sŭs) [L., a braid] An intertwining of nerves or blood vessels.

PMS See *premenstrual syndrome.*

pneumothorax (noo-mō-thōr′aks) The presence of air in the pleural cavity.

podocyte (pod′ō-sīt) [fr. *pous, podos,* foot + G. *kytos,* a hollow (cell)] Epithelial cell of Bowman's capsule attached to the outer surface of the glomerular capillary basement membrane; forms part of the filtration membrane.

polar body The oocyte receiving little cytoplasm; results from the first and the second meiotic division.

polar covalent bond Chemical bond in which electrons are shared unequally between two atoms.

polarize Development of a difference in potential (charge) between two points, as between the inside and outside of a cell membrane.

polycythemia (pol′ē-sī-thē′mē-ă) [G. *polys,* many + *kytos,* cell] Increase in red blood cell numbers above the normal value.

polysaccharide (pol-ē-sak′ă-rīd) [many sugars] Many monosaccharides chemically bound together, such as glycogen and starch.

pons (ponz) [L., bridge] The part of the brainstem between the medulla oblongata and midbrain; contains nerve tracts between the cerebrum and cerebellum, as well as ascending and descending tracts.

portal system (pōr′tăl) System of vessels in which blood, after passing through one capillary bed, is conveyed through a second capillary network.

positive feedback Mechanism by which any deviation from an ideal normal value or set point is made greater.

posterior (pos-tēr′ē-ŏr) [L. *posterus,* following] That which follows; in humans, toward the back.

posterior horn The posterior extension of spinal cord gray matter; contains neuron cell bodies that receive input from primary sensory neurons and relay that input to the brain; also called the dorsal horn.

posterior pituitary gland The posterior portion of the pituitary gland, which consists of processes of nerve cells that have their cell bodies located in the hypothalamus; secretes oxytocin and antidiuretic hormone.

postganglionic (pōst′gang-glē-on′ik) Autonomic neurons whose cell bodies are located outside the central nervous system and that receive synaptic stimulation from preganglionic autonomic neurons.

preganglionic (prē′gang-glē-on′ik) Autonomic neurons whose cell bodies are located in the central nervous system and that synapse with postganglionic neurons.

preload (prē′lōd) The degree to which the ventricular wall is stretched at the end of diastole; increases as the venous return increases.

premenstrual syndrome (PMS) (prē-men′stroo-al sin′drōm) In some women of reproductive age, the regular monthly experience of physiological and emotional distress, usually during the few days preceding menses, typically involving fatigue, edema, irritability, tension, anxiety, and depression.

prenatal period (prē-nā′tăl) [L. *prae,* before + *natalis,* relating to birth] The period before birth.

prepuce (prē′poos) In the male, a free fold of skin that almost completely covers the glans penis; the foreskin; in the female, a fold of mucous membrane that covers the clitoris.

primary response Immune response that occurs as a result of the first exposure to an antigen; results in the production of antibodies and memory cells.

prime mover Muscle that plays the principal role in accomplishing a movement.

primitive streak A shallow groove in the ectodermal surface of the embryonic disk; cells migrating through the streak become mesoderm.

process (pros′es, prō′ses) Projection on a bone.

product (prod′ŭkt) Substance produced in a chemical reaction.

progesterone (prō-jes′ter-ōn) Hormone secreted primarily by the corpus luteum and the placenta; aids in growth and development of female reproductive organs and secondary sexual characteristics; causes growth and maturation of the endometrium of the uterus during the menstrual cycle.

prolactin (prō-lak'tin) [L. *pro*, precursor + *lact*, milk] Hormone of the anterior pituitary gland that stimulates the secretion of milk.

pronation (prō-nā'shŭn) [L. *pronare*, to bend forward] Rotation, as of the forearm, starting in the anatomical position, so that the anterior surface faces posteriorly.

proprioceptive neurons (prō'prē-ō-sep'tiv) [L. *proprius*, one's own + *capio*, to take] Nerves that innervate the joints and tendons and provide information about the position of the body and its various parts.

prophase (prō'fāzs) [G. *prophasis*, to foreshadow] The first stage of mitosis or meiosis consisting of contraction and increase in thickness of the chromosomes.

prostaglandin (pros-tă-glan'din) Class of physiologically active substances present in many tissues; effects include vasodilation, stimulation and contraction of uterine smooth muscle, and promotion of inflammation and pain.

prostate gland (pros'tāt) [G. *prostates*, one standing before] Gland that surrounds the beginning of the urethra in the male. The secretion of the gland is a milky fluid that is discharged into the urethra as part of the semen.

protein (prō'tēn) [G. *proteios*, primary] Large molecule consisting of long sequences of amino acids (polypeptides) linked by peptide bonds.

proteoglycan (prō'tē-ō-glī'kan) [G. *proteo*, protein + *glycan*, polysaccharide] Macromolecule consisting of numerous polysaccharides attached to a common protein core, attract and retain large amounts of water.

proteolytic (prō'tē-ō-lit'ik) An enzyme capable of digesting proteins or polypeptides.

proton (prō'ton) [G. *protos*, first] Positively charged particle found in the nuclei of atoms.

provitamin (prō-vīt'ă-min) Substance that can be converted into a vitamin.

proximal (prok'si-măl) [L. *proximus*, nearest] Closer to the point of attachment to the body than another structure.

proximal tubule Convoluted portion of the nephron that extends from Bowman's capsule to the descending limb of Henle's loop.

pterygoid (ter'ĭ-goyd) [G. *pteryx*, wing] Wing-shaped structure; two of the muscles of mastication, attached to wing-shaped bony projections.

PTH See *parathyroid hormone*.

puberty (pū'ber-tē) [L. *pubertas*, grown up] Series of events that transform a child into a sexually mature adult; involves an increase in the secretion of all reproductive hormones.

pudendal cleft (pū-den'dal) [L. *pudeo*, to feel ashamed] Cleft between the labia majora.

pudendum (pū-den'dŭm), pl. **pudenda** (pū-den'da) The external genitals, especially the female genitals. See *vulva*.

pulmonary capacity (pŭl'mō-nār-ē) The sum of two or more pulmonary volumes.

pulmonary circulation Blood flow through the system of blood vessels that carry blood from the right ventricle of the heart to the lungs and back from the lungs to the left atrium.

pulmonary semilunar valve The semilunar valve found at the base of the pulmonary trunk where it exits from the right ventricle.

pulmonary trunk Large elastic artery that carries blood from the right ventricle of the heart to the right and left pulmonary arteries.

pulmonary volume Lung volume, measured by spirometry; deviations from a normal value can be used to diagnose certain lung diseases; the pulmonary volumes are the tidal volume, inspiratory reserve volume, expiratory reserve volume, and residual volume.

pulp (pŭlp) [L. *pulpa*, flesh] The soft tissue inside a tooth, consisting of connective tissue, blood vessels, nerves, and lymphatic vessels.

pulse (pŭls) A pressure wave that travels rapidly along the arteries when blood is ejected from the left ventricle into the aorta.

pulse pressure The difference between systolic and diastolic pressure.

pupil (pū'pĭl) [L. *pupa*, a doll, because you can see a little reflection or doll in the pupil of another person's eye] Opening in the iris of the eye through which light enters.

Purkinje fiber (pŭr-kĭn'jē) Named for the Bohemian anatomist/physiologist, Johannes Purkinje (1787–1869). A specialized cardiac muscle fiber that conducts action potentials through cardiac muscle; forms part of the conduction system of the heart.

pus (pŭs) Product of inflammation consisting of a liquid containing white blood cells, dead cells, and cell fragments.

pyloric sphincter (pī-lōr'ik) [G., gatekeeper] A thickened ring of smooth muscle at the distal end of the stomach.

pyrogen (pī'rō-jen) Chemical released by microorganisms, neutrophils, monocytes, and other cells that stimulates fever production by acting on the hypothalamus.

pyruvic acid (pī-roo'vik) Three-carbon end product of glycolysis; two pyruvic acid molecules are produced from each glucose molecule.

Q

quadrant (kwäh'drant) [L. *quadrans*, a quarter] One-quarter of a circle; the abdomen is divided into right upper, right lower, left upper, and left lower quadrants by a horizontal and a vertical line intersecting at the umbilicus.

R

RBC See *red blood cell*.

reactant (rē-ak'tant) Substance taking part in a chemical reaction.

receptor (rē-sĕp'tōr) [L., receiver] A protein molecule on the cell surface or within the cytoplasm that binds to a specific factor such as a drug, hormone, antigen, or neurotransmitter; one of the sensory nerve endings in the skin, deep tissues, viscera, and special sense organs.

recessive (rē-ses'iv) In genetics, a gene that may not be expressed phenotypically because of the expression of a contrasting dominant gene.

rectum (rek'tŭm) [L. *rectus*, straight] The last, straight part of the large intestine; between the colon and the anal canal.

rectus (rek'tŭs) Straight.

red blood cell (RBC) Biconcave disk that contains hemoglobin, which transports oxygen and carbon dioxide; red blood cells do not have a nucleus.

reflex (rē'fleks) Automatical responses to stimuli; does not require conscious thought.

reflex arc Consists of a sensory receptor, afferent (sensory) neuron, association neuron, efferent (motor) neuron, and effector organ.

regeneration (rē'jen-er-ā'shŭn) Tissue repair in which the damaged cells are replaced by cells of the same type as those damaged.

releasing hormone Hormone that is released from neurons in the hypothalamus and flows through the hypothalamic–pituitary portal system to the anterior pituitary gland; functions to regulate the secretion of hormones from the cells of the anterior pituitary gland.

renal capsule (rē'năl) The connective tissue capsule that surrounds each kidney.

renal corpuscle (kōr'pŭs-l) The structure composed of a Bowman's capsule and its glomerulus.

renal pyramid (pir'ă-mid) Cone-shaped structure that extends from the renal sinus, where the apex is located, into the cortex of the kidney, where the base is located.

renal sinus (sī'nŭs) The cavity central to the medulla of the kidney that is filled with adipose tissue and contains the renal pelvis.

renin (rē'nin, ren'in) Enzyme secreted by the kidney that converts the plasma protein, angiotensinogen to angiotensin I.

replacement Tissue repair in which the damaged cells are replaced by cells of a type different from those damaged.

retinal (rĕt'i-năl) Relating to the retina; retinaldehyde most commonly referring to the all-trans form (all-trans-retinal).

respiration (res-pi-rā'shŭn) [L. *respiratio*, to breathe] Process in which oxygen is used to oxidize organic fuel molecules, providing a source of energy as well as carbon dioxide and water; includes venti-lation, gas exchange, transport of oxygen and carbon dioxide in the blood, gas exchange between the blood and the tissues, and cell metabolism.

respiratory membrane (res'pi-ră-tōr-ē, rĕ-spīr'ă-tōr-ē) Membrane in the lungs across which gas exchange occurs with blood; consists of a thin layer of fluid, the alveolar epithelium, a base-ment membrane of the alveolar epithelium, interstitial space, the basement membrane of the capillary endothelium, and the capillary endothelium.

respiratory system Includes the nose, nasal cavity, pharynx, larynx, trachea, bronchi, and lungs.

resting membrane potential The charge difference across the membrane of a resting cell (i.e., a cell that has not been stimulated to produce an action potential).

rete testis (rē'tē) The network of canals at the termination of the straight portion of the seminiferous tubules.

reticular formation (rē-tik'ū-lăr) [L. *rete*, net] A loose network of neuron cell bodies scattered throughout the brainstem; involved in the regulation of cycles such as the sleep–wake cycle.

retina (ret'i-nă) [L. *rete*, a net] The inner, light-sensitive tunic of the eye; nervous tunic.

retinaculum (ret-i-nak'ū-lŭm) [L., band, bracelet, halter, to hold back] Dense, regular connective tissue sheath holding down the tendons at the wrist, ankle, or other sites.

retinal (rĕt'i-năl) Relating to the retina; Retinaldehyde most commonly referring to the all-trans form (all-trans-retinal).

retroperitoneal (re'trō-per'i-tō-nē'ăl) Located behind the parietal peritoneum; includes the kidneys, adrenal glands, pancreas, portions of the intestines, and urinary bladder.

reversible reaction (rē-ver'si-bl) Chemical reaction in which the reaction can proceed from reactants to products, or from products to reactants; the amount of reactants relative to products is constant at equilibrium.

rhodopsin (rō-dop'sin) [G. *rhodon*, rose or red color + *opsin*, protein portion of rhodopsin] A purplish red molecule found in the external segment of the rods of the retina. Action of light converts it to opsin and all-trans-retinal.

ribonucleic acid (rī'bō-noo-klē'ik) **(RNA)** Type of nucleic acid containing the sugar ribose; involved in protein synthesis.

ribosomal RNA (rRNA) (ri'bō-sōm-ăl) RNA that is associated with certain proteins to form ribosomes.

ribosome (ri'bō-sōm) [ribose, a specific sugar] Small, spherical, cytoplasmic organelle where protein synthesis occurs.

right lymphatic duct Lymphatic duct that empties into the right subclavian vein; drains the right side of the head and neck, the right upper thorax, and the right upper limb.

RNA See *ribonucleic acid.*

rod Photoreceptor cell in the retina of the eye with a rod-shaped photoreceptive process; very light-sensitive cell that is important in dim light.

rotator cuff (rō-tā′tōr, rō-tā′tōr) Four deep muscles that attach the humerus to the scapula.

rRNA See *ribosomal RNA.*

ruga (roo′gă) Ridge or fold in the mucous membrane of the stomach.

S

SA See *sinoatrial.*

sagittal plane (saj′i-tăl) [L. *sagitta,* the flight of an arrow] Plane running vertically through the body and dividing it into right and left parts.

saliva (să-lī′vă) A fluid containing enzymes and mucus, produced by the salivary glands and released into the oral cavity.

salivary gland (sal′i-văr-rē) Gland opening into the mouth and producing saliva.

salt Molecule consisting of a positively charged ion other than hydrogen, and a negatively charged ion other than hydroxide.

sarcolemma (sar′kō-lem′ă) [G. *sarx,* flesh, means muscle; *lemma,* husk] The cell membrane of a muscle fiber.

sarcomere (sar′kō-mēr) [*sarco-* + G. *meros,* part] The part of a myofibril formed of actin and myosin myofilaments, extending from Z disk to Z disk; the structural and functional unit of a muscle.

sarcoplasm (sar′kō-plazm) [*sarco-* + *plasma,* a thing formed] The cytoplasm of a muscle fiber.

sarcoplasmic reticulum (sar′kō-plaz′mik re-tik′ū-lŭm) The endoplasmic reticulum of a muscle fiber.

scapula (skap′ū-lă) The shoulder blade.

Schwann cell Named for the German histologist/physiologist, Theodor Schwann (1810–1882). Neuroglial cell forming myelin sheaths around axons in the peripheral nervous system.

sciatic (sī-at′ik) [fr. *ischion,* the hip joint] The ischiadic or sciatic nerve.

sclera (sklēr′ă) [L. *skleros,* hard] The dense, white, opaque posterior four-fifths of the fibrous tunic of the eye; white of the eye.

scoliosis (skō-lē-ō′sis) [G., a crookedness] An abnormal lateral curvature of the spine.

scrotum (skrō′tum) Musculocutaneous sac containing the testes.

sebaceous gland (sē-bā′shŭs) [L. *sebum,* tallow] Gland of the skin that produces sebum; usually associated with a hair follicle.

sebum (sē′bŭm) [L., tallow] Oily, white, fatty substance produced by the sebaceous glands; oils hair and the surface of the skin.

secondary response See *memory response.*

secretin (se-krē′tin) Hormone released from the epithelium of the duodenum; inhibits gastric secretion.

sella turcica (sel′ătŭr′sĭ-kă) [L., saddle, Turkish] The saddle-shaped depression in the inner surface of the skull where the pituitary gland is located.

semen (sē′men) [L., seed] Penile ejaculate; thick, yellowish white, viscous fluid containing sperm cells and secretions of the testes, seminal vesicles, prostate gland, and bulbourethral glands.

semicircular canal (sem′ē-sir′kū-lăr) One of three canals in each temporal bone; involved in the detection of motion.

semilunar valve (sem-ē-loo′năr) One of two valves in the heart composed of three crescent-shaped cusps that prevent flow of blood into the ventricles following ejection; located at the beginning of the aorta and pulmonary trunk.

seminal vesicle (sem′i-năl ves′i-kl) One of two glandular structures that empty into the ejaculatory ducts; its secretion is one of the components of semen.

seminiferous tubule (sem′i-nif′er-ŭs) Tubule in the testis in which sperm cells develop.

sensory neuron Neuron that extends from sensory receptors in the periphery to the central nervous system.

septum (sep′tŭm) [L. *saeptum,* a partition or wall] A thin wall dividing two cavities or masses of softer tissue.

serosa (se-rō′să) The smooth, outermost covering of an organ where it faces a cavity and is not sur-rounded by connective tissue.

serous membrane (sēr′ŭs) Thin sheet consisting of epithelium and connective tissue that lines cavities not opening to the outside of the body; does not contain glands but does secrete serous fluid.

Sertoli cell (ser-tō′lē) Named for the Italian histologist, Enrico Sertoli (1842–1910). Cell in the wall of the seminiferous tubules to which spermatogonia and spermatids are attached.

serum (sēr′ŭm) Fluid portion of blood after the removal of fibrin and formed elements.

sex chromosome A chromosome other than an autosome; responsible for sex determination.

sinoatrial (SA) node (sī′nō-ā′trē-ăl) Mass of specialized cardiac muscle fibers located in the right atrium near the opening of the superior vena cava that acts as the "pacemaker" of the cardiac conduction system.

sliding filament model Mechanism by which actin and myosin myofilaments slide over one another during muscle contraction.

solute (sol′ūt, sō′loot) [L. *solutus,* dissolved] Dissolved substance in a solution.

solution (sō-loo′shŭn) Homogeneous mixture formed when a solute dissolves in a solvent (liquid).

solvent (sol′vent) [L. *solvens,* dissolve] Liquid that holds another substance in solution.

somatic motor (sō-mat′ik) [G. *soma,* body or bodily] A type of motor (efferent) neuron of the peripheral nervous system that innervates skeletal muscle.

somesthetic (sō′mes-thet′ik) [G. *soma,* body + *aisthesis,* sensation] Consciously perceived body sensations.

somesthetic cortex That part of the cerebral cortex involved with the conscious perception and localization of general body sensations.

spermatid (sper′mă-tid) A cell in the late stage of the development of the sperm cell (male sex cell). It is haploid and is derived from the secondary spermatocyte.

spermatocyte (sper′mă-tō′sīt) Cell arising from a spermatogonium and destined to give rise to spermatozoa.

spermatogenesis (sper′mă-tō-jen′ĕ-sis) Formation and development of sperm cells.

spermatogonium (sper′mă-tō-gō′nē-ŭm), pl. **spermatogonia** (sper′mă-tō-gō′nē-ă) The most peripheral germ cells in the seminiferous tubules scattered between the Sertoli cells; divide by mitosis and some form primary spermatocytes.

spermatozoon (sper′mă-tō-zō′on), pl. **spermatazoa** (sper′mă-tō-zō′ă) [G. *sperma,* seed + *zoon,* animal] Male gamete, or sex cell, composed of a head, midpiece, and tail; contains the genetic information transmitted by the male; sperm cell.

sperm cell [G. *sperma,* seed] Male reproductive cell; see *spermatozoon.*

sphenoid (sfē′noyd) [G. *sphenoeides,* wedge + *eidos,* resemblance] Sphenoid bone or relating to the sphenoid bone.

sphygmomanometer (sfig′mō-mă-nom′ĕ-ter) [G. *sphygmos,* pulse; *manos* + *metron,* measure] An instrument for measuring blood pressure consisting of an arm sleeve and inflating bulb with a device attached for measuring pressure in the arm sleeve.

spina bifida (spī′nă bif′ĭ-dă, bī′fĭ-dă) A defect in the spinal column, consisting in absence of the vertebral arches, through which the spinal membranes, with or without spinal cord tissue, may protrude.

spinal cord Portion of the central nervous system extending from the foramen magnum at the base of the skull to the second lumbar vertebra; consists of a central gray portion and a peripheral white portion.

spinal nerve Peripheral nerve exiting from the spinal cord.

spirometer (spī-rom′ĕ-ter) [L. *spiro,* to breathe + G. *metron,* measure] Meter used for measuring the volume of respiratory gases; usually consisting of a counterbalanced cylindrical bell sealed by dipping into a circular trough of water.

spirometry (spī-rom′ĕ-trē) Process of making pulmonary measurements with a spirometer.

spleen (splēn) Large lymphatic organ in the left upper part of the abdominal cavity, between the stomach and diaphragm; composed of white and red pulp; responds to foreign substances in the blood, destroys worn out red blood cells, and is a reservoir for blood.

squamous (skwā′mŭs) [L. *squama,* a scale] Scalelike, flat.

stapes (stā′pēz) [L., stirrup] The third of the three middle ear bones; attached to the oval window; the stirrup.

Starling's law of the heart Named for the English physiologist, Ernest Starling (1866–1927). Force of contraction of cardiac muscle is a function of the length of its muscle fibers at the end of dias-tole; the greater the degree of filling of the heart (the greater the venous return), the greater the force of contraction of the cardiac muscle.

stem cell Single population of cells that differentiate to give rise to the formed elements of blood.

stenosed valve (sten′ōzd) A valve that has its opening narrowed or partially closed.

sternum (ster′nŭm) [L. *sternon,* chest] Breastbone.

steroid (stēr′oyd, ster′oyd) Large family of lipids, including some hormones, vitamins, and cholesterol.

stethoscope (steth′ō-skōp) [G. *stetho-,* chest + *skopeo,* to view] An instrument originally devised for aid in hearing the respiratory and cardiac sounds in the chest and now used in hearing other sounds in the body as well.

strabismus (stra-biz′mŭs) [G. *strabismos,* a squinting] Lack of parallelism of the visual axes of the eyes.

stratum (strat′ŭm), pl. **strata** (strat′ă) [L., bed cover, layer] Layer of tissue.

stratum basale (bă-săl′ē) The deepest layer of the epidermis; consists of columnar cells that undergo mitotic divisions.

stratum corneum (kōr′nē-ŭm) The most superficial layer of the epidermis; consists of dead, squamous cornified cells that have undergone keratinization.

stria, pl., **striae** (stri′e) [L., channel, furrow] Bands of thin, wrinkled skin, becoming red and white, that occur commonly on the abdomen, buttocks, and thighs, at puberty, and/or during and following pregnancy and result from over-extension of the skin.

stroke volume The volume of blood ejected from either the right or left ventricle during each heartbeat.

styloid (stī′loyd) [G. *stylos,* a stake or pen] A slender, pencil-shaped process.

subarachnoid space (sŭb-ă-rak′noyd) The fluid-filled space below the arachnoid layer covering the brain and spinal cord; contains cerebrospinal fluid.

subcutaneous (sŭb-koo-tā′nē-ŭs) [L. *sub*, under + *cutis*, skin] Under the skin; same tissue as the hypodermis.

sublingual gland (sŭb-ling′gwăl) One of a pair of salivary glands located below the tongue.

submandibular gland (sŭb-man-dib′ū-lăr) One of a pair of salivary glands located below the mandible.

submucosa (sŭb-moo-kō′să) The layer of connective tissue deep to the mucous membrane.

sulcus (soo′kŭs), pl. **sulci** (sŭl′sī) [L., ditch] A groove on the surface of the brain between gyri.

superficial (soo-per-fish′ăl) [L. *superficialis*, surface] Toward or on the surface.

superior (soo-pēr′ē-ōr) [L., higher] Up, or higher, with reference to the anatomical position.

superior vena cava (vē′nă kā′vă) Receives blood from the head, neck, and upper limbs and empties into the right atrium of the heart.

supination (soo′pi-nā′shŭn) [L. *supino*, to place something on its back] Rotation of the forearm so that the anterior surface is anterior; that is, the forearm is in the anatomical position.

surfactant (ser-fak′tănt) A mixture of lipoprotein molecules produced by the secretory cells of the alveolar epithelium of the lung; reduces water surface tension.

suture (soo′choor) [L. *surtura*, a seam] Fibrous joint between flat bones of the skull.

sweat gland (swet) Usually a secretory organ that produces a watery secretion called sweat that is released onto the surface of the skin; some sweat glands, however, produce an organic secretion.

sympathetic (sim-pă-thet′ik) [G. *sympatheo*, to feel with + *pathos*, suffering] Subdivision of the autonomic nervous system with preganglionic nerve cell bodies located in the thoracic and lumbar regions of the spinal cord; generally involved in preparing the body for immediate physical activity.

synapse (sin′aps) [G. *syn*, together + *haptein*, to clasp] Junction between a nerve cell and another nerve cell, muscle cell, or gland cell; in a chemical synapse, chemicals are released from the nerve cell as a result of an action potential in the nerve cell, the chemicals cross the cleft between the cells, and they cause some response in the postsynaptic cell.

synapsis (si-nap′sis) The pairing of homologous chromosomes during prophase of the first meiotic division.

synergist (sin′er-jist) A muscle that works with another muscle to cause a movement.

synovial cavity (si-nō′vē-ăl) [G. *syn*, coming together + *ovia*, resembling egg albumin] Cavity surrounding articulating bones of a freely movable or synovial joint; contains synovial fluid.

synovial fluid A somewhat viscous substance serving as a lubricant in movable joints, tendon sheaths, and bursae.

synovial joint Freely movable joint.

synovial membrane Lines the inside of a joint cavity; produces synovial fluid.

synthesis reaction (sin′thĕ-sis) The combination of atoms, ions, or molecules to form a new and larger molecule.

systemic circulation (sis-tem′ik) Blood flow through the system of blood vessels that carry blood from the left ventricle of the heart to the tissues of the body and back from the body to the right atrium.

systemic inflammation Inflammation that occurs in many areas of the body; in addition to the symptoms of local inflammation, can include increased neutrophil numbers in the blood, fever, and shock.

systole (sis′tō-lē) [G., a contracting] Contraction of the heart chambers during which blood leaves the chambers; usually refers to ventricular contraction.

systolic pressure (sis-tol′ik) The maximum arterial blood pressure reached during ventricular systole.

T

target tissue Tissue on which a hormone acts.

tarsal (tar′săl) [G. *tarsos*, sole of foot] Bone of the instep of the foot.

taste bud Sensory structure that is found mostly on the tongue and functions as a taste receptor.

tectorial membrane (tek-tōr′ē-ăl) [L., a covering] Membrane attached to the spiral lamina and extending over the hair cells; hairs of the hair cells have their tips embedded in the membrane.

telophase (tel′ō-fāz) [G. *telos*, end + *phasis*, an appearance] The final stage of mitosis or meiosis that begins when migration of chromosomes to the poles of the cells has been completed.

temporal (tem′pŏ-răl) [L. *tempus*, time] Indicating the temple; the temple of the head is so named because it is there that the hair first begins turning white, indicating the passage of time.

tendinous intersection (ten′di-nŭs) One of the bands of connective tissue crossing the rectus abdominus muscle subdividing it and attaching it to adjacent connective tissue.

tendon (ten′dŏn) A tough connective tissue band connecting a muscle to bone.

teniae coli (tē′nē-ē kō′lī) [G. *tainia*, band, tapeworm + *coli*, colon] The segmented, longitudinal smooth muscle layer of the colon.

testis (tes′tis), pl. **testes** (test′tēz) One of two male reproductive glands located in the scrotum; produces testosterone and sperm cells.

testosterone (tes′tos′tĕ-rōn) Steroid hormone secreted primarily by the testes; aids in spermatogenesis, controls maintenance and development of male reproductive organs and secondary sexual characteristics, and influences sexual behavior.

tetanus (tet′ă-nŭs) [L. *tetanus*, convulsive tension] A sustained muscular contraction caused by a series of nerve stimuli repeated so rapidly that the individual contractions are fused, producing as sustained tetanic contraction. A disease marked by painful tonic muscular contractions, caused by the neurotoxin of *Clostidium tetani* action on the central nervous system.

tetany (tet′ă-nē) A condition in muscle contraction in which there is no relaxation between muscle twitches.

tetraiodothyronine (tet′ră-ī-ō-dō-thī′rō-nēn) (T4) One of the thyroid hormones that contains four iodine atoms; also called thyroxine.

thalamus (thal′ă-mŭs) [G., a bedroom] A large mass of gray matter making up the bulk of the diencephalon; involved in the relay of sensory input to the cerebrum.

thoracic cavity (thō-ras′ik) Space bounded by the neck, the thoracic wall, and the diaphragm.

thoracic duct Largest lymph vessel in the body; drains the left side of the head and neck, the left upper thorax, the left upper limb, and the inferior half of the body into the left subclavian vein.

thorax (thō′raks) [G., breastplate] The chest; the upper part of the trunk between the neck and the abdomen.

thrombocyte (throm′bō-sīt) [*thrombos-*, clot + G. *kytos*, cell] A cell fragment involved in platelet plug and clot formation; also called a platelet.

thrombosis (throm′bō′sis) [G. *thrombos*, clot] The formation or presence of a clot (thrombus) inside of a blood vessel.

thrombus (throm′bŭs) [G. *thrombos*, clot] A clot within the cardiovascular system.

thymosin (thī′mō-sin) A hormone secreted from the thymus gland that helps activate the immune system.

thymus (thī′mŭs) [G. *thymos*, sweetbread] Bilobed lymphatic organ located in the inferior neck and superior mediastinum; involved with the maturation of T cells.

thyroid cartilage (thī′royd) [G. *thyreoeides*, shield] Largest laryngeal cartilage; forms the laryngeal prominence, or Adam's apple.

thyroid follicle One of many small spheres with walls consisting of cuboidal epithelial cells in the thyroid gland, and which is filled with proteins to which thyroid hormones are attached until they are secreted.

thyroid gland Endocrine gland located inferior to the larynx and consisting of two lobes connected by a narrow band; secretes the thyroid hormones.

thyroid hormone Any hormone secreted by the thyroid gland; especially those such as thyroxine that contain iodine and function to regulate metabolism and maturation of tissues.

thyroid-stimulating hormone (TSH) Hormone released from the hypothalamus that stimulates thyroid hormone secretion from the thyroid gland.

thyroxine (thī-rok′sēn, thī-rok′sin) See *tetraiodothyronine*.

tissue (tish′ū) [L. *texo*, to weave] Collection of cells with similar structure and function and the substances between the cells.

tissue repair Substitution of viable cells for damaged or dead cells by regeneration or replacement.

tonsil (ton′sil) Any collection of lymphoid tissue; usually refers to large collections of lymphoid tissue beneath mucous membranes of the oral cavity and pharynx; lingual, pharyngeal, and palatine tonsils.

trabecula (tră-bek′ū-lă) [L. *trabs*, beam] A beam or plate of cancellous bone or other tissue.

trachea (trā′kē-ă) [G. *tracheia arteria*, rough artery] Air tube extending from the larynx into the thorax, where it divides to form bronchi; has 16–20 C-shaped pieces of cartilage in its walls.

tracheostomy (trā′kē-os′tō-mē) An incision into the trachea.

tract (trakt) Nerve tract; a bundle of neuron cell processes (axons) in the central nervous system, usually sharing a common function.

transfer RNA (tRNA) RNA that attaches to individual amino acids and transports them to the ribosomes, where they are connected to form a protein polypeptide chain.

transverse plane (trans-vers′) Plane separating the body into superior and inferior parts.

transverse section A cut made at right angles to the long axis of an organ.

trapezius (tra-pē′zē-ŭs) Back muscle, shaped like a trapezium (a four-sided geometric figure in which no two sides are parallel), that rotates the scapula.

triacylglycerol (trī-as′il-glis′er-ol) See *triglyceride*.

triceps brachii (trī′seps brā′kē-ī) A three-headed muscle in the posterior arm that extends the forearm.

tricuspid valve (trī-kŭs′pid) Valve closing the opening between the right atrium and right ventricle of the heart.

triglyceride (trī-glis′er-īd) A common type of lipid, or fat, with three fatty acids bound to a glycerol molecule; also called a triacylglycerol.

trigone (trī′gōn) [L. *trigonium*, triangle] A triangular smooth area at the base of the urinary bladder between the openings of the two ureters and that of the urethra.

triiodothyronine (trī-ī′ō-dō-thī′rō-nēn) **(T3)** One of the thyroid hormones that contains three iodine atoms.

tRNA See *transfer RNA.*

trochanter (trō′kanter) [G., a runner] One of the large tubercles of the proximal femur.

trophoblast (trō′fō-blast) [G. *trophe*, nourishment + *blastos*, germ] The outer part of the blastocyst; enters the uterus and becomes the embryonic portion of the placenta.

trypsin (trip′sin) An enzyme released from the pancreas that digests proteins.

TSH See *thyroid-stimulating hormone.*

tubercle (too′ber-kl) A lump or knob on a bone.

tuberosity (too′ber-os′i-tē) Lump on a bone, usually larger than a tubercle.

tubular reabsorption Movement of materials, by means of diffusion or active transport, from the filtrate within a nephron into the blood.

tubular secretion Movement of materials, by means of active transport, from the blood into the filtrate of a nephron.

tumor (too′mŏr) Any swelling, one of the cardinal signs of inflammation, or a new growth of tissue in which the multiplication of cells is uncontrolled and progressive; see also *neoplasm.*

tunic (too′nik) [L., coat] A layer or coat; one of the three enveloping layers of the wall of the eye; the three tunics are the fibrous, vascular, and nervous tunics; one of the three layers of blood vessels: tunica intimima, tunica media, and tunica adventitia.

tunica adventitia (too′ni-kă ad-ven-tish′ă) Outermost fibrous coat of a vessel or an organ that is derived from the surrounding connective tissue.

tunica intima (in′ti-mă) Innermost layer of a blood or lymphatic vessel; consists of endothelium and a small amount of connective tissue.

tunica media Middle, usually muscular, coat of an artery or other tubular structure.

tympanic membrane (tim-pan′ik) [drumlike] Cellular membrane that covers the inner opening of the external auditory meatus and separates the middle and external ears; vibrates in response to sound waves; the eardrum.

U

ulcer (ŭl′ser) [L. *ulcus,* a sore] A lesion on the surface of the skin or a mucous membrane such as in the stomach or intestine caused by a superficial loss of tissue, usually with inflammation.

umbilical cord (ŭm-bil′i-kăl) [L., naval] Cord connecting the fetus to the placenta; contains two umbilical arteries, which originate from the embryo's internal iliac arteries, that carry blood from the embryo to the placenta, and one umbilical vein that carries blood back to the fetus.

umbilical vein Vein in the umbilical cord of the fetus by which the fetus receives nourishment from the placenta; becomes the round ligament of the liver in the adult.

upper motor neuron A motor neuron located in the cerebral cortex and synapsing with a lower motor neuron in the brainstem or spinal cord.

ureter (ū-rē′ter, ū′re-ter) [G. *oureter,* urinary canal] Tube conducting urine from the kidney to the urinary bladder.

urethra (ū-rē′thră) A duct leading from the bladder, discharging the urine externally.

uterus (ū′ter-ŭs) Hollow muscular organ in which the fertilized oocyte develops into a fetus.

utricle (ū′trĭ-kl) The larger of the two membranous sacs in the vestibule of the labyrinth. The semi-circular canals arise from this.

uvula (ū′vū-lă) [L. *uva,* grape] Small, grapelike appendage at the posterior margin of the soft palate.

V

vaccine (vak′sēn, vak-sēn′) Preparation of killed microorganisms, altered microorganisms, or derivatives of microorganisms intended to produce immunity; usually administered by injection, but sometimes ingestion is preferred.

vagina (vă-jī′nă) [L., sheath] Genital canal in the female, extending from the uterus to the vulva.

variable region Part of an antibody that combines with an antigen; responsible for the specificity of the antibody.

varicose (văr′ĭ-kōs) **vein** A vein that is so dilated that the cusps of the valves are no longer capable of preventing backflow of blood; usually the veins in the lower legs or the hemorrhoidal veins.

vasoconstriction (vā′sō-kon-strik′shŭn) Decreased diameter of blood vessels.

vasodilation (vā′sō-dī-lā′shŭn) Increased diameter of blood vessels.

vasomotor center (vā-sō-mō′ter) An area of the lower pons and upper medulla oblongata that continually transmits a low frequency of action potentials through sympathetic neurons to smooth muscle in blood vessels; can cause vasoconstriction and vasodilation.

vasomotor tone Partial constriction of blood vessels in the periphery, which results from relatively constant sympathetic stimulation.

vasopressin (vā-sō-pres′in) [L. *vaso,* blood vessel + *pressum,* to press down] A peptide hormone related to oxytocin secreted from the posterior pituitary gland. In large doses causes contraction of blood vessel smooth muscle; see *antidiuretic hormone.*

vein (vān) Blood vessel that carries blood toward the heart.

venous return (vē′nŭs) Volume of blood returning to the heart.

ventilation (ven-ti-lā′shŭn) Movement of air in and out of the lungs.

ventral (ven′trăl) [L. *venter,* belly] In humans, synonymous with anterior.

ventral root Motor (efferent) root of a spinal nerve.

ventricle (ven′tri-kl) [L. *venter,* belly] A cavity; in the brain, one of four cavities filled with cerebrospinal fluid; one of two chambers of the heart that pump blood into arteries; there is a left and a right ventricle.

vernix caseosa (ver′niks kā′sē-ō′să) Epithelial cells and sebaceous matter that cover the skin of the fetus.

vesicle (ves′i-kl) [L. *vesicula,* blister or bladder] A small, membrane-bound sac containing material to be transported across the cell membrane.

vestibular fold (ves-tib′ū-lăr) [L., entrance hall] A false vocal fold.

vestibule (ves′ti-bool) A small cavity or a space at the entrance of a canal; see also *vulva.*

villus (vil′ŭs), pl. **villi** (vil′ī) [L., shaggy hair] Projection of the mucous membrane in the small intestine that increases surface area.

visceral (vis′er-ăl) [L. *viscus,* the soft parts, internal organs] Relating to the internal organs.

visceral peritoneum (per′i-tō-nē′ŭm) [L., organ] That part of the serous membrane in the abdominal cavity covering the surface of some abdominal organs.

vitamin (vīt′ă-min) [L. *vita,* life + *amine,* from ammonia] One of a group of organic substances, present in minute amounts in natural foods, that are essential to normal metabolism; insufficient amounts in the diet may cause deficiency diseases.

vitamin D Fat-soluble vitamin produced from a pre-cursor molecule in skin exposed to ultraviolet light; increases calcium and phosphate uptake in the intestine.

vitreous humor (vit′rē-ŭs) Transparent, jellylike substance that fills the posterior compartment of the eye; helps maintain pressure within the eye and holds the lens and retina in place.

vocal fold (vō′kăl) One of the ligaments that extends from the posterior surface of the thyroid cartilage to the paired cartilages of the larynx; the superior pair are the false vocal folds, and the inferior pair are the true vocal folds.

vulva (vŭl′vă) [L., a wrapper or covering, seed covering, womb] The external genitalia of the female; the mons pubis, labia majora and minora, the clitoris, the vestibule and its glands, the opening of the urethra, and the opening of the vagina.

W

WBC See *white blood cell.*

white blood cell (WBC) Round nucleated blood cell involved in immunity; includes neutrophils, basophils, eosinophils, lymphocytes, and monocytes. Also called leukocyte.

X

X-linked A trait caused by a gene on the X chromosome.

Y

yolk sac (yōk, yōlk) Highly vascular endodermal layer surrounding the yolk of an embryo.

Z

zona pellucida (zō′nă pellu′sida) [L. *zone,* girdle + *pellucidus,* passage of light] An extracellular coat surrounding the oocyte; appears translucent.

zygomatic (zī′gō-mat′ik) [G. *zygon,* yoke] Referring to the zygomatic, or cheek, bone; the zygomatic arch is a bony arch created by the junction of the zygomatic and temporal bones.

zygomaticus muscle (zī′gō-mat′i-kŭs) A muscle originating on the zygomatic bone and inserting onto the corner of the mouth, involved in smiling.

zygote (zī′gōt) [G. *zygotos,* yoked] The single-celled, diploid product of fertilization, resulting from the union of the sperm cell and an oocyte.

Credits

Photographs

Chapter 1
Opener: © Quest/Science Photo Library/Photo Researchers, Inc.; **1.1b:** © Bart Harris/CORBIS; **1.7a & b:** © McGraw-Hill Higher Education, Inc./Eric Wise, photographer; **1.8a & b:** © McGraw-Hill Higher Education, Inc./Eric Wise, photographer; **1.10a:** © McGraw-Hill Higher Education, Inc./Eric Wise, photographer; **1.10b-d:** © R. T Hutchings.

Chapter 2
Opener: © Prof. P. Motta/Dept. of Anatomy/University "La Sapienza", Rome/Science Photo Library/Custom Medical Stock Photo; **2.3c:** © Trent Stephens; **2A:** © ISM/Sovereign/Phototake, Inc.; **2B:** © Simon Fraser/Royal Victoria Infirmary/Photo Researchers, Inc.; **2.11c:** © Barry King/Tom Stack & Assocs.

Chapter 3
Opener: © Dennis Kunkel Microscopy, Inc.; **3.2b:** © J. David Robertson, from Charles Flickinger, Medical Cell Biology, Philadelphia; **3.12b:** © Don Fawcett/Photo Researchers, Inc.; **3.12c:** © Bernard Gilula/Photo Researchers, Inc.; **3.15b:** © J. David Robertson, from Charles Flickinger, Medical Cell Biology, Philadelphia; **3.16b:** © Robert Bollender/Don Fawcett/Visuals Unlimited; **3.18b:** © Don Fawcett/Visuals Unlimited; **3.19b:** © Don Fawcett/Photo Researchers, Inc.; **3.20b:** © Biology Media/Photo Researchers, Inc.; **3.26a-f:** © Ed Reschke.

Chapter 4
Opener: © Prof. P. Motta & E. Vizza/Science Photo Library/Photo Researchers, Inc.; **4.1b:** © Victor Eroschenko; **4.1c:** © Ed Reschke; **4.2a (2):** © McGraw-Hill Higher Education, Inc./Dennis Strete, photographer; **4.2b (2):** © Victor Eroschenko; **4.2c (2):** © Victor Eroschenko; **4.2d (2):** © Victor Eroschenko; **4.3a (2):** © Victor Eroschenko; **4.3b (2):** © Victor Eroschenko; **4.3b (3):** © Victor Eroschenko; **4.5 (2):** © Ed Reschke; **4.6 (2):** © Ed Reschke; **4.7a (2):** © Victor Eroschenko; **4.7b (2):** © Victor Eroschenko; **4.8a (2):** © Carolina Biological Supply/Phototake; **4.8b (2):** © Victor Eroschenko; **4.8c (2):** © Victor Eroschenko; **4.9 (2):** © Trent Stephens; **4.10 (2):** © Ed Reschke; **4.11a (2):** © Ed Reschke; **4.11b (2):** © Ed Reschke; **4.11c (2):** © Victor Eroschenko; **4.12:** © Trent Stephens.

Chapter 5
Opener: © Dennis Kunkel Microscopy, Inc.; **5.3b:** © John Cunningham/Visuals Unlimited; **5.10a-c:** © Thomas B. Habif; **5.Aa:** © James Stevenson/Science Photo Library/Photo Researchers, Inc.; **5.Ab:** © Stan Levy/Photo Researchers, Inc.

Chapter 6
Opener: © Prof. P. Motta/Dept. of Anatomy/University "La Sapienza", Rome/Science Photo Library/Photo Researchers, Inc.; **6.3b:** © Trent Stephens; **6.5b:** © Visuals Unlimited; **6.7a** © Ed Reschke/Peter Arnold, Inc.; **6.7b:** © Bio-Photo Assocs/Photo Researchers, Inc.; **6.A:** © Ewing Galloway, Inc./Index Stock; **6.12b:** © McGraw-Hill Higher Education, Inc./Eric Wise, photographer; **6.23d:** © Trent Stephens; **6.24a:** © McGraw-Hill Companies/Andy Resek, Photographer; **6.24b, 6.28, 6.36, 6.41a-f:** © McGraw-Hill Higher Education, Inc./Eric Wise, photographer; **6.Da:** © James Stevenson/Science Photo Library/Photo Researchers, Inc.; **6.Db:** © CNRI/Science Photo Library/Photo Researchers, Inc. **6.Ea:** © Princess Margaret Rose Orthopaedic Hospital/Science Photo Library/Photo Researchers, Inc.; **6.Eb:** © Dr. Michael Klein/Peter Arnold, Inc.

Chapter 7
Opener: © Professors P. M. Motta, P. M. Andrews, K. R. Porter & J. Vial/Science Photo Library/Photo Researchers, Inc.; **7.3a:** © Richard Rodewald; **7.5b:** © Fred Hossler/Visuals Unlimited; **7.14a & b:** © Steve Barden/Action Plus; **7.24a & b, 7.26a & b:** © McGraw-Hill Higher Education, Inc./Eric Wise, photographer; **7.A:** © Phillippe Plailly/Science Photo Library/Photo Researchers, Inc.

Chapter 8
Opener: © Quest/Science Photo Library/Photo Researchers, Inc.: **8.22:** © Courtesy of Branislav Vidic; **8.25a:** © R. T. Hutchings; **8.25b:** © McGraw-Hill Higher Education, Inc./Rebecca Gray, photographer/Don Kincaid, dissections; **8.33a:** © Deep Light Productions/SPL/Photo Researchers, Inc.; **8.A1:** © Howard J. Radzyner/Phototake; **8.A2:** © CNRI/Science Photo Library/Photo Researchers, Inc.

Chapter 9
Opener: © Mireille Lavigne-Rebillard, INSERM Unit 254, Montpellier ("from the site Promenade around the cochlea (http://www.iurc.montp.inserm.fr/cric/audition/fran%E7ais/cochlea/fcochlea.htm)" by R. Pujol, S Blatrix and T. Pujol, CRIC, University of Montpellier /INSERM; **9.5:** © McGraw-Hill Higher Education, Inc./Eric Wise, photographer; **9.14:** © A. L. Blum/Visuals Unlimited; **9.16b:** © McGraw-Hill Higher Education, Inc./Rebecca Gray, photographer/Don Kincaid, dissections; **9.Ba & b:** Reproduced from Isihara's *Tests for Color Blindness* published by Kanehara & Co., Ltd., Tokyo, Japan, but tests for color blindness cannot be conducted with this material. For accurate testing, the original plates should be used.; **9.21 & b2:** © Trent Stephens; **9.23a:** © Jerry Wachter/Photo Researchers, Inc.

Chapter 10
Opener: © Dr. L. Orci, University of geneva/Science Photo Library/Photo Researchers, Inc.; **10.13d, 10.16b:** © Victor Eroschenko; **10.20:** © Bio-Photo Assocs/Photo Researchers, Inc.; **10.Aa & b:** © Ken Greer/Visuals Unlimited.

Chapter 11
Opener: © Dennis Kunkel Microscopy, Inc.; **11.1b:** © McGraw-Hill Higher Education, Inc./Eric Wise, photographer; **11.3a:** © National Cancer Institute/Science Photo Library/Photo Researchers, Inc.; **11.6:** © Ed Reschke; **11.7a-e:** © Victor Eroschenko.

Chapter 12
Opener: © David M. Phillips/Photo Researchers, Inc.; **12.5b:** © R. T. Hutchings; **12.7a & b:** © McMinn & Hutchings, *Color Atlas of Human Anatomy*/Mosby; **12.13b:** © Ed Reschke; **12.19:** © Terry Cockerham/Cynthia Alexander/Synapse Media Productions; **12.A:** © Hank Morgan/Science Source/Photo Researchers, Inc.

Chapter 13
Opener: © Prof. P. Motta/Dept. of Anatomy/University "La Sapienza", Rome/Science Photo Library/Photo Researchers, Inc. **13.2:** © Visuals Unlimited.

Chapter 14
Opener: © Dennis Kunkel Microscopy, Inc.; **14.6b:** © Trent Stephens; **14.A:** © Kenneth Greer/Visuals Unlimited.

Chapter 15
Opener: © David Phillips/Photo Researchers, Inc.; **15.2b:** © R. T. Hutchings; **15.4b:** © CNRI/Phototake; **15.6a & b:** © Branislav Vidic; **15.14a:** © Custom Medical Stock Photo; **15.A:** © Deni McIntyre/Photo Researchers, Inc.

Chapter 16
Opener: © Dennis Kunkel Microscopy, Inc.; **16.14d:** © David M. Phillips/Visuals Unlimited; **16.21b:** © CNRI/Science Photo Library/Photo Researchers, Inc.; **16.A:** © Maxine Cass, 2003.

Chapter 17
Opener: © Professors P. Motta & T. Naguro/Science Photo Library/Photo Researchers, Inc; **17.11:** © Kyle Rothenborg/Pacific Stock.

Chapter 18
Opener: © Dennis Kunkel Microscopy, Inc.; **18.3b:** © McGraw-Hill Higher Education, Inc./Rebecca Gray, photographer/Don Kincaid, dissections.

Chapter 19
Opener: © Prof. P. Motta & S. Makabe/Science Photo Library/Photo Researchers, Inc.; **19.Ba-c:** © The McGraw-Hill Companies, Inc./Jill Braaten, Photographer.

Chapter 20
Opener: © Dr. Yorgus Nikas/Jake Burns/Phototake; **20.8:** © John Giannicchi/Photo Researchers, Inc.; **20.13a & b:** © Petit Format/Nestle/Photo Researchers, Inc.; **20.18:** © CNRI/Science Photo Library/Photo Researchers, Inc.

Index

Prefixes, Suffixes, and Combining Forms

The ability to break down medical terms into separate components or to recognize a complete word depends on mastery of the combining forms (roots or stems) and the prefixes and suffixes that alter or modify their meanings. Common prefixes, suffixes, and combining forms are listed below in boldface type, followed by the meaning of each form and an example illustrating its use.

a-, an- without, lack of: *a*phasia (lack of speech), *an*aerobic (without oxygen)

ab- away from: *ab*ductor (leading away from)

-able capable: vi*able* (capable of living)

acou- hearing: *acou*stics (science of sound)

acr- extremity: *acr*omegaly (large extremities)

ad- to, toward, near to: *ad*renal (near the kidney)

adeno- gland: *adeno*ma (glandular tumor)

-al expressing relationship: neur*al* (referring to nerves)

-algia pain: gastr*algia* (stomach pain)

angio- vessel: *angio*graphy (radiography of blood vessels)

ante- before, forward: *ante*cubital (before elbow)

anti- against, reversed: *anti*peristalsis (reversed peristalsis)

arthr- joint: *arthr*itis (inflammation of a joint)

-ary associated with: urin*ary* (associated with urine)

-asis condition, state of: homeost*asis* (state of staying the same)

auto- self: *auto*lysis (self breakdown)

bi- twice, double: *bi*cuspid (two cusps)

bio- live: *bio*logy (study of living)

-blast bud, germ: fibro*blast* (fiber-producing cell)

brady- slow: *brady*cardia (slow heart rate)

-c expressing relationship: cardia*c* (referring to heart)

carcin- cancer: *carcin*ogenic (causing cancer)

cardio- heart: *cardio*pathy (heart disease)

cata- down, according to: *cata*bolism (breaking down)

cephal- head: *cephal*ic (toward the head)

-cele hollow: blasto*cele* (hollow cavity inside a blastocyst)

cerebro- brain: *cerebro*spinal (referring to brain and spinal cord)

chol- bile: a*chol*ic (without bile)

cholecyst- gallbladder: *cholecyst*okinin (hormone causing the gallbladder to contract)

chondr- cartilage: *chondr*ocyte (cartilage cell)

-cide kill: bacteri*cide* (agent that kills bacteria)

circum- around, about: *circum*duction (circular movement)

-clast smash, break: osteo*clast* (cell that breaks down bone)

co-, com-, con- with, together: *co*enzyme (molecule that functions with an enzyme), *com*misure (coming together), *con*vergence (to incline together)

contra- against, opposite: *contra*lateral (opposite side)

crypto- hidden: *crypto*rchidism (undescended or hidden testes)

cysto- bladder, sac: *cysto*cele (hernia of a bladder)

-cyte-, cyto- cell: erythro*cyte* (red blood cell), *cyto*skeleton (supportive fibers inside a cell)

de- away from: *de*hydrate (remove water)

derm- skin: *derm*atology (study of the skin)

di- two: *di*ploid (two sets of chromosomes)

dia- through, apart, across: *dia*pedesis (ooze through)

dis- reversal, apart from: *dis*sect (cut apart)

-duct- leading, drawing: ab*duct* (lead away from)

-dynia pain: masto*dynia* (breast pain)

dys- difficult, bad: *dys*mentia (bad mind)

e- out, away from: *e*viscerate (take out viscera)

ec- out from: *ec*topic (out of place)

ecto- on outer side: *ecto*derm (outer skin)

-ectomy cut out: append*ectomy* (cut out the appendix)

-edem- swell: myo*edem*a (swelling of a muscle)

em-, en- in: *em*pyema (pus in), *en*cephalon (in the brain)

-emia blood: an*emia* (deficiency of blood)

endo- within: *endo*metrium (within the uterus)

entero- intestine: *entero*itis (inflammation of the intestine)

epi- upon, on: *epi*dermis (on the skin)

erythro- red: *erythro*cyte (red blood cell)

eu- well, good: *eu*phoria (well-being)

ex- out, away from: *ex*halation (breathe out)

exo- outside, on outer side: *exo*genous (originating outside)

extra- outside: *extra*cellular (outside the cell)

-ferent carry: af*ferent* (carrying to the central nervous system)

-form expressing resemblance: fusi*form* (resembling a fusion)

gastro- stomach: *gastro*dynia (stomach ache)

-genesis produce, origin: patho*genesis* (origin of disease)

gloss- tongue: hypo*gloss*al (under the tongue)

glyco- sugar, sweet: *glyco*lysis (breakdown of sugar)

-gram a drawing: myo*gram* (drawing of a muscle contraction)

-graph instrument that records: myo*graph* (instrument for measuring muscle contraction)

hem- blood: *hem*opoiesis (formation of blood)

hemi- half: *hemi*plegia (paralysis of half of the body)

hepato- liver: *hepato*itis (inflammation of the liver)

hetero- different, other: *hetero*zygous (different genes for a trait)

hist- tissue: *hist*ology (study of tissues)

homeo-, homo- same: *homeo*stasis (state of staying the same), *homo*logous (alike in structure or origin)

hydro- wet, water: *hydro*cephalus (fluid within the head)

hyper- over, above, excessive: *hyper*trophy (overgrowth)

hypo- under, below, deficient: *hypo*tension (low blood pressure)

-ia, -id expressing condition: neuralg*ia* (pain in nerve), flacc*id* (state of being weak)

-iatr- treat, cure: ped*iatr*ics (treatment of children)

-im not: *im*permeable (not permeable)

in- in, into: *in*jection (forcing fluid into)

infra- below, beneath: *infra*orbital (below the eye)

inter- between: *inter*costal (between the ribs)

intra- within: *intra*ocular (within the eye)

-ism condition, state of: dimorph*ism* (condition of two forms)